CATEGORICAL LOGIC
AND TYPE THEORY

STUDIES IN LOGIC

AND

THE FOUNDATIONS OF MATHEMATICS

VOLUME 141

Honorary Editor:

P. SUPPES

ELSEVIER
AMSTERDAM • LAUSANNE • NEW YORK • OXFORD • SHANNON • SINGAPORE • TOKYO

CATEGORICAL LOGIC
AND TYPE THEORY

Bart JACOBS
Research Fellow of the
Royal Netherlands Academy of Arts and Sciences

Computing Science Institute,
University of Nijmegen, P.O. Box 9010, 6500 GL Nijmegen,
The Netherlands

ELSEVIER
AMSTERDAM • LAUSANNE • NEW YORK • OXFORD • SHANNON • SINGAPORE • TOKYO

ELSEVIER SCIENCE B.V.
Sara Burgerhartstraat 25
P.O. Box 211, 1000 AE Amsterdam, The Netherlands

First edition: 1999
Paperback edition 2001

Library of Congress Cataloging in Publication Date
A catalog record from the Library of Congress has been applied for.

ISBN: 0 444 50170 3 (Hardbound)
ISBN: 0 444 50853 8 (Paperback)

Transferred to digital printing 2005.

Preface

This book has its origins in my PhD thesis, written during the years 1988 – 1991 at the University of Nijmegen, under supervision of Henk Barendregt. The thesis concerned categorical semantics of various type theories, using fibred categories. The connections with logic were not fully exploited at the time. This book is an attempt to give a systematic presentation of both logic and type theory from a categorical perspective, using the unifying concept of a fibred category. Its intended audience consists of logicians, type theorists, category theorists and (theoretical) computer scientists.

The main part of the book was written while I was employed by NWO, the National Science Foundation in The Netherlands. First, during 1992 – 1994 at the Mathematics Department of the University of Utrecht, and later during 1994 – 1996 at CWI, Center for Mathematics and Computer Science, in Amsterdam. The work was finished in Nijmegen (where it started): currently, I am employed at the Computing Science Institute of the University of Nijmegen as a Research Fellow of the Royal Netherlands Academy of Arts and Sciences.

This book could not have been written without the teaching, support, encouragement, advice, criticism and help of many. It is a hopeless endeavour to list them all. Special thanks go to my friends and (former) colleagues at Nijmegen, Cambridge (UK), Utrecht and Amsterdam, but also to many colleagues in the field. The close cooperation with Thomas Streicher and Claudio Hermida during the years is much appreciated, and their influence can be felt throughout this work. The following persons read portions of the manuscript and provided critical feedback, or contributed in some other way: Lars Birkedal, Zinovy Diskin, Herman Geuvers, Claudio Hermida, Peter Lietz, José Meseguer, Jaap van Oosten, Wesley Phoa, Andy Pitts, Thomas Streicher,

Hendrik Tews and Krzysztof Worytkiewicz. Of course, the responsibility for mistakes remains entirely mine.

The diagrams in this book have been produced with Kristoffer Rose's Xγ-pic macros, and the proof trees with Paul Taylor's macros. The style files have been provided by the publisher.

Bart Jacobs,
Nijmegen, August 1998.

Contents

Preliminaries

A brief account will be given of the organisation of this book, of what is presupposed, and of some of the notions and notations that will be used.

Organisation of the book

The contents form a mixture of logic, type theory and category theory. There are three Chapters 1, 7 and 9 dealing explicitly with (fibred) category theory. The other chapters have mixed contents. Chapter 1 starts off with an introduction to the basic concepts of fibred category theory. This material will subsequently be used in the Chapters 2, 3, 4, 5, and 8, respectively on simple type theory, equational logic, first and higher order predicate logic (over simple type theory) and on polymorphic type theory. Only basic fibred category theory is needed here, since there is no type dependency. The first few sections (of these chapters) give introductions to the relevant systems of logic and type theory. It should be possible to skip the first chapter and start reading the beginning of these subsequent Chapters 2, 3, 4, 5, and 8. A return to Chapter 1 may then take place on a call-by-need basis. In such a way, the reader may oscillate between logical and type theoretical expositions on the one hand, and categorical expositions on the other. Towards the end of Chapter 8 on polymorphic type theory some extra material on the effective topos (from Chapter 6) and on internal categories (from Chapter 7) is used.

In the last two Chapters 10 and 11 on first and higher order dependent type theory the distinction between logical and type theoretical elements on the one side and categorical elements on the other, becomes less pronounced.

Familiarity with fibred category theory (from Chapters 1, 7 and 9) is assumed at this stage.

The essential dependencies between the various chapters are sketched in the following diagram.

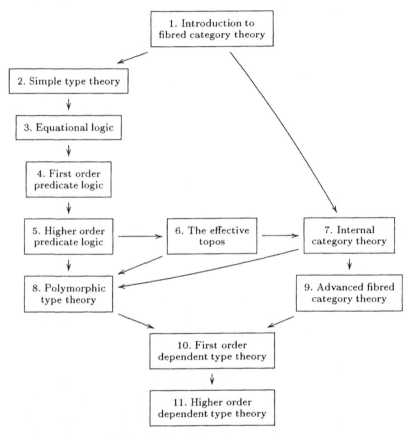

Prerequisites

The reader is assumed to be familiar with the basic notions of category theory, such as adjunctions, (co)limits and Cartesian closed categories (CCCs). Familiarity with predicate logic is assumed, and also some nodding acquaintance with type theory is presupposed. But this is not essential, for example for readers who are mainly interested in categorical aspects. Also, in examples of models we shall use some basic notions and results from domain theory and recursion theory.

We shall be more specific, especially about notational conventions.

Category theory
Arbitrary categories will be written as $\mathbb{A}$, $\mathbb{B}$, $\mathbb{C}$, ... in open face. Specific categories like **Sets**, **PoSets** and **Eff** will be written in bold face. (But also arbitrary *internal* categories **A**, **B**, **C**, ... will be in bold face.) We use capital letters for objects, and write $X \in \mathbb{C}$ to express that X is an object of the category $\mathbb{C}$. Small, non-capitalised letters are used for morphisms in a category (also called maps, or arrows). The homset (or class) $\mathbb{C}(X, Y)$ is the collection of morphisms from X to Y in a category $\mathbb{C}$. The notations $f \colon X \to Y$ and $X \xrightarrow{f} Y$ are also used for $f \in \mathbb{C}(X, Y)$. We use special arrows $X \rightarrowtail Y$ for monic maps (also called monos) and $X \twoheadrightarrow Y$ for epic maps (or epis). We recall that a category $\mathbb{C}$ is called locally small if all its collections $\mathbb{C}(X, Y)$ of homomorphisms are small sets (as opposed to proper classes). And $\mathbb{C}$ is called small if additionally its collection of objects is a small set. The opposite of a category $\mathbb{C}$ will be written as $\mathbb{C}^{\mathrm{op}}$. In the context of a fibration we generally use letters like I, J, K and u, v, w for objects and morphisms of the "base" category and letters like X, Y, Z and f, g, h for objects and morphisms of the "total" category.

The identity morphism on an object X is $\mathrm{id}_X \colon X \to X$, or simply $\mathrm{id} \colon X \to X$. Composition of morphisms $f \colon X \to Y$ and $g \colon Y \to Z$ is usually written as $g \circ f \colon X \to Z$. Sometimes we write $gf \colon X \to Z$ for this composite, especially when f and g are functors. Occasionally we use a double arrow notation $f, g \colon X \rightrightarrows Y$ to express that f and g are two parallel morphisms. A natural transformation α between functors $F, G \colon \mathbb{A} \rightrightarrows \mathbb{B}$ is usually written with double arrow $\Rightarrow$ as $\alpha \colon F \Rightarrow G$, for example, in a diagram:

$$F \left(\underset{\Longrightarrow}{\overset{\mathbb{A}}{}} \alpha \right) G$$

$$\mathbb{B}$$

This $\Rightarrow$ notation will, more generally, be used for 2-cells in a 2-category. And it will sometimes also occur as alternative $X \Rightarrow Y = Y^X$ for an exponent object Y^X in a Cartesian closed category (CCC). We generally use 1 for a terminal object (also called final object or empty product) in a category. Binary Cartesian products are written as $X \times Y$ with projections $\pi \colon X \times Y \to X$, $\pi' \colon X \times Y \to Y$ and tuples $\langle f, g \rangle \colon Z \to X \times Y$ for $f \colon Z \to X$ and $g \colon Z \to Y$. As a special case of tupleing, we often write δ or $\delta(X)$ for the diagonal $\langle \mathrm{id}, \mathrm{id} \rangle \colon X \to X \times X$ on X, and δ or $\delta(I, X)$ for the "parametrised" diagonal $\langle \mathrm{id}, \pi' \rangle \colon I \times X \to (I \times X) \times X$, which duplicates X, with parameter I.

Associated with the abovementioned exponent object Y^X in a CCC there are evaluation and abstraction maps, which will be written as $\text{ev}\colon Y^X \times X \to Y$ and $\Lambda(f)\colon Z \to Y^X$, for $f\colon Z \times X \to Y$.

An initial object (or empty coproduct) is usually written as 0. For binary coproducts we write $X + Y$ with coprojections $\kappa\colon X \to X + Y$, $\kappa'\colon Y \to X + Y$ and cotuples $[f, g]\colon X + Y \to Z$, where $f\colon X \to Z$ and $g\colon Y \to Z$. The codiagonal $\nabla = [\text{id}, \text{id}]\colon X + X \to X$ is an example of a cotuple.

For functors $F\colon \mathbb{A} \to \mathbb{B}$ and $G\colon \mathbb{B} \to \mathbb{A}$ in an adjunction $(F \dashv G)$ the homset isomorphism $\mathbb{B}(FX, Y) \cong \mathbb{A}(X, GY)$ is often written as a bijective correspondence between morphisms $f\colon FX \to Y$ and $g\colon X \to GY$ via double lines:

$$
\frac{FX \xrightarrow{\;\;f\;\;} Y}{X \xrightarrow[g]{} GY} \qquad \textit{e.g.} \text{ for exponents:} \qquad \frac{Z \times X \longrightarrow Y}{Z \longrightarrow Y^X = X \Rightarrow Y}
$$

In such a situation, transposition is sometimes written as $(f\colon FX \to Y) \mapsto (f^{\vee}\colon X \to GY)$ and $(g\colon X \to GY) \mapsto (g^{\wedge}\colon FX \to Y)$, or more ambiguously, as $f \mapsto \bar{f}$ and $g \mapsto \bar{g}$. We reserve the symbols η for the unit natural transformation $\text{id} \Rightarrow GF$, and ε for the counit natural transformation $FG \Rightarrow \text{id}$ of an adjunction $(F \dashv G)$. We recall that these natural transformations have components $\eta_X = (\text{id}_{FX})^{\vee}$ and $\varepsilon_Y = (\text{id}_{GY})^{\wedge}$. In case both η and ε are (natural) isomorphisms, the categories $\mathbb{A}$ and $\mathbb{B}$ are called equivalent. This is written as $\mathbb{A} \simeq \mathbb{B}$.

For the rest, we generally follow usual categorical notation, *e.g.* as in the standard reference [187]. Another (more recent) reference text is [36]. And [186, 19, 61] may be used as introductions.

Logic

Logic as presented in this book differs from traditional accounts in three aspects. (1) We standardly use many-typed (predicate) logic, in which variables occurring in predicates need not be restricted to a single type (or, in more traditionally terminology, to a single sort). (2) We do not restrict ourselves to logic over simple type theory, but also allow logics over polymorphic and dependent type theories. (3) Contexts of variable declarations will be explicitly written at all times.

Hence a logical entailment

$$
n + 5 = 7 \vdash n = 2
$$

is seen as incomplete, and will be written with explicit variable declaration as

$$
n\colon \mathsf{N} \mid n + 5 = 7 \vdash n = 2.
$$

The sign '|' is used to separate the type theoretic context $n: \mathsf{N}$ from the logical context $n + 5 = 7$. These contexts will also be present in derivation rules. The reason for carrying contexts explicitly along comes from their important categorical rôle as indices.

We use as propositional connectives $\perp$ for *falsum* (or absurdity), $\vee$ for disjunction, $\top$ for truth, $\wedge$ for conjunction and $\supset$ for implication. Negation $\neg$ will be defined as $\neg\varphi \equiv \varphi \supset \perp$. Existential $\exists$ and universal $\forall$ quantification will be written in typed form as $\exists x: \sigma. \varphi$ and $\forall x: \sigma. \varphi$. And we similarly use the notation $=_\sigma$ for typed equality (on type σ). All these proposition formers will be used with their standard rules. (But for $\exists, \forall$ and $=_\sigma$ we also use the equivalent—but less standard—adjoint rules, see Lemmas 4.1.7 and 4.1.8.) Higher order logic will be described via a distinguished (constant) type $\mathsf{Prop}: \mathsf{Type}$, which enables quantification over propositions, like in $\forall \alpha: \mathsf{Prop}. \varphi$.

By default, logic will be constructive logic. Non-constructive, classical logic (with the additional double negation rule: $\neg\neg\varphi$ entails φ) will not be very important, since the logic of most of the models that we consider is constructive. See [67, 23, 335] for more information on constructive logic.

Type theory

Mostly, standard type theoretical notation will be used. For example, exponent types are written as $\sigma \to \tau$ and (dependent) product types as $\Pi x: \sigma. \tau$. The associated introduction and elimination operations are lambda-abstraction $\lambda x: \sigma. M$ and application $M \cdot N$, or simply MN. (Sometimes we also use "meta-lambda-abstraction" $\lambda x. f(x)$ for the actual function $x \mapsto f(x)$, not in some formal calculus.) We standardly describe besides "limit types" also "colimit types" like coproduct (disjoint union) $\sigma + \tau$, dependent sum $\Sigma x: \sigma. \tau$, equality $\mathrm{Eq}_\sigma(x, x')$ and quotient σ/R. There is no established notation for the introduction and elimination operations associated with these type formers. The notation that we shall use is given in Figure 0.1. The precise rules will be given later. For these "colimit" type formers there are typical "commutation conversions" (involving substitution of elimination terms) and "Frobenius properties" (describing commutation with products). We write (*e.g.* in the above table) $M[N/x]$ for the result of substituting N for all free occurrences of x in M. This applies to terms, types or kinds M, N. In a type theoretic context, an equation $M = N$ between terms usually describes convertibility. We shall use $\equiv$ to denote syntactic equality (following [13]).

Familiarity with the propositions-as-types correspondence (between derivability in logic and inhabitation in type theory) will be convenient, but not necessary. For basic information on type theory we refer to [14, 98]. Also the standard textbook [13] on the untyped lambda calculus is relevant, since many

	introduction	elimination
$\sigma + \tau$	$\kappa M\colon \sigma + \tau,\ \kappa'N\colon \sigma + \tau$ (for $M\colon \sigma, N\colon \tau$)	unpack P as $[\kappa x$ in $Q, \kappa'y$ in $R]$ (for $P\colon \sigma + \tau, Q(x), R(y)\colon \rho$ where $x\colon \sigma, y\colon \tau$)
$\Sigma x\colon \sigma.\,\tau$	$\langle M, N\rangle\colon \Sigma x\colon \sigma.\,\tau$ (for $M\colon \sigma, N\colon \tau[M/x]$)	unpack P as $\langle x, y\rangle$ in $Q\colon \rho$ (for $P\colon \Sigma x\colon \sigma.\,\tau, Q(x, y)\colon \rho$ where $x\colon \sigma, y\colon \tau$)
$\mathrm{Eq}_\sigma(x, x')$	$\mathsf{r}_\sigma(M)\colon \mathrm{Eq}_\sigma(M, M)$ (for $M\colon \sigma$)	Q with $x' = x$ via $P\colon \rho$ (for $P\colon \mathrm{Eq}_\sigma(x, x'), Q(x)\colon \rho[x/x']$ where $x, x'\colon \sigma$)
σ/R	$[M]_R\colon \sigma/R$ (for $M\colon \sigma$)	pick x from P in $Q\colon \rho$ (for $P\colon \sigma/R, Q(x)\colon \rho$ where $x\colon \sigma$)

Fig. 0.1. Introduction and elimination terms for "colimit" types

of the typed notions stem from the untyped setting.

Order theory

We briefly mention some of the ordered sets that will be used. A set $X = \langle X, \leq \rangle$ carrying an "order" relation $\leq\, \subseteq X \times X$ which is reflexive and transitive is called a preorder. And it is a partially ordered set (or poset, for short) if the order is additionally anti-symmetric. A function $f\colon X \to Y$ between the underlying sets of two preorders or posets X, Y is called monotone if it satisfies $x \leq x' \Rightarrow f(x) \leq f(x')$ for all $x, x' \in X$. Posets with monotone functions form a category **PoSets**. A poset is a lattice if it contains a bottom element $\perp\, \in X$, a top element $\top \in X$, a meet $x \wedge y \in X$ and a join $x \vee y \in X$ for all elements $x, y \in X$. Such a lattice is a Heyting algebra (HA) if it additionally admits an operation $\supset\colon X \times X \to X$ with $z \leq x \supset y$ if and only if $z \wedge x \leq y$. Hence a Heyting algebra is a poset bicartesian closed category. A Boolean algebra (BA) is a Heyting algebra in which $\neg\neg x \leq x$ holds, where $\neg x = x \supset \perp$. Heyting algebras and Boolean algebras form models of constructive and classical propositional logic (respectively).

A poset X is called a *complete* lattice if every subset $a \subseteq X$ has a join $\bigvee a \in X$. Every subset $a \subseteq X$ then also has a meet given by $\bigwedge a = \bigvee\{x \mid x$ is a lower bound of $a\}$. A complete Heyting algebra (CHA)—also called a frame, or a locale—is a Heyting algebra, which is complete as a poset. A poset X is a directed complete partial order (dcpo) if every *directed* subset

$a \subseteq X$ has a join $\bigvee a \in X$, where a subset $a \subseteq X$ is directed if a is non-empty and satisfies: for all $x, y \in a$ there is a $z \in a$ with $x \leq z$ and $y \leq z$. For emphasis we sometimes write $\bigvee^{\uparrow}$ instead of $\bigvee$ for a join of a directed subset. A function $f: X \to Y$ between dcpos is (Scott-) continuous if it is monotone and preserves suprema of directed subsets. Dcpos with continuous functions form a category **Dcpo**, which is Cartesian closed. Also complete lattices with continuous functions from a CCC. For more information, see *e.g.* [69, 3, 170].

Recursion theory
The categories of PERs and of ω-sets (and also the effective topos) will occur in many examples. They involve some basic recursion theory. We assume some coding $(\varphi_n)_{n \in \mathbb{N}}$ of the partial recursive functions, and use it to describe what is called Kleene application on natural numbers:

$$n \cdot m = \begin{cases} \varphi_n(m) & \text{if } \varphi_n(m) \downarrow \ (i.e. \text{ if } \varphi_n(m) \text{ is defined}) \\ \uparrow & \text{otherwise } (i.e. \text{ undefined, otherwise}). \end{cases}$$

For a partial recursive function $f: \mathbb{N}^n \times \mathbb{N} \to \mathbb{N}$ we let $\vec{x} \mapsto \Lambda y. f(\vec{x}, y)$ be the partial recursive function $s_1^n(e, -): \mathbb{N}^n \to \mathbb{N}$ that is obtained from the "*s-m-n*-theorem" by writing

$$f(\vec{x}, y) = \varphi_e(\vec{x}, y) = \varphi_{s_1^n(e, \vec{x})}(y).$$

Then $(\Lambda y. f(\vec{x}, y)) \cdot z \simeq f(\vec{x}, z)$, where $\simeq$ is Kleene equality; it expresses that the left hand side is defined if and only if the right hand side is defined, and in that case both sides are equal. We further use a recursive bijection $\langle -, - \rangle: \mathbb{N} \times \mathbb{N} \xrightarrow{\cong} \mathbb{N}$ with recursive projection functions $\mathbf{p}, \mathbf{p}': \mathbb{N} \rightrightarrows \mathbb{N}$. See *e.g.* [66, 294, 236] for more information.

Chapter 0

Prospectus

This introductory chapter is divided into two parts. It first discusses some generalities concerning logic, type theory and category theory, and describes some themes that will be developed in this book. It then continues with a description of the (standard) logic and type theory of ordinary sets, from the perspective of fibred category theory—typical of this book. This description focuses on the fundamental adjunctions that govern the various logical and type theoretic operations.

0.1 Logic, type theory, and fibred category theory

A logic is always a logic over a type theory. This statement sums up our approach to logic and type theory, and forms an appropriate starting point. It describes a type theory as a "theory of sorts", providing a domain of reasoning for a logic. Roughly, types are used to classify values, so that one can distinguish between zero as a natural number $0: \mathbb{N}$ and zero as a real number $0: \mathbb{R}$, and between addition $+: \mathbb{N} \times \mathbb{N} \to \mathbb{N}$ on natural numbers and addition $+: \mathbb{R} \times \mathbb{R} \to \mathbb{R}$ on real numbers. In these examples we use atomic types $\mathbb{N}$ and $\mathbb{R}$ and composite types $\mathbb{N} \times \mathbb{N} \to \mathbb{N}$ and $\mathbb{R} \times \mathbb{R} \to \mathbb{R}$ obtained with the type constructors $\times$ for Cartesian product, and $\to$ for exponent (or function space). The relation ':' as in $0: \mathbb{N}$, is the inhabitation relation of type theory. It expresses that 0 is of type $\mathbb{N}$, *i.e.* that 0 inhabits $\mathbb{N}$. It is like membership $\in$ in set theory, except that $\in$ is untyped, since everything is a set. But a string is something which does not inhabit the type of natural numbers. Hence we

1

shall have to deal with rules regulating inhabitation, like

$$\frac{}{0:\mathbb{N}} \quad \text{and} \quad \frac{n:\mathbb{N}}{\mathsf{succ}(n):\mathbb{N}}$$

The first rule is unconditional: it has no premises and simply expresses that the term 0 inhabits the type $\mathbb{N}$. The second rule tells that if we know that n inhabits $\mathbb{N}$, then we may conclude that $\mathsf{succ}(n)$ also inhabits $\mathbb{N}$, where $\mathsf{succ}(-)$ may be read as successor operation. In this way one can generate terms, like $\mathsf{succ}(\mathsf{succ}(0)):\mathbb{N}$ inhabiting the type $\mathbb{N}$.

In predicate logic one reasons about such terms in a type theory, like in

$$\forall x:\mathbb{N}.\ \exists y:\mathbb{N}.\ y > \mathsf{succ}(x).$$

This gives an example of a proposition. The fact that this expression is a proposition may also be seen as an inhabitation statement, so we can write

$$(\forall x:\mathbb{N}.\ \exists y:\mathbb{N}.\ y > \mathsf{succ}(x)) : \mathsf{Prop}$$

using a type Prop of propositions. In this particular proposition there are no free variables, but in predicate logic an arbitrary proposition $\varphi:\mathsf{Prop}$ may contain free variables. These variables range over types, like in:

$$x > 5 : \mathsf{Prop}, \text{ where } x:\mathbb{N} \qquad \text{or} \qquad x > 5 : \mathsf{Prop}, \text{ where } x:\mathbb{R}.$$

We usually write these free variables in a "context", which is a sequence of variable declarations. In the examples the sequence is a singleton, so we write

$$x:\mathbb{N} \vdash x > 5 : \mathsf{Prop} \qquad \text{and} \qquad x:\mathbb{R} \vdash x > 5 : \mathsf{Prop}.$$

The turnstile symbol $\vdash$ separates the context from the conclusion: we read the sequent $x:\mathbb{N} \vdash x > 5:\mathsf{Prop}$ as: in the context where the variable x is of type $\mathbb{N}$, the expression $x > 5$ is a proposition. Well-typedness is of importance, since if x is a string, then the expression $x > 5$ does not make sense (unless one has a different operation $>$ on strings, and one reads '5' as a string).

This explains what we mean with: a logic is always a logic over a type theory. Underlying a logic there is always a calculus of typed terms that one reasons about. But one may ask: what about single-sorted logic (*i.e.* single-typed, or untyped, logic) in which variables are thought of as ranging over a single domain, so that types do not really play a rôle? Then one still has a type theory, albeit a very primitive one with only one type (namely the type of the domain), and no type constructors. In such situations one often omits the (sole) type, since it has no rôle. But formally, it is there. And what about propositional logic? It is included as a border case: it can be seen as a degenerate predicate logic in which all predicates are closed (*i.e.* do not contain term variables), so one can see propositional logic as a logic over the empty type theory.

We distinguish three basic kinds of type theory:

- simple type theory (STT);
- dependent type theory (DTT);
- polymorphic type theory (PTT).

In *simple* type theory there are types built up from atomic types (like $\mathbb{N}$, $\mathbb{R}$ above) using type constructors like exponent $\rightarrow$, Cartesian product $\times$ or coproduct (disjoint union) $+$. Term variables $x: \sigma$ are used to build up terms, using atomic terms and introduction and elimination operations associated with the type constructors (like tuples and projections for products $\times$). Types in simple type theory may be seen as sets, and (closed) terms inhabiting types as elements of these sets. In *dependent* type theory, one allows a term variable $x: \sigma$ to occur in another type $\tau(x)$: Type. This increases the expressive power, for example because one can use in DTT the type Matrix(n, m) of $n \times m$ matrices (say over some fixed field), for $n: \mathbb{N}$ and $m: \mathbb{N}$ terms of type $\mathbb{N}$. If one thinks of types as sets, this type dependency is like having for each element $i \in I$ of a set I, another set $X(i)$. One usually writes $X_i = X(i)$ and sees $(X_i)_{i \in I}$ as an I-indexed family of sets. Thus, in dependent type theory one allows type-indexed-types, in analogy with set-indexed-sets. Finally, in *polymorphic* type theory, one may use additional type variables α to build up types. So type variables α may occur inside a type $\sigma(\alpha)$, like in the type list(α) of lists of type α. This means that one has types, indexed by (or parametrised by) the universe Type of all types. In a set theoretic picture this involves a set $X_A = X(A)$ for each set A. One gets indexed collections $(X_A)_{A \in \mathbf{Sets}}$ of sets X_A.

These three type theories are thus distinguished by different forms of indexing of types: no indexing in simple type theory, indexing by term variables $x: \sigma$ in dependent type theory, and indexing by type variables α: Type in polymorphic type theory. One can also combine dependent and polymorphic type theory, into more complicated type theories, for example, into what we call polymorphic dependent type theory (PDTT) or full higher order dependent type theory (FhoDTT).

What we have sketched in the beginning of this section is predicate logic over simple type theory. We shall call this simple predicate logic (SPL). An obvious extension is to consider predicate logic over dependent type theory, so that one can reason about terms in a dependent type theory. Another extension is logic over polymorphic type theory. This leads to dependent predicate logic (DPL) and to polymorphic predicate logic (PPL). If one sees a typed calculus as a (rudimentary) programming language, then these logics may be used as program logics to reason about programs written in simple, dependent, or polymorphic type theory. This describes logic as a "module" that one can plug onto a type theory.

This book focuses on such structural aspects of logic and type theory. The language and techniques of category theory will be essential. For example, we talked about a logic *over* a type theory. Categorically this will correspond to one ("total") category, capturing the logic, being *fibred over* another ("base") category, capturing the type theory. Indeed, we shall make special use of tools from fibred category theory. This is a special part of category theory, stemming from the work of Grothendieck in algebraic geometry, in which (continuous) indexing of categories is studied. As we already mentioned, the various forms of type theoretic indexing distinguish varieties of type theory. And also, putting a logic on top of some type theory (in order to reason about it) will be described by putting a fibration on top of the categorical structure corresponding to the type theory. In this way we can put together complicated structures in a modular way.

Fibred category theory is ordinary category theory with respect to a base category. Also, one can say, it is ordinary category theory *over* a base category. Such a base category is like a universe. For example, several concepts in category theory are defined in terms of sets. One says that a category $\mathbb{C}$ has arbitrary products if for each *set* I and each I-indexed collection $(X_i)_{i \in I}$ of objects $X_i \in \mathbb{C}$ there is a product object $\prod_{i \in I} X_i \in \mathbb{C}$ together with projection morphisms $\pi_j : (\prod_{i \in I} X_i) \to X_j$, which are suitably universal. In category theory one is not very happy with this privileged position of sets and so the question arises: is there a way to make sense of such products with respect to an object I of a 'universe' or 'base category' $\mathbb{B}$, more general than the category **Sets** of sets and functions? This kind of generality is needed to interpret logical products $\forall x : \sigma. \varphi$ or type theoretic products $\Pi x : \sigma. \tau$ when the domain of quantification σ is not interpreted as a set (but as some ordered set, or algebra, for example).

Another example is local smallness. A category $\mathbb{C}$ is locally small if for each pair of objects $X, Y \in \mathbb{C}$ the morphisms $X \to Y$ in $\mathbb{C}$ form a *set* (as opposed to a proper class). That is, if one has homsets $\mathbb{C}(X, Y) \in$ **Sets** as objects in the category of sets. Again the question arises whether there is a way of saying that $\mathbb{C}$ is locally small with respect to an arbitrary universe or base category $\mathbb{B}$ and not just with respect to **Sets**.

Fibred category theory provides answers to such questions. It tells what it means for a category $\mathbb{E}$ to be 'fibred over' a base category $\mathbb{B}$. In that case we write $\begin{smallmatrix} \mathbb{E} \\ \downarrow \\ \mathbb{B} \end{smallmatrix}$, where the arrow $\mathbb{E} \to \mathbb{B}$ is a functor which has a certain property that makes it into a fibration. And in such a situation one can answer the above questions: one can define quantification with respect to objects $I \in \mathbb{B}$ and say when one has appropriate hom-objects $\underline{\mathrm{Hom}}(X, Y) \in \mathbb{B}$ for $X, Y \in \mathbb{E}$.

The ways of doing this will be explained in this book. And for a category $\mathbb{C}$ there is always a 'family fibration' $\begin{smallmatrix} \mathrm{Fam}(\mathbb{C}) \\ \downarrow \\ \mathbf{Sets} \end{smallmatrix}$ of set-indexed families in $\mathbb{C}$. The fibred notions of quantification and local smallness, specialised to this family fibration, are the ordinary notions described above. Thus, in the family fibration we have our standard universe (or base category) of sets.

There are many categorical notions arising naturally in logic and type theory (see the list below). And many arguments in category theory can be formulated conveniently using logic and type theory as "internal" language (sometimes called the "Mitchell-Bénabou" language, in the context of topos theory). These fields however, have different origins: category theory arose in the work of Eilenberg and Mac Lane in the 1940s within mathematics, and was in the beginning chiefly used in algebra and topology. Later it found applications in almost all areas of mathematics (and computer science as well, more recently). Type theory is also from this century, but came up earlier in foundational work by Russell in logic (to avoid paradoxes). Recently, type theory has become important in various (notably functional) programming languages, and in computer mathematics: many type theories have been used during the last two decades as a basis for so-called proof-assistants. These are special computer programs which assist in the verification of mathematical statements, expressed in the language of some (typed) logic. The use of types in these areas imposes certain restrictions on what can be expressed, but facilitates the detection of various errors. We think it is in a sense remarkable that two such fundamental fields (of category theory and of type theory)— with their apparent differences and different origins—are so closely related. This close relationship may be beneficial in the use and further development of both these fields.

We shall be especially interested in categorical phenomena arising within logic and type theory. Among these we mention the following.

(i) Every context of variable declarations (in type theory) or of premises (in logic) is an index. It is an index for a 'fibre' category which captures the logic or type theory that takes place within that context—with the declared variables, or under the assumptions. The importance of this categorical rôle of contexts is our motivation for paying more than usual attention to contexts in our formulations of type theory and logic.

(ii) Appropriately typed sequences of terms give rise to morphisms between contexts. This is the canonical way to produce a category from types and terms. These context morphisms induce substitution functors between fibre categories. The structural operations of weakening (adding a dummy assumption) and contraction (replacing two assumptions of the same kind by a single one) appear as special cases of these substitution functors: weakening is substitution along a projection π, and contraction is substitution along a diagonal δ. These π and δ may be Cartesian projections and diagonals in simple and polymorphic type theories, or 'dependent' projections and diagonals in dependent type theory.

(iii) The basic operations of logic and type theory can be described as adjoints in category theory. Such operations standardly come with an introduction and an elimination operation, which are each other's inverses (via the so-called (β)- and (η)-conversions). Adjoint correspondences capture such situations. This may be familiar for the (simple) type theoretic constructors 1, $\times$, 0, $+$ and $\rightarrow$ (and for their propositional counterparts $\top$, $\wedge$, $\bot$, $\vee$ and $\supset$), since these are the operations of bicartesian closed categories (which can be described via standard adjunctions). But also existential $\exists x\colon \sigma.\,(-)$ and universal $\forall x\colon \sigma.\,(-)$ quantification in predicate logic over a type σ, dependent sum $\Sigma x\colon \sigma.\,(-)$ and product $\Pi x\colon \sigma.\,(-)$ in dependent type theory over a type σ, and polymorphic sum $\Sigma\alpha\colon \mathsf{Type}.\,(-)$ and product $\Pi\alpha\colon \mathsf{Type}.\,(-)$ in polymorphic type theory over the universe Type of types, are characterised as left and right adjoints, namely to the weakening functor which adds an extra dummy assumption $x\colon \sigma$, or $\alpha\colon \mathsf{Type}$. Moreover, equality $=_\sigma$ on a type σ is characterised as left adjoint to the contraction functor which replaces two variables $x, y\colon \sigma$ by a single one (by substituting x for y). By 'being characterised' we mean that the standard logical and type-theoretical rules for these operations are (equivalent to) the rules that come out by describing these operations as appropriate adjoints.

The most important adjunctions are:

$$
\begin{array}{rcl}
\text{existential } \exists,\text{ sum } \Sigma & \dashv & \text{weakening} \\
\text{weakening} & \dashv & \text{universal } \forall,\text{ product } \Pi \\
\text{equality} & \dashv & \text{contraction} \\
\text{truth} & \dashv & \text{comprehension (or 'subsets types')} \\
\text{(but also: equality} & \dashv & \text{comprehension, via a different functor)} \\
\text{quotients} & \dashv & \text{equality.}
\end{array}
$$

The first four of these adjoints were recognised by Lawvere (and the last two are identified in this book). Lawvere first described the quantifiers $\exists, \forall$ as left and right adjoints to arbitrary substitution functors. The above picture with separate adjoints to weakening and to contraction functors is a refinement, since, as we mentioned in (ii), weakening and contraction functors are special cases of substitution functors. (These operations of weakening and contraction can be suitably organised as a certain comonad; we shall define quantification and equality abstractly with respect to such comonads.)

(iv) As we mentioned above, the characteristic aspect of dependent type theory is that types may depend on types, in the sense that term variables inhabiting types may occur in other types. And the characteristic aspect of polymorphic type theory is that type variables may occur in types. Later we shall express this as: types may depend on kinds. These dependencies amount to certain forms of indexing. They are described categorically by fibred (or indexed) categories. Thus, if one knows the dependencies in a type theory, then one knows its underlying categorical structure. The additional type theoretic structure may be described via certain adjunctions, as in the previous point.

(v) Models of logics and type theories are (structure preserving) functors. From a specific system in logic or type theory one can syntactically build a so-called 'classifying' (fibred) category, using a term model—or generalised Lindenbaum-Tarski—construction. A model of this system is then a (fibred) functor with this classifying (fibred) category as domain, preserving appropriate structure. We shall make systematic use of this *functorial semantics*. It was introduced by Lawvere for single-typed simple type theories. And it extends to other logics and type theories, and thus gives a systematic description of models of (often complicated) logics and type theories.

(vi) If $\sigma = \sigma(\alpha)$ is a type (in polymorphic type theory) in which a free type variable α occurs, then, under reasonable assumptions about type formation, the operation $\tau \mapsto \sigma[\tau/\alpha]$ of substituting a type τ for α, is functorial. This functoriality is instrumental in describing the rules of (co-)inductively defined data types in terms of (co-)algebras of this functor. And the reasoning principles (or logic) associated with such data types can also be captured in terms of (co-)algebras (but for a different functor, obtained by lifting the original functor to the logical world of predicates and relations).

(vii) A logical framework is a type theory $\mathcal{T}$ which is expressive enough so that one can formulate other systems $\mathcal{S}$ of logic or of type theory inside $\mathcal{T}$. Categorically one may then describe (the term model of) $\mathcal{S}$ as an internal category in (the term model of) $\mathcal{T}$. We briefly discuss dependent type theory as a logical framework in Section 10.2, but we refer to [87] for this connection with internal categories.

This is not a book properly on logic or on type theory. Many logical and type theoretical calculi are described and some illustrations of their use are given, but there is nothing about specific proof-theoretic properties like cut-elimination, Church-Rosser or strong normalisation. Therefore, see [14]. The emphasis here lies on categorical semantics. This is understood as follows. Category theory provides means to say what a model of, say predicate logic, should look like. It gives a specification, or a hollow structure, which captures the essentials. A proper model is something else, namely an instance of such a structure. We shall describe both these hollow structures, and some instances of these. (But we do not investigate the local structure or theories of the example models, like for example in [197] or in [13, Chapter 19].)

So what, then, is the advantage of knowing what the categorical structures are, corresponding to certain logics and type theories? Firstly, it enables us to easily and quickly recognise that certain mathematical structures are models of some logical or type theoretical calculus, without having to write out an interpretation in detail. The latter can be given for the 'hollow categorical structure', and need not be repeated for the particular instances. One only has to check that the particular structure is an instance of the general categorical structure. For example, knowing that a particular category (of domains, say) is Cartesian closed yields the information that we can interpret simple type theory. Secondly, once this is realised, we can turn things around, and start using our calculus (suitably incorporating the constants in a signature) to reason directly and conveniently about a (concrete or abstract categorical) model. This is the logician's view of the mathematician's use of language: when reasoning about a particular mathematical structure (say a group G), one formally adds the elements $a \in G$ as constants $\underline{a}$ to the language, and one uses the resulting "internal" language to reason directly about G. The same approach applies to more complex mathematical structures, like a fibred category of domains: one then needs a suitable type theoretic language to reason about such a complex (indexed) structure. The third advantage is that a clear (categorical) semantics provides a certain syntactic hygiene, and deepens the understanding of the various logical and type theoretical systems. For example, the principle that a (possibly new) operation in logic or type theory should correspond to an adjoint gives certain canonical introduction, elimination and conversion rules for the constructor. Fourthly, models can be used to obtain new results about one's logical or type theoretical system. Consistency, conservativity and independence results are often obtained in this manner. Finally, and maybe most importantly, models provide meaning to one's logical or type theoretical language, resulting in a better understanding of the syntax.

There are so many systems of logic and type theory because there are certain "production rules" which generate new systems from given ones.

(i) There are three basic type theories: simple type theory (STT), dependent type theory (DTT) and polymorphic type theory (PTT).

(ii) Given a certain type theory, one can construct a logic over this type theory with predicates $\varphi(\vec{x})$: **Prop** containing free variables $\vec{x}$ inhabiting types. This allows us to reason about (terms in) the given type theory.

(iii) Given a logic (over some type theory), one can construct a new type theory (extending the given one) by a propositions-as-types upgrade: one considers the propositions φ in the logic as types in the new type theory, and derivations in the logic as terms in the new type theory.

This modularity is reflected categorically in the following three points.

(i) There are three basic categorical structures: for STT (Cartesian closed categories), for DTT (what we call closed comprehension categories) and for PTT (certain fibred Cartesian closed categories).

(ii) Putting a logic on a type theory corresponds to putting a preorder fibration on top of the structure describing the type theory. For logic one uses preorder structures, since in logic one is interested in provability and not in explicit proofs (or proof-terms, as in type theory), which are described as non-trivial morphisms.

(iii) Under a propositions-as-types upgrade one replaces a preorder fibration by an ordinary fibration (with proper fibre categories), thus making room for proof-terms as proper morphisms.

(Both second points are not as unproblematic as they may seem, because one may have complicated type theories, say with two syntactic universes of types and of kinds, in which there are many ways of putting a logic on top of such a type theory: one may wish to reason about types, or about kinds, or about both in the same logic. Categorically, there are similarly different ways in which a preorder fibration can be imposed.)

By the very nature of its contents, this book is rather descriptive. It contains few theorems with deep mathematical content. The influence of computer science may be felt here, in which much emphasis is put on the description of various languages and formalisms.

Also, it is important to stress that this is not a book properly on fibred category theory. And it is not intended as such. It does contain the basic concepts and results from fibred category theory, but only as far as they are directly useful in logic or type theory (and not in topology, for example). Some of these basic results have not been published previously, but have been folklore for some time already. They have been discovered and rediscovered by various people, and the precise flow of ideas is hard to track in detail. What we present in this book is not a detailed historical account, and we therefore apologise in advance for any misrepresentation of history.

We sketch what we see as the main lines. In the development of fibred category and categorical logic one can distinguish an initial French period starting in the 1960s with Grothendieck's definition of a fibration (*i.e.* a fibred category), published in [107]. It was introduced in order to study descent. The ensuing theory was further developed by Grothendieck and (among others) Giraud [100] and Bénabou. The latter's work is more logical and foundational in spirit than Grothendieck's (involving for example suitable fibred notions of local smallness and definability), and is thus closest to the current work. Many of the basic notions and results stem from this period.

In the late 1960s Lawvere first applied indexed categories in the study of logic. Especially, he described quantification and equality in terms of adjoints to substitution functors, and showed that also comprehension involves an adjunction. This may be seen as the start of categorical logic (explicitly, in his influential "Perugia Lecture Notes" and also in [192, 193]). At about the same time, the notion of elementary topos was formulated, by Lawvere and Tierney. This resulted in renewed attention for indexed (and internal) categories, to study phenomena over (and inside) toposes. See for example [173, 169] and the references there.

Then, in the 1980s there is the start of a type theoretic boom, in which indexed and fibred categories are used in the semantics of polymorphic and dependent type theories, see the basic papers [306, 307, 148] and the series of PhD theses [45, 330, 75, 185, 318, 252, 260, 7, 154, 89, 217, 86, 60, 289, 125, 4, 198, 133]. This book collects much material from this third phase. Explicitly, the connection between simple type theory and Cartesian closed categories was first established by Lawvere and Lambek. Later, dependent type theory was related to locally Cartesian closed categories by Seely, and to the more general "display map categories" by Taylor. The relation between polymorphic type theory and certain fibred (or indexed, or internal) Cartesian closed categories is due to Seely, Lamarche and Moggi. Finally, more complicated systems combining polymorphic and dependent systems (like the calculus of constructions) were described categorically by Hyland, Pitts, Streicher, Ehrhard, Curien, Pavlović, Jacobs and Dybjer. This led to the (surprising) discovery of complete internal categories by Moggi and Hyland (and to the subsequent development of 'synthetic' domain theory in abstract universes).

Interestingly, fibred categories are becoming more and more important in various other areas of (theoretical) computer science, precisely because the aspects of indexing and substitution (also called renaming, or relabelling) are so fundamental. Among these areas we mention (without pretension to be in any sense complete): database theory [295, 151, 9], rewriting [12], automata theory [175, 10], abstract environments [279], data flow networks [310], constraint programming [219], concurrency theory [345, 131], program analysis [230, 25], abstract domain theory [146] and specification [152, 327, 48, 159].

Many topics in the field of categorical logic and type theory are not discussed in this book. Sometimes because the available material is too recent (and unsettled), sometimes because the topic deviates too much from the main line, but mostly simply because of lack of space. Among these topics we mention (with a few references): inductively and co-inductively defined types in dependent type theory [70, 71], categorical combinators [63, 290, 116], categorical normalisation proofs [147, 238, 5], fixed points [16], rewriting and 2-categorical structure [308, 278], modal logic [93], μ-calculi [313], synthetic domain theory [144, 331, 264], a fibred Giraud theorem [229], a fibred adjoint functor theorem [47, 246], descent theory [168] (especially with its links to Beth definability [208]), fibrations in bi-categories [315, 317], 2-fibrations [127], and the theory of stacks [100].

The choice has been made to present details of interpretation functions for simple type theory in full detail in Chapter 2, together with the equivalent functorial interpretation. In later chapters interpretations will occur mostly in the more convenient functorial form. For detailed information about interpretation functions in polymorphic and (higher order) dependent type theories we refer to [319, 61]. As we proceed we will be increasingly blurring the distinction between certain type theories and certain fibred categories, thus decreasing the need for explicit interpretations

0.2 The logic and type theory of sets

We shall now try to make the fibred perspective more concrete by describing the (familiar) logic and type theory of ordinary sets in fibred form. Therefore we shall use the fibrations of predicates over sets and of families of sets over sets, without assuming knowledge of what precisely constitutes a fibration. In a well-known situation we thus describe some of the structures that will be investigated in more abstract form in the course of this book. We shall write **Sets** for the category of (small) sets and ordinary functions between them.

Predicates on sets can be organised in a category, that will be called **Pred**, as follows.

objects   pairs (I, X) where $X \subseteq I$ is a subset of a set I; in this situation we consider X as a predicate on a type I, and write $X(i)$ for $i \in X$ to emphasise that an element $i \in I$ may be understood as a free variable in X. When I is clear from the context, we sometimes write X for the object $(X \subseteq I)$.

morphisms $(I, X) \to (J, Y)$ are functions $u: I \to J$ between the underlying sets satisfying

$$X(i) \text{ implies } Y(u(i)), \quad \text{for each } i \in I.$$

Diagrammatically, this condition on such a function

$u: I \to J$ amounts to the existence of a necessarily unique (dashed) map

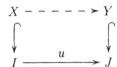

indicating that u restricts appropriately.

There is an obvious forgetful functor **Pred** $\to$ **Sets** sending a predicate to its underlying set (or type): $(I, X) \mapsto I$. This functor is a "fibration". And although it plays a crucial rôle in this situation, we do not give it a name, but simply write it vertically as $\begin{smallmatrix} \mathbf{Pred} \\ \downarrow \\ \mathbf{Sets} \end{smallmatrix}$ to emphasise that it describes predicates as living over sets.

For a specific set I, the "fibre" category **Pred**$_I$ is defined as the subcategory of **Pred** of predicates $(X \subseteq I)$ on I and of morphisms that are mapped to the identity function on I. This category **Pred**$_I$ may be identified with the poset category $\langle P(I), \subseteq \rangle$ of subsets of I, ordered by inclusion. For a function $u: I \to J$ there is "substitution" functor $u^*: P(J) \to P(I)$ in the reverse direction, by

$$(Y \subseteq J) \longmapsto (\{ i \mid u(i) \in Y \} \subseteq I).$$

Clearly we have $Y \subseteq Y' \Rightarrow u^*(Y) \subseteq u^*(Y')$, so that u^* is indeed a functor. Two special cases of substitution are weakening and contraction. Weakening is substitution along a Cartesian projection $\pi: I \times J \to I$. It consists of a functor

$$P(I) \xrightarrow{\pi^*} P(I \times J) \qquad \text{sending} \qquad X \mapsto \{ (i, j) \mid i \in X \text{ and } j \in J \}$$

by adding a dummy variable $j \in J$ to a predicate X. Contraction is substitution along a Cartesian diagonal $\delta: I \to I \times I$. It is a functor

$$P(I \times I) \xrightarrow{\delta^*} P(I) \qquad \text{given by} \qquad Y \mapsto \{ i \in I \mid (i, i) \in Y \}.$$

It replaces two variables of type I by a single variable.

Each fibre category $P(I)$ is a Boolean algebra, with the usual set theoretic operations of intersection $\cap$, top element $(I \subseteq I)$, union $\cup$, bottom element $(\emptyset \subseteq I)$, and complement $I \backslash (-)$. These operations correspond to the propositional connectives $\wedge, \top, \vee, \bot, \neg$ in (Boolean) logic. They are preserved by substitution functors u^* between fibre categories.

The categorical description of the quantifiers $\exists, \forall$ is less standard (than the propositional structure of subsets). These quantifiers are given by operations between the fibres—and not inside the fibres, like the propositional connectives—since they bind free variables in predicates (and thus change the underlying types). They turn out to be adjoints to weakening, as expressed by the fundamental formula:

$$\exists \dashv \pi^* \dashv \forall.$$

In more detail, we define for a predicate $Y \subseteq I \times J$,

$$\exists(Y) = \{i \in I \mid \exists j \in J. (i, j) \in Y\}$$
$$\forall(Y) = \{i \in I \mid \forall j \in J. (i, j) \in Y\}.$$

These assignments $Y \mapsto \exists(Y)$ and $Y \mapsto \forall(Y)$ are functorial $P(I \times J) \rightrightarrows P(I)$. And they are left and right adjoints to the above weakening functor $\pi^* : P(I) \to P(I \times J)$ because there are the following basic adjoint correspondences.

$$\frac{Y \subseteq \pi^*(X) \quad \text{over } I \times J}{\exists(Y) \subseteq X \quad \text{over } I} \qquad \text{and} \qquad \frac{\pi^*(X) \subseteq Y \quad \text{over } I \times J}{X \subseteq \forall(Y) \quad \text{over } I}$$

(Where the double line means: if and only if.)

For a set (or type) I, equality $i = i'$ for elements $i, i' \in I$ forms a predicate on $I \times I$. Such equality can also be captured categorically, namely as left adjoint to the contraction functor $\delta^* : P(I \times I) \to P(I)$. One defines for a predicate $X \subseteq I$ the predicate $\mathrm{Eq}(X)$ on $I \times I$ by

$$\mathrm{Eq}(X) = \{(i, i') \in I \times I \mid i = i' \text{ and } i \in X\}.$$

Then there are adjoint correspondences

$$\frac{\mathrm{Eq}(X) \subseteq Y \quad \text{over } I \times I}{X \subseteq \delta^*(Y) \quad \text{over } I}$$

Notice that the predicate $\mathrm{Eq}(X)$ is equality on I for the special case where X is the top element I. See also Exercise 0.2.2 below for a description of a right adjoint to contraction, in terms of inequality.

The operations of predicate logic can thus be identified as certain structure in this fibration $\begin{smallmatrix}\textbf{Pred}\\\downarrow\\\textbf{Sets}\end{smallmatrix}$, namely as structure in and between its fibres. Moreover, it is a property of the fibration that this logical structure exists, since it can be characterised in a universal way—via adjoints—and is thus given uniquely up-to-isomorphism. The same holds for the other logical and type theoretical operations that we identify below.

Comprehension is the assignment of a set to a predicate, or, as we shall say more generally later on, of a type to a predicate. This assignment takes a predicate to the set of elements for which the predicate holds. It also has a universal property. Therefore we first need the "truth" functor $1: \textbf{Sets} \to$

Pred, which assigns to a set I the truth predicate $1(I) = (I \subseteq I)$ on I; it is the terminal object in the fibre over I. Comprehension (or subset types, as we shall also say) is then given by a functor $\{-\}: \textbf{Pred} \to \textbf{Sets}$, namely

$$\{(Y \subseteq J)\} = \{j \in J \mid Y(j)\} = Y.$$

Hence $\{-\}: \textbf{Pred} \to \textbf{Sets}$ is simply $(Y \subseteq J) \mapsto Y$. It is right adjoint to the truth functor $1: \textbf{Sets} \to \textbf{Pred}$ since there is a bijective correspondence between functions u and v in a situation:

$$\frac{1(I) \xrightarrow{\;u\;} (Y \subseteq J) \quad \text{in } \textbf{Pred}}{I \xrightarrow[v]{} \{(Y \subseteq J)\} \quad \text{in } \textbf{Sets}}$$

In essence this correspondence tells us that $Y(j)$ holds if and only if $j \in \{(Y \subseteq J)\}$.

Quotient sets can also be described using the fibration of predicates over sets. We first form the category **Rel** of (binary) relations on sets by pullback:

$$\begin{array}{ccc}
\textbf{Rel} & \longrightarrow & \textbf{Pred} \\
\downarrow & \lrcorner & \downarrow \\
\textbf{Sets} & \longrightarrow & \textbf{Sets} \\
& {\scriptstyle I \,\mapsto\, I \times I} &
\end{array}$$

Via this pullback we restrict ourselves to predicates with underlying sets of the form $I \times I$. Explicitly, the category **Rel** has

objects	pairs (I, R) where $R \subseteq I \times I$ is a (binary) relation on $I \in \textbf{Sets}$.
morphisms	$(I, R) \to (J, S)$ are functions $u: I \to J$ between the underlying sets with the property

$$R(i, i') \text{ implies } S(u(i), u(i')), \quad \text{for all } i, i' \in I.$$

The functor $\textbf{Rel} \to \textbf{Sets}$ in the diagram is then $(I, R) \mapsto I$. It will turn out to be a fibration by construction. The abovementioned equality predicate yields an equality functor $\text{Eq}: \textbf{Sets} \to \textbf{Rel}$, namely

$$J \mapsto \text{Eq}(J) = \{(j, j) \mid j \in J\}.$$

Quotients in set theory can then be described in terms of a left adjoint Q to this equality functor Eq: a relation $R \subseteq I \times I$ is mapped to the quotient set $I/\overline{R}$, where $\overline{R} \subseteq I \times I$ is the least equivalence relation containing R. Indeed there is an adjoint correspondence between functions v and u in:

$$Q(I, R) = I/\overline{R} \xrightarrow{\ v\ } J \quad \text{in } \mathbf{Sets}$$

$$\overline{\qquad\qquad\qquad\qquad\qquad}$$

$$R \xrightarrow{\ u\ } \mathrm{Eq}(J) \quad \text{in } \mathbf{Rel}$$

This correspondence can be reformulated as: for each functon $u: I \to J$ with $u(i) = u(i')$ for all $i, i' \in I$ for which $R(i, i')$ holds, there is a unique function $v: I/\overline{R} \dashrightarrow J$ in a commuting triangle

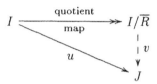

Finally we mention that predicates over sets give us higher order logic. There is a distinguished set $2 = \{0, 1\}$ of propositions, with special predicate $(\{1\} \subseteq 2)$ for truth: for every predicate $(X \subseteq I)$ on a set I, there is a unique function $\mathrm{char}(X \subseteq I): I \to 2$ with

$$(X \subseteq I) = \mathrm{char}(X \subseteq I)^*(\{1\} \subseteq 2).$$

This existence of "characteristic morphisms" is what makes the category of sets a topos. It allows us to quantify via this set 2 over propositions.

This completes our first glance at the fibred structure of the logic of sets. In the remainder of this section we sketch some of the type theoretic structure of sets, again in terms of a fibration, namely in terms of the "family" fibration $\begin{smallmatrix}\mathrm{Fam}(\mathbf{Sets})\\ \downarrow\\ \mathbf{Sets}\end{smallmatrix}$ of set-indexed-sets. It captures the dependent type theory (with type-indexed-types) of sets.

The category $\mathrm{Fam}(\mathbf{Sets})$ of families of sets has

objects   pairs (I, X) consisting of an index set I and a family $X = (X_i)_{i \in I}$ of I-indexed sets X_i.

morphisms $(I, X) \to (J, Y)$ are pairs (u, f) consisting of functions

$$I \xrightarrow{\ u\ } J \quad \text{and} \quad f = \left(X_i \xrightarrow{\ f_i\ } Y_{u(i)} \right)_{i \in I}$$

There is a projection functor $\mathrm{Fam}(\mathbf{Sets}) \to \mathbf{Sets}$ sending an indexed family to its underlying set index set: $(I, X) \mapsto I$. It will turn out to be a fibration. Essentially this will mean that there are (appropriate) substitution or reindexing functors: for a function $u: I \to J$ between index sets, we can map a family $Y = (Y_j)_{j \in J}$ over J to a family over I via:

$$(Y_j)_{j \in J} \longmapsto (Y_{u(i)})_{i \in I}.$$

We shall write u^* for this operation. It extends to a functor between "fibre" categories: for an arbitrary set K, let $\mathrm{Fam}(\mathbf{Sets})_K$ be the "fibre" subcategory of $\mathrm{Fam}(\mathbf{Sets})$ of those families (K, X) with K as index set, and with morphisms (id_K, f) with the identity on K as underlying function. Then $u \colon I \to J$ yields a substitution functor $u^* \colon \mathrm{Fam}(\mathbf{Sets})_J \to \mathrm{Fam}(\mathbf{Sets})_I$.

Notice that there is an inclusion functor $\mathbf{Pred} \hookrightarrow \mathrm{Fam}(\mathbf{Sets})$ of predicates into families, since every predicate $(X \subseteq I)$ yields an I-indexed family $(X_i)_{i \in I}$ with

$$X_i = \begin{cases} \{*\} & \text{if } i \in X \\ \emptyset & \text{otherwise.} \end{cases}$$

It is not hard to see that this yields a full and faithful functor $\mathbf{Pred} \hookrightarrow \mathrm{Fam}(\mathbf{Sets})$, which commutes with substitution. It is a 'morphism of fibrations'.

Our aim is to describe the dependent coproduct $\coprod$ and product $\prod$ of families of sets as adjoints to weakening functors, in analogy with the situation for existential $\exists$ and universal $\forall$ quantification in the logic of sets. But in this situation of families of sets we have weakening functors π^* induced not by Cartesian projections $\pi \colon I \times J \to I$, but by "dependent" projections $\pi \colon \{I \mid X\} \to I$, with domain $\{I \mid X\}$ given by the disjoint union:

$$\{I \mid X\} = \{(i, x) \mid i \in I \text{ and } x \in X_i\}$$

which generalises the Cartesian product. The weakening functor π^* associated with this dependent projection $\pi \colon \{I \mid X\} \to I$ sends a family $Y = (Y_i)_{i \in I}$ over I to a family $\pi^*(Y)$ over $\{I \mid X\}$ by vacuously adding an extra index x, as in:

$$\pi^*(Y) = (Y_i)_{(i \in I, x \in X_i)}.$$

(As we shall see later, the projection $\pi \colon \{I \mid X\} \to I$ arises in a canonical way, since the assignment $(I, X) \mapsto \{I \mid X\}$ yields a functor $\mathrm{Fam}(\mathbf{Sets}) \to \mathbf{Sets}$, which is right adjoint to the terminal object functor $1 \colon \mathbf{Sets} \to \mathrm{Fam}(\mathbf{Sets})$, sending a set J to the J-indexed collection $(\{*\})_{j \in J}$ of singletons. The counit of this adjunction has the projection π as underlying map. Thus, the operation $(I, X) \mapsto \{I \mid X\}$ is like comprehension for predicates, as described above.)

The claim is that the dependent coproduct $\coprod$ and product $\prod$ for set-indexed sets are left and right adjoints to the weakening functor π^*. Therefore we have to define coproduct $\coprod$ and product $\prod$ as functors

$$\mathbf{Fam(Sets)}_{\{I \mid X\}} \xleftarrow{\quad \pi^* \quad} \mathbf{Fam(Sets)}_I$$

with $\coprod$ above and $\prod$ below the arrow.

$$\{I \mid X\} \xrightarrow{\quad \pi \quad} I$$

acting on an $\{I \mid X\}$-indexed family $Z = (Z_{(i,x)})_{i \in I, x \in X}$, and producing an I-indexed family. These functors are given by

$$\coprod(Z)_i = \{(x, z) \mid x \in X_i \text{ and } z \in Z_{(i,x)}\}$$
$$\prod(Z)_i = \{\varphi \colon X_i \to \bigcup_{x \in X,} Z_{(i,x)} \mid \forall x \in X_i.\, \varphi(x) \in Z_{(i,x)}\}.$$

We then get the fundamental relation

$$\coprod \dashv \pi^* \dashv \prod$$

since there are bijective adjoint correspondences between families of functions f and g in:

$$\frac{Z \xrightarrow{\ f\ } \pi^*(Y) \quad \text{over } \{I \mid X\}}{\coprod(Z) \xrightarrow{\ g\ } Y \quad \text{over } I} \quad \text{and} \quad \frac{\pi^*(Y) \xrightarrow{\ f\ } Z \quad \text{over } \{I \mid X\}}{Y \xrightarrow{\ g\ } \prod(Z) \quad \text{over } I}$$

Also in this situation, there are adjoints to contraction functors δ^* (induced by *dependent* diagonals), given by equality and inequality. But we do not further pursue this matter, and conclude our introduction at this point. What we have sketched is that families of sets behave like dependent types, and that subsets behave like predicates, yielding a logic over (dependent) type theory. We have shown that the basic operations of this logic and of this type theory can be described by adjunctions, in a fibred setting. In the course of this book we shall (among many other things) be more precise about what it means to have such a logic over a type theory and we shall axiomatise all of the structure found above, and identify it in many other situations.

Finally, the next few exercises may help the reader to become more familiar with the structure described above.

Exercises

0.2.1.   Define a left adjoint $F: \mathrm{Fam}(\mathbf{Sets}) \to \mathbf{Pred}$ to the inclusion functor

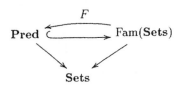

such that: (1) F makes the triangle commute (so it does not change the index set), and (2) F commutes with substitution.

0.2.2.   Define for a subset $X \subseteq I$ the relation $\mathrm{nEq}(X) \subseteq I \times I$ by

$$\mathrm{nEq}(X) = \{(i, i') \mid i \neq i' \text{ or } i \in X\}$$

and show that the assignment $X \mapsto \mathrm{nEq}(X)$ is right adjoint to contraction $\delta^*: P(I \times I) \to P(I)$. Notice that $\mathrm{nEq}(X)$ at the bottom element $X = \emptyset$ is inequality on I.

0.2.3.   Show that the equality functor $\mathrm{Eq}: \mathbf{Sets} \to \mathbf{Rel}$ also has a right adjoint.

0.2.4.   Check that the operation $(I, X) \mapsto \{I \mid X\}$ yields a functor $\mathrm{Fam}(\mathbf{Sets}) \to \mathbf{Sets}$, and show that it is right adjoint to the terminal object functor $\mathbf{Sets} \to \mathrm{Fam}(\mathbf{Sets})$, mapping a set J to the family of singletons $(\{*\})_{j \in J}$. Describe the unit and counit of the adjunction explicitly.

Chapter 1

Introduction to fibred category theory

This first proper chapter starts with the basics of fibred category theory; it provides the foundation for much of the rest of this book. A fibration, or fibred category, is designed to capture collections $(\mathbb{C}_I)_{I \in \mathbb{B}}$ of categories $\mathbb{C}_I$ varying over a base category $\mathbb{B}$, generalising for example collections of sets $(X_i)_{i \in I}$ varying over a base, or index, set I. The main categorical examples are the indexed collections of categories

$$\left(\mathbb{B}/I\right)_{I \in \mathbb{B}} \qquad \left(\mathrm{Sub}(I)\right)_{I \in \mathbb{B}} \qquad \left(\mathbb{B}/\!\!/I\right)_{I \in \mathbb{B}}$$

consisting of slice categories $\mathbb{B}/I$ over I, posets $\mathrm{Sub}(I)$ of subobjects of I, and what we call 'simple slice categories' $\mathbb{B}/\!\!/I$ over I. The ordinary slice categories will be used for dependent type theory, the posets of subobjects for predicate logic, and the simple slice categories for simple type theory (whence the name). The slice categories $\mathbb{B}/I$ will be used as leading example in the first section when we introduce fibrations. The other examples $\mathrm{Sub}(I)$ and $\mathbb{B}/\!\!/I$ will be introduced soon afterwards, in Section 1.3.

In all of these cases, a morphism $u \colon I \to J$ in the base category $\mathbb{B}$ induces a substitution functor, commonly written as u^*, acting in the reverse direction. That is, there are substitution functors:

$$\mathbb{B}/J \xrightarrow{\;u^*\;} \mathbb{B}/I \qquad \mathrm{Sub}(J) \xrightarrow{\;u^*\;} \mathrm{Sub}(I) \qquad \mathbb{B}/\!\!/J \xrightarrow{\;u^*\;} \mathbb{B}/\!\!/I$$

Weakening functors and contraction functors arise as special cases of substitution functors u^*, namely (respectively) as π^*, where π is a projection morphism in $\mathbb{B}$, and as δ^*, where δ is a diagonal morphism in $\mathbb{B}$.

These two aspects—indexing and substitution—will be studied systematically in this first chapter, in terms of fibrations. The notion of 'fibred category', or 'fibration', is due to Grothendieck [107].

This chapter develops the basic theory of fibrations and shows how various notions from ordinary category theory—such as adjunctions, products and coproducts—make sense for fibred categories as well. In the last section 1.10 we describe the notion of 'indexed category', a common alternative formulation of variable category, and explain why an indexed category should be regarded as simply a particular kind of fibrations (namely as a 'cloven' one). Chapter 7 describes internal categories, which also correspond to certain fibrations, namely to so-called 'small' fibrations.

The ten sections which together form this chapter contain the essentially standard, first part of the theory of fibrations, geared towards use in categorical logic and type theory. The main notions are: Cartesian morphism, substitution functor, change-of-base, fibred adjunction, fibred (co)product and indexed category. These will be introduced together with many examples. Sometimes the theory is further developed in exercises, but mostly, the exercises of a section serve to familiarise the reader with the new material in that section. There is a later chapter (Chapter 9) which continues the study of fibrations.

1.1 Fibrations

Basically, a fibration is a categorical structure which captures *indexing* and *substitution*. Since the formal definition of a fibration is a bit technical—see Definition 1.1.3 below—we start with the following introductory observations. These focus on the special case of a codomain fibration, and will lead to the general definition of fibration towards the end of this section. The exercises contain many elementary results about fibrations, which should help the reader to get acquainted with the concepts involved.

Indexing

Suppose we wish to consider a family of sets, ranging over some index set I. There are two ways of doing so.

(a) **Pointwise** (or **split**) **indexing**: as a collection $(X_i)_{i \in I}$, where each X_i is a set. Probably this way is most elementary and comes first to one's mind. One can think of this collection as being given by a function (or functor) $I \to \mathbf{Sets}$, namely $i \mapsto X_i$.

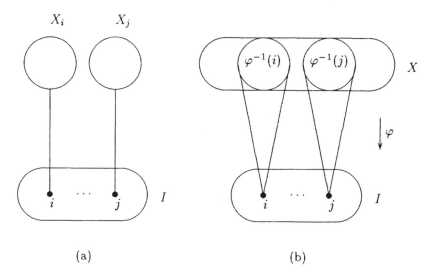

Fig. 1.1. Pointwise indexing (a) and display indexing (b) of set-indexed-sets

(b) **Display indexing**: as a function $\varphi: X \to I$. The sets in the family then appear as **fibres** "over i"

$$\varphi^{-1}(i) = \{x \in X \mid \varphi(x) = i\}$$

for each $i \in I$.

A picture suggesting the difference between these ways of indexing is presented in Figure 1.1.

These descriptions are equivalent: given a collection $(X_i)_{i \in I}$ as in (a), take X to be the disjoint union $\coprod_{i \in I} X_i = \{(i, x) \mid i \in I \text{ and } x \in X_i\}$; it comes equipped with a projection function $\pi: \coprod_{i \in I} X_i \to I$ sending $(i, x) \mapsto i$. Up-to-isomorphism, the fibre $\pi^{-1}(i)$ over i is the original X_i. Conversely, given a function $\varphi: X \to I$ as in (b), put $X_i = \varphi^{-1}(i)$. This yields a collection $(X_i)_{i \in I}$ as in (a), together with an isomorphism $\coprod_{i \in I} X_i \cong X$.

(For the set theoretic purist we remark that the passage from (a) to (b) relies on the Axiom of Replacement. Also we should mention that the fibres $\varphi^{-1}(i)$ in (b) are necessarily disjoint, whereas the sets X_i in (a) need not be disjoint. But that is not essential at this stage.)

Although pointwise indexing (a) seems more natural at first, display indexing (b) has the great advantage that it generalises to arbitrary categories, since it only involves the notion of a morphism, see Definition 1.1.5 below.

Hence in the sequel we often describe a family of sets as a function $\varphi: X \to I$ as in (b). We then loosely speak about the fibres $X_i = \varphi^{-1}(i)$ and say that X is a **family over** I and that φ **displays** the family (X_i). In order to emphasise that we think of such a map φ as a family, we often write it vertically as $\begin{pmatrix} X \\ \downarrow\varphi \\ I \end{pmatrix}$. A **constant family** is one of the form $\begin{pmatrix} I \times X \\ \downarrow\pi \\ I \end{pmatrix}$, where π is the Cartesian product projection; often it is written simply as $I^*(X)$. Notice that all fibres of this constant family are (isomorphic to) X.

Such families $\begin{pmatrix} X \\ \downarrow\varphi \\ I \end{pmatrix}$ of sets give rise to two categories: the **slice category Sets/I** and the **arrow category Sets$^{\to}$**. The objects of **Sets/I** are the I-indexed families, for a fixed set I; the objects of **Sets$^{\to}$** are all the I-indexed families, for all possible I. Here are the definitions.

Sets/I   **objects**   families $\begin{pmatrix} X \\ \downarrow\varphi \\ I \end{pmatrix}$.

   morphisms $\begin{pmatrix} X \\ \downarrow\varphi \\ I \end{pmatrix} \xrightarrow{\ f\ } \begin{pmatrix} Y \\ \downarrow\psi \\ I \end{pmatrix}$ are functions $f: X \to Y$
   making the following diagram commute.

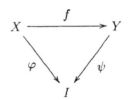

Notice that f can thus be seen as a collection of functions $f_i: X_i \to Y_i$—where $X_i = \varphi^{-1}(i)$ and $Y_i = \psi^{-1}(i)$ are the fibres involved (for $i \in I$). Composition and identities in **Sets/I** are inherited from **Sets**.

Sets$^{\to}$ **objects**   families $\begin{pmatrix} X \\ \downarrow\varphi \\ I \end{pmatrix}$, for arbitrary sets I.

   morphisms $\begin{pmatrix} X \\ \downarrow\varphi \\ I \end{pmatrix} \xrightarrow{(u,\,f)} \begin{pmatrix} Y \\ \downarrow\psi \\ J \end{pmatrix}$ are pairs of functions
   $u: I \to J$ and $f: X \to Y$ for which the following

diagram commutes.

$$
\begin{array}{ccc}
X & \xrightarrow{\;f\;} & Y \\
{\scriptstyle \varphi}\downarrow & & \downarrow{\scriptstyle \psi} \\
I & \xrightarrow{\;u\;} & J
\end{array}
$$

Hence, objects in the arrow category **Sets$^{\rightarrow}$** involve an extra function u between the index sets. Notice that one can now view f as a collection of functions $f_i \colon X_i \to Y_{u(i)}$, since for $x \in \varphi^{-1}(i)$, $f(x)$ lands in $\psi^{-1}(u(i))$. Composition and identities in **Sets$^{\rightarrow}$** are component-wise inherited from **Sets**.

We further remark that there is a **codomain functor** cod: **Sets$^{\rightarrow}$** $\to$ **Sets**; it maps

$$
\left(\begin{array}{c} X \\ \downarrow{\scriptstyle \varphi} \\ I \end{array} \right) \mapsto I \qquad \text{and} \qquad (u, f) \mapsto u.
$$

Also, for each I, there is a (non-full) inclusion functor **Sets**$/I \hookrightarrow$ **Sets$^{\rightarrow}$**.

Substitution

Suppose a family 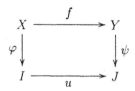 over a set J is given. Substitution involves changing the index set J. More specifically substitution along a function $u \colon I \to J$ involves creating a family of sets with the domain I of u as new index set and with fibres $Y_{u(i)}$ for $i \in I$. Thus the family $(Y_j)_{j \in J}$ is turned into a family $(X_i)_{i \in I}$ with $X_i = Y_{u(i)}$. This family $(X_i)_{i \in I}$ can be obtained in the following way. Form the pullback of ψ against u:

$$
u^*(\psi) = \varphi \left\downarrow \begin{array}{ccc} X & \xrightarrow{\;f\;} & Y \\ & & \\ I & \xrightarrow{\;u\;} & J \end{array} \right\downarrow \psi \qquad (*)
$$

That is, form the set $X = \{(i, y) \in I \times Y \mid u(i) = \psi(y)\}$ with obvious projection functions $I \xleftarrow{\varphi} X \xrightarrow{f} Y$. One obtains a new family $\left(\begin{array}{c} X \\ \downarrow{\scriptstyle \varphi} \\ I \end{array} \right)$ over I with fibres

$$
X_i = \varphi^{-1}(i) \cong \{y \in Y \mid \psi(y) = u(i)\} = \psi^{-1}(u(i)) = Y_{u(i)}.
$$

One normally writes $u^*(\psi)$ for the result φ of substituting ψ along u.

1.1.1. Examples. The following four special cases of substitution along a map u are worth mentioning separately.

(i) Suppose u is an element $j \in J$, that is, u is of the form $j: 1 \to J$ where $1 = \{*\}$ is a one-element (terminal) set. Then $u^*(\psi) = j^*(\psi)$ becomes the family $\begin{pmatrix} X_j \\ \downarrow \\ 1 \end{pmatrix}$. It can be identified with the fibre X_j. Thus, substituting along a specific element j yields the fibre X_j over this element j.

(ii) Substitution of an ordinary (non-indexed) set X, described as a family $\begin{pmatrix} X \\ \downarrow \\ 1 \end{pmatrix}$ over a singleton set 1, along the unique map $I \dashrightarrow 1$ yields the constant family $I^*(X) = \begin{pmatrix} I \times X \\ \downarrow \\ I \end{pmatrix}$. This is because the pullback of two maps $I \to 1$ and $X \to 1$ with the terminal object 1 as common codomain, is the Cartesian product $I \times X$ of their domains.

(iii) In case u is a projection $\pi: J \times I \to J$, then $\pi^*(\psi)$ is $\psi \times \mathrm{id}$, since the following diagram is a pullback square.

$$
\begin{array}{ccc}
Y \times I & \xrightarrow{\ \pi\ } & Y \\
{\scriptstyle \psi \times \mathrm{id}}\downarrow\ \ \lrcorner & & \ \ \downarrow{\scriptstyle \psi} \\
J \times I & \xrightarrow[\ \pi\ ]{} & J
\end{array}
$$

One obtains as fibre over $(j, i) \in J \times I$

$$\pi^*(\varphi)^{-1}(j, i) = (\psi \times \mathrm{id})^{-1}(j, i) = \{(y, j, i) \mid \psi(y) = j\} \cong \psi^{-1}(j) \times I$$

which shows that there is an extra "dummy" index variable i in the family $\pi^*(\varphi)$ which plays no rôle. Later in Section 3.1 (explicitly in Example 3.1.1) we shall see that in logical terms, substitution along a projection is **weakening** (*i.e.* adding an extra assumption).

(iv) The dual (in some sense) of (iii) is substitution along a diagonal $\delta: J \to J \times J$. For a family $\begin{pmatrix} Y \\ \downarrow{\scriptstyle \psi} \\ J \times J \end{pmatrix}$ the fibre of $\delta^*(\psi)$ over $j \in J$ is

$$\delta^*(\psi)_j = \{y \in Y \mid \psi(y) = (j, j)\} = Y_{(j,j)}$$

which is the family $Y_{(j,j')}$ restricted to $j = j'$. This is **contraction**: replacing two variables j, j' by a single one via substituting $[j/j']$.

Notice that the pair (u, f) in the pullback diagram $(*)$ above is a morphism $u^*(\psi) \to \psi$ in the arrow category $\mathbf{Sets}^{\to}$. For a moment let us call this pair (u, f) the "substitution morphism" (later it will be called a Cartesian morphism). This substitution morphism has a universal property: suppose we have another morphism,

$$\left(\begin{array}{c} Z \\ \downarrow \chi \\ K \end{array} \right) \xrightarrow{(v,g)} \left(\begin{array}{c} Y \\ \downarrow \psi \\ J \end{array} \right)$$

in **Sets**$^{\rightarrow}$ such that $v: K \to J$ factors through $u: I \to J$, say via $w: K \to I$ with $v = u \circ w$, as in

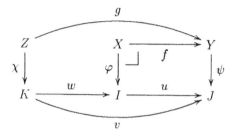

Then there is a unique morphism

$$\left(\begin{array}{c} Z \\ \downarrow \chi \\ K \end{array} \right) \xrightarrow{(w,h)}_{- - \succ} \left(\begin{array}{c} X \\ \downarrow \varphi \\ I \end{array} \right)$$

in **Sets**$^{\rightarrow}$ which is sent to w by the codomain functor cod: **Sets**$^{\rightarrow} \to$ **Sets**, and for which the composite

$$\left(\begin{array}{c} Z \\ \downarrow \\ K \end{array} \right) \xrightarrow{(w,h)}_{- - \succ} \left(\begin{array}{c} X \\ \downarrow \\ I \end{array} \right) \xrightarrow{(u,f)} \left(\begin{array}{c} Y \\ \downarrow \\ J \end{array} \right) \qquad \text{is} \qquad \left(\begin{array}{c} Z \\ \downarrow \\ K \end{array} \right) \xrightarrow{(v,g)} \left(\begin{array}{c} Y \\ \downarrow \\ J \end{array} \right).$$

This holds because X was constructed as a pullback:

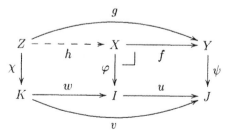

The presence of such 'best possible substitution morphisms' $u^*(\psi) \to \psi$ is the cardinal property of the codomain functor cod: **Sets**$^{\rightarrow} \to$ **Sets** that makes it a fibration. Definition 1.1.3 below captures this property abstractly in purely categorical terms. And in Section 1.4 we shall see how this property induces—by choosing substitution morphisms—substitution functors u^*.

We introduce some notation and terminology. Let $p: \mathbb{E} \to \mathbb{B}$ be a functor. It can be seen as a (display) family $\left(\begin{smallmatrix} \mathbb{E} \\ \downarrow p \\ \mathbb{B} \end{smallmatrix} \right)$ of categories: for an object $I \in \mathbb{B}$, the **fibre** or **fibre category** $\mathbb{E}_I = p^{-1}(I)$ over I is the category with

objects	$X \in \mathbb{E}$ with $pX = I$.
morphisms	$X \to Y$ in $\mathbb{E}_I$ are morphisms $f: X \to Y$ in $\mathbb{E}$ for which pf is the identity map on I in $\mathbb{B}$.

An object $X \in \mathbb{E}$ such that $pX = I$ (*i.e.* an $X \in \mathbb{E}_I$) is said to be **above** I; similarly, a morphism f in $\mathbb{E}$ with $pf = u$ is said to be **above** u. This terminology is in accordance with our 'vertical' notation $\begin{smallmatrix} \mathbb{E} \\ \downarrow p \\ \mathbb{B} \end{smallmatrix}$. A morphism in $\mathbb{E}$ will be called **vertical** if it is above some identity morphism in $\mathbb{B}$, that is, when it is in a fibre category. For $X, Y \in \mathbb{E}$ and $u: pX \to pY$ in $\mathbb{B}$ we sometimes write

$$\mathbb{E}_u\big(X, Y\big) = \{f: X \to Y \text{ in } \mathbb{E} \mid f \text{ is above } u\} \subseteq \mathbb{E}\big(X, Y\big).$$

When considering such a family of categories $\begin{smallmatrix} \mathbb{E} \\ \downarrow p \\ \mathbb{B} \end{smallmatrix}$, we call $\mathbb{B}$ the **base category** and $\mathbb{E}$ the **total category**.

1.1.2. Examples. (i) Consider the codomain functor $\begin{smallmatrix} \mathbf{Sets}^{\to} \\ \downarrow \\ \mathbf{Sets} \end{smallmatrix}$. An object above $I \in \mathbf{Sets}$ is a family $\left(\begin{smallmatrix} X \\ \downarrow \varphi \\ I \end{smallmatrix} \right)$ over I; a vertical morphism has the form

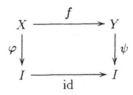

Thus the fibre category above $I \in \mathbf{Sets}$ can be identified with the slice category $\mathbf{Sets}/I$ of families over I and commuting triangles. Notice that the fibre $\mathbf{Sets}/1$ (or slice) over a singleton (terminal) set 1 can be identified with the base category $\mathbf{Sets}$ itself.

(ii) For a functor $p: \mathbb{E} \to \mathbb{B}$, the fibre category $\mathbb{E}_I$ over $I \in \mathbb{B}$ can be constructed via a pullback: one has a pullback of categories

is a pullback of sets, as described in Example 1.1.1 (i).

Finally, we come to the definition of 'fibration'.

1.1.3. Definition. Let $p: \mathbb{E} \to \mathbb{B}$ be a functor.

(i) A morphism $f: X \to Y$ in $\mathbb{E}$ is **Cartesian over** $u: I \to J$ in $\mathbb{B}$ if $pf = u$ and every $g: Z \to Y$ in $\mathbb{E}$ for which one has $pg = u \circ w$ for some $w: pZ \to I$, uniquely determines an $h: Z \to X$ in $\mathbb{E}$ above w with $f \circ h = g$. In a diagram:

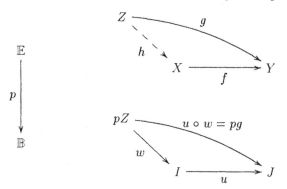

We call $f: X \to Y$ in the total category $\mathbb{E}$ Cartesian if it is Cartesian over its underlying map pf in $\mathbb{B}$.

(ii) The functor $p: \mathbb{E} \to \mathbb{B}$ is a **fibration** if for every $Y \in \mathbb{E}$ and $u: I \to pY$ in $\mathbb{B}$, there is a Cartesian morphism $f: X \to Y$ in $\mathbb{E}$ above u. Sometimes a fibration will be called a **fibred category** or a **category (fibred) over** $\mathbb{B}$.

We often say that a Cartesian morphism $f: X \to Y$ above $u: I \to pY$ is a **terminal** or **Cartesian** lifting of u in a situation:

$$X \dashrightarrow^{\ f\ } Y$$

$$I \xrightarrow{\ u\ } J$$

(Later, in Section 9.1, we shall describe 'opfibrations' as functors $p: \mathbb{E} \to \mathbb{B}$ in which one has 'initial' or 'opcartesian' liftings of maps $pX \to J$ in $\mathbb{B}$.)

The previous two diagrams embody a convention that will be used throughout: if a diagram is drawn in two parts, one above the other, then "above" in the diagram means "above" in the categorical sense described before Example 1.1.2. Further, a fibration is written vertically as $\begin{smallmatrix}\mathbb{E}\\\downarrow\\\mathbb{B}\end{smallmatrix}$ and is pronounced as '$\mathbb{E}$ over $\mathbb{B}$'. Often the name of the functor is omitted if it is clear what 'over' means.

1.1.4. Proposition. *Cartesian liftings are unique up-to-isomorphism (in a slice): if f and f' with $\operatorname{cod} f = \operatorname{cod} f'$ are both Cartesian over the same map, then there is a unique vertical isomorphism $\varphi\colon X \overset{\cong}{\Rightarrow} X'$ with $f' \circ \varphi = f$.* $\square$

(The proof is left as Exercise 1.1.1 (i) below.)

The reader is now invited to check that with respect to the codomain functor $\operatorname{cod}\colon \mathbf{Sets}^{\rightarrow} \to \mathbf{Sets}$ the Cartesian morphisms in $\mathbf{Sets}^{\rightarrow}$ are precisely the pullback squares in $\mathbf{Sets}$ and that the functor cod is a fibration.

The following is a mild generalisation of what has been considered above for the category of sets.

1.1.5. Definition. For an arbitrary category $\mathbb{B}$, the **arrow category** $\mathbb{B}^{\rightarrow}$ has families $\begin{pmatrix}X\\\downarrow\varphi\\I\end{pmatrix}$ as objects; thus maps $\varphi\colon X \to I$ in $\mathbb{B}$ are objects of $\mathbb{B}^{\rightarrow}$.

A morphism $\begin{pmatrix}X\\\downarrow\varphi\\I\end{pmatrix} \to \begin{pmatrix}Y\\\downarrow\psi\\J\end{pmatrix}$ in $\mathbb{B}^{\rightarrow}$ consists of a pair of morphisms $u\colon I \to J$, $f\colon X \to Y$ in $\mathbb{B}$ such that $\psi \circ f = u \circ \varphi$.

For an object $I \in \mathbb{B}$ the **slice category** $\mathbb{B}/I$ is the subcategory of $\mathbb{B}^{\rightarrow}$ of families over I (*i.e.* with codomain I) and morphisms (u, f) where $u = \operatorname{id}_I$. Sometimes, a slice category is simply called a slice.

1.1.6. Proposition. *Consider the codomain functor $\operatorname{cod}\colon \mathbb{B}^{\rightarrow} \to \mathbb{B}$.*

(i) *The fibre category over $I \in \mathbb{B}$ is the slice category $\mathbb{B}/I$.*

(ii) *Cartesian morphisms in $\mathbb{B}^{\rightarrow}$ coincide with pullback squares in $\mathbb{B}$.*

(iii) *The functor cod is a fibration if and only if $\mathbb{B}$ has pullbacks. In that case it called the* **codomain fibration** *on $\mathbb{B}$.* $\square$

(The proof is also left as an exercise.)

The notation $\mathbb{B}^{\rightarrow}$ for the category of arrows of $\mathbb{B}$ comes from the fact that $\mathbb{B}^{\rightarrow}$ can be seen as the category of functors from $\cdot \to \cdot$ to $\mathbb{B}$, and natural transformations between them. Similarly we write $\mathbb{B}^{\rightarrow\rightarrow}$ for the category of functors from $\cdot \to \cdot \to \cdot$ to $\mathbb{B}$. Notice that $\mathbb{B}^{\rightarrow\rightarrow}$ is not the same as $\left(\mathbb{B}^{\rightarrow}\right)^{\rightarrow}$.

Alternatively, one can see $\mathbb{B}^{\rightarrow}$ as the comma category $(\mathbb{B} \downarrow \mathbb{B})$, see [187]. In writing $\begin{smallmatrix}\mathbb{B}^{\rightarrow}\\\downarrow\\\mathbb{B}\end{smallmatrix}$ we always refer to the codomain fibration on $\mathbb{B}$ (and not to the domain fibration described in Exercise 1.1.8 below).

We started this section by describing set-indexed families of sets, either as

(a) pointwise $(X_i)_{i \in I}$ or as   (b) display $\begin{array}{c} X \\ \downarrow \\ I \end{array}$.

We emphasise that it is important to have both pictures in mind "at the same time". There is a great similarity with indexed families of categories; they can be presented either as

(a) $(\mathbb{E}_I)_{I \in \mathbb{B}}$ or as   (b) $\begin{array}{c} \mathbb{E} \\ \downarrow \\ \mathbb{B} \end{array}$.

In (b) one gets a picture as given by fibrations, and in (a) as given by so-called 'indexed categories'. It turns out that there is also a way of switching between (a) and (b) for categories, given by the 'Grothendieck construction', which is an extension of what we have for sets. The details are in the last section of this chapter, together with a short discussion on fibrations versus indexed categories.

For the time being however, we concentrate on (b) for categories, in order to become more familiar with fibred categories. But it is good to keep (a) in mind. For example, when confronted with a fibration, always ask what the fibres are.

Exercises 1–4 collect some useful facts about Cartesian morphisms and fibrations. We will often make use of them.

Exercises

1.1.1.   (i)   Prove Proposition 1.1.4.
  (ii)   Suppose f is Cartesian and g and h are above the same map. Show that $f \circ g = f \circ h$ implies $g = h$.
1.1.2.   Let $p: \mathbb{E} \to \mathbb{B}$ be a functor; assume $f: X \to Y$ is in $\mathbb{E}$ and put $u = pf$. Show that f is Cartesian if and only if for each $Z \in \mathbb{E}$ and $v: pZ \to pX$ in $\mathbb{B}$, the function

$$\mathbb{E}_v \big(Z, X \big) \xrightarrow{\ f \, \circ \, (-)\ } \mathbb{E}_{u \circ v} \big(Z, Y \big)$$

is an isomorphism.
1.1.3.   Consider the total category of a fibration. Show that
  (i)   every morphism factors as a vertical map followed (diagrammatically) by a Cartesian one;
  (ii)   a Cartesian map above an isomorphism is an isomorphism. Especially a vertical Cartesian map is an isomorphism.
1.1.4.   Let $\begin{array}{c} \mathbb{E} \\ \downarrow p \\ \mathbb{B} \end{array}$ be a fibration. Prove that
  (i)   all isomorphisms in $\mathbb{E}$ are Cartesian;

(ii) if $\xrightarrow{f}\xrightarrow{g}$ is a composable pair of Cartesian morphisms in $\mathbb{E}$, then also
their composite $\xrightarrow{g\circ f}$ is Cartesian.
Hence it makes sense to talk about the subcategory $\mathrm{Cart}(\mathbb{E}) \hookrightarrow \mathbb{E}$ having
all objects from $\mathbb{E}$ but only the Cartesian arrows. We write $|p|$ for the
composite $\mathrm{Cart}(\mathbb{E}) \hookrightarrow \mathbb{E} \to \mathbb{B}$.

(iii) Let $\xrightarrow{f}\xrightarrow{g}$ be a composable pair in $\mathbb{E}$ again. Show now that if g and
$g \circ f$ are Cartesian, then f is Cartesian as well.

(iv) Verify that a consequence of (iii) is that the functor $|p|: \mathrm{Cart}(\mathbb{E}) \to \mathbb{B}$
is a fibration. From (ii) in the previous exercise it follows that all fibres
of $|p|$ are groupoids (*i.e.* that all maps in the fibres are isomorphisms).
[This $|p|$ will be called the **fibration of objects of** p.]

1.1.5. Verify that the following two results—known as the Pullback Lemmas—are
a consequence of (ii) and (iii) in the previous exercise. Consider

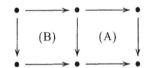

(i)   If (A) and (B) are pullback squares, then the outer rectangle is also a
pullback square.

(ii) If the outer rectangle and (A) are pullback squares, then (B) is a pullback square as well.

1.1.6. Consider a functor $p: \mathbb{E} \to \mathbb{B}$. We describe a slightly weaker notion of Cartesianness, than the one above. Call a morphism $f: X \to Y$ in $\mathbb{E}$ **weak Cartesian** if for each $g: Z \to Y$ with $pf = pg$ there is a unique vertical $h: Z \to X$
with $f \circ h = g$. Show that the functor p is a fibration if and only if both
(a) every morphism $u: I \to pY$ in $\mathbb{B}$ has a weak Cartesian lifting $f: X \to Y$;
(b) the composition of two weak Cartesian morphisms is again weak Cartesian.

1.1.7. Check that the following are (trivial) examples of fibrations

$$\begin{array}{cccc} \mathbb{B}\times\mathbb{C} & \mathbb{B} & \mathbb{B} & X \\ \downarrow{\scriptstyle\mathrm{fst}} & \downarrow{\scriptstyle\mathrm{id}} & \downarrow & \downarrow \\ \mathbb{B} & \mathbb{B} & 1=\{*\} & I \end{array}$$

where X, I are sets (*i.e.* discrete categories).

1.1.8. For an arbitrary category $\mathbb{B}$, consider the domain functor $\mathrm{dom}: \mathbb{B}^{\to} \to \mathbb{B}$.

(i)   Describe the fibre category above $I \in \mathbb{B}$. It is usually called the **opslice
category** or simply **opslice** and written as $I\backslash\mathbb{B}$.

(ii) Show that dom is a fibration (without any assumptions about $\mathbb{B}$).

(iii) Show also that for each $I \in \mathbb{B}$ the domain functor $\mathrm{dom}_I: \mathbb{B}/I \to \mathbb{B}$ is a
fibration.

1.1.9. Assume $\mathbb{B}$ is a category with pullbacks. Show that the functor $\mathbb{B}^{\to\to} \longrightarrow \mathbb{B}^{\to}$
sending $\xrightarrow{f}\xrightarrow{g}$ to $\xrightarrow{g}$ is a fibration. Is the composite $\mathbb{B}^{\to\to} \longrightarrow \mathbb{B}^{\to} \xrightarrow{\mathrm{cod}} \mathbb{B}$
also a fibration?

1.1.10. Show that the object functor **Cat** → **Sets** is a fibration. Also that the forgetful functor **Sp** → **Sets** is a fibration—where **Sp** is the category of topological spaces and continuous functions.

1.1.11. Let **Fld** be the category of fields and field homomorphisms; **Vect** is the category of vector spaces: objects are triples $(K, V, \cdot)$ where K is a field of scalars, V is an Abelian group of vectors and $\cdot : K \times V \to V$ is an action of scalar multiplication (which distributes both over scalar and over vector addition). A morphism $(K, V, \cdot) \to (L, W, \cdot)$ in **Vect** is a pair (u, f) where $u : K \to L$ is a field homomorphism and $f : V \to W$ a group homomorphism such that $f(a \cdot x) = u(a) \cdot f(x)$ for all $a \in K$ and $x \in V$.

Check that the obvious forgetful functor **Vect** → **Fld** is a fibration. What are the fibres? Which maps are Cartesian?

1.2 Some concrete examples: sets, ω-sets and PERs

In this section we shall describe some specific fibred categories which will be used as leading examples. They involve firstly families indexed over sets and secondly the categories of ω-sets and of partial equivalence relations (PERs). The latter will provide important examples of models of various type theories. Later we shall describe the three categories of sets, ω-sets and PERs as (reflective) subcategories of the effective topos **Eff**. This topos thus provides a framework for studying them together. The subsections about ω-sets and PERs contain little fibred category theory; they only contain the basic definitions and properties of ω-sets and PERs. Fibred aspects will be studied later.

Set-indexed families

Assume $\mathbb{C}$ is an arbitrary category. We will describe a category Fam($\mathbb{C}$) of set-indexed families of objects and arrows of $\mathbb{C}$. As objects of Fam($\mathbb{C}$) we take collections $(X_i)_{i \in I}$ where for each element i of the index set I, X_i is an object of $\mathbb{C}$. Objects of Fam($\mathbb{C}$) may thus be seen as pairs (I, X) with I a set and X a function $X : I \to \mathbb{C}_0$—where $\mathbb{C}_0$ is the collection of objects of $\mathbb{C}$.

What, then, is a map $(X_i)_{i \in I} \to (Y_j)_{j \in J}$? We take it to consist of a function $u : I \to J$ between the index sets together with a collection of morphisms $f_i : X_i \to Y_{u(i)}$ in $\mathbb{C}$, for $i \in I$. Composition in Fam($\mathbb{C}$) is done as follows. Given two morphisms

$$(X_i)_{i \in I} \xrightarrow{\;(u, (f_i)_{i \in I})\;} (Y_j)_{j \in J} \xrightarrow{\;(v, (g_j)_{j \in J})\;} (Z_k)_{k \in K}$$

involving for $i \in I$ and $j \in J$ maps in $\mathbb{C}$:

$$X_i \xrightarrow{f_i} Y_{u(i)} \qquad \text{and} \qquad Y_j \xrightarrow{g_j} Z_{v(j)}$$

Thus, for each $i \in I$, we can form a composite in $\mathbb{C}$

$$X_i \xrightarrow{f_i} Y_{u(i)} \xrightarrow{g_{u(i)}} Z_{v(u(i))}.$$

So that we obtain a composite morphism $(v \circ u, (g_{u(i)} \circ f_i)_{i \in I})$ in $\mathrm{Fam}(\mathbb{C})$ from the family $(X_i)_{i \in I}$ to the family $(Z_k)_{k \in K}$.

There is a projection functor $p \colon \mathrm{Fam}(\mathbb{C}) \to \mathbf{Sets}$ which maps families to their index sets:

$$(X_i)_{i \in I} \mapsto I \qquad \text{and} \qquad (u, (f_i)_{i \in I}) \mapsto u.$$

From what we have seen in the previous section we may expect that such a functor from indexed collections to index sets is a fibration. And indeed p is a fibration: given a function $u \colon I \to J$ and an indexed collection $(Y_j)_{j \in J}$ above J we can find a Cartesian lifting in a diagram

$$(??_i)_{i \in I} \dashrightarrow^{?} (Y_j)_{j \in J}$$

$$I \xrightarrow{\quad u \quad} J$$

The obvious choice is to take $??_i = Y_{u(i)}$. Then as map $\dashrightarrow^{?}$ one can take $(u, (\mathrm{id}_{Y_{u(i)}})_{i \in I})$, which is above u. The verification of the required universal property of this lifting is left to the reader.

1.2.1. Definition. The above fibration $\begin{smallmatrix} \mathrm{Fam}(\mathbb{C}) \\ \downarrow \\ \mathbf{Sets} \end{smallmatrix}$ will be called the **family fibration of** $\mathbb{C}$. The fibre over $I \in \mathbf{Sets}$ is the (functor) category $\mathbb{C}^I$ of I-indexed families of objects and arrows in $\mathbb{C}$.

Recall that the category $\mathbb{C}$ is a parameter in this construction. Especially we can take $\mathbb{C} = \mathbf{Sets}$ (like in the Prospectus). The resulting family fibration $\begin{smallmatrix} \mathrm{Fam}(\mathbf{Sets}) \\ \downarrow \\ \mathbf{Sets} \end{smallmatrix}$ of set-indexed sets gives a precise description of pointwise indexing of families of sets as in (a) in the beginning of the previous section. On the other hand, the arrow fibration $\begin{smallmatrix} \mathbf{Sets}^{\to} \\ \downarrow \\ \mathbf{Sets} \end{smallmatrix}$ captures display indexing as considered under (b). The fact that pointwise indexing of sets is essentially the same as display indexing finds its precise categorical formulation in the statement that the categories $\mathrm{Fam}(\mathbf{Sets})$ and $\mathbf{Sets}^{\to}$ are equivalent. In fact, the *fibrations* $\begin{matrix} \text{Fam(Sets)} \\ \downarrow \\ \textbf{Sets} \end{matrix}$ and $\begin{matrix} \textbf{Sets}^{\rightarrow} \\ \downarrow \\ \textbf{Sets} \end{matrix}$ are equivalent in a sense appropriate to fibred category theory, see Section 1.7. For the time being, we formulate this as follows.

1.2.2. Proposition. *There is an equivalence of categories in (the top line of) a commuting triangle*

$$
\begin{array}{ccc}
\text{Fam(Sets)} & \xrightarrow{\;\simeq\;} & \textbf{Sets}^{\rightarrow} \\
& \searrow \qquad \swarrow \text{cod} & \\
& \textbf{Sets} &
\end{array}
$$

where the functor Fam(**Sets**) $\rightarrow$ **Sets**$^{\rightarrow}$ *sends*

$$
(X_i)_{i \in I} \mapsto \text{the projection} \left(\begin{matrix} \coprod_{i \in I} X_i \\ \downarrow \pi \\ I \end{matrix} \right).
$$

$\square$

ω-Sets

Our next example in this section involves the category ω-**Sets** of so-called **omega sets**. It combines the set-theoretic with the recursion-theoretic and will play an important rôle in the sequel. An informative source is [143], but see also [199] and later sections in this book.

Recall that we write $e \cdot n$ for Kleene application: $e \cdot n$ is the outcome $\varphi_e(n)$ of applying the e-th partial recursive function φ_e to n. A code or index for a partial recursive function f will be written as $\Lambda x. f(x)$.

1.2.3. Definition. An ω-*set* (*i.e.* an object of the category ω-**Sets** that we are about to describe) is a set X together with for each element $x \in X$ a non-empty set of natural numbers, written as

$$
E(x) \subseteq \mathbb{N}
$$

One calls E the **existence predicate** of the ω-set. We then write (X, E)—or sometimes (X, E_X)—for the object itself. A morphism $f: (X, E) \rightarrow (Y, E)$ in ω-**Sets** is a function $f: X \rightarrow Y$ between the underlying sets, for which there is a **code** $e \in \mathbb{N}$ which **tracks** f in the sense that

for $x \in X$ and $n \in E_X(x)$ one has: $e \cdot n$ is defined and $e \cdot n \in E_Y(f(x))$.

Notice that only the existence of such a code and not the code itself, is part of the definition of a morphism. The identity function $(X, E) \rightarrow (X, E)$ is then tracked by a code $\Lambda x. x$ for the identity function on $\mathbb{N}$. And for morphisms $(X, E) \xrightarrow{f} (Y, E) \xrightarrow{g} (Z, E)$ in ω-**Sets**, say with f tracked by e and g by d, the composite $(X, E) \xrightarrow{g \circ f} (Z, E)$ is tracked by a code $\Lambda x.\, d \cdot (e \cdot x)$ for the function $x \mapsto d \cdot (e \cdot x)$. This constitutes a category which will be denoted by ω-**Sets**. It comes with an obvious forgetful functor ω-**Sets** $\to$ **Sets** which forgets the existence predicate.

In the future, in writing $e \cdot n \in E(f(x))$ as above, we implicitly assume that $e \cdot n$ is defined.

1.2.4. Proposition. *The category ω-**Sets** has finite limits and exponents.*

Proof. The constructions on the underlying sets are as for sets. Some extra care is needed to deal with the codes. For example, one has a Cartesian product

$$(X, E) \times (Y, E) = (X \times Y, E)$$

with

$$E(x, y) = \{\langle n, m\rangle \in \mathbb{N} \mid n \in E(x) \text{ and } m \in E(y)\}$$

where $\langle -, -\rangle$ is an effective coding of $\mathbb{N} \times \mathbb{N}$ into $\mathbb{N}$. The projections in ω-**Sets** are the projections $X \leftarrow X \times Y \to Y$ in **Sets** tracked by codes for the projection functions associated with the effective coding. The exponent is given by

$$(X, E) \Rightarrow (Y, E) = (\{f \in Y^X \mid f \text{ is tracked by some } e \in \mathbb{N}\}, E)$$

with

$$E(f) = \{e \in \mathbb{N} \mid e \text{ tracks } f\}. \qquad \square$$

Since the category ω-**Sets** has pullbacks, the codomain functor $\begin{smallmatrix} \omega\text{-}\mathbf{Sets}^{\to} \\ \downarrow \\ \omega\text{-}\mathbf{Sets} \end{smallmatrix}$ is a fibration, see Proposition 1.1.6. This yields display families $\begin{pmatrix} (X,E) \\ \downarrow \\ (I,E) \end{pmatrix}$ of ω-sets indexed by ω-sets, as described in (b) in the beginning of the previous section. At the end of Section 1.4 it will be shown how to describe ω-set-indexed ω-sets pointwise as in (a), and how to get an equivalence result like Proposition 1.2.2 for ω-sets.

Next we describe the relation between sets and ω-sets as: **Sets** is a reflective subcategory of ω-**Sets**. Obviously any set X can be turned into an ω-set (X, E) with $E(x) = \mathbb{N}$ for each $x \in X$. One obtains a functor $\nabla\colon \mathbf{Sets} \to \omega$-**Sets** in this way, since for a function $f\colon X \to Y$ any code of a total recursive function can be used to get $f\colon \nabla X \to \nabla Y$ in ω-**Sets**. Thus ∇ is full and faithful.

This functor ∇ turns out to be right adjoint to the forgetful functor

ω-**Sets** $\to$ **Sets**, since there is a bijective correspondence

$$\frac{(X, E) \xrightarrow{\quad f \quad} \nabla Y \quad \text{in } \omega\text{-Sets}}{X \xrightarrow[f]{\quad\quad} Y \quad \text{in Sets}}$$

One uses the fact that the code of f in ω-**Sets** is irrelevant in this case. Thus **Sets** is a reflective subcategory of ω-**Sets**, in a situation,

$$\textbf{Sets} \underset{\nabla}{\overset{\text{forget}}{\rightleftarrows}} \omega\text{-}\textbf{Sets} \qquad \text{with} \qquad \text{forget} \dashv \nabla.$$

full and faithful

Later in Section 6.2 we shall see that the categories of **Sets** and ω-**Sets** can be described as the categories of sheaves and separated objects for the double negation nucleus in the effective topos **Eff**. It explains the reflection **Sets** $\leftrightarrows$ ω-**Sets**.

Partial equivalence relations

Next we introduce the category **PER** of partial equivalence relations (on the natural numbers) and show how it forms a reflective subcategory of the above category ω-**Sets**. PERs were first introduced in [302], and have since then been used extensively in the semantics of various type theories, see *e.g.* [143, 41, 199, 81, 31, 26, 197], or [33] for a recent reference—where categories of PERs are identified within exact completions—with many pointers to the literature.

1.2.5. Definition. A **partial equivalence relation** (abbreviated as 'PER') on $\mathbb{N}$ is a subset $R \subseteq \mathbb{N} \times \mathbb{N}$ which, as a relation, is symmetric and transitive. For such a PER R one writes

$$|R| = \{n \in \mathbb{N} \mid nRn\} \qquad \text{for \textbf{domain}}$$
$$[n] = [n]_R = \{m \in \mathbb{N} \mid mRn\}$$
$$\mathbb{N}/R = \{[n] \mid n \in |R|\} \qquad \text{for \textbf{quotient}}$$
$$\text{PER} = \{R \subseteq \mathbb{N} \times \mathbb{N} \mid R \text{ is a PER}\}.$$

Notice that a PER R is an equivalence relation on its domain $|R|$, so formally we should write $|R|/R$ instead of $\mathbb{N}/R$ for the quotient. But the latter notation is clearer. Every equivalence relation S on a subset of $\mathbb{N}$ forms a PER $S \subseteq \mathbb{N} \times \mathbb{N}$, see Exercise 1.2.5.

It is easy to see that PERs are closed under arbitrary intersections. Hence, ordered by inclusion, they form a complete lattice with joins

$$\bigvee_i S_i = \bigcap \{R \in \text{PER} \mid R \supseteq \bigcup_i S_i\}.$$

A category of PERs is formed with

objects $R \in \text{PER}$.

morphisms $R \to S$ are functions $f: \mathbb{N}/R \to \mathbb{N}/S$ between the quotient sets, which are **tracked** (or, **have a code**). That is, for some code $e \in \mathbb{N}$, one has

$$\forall n \in |R|. \, f([n]_R) = [e \cdot n]_S.$$

We shall write **PER** to denote this category.

1.2.6. Proposition. *The category* **PER** *has finite limits and exponents.*

Proof. As terminal PER one can take $\{(0,0)\}$, or $\mathbb{N} \times \mathbb{N}$. For the product of R and S one can use the relation

$$R \times S = \{(n,m) \mid \mathsf{p}nR\mathsf{p}m \text{ and } \mathsf{p}'nS\mathsf{p}'m\}$$

where p, p' are the recursive projection functions associated with the effective pairing $\langle -, - \rangle: \mathbb{N} \times \mathbb{N} \overset{\cong}{\to} \mathbb{N}$. The equaliser of $f, g: R \rightrightarrows S$ is $R' \rightarrowtail R$ where

$$R' = \{(n,m) \in R \mid f([n]) = g([m])\}.$$

And the exponent of R, S is

$$R \Rightarrow S = \{(n,n') \mid \forall m, m' \in \mathbb{N}. \, mRm' \Rightarrow n \cdot mSn' \cdot m'\}. \qquad \square$$

Since the category **PER** has finite limits, we have a codomain fibration $\begin{smallmatrix} \mathbf{PER}^{\rightarrow} \\ \downarrow \\ \mathbf{PER} \end{smallmatrix}$ of PER-indexed PERs in display style (like in (b) in the beginning of the previous section). As for sets and for ω-sets, there is also pointwise indexing as in (a) for PERs, see Proposition 1.5.3.

An important point to note at this stage is that the category **PER** is a full subcategory of the category ω-**Sets** of ω-sets introduced earlier in this section. The inclusion **PER** $\hookrightarrow \omega$-**Sets** is given by

$$R \mapsto (\mathbb{N}/R, \in)$$

where $\in$ is the existence predicate $\in ([n]) = [n] = \{m \in \mathbb{N} \mid nRm\}$. Indeed a morphism $f: (\mathbb{N}/R, \in) \to (\mathbb{N}/S, \in)$ in ω-**Sets** is a function $f: \mathbb{N}/R \to \mathbb{N}/S$ for which there is a code $e \in \mathbb{N}$ such that

$$\forall [n] \in \mathbb{N}/R. \, \forall m \in [n]. \, e \cdot m \in f([m])$$

But this is equivalent to

$$\forall m \in |R|.\, [e \cdot m] = f([m])$$

and this precisely says that f is a morphism $R \to S$ in **PER**, tracked by e. Thus we have a full and faithful functor **PER** $\hookrightarrow$ ω-**Sets**, which is the identity on morphisms.

One can show that this functor **PER** $\hookrightarrow$ ω-**Sets** preserves finite limits and exponents. This is left to the reader. What we will describe is a left adjoint $\mathbf{r}(-)\colon$ ω-**Sets** $\to$ **PER** to this inclusion, which is obtained by forcing the existence sets $E(x) \subseteq \mathbb{N}$ to be disjoint, see explicitly in Exercise 1.2.9. For an ω-set (X, E) with elements $x, x' \in X$, put

$$x \smile x' \;\Leftrightarrow\; E(x) \cap E(x') \neq \emptyset$$

and write $\sim$ for the transitive closure of $\smile$ in X. Define

$$\mathbf{r}(X, E) = \{(m, m') \mid \exists x, x' \in X.\, m \in E(x) \text{ and } m' \in E(x') \text{ and } x \sim x'\}.$$

There is then a bijective correspondence in

$$\frac{\mathbf{r}(X, E) \xrightarrow{\;f\;} R \quad \text{in } \mathbf{PER}}{(X, E) \xrightarrow{\;g\;} (\mathbb{N}/R, \in) \quad \text{in } \omega\text{-}\mathbf{Sets}}$$

given as follows.

- Assume $f\colon \mathbf{r}(X, E) \to R$ in **PER**, say tracked by e. Define a transpose $f^{\vee}\colon (X, E) \to (\mathbb{N}/R, \in)$ in ω-**Sets** by

$$f^{\vee}(x) \;=\; f([m]), \quad \begin{array}{l} \text{where } m \in E(x) \text{ is arbitrary} \\ (\text{recall } E(y) \neq \emptyset \text{ for each } y \in X) \\ \text{and where } [m] \text{ is the class of } m \text{ in } \mathbf{r}(X, E). \end{array}$$

  Then e also tracks $f^{\vee}$ in ω-**Sets**.
- Conversely, given $g\colon (X, E) \to (\mathbb{N}/R, \in)$ in ω-**Sets**, say tracked by d, then one easily checks that g is constant on $\sim$-equivalence classes, *i.e.* that:

$$x \sim x' \;\Rightarrow\; g(x) = g(x').$$

But then we may define a transpose $g^{\wedge}\colon \mathbf{r}(X, E) \to R$ in **PER** by

$$g^{\wedge}([m]) \;=\; g(x), \quad \text{where } m \in E(x).$$

This yields a well-defined function, which is tracked in **PER** by d.

It is easy to see that the passages $f \mapsto f^{\vee}$ and $g \mapsto g^{\wedge}$ are each others inverses. Thus, also **PER** is a reflective subcategory of ω-**Sets**:

$$\textbf{PER} \quad \xleftrightarrow[\mathbb{N}/(-)]{\textbf{r}} \quad \omega\textbf{-Sets} \qquad \text{with} \qquad \textbf{r} \dashv \mathbb{N}/(-).$$

full and faithful

The relations that we have established between sets, ω-sets and PERs are summarised in the following result.

1.2.7. Proposition. *There is a diagram of functors,*

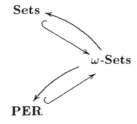

in which the $\hookrightarrow$'s are full and faithful functors (preserving finite limits and exponents), with the arrows in opposite direction as left adjoint. Thus both **Sets** *and* **PER** *are reflective subcategories of ω-**Sets**.* $\square$

Exercises

1.2.1. Prove that a morphism $(u, (f_i)_{i \in I})$ in Fam($\mathbb{C}$) is Cartesian if and only if each f_i is an isomorphism in $\mathbb{C}$.

1.2.2. Consider a map $f = (f_i: X_i \to Y_i)_{i \in I}$ in the fibre Fam($\mathbb{C}$)$_I = \mathbb{C}^I$ over $I \in$ **Sets**. Prove that f is a mono in this fibre if and only if each f_i is a mono in $\mathbb{C}$.

1.2.3. For an arbitrary category $\mathbb{B}$, let $\mathbb{B}^{\rightarrow}_{\bullet}$ be the category with **pointed families** as objects; these are pairs $\langle \left(\begin{smallmatrix} X \\ \downarrow \varphi \\ I \end{smallmatrix} \right), s \rangle$ where s is a section of φ (*i.e.* a map $s: I \to X$ with $\varphi \circ s = $ id). A morphism $\langle \left(\begin{smallmatrix} X \\ \downarrow \varphi \\ I \end{smallmatrix} \right), s \rangle \longrightarrow \langle \left(\begin{smallmatrix} Y \\ \downarrow \psi \\ J \end{smallmatrix} \right), t \rangle$ in $\mathbb{B}^{\rightarrow}_{\bullet}$ consists of a pair of morphisms $u: I \to J$, $f: X \to Y$ in $\mathbb{B}$ with $\psi \circ f = u \circ \varphi$ and also $f \circ s = t \circ u$. Thus morphisms of pointed families preserve the points (*i.e.* sections) of the families. Prove that

(i) if the category $\mathbb{B}$ has pullbacks then the functor $\mathbb{B}_\bullet^\to \to \mathbb{B}$ sending

$$\left\langle \begin{pmatrix} X \\ \downarrow\varphi \\ I \end{pmatrix}, s \right\rangle \text{ to the index object } I \text{ is a fibration;}$$

(ii) for $\mathbb{B} = \mathbf{Sets}$, there is an equivalence of categories $\mathrm{Fam}(\mathbf{Sets}_\bullet) \xrightarrow{\approx} \mathbf{Sets}_\bullet^\to$ like in Proposition 1.2.2, where $\mathbf{Sets}_\bullet$ is the category of **pointed** sets: objects are sets containing a distinguished base point, morphisms are functions preserving such points.

1.2.4. Check that for a PER R one has $R \subseteq |R|^2$.

1.2.5. Let I be a set. A **partition** of I is a collection $Q \subseteq P(I)$ of subsets of I satisfying (1) every set in Q is non-empty (2) if $a, b \in Q$ and $a \cap b \neq \emptyset$, then $a = b$ (3) $\bigcup Q = I$. A **partial partition** of I is a subset $Q \subseteq P(I)$ satisfying (1) and (2) but not necessarily (3). Show that
   (i) there is a bijective correspondence between partitions and equivalence relations and between partial partitions and partial equivalence relations (on I);
   (ii) there is a bijective correspondence between partial equivalence relations on I and equivalence relations on subsets of I.

1.2.6. Notice that for $R \in \mathbf{PER}$, the "global sections" or "global elements" homset $\mathbf{PER}(1, R)$ is isomorphic to the quotient $\mathbb{N}/R$. And also that all homsets in $\mathbf{PER}$ and in ω-$\mathbf{Sets}$ are countable.

1.2.7. (i) Prove that for each ω-set (I, E) the slice category ω-$\mathbf{Sets}/(I, E)$ is Cartesian closed, *i.e.* that ω-$\mathbf{Sets}$ is a locally Cartesian closed category (LCCC).
   (ii) Show that also $\mathbf{PER}$ is an LCCC.

1.2.8. Show that a map $\nabla X \to (\mathbb{N}/R, \in)$ in ω-$\mathbf{Sets}$ is constant (where X is a set and R is a PER).

1.2.9. (i) Prove that the unit $\eta_{(X,E)}$ of the reflection $\mathbf{PER} \leftrightarrows \omega$-$\mathbf{Sets}$ at $(X, E) \in$ ω-$\mathbf{Sets}$ is an isomorphism if and only if the existence predicate $E \colon X \to P\mathbb{N}$ has disjoint images (*i.e.* satisfies $E(x) \cap E(y) \neq \emptyset \Rightarrow x = y$).
   Conclude that $\mathbf{PER}$ is equivalent to the full subcategory of ω-$\mathbf{Sets}$ on these objects with such disjoint images. These ω-sets are also called **modest sets** (after D. Scott). In this situation the existence predicate $E \colon X \to P\mathbb{N}$ may equivalently be described via a surjective function $U \twoheadrightarrow X$, where $U \subseteq \mathbb{N}$ (*i.e.* via a subquotient of $\mathbb{N}$), see *e.g.* [143, Definition 1.1].
   (ii) In view of (i), describe the reflector $\mathbf{r} \colon \omega$-$\mathbf{Sets} \to \mathbf{PER}$ as 'forcing images to be disjoint', by taking a suitable quotient

$$(X \to P\mathbb{N}) \mapsto (X/\sim \to P\mathbb{N}).$$

1.2.10. (i) Let $\mathrm{Eq}(\mathbb{N}) = \{(n, n) \mid n \in \mathbb{N}\} \subseteq \mathbb{N} \times \mathbb{N}$ be the 'diagonal' PER. Show that it is a natural numbers object (NNO) in $\mathbf{PER}$. Also that the resulting ω-set $N = (\mathbb{N}, \in)$ with $\in(n) = \{n\}$ is NNO in ω-$\mathbf{Sets}$.
   (ii) Check that the maps $\mathrm{Eq}(\mathbb{N}) \to \mathrm{Eq}(\mathbb{N})$ in $\mathbf{PER}$, *i.e.* the maps $N \to N$ in ω-$\mathbf{Sets}$, can be identified with the (total) recursive functions $\mathbb{N} \to \mathbb{N}$.

1.2.11. Show that the category ω-**Sets** has finite colimits. And conclude, from the reflection **PER** $\leftrightarrows$ ω-**Sets** that **PER** also has finite colimits.

1.2.12. Prove that the reflector $\mathbf{r} \colon \omega$-**Sets** $\to$ **PER** preserves finite products, but does *not* preserve equalisers.

[*Hint.* For a counter example, consider in **Sets** on the two-element set $2 = \{0, 1\}$ the identity and twist maps $\mathrm{id}, \neg \colon 2 \rightrightarrows 2$, with empty set $\emptyset \rightarrowtail 2$ as equaliser. By applying $\nabla \colon$ **Sets** $\to$ ω-**Sets** we get an equaliser diagram in ω-**Sets** (since ∇ is right adjoint). But it is not preserved by the reflector $\mathbf{r}$, since $\mathbf{r}(\nabla 2)$ is terminal, and $\mathbf{r}(\nabla \emptyset)$ is initial.]

1.2.13. (i) Prove that $\mathrm{Fam}(-) \colon$ **Cat** $\to$ **Cat** is a (2-)functor. (One has to ignore aspects of size here, because categories $\mathrm{Fam}(\mathbb{C})$ are not small; for example, $\mathrm{Fam}(\mathbf{1})$ is isomorphic to **Sets**.)

(ii) Show that $\mathrm{Fam}(\mathbb{C})$ is the free completion of $\mathbb{C}$ with respect to set-indexed coproducts. This means that $\mathrm{Fam}(\mathbb{C})$ has set-indexed coproducts and that there is a unit $\mathbb{C} \to \mathrm{Fam}(\mathbb{C})$ which is universal among functors from $\mathbb{C}$ to categories $\mathbb{D}$ with set-indexed coproducts $\coprod_{i \in I} X_i$.

(iii) Prove that a category $\mathbb{C}$ has arbitrary coproducts if and only if the unit $\mathbb{C} \to \mathrm{Fam}(\mathbb{C})$ has a left adjoint.

[The $\mathrm{Fam}(-)$ operation forms a so-called 'KZ-doctrine', see [180].]

1.3 Some general examples

So far we have seen codomain fibrations $\begin{smallmatrix} \mathbb{B}^{\to} \\ \downarrow \\ \mathbb{B} \end{smallmatrix}$ for categories $\mathbb{B}$ with pullbacks (in Proposition 1.1.6), and family fibrations $\begin{smallmatrix} \mathrm{Fam}(\mathbb{C}) \\ \downarrow \\ \mathbf{Sets} \end{smallmatrix}$ (in Definition 1.2.1) for arbitrary categories $\mathbb{C}$. In this section we shall introduce 'simple fibrations' $\begin{smallmatrix} s(\mathbb{B}) \\ \downarrow \\ \mathbb{B} \end{smallmatrix}$, for categories $\mathbb{B}$ with Cartesian products, and fibrations $\begin{smallmatrix} \mathrm{Sub}(\mathbb{B}) \\ \downarrow \\ \mathbb{B} \end{smallmatrix}$ and $\begin{smallmatrix} \mathrm{Rel}(\mathbb{B}) \\ \downarrow \\ \mathbb{B} \end{smallmatrix}$ of subobjects and relations (for categories $\mathbb{B}$ with pullbacks).

Simple fibrations

This first construction will be of central importance in the next chapter on simple type theory. Let $\mathbb{B}$ be an arbitrary category with Cartesian products $\times$. We write $s(\mathbb{B})$ for the category having

objects   pairs (I, X) of objects of $\mathbb{B}$.

morphisms $(I, X) \to (J, Y)$ are pairs of morphisms (u, f) in $\mathbb{B}$ with $u \colon I \to J$ and $f \colon I \times X \to Y$.

The composite of $(I, X) \xrightarrow{(u,f)} (J, Y) \xrightarrow{(v,g)} (K, Z)$ is $(v \circ u,\ g \circ \langle u \circ \pi, f \rangle)$, where the second component is obtained as composite

$$I \times X \xrightarrow{\langle u \circ \pi, f \rangle} J \times Y \xrightarrow{g} Z$$

And the identity on (I, X) is the pair (id_I, π') with π' the second projection $I \times X \to X$. There is then an obvious projection functor $\mathrm{s}(\mathbb{B}) \to \mathbb{B}$ given by

$$(I, X) \mapsto I \qquad \text{and} \qquad (u, f) \mapsto u.$$

Intuitively, maps $f: X \to Y$ in the fibre $\mathrm{s}(\mathbb{B})_I$ over $I \in \mathbb{B}$ are I-indexed families $f_i: X \to Y$, for $i \in I$, where the objects are kept fixed. Remember the family fibration from the previous section where we had maps $f_i: X_i \to Y_i$ over I.

The above functor will be written as $\mathrm{s}_\mathbb{B}: \mathrm{s}(\mathbb{B}) \to \mathbb{B}$ and called the **simple fibration** on $\mathbb{B}$. It is a fibration indeed, since for $(J, Y) \in \mathrm{s}(\mathbb{B})$ and $u: I \to J$ in $\mathbb{B}$ one finds a Cartesian lifting of u as:

$$(I, Y) \xdashrightarrow{(u, \pi')} (J, Y)$$

$$I \xrightarrow{u} J$$

1.3.1. Definition. For a category $\mathbb{B}$ with Cartesian products, the **simple fibration** on $\mathbb{B}$ is the above projection functor $\begin{smallmatrix} \mathrm{s}(\mathbb{B}) \\ \downarrow \\ \mathbb{B} \end{smallmatrix}$.

The fibre $\mathrm{s}(\mathbb{B})_I$ over $I \in \mathbb{B}$ will often be written as $\mathbb{B}/\!\!/ I$ and called the **simple slice** over I. (Its objects are $X \in \mathbb{B}$ and its maps $X \to Y$ are $I \times X \to Y$ in $\mathbb{B}$.)

Notice that all these simple slices $\mathbb{B}/\!\!/ I$ have the same objects, namely the objects from $\mathbb{B}$. There is an obvious functor $I^*: \mathbb{B} \to \mathbb{B}/\!\!/ I$ by $X \mapsto X$ and $f \mapsto f \circ \pi'$. There is a similar functor $I^*: \mathbb{B} \to \mathbb{B}/I$ given by $X \mapsto \left(\begin{smallmatrix} I \times X \\ \downarrow \pi \\ I \end{smallmatrix} \right)$, and $f \mapsto \mathrm{id}_I \times f$, as used earlier in Section 1. We write I^* for both these functors $\mathbb{B} \to \mathbb{B}/\!\!/ I$ and $\mathbb{B} \to \mathbb{B}/I$. These simple and ordinary slices have much in common (see Exercises 1.3.2 – 1.3.4 below, and Corollary 1.10.16). For example, if $\mathbb{B}$ additionally has a terminal object 1, then for both the simple and the ordinary slice there are isomorphisms of categories

$$\mathbb{B} \xrightarrow[\cong]{1^*} \mathbb{B}/\!\!/ 1 \qquad \text{and} \qquad \mathbb{B} \xrightarrow[\cong]{1^*} \mathbb{B}/1$$

It is useful to present an immediate generalisation of this 'simple' construction. It is based on the following notion from [156].

1.3.2. Definition. (i) A **CT-structure** is a pair $(\mathbb{B}, T)$ where $\mathbb{B}$ is a category with finite products and T is a non-empty collection of objects from $\mathbb{B}$. Such a CT-structure will be called **non-trivial** if there is at least one object $X \in T$ equipped with an arrow $1 \to X$ from the terminal object to X.

(ii) A **morphism of CT-structures** from $(\mathbb{B}, T)$ to $(\mathbb{A}, S)$ is a finite product preserving functor $K \colon \mathbb{B} \to \mathbb{A}$ which satisfies $K[T] \subseteq S$ (*i.e.* $X \in T$ implies $KX \in S$).

This condition of non-triviality for CT-structures usually expresses that some domain is non-empty and can be seen as a non-degeneracy condition. The 'C' and the 'T' in 'CT-structure' stand for 'context' and 'type'. As will be explained in the next chapter, in such a CT-structure $(\mathbb{B}, T)$ one can view $\mathbb{B}$ as a category of contexts and T as a collection of types; the inclusion $T \subseteq \mathrm{Obj}\,\mathbb{B}$ can then be seen as identification of a type σ with the corresponding singleton context $(x \colon \sigma)$. The two extreme cases are T is $\mathrm{Obj}\,\mathbb{B}$ and T is a singleton.

An example of a CT-structure is $\mathbb{B} = \omega\text{-}\mathbf{Sets}$ and $T = $ objects of the form ∇X (where X is a set).

1.3.3. Definition. Suppose $(\mathbb{B}, T)$ is a CT-structure. Let $\mathrm{s}(T)$ be the category with

objects	pairs (I, X) with $I \in \mathbb{B}$ and $X \in T$.
morphisms	$(I, X) \to (J, Y)$ are pairs (u, f) in $\mathbb{B}$ with $u \colon I \to J$ and $f \colon I \times X \to Y$.

This generalises the earlier definition of $\mathrm{s}(\mathbb{B})$ by restricting the second component of objects to the types T.

As before, one obtains a fibration $\begin{smallmatrix} \mathrm{s}(T) \\ \downarrow s_T \\ \mathbb{B} \end{smallmatrix}$. It will be called the **simple fibration** associated with the CT-structure $(\mathbb{B}, T)$.

Notice that the original construction $s_\mathbb{B} \colon \mathrm{s}(\mathbb{B}) \to \mathbb{B}$ is the special case $s_T \colon \mathrm{s}(T) \to \mathbb{B}$ where T consists of all objects of $\mathbb{B}$. The other extreme is where T is a singleton, say $T = \{\Omega\}$. We then omit the curly braces $\{-\}$ and write $s_\Omega \colon \mathrm{s}(\Omega) \to \mathbb{B}$ for the resulting simple fibration. CT-structures with one type (*i.e.* of the form $(\mathbb{B}, \Omega)$) will be used for the semantics of the *untyped* lambda calculus—because 'untyped' is the same as 'typed with a single type', see Section 2.5.

The generalised "CT" version of a simple fibration involves a restriction to a subset of the *objects*. A similar generalisation exists for codomain fibrations, involving a restriction to a subset of the *arrows*. This leads to the notion of a display map category and—in a further generalisation—to the notion of a comprehension category. The details are in Chapter 10 on dependent type theory, especially in Section 10.4.

Monos and subobjects

In the codomain fibration $\begin{smallmatrix} \mathbb{B}^{\to} \\ \downarrow \\ \mathbb{B} \end{smallmatrix}$ we took all maps of $\mathbb{B}$ as families. An obvious restriction is to consider monic maps $X \rightarrowtail I$ only. In the case $\mathbb{B} = \mathbf{Sets}$, a fibre of such a monic (injective) map can have at most one element, so it is either empty or a singleton. We write $\mathrm{Mono}(\mathbb{B})$ for the full subcategory of $\mathbb{B}^{\to}$ consisting of monic families. If the category $\mathbb{B}$ has pullbacks, then the (restricted) codomain functor $\begin{smallmatrix} \mathrm{Mono}(\mathbb{B}) \\ \downarrow \\ \mathbb{B} \end{smallmatrix}$ is a fibration; it will be called the **fibration of monos** (of $\mathbb{B}$). This functor is a fibration because a pullback of a mono along an arbitrary map is a mono again. Notice that the fibres of this fibration $\begin{smallmatrix} \mathrm{Mono}(\mathbb{B}) \\ \downarrow \\ \mathbb{B} \end{smallmatrix}$ are all preordered categories. Such a fibration will be called **preordered**, or a **fibred preorder**. But notice that the total category $\mathrm{Mono}(\mathbb{B})$ itself, is not a preorder.

The preorder $\sqsubseteq$ of monos, say in the fibre over I, is given as follows. For $X \overset{m}{\rightarrowtail} I$ and $Y \overset{n}{\rightarrowtail} I$ one has $m \sqsubseteq n$ if and only if there is a (necessarily unique, monic) map $f \colon X \to Y$ with $n \circ f = m$. One can then form the "poset reflection" of this preorder $\sqsubseteq$ on the monos over I. It yields a poset with equivalence classes of monos as elements (where $m \sim n$ if and only if both $m \sqsubseteq n$ and $n \sqsubseteq m$, if and only if there is an isomorphism $\varphi \colon X \overset{\cong}{\to} Y$ with $n \circ \varphi = m$). These equivalence classes are called subobjects (of I); the resulting poset will be written as $\mathrm{Sub}(I)$.

Usually one does not distinguish notationally between a mono and the corresponding subobject. We write $\mathrm{Sub}(\mathbb{B})$ for the category obtained from $\mathrm{Mono}(\mathbb{B})$ by taking subobjects as objects. One gets the **fibration** $\begin{smallmatrix} \mathrm{Sub}(\mathbb{B}) \\ \downarrow \\ \mathbb{B} \end{smallmatrix}$ **of subobjects** in $\mathbb{B}$. The fibres $\mathrm{Sub}(I)$ are partial orders. For $\mathbb{B} = \mathbf{Sets}$ the subobject fibration $\begin{smallmatrix} \mathrm{Sub}(\mathbf{Sets}) \\ \downarrow \\ \mathbf{Sets} \end{smallmatrix}$ was written as $\begin{smallmatrix} \mathrm{Pred} \\ \downarrow \\ \mathbf{Sets} \end{smallmatrix}$ in the Prospectus. For a specific set $I \in \mathbf{Sets}$, the fibre $\mathrm{Sub}(I)$ above I is the (partially ordered) category $(\mathcal{P}I, \subseteq)$ of subsets of I.

1.3.4. Remark. At this stage we have already seen the three fibrations that will play a crucial rôle in this book. They are the simple fibration $\begin{smallmatrix} s(\mathbb{B}) \\ \downarrow \\ \mathbb{B} \end{smallmatrix}$, the codomain fibration $\begin{smallmatrix} \mathbb{B}^{\to} \\ \downarrow \\ \mathbb{B} \end{smallmatrix}$ and the subobject fibration $\begin{smallmatrix} \mathrm{Sub}(\mathbb{B}) \\ \downarrow \\ \mathbb{B} \end{smallmatrix}$. The last fibration will be used to describe the so-called internal (predicate) logic of $\mathbb{B}$;

this will become clear in Chapters 3 and 4. The first two will be used in the categorical description of type theories. The simple fibration will be used for simple type theory and the codomain fibration for dependent type theory. We therefore often jointly refer to these two fibrations as the **type theoretic fibrations**.

Relations

A (binary) **relation** on an object I in a category $\mathbb{B}$ with finite limits is a subobject $R \rightarrowtail I \times I$. The category $\text{Rel}(\mathbb{B})$ has such relations as objects; a morphism from $R \rightarrowtail I \times I$ to $S \rightarrowtail J \times J$ in $\text{Rel}(\mathbb{B})$ is a map $u : I \to J$ in $\mathbb{B}$ giving rise to a commuting diagram

$$
\begin{array}{ccc}
R & \dashrightarrow & S \\
\downarrow & & \downarrow \\
I \times I & \xrightarrow{\ u \times u\ } & J \times J
\end{array}
$$

Notice that there is no need to mention the (name of the) top dashed arrow, because there can be only one such map. Set-theoretically the diagram expresses that iRi' implies $u(i)Su(i')$. The functor $\text{Rel}(\mathbb{B}) \to \mathbb{B}$ sending a relation $R \rightarrowtail I \times I$ to its carrier I is then a fibration—again, since monos are stable under pullback.

Often we are interested in special relations. Categorically, a relation $\langle r_1, r_2 \rangle : R \rightarrowtail I \times I$ is called

(i) **reflexive** if the diagonal $\delta_I = \langle \text{id}, \text{id} \rangle : I \rightarrowtail I \times I$ factors through $R \rightarrowtail I \times I$, *i.e.* if there is a map

(ii) **symmetric** if there is a 'swap' map

(iii) **transitive** if, after forming the pullback T of triples (in which both the first two and the last two components are related by R) as on the left

where $t = \langle r_1 \circ r_{12}, r_2 \circ r_{23} \rangle : T \to I \times I$. It is not hard to see that in **Sets** these definitions coincide with the usual formulations.

Then, a relation $R \rightarrowtail I \times I$ is called an **equivalence relation** if it is reflexive, symmetric and transitive. It is a **partial equivalence relation** if it is symmetric and transitive, but not necessarily reflexive. One obtains corresponding fibrations $\begin{smallmatrix} \mathrm{ERel}(\mathbb{B}) \\ \downarrow \\ \mathbb{B} \end{smallmatrix}$ and $\begin{smallmatrix} \mathrm{Per}(\mathbb{B}) \\ \downarrow \\ \mathbb{B} \end{smallmatrix}$.

Exercises

1.3.1. (i) Show that in the total category $s(\mathbb{B})$ of a simple fibration $\begin{smallmatrix} s(\mathbb{B}) \\ \downarrow \\ \mathbb{B} \end{smallmatrix}$ a morphism $(u, f) : (I, X) \to (J, Y)$ is Cartesian if and only if there is an isomorphism $h : I \times X \overset{\cong}{\to} I \times Y$ in $\mathbb{B}$ such that $\pi \circ h = \pi$ and $\pi' \circ h = f$.

    (ii) Show that the assignment $(I, X) \mapsto I^*(X) = \begin{pmatrix} I \times X \\ \downarrow \pi \\ I \end{pmatrix}$ extends to a full and faithful functor $s(\mathbb{B}) \to \mathbb{B}^\to$. Prove that it maps Cartesian morphisms to pullback squares.

    [This functor restricts to a full and faithful functor $\mathbb{B}/\!\!/ I \to \mathbb{B}/I$.]

1.3.2. Consider a simple fibration $\begin{smallmatrix} s(\mathbb{B}) \\ \downarrow \\ \mathbb{B} \end{smallmatrix}$ for a category $\mathbb{B}$ with finite products $(1, \times)$. Prove that

    (i) each fibre $\mathbb{B}/\!\!/ I$ has finite products, and $I^* : \mathbb{B} \to \mathbb{B}/\!\!/ I$ preserves these products;

    (ii) the following are equivalent:
        (a) $\mathbb{B}$ is Cartesian closed;
        (b) each fibre $\mathbb{B}/\!\!/ I$ is Cartesian closed;
        (c) each functor $I^* : \mathbb{B} \to \mathbb{B}/\!\!/ I$ has a right adjoint $I \Rightarrow (-)$.

1.3.3. In case a category $\mathbb{B}$ has finite limits (*i.e.* additionally has equalisers with respect to the previous exercise), prove that $\mathbb{B}$ is Cartesian closed if and only if for each $I \in \mathbb{B}$, the functor $I^* : \mathbb{B} \to \mathbb{B}/I$ (to the ordinary slice) mapping X to $I^*(X) = \begin{pmatrix} I \times X \\ \downarrow \pi \\ I \end{pmatrix}$, has a right adjoint $\prod_I$.

[*Hint.* One obtains a right adjoint $\prod_I$ by mapping a family $\begin{pmatrix} X \\ \downarrow \varphi \\ I \end{pmatrix}$ to the domain of the equaliser e in

$$
\prod_I(\varphi) \xmapsto{\quad e \quad} (I \Rightarrow X) \underset{\Lambda(\pi)}{\overset{I \Rightarrow \varphi}{\rightrightarrows}} (I \Rightarrow I).
$$
]

1.3.4.   Let $\mathbb{B}$ be a category with finite products and I an object in $\mathbb{B}$.

(i)   Show that the functor $I \times (-) \colon \mathbb{B} \to \mathbb{B}$ forms a comonad on $\mathbb{B}$.

(ii)   Show that the simple slice $\mathbb{B} /\!/ I$ is the Kleisli category of this comonad $I \times (-)$ and that the ordinary slice $\mathbb{B}/I$ is its Eilenberg-Moore category.

1.3.5.   Let $\mathbb{B}$ have finite limits. Prove that a map of families $\begin{pmatrix} X \\ \downarrow \\ I \end{pmatrix} \to \begin{pmatrix} Y \\ \downarrow \\ J \end{pmatrix}$ is a mono in $\mathbb{B}^{\to}$ if and only if both its components $I \to J$ and $X \to Y$ are monos in $\mathbb{B}$.

1.3.6.   A **regular mono** is a mono that occurs as an equaliser. Write $\mathrm{RegSub}(\mathbb{B})$ for the full subcategory $\mathrm{Sub}(\mathbb{B})$ consisting of (equivalence classes of) regular monos. Show that the codomain functor $\begin{array}{c} \mathrm{RegSub}(\mathbb{B}(\\ \downarrow \\ \mathbb{B} \end{array}$ is a fibration.

1.3.7.   Let $\mathbb{B}$ be a category with finite limits.

(i)   Show that if $\mathbb{B}$ is a CCC then also $\mathbb{B}^{\to}$ is a CCC and $\begin{array}{c} \mathbb{B}^{\to} \\ \downarrow \\ \mathbb{B} \end{array}$ is a functor which strictly preserves the CCC-structure.

(ii)   Show that the same holds for $\mathrm{Sub}(\mathbb{B})$ instead of $\mathbb{B}^{\to}$.

[*Hint.* For families $\begin{pmatrix} X \\ \downarrow \varphi \\ I \end{pmatrix}$ and $\begin{pmatrix} Y \\ \downarrow \psi \\ J \end{pmatrix}$ construct the exponent family $\varphi \Rightarrow \psi$ over the exponent object $I \Rightarrow J$ in $\mathbb{B}$ as in the following pullback diagram.

$$
\begin{array}{ccc}
U & \longrightarrow & (X \Rightarrow Y) \\
{\scriptstyle \varphi \Rightarrow \psi} \downarrow\; \llcorner & & \downarrow {\scriptstyle X \Rightarrow \psi} \\
(I \Rightarrow J) & \underset{\varphi \Rightarrow J}{\longrightarrow} & (X \Rightarrow J)
\end{array}
$$
]

1.3.8.   Give a categorical formulation of anti-symmetry of a relation $R \rightarrowtail I \times I$.

1.3.9. Verify in detail that the following functors are fibrations.

$$\mathrm{Sub}(\mathbb{B}) \qquad \mathrm{Rel}(\mathbb{B}) \qquad \mathrm{Per}(\mathbb{B}) \qquad \mathrm{ERel}(\mathbb{B})$$
$$\downarrow \qquad\qquad \downarrow \qquad\qquad \downarrow \qquad\qquad \downarrow$$
$$\mathbb{B} \qquad\qquad \mathbb{B} \qquad\qquad \mathbb{B} \qquad\qquad \mathbb{B}$$

1.3.10. Define an alternative category of relations $R \rightarrowtail I \times J$ on two possibly different objects in a category $\mathbb{B}$, which is fibred over $\mathbb{B} \times \mathbb{B}$.

1.3.11. Let $\begin{smallmatrix}\mathbb{E}\\\downarrow p\\\mathbb{B}\end{smallmatrix}$ be a fibration. Prove that p is preordered (*i.e.* all its fibre categories are preorders) if and only if above each map $pX \to pY$ in $\mathbb{B}$ there is at most one arrow $X \to Y$ in $\mathbb{E}$ (i.e. if p is faithful). Conclude that in the total category of a preorder fibration, a vertical morphism is monic.

1.4 Cloven and split fibrations

The definition of a fibration is of the form "for every x and y there is a z such that ...". This does not imply that we are given for each pair x, y an explicit z, unless we make use of the Axiom of Choice. The differences in the way the structure of a fibration may be given will concern us in this section. Briefly, a fibration is called **cloven** if it comes together with a choice of Cartesian liftings; and it is called **split** if it is cloven and the given liftings are well-behaved in the sense that they satisfy certain functoriality conditions. These fibrations behave more pleasantly, and therefore we prefer to work with fibrations in split form (if this is possible). Cloven and split fibrations give rise to so-called indexed categories $\mathbb{B}^{\mathrm{op}} \to \mathbf{Cat}$. These generalise set-valued functors (or presheaves) $\mathbb{B}^{\mathrm{op}} \to \mathbf{Sets}$.

We recall that a functor $\begin{smallmatrix}\mathbb{E}\\\downarrow p\\\mathbb{B}\end{smallmatrix}$ is a fibration if for every map $u \colon I \to J$ in the base category $\mathbb{B}$ and every object $X \in \mathbb{E}$ above J in the total category, there is a Cartesian lifting $\bullet \to X$ in $\mathbb{E}$. Assume now we *choose* for every such u and X a specific Cartesian lifting and write it as

$$u^*(X) \xrightarrow{\quad \overline{u}(X) \quad} X$$

(By Proposition 1.1.4 we can only choose up-to vertical isomorphisms.)

We claim that, having made such choices, every map $u \colon I \to J$ in $\mathbb{B}$ determines a functor u^* from the fibre $\mathbb{E}_J$ over J to the fibre $\mathbb{E}_I$ over I. (Note the direction!) The recipe for $u^* \colon \mathbb{E}_J \to \mathbb{E}_I$ is as follows.

- for an object $X \in \mathbb{E}_J$ one has $pX = J$ and so we take $u^*(X) \in \mathbb{E}_I$ to be the domain of the previously determined Cartesian lifting $\overline{u}(X) \colon u^*(X) \to X$;

- for a map $f: X \to Y$ in $\mathbb{E}_J$, consider the following diagram in $\mathbb{E}$.

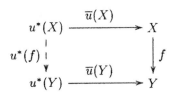

$$I \xrightarrow{\quad u \quad} J$$

The composite $f \circ \overline{u}(X): u^*(X) \to Y$ is above u, since f is vertical. Because $\overline{u}(Y)$ is by definition the *terminal* lifting of u with codomain Y, there is a unique map $u^*(X) \dashrightarrow u^*(Y)$, call it $u^*(f)$, with $\overline{u}(Y) \circ u^*(f) = f \circ \overline{u}(X)$.

By uniqueness, u^* preserves identities and composition. Thus one obtains a functor $u^*: \mathbb{E}_J \to \mathbb{E}_I$. Such functors u^* are known under various names: as **reindexing** functors, **substitution** functors, **relabelling** functors, **inverse image** functors or sometimes also as **change-of-base** or as **pullback functors**. We mostly use the first two names.

1.4.1. Convention. An unlabelled arrow $u^*(X) \to X$ in a diagram is always a (chosen) Cartesian morphism $\overline{u}(X): u^*(X) \to X$ as above. Omitting these labels makes diagrams more readable. Choosing an object $u^*(X)$ will often be called substitution or reindexing (along u).

1.4.2. Example. Assume $\mathbb{B}$ is a category with *chosen* pullbacks and consider the codomain fibration $\begin{smallmatrix} \mathbb{B}^{\to} \\ \downarrow \\ \mathbb{B} \end{smallmatrix}$. Recall that the fibre over $I \in \mathbb{B}$ can be identified with the slice $\mathbb{B}/I$. A morphism $u: I \to J$ induces by the above recipe a substitution functor $u^*: \mathbb{B}/J \to \mathbb{B}/I$ by pullbacks (as described before Example 1.1.1). Usually it is called the pullback functor induced by u. As a special case we have $I^*: \mathbb{B} \cong \mathbb{B}/1 \to \mathbb{B}/I$ resulting from the unique map $!_I: I \to 1$ from I to the terminal object $1 \in \mathbb{B}$. It sends an object $X \in \mathbb{B}$ to the Cartesian projection $\left(\begin{smallmatrix} I \times X \\ \downarrow \\ I \end{smallmatrix} \right)$.

We return to our general fibration $\begin{smallmatrix} \mathbb{E} \\ \downarrow p \\ \mathbb{B} \end{smallmatrix}$ (with chosen liftings). A good question is the following: given two composable morphisms

$$I \xrightarrow{\ u\ } J \xrightarrow{\ v\ } K$$

in $\mathbb{B}$, are the two resulting functors $\mathbb{E}_K \rightrightarrows \mathbb{E}_I$, namely

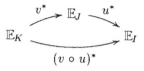

equal? It turns out that in general they are not equal but naturally isomorphic: one gets a unique mediating map as in the diagram on the left below.

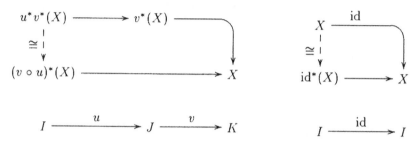

This isomorphism $u^*v^*(X) \cong (v \circ u)^*(X)$ arises because Cartesian morphisms are closed under composition, so that there are two Cartesian liftings of $v \circ u$ at X, as indicated.

There is a similar phenomenon for identities, as in the diagram on the right, since identities are Cartesian. Hence, in general the substitution functor $(\mathrm{id}_I)^*$ induced by the identity on $I \in \mathbb{B}$, is only naturally isomorphic—and not equal—to the identity functor on the fibre category $\mathbb{E}_I$.

Sometimes these morphisms $u^*v^*(X) \xrightarrow{\cong} (v \circ u)^*(X)$ and $X \xrightarrow{\cong} \mathrm{id}^*(X)$ are identities—we then call the fibration **split**—but often this does not happen (*e.g.* in the above pullback example with $\mathbb{B} = \mathbf{Sets}$ and with canonical pullbacks in **Sets**).

(The case of identities is not problematic since we can always choose $(\mathrm{id}_I)^* = \mathrm{id}_{\mathbb{E}_I}$.)

It is not hard to check that the maps determined in the diagrams (for every X) yield *natural* isomorphisms $\mathrm{id} \xRightarrow{\cong} (\mathrm{id})^*$ and $u^*v^* \xRightarrow{\cong} (v \circ u)^*$. Moreover, they satisfy certain coherence conditions, which will be given below.

Thus, when we work with reindexing functors, lots of (coherent) isomorphisms crop up. It is time to sum up the above discussion in a few definitions and results.

1.4.3. Definition. (i) A fibration is called **cloven** if it comes equipped with a **cleavage**; that is, with a choice of Cartesian liftings. This cleavage then induces substitution functors u^* between the fibres.

(ii) A **split** fibration is a cloven fibration for which the induced substitution functors are such that the canonical natural transformations are identities:

$$\text{id} \overset{=}{\Longrightarrow} (\text{id})^* \qquad \text{and} \qquad u^* v^* \overset{=}{\Longrightarrow} (v \circ u)^*.$$

The cleavage involved is then often called a **splitting**.

The family fibration $\begin{smallmatrix} \text{Fam}(\mathbb{C}) \\ \downarrow \\ \textbf{Sets} \end{smallmatrix}$ is an example of a fibration which can be equipped with a splitting. In fact the choice of lifting described in the beginning of Section 1.2 makes the fibration split.

By using a version of the Axiom of Choice of suitable strength in the meta-theory, one can always provide a fibration with a cleavage. We usually indicate explicitly when we do so. Later in Corollary 5.2.5 it will be shown that every fibration is equivalent (in a fibred sense) to a split one. The construction used there involves the fibred Yoneda Lemma. For some codomain examples, like for set-indexed sets in Proposition 1.2.2, and similarly for ω-sets and PERs below, we can give an elementary equivalent split description, see Propositions 1.4.9 and 1.5.3 below.

By choosing substitution functors one obtains from a cloven fibration $\begin{smallmatrix} \mathbb{E} \\ \downarrow \\ \mathbb{B} \end{smallmatrix}$ an assignment $I \mapsto \mathbb{E}_I$ which is almost a functor $\mathbb{B}^{\text{op}} \to \textbf{Cat}$. It yields a so-called 'pseudo-functor'.

1.4.4. Definition. (i) An **indexed category** —or, to be more precise, a $\mathbb{B}$**-indexed category**—is a **pseudo functor** $\Psi \colon \mathbb{B}^{\text{op}} \to \textbf{Cat}$. It consists of a mapping which assigns to each object $I \in \mathbb{B}$ a category $\Psi(I)$ and to each morphism $u \colon I \to J$ a functor $\Psi(u) \colon \Psi(J) \to \Psi(I)$; note the reverse direction. Such a functor $\Psi(u)$ is often simply denoted by u^* when no confusion arises. Additionally, a pseudo functor involves natural isomorphisms

$$\eta_I \colon \text{id} \overset{\cong}{\Longrightarrow} (\text{id}_I)^* \qquad \text{for } I \in \mathbb{B}$$

$$\mu_{u,v} \colon u^* v^* \overset{\cong}{\Longrightarrow} (v \circ u)^* \qquad \text{for } I \overset{u}{\longrightarrow} J \overset{v}{\longrightarrow} K \text{ in } \mathbb{B}$$

which satisfy certain coherence conditions:

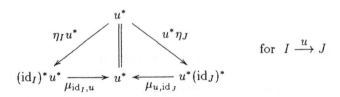

for $I \overset{u}{\longrightarrow} J$

$$
\begin{array}{ccc}
u^* v^* w^* & \xrightarrow{\;u^* \mu_{v,w}\;} & u^*(w \circ v)^* \\[2mm]
{\scriptstyle \mu_{u,v} w^*} \Big\downarrow & & \Big\downarrow {\scriptstyle \mu_{u,w \circ v}} \\[2mm]
(v \circ u)^* w^* & \xrightarrow[\;\mu_{v \circ u,w}\;]{} & (w \circ v \circ u)^*
\end{array}
\qquad \text{for } I \xrightarrow{u} J \xrightarrow{v} K \xrightarrow{w} L
$$

There is a formal similarity with the diagrams for a monad. It is made explicit in Exercise 1.10.7.

(ii) **A split indexed category** is a functor $\Psi \colon \mathbb{B}^{\mathrm{op}} \to \mathbf{Cat}$; it is an indexed category for which the η's and μ's in (i) are identities.

What we have said in the beginning of this section can now be summarised in the following result.

1.4.5. Proposition. *Let* $\;\begin{smallmatrix}\mathbb{E}\\\downarrow p\\\mathbb{B}\end{smallmatrix}\;$ *be a fibration with a cleavage. The assignment*

$$I \mapsto \mathbb{E}_I \qquad and \qquad u \mapsto (\textit{the substitution functor } u^*)$$

determines a $\mathbb{B}$-*indexed category. This indexed category is split whenever the cleavage of* p *is a splitting.* $\qquad\square$

Notice that in this definition of indexed category, the coherent isomorphisms η, μ are part of the *structure*. In fibrations, one does not have such structure, but it follows from the universal *property* of lifting, once a choice of liftings is made. For a more detailed discussion on fibrations versus indexed categories, see 1.10.4.

We have seen in Proposition 1.2.2 that the non-split codomain fibration $\begin{smallmatrix}\mathbf{Sets}^{\to}\\\downarrow\\\mathbf{Sets}\end{smallmatrix}$ can equivalently be described as a split family fibration $\begin{smallmatrix}\mathsf{Fam}(\mathbf{Sets})\\\downarrow\\\mathbf{Sets}\end{smallmatrix}$ Remember that the codomain fibration captures display indexing with substitution by pullback, whereas the family fibration captures pointwise indexing with substitution by composition. The latter gives a split fibration. There are similar phenomena for ω-sets and for PERs.

(Later, in Section 1.7 we will see that the equivalence below is an equivalence in a sense appropriate for *fibred* categories.)

1.4.6. Definition. Let $\mathrm{UFam}(\omega\text{-}\mathbf{Sets})$ be the category of "uniform families" of ω-**Sets**. It has:

objects	omega-sets (I, E) together with for each element $i \in I$ an ω-set (X_i, E_i). We shall often simply write $(X_i, E_i)_{i \in (I,E)}$ for such objects.

morphisms $(X_i, E_i)_{i \in (I,E)} \;\to\; (Y_j, E_j)_{j \in (J,E)}$ are pairs $(u, (f_i)_{i \in I})$ where $u \colon (I, E) \to (J, E)$ is a morphism in ω-**Sets** between the underlying index objects and $(f_i \colon X_i \to Y_{u(i)})_{i \in I}$ is a collection of functions between the fibres, which is 'tracked uniformly': there is a code $e \in \mathbb{N}$ such that for every $i \in I$ and $n \in E(i)$ one has that $e \cdot n$ tracks f_i. Explicitly, for some $e \in \mathbb{N}$, we have

$$\forall i \in I. \, \forall n \in E(i). \, \forall x \in X_i. \, \forall m \in E_i(x).$$
$$e \cdot n \cdot m \in E_{u(i)}(f_i(x)).$$

We leave it to the reader to verify that one obtains a category. There is a first projection functor UFam(ω-**Sets**) $\to \omega$-**Sets**—which is a split fibration, much in the same way as for **Sets**. This category UFam(ω-**Sets**) captures ω-sets pointwise indexing ω-sets. It is related to the arrow category ω-**Sets**$^{\to}$, capturing display indexing, in the following manner.

1.4.7. Proposition. *The projection functor* $\begin{smallmatrix} \text{UFam}(\omega\text{-}\mathbf{Sets}) \\ \downarrow \\ \omega\text{-}\mathbf{Sets} \end{smallmatrix}$ *given by the mapping* $(X_i, E_i)_{i \in (I,E)} \mapsto (I, E)$, *is a split fibration. Moreover, there is an equivalence of categories in a commuting triangle.*

$$
\begin{array}{ccc}
\text{UFam}(\omega\text{-}\mathbf{Sets}) & \xrightarrow{\;\simeq\;} & \omega\text{-}\mathbf{Sets}^{\to} \\
& \searrow \quad \swarrow \text{cod} & \\
& \omega\text{-}\mathbf{Sets} &
\end{array}
$$

where the functor UFam(ω-**Sets**) $\to \omega$-**Sets**$^{\to}$ *sends*

$$(X_i, E_i)_{i \in (I,E)} \mapsto \left(\begin{array}{c} (\coprod_{i \in I} X_i, E) \\ \downarrow \pi \\ (I, E) \end{array} \right),$$

with the existence predicate E on the disjoint union $\coprod_{i \in I} X_i$ given by

$$E(i, x) = \{\langle n, m \rangle \mid n \in E_I(i) \text{ and } m \in E_i(x)\}.$$

Proof. The functor $\begin{smallmatrix} \text{UFam}(\omega\text{-}\mathbf{Sets}) \\ \downarrow \\ \omega\text{-}\mathbf{Sets} \end{smallmatrix}$ is a fibration because for $u \colon (I, E) \to (J, E)$ in ω-**Sets**, and a family $(Y_j, E_j)_{j \in (J,E)}$ over (J, E), we can form a family $u^*((Y_j, E_j)_{j \in (J,E)})$ over (I, E) as $(Y_{u(i)}, E_i)$, where $E_i(y) = E_{u(i)}(y)$. There is then an associated Cartesian lifting $(u, (\mathrm{id})) \colon u^*((Y_j, E_j)_{j \in (J,E)}) \to (Y_j, E_j)_{j \in (J,E)}$ over u. This choice of liftings forms a splitting.

The projection π is well-defined, since it is tracked by a code for the first projection $\langle n, m \rangle \mapsto n$. We get a functor $\mathcal{P} \colon \text{UFam}(\omega\text{-}\mathbf{Sets}) \to \omega\text{-}\mathbf{Sets}^{\to}$ by sending a morphism

$$(X_i, E_i)_{i \in (I,E)} \xrightarrow{\quad (u,f) \quad} (Y_j, E_j)_{j \in (J,E)}$$

to the square

$$
\begin{array}{ccc}
(\coprod_{i \in I} X_i, E) & \xrightarrow{\quad \{u,f\} \quad} & (\coprod_{j \in J} Y_j, E) \\
\pi \downarrow & & \downarrow \pi \\
(I, E) & \xrightarrow{\quad u \quad} & (J, E)
\end{array}
$$

where $\{u, f\}$ is the function $(i, x) \mapsto (u(i), f_i(x))$, tracked by $\Lambda z. \langle e \cdot (\mathbf{p}z), d \cdot (\mathbf{p}z) \cdot (\mathbf{p}'z) \rangle$, in which e is a code for u and d is a code for the family of functions $f = (f_i)_{i \in I}$. We leave it to the reader to verify that $\mathcal{P}$ is a full and faithful functor.

In the reverse direction, one maps a family $\left(\begin{array}{c} (X,E) \\ \downarrow \varphi \\ (I,E) \end{array} \right)$ in ω-**Sets**$^{\rightarrow}$ to the family (X_i, E_i), where for $i \in I$ the set X_i is the fibre $\varphi^{-1}(i)$ over $i \in I$, and $E_i(x) = E_X(x)$. This is evidently functorial, and yields an equivalence $\mathrm{UFam}(\omega\text{-}\mathbf{Sets}) \xrightarrow{\simeq} \omega\text{-}\mathbf{Sets}^{\rightarrow}$, commuting with the functors to ω-**Sets**. $\qquad \square$

Notice that in the split fibration $\begin{array}{c} \mathrm{UFam}(\omega\text{-}\mathbf{Sets}) \\ \downarrow \\ \omega\text{-}\mathbf{Sets} \end{array}$ one has substitution by composition, whereas in $\begin{array}{c} \omega\text{-}\mathbf{Sets}^{\rightarrow} \\ \downarrow \\ \omega\text{-}\mathbf{Sets} \end{array}$ one has substitution by pullbacks. The former is evidently functorial, whereas the latter is only 'pseudo-functorial'. This is precisely as for $\begin{array}{c} \mathrm{Fam}(\mathbf{Sets}) \\ \downarrow \\ \mathbf{Sets} \end{array}$ and $\begin{array}{c} \mathbf{Sets}^{\rightarrow} \\ \downarrow \\ \mathbf{Sets} \end{array}$ in Proposition 1.2.2.

Recall that there is a full subcategory **PER** $\hookrightarrow \omega$-**Sets** of partial equivalence relations inside the category of ω-sets. One may thus restrict the indexed objects in the definition of the category $\mathrm{UFam}(\omega\text{-}\mathbf{Sets})$ to PERs. This yields a category $\mathrm{UFam}(\mathbf{PER})$ of ω-set-indexed-PERs, instead of ω-set-indexed-ω-sets. We get another example of a split fibration $\begin{array}{c} \mathrm{UFam}(\mathbf{PER}) \\ \downarrow \\ \omega\text{-}\mathbf{Sets} \end{array}$, which will play an important rôle in the sequel. Therefore we spell out its definition in detail.

1.4.8. Definition. Let $\mathrm{UFam}(\mathbf{PER})$ be the category with

objects   collections $(R_i)_{i \in I}$ of PERs R_i indexed by an ω-set (I, E). As above, these are often written as $(R_i)_{i \in (I,E)}$.

morphisms $(R_i)_{i \in (I,E)} \to (S_j)_{j \in (J,E)}$ are pairs (u, f) where $u \colon (I, E) \to (Y, E)$ is a morphism in ω-**Sets** and $f = (f_i \colon R_i \to S_{u(i)})_{i \in I}$ is a collection of functions between the fibres, which is tracked uniformly: there is an $e \in \mathbb{N}$ such that for every $i \in I$ and $n \in E(i)$ the code $e \cdot n$ tracks f_i in **PER**.

1.4.9. Proposition. *The first projection* $\begin{array}{c} \text{UFam}(\textbf{PER}) \\ \downarrow \\ \omega\text{-}\textbf{Sets} \end{array}$ *mapping* $(R_i)_{i \in (I,E)} \mapsto (I, E)$ *is a split fibration. Substitution is by composition, precisely as above.* □

In the next section we shall see how one can further restrict the index objects to **PER** $\hookrightarrow$ ω-**Sets** via what is called change-of-base.

We close this section with a simple lemma which turns out to be very useful in calculating with fibrations. It is essentially a reformulation of Exercise 1.1.2, and tells us that a morphism in a total category corresponds to a morphism in the basis together with a vertical map. It enables us to switch smoothly between global structure in the total category and local structure in the fibres.

1.4.10. Lemma. *Let* $\begin{array}{c} \mathbb{E} \\ \downarrow p \\ \mathbb{B} \end{array}$ *be a fibration. For every cleavage one has an iso-morphism of sets (or classes)*

$$\mathbb{E}(X, Y) \cong \coprod_{u \colon pX \to pY} \mathbb{E}_{pX}\big(X, u^*(Y)\big)$$

where $\coprod$ *is disjoint union. The isomorphism is natural in X and Y, between functors* $\mathbb{E}^{\mathrm{op}} \times \mathbb{E} \rightrightarrows$ **Sets**.

Proof. Given $f \colon X \to Y$ in $\mathbb{E}$ take $u = pf \colon pX \to pY$ and $f' \colon X \to u^*(Y)$ to be the vertical part of f, *i.e.* the unique vertical map with $\overline{u}(Y) \circ f' = f$. Conversely given $u \colon pX \to pY$ and $f' \colon X \to u^*(Y)$ above pX one obtains $f = \overline{u}(Y) \circ f' \colon X \to Y$. Naturality is left as exercise. □

Finally, there is a principle of mathematical purity that deserves attention. One should not define a property for fibrations in terms of a specific cleavage; definitions should be 'cleavage-free' or 'intrinsic'. Sometimes it can be subtle that a certain property is intrinsic: consider as example,

"every substitution functor u^* has a left adjoint $\coprod_u$"

This property does not depend on a cleavage: two different cleavages induce naturally isomorphic substitution functors (see Exercise 1.4.3 below); so one of them has an adjoint if and only if the other has an adjoint.

Exercises

1.4.1. Describe weakening functors π^* and contraction functors δ^* (both on objects and on morphisms) for projections $\pi \colon I \times J \to I$ and diagonals

$\delta = \langle \mathrm{id}, \pi' \rangle : I \times J \to (I \times J) \times J$ in a family fibration, a codomain fibration, a subobject fibration, and in a simple fibration.

1.4.2. Prove that if a map $u: I \to J$ in the base category of a fibration is an isomorphism, then so is a Cartesian lifting $\overline{u}(X): u^*(X) \to X$, for each X above J.

1.4.3. Given a fibration with two cleavages. Show that for each morphism in the basis the two induced substitution functors are naturally isomorphic.

1.4.4. Let $\begin{smallmatrix} \mathbb{E} \\ \downarrow p \\ \mathbb{B} \end{smallmatrix}$ be a fibration, and consider the squares (a) in $\mathbb{E}$ over (b) in $\mathbb{B}$.

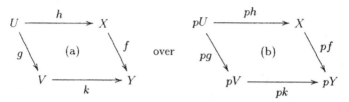

where h and k are both Cartesian. Prove that (a) is a pullback square in $\mathbb{E}$ if and only if (b) is a pullback square in $\mathbb{B}$.

[Notice that as a result, the square defining u^* on morphisms in the beginning of this section, is a pullback in the total category.]

1.4.5. Show that any poset fibration (all of whose fibre categories are posets) is split.

1.4.6. Assume functors $\mathbb{B} \xrightarrow{K} \mathbb{C} \xleftarrow{L} \mathbb{A}$ and form the comma category $(K \downarrow L)$. Show that the (first) projection functor $\begin{smallmatrix} (K \downarrow L) \\ \downarrow \\ \mathbb{B} \end{smallmatrix}$ is a split fibration.

[The second projection is an "opfibration", see Lemma 9.1.6.]

1.4.7. Show that the split indexed category induced by the family fibration $\begin{smallmatrix} \mathrm{Fam}(\mathbb{C}) \\ \downarrow \\ \mathbf{Sets} \end{smallmatrix}$ sends a set I to the (functor) category $\mathbb{C}^I$ of I-indexed families of objects and morphisms of $\mathbb{C}$. What is the morphism part of this functor?

1.4.8. The following tells that choosing a cleavage is functorial—in a suitable sense. Let $p: \mathbb{E} \to \mathbb{B}$ be a functor. Form the pullback in **Cat**

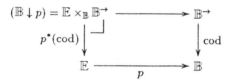

and define a functor $F: \mathbb{E}^{\to} \longrightarrow \mathbb{E} \times_{\mathbb{B}} \mathbb{B}^{\to}$ by $(f: X' \to X) \mapsto (X, pf)$. Prove that p is a cloven fibration if and only if F has a 'right-adjoint-right-inverse', *i.e.* a right adjoint with identity $FG \xrightarrow{=} \mathrm{id}$ as counit.

[This result may be found in [105], where it is attributed to Chevalley. It shows the 'algebraic nature' of the concept of (cloven) fibration; it is comparable to the result that chosen products $\times$ may be given by an adjoint functor. The result forms the basis for a 2-categorical formulation of the concept of fibration in [315, 317], see Definition 9.4.1 later on.]

1.4.9.   Check that the natural isomorphisms $\eta\colon \mathrm{id} \overset{\cong}{\Longrightarrow} \mathrm{id}^*$ and $\mu\colon u^* v^* \overset{\cong}{\Longrightarrow} (v \circ u)^*$ in a cloven fibration make the diagrams in Definition 1.4.4 commute.

1.4.10.   Later in Section 1.10 it will be shown that each indexed category gives rise to a cloven fibration; the latter is split whenever the indexed category is split. Try to find this construction now already.
[*Hint.* Have another look at the first part of Section 1.1 and try to generalise the disjoint union which is used to go from pointwise to display indexing.]

1.5 Change-of-base and composition for fibrations

So far we have seen several examples of fibrations. In this section we introduce two basic techniques for constructing new fibrations from old, namely change-of-base (or pullback) and composition. This will give rise to new examples of fibrations, but also to a rediscovery of some old ones.

1.5.1. Lemma (Change-of-base). *Let* $\begin{smallmatrix}\mathbb{E}\\\downarrow p\\\mathbb{B}\end{smallmatrix}$ *be a fibration and* $K\colon \mathbb{A} \to \mathbb{B}$ *be a functor. Form the pullback in* **Cat**

$$
\begin{array}{ccc}
\mathbb{A} \times_{\mathbb{B}} \mathbb{E} & \longrightarrow & \mathbb{E} \\
K^*(p)\Big\downarrow \;\ulcorner & & \Big\downarrow p \\
\mathbb{A} & \underset{K}{\longrightarrow} & \mathbb{B}
\end{array}
$$

In this situation, the functor $K^*(p)$ *is also a fibration. It is cloven or split in case* p *is cloven or split.*

We should point out that we are using the ordinary pullback of categories here: $\mathbb{A} \times_{\mathbb{B}} \mathbb{E}$ has pairs $I \in \mathbb{A}$, $X \in \mathbb{E}$ with $KI = pX$ as objects. So we use equalities between objects, instead of isomorphisms.

Proof. Given an object $(J, Y) \in \mathbb{A} \times_{\mathbb{B}} \mathbb{E}$ (so $KJ = pY$) and a morphism $u\colon I \to J$ in $\mathbb{A}$. Let $f\colon X \to Y$ in $\mathbb{E}$ be a Cartesian lifting of $Ku\colon KI \to KJ$ in $\mathbb{B}$. The pair (u, f) is then $K^*(p)$-Cartesian over u:

$$
\begin{array}{ccccc}
\mathbb{A} \times_{\mathbb{B}} \mathbb{E} & & & (I, X) \overset{(u,f)}{-\,-\,-\,-\,\blacktriangleright} (J, Y) \\
K^*(p)\Big\downarrow & & & \\
\mathbb{A} & & & I \underset{u}{\longrightarrow} J
\end{array}
$$

$\square$

1.5.2. Examples. In general change-of-base is a useful tool for defining (fibred) categories. For example, it can be used to take out a certain part of a fibration.

(i) Let **FinSets** $\hookrightarrow$ **Sets** be the category of finite sets. Change-of-base

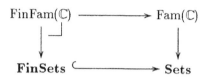

yields the fibration $\begin{smallmatrix} \mathrm{FinFam}(\mathbb{C}) \\ \downarrow \\ \mathbf{FinSets} \end{smallmatrix}$ of finite families of objects and arrows in $\mathbb{C}$. Such a diagram will often be called a **change-of-base situation**.

(ii) Let $\mathbb{C}$ be a locally small category with terminal object 1. By change-of-base along the global sections functor $\Gamma = \mathbb{C}(1, -) \colon \mathbb{C} \to \mathbf{Sets}$ one obtains the so-called **scone** $\mathrm{Sc}(\mathbb{C})$ and the **injective scone** $\mathrm{iSc}(\mathbb{C})$ in

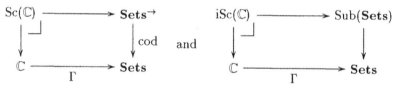

The previous lemma yields that the scone and injective scone of $\mathbb{C}$ are fibred over $\mathbb{C}$. Sometimes the scone $\mathrm{Sc}(\mathbb{C})$ is called the **Freyd cover** of $\mathbb{C}$, see [85] or [186]. It can also be described as the comma category $(\mathbf{Sets} \downarrow \Gamma)$.

Next we show how two specific fibrations that we already know can be reconstructed via change-of-base. For the first example, recall from the previous section the fibration $\begin{smallmatrix} \mathrm{UFam}(\mathbf{PER}) \\ \downarrow \\ \omega\text{-}\mathbf{Sets} \end{smallmatrix}$ of ω-set-indexed-PERs. This fibration can be turned into a fibration of PER-indexed PERs by restricting the index objects to $\mathbf{PER} \hookrightarrow \omega\text{-}\mathbf{Sets}$ via change-of-base. What we get is a split fibration, which turns out to be equivalent to the codomain fibration on PERs.

1.5.3. Proposition. *Form the split fibration* $\begin{smallmatrix} \mathrm{UFam}(\mathbf{PER}) \\ \downarrow \\ \mathbf{PER} \end{smallmatrix}$ *of PER-indexed PERs in the following change-of-base situation.*

There is then an equivalence of categories in a commuting triangle:

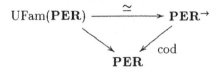

relating pointwise and display indexing of PERs.

Using the same notation UFam(**PER**) for different total categories here may seem confusing, but is rather convenient, and should not lead to problems as long as we use the entire fibration, as in $\begin{smallmatrix} \text{UFam}(\mathbf{PER}) \\ \downarrow \\ \mathbf{PER} \end{smallmatrix}$ and $\begin{smallmatrix} \text{UFam}(\mathbf{PER}) \\ \downarrow \\ \omega\text{-Sets} \end{smallmatrix}$, *i.e.* the total category together with the base category; then one can still see the difference.

Proof. One maps a family of PERs $(R_{[n]})_{[n] \in \mathbb{N}/A}$ indexed by a PER A, to the projection $\pi \colon \{A \mid R\} \to A$, where $\{A \mid R\}$ is the PER

$$\{A \mid R\} = \{\langle n, m \rangle \mid \mathbf{p}n A \mathbf{p}m \text{ and } \mathbf{p}'n R_{[\mathbf{p}n]} \mathbf{p}'m\}$$

and π is the projection given by $[\langle n, m \rangle] \mapsto [n]$.

In the reverse direction, one maps a family $\left(\begin{smallmatrix} R \\ \downarrow \varphi \\ A \end{smallmatrix} \right)$ in **PER**$^{\rightarrow}$ to the collection $R_{[n]}$ for $n \in |A|$ where $R_{[n]}$ is the fibre

$$R_{[n]} = \{(m, m') \in R \mid \varphi([m]) = [n]\}.$$

Further details of the equivalence are left to the reader.   □

Notice that we have obtained a fibration $\begin{smallmatrix} \text{UFam}(\mathbf{PER}) \\ \downarrow \\ \mathbf{PER} \end{smallmatrix}$ by change-of-base which is equivalent to the fibration $\begin{smallmatrix} \mathbf{PER}^{\rightarrow} \\ \downarrow \\ \mathbf{PER} \end{smallmatrix}$. In such a situation we also say that there is a change-of-base situation

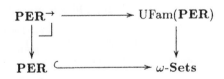

But notice also that this diagram is not a pullback in **Cat**, in the sense that we used before. We shall be similarly sloppy in the next result.

But first recall (*e.g.* from [42]) that in a category $\mathbb{C}$ with finite limits and finite coproducts $(0, +)$ one says that coproducts are **universal** if in a diagram,

$$
\begin{array}{ccccc}
X' & \longrightarrow & Z & \longleftarrow & Y' \\
\downarrow & \lrcorner & \downarrow & \llcorner & \downarrow \\
X & \xrightarrow{\ \kappa\ } & X+Y & \xleftarrow{\ \kappa'\ } & Y
\end{array}
\qquad (*)
$$

the left and right squares are pullbacks, then the top row is (also) a coproduct diagram (*i.e.* the induced cotuple $X' + Y' \to Z$ is an isomorphism). And coproducts are called **disjoint** if the coprojections κ, κ' are monos and form a pullback square

$$
\begin{array}{ccc}
0 & \longrightarrow & Y \\
\downarrow & \lrcorner & \downarrow {\scriptstyle \kappa'} \\
X & \xrightarrow{\ \kappa\ } & X+Y
\end{array}
$$

Below we use the fact that if coproducts are universal, then the initial object 0 is **strict**: every map $X \to 0$ is an isomorphism, see Exercise 1.5.6.

Notice that these notions easily extend to coproducts $\coprod_{i \in I} X_i$ indexed by an arbitrary set I. This is what we shall use.

1.5.4. Proposition. *Let $\mathbb{C}$ be a category with finite limits and set-indexed coproducts $\coprod_{i \in I} X_i$, which are universal and disjoint. There is then a copower functor*

$$
\mathbf{Sets} \xrightarrow{\ \ \coprod\ \ } \mathbb{C} \qquad by \qquad I \longmapsto I \cdot 1 = \coprod_{i \in I}(1)
$$

where $1 \in \mathbb{C}$ is the terminal object. This yields an equivalence of categories

$$
\mathbb{C}/\coprod(I) \xrightarrow{\ \simeq\ } \mathbb{C}^I, \quad natural \ in \ I \in \mathbf{Sets}.
$$

Then we can obtain the family fibration on $\mathbb{C}$ in a change-of-base situation,

$$
\begin{array}{ccc}
\mathrm{Fam}(\mathbb{C}) & \longrightarrow & \mathbb{C}^{\to} \\
\downarrow & \lrcorner & \downarrow {\scriptstyle \mathrm{cod}} \\
\mathbf{Sets} & \xrightarrow{\ \coprod\ } & \mathbb{C}
\end{array}
$$

By this result, we have a correspondence between pointwise indexing $(X_i)_{i \in I}$ of an I-indexed family in $\mathbb{C}$ and display indexing $\left(\begin{smallmatrix} X \\ \downarrow \\ \coprod(I) \end{smallmatrix} \right)$, over index objects

$\coprod(I) = \coprod_{i \in I} 1 \in \mathbb{C}$—assuming that $\mathbb{C}$ has a suitably rich coproduct-structure.

Proof. For a function $u: I \to J$ one takes $\coprod(u): \coprod(I) \to \coprod(J)$ to be the unique map with $\coprod(u) \circ \kappa_i = \kappa_{u(i)}: 1 \to \coprod_{j \in J} 1 = \coprod(J)$.

One can define two functors

$$\mathbb{C}^I \underset{G_I}{\overset{F_I}{\rightleftarrows}} \mathbb{C}/\coprod(I)$$

as follows. For a collection $X = (X_i)_{i \in I}$ in $\mathbb{C}^I$, take

$$F\big((X_i)_{i \in I}\big) \overset{\text{def}}{=} \left(\begin{array}{c} \coprod_{i \in I} X_i \\ \downarrow \pi_X \\ \coprod(I) \end{array} \right)$$

where π_X is the unique map with $\pi_X \circ \kappa_i = \kappa_i \circ {!}_{X_i}$. In the reverse direction, one takes

$$G\left(\begin{array}{c} Z \\ \downarrow \varphi \\ \coprod(I) \end{array} \right) \overset{\text{def}}{=} (Z_i)_{i \in I},$$

where each Z_i (for $i \in I$) is obtained in a pullback square:

$$\begin{array}{ccc} Z_i & \longrightarrow & Z \\ \downarrow & \lrcorner & \downarrow \varphi \\ 1 & \underset{\kappa_i}{\longrightarrow} & \coprod(I) \end{array}$$

There is a natural isomorphism $FG \cong \text{id}$ since by universality the maps $Z_i \to Z$ in this diagram yield a cotuple isomorphism $\coprod_i Z_i \overset{\cong}{\to} Z$. In order to prove $GF \cong \text{id}$ one applies the definition of G to $F((X_i)_{i \in I})$, leading to pullbacks of π_X (as above) along κ_i. This gives us the original collection $(X_i)_{i \in I}$, since there are pullback squares

$$\begin{array}{ccc} X_i & \longrightarrow & \coprod_{i \in I} X_i \\ \downarrow & \lrcorner & \downarrow \pi_X \\ 1 & \underset{\kappa_i}{\longrightarrow} & \coprod(I) \end{array}$$

by the following argument. Assume $u\colon K \to \coprod_{i \in I} X_i$ with $\pi_X \circ u = \kappa_i \circ !_K$. Then for each $j \in I$, one can form the pullback square as on the left

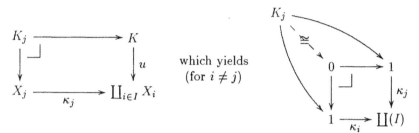

which yields
(for $i \neq j$)

so that we get an isomorphism $K_j \cong 0$ (since the initial object 0 is strict). Hence $K \cong \coprod_{j \in I} K_j \cong K_i$, which yields the required mediating map $K \to X_i$. □

The second way of constructing new fibrations is simply by composition. This shows that repeated indexing is a form of indexing.

1.5.5. Lemma. *Let* $\begin{smallmatrix} \mathbb{E} \\ \downarrow p \\ \mathbb{B} \end{smallmatrix}$ *and* $\begin{smallmatrix} \mathbb{B} \\ \downarrow r \\ \mathbb{A} \end{smallmatrix}$ *be fibrations.*

(i) *The composite* $\begin{smallmatrix} \mathbb{E} \\ \downarrow rp \\ \mathbb{A} \end{smallmatrix}$ *is then also a fibration, in which*

f *in* $\mathbb{E}$ *is* rp-*Cartesian* $\Leftrightarrow$ f *is* p-*Cartesian and* pf *is* r-*Cartesian.*

In case both p *and* r *are cloven (or split), then the composite fibration* rp *is also cloven (or split).*

(In such a situation one often calls p *a* **fibration over** r, *see also Section 9.4.)*

(ii) *For each object* $I \in \mathbb{A}$ *one obtains a functor* p_I *from* $\mathbb{E}_I = (rp)^{-1}(I)$ *to* $\mathbb{B}_I = r^{-1}(I)$ *by restriction. All of these* p_I's *are fibrations.*

Proof. (i) Given $Y \in \mathbb{E}$ and $u\colon I \to rp(Y)$ in $\mathbb{A}$. Let f be an r-Cartesian lifting of u and g a p-Cartesian lifting of f; one obtains that g is rp-Cartesian over u:

$$\bullet \,\,-\,-\,-\,\overset{g}{-}\,-\,-\,\twoheadrightarrow Y$$

$$\bullet \,\,-\,-\,-\,\overset{f}{-}\,-\,\twoheadrightarrow p(Y)$$

$$I \xrightarrow{\quad u \quad} rp(Y)$$

(ii) Left to the reader.   □

1.5.6. Example. Let $\mathbb{B}$ be a category with pullbacks, and write

$$\longrightarrow \cdots \longrightarrow \quad (n \text{ times})$$

for the linear order of length n, considered as a category. Consider then the sequence of functor categories and generalised codomain functors:

$$\cdots \longrightarrow \mathbb{B}^{\rightarrow\rightarrow\rightarrow} \longrightarrow \mathbb{B}^{\rightarrow\rightarrow} \longrightarrow \mathbb{B}^{\rightarrow} \longrightarrow \mathbb{B} \longrightarrow 1$$

sending

$$\left(\xrightarrow{f_n} \xrightarrow{f_{n-1}} \cdots \xrightarrow{f_1} \right) \longmapsto \left(\xrightarrow{f_{n-1}} \cdots \xrightarrow{f_1} \right).$$

Then each of these functors is a fibration, and all (finite) composites are fibrations.

The last two exercises 1.5.6 and 1.5.7 below contain some useful facts about universal and disjoint coproducts, which are there for future reference. For more information, see [42, 51]. There, a category $\mathbb{C}$ with coproducts is called extensive if the canonical functors $\mathbb{C}/X \times \mathbb{C}/Y \to \mathbb{C}/(X+Y)$ are equivalences. This definition does not require $\mathbb{C}$ to have pullbacks: it can be shown that the relevant pullbacks for universality and disjointness are induced.

(A comparable property for 'extensive fibrations' may be found in Exercise 9.2.13 (iii))

Exercises

1.5.1. See the difference between the (total) categories FinFam(**Sets**) and Fam(**FinSets**).

1.5.2. Define a split fibration of PER-indexed-ω-sets by change-of-base.

1.5.3. Consider the fibrations $\begin{smallmatrix}\mathbb{E}\\\downarrow p\\\mathbb{B}\end{smallmatrix}$ and $\begin{smallmatrix}\mathbb{B}\\\downarrow r\\\mathbb{A}\end{smallmatrix}$ in the 'composition' Lemma 1.5.5.
   (i) Prove Lemma 1.5.5 (ii).
   (ii) Let $f: X \to Y$ be a morphism in $\mathbb{E}$ and write $I = pX \in \mathbb{B}$. Show that f is p-Cartesian if and only if it can be written as $g \circ h$ with g rp-Cartesian and h p_I-Cartesian.

1.5.4. Consider the scone construction from Example 1.5.2 (ii), and prove: if $\mathbb{C}$ is Cartesian closed, then so is Sc($\mathbb{C}$), and the functor Sc($\mathbb{C}$) $\to \mathbb{C}$ preserves this structure.
   [This result can also be proved via more advanced fibred techniques, see Example 9.2.5 (i).]

1.5.5. Let A be a complete lattice. Check that coproducts $\bigvee_{i \in I} x_i$ in A are universal if and only if A is a frame, *i.e.* satisfies $y \wedge (\bigvee_{i \in I} x_i) = \bigvee_{i \in I} (y \wedge x_i)$. And that coproducts are disjoint if and only if A has at most two elements.

1.5.6. (Cockett, see *e.g.* [51]) Let $\mathbb{B}$ be a **distributive category**, *i.e.* a category with finite products $(1, \times)$ and coproducts $(0, +)$ which are distributive: the canonical maps $(Z \times X) + (Z \times Y) \to Z \times (X + Y)$ are isomorphisms. (Alternatively, universality of coproducts as in diagram (∗) on page 59 holds for the special case where $Z \to X + Y$ is a Cartesian projection $Z \times (X + Y) \to X + Y$.)

(i) Use distributivity to show that morphisms of the form

$$X \times X \xrightarrow{\;\kappa \times \mathrm{id}\;} (X + Y) \times X$$

are split monos.

(ii) Prove that coprojections $X \xrightarrow{\kappa} X + Y \xleftarrow{\kappa'} Y$ are monos.

[*Hint.* For $f, g \colon Z \rightrightarrows X$ with $\kappa \circ f = \kappa \circ g$, consider the diagram:

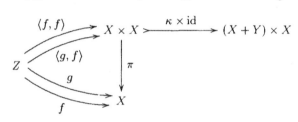

This slightly simplifies the argument in the proof of [51, Lemma 3.1].]

(iii) Show that the canonical maps $0 \to 0 \times Z$ are isomorphisms. (Hence a distributive category is characterised by: functors $(-) \times Z$ preserve finite coproducts.)

[*Hint.* Notice that the codiagonal $\nabla = [\mathrm{id}, \mathrm{id}] \colon (0 \times Z) + (0 \times Z) \to 0 \times Z$ is an isomorphism. Hence the two coprojections $\kappa, \kappa' \colon 0 \times Z \rightrightarrows (0 \times Z) + (0 \times Z)$ are equal, and so any two maps $0 \times Z \rightrightarrows Y$ are equal.]

(iv) Conclude from (iii) that 0 is a **strict** initial object: every map $Z \to 0$ is an isomorphism.

1.5.7. Prove that in a category $\mathbb{B}$ with disjoint and universal coproducts, diagrams of the form

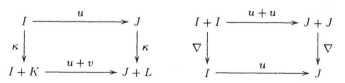

are pullback squares. Show also that the coproduct functor $+ \colon \mathbb{B} \times \mathbb{B} \to \mathbb{B}$ preserves pullbacks.

1.6 Fibrations of signatures

Signatures will be used in this book as the basic structures that generate a logic or a type theory. They contain the basic types and function symbols (possibly also predicate symbols) which are used to build a logic or type theory on. The aim of this section is twofold: first to define signatures and organise them in suitable (fibred) categories. It turns out that these categories of signatures can be introduced most conveniently by change-of-base. The second aim is to use signatures, together with categories of models, to illustrate the organisational power of fibrations.

In universal algebra and traditional logic one uses 'sort' for what we prefer to call 'type'. A typical signature consists of a set of basic types, say $\{N, B, \ldots\}$, together with a set of typed function symbols, containing for example

$$
\begin{array}{rcl}
+ & : & N, N \longrightarrow N \\
\text{succ} & : & N \longrightarrow N \\
\wedge & : & B, B \longrightarrow B \\
= & : & N, N \longrightarrow B \\
& \vdots &
\end{array}
$$

A signature is called **single-typed** if it has only one basic type and **many-typed** otherwise. Many-typed signatures are of fundamental importance for algebraic data types and specifications, see *e.g.* [77]. Here (and in the next chapter) we investigate 'pure' signatures without equations. The latter are included in Chapter 3 on equational logic. And in Chapter 4 on first order logic we shall have signatures with (many-typed) predicate symbols. Signatures underlying higher order logic in Chapter 5 have a distinguished type Prop for propositions.

Alternative, older names for 'many-typed' are 'many-sorted' and 'heterogeneous' (as used for example in [34]), which are in contrast with 'single-sorted' and 'homogeneous'. Mathematical interest has been focussed mainly on single-typed signatures, but the more general many-typed signatures are standard in computer science.

Formally, a **many-typed signature** Σ is a pair $(T, \mathcal{F})$ where T is a set of (basic) types and $\mathcal{F} \colon T^\star \times T \to \mathbf{Sets}$ is a mapping which assigns to every sequence of types $\langle \sigma_1, \ldots, \sigma_n \rangle \in T^\star$ and $\sigma_{n+1} \in T$ a set $\mathcal{F}(\langle \sigma_1, \ldots, \sigma_n \rangle, \sigma_{n+1})$ of function symbols taking inputs of type $\sigma_1, \ldots, \sigma_n$ and yielding an output of type σ_{n+1}. In order to simplify the notation, we shall write for a signature $\Sigma = (T, \mathcal{F})$,

$$
|\Sigma| = T
$$

for the underlying set of types, and

$$F: \sigma_1, \ldots, \sigma_n \longrightarrow \sigma_{n+1} \qquad \text{if} \qquad F \in \mathcal{F}(\langle \sigma_1, \ldots, \sigma_n \rangle, \sigma_{n+1}).$$

(Notice that these sets $\mathcal{F}(\alpha)$ for $\alpha \in T^\star \times T$ need not be disjoint, so we may have overloading of function symbols, like in:

$$+: \mathsf{N}, \mathsf{N} \longrightarrow \mathsf{N} \qquad \text{and} \qquad +: \mathsf{R}, \mathsf{R} \longrightarrow \mathsf{R}.$$

See also Exercise 1.6.3 below.)

A morphism $\Sigma \to \Sigma'$ of many-typed signatures consists of a function $u: |\Sigma| \to |\Sigma'|$ between the underlying sets of types together with a family of functions (f_α) between sets of function symbols such that

$$F: \sigma_1, \ldots, \sigma_n \longrightarrow \sigma_{n+1} \Rightarrow f_\alpha(F): u(\sigma_1), \ldots, u(\sigma_n) \longrightarrow u(\sigma_{n+1})$$

where the subscript α is $\langle \langle \sigma_1, \ldots, \sigma_n \rangle, \sigma_{n+1} \rangle$. Thus one obtains a category **Sign** together with a forgetful functor **Sign** $\to$ **Sets** sending a signature Σ to its underlying set of types $|\Sigma|$. It is a split fibration because for a signature Σ and a function $u: S \to |\Sigma|$ one can form a many-typed signature over S with function symbols,

$$F: \sigma_1, \ldots, \sigma_n \longrightarrow \sigma_{n+1} \overset{\text{def}}{\Leftrightarrow} F: u(\sigma_1), \ldots, u(\sigma_n) \longrightarrow u(\sigma_{n+1}) \text{ in } \Sigma.$$

This is captured all at once in the following definition of the category of signatures. (We hope the reader will appreciate its conciseness.)

1.6.1. Definition. The category **Sign** of **many-typed signatures** is defined in the change-of-base situation

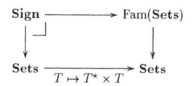

where $T^\star$ is the free monoid of finite sequences on T (the "Kleene star"). As an immediate consequence of Lemma 1.5.1 we get that $\begin{smallmatrix} \textbf{Sign} \\ \downarrow \\ \textbf{Sets} \end{smallmatrix}$ is a split fibration.

Often we simply use 'signature' for 'many-typed signature'.

1.6.2. Convention. A morphism $\phi: \Sigma \to \Sigma'$ of signatures consists of a pair $(u, (f))$ as describe above. We usually write ϕ both for u and for all of the f's. Thus we get

$$F: \sigma_1, \ldots, \sigma_n \longrightarrow \sigma_{n+1} \text{ in } \Sigma$$
$$\Rightarrow \phi(F): \phi(\sigma_1), \ldots, \phi(\sigma_n) \longrightarrow \phi(\sigma_{n+1}) \text{ in } \Sigma'.$$

This is notationally rather convenient and not likely to cause much confusion.

Terms

The description below of the terms associated with a signature is as in universal algebra: it is based on indexed sets of term variables. In the next chapter we shall give a more type theoretic description (which will be used in the subsequent remainder of the book), based on a fixed infinite set of term variables $\{v_0, v_1, \ldots\}$ which are linked to a type in a context containing type declarations of the form $v_i : \sigma_i$.

Suppose Σ is a signature with $T = |\Sigma|$ as underlying set of types. A T-indexed collection of sets $X = (X_\sigma)_{\sigma \in T}$ can be seen as providing a set of variables X_σ for every type $\sigma \in T$. One can form a new T-indexed collection

$$\left(\mathrm{Terms}_\tau(X) \right)_{\tau \in T}$$

where $\mathrm{Terms}_\tau(X)$ is the set of terms of type τ. These collections are defined as follows.

- $X_\tau \subseteq \mathrm{Terms}_\tau(X)$;
- if $F : \tau_1, \ldots, \tau_n \longrightarrow \tau_{n+1}$ in Σ and $M_1 \in \mathrm{Terms}_{\tau_1}(X), \ldots, M_n \in \mathrm{Terms}_{\tau_n}(X)$ then $F(M_1, \ldots, M_n) \in \mathrm{Terms}_{\tau_{n+1}}(X)$.

Hence a term is a (well-typed) string consisting of variables $x \in \bigcup_{\sigma \in T} X_\sigma$ and function symbols F from Σ. There are associated notions of **free variable** and **substitution**:

$$\mathrm{FV}(x) = \{x\}$$
$$\mathrm{FV}(F(M_1, \ldots, M_n)) = \mathrm{FV}(M_1) \cup \cdots \cup \mathrm{FV}(M_n)$$

and for $y \in X_\tau$ and $N \in \mathrm{Terms}_\tau(X)$,

$$x[N/y] = \begin{cases} N & \text{if } x = y \\ x & \text{else} \end{cases}$$
$$F(M_1, \ldots, M_n)[N/y] = F(M_1[N/y], \ldots, M_n[N/y]).$$

In a similar way one defines simultaneous substitution $M[\vec{N}/\vec{y}]$. Notice that the dependence on the signature Σ is left implicit in the above definition of terms.

(Set theoretic) semantics

Let Σ be a signature, once again with $T = |\Sigma|$ as its underlying set of types. A **model** or **algebra** for Σ consists of a T-indexed collection $(A_\sigma)_{\sigma \in T}$ of **carrier** sets together with a collection of suitably typed functions: for each function symbol $F : \sigma_1, \ldots, \sigma_n \longrightarrow \sigma_{n+1}$ in Σ, an actual function $[\![F]\!] : A_{\sigma_1} \times \cdots \times A_{\sigma_n} \to A_{\sigma_{n+1}}$ between the corresponding carrier sets.

Thus a model consists of a pair $((A_\sigma)_{\sigma \in T}, [\![\_]\!])$.

1.6.3. Example. An obvious way to model a signature containing one function symbol

$$\text{if: B, N, N} \longrightarrow \text{N}$$

for an if-conditional on the natural numbers, is to use carriers

$$A_\text{B} = \{0, 1\} \qquad A_\text{N} = \mathbb{N}, \quad \text{the set of natural numbers}$$

and a function

$$[\![\text{if}]\!] : A_\text{B} \times A_\text{N} \times A_\text{N} \longrightarrow A_\text{N}$$

$$(b, n, m) \quad \mapsto \quad \begin{cases} n & \text{if } b = 1 \\ m & \text{otherwise.} \end{cases}$$

Of course, one can more generally interpret 'if' in a distributive category (with natural numbers object N and $B = 1 + 1$, see Section 2.6), but here we restrict ourselves to set theoretic models.

Such a model $((A_\sigma)_{\sigma \in T}, [\![\_]\!])$ for Σ can be used to interpret Σ-terms: suppose we have a collection of variable sets $X = (X_\sigma)_{\sigma \in T}$ together with a **valuation**

$$\left(\rho_\sigma \colon X_\sigma \longrightarrow A_\sigma \right)_{\sigma \in T}.$$

Such a valuation consists of functions assigning values in the model to the variables. Then there is an **interpretation** consisting of functions

$$\left([\![\_]\!]_\rho^\tau \colon \text{Terms}_\tau(X) \longrightarrow A_\tau \right)_{\tau \in T}$$

given by

$$[\![x]\!]_\rho = \rho_\tau(x) \qquad\qquad \text{for } x \in X_\tau$$
$$[\![F(M_1, \ldots, M_n)]\!]_\rho = [\![F]\!]([\![M_1]\!]_\rho, \ldots, [\![M_n]\!]_\rho)$$

For readability's sake we have omitted the superscripts τ in $[\![-]\!]^\tau$. One obtains a bijective correspondence between valuations and interpretations:

$$\frac{\left(X_\sigma \xrightarrow{\rho_\sigma} A_\sigma \right)_{\sigma \in T}}{\left(\text{Terms}_\tau(X) \xrightarrow[{[\![-]\!]_\rho^\tau}]{} A_\tau \right)_{\tau \in T}}$$

For a valuation $(\rho_\sigma \colon X_\sigma \to A_\sigma)$ together with elements $x \in X_\sigma$ and $a \in A_\sigma$ one defines a new valuation $\rho(x \mapsto a)$ by

$$\rho(x \mapsto a)(y) = \begin{cases} a & \text{if } y = x \\ \rho(y) & \text{else.} \end{cases}$$

A term $M \in \text{Terms}_\tau(X)$ contains only finitely many variables, say $x_1 \in X_{\sigma_1}, \ldots, x_n \in X_{\sigma_n}$. Such a term thus induces a function

$$A_{\sigma_1} \times \cdots \times A_{\sigma_n} \xrightarrow{\; [\![\, M \,]\!] \;} A_\tau$$

by

$$(a_1, \ldots, a_n) \longmapsto [\![\, M \,]\!]_{\rho(x_1 \mapsto a_1, \ldots, x_n \mapsto a_n)}.$$

In the expression on the right hand side, the valuation ρ does not play a rôle anymore. Interpreting a term as such a map (without valuations) gives a more categorical description.

1.6.4. Definition. The category **S-Model** of (set theoretic) models of many-typed signatures has

objects $(\Sigma, (A_\sigma), [\![\,_-\,]\!])$ where $((A_\sigma), [\![\,_-\,]\!])$ is a model for Σ.

morphisms $(\phi, (H_\sigma)) : (\Sigma, (A_\sigma), [\![\,_-\,]\!]) \longrightarrow (\Sigma', (A'_\sigma), [\![\,_-\,]\!]')$ consist of

- a morphism of signatures $\phi : \Sigma \to \Sigma'$
- a $|\Sigma|$-indexed collection of functions

$$H_\sigma : A_\sigma \longrightarrow A'_{\phi(\sigma)}$$

such that for each function symbol

$$F : \sigma_1, \ldots, \sigma_n \longrightarrow \sigma_{n+1} \text{ in } \Sigma$$

the following diagram commutes.

$$
\begin{array}{ccc}
A_{\sigma_1} \times \cdots \times A_{\sigma_n} & \xrightarrow{\; H_{\sigma_1} \times \cdots \times H_{\sigma_n} \;} & A'_{\phi(\sigma_1)} \times \cdots \times A'_{\phi(\sigma_n)} \\
{\scriptstyle [\![\, F \,]\!]} \downarrow & & \downarrow {\scriptstyle [\![\, \phi(F) \,]\!]'} \\
A_{\sigma_{n+1}} & \xrightarrow{\; H_{\sigma_{n+1}} \;} & A'_{\phi(\sigma_{n+1})}
\end{array}
$$

Such set theoretic models of many-typed signatures and their morphisms are studied in some detail in [34].

There are a projection functors

$$\textbf{S-Model} \longrightarrow \textbf{Sign} \longrightarrow \textbf{Sets}$$
$$(\Sigma, (A_\sigma), [\![\,_-\,]\!]) \longmapsto \Sigma \longmapsto |\Sigma|.$$

They will play a rôle below; but first we describe syntactically constructed models.

1.6.5. Example. Let Σ be a signature with $T = |\Sigma|$ as set of basic types, and let $X = (X_\sigma)_{\sigma \in T}$ be a collection of typed variables. The sets of terms $\mathrm{Terms}_\tau(X)$ (for $\tau \in T$) form carriers for the so-called **term model** of Σ, with variables from X. A function symbol $F: \tau_1, \ldots, \tau_n \longrightarrow \tau_{n+1}$ has an interpretation as a function

$$\mathrm{Terms}_{\tau_1}(X) \times \cdots \times \mathrm{Terms}_{\tau_n}(X) \longrightarrow \mathrm{Terms}_{\tau_{n+1}}(X)$$

described by

$$(M_1, \ldots, M_n) \longmapsto F(M_1, \ldots, M_n).$$

The term model on the empty collection of variables $(\emptyset)_{\sigma \in T}$ is usually called the **initial model** of Σ. It is initial object in the fibre category over Σ of the fibration described in (i) below.

1.6.6. Lemma. (i) *The functor* $\begin{smallmatrix} \textbf{S-Model} \\ \downarrow \\ \textbf{Sign} \end{smallmatrix}$ *sending a model to its underlying signature is a split fibration. The fibre over $\Sigma \in$ **Sign** is the category of models with signature Σ.*

(ii) *The functor* $\begin{smallmatrix} \textbf{S-Model} \\ \downarrow \\ \textbf{Sets} \end{smallmatrix}$ *which sends a model to its underlying set of types is a split fibration. The fibre over $T \in$ **Sets** is the category of models of signatures with T as set of types.*

(iii) *For every set of types T, the fibre category* (models of signatures over T) *is fibred over the category* (signatures over T).

Proof. (i) Given a model $(\Sigma', (A'), [\![_-]\!]')$ and a signature morphism $\phi: \Sigma \to \Sigma'$ one obtains a model $((A_\sigma), [\![_-]\!])$ over Σ by putting

$$A_\sigma \stackrel{\mathrm{def}}{=} A'_{\phi(\sigma)} \qquad \text{and} \qquad [\![F]\!] \stackrel{\mathrm{def}}{=} [\![\phi(F)]\!]'.$$

(ii) + (iii) Directly by Lemma 1.5.5, using that $\begin{smallmatrix} \textbf{Sign} \\ \downarrow \\ \textbf{Sets} \end{smallmatrix}$ is a split fibration. $\square$

This lemma exhibits a general pattern which can be described roughly as follows. Given a notion $\mathcal{P}$ and another notion $\mathcal{Q}(\alpha)$ involving a parameter α of type $\mathcal{P}$, then, in general, the category of $\mathcal{Q}(\alpha)$'s is fibred over the category of $\mathcal{P}$'s, provided the $\mathcal{Q}(\alpha)$'s are suitably closed under substitution along morphisms of $\mathcal{P}$'s. To put it more concisely as a slogan:

if $\mathcal{Q}$'s depend on $\mathcal{P}$'s then $\mathcal{Q}$'s are fibred over $\mathcal{P}$'s

In the above lemma, we have models—involving signatures and thus sets of types—fibred over signatures and thus fibred over sets. A similar example is given by the vector spaces which involve fields and are fibred over fields, see Exercise 1.1.11.

However, this slogan is not entirely correct since the $\mathcal{Q}$'s can also be 'op-fibred' over the $\mathcal{P}$'s, which happens in case substitution acts covariantly (see Section 9.1).

But the point is that fibrations have a great organisational strength. They provide appropriate ways of layering mathematical structures, by making explicit what depends on what. This is the reason which makes elementary lemmas like the above one important.

In later chapters, this aspect will be crucial in modelling logics and type theories: for a type theory with, say, propositions depending on types (in a sense to be made precise in Section 11.5) the underlying structure involves a category of propositions fibred over a category of types.

Single-typed signatures

1.6.7. Definition. (i) We recall that a signature Σ is called **single-typed** if its underlying set of types $|\Sigma|$ is a singleton.

(ii) The category **Sign$_{ST}$** of single-typed signatures is defined by the change-of-base situation

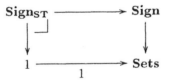

where 1 is the (one-object one-arrow) terminal category and the functor $1\colon 1 \to$ **Sets** points to a singleton set.

(iii) The category **S-Model$_{ST}$** of (set-theoretic) models for single-typed signatures arises in the change-of-base situation

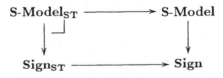

where the functor **Sign$_{ST}$** $\to$ **Sign** comes from (ii). Thus also models of single-typed signatures are fibred over their underlying signatures.

As we mentioned earlier, many mathematical texts on signatures are restricted to the single-typed case. A signature for a monoid acting on a set is then not described by function symbols

$$m\colon \mathsf{M}, \mathsf{M} \longrightarrow \mathsf{M}, \qquad e\colon () \longrightarrow \mathsf{M}, \qquad a\colon \mathsf{M}, \mathsf{X} \longrightarrow \mathsf{X}$$

but by a collection of function symbols

$$a_z : X \longrightarrow X$$

one for each element z in the carrier A_M of M. Such a single-typed description is not only artificial but it also involves a mixture of syntax and semantics (namely M and A_M). Such practices which have arisen in mathematics are not necessarily well-fitted for applications in computer science.

Exercises

1.6.1. Write down a (single-typed) signature for groups and also a (many-typed) signature for vector spaces.

1.6.2. A many-typed signature is called **finite** if it has only finitely many types and function symbols. Define the subcategory of **FinSign** $\hookrightarrow$ **Sign** consisting of finite signatures by change-of-base.

1.6.3. Many-typed signatures are sometimes defined (like in [343] or in [282, 2.2.1]) as objects of the category **Sign'** which arises in the following change-of-base situation.

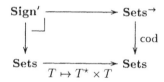

(i) Describe the category **Sign'** in elementary terms.

(ii) Show that the categories **Sign** and **Sign'** are equivalent.

(iii) One often prefers **Sign** to **Sign'** because signatures in **Sign** allow **overloading** of function symbols: for example the use of $+$ both for addition of integers and for addition of reals. Explain.

[Another advantage of **Sign** is that $\begin{smallmatrix}\textbf{Sign}\\\downarrow\\\textbf{Sets}\end{smallmatrix}$ is a *split* fibration.]

1.6.4. Describe the category **S-Model$_{ST}$** of models of single-typed signatures in detail.

1.6.5. The category **Sign** captures signatures of functions. A **signature of predicates** consists of a set of types T together with predicate symbols $R : \sigma_1, \ldots, \sigma_n$ where each σ_i is a type (element of T). Define an appropriate category of such signatures of predicates by change-of-base. Define also a category with both function and predicate symbols by change-of-base. [Such a category will be introduced in Definition 4.1.1.]

1.6.6. Let Σ be a signature and $T = |\Sigma|$ its set of types. We write **S-Model**(Σ) for the fibre category over Σ of the fibration $\begin{smallmatrix}\textbf{S-Model}\\\downarrow\\\textbf{Sign}\end{smallmatrix}$. This is the category of Σ-models.

(i)   Show that the assignment

$$X = (X_\sigma)_{\sigma \in T} \mapsto \Big(\mathrm{Terms}_\tau(X) \Big)_{\tau \in T}$$

extends to a functor from the category $\mathrm{Fam}(\mathbf{Sets})_T = \mathbf{Sets}^T$ of T-indexed families of sets to $\mathbf{S\text{-}Model}(\Sigma)$.

(ii) Assume $(A_\sigma)_{\sigma \in T}, [\![\_]\!])$ is a Σ-model. Verify that an interpretation $[\![\_]\!]^\tau_\rho \colon \mathrm{Terms}_\tau(X) \to A_\tau$ is a morphism of Σ-models.

1.6.7.   Let $((A_\sigma), [\![\_]\!])$ be a Σ-model and ρ a valuation $(X_\sigma \to A_\sigma)$.

(i)   Show that

$$[\![M[N/x]]\!]_\rho = [\![M]\!]_{\rho(x \mapsto [\![N]\!]_\rho)}.$$

(ii) Let $((B_\sigma), [\![\_]\!])$ be another Σ-model and $(H_\sigma \colon A_\sigma \to B_\sigma)$ be a morphism of Σ-models (*i.e.* a morphism in the fibre of $\mathbf{S\text{-}Model}$ over Σ). Show that

$$H\big([\![M]\!]_\rho \big) = [\![M]\!]_{H \circ \rho}.$$

1.7 Categories of fibrations

In this section we shall introduce and study "fibred functors" as appropriate morphisms between fibred categories (preserving the relevant structure). Also we shall describe "fibred natural transformations" between such fibred functors—just like ordinary natural transformations are morphisms between morphisms of ordinary categories (*i.e.* functors). We shall describe four (2-)categories of fibrations according to the following table.

		over a fixed basis $\mathbb{B}$	over arbitrary bases
split		$\mathbf{Fib_{split}}(\mathbb{B})$	$\mathbf{Fib_{split}}$
not necessarily split		$\mathbf{Fib}(\mathbb{B})$	$\mathbf{Fib}$

By laying down what appropriate morphisms of fibrations are, we can use categorical language to talk about fibrations as objects. This enables us to express some elementary facts about fibrations. Also, we say what fibred natural transformations (2-cells) are. Then we can apply various 2-categorical notions in the context of fibrations, like equivalence, and adjointness; the latter is studied in the next section.

We start with the category named $\mathbf{Fib}$, because it is most general among the categories in the table: it contains the other three as subcategories.

1.7.1. Definition. (i) A **morphism** $\left(\begin{smallmatrix} \mathbb{E} \\ \downarrow p \\ \mathbb{B} \end{smallmatrix} \right) \to \left(\begin{smallmatrix} \mathbb{D} \\ \downarrow q \\ \mathbb{A} \end{smallmatrix} \right)$ **of fibrations** consists of a pair of functors $K: \mathbb{B} \to \mathbb{A}$ and $H: \mathbb{E} \to \mathbb{D}$ such that the diagram

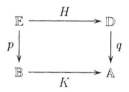

commutes and H sends Cartesian morphisms in $\mathbb{E}$ to Cartesian morphisms in $\mathbb{D}$. Such a functor H will be called **fibred**. This yields a category which will be written as **Fib**.

(ii) The subcategory $\mathbf{Fib_{split}} \hookrightarrow \mathbf{Fib}$ has split fibrations as objects and morphisms (K, H) as above where H preserves the splitting on-the-nose (that is, up-to-equality and not up-to-isomorphism).

Notice that in (i) we require the square to commute on-the-nose, not up-to-isomorphism. As it stands, the notion of morphism of fibrations is easy to work with and does what we want. For a more abstract approach, see [317].

Here is a first result that we can now express.

1.7.2. Lemma. *The functors*

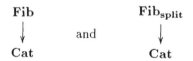

sending a (split) fibration to its base category are fibrations themselves. Reindexing is done by change-of-base, see Lemma 1.5.1. □

1.7.3. Definition. (i) For a fixed category $\mathbb{B}$, the category $\mathbf{Fib}(\mathbb{B})$ of fibrations with $\mathbb{B}$ as base category is defined to be the fibre over $\mathbb{B}$ of the above fibration $\begin{smallmatrix} \mathbf{Fib} \\ \downarrow \\ \mathbf{Cat} \end{smallmatrix}$. It thus has fibrations with basis $\mathbb{B}$ as objects. A morphism $\left(\begin{smallmatrix} \mathbb{E} \\ \downarrow p \\ \mathbb{B} \end{smallmatrix} \right) \to \left(\begin{smallmatrix} \mathbb{D} \\ \downarrow q \\ \mathbb{B} \end{smallmatrix} \right)$ in $\mathbf{Fib}(\mathbb{B})$ is then determined by a functor $H: \mathbb{E} \to \mathbb{D}$ making the triangle

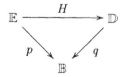

commute and preserving Cartesian morphisms. We call such a functor a **fibred functor** (as before) or a **functor over** $\mathbb{B}$.

(ii) Similarly the category $\mathbf{Fib}_{\mathbf{split}}(\mathbb{B})$ is defined to be the fibre over $\mathbb{B}$ of the fibration $\begin{smallmatrix} \mathbf{Fib}_{\mathbf{split}} \\ \downarrow \\ \mathbf{Cat} \end{smallmatrix}$. Morphisms in $\mathbf{Fib}_{\mathbf{split}}(\mathbb{B})$ are fibred functors H as in the triangle, which preserve the splitting on-the-nose. They will be called **split functors**.

(iii) If $H \colon \mathbb{E} \to \mathbb{D}$ is a fibred or split functor as in (i) or (ii), then for each object $I \in \mathbb{B}$ one obtains by restriction a functor $\mathbb{E}_I \to \mathbb{D}_I$ between the fibres over I; it will be written as H_I.

Often the name 'Cartesian functor' is used for what is called a 'fibred functor' here. This predicate 'Cartesian' is not very appropriate, because such functors are not Cartesian morphisms for some fibration.

Notice that the category $\mathbf{Fib}(1)$ of fibrations on the terminal category 1 can be identified with the category $\mathbf{Cat}$ of categories.

1.7.4. Lemma. *The categories* $\mathbf{Fib}(\mathbb{B})$ *and* $\mathbf{Fib}_{\mathbf{split}}(\mathbb{B})$ *have finite products; these are preserved by change-of-base.*

Proof. The identity functor $\begin{smallmatrix} \mathbb{B} \\ \downarrow \\ \mathbb{B} \end{smallmatrix}$ is terminal object, and the Cartesian product of two fibrations $\begin{smallmatrix} \mathbb{E} \\ \downarrow p \\ \mathbb{B} \end{smallmatrix}$ and $\begin{smallmatrix} \mathbb{D} \\ \downarrow q \\ \mathbb{B} \end{smallmatrix}$ on $\mathbb{B}$ is defined in:

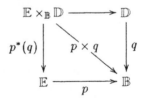

This yields a fibration $p \times q$ by Lemmas 1.5.1 and 1.5.5 (since it is obtained by change-of-base and composition). □

The next two lemmas give examples of morphisms of fibrations.

1.7.5. Lemma. *Let* $\mathbb{A}$ *and* $\mathbb{B}$ *be categories with pullbacks and let* $K \colon \mathbb{A} \to \mathbb{B}$ *be a pullback preserving functor. There are then extensions of* K *to morphisms*

$$\left(\begin{smallmatrix} \mathrm{Sub}(\mathbb{A}) \\ \downarrow \\ \mathbb{A} \end{smallmatrix} \right) \longrightarrow \left(\begin{smallmatrix} \mathrm{Sub}(\mathbb{B}) \\ \downarrow \\ \mathbb{B} \end{smallmatrix} \right) \qquad \text{and} \qquad \left(\begin{smallmatrix} \mathbb{A}^{\to} \\ \downarrow \\ \mathbb{A} \end{smallmatrix} \right) \longrightarrow \left(\begin{smallmatrix} \mathbb{B}^{\to} \\ \downarrow \\ \mathbb{B} \end{smallmatrix} \right) .$$

between the corresponding subobject and codomain fibrations.

Proof. The functor K preserves monos, since $m: X \to I$ is a mono if and only if the following diagram is a pullback.

$$
\begin{array}{ccc}
X & \xrightarrow{\;\mathrm{id}\;} & X \\
{\scriptstyle \mathrm{id}}\big\downarrow & & \big\downarrow{\scriptstyle m} \\
X & \xrightarrow[m]{} & I
\end{array}
$$

Thus one can define a functor $\mathrm{Sub}(\mathbb{A}) \to \mathrm{Sub}(\mathbb{B})$ by

$$
\left(X \xrightarrowtail{\;m\;} I \right) \longmapsto \left(KX \xrightarrowtail{\;Km\;} KI \right).
$$

It preserves Cartesian morphisms because K preserves pullbacks.

The extension to codomain fibrations is obvious. $\square$

1.7.6. Lemma. *Let $K: (\mathbb{A}, S) \to (\mathbb{B}, T)$ be a morphism of CT-structures (see Definition 1.3.2). One obtains an extension of K to a morphism between the corresponding simple fibrations $\begin{smallmatrix} s(S) \\ \downarrow \\ \mathbb{A} \end{smallmatrix} \to \begin{smallmatrix} s(T) \\ \downarrow \\ \mathbb{B} \end{smallmatrix}$ which preserves the splitting on-the-nose.*

Proof. By definition K preserves finite products, so let $\gamma_{I,J}: KI \times KJ \xrightarrow{\;\cong\;} K(I \times J)$ be the inverse of the canonical map $\langle K\pi, K\pi' \rangle$. One can define a functor $s(K): s(S) \to s(T)$ on objects by $(I, X) \mapsto (KI, KX)$ and on arrows $(u, f): (I, X) \to (J, Y)$—where $u: I \to J$ and $f: I \times X \to Y$—by $(u, f) \mapsto (Ku, Kf \circ \gamma_{I,X})$. The splitting is preserved since

$$
s(K)(u, \pi') = (K(u), K(\pi') \circ \gamma) = (K(u), \pi').   \square
$$

As special case, a finite product preserving functor $\mathbb{A} \to \mathbb{B}$ induces a morphism $\begin{smallmatrix} s(\mathbb{A}) \\ \downarrow \\ \mathbb{A} \end{smallmatrix} \to \begin{smallmatrix} s(\mathbb{B}) \\ \downarrow \\ \mathbb{B} \end{smallmatrix}$ between the corresponding simple fibrations.

2-categorical structure

It turns out that the homsets

$$
\mathbf{Fib}(p, q) \qquad \text{and} \qquad \mathbf{Fib}(\mathbb{B})(p, q)
$$

(and their split versions) are categories themselves. One thus gets 2-categories of fibrations. This extra structure enables us to express various 2-categorical notions—like adjunctions, equivalences or (co)monads—for fibred categories. In general, these notions will be quite different in **Fib** and in **Fib**($\mathbb{B}$), see [125–127] for an investigation. We shall not make deep use of the 2-categorical aspects. And we usually spell out the details of the 2-categorical notions that we use for fibrations. But we do find it convenient to have the language of 2-categories at hand.

1.7.7. Definition. Assume (K, H) and (L, G) are morphisms $\left(\begin{smallmatrix} \mathbb{E} \\ \downarrow p \\ \mathbb{B} \end{smallmatrix} \right) \Rightarrow$ $\left(\begin{smallmatrix} \mathbb{D} \\ \downarrow q \\ \mathbb{A} \end{smallmatrix} \right)$ in **Fib** (*i.e.* 1-cells) as below. A **2-cell** $(K, H) \Rightarrow (L, G)$ in **Fib** consists of a pair of natural transformations $\sigma \colon K \Rightarrow L$ and $\tau \colon H \Rightarrow G$ in a diagram:

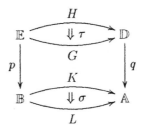

such that τ is above σ; that is, for $X \in \mathbb{E}$, the component τ_X is above the component σ_{pX}. This may be expressed as: the two 2-cells $qH \Rightarrow qG$ and $Kp \Rightarrow Lp$ in the diagram are equal. One obtains that **Fib** is a 2-category, with identities and composition of 2-cells inherited from **Cat**.

The 2-cells in the category **Fib**$_{\text{split}}$ are as in **Fib**. And a 2-cell in **Fib**($\mathbb{B}$) or **Fib**$_{\text{split}}$($\mathbb{B}$) is given by a diagram

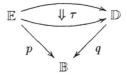

in which every component of τ is vertical. Such a 2-cell in **Fib**($\mathbb{B}$) or **Fib**$_{\text{split}}$($\mathbb{B}$) is often called a **vertical** or **fibred** natural transformation.

Since 'equivalence' is a 2-categorical notion we have that two fibrations $\begin{smallmatrix} \mathbb{E} \\ \downarrow p \\ \mathbb{B} \end{smallmatrix}$ and $\begin{smallmatrix} \mathbb{D} \\ \downarrow q \\ \mathbb{B} \end{smallmatrix}$ with the same basis $\mathbb{B}$ are **equivalent** (formally: equivalent *in* **Fib**($\mathbb{B}$), or *over* $\mathbb{B}$) if there are fibred functors $F \colon \mathbb{E} \to \mathbb{D}$ and $G \colon \mathbb{D} \to \mathbb{E}$ with vertical natural isomorphisms $GF \cong \text{id}_{\mathbb{E}}$ and $FG \cong \text{id}_{\mathbb{D}}$. Several of the equivalences between total categories that we have seen before (see Propositions 1.2.2, 1.4.7, 1.5.3 and Exercise 1.2.3) are actually *fibred* equivalences.

1.7.8. Proposition. *There are fibred equivalences over* **Sets**:

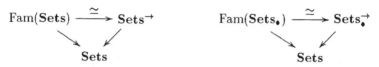

and over ω-**Sets** *and* **PER**:

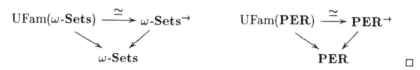

Notice that all the fibrations on the left of $\xrightarrow{\simeq}$ are split, because they involve pointwise indexing.

We mention two lemmas involving fibred 2-cells. The first one is easy.

1.7.9. Lemma. *Let* $K: \mathbb{A} \to \mathbb{B}$ *be a functor. Change-of-base along* K *yields a 2-functor* $K^*: \mathbf{Fib}(\mathbb{B}) \to \mathbf{Fib}(\mathbb{A})$.

It restricts to $\mathbf{Fib}_{\mathbf{split}}(\mathbb{B}) \to \mathbf{Fib}_{\mathbf{split}}(\mathbb{A})$. $\qquad\qquad\qquad$ □

The second lemma is more involved and may be skipped at first reading. The essential point about fibrations is that (single) morphisms in the base category can be lifted. By the universal property of such liftings one can also lift a natural transformation. This is the content of the next result. Since a natural transformation consists of a family of arrows, one needs to lift many maps at the same time, and so we require a cleavage.

1.7.10. Lemma. *Assume that two functors* $K, L: \mathbb{A} \rightrightarrows \mathbb{B}$ *are given with a natural transformation* $\sigma: K \Rightarrow L$ *between them. Let* $\begin{smallmatrix} \mathbb{E} \\ \downarrow p \\ \mathbb{B} \end{smallmatrix}$ *be a cloven fibration; then there is a lifting* $\overline{\sigma}: K'\langle\sigma\rangle \Rightarrow L'$ *of* $\sigma: K \Rightarrow L$ *in a diagram,*

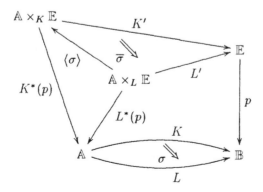

where $\langle \sigma \rangle$ is the functor which sends (I, X) to $(I, \sigma_I^*(X))$. The pair $(\sigma, \overline{\sigma})$ is a 2-cell in **Fib** *from $(K, K'\langle \sigma \rangle)$ to (L, L'). All components of the lifted natural transformation $\overline{\sigma}$ are Cartesian.*

This lifting of σ to $\overline{\sigma}$ enjoys a certain universal property, which will not be made explicit here. But the reader may consult [171] (or also [252, II, 1.7]). In [171] such lifting of natural transformations is described as lifting of 2-cells in a 2-category, and used to give a definition of when a 1-cell $E \to B$ is a fibration (in this 2-category). This yields an alternative to the (2-categorical) definition based on Exercise 1.4.8.

(Later in Exercise 9.3.8 we shall relate (families of) adjoints to reindexing functors $\sigma_I^*: \mathbb{E}_{LI} \to \mathbb{E}_{KI}$ between fibres to adjoints to the above functor $\langle \sigma \rangle: \mathbb{A} \times_K \mathbb{E} \to \mathbb{A} \times_L \mathbb{E}$ between total categories.)

Proof. The component of $\overline{\sigma}$ at $(I, X) \in \mathbb{A} \times_L \mathbb{E}$ is obtained from the cleavage, as:

$$\left(\overline{\sigma}\right)_{(I,X)} \overset{\text{def}}{=} \left(K'\langle \sigma \rangle(I, X) = K'(I, \sigma_I^*(X)) = \sigma_I^*(X) \xrightarrow{\overline{\sigma}_I(X)} X = L'(I, X) \right)$$

using that $X \in \mathbb{E}$ is above the codomain of $\sigma_I: KI \to LI = pX$ in $\mathbb{B}$. This $\overline{\sigma}$ is a natural transformation since for a morphism $(u, f): (I, X) \to (J, Y)$ in $\mathbb{A} \times_L \mathbb{E}$ where $u: I \to J$ in $\mathbb{A}$ and $f: X \to Y$ in $\mathbb{E}$ with $pf = Lu$—one has a naturality square in $\mathbb{B}$:

$$
\begin{array}{ccc}
KI & \xrightarrow{\quad \sigma_I \quad} & LI = pX \\
\ \ \ \searrow Ku & & \searrow Lu = pf \\
& KJ \xrightarrow{\quad \sigma_J \quad} & LJ = pY
\end{array}
$$

And above this diagram in $\mathbb{E}$:

$$
\begin{array}{ccc}
\sigma_I^*(X) & \xrightarrow{\ (\overline{\sigma})_{(I,X)}\ } & X \\
K'\langle\sigma\rangle(u,f) = Ku \ \searrow & & \searrow f = L'(u,f) \\
& \sigma_J^*(Y) \xrightarrow{\ (\overline{\sigma})_{(J,Y)}\ } & Y
\end{array}
$$

where the dashed arrow is the unique one above Ku making the square commute (because $(\overline{\sigma})_{(J,Y)}$ is Cartesian). Thus, basically, $\overline{\sigma}$ is a natural transformation by definition of $\langle \sigma \rangle$. □

Exercises

1.7.1. Show that the categories **Fib** and **Fib$_{\text{split}}$** both have finite products.

1.7.2. Let $\begin{smallmatrix}\mathbb{E}\\\downarrow p\\\mathbb{B}\end{smallmatrix}$ and $\begin{smallmatrix}\mathbb{D}\\\downarrow q\\\mathbb{B}\end{smallmatrix}$ be fibrations and $H\colon \mathbb{E} \to \mathbb{D}$ a functor with $qH = p$.

   (i) Assume H is full and faithful; prove that H reflects Cartesianness, *i.e.* that Hf is Cartesian implies that f is Cartesian.
   [*Hint.* Use Exercise 1.1.2]

   (ii) Assume now that H is a fibred functor, *i.e.* that it preserves Cartesianness. Show that

$$H\colon \mathbb{E} \to \mathbb{D} \text{ is full} \quad \Leftrightarrow \quad \text{every } H_I\colon \mathbb{E}_I \to \mathbb{D}_I \text{ is full.}$$

   And that the same holds for 'faithful' instead of 'full'.

1.7.3. Let **2** be the two-element poset category $\{\bot, \top\}$ with $\bot \leq \top$. Describe an isomorphism of fibrations $\begin{smallmatrix}\text{Fam}(\mathbf{2})\\\downarrow\\\mathbf{Sets}\end{smallmatrix} \cong \begin{smallmatrix}\text{Sub}(\mathbf{Sets})\\\downarrow\\\mathbf{Sets}\end{smallmatrix}$ in **Fib(Sets)**.

1.7.4. Verify that the assignment $\mathbb{C} \mapsto \left(\begin{smallmatrix}\text{Fam}(\mathbb{C})\\\downarrow\\\mathbf{Sets}\end{smallmatrix} \right)$ extends to a (2-)functor

   $\mathbf{Cat} \to \mathbf{Fib}_{\text{split}}(\mathbf{Sets})$.

1.7.5. Check that the assignment $I \mapsto \left(\begin{smallmatrix}\mathbb{B}/I\\\downarrow \text{dom}_I\\\mathbb{B}\end{smallmatrix} \right)$ yields a functor $\mathbb{B} \to$

   Fib$_{\text{split}}(\mathbb{B})$ which preserves finite products.

1.7.6. Let $\mathbb{A}, \mathbb{B}$ be categories with finite products and let $K\colon \mathbb{A} \to \mathbb{B}$ be a functor preserving these. Lemma 1.7.6 yields a map $(K, s(K))\colon \begin{smallmatrix}s(\mathbb{A})\\\downarrow\\\mathbb{A}\end{smallmatrix} \to \begin{smallmatrix}s(\mathbb{B})\\\downarrow\\\mathbb{B}\end{smallmatrix}$ between the associated simple fibrations. Show that the functor $s(K)\colon s(\mathbb{A}) \to s(\mathbb{B})$ between the total categories, restricted to a fibre $\mathbb{A}/\!\!/I \to \mathbb{B}/\!\!/KI$, preserves finite products (see also Exercise 1.3.2).

1.7.7. (See [105, Theorem 3.9].) Notice that (as a special case of Exercise 1.4.6), for every functor $F\colon \mathbb{A} \to \mathbb{B}$, the projection functor $\begin{smallmatrix}(\mathbb{B}\downarrow F)\\\downarrow\\\mathbb{B}\end{smallmatrix}$ from the comma category to $\mathbb{B}$ is a split fibration. Prove that the assignment

$$\left(\begin{smallmatrix}\mathbb{A}\\\downarrow F\\\mathbb{B}\end{smallmatrix} \right) \mapsto \left(\begin{smallmatrix}(\mathbb{B}\downarrow F)\\\downarrow\\\mathbb{B}\end{smallmatrix} \right)$$

   yields a functor $\mathbf{Cat}/\mathbb{B} \to \mathbf{Fib}_{\text{split}}(\mathbb{B})$, which is left adjoint to the inclusion (in the reverse direction). Describe concretely how each functor factors through a split fibration.

1.7.8. Verify that $\langle \sigma \rangle$ in Lemma 1.7.10 is a *fibred* functor $L^*(p) \to K^*(p)$.

1.7.9. Let $\begin{smallmatrix}\mathbb{E}\\\downarrow p\\\mathbb{B}\end{smallmatrix}$ be a fibration. A **fibred monad** on p is a monad on p in the 2-category **Fib$(\mathbb{B})$**. It is thus given by a fibred functor $T\colon \mathbb{E} \to \mathbb{E}$ together with vertical unit $\eta\colon \text{id}_{\mathbb{C}} \Rightarrow T$ and vertical multiplication $\mu\colon T^2 \Rightarrow T$, satisfying $\mu \circ T\eta = \text{id} = \mu \circ \eta_T$ and $\mu \circ T\mu = \mu \circ \mu T$ as usual.

(i)   Show that the (ordinary) Kleisli category $\mathbb{E}_T$ is fibred over $\mathbb{B}$.
(ii) Show also that the Eilenberg-Moore category of algebras $\mathbb{E}^T$ of the
      monad T is fibred over $\mathbb{B}$. (Note that every algebra is automatically
      vertical.)

[These fibrations $\begin{smallmatrix} \mathbb{E}_T \\ \downarrow \\ \mathbb{B} \end{smallmatrix}$ and $\begin{smallmatrix} \mathbb{E}^T \\ \downarrow \\ \mathbb{B} \end{smallmatrix}$ are the Kleisli- and Eilenberg-Moore-objects in the 2-category **Fib**($\mathbb{B}$) (see [314] for what this means). The constructions in **Fib** are quite different, see [129].]

1.8 Fibrewise structure and fibred adjunctions

In ordinary categories one can describe binary products × or coproducts + in familiar ways, for example in terms of their universal properties. The question arises whether such structure also makes sense for fibred categories, and if so, what does it mean. One answer here will be: products × in every fibre, preserved by reindexing functors u^* between these fibres. This gives "fibrewise structure". It will be our first concern in this section.

In a next step one notices that (chosen) products × for ordinary categories can equivalently be described in terms of ordinary adjunctions; that is, in terms of adjunctions in the 2-category **Cat** of categories. It turns out that such fibrewise structure can similarly be described in terms of suitable adjunctions between fibrations. Formally, such "fibred" adjunctions are adjunctions in a 2-category of fibrations **Fib**($\mathbb{B}$) over a fixed base category $\mathbb{B}$. This will be our second concern.

(There is also an alternative answer which is of a global nature and will be of less interest here. It involves structure defined by adjunctions in the 2-category **Fib** of fibrations over arbitrary bases. See for example Exercises 1.8.10 and 1.8.11. In the latter one finds how adjunctions in **Fib** reduce to adjunctions over a fixed basis.)

1.8.1. Definition. Let ◇ be some categorical property or structure (for example some limit or colimit or exponent)

(i)   We say a fibration has **fibred** ◇'s (or also, **fibrewise** ◇'s) if all fibre categories have ◇'s and reindexing functors preserve ◇'s. A split fibration has **split fibred** ◇'s if all fibres have (chosen) ◇'s and the reindexing functors induced by the splitting preserve ◇'s on-the-nose.

(The predicate 'fibred' is sometimes omitted, when it is clear that we talk about fibred categories.)

(ii) A morphism $\left(\begin{smallmatrix} \mathbb{E} \\ \downarrow p \\ \mathbb{B} \end{smallmatrix} \right) \xrightarrow{(K,L)} \left(\begin{smallmatrix} \mathbb{D} \\ \downarrow q \\ \mathbb{A} \end{smallmatrix} \right)$ of fibrations with ◇'s preserves (fibred) ◇'s if for each object $I \in \mathbb{B}$ the functor $L_I \colon \mathbb{E}_I \to \mathbb{D}_{KI}$ preserves ◇'s.

For the split version, one requires preservation on-the-nose.

The following notion deserves explicit attention because of its frequent use.

1.8.2. Definition. A (split) **fibred CCC** or **Cartesian closed fibration** is a fibration with (split) fibred finite products and exponents.

1.8.3. Examples. (i) Usually, ordinary categorical structure exists in a category $\mathbb{C}$ if and only if the corresponding fibred structure exists in the family fibration $\begin{smallmatrix} \text{Fam}(\mathbb{C}) \\ \downarrow \\ \textbf{Sets} \end{smallmatrix}$. For example:

$$\mathbb{C} \text{ is a CCC (with chosen structure)} \quad \Leftrightarrow \quad \begin{smallmatrix} \text{Fam}(\mathbb{C}) \\ \downarrow \\ \textbf{Sets} \end{smallmatrix} \text{ is a split fibred CCC.}$$

The implication ($\Rightarrow$) follows from a pointwise construction: *e.g.* the Cartesian product of families $(X_j)_{j \in J}$ and $(Y_j)_{j \in J}$ in the fibre over J is $(X_j \times Y_j)_{j \in J}$. Reindexing preserves this structure on-the-nose: for $u: I \to J$ in **Sets** we get:

$$
\begin{aligned}
u^*\big((X_j)_{j \in J} \times (Y_j)_{j \in J}\big) &= u^*\big((X_j \times Y_j)_{j \in J}\big) \\
&= (X_{u(i)} \times Y_{u(i)})_{i \in I} \\
&= (X_{u(i)})_{i \in I} \times (Y_{u(i)})_{i \in I} \\
&= u^*\big((X_j)_{j \in J}\big) \times u^*\big((Y_j)_{j \in J}\big).
\end{aligned}
$$

The implication ($\Leftarrow$) in the reverse direction follows from the fact that the category $\mathbb{C}$ is isomorphic to the fibre $\text{Fam}(\mathbb{C})_1$ above the terminal object 1— which is a CCC, by assumption.

(ii) Exercise 1.3.1 almost contains the result that for a category $\mathbb{B}$ with finite products, the simple fibration $\begin{smallmatrix} s(\mathbb{B}) \\ \downarrow \\ \mathbb{B} \end{smallmatrix}$ has split finite products. The only requirement that should still be verified is that reindexing functors preserve the fibrewise structure. This is easy. Moreover, this result can be extended to: $\begin{smallmatrix} s(\mathbb{B}) \\ \downarrow \\ \mathbb{B} \end{smallmatrix}$ is a split fibred CCC if and only if $\mathbb{B}$ is a CCC.

(iii) For a category $\mathbb{B}$ with finite limits, the codomain fibration $\begin{smallmatrix} \mathbb{B}^{\to} \\ \downarrow \\ \mathbb{B} \end{smallmatrix}$ on $\mathbb{B}$ always has fibred finite limits. The same holds for the subobject fibration $\begin{smallmatrix} \text{Sub}(\mathbb{B}) \\ \downarrow \\ \mathbb{B} \end{smallmatrix}$ on $\mathbb{B}$. And for a finite limit preserving functor $F: \mathbb{A} \to \mathbb{B}$ between categories $\mathbb{A}, \mathbb{B}$ with finite limits, the induced morphisms of fibrations $\begin{smallmatrix} \mathbb{A}^{\to} \\ \downarrow \\ \mathbb{A} \end{smallmatrix} \to \begin{smallmatrix} \mathbb{B}^{\to} \\ \downarrow \\ \mathbb{B} \end{smallmatrix}$ and $\begin{smallmatrix} \text{Sub}(\mathbb{A}) \\ \downarrow \\ \mathbb{A} \end{smallmatrix} \to \begin{smallmatrix} \text{Sub}(\mathbb{B}) \\ \downarrow \\ \mathbb{B} \end{smallmatrix}$ (see Proposition 1.7.5) preserve fibred finite limits.

(iv) A category $\mathbb{B}$ with finite limits is a **locally Cartesian closed category** (LCCC)—*i.e.* all its slice categories $\mathbb{B}/I$ are Cartesian closed—if and only if the codomain fibration $\begin{smallmatrix} \mathbb{B}^{\rightarrow} \\ \downarrow \\ \mathbb{B} \end{smallmatrix}$ is a fibred CCC. The (if)-part of the statement is obvious by definition of LCCC. For (only if), it remains to verify that the reindexing functors (given by pullback) preserve the exponents in the fibres (often called **local exponentials**). This will be postponed until Exercise 1.9.4 (iii).

(v) Recall from Section 1.2 that the categories **PER** and ω-**Sets** have finite limits and are Cartesian closed. By a pointwise construction (as in (i) above) this structure lifts to *split* fibrewise finite limits and exponents for the fibrations $\begin{smallmatrix} \mathrm{UFam}(\omega\text{-}\mathbf{Sets}) \\ \downarrow \\ \omega\text{-}\mathbf{Sets} \end{smallmatrix}$, $\begin{smallmatrix} \mathrm{UFam}(\mathbf{PER}) \\ \downarrow \\ \omega\text{-}\mathbf{Sets} \end{smallmatrix}$ and $\begin{smallmatrix} \mathrm{UFam}(\mathbf{PER}) \\ \downarrow \\ \mathbf{PER} \end{smallmatrix}$ of ω-sets and PERs over ω-sets, and of PERs over PERs.

The following result is often useful.

1.8.4. Lemma. *Let $\Diamond$ be as in Definition 1.8.1. If a (split) fibration p has (split) fibred $\Diamond$'s, then so has a fibration $K^*(p)$ obtained by change-of-base. Moreover, the associated morphism of fibrations $K^*(p) \to p$ preserves $\Diamond$'s.*

Proof. Suppose p has fibred $\Diamond$'s. The fibre of $K^*(p)$ above I is isomorphic to the fibre of p above KI. Hence $K^*(p)$ has $\Diamond$'s in its fibre categories. They are preserved under reindexing, since the reindexing functors of $K^*(p)$ are obtained from those of p. $\qquad\square$

1.8.5. Example. The category **Sets** has all (small) limits and colimits. Hence by Example 1.8.3 (i) the family fibration $\begin{smallmatrix} \mathrm{Fam}(\mathbf{Sets}) \\ \downarrow \\ \mathbf{Sets} \end{smallmatrix}$ of set-indexed sets has these limits and colimits in split form. Recall from Definition 1.6.1 that the fibration $\begin{smallmatrix} \mathbf{Sign} \\ \downarrow \\ \mathbf{Sets} \end{smallmatrix}$ of many-typed signatures is obtained by change-of-base from this family fibration. Hence the fibration of signatures has split limits and colimits. Moreover, the morphism of fibrations $\begin{smallmatrix} \mathbf{Sign} \\ \downarrow \\ \mathbf{Sets} \end{smallmatrix} \to \begin{smallmatrix} \mathrm{Fam}(\mathbf{Sets}) \\ \downarrow \\ \mathbf{Sets} \end{smallmatrix}$ preserves these.

Adjunctions between fibred categories

We begin the study of fibred adjunctions with an example. Recall that an ordinary category $\mathbb{C}$ has a terminal object if and only if the unique functor $\mathbb{C} \dashrightarrow \mathbf{1}$ from $\mathbb{C}$ to the terminal category $\mathbf{1}$ has a right adjoint (written as $1\colon \mathbf{1} \to \mathbb{C}$). The situation is similar for fibred categories. Consider for example a codomain fibration $\begin{smallmatrix} \mathbb{B}^{\rightarrow} \\ \downarrow \\ \mathbb{B} \end{smallmatrix}$. Every fibre $\mathbb{B}/J$ has a terminal object, namely the identity family $1J = \left(\begin{smallmatrix} J \\ \downarrow \mathrm{id} \\ J \end{smallmatrix} \right)$. The assignment $J \mapsto 1J$ then extends to a functor $1\colon \mathbb{B} \to \mathbb{B}^{\rightarrow}$. It has the following properties.

(i) This functor 1 can be described as a fibred functor $\mathrm{id}_{\mathbb{B}} \to \mathrm{cod}$ as in

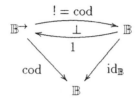

where the identity functor $\mathrm{id}_{\mathbb{B}}$ is the terminal object in the category **Fib**($\mathbb{B}$) of fibrations over $\mathbb{B}$.

(ii) The functor 1 is right adjoint to the unique morphism $\mathrm{cod} \dashrightarrow \mathrm{id}_{\mathbb{B}}$ in **Fib**($\mathbb{B}$): there are obvious adjoint correspondences

$$
\left(\begin{matrix} X \\ \downarrow\varphi \\ I \end{matrix} \right) \longrightarrow \left(\begin{matrix} J \\ \downarrow\mathrm{id} \\ J \end{matrix} \right) = 1J \quad \text{in } \mathbb{B}^{\to}
$$

$$
\mathrm{cod}\left(\begin{matrix} X \\ \downarrow\varphi \\ I \end{matrix} \right) = I \longrightarrow J \quad \text{in } \mathbb{B}
$$

Moreover, the unit and counit of this adjunction are vertical in the above triangle.

These two points establish that the fibred terminal object functor $1 \colon \mathbb{B} \longrightarrow \mathbb{B}^{\to}$ obtained by taking fibrewise terminal objects, is a 'fibred right adjoint' to the functor $\mathrm{cod} \dashrightarrow \mathrm{id}_{\mathbb{B}}$—just like in the case of ordinary categories a terminal object in $\mathbb{C}$ is given by a right adjoint to the functor $\mathbb{C} \dashrightarrow 1$.

Below we present the general formulation of the notion of fibred adjunction. Formally, it is an adjunction in a 2-category of fibrations over a fixed base category.

1.8.6. Definition. (i) Let $\begin{smallmatrix}\mathbb{E}\\\downarrow p\\\mathbb{B}\end{smallmatrix}$ and $\begin{smallmatrix}\mathbb{D}\\\downarrow q\\\mathbb{B}\end{smallmatrix}$ be fibrations with the same base category $\mathbb{B}$. A **fibred adjunction** over $\mathbb{B}$ is given by fibred functors F, G in

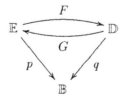

together with *vertical* natural transformations $\eta \colon \mathrm{id}_{\mathbb{E}} \Rightarrow GF$ and $\varepsilon \colon FG \Rightarrow \mathrm{id}_{\mathbb{D}}$ satisfying the usual triangular identities $G\varepsilon \circ \eta G = \mathrm{id}$ and $\varepsilon F \circ F\eta = \mathrm{id}$. This is an adjunction in the 2-category **Fib**($\mathbb{B}$); it obviously involves an ordinary adjunction $(F \dashv G)$.

(ii) A **split fibred adjunction** over $\mathbb{B}$ is an adjunction in the 2-category **Fib**$_{\mathbf{split}}(\mathbb{B})$; it consists a fibred adjunction as above in which the fibrations p and q are split and also the functors F and G are split (*i.e.* preserve the splitting).

Notice that verticality of the unit η of an adjunction $(F \dashv G)$ between fibred functors as above implies verticality of the counit, and vice-versa.

1.8.7. Examples. (i) Every ordinary adjunction $(F \dashv G)$ in:

$$\mathbb{C} \xleftarrow[\;\;G\;\;]{\;\;F\;\;} \perp \;\;\mathbb{D}$$

lifts to a split fibred adjunction $(\mathrm{Fam}(F) \dashv \mathrm{Fam}(G))$ over **Sets** in:

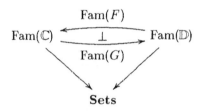

by a pointwise construction. Essentially this follows from the 2-functoriality of $\mathrm{Fam}(-)$ in Exercise 1.7.4.

(ii) In a similar way, the reflection

$$\mathbf{PER} \xhookrightarrow{\;\;\;\;} \omega\text{-}\mathbf{Sets}$$

from Proposition 1.2.7 lifts to a fibred reflection

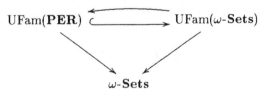

again by a pointwise construction.

(But this lifting over ω-**Sets** is less trivial than over **Sets** in the previous example, since one needs to check that the (pointwise defined) units and counits have uniform realisers.)

The earlier example involving fibred terminal objects for a codomain fibration can now be described for arbitrary fibrations.

1.8.8. Lemma. *A fibration $\begin{smallmatrix}\mathbb{E}\\\downarrow p\\\mathbb{B}\end{smallmatrix}$ has a fibred terminal object if and only if the unique morphism from p to the terminal object in* **Fib**$(\mathbb{B})$ *has a fibred right adjoint, say 1, in*

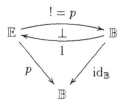

Proof. Assume that each fibre category $\mathbb{E}_I$ has a terminal object $1I$, and that these terminal objects are preserved by reindexing functors: for $u: I \rightarrow J$ in $\mathbb{B}$ one has $u^*(1J) \stackrel{\cong}{\longrightarrow} 1I$ over I. Then one gets a functor $1: \mathbb{B} \rightarrow \mathbb{E}$, since a morphism $u: I \rightarrow J$ in $\mathbb{B}$ can be mapped to the composite $1I \cong u^*(1J) \rightarrow 1J$ over u. Thus $p \circ 1 = \mathrm{id}_\mathbb{B}$. Moreover, 1 is a fibred functor in the above diagram, since each map $1u$ is Cartesian by construction. Further, there are adjoint correspondences

$$\frac{pX \xrightarrow{\;\;u\;\;} J \quad \text{in } \mathbb{B}}{X \xrightarrow[f]{} 1J \quad \text{in } \mathbb{E}}$$

given by $u \mapsto (X \stackrel{!}{\rightarrow} 1pX \stackrel{1u}{\rightarrow} 1J)$ and $f \mapsto pf$. The resulting unit is the unique map $!: X \dashrightarrow 1pX$ (which is p-vertical) and counit is the identity $p1J \stackrel{\cong}{\Rightarrow} J$ (which is $\mathrm{id}_\mathbb{B}$-vertical).

Conversely, if the above functor $p: p \dashrightarrow \mathrm{id}_\mathbb{B}$ has a fibred right adjoint $1: \mathbb{B} \rightarrow \mathbb{E}$, then for each object $I \in \mathbb{B}$ the object $1I$ is terminal in the fibre $\mathbb{E}_I$ over I: the counit component ε_I is $\mathrm{id}_\mathbb{B}$-vertical and therefore an identity $p1I \stackrel{\cong}{\Rightarrow} I$. Hence the transpose of a map $f: X \rightarrow 1I$ is $pf: pX \rightarrow I$, so that there is precisely one vertical map $X \rightarrow 1I$.

Further, reindexing functors preserve these fibred terminal objects: a map $u: J \rightarrow I$ in $\mathbb{B}$ is $\mathrm{id}_\mathbb{B}$-Cartesian over itself; hence $1u: 1J \rightarrow 1I$ is p-Cartesian over u, since 1 is by assumption a fibred functor. But by definition, also the lifting $\overline{u}(1I): u^*(1I) \rightarrow 1I$ is Cartesian over u. This yields an isomorphism $u^*(1I) \stackrel{\cong}{\rightarrow} 1J$, since Cartesian liftings are unique, up-to-isomorphism. $\quad\square$

Having seen this lemma, one expects that in general the structure induced by a fibred adjunction is induced fibrewise and is preserved under reindexing. The following result states that this is indeed the case. The preservation is expressed by a so-called 'Beck-Chevalley condition', which may be a bit puzzling at first sight. We elaborate later on.

(We should emphasise that not all fibrewise structure comes from fibred adjunctions. For example, a fibration may have fibrewise a monoidal structure.)

1.8.9. Lemma. *Let* $\begin{smallmatrix} \mathbb{E} \\ \downarrow p \\ \mathbb{B} \end{smallmatrix}$ *and* $\begin{smallmatrix} \mathbb{D} \\ \downarrow q \\ \mathbb{B} \end{smallmatrix}$ *be fibrations and let* $H\colon \mathbb{E} \to \mathbb{D}$ *be a fibred functor. This functor* H *has a fibred left (resp. right) adjoint if and only if both*

(a) For each object $I \in \mathbb{B}$ *the functor* $H_I\colon \mathbb{E}_I \to \mathbb{D}_I$ *restricted to the fibres over* I *has a left (resp. right) adjoint* $K(I)$.
(b) The **Beck-Chevalley condition** *holds, i.e. for every map* $u\colon I \to J$ *in* $\mathbb{B}$ *and for every pair of reindexing functors*

$$\mathbb{E}_J \xrightarrow{\ u^* \ } \mathbb{E}_I \qquad\qquad \mathbb{D}_J \xrightarrow{\ u^\# \ } \mathbb{D}_I$$

the canonical natural transformation

$$K(I)u^\# \Longrightarrow u^* K(J) \qquad\qquad (resp.\ u^* K(J) \Longrightarrow K(I)u^\#)$$

is an isomorphism.

The lemma describes global adjunctions $K \dashv H$ (or $H \dashv K$) in terms of local adjunctions $K(I) \dashv H_I$ (or $H_I \dashv K(I)$) which are suitably preserved by reindexing functors. In the local left adjoint situation:

$$
\begin{array}{ccc}
\mathbb{E}_J & \xrightarrow{\ \ \ u^* \ \ \ } & \mathbb{E}_I \\[2pt]
K(J) \left\uparrow\downarrow\right\dashv H_J & & K(I) \left\uparrow\downarrow\right\dashv H_I \\[2pt]
\mathbb{D}_J & \xrightarrow[\ \ \ u^\# \ \ \ ]{} & \mathbb{D}_I
\end{array}
$$

the canonical map $K(I)u^\# \Rightarrow u^* K(J)$ arises as the transpose of

$$u^\# \xrightarrow{\ u^\#(\eta)\ } u^\# H_J K(J) \xrightarrow{\ \cong\ } H_I u^* K(J)$$

Alternatively, it may be described as the following (pasting) composite.

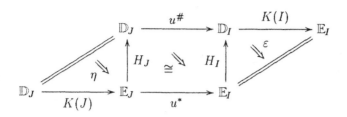

Proof. First, in case $K:\mathbb{D} \to \mathbb{E}$ is a fibred left or right adjoint to H, then one obtains adjunctions between the fibres since the unit and counit of a fibred adjunction are vertical. For a morphism $u:I \to J$ in $\mathbb{B}$ and an object $Y \in \mathbb{D}$ over $J \in \mathbb{B}$, we get two Cartesian liftings of u at Y in a situation:

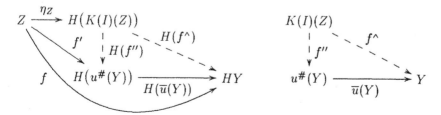

$(*)$

An appropriate diagram chase shows that this map $K(u^{\#}(Y)) \xrightarrow{\cong} u^{*}(KY)$ is the canonical isomorphism induced by the adjunction.

Conversely, assume local adjunctions satisfying Beck-Chevalley. We shall do the left adjoint case. We claim that for each object $Z \in \mathbb{D}$, say above $I \in \mathbb{B}$, the (vertical) unit component $\eta_Z: Z \to H(K(I)(Z))$ is a (global) universal map from Z to H. Indeed, for a morphism $f:Z \to HY$ in $\mathbb{D}$, say above $u:I \to J$ in $\mathbb{B}$, write $f = H(\overline{u}(Y)) \circ f':Z \to H(u^{\#}(Y)) \to HY$. By the local adjunctions $K(I) \dashv H_I$ we get a unique vertical map $f'':K(I)(Z) \to u^{\#}(Y)$ with $H(f'') \circ \eta_Z = f'$. Then $f^{\wedge} = \overline{u}(Y) \circ f'':K(I)(Z) \to Y$ is the required unique map with $H(f^{\wedge}) \circ \eta_Z = f$ in:

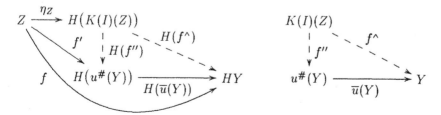

The assignment $Z \mapsto K(pZ)(Z)$ now extends to a functor $K: \mathbb{D} \to \mathbb{E}$, which is left adjoint to H (see *e.g.* [187, IV, 1, Theorem 2]). What remains is to show that K is a fibred functor. This follows because, by universality of η, the triangle of the above diagram (∗) commutes.   □

There is a similar result for *split* fibred adjunctions.

1.8.10. Lemma. *Let* $\left(\begin{smallmatrix} \mathbb{E} \\ \downarrow p \\ \mathbb{B} \end{smallmatrix} \right) \xrightarrow{H} \left(\begin{smallmatrix} \mathbb{D} \\ \downarrow q \\ \mathbb{B} \end{smallmatrix} \right)$ *be a split functor between split fibrations. Then H has a* split *fibred left/right adjoint if and only if one has like in the previous lemma, (a) and the Beck-Chevalley condition (b), but this time with the canonical map being an* identity.   □

1.8.11. Excurs on the Beck-Chevalley condition. The above lemmas express that a (split) fibred adjunction corresponds to fibrewise adjunctions, involving adjunctions between fibres and reindexing functors preserving this structure. The latter is formulated by a Beck-Chevalley condition, which requires a certain natural transformation to be an isomorphism. We shall have a closer look at this condition via an example.

Let $\mathbb{C}$ be an ordinary category with Cartesian products, given by a right adjoint $\times: \mathbb{C} \times \mathbb{C} \to \mathbb{C}$ in **Cat** to the diagonal $\Delta: \mathbb{C} \to \mathbb{C} \times \mathbb{C}$. The unit η_Z is usually described as the diagonal $\langle \mathrm{id}_Z, \mathrm{id}_Z \rangle: Z \to Z \times Z$ and the counit $\varepsilon_{(X,Y)}$ as the pair $(\pi, \pi'): (X \times Y, X \times Y) \to (X, Y)$ of projections in $\mathbb{C} \times \mathbb{C}$. If $\mathbb{D}$ is another category with Cartesian products then one says that a functor $F: \mathbb{C} \to \mathbb{D}$ preserves these products if the pair $F(\pi_{X,Y})$, $F(\pi'_{X,Y})$ forms a Cartesian product diagram in $\mathbb{D}$. Put a bit differently, one requires that the canonical map

$$F(X \times Y) \xrightarrow{\langle F(\pi_{X,Y}), F(\pi'_{X,Y}) \rangle} FX \times FY \qquad (*)$$

is an isomorphism. It arises as transpose of the pair

$$(F(X \times Y), F(X \times Y)) \xrightarrow{(F(\pi_{X,Y}), F(\pi'_{X,Y}))} (FX, FY)$$

in $\mathbb{D} \times \mathbb{D}$. That is, of

$$(F \times F)(X \times Y, X \times Y) \xrightarrow{(F \times F)(\varepsilon_{(X,Y)})} (F \times F)(X, Y)$$

which is a specific case of the above general description of canonical map. We have thus shown that the canonical map formulation as used in the Beck-Chevalley condition corresponds to the usual formulation of preservation for

Cartesian products. The correspondence is a general phenomenon, which is described in more detail in Exercise 1.8.7 below.

In the above Lemma 1.8.10, dealing with split fibred adjunctions and adjunctions between fibres, it is required that the canonical map is the *identity*. In the example of Cartesian products, the requirement that the map (*) is the identity contains much more information than merely $F(X \times Y) = FX \times FY$: it implies that $(F \times F)(\varepsilon_{X,Y}) = \varepsilon'_{FX,FY}$, where ε' is the counit of the adjunction associated with the Cartesian products on $\mathbb{D}$. It also implies $F(\eta_Z) = \eta'_{FZ}$, since

$$
\begin{aligned}
\eta'_{FZ} &= \langle \pi_{FZ,FZ}, \pi_{FZ,FZ} \rangle \circ \langle F(\mathrm{id}_Z), F(\mathrm{id}_Z) \rangle \\
&= \langle F(\pi_{Z,Z}), F(\pi'_{Z,Z}) \rangle \circ F\langle \mathrm{id}_Z, \mathrm{id}_Z \rangle \\
&= F\langle \mathrm{id}_Z, \mathrm{id}_Z \rangle \\
&= F(\eta_Z).
\end{aligned}
$$

Thus, the requirement that the canonical map (*) is an identity morphism leads to a so-called "map of adjunctions" (see [187]), namely from the adjunction ($\Delta \dashv \times$) on $\mathbb{C}$ to the adjunction ($\Delta \dashv \times$) on $\mathbb{D}$, as in the following diagram.

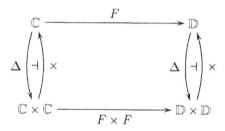

Conversely one easily establishes that if this diagram forms a map of adjunctions, then the canonical map (*) is an identity. Again, this holds more generally, as made explicit by the next lemma below. The proof is easy and left to the reader.

1.8.12. Lemma. *Let* $\left(\begin{smallmatrix} \mathbb{E} \\ \downarrow p \\ \mathbb{B} \end{smallmatrix} \right) \xrightarrow{H} \left(\begin{smallmatrix} \mathbb{D} \\ \downarrow q \\ \mathbb{B} \end{smallmatrix} \right)$ *be a split functor between split fibrations p and q. Then H has a split fibred left/right adjoint if and only if both*

(a) For each object $I \in \mathbb{B}$ the functor $H_I : \mathbb{E}_I \to \mathbb{D}_I$ restricted to the fibres over I has a left (resp. right) adjoint $K(I)$.

(b) for every map $u: I \to J$ in $\mathbb{B}$, the pair of reindexing functors $u^*: \mathbb{E}_J \to \mathbb{E}_I$, $u^\#: \mathbb{D}_J \to \mathbb{D}_I$ induced by the splitting, forms a map of adjunctions in

$$
\begin{array}{ccc}
\mathbb{E}_J & \xrightarrow{\quad u^* \quad} & \mathbb{E}_I \\
H_J \left(\ \right) K(J) & & H_I \left(\ \right) K(I) \\
\mathbb{D}_J & \xrightarrow[\quad u^\# \quad]{} & \mathbb{D}_I
\end{array}
$$
□

In the sequel we shall often describe a specific fibred adjunction by a collection of fibrewise adjunctions and leave verification of the Beck-Chevalley condition as an exercise. It usually follows in a straightforward way when the adjunctions between the fibres are defined in a suitably uniform manner.

Since a fibred adjunction involves ordinary adjunctions between fibre categories it is immediate that a fibred right adjoint preserves fibred limits, and that a fibred left adjoint preserves fibred colimits (see *e.g.* [187, V, 5, Theorem 1]). There are also fibred versions of the adjoint functor theorems, but we shall not need them and we refer the interested reader to [47] and [246]. They involve suitable fibred notions of generators and well-poweredness.

Exercises

1.8.1.   Explain in detail what a 'fibred LCCC' is.

1.8.2.   Let $(\mathbb{B}, T)$ be a non-trivial CT-structure. Prove that the associated simple fibration $\begin{smallmatrix} s(T) \\ \downarrow \\ \mathbb{B} \end{smallmatrix}$ has a fibred terminal object if and only if the collection of types T contains a terminal object (in $\mathbb{B}$).

1.8.3.   (i) Prove that a category with Cartesian products has distributive coproducts (see Exercise 1.5.6) if and only if its simple fibration has fibred (distributive) coproducts.

        (ii) And similarly, that a category with pullbacks has (finite) universal coproducts if and only if its codomain fibration has fibred (universal) coproducts.

1.8.4.   Show in detail (as in Lemma 1.8.8) that a fibration $\begin{smallmatrix} \mathbb{E} \\ \downarrow p \\ \mathbb{B} \end{smallmatrix}$ has fibred Cartesian products $\times$ if and only if the diagonal $\Delta: p \to p \times p$ in $\mathbf{Fib}(\mathbb{B})$ has a fibred right adjoint.

1.8.5.   Consider fibrations $\begin{smallmatrix} \mathbb{E} \\ \downarrow p \\ \mathbb{B} \end{smallmatrix}$ and $\begin{smallmatrix} \mathbb{D} \\ \downarrow q \\ \mathbb{B} \end{smallmatrix}$ together with a (not necessarily fibred) functor $F: \mathbb{E} \to \mathbb{D}$ with right adjoint G such that (a) $qF = p$ and $pG = q$, and (b) the unit and counit of the adjunction $(F \dashv G)$ are vertical. Prove that G is then a fibred functor.

[*Hint.* For a short proof, use Exercise 1.1.2, but see also [344, Lemma 4.5].]

1.8.6.   Let **SignVar** be the category of 'signatures with variables' obtained in the following change-of-base situation,

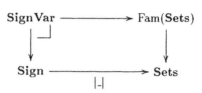

Thus an object of **SignVar** is a many-typed signature Σ together with a $|\Sigma|$-indexed collection $X = (X_\sigma)_{\sigma \in |\Sigma|}$ of sets (of variables). Show that the term model assignment

$$(\Sigma, X) \mapsto (\mathrm{Terms}_\tau(X))$$

described in Example 1.6.5 extends to a left adjoint to the forgetful functor **S-Model** → **SignVar** which sends a model $(\Sigma, (A_\sigma), [\![\_]\!])$ to $(\Sigma, (A_\sigma))$ in

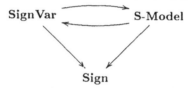

Check that it is *not* a fibred adjunction (as noted by Meseguer).

1.8.7.   Consider two adjunctions in the following (non-commuting) diagram.

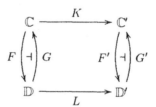

Following [157] we say that a **pseudo-map of adjunctions** from $(F \dashv G)$ to $(F' \dashv G')$ consists of a pair of functors $K : \mathbb{C} \to \mathbb{C}'$, $L : \mathbb{D} \to \mathbb{D}'$ together with natural isomorphisms $\varphi : F'K \overset{\cong}{\Rightarrow} LF$ and $\psi : G'L \overset{\cong}{\Rightarrow} KG$ satisfying $\psi F \circ G' \varphi \circ \eta' K = K\eta$ and $L\varepsilon \circ \varphi G \circ F'\psi = \varepsilon' L$, where η, ε and η', ε' are the unit and counit of the adjunctions $(F \dashv G)$ and $(F' \dashv G')$.

[A map of adjunctions, as defined in [187], has $\varphi = \mathrm{id}$ and $\psi = \mathrm{id}$. But see also *loc. cit.* Exercise IV 7 4, where there is a weaker notion (due to Kelly—with natural transformation φ and ψ^{-1} as above, except that they need not be isomorphisms.]

(i)   These isomorphisms φ and ψ turn out to determine each other: given an isomorphism $F'K \cong LF$, show that one obtains a pseudo-map of adjunctions if and only if the canonical map $KG \Rightarrow G'L$ is an isomorphism. The latter is obtained by transposing $F'KG \cong LFG \Rightarrow L$.

(ii) Formulate and prove a dual version of (i).

(iii) Show that a result like Lemma 1.8.12 can be obtained for arbitrary (non-split) fibrations with 'map of adjunctions' in (b) replaced by 'pseudo-map of adjunctions'.

1.8.8.   Let $\mathbb{J}$ be a category (thought of as index) and $\begin{smallmatrix}\mathbb{E}\\\downarrow p\\\mathbb{B}\end{smallmatrix}$ be a fibration.

(i)   Show that the composition functor $(p \circ -): \mathbb{E}^{\mathbb{J}} \to \mathbb{B}^{\mathbb{J}}$ between functor categories is a fibration.

Let $\delta: \mathbb{B} \to \mathbb{B}^{\mathbb{J}}$ be the diagonal functor which maps $I \in \mathbb{B}$ to the constant functor $\mathbb{J} \to \mathbb{B}$ that maps everything to I (*i.e.* the exponential transpose of the projection $\mathbb{B} \times \mathbb{J} \to \mathbb{B}$). Form the exponent fibration $p^{\mathbb{J}}$ by change-of-base,

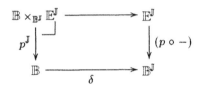

(ii)   Describe the resulting fibred diagonal functor $\Delta: p \to p^{\mathbb{J}}$ over $\mathbb{B}$.

(iii) Show that the fibration p has fibred limits (resp. colimits) of shape $\mathbb{J}$ if and only if this Δ has a fibred right (resp. left) adjoint.

(iv) Give a similar analysis for *split* (co)limits.

1.8.9.   Exponents in an ordinary category can be described in terms of adjunctions involving a parameter, see [187]. This approach does not generalise readily to fibred categories. We sketch an alternative approach, as taken in [157].

Let $\mathbb{C}$ be a category with Cartesian products, say described by the functor $\times: \mathbb{C} \times \mathbb{C} \to \mathbb{C}$. Write $|\mathbb{C}|$ for the discrete category underlying $\mathbb{C}$, of objects only. We extend the Cartesian products to a functor $\mathrm{prod}: |\mathbb{C}| \times \mathbb{C} \to |\mathbb{C}| \times \mathbb{C}$ by $(X, X') \mapsto (X, X \times X')$.

(i)   Check that the category $\mathbb{C}$ has (chosen) exponents if and only if this functor prod has a right adjoint.

(ii)   Show that for a fibration $\begin{smallmatrix}\mathbb{E}\\\downarrow p\\\mathbb{B}\end{smallmatrix}$ with Cartesian products, one can define in a similar way a *fibred* functor $\mathrm{prod}: |p| \times p \to |p| \times p$, where $|p|$ is the object fibration associated with p, as introduced in Exercise 1.1.4.

(iii) Prove now that such a fibration p has fibred exponents if and only if this functor prod has a fibred right adjoint.

(iv) Assume next that $\begin{smallmatrix}\mathbb{E}\\\downarrow p\\\mathbb{B}\end{smallmatrix}$ is a *split* fibration with split Cartesian products. Write $\mathrm{Split}(\mathbb{E})$ for the subcategory of $\mathbb{E}$ with arrows obtained from the splitting and $\overset{\text{Split}(\mathbb{E})}{\underset{\mathbb{B}}{\downarrow\|p\|}}$ for the resulting split fibration. Show that p has split exponents if and only if the split functor prod: $\|p\| \times p \to \|p\| \times p$ has a split fibred right adjoint.

[For a split fibration p, $\|p\|$ (instead of $|p|$) is the appropriate fibration of objects of p.]

1.8.10. Definition 1.8.6 describes adjunctions in the 2-category $\mathbf{Fib}(\mathbb{B})$ for a fixed base category $\mathbb{B}$. One can also consider adjunctions in the 2-category $\mathbf{Fib}$ of fibrations over arbitrary bases.

(i) Describe such adjunctions in $\mathbf{Fib}$ in detail.

(ii) Recall from Exercise 1.7.1 that the category $\mathbf{Fib}$ has Cartesian products. Show that a fibration p has fibred Cartesian products plus Cartesian products in its base category if and only if the diagonal $\Delta: p \to p \times p$ in $\mathbf{Fib}$ has a right adjoint in $\mathbf{Fib}$, *i.e.* if p has Cartesian products in $\mathbf{Fib}$.

1.8.11. In this exercise we relate adjunctions in $\mathbf{Fib}(-)$ and adjunctions in $\mathbf{Fib}$, following [125–127]. Consider a fibration $\overset{\mathbb{E}}{\underset{\mathbb{B}}{\downarrow p}}$ and a functor $F: \mathbb{A} \to \mathbb{B}$.

(i) Show that a (ordinary) right adjoint $G: \mathbb{B} \to \mathbb{A}$ to F induces a right adjoint in $\mathbf{Fib}$ to $F^*(p) \to p$.

[*Hint.* For $X \in \mathbb{E}$ above $I \in \mathbb{B}$, consider the pair $(GI, \varepsilon_I^*(X))$ in the total category $\mathbb{A} \times_{\mathbb{B}} \mathbb{E}$ of $F^*(p)$, where ε is the counit of the adjunction $(F \dashv G)$.]

(ii) Assume now that F has a right adjoint G. Let $\overset{\mathbb{D}}{\underset{\mathbb{A}}{\downarrow q}}$ also be a fibration and let $F': \mathbb{D} \to \mathbb{E}$ form together with $F: \mathbb{A} \to \mathbb{B}$ a morphism $(F, F'): q \to p$ in $\mathbf{Fib}$. Show that there is a right adjoint $(G, G'): p \to q$ in $\mathbf{Fib}$ to (F, F') if and only if there is a right adjoint in $\mathbf{Fib}(\mathbb{A})$ to the induced functor $q \to F^*(p)$.

1.9 Fibred products and coproducts

In the previous section we have studied structure *inside* the fibres of a fibration. Now we move to structure *between* the fibres, given by adjoints to (certain) substitution functors. It will be described as fibred products $\prod$ and coproducts $\coprod$.

Two forms of such quantification $\prod, \coprod$ will be discussed in this section: the first one is "simple" quantification along Cartesian projections in a base category, and the second one is quantification along arbitrary morphisms (in a sense to be made precise). These two forms of quantification will turn out to be instances of a general notion, to be described in Section 9.3. For the moment we are satisfied with elementary descriptions.

Recall that an ordinary category $\mathbb{C}$ has set-indexed products if for every set I and every I-indexed collection $(Y_i)_{i \in I}$ of objects in $\mathbb{C}$, there is a product object $\prod_{i \in I} Y_i$ in $\mathbb{C}$. Put differently, if each diagonal $\Delta_I: \mathbb{C} \to \mathbb{C}^I$ has a right adjoint $\prod_I$ (using the Axiom of Choice). We can express this also in terms of the family fibration $\begin{smallmatrix} \mathrm{Fam}(\mathbb{C}) \\ \downarrow \\ \mathbf{Sets} \end{smallmatrix}$ on $\mathbb{C}$. The category $\mathbb{C}^I$ is isomorphic to the fibre $\mathrm{Fam}(\mathbb{C})_I$ over I and the diagonal Δ_I is the composite

$$\mathbb{C} \cong \mathrm{Fam}(\mathbb{C})_1 \xrightarrow{\; !_I^* \;} \mathrm{Fam}(\mathbb{C})_I \cong \mathbb{C}^I$$

where $!_I^*$ is the reindexing functor associated with the unique map $!_I: I \dashrightarrow 1$. Thus, set-indexed products in $\mathbb{C}$ can be described in terms of right adjoints to certain reindexing functors of the family fibration $\begin{smallmatrix} \mathrm{Fam}(\mathbb{C}) \\ \downarrow \\ \mathbf{Sets} \end{smallmatrix}$ on $\mathbb{C}$. It is precisely this aspect which is generalised in the present section: in a fibration $\begin{smallmatrix} \mathbb{E} \\ \downarrow p \\ \mathbb{B} \end{smallmatrix}$ the objects and morphism in the total category $\mathbb{E}$ are understood as indexed by $\mathbb{B}$. Thus right adjoints to reindexing functors $!_I^*$ (for $I \in \mathbb{B}$) will yield suitably generalised products of an I-indexed collection $X \in \mathbb{E}_I$ in the fibre over I. In this way one defines quantification with respect to an arbitrary base category $\mathbb{B}$—and not just with respect to **Sets**. This leads to a truly general theory of quantification, which finds applications later on in describing $\forall, \exists$ in logic and Π, Σ in type theory.

Actually, it will be more appropriate to describe quantification in terms of adjoints to reindexing functors π^* induced by Cartesian projections $\pi: I \times J \to I$, instead of just to $!_I^*$. The latter then appear via projections $\pi: 1 \times I \dashrightarrow 1$. Such a description involves quantification with a parameter.

1.9.1. Definition. Let $\mathbb{B}$ be a category with Cartesian products $\times$ and $\begin{smallmatrix} \mathbb{E} \\ \downarrow p \\ \mathbb{B} \end{smallmatrix}$ be a fibration. We say that p has **simple products** (resp. **simple coproducts**) if both

- for every pair of objects $I, J \in \mathbb{B}$, every "weakening functor"

$$\mathbb{E}_I \xrightarrow{\;\; \pi_{I,J}^* \;\;} \mathbb{E}_{I \times J}$$

  induced by the Cartesian projection $\pi_{I,J}: I \times J \to I$, has a right adjoint $\prod_{(I,J)}$ (resp. a left adjoint $\coprod_{(I,J)}$);
- the Beck-Chevalley condition holds: for every $u: K \to I$ in $\mathbb{B}$ and $J \in \mathbb{B}$, in

the diagram

$$
\begin{array}{ccc}
\mathbb{E}_I & \xrightarrow{\;\;u^*\;\;} & \mathbb{E}_K \\
\pi^*_{I,J} \Big\downarrow\;\;\Big(\;\;\Big) & & \pi^*_{K,J} \Big\downarrow\;\;\Big(\;\;\Big) \\
\mathbb{E}_{I\times J} & \xrightarrow[(u \times \mathrm{id})^*]{} & \mathbb{E}_{K\times J}
\end{array}
$$

the canonical natural transformation

$$
u^* \textstyle\prod_{(I,J)} \Longrightarrow \prod_{(K,J)} (u \times \mathrm{id})^*
$$
$$
(\text{resp. } \textstyle\coprod_{(K,J)} (u \times \mathrm{id})^* \Longrightarrow u^* \coprod_{(I,J)})
$$

is an isomorphism.

Later, in Section 9.3, this form of quantification will be described in terms of simple fibrations. That is why we call this 'simple' quantification. As in the previous section, the Beck-Chevalley condition guarantees that the induced structure is preserved by reindexing functors (and hence that it is essentially the same in all fibres). This Beck-Chevalley condition is not a formality: it may fail, see Exercise 1.9.10 below. Recall from Example 1.1.1 (iii) that we call functors of the form π^* 'weakening functors' because they add a dummy variable.

One can formulate appropriate versions of quantification (in Definition 1.9.1 above and also in Definition 1.9.4 below) for *split* fibrations. The canonical isomorphism mentioned in the Beck-Chevalley condition is then required to be an identity (for the adjoints to the reindexing functors induced by the splitting).

The next result shows that the above simple quantification gives us what we expect in the situation of the standard fibration over sets.

1.9.2. Lemma. *For an arbitrary category $\mathbb{C}$ one has:*

$$
\textit{the family fibration } \begin{array}{c} \mathrm{Fam}(\mathbb{C}) \\ \downarrow \\ \mathbf{Sets} \end{array} \textit{ has (split) simple products/coproducts}
$$
$$
\Leftrightarrow \mathbb{C} \textit{ has set-indexed products/coproducts.}
$$

Proof. We shall do the case of products.

($\Leftarrow$) For sets I, J one defines a product functor $\prod_{(I,J)} \colon \mathrm{Fam}(\mathbb{C})_{I\times J} \to \mathrm{Fam}(\mathbb{C})_I$ by

$$
\big(Y_{(i,j)}\big)_{(i,j)\in I\times J} \mapsto \big(\textstyle\prod_{j\in J} Y_{(i,j)}\big)_{i\in I}.
$$

Then one obtains the following isomorphisms, establishing an adjunction

$\pi^*_{I,J} \dashv \prod_{(I,J)}.$

$$\text{Fam}(\mathbb{C})_{I \times J}\left(\pi^*_{I,J}((X_i)_{i \in I}),\ (Y_{(i,j)})_{(i,j) \in I \times J}\right)$$

$$= \text{Fam}(\mathbb{C})_{I \times J}\left((X_i)_{(i,j) \in I \times J},\ (Y_{(i,j)})_{(i,j) \in I \times J}\right)$$

$$\cong \prod_{(i,j) \in I \times J} \mathbb{C}\left(X_i,\ Y_{(i,j)}\right)$$

$$\cong \prod_{i \in I} \prod_{j \in J} \mathbb{C}\left(X_i,\ Y_{(i,j)}\right)$$

$$\cong \prod_{i \in I} \mathbb{C}\left(X_i,\ \textstyle\prod_{j \in J} Y_{(i,j)}\right)$$

$$\cong \text{Fam}(\mathbb{C})_I\left((X_i)_{i \in I},\ \textstyle\prod_{(I,J)}((Y_{(i,j)})_{(i,j) \in I \times J})\right).$$

Beck-Chevalley holds, by an easy calculation.

($\Rightarrow$) Let 1 be a one-element, terminal set. For each set I, the diagonal functor $\Delta_I \colon \mathbb{C} \to \mathbb{C}^I$ is the composite,

$$\mathbb{C} \cong \text{Fam}(\mathbb{C})_1 \xrightarrow{\ \pi^*_{1,I}\ } \text{Fam}(\mathbb{C})_{1 \times I} \cong \mathbb{C}^I.$$

Since this weakening functor $\pi^*_{1,I}$ has a right adjoint $\prod_{(1,I)}$, also the diagonal Δ_I has a right adjoint. Thus $\mathbb{C}$ has I-indexed products, for each set I.   □

1.9.3. Proposition. *Let $\mathbb{B}$ be a category with finite products.*

(i) *The simple fibration* $\begin{smallmatrix} s(\mathbb{B}) \\ \downarrow \\ \mathbb{B} \end{smallmatrix}$ *on $\mathbb{B}$ always has simple coproducts.*

(ii) *And it has simple products if and only if $\mathbb{B}$ is Cartesian closed.*

Proof. (i) For a projection $\pi \colon I \times J \to I$ we can define a coproduct functor

$$s(\mathbb{B})_{I \times J} = \mathbb{B}/\!\!/(I \times J) \xrightarrow{\ \coprod_{(I,J)}\ } \mathbb{B}/\!\!/I = s(\mathbb{B})_I$$

between the corresponding simple slices by $X \mapsto J \times X$, since:

$$\mathbb{B}/\!\!/(I \times J)\left(X,\ \pi^*(Y)\right) \cong \mathbb{B}\left((I \times J) \times X,\ Y\right)$$

$$\cong \mathbb{B}\left(I \times (J \times X),\ Y\right)$$

$$\cong \mathbb{B}/\!\!/I\left(\textstyle\coprod_{(I,J)}(X),\ Y\right).$$

(ii) If the category $\mathbb{B}$ is Cartesian closed, we can define a product functor $\prod_{(I,J)} \colon \mathbb{B}/\!\!/(I \times J) \to \mathbb{B}/\!\!/I$ by $X \mapsto J \Rightarrow X$. This yields simple products. And conversely, if the simple fibration has simple products, then $\mathbb{B}$ is Cartesian closed by Exercise 1.3.2 (ii), because each functor $I^* \colon \mathbb{B} \to \mathbb{B}/\!\!/I$ has a right adjoint (since it can be written as composite $\mathbb{B} \cong \mathbb{B}/\!\!/1 \xrightarrow{\ \pi^*\ } \mathbb{B}/\!\!/I$).   □

We turn to the second "non-simple" form of quantification; it does not deal with quantification solely along Cartesian projections, but along all morphisms in a base category.

1.9.4. Definition. Let $\mathbb{B}$ be a category with pullbacks and $\begin{smallmatrix} \mathbb{E} \\ \downarrow p \\ \mathbb{B} \end{smallmatrix}$ a fibration on $\mathbb{B}$. One says that p has **products** (resp. **coproducts**) if both

- for every morphism $u: I \to J$ in $\mathbb{B}$, every substitution functor $u^*: \mathbb{E}_J \to \mathbb{E}_I$ has a right adjoint $\prod_u$ (resp. a left adjoint $\coprod_u$);
- the Beck-Chevalley condition holds: for every pullback in $\mathbb{B}$ of the form

$$
\begin{array}{ccc}
K & \xrightarrow{\ v\ } & L \\
{\scriptstyle r}\downarrow & \lrcorner & \downarrow {\scriptstyle s} \\
I & \xrightarrow[\ u\ ]{} & J
\end{array}
$$

the canonical natural transformation

$$s^* \prod_u \Longrightarrow \prod_v r^* \qquad (\text{resp. } \coprod_v r^* \Longrightarrow s^* \coprod_u)$$

is an isomorphism.

It is easy to see that this second form of quantification is really an extension of the earlier 'simple' one. If one has quantification along all morphisms, then in particular along Cartesian projections; and the Beck-Chevalley condition holds since for every $u: K \to I$ and $J \in \mathbb{B}$ the following diagram is a pullback.

$$
\begin{array}{ccc}
K \times J & \xrightarrow{\ u \times \mathrm{id}\ } & I \times J \\
{\scriptstyle \pi}\downarrow & \lrcorner & \downarrow {\scriptstyle \pi} \\
K & \xrightarrow[\ u\ ]{} & I
\end{array}
$$

We emphasise that the simple form of quantification is described in terms of adjoints to *weakening* functors π^* (induced by Cartesian projections π) and the subsequent one in terms of adjoints to arbitrary *substitution* functors u^*. The latter is the formulation first identified by Lawvere in [192]. For the quantifiers $\forall, \exists$ in logic and Π, Σ in simple or polymorphic type theory, it suffices to have quantification along projections. But in dependent type theory the above Cartesian projections will have to be generalised in a suitable way to 'dependent' projections, see Section 10.3.

Equality can be captured in terms of (left) adjoints to *contraction* functors δ^* induced by diagonals δ, see Chapter 3.

It is probably worth noting the following. Adjoints are determined up-to-isomorphism, so the left and right adjoints $\coprod_{\mathrm{id}}$ and $\prod_{\mathrm{id}}$ to an identity substitution function $\mathrm{id}^* \cong \mathrm{id}$ are themselves (naturally) isomorphic to the identity: $\coprod_{\mathrm{id}} \cong \mathrm{id} \cong \prod_{\mathrm{id}}$. For composable maps v, u in the base category, there is an isomoporphism $(v \circ u)^* \cong u^* \circ v^*$, see Section 1.4. It leads to isomorphisms $\coprod_{v \circ u} \cong \coprod_v \circ \coprod_u$ and $\prod_{v \circ u} \cong \prod_v \circ \prod_u$ since adjunctions can be composed, see [187, Chapter IV, ß 8].

Our first example of this second form of quantification again involves family fibrations. It extends Lemma 1.9.2. Notice the explicit use of equality in the definition of $\prod_u$ in the proof. It returns in more abstract form in Example 4.3.7.

1.9.5. Lemma. *Let $\mathbb{C}$ be an arbitrary category. Then:*

$$\text{the family fibration} \quad \begin{array}{c} \mathrm{Fam}(\mathbb{C}) \\ \downarrow \\ \mathbf{Sets} \end{array} \quad \text{has (split) products/coproducts}$$

$$\Leftrightarrow \mathbb{C} \text{ has set-indexed products/coproducts.}$$

Proof. The interesting part is the implication ($\Leftarrow$). For $u : I \to J$ in **Sets** define product and coproduct functors $\prod_u, \coprod_u : \mathrm{Fam}(\mathbb{C})_I \rightrightarrows \mathrm{Fam}(\mathbb{C})_J$ by

$$(Y_i)_{i \in I} \mapsto \left(\prod \{Y_i \mid u(i) = j\} \right)_{j \in J}$$

$$(Y_i)_{i \in I} \mapsto \left(\coprod \{Y_i \mid u(i) = j\} \right)_{j \in J}. \qquad \square$$

1.9.6. Lemma. *The fibration* $\begin{array}{c} \mathrm{UFam}(\mathbf{PER}) \\ \downarrow \\ \omega\text{-}\mathbf{Sets} \end{array}$ *of PERs over ω-sets has both products and coproducts (along all maps in ω-Sets).*

Proof. This follows in fact from the fibred reflection $\mathrm{UFam}(\mathbf{PER}) \leftrightarrows \mathrm{UFam}(\omega\text{-}\mathbf{Sets}) \simeq \omega\text{-}\mathbf{Sets}^{\to}$ over ω-**Sets** in Proposition 1.8.7 (ii), using the reflection lemma 9.3.9 later on. Here we give the explicit formulas: for a morphism $u : (I, E) \to (J, E)$ in ω-**Sets** and a family $R = (R_i)_{i \in (I,E)}$ over (I, E) we get a product and coproduct over (J, E) by

$$\prod_u(R)_j = \{(n, n') \mid \forall i \in I. u(i) = j \Rightarrow \forall m, m' \in E(i). n \cdot m R_i n' \cdot m'\}$$

$$\coprod_u(R)_j = \mathrm{r}\left(\coprod_{u(i)=j} \mathbb{N}/R_i, E \right)$$

where r is the left adjoint to the inclusion **PER** $\hookrightarrow \omega$-**Sets**, and E is the existence predicate on the disjoint union $\coprod_{u(i)=j} \mathbb{N}/R_i$ given by $E(i, [n]_{R_i}) = \{\langle n, n' \rangle \mid n \in E(i) \text{ and } n' \in [n]_{R_i}\}$. $\qquad \square$

The following result is often quite useful. The proof is left as an exercise.

1.9.7. Lemma. *Consider a fibration for which each reindexing functor has both a left $\coprod$ and a right $\prod$ adjoint. Then Beck-Chevalley holds for coproducts $\coprod$ if and only if it holds for products $\prod$.* $\qquad \square$

The next result for codomain fibrations is the analogue of Proposition 1.9.3 for simple fibrations. The third point is due to Freyd [83].

1.9.8. Proposition. *For a category $\mathbb{B}$ with finite limits, the codomain fibration $\begin{smallmatrix} \mathbb{B}^{\to} \\ \downarrow \\ \mathbb{B} \end{smallmatrix}$ on $\mathbb{B}$ has*

(i) *coproducts $\coprod_u$; they are given by composition;*
(ii) *simple products $\prod_{(I,J)}$ if and only if $\mathbb{B}$ is Cartesian closed;*
(iii) *products $\prod_u$ if and only if $\mathbb{B}$ is locally Cartesian closed.*

Proof. (i) For $u: I \to J$ one defines a coproduct functor $\coprod_u: \mathbb{B}/I \to \mathbb{B}/J$ by

$$\left(X \xrightarrow{\varphi} I \right) \mapsto \left(X \xrightarrow{u \circ \varphi} J \right) \quad \text{and} \quad \left(\varphi \xrightarrow{f} \psi \right) \mapsto \left((u \circ \varphi) \xrightarrow{f} (u \circ \psi) \right).$$

The adjunction $(\coprod_u \dashv u^*)$ then follows from the bijective correspondence between maps $f: \coprod_u \varphi \to \psi$ over J and $g: \varphi \to u^*(\psi)$ over I in:

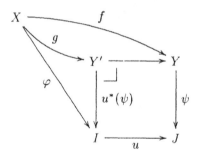

Beck-Chevalley follows from the Pullback Lemma (see Exercise 1.1.5).

(ii) The proof is essentially as in Exercise 1.3.3, except that we have to deal with an extra parameter object. In case $\mathbb{B}$ is Cartesian closed we can form a simple product of a family $\begin{pmatrix} X \\ \downarrow \varphi \\ I \times J \end{pmatrix}$ over $I \times J$ along a projection $\pi: I \times J \to I$ as the family $\begin{pmatrix} P \\ \downarrow \\ I \end{pmatrix}$ over I, in the pullback diagram:

$$\begin{array}{ccc} P & \longrightarrow & J \Rightarrow X \\ {\scriptstyle \prod_{(I,J)}(\varphi)} \downarrow \; \lrcorner & & \downarrow {\scriptstyle J \Rightarrow \varphi} \\ I & \xrightarrow{\;\;\Lambda(\mathrm{id}_{I \times J})\;\;} & J \Rightarrow (I \times J) \end{array}$$

Informally, P consists of the pairs (i, f) with $\varphi(f(j)) = \langle i, j \rangle$, for all j.

Conversely, if the codomain fibration has simple products, then in particular each functor $I^*: \mathbb{B} \to \mathbb{B}/I$ has a right adjoint. Hence $\mathbb{B}$ is Cartesian closed by Exercise 1.3.3.

(iii) If $\mathbb{B}$ has finite limits, then each slice $\mathbb{B}/I$ has finite products. Hence

$\mathbb{B}$ is an LCCC $\Leftrightarrow$ each slice $\mathbb{B}/I$ is Cartesian closed

$\Leftrightarrow$ for each object $u: J \to I$ in $\mathbb{B}/I$, the functor

$$u^*: \mathbb{B}/I \longrightarrow (\mathbb{B}/I)/u = \mathbb{B}/J$$

has a right adjoint $\prod_u$ (see Exercise 1.3.3)

$\Leftrightarrow$ the codomain fibration $\begin{smallmatrix} \mathbb{B}^{\to} \\ \downarrow \\ \mathbb{B} \end{smallmatrix}$ has products $\prod_u$.

This last step is justified by the fact that Beck-Chevalley always holds by the previous lemma. $\square$

For an explicit formulation of the Cartesian products and exponents in the slices $\mathbb{B}/I$ in terms of $\coprod$ and $\prod$, see Exercise 1.9.2 below.

1.9.9. Corollary. *If a category $\mathbb{B}$ with finite limits is Cartesian closed/locally Cartesian closed, then its subobject fibration $\begin{smallmatrix} \mathrm{Sub}(\mathbb{B}) \\ \downarrow \\ \mathbb{B} \end{smallmatrix}$ has simple/ordinary products $\prod$.*

Proof. Since right adjoints $\prod$ preserve monos, they restrict to functors between (posets of) subobjects. $\square$

The following result tells how simple and ordinary products are related. It shows that ordinary products are simple products relativised to all slices of the base category. This is sometimes called localisation, see *e.g.* [246].

1.9.10. Theorem. *Let $\begin{smallmatrix} \mathbb{E} \\ \downarrow p \\ \mathbb{B} \end{smallmatrix}$ be a fibration on a base category $\mathbb{B}$ with pullbacks. For each object $I \in \mathbb{B}$, write $I^*(p)$ for the fibration obtained by change-of-base in*

$$
\begin{array}{ccc}
\mathbb{B}/I \times_{\mathbb{B}} \mathbb{E} & \longrightarrow & \mathbb{E} \\
\scriptstyle I^*(p) = \mathrm{dom}_I^*(p) \downarrow \;\lrcorner & & \downarrow \scriptstyle p \\
\mathbb{B}/I & \xrightarrow{\;\mathrm{dom}_I\;} & \mathbb{B}
\end{array}
$$

Then p has (ordinary) products $\prod_u$ if and only if each fibration $I^*(p)$ has simple products $\prod_{(v,w)}$.

A similar result holds for coproducts $\coprod$.

Proof. Assume p has products $\prod_u$ along an arbitrary morphism u in $\mathbb{B}$. Let $v\colon K \to I$ and $w\colon L \to I$ be objects of the slice $\mathbb{B}/I$ and consider their pullback

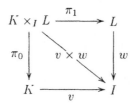

A simple product $\prod_{(v,w)}$ along the Cartesian projection $\pi_0\colon v \times w \to v$ in $\mathbb{B}/I$ is then given by

$$X \mapsto \prod_{\pi_0}(X).$$

As a result of the Beck-Chevalley condition for products in p one gets in $I^*(p)$ that for $u\colon J \to K$,

$$u^* \prod_{\pi_0} \cong \prod_{\lambda_0}(u \times \mathrm{id})^*$$

where λ_0 is the first projection $(v \circ u) \times w \to (v \circ u)$ in $\mathbb{B}/I$. This is the appropriate formulation of Beck-Chevalley for simple products.

Conversely, assume that each fibration $I^*(p)$ has simple products. Then for a map $u\colon J \to I$ in $\mathbb{B}$, the fibration $I^*(p)$ has a product $\prod_u$ along the 'projection' $!_u = u\colon u \dashrightarrow \mathrm{id}_I$ in $\mathbb{B}/I$. Beck-Chevalley also holds: for $v\colon K \to I$ consider the following pullback square in $\mathbb{B}/I$.

$$
\begin{array}{ccc}
v \times u & \xrightarrow{\;v' = !_v \times \mathrm{id}\;} & u \\
{\scriptstyle v^*(u)}\downarrow\;\lrcorner & & \downarrow\;{\scriptstyle !_u = u} \\
v & \xrightarrow[\;!_v = v\;]{} & \mathrm{id}_I
\end{array}
$$

It yields $v^* \prod_u \cong \prod_{v^*(u)} v'^*$ as required. $\qquad\square$

1.9.11. Definition. A fibration is called **complete** if it has products $\prod_u$ and fibred finite limits. Dually, a fibration is **cocomplete** if it has coproducts $\coprod_u$ and fibred finite colimits.

The codomain fibration associated with a locally Cartesian closed category is thus complete. And fibrations that we know to be equivalent to codomain fibrations of LCCCs are complete, like $\begin{array}{c}\text{Fam}(\mathbf{Sets})\\\downarrow\\\mathbf{Sets}\end{array}$, $\begin{array}{c}\text{UFam}(\omega\text{-}\mathbf{Sets})\\\downarrow\\\omega\text{-}\mathbf{Sets}\end{array}$ and $\begin{array}{c}\text{UFam}(\mathbf{PER})\\\downarrow\\\mathbf{PER}\end{array}$. Also $\begin{array}{c}\text{UFam}(\mathbf{PER})\\\downarrow\\\omega\text{-}\mathbf{Sets}\end{array}$ is complete, see Lemma 1.9.6. And the family fibration of a complete category is complete.

Ordinary categories are complete in case they have arbitrary products and equalisers. Above we have required all finite limits instead of just equalisers. Under certain technical assumptions, it is possible to obtain fibred finite products from products $\prod_u$ and fibred equalisers, so that we get all fibred finite limits, see Exercise 9.5.11. But in general it is more convenient to require explicitly the presence of all fibred finite limits.

In Section 7.4 it will be shown how every small diagram in a complete fibration has a limit. The difficulty in getting such a result lies in saying what a small diagram in a fibred category is.

The following technical result will be used frequently in the categorical description of logics and type theories. It deals with distribution of coproducts $\coprod$ over Cartesian products $\times$ in the fibres. It is a generalisation of the distribution of $\bigvee$ over $\wedge$ in a frame, see Exercise 1.9.6. In logic it corresponds to the equivalence of $\exists x \colon \sigma. \, (\varphi \wedge \psi(x))$ and $\varphi \wedge \exists x \colon \sigma. \, \psi(x)$, if x does not occur free in φ. It also occurs as an equivalence between $\nu x. \, (P \,\|\, \pi^*(Q))$ and $(\nu x. \, P) \,\|\, Q$ in process theory, where ν is restriction and $\|$ is parallel composition, see [219].

1.9.12. Lemma (Frobenius). *Let* $\begin{smallmatrix} \mathbb{E} \\ \downarrow p \\ \mathbb{B} \end{smallmatrix}$ *be a fibred CCC.*

(i) *Suppose p has simple coproducts. For each pair of objects $I, J \in \mathbb{B}$ in the basis and each pair of objects $Y \in \mathbb{E}_I$, $Z \in \mathbb{E}_{I \times J}$ in appropriate fibres, the canonical morphism*

$$\coprod_{(I,J)} (\pi^*_{I,J}(Y) \times Z) \longrightarrow Y \times \coprod_{(I,J)}(Z)$$

is an isomorphism.

(ii) *Suppose now p has coproducts. Then for each $u \colon I \to J$ in $\mathbb{B}$, $Y \in \mathbb{E}_J$ and $Z \in \mathbb{E}_I$, the canonical morphism*

$$\coprod_u (u^*(Y) \times Z) \longrightarrow Y \times \coprod_u(Z)$$

is an isomorphism.

Proof. We do only (i). First of all, the Frobenius map is obtained as transpose of the composite

$$\pi^*_{I,J}(Y) \times Z \xrightarrow{\ \mathrm{id} \times \eta_Z\ } \pi^*_{I,J}(Y) \times \pi^*_{I,J}(\coprod_{(I,J)}(Z))$$
$$\downarrow \cong$$
$$\pi^*_{I,J}(Y \times \coprod_{(I,J)}(Z)).$$

It is an isomorphism by Yoneda:

$$\frac{\coprod_{(I,J)}(\pi^*_{I,J}(Y) \times Z) \longrightarrow W}{\frac{\pi^*_{I,J}(Y) \times Z \longrightarrow \pi^*_{I,J}(W)}{\frac{Z \longrightarrow \pi^*_{I,J}(Y) \Rightarrow \pi^*_{I,J}(W) \cong \pi^*_{I,J}(Y \Rightarrow W)}{\frac{\coprod_{(I,J)}(Z) \longrightarrow Y \Rightarrow W}{Y \times \coprod_{(I,J)}(Z) \longrightarrow W}}}}$$

$\square$

Notice that the Frobenius map is an isomorphism because reindexing functors preserve exponents. Even if there are no fibred exponents around, the Frobenius map can still be an isomorphism. In that case we shall speak of **(simple) coproducts with the Frobenius property**, or briefly, of **(simple) coproducts satisfying Frobenius**.

Finally we should also say what it means for a morphism of fibrations to preserve the above (simple) products and coproducts.

1.9.13. Definition. Let $\left(\begin{smallmatrix} \mathbb{E} \\ \downarrow p \\ \mathbb{B} \end{smallmatrix} \right) \xrightarrow{(K,L)} \left(\begin{smallmatrix} \mathbb{D} \\ \downarrow q \\ \mathbb{A} \end{smallmatrix} \right)$ be a morphism of fibrations.

(i) Assume that p and q have simple products (resp. coproducts). Then $(K, L): p \to q$ **preserves simple products** (resp. **coproducts**) if both

- $K: \mathbb{B} \to \mathbb{A}$ preserves binary products, say with $\gamma_{I,J}$ as inverse of the canonical map $K(I \times J) \to KI \times KJ$;
- for each pair $I, J \in \mathbb{B}$, in

the canonical natural transformation

$$H \prod_{(I,J)} \Longrightarrow \prod_{(KI,KJ)} \gamma^*_{I,J} H$$
$$(\text{resp. } \coprod_{(KI,KJ)} \gamma^*_{I,J} H \Longrightarrow H \coprod_{(I,J)})$$

is an isomorphism.

(ii) Assume now that p and q have products (resp. coproducts). The map $(K, L): p \to q$ **preserves products** (resp. **coproducts**) if both

- $K: \mathbb{B} \to \mathbb{A}$ preserves pullbacks;
- for every $u: I \to J$ in $\mathbb{B}$, the canonical natural transformation,

$$H \prod_u \Longrightarrow \prod_{Ku} H \qquad (\text{resp.} \coprod_{Ku} H \Longrightarrow H \coprod_u)$$

is an isomorphism.

This notion occurs in the following useful lemma on quantification and change-of-base.

1.9.14. Lemma. *Let $\begin{smallmatrix} \mathbb{E} \\ \downarrow p \\ \mathbb{B} \end{smallmatrix}$ be a fibration and $K: \mathbb{A} \to \mathbb{B}$ a finite limit (product) preserving functor. Then*

$$p \text{ has (simple) products/coproducts}$$
$$\Rightarrow K^*(p) \text{ has (simple) products/coproducts.}$$

Moreover, the morphism of fibrations $K^*(p) \to q$ preserves these. □

1.9.15. Example. By Lemma 1.9.5 the family fibration $\begin{smallmatrix} \textbf{Fam}(\textbf{Sets}) \\ \downarrow \\ \textbf{Sets} \end{smallmatrix}$ has both products and coproducts. Hence also the fibration $\begin{smallmatrix} \textbf{Sign} \\ \downarrow \\ \textbf{Sets} \end{smallmatrix}$ of many-typed signatures has products and coproducts, because it is obtained by change-of-base, see Definition 1.6.1, and because the functor $T \mapsto T^* \times T$ preserves pullbacks. The morphism of fibrations $\begin{smallmatrix} \textbf{Sign} \\ \downarrow \\ \textbf{Sets} \end{smallmatrix} \to \begin{smallmatrix} \textbf{Fam}(\textbf{Sets}) \\ \downarrow \\ \textbf{Sets} \end{smallmatrix}$ then preserves these induced products and coproducts.

Exercises

1.9.1. Fill in the details of the proof of Lemma 1.9.5 and pay special attention to the Beck-Chevalley condition.

1.9.2. Assume $\mathbb{B}$ is a category whose codomain fibration $\begin{smallmatrix} \mathbb{B}^{\to} \\ \downarrow \\ \mathbb{B} \end{smallmatrix}$ is complete. By Proposition 1.9.8 (iii) we know that $\mathbb{B}$ is then locally Cartesian closed. Show that for objects φ, ψ in the slice $\mathbb{B}/I$, the Cartesian product and exponent are given by the formulas:

$$\varphi \times \psi = \coprod_\varphi \varphi^*(\psi) \qquad \text{and} \qquad \varphi \Rightarrow \psi = \prod_\varphi \varphi^*(\psi).$$

1.9.3. Conclude from Propositions 1.9.3 (ii) and 1.9.8 (iii) that (finite) coproducts are automatically distributive in a CCC, and universal in an LCCC.

1.9.4. (i) (Lawvere [193], p.6) Show that Lemma 1.9.12 (ii) can be strengthened in the following way. Consider a fibration with fibred finite products and coproducts $\coprod_u$, in which each fibre category has exponents. Then:

the Frobenius property holds
$\Leftrightarrow$ reindexing functors preserve exponents.

(ii) Assume $\mathbb{B}$ has finite limits; show that the codomain fibration $\begin{smallmatrix} \mathbb{B}^{\rightarrow} \\ \downarrow \\ \mathbb{B} \end{smallmatrix}$ has coproducts satisfying the Frobenius property.

(iii) Conclude from (i) and (ii) that if $\mathbb{B}$ is locally Cartesian closed (*i.e.* every slice is Cartesian closed), then the codomain fibration on $\mathbb{B}$ is Cartesian closed. This fills the gap in Example 1.8.3 (iv).

1.9.5. Let a fibration $\begin{smallmatrix} \mathbb{E} \\ \downarrow p \\ \mathbb{B} \end{smallmatrix}$ have coproducts $\coprod_u$. Prove that for a mono $m\colon I' \rightarrowtail I$ in $\mathbb{B}$ the coproduct functor $\coprod_m\colon \mathbb{E}_{I'} \to \mathbb{E}_I$ is full and faithful.

[*Hint.* Write the mono in a pullback square, and use Beck-Chevalley.]

1.9.6. Let $\mathbb{C}$ be a category with finite products and set-indexed coproducts. The family fibration $\begin{smallmatrix} \text{Fam}(\mathbb{C}) \\ \downarrow \\ \textbf{Sets} \end{smallmatrix}$ then has fibred finite products and (simple) coproducts. Show that the Frobenius property holds if and only if the coproducts in $\mathbb{C}$ are distributive (*i.e.* if functors $X \times (-)\colon \mathbb{C} \to \mathbb{C}$ preserve coproducts: the canonical maps $\coprod_i (X \times Y_i) \to X \times (\coprod_i Y_i)$ are isomorphisms).

[Especially, for every frame A the family fibration $\begin{smallmatrix} \text{Fam}(A) \\ \downarrow \\ \textbf{Sets} \end{smallmatrix}$ has fibred finite products and coproducts satisfying Frobenius.]

1.9.7. Let $\begin{smallmatrix} \mathbb{E} \\ \downarrow p \\ \mathbb{B} \end{smallmatrix}$ be a fibred CCC with simple products and coproducts.

(i) Show that for $X \in \mathbb{E}_{I \times J}$ and $Y \in \mathbb{E}_I$ there is an isomorphism over I

$$(\coprod_{(I,J)} X) \Rightarrow Y \cong \prod_{(I,J)} (X \Rightarrow \pi^*(Y))$$

[Recall from logic the equivalence of $((\exists x\colon \sigma. \varphi) \supset \psi)$ and $(\forall x\colon \sigma. (\varphi \supset \psi))$ if x is not free in ψ.]

(ii) Formulate and prove a similar result for non-simple coproducts $\coprod_u$ and products $\prod_u$.

1.9.8. Let $\mathbb{B}$ be an LCCC. Describe the associated coproduct $\coprod$ and product $\prod$ along morphisms in $\mathbb{B}$ as fibred functors in a situation:

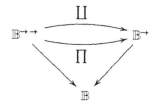

1.9.9. Let $\mathbb{B}$ be an LCCC. Show that complete distributivity (or the Axiom of

Choice, see Exercise 10.2.1) holds: the canonical map

$$\prod_u \coprod_\varphi(\psi) \longleftarrow \coprod_{\prod_u(\varphi)} \prod_{u'} \varepsilon^*(\psi)$$

is an isomorphism—where u' is the pullback of u along $\prod_u(\varphi)$ and ε is the counit of the adjunction $(u^* \dashv \prod_u)$ at φ.

1.9.10. Let **Dcpo** be the category of directed complete partial orders (dcpos) and (Scott-)continuous functions. A subset A of a dcpo X is Scott-closed if A is a lower set closed under directed joins. (This means that A is closed in the Scott topology on X.)

(i) Define a fibration of Scott-closed subsets (ordered by inclusion) over **Dcpo**.

(ii) Show that a left adjoint $\coprod_{(X,Y)}$ along a projection $\pi\colon X \times Y \to X$ exists and is given by

$$A \mapsto \overline{\{x \in X \mid \exists y \in Y.\,(x,y) \in A\}}$$

where $\overline{(\cdot)}$ is Scott-closure.

(iii) Show that in case Beck-Chevalley would hold, one would get

$$x \in \coprod_{(X,Y)}(A) \;\Leftrightarrow\; \exists y \in Y.\,(x,y) \in A$$

i.e. that $\{x \in X \mid \exists y \in Y.\,(x,y) \in A\}$ is already Scott-closed.
[*Hint.* Consider the pullback of $\pi\colon X \times Y \to X$ and $x\colon 1 \to X$.]

(iv) Check that the latter is not the case: consider $X = \mathbb{N} \cup \{\infty\}$ with the usual total order of $\mathbb{N}$ plus a top element, and $Y = \mathbb{N} \cup \{\infty\}$ with discrete order. Take $A = \{(x,y) \in \mathbb{N} \times \mathbb{N} \mid x \leq_\mathbb{N} y\} \subseteq X \times Y$, where $\leq_\mathbb{N}$ is the usual order on $\mathbb{N}$.

[This gives an example where one has left adjoints to π^*'s but no Beck-Chevalley. This example (or counter example) is due to Pitts (see also [61, Chapter 1, Exercise (7)]).]

1.9.11. In Exercise 1.2.13 one finds that a category $\mathbb{C}$ has set-indexed coproducts if and only if the unit $\mathbb{C} \to \mathrm{Fam}(\mathbb{C})$ has a left adjoint. We describe an analogue of this result (due to Bénabou) for fibred categories.

Let $\begin{smallmatrix}\mathbb{E}\\\downarrow{\scriptstyle p}\\\mathbb{B}\end{smallmatrix}$ be a fibration, where $\mathbb{B}$ has pullbacks. Define the fibration $\mathrm{Fam}(p)$ to be the composite $\mathrm{cod} \circ \mathrm{dom}^*(p)$ in

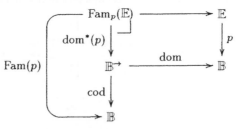

and define $\eta_p\colon \mathbb{E} \to \mathrm{Fam}_p(\mathbb{E}) = \mathbb{B}^\to \times_\mathbb{B} \mathbb{E}$ by $X \mapsto (\mathrm{id}_{pX}, X)$.

(i) Show that η_p is a fibred functor $p \to \text{Fam}(p)$.

(ii) Prove that p has coproducts if and only if η_p has a fibred left adjoint.

1.10 Indexed categories

We recall from the first section 1.1 that there are two ways of describing I-indexed families of sets: (a) pointwise indexing via indexed collections $(X_i)_{i \in I}$, or (b) display indexing via functions $\left(\begin{smallmatrix} X \\ \downarrow \\ I \end{smallmatrix} \right)$. The collection in (a) may be described as a functor from the discrete category I to **Sets**. Equivalently as a functor $I^{\text{op}} \to$ **Sets**. It has been shown that (a) and (b) are essentially the same for sets.

The reader may already have noticed that similar descriptions exist for indexing of categories: one has (a) indexed categories $\mathbb{B}^{\text{op}} \to$ **Cat** and (b) fibrations $\begin{smallmatrix} \mathbb{E} \\ \downarrow p \\ \mathbb{B} \end{smallmatrix}$, giving pointwise and display indexing for categories. Proposition 1.4.5 describes how to go from (b) to (a)—for a cloven fibration—by mapping an object I of the base category $\mathbb{B}$ to the fibre category $\mathbb{E}_I$ over I. In this section one finds the so-called 'Grothendieck construction' which establishes a passage in the reverse direction from (a) to (b). It occurs in Grothendieck's original paper [107] on fibred categories.

A discussion on fibrations versus indexed categories is included.

1.10.1. Definition (Grothendieck construction). Let $\Psi \colon \mathbb{B}^{\text{op}} \to$ **Cat** be an indexed category. The **Grothendieck completion** $\int_{\mathbb{B}}(\Psi)$ (or simply $\int \Psi$) of Ψ is the category with

objects (I, X) where $I \in \mathbb{B}$ and $X \in \Psi(I)$.

morphisms $(I, X) \to (J, Y)$ are pairs (u, f) with $u \colon I \to J$ in $\mathbb{B}$ and $f \colon X \to u^*(Y) = \Psi(u)(Y)$ in $\Psi(I)$.

Composition and identities in $\int_{\mathbb{B}}(\Psi)$ involve the isomorphisms η and μ from Definition 1.4.4. The identity $(I, X) \to (I, X)$ in $\int \Psi$ is the pair $(\text{id}, \eta_I(X))$, where η_I is the natural isomorphism $\text{id}_{\Psi(I)} \xRightarrow{\cong} (\text{id}_I)^*$. And composition in $\int \Psi$ of

$$(I, X) \xrightarrow{\ (u, f)\ } (J, Y) \xrightarrow{\ (v, g)\ } (K, Z)$$

i.e. of

$$\left\{ \begin{array}{c} I \xrightarrow{\ u\ } J \\ X \xrightarrow[f]{} u^*(Y) \end{array} \right. \qquad \text{and} \qquad \left\{ \begin{array}{c} J \xrightarrow{\ v\ } K \\ Y \xrightarrow[g]{} v^*(Z) \end{array} \right.$$

is defined as

$$
\left\{
\begin{array}{l}
I \xrightarrow{\ u\ } J \xrightarrow{\ v\ } K \\[2mm]
X \xrightarrow[\ f\ ]{\qquad} u^*(Y) \xrightarrow[\ u^*(g)\ ]{\qquad} u^*v^*(Z) \xrightarrow[\ \mu_{u,v}(Z)\ ]{\ \cong\ } (v \circ u)^*(Z)
\end{array}
\right.
$$

The required equalities for identity and composition follow from the coherence diagrams in Definition 1.4.4. In fact, these coherence conditions capture precisely what is required for $\int \Psi$ to be a category.

1.10.2. Proposition. (i) *The first projection* $\begin{smallmatrix} \int^{\Psi} \\ \downarrow \\ \mathbb{B} \end{smallmatrix}$ *is a cloven fibration. It is split whenever the indexed category* Ψ *is split.*

(ii) *Turning a cloven fibration first into an indexed category (as in Proposition 1.4.5) and then again into a fibration yields a fibration which is equivalent to the original one.*

(iii) *Also, turning an indexed category first into a fibration and then into an indexed category yields a result which is "essentially the same" as the original.*

Proof. (i) For $u \colon I \to J$ in $\mathbb{B}$ and $Y \in \Psi(J)$ there is a cleavage

$$
(I, u^*(Y)) \ \ -\ -\ \xrightarrow{\ (u,\mathrm{id})\ }\ -\ -\ \blacktriangleright\ (J, Y)
$$

$$
I \xrightarrow{\qquad u \qquad} J
$$

(ii) Easy.

(iii) In order to make the statement precise one first has to introduce a notion of equivalence for indexed categories. We leave this to the meticulous reader. Below one does find the appropriate notions for *split* indexed categories. $\qquad\square$

1.10.3. Examples. (i) The family fibration $\begin{smallmatrix} \mathrm{Fam}(\mathbb{C}) \\ \downarrow \\ \mathbf{Sets} \end{smallmatrix}$ arises by applying the Grothendieck construction to the split indexed category $\mathbf{Sets}^{\mathrm{op}} \to \mathbf{Cat}$ given by $I \mapsto \mathbb{C}^I$.

(ii) The previous example can be extended to categories in the following way. For a fixed category $\mathbb{C}$ one obtains a split indexed category $\mathbf{Cat}^{\mathrm{op}} \to \mathbf{Cat}$ by $\mathbb{A} \mapsto$ [the functor category $\mathbb{C}^{\mathbb{A}}$]. The resulting split fibration will be written as $\begin{smallmatrix} \mathrm{Fam}(\mathbb{C}) \\ \downarrow \\ \mathbf{Cat} \end{smallmatrix}$ and called the **family fibration over Cat.**

(iii) For a category $\mathbb{B}$ with finite products, one gets a split indexed category $\mathbb{B}^{op} \to \mathbf{Cat}$ by mapping $I \in \mathbb{B}$ to the simple slice category $\mathbb{B} /\!/ I$. The resulting fibration is the simple fibration $\begin{smallmatrix} s(\mathbb{B}) \\ \downarrow \\ \mathbb{B} \end{smallmatrix}$ on $\mathbb{B}$.

(iv) For a category $\mathbb{B}$ with explicitly given pullbacks, the assignment $I \mapsto \mathbb{B}/I$ extends to an (in general non-split) indexed category $\mathbb{B}^{op} \to \mathbf{Cat}$. Its associated fibration is the cloven codomain fibration $\begin{smallmatrix} \mathbb{B}^{\to} \\ \downarrow \\ \mathbb{B} \end{smallmatrix}$.

Later on in this section, these latter two type theoretic fibrations will reappear in connection with indeterminates.

1.10.4. Discussion. We have seen that the two ways (a) (= pointwise) and (b) (= display) of indexing sets (and the associated pictures) as described in Section 1.1 extend to categories: indexed categories correspond to pointwise indexing (a) and (cloven) fibrations to display indexing (b). In the following comparison of these two forms of indexing (for categories), we shall discuss one conceptual difference and a number of technical differences.

(i) The notion of indexed category involves some explicit structure (namely reindexing functors and mediating isomorphisms id $\overset{\cong}{\Rightarrow}$ id* and $u^* \circ v^* \overset{\cong}{\Rightarrow} (v \circ u)^*$ in Definition 1.4.4), which is left implicit in fibrations. So an indexed category has a *structure* where a fibration has a *property*. The defining property of a fibration determines such structure once a choice of cleavage has been made, see Section 1.7. In general in category theory one prefers properties to structures.

We mention the following two disadvantages of working with explicit reindexing functors and mediating isomorphisms.

(a) It means that one has to check every time explicitly whether a property is *intrinsic* or not, *i.e.* whether or not it depends on the specific structure. For instance, in the indexed category $I \mapsto \mathbb{B}/I$ in Example 1.10.3 (iv) above, each reindexing functor (given by pullback) has a left adjoint (by composition). This property does not just hold for the given indexed category arising from the explicitly given pullbacks, but for all such indexed categories arising from all possible choices of pullbacks. In fibred category theory one leaves this structure implicit, which enables a natural and intrinsic formulation of this property of the codomain fibration $\begin{smallmatrix} \mathbb{B}^{\to} \\ \downarrow \\ \mathbb{B} \end{smallmatrix}$, see Proposition 1.9.8 (i).

(b) Dealing explicitly with the mediating isomorphisms $\eta : \text{id} \overset{\cong}{\Rightarrow} \text{id}^*$ and $\mu : (u^* \circ v^*) \overset{\cong}{\Rightarrow} (v \circ u)^*$ (and the associated coherence conditions) is cumbersome. Of course one can ignore them, but that means pretending there is no problem. This is dangerous, because coherence conditions may fail.

(ii) In Section 1.6 we saw that fibrations are closed under composition. Of course a similar result can be formulated for indexed categories (try it!), but it lacks the smoothness and clarity that one has with fibrations. Thus simple and clarifying results like Lemma 1.6.6 fall outside the direct scope of indexed categories. Later we shall make crucial use of this closedness under composition in the categorical description of logics and type theories which involve different levels of indexing, see Sections 8.6, 9.4, 8.6, 11.2 and 11.3.

(iii) This last point is related to another advantage of fibrations over indexed categories, namely that the notion of fibration makes sense in a 2-category, see *e.g.* [317, 171]. This is like display indexing of families, which makes sense in any category.

(iv) Some constructions are easier for indexed categories. Change-of-base is slightly simpler (for indexed categories) because it is done by composition (see Proposition 1.10.6 below). Considerably more elementary is the construction which yields the opposite: for an indexed category Ψ one takes the opposite of the fibre categories $\Psi(I)$. Probably the easiest way to understand the opposite of a fibration is: first turn it into an indexed category, take the opposite fibre-wise and turn the result back into a fibration. An explicit fibred construction is described in Definition 1.10.11 below.

(v) The categorical semantics of simple and polymorphic lambda calculi can easily be described in terms of indexed categories, as in [307, 156, 61]. But, if one wishes to use indexed categories also for the semantics of type dependency, one ends up describing the relevant structure in terms of the associated Grothendieck completions. In general, if one uses the Grothendieck completion all the time, one might as well use fibrations from the beginning.

Finally one sometimes hears that indexed categories are more elementary and easier to understand and use than fibrations. We disagree at this point. Properly explained and exemplified, fibrations give a clearer picture of indexing and are more convenient to use. Eventually, of course, one tends to think of indexed categories and (cloven) fibrations interchangeably. But, and here we quote Bénabou [29, p. 31, in (v)]: "An indexed category is just a presentation of a fibred category".

On a more practical note, we shall often use *split* indexed categories—in particular, as a means to introduce a split fibration—but hardly ever non-split ones. In those situations we prefer to use fibrations. In line with this approach we will describe 1- and 2-cells for split indexed categories only.

1.10.5. Definition. A **morphism of split indexed categories** from $\Psi \colon \mathbb{B}^{\mathrm{op}} \to \mathbf{Cat}$ to $\Phi \colon \mathbb{A}^{\mathrm{op}} \to \mathbf{Cat}$ is a pair (K, α) where $K \colon \mathbb{B} \to \mathbb{A}$ is a functor and $\alpha \colon \Psi \Rightarrow \Phi K^{\mathrm{op}}$ is a natural transformation. Notice that the components of α are functors $\alpha_I \colon \Psi(I) \to \Phi(KI)$. This determines a category **ICat**.

1.10.6. Proposition. *The functor* $\begin{smallmatrix} \mathbf{ICat} \\ \downarrow \\ \mathbf{Cat} \end{smallmatrix}$ *, which sends an indexed category to its base, is a split fibration. The fibre above a category* $\mathbb{B}$ *will be denoted by*

ICat($\mathbb{B}$); *it contains split indexed categories with base $\mathbb{B}$ and natural transformations between them.*

Proof. For an indexed category $\Psi\colon \mathbb{B}^{\mathrm{op}} \to \mathbf{Cat}$ and an arbitrary functor $K\colon \mathbb{A} \to \mathbb{B}$, put $K^*(\Psi) = \Psi \circ K^{\mathrm{op}}\colon \mathbb{A}^{\mathrm{op}} \to \mathbf{Cat}$ and $\overline{K}(\Psi) = (K, (\mathrm{id}_{\Psi(KI)})_{I \in \mathbb{A}})$ in **ICat**. $\qquad\square$

1.10.7. Theorem (Grothendieck). *The Grothendieck construction from Definition 1.10.1 gives an equivalence of fibrations:*

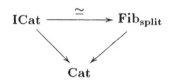

Proof. The Grothendieck construction determines the object part of a functor $\mathcal{G}\colon \mathbf{ICat} \to \mathbf{Fib}_{\mathrm{split}}$. For a morphism $(K, \alpha)\colon (\Psi\colon \mathbb{B}^{\mathrm{op}} \to \mathbf{Cat}) \longrightarrow (\Phi\colon \mathbb{A}^{\mathrm{op}} \to \mathbf{Cat})$ in **ICat**, one defines $\mathcal{G}(K, \alpha) = (K, \int \alpha)$, where $\int \alpha\colon \int_{\mathbb{B}} \Psi \to \int_{\mathbb{A}} \Phi$ is layed down by $(I, X) \mapsto (KI, \alpha_I(X))$ and $(u, f) \mapsto (Ku, \alpha_I(f))$—with I the domain of u.

In the reverse direction there is a functor $\mathcal{I}\colon \mathbf{Fib}_{\mathrm{split}} \to \mathbf{ICat}$; it maps a split fibration $\begin{smallmatrix}\mathbb{E}\\ \downarrow{\scriptstyle p}\\ \mathbb{B}\end{smallmatrix}$ to the functor $\mathcal{I}(p)\colon \mathbb{B}^{\mathrm{op}} \to \mathbf{Cat}$ which maps $I \mapsto \mathbb{E}_I$ and $u \mapsto u^*$, as described in Proposition 1.4.5. Clearly, for a morphism $\left(\begin{smallmatrix}\mathbb{E}\\ \downarrow{\scriptstyle p}\\ \mathbb{B}\end{smallmatrix} \right) \xrightarrow{(K,H)}$ $\left(\begin{smallmatrix}\mathbb{D}\\ \downarrow{\scriptstyle q}\\ \mathbb{A}\end{smallmatrix} \right)$ in **Fib**$_{\mathrm{split}}$, one takes $\mathcal{I}(K, H) = (K, (H_I)_{I \in \mathbb{A}})$, where $H_I\colon \mathbb{E}_I \to \mathbb{D}_{KI}$ is the restriction to the fibres. Naturality in I is obtained from the fact that H preserves the splitting on-the-nose. The required fibred equivalence follows readily. $\qquad\square$

Notice that the above result gives a categorical version of the equivalence $\mathbf{Fam}(\mathbf{Sets}) \xrightarrow{\simeq} \mathbf{Sets}^{\to}$ in Proposition 1.2.2 involving set-indexed families of sets.

Next we mention 2-cells for split indexed categories. In order not to complicate matters too much, we restrict ourselves to a fixed base category $\mathbb{B}$ (and so we allow only $K = \mathrm{id}\colon \mathbb{B} \to \mathbb{B}$ as functor between base categories in Definition 1.10.5).

1.10.8. Definition. Let Ψ and Φ be split indexed categories $\mathbb{B}^{\mathrm{op}} \rightrightarrows \mathbf{Cat}$ and $\alpha, \beta\colon \Psi \rightrightarrows \Phi$ natural transformations (*i.e.* 1-cells in **ICat**($\mathbb{B}$)). A **2-cell** $\sigma\colon \alpha \Rightarrow \beta$ in **ICat**($\mathbb{B}$) is defined to be a **modification** (see *e.g.* [176]), *i.e.* a family $\sigma_I\colon \alpha_I \Rightarrow \beta_I$ of natural transformations such that for each $u\colon I \to J$ in

$\mathbb{B}$ one has

$$\sigma_I \, \Psi(u) = \Phi(u) \, \sigma_J.$$

In a diagram:

$$
\begin{array}{ccc}
\Psi(J) & \xrightarrow{\quad\Psi(u)\quad} & \Psi(I) \\
\alpha_J \left(\overset{\sigma_J}{\Rightarrow} \right) \beta_J & & \alpha_I \left(\overset{\sigma_I}{\Rightarrow} \right) \beta_I \\
\Phi(J) & \xrightarrow[\quad\Phi(u)\quad]{} & \Phi(I)
\end{array}
$$

The proof of the following result is left to the interested reader.

1.10.9. Proposition. *The fibred equivalence* **ICat** $\overset{\simeq}{\Rightarrow}$ **Fib**$_{\mathrm{split}}$ *in Theorem 1.10.7 gives rise to an equivalence*

$$\mathbf{ICat}(\mathbb{B}) \xrightarrow{\quad\simeq\quad} \mathbf{Fib}_{\mathrm{split}}(\mathbb{B})$$

of 2-categories, for each category $\mathbb{B}$. □

As we already mentioned in the discussion in 1.10.4, there is considerable difference between taking opposites for (split) indexed categories and for fibrations. We shall briefly describe both constructions.

1.10.10. Definition. For a split indexed category $\Psi \colon \mathbb{B}^{\mathrm{op}} \to \mathbf{Cat}$ with basis $\mathbb{B}$, define the opposite split indexed category Ψ^{op} as the "fibrewise opposite" $\Psi^{\mathrm{op}} \colon \mathbb{B}^{\mathrm{op}} \to \mathbf{Cat}$ given by

$$I \mapsto \Psi(I)^{\mathrm{op}} \qquad \text{and} \qquad \left(I \xrightarrow{u} J \right) \mapsto \left(\Psi(J)^{\mathrm{op}} \xrightarrow{u^*} \Psi(I)^{\mathrm{op}} \right).$$

This definition of opposite for indexed categories is nice and simple. In contrast, the opposite for fibred categories is rather involved. The opposite p^{op} of a fibration p should mean: fibrewise the opposite. The requirement that such a construction be intrinsic makes the definition below somewhat complicated. For *split* fibrations it is much simpler (see Exercise 1.10.9), since it is essentially as for split indexed categories.

Recall that an arrow in the total category of a fibration factors as a vertical morphism followed by a Cartesian one. The intuition behind the definition of the opposite is that all vertical maps in such composites are reversed.

1.10.11. Definition (Bénabou [28]). Let $\begin{smallmatrix}\mathbb{E}\\\downarrow p\\\mathbb{B}\end{smallmatrix}$ be a fibration. A new fibration (with the same basis) written as $\begin{smallmatrix}\mathbb{E}^{(\mathrm{op})}\\\downarrow p^{\mathrm{op}}\\\mathbb{B}\end{smallmatrix}$ will be described which is fibrewise the opposite of p. Let

$$CV = \{(f_1, f_2) \mid f_1 \text{ is Cartesian, } f_2 \text{ is vertical and } \mathrm{dom}(f_1) = \mathrm{dom}(f_2)\}.$$

An equivalence relation is defined on the collection CV by

$$(f_1, f_2) \sim (g_1, g_2)$$
$$\Leftrightarrow \text{ there is a vertical } h \text{ with } g_1 \circ h = f_1 \text{ and } g_2 \circ h = f_2, \text{ as in}$$

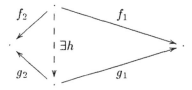

The equivalence class of (f_1, f_2) will be written as $[f_1, f_2]$.

The total category $\mathbb{E}^{(\mathrm{op})}$ of p^{op} has $X \in \mathbb{E}$ as objects. Its morphisms $X \to Y$ are equivalence classes $[f_1, f_2]$ of maps f_1, f_2 as in:

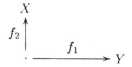

Composition is described by the following diagram

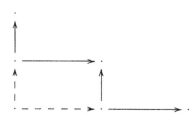

in which the horizontal dashed arrow is a Cartesian lifting, and the vertical dashed arrow is induced. The functor $p^{\mathrm{op}} \colon \mathbb{E}^{(\mathrm{op})} \to \mathbb{B}$ is then defined by $X \mapsto pX$ and $[f_1, f_2] \mapsto p(f_1)$.

The proof of the following result is left to the reader.

1.10.12. Lemma. *Let* $\begin{smallmatrix}\mathbb{E}\\\downarrow p\\\mathbb{B}\end{smallmatrix}$ *be a fibration. One has that*

(i) *the functor* $\begin{smallmatrix} \mathbb{E}^{(\mathrm{op})} \\ \downarrow p^{\mathrm{op}} \\ \mathbb{B} \end{smallmatrix}$ *is a fibration; and a morphism* $[f_1, f_2]$ *in* $\mathbb{E}^{(\mathrm{op})}$ *is Cartesian if and only if the vertical map* f_2 *is an isomorphism;*

(ii) *this fibration* p^{op} *is the fibrewise opposite of p, that is, for each object* $I \in \mathbb{B}$ *there is an isomorphism of fibre categories*

$$\left(\mathbb{E}^{(\mathrm{op})}\right)_I \cong \left(\mathbb{E}_I\right)^{\mathrm{op}}, \qquad \textit{natural in I;}$$

(iii) *there is an isomorphism of fibrations* $(p^{\mathrm{op}})^{\mathrm{op}} \cong p$ *over* $\mathbb{B}$. $\qquad\square$

In ordinary category theory, a category $\mathbb{C}$ has limits of shape $\mathbb{J}$ if and only if the opposite category $\mathbb{C}^{\mathrm{op}}$ has colimits of shape $\mathbb{J}$. Similar results exist for fibred categories, because the opposite of a fibration is taken fibrewise.

1.10.13. Lemma. *For a fibration p one has:*

$$p \text{ has fibred limits of shape } \mathbb{J} \quad\Leftrightarrow\quad p^{\mathrm{op}} \text{ has fibred colimits of shape } \mathbb{J}$$
$$p \text{ has simple products} \quad\Leftrightarrow\quad p^{\mathrm{op}} \text{ has simple coproducts;}$$
$$p \text{ has products} \quad\Leftrightarrow\quad p^{\mathrm{op}} \text{ has coproducts.} \qquad\square$$

We close this section with an investigation of **adjoining indeterminates** (or adding elements) to a category. It gives rise to indexed categories, and shows in particular how the type theoretic simple and codomain fibrations arise from the same pattern.

Let $\mathbb{B}$ be a category with terminal object 1 and let I be an object of $\mathbb{B}$. One can form a new category $\mathbb{B}[x : I]$—or $\mathbb{B}[x : 1 \to I]$ in the notation of [186]—by adding an indeterminate x of type I as follows. Consider the underlying graph of $\mathbb{B}$ and add an extra edge $x : 1 \to I$, where x is a new symbol. Let $\mathbb{B}[x : I]$ be the free category with terminal object 1, generated by this extended graph, incorporating the terminal object $1 \in \mathbb{B}$ and the equations that hold in $\mathbb{B}$. It comes equipped with an inclusion $\eta_I : \mathbb{B} \to \mathbb{B}[x : I]$ which preserves the terminal object. This functor together with $x : 1 \to I$ in $\mathbb{B}[1 : I]$ is universal in the following way. For any terminal object preserving functor $F : \mathbb{B} \to \mathbb{C}$ together with a morphism $a : 1 \to FI$ in $\mathbb{C}$, there is an (up-to-unique-isomorphism) unique terminal object preserving functor $\overline{F} : \mathbb{B}[x : I] \to \mathbb{C}$ such that $\overline{F}x = a$ in

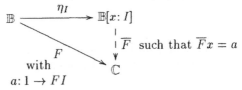

Below we are particularly interested in the case where the category $\mathbb{B}$ has finite products or finite limits. This structure can then be extended to $\mathbb{B}[x : I]$ and the universal property holds for functors preserving such structure.

1.10.14. Proposition. *The assignment $I \mapsto \mathbb{B}[x : I]$ extends to an indexed category $\mathbb{B}^{\mathrm{op}} \to \mathbf{Cat}$.*

Proof. For $u : I \to J$ in $\mathbb{B}$ one obtains a functor $u^* : \mathbb{B}[y : J] \to \mathbb{B}[x : I]$ by the universal property applied to

In the particular cases where the category $\mathbb{B}$ has finite products or finite limits, there are simpler ways to describe the category $\mathbb{B}[x : I]$. The following can be found in [186]: (i) in Section I, 7 and (ii) in Exercise 2 in II, 16.

1.10.15. Proposition. (i) *In case $\mathbb{B}$ has finite products, $\mathbb{B}[x : I]$ is equivalent to the simple slice category $\mathbb{B}/\!\!/ I$.*

(ii) *In case $\mathbb{B}$ has finite limits, $\mathbb{B}[x : I]$ is equivalent to the ordinary slice category $\mathbb{B}/I$.*

Proof. It is not hard to verify that the functors $I^* : \mathbb{B} \to \mathbb{B}/\!\!/ I$ and $I^* : \mathbb{B} \to \mathbb{B}/I$ satisfy the appropriate universal properties. $\qquad\square$

1.10.16. Corollary. *Applying the Grothendieck construction to the indexed category $I \mapsto \mathbb{B}[x : I]$ from Proposition 1.10.14 yields*

(i) *the simple fibration* $\begin{smallmatrix} \mathrm{s}(\mathbb{B}) \\ \downarrow \\ \mathbb{B} \end{smallmatrix}$ *in case $\mathbb{B}$ has finite products;*

(ii) *the codomain fibration* $\begin{smallmatrix} \mathbb{B}^{\to} \\ \downarrow \\ \mathbb{B} \end{smallmatrix}$ *in case $\mathbb{B}$ has finite limits.* $\qquad\square$

The 'type theoretic' simple and codomain fibrations thus arise by the same procedure of adjoining indeterminates. For more information, see [125] or [129]. There one finds a description of adding indeterminates to fibred categories.

Exercises

1.10.1. (i) Give the split indexed category yielding the fibration $\begin{smallmatrix} \mathrm{UFam}(\omega\text{-}\mathbf{Sets}) \\ \downarrow \\ \omega\text{-}\mathbf{Sets} \end{smallmatrix}$ from Section 1.2 upon application of the Grothendieck construction.

(ii) Do the same for the fibration $\begin{smallmatrix} \mathbf{S\text{-}Model} \\ \downarrow \\ \mathbf{Sign} \end{smallmatrix}$ from Section 1.6.

1.10.2. (i) Show that the Grothendieck construction applied to a (representable) functor $\mathbb{B}(-, I)\colon \mathbb{B}^{\mathrm{op}} \to \mathbf{Cat}$—which is a split indexed category with discrete fibre categories—yields the domain fibration $\begin{smallmatrix} \mathbb{B}/I \\ \downarrow {\scriptstyle \mathrm{dom}_I} \\ \mathbb{B} \end{smallmatrix}$.

In line with this result, one calls a fibration $\begin{smallmatrix} \mathbb{E} \\ \downarrow p \\ \mathbb{B} \end{smallmatrix}$ **representable** if it is equivalent (as a fibration) to dom_I for some $I \in \mathbb{B}$.

(ii) Show that a presheaf $H\colon \mathbb{B}^{\mathrm{op}} \to \mathbf{Sets}$ is representable (in the ordinary sense) if and only if the associated Grothendieck fibration is representable (as a fibration).

[Representability will be further investigated in Section 5.2.]

1.10.3. Show that the category $\mathbf{ICat}(\mathbb{B})$ of indexed categories with basis $\mathbb{B}$ has finite products.

1.10.4. Show that a (split) indexed category $\Psi\colon \mathbb{B}^{\mathrm{op}} \to \mathbf{Cat}$ and a functor $K\colon \mathbb{A} \to \mathbb{B}$ give rise to a pullback diagram of categories:

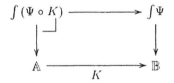

1.10.5. Say that a split indexed category $\Psi\colon \mathbb{B}^{\mathrm{op}} \to \mathbf{Cat}$ has (**indexed**) **Cartesian products** $\times$ if the diagonal $\Delta\colon \Psi \to \Psi \times \Psi$ in $\mathbf{ICat}(\mathbb{B})$ has a right adjoint in the 2-category $\mathbf{ICat}(\mathbb{B})$.

(i) Describe what this means concretely.

(ii) Verify that Ψ has indexed Cartesian products if and only if its associated Grothendieck fibration has split fibred Cartesian products.

1.10.6. For two fibrations $\begin{smallmatrix} \mathbb{E} \\ \downarrow p \\ \mathbb{B} \end{smallmatrix}$ and $\begin{smallmatrix} \mathbb{D} \\ \downarrow q \\ \mathbb{B} \end{smallmatrix}$ over $\mathbb{B}$ we write $\mathbf{Fib}(\mathbb{B})(p, q)$ for the hom-category of fibred functors $p \to q$ over $\mathbb{B}$ and vertical natural transformations between them. For $I \in \mathbb{B}$ consider the assignment

$$I \mapsto \mathbf{Fib}(\mathbb{B})\Big(\mathrm{dom}_I \times p,\ q\Big).$$

(i) Show that it extends to a split indexed category $\mathbb{B}^{\mathrm{op}} \to \mathbf{Cat}$.

(ii) Write $p \Rightarrow q$ for the resulting split fibration. Show that there is an *equivalence* of categories

$$\mathbf{Fib}(\mathbb{B})\Big(r \times p,\ q\Big) \simeq \mathbf{Fib}(\mathbb{B})\Big(r,\ p \Rightarrow q\Big).$$

(iii) Now assume that the above p, q are *split* fibrations and consider the split fibration (for which we also write $p \Rightarrow q$) resulting from the indexed category,

$$I \mapsto \mathbf{Fib}_{\mathbf{split}}(\mathbb{B})\Big(\mathrm{dom}_I \times p,\ q\Big).$$

of split fibred functors. Show that one now gets an isomorphism of hom-categories,

$$\mathbf{Fib}_{\mathbf{split}}(\mathbb{B})\big(r \times p,\, q\big) \cong \mathbf{Fib}_{\mathbf{split}}(\mathbb{B})\big(r,\, p \Rightarrow q\big).$$

[Thus $p \Rightarrow q$ behaves like an exponent, see also [36, II, Lemma 8.4.4]. Its definition can be understood in terms of the Fibred Yoneda Lemma 5.2.4. If p and q are presheaves (*i.e.* discrete fibrations), then $p \Rightarrow q$ is the usual exponent of presheaves (see Example 5.4.2).]

1.10.7. (i) Check that one gets a category $\int \Psi$ as described in Definition 1.10.1 in case η and μ are natural transformations satisfying the coherence conditions (but are not necessarily isomorphisms as in Definition 1.4.4), but that the result need not be fibred over $\mathbb{B}$.

(ii) Show that a monad (T, η, μ) on a category $\mathbb{A}$ corresponds to a "pseudo-functor" $\mathbb{A}\colon 1^{\mathrm{op}} \to \mathbf{Cat}$, without the requirement that the maps η and μ are isomorphisms.

(iii) Show that in the situation of (ii), the Grothendieck construction as in (i) corresponds to taking the Kleisli category of the monad (T, η, μ).

1.10.8. Let $\mathbb{B}$ be a category with finite limits. We write $\begin{smallmatrix} \mathrm{Inv}(\mathbb{B}) \\ \downarrow \\ \mathbb{B} \end{smallmatrix}$ for the opposite of the codomain fibration $\begin{smallmatrix} \mathbb{B}^{\to} \\ \downarrow \\ \mathbb{B} \end{smallmatrix}$. The category $\mathrm{Inv}(\mathbb{B})$ is sometimes called the inverse arrow category of $\mathbb{B}$.

(i) Describe the fibration $\begin{smallmatrix} \mathrm{Inv}(\mathbb{B}) \\ \downarrow \\ \mathbb{B} \end{smallmatrix}$ in detail.

(ii) Show that its fibre above the terminal object is (isomorphic to) $\mathbb{B}^{\mathrm{op}}$.

1.10.9. Let p be a split fibration. Show that the opposite fibration p^{op} can also be obtained by turning p first into a split indexed category, taking the opposite of all fibres and changing it back into a fibration.

1.10.10. Let p be a Cartesian closed fibration. Describe the (fibred) exponents via a fibred functor $\Rightarrow\colon p^{\mathrm{op}} \times p \to p$.

1.10.11. Let $\mathbb{C}$ be a category with pullbacks; consider the simple fibration $\begin{smallmatrix} s(\mathbb{C}) \\ \downarrow s_{\mathbb{C}} \\ \mathbb{C} \end{smallmatrix}$. There is a functor $\{-\}\colon s(\mathbb{C}) \to \mathbb{C}$ given by $(I, X) \mapsto I \times X$ and $(u, f) \mapsto \langle u \circ \pi, f \rangle$. Form the fibration q by change-of-base

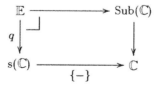

and let r be the fibration $\big(s_{\mathbb{C}} \circ q^{\mathrm{op}}\big)^{\mathrm{op}}\colon \mathbf{D}\mathbb{C} \to \mathbb{C}$. Describe the total category $\mathbf{D}\mathbb{C}$ in detail.

[It is the **dialectica category** of de Paiva [243].]

Chapter 2

Simple type theory

In this chapter we introduce the first and most elementary type theory, namely simple type theory (STT), which goes back to Church [49]. Here we use the terminology *simply typed* for type theories without type variables and *polymorphically typed* for type theories which do have such type variables. Chapter 8 is devoted to these polymorphic type theories (PTTs). Although there are no type variables α: Type in STT, term variables v: σ inhabiting types σ: Type do exist. But these are allowed to occur only in terms—and not in types, like in dependent type theories (DTTs) (see Chapter 10).

In the present chapter we give categorical semantics of simple type theory, both in (traditional) terms of ordinary categories, and also in terms of fibred categories. We begin with the syntax of calculi of types and terms, starting from a many-typed signature as defined in Section 1.6. From now on, terms will be described systematically in contexts. These are finite sequences of variable declarations v_i: σ_i, describing the types σ_i of free variables v_i. The rules for term formation will guarantee that variables only appear free in a term, if they occur in the context of the term. This calculus types and terms gives in a canonical way rise to a category, which is commonly called its classifying category. A model of a calculus can conveniently be described in terms of a suitable structure preserving functor from its classifying category into some other 'receiving' category. This is the essence of Lawvere's functorial semantics. In Section 2.2 it will first be described for ordinary categories. Later on, this functorial semantics will also be used for fibred categories.

STT is commonly studied with the following constructors for the formation of new types: exponents $\rightarrow$ (or function spaces), finite products $(1, \times)$ and fi-

119

nite coproducts $(0, +)$ (or disjoint unions). Models of calculi with $(\to, 1, \times)$ are easily described in terms of Cartesian closed categories, see *e.g.* [186, 63, 61]. Additional coproduct types $(0, +)$ can be described with categories having additionally finite coproducts. Under the propositions-as-types perspective, the study of these type formers $(\to, 1, \times, 0, +)$ amounts to the study of the proof theory of the propositional connectives $(\supset, \top, \wedge, \perp, \vee)$. It turns out that the minimal calculus, with exponent types only, is most difficult to capture categorically. This is because categorical exponents are not described in isolation, but require (binary) products.

Using fibred categories one can resolve this difficulty. In a fibred description of a type theory (or of a logic), contexts form objects of a base category. The fibre above such a context contains what happens in that context. This view is fundamental. For simple type theory it suffices to consider simple fibrations (introduced in Section 1.3), since types do not contain any (term or type) variables and hence do not depend on a context. It will turn out that exponent types $\to$ can then be described by right adjoints to weakening functors π^*, *i.e.* by what were called simple products in Section 1.9. This will be done in Section 2.4. Additionally, Cartesian product types $\times$ are captured as left adjoints to these weakening functors π^*. This description of $\to$ and $\times$ is in fact a special case of Π and Σ in a situation where Π becomes $\to$ and Σ becomes $\times$ because there is no type dependency (see also Example 10.1.2 later on).

Historically, Church's *untyped* λ-calculus came before his simple type theory. In this untyped λ-calculus there is no typing discipline and each term may be applied to every other term (including itself, which gives self-application, like in the term $\lambda x. xx$). But the untyped λ-calculus may be understood as a simply typed λ-calculus with only one type, say Ω, with $\Omega \to \Omega = \Omega$. Specialising the fibred description of exponents to this particular case with one type, naturally gives us a notion of model for the untyped λ-calculus. This will be done in Section 2.5. In our fibred approach we thus get (the semantics of) untyped λ-calculi as a special case of (the semantics of) simply typed calculi. It comes almost for free.

In the final section 2.6 of this chapter one can find how simple fibrations may also be used to give a suitable description of data types with (simple) parameters.

2.1 The basic calculus of types and terms

Starting from a many-typed signature we will define various simply typed calculi: in this section we introduce the basic calculus which gives a detailed description of the terms associated with a signature. Later, in Section 2.3 we will define the calculus $\lambda 1$ by adding exponent types via a type constructor $\rightarrow$, and extensions of this $\lambda 1$ with finite product types $(1, \times)$ and finite coproduct types $(0, +)$. With all these calculi one associates in a canonical way a classifying category: it is obtained as a term model (or generalised Lindenbaum-Tarski) construction. It will play a crucial rôle in the Lawvere's functorial semantics in the next section.

Let Σ be a many-typed signature with $T = |\Sigma|$ as its underlying set of types. We assume a denumerably infinite set $\text{Var} = \{v_1, v_2, \ldots\}$, elements of which will be called (term) variables. A **context** Γ is then a finite sequence of variable declarations written as

$$\Gamma = (v_1 : \sigma_1, \ldots, v_n : \sigma_n).$$

By convention, we list the variables in a context starting with v_1. We can concatenate contexts $\Gamma = (v_1 : \sigma_1, \ldots, v_n : \sigma_n)$ and $\Delta = (v_1 : \tau_1, \ldots, v_n : \tau_m)$ as

$$\Gamma, \Delta = (v_1 : \sigma_1, \ldots, v_n : \sigma_n, v_{n+1} : \tau_1, \ldots, v_{n+m} : \tau_m).$$

This precise use of variables v_i has two advantages: it prevents name clashes of variables and is fairly close to a categorical description. There is nothing deep to it since variables are merely placeholders. The extra book-keeping which it requires is bearable. And in situations where it does not matter which of the variables v_i is being used, we freely use meta-variables $x, y, z, \ldots$ Especially in later chapters we shall use mostly these meta-variables, but for the moment it is better to be precise.

Terms are thus described with respect to a fixed collection of variables, which receive their types in contexts. And not, as in universal algebra (see Section 1.6), with respect to various collections $(X_\sigma)_{\sigma \in T}$ of sets X_σ of variables which are already typed.

In type theory one uses the notation

$$\Gamma \vdash M : \tau$$

to express that M is a term of type τ in context Γ. In such a situation one sometimes says that M inhabits τ, or just that τ is inhabited (by M, in context Γ). A typical example of such an inhabitation sequent is

$$n : \mathsf{N}, m : \mathsf{N} \vdash \mathsf{plus}(\mathsf{times}(m, n), m) : \mathsf{N}.$$

Such a typing sequent can be obtained by successive applications of the following basic rules.

identity

$$\overline{\rule{0pt}{1em}\quad v_1 : \sigma \vdash v_1 : \sigma \quad}$$

function symbol

$$\frac{\Gamma \vdash M_1 : \sigma_1 \quad \cdots \quad \Gamma \vdash M_n : \sigma_n}{\Gamma \vdash F(M_1, \ldots, M_n) : \sigma_{n+1}} \text{ (for } F : \sigma_1, \ldots, \sigma_n \longrightarrow \sigma_{n+1} \text{ in } \Sigma)$$

Plus the following **structural rules**:

weakening

$$\frac{v_1 : \sigma_1, \ldots, v_n : \sigma_n \vdash M : \tau}{v_1 : \sigma_1, \ldots, v_n : \sigma_n, v_{n+1} : \sigma_{n+1} \vdash M : \tau}$$

contraction

$$\frac{\Gamma, v_n : \sigma, v_{n+1} : \sigma \vdash M : \tau}{\Gamma, v_n : \sigma \vdash M[v_n / v_{n+1}] : \tau}$$

exchange

$$\frac{\Gamma, v_i : \sigma_i, v_{i+1} : \sigma_{i+1}, \Delta \vdash M : \tau}{\Gamma, v_i : \sigma_{i+1}, v_{i+1} : \sigma_i, \Delta \vdash M[v_i / v_{i+1}, v_{i+1} / v_i] : \tau}$$

These last three rules allow us to add an extra variable declaration, to replace two variables of the same type by a single one and to permute assumptions. Often, these structural rules are not listed explicitly. But here we emphasise them, because weakening and contraction play an important rôle in the categorical description of type constructors. (Also it is good to be explicitly aware of such rules, because their use may be restricted, as in linear logic, see [97, 98].)

We thus have rules for deriving inhabitation sequents $\Gamma \vdash M : \sigma$. Formally we say that such a sequent is **derivable** if there is a derivation tree consisting of the above rules, with the sequent $\Gamma \vdash M : \sigma$ as conclusion. In that case we sometimes write

$$\blacktriangleright \Gamma \vdash M : \sigma$$

for: $\Gamma \vdash M : \sigma$ is derivable.

As an example, consider a signature with two function symbols:

$$\mathsf{plus} : \mathsf{N}, \mathsf{N} \longrightarrow \mathsf{N}, \qquad \mathsf{if} : \mathsf{B}, \mathsf{N}, \mathsf{N} \longrightarrow \mathsf{N}.$$

Then one can derive an inhabitation statement

$$v_1 : \mathsf{B}, v_2 : \mathsf{N} \vdash \mathsf{if}(v_1, v_2, \mathsf{plus}(v_2, v_2)) : \mathsf{N}.$$

Formally, this is done as follows.

$$
\cfrac{
  \cfrac{
    \cfrac{v_1\colon \mathsf{B} \vdash v_1\colon \mathsf{B}}{v_1\colon \mathsf{B}, v_2\colon \mathsf{N} \vdash v_1\colon \mathsf{B}}\ (\mathrm{W})
    \qquad
    \cfrac{\cfrac{v_1\colon \mathsf{N} \vdash v_1\colon \mathsf{N}}{v_1\colon \mathsf{N}, v_2\colon \mathsf{N} \vdash v_1\colon \mathsf{N}}\ (\mathrm{W})}{v_1\colon \mathsf{N}, v_2\colon \mathsf{N} \vdash v_2\colon \mathsf{N}}\ (\mathrm{E})
    \qquad
    \cfrac{\cfrac{v_1\colon \mathsf{N} \vdash v_1\colon \mathsf{N} \qquad v_1\colon \mathsf{N} \vdash v_1\colon \mathsf{N}}{\cfrac{v_1\colon \mathsf{N} \vdash \mathsf{plus}(v_1,v_1)\colon \mathsf{N}}{v_1\colon \mathsf{N}, v_2\colon \mathsf{N} \vdash \mathsf{plus}(v_1,v_1)\colon \mathsf{N}}}}{v_1\colon \mathsf{N}, v_2\colon \mathsf{N} \vdash \mathsf{plus}(v_2,v_2)\colon \mathsf{N}}\ (\mathrm{E})
  }{}
}{v_1\colon \mathsf{B}, v_2\colon \mathsf{N} \vdash \mathsf{if}(v_1, v_2, \mathsf{plus}(v_2,v_2))\colon \mathsf{N}}
$$

The annotations (W) and (E) indicate applications of the Weakening and Exchange rules. In similar fashion, one can write a derivation tree for

$$v_1\colon \mathsf{N},\, v_2\colon \mathsf{N},\, v_3\colon \mathsf{B},\, v_4\colon \mathsf{N},\, v_5\colon \mathsf{B} \vdash \mathsf{if}(v_5, \mathsf{plus}(v_1, v_4), v_2)\colon \mathsf{N}.$$

Intuitively, this may be clear, but the formal derivation is involved. In Exercise 2.1.1 below, we present some extra (derivable) rules which make it easier to form such terms.

This calculus of types and terms may be called the **term calculus** of a signature Σ.

Substitution $M[N/v_n]$ of a term N for a variable v_n in M is best defined on 'raw' terms (*i.e.* not necessarily well-typed terms), as

$$
\begin{aligned}
v_m[N/v_n] &= \begin{cases} N & \text{if } m = n \\ v_m & \text{else} \end{cases} \\
F(M_1, \ldots, M_n)[N/v_n] &= F(M_1[N/v_n], \ldots, M_n[N/v_n]).
\end{aligned}
$$

As a derived rule one then has

substitution

$$
\cfrac{\Gamma, v_n\colon \sigma \vdash M\colon \tau \qquad \Gamma \vdash N\colon \sigma}{\Gamma \vdash M[N/v_n]\colon \tau}
$$

This rule expresses that if the term N and variable v_n have the same type, then performing substitution $[N/v_n]$ transforms well-typed terms into well-typed terms. The rule is consequence of a much more general substitution result, which is presented as Exercise 2.1.2.

It is useful to emphasise once again the difference between the above term calculus and the sets of terms $\mathrm{Terms}_\tau(X)$ of type τ, built upon a T-indexed collection of sets of variables $X = (X_\sigma)_{\sigma \in T}$, as described in Section 1.6. The main difference lies in the fact that in the latter approach the sets of variables $(X_\sigma)_{\sigma \in T}$ form a parameter. This is usual in universal algebra. In the type theoretic approach in this section (and in the rest of this book) we fix in advance the set from which variables can be taken.

There is a way to switch for individual terms between these descriptions. If $\Gamma = (v_1 : \sigma_1, \ldots, v_n : \sigma_n)$ and $\Gamma \vdash M : \tau$ in type theory are given, then one can form sets $X_\sigma = \{v \in \text{Var} \mid v : \sigma \text{ occurs in } \Gamma\}$. This yields a collection $X = (X_\sigma)_{\sigma \in T}$ of term variables and the term M can now be described as a term $M \in \text{Terms}_\tau(X)$. In particular, in a Σ-algebra $(A_\sigma, [\![-]\!])$ as described in Section 1.6 the term M yields a function

$$A_{\sigma_1} \times \cdots \times A_{\sigma_n} \xrightarrow{\quad [\![\Gamma \vdash M : \tau]\!] = \quad} A_\tau$$
$$\vec{\lambda}\vec{a}. [\![M]\!]_{\rho(\vec{v} \mapsto \vec{a})}$$

as described before Definition 1.6.4.

Conversely, assume an arbitrary collection $X = (X_\sigma)_{\sigma \in T}$ of term variables and a term M in $\text{Terms}_\tau(X)$, as in the alternative description. We know that M is formed in a finite number of steps and can thus contain only a finite number of variables $x_i \in X_{\sigma_i}$, say, with $1 \leq i \leq n$. Replacing these $x_i \in X_{\sigma_i}$ by $v_i : \sigma_i$, one gets a term $v_1 : \sigma_1, \ldots, v_n : \sigma_n \vdash M[\vec{v}/\vec{x}] : \tau$ as in type theory.

We close this section by showing how contexts and appropriately typed (sequences of) terms form a category. The intuition behind terms-as-morphisms is the following. A term in context $v_1 : \sigma_1, \ldots, v_n : \sigma_n \vdash M : \tau$ may be seen as an operation which maps inputs $a_i : \sigma_i$ on the left of the turnstile $\vdash$ to an output $M[\vec{a}/\vec{v}] : \tau$ on the right, via substitution. Thus one expects such a term to form a morphism

$$\sigma_1 \times \cdots \times \sigma_n \xrightarrow{\quad M \quad} \tau$$

in a suitable category, so that, roughly, $\vdash$ becomes $\to$. This is formalised in the next definition. Morphisms will not be individual terms, but sequences of terms. Such sequences are often called context morphisms.

2.1.1. Definition. The above term calculus on a signature Σ will serve as a basis for the **classifying category** (or **term model**) $\mathcal{Cl}(\Sigma)$ of Σ. Its objects are contexts Γ of variable declarations. And its morphisms $\Gamma \to \Delta$—where $\Delta = (v_1 : \tau_1, \ldots, v_m : \tau_m)$—are m-tuples $(M_1, \ldots, M_m)$ of terms for which we can derive $\Gamma \vdash M_i : \tau_i$, for each $1 \leq i \leq m$.

The identity on an object $\Gamma = (v_1 : \sigma_1, \ldots, v_n : \sigma_n)$ in $\mathcal{Cl}(\Sigma)$ is the n-tuple of variables

$$\Gamma \xrightarrow{\quad (v_1, \ldots, v_n) \quad} \Gamma$$

And the composite of context morphisms

$$\Gamma \xrightarrow{\quad (M_1, \ldots, M_m) \quad} \Delta \xrightarrow{\quad (N_1, \ldots, N_k) \quad} \Theta$$

is the k-tuple $(L_1, \ldots, L_k)$ defined by simultaneous substitution:

$$L_i = N_i[M_1/v_1, \ldots, M_m/v_m].$$

It is then almost immediate that identities are identities indeed. Associativity requires a suitable substitution lemma, see Exercise 2.1.3 below.

We notice that the construction of a classifying category of a type theory is like the construction of the Lindenbaum algebra of a (propositional) logic. In the first case a turnstile $\vdash$ in type theory becomes an arrow $\to$ in a category, and in the second case a turnstile $\vdash$ in logic becomes an inequality $\leq$ in a preorder (or poset). Under a propositions-as-types reading, the preorder that one obtains is the underlying preorder of the classifying category.

2.1.2. Proposition. *The classifying category* $\mathcal{C}\ell(\Sigma)$ *of a signature* Σ *has finite products.*

Proof. The empty context $\emptyset$ is terminal object, since for any context Γ there is precisely one morphism $\Gamma \dashrightarrow \emptyset$, namely the empty sequence (). The Cartesian product of contexts $\Gamma = (v_1 : \sigma_1, \ldots, v_n : \sigma_n)$ and $\Delta = (v_1 : \tau_1, \ldots, v_m : \tau_m)$ is their concatenation Γ, Δ with projection morphisms:

$$\Gamma \xleftarrow{\quad (v_1, \ldots, v_n) \quad} (\Gamma, \Delta) \xrightarrow{\quad (v_{n+1}, \ldots, v_{n+m}) \quad} \Delta \qquad \square$$

Exercises

2.1.1.   Prove that the following 'extended' structural rules are derivable.

(i)
$$\frac{\Gamma, v_n : \sigma, \Delta, v_{n+m} : \rho, \Theta \vdash M : \tau}{\Gamma, v_n : \rho, \Delta, v_{n+m} : \sigma, \Theta \vdash M[v_n/v_{n+m}, v_{n+m}/v_n] : \tau}$$

(ii)
$$\frac{\Gamma, v_n : \sigma, \Delta \vdash M : \tau}{\Gamma, \Delta, v_{n+m} : \sigma \vdash M[v_{n+m}/v_n, v_n/v_{n+1}, \ldots, v_{n+m-1}/v_{n+m}] : \tau}$$

(iii)
$$\frac{\Gamma, v_n : \sigma, \Delta, v_{n+m} : \sigma, \Theta \vdash M : \tau}{\Gamma, v_n : \sigma, \Delta, \Theta \vdash \\ M[v_n/v_{n+m}, v_{n+m}/v_{n+m+1}, \ldots, v_{n+m+k-1}/v_{n+m+k}] : \tau}$$

(iv)
$$\frac{\Gamma, \Gamma \vdash M : \tau}{\Gamma \vdash M[v_1/v_{n+1}, \ldots, v_n/v_{2n}] : \tau}$$

2.1.2.   Derive the following substitution rule.

$$\frac{\Gamma, v_n : \sigma, \Delta \vdash M : \tau \qquad \Theta \vdash N : \sigma}{\Gamma, \Theta, \Delta \vdash M^*[N^\#/v_n] : \tau}$$

where, assuming Δ to be of length m and Θ of length k,

$$M^* = M[v_{n+k}/v_{n+1}, \ldots, v_{n+k+m-1}/v_{n+m}]$$
$$N^\# = N[v_n/v_1, \ldots, v_{n+k-1}/v_k].$$

2.1.3.   (i) Prove the following **substitution lemma** (see also [13], 2.1.16) for 'raw' terms.

$$M[N/v_n][L/v_m] \equiv M[N[L/v_m]/v_n], \qquad \text{if } v_m \text{ is not free in } M.$$

(The sign $\equiv$ is used for syntactical identification, as opposed to conversion later on.)

(ii) Show that—as a result—composition in classifying categories $\mathcal{C}\ell(\Sigma)$ is associative.

2.2 Functorial semantics

In Section 1.6 we have seen the notion of model (or algebra) for a many-typed signature Σ. It consists of a collection of sets A_σ for types σ in Σ, and of an actual function $[\![F]\!]: A_{\sigma_1} \times \cdots \times A_{\sigma_n} \longrightarrow A_{\sigma_{n+1}}$ for each function symbol $F: \sigma_1, \ldots, \sigma_n \longrightarrow \sigma_{n+1}$ in Σ. Below we shall re-describe, following Lawvere [191], such a model as a finite product preserving functor $\mathcal{C}\ell(\Sigma) \to$ **Sets** from the classifying category of Σ to sets. This alternative formulation of *model of a signature* admits generalisation to model functors $\mathcal{C}\ell(\Sigma) \to \mathbb{B}$ into receiving categories $\mathbb{B}$ other than **Sets**.

But first we re-describe set-theoretic models. Recall the fibration $\begin{smallmatrix} \textbf{S-Model} \\ \downarrow \\ \textbf{Sign} \end{smallmatrix}$ of set theoretic models of many-typed signatures over their signatures from Section 1.6. The fibre category of Σ-models will be written as **S-Model**(Σ).

2.2.1. Theorem. *For each many-typed signature Σ, there is an equivalence of categories*

$$\textbf{S-Model}(\Sigma) \simeq \textbf{FPCat}\big(\mathcal{C}\ell(\Sigma), \textbf{Sets}\big)$$

where the right hand side denotes the hom-category of finite product preserving functors and natural transformations between them.

Thus, set-theoretic Σ-models correspond to finite product preserving functors from the classifying category of Σ to **Sets** and morphisms of Σ-models to natural transformations between the corresponding functors. Above, **FPCat** stands for the 2-category which has categories with finite products as objects and functors preserving such structure as morphisms; 2-cells are ordinary natural transformations.

Proof. For the passage **S-Model**$(\Sigma) \longrightarrow \textbf{FPCat}\big(\mathcal{C}\ell(\Sigma), \textbf{Sets}\big)$, let $(A_\sigma)_{\sigma \in T}$ be a Σ-model, where $T = |\Sigma|$ is the set of types underlying Σ. One defines an associated model functor $\mathcal{A}: \mathcal{C}\ell(\Sigma) \to \mathbf{Sets}$ by

$$\Gamma = (v_1 : \sigma_1, \ldots, v_n : \sigma_n) \mapsto A_{\sigma_1} \times \cdots \times A_{\sigma_n}$$
$$(M_1, \ldots, M_m) : \Gamma \to \Delta \mapsto \lambda(a_1, \ldots, a_n). (\llbracket M_1 \rrbracket_{\rho(\vec{v} \mapsto \vec{a})}, \ldots, \llbracket M_m \rrbracket_{\rho(\vec{v} \mapsto \vec{a})})$$
$$= \langle \llbracket \Gamma \vdash M_1 : \tau_1 \rrbracket, \ldots, \llbracket \Gamma \vdash M_m : \tau_m \rrbracket \rangle,$$

where $\llbracket \Gamma \vdash M_i : \tau_i \rrbracket$ is the interpretation of the term $\Gamma \vdash M_i : \tau_i$ as a function $A_{\sigma_1} \times \cdots \times A_{\sigma_n} \to A_{\tau_i}$ (as we described before Definition 2.1.1).

As an example we show that $\mathcal{A}: \mathcal{C}\ell(\Sigma) \to \mathbf{Sets}$ preserves identities.

$$\begin{aligned}
\mathcal{A}(\mathrm{id}_\Gamma) &= \mathcal{A}(v_1, \ldots, v_n) \\
&= \lambda(a_1, \ldots, a_n). (\llbracket v_1 \rrbracket_{\rho(\vec{v} \mapsto \vec{a})}, \ldots, \llbracket v_n \rrbracket_{\rho(\vec{v} \mapsto \vec{a})}) \\
&= \lambda(a_1, \ldots, a_n). (a_1, \ldots, a_n) \\
&= \mathrm{id}_{\mathcal{A}(\Gamma)}.
\end{aligned}$$

From Exercise 1.6.7 (i) it follows that $\mathcal{A}$ preserves composition. It is almost immediate that $\mathcal{A}$ preserves finite products. But note that products are not preserved *on-the-nose*, due to an implicit use of bracketing in $A_{\sigma_1} \times \cdots \times A_{\sigma_n}$.

A morphism of Σ-models $(H_\sigma : A_\sigma \to B_\sigma)_{\sigma \in T}$ induces a natural transformation $\mathcal{A} \Rightarrow \mathcal{B}$ between the corresponding functors, with component at $\Gamma = (v_1 : \sigma_1, \ldots, v_n : \sigma_n)$ given by

$$\lambda(a_1, \ldots, a_n). (H_{\sigma_1}(a_1), \ldots, H_{\sigma_n}(a_n)) = H_{\sigma_1} \times \cdots \times H_{\sigma_n}.$$

Naturality follows from Exercise 1.6.7 (ii).

In the reverse direction, a finite product preserving functor $\mathcal{M}: \mathcal{C}\ell(\Sigma) \to \mathbf{Sets}$ determines a set-theoretic model of Σ with carrier sets

$$\left(\mathcal{M}_\sigma \stackrel{\mathrm{def}}{=} \mathcal{M}(v_1 : \sigma) \right)_{\sigma \in T}$$

and interpretation of Σ-function symbol $F: \sigma_1, \ldots, \sigma_n \longrightarrow \sigma_{n+1}$,

$$\llbracket F \rrbracket \stackrel{\mathrm{def}}{=} \mathcal{M}\left(F(v_1, \ldots, v_n) : (v_1 : \sigma_1, \ldots, v_n : \sigma_n) \to (v_1 : \sigma_{n+1}) \right) \circ \varphi,$$

where φ is the isomorphism in

$$\mathcal{M}\left(v_1 : \sigma_1, \ldots, v_n : \sigma_n \right) \xleftarrow[\cong]{\varphi} \mathcal{M}\left(v_1 : \sigma_1 \right) \times \cdots \times \mathcal{M}\left(v_1 : \sigma_n \right)$$
$$= \mathcal{M}_{\sigma_1} \times \cdots \times \mathcal{M}_{\sigma_n}$$

making $\llbracket F \rrbracket$ a well-typed function. A natural transformation $\alpha: \mathcal{M} \Rightarrow \mathcal{N}$ between functors $\mathcal{M}, \mathcal{N}: \mathcal{C}\ell(\Sigma) \rightrightarrows \mathbf{Sets}$ determines a morphism of models, with functions

$$\left(\mathcal{M}_\sigma \longrightarrow \mathcal{N}_\sigma \right)_{\sigma \in T},$$

given by

$$\lambda a \in \mathcal{M}_\sigma. \alpha_{(v_1 : \sigma)}(a). \qquad \square$$

The next definition embodies a crucial step in functorial semantics; it generalises models of a signature in **Sets** to an arbitrary category with finite products. The previous theorem suggests to define such models simply as finite product preserving functors with the classifying category of the signature as domain.

2.2.2. Definition. Let Σ be a many-typed signature and $\mathbb{B}$ a category with finite products. A **model** of Σ in $\mathbb{B}$ consists of a functor

$$\mathcal{C}\ell(\Sigma) \xrightarrow{\ \ \mathcal{M}\ \ } \mathbb{B}$$

preserving finite products. A **morphism** between two such Σ-models $\mathcal{M}, \mathcal{N}: \mathcal{C}\ell(\Sigma) \rightrightarrows \mathbb{B}$ in $\mathbb{B}$ is then a natural transformation $\mathcal{M} \Rightarrow \mathcal{N}$. Hence the category of Σ-models in $\mathbb{B}$ is defined to be the hom-category $\mathbf{FPCat}\big(\mathcal{C}\ell(\Sigma),\, \mathbb{B}\big)$.

More explicitly, a model of a signature Σ in a category $\mathbb{B}$ is given by an object

$$[\![\, \sigma \,]\!] \in \mathbb{B}$$

for every type $\sigma \in |\Sigma|$ and a morphism

$$[\![\, F \,]\!]: [\![\, \sigma_1 \,]\!] \times \cdots \times [\![\, \sigma_n \,]\!] \longrightarrow [\![\, \sigma_{n+1} \,]\!] \quad \text{in } \mathbb{B},$$

for every function symbol $F: \sigma_1, \ldots, \sigma_n \longrightarrow \sigma_{n+1}$ in Σ. The force of the above definition lies in the fact that it tells us what a model of a signature is in an arbitrary category with finite products. It is completely general. For example, a **continuous Σ-algebra** is defined in [101] as a Σ-algebra whose carriers are directed complete partial orders (dcpos) (posets with joins of directed subsets) and whose interpretations of function symbols are continuous functions (preserving these joins). Thus, such a continuous Σ-algebra is a model $\mathcal{C}\ell(\Sigma) \to \mathbf{Dcpo}$ of Σ in the category $\mathbf{Dcpo}$ of dcpos and continuous functions. Another example (involving partial functions) may be found in Exercise 2.2.1 below.

2.2.3. Example. Among all the models a signature Σ can have there is one very special: it is simply the identity functor $\mathcal{C}\ell(\Sigma) \to \mathcal{C}\ell(\Sigma)$. This model of Σ in $\mathcal{C}\ell(\Sigma)$ is called the **generic model** of Σ. It is the categorical version of the term model constructed in Example 1.6.5 in the style of universal algebra.

In a category of models $\mathbf{FPCat}\big(\mathcal{C}\ell(\Sigma),\, \mathbb{B}\big)$—like in any category—one may have initial and terminal objects. These are initial or terminal models of Σ in $\mathbb{B}$. They play a distinguished rôle in the semantics of data types.

The following two results gives a clearer picture of what such categorical models are.

2.2.4. Lemma. (i) *There is a forgetful functor*

$$\mathbf{FPCat} \xrightarrow{\quad \mathrm{Sign}(-) \quad} \mathbf{Sign}$$

given as follows. For a category $\mathbb{B} \in \mathbf{FPCat}$ *the* **underlying signature** $\mathrm{Sign}(\mathbb{B})$ *of* $\mathbb{B}$ *has objects from* $\mathbb{B}$ *as types and function symbols given by*

$$F \colon X_1, \ldots, X_n \longrightarrow X_{n+1} \quad in \quad \mathbf{Sign}(\mathbb{B})$$
$$\overset{\mathrm{def}}{\Leftrightarrow} F \ is \ a \ morphism \ X_1 \times \cdots \times X_n \to X_{n+1} \ in \ \mathbb{B}.$$

(ii) *In the reverse direction, taking classifying categories yields a functor*

$$\mathcal{C}\ell(-) \colon \mathbf{Sign} \longrightarrow \mathbf{FPCat}.$$

For a morphism of signatures $\phi \colon \Sigma \to \Sigma'$ *in* $\mathbf{Sign}$ *one obtains a functor* $\mathcal{C}\ell(\phi) \colon \mathcal{C}\ell(\Sigma) \to \mathcal{C}\ell(\Sigma')$ *by replacing every* Σ*-type and function symbol by its image under* ϕ*. For a term* M *we shall often write* ϕM *for* $\mathcal{C}\ell(\phi)(M)$*.*

Proof. (i) For a morphism $K \colon \mathbb{A} \to \mathbb{B}$ in **FPCat**—*i.e.* for a finite product preserving functor—one has a signature morphism $\mathrm{Sign}(K) \colon \mathrm{Sign}(\mathbb{A}) \to \mathrm{Sign}(\mathbb{B})$ which sends $X \in \mathbb{A}$ to $KX \in \mathbb{B}$ and a map $X_1 \times \ldots \times X_n \to X_{n+1}$ in $\mathbb{A}$ to the composite $KX_1 \times \ldots \times KX_n \cong K(X_1 \times \ldots \times X_n) \to KX_{n+1}$ in $\mathbb{B}$.
(ii) Easy. $\qquad\square$

(Here, we should allow signatures with classes (as opposed to sets) of types and function symbols if we wish to define **Sign**($\mathbb{B}$) for a non-small category $\mathbb{B}$—with finite products.)

2.2.5. Theorem. *For a signature* Σ *and a category* $\mathbb{B}$ *with finite products, there is a bijective correspondence (up-to-isomorphism) between morphisms of signatures* ϕ *and models* $\mathcal{M}$ *as in*

$$\frac{\Sigma \xrightarrow{\quad \phi \quad} \mathrm{Sign}(\mathbb{B})}{\mathcal{C}\ell(\Sigma) \xrightarrow[\mathcal{M}]{\quad} \mathbb{B}}$$

We do not obtain a precise correspondence (but only "up-to-isomorphism") between the ϕ and $\mathcal{M}$ in the theorem because first translating a model $\mathcal{M}$ into a morphism of signatures $\phi_{\mathcal{M}}$ and then back into a model $\mathcal{M}_{\phi_{\mathcal{M}}}$ does not precisely return $\mathcal{M}$, because $\mathcal{M}$ preserves finite products only up-to-isomorphism. Thus, in a suitable 2-categorical sense, the functor $\mathcal{C}\ell(-)$ is left adjoint to the forgetful functor $\mathrm{Sign}(-) \colon \mathbf{FPCat} \to \mathbf{Sign}$, and $\mathcal{C}\ell(\Sigma)$ is the **free category with finite products** generated by the signature Σ.

Proof. The proof is essentially a reformulation of the proof of Theorem 2 2.1 with the receiving category **Sets** replaced by the category $\mathbb{B}$. For a morphism of signatures $\phi: \Sigma \to \mathrm{Sign}(\mathbb{B})$, one defines a model $\mathcal{M}: \mathcal{Cl}(\Sigma) \to \mathbb{B}$ by

$$\Gamma = (v_1:\sigma_1,\ldots,v_n:\sigma_n) \;\mapsto\; \phi(\sigma_1) \times \cdots \times \phi(\sigma_n)$$
$$(M_1,\ldots,M_m):\Gamma \to \Delta \;\mapsto\; \langle \mathcal{M}(\Gamma \vdash M_1:\tau_1),\ldots,\mathcal{M}(\Gamma \vdash M_m:\tau_m)\rangle$$

where $\Delta = v_1:\tau_1,\ldots,v_m:\tau_m$. This operation $\mathcal{M}(-)$ is a mapping which interprets a term $\Gamma \vdash M:\tau$ in context $\Gamma = v_1:\sigma,\ldots,v_n:\sigma_n$ as a morphism in $\mathbb{B}$:

$$\phi(\sigma_1) \times \cdots \times \phi(\sigma_n) = \mathcal{M}(\Gamma) \xrightarrow{\;\;\mathcal{M}(\Gamma \vdash M:\tau)\;\;} \mathcal{M}(v_1:\tau) = \phi(\tau)$$

It is defined by induction on the derivation of $\Gamma \vdash M:\tau$ as:

- **identity.**
$$\mathcal{M}(v_1:\sigma \vdash v_1:\sigma) = \mathrm{id}:\phi(\sigma) \to \phi(\sigma).$$

- **function symbol.** For $F:\tau_1,\ldots,\tau_m \longrightarrow \tau_{n+1}$,
$$\mathcal{M}(\Gamma \vdash F(M_1,\ldots,M_m):\tau_{m+1})$$
$$= \phi(F) \circ \langle \mathcal{M}(\Gamma \vdash M_1:\tau_1),\ldots,\mathcal{M}(\Gamma \vdash M_m:\tau_m)\rangle.$$

- **weakening.** Suppose $\Gamma \vdash M:\tau$. Then
$$\mathcal{M}(\Gamma,v_n:\sigma \vdash M:\tau) = \mathcal{M}(\Gamma \vdash M:\tau) \circ \pi.$$

- **contraction.** Suppose $\Gamma,v_n:\sigma,v_{n+1}:\sigma \vdash M:\tau$. Then
$$\mathcal{M}(\Gamma,v_n:\sigma \vdash M[v_n/v_{n+1}]:\tau)$$
$$= \mathcal{M}(\Gamma,v_n:\sigma,v_{n+1}:\sigma \vdash M:\tau) \circ \langle \mathrm{id},\pi'\rangle.$$

- **exchange.** Suppose $\Gamma,v_i:\sigma_i,v_{i+1}:\sigma_{i+1},\Gamma' \vdash M:\tau$. Then
$$\mathcal{M}(\Gamma,v_i:\sigma_{i+1},v_{i+1}:\sigma_i,\Gamma' \vdash M[v_i/v_{i+1},v_{i+1}/v_i]:\tau)$$
$$= \mathcal{M}(\Gamma,v_i:\sigma_i,v_{i+1}:\sigma_{i+1},\Gamma' \vdash M:\tau) \circ \mathrm{id} \times \langle \pi',\pi\rangle \times \mathrm{id}.$$

Further details are left to the reader in Exercise 2.2.2 below.

In the reverse direction, given a model $\mathcal{M}: \mathcal{Cl}(\Sigma) \to \mathbb{B}$ one obtains a morphism of signatures $\Sigma \to \mathrm{Sign}(\mathbb{B})$ by $\sigma \mapsto \mathcal{M}(v_1:\sigma)$ and $F \mapsto \mathcal{M}(F) \circ \varphi$, where φ is a mediating isomorphism, like in the proof of Theorem 2.2.1. $\square$

Notice in the above proof the importance of *projections* $\pi: I \times J \to I$ for the interpretation of weakening and of (*parametrised*) *diagonals* $\delta = \langle \mathrm{id},\pi'\rangle: I \times J \to (I \times J) \times J$ for contraction.

2.2.6. Definition. The adjunction $\mathcal{Cl}(-) \dashv \mathrm{Sign}(-)$ in the previous theorem gives rise to a monad $T = \mathrm{Sign}(\mathcal{Cl}(-))$ on the category **Sign** of signatures. The resulting Kleisli category—written as $\mathbf{Sign}_{\mathrm{tr}}$—will be called the **category of signatures and translations**. Thus a translation $\phi: \Sigma \to \Sigma'$ is understood as a mapping of types to contexts and of function symbols to terms (instead of: types to types and function symbols to function symbols, as in the category **Sign**). Formally, such a translation ϕ is a morphism of signatures $\Sigma \to \text{Sign}(\mathcal{C}l(\Sigma'))$.

The category **Sign**$_{\text{tr}}$ is in fact more useful than **Sign**: translations occur more naturally than morphisms of signatures, as the following examples illustrate.

2.2.7. Examples. (i) The classic example of a translation of signatures involves two (single-typed) signatures for groups, see [212], Definitions 1.1 and 1.2. For reasons of clarity we provide the following two signatures with equations; but they do not play a rôle at this stage.

(1) Let Σ_1 be the signature with one type G and three function symbols

$$\text{m: G, G} \longrightarrow \text{G}, \qquad \text{e: ()} \longrightarrow \text{G}, \qquad \text{i: G} \longrightarrow \text{G}$$

giving a multiplication, unit and inverse operation. The equations are the familiar ones for groups:

$$v_1 \colon \text{G} \ \vdash \ \text{m}(\text{e}, v_1) =_G v_1 \qquad\qquad v_1 \colon \text{G} \ \vdash \ \text{m}(\text{i}(v_1), v_1) =_G \text{e}$$
$$v_1 \colon \text{G} \ \vdash \ \text{m}(v_1, \text{e}) =_G v_1 \qquad\qquad v_1 \colon \text{G} \ \vdash \ \text{m}(v_1, \text{i}(v_1)) =_G \text{e}$$
$$v_1 \colon \text{G}, v_2 \colon \text{G}, v_3 \colon \text{G} \ \vdash \ \text{m}(v_1, \text{m}(v_2, v_3)) =_G \text{m}(\text{m}(v_1, v_2), v_3).$$

Such equations will be studied systematically in the next chapter.

(2) Less standard is the following signature Σ_2 for groups. It has again one type G but only two function symbols,

$$\text{d: G, G} \longrightarrow \text{G} \qquad \text{and} \qquad \text{a: ()} \longrightarrow \text{G}$$

satisfying a single equation

$$v_1 \colon \text{G}, v_2 \colon \text{G}, v_3 \colon \text{G} \vdash$$
$$\text{d}(\text{d}(\text{d}(v_3, \text{d}(v_1, \text{d}(v_1, v_1))), \text{d}(v_3, \text{d}(v_2, \text{d}(v_1, v_1)))), v_1) =_G v_2.$$

Notice that the second function symbol (or constant) a does not occur in this equation; its sole rôle is to ensure that groups have at least one element. (It is not present in [212], so that groups may be empty there.)

There is a *translation* $\Sigma_2 \to \Sigma_1$ which maps the type G to itself, and the function symbol d to the Σ_1-term

$$v_1 \colon \text{G}, v_2 \colon \text{G} \vdash \text{m}(\text{i}(v_1), v_2) \colon \text{G}$$

and a to an arbitrary term in G, *e.g.* e. This is a translation—and not a morphism of signatures—because the function symbol d of Σ_2 is mapped to a *term* of Σ_1—and not to a function symbol of Σ_1. For more details, see Exercise 2.2.3 below.

(ii) Boolean logic can be described by the (functionally complete) pair of connectives

$$\neg: B \longrightarrow B \qquad \text{and} \qquad \wedge: B, B \longrightarrow B$$

of negation and conjunction. Alternatively, negation and implication can be used:

$$\neg: B \longrightarrow B \qquad \text{and} \qquad \supset: B, B \longrightarrow B$$

Or also the Sheffer stroke

$$|: B, B \longrightarrow B.$$

The standard definitions

$$v_1 \supset v_2 = \neg(v_1 \wedge \neg v_2) \qquad \text{and} \qquad v_1 | v_2 = \neg(v_1 \wedge v_2)$$

yield translations from the last two signatures into the first one.

Exercises

2.2.1.   (i)   Let **Sets.** be the category of pointed sets as described in Exercise 1.2.3. It can be seen as the category of sets and partial functions. Show that **Sets.** has finite products.

(ii)   Let Σ be a signature. Define a **partial Σ-algebra** (or **model**) to be a finite product preserving functor $\mathcal{C}l(\Sigma) \to$ **Sets.**. Describe such a partial algebra in detail.

2.2.2.   Consider the interpretation $\mathcal{M}$ associated with a morphism of signatures $\phi: \Sigma \to \mathbf{Sign}(\mathbb{B})$ in the proof of Theorem 2.2.5.

(i)   Let $\Gamma = (v_1: \sigma_1, \ldots, v_n: \sigma_n)$ be a context with a term $\Gamma \vdash M: \tau$ such that the variable v_k in Γ does not occur (free) in M. Prove that

$$\mathcal{M}(\Gamma \vdash M: \sigma) = \mathcal{M}(\Gamma^k \vdash M^k: \sigma) \circ \langle \pi_1, \ldots, \pi_{k-1}, \pi_{k+1}, \ldots, \pi_n \rangle,$$

where

$$\Gamma^k = v_1: \sigma_1, \ldots, v_{k-1}: \sigma_k, v_k: \sigma_{k+1}, \ldots, v_{n-1}: \sigma_n$$
$$M^k = M[v_k/v_{k+1}, \ldots, v_{n-1}/v_n],$$

and π_i is the obvious projection map $\mathcal{M}(\sigma_1) \times \cdots \times \mathcal{M}(\sigma_n) \to \mathcal{M}(\sigma_i)$.

(ii)   Next, for $\Gamma = v_1: \sigma_1, \ldots, v_n: \sigma_n$, consider a term $\Gamma \vdash N: \tau$, and a context morphism $\vec{M}: \Delta \to \Gamma$. Prove that

$$\mathcal{M}(\Delta \vdash N[\vec{M}/\vec{v}]: \tau) = \mathcal{M}(\Gamma \vdash N: \tau) \circ$$
$$\langle \mathcal{M}(\Delta \vdash M_1: \sigma_1), \ldots, \mathcal{M}(\Delta \vdash M_n: \sigma_n) \rangle.$$

(iii)   Conclude from (i) and (ii) that $\mathcal{M}$ preserves identities and composition.

2.2.3.   (i)   Check that the translation in Example 2.2.7 (i) of d as $m(i(v_1), v_2)$ satisfies the equation for d.

(ii)   Find also a translation of signatures $\Sigma_1 \to \Sigma_2$.

(iii) In (ii) of the same example, define a translation from the first signature with ¬ and ∧ into the last one with the Sheffer stroke |.

2.2.4. We describe a category of **categorical models of signatures**. Let **C-Model** be the category with

objects $(\Sigma, \mathbb{A}, \mathcal{M})$ where $\mathcal{M}: \mathcal{Cl}(\Sigma) \to \mathbb{A}$ is a finite product preserving functor.

morphisms $(\Sigma, \mathbb{A}, \mathcal{M}) \to (\Sigma', \mathbb{A}', \mathcal{M}')$ consist of a triple (ϕ, K, α) where $\phi: \Sigma \to \Sigma'$ is a morphism of signatures, $K: \mathbb{A} \to \mathbb{A}'$ is a finite product preserving functor and α is a natural transformation $K\,\mathcal{M} \Rightarrow \mathcal{M}'\,\mathcal{Cl}(\phi)$ in

$$
\begin{array}{ccc}
\mathcal{Cl}(\Sigma) & \xrightarrow{\;\;\mathcal{Cl}(\phi)\;\;} & \mathcal{Cl}(\Sigma') \\
\mathcal{M} \downarrow & {\overset{\alpha}{\nearrow}} & \downarrow \mathcal{M}' \\
\mathbb{A} & \xrightarrow[\;\;\;K\;\;\;]{} & \mathbb{A}'
\end{array}
$$

(i) Show that the projection $\begin{smallmatrix}\textbf{C-Model}\\ \downarrow \\ \textbf{Sign}\end{smallmatrix}$ is a fibration.

(ii) Verify that this fibration has fibred finite products.

[In Section 9.1 it will be shown that the fibration $\begin{smallmatrix}\textbf{C-Model}\\ \downarrow \\ \textbf{Sign}\end{smallmatrix}$ comes from a canonical construction as one leg of a fibred span. There is also a projection functor $\begin{smallmatrix}\textbf{C-Model}\\ \downarrow \\ \textbf{FPCat}\end{smallmatrix}$, which is an 'opfibration', since reindexing works in the other direction.]

2.2.5. Let $\mathcal{M}: \mathcal{Cl}(\Sigma) \to \mathbb{B}$ be a Σ-model. For terms $\Gamma \vdash N, N': \sigma$ write

$$
\mathcal{M} \models N =_\sigma N' \qquad \text{for} \qquad \mathcal{M}(N) = \mathcal{M}(N')
$$

where on the right hand side, N and N' are treated as morphisms $\Gamma \rightrightarrows \sigma$ in $\mathcal{Cl}(\Sigma)$. Let $\phi: \Sigma \to \Sigma'$ be a morphism of signatures. Show that the 'satisfaction lemma'

$$
\mathcal{M} \models \phi N =_{\phi\sigma} \phi N' \;\Leftrightarrow\; \phi^*(\mathcal{M}) \models N =_\sigma N'
$$

boils down to a tautology—where $\phi^*(\mathcal{M})$ is the outcome of reindexing along ϕ, see (i) in the previous exercise.

[This property is fundamental in the definition of an institution [152].]

2.3 Exponents, products and coproducts

In this section we discuss three simply typed λ-calculi, which will be written as $\lambda 1$, $\lambda 1_\times$ and $\lambda 1_{(\times, +)}$. The calculus $\lambda 1$ has exponent (or arrow) types $\sigma \to \tau$; $\lambda 1_\times$ additionally has finite product types $1, \sigma \times \tau$ which allow one to form finite tuples of terms (including the empty tuple); and in $\lambda 1_{(\times,+)}$ one has finite coproduct types $0, \sigma + \tau$. With these one can form finite cotuples. These calculi are built on top of a many-typed signature. A brief discussion of the propositions-as-types analogy is included.

(We shall not discuss the rewriting properties of these type theories. We refer to [186] for proofs of Church-Rosser and strong normalisation for type theories with $\rightarrow$ and $\times$—building on ideas of Tait and de Vrijer. A singleton type is included in [64].)

At this stage we begin to be more sloppy in the use of variables: instead of the formally numbered variables $\{v_n \mid n \in \mathbb{N}\}$ we now start using meta-variables u, v, w, x, y, z. This is more convenient for human beings (as opposed to computers). We shall require that no two variables occurring in a context Γ are equal. In particular, in writing an extended context $\Gamma, x : \sigma$ it is assumed that x does not occur in Γ.

$\lambda 1$-calculi

Let Σ be a many-typed signature with $T = |\Sigma|$ as its underlying set of types. Let T_1 be the least set containing T, which is formally closed under $\rightarrow$, *i.e.* $T \subseteq T_1$ and $\sigma, \tau \in T_1 \Rightarrow (\sigma \rightarrow \tau) \in T_1$. We now call elements of T_1 types, and if we wish to stress that $\sigma \in T_1$ actually is a member of T, then we call it an **atomic** or **basic** type. In order to spare on parentheses one usually writes

$$\sigma_1 \rightarrow \sigma_2 \rightarrow \cdots \rightarrow \sigma_{n-1} \rightarrow \sigma_n$$

for

$$\sigma_1 \rightarrow (\sigma_2 \rightarrow \cdots \rightarrow (\sigma_{n-1} \rightarrow \sigma_n) \cdots).$$

Instead of extending T to T_1 we can also say that there are the following two type formation rules.

$$\frac{}{\vdash \sigma : \mathsf{Type}} \; (\text{for } \sigma \in T) \qquad \frac{\vdash \sigma : \mathsf{Type} \qquad \vdash \tau : \mathsf{Type}}{\vdash \sigma \rightarrow \tau : \mathsf{Type}}$$

Notice that in these type formation statements of the form $\vdash \sigma : \mathsf{Type}$ we have an empty context because types in STT are not allowed to contain any variables. This will be different in calculi with polymorphic or dependent types.

The **simply typed λ-calculus** $\lambda 1(\Sigma)$ built on top of Σ has all the rules of the term calculus of Σ—described in Section 2.1—plus the following introduction and elimination rules for abstraction and application.

$$\frac{\Gamma, v : \sigma \vdash M : \tau}{\Gamma \vdash \lambda v : \sigma.\, M : \sigma \rightarrow \tau} \qquad \frac{\Gamma \vdash M : \sigma \rightarrow \tau \qquad \Gamma \vdash N : \sigma}{\Gamma \vdash M N : \tau}$$

Intuitively one thinks of the abstraction term $\lambda v : \sigma. M$ as the function $a \mapsto M[a/v]$, so that $\sigma \to \tau$ is the type of functions taking inputs of type σ and returning an output of type τ. The term former $\lambda v : \sigma. (-)$ binds the variable v. The application term MN (sometimes written as $M \cdot N$) describes the application of a function $M : \sigma \to \tau$ to an argument $N : \sigma$. Notice that this application is required to be well-typed in an obvious sense.

This explains the associated two conversions

$$\frac{\Gamma, v : \sigma \vdash M : \tau \quad \Gamma \vdash N : \sigma}{\Gamma \vdash (\lambda v : \sigma. M)N = M[N/v] : \tau} \qquad \frac{\Gamma \vdash M : \sigma \to \tau}{\Gamma \vdash \lambda v : \sigma. Mv = M : \sigma \to \tau}$$

where in the latter case it is assumed that v is not free in M. The first of these rules describes what is called (β)-conversion, and the second describes (η)-conversion. This (β) is evaluation of a function on an argument, and (η) is extensionality of functions. Here we have written these conversions as rules, with all types explicitly present. Often they are simply written as

$$(\lambda v : \sigma. M)N = M[N/v] \qquad \text{and} \qquad \lambda v : \sigma. Mv = M$$

like in:

$$(\lambda v : \mathsf{N}. \, \mathsf{plus}(v, 3)) \, 4 = \mathsf{plus}(4, 3).$$

Substitution is extended to these new abstraction and application terms by

$$(\lambda v : \sigma. M)[L/w] \equiv \lambda v : \sigma. (M[L/w])$$
$$(MN)[L/w] \equiv (M[L/w])(N[L/w])$$

under the (usual) proviso that v is not free in L (to avoid that a variable which is free in L becomes bound after substitution; this can always be avoided by a change of name of the bound variable v in $\lambda v : \sigma. M$). We write $\equiv$ to indicate that this involves a syntactic identification. One further extends the conversion relation $=$ to become an equivalence relation which is compatible with abstraction and application in the sense that

$$\frac{\Gamma, v : \sigma \vdash M = M' : \tau}{\Gamma \vdash \lambda v : \sigma. M = \lambda v : \sigma. M' : \sigma \to \tau}$$

$$\frac{\Gamma \vdash M = M' : \sigma \to \tau \quad \Gamma \vdash N = N' : \sigma}{\Gamma \vdash MN = M'N' : \tau}$$

see [13]. The first of these rules is often called the (ξ)-rule.

Thus, $\lambda 1(\Sigma)$ extends the signature Σ with means for introducing functions and applying them to arguments. This calculus gives rise to a syntactically constructed category $\mathcal{C}l1(\Sigma)$, called the $\lambda 1$-**classifying category of** Σ. Its objects are contexts

$$\Gamma = (v_1 : \sigma_1, \ldots, v_n : \sigma_n) \qquad \text{with} \qquad \sigma_i \in T_1.$$

Note that the $\lambda 1$-classifying category has arrow types occurring in its objects. A morphism $\Gamma \to \Delta$ in $\mathcal{C}\ell 1(\Sigma)$, where $\Delta = (v_1 : \tau_1, \ldots, v_m : \tau_m)$, is an m-tuple of equivalence classes (with respect to conversion $=$) of terms

$$([M_1], \ldots, [M_m]) \qquad \text{with} \qquad \Gamma \vdash M_i : \tau_i \text{ in } \lambda 1(\Sigma).$$

Thus a second difference between the $\lambda 1$-classifying category $\mathcal{C}\ell 1(\Sigma)$ and the classifying category $\mathcal{C}\ell(\Sigma)$ described Section 2.1, is that in the former one takes equivalence classes of terms—instead of terms themselves—as constituents of context morphisms.

2.3.1. Proposition. *The $\lambda 1$-classifying category $\mathcal{C}\ell 1(\Sigma)$ of a signature Σ has finite products. If $T = |\Sigma|$ is the underlying set of types of Σ, then $\mathcal{C}\ell 1(\Sigma)$ together with the set of types T_1 (obtained by closing the set of basic types T under $\to$) is a CT-structure (see Definition 1.3.2); it is non-trivial if and only if T is non-empty.*

Proof. Finite products in $\mathcal{C}\ell 1(\Sigma)$ are as in $\mathcal{C}\ell(\Sigma)$: the empty context is terminal and concatenation of contexts yields Cartesian products. The inclusion $T_1 \hookrightarrow \mathrm{Obj}\, \mathcal{C}\ell 1(\Sigma)$ involves identification of a type σ with the corresponding singleton context $(v_1 : \sigma)$. $\qquad\square$

The identification used in the proof is very handy. We shall freely make use of it and consider a type σ as an object of a classifying category by identifying it with the singleton context $(v_1 : \sigma)$.

The above proposition describes the context structure in $\lambda 1$-classifying categories in terms of finite products. An appropriate categorical description of the structure induced by the exponent types $\sigma \to \tau$ may be found in the next section. It uses that the pair $(\mathcal{C}\ell(\Sigma), T_1)$ is a CT-structure.

Propositions as types

Let T be a non-empty set, elements of which will be seen as propositional constants. And let T_1 be the formal closure of T under $\to$, as above. The elements of T_1 may be seen as propositions of **minimal intuitionistic logic** (MIL, for short), since they are built up from constants using only $\to$ (or $\supset$) for implication. For $\sigma_1, \ldots, \sigma_n, \tau \in T_1$ we can write

$$\blacktriangleright \quad \sigma_1, \ldots, \sigma_n \vdash_{\text{MIL}} \tau$$

if τ is derivable from assumptions $\sigma_1, \ldots, \sigma_n$ in minimal intuitionistic logic. The (non-structural) rules of MIL are $\to$-introduction and $\to$-elimination.

Let $\mathcal{A}$ now be a collection of such sequents $\sigma_1, \ldots, \sigma_n \vdash \tau$, which we regard as axioms (with $\sigma_i, \tau \in T$). That is, for each sequent $S \in \mathcal{A}$, we have a rule

$$\frac{}{S}$$

expressing that S is derivable without further ado. We wish to consider which other sequents are derivable, assuming these axioms in $\mathcal{A}$. For example, if $\mathcal{A}$ contains the sequent $\varphi \vdash \psi$, then we can derive $\psi \to \chi \vdash \varphi \to \chi$ as follows.

$$\cfrac{\cfrac{\psi \to \chi \vdash \psi \to \chi}{\varphi, \psi \to \chi \vdash \psi \to \chi} \qquad \cfrac{\varphi \vdash \chi}{\varphi, \psi \to \chi \vdash \psi}}{\cfrac{\varphi, \psi \to \chi \vdash \chi}{\psi \to \chi \vdash \varphi \to \chi}}$$

Let $\Sigma_{\mathcal{A}}$ be the signature constructed from a set $\mathcal{A}$ of axioms in the following way. Take the set T of atomic propositions as atomic types in $\Sigma_{\mathcal{A}}$, and choose for every sequent $\sigma_1, \ldots, \sigma_n \vdash \tau$ in $\mathcal{A}$ a new function symbol $F \colon \sigma_1, \ldots, \sigma_n \longrightarrow \tau$. Think of F as an atomic proof-object for the axiom.

If we assume in the above example a function symbol $F \colon \varphi \longrightarrow \psi$ corresponding to the axiom $\varphi \vdash \psi$, then there is a λ-term which codes the derivation, namely

$$v \colon \psi \to \chi \vdash \lambda w \colon \varphi.\, v(Fw) \colon \varphi \to \chi.$$

More generally, one can prove that

$\sigma_1, \ldots, \sigma_n \vdash_{\text{MIL}} \tau$ is derivable from $\mathcal{A}$

$\Leftrightarrow$ there is a term M with $v_1 \colon \sigma_1, \ldots, v_n \colon \sigma_n \vdash M \colon \tau$ in $\lambda 1(\Sigma_{\mathcal{A}})$.

This gives an example of what is known as the paradigm of **propositions-as-types** or better as **propositions-as-types** and **proofs-as-terms**. This perspective was first brought forward clearly in Howard [140], but goes back to Curry and Feys [65]. The above bi-implication $\Leftrightarrow$ depends on the fact that derivations in MIL correspond directly to $\lambda 1$-terms. In particular, the introduction and elimination rules for implication in logic have the same form as the introduction and implication rules for exponents in type theory.

As a result, *provability* in logic corresponds to *inhabitation* in type theory. A term $M \colon \sigma \to \tau$ may be seen as a proof of the proposition $\sigma \to \tau \colon M$ transforms each proof $N \colon \sigma$ of σ into a proof $MN \colon \tau$ of τ. This is the so-called Brouwer-Heyting-Kolmogorov interpretation of the $\to$-connective in constructive logic, see [140, 335]. This interpretation extends to finite conjunctions $(\top, \wedge)$ and disjunctions $(\bot, \vee)$, which, by including proofs, may be read as finite products $(1, \times)$ and coproducts $(0, +)$. Later in Section 8.1 we shall see how the quantifiers $\forall, \exists$ in predicate logic correspond to product $\prod$ and sum $\coprod$ of types over kinds in polymorphic type theory.

The analogy between derivations and terms goes even further: the (β)- and (η)-conversions for $\lambda 1$-terms correspond to certain identifications on derivations, namely to:

$$\left(\begin{array}{cc} \dfrac{\begin{array}{c} \vdots \\ \sigma \vdash \tau \end{array}}{\dfrac{\vdash \sigma \to \tau \qquad \vdash \sigma}{\vdash \tau}} \end{array} \right) \overset{\beta}{=} \left(\dfrac{\vdash \sigma \quad \begin{array}{c} \vdots \\ \sigma \vdash \tau \end{array}}{\vdash \tau} \right)$$

$$\left(\dfrac{\dfrac{\begin{array}{c} \vdots \\ \vdash \sigma \to \tau \end{array} \quad \sigma \vdash \sigma}{\sigma \vdash \tau}}{\vdash \sigma \to \tau} \right) \overset{\eta}{=} \left(\dfrac{\begin{array}{c} \vdots \end{array}}{\vdash \sigma \to \tau} \right)$$

Via (β)-conversion one can thus remove an introduction step which is immediately followed by an elimination step. And (η) does the same for elimination immediately followed by introduction. For more details, see [280, 140].

(We have been overloading the arrow $\to$ by using it both for exponent types $\sigma \to \tau$ and for implication propositions $\varphi \to \psi$. This is convenient in explaining the idea of propositions-as-types. But from now on we shall be using $\supset$ instead of $\to$ for implication in logic.)

From the way the $\lambda 1$-classifying category $\mathcal{C}\ell 1(\Sigma)$ was constructed, we immediately get another correspondence:

$$\sigma_1, \ldots, \sigma_n \vdash_{\text{MIL}} \tau \text{ is derivable from } \mathcal{A}$$
$$\Leftrightarrow \text{ there is a morphism } \sigma_1 \times \cdots \times \sigma_n \to \tau \text{ in } \mathcal{C}\ell 1(\Sigma_{\mathcal{A}}).$$

where we identify σ with the singleton context $(v_1 \colon \sigma)$ in $\mathcal{C}\ell 1(\Sigma)$ as discussed after Proposition 2.3.1. Here we have an elementary example of **propositions-as-objects** and **proofs-as-morphisms**. This basic correspondence forms the heart of categorical logic, as often emphasised by Lawvere and Lambek.

$\lambda 1_\times$-calculi

Let Σ be a many-typed signature. The calculus $\lambda 1_\times(\Sigma)$ will be introduced as $\lambda 1(\Sigma)$ extended with finite product types. This new calculus $\lambda 1_\times(\Sigma)$ has all the rules of $\lambda 1(\Sigma)$ plus the following type formation rules.

$$\dfrac{}{\vdash 1 \colon \mathsf{Type}} \qquad\qquad \dfrac{\vdash \sigma \colon \mathsf{Type} \qquad \vdash \tau \colon \mathsf{Type}}{\vdash \sigma \times \tau \colon \mathsf{Type}}$$

We use 1 as a new symbol (not occurring in $|\Sigma|$) for singleton (or unit) type (empty product). Additionally, there are in $\lambda 1_\times(\Sigma)$ the following associated introduction and elimination rules for tupleing and projecting.

$$\frac{}{\vdash \langle \rangle : 1}$$

$$\frac{\Gamma \vdash M : \sigma \quad \Gamma \vdash N : \tau}{\Gamma \vdash \langle M, N \rangle : \sigma \times \tau}$$

$$\frac{\Gamma \vdash P : \sigma \times \tau}{\Gamma \vdash \pi P : \sigma}$$

$$\frac{\Gamma \vdash P : \sigma \times \tau}{\Gamma \vdash \pi' P : \tau}$$

Formally it would be better to give insert appropriate indices, as in $\pi_{\sigma,\tau} P, \pi'_{\sigma,\tau} P$, but that would make the notation rather heavy.

Associated with these introduction and elimination rules there are the following conversions.

$$\frac{\Gamma \vdash M : 1}{\Gamma \vdash M = \langle \rangle : 1}$$

$$\frac{\Gamma \vdash M : \sigma \quad \Gamma \vdash N : \tau}{\Gamma \vdash \pi \langle M, N \rangle = M : \sigma}$$

$$\frac{\Gamma \vdash M : \sigma \quad \Gamma \vdash N : \tau}{\Gamma \vdash \pi' \langle M, N \rangle = N : \tau}$$

$$\frac{\Gamma \vdash P : \sigma \times \tau}{\Gamma \vdash \langle \pi P, \pi' P \rangle = P : \sigma \times \tau}$$

Substitution is extended to the new terms by

$$\langle \rangle [L/v] \equiv \langle \rangle \qquad \langle M, N \rangle [L/v] \equiv \langle M[L/v], N[L/v] \rangle$$
$$(\pi P)[L/v] \equiv \pi(P[L/v]) \qquad (\pi' P)[L/v] \equiv \pi'(P[L/v]).$$

We continue to write $=$ for the compatible equivalence relation generated by the above conversions plus the (β)- and (η)-conversions of $\lambda 1$.

The advantage of having finite product types around is that one no longer needs to distinguish between contexts and types. Terms with multiple variables

$$v_1 : \sigma_1, \ldots, v_n : \sigma_n \vdash M : \tau$$

correspond bijectively to terms with a single variable

$$v_1 : \sigma_1 \times \cdots \times \sigma_n \vdash N : \tau$$

where the product type $\sigma_1 \times \cdots \times \sigma_n$ is the singleton type 1 if $n = 0$.

We thus define the $\lambda 1_\times$-**classifying category** $\mathcal{C}\ell 1_\times(\Sigma)$ of Σ with

objects   types σ, built up form atomic types and $(1, \times, \rightarrow)$.

morphisms $\sigma \rightarrow \tau$ are equivalence classes (with respect to conversion) $[M]$ of terms $v_1 : \sigma \vdash M : \tau$.

The identity on σ is the equivalence class of the term $v_1 : \sigma \vdash v_1 : \sigma$ and composition involves substitution.

2.3.2. Proposition. *The $\lambda 1_\times$-classifying category $\mathcal{C}\ell 1_\times(\Sigma)$ of a signature Σ is Cartesian closed.*

Proof. It is easy to see that the type 1 is terminal and that the product type $\sigma \times \tau$ is a Cartesian product of σ and τ in $\mathcal{Cl}1_\times(\Sigma)$. This holds, almost by definition of 1, $\times$. It will be shown that the exponent type $\sigma \to \tau$ is the exponent object in $\mathcal{Cl}1_\times(\Sigma)$.

Assume an arrow $\rho \times \sigma \to \tau$ in $\mathcal{Cl}1_\times(\Sigma)$, say given by a term

$$z \colon \rho \times \sigma \vdash M \colon \tau.$$

Then one can form the abstraction term

$$x \colon \rho \vdash \lambda y \colon \sigma.\, M[\langle x, y\rangle/z] \colon \sigma \to \tau$$

which supports the following definition of categorical abstraction.

$$\Lambda([M]) = [\lambda y \colon \sigma.\, M[\langle x, y\rangle/z]] \colon \rho \longrightarrow (\sigma \to \tau).$$

Remember that the outer square braces $[-]$ denote the equivalence class with respect to conversion and that the inner ones are part of the notation for substitution. In a similar way, the term

$$w \colon (\sigma \to \tau) \times \sigma \vdash (\pi w)(\pi' w) \colon \tau$$

gives rise to the evaluation morphism

$$\mathrm{ev} = [(\pi w)(\pi' w)] \colon (\sigma \to \tau) \times \sigma \longrightarrow \tau$$

The categorical (β)- and (η)-conversions follow from the syntactical ones: first, for $z \colon \rho \times \sigma$,

$$
\begin{aligned}
\mathrm{ev} \circ \Lambda([M]) \times \mathrm{id} &= \left[(\pi w)(\pi' w)[\langle (\lambda y \colon \sigma.\, M[\langle x, y\rangle/z])[\pi z/x], \pi' z\rangle/w]\right] \\
&= \left[(\lambda y \colon \sigma.\, M[\langle \pi z, y\rangle/z])(\pi' z)\right] \\
&= \left[M[\langle \pi z, \pi' z\rangle/z]\right] \\
&= [M] \qquad\qquad \text{and}
\end{aligned}
$$

$$
\begin{aligned}
\Lambda(\mathrm{ev} \circ [N] \times \mathrm{id}) &= \left[\lambda y \colon \sigma.\, (\pi w)(\pi' w)[\langle N, y\rangle/w]\right] \\
&= \left[\lambda y \colon \sigma.\, Ny\right] \\
&= [N]. \qquad\qquad\qquad\qquad\qquad\quad \square
\end{aligned}
$$

$\lambda 1_{(\times, +)}$-*calculi*

In a next step we form on top of a signature Σ a calculus $\lambda 1_{(\times, +)}(\Sigma)$ which has exponent and finite product types as in $\lambda 1_\times(\Sigma)$, but additionally $\lambda 1_{(\times, +)}(\Sigma)$ has finite coproduct types (also called disjoint union types). This means that there are additional type formation rules:

$$\frac{}{\vdash 0 \colon \mathsf{Type}} \qquad\qquad \frac{\vdash \sigma \colon \mathsf{Type} \qquad \vdash \tau \colon \mathsf{Type}}{\vdash \sigma + \tau \colon \mathsf{Type}}$$

where 0 is a new symbol for the empty type (or, empty coproduct). There are the following introduction and elimination rules for these coproduct types $(0, +)$.

$$\frac{\Gamma \vdash M : \sigma}{\Gamma \vdash \kappa M : \sigma + \tau} \qquad\qquad \frac{\Gamma \vdash N : \tau}{\Gamma \vdash \kappa' N : \sigma + \tau}$$

$$\frac{\Gamma \vdash P : \sigma + \tau \quad \Gamma, x : \sigma \vdash Q : \rho \quad \Gamma, y : \tau \vdash R : \rho}{\Gamma \vdash \text{unpack } P \text{ as } [\kappa x \text{ in } Q, \kappa' y \text{ in } R] : \rho} \qquad \frac{}{\Gamma, z : 0 \vdash \{\} : \rho}$$

Thus, instead of projections π, π' for products, we now have coprojections κ, κ' for coproducts. The variables x in Q and y in R become bound in the "unpack" or "case" term unpack P as $[\kappa x \text{ in } Q, \kappa' y \text{ in } R]$. It can be understood as follows. Look at $P : \sigma + \tau$; if P is in σ, then do Q with P for x; else if P is in τ, do R with P for y. This explains the conversions:

$$\frac{\Gamma \vdash M : \sigma \quad \Gamma, x : \sigma \vdash Q : \rho \quad \Gamma, y : \tau \vdash R : \rho}{\Gamma \vdash \text{unpack } \kappa M \text{ as } [\kappa x \text{ in } Q, \kappa' y \text{ in } R] = Q[M/x] : \rho}$$

$$\frac{\Gamma \vdash N : \tau \quad \Gamma, x : \sigma \vdash Q : \rho \quad \Gamma, y : \tau \vdash R : \rho}{\Gamma \vdash \text{unpack } \kappa' N \text{ as } [\kappa x \text{ in } Q, \kappa' y \text{ in } R] = R[N/y] : \rho}$$

$$\frac{\Gamma \vdash P : \sigma + \tau \quad \Gamma, z : \sigma + \tau \vdash R : \rho}{\Gamma \vdash \text{unpack } P \text{ as } [\kappa x \text{ in } R[(\kappa x)/z], \kappa' y \text{ in } R[(\kappa' y)/z]] = R[P/z] : \rho}$$

$$\frac{\Gamma, z : 0 \vdash M : \rho}{\Gamma, z : 0 \vdash M = \{\} : \rho}$$

The latter rule tells that in a context in which the empty type 0 is inhabited, each term M must be convertible to the empty cotuple $\{\}$.

The following commutation result is often useful in calculations.

2.3.3. Lemma. *In the above calculus with coproduct types $+$ one has the following* **commutation conversion.**

$$\frac{\Gamma \vdash P : \sigma + \tau \quad \Gamma, x : \sigma \vdash Q : \rho \quad \Gamma, y : \tau \vdash R : \rho \quad \Gamma, z : \rho \vdash L : \mu}{\Gamma \vdash L[(\text{unpack } P \text{ as } [\kappa x \text{ in } Q, \kappa' y \text{ in } R])/z]}$$
$$= \text{unpack } P \text{ as } [\kappa x \text{ in } L[Q/z], \kappa' y \text{ in } L[R/z]] : \mu$$

It tells that cotupleing (or unpacking) commutes with substitution.

Proof. Because

$$L[(\text{unpack } P \text{ as } [\kappa x \text{ in } Q, \kappa' y \text{ in } R])/z]$$
$$= L[(\text{unpack } w \text{ as } [\kappa x \text{ in } Q, \kappa' y \text{ in } R])/z][P/w]$$
$$= \text{unpack } P \text{ as } [\ \kappa x \text{ in } L[(\text{unpack } \kappa x \text{ as } [\kappa x \text{ in } Q, \kappa' y \text{ in } R])/z],$$
$$\kappa' y \text{ in } L[(\text{unpack } \kappa' y \text{ as } [\kappa x \text{ in } Q, \kappa' y \text{ in } R])/z]\ ]$$
$$= \text{unpack } P \text{ as } [\kappa x \text{ in } L[Q/z], \kappa' y \text{ in } L[R/z]]. \qquad \square$$

Such a commutation conversion is typical for "colimit" types, like $+$, Σ, quotients and equality (which are described categorically by left adjoints). We use it for example in the proof of the next result, establishing distributivity of $\times$ over $+$ in type theory. Essentially, this follows from the presence of the "parameter" context Γ in the above $+$-elimination rule. The second point gives a type theoretic version of the argument sketched in Exercise 1.5.6 (i) and (ii).

2.3.4. Proposition. (i) *Type theoretic coproducts $+$ are automatically distributive: the canonical term*

$$u\colon (\sigma \times \tau) + (\sigma \times \rho) \vdash P(u) \stackrel{\text{def}}{=}$$
$$\text{unpack } u \text{ as } [\kappa x \text{ in } \langle \pi x, \kappa(\pi' x)\rangle, \kappa' y \text{ in } \langle \pi y, \kappa'(\pi' y)\rangle)]\colon \sigma \times (\tau + \rho)$$

is invertible—without assuming exponent types.

(ii) *Type theoretic coprojections κ, κ' are automatically "injective": the rules*

$$\frac{\Gamma \vdash \kappa M = \kappa M'\colon \sigma + \tau}{\Gamma \vdash M = M'\colon \sigma} \quad and \quad \frac{\Gamma \vdash \kappa' N = \kappa' N'\colon \sigma + \tau}{\Gamma \vdash N = N'\colon \tau}$$

are derivable, where $=$ denotes conversion.

Proof. (i) We have to produce a term "in the reverse direction":

$$v\colon \sigma \times (\tau + \rho) \vdash Q(v)\colon (\sigma \times \tau) + (\sigma \times \rho)$$

with conversions $P[Q(v)/u] = v$ and $Q[P(u)/v] = u$. First we notice that there is a term

$$x\colon \sigma, w\colon \tau + \rho \vdash \text{unpack } w \text{ as } [\kappa y \text{ in } \kappa\langle x, y\rangle, \kappa' z \text{ in } \kappa'\langle x, z\rangle]\colon (\sigma \times \tau) + (\sigma \times \rho).$$

Hence if we have a variable $v\colon \sigma \times (\tau + \rho)$, then we can substitute $[\pi v/x]$ and $[\pi' v/w]$ in this term, so that we can define

$$Q(v) \stackrel{\text{def}}{=} \text{unpack } \pi' v \text{ as } [\kappa y \text{ in } \kappa\langle \pi v, y\rangle, \kappa' z \text{ in } \kappa'\langle \pi v, z\rangle].$$

Using the commutation conversions from the previous lemma, we show that these terms P and Q are each other's inverses:

$$
\begin{aligned}
P[Q(v)/u] &= \text{unpack } \pi'v \text{ as } [\ \kappa y \text{ in } P[\kappa\langle\pi v, y\rangle/u], \\
&\qquad\qquad\qquad \kappa'z \text{ in } P[\kappa'\langle\pi v, z\rangle/u]\ ] \\
&= \text{unpack } \pi'v \text{ as } [\kappa y \text{ in } \langle\pi v, \kappa y\rangle, \kappa'z \text{ in } \langle\pi v, \kappa'z\rangle] \\
&= \langle\pi v, \pi'v\rangle \\
&= v.
\end{aligned}
$$

$$
\begin{aligned}
Q[P(u)/v] &= \text{unpack } u \text{ as } [\ \kappa x \text{ in } Q[\langle\pi x, \kappa(\pi'x)\rangle/v], \\
&\qquad\qquad\qquad \kappa'y \text{ in } Q[\langle\pi y, \kappa'(\pi'y)\rangle/v]\ ] \\
&= \text{unpack } u \text{ as } [\kappa x \text{ in } \kappa\langle\pi x, \pi'x\rangle, \kappa'y \text{ in } \kappa'\langle\pi y, \pi'y\rangle] \\
&= \text{unpack } u \text{ as } [\kappa x \text{ in } \kappa x, \kappa'y \text{ in } \kappa'y] \\
&= u.
\end{aligned}
$$

(ii) For a variable $w\colon (\sigma + \tau) \times \sigma$ we define a term

$$
L(w) \stackrel{\text{def}}{=} \text{unpack } \pi w \text{ as } [\kappa x \text{ in } \langle x, \pi'w\rangle, \kappa'y \text{ in } \langle\pi'w, \pi'w\rangle]\colon \sigma \times \sigma. \qquad .
$$

Then for $x, z\colon \sigma$ we get a conversion

$$
L[\langle\kappa x, z\rangle/w] = \text{unpack } \kappa x \text{ as } [\kappa x \text{ in } \langle x, z\rangle, \kappa'y \text{ in } \langle z, z\rangle] = \langle x, z\rangle.
$$

Hence we reason as follows. For terms $\Gamma \vdash M, M'\colon \sigma$,

$$
\begin{aligned}
\kappa M = \kappa M' &\Rightarrow \langle\kappa M, M\rangle = \langle\kappa M', M\rangle \\
&\Rightarrow \langle M, M\rangle = L[\langle\kappa M, M\rangle/w] = L[\langle\kappa M', M\rangle/w] = \langle M', M\rangle \\
&\Rightarrow M = \pi\langle M, M\rangle = \pi\langle M', M\rangle = M'. \qquad\qquad \square
\end{aligned}
$$

We also describe classifying categories $\mathcal{C}\ell_{1(\times,+)}(\Sigma)$ involving type theoretic exponents, finite products and coproducts. The definition is as for $\mathcal{C}\ell_{1\times}(\Sigma)$ above: types—but this time also with finite coproducts—are objects and morphisms $\sigma \to \tau$ are equivalence classes $[M]$ of terms $v_1\colon \sigma \vdash M\colon \tau$—where the conversion relation of course includes the above conversions for finite coproducts.

2.3.5. Proposition. *Categories $\mathcal{C}\ell_{1(\times,+)}(\Sigma)$ are Cartesian closed and have finite coproducts.*

Such a Cartesian closed category with finite coproducts is sometimes called a bicartesian closed category (BiCCC).

Proof. Cartesian closure is obtained as in the proof of Proposition 2.3.2. We concentrate on finite coproducts.

The empty type 0 is initial object in $\mathcal{C}\ell_{1(\times,+)}(\Sigma)$, since for every type σ we have a term $z\colon 0 \vdash \{\}\colon \sigma$. And if there is another term $z\colon 0 \vdash M\colon \sigma$, then $z\colon 0 \vdash M = \{\}\colon \sigma$, so that $[M] = [\{\}]\colon 0 \to \sigma$ in $\mathcal{C}\ell_{1(\times,+)}(\Sigma)$.

The coproduct type $\sigma + \tau$ is also the coproduct object: there are coprojection maps $\sigma \to (\sigma + \tau) \leftarrow \tau$ given by terms

$$x: \sigma \vdash \kappa x: \sigma + \tau \qquad \text{and} \qquad y: \tau \vdash \kappa' y: \sigma + \tau$$

And for each pair of morphisms $\sigma \to \rho$, $\tau \to \rho$, say described by

$$x: \sigma \vdash Q: \rho \qquad \text{and} \qquad y: \tau \vdash R: \rho$$

we have a cotuple $\sigma + \tau \to \rho$ given by the unpack term

$$z: \sigma + \tau \vdash \text{unpack } z \text{ as } [\kappa x \text{ in } Q, \kappa' y \text{ in } R]: \rho. \qquad \square$$

With this proposition in mind, the above Lemma 2.3.3 can be understood as saying that for $f: \sigma \to \rho$, $g: \tau \to \rho$ and $h: \rho \to \mu$ one has $h \circ [f, g] = [h \circ f, h \circ g]$—where the square braces $[-, -]$ are used for (categorical) cotupleing.

Distributive signatures

In this section we have seen various ways of forming new types starting from a set of atomic types. Signatures as first introduced in Definition 1.6.1 involve function symbols $F: \sigma_1, \ldots, \sigma_n \longrightarrow \sigma_{n+1}$ where each σ_i is an atomic type. This restriction to atomic types σ_i is not really practical. For example, a signature for natural numbers with zero, addition and predecessor may be given by an atomic type N and function symbols

$$0: 1 \longrightarrow N, \qquad \text{plus}: N \times N \longrightarrow N, \qquad \text{pred}: N \longrightarrow 1 + N$$

involving the derived (non-atomic) type $N \times N$ and $1 + N$. Here, the construction $1 + (-)$ is used to deal with partial operations: the predecessor pred yields an outcome in 1 if applied to zero, and in N otherwise. Similarly one can describe a subtraction function symbol min: $N \times N \longrightarrow 1 + N$ via coproducts.

In order to get this kind of expressiveness, one needs to have functions symbols $F: \sigma_1, \ldots, \sigma_n \longrightarrow \sigma_{n+1}$, where the σ_i may be formed from atomic types using finite products and coproducts. This leads to what we call distributive signatures, see [160]; they are called distributive graphs in [342]. Notice that in the presence of finite product types $(1, \times)$ we may restrict ourselves to function symbols $F: \sigma \longrightarrow \tau$ with precisely one input type. This leads to the following description, which is much like Definition 1.6.1.

2.3.6. Definition. For a set T (of atomic types) let us write $\overline{T}$ for the closure of T under finite products $(1, \times)$ and coproducts $(0, +)$. A category **DistrSign** of **distributive signatures** is defined by the following change-of-base situation.

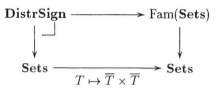

There are special kinds of distributive signatures which are useful to describe inductively and co-inductively defined types. These will be called **Hagino signatures**, after [111, 110, 112]. They occur in two forms, namely the inductive form and the dual co-inductive form. Here we only give the description of these Hagino signatures. They will be further investigated in Section 2.6 in STT, and also in Section 8.2 in PTT.

2.3.7. Definition. Let T be a set (of atomic types) and X a fresh symbol which is not in T. It serves as a type variable. A **Hagino signature** is a distributive signature with one single function symbol, which is either of the form

$$\sigma \xrightarrow{\text{constr}} X \qquad \text{or} \qquad X \xrightarrow{\text{destr}} \sigma$$

where σ is a "Hagino" type in the closure $\overline{T \cup \{X\}}$ of $T \cup \{X\}$ under finite products and coproducts. Sometimes we shall write $\sigma(X)$ for σ to emphasise the possible occurrence of X in σ.

In case this $\sigma(X)$ is of the form $\sigma_1(X) + \cdots + \sigma_n(X)$ the constructor constr may be understood as an n-tuple of function symbols $\text{constr}_i(X) : \sigma_i \to X$. Dually, if $\sigma(X)$ is of the form $\sigma_1(X) \times \cdots \times \sigma_n(X)$ the destructor destr corresponds to an n-tuple $\text{destr}_i : X \to \sigma_i(X)$.

Examples of Hagino signatures are

$$1 + X \longrightarrow X \qquad \text{for natural numbers}$$
$$1 + \alpha \times X \longrightarrow X \qquad \text{for finite lists of type } \alpha$$
$$X \longrightarrow \alpha \times X \qquad \text{for streams (or infinite lists) of type } \alpha.$$

In the first case the constructor is understood as the cotuple $[0, S] : 1 + X \to X$ of zero and successor. And for the finite lists one has a constructor nil: $1 \to X$ for the empty list and a constructor cons: $\alpha \times X \to X$ which turns an element of type α together with a list into a new list. In the third example one has two destructors: one for the head and one for the tail of an infinite list.

As another example, one can use these Hagino signatures to describe the connectives in propositional logic, for example with (cotupled) constructors:

$$1 + (X \times X) + 1 + (X \times X) + (X \times X) \xrightarrow{[\top, \wedge, \bot, \vee, \supset]} X$$

The idea behind a Hagino signature of the form $\sigma \rightarrow X$ is that X is the free type generated by the constructor. Later in Section 2.6 it will be described as a suitably initial fixed point $\sigma(X) \overset{\cong}{\rightarrow} X$ of an associated polynomial functor

$$X \mapsto \sigma(X)$$

where we have written the occurrences of X in σ explicitly. And in the dual case $X \rightarrow \sigma$ one thinks of X as the cofree type generated by the destructor. It corresponds to a fixed point $X \overset{\cong}{\rightarrow} \sigma(X)$, which is terminal in a suitable sense.

Finally, there is no need to restrict oneself to the finite product and coproduct type constructors in defining a category of signatures. One can also use exponent types; this leads to so-called higher type signatures, see *e.g.* [275].

Exercises

2.3.1.   Let Σ be a signature. Define finite product preserving functors

$$\mathcal{Cl}(\Sigma) \longrightarrow \mathcal{Cl}1(\Sigma) \longrightarrow \mathcal{Cl}1_\times(\Sigma) \longrightarrow \mathcal{Cl}1_{(\times,+)}(\Sigma).$$

2.3.2.   (i)   Give a proof of the above propositions-as-types bi-implication relating provability in MIL and inhabitation in $\lambda 1$.
          (ii) Formulate and prove similar results for $\lambda 1_\times$ and $\lambda 1_{(\times,+)}$.

2.3.3.   Give the precise correspondence in $\lambda 1_\times$ between terms $v_1 : \sigma_1, \ldots, v_n : \sigma_n \vdash M : \tau$ in contexts of arbitrary length and terms $v_1 : (\sigma_1 \times \cdots \times \sigma_n) \vdash N : \tau$ in contexts of length one.

2.3.4.   Give a concrete description of the category **DistrSign** of distributive signatures.

2.3.5.   Consider the following alternative description of a classifying category, say $\mathcal{Cl}1'_{(\times,+)}(\Sigma)$, with prime $'$. Objects are types σ as in $\mathcal{Cl}1_{(\times,+)}(\Sigma)$, but as morphisms $\sigma \rightarrow \tau$ we now take equivalence classes $[M]$ of closed terms $\vdash M : \sigma \rightarrow \tau$. Show that one gets a category $\mathcal{Cl}1'_{(\times,+)}(\Sigma)$ in this way and that it is isomorphic to the category $\mathcal{Cl}1_{(\times,+)}(\Sigma)$ described above.

2.4 Semantics of simple type theories

In the previous section we introduced firstly $\lambda 1$-calculi with exponent types, and secondly the (slightly) more complicated calculi $\lambda 1_\times$ and $\lambda 1_{(\times,+)}$, which are obtained by adding finite product and coproduct types. The (categorical) semantics of such calculi will be described in reverse order. The $\lambda 1_\times$-calculi have a straightforward (functorial) interpretation in Cartesian closed categories, as described *e.g.* in [186, 63, 61]. The finite product and exponent types in the calculus can be interpreted simply as finite product and exponent objects in a Cartesian closed category. Similarly, $\lambda 1_{(\times,+)}$-calculi can be interpreted in bicartesian closed categories (which additionally have finite co-products).

The semantics of $\lambda 1$-calculi—with exponent types only—is more subtle, since there is no identification of contexts and types involved. As a result there is no straightforward way to describe exponent types $\sigma \to (-)$ as right adjoints to product type functors $\sigma \times (-)$. Whereas there are no finite products of *types* in $\lambda 1$, one does have finite products of *contexts* (given by concatenation). Especially there are context projections $\pi \colon (\Gamma, v \colon \sigma) \to \Gamma$, inducing weakening functors π^*, which add an extra dummy variable $v \colon \sigma$. It turns out that exponent types $\sigma \to (-)$ can be captured categorically as right adjoints to such a weakening functors π^* in simple fibrations. This approach does not rely on product types $\sigma \times \tau$. Actually, these product types $\sigma \times (-)$ can be captured dually as *left* adjoints to these π^*'s. This view on exponents and products in the simple type theory comes from [156].

Unravelling the structure induced by right adjoints to π^*'s leads to an elementary formulation in Lemma 2.4.7 of what a '$\lambda 1$-category' is. It will be useful in the next section on the untyped λ-calculus.

But, as promised, we start with the calculi $\lambda 1_\times$ and $\lambda 1_{(\times,+)}$. Let **CCC** denote the category of Cartesian closed categories and functors preserving this structure. Similarly, let **BiCCC** $\hookrightarrow$ **CCC** be the subcategory of Cartesian closed categories with finite coproducts, and functors preserving all this structure. Recall from Propositions 2.3.2 and 2.3.5 that the classifying categories $\mathcal{C}\ell 1_\times(\Sigma)$ and $\mathcal{C}\ell 1_{(\times,+)}$ are objects of **CCC** and **BiCCC** respectively.

2.4.1. Definition. Let Σ be a signature. A **model for the calculus** $\lambda 1_\times(\Sigma)$ in a Cartesian closed category $\mathbb{A}$ is a functor

$$\mathcal{C}\ell 1_\times(\Sigma) \xrightarrow{\;\;\mathcal{M}\;\;} \mathbb{A} \qquad \text{in } \mathbf{CCC}.$$

Similarly, a model of $\lambda 1_{(\times,+)}(\Sigma)$ in a bicartesian closed category $\mathbb{A}$ is a functor

$$\mathcal{C}\ell 1_{(\times,+)}(\Sigma) \xrightarrow{\;\;\mathcal{M}\;\;} \mathbb{A} \qquad \text{in } \mathbf{BiCCC}.$$

We have a closer look at what a model is in the bicartesian closed category **Sets** of sets and ordinary functions. Suppose $[\![-]\!] \colon \mathcal{C}\ell 1_{(\times,+)}(\Sigma) \to \mathbf{Sets}$ is such a model. It involves

- a model of the signature Σ in **Sets**. Formally it is obtained by precomposition with the inclusion functor $\mathcal{C}\ell(\Sigma) \hookrightarrow \mathcal{C}\ell 1_{(\times,+)}(\Sigma)$ described in Exercise 2.3.1.
- a one-element, terminal set $[\![1]\!] = \{*\}$ and the empty set $[\![0]\!] = \emptyset$.
- binary products $[\![\sigma \times \tau]\!] \cong [\![\sigma]\!] \times [\![\tau]\!]$ and coproducts $[\![\sigma + \tau]\!] \cong [\![\sigma]\!] + [\![\tau]\!]$

of sets (where the coproduct + of set is given by disjoint union).

- function spaces $[\![\, \sigma \to \tau\,]\!] \cong [\![\, \tau\,]\!]^{[\![\, \sigma\,]\!]}$.

But Definition 2.4.1 covers models in any bicartesian closed category—and not just in **Sets**. The last three points are then modified according to the particular BiCCC-structure of the category involved.

There are results (similar to Theorem 2.2.5) for $\lambda 1_\times$ and $\lambda 1_{(\times, +)}$ which give a correspondence between functorial models and morphisms of signatures. We merely state these results here and leave the proof to the meticulous reader. Recall that for a category $\mathbb{B}$ with finite products there is an associated signature $\mathrm{Sign}(\mathbb{B})$, see Lemma 2.2.4.

2.4.2. Theorem. *Let Σ be a many-typed signature.*

(i) *For a Cartesian closed category $\mathbb{B}$ there is a bijective correspondence (up-to-isomorphism) between morphisms of signatures and models in*

$$\frac{\Sigma \xrightarrow{\ \phi\ } \mathrm{Sign}(\mathbb{B}) \quad in\ \mathbf{Sign}}{\mathcal{C}\ell 1_\times(\Sigma) \xrightarrow[\ \mathcal{M}\ ]{} \mathbb{B} \quad in\ \mathbf{CCC}}$$

(ii) *Similarly, for a Cartesian closed category $\mathbb{C}$ with finite coproducts there is a correspondence*

$$\frac{\Sigma \xrightarrow{\ \phi\ } \mathrm{Sign}(\mathbb{C}) \quad in\ \mathbf{Sign}}{\mathcal{C}\ell 1_{(\times, +)}(\Sigma) \xrightarrow[\ \mathcal{M}\ ]{} \mathbb{C} \quad in\ \mathbf{BiCCC}}$$ □

We turn to $\lambda 1$-calculi. Their categorical semantics will be described in terms of a simple fibration $\begin{smallmatrix} s(T) \\ \downarrow \\ \mathbb{B} \end{smallmatrix}$ associated with a CT-structure $(\mathbb{B}, T)$. We recall from Section 1.3 that the latter consists of a category $\mathbb{B}$ (of contexts) with finite products and a collection (of types) $T \subseteq \mathrm{Obj}\,\mathbb{B}$. Such a CT-structure is non-trivial if there is a type $X \in T$ and an arrow $1 \to X$ from the terminal object $1 \in \mathbb{B}$ to X.

First we describe how simple quantification, as described in Section 1.9, extends to these CT-structures, yielding quantification over types (*i.e.* over objects in T).

2.4.3. Definition. Let $(\mathbb{B}, T)$ be a CT-structure and $\begin{smallmatrix} \mathbb{E} \\ \downarrow p \\ \mathbb{B} \end{smallmatrix}$ a fibration. We say that p has **simple T-products** if for each $I \in \mathbb{B}$ and $X \in T$, every weakening functor $\pi^*_{I,X} \colon \mathbb{E}_I \to \mathbb{E}_{I \times X}$ induced by the projection $\pi_{I,X} \colon I \times X \to I$ has a right adjoint $\prod_{(I,X)}$—plus a Beck-Chevalley condition as in Definition 1.9.1.

Similarly one defines **simple T-coproducts** in terms of adjunctions $\coprod_{(I,X)} \dashv \pi^*_{I,X}$.

So in defining products and coproducts with respect to a CT-structure $(\mathbb{B}, T)$ we restrict the projections $\pi_{I,X}: I \times X \to I$ along which one has quantification, to those with $X \in T$, the set of types. Thus we quantify over types only, and not over all contexts. The simple products and coproducts as described in Definition 1.9.1 come out as special case, namely where $T = \text{Obj}\,\mathbb{B}$. The other extreme, where T is a singleton, will also be of importance, namely for the untyped λ-calculus and also for the second order polymorphic λ-calculus $\lambda 2$.

We have prepared the grounds for a categorical description of exponent types $\to$, without assuming Cartesian product types $\times$. Notice that we do assume Cartesian products in our base categories, but these correspond to context concatenation. For convenience, we restrict ourselves to the split case.

2.4.4. Definition ([156]). (i) A **$\lambda 1$-category** is a non-trivial CT-structure $(\mathbb{B}, T)$ for which the associated simple fibration $\begin{smallmatrix} \text{s}(T) \\ \downarrow \\ \mathbb{B} \end{smallmatrix}$ has split simple T-products.

(ii) A **morphism of $\lambda 1$-categories** from $(\mathbb{B}, T)$ to $(\mathbb{B}', T')$ consists of a morphism of CT-structures $K: (\mathbb{B}, T) \to (\mathbb{B}', T')$ whose extension to a morphism of fibrations $\begin{smallmatrix} \text{s}(T) \\ \downarrow \\ \mathbb{B} \end{smallmatrix} \to \begin{smallmatrix} \text{s}(T') \\ \downarrow \\ \mathbb{B}' \end{smallmatrix}$ (see Lemma 1.7.6) preserves simple products.

The content of this definition is that the exponent types of a $\lambda 1$-calculus are simple products (with respect to the set of types). Before going on, let us check that this works for syntactically constructed classifying categories.

2.4.5. Example. Let Σ be a signature with non-empty underlying set $T = |\Sigma|$ of atomic types. The latter ensures that the associated CT-structure $(\mathcal{Cl}1(\Sigma), T_1)$ is non-trivial, see Proposition 2.3.1. Consider the resulting fibration $\begin{smallmatrix} \text{s}(T_1) \\ \downarrow \\ \mathcal{Cl}1(\Sigma) \end{smallmatrix}$. For each context $\Gamma \in \mathcal{Cl}1(\Sigma)$ and type $\sigma \in T_1$, the projection morphism $\pi: \Gamma \times \sigma \to \Gamma$ in $\mathcal{Cl}1(\Sigma)$ gives rise to the weakening functor between the fibres:

$$\text{s}(T_1)_\Gamma \xrightarrow{\quad \pi^* \quad} \text{s}(T_1)_{\Gamma \times \sigma}$$

given by

$$\begin{cases} (\Gamma, \rho) \mapsto (\Gamma \times \sigma, \rho) \\ \left(\Gamma \times \rho_1 \xrightarrow{[M]} \rho_2\right) \mapsto \left(\Gamma \times \sigma \times \rho_1 \xrightarrow{[M]} \rho_2\right). \end{cases}$$

i.e. by

$$\begin{cases} \quad (\vdash \rho \colon \mathsf{Type}) \;\mapsto\; (\vdash \rho \colon \mathsf{Type}) \\ (\Gamma, z \colon \rho_1 \vdash M \colon \rho_2) \;\mapsto\; (\Gamma, x \colon \sigma, z \colon \rho_1 \vdash M \colon \rho_2). \end{cases}$$

This π^* adds an extra hypothesis of type σ. We should define a right adjoint

$$s(T_1)_{\Gamma \times \sigma} \xrightarrow{\;\;\prod_{(\Gamma, \sigma)}\;\;} s(T_1)_{\Gamma}$$

in the reverse direction. It naturally suggests itself as

$$(\Gamma \times \sigma, \tau) \mapsto (\Gamma, \sigma \to \tau) \qquad \textit{i.e. as} \qquad (\vdash \tau \colon \mathsf{Type}) \mapsto (\vdash \sigma \to \tau \colon \mathsf{Type}).$$

We then have to establish a bijective correspondence

$$\frac{\pi^*(\Gamma, \rho) = (\Gamma \times \sigma, \rho) \xrightarrow{\;\;[M]\;\;} (\Gamma \times \sigma, \tau)}{(\Gamma, \rho) \xrightarrow[\;\;[N]\;\;]{} (\Gamma, \sigma \to \tau) = \prod_{(\Gamma, \sigma)}(\Gamma \times \sigma, \tau)}$$

between terms M, N in

$$\frac{\Gamma, x \colon \sigma, z \colon \rho \vdash M \colon \tau}{\Gamma, z \colon \rho \vdash N \colon \sigma \to \tau}$$

It is given by abstraction and application:

$$M \mapsto \lambda x \colon \sigma.\, M \qquad \text{and} \qquad N \mapsto N x.$$

The fact that these operations are each others inverse corresponds precisely to the (β)- and (η)-conversions described in the previous section. We conclude that $(\mathcal{C}\ell 1(\Sigma), T_1)$ is a $\lambda 1$-category.

In view of this example, and in analogy with Definition 2.2.2, the following definition describes functorial semantics for $\lambda 1$.

2.4.6. Definition. Let Σ be a signature with $S = |\Sigma|$ as set of atomic types. A **$\lambda 1$-model** is a morphism of $\lambda 1$-categories $\mathcal{M} \colon (\mathcal{C}\ell 1(\Sigma), S_1) \to (\mathbb{B}, T)$.

Next we give a more amenable description of $\lambda 1$-categories.

2.4.7. Lemma. *Let $(\mathbb{B}, T)$ be a non-trivial CT-structure. The following two statements are equivalent.*

(i) *The pair $(\mathbb{B}, T)$ forms a $\lambda 1$-category.*

(ii) *The collection $T \subseteq \mathrm{Obj}\,\mathbb{B}$ is closed under exponents. That is, for types $X, Y \in T$ there is an exponent type $X \Rightarrow Y \in T$ together with an evaluation morphism $\mathrm{ev} \colon (X \Rightarrow Y) \times X \to Y$ such that for each object $I \in \mathbb{B}$ and map*

$f: I \times X \to Y$ in $\mathbb{B}$ *there is a unique abstraction map* $\Lambda(f): I \to X \Rightarrow Y$ *with* $\mathrm{ev} \circ \Lambda(f) \times \mathrm{id} = f$.

Proof. (ii) $\Rightarrow$ (i). For $I \in \mathbb{B}$ and $X \in T$ we can define a product functor $\prod_{(I,X)}: \mathrm{s}(T)_{I \times X} \to \mathrm{s}(T)_I$ by $Y \mapsto X \Rightarrow Y$. Then we get correspondences in the fibres:

$$\frac{\frac{\frac{\pi_{I,X}^*(Z) = Z \longrightarrow Y \quad \text{over } I \times X}{(I \times X) \times Z \longrightarrow Y \quad \text{in } \mathbb{B}}}{I \times Z \longrightarrow X \Rightarrow Y \quad \text{in } \mathbb{B}}}{Z \longrightarrow \prod_{(I,X)}(Y) \quad \text{over } I}$$

This describes (simple product) adjunctions $\pi_{I,X}^* \dashv \prod_{(I,X)}$.

(i) $\Rightarrow$ (ii). For types $X, Y \in T$, we consider Y as an object $(1 \times X, Y)$ of the fibre over $1 \times X$ and thus we can take

$$X \Rightarrow Y \stackrel{\mathrm{def}}{=} \prod_{(1,X)}(Y) \in T.$$

Notice that for an object $I \in \mathbb{B}$, reindexing along $!_I: I \dashrightarrow 1$ in $\mathbb{B}$ yields

$$
\begin{aligned}
X \Rightarrow Y &= !_I^*(X \Rightarrow Y) \\
&= !_I^*(\textstyle\prod_{(1,X)}(Y)) \\
&= \textstyle\prod_{(I,X)}((!_I \times \mathrm{id})^*(Y)) \quad \text{by Beck-Chevalley} \\
&= \textstyle\prod_{(I,X)}(Y).
\end{aligned}
$$

The counit (at Y) of the adjunction $\pi_{I,X}^* \dashv \prod_{(I,X)}$ is a morphism

$$\prod_{(I,X)}(Y) \xrightarrow{\;\varepsilon_Y^{(I,X)}\;} Y$$

in the fibre over $I \times X$. For $I = 1$, it forms a map in $\mathbb{B}$

$$(1 \times X) \times \prod_{(1,X)}(Y) \xrightarrow{\;\varepsilon_Y^{(I,X)}\;} Y$$

Hence we can define an evaluation map

$$\mathrm{ev}_{X,Y} \stackrel{\mathrm{def}}{=} \left((X \Rightarrow Y) \times X \xrightarrow[\cong]{\;\langle\langle !, \pi'\rangle, \pi\rangle\;} (1 \times X) \times (X \Rightarrow Y) \xrightarrow{\;\varepsilon_Y^{(1,X)}\;} Y \right)$$

The definition of abstraction is a bit tricky. Since $(\mathbb{B}, T)$ is by assumption non-trivial, we may assume an object $Z \in T$ with an arrow $z_0: 1 \to Z$.

For a map $f: I \times X \to Y$ in $\mathbb{B}$ (with $I \in \mathbb{B}$ and $X, Y \in T$) one has $f \circ \pi: (I \times X) \times Z \to Y$ in $\mathbb{B}$ and thus $f \circ \pi: \pi^*_{I,X}(Z) \to Y$ in the fibre over $I \times X$. Taking the transpose across the adjunction $\pi^*_{I,X} \dashv \prod_{(I,X)}$ yields a morphism $(f \circ \pi)^{\vee}: Z \to \prod_{(I,X)}(Y)$ in $s(T)_I$. As noticed above, $X \Rightarrow Y = \prod_{(I,X)}(Y)$ and thus $(f \circ \pi)^{\vee}: I \times Z \to X \Rightarrow Y$ in $\mathbb{B}$. Hence we take

$$\Lambda(f) \stackrel{\text{def}}{=} \left(I \xrightarrow{\ \langle \text{id}, z_0 \circ \, ! \rangle\ } I \times Z \xrightarrow{\ (f \circ \pi)^{\vee}\ } X \Rightarrow Y \right)$$

(The auxiliary type Z is first used to introduce a dummy variable which is later removed by substituting z_0.)

The validity of the categorical (β)- and (η)-equations follows from computations in the fibres. We shall do (β) and write $\bullet$ for composition in the fibres.

$$\text{ev} \circ \Lambda(f) \times \text{id}$$

$$= \varepsilon_Y^{(1,X)} \circ \langle \langle !, \pi' \rangle, \pi \rangle \circ \Lambda(f) \times \text{id}$$

$$= \varepsilon_Y^{(1,X)} \circ (!_I \times \text{id}) \times \text{id} \circ \langle \text{id}, \Lambda(f) \circ \pi \rangle$$

$$= (!_I \times \text{id})^* (\varepsilon_Y^{(1,X)}) \circ \langle \text{id}, \Lambda(f) \circ \pi \rangle$$

$$= \varepsilon_Y^{(I,X)} \circ \langle \text{id}, \Lambda(f) \circ \pi \rangle$$

$$= \varepsilon_Y^{(I,X)} \circ \langle \text{id}, \textstyle\prod_{(I,X)}(f \circ \pi) \circ \langle \pi, \eta_Z^{(I,X)} \rangle \circ \langle \text{id}, z_0 \circ \, ! \rangle \circ \pi \rangle$$

$$= \varepsilon_Y^{(I,X)} \circ \langle \pi, \textstyle\prod_{(I,X)}(f \circ \pi) \circ \pi \times \text{id} \rangle \circ \langle \pi, \eta_Z^{(I,X)} \circ \pi \times \text{id} \rangle$$
$$\qquad \circ \langle \text{id}, z_0 \circ \, ! \rangle$$

$$= \varepsilon_Y^{(I,X)} \bullet \pi^*_{I,X} \textstyle\prod_{(I,X)}(f \circ \pi) \bullet \pi^*_{I,X}(\eta_Z^{(I,X)}) \circ \langle \text{id}, z_0 \circ \, ! \rangle$$

$$= (f \circ \pi) \bullet \varepsilon_{\pi^*_{I,X}(Z)}^{(I,X)} \bullet \pi^*_{I,X}(\eta_Z^{(I,X)}) \circ \langle \text{id}, z_0 \circ \, ! \rangle$$

$$= (f \circ \pi) \bullet \text{id} \circ \langle \text{id}, z_0 \circ \, ! \rangle$$

$$= f \circ \pi \circ \langle \text{id}, z_0 \circ \, ! \rangle$$

$$= f. \qquad\qquad\qquad\qquad \square$$

2.4.8. Corollary (Proposition 1.9.3 (ii)). *A category* $\mathbb{B}$ *with finite products is Cartesian closed if and only if the simple fibration* $\begin{smallmatrix} s(\mathbb{B}) \\ \downarrow \\ \mathbb{B} \end{smallmatrix}$ *on* $\mathbb{B}$ *has simple products (i.e. forms a $\lambda1$-category).*

Proof. Take $T = \text{Obj}\,\mathbb{B}$ in Lemma 2.4.7. $\qquad\qquad\qquad\qquad \square$

The semantics of $\lambda1_{\times}$-calculi can thus be seen as a special case of $\lambda1$-calculi (where the collection of types T contains all objects). The other extreme where T is a singleton describes the semantics of the untyped λ-calculus; this will be the subject of the next section.

We close this section with an example of a $\lambda 1$-category, involving Scott-closed subsets as types.

2.4.9. Example. Let D be a directed complete partial order (dcpo). Non-empty closed subsets $X \subseteq D$ (with respect to the Scott topology) are often called **ideals**.. They are the non-empty directed lower sets X, satisfying (i) $X \neq \emptyset$, or equivalently, $\perp \in X$, (ii) $x \leq y \in X \Rightarrow x \in X$, and (iii) directed $a \subseteq X \Rightarrow \bigvee^{\uparrow} a \in X$. We show that ideals can be used (as types) to model simply typed calculi with exponents (provided one has an interpretation of the signature). Therefore, we form a base category $\mathbb{B}$ with finite sequences $\langle X_1, \ldots, X_n \rangle$ of such ideals X_i as objects. These sequences may be understood as contexts. A morphism $\langle X_1, \ldots, X_n \rangle \to \langle Y_1, \ldots, Y_m \rangle$ in $\mathbb{B}$ is given by a sequence $\langle f_1, \ldots, f_m \rangle$ of continuous functions $f_j \colon D^n \to D$ satisfying $f_j[\vec{X}] \subseteq Y_j$. That is, for all $x_i \in X_i$, one has $f_j(\vec{x}) \in Y_j$. The empty sequence is then terminal object in $\mathbb{B}$ and concatenation of sequences yields Cartesian products in $\mathbb{B}$.

Now let us assume that D is reflexive, *i.e.* that it isomorphic to its own space of continuous functions $[D \to D]$, via continuous maps $F \colon D \to [D \to D]$ and $G \colon [D \to D] \to D$ satisfying $F \circ G = \text{id}$ and $G \circ F = \text{id}$. An example of such a dcpo is D. Scott's D_∞, see [301, 13]; it forms a model of the untyped λ-calculus, as will be explained in the next section. In a standard way one forms an exponent of ideals $X, Y \subseteq D$ by

$$X \Rightarrow Y = \{z \in D \mid \forall x \in X.\, F(z)(x) \in Y\}.$$

One easily verifies that $X \Rightarrow Y$ is an ideal again.

Let $T \subseteq \text{Obj}\,\mathbb{B}$ be the collection of ideals (*i.e.* of sequences of length one). One obtains a CT-structure $(\mathbb{B}, T)$. We claim that it is a $\lambda 1$-category. Using the above Lemma 2.4.7 this is readily established: one has an evaluation map

$$\text{ev} = \left(\langle X \Rightarrow Y, X \rangle \xrightarrow{\lambda xy.\, F(x)(y)} Y \right)$$

And for $f \colon \langle \vec{Z}, X \rangle \to Y$ one takes as abstraction map

$$\Lambda(f) = \left(\vec{Z} \xrightarrow{\lambda \vec{z}.\, G(\lambda x.\, f(\vec{z}, x))} (X \Rightarrow Y) \right)$$

Exercises

2.4.1. Verify that the Beck-Chevalley condition in Example 2.4.5 corresponds precisely to the proper distribution of substitution over abstraction and application (as described in the previous section).

2.4.2.   Check the (η)-conversion $\Lambda(\mathrm{ev} \circ g \times \mathrm{id}) = g$ in the proof of Lemma 2.4.7.

2.4.3.   Extend Lemma 2.4.7 to morphisms, in the sense that a morphism of $\lambda 1$-categories corresponds to a morphism of CT-structures which preserves the relevant exponents.

2.4.4.   Show that the inclusion functor $\mathcal{C}\ell 1(\Sigma) \hookrightarrow \mathcal{C}\ell 1_\times(\Sigma)$ extends to a morphism of $\lambda 1$-categories. Conclude that every $\lambda 1_\times(\Sigma)$-model is a $\lambda 1(\Sigma)$-model.

2.4.5.   Consider a model $\mathcal{M}$ of a propositional logic $\mathcal{L}$ in a certain poset $(X, \leq)$, for example in a Heyting algebra. Show that such an $\mathcal{M}$ can also be understood as a functorial model $\mathcal{M}: \mathrm{LA}(\mathcal{L}) \to X$, from the Lindenbaum algebra $\mathrm{LA}(\mathcal{L})$ of propositions (modulo $\varphi \sim \psi \Leftrightarrow \varphi \vdash \psi$ and $\psi \vdash \varphi$) into X. Check that interpretation of the logical connectives $\top, \wedge, \supset$ etc. corresponds to preservation of this structure by $\mathcal{M}$.

2.4.6.   Let $(\mathbb{B}, T)$ be a $\lambda 1$-category. Define $\widehat{T}$ to be the smallest collection containing T which satisfies

$$1 \in \widehat{T} \qquad \text{and} \qquad X, Y \in \widehat{T} \Rightarrow X \times Y \in \widehat{T}.$$

Hence $\widehat{T}$ is obtained by closing T under finite products. Let $\widehat{T}$ also denote the full subcategory of $\mathbb{B}$ with objects in this collection.

(i)   Show that $\widehat{T}$ is Cartesian closed.

(ii) Show that—as a result—$\lambda 1$-classifying categories $\mathcal{C}\ell 1(\Sigma)$ are Cartesian closed. Describe an exponent $(v_1: \sigma_1, v_2: \sigma_2) \Rightarrow (v_1: \tau_1, v_2: \tau_2, v_3: \tau_3)$ explicitly.

2.4.7.   Let $(\mathbb{B}, T)$ be a CT-structure. Show that the associated simple fibration
$$\begin{array}{c} \mathrm{s}(T) \\ \downarrow \\ \mathbb{B} \end{array}$$
has simple T-coproducts $\coprod$ if and only if the collection of types T is closed under binary products $\times$. Hence binary product types are described by *left* adjoints to weakening functors.
[Details of the proof may be found in [156].]

2.4.8.   Formulate and prove a result like Theorem 2.4.2 for $\lambda 1$-categories.

2.5 Semantics of the untyped lambda calculus as a corollary

As a general point we observe that *untyped* can be identified with *typed in a universe with only one type*. In the untyped λ-calculus (see [13]) one can build terms from variables v via application MN and abstraction $\lambda v. M$, without any type restrictions (because there are no types). We can see this untyped λ-calculus as a (typed) $\lambda 1$-calculus with a single type Ω satisfying $\Omega = \Omega \to \Omega$. Every untyped term $M(\vec{v})$ can then be typed as $v_1: \Omega, \ldots, v_n: \Omega \vdash M: \Omega$.

The notions and results developed for the simply typed λ-calculus in the previous section are based on CT-structures. Specialising to such structures with only one type (*i.e.* to the single-typed case) yields appropriate notions for the untyped λ-calculus. This constitutes a precise mathematical elaboration of the point of view—stressed by D. Scott—that the untyped λ-calculus should be seen as a special case of the (simply) typed one. The notion of 'λ-category' that we arrive at through this analysis, is in fact a mild generalisation of an early notion of Obtułowicz, see [233, 235]. More information on the semantics of the untyped λ-calculus can be found in [301, 303, 304, 221, 181, 13, 186, 63, 156, 158, 274].

2.5.1. Definition. A **λ-category** is a category $\mathbb{B}$ with finite products containing a distinguished object $\Omega \in \mathbb{B}$, such that the (single-typed) CT-structure $(\mathbb{B}, \{\Omega\})$ is a λ1-category, *i.e.* such that simple fibration $\begin{smallmatrix} s(\Omega) \\ \downarrow \\ \mathbb{B} \end{smallmatrix}$ has simple Ω-products.

The following is then a special case of Lemma 2.4.7.

2.5.2. Lemma. *Let $\mathbb{B}$ be a category with finite products and let $\Omega \in \mathbb{B}$ be a non-empty object (i.e. with non-empty hom-set $\mathbb{B}(1, \Omega)$). The pair $(\mathbb{B}, \Omega)$ is then a λ-category if and only if there is a map* app$: \Omega \times \Omega \to \Omega$ *such that for each $f: I \times \Omega \to \Omega$ there is precisely one $\lambda(f): I \to \Omega$ with* app $\circ \lambda(f) \times$ id $= f$.

Proof. Lemma 2.4.7 requires the singleton set $\{\Omega\}$ of types to be closed under exponents. This is the case if and only if Ω itself is the exponent $\Omega \Rightarrow \Omega$. The result follows easily by reading app for ev and $\lambda(f)$ for $\Lambda(f)$ in the formulation of Lemma 2.4.7.   □

2.5.3. Examples. (i) Consider a signature with one atomic type Ω and no function symbols. Identify the exponent type $\Omega \to \Omega$ with Ω. In the resulting λ1-calculus on this signature we can provide every untyped term $M(\vec{v})$ with a typing $v_1: \Omega, \ldots, v_n: \Omega \vdash M: \Omega$. The classifying λ1-category can then be described as follows.

objects $n \in \mathbb{N}$.

morphisms $n \to m$ are m-tuples $([M_1], \ldots, [M_m])$ of $\beta\eta$-equivalence classes of untyped λ-terms M_i with free variables among $v_1, \ldots, v_n$.

The object 0 is then terminal and $n + m$ is the Cartesian product of the objects n and m. The object 1 plays the rôle of Ω in the above lemma: there is an application map

$$\text{app} = \left(2 \xrightarrow{\ [v_1 v_2]\ } 1\right)$$

And for each morphism $[M]: n + 1 \to 1$ there is an associated abstraction map

$$\lambda([M]) = \left(n \xrightarrow{\ [\lambda v_{n+1}.\, M]\ } 1\right)$$

satisfying the required properties.

The result is a categorical version of what is called the **closed term model** in [13]. It is the (pure) λ-classifying category; 'pure', because there are no function symbols involved.

(ii) Let D be a dcpo which is "reflexive", *i.e.* which is isomorphic to the dcpo of its own continuous endofunctions, *i.e.* $D \cong [D \to D]$, say via continuous $F: D \to [D \to D]$ and $G: [D \to D] \to D$ with $F \circ G = \text{id}$ and $G \circ F = \text{id}$, as in Example 2.4.9. The first example of such a D is D. Scott's D_∞, see [301, 13]. It will be described as a $\lambda 1$-category. A base category $\mathbb{D}$ is formed with $n \in \mathbb{N}$ as objects; the object n is the context consisting of n variables. Morphism $n \to m$ are sequences $(f_1, \ldots, f_m)$ where each f_i is a continuous function $D^n \to D$. Composition in $\mathbb{D}$ is done in the obvious way and identities are sequences of projections. The object $0 \in \mathbb{D}$ is terminal and $n + m$ is a Cartesian product of n, m. As distinguished object ("Ω") we take $1 \in \mathbb{D}$. Notice that 1 is a non-empty object since the set D is non-empty: it contains, for example, the identity combinator $\boldsymbol{I} = G(\text{id}_D)$.

One has app: $1 + 1 \to 1$ as a continuous function $D \times D \to D$ described by $(x, y) \mapsto F(x)(y)$. For $f: n + 1 \to 1$ in $\mathbb{D}$ one takes $\lambda(f)(\vec{x}) = G(\lambda y. f(\vec{x}, y))$, which yields a morphism $n \to 1$. Then

$$\big(\text{app} \circ \lambda(f) \times \text{id}\big)(\vec{x}, z) = F\big(G(\lambda y. f(\vec{x}, y))\big)(z) = f(\vec{x}, z).$$

It is easy to see that $\lambda(f)$ is unique in satisfying this equation.

(iii) The previous example can be generalised in the following sense. Let $\mathbb{B}$ be a Cartesian closed category containing an (extensional) **reflexive object** Ω. This means that there is an isomorphism $\Omega \cong (\Omega \Rightarrow \Omega)$, say via maps $F: \Omega \to (\Omega \Rightarrow \Omega)$ and $G: (\Omega \Rightarrow \Omega) \to \Omega$ with $F \circ G = \text{id}$ and $G \circ F = \text{id}$. Then we can define application and abstraction operations, namely:

$$\text{app} = \Big(\Omega \times \Omega \xrightarrow{\quad \text{ev} \circ F \times \text{id} \quad} \Omega\Big)$$

And for $f: I \times \Omega \to \Omega$ there is:

$$\lambda(f) = \Big(I \xrightarrow{\quad G \circ \Lambda(f) \quad} \Omega\Big)$$

One obtains a λ-category as described in Lemma 2.5.2.

The notion of a *CCC with reflexive object* was used by D. Scott as a categorical model of the untyped λ-calculus. The above notion of λ-category is more economical in the sense that it does not require all exponents in the ambient category, but only the relevant one, namely $\Omega \Rightarrow \Omega$. But a λ-category can be described as a reflexive object in a richer (presheaf) environment, see Exercise 2.5.1 below.

Obtułowicz [233, 235] introduced what he called a *Church algebraic theory*. It is a λ-category $(\mathbb{B}, \Omega)$ in which the collection of objects of $\mathbb{B}$ is of the form $\{\Omega^n \mid n \in \mathbb{N}\}$—as in Examples (i) and (ii) above. In fact, Obtułowicz defined a non-extensional version, as in Exercise 2.5.2 below.

Exercises

2.5.1. Let $\mathbb{B}$ be a category with finite products and $\Omega \in \mathbb{B}$ be a non-empty object. Show that $(\mathbb{B}, \Omega)$ is a λ-category if and only if the associated representable presheaf $\mathbb{B}(-, \Omega) : \mathbb{B}^{\mathrm{op}} \to \mathbf{Sets}$ is a reflexive object in the (Cartesian closed) ("topos") category $\mathbf{Sets}^{\mathbb{B}^{\mathrm{op}}}$ of presheaves.

[Familiarity with the Cartesian closed structure of $\mathbf{Sets}^{\mathbb{B}^{\mathrm{op}}}$ is assumed here; see Example 5.4.2. Especially with the fact that the Yoneda embedding $X \mapsto \mathbb{B}(-, X)$ preserves exponents.]

2.5.2. The formulation below is based on [156] and uses *semi-adjunctions* from [119]. These provide general categorical means to describe non-extensionality.

A **non-extensional λ-category** is given by a non-trivial CT-structure $(\mathbb{B}, \Omega)$ such that the associated simple fibration $\begin{smallmatrix} s(\Omega) \\ \downarrow \\ \mathbb{B} \end{smallmatrix}$ has 'semi-products'; that is, every $\pi^*_{I,\Omega}$ has a right semi-adjoint and for every $u : I \to J$ in $\mathbb{B}$ the pair $(u^*, (u \times \mathrm{id})^*)$ forms a morphism of semi-adjunctions. See [119] for the details of these 'semi' notions.

(i) Show that a (non-trivial) CT-structure $(\mathbb{B}, \Omega)$ is a non-extensional λ-category if and only if there is an application map $\mathrm{app} : \Omega \times \Omega \to \Omega$ such that for each $f : I \times \Omega \to \Omega$ there is a (not necessarily unique) abstraction map $\lambda(f) : I \to \Omega$ subject to the equations

$$\mathrm{app} \circ \lambda(f) \times \mathrm{id} = f \qquad \text{and} \qquad \lambda(f \circ g \times \mathrm{id}) = \lambda(f) \circ g.$$

(ii) Let D be a dcpo such that the continuous endofunctions $[D \to D]$ form a **retract** of D, say via $F : D \to [D \to D]$ and $G : [D \to D] \to D$ with $F \circ G = \mathrm{id}$, but not necessarily $G \circ F = \mathrm{id}$. An example of such a dcpo is $P\omega$, see [13]. Show that the construction in Example 2.5.3 (ii) applied to such a dcpo yields an example of a non-extensional λ-category.

2.6 Simple parameters

In the preceding two sections we have been using simple fibrations for the semantics of simple type theory. Here we show how these simple fibrations can also be used to systematically describe data types with *simple* parameters. We shall first briefly describe finite coproducts with simple parameters, next natural numbers with simple parameters, and finally arbitrary inductively defined data types (as given by Hagino signatures) with simple parameters. For the latter we shall make essential use of so-called strong functors. This approach comes from [160], where it is presented in terms of simple slice categories (instead of simple fibrations).

Recall that for a category $\mathbb{B}$ with finite products there is a simple fibration $\begin{smallmatrix} s(\mathbb{B}) \\ \downarrow \\ \mathbb{B} \end{smallmatrix}$ on $\mathbb{B}$, with fibred finite products. The fibre over $I \in \mathbb{B}$ is written as $\mathbb{B}/\!\!/I$ and is called the simple slice over I. Its objects are $X \in \mathbb{B}$, and its morphisms $X \to Y$ are maps $I \times X \to Y$ in $\mathbb{B}$.

Distributive coproducts

A coproduct object $X + Y$ comes, by definition, equipped with (natural) bijective correspondences

$$\frac{X \longrightarrow Z \qquad Y \longrightarrow Z}{X + Y \longrightarrow Z}$$

Say we have **coproducts with simple parameters** if for each parameter object $I \in \mathbb{B}$ there are bijective correspondences

$$\frac{I \times X \longrightarrow Z \qquad I \times Y \longrightarrow Z}{I \times (X + Y) \longrightarrow Z} \qquad (*)$$

natural in X, Y, Z. Then we have the following result.

2.6.1. Proposition. *Let $\mathbb{B}$ be a category with binary products $\times$ and coproducts $+$. The following statements are then equivalent.*

(i) *$\mathbb{B}$ has coproducts with simple parameters (as described above).*

(ii) *The simple fibration $\begin{smallmatrix} s(\mathbb{B}) \\ \downarrow \\ \mathbb{B} \end{smallmatrix}$ on $\mathbb{B}$ has fibred coproducts.*

(iii) *$\mathbb{B}$ has distributive coproducts: the canonical maps*

$$(I \times X) + (I \times Y) \longrightarrow I \times (X + Y)$$

are isomorphisms.

Proof. (i) $\Leftrightarrow$ (ii). Almost immediate: the correspondence $(*)$ above precisely says that each simple slice $\mathbb{B}/\!\!/I$ (*i.e.* fibre over I) has coproducts. Preservation by reindexing functors is obvious. And (ii) $\Leftrightarrow$ (iii) is Exercise 1.8.3 (i).   □

This correspondence between data types (coproducts in this case) with simple parameters and a fibred version of such data types in a simple fibration will be elaborated further.

Natural numbers

Recall that in a category with finite products a **natural numbers object** (NNO) consists of a zero and successor diagram

$$1 \xrightarrow{\ 0\ } N \xrightarrow{\ S\ } N$$

which is initial in the sense that for an arbitrary diagram of the form $1 \xrightarrow{x} X \xrightarrow{g} X$ there is a unique $h: N \dashrightarrow X$ making the following diagram commute.

$$
\begin{array}{ccccc}
1 & \xrightarrow{\quad 0 \quad} & N & \xrightarrow{\quad S \quad} & N \\
\big\| & & \downarrow h & & \downarrow h \\
1 & \xrightarrow[\quad x \quad]{} & X & \xrightarrow[\quad g \quad]{} & X
\end{array}
$$

In functional notation, this is written as:

$$h\,0 = x \qquad \text{and} \qquad h\,(S\,n) = g\,(h\,n).$$

Recall that in **Sets** this mediating map $h: N \dashrightarrow X$ is obtained by iteration as:

$$h(n) = g^{(n)}(x) \qquad \text{where} \qquad \begin{cases} g^{(0)}(x) &= x \\ g^{(n+1)}(x) &= g(g^{(n)}(x)) \end{cases}$$

We say that $1 \xrightarrow{\ 0\ } N \xrightarrow{\ S\ } N$ is an **NNO with simple parameters** if for each parameter object I and pair of maps $f: I \times 1 \to X$ and $g: I \times X \to X$, there is a unique $h: I \times N \dashrightarrow X$ making the following diagram commute.

$$
\begin{array}{ccccc}
I \times 1 & \xrightarrow{\ \text{id} \times 0\ } & I \times N & \xrightarrow{\ \text{id} \times S\ } & I \times N \\
\big\| & & \downarrow \langle \pi, h \rangle & & \downarrow \langle \pi, h \rangle \\
I \times 1 & \xrightarrow[\ \langle \pi, x \rangle\ ]{} & I \times X & \xrightarrow[\ \langle \pi, g \rangle\ ]{} & I \times X
\end{array}
$$

where we have written $f: I \times 1 \to X$ instead of $f: I \to X$ for purely formal reasons. In functional notation we now have equations:

$$h\,(i,\, 0) = f\,i \qquad \text{and} \qquad h\,(i,\, S\,n) = g\,(i,\, h(i,\, n)).$$

They emphasise that such an NNO involves an extra parameter i. By taking the terminal object 1 as parameter object one sees that an NNO with simple parameters is an ordinary NNO. The reverse direction can be obtained in Cartesian closed categories. Below we give alternative descriptions of such

NNOs with simple parameters: they are fibred NNOs in simple fibrations. Therefore we need the following *fibrewise* notion. It is in fact a special case of Definition 1.8.1.

2.6.2. Definition. A fibration with a fibred terminal object has a **fibred natural numbers object** if each fibre has an NNO and reindexing functors preserve NNOs (*i.e.* if $0, S$ form an NNO, then so do $u^*(0), u^*(S)$).

2.6.3. Proposition. *For a category $\mathbb{B}$ with finite products, the following statements are equivalent.*

(i) $\mathbb{B}$ *has an NNO with simple parameters.*

(ii) $\mathbb{B}$ *has an NNO $0, S$ and for each $I \in \mathbb{B}$, the functor $I^*: \mathbb{B} \to \mathbb{B}/\!/I$ applied to $0, S$ yields an NNO $I^*(0), I^*(S)$ in the simple slice $\mathbb{B}/\!/I$ over I.*

(iii) *The simple fibration $\begin{smallmatrix} s(\mathbb{B}) \\ \downarrow \\ \mathbb{B} \end{smallmatrix}$ on $\mathbb{B}$ has a fibred NNO.*

Proof. (i) $\Leftrightarrow$ (ii). By definition of NNO with simple parameters.

(ii) $\Rightarrow$ (iii). Each fibre $\mathbb{B}/\!/I$ has an NNO $I^*(0), I^*(S)$. These are preserved under reindexing, since for $u: I \to J$ in $\mathbb{B}$ one has $u^* \circ J^* \cong I^*$.

(iii) $\Rightarrow$ (ii). Assume the simple fibration on $\mathbb{B}$ has a fibred NNO. Then $\mathbb{B}$ has an NNO $0, S$, since $\mathbb{B}$ is isomorphic to the simple slice $\mathbb{B}/\!/1$ over 1. Moreover, the pair $I^*(0), I^*(S)$ is an NNO in $\mathbb{B}/\!/I$, since reindexing functors preserves NNOs. □

Hagino signatures and strong functors

Recall from Definition 2.3.7 that a Hagino signatureinvolves a set S of atomic types, a type variable X and either a *constructor* function symbol constr: $\sigma \to X$ (in the inductive case) or a *destructor* function symbol destr: $X \to \sigma$ (in the co-inductive case), where σ is a type in the closure $\overline{S \cup \{X\}}$ of the set $S \cup \{X\}$ under finite products $(1, \times)$ and finite coproducts $(0, +)$. A model of the set (or subsignature) S in a category $\mathbb{B}$ consists of a functor $A: S \to \mathbb{B}$, *i.e.* of a collection $(A_s)_{s \in S}$ of objects $A_s \in \mathbb{B}$. Such a model assigns values in $\mathbb{B}$ to the atomic types $s \in S$. The category of models of S in $\mathbb{B}$ is the functor category $\mathbb{B}^S$, in which a morphism $f: (A_s)_{s \in S} \to (B_s)_{s \in S}$ consists of a collection $f = (f_s: A_s \to B_s)_{s \in S}$ of morphisms in $\mathbb{B}$.

Models of a Hagino signature can be described conveniently in terms of associated polynomial functors. This will be done first.

2.6.4. Definition. Each model $A: S \to \mathbb{B}$ of a set of atomic types S in a distributive category $\mathbb{B}$ together with a type $\sigma \in \overline{S \cup \{X\}}$ determines a **poly-**

nomial functor $T(A)_\sigma \colon \mathbb{B} \to \mathbb{B}$ which follows the structure of σ:

$$T(A)_\sigma \stackrel{\text{def}}{=} \begin{cases} \text{the constant functor } A_s & \text{if } \sigma \equiv s \in S \\ \text{the identity functor} & \text{if } \sigma \equiv X \\ \text{the constant functor } 0 & \text{if } \sigma \equiv 0 \\ \text{the constant functor } 1 & \text{if } \sigma \equiv 1 \\ Y \mapsto T(A)_{\sigma_1}(Y) + T(A)_{\sigma_2}(Y) & \text{if } \sigma \equiv \sigma_1 + \sigma_2 \\ Y \mapsto T(A)_{\sigma_1}(Y) \times T(A)_{\sigma_2}(Y) & \text{if } \sigma \equiv \sigma_1 \times \sigma_2. \end{cases}$$

For an arbitrary endofunctor $T \colon \mathbb{B} \to \mathbb{B}$ an **algebra** (or T-**algebra**) consists of a "carrier" object $Y \in \mathbb{B}$ together with a morphism $\varphi \colon T(Y) \to Y$. Dually, a **co-algebra** is a pair (Z, ψ) consisting of a carrier object Z and a map $\psi \colon Z \to T(Z)$ pointing in the reverse direction. In both the algebraic and in the co-algebraic case one can understand the functor T as describing a signature of operations. For instance, if $T(X) = 1 + X \times X + X$, then a T-algebra $T(Y) \to Y$ consists of a carrier Y on which we have three operations $1 \to Y$, $Y \times Y \to Y$ and $Y \to Y$. Every group G carries such a T-algebra structure $T(G) \to G$ consisting of the cotuple of unit, multiplication and inverse operations. Co-algebras $Z \to T(Z)$ generally describe "dynamical systems" (in an abstract sense), where Z is the state space, and the map $Z \to T(Z)$ is the dynamics, or transition function, acting on the state space (see *e.g.* [167]). Typical examples arise from automata: if Σ is a finite alphabet, then the functor $T(X) = (1 + X)^\Sigma$ is polynomial. A co-algebra $Z \to T(Z)$ may be described as a transition function $Z \times \Sigma \to 1 + Z$ which yields for every state $z \in Z$ and input symbol $a \in \Sigma$ either an outcome in 1, if the computation is unsuccessful, or a new state in Z. It is a certain automaton.

One forms a category $\mathbf{Alg}(T)$ with T-algebras as objects and as morphisms

$$\left(T(Y) \xrightarrow{\ \varphi\ } Y \right) \xrightarrow{\hspace{2.5cm} h \hspace{2.5cm}} \left(T(Z) \xrightarrow{\ \psi\ } Z \right)$$

maps $h \colon Y \to Z$ in the underlying category $\mathbb{B}$ between the carriers for which the following diagram commutes.

$$\begin{array}{ccc} T(Y) & \xrightarrow{\ T(h)\ } & T(Z) \\ {\scriptstyle\varphi}\downarrow & & \downarrow{\scriptstyle\psi} \\ Y & \xrightarrow[\ h\ ]{} & Z \end{array}$$

Dually, there is a category $\mathbf{CoAlg}(T)$ of co-algebras and similar, structure preserving morphisms between carriers. In these categories of algebras and co-algebras one can study initial and terminal objects. An initial algebraindexSInitial!– algebra for a functor $T: \mathbb{B} \to \mathbb{B}$ is a terminal co-algebraindexالسTerminal!– coalgebra for $T^{\mathrm{op}}: \mathbb{B}^{\mathrm{op}} \to \mathbb{B}^{\mathrm{op}}$. Notice that an initial algebra of the functor $X \mapsto 1 + X$ is a natural numbers object. In terms of these (co-)algebras one can describe many more data types than just natural numbers.

But first we mention the following basic result.

2.6.5. Lemma (Lambek). *An initial T-algebra $\varphi: T(Y) \to Y$ is an isomorphism.*

Thus initial algebras are fixed points $T(Y) \stackrel{\cong}{\to} Y$ of functors. By duality, a similar result holds for terminal co-algebras. In Exercise 2.6.4 below, we sketch the standard construction of such fixed points, generalising Tarski's fixed point construction in posets.

Proof. Considering the T-algebra $T(\varphi): T^2(Y) \to T(Y)$. One obtains by initiality an algebra map $f: \varphi \dashrightarrow T(\varphi)$, *i.e.* a morphism $f: Y \to T(Y)$ in $\mathbb{B}$ with $f \circ \varphi = T(\varphi) \circ T(f)$. But then, $\varphi \circ f$ is an algebra map $\varphi \to \varphi$ and must be the identity. Thus also $f \circ \varphi = T(\varphi) \circ T(f) = T(\varphi \circ f) = T(\mathrm{id}) = \mathrm{id}$. $\square$

2.6.6. Definition. Consider a type $\sigma(X)$ built with finite product and co-products from $S \cup \{X\}$ and a model $S: A \to \mathbb{B}$ of the atomic types in a distributive category $\mathbb{B}$. A **(initial) model**n $\mathbb{B}$ of an inductive Hagino signature

$$\sigma(X) \xrightarrow{\quad \mathsf{constr} \quad} X$$

is an (initial) $T(A)_\sigma$-algebra, written for convenience with the same name, as:

$$T(A)_\sigma(X) \xrightarrow{\quad \mathsf{constr} \quad} X$$

A **(terminal) model** of a co-inductive Hagino signature $\mathsf{destr}: X \to \sigma(X)$ is a (terminal) co-algebra $\mathsf{destr}: X \to T(A)_\sigma(X)$ of the associated functor.

Hagino signatures $\sigma \to X$ or $X \to \sigma$ are often used in programming languages to define a new type X recursively. The inductive case, say of the form $(\sigma_1 + \cdots + \sigma_n) \to X$ occurs in the functional programming language ML [224, 251] with syntax

$$\mathsf{datatype}\ X\ =\ C_1\ \mathsf{of}\ \sigma_1\ |\ \cdots\ |\ C_n\ \mathsf{of}\ \sigma_n$$

where $C_1, \ldots, C_n$ are constructors. Categorically, one combines these C_i into a single constructor $\mathsf{constr} = [C_1, \ldots, C_n]: (\sigma_1 + \cdots + \sigma_n) \to X$ via a cotuple. Describing constr as initial algebra of the functor associated with the type $(\sigma_1 + \cdots + \sigma_n)(X)$ provides appropriate elimination rules for such data types, which are used to define operations on them. Initiality tells us that it is the freely generated structure, and hence how it behaves with respect to arbitrary such structures. Co-algebras can be used to describe infinite data structures (and more generally, dynamical systems [167]), for example in object-oriented languages, see [283, 162, 164]. Terminal co-algebras are minimal realisations, in which all behaviourally indistinguishable (bisimilar) states are identified, see also [298].

In the programming language CHARITY, see [52], one can define both these initial and terminal types. Thus one can define for example a type of trees of finite depth with nodes having infinitely many branches.

These recursively defined types with initial or terminal characterisations occur already in [6], but were first investigated systematically from a type theoretic perspective by Hagino [111, 110, 112].

The above is standard theory. Here we show how we can use simple fibrations in order to get appropriate versions with parameters of such data types. The approach comes from [160], but there, the language of fibred categories is not used. What we need first is the notion of a strong functor.

2.6.7. Definition. Let $\mathbb{B}$ be a category with finite products. A functor $T: \mathbb{B} \to \mathbb{B}$ is called **strong** is it comes equipped with a **strength** natural transformation st with components $\text{st}_{I,X}: I \times TX \to T(I \times X)$ making the following two diagrams commute.

$$
\begin{array}{ccc}
I \times TX & \xrightarrow{\ \text{st}\ } & T(I \times X) \\
& {}_{\pi'}\searrow & \downarrow{\scriptstyle T(\pi')} \\
& & TX
\end{array}
$$

$$
\begin{array}{ccccc}
I \times (J \times TX) & \xrightarrow{\ \text{id} \times \text{st}\ } & I \times T(J \times X) & \xrightarrow{\ \text{st}\ } & T(I \times (J \times X)) \\
{\scriptstyle\cong}\downarrow & & & & \downarrow{\scriptstyle\cong} \\
(I \times J) \times TX & & \xrightarrow{\hspace{3cm}\text{st}\hspace{3cm}} & & T((I \times J) \times X)
\end{array}
$$

2.6.8. Examples. (i) Every functor $T: \mathbf{Sets} \to \mathbf{Sets}$ is strong with strength $I \times TX \to T(I \times X)$ given by

$$(i, a) \mapsto T\big(\lambda x \in X. \langle i, x \rangle\big)(a).$$

For example, for a set A, let $\text{list}(A)$ (or A^*) be the set of finite sequences of elements of A. The assignment $A \mapsto \text{list}(A)$ forms a functor on **Sets** with strength $I \times \text{list}(A) \to \text{list}(I \times A)$ given by

$$(i, \langle a_1, \ldots, a_n \rangle) \mapsto \langle (i, a_1), \ldots, (i, a_n) \rangle.$$

(ii) On a distributive category $\mathbb{B}$, identity functors and constant functors are strong. Moreover, if $T, S: \mathbb{B} \to \mathbb{B}$ are strong, then so are

$$Y \mapsto T(Y) \times S(Y) \qquad \text{and} \qquad Y \mapsto T(Y) + S(Y).$$

Hence every polynomial functor $T(A)_\sigma: \mathbb{B} \to \mathbb{B}$ in Definition 2.6.4 is strong.

The following basic result, due to Plotkin, gives an alternative description of strong functors in terms of simple fibrations.

2.6.9. Proposition. *Let $\mathbb{B}$ be a category with finite products. There is a bijective correspondence*

$$
\begin{array}{c}
\text{strong functors } \mathbb{B} \longrightarrow \mathbb{B} \\
\hline\hline
\end{array}
$$

$$
\begin{array}{c}
\text{split functors} \quad
\begin{array}{c}
s(\mathbb{B}) \longrightarrow s(\mathbb{B}) \\
\searrow \quad \swarrow \\
\mathbb{B}
\end{array}
\end{array}
$$

Given this correspondence, we shall write $T/\!\!/ I: \mathbb{B}/\!\!/ I \to \mathbb{B}/\!\!/ I$ for the endofunctor on the simple slice over I, associated with a strong functor $T: \mathbb{B} \to \mathbb{B}$.

Proof. Let (T, st) be a strong functor on $\mathbb{B}$. We define a split functor $\overline{T}: s(\mathbb{B}) \to s(\mathbb{B})$ by

$$(I, X) \mapsto (I, T(X)) \qquad \text{and} \qquad (u, f) \mapsto (u, T(f) \circ \mathrm{st}).$$

Conversely, let $R: s(\mathbb{B}) \to s(\mathbb{B})$ be a split endofunctor on $\begin{smallmatrix} s(\mathbb{B}) \\ \downarrow \\ \mathbb{B} \end{smallmatrix}$. It leads by restriction to functors $R_I: \mathbb{B}/\!\!/ I \to \mathbb{B}/\!\!/ I$ on the fibres. Hence we get a functor $\overline{R}$ on $\mathbb{B}$, via the functor R_1 over the terminal object 1:

$$\overline{R} = \left(\mathbb{B} \xrightarrow{\;\cong\;} \mathbb{B}/\!\!/ 1 \xrightarrow{\;R_1\;} \mathbb{B}/\!\!/ 1 \xrightarrow{\;\cong\;} \mathbb{B} \right)$$

It satisfies $I^* \circ \overline{R} = R_I \circ I^*$ and hence in particular for $X \in \mathbb{B}$, $\overline{R}(X) = R_I(X)$. A strength map $\mathrm{st}: I \times \overline{R}(X) \to \overline{R}(I \times X)$ is obtained as follows. The identity map $I \times X \to I \times X$ in $\mathbb{B}$ forms a morphism $I \to I \times X$ in $\mathbb{B}/\!\!/ I$. Thus by applying the functor R_I one obtains a morphism $R_I(X) \to R_I(I \times X)$ in $\mathbb{B}/\!\!/ I$. It corresponds to a map $I \times \overline{R}(X) \to \overline{R}(I \times X)$ in $\mathbb{B}$.

We leave it to the reader to verify that $\overline{\overline{T}} = T$ and $\overline{\overline{R}} = R$. $\qquad\square$

The following definition contains a compact reformulation of a notion used by Cockett and Spencer [52, 53] in their description of initial models of Hagino signatures with parameters.

2.6.10. Definition. An algebra $\varphi: TX \to X$ for a strong functor $T: \mathbb{B} \to \mathbb{B}$ is called **initial with simple parameters** if for each object $I \in \mathbb{B}$, the functor

$I^*: \mathbb{B} \to \mathbb{B}/\!/I$ maps φ to an initial algebra $I^*(\varphi)$ for the functor $T/\!/I: \mathbb{B}/\!/I \to \mathbb{B}/\!/I$ on the simple slice category over I.

Notice that this really is a fibrewise definition: it essentially says that in each fibre $\mathbb{B}/\!/I$ of the simple fibration $\begin{smallmatrix} s(\mathbb{B}) \\ \downarrow \\ \mathbb{B} \end{smallmatrix}$ the associated functor $T/\!/I: \mathbb{B}/\!/I \to \mathbb{B}/\!/I$ has an initial algebra $\varphi^I = I^*(\varphi)$, and that these initial algebras are preserved under reindexing. Since the category $\mathbb{B}$ can be identified with the fibre $\mathbb{B}/\!/1$ over 1, it suffices to have an algebra there, which is preserved by each reindexing functor $I^*: \mathbb{B} \to \mathbb{B}/\!/I$ associated with the map $I \to 1$ (as in Proposition 2.6.3).

If we spell out initiality with simple parameters of $\varphi: T(X) \to X$ as described above, then we come to the formulation used by Cockett and Spencer. It says that for each parameter object $I \in \mathbb{B}$ and for each "algebra with parameter" $\psi: I \times T(Y) \to Y$ in $\mathbb{B}$, there is a unique map $h: I \times X \dashrightarrow Y$ making the following diagram commute.

$$
\begin{array}{ccccc}
I \times T(X) & \xrightarrow{\langle \pi, \mathrm{st} \rangle} & I \times T(I \times X) & \xrightarrow{\mathrm{id} \times Th} & I \times T(Y) \\
{\scriptstyle \mathrm{id} \times \varphi} \downarrow & & & & \downarrow {\scriptstyle \psi} \\
I \times X & & \xrightarrow{\hspace{3cm} h \hspace{3cm}} & & Y
\end{array}
$$

The reader may wish to check that an algebra of the functor $X \mapsto 1 + X$ which is initial with simple parameters, is an NNO with simple parameters, as explicitly described in the beginning of this section.

We have only sketched the basics of the theory of (co-)inductively defined types, with emphasis on simple parameters. If one replaces the simple fibration by the codomain fibration, then one gets a theory with **dependent parameters**. For example, one can say that an NNO with dependent parameters $0, S$ in a category $\mathbb{B}$ is an NNO $0, S$ in $\mathbb{B}$ such that for each parameter object $I \in \mathbb{B}$, the functor $I^*: \mathbb{B} \to \mathbb{B}/I$ (to the *ordinary* slice category) maps $0, S$ to an NNO $I^*(0), I^*(S)$ in $\mathbb{B}/I$. This can alternatively be described as a fibred NNO for the codomain fibration on $\mathbb{B}$. In this dependent setting there is the following analogue of Proposition 2.6.9. It stems from unpublished work of Paré, see also [172, Proposition 3].

2.6.11. Proposition. *Let $\mathbb{B}$ be a category with finite limits. There is a bijec-*

tive correspondence (up-to-isomorphism) between

$$\frac{\text{strong, pullback preserving functors } \mathbb{B} \longrightarrow \mathbb{B}}{\text{fibred, fibred pullback preserving functors}} \quad \begin{array}{c} \mathbb{B}^{\rightarrow} \longrightarrow \mathbb{B}^{\rightarrow} \\ \searrow \quad \swarrow \\ \mathbb{B} \end{array}$$

Proof. Assuming a strong, pullback preserving functor $T: \mathbb{B} \to \mathbb{B}$, we define a functor $\overline{T}: \mathbb{B}^{\rightarrow} \to \mathbb{B}^{\rightarrow}$ by sending a family $\left(\begin{smallmatrix} X \\ \downarrow \varphi \\ I \end{smallmatrix} \right)$ to the composite $\left(\begin{smallmatrix} X' \\ \downarrow \\ I \end{smallmatrix} \right)$ in the following diagram.

$$\overline{T}(\varphi) \left(\begin{array}{ccc} X' & \xrightarrow{\hspace{3cm}} & T(X) \\ \downarrow \quad \lrcorner & & \downarrow T(\varphi) \\ I \times T(1) \xrightarrow{\text{st}} T(I \times 1) \xrightarrow[T(\pi)]{\cong} & T(I) \\ \downarrow \pi & & \\ I & & \end{array} \right)$$

It is not hard to see that because T preserves pullbacks, this $\overline{T}$ is a fibred functor, which preserves fibred pullbacks.

Conversely, given a fibred functor $R: \mathbb{B}^{\rightarrow} \to \mathbb{B}^{\rightarrow}$ preserving fibred pullbacks, we get a pullback preserving functor

$$\overline{R} = \left(\mathbb{B} \xrightarrow{\cong} \mathbb{B}/1 \xrightarrow{R_1} \mathbb{B}/1 \xrightarrow{\cong} \mathbb{B} \right)$$

on $\mathbb{B}$. Because R is a fibred functor, the Cartesian morphism in $\mathbb{B}^{\rightarrow}$ on the left below, is sent to the Cartesian morphism on the right.

$$\left(\begin{array}{ccc} I \times X & \longrightarrow & X \\ \pi \downarrow \quad \lrcorner & & \downarrow \\ I & \longrightarrow & 1 \end{array} \right) \xrightarrow{R} \left(\begin{array}{ccc} I \times \overline{R}(X) & \longrightarrow & \overline{R}(X) \\ \pi \downarrow \quad \lrcorner & & \downarrow \\ I & \longrightarrow & 1 \end{array} \right)$$

As a result, the functor $R: \mathbb{B}^{\rightarrow} \to \mathbb{B}^{\rightarrow}$ restricts to a split functor $R': s(\mathbb{B}) \to s(\mathbb{B})$, since the full subcategory $s(\mathbb{B}) \hookrightarrow \mathbb{B}^{\rightarrow}$ consists of Cartesian projections. By Proposition 2.6.9, the restriction of R' to the fibre over 1 is then strong. But this is $\overline{R}$, as described above.   □

A different extension of the basic theory, to be elaborated in Section 9.2, goes as follows. Given a polynomial functor $T: \mathbb{B} \to \mathbb{B}$ on the base category of a fibration $\overset{\mathbb{E}}{\underset{\mathbb{B}}{\downarrow}}$, then, under suitable assumptions, one can lift $T \colon \mathbb{B} \to \mathbb{B}$ to a fibred functor $\mathrm{Pred}(T) \colon \mathbb{E} \to \mathbb{E}$ on the total category of the fibration. It turns out that algebras of this lifted functor $\mathrm{Pred}(T)$ capture the induction principles which are needed to reason about the (initial) data type associated with T. And dually, co-algebras of $\mathrm{Pred}(T)$ may be used to reason about (terminal) co-algebras of T. This approach exploits a fibration $\overset{\mathbb{E}}{\underset{\mathbb{B}}{\downarrow}}$ as providing a logic of predicates in $\mathbb{E}$ to reason about types in the base category $\mathbb{B}$. This view on fibrations will be developed in the next three chapters.

Exercises

2.6.1. Consider a distributive category and define $\underline{n} = 1 + \cdots + 1$ (n times).
   (i) Prove that $\underline{n} + \underline{m} \cong \underline{n + m}$ and $\underline{n} \times \underline{m} \cong \underline{n \times m}$.
   (ii) Show that $\underline{2}$ carries the structure of a Boolean algebra.
       [*Hint.* Use $\underline{2} \times \underline{2} \cong \underline{2} + \underline{2}$ to define conjunction $\wedge \colon \underline{2} \times \underline{2} \to \underline{2}$ via the cotuple of cotuples $[[\kappa, \kappa], [\kappa, \kappa']] \colon \underline{2} + \underline{2} \to \underline{2}$.]
   (iii) Define a choice operation if: $\underline{2} \times X \times Y \to X + Y$.
   [For more such programming in distributive categories, see [341, 52, 53].]

2.6.2. Show that in a poset category with finite products $(\top, \wedge)$ and finite coproducts $(\bot, \vee)$, distributivity of $\wedge$ over $\vee$ implies distributivity of $\vee$ over $\wedge$ and vice-versa. In that case one has a distributive lattice. Note that this correspondence between distributivities does not hold for arbitrary categories.

2.6.3. Show that the assignment $A \mapsto T(A)_\sigma$ in Definition 2.6.4 extends to a functor $\mathbb{B}^S \to \mathbb{B}^{\mathbb{B}}$.

2.6.4. Show that the initial algebra of an endofunctor $T \colon \mathbb{B} \to \mathbb{B}$ can be constructed from the colimit X of the ω-chain,

$$0 \overset{!}{\longrightarrow} T(0) \overset{T(!)}{\longrightarrow} T^2(0) \overset{T^2(!)}{\longrightarrow} T^3(0) \longrightarrow \cdots \longrightarrow X$$

in case this colimit exists in $\mathbb{B}$ and is preserved by T—where $0 \in \mathbb{B}$ is initial object. This is as in [309].
   Prove that, dually a terminal co-algebra can be constructed as limit Y of the ω-chain,

$$1 \overset{!}{\longleftarrow} T(1) \overset{T(!)}{\longleftarrow} T^2(1) \overset{T^2(!)}{\longleftarrow} T^3(1) \longleftarrow \cdots \longleftarrow Y$$

provided T preserves such ω-limits.

2.6.5. Prove that, on a distributive category $\mathbb{B}$, the assignments $Y \mapsto T(Y) \times S(Y)$ and $Y \mapsto T(Y) + S(Y)$ are strong functors $\mathbb{B} \to \mathbb{B}$, assuming that both S and T are strong functors (as claimed in Example 2.6.8 (ii)).

2.6.6. Following [50] we say that a category $\mathbb{B}$ with finite products has **list objects** if for each $A \in \mathbb{B}$ there is an object $\mathrm{list}(A)$ equipped with a pair of maps

$$\mathrm{nil}\colon 1 \longrightarrow \mathrm{list}(A) \qquad \text{and} \qquad \mathrm{cons}\colon A \times \mathrm{list}(A) \longrightarrow \mathrm{list}(A)$$

such that for each $X \in \mathbb{B}$ which comes together with maps $x\colon 1 \longrightarrow X$ and $g\colon A \times X \longrightarrow X$, there is a unique morphism $h\colon \mathrm{list}(A) \dashrightarrow X$ with $h \circ \mathrm{nil} = x$ and $h \circ \mathrm{cons} = g \circ \mathrm{id} \times h$.

(i) Formulate appropriate fibrewise list objects and list objects with simple parameters such that a result like Proposition 2.6.3 can be obtained.

(ii) Show that a list object on A is an initial algebra of the functor $X \mapsto 1 + (A \times X)$.

. (iii) Check that the formulation with simple parameters from (i) coincides with the one in Definition 2.6.10.

2.6.7. Show that in a Cartesian closed category an initial algebra is always initial with simple parameters.

2.6.8. Define what a terminal co-algebra with simple parameters is. Show that each terminal co-algebra is automatically terminal with simple parameters.

2.6.9. Consider a comonad $G\colon \mathbb{C} \to \mathbb{C}$ and a functor $T\colon \mathbb{C} \to \mathbb{C}$ with a natural transformation $\sigma\colon GT \Rightarrow TG$.

(i) Say what it means that (T, σ) forms a map of comonads $G \to G$, *i.e.* that σ commutes appropriately with the comonads counit ε and comultiplication δ.

(ii) Assume that $\mathbb{C}$ has Cartesian products $\times$. Prove that a natural transformation $\mathrm{st}\colon (\times \circ T \times \mathrm{id}) \Rightarrow (T \circ \times)$ makes the functor T strong if and only if for each object $I \in \mathbb{C}$, the induced natural transformation $\mathrm{st}^I\colon T(-) \times I \Rightarrow T((-) \times I)$ forms a map of comonads.
[Recall from Exercise 1.3.4 that the functor $(-) \times I\colon \mathbb{C} \to \mathbb{C}$ carries a comonad structure.]

2.6.10. Let $\mathbb{B}$ be a category with finite products. A **strong monad** on $\mathbb{B}$ is given by a 4-tuple $(T, \eta, \mu, \mathrm{st})$, where (T, η, μ) is a monad on $\mathbb{B}$ and (T, st) is a strong functor. Additionally, the following two diagrams are required to commute.

$$
\begin{array}{ccc}
I \times X & \xrightarrow{\ \mathrm{id} \times \eta\ } & I \times TX \\
& {\scriptstyle \eta}\searrow & \ \ \downarrow {\scriptstyle \mathrm{st}} \\
& & T(I \times X)
\end{array}
\qquad
\begin{array}{ccc}
I \times T^2 X \xrightarrow{\ \mathrm{st}\ } T(I \times TX) \xrightarrow{\ T(\mathrm{st})\ } T^2(I \times X) \\
{\scriptstyle \mathrm{id} \times \mu}\downarrow \hspace{5.5cm} \downarrow {\scriptstyle \mu} \\
I \times TX \xrightarrow{\hspace{4cm}\mathrm{st}\hspace{4cm}} T((I \times X))
\end{array}
$$

(i) Show that the (finite) lists and powerset operations $X \mapsto \mathrm{list}(X)$ and $X \mapsto P(X)$ are strong monads on **Sets**.

(ii) Show in line with Proposition 2.6.9 that there is a bijective correspondence between strong monads on $\mathbb{B}$ and split monads on the simple fibration $\begin{smallmatrix} s(\mathbb{B}) \\ \downarrow \\ \mathbb{B} \end{smallmatrix}$ (see Exercise 1.7.9).

Chapter 3

Equational Logic

At this point we start the categorical investigation of logic. This chapter will be about a logic of equations between terms in simple type theory (STT). First order logic with more general predicates on terms (than equations) may be found in the next chapter. And the subsequent chapter 5 deals with higher order logic, in which there is a special type **Prop** of propositions. This leads to higher order quantification. All these logics are many-typed logics with types (and terms) as in STT. Or, as we like to put it, these are "simple" logics (fibred) over STT. Later we shall also see logic over polymorphic type theory (PTT) and over dependent type theory (DTT). These have greater expressive power at the level of types.

But for the moment we restrict ourselves to equational logic over simple types. There, one has equations between terms as propositions. Propositions form a new syntactic universe (besides types). They are the entities that one reasons about, and occur in the relation ⊢ of logical entailment. We start this series of chapters on logic (3, 4, 5) with a few remarks on logics in general and with an explanation of the (logical) terminology and notation that we shall be using in the rest of this book. Starting from these generalities we can already construct a fibration from a logic as a term model (or classifying fibration), capturing the essential context structure of the logic. The subsequent sections in this chapter contain an exposition of the traditional approach to the semantics of non-conditional equational logic in terms of categories with finite products, and also an exposition of the fibred approach. The latter makes use of Lawvere's description of equality via left adjoints to contraction functors. This fibred approach presents equality as an "internal" notion, in the logic of a fibration. It is very general and close to syntax. And it fits nicely into a uniform categorical description of logics. This fibred line will be pursued in subsequent chapters.

The way in which a fibred category $\begin{smallmatrix} \mathbb{E} \\ \downarrow \\ \mathbb{B} \end{smallmatrix}$ provides us with means to reason about what happens in the base category $\mathbb{B}$, is described in Section 3.5. In particular, in Definition 3.5.3, validity of equations in a fibration (admitting equality) is introduced. This shows how $\mathbb{E}$ gives us a logic over $\mathbb{B}$. We will show how choosing different fibred categories on the same base category gives different logics (with different notions of equality) to reason about this base category, see Examples 3.5.4 and 3.5.5. In the subsequent and final section 3.6 the functorial semantics from the previous chapter is extended from ordinary categories to fibred categories. It enables us to capture models of logics as certain structure preserving morphisms of fibrations.

3.1 Logics

A logic is a formal system for reasoning. There are various such systems, with variation determined by, for example:

- what to reason about; this determines the form of the atomic propositions;
- which means to use; this determines the logical connectives used to build compound propositions;
- which rules to follow; for example whether to follow the constructive or classical rules for negation.

In this chapter we study many-typed equational logic. It has equations between terms from STT as atomic propositions, and so it may be called *simple* equational logic (in contrast to polymorphic or dependent equational logic, for example). Our (categorical) account of equational logic does not involve any connectives. These can be added later and studied separately, see the next chapter. In order to describe a (not necessarily equational) logic over STT, we start with a (many-typed) signature, containing the atomic types and function symbols, that will generate an underlying simply typed calculus (as in the previous chapter). In predicate logic the signature may contain atomic predicate symbols, but in equational logic one restricts oneself to equations as (atomic) propositions. In general, a signature together with a collection of propositions (serving as axioms) will be called a **specification**. And a specification in which the collection of axioms is closed under derivability will be called a **theory**.

Usually, a statement in a logic is written as

$$\varphi_1, \ldots, \varphi_n \vdash \psi$$

where $\varphi_1, \ldots, \varphi_n$ and ψ are propositions. Such a sequent expresses that ψ follows (as conclusion) from the assumptions $\varphi_1, \ldots, \varphi_n$. These propositions $\varphi_1, \ldots, \varphi_n, \psi$ may contain (free) variables of certain types. The context in which these variables are declared is left implicit in the above formulation. Contexts are very important in a categorical description of logic—since they are indices—and therefore we prefer to use statements of the form

$$\Gamma \mid \varphi_1, \ldots \varphi_n \vdash \psi$$

in which the context Γ containing all the free variables of $\varphi_1, \ldots, \varphi_n$ and ψ, is written explicitly. The sign '$\mid$' is used as a separator and has no logical meaning. Its rôle is to separate the **type context** Γ from the **proposition context** $\varphi_1, \ldots, \varphi_n$, much like '$\mid$' in the standard notation $\{i \in I \mid \varphi(i)\}$ for comprehension separates the set-theoretical from the logical. In [186] the type context Γ is written as a subscript of the turnstile $\vdash_\Gamma$. It leads to sequents of the form $\varphi_1, \ldots \varphi_n \vdash_\Gamma \psi$. But this notation is not very convenient when we deal with rules (like for $\forall$ or $\exists$) that change the type context Γ. So we put the type context Γ at the beginning of the sequent.

As an example, in equational logic one can have a sequent

$$v_1 : \mathsf{N}, v_2 : \mathsf{N} \mid v_1 =_\mathsf{N} 3, v_2 + v_1 =_\mathsf{N} 5 \vdash v_2 =_\mathsf{N} 2.$$

with type context $v_1 : \mathsf{N}, v_2 : \mathsf{N}$, proposition context $v_1 =_\mathsf{N} 3, v_2 + v_1 =_\mathsf{N} 5$ and conclusion $v_2 =_\mathsf{N} 2$. Such a sequent involves ingredients (such as $\mathsf{N}, +, 3, 5, 2$ in this case) which come from an underlying signature as in STT, describing the basic types and function symbols that we use. This signature determines which terms (like $v_2 + v_1$ above) can be formed, and hence also which equations (between terms) can be used. Additionally, we may wish to have certain equations as axioms in equational logic. For example, the monoid equations in reasoning about monoids. An equational specification consists of a signature Σ together with a set $\mathcal{A}$ (for axioms) of equations between Σ-terms. A precise definition will be given in the next section. Later on, in predicate logic, we will use slightly different specifications, consisting of a triple $(\Sigma, \Pi, \mathcal{A})$, where Σ is a signature, Π is an additional set of typed predicate symbols $P : \sigma_1, \ldots, \sigma_n$, and $\mathcal{A}$ is a set of axioms.

Context rules

In Figure 3.1 we list the context rules which will be used in all of the logics that we consider. We write Γ for a type context of the form $x_1 : \sigma_1, \ldots, x_n : \sigma_n$

axiom

$$\frac{}{\Gamma \mid \Theta \vdash \psi} \text{ (if } (\Gamma \mid \Theta \vdash \psi) \in \mathcal{A})$$

identity

$$\frac{\Gamma \vdash \psi \colon \mathsf{Prop}}{\Gamma \mid \psi \vdash \psi}$$

cut

$$\frac{\Gamma \mid \Theta \vdash \varphi \qquad \Gamma \mid \Theta', \varphi \vdash \psi}{\Gamma \mid \Theta, \Theta' \vdash \psi}$$

weakening for propositions

$$\frac{\Gamma \mid \Theta \vdash \psi \qquad \Gamma \vdash \varphi \colon \mathsf{Prop}}{\Gamma \mid \Theta, \varphi \vdash \psi}$$

contraction for propositions

$$\frac{\Gamma \mid \Theta, \varphi, \varphi \vdash \psi}{\Gamma \mid \Theta, \varphi \vdash \psi}$$

exchange for propositions

$$\frac{\Gamma \mid \Theta, \varphi, \chi, \Theta' \vdash \psi}{\Gamma \mid \Theta, \chi, \varphi, \Theta' \vdash \psi}$$

weakening for types

$$\frac{\Gamma \mid \Theta \vdash \psi}{\Gamma, x \colon \sigma \mid \Theta \vdash \psi}$$

contraction for types

$$\frac{\Gamma, x \colon \sigma, y \colon \sigma \mid \Theta \vdash \psi}{\Gamma, x \colon \sigma \mid \Theta[x/y] \vdash \psi[x/y]}$$

exchange for types

$$\frac{\Gamma, x \colon \sigma_i, y \colon \sigma_{i+1}, \Gamma' \mid \Theta \vdash \psi}{\Gamma, y \colon \sigma_{i+1}, x \colon \sigma_i, \Gamma' \mid \Theta \vdash \psi}$$

substitution

$$\frac{\Gamma \vdash M \colon \sigma \qquad \Delta, x \colon \sigma, \Delta' \mid \Theta \vdash \psi}{\Delta, \Gamma, \Delta' \mid \Theta[M/x] \vdash \psi[M/x]}$$

Fig. 3.1. Context rules in logic

in which (term) variables x_i are declared of type σ_i, and Θ for a proposition context consisting of a sequence $\varphi_1, \ldots, \varphi_m$ of propositions. In combined contexts $\Gamma \mid \Theta$ we ensure that all the free variables occurring in (the propositions in) the proposition context Θ are declared in Γ. Further, sometimes we apply substitution $\Theta[M/v]$ to proposition contexts. It means substitution $\varphi[M/v]$ applied to all the propositions φ in Θ.

In the "axiom" rule in Figure 3.1 it is assumed that there is a given set of axioms $\mathcal{A}$. If there is no such set specified, the rule does not apply.

Notice that the "identity" rule starts from the assumption $\Gamma \vdash \psi \colon \mathsf{Prop}$ that ψ is a well-formed proposition in type context Γ. How such statements are obtained depends on the specific logic that we are using. For example, in equational logic one only has equational propositions $\Gamma \vdash \psi \colon \mathsf{Prop}$ with ψ of the form $M =_\sigma M'$, where M, M' are terms of type σ in context Γ.

As in STT, in concatenated contexts Γ, Δ we always assume that the variables in Γ and Δ are distinct. Especially, in writing $\Gamma, x : \sigma$ it is implicitly assumed that the variable x does not occur in Γ.

We sometimes write

$$\blacktriangleright \Gamma \mid \Theta \vdash \varphi$$

to express that the sequent $\Gamma \mid \Theta \vdash \varphi$ is derivable. This means that there is a derivation tree regulated by the above rules (and possibly some extra rules specific to the logic) with $\Gamma \mid \Theta \vdash \varphi$ as conclusion. Notice that in the formalism that we use all assumptions are explicitly present at every stage of the derivation in the type and proposition contexts.

The following rule is in general *not* valid.

strengthening

$$\frac{\Gamma, x : \sigma \mid \Theta \vdash \psi}{\Gamma \mid \Theta \vdash \psi} \; (x \text{ not free in } \Theta, \psi)$$

The problem lies in the fact that (the interpretation of) the type σ may be empty. The (then absurd) assumption $x : \sigma$ that σ is inhabited may lead to conclusions, which can not be obtained without the assumption $x : \sigma$. See Exercise 3.1.3 for more details.

Fibrations of contexts in logic

The above rules suffice to describe the basic categorical structure in a logic over a simple type theory. Let $(\Sigma, \Box)$ be a specification for some system of logic, where Σ is a many-typed signature and $\Box$ is something extra, determined by the specific logic; it may consist of collections of additional atomic symbols and/or axioms. For example, for equational logic, $\Box$ will be a set of equations which serve as axioms. And in predicate logic it will consist of predicate symbols plus axioms. Given Σ and $\Box$ we can start forming sequents. The categorical way to understand these sequents is as follows.

$$\overbrace{x_1 : \sigma_1, \ldots, x_n : \sigma_n}^{\substack{\text{index object in} \\ \text{the base category}}} \mid \underbrace{\varphi_1, \ldots, \varphi_m \vdash \psi}_{\substack{\text{inequality } \leq \text{ in the} \\ \text{fibre over the index}}}$$

A type context $\Gamma = (x_1 : \sigma_1, \ldots, x_n : \sigma_n)$ is thus an index for a logic describing what happens in this context. This is a basic theme.

We can formalise this view. The specific logic that we have gives rise to a fibration of contexts:

$$\mathcal{L}(\Sigma, \square)$$
$$\downarrow$$
$$\mathcal{Cl}(\Sigma)$$

with the classifying category $\mathcal{Cl}(\Sigma)$ of the signature Σ as basis. This fibration has the following properties.

(a) The fibre over a type context $\Gamma \in \mathcal{Cl}(\Sigma)$ contains the logic in context Γ: its objects are sequents of the form $\Gamma \mid \Theta$; a morphism $(\Gamma \mid \Theta) \to (\Gamma \mid \Theta')$ exists if each proposition ψ in Θ' is derivable from Θ in type context Γ, *i.e.* if
▶ $\Gamma \mid \Theta \vdash \psi$ for each ψ in Θ'.

(b) The fibration is a fibred preorder, *i.e.* all fibre categories are preorders. This is typical for 'logical' fibrations (in contrast to 'type theoretic' fibrations), because in logic one does not distinguish between different proofs of the same proposition: there are no explicit proof-objects or proof-terms, which can serve as (proper) morphisms.

(c) The base category $\mathcal{Cl}(\Sigma)$ has finite products; as we have seen in the previous chapter, these are given by concatenation of type contexts.

(d) The fibration has fibred finite products; this structure is obtained from concatenation of proposition contexts.

We proceed to describe the total category $\mathcal{L}(\Sigma, \square)$ in detail.

objects	pairs $\Gamma \mid \Theta$ consisting of a type context Γ and a propositions context Θ, such that all free (term) variables in propositions in Θ are declared in Γ.
morphisms	$(\Gamma \mid \Theta) \to (\Gamma' \mid \Theta')$ are context morphisms $\vec{M} \colon \Gamma \to \Gamma'$ in $\mathcal{Cl}(\Sigma)$ such that for each proposition ψ in Θ' one can derive

$$\Gamma \mid \Theta \vdash \psi[\vec{M}/\vec{v}]$$

where $[\vec{M}/\vec{v}]$ is simultaneous substitution for the variables v_i declared in Γ'.

Identities in $\mathcal{L}(\Sigma, \square)$ are identities in $\mathcal{Cl}(\Sigma)$, by the identity rule. Also composition is inherited from $\mathcal{Cl}(\Sigma)$: for morphisms in $\mathcal{L}(\Sigma, \square)$,

$$(\Gamma \mid \Theta) \xrightarrow{\vec{M}} (\Gamma' \mid \Theta') \xrightarrow{\vec{N}} (\Gamma'' \mid \Theta'')$$

let $\vec{L} = \vec{N} \circ \vec{M}$ be the composite in the base category $\mathcal{Cl}(\Sigma)$—which means $L_i = N_i[\vec{M}/\vec{v}]$. This map $\vec{L}$ is also a morphism $(\Gamma \mid \Theta) \to (\Gamma'' \mid \Theta'')$ in the total category $\mathcal{L}(\Sigma, \square)$. This follows from a combination of the cut and substitution rule: for each proposition ψ in Θ'' one can derive

$$\Gamma' \mid \Theta' \vdash \psi[\vec{N}/\vec{w}]$$

(with $\vec{w}$ declared in Γ'') and thus by substituting $\vec{M}$ one can also derive

$$\Gamma \mid \Theta'[\vec{M}/\vec{v}] \vdash \psi[\vec{L}/\vec{w}].$$

But for each φ in Θ'

$$\Gamma \mid \Theta \vdash \varphi[\vec{M}/\vec{v}]$$

is derivable, which yields, by repeated application of the cut and contraction (for propositions) rules, that

$$\Gamma \mid \Theta \vdash \psi[\vec{L}/\vec{w}].$$

is derivable. This makes $\vec{L}$ a morphism in $\mathcal{L}(\Sigma, \square)$.

The projection functor $\begin{smallmatrix} \mathcal{L}(\Sigma,\square) \\ \downarrow \\ \mathcal{C}l(\Sigma) \end{smallmatrix}$ given as $(\Gamma \mid \Theta) \mapsto \Gamma$ is a split fibration. The fibre over $\Gamma \in \mathcal{C}l(\Sigma)$ indeed contains the logic in Γ, as stated in (a)+(b) above. As to (d), the terminal object in the fibre over type context Γ is $\Gamma \mid \emptyset$ (Γ with empty proposition context) and the Cartesian product of $\Gamma \mid \Theta$ and $\Gamma \mid \Theta'$ is $\Gamma \mid \Theta, \Theta'$ (Γ with concatenated proposition contexts).

3.1.1. Example. Consider type contexts

$$(\Gamma = x_1 : \sigma_1, \ldots, x_n : \sigma_n) \qquad \text{and} \qquad \Delta = (y_1 : \tau_1, \ldots, y_m : \tau_m)$$

together with a context morphism $\vec{M} \colon \Gamma \to \Delta$, so that $\Gamma \vdash M_j : \tau_j$. The (categorical) substitution functor $(\vec{M})^*$ associated with $\vec{M}$ is a functor from the fibre over Δ to the fibre over Γ. It maps a proposition context Θ in type context Δ to a proposition context in context Γ by performing substitution $[\vec{M}/\vec{y}]$ in syntax:

$$(\Delta \mid \Theta) \mapsto (\Gamma \mid \Theta[\vec{M}/\vec{y}]).$$

There are two special cases of this general description of substitution which should be singled out, namely *weakening* and *contraction* (see also Example 1.1.1).

(i) Let us write

$$(\Gamma, x : \sigma) \xrightarrow{\pi} \Gamma$$

for the context morphism $(x_1, \ldots, x_n)$ consisting of the variables in Γ. Then we get an associated substitution functor π^* which performs weakening. It acts as follows.

$$(\Gamma \mid \Theta) \mapsto (\Gamma, x : \sigma \mid \Theta).$$

That is, it adds a dummy variable declaration (or assumption) $x : \sigma$. In syntax this is not an explicit operation, since there is no notation for weakening. Only when one moves to a categorical level, it becomes an explicit operation. It makes things more cumbersome, but it better brings forward the structural aspects. For example, in the next chapter on predicate logic we shall see how one can capture existential $\exists$ and universal $\forall$ quantification as left and right adjoints to these weakening functors π^*.

(Pavlović [256] proposes explicit notation in syntax for weakening: given a proposition $\Gamma \vdash \varphi : \mathsf{Prop}$, he writes $\Gamma, x : \sigma \vdash \varphi(\sharp) : \mathsf{Prop}$ for φ with this dummy variable x added by weakening.)

(ii) Now write

$$(\Gamma, x : \sigma) \xrightarrow{\ \delta\ } (\Gamma, x : \sigma, y : \sigma)$$

for the diagonal context morphism $(x_1, \ldots, x_n, x, x)$. The associated substitution functor δ^* performs contraction:

$$(\Gamma, x : \sigma, y : \sigma \mid \Theta) \mapsto (\Gamma, x : \sigma \mid \Theta[x/y]).$$

It replaces two variables x, y of the same type by a single variable occurring in both places via substitution $[x/y]$ of x for y. This is an operation which can be described explicitly in syntax. Later in this chapter we shall capture equality via left adjoints to such contraction functors δ^*.

Exercises

3.1.1. Prove that a morphism $\vec{M} : (\Gamma \mid \Theta) \to (\Gamma' \mid \Theta')$ in $\mathcal{L}(\Sigma, \square)$ is Cartesian if and only if for each φ in Θ one can derive $\Gamma \mid \Theta'[\vec{M}/\vec{v}] \vdash \varphi$.

3.1.2. Verify in detail that the fibration $\begin{smallmatrix} \mathcal{L}(\Sigma, \square) \\ \downarrow \\ \mathcal{C}\ell(\Sigma) \end{smallmatrix}$ has fibred finite products. Show that the total category $\mathcal{L}(\Sigma, \square)$ also has finite products: $\emptyset \mid \emptyset$ is terminal object and the Cartesian product of $\Gamma \mid \Theta$ and $\Gamma' \mid \Theta'$ is $\Gamma, \Gamma' \mid \Theta, \Theta'$.

3.1.3. Check that the following is an example showing that the strengthening rule is not valid. Consider in **Sets** two functions $f, g : X \rightrightarrows Y$. Since **Sets** is a distributive category, we have $X \times \emptyset \cong \emptyset$, so that $f \circ \pi = g \circ \pi : X \times \emptyset \to Y$. This means that we have validity of

$$x : X, z : 0 \mid \emptyset \vdash f(x) =_Y g(x),$$

with $z : 0$ not occurring on the right of $\mid$. But evidently, me may not conclude

$$x : X \mid \emptyset \vdash f(x) =_Y g(x)$$

since f and g are arbitrary functions.

3.2 Specifications and theories in equational logic

The present section deals with the syntactic aspects of equational logic over simple type theory (STT). We investigate the rules which are specific for reasoning with equations between terms (in STT). The main result in this section is the reformulation (in Lemma 3.2.3) of the standard rules of equational logic in a single 'mate' rule (after Lawvere). It prepares the ground for a categorical description of equality in terms of left adjoints to contraction functors δ^* in Section 3.4.

First we have to make precise what kind of atomic propositions may be used in equational logic. These will be equations of the form $M =_\sigma M'$, for terms M, M' of the same type σ in STT. Formally:

Equational proposition formation

$$\frac{\Gamma \vdash M : \sigma \qquad \Gamma \vdash M' : \sigma}{\Gamma \vdash M =_\sigma M' : \mathsf{Prop}}$$

The type subscript σ in $=_\sigma$ is used to emphasise that we are dealing with equality of terms of the same type σ. But more importantly, to distinguish propositional equality $M =_\sigma M'$ from conversion $M = M'$, as we have seen in the previous chapter, which comes with the type formers $\to, \times, 1, +, 0$ in STT. These should not be confused: conversion $=$ belongs to type theory, whereas propositional equality $=_\sigma$ is part of logic. Sometimes we call conversion **external** equality and propositional equality **internal** equality. The latter because $=_\sigma$ can only be established within formal logic. This is in line with categorical terminology.

Internal equality contains external equality via the following rule.

From external to internal equality

$$\frac{\Gamma \vdash M : \sigma \qquad \Gamma \vdash M' : \sigma \qquad \Gamma \vdash M = M' : \sigma}{\Gamma \vdash M =_\sigma M'}$$

It says that convertible terms are (propositionally) equal in logic. As a consequence, in logic, terms are considered up-to-conversion. One may also postulate a rule in the reverse direction (so that internal and external equality become the same, in what is sometimes called "extensional" logic), but we shall not do so in general.

In equational logic we shall only use atomic propositions $M =_\sigma M'$ and no compound propositions with connectives, like $\wedge, \vee, \supset$. The reason is that we wish to study equality in isolation. The sequents in our logic thus have the following form.

3.2.1. Definition. Let Σ be a signature.

(i) A **Σ-equation** is a sequent of the form

$$\Gamma \mid M_1 =_{\sigma_1} M_1', \ldots, M_n =_{\sigma_n} M_n' \vdash M_{n+1} =_{\sigma_{n+1}} M_{n+1}'$$

where for each i, both M_i and M_i' are Σ-terms of type σ_i in context Γ, so that $M_i =_{\sigma_i} M_i'$ is a well-formed proposition.

Such an equation will be called **non-conditional** or **algebraic** if $n = 0$, that is, if its proposition context is empty. We then write the sequent as

$$\Gamma \mid \emptyset \vdash M =_\sigma M' \qquad \text{or simply as} \qquad \Gamma \vdash M =_\sigma M'.$$

(ii) An **equational specification** is a pair $(\Sigma, \mathcal{H})$ where Σ is a signature and $\mathcal{H}$ (for Horn) is a collection of Σ-equations. An **algebraic specification** is a pair $(\Sigma, \mathcal{A})$ where $\mathcal{A}$ is a collection of algebraic equations.

Notice, by the way, that the notation $\Gamma \vdash M =_\sigma M'$ in this definition was already used in the earlier rule that described internal equality resulting from external equality.

3.2.2. Examples. (i) In Example 2.2.7 (i) one finds two algebraic specifications for groups: Σ_1 with five axioms and Σ_2 with one axiom.

(ii) The classical example in algebra of a conditional specification is of a **torsion free** group, *i.e.* of a group G without elements with finite period, except its unit. This specification has infinitely many conditional axioms, one for each $n \in \mathbb{N}$:

$$x \colon \mathsf{G} \mid \underbrace{x \bullet \cdots \bullet x}_{n \text{ times}} =_\mathsf{G} \mathsf{e} \vdash x =_\mathsf{G} \mathsf{e},$$

where $\bullet$ is the multiplication of the group G, and $\mathsf{e} \colon \mathsf{G}$ its unit.

(iii) Assume a signature in which for a type σ one also has a type $P\sigma$ intended as type of finite subsets of σ. For a cardinality operation $\mathsf{card} \colon P\sigma \to \mathsf{N}$ one may expect a conditional equation

$$x \colon \sigma, y \colon P\sigma \mid \mathsf{elem}\,(x, y) =_\mathsf{B} \mathsf{ff} \vdash \mathsf{card}\,(\mathsf{add}\,(x, y)) =_\mathsf{N} \mathsf{card}\,(y) + 1$$

where $\mathsf{ff} \colon \mathsf{B}$ is the boolean constant 'false' and **elem** and **add** are the obvious set theoretic operations.

Next we describe the typical rules of equational logic—besides the standard context rules from the previous section. They will also be used in any of the later logics with equality. We start with substitution; one puts

$$\left(N =_\tau N'\right)[M/x] \equiv \left(N[M/x] =_\tau N'[M/x]\right)$$

where $\equiv$ means syntactic identification. Categorically, this distribution of substitution over equations will be captured—as always—by a Beck-Chevalley condition, see Definition 3.4.1. The corresponding substitution rule is

substitution

$$\frac{\Gamma \vdash M : \sigma \qquad \Delta, x : \sigma, \Delta' \mid \overrightarrow{N} =_{\vec{\tau}} \overrightarrow{N'} \vdash L =_{\rho} L'}{\Delta, \Gamma, \Delta' \mid \overrightarrow{N[M/x]} =_{\vec{\tau}} \overrightarrow{N'[M/x]} \vdash L[M/x] =_{\rho} L'[M/x]}$$

which we explicitly mention as a special case of the substitution rule in the previous section. The vector notation $\overrightarrow{N} =_{\vec{\tau}} \overrightarrow{N'}$ is a shorthand for a sequence of assumptions $N_1 = _{\tau_1} N_1', \ldots, N_k = _{\tau_k} N_k'$.

The next four rules are the **basic rules of equational logic.**

reflexivity   **symmetry**

$$\frac{\Gamma \vdash M : \sigma}{\Gamma \mid \Theta \vdash M =_{\sigma} M} \qquad\qquad \frac{\Gamma \mid \Theta \vdash M =_{\sigma} M'}{\Gamma \mid \Theta \vdash M' =_{\sigma} M}$$

transitivity

$$\frac{\Gamma \mid \Theta \vdash M =_{\sigma} M' \qquad \Gamma \mid \Theta \vdash M' =_{\sigma} M''}{\Gamma \mid \Theta \vdash M =_{\sigma} M''}$$

replacement

$$\frac{\Gamma \mid \Theta \vdash M =_{\sigma} M' \qquad \Gamma, x : \sigma \vdash N : \tau}{\Gamma \mid \Theta \vdash N[M/x] =_{\tau} N[M'/x]}$$

The next lemma gives a more concise formulation of these rules, and paves the way for Lawvere's categorical account of equality—which is in Section 3.4. It describes equality via left adjoints to contraction functors. Remember that the latter replace two variables $x, y : \sigma$ of the same type by a single one, using substitution $[x/y]$, see Example 3.1.1 (ii).

3.2.3. Lemma. *Consider for terms $\Gamma, x : \sigma, y : \sigma \vdash N, N' : \tau$ the following rule.*

Lawvere equality

$$\frac{\Gamma, x : \sigma \mid \Theta \vdash N[x/y] =_{\tau} N'[x/y]}{\Gamma, x : \sigma, y : \sigma \mid \Theta, x =_{\sigma} y \vdash N =_{\tau} N'} \text{ (=-mate)}$$

Under the assumption of the substitution rule, the above four basic rules of equational logic are equivalent to this equality rule of Lawvere.

The double line indicates that the rule may be applied in both directions. Note that it is implicit in the notation that the variable y does not occur in the proposition context Θ.

Proof. Assume Lawvere's rule. Reflexivity follows by applying it upwards:

$$\frac{x\colon\sigma, y\colon\sigma \mid x =_\sigma y \vdash x =_\sigma y}{x\colon\sigma \mid \emptyset \vdash (x[x/y] =_\sigma y[x/y]) \equiv (x =_\sigma x)}$$

We can immediately use reflexivity to obtain symmetry in:

$$\frac{x\colon\sigma \mid \emptyset \vdash (x =_\sigma x) \equiv (y[x/y] =_\sigma x[x/y])}{x\colon\sigma, y\colon\sigma \mid x =_\sigma y \vdash y =_\sigma x}$$

And transitivity is got by taking:

$$\frac{x\colon\sigma, y\colon\sigma \mid x =_\sigma y \vdash (x =_\sigma y) \equiv (x[y/z] =_\sigma z[y/z])}{x\colon\sigma, y\colon\sigma, z\colon\sigma \mid x =_\sigma y,\, y =_\sigma z \vdash x =_\sigma z}$$

which is an instantiation of Lawvere's rule with $\Gamma = (x\colon\sigma)$ and $\Theta = (x =_\sigma y)$. Finally, in order to derive replacement, assume

$$\Gamma \mid \Theta \vdash M =_\sigma M' \qquad \text{and} \qquad \Gamma, x\colon\sigma \vdash N\colon\tau.$$

Let $N' = N[y/x]$. Then,

$$\frac{\Gamma \mid \Theta \vdash M =_\sigma M' \qquad \dfrac{\dfrac{\dfrac{\dfrac{\Gamma, x\colon\sigma \vdash N\colon\tau}{\Gamma, x\colon\sigma \mid \Theta \vdash N[x/y] =_\tau N'[x/y]}\ (\text{refl})}{\Gamma, x\colon\sigma, y\colon\sigma \mid \Theta, x =_\sigma y \vdash N =_\tau N'}\ (\text{=-mate})}{\Gamma \mid \Theta, M =_\sigma M' \vdash N[M/x] =_\tau N'[M'/y]}\ (\text{subst})\qquad \Gamma \mid \Theta, M =_\sigma M' \vdash N[M/x] =_\tau N[M'/x]}{\Gamma \mid \Theta \vdash N[M/x] =_\tau N[M'/x]}\ (\text{cut})$$

In the reverse direction, assume the four basic rules (plus substitution), and consider two terms $\Gamma, x\colon\sigma, y\colon\sigma \vdash N, N'\colon\tau$. Lawvere's rule downwards is then obtained as follows. First one deduces

$$\frac{\Gamma, x\colon\sigma, y\colon\sigma \mid \Theta, x =_\sigma y \vdash x =_\sigma y \qquad \Gamma, x\colon\sigma, y\colon\sigma, z\colon\sigma \vdash N[z/x]\colon\tau}{\Gamma, x\colon\sigma, y\colon\sigma \mid \Theta, x =_\sigma y \vdash N[z/x][x/z] =_\tau N[z/x][y/z]}\ (\text{repl})$$
$$\equiv (N =_\tau N[y/x]).$$

Similarly one gets

$$\Gamma, x\colon\sigma, y\colon\sigma \mid \Theta, x =_\sigma y \vdash N' =_\tau N'[x/y].$$

But then, using the assumption $N[x/y] =_\tau N'[x/y]$, we are done by symmetry and transitivity.

And Lawvere's rule in upward direction is deduced as follows.

$$
\cfrac{
\cfrac{\Gamma, x{:}\,\sigma \vdash x{:}\,\sigma}{\Gamma, x{:}\,\sigma \mid \Theta \vdash x =_\sigma x}
\quad
\cfrac{\Gamma, x{:}\,\sigma, y{:}\,\sigma \mid \Theta, x =_\sigma y \vdash N =_\tau N'}{\Gamma, x{:}\,\sigma \mid \Theta, x =_\sigma x \vdash N[x/y] =_\tau N'[x/y]} \text{(subst)}
}{\Gamma, x{:}\,\sigma \mid \Theta \vdash N[x/y] =_\tau N'[x/y]} \text{(cut)}
\qquad \square
$$

The above formulation of Lawvere's rule can be strengthened a bit, so that the variable y is allowed to occur in the proposition context Θ. This will be relevant later in connection with the Frobenius property.

3.2.4. Lemma. *The above equality rule of Lawvere is equivalent to the following rule.*

Lawvere equality with Frobenius

$$
\cfrac{\Gamma, x{:}\,\sigma \mid \Theta[x/y] \vdash N[x/y] =_\tau N'[x/y]}{\Gamma, x{:}\,\sigma, y{:}\,\sigma \mid \Theta, x =_\sigma y \vdash N =_\tau N'} \text{(=-mate)}
$$

Proof. This extended equality rule in upward direction follows simply by substituting $[x/y]$ and using reflexivity (which follows from the earlier Lawvere rule). Downwards, it suffices to prove for terms $\Gamma, x{:}\,\sigma, y{:}\,\sigma \vdash L, L'{:}\,\rho$ that the following sequent is derivable.

$$\Gamma, x{:}\,\sigma, y{:}\,\sigma \mid L[x/y] =_\rho L'[x/y], x =_\sigma y \vdash L =_\rho L'.$$

Since then one can apply the cut rule to all equations $L =_\rho L'$ in Θ. One derives this sequent via an immediate application of Lawvere's rule:

$$
\cfrac{\Gamma, x{:}\,\sigma \mid L[x/y] =_\rho L'[x/y] \vdash L[x/y] =_\rho L'[x/y]}{\Gamma, x{:}\,\sigma, y{:}\,\sigma \mid L[x/y] =_\rho L'[x/y], x =_\sigma y \vdash L =_\rho L'}
\qquad \square
$$

3.2.5. Definition. (i) An equational specification $(\Sigma, \mathcal{H})$ will be called a **theory** if its set of equations $\mathcal{H}$ is closed under derivability. This means that if there is a derivation of an equation $E = (\Gamma, \Theta \vdash M =_\sigma M')$ from assumptions $E_1, \ldots, E_n \in \mathcal{H}$, then E must be in $\mathcal{H}$. The rules which can be used in such a derivation are the context rules of the previous section plus the above four basic rules of equational logic (or, equivalently plus Lawvere's equality rule).

Similarly, an algebraic specification $(\Sigma, \mathcal{A})$ is called a theory if the set $\mathcal{A}$ of algebraic (non-conditional) equations is closed under derivability—where the same rules as above may be used, but with empty proposition context Θ.

(ii) Every equational specification $(\Sigma, \mathcal{H})$ gives rise to a theory by closing $\mathcal{H}$ under derivability: one takes $\overline{\mathcal{H}}$ to be the least collection satisfying

- $\mathcal{H} \subseteq \overline{\mathcal{H}}$;
- if equation E is derivable from $E_1, \ldots, E_n \in \overline{\mathcal{H}}$, then $E \in \overline{\mathcal{H}}$.

We write $Th(\Sigma, \mathcal{H}) = (\Sigma, \overline{\mathcal{H}})$ for the **theory associated with** $(\Sigma, \mathcal{H})$. Sometimes we write $E \in Th(\Sigma, \mathcal{H})$ instead of $E \in \overline{\mathcal{H}}$.

Similarly there is a theory $Th(\Sigma, \mathcal{A})$ associated with an algebraic specification $(\Sigma, \mathcal{A})$.

3.2.6. Definition. (i) The category **EqSpec** has

> **objects**   equational specifications $(\Sigma, \mathcal{H})$.
>
> **morphisms** $(\Sigma, \mathcal{H}) \to (\Sigma', \mathcal{H}')$ are morphisms $\phi \colon \Sigma \to \Sigma'$ of signatures such that
> $$E \in \mathcal{H} \;\Rightarrow\; \phi E \in Th(\Sigma', \mathcal{H}'),$$
> where ϕE is obtained from E by applying ϕ to all types and terms in E.

Such a morphism ϕ in **EqSpec** will be called a **morphism of equational specifications**.

(ii) In the same vain there is a subcategory **AlgSpec** $\hookrightarrow$ **EqSpec**, objects of which are algebraic specifications; its morphisms are morphisms of signatures which map non-conditional equations to derivable non-conditional equations (using only the rules with empty proposition context).

Exercises

3.2.1.   Show that the following rule is derivable (or: *admissible*) in equational logic.
$$\frac{\Gamma \mid \Theta \vdash M =_\sigma M' \qquad \Gamma, x \colon \sigma \mid \Theta \vdash N =_\tau N'}{\Gamma \mid \Theta \vdash N[M/x] =_\tau N'[M'/x]}$$

3.2.2.   Consider the first equational signature for groups in Examples 2.2.7 (i). Give a formal derivation of the following basic result about groups.
$$x \colon \mathsf{G}, y \colon \mathsf{G} \vdash \mathsf{i}(\mathsf{m}(x, y)) =_{\mathsf{G}} \mathsf{m}(\mathsf{i}(y), \mathsf{i}(x)).$$

3.2.3.   Check that the projections

$$\begin{array}{ccc} \textbf{EqSpec} & & \textbf{AlgSpec} \\ \downarrow & \text{and} & \downarrow \\ \textbf{Sign} & & \textbf{Sign} \end{array}$$

are split fibrations.
[Recall the discussion after Lemma 1.6.6 about the organisational power of fibrations.]

3.3 Algebraic specifications

In Section 2.2 we have described the semantics of a many-typed signature Σ in terms of finite product preserving functors $\mathcal{M}: \mathcal{C}\ell(\Sigma) \to \mathbb{B}$, where $\mathcal{C}\ell(\Sigma)$ is the classifying category of Σ. In this section we investigate how to model algebraic equations in a similar fashion, using ordinary categories. In the subsequent three sections of this chapter we use fibred categories to model arbitrary, conditional equations in a systematic manner.

Suppose we have a model $\mathcal{M}: \mathcal{C}\ell(\Sigma) \to \mathbb{B}$ of Σ in a category $\mathbb{B}$ and two terms $\Gamma \vdash N, N': \sigma$ in the term calculus of Σ. An algebraic Σ-equation

$$\Gamma \vdash N =_\sigma N'$$

is said to be **valid** (or to **hold**) in $\mathcal{M}$ if the two resulting maps

$$\mathcal{M}(\Gamma) \underset{\mathcal{M}(N')}{\overset{\mathcal{M}(N)}{\rightrightarrows}} \mathcal{M}(\sigma)$$

are equal in $\mathbb{B}$. Thus, an equation holds under an interpretation, if the two terms are interpreted as equal maps.

For a set $\mathcal{A}$ of algebraic equations, we write $\mathcal{M} \models \mathcal{A}$ if all equations in $\mathcal{A}$ are valid in $\mathcal{M}$.

3.3.1. Example. Consider the first specification of groups as in Example 2.2.7 with one type G and three function symbols $m: G, G \longrightarrow G, e: () \longrightarrow G, i: G \longrightarrow G$ and the familiar equations:

$$v_1: G \vdash m(e, v_1) =_G v_1 \qquad\qquad v_1: G \vdash m(i(v_1), v_1) =_G e$$
$$v_1: G \vdash m(v_1, e) =_G v_1 \qquad\qquad v_1: G \vdash m(v_1, i(v_1)) =_G e$$
$$v_1: G, v_2: G, v_3: G \vdash m(v_1, m(v_2, v_3)) =_G m(m(v_1, v_2), v_3).$$

A model $\mathcal{M}$ of this specification in a category $\mathbb{B}$ with finite products, then consists of an object $G = \mathcal{M}(G) \in \mathbb{B}$ together with maps,

$$G \times G \xrightarrow{\;m = \mathcal{M}(m)\;} G \qquad 1 \xrightarrow{\;e = \mathcal{M}(e)\;} G \qquad G \xrightarrow{\;i = \mathcal{M}(i)\;} G$$

such that the above equations hold. This means explicitly that, for example,

$$m \circ \langle e \circ !, \mathrm{id} \rangle = m \circ \langle \mathrm{id}, e \circ ! \rangle = \mathrm{id}_G.$$

In a similar way one can describe the other three equations above as equations in $\mathbb{B}$. One gets precisely the diagrams of an **internal group** in $\mathbb{B}$:

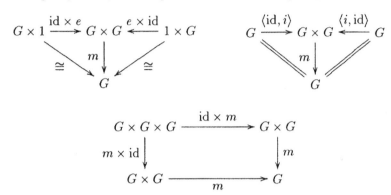

We briefly mention validity of equations with conditions. This can be expressed in case the receiving category $\mathbb{B}$ additionally has equalisers. We write $\mathrm{Eq}(u, v)$ for the (monic) equaliser map of u, v in

$$I' \overset{\mathrm{Eq}(u,v)}{\rightarrowtail} I \underset{v}{\overset{u}{\rightrightarrows}} J$$

Whenever convenient, we also use $\mathrm{Eq}(u, v)$ for the corresponding subobject. Recall that for a category $\mathbb{B}$ with finite limits, the posets $\mathrm{Sub}(I)$ of subobjects of an object $I \in \mathbb{B}$ have finite products (*i.e.* intersections). These will be denoted by $\wedge$ and $\top$.

Let E be a conditional Σ-equation,

$$\Gamma \mid M_1 =_{\sigma_1} M_1', \ldots, M_n =_{\sigma_n} M_n' \vdash N =_{\tau} N'$$

We say that E **holds in** a model $\mathcal{M} \colon \mathcal{Cl}(\Sigma) \to \mathbb{B}$, or that $\mathcal{M}$ **validates** E, in case the intersection of the equalisers of the assumptions is contained in the equaliser of the conclusion:

$$\mathrm{Eq}(\mathcal{M}(M_1), \mathcal{M}(M_1')) \wedge \cdots \wedge \mathrm{Eq}(\mathcal{M}(M_n), \mathcal{M}(M_n')) \leq \mathrm{Eq}(\mathcal{M}(N), \mathcal{M}(N'))$$

where $\leq$ is the order of the poset $\mathrm{Sub}(\mathcal{M}(\Gamma))$.

Since $\mathcal{M}$ preserves finite products, the left hand side is isomorphic to a single equaliser, namely to the equaliser of the two context morphisms

$$\mathcal{M}(\Gamma) \underset{\mathcal{M}(M_1', \ldots, M_n')}{\overset{\mathcal{M}(M_1, \ldots, M_n)}{\rightrightarrows}} \mathcal{M}(\sigma_1 \times \cdots \times \sigma_n)$$

Notice that this definition restricts to the earlier one for non-conditional equations—since for $n = 0$ the empty meet is $\top$ (and $\top \leq \mathrm{Eq}(u, v) \Leftrightarrow u = v$). We call a rule

$$\frac{S_1}{S_2}$$

sound if validity of the sequent S_1 implies validity of the sequent S_2. At this stage we only know what it means for an equational sequent $\Gamma \mid \vec{M} =_{\vec{\sigma}} \vec{M}' \vdash N =_\tau N'$ to be valid, but in the next chapter we shall see validity for more general sequents.

3.3.2. Lemma (Soundness). *Let $(\Sigma, \mathcal{A})$ be an algebraic specification and let $\mathcal{M} : \mathcal{C}\ell(\Sigma) \to \mathbb{B}$ be a model of $\mathcal{A}$. Then every (algebraic) equation derivable from $\mathcal{A}$ holds in $\mathcal{M}$. Thus $\mathcal{M}$ is a model of the theory of $(\Sigma, \mathcal{A})$.*

Proof. One shows that all derivation rules are sound. Reflexivity, symmetry and transitivity are obvious. For replacement assume validity of $\Gamma \vdash M =_\sigma M'$; then for each term $\Gamma, x{:}\sigma \vdash N{:}\tau$ we get validity of $\Gamma \vdash N[M/x] = N[M'/x]{:}\tau$ from

$$\begin{aligned} \mathcal{M}(N[M/x]) &= \mathcal{M}(N) \circ \langle \mathrm{id}, \mathcal{M}(M) \rangle \quad \text{by Exercise 2.2.2} \\ &= \mathcal{M}(N) \circ \langle id, \mathcal{M}(M') \rangle \\ &= \mathcal{M}(N[M'/x]). \end{aligned}$$

In a similar way one obtains soundness of the substitution rule (using Exercise 2.2.2 again). $\qquad\Box$

We next describe classifying categories for algebraic specifications. They behave like classifying categories for signatures—and are constructed as suitable quotients of these.

3.3.3. Definition. Let $(\Sigma, \mathcal{A})$ be an algebraic specification. We say that Σ-terms $\Gamma \vdash N, N'{:}\sigma$ are **equivalent modulo** $\mathcal{A}$ if the equation $\Gamma \vdash N =_\sigma N'$ is derivable using the equations from $\mathcal{A}$ as axioms. We define a classifying category $\mathcal{C}\ell(\Sigma, \mathcal{A})$ with

> **objects**   contexts Γ.
>
> **morphisms** $\Gamma \to \Delta$ are sequences $(|M_1|, \ldots, |M_n|)$ of equivalence classes (modulo $\mathcal{A}$) of terms $\vec{M}{:}\Gamma \to \Delta$ in the classifying category $\mathcal{C}\ell(\Sigma)$ of Σ.

(Notice the following subtlety of notation: we use $|M|$ for the equivalence class of M modulo propositional (or internal) equality, where we used $[M]$ for the equivalence class modulo conversion (or external equality) in the previous chapter.)

Thus the classifying category $\mathcal{Cl}(\Sigma, \mathcal{A})$ of an algebraic specification $(\Sigma, \mathcal{A})$ is obtained by making certain identifications (induced by the axioms $\mathcal{A}$) in the classifying category $\mathcal{Cl}(\Sigma)$ of the signature Σ. As a result, there is a canonical quotient functor $\mathcal{Cl}(\Sigma) \twoheadrightarrow \mathcal{Cl}(\Sigma, \mathcal{A})$.

With this definition of classifying category of an algebraic specification $(\Sigma, \mathcal{A})$, we can understand a model of $(\Sigma, \mathcal{A})$ in a category $\mathbb{B}$ functorially, namely as a finite product preserving functor $\mathcal{Cl}(\Sigma, \mathcal{A}) \to \mathbb{B}$. This is the content of the following result.

3.3.4. Theorem. *A classifying category $\mathcal{Cl}(\Sigma, \mathcal{A})$ has finite products. Moreover, there is a bijective correspondence between*

$$\frac{\mathcal{Cl}(\Sigma, \mathcal{A}) \xrightarrow{\ \mathcal{M}\ } \mathbb{B} \quad in\ \mathbf{FPCat}}{\mathcal{Cl}(\Sigma) \xrightarrow[\mathcal{N}]{} \mathbb{B} \quad in\ \mathbf{FPCat}\ with\ \mathcal{N} \models \mathcal{A}.}$$

Proof. The finite product structure in $\mathcal{Cl}(\Sigma, \mathcal{A})$ is given by concatenation of contexts as in $\mathcal{Cl}(\Sigma)$. For a functor $\mathcal{M}: \mathcal{Cl}(\Sigma, \mathcal{A}) \to \mathbb{B}$ in $\mathbf{FPCat}$ one obtains a functor $\overline{\mathcal{M}}$ as composite $\mathcal{Cl}(\Sigma) \twoheadrightarrow \mathcal{Cl}(\Sigma, \mathcal{A}) \to \mathbb{B}$ satisfying $\overline{\mathcal{M}} \models \mathcal{A}$ because for every equation $\Gamma \vdash N =_\sigma N'$ in $\mathcal{A}$, the terms N and N' are equivalent modulo $\mathcal{A}$, and thus give rise to the same morphism in $\mathcal{Cl}(\Sigma, \mathcal{A})$. This is because the functor $\mathcal{Cl}(\Sigma) \twoheadrightarrow \mathcal{Cl}(\Sigma, \mathcal{A})$ maps context morphisms $\vec{M}$ to their equivalence classes $\overrightarrow{[M]}$.

In the reverse direction, for a model $\mathcal{N}: \mathcal{Cl}(\Sigma) \to \mathbb{B}$ of Σ with $\mathcal{N} \models \mathcal{A}$ one has by soundness that if N, N' are equivalent modulo $\mathcal{A}$, then $\mathcal{N}(N) = \mathcal{N}(N')$. Thus $\mathcal{N}$ restricts to a well-defined functor $\mathcal{Cl}(\Sigma, \mathcal{A}) \to \mathbb{B}$. □

By this result, we can take a model of an algebraic specification $(\Sigma, \mathcal{A})$ in a category $\mathbb{B}$ (with finite products) to be a finite product preserving functor $\mathcal{Cl}(\Sigma, \mathcal{A}) \to \mathbb{B}$.

3.3.5. Corollary (Completeness). *Let $(\Sigma, \mathcal{A})$ be an algebraic specification. An (algebraic) equation is derivable from $\mathcal{A}$ if and only if it holds in all models of $(\Sigma, \mathcal{A})$.*

Proof. The (only-if) follows from the soundness Lemma 3.3.2. For (if), there is the 'generic' model id: $\mathcal{Cl}(\Sigma, \mathcal{A}) \to \mathcal{Cl}(\Sigma, \mathcal{A})$ of $(\Sigma, \mathcal{A})$ in its own classifying category. If an equation holds in all models, then it certainly holds in this particular model. Then, by the previous result, it holds in $\mathcal{Cl}(\Sigma) \twoheadrightarrow \mathcal{Cl}(\Sigma, \mathcal{A})$. But this latter model validates precisely the equations which are derivable from $\mathcal{A}$. □

In Lemma 2.2.4 we saw how every category with finite products induces a many-typed signature. Below we show that it actually induces an algebraic specification: the equations that one gets are precisely those that hold in the category.

3.3.6. Definition. Let $\mathbb{B}$ be a category with finite products and let $\mathrm{Sign}(\mathbb{B})$ be its associated signature (as in Lemma 2.2.4). Then one can form terms $\Gamma \vdash M\colon X$ and equations $\Gamma \vdash M =_X M'$ using this signature. Recall from Theorem 2.2.5 that there is a model $\varepsilon\colon \mathcal{Cl}(\mathrm{Sign}(\mathbb{B})) \to \mathbb{B}$ of the signature of $\mathbb{B}$ in itself.

We write $\mathcal{A}(\mathbb{B})$ for the set of non-conditional $\mathrm{Sign}(\mathbb{B})$-equations which hold in ε (as described in the beginning of this section). The pair $(\mathrm{Sign}(\mathbb{B}), \mathcal{A}(\mathbb{B}))$ is the algebraic specification associated with $\mathbb{B}$. By the previous theorem we get a model $\mathcal{Cl}(\mathrm{Sign}(\mathbb{B}), \mathcal{A}(\mathbb{B})) \to \mathbb{B}$, which we also denote by ε.

3.3.7. Example. The underlying signature $\mathrm{Sign}(\mathbb{B})$ of a category $\mathbb{B}$ with finite products has function symbols

$$X_1, X_2 \xrightarrow{\ \mathsf{pair}\ } X_1 \times X_2 \quad X_1 \times X_2 \xrightarrow{\ \mathsf{proj}\ } X_1 \quad X_1 \times X_2 \xrightarrow{\ \mathsf{proj}'\ } X_2$$

which arise from the following maps in $\mathbb{B}$, see Definition 2.2.5 (i).

$$X_1, X_2 \xrightarrow[=\ \mathrm{id}]{\varepsilon(\mathsf{pair})} X_1 \times X_2 \quad X_1 \times X_2 \xrightarrow[=\ \pi]{\varepsilon(\mathsf{proj})} X_1 \quad X_1 \times X_2 \xrightarrow[=\ \pi']{\varepsilon(\mathsf{proj}')} X_2$$

These function symbols come equipped with equations in $\mathcal{A}(\mathbb{B})$,

$$x\colon X_1, y\colon X_2 \ \vdash\ \mathsf{proj}\,(\mathsf{pair}\,(x,y)) =_{X_1} x$$
$$x\colon X_1, y\colon X_2 \ \vdash\ \mathsf{proj}'\,(\mathsf{pair}\,(x,y)) =_{X_2} y$$
$$z\colon X_1 \times X_2 \ \vdash\ \mathsf{pair}\,(\mathsf{proj}\,(z), \mathsf{proj}'\,(z)) =_{X_1 \times X_2} z$$

Similarly, there is an 'empty tuple' function symbol in $\mathrm{Sign}(\mathbb{B})$

$$() \xrightarrow{\ \langle\rangle\ } 1 \qquad \text{with equation} \qquad \Gamma \vdash M =_1 \langle\rangle.$$

Combining these we obtain an isomorphism of context objects in the classifying category $\mathcal{Cl}(\mathrm{Sign}(\mathbb{B}), \mathcal{A}(\mathbb{B}))$, namely

$$(x_1\colon X_1, \ldots, x_n\colon X_n) \cong (z\colon X_1 \times \cdots \times X_n).$$

The latter isomorphism will be used in the proof of the next result. It states that every category with finite products can be understood as a classifying category, namely of its own algebraic specification. Hence one can identify (following Lawvere) and algebraic theory with a category with finite products.

3.3.8. Theorem. *A category $\mathbb{B}$ with finite products is equivalent to the classifying category $\mathcal{Cl}(\mathrm{Sign}(\mathbb{B}), \mathcal{A}(\mathbb{B}))$ of its own theory of algebraic equations.*

Proof. One can define a functor $\theta \colon \mathbb{B} \to \mathcal{C}\ell(\mathrm{Sign}(\mathbb{B}), \mathcal{A}(\mathbb{B}))$ by mapping an object to the associated singleton context: $X \mapsto (x \colon X)$. Then

$$(\varepsilon \circ \theta)(X) \; = \; \varepsilon(x \colon X)$$
$$= \; X.$$
$$(\theta \circ \varepsilon)(x_1 \colon X_1, \ldots, x_n \colon X_n) \; = \; \theta(X_1 \times \cdots \times X_n)$$
$$= \; (z \colon X_1 \times \cdots \times X_n)$$
$$\cong \; (x_1 \colon X_1, \ldots, x_n \colon X_n). \qquad \square$$

This is a useful result; it shows that instead of the diagrammatic categorical language one can use a type theoretic "internal" language to establish certain results in a category $\mathbb{B}$ with finite products. Explicitly, if we wish to prove that two arrows in $\mathbb{B}$ that we can describe as terms are equal, then it suffices to prove the equality between these terms in the equational logic with specification $(\mathrm{Sign}(\mathbb{B}), \mathcal{A}(\mathbb{B}))$ associated with $\mathbb{B}$. The weakness of this result however, lies in the fact that the terms that occur in our equational logic are of very simple form. For example, if we have a group object in $\mathbb{B}$—as described in Example 3.3.1—then we can use the language of types and terms and the associated equational logic to prove things about such an object (living in an arbitrary universe $\mathbb{B}$). This is what is usually done in mathematics (form a logician's point of view): one uses a suitable internal language to reason directly in a particular structure—but usually with a language which is more expressive than the one we consider so far.

Also, one can understand every finite product preserving functor $F \colon \mathbb{B} \to \mathbb{C}$ as a functorial model of the specification $(\mathrm{Sign}(\mathbb{B}), \mathcal{A}(\mathbb{B}))$ of $\mathbb{B}$ in the category $\mathbb{C}$. Thus $F \colon \mathbb{B} \to \mathbb{C}$ is a model of (the theory of) $\mathbb{B}$ in $\mathbb{C}$.

Similar correspondences between certain kinds of categories and certain kinds of theories have been established. Most famous is the correspondence between categories with finite limits and "essentially algebraic" theories (see [83]). In these essentially algebraic theories one has (hierarchies of) partial operations, with the domain of an operation described by the extension of a finite conjunction of equations involving operations which are lower in the hierarchy.

Exercises

3.3.1.   Check in detail that the equations in Example 3.3.1 lead to the diagrams describing an internal group.

3.3.2.   The following is based on Exercise 1.2.3.

    (i)   Show that the category **Sets.** of pointed sets (or of sets and partial functions) has finite limits.

(ii) Let Σ be a many-sorted signature and $\mathcal{M}: \mathcal{Cl}(\Sigma) \to \mathbf{Sets}_\bullet$ be a finite product preserving functor (*i.e.* a partial Σ-algebra). Find out what it means for a conditional Σ-equation to hold in $\mathcal{M}$; pay special attention to undefinedness.

3.3.3. Consider a category $\mathbb{B}$ with finite products. Show that the following $\mathrm{Sign}(\mathbb{B})$-equations hold.

(i) For an object $X \in \mathbb{B}$

$$x: X \vdash \mathrm{id}_X(x) =_X x.$$

(ii) For composable maps $X \xrightarrow{f} Y \xrightarrow{g} Z$ in $\mathbb{B}$

$$x: X \vdash (g \circ f)(x) =_Z g(y)[f(x)/y].$$

3.3.4. Let $(\Sigma, \mathcal{A})$ be an equational signature. For a category $\mathbb{B}$ with finite products, let $\mathrm{Mod}\big((\Sigma, \mathcal{A}), \mathbb{B}\big)$ be the category of models of $(\Sigma, \mathcal{A})$ in $\mathbb{B}$ consisting of finite product preserving functors $\mathcal{Cl}(\Sigma, \mathcal{A}) \to \mathbb{B}$ and natural transformations between them.

(i) Show that each functor $K: \mathbb{B} \to \mathbb{A}$ in **FPCat** induces a functor

$$\mathrm{Mod}\big((\Sigma, \mathcal{A}), \mathbb{B}\big) \longrightarrow \mathrm{Mod}\big((\Sigma, \mathcal{A}), \mathbb{A}\big)$$

by composition with K.

(ii) Show also that each morphism $\phi: (\Sigma', \mathcal{A}') \to (\Sigma, \mathcal{A})$ of algebraic specifications induces a functor

$$\mathrm{Mod}\big((\Sigma, \mathcal{A}), \mathbb{B}\big) \longrightarrow \mathrm{Mod}\big((\Sigma', \mathcal{A}'), \mathbb{B}\big).$$

[Thus morphisms of receiving categories and of algebraic specifications act in opposite directions on models. This gives rise to a "fibred span", see Definition 9.1.5.]

3.3.5. Let $(\Sigma, \mathcal{A})$ and $(\Sigma', \mathcal{A}')$ be algebraic specifications and consider a functor $\mathcal{Cl}(\Sigma, \mathcal{A}) \to \mathcal{Cl}(\Sigma', \mathcal{A}')$ in **FPCat**. Explain how (the categorical notion of) faithfulness of this functor corresponds to (the logical notion of) conservativity: if an equation holds after translation, then it must already hold before the translation.

3.3.6. For an algebraic specification $(\Sigma, \mathcal{A})$, let $\mathcal{Cl}1_\times(\Sigma, \mathcal{A})$ be the (Cartesian closed) category formed as follows. Its types are obtained by closing the atomic types in Σ under $1, \times, \to$. And its morphisms $|M|: \sigma \to \tau$ are equivalence classes $|M|$ of terms $x: \sigma \vdash M: \tau$, where two terms are equivalent if one can prove from the axioms in $\mathcal{A}$ that they are (propositionally) equal. (In this case the conversions associated with $1, \times, \to$ are included in the internal equality, via the rule (from external to internal equality), described in the beginning of the previous section.)

(i) Check that $\mathcal{Cl}1_\times(\Sigma, \mathcal{A})$ is a CCC, and that for an arbitrary CCC $\mathbb{C}$

there is a bijective correspondence

$$\frac{\mathcal{C}l_\times(\Sigma, \mathcal{A}) \longrightarrow \mathbb{C} \quad \text{in } \mathbf{CCC}}{(\Sigma, \mathcal{A}) \longrightarrow (\mathrm{Sign}(\mathbb{C}), \mathcal{A}(\mathbb{C})) \quad \text{in } \mathbf{AlgSpec}}$$

(ii) Let $\mathbb{B}$ be a category with finite products, and $\mathbb{C}$ be a Cartesian closed category. Establish a correspondence

$$\frac{\mathcal{C}l_\times(\mathrm{Sign}(\mathbb{B}), \mathcal{A}(\mathbb{B})) \longrightarrow \mathbb{C} \quad \text{in } \mathbf{CCC}}{\mathbb{B} \longrightarrow \mathbb{C} \quad \text{in } \mathbf{FPCat}}$$

[A standard gluing argument shows that the resulting functor $\mathbb{B} \to \mathcal{C}l_\times(\mathrm{Sign}(\mathbb{B}), \mathcal{A}(\mathbb{B}))$ is full and faithful, see *e.g.* [183, *Annexe C*] or [61, 4.10]. This means that adding exponents to an algebraic theory does not introduce new terms between old types, or new equations between old terms.]

3.4 Fibred equality

We start with a categorical description of equality in terms of adjunctions; to be more precise, in terms of left adjoints to **contraction functors** δ^*. It was first put forward by Lawvere in [193]. This approach captures the mate rule for equality in Lemma 3.2.3 categorically. The present section contains the technical prerequisites, and the next section shows how this fibred equality is used for the semantics of conditional equations. The goal is the fundamental Definition 3.5.3 of validity of an equation in a fibration.

In a (base) category with Cartesian products $\times$ we shall write for objects I, J

$$I \times J \xrightarrow{\quad \delta = \delta(I, J) = \langle \mathrm{id}, \pi' \rangle \quad} (I \times J) \times J$$

for the 'parametrised' diagonal which duplicates J, with parameter I. It is used to interpret contraction for types, see for example in the proof of Theorem 2.2.5 (iii). Notice that such a diagonal is a split mono: it is a section of the two projections $(I \times J) \times J \rightrightarrows I \times J$.

3.4.1. Definition. Let $\begin{smallmatrix}\mathbb{E}\\\downarrow p\\\mathbb{B}\end{smallmatrix}$ be a fibration on a base category $\mathbb{B}$ with Cartesian products.

(i) This p is said to have (**simple**) **equality** if both

• for every pair $I, J \in \mathbb{B}$, each contraction functor $\delta(I, J)^*$ has a left adjoint

$$\mathbb{E}_{I \times J} \xrightarrow{\quad \mathrm{Eq}_{I,J} = \coprod_{\delta(I,J)} \quad} \mathbb{E}_{(I \times J) \times J}.$$

- the Beck-Chevalley condition holds: for each map $u \colon K \to I$ in $\mathbb{B}$ (between the parameter objects) the canonical natural transformation

$$\mathrm{Eq}_{K,J}(u \times \mathrm{id})^* \Longrightarrow ((u \times \mathrm{id}) \times \mathrm{id})^* \mathrm{Eq}_{I,J}$$

is an isomorphism.

(ii) If p is a fibration with fibred finite products $\times$, then we say that p has **equality with the Frobenius property** (or briefly, **equality satisfying Frobenius**) if it has equality as described above in such a way that for all objects $X \in \mathbb{E}_{(I \times J) \times J}$ and $Y \in \mathbb{E}_{I \times J}$, the canonical map

$$\mathrm{Eq}_{I,J}(\delta^*(X) \times Y) \longrightarrow X \times \mathrm{Eq}_{I,J}(Y)$$

is an isomorphism.

The canonical Beck-Chevalley map is obtained in the standard way by transposing the composite

$$(u \times \mathrm{id})^* \xrightarrow{\;\;(u \times \mathrm{id}^*(\eta)\;\;} (u \times \mathrm{id})^* \delta_{I,J} \mathrm{Eq}_{I,J} \cong \delta_{K,J}^* ((u \times \mathrm{id}) \times \mathrm{id})^* \mathrm{Eq}_{I,J}$$

With this notion of equality we will be able to define validity of an equation between morphisms (terms) in a base category, see Definition 3.5.3 in the next section. In this section we concentrate on the technicalities of such fibred equality.

Note that the above definition speaks of *simple* equality. This is to distinguish it from other forms of equality, to be described later in Section 9.3. The name 'simple' refers to an involvement of simple fibrations, see Exercise 3.4.1 below. In this and the next few chapters we shall only use simple equality and therefore we can safely omit the word 'simple' for the time being.

Above, we only consider *left* adjoints to the contraction functors δ^*. In presence of fibred exponents, these left adjoints induce right adjoints to δ^*, see Exercise 3.4.2.

3.4.2. Notation. Let $\begin{smallmatrix} \mathbb{E} \\ \downarrow p \\ \mathbb{B} \end{smallmatrix}$ be a fibration with equality as described before. Assume p has a terminal object functor $1 \colon \mathbb{B} \to \mathbb{E}$, see Lemma 1.8.8. For parallel maps $u, v \colon I \rightrightarrows J$ in $\mathbb{B}$ we write

$$\mathrm{Eq}(u, v) \stackrel{\mathrm{def}}{=} \langle \langle \mathrm{id}, u \rangle, v \rangle^* \bigl(\mathrm{Eq}_{I,J}(1) \bigr) \in \mathbb{E}_I$$

in a situation:

$$\text{Eq}(u, v) \xrightarrow{\hspace{3cm}} \text{Eq}_{I,J}(1) \hspace{3cm} 1$$

$$I \xrightarrow{\;\langle\langle\text{id}, u\rangle, v\rangle\;} (I \times J) \times J \xleftarrow{\;\delta\;} I \times J$$

where $1 = 1(I \times J)$ is the terminal object in the fibre over $I \times J$. This yields an equality predicate $\text{Eq}(u, v)$ in the fibre over the domain I of the maps u, v. One thinks of the predicate $\text{Eq}(u, v)$ at $i \in I$ as expressing the truth of $u(i) =_J v(i)$ in what may be called the "internal logic of the fibration", *i.e.* in the logic which is based on what holds in this fibration, see the next chapter. We thus say that $u, v \colon I \rightrightarrows J$ are **internally equal** if there is a "proof" $1 \to \text{Eq}(u, v)$ over I. This need not be the same as **external equality** of $u, v \colon I \rightrightarrows J$, which simply means equality $u = v$ of u, v as morphisms of the base category. Below, in Lemma 3.4.5 we shall formally prove that internal equality is reflexive, so that external equality implies internal equality. The converse need not be the case—see the next section for examples. In case internal equality in a fibration does imply external equality we will say that the fibration has **very strong** equality. The logic then often called **extensional**. This terminology using strength is borrowed from type theory where "strong" and "very strong" forms of equality exist, see Section 11.4 later on.

Substitution in such equality predicates $\text{Eq}(u, v)$ is done by composition:

$$
\begin{aligned}
& w^*\big(\text{Eq}(u, v)\big) \\
&\cong\; (\langle\langle\text{id}, u\rangle, v\rangle \circ w)^*\,\text{Eq}(1) \\
&\cong\; ((w \times \text{id}) \times \text{id}) \circ \langle\langle\text{id}, u \circ w\rangle, v \circ w\rangle)^*\,\text{Eq}(1) \\
&\cong\; \langle\langle\text{id}, u \circ w\rangle, v \circ w\rangle^*\,((w \times \text{id}) \times \text{id})^*\,\text{Eq}(1) \\
&\cong\; \langle\langle\text{id}, u \circ w\rangle, v \circ w\rangle^*\,\text{Eq}(w \times \text{id})^*(1)) \qquad \text{by Beck-Chevalley} \\
&\cong\; \langle\langle\text{id}, u \circ w\rangle, v \circ w\rangle^*\,\text{Eq}(1) \\
&\cong\; \text{Eq}(u \circ w, v \circ w).
\end{aligned}
$$

As a special case of Frobenius one obtains for the projection morphism $\pi \colon (I \times J) \times J \to I \times J$ that

$$\text{Eq}_{I,J}(X \times Y) \cong \text{Eq}_{I,J}(\delta^* \pi^*(X) \times Y) \cong \pi^*(X) \times \text{Eq}_{I,J}(Y).$$

And so in particular for $Y = 1$ we get

$$\text{Eq}_{I,J}(X) \cong \pi^*(X) \times \text{Eq}_{I,J}(1).$$

This latter isomorphism is often useful. Informally it says that

$$\text{Eq}(X)_{(i,j,j')} = X_{(i,j)} \wedge (j =_J j').$$

We continue with a basic observation.

3.4.3. Lemma. *A fibration with coproducts $\coprod_u$ (satisfying Frobenius) has equality Eq (satisfying Frobenius).*

Proof. Suppose $\begin{smallmatrix}\mathbb{E}\\\downarrow p\\\mathbb{B}\end{smallmatrix}$ has coproducts. Since every reindexing functor u^* has a left adjoint $\coprod_u$, we especially have left adjoints ($\coprod_\delta \dashv \delta^*$) to contraction functors δ^*. Beck-Chevalley holds, since for $u \colon K \to I$ the following is a pullback diagram in $\mathbb{B}$.

$$
\begin{array}{ccc}
K \times J & \xrightarrow{\;\; u \times \mathrm{id} \;\;} & I \times J \\
{\scriptstyle \delta(K,J)=\delta}\downarrow \lrcorner & & \downarrow {\scriptstyle \delta = \delta(I,J)} \\
(K \times J) \times J & \xrightarrow[\;\; (u \times \mathrm{id}) \times \mathrm{id} \;\;]{} & (I \times J) \times J
\end{array}
$$

In case p has fibred finite products and the coproducts of p satisfy the Frobenius property, then Frobenius obviously holds for equality as well. $\square$

3.4.4. Examples. (i) By the previous result (plus Proposition 1.9.8 and Lemma 1.9.7), each codomain fibration $\begin{smallmatrix}\mathbb{B}^{\rightarrow}\\\downarrow\\\mathbb{B}\end{smallmatrix}$ has equality satisfying Frobenius. For parallel arrows $u, v \colon I \rightrightarrows J$ in $\mathbb{B}$ one has, following 3.4.2, an equality predicate,

$$
\mathrm{Eq}(u, v) = \langle\langle \mathrm{id}, u\rangle, v\rangle^* \big(\coprod_\delta (1) \big) = \langle\langle \mathrm{id}, u\rangle, v\rangle^*(\delta)
$$

in a pullback situation:

$$
\begin{array}{ccc}
K & \longrightarrow & I \times J \\
{\scriptstyle \mathrm{Eq}(u,v)}\downarrow \lrcorner & & \downarrow {\scriptstyle \delta = \delta(I,J)} \\
I & \xrightarrow[\;\; \langle\langle \mathrm{id}, u\rangle, v\rangle \;\;]{} & (I \times J) \times J
\end{array}
$$

It is easily established that $\mathrm{Eq}(u, v)$ is then the equaliser of u and v: this is in fact the standard way to get equalisers via pullbacks and products. Thus the notation $\mathrm{Eq}(u, v)$ for the equaliser of u, v (as used for example in the previous section) coincides with the notation introduced in 3.4.2 above.

(ii) The situation for a subobject fibration $\begin{smallmatrix}\mathrm{Sub}(\mathbb{B})\\\downarrow\\\mathbb{B}\end{smallmatrix}$ is similar: since monos are closed under composition and the diagonal $\delta = \langle \mathrm{id}, \pi'\rangle$ is monic, each pullback functor δ^* has a left adjoint by composition. Hence equality comes for free in subobject fibrations. It is easily verified that Frobenius holds.

(iii) Suppose $\mathbb{C}$ is a category with initial object 0. The family fibration
$$\begin{array}{c}\text{Fam}(\mathbb{C})\\\downarrow\\\textbf{Sets}\end{array}$$ then has equality: for a family $X = (X_{(i,j)})_{(i,j)\in I\times J}$ of $\mathbb{C}$-objects over $I \times J$ one defines a family over $(I \times J) \times J$ by

$$\mathrm{Eq}(X)_{(i,j,j')} = \begin{cases} X_{(i,j)} & \text{if } j = j' \\ 0 & \text{else.} \end{cases}$$

We get a bijective correspondence

$$\frac{(\mathrm{Eq}(X)_{(i,j,j')}) \longrightarrow (Y_{(i,j,j')})}{(X_{(i,j)}) \longrightarrow (Y_{(i,j,j)}) = \delta^*(Y_{(i,j,j')})}$$

In case the category $\mathbb{C}$ additionally has finite products in such a way that functors $Z \times (-)\colon \mathbb{C} \to \mathbb{C}$ preserve the initial object (which simply means $0 \dashrightarrow Z \times 0$ is an isomorphism), then the family fibration has finite products as well (by Example 1.8.3 (i)) and equality satisfies the Frobenius property: for a family $Y = (Y_{(i,j,j')})$ over $(I \times J) \times J$,

$$\mathrm{Eq}(\delta^*(Y) \times X)_{(i,j,j')} = \begin{cases} Y_{(i,j,j)} \times X_{(i,j)} & \text{if } j = j' \\ 0 & \text{otherwise} \end{cases}$$

$$\big(Y \times \mathrm{Eq}(X)\big)_{(i,j,j')} = \begin{cases} Y_{(i,j,j)} \times X_{(i,j)} & \text{if } j = j' \\ Y_{(i,j,j')} \times 0 \cong 0 & \text{otherwise.} \end{cases}$$

The Frobenius property is thus a distributivity condition (like in Exercise 1.9.6).

Notice that for functions $u, v\colon I \rightrightarrows J$ the family $\mathrm{Eq}(u, v)$ over I (see 3.4.2) is given by

$$\mathrm{Eq}(u, v)_i = \begin{cases} 1 & \text{if } u(i) = v(i) \\ 0 & \text{else.} \end{cases}$$

(where $1, 0$ are terminal and initial object in $\mathbb{C}$).

(iv) Let $(\Sigma, \mathcal{H})$ be an equational specification, consisting of a signature Σ and a set $\mathcal{H}$ of possibly conditional equations between Σ-terms. In the first section of this chapter we outlined a general construction which produces a term model fibration that captures the logic involved. We claim that this fibration $\begin{array}{c}\mathcal{L}(\Sigma,\mathcal{H})\\\downarrow\\\mathcal{C}\ell(\Sigma)\end{array}$ thus associated with this equational specification $(\Sigma, \mathcal{H})$ admits equality satisfying Frobenius. For contexts $\Gamma, \Gamma' \in \mathcal{C}\ell(\Sigma)$ we must exhibit a left adjoint to δ^*, where δ is the parametrised diagonal $\Gamma, \Gamma' \to \Gamma, \Gamma', \Gamma'$ in the base category $\mathcal{C}\ell(\Sigma)$. For convenience we suppose Γ' to be $(x\colon \sigma)$ of length one. We then define an equality functor, using propositional equality $=_\sigma$ from equational logic:

$$\mathrm{Eq}(\Gamma, x{:}\sigma \mid \Theta) \stackrel{\text{def}}{=} (\Gamma, x{:}\sigma, y{:}\sigma \mid \Theta, x =_\sigma y).$$

The required adjunction boils down to a bijective correspondence

$$\frac{\Gamma, x{:}\sigma, y{:}\sigma \mid \Theta, x =_\sigma y \vdash \Theta'}{\Gamma, x{:}\sigma \mid \Theta \vdash \Theta'[x/y] = \delta^*(\Theta')}$$

which is (essentially) Lawvere's equality rule as described in Lemma 3.2.3.

The Frobenius property holds because

$$
\begin{aligned}
&\mathrm{Eq}((\Gamma, x{:}\sigma \mid \Theta[x/y]) \times (\Gamma, x{:}\sigma \mid \Theta')) \\
&= \ \mathrm{Eq}(\Gamma, x{:}\sigma \mid \Theta[x/y], \Theta') \\
&= \ (\Gamma, x{:}\sigma, y{:}\sigma \mid \Theta[x/y], \Theta', x =_\sigma y) \\
&\cong \ (\Gamma, x{:}\sigma, y{:}\sigma \mid \Theta, \Theta', x =_\sigma y) \\
&= \ (\Gamma, x{:}\sigma, y{:}\sigma \mid \Theta) \times (\Gamma, x{:}\sigma, y{:}\sigma \mid \Theta', x =_\sigma y) \\
&= \ (\Gamma, x{:}\sigma, y{:}\sigma \mid \Theta) \times \mathrm{Eq}(\Gamma, x{:}\sigma \mid \Theta')
\end{aligned}
$$

where the isomorphism $\cong$ follows from Lawvere's extended equality rule in Lemma 3.2.4. The Frobenius property is thus a result of the parametrised formulation of this rule involving a proposition context Θ.

One easily verifies that for parallel context morphisms $\vec{M}, \vec{N} \colon \Gamma \rightrightarrows \Delta$ in $\mathcal{C}\!\ell(\Sigma)$ equality is given by the proposition context

$$\mathrm{Eq}(\vec{M}, \vec{N}) = (\Gamma \mid \vec{M} =_{\vec{\sigma}} \vec{N}).$$

Hence these morphisms $\vec{M}, \vec{N}$ in the base category are internally equal in the fibration if one can prove (using the axioms from $\mathcal{H}$) that

$$\Gamma \mid \emptyset \vdash M_i =_{\sigma_i} N_i.$$

for each i. Hence by using a different set of axioms $\mathcal{H}'$ one gets a different fibration $\begin{smallmatrix} \mathcal{L}(\Sigma, \mathcal{H}') \\ \downarrow \\ \mathcal{C}\!\ell(\Sigma) \end{smallmatrix}$ on the same base category, in which other internal equalities hold. This gives us a different logic to reason about morphisms in the base category $\mathcal{C}\!\ell(\Sigma)$, *i.e.* about Σ-terms.

This concludes the series of examples.

The following lemma gives some standard combinators for equality.

3.4.5. Lemma. *Let* $\begin{smallmatrix} \mathbb{E} \\ \downarrow p \\ \mathbb{B} \end{smallmatrix}$ *be a fibration with fibred finite products and equality satisfying Frobenius. Then, for parallel morphisms* $u, v, w \colon I \rightrightarrows J$ *and*

$t: I \times J \to K$ *in* $\mathbb{B}$ *there are the following vertical combinators in* $\mathbb{E}$.

$$1 \xrightarrow{\quad\text{refl}\quad} \text{Eq}(u, u)$$

$$\text{Eq}(u, v) \xrightarrow[\cong]{\quad\text{sym}\quad} \text{Eq}(v, u)$$

$$\text{Eq}(u, v) \times \text{Eq}(v, w) \xrightarrow{\quad\text{trans}\quad} \text{Eq}(u, w)$$

$$\text{Eq}(u, v) \xrightarrow{\quad\text{repl}\quad} \text{Eq}(t \circ \langle \text{id}, u \rangle, t \circ \langle \text{id}, v \rangle)$$

$$u^*(X) \times \text{Eq}(u, v) \xrightarrow{\quad\text{subst}\quad} v^*(X)$$

These are preserved under reindexing and make some 'obvious' diagrams commute, e.g.,

Proof. By reindexing the unit $\eta: 1 \to \delta^* \text{Eq}(1)$ above $I \times J$ along $\langle \text{id}, u \rangle: I \to I \times J$, one obtains the reflexivity combinator refl as composite

$$1 \cong \langle \text{id}, u \rangle^*(1) \xrightarrow{\langle \text{id}, u \rangle^*(\eta)} \langle \text{id}, u \rangle^* \delta^* \text{Eq}(1) \cong \langle\langle \text{id}, u \rangle, u \rangle^* \text{Eq}(1) = \text{Eq}(u, u).$$

Let γ be the parametrised twist map $\langle \pi \times \text{id}, \pi' \circ \pi \rangle: (I \times J) \times J \overset{\cong}{\to} (I \times J) \times J$ which exchanges the first and second J. Then

$$
\begin{aligned}
\text{Eq}(u, v) &= \langle\langle \text{id}, u \rangle, v \rangle^* \text{Eq}(1) \\
&\cong \langle\langle \text{id}, u \rangle, v \rangle^* \gamma^* \text{Eq}(1) \\
&\cong \langle\langle \text{id}, v \rangle, u \rangle^* \text{Eq}(1) \\
&= \text{Eq}(v, u)
\end{aligned}
$$

This yields the symmetry combinator sym. The transitivity combinator trans arises as follows. Consider above $(I \times J) \times J$ the first projection

$$\text{Eq}(1) \times 1 \longrightarrow \text{Eq}(1) \cong \delta^*(\pi \times \text{id})^* \text{Eq}(1).$$

By transposing across $(\text{Eq} \dashv \delta^*)$ and using that $1 \cong (\pi \times \text{id})^*(1)$ on the left hand side, one obtains

$$\text{Eq}\big(\text{Eq}(1) \times (\pi \times \text{id})^*(1)\big)$$
$$\cong\ \pi^*\text{Eq}(1) \times \text{Eq}((\pi \times \text{id})^*(1)) \qquad \text{by Frobenius}$$
$$\cong\ \pi^*\text{Eq}(1) \times ((\pi \times \text{id}) \times \text{id})^*\text{Eq}(1) \quad \text{by Beck-Chevalley.}$$

Thus we have a map

$$\pi^*\text{Eq}(1) \times ((\pi \times \text{id}) \times \text{id})^*\text{Eq}(1) \longrightarrow (\pi \times \text{id})^*\text{Eq}(1)$$

above $((I \times J) \times J) \times J$. By reindexing along the 4-tuple $\langle\langle\langle\text{id}, u\rangle, v\rangle, w\rangle\colon I \to ((I \times J) \times J) \times J$ one gets the required transitivity combinator.

For the replacement combinator repl, assume a map $t\colon I \times J \to K$ in $\mathbb{B}$, and consider above $I \times J$ the composite

$$1 \xrightarrow{\ \text{refl}\ } \text{Eq}(t, t) = \text{Eq}(t \circ \pi \circ \delta, t \circ \pi \times \text{id} \circ \delta)$$
$$\cong\ \delta^*(\text{Eq}(t \circ \pi, t \circ \pi \times \text{id}))$$

It yields a morphism above $(I \times J) \times J$ by transposition:

$$\text{Eq}(1) \longrightarrow \text{Eq}(t \circ \pi, t \circ \pi \times \text{id})$$

Hence by reindexing along $\langle\langle\text{id}, u\rangle, v\rangle\colon I \to (I \times J) \times J$ one obtains the required map

$$\text{Eq}(u, v) \longrightarrow \text{Eq}(t \circ \langle\text{id}, u\rangle, t \circ \langle\text{id}, v\rangle)$$

Finally for the substitution combinator subst notice that $\pi'^*(X) \cong \delta^*\pi'^*(X)$, so we have over $I \times J$ a projection map

$$\pi'^*(X) \times 1 \longrightarrow \delta^*\pi'^*(X)$$

By transposing and using Frobenius we get

$$\pi^*\pi'^*(X) \times \text{Eq}(1) \longrightarrow \pi'^*(X)$$

The subst combinator arises by reindexing along $\langle\langle\text{id}, u\rangle, v\rangle$. $\qquad\square$

The next result gives an application of these combinators; the proof requires some elementary, but non-trivial, categorical manipulations. The result states that two tuples are equal in the internal logic of a fibration if and only if their components are equal. It also occurs in Lawvere's paper [193] (as the second corollary on page 10), but some stronger form of Beck-Chevalley is used there. See Exercise 3.4.7 below.

For convenience, we present the result for fibred preorders.

3.4.6. Proposition. *Consider a fibred preorder with fibred finite products and equality satisfying Frobenius. Then there is a vertical isomorphism*

$$\text{Eq}(\langle u_1, u_2 \rangle, \langle v_1, v_2 \rangle) \cong \text{Eq}(u_1, v_1) \wedge \text{Eq}(u_2, v_2).$$

Proof. Assume the morphisms u_1, v_1, u_2, v_2 in the base category are given as follows.

$$J \overset{u_1}{\underset{v_1}{\rightleftarrows}} I \overset{u_2}{\underset{v_2}{\rightrightarrows}} K$$

The ($\leq$)-part of the result is easy, since by applying the above replacement combinator one obtains

$$\text{Eq}(\langle u_1, u_2 \rangle, \langle v_1, v_2 \rangle) \leq \text{Eq}(\pi \circ \pi' \circ \langle \text{id}, \langle u_1, u_2 \rangle \rangle, \pi \circ \pi' \circ \langle \text{id}, \langle v_1, v_2 \rangle \rangle)$$
$$\cong \text{Eq}(u_1, v_1)$$

and similarly for $\text{Eq}(u_2, v_2)$.

The ($\geq$)-part requires more work. Our first aim is to prove

$$\text{Eq}(u_1 \circ \pi, v_1 \circ \pi) \leq \text{Eq}(u_1 \times \text{id}, v_1 \times \text{id}). \qquad (*)$$

Consider therefore the diagram

$$
\begin{array}{ccc}
(I \times K) \times J & \overset{\alpha}{\longrightarrow} & (I \times K) \times (J \times K) \\
{\scriptstyle \delta} \downarrow & & \downarrow {\scriptstyle \delta} \\
((I \times K) \times J) \times J & \overset{\beta}{\longrightarrow} & ((I \times K) \times (J \times K)) \times (J \times K)
\end{array}
$$

which commutes for the "obvious" maps

$$\alpha = \langle \pi, \langle \pi', \pi' \circ \pi \rangle \rangle$$
$$\beta = \langle \langle \pi \circ \pi, \langle \pi' \circ \pi, \pi' \circ \pi \circ \pi \rangle \rangle, \langle \pi', \pi' \circ \pi \circ \pi \rangle \rangle.$$

The terminal object 1 above $(I \times K) \times J$ comes together with a morphism

$$1 \cong \alpha^*(1) \overset{\alpha^*(\eta)}{\leq} \alpha^* \delta^* \text{Eq}(1) \cong \delta^* \beta^* \text{Eq}(1)$$

which yields by transposition

$$\text{Eq}(1) \leq \beta^* \text{Eq}(1) \quad \text{above } ((I \times K) \times J) \times J.$$

Reindexing along $\langle \langle \text{id}, u_1 \circ \pi \rangle, v_1 \circ \pi \rangle : I \times K \to ((I \times K) \times J) \times J$ yields the required map $(*)$.

Using the inequality $(*)$ we get:

$$
\begin{aligned}
\mathrm{Eq}(u_1, v_1) &\cong \langle \mathrm{id}, v_2 \rangle^* \pi^* \mathrm{Eq}(u_1, v_1) \\
&\cong \langle \mathrm{id}, v_2 \rangle^* \mathrm{Eq}(u_1 \circ \pi, v_1 \circ \pi) \\
&\leq \langle \mathrm{id}, v_2 \rangle^* \mathrm{Eq}(u_1 \times \mathrm{id}, v_1 \times \mathrm{id}) \\
&= \mathrm{Eq}(\langle u_1, v_2 \rangle, \langle v_1, v_2 \rangle).
\end{aligned}
$$

Further, from replacement we obtain

$$
\begin{aligned}
\mathrm{Eq}(u_2, v_2) &\leq \mathrm{Eq}(u_1 \times \mathrm{id} \circ \langle \mathrm{id}, u_2 \rangle, u_1 \times \mathrm{id} \circ \langle \mathrm{id}, v_2 \rangle) \\
&= \mathrm{Eq}(\langle u_1, u_2 \rangle, \langle u_1, v_2 \rangle).
\end{aligned}
$$

But then

$$
\begin{aligned}
\mathrm{Eq}(u_1, v_1) \wedge \mathrm{Eq}(u_2, v_2) &\cong \mathrm{Eq}(u_2, v_2) \wedge \mathrm{Eq}(u_1, v_1) \\
&\leq \mathrm{Eq}(\langle u_1, u_2 \rangle, \langle u_1, v_2 \rangle) \wedge \mathrm{Eq}(\langle u_1, v_2 \rangle, \langle v_1, v_2 \rangle) \\
&\leq \mathrm{Eq}(\langle u_1, u_2 \rangle, \langle v_1, v_2 \rangle),
\end{aligned}
$$

the latter by transitivity. $\qquad\square$

For future use, we mention at the end of this section what it means for a morphism of fibrations to preserve equality Eq.

3.4.7. Definition. Let $\begin{smallmatrix} \mathbb{E} \\ \downarrow p \\ \mathbb{B} \end{smallmatrix} \xrightarrow{(K,L)} \begin{smallmatrix} \mathbb{E}' \\ \downarrow p' \\ \mathbb{B}' \end{smallmatrix}$ be a morphism of fibrations. We say that (K, L) **preserves equality** (or, is a **morphism of fibrations with equality**) if $K \colon \mathbb{B} \to \mathbb{B}'$ preserves finite products and for each pair of objects $I, J \in \mathbb{B}$, the canonical natural transformation

$$
\mathrm{Eq}'_{(KI, KJ)} \gamma_1^* \, L \Longrightarrow \gamma_2^* \, L \, \mathrm{Eq}_{(I,J)}
$$

is an isomorphism—where $\gamma_1 \colon KI \times KJ \xrightarrow{\cong} K(I \times J)$ and $\gamma_2 \colon (KI \times KJ) \times KJ \xrightarrow{\cong} K((I \times J) \times J)$ are the canonical isomorphisms.

Exercises

3.4.1. Let $\mathbb{B}$ be a category with finite products.
   (i) Extend the assignment $(I, J) \mapsto \delta(I, J) = \langle \mathrm{id}, \pi' \rangle \colon I \times J \to (I \times J) \times J$ to a functor $\delta \colon s(\mathbb{B}) \to \mathbb{B}^{\to}$.
   (ii) Show that δ sends Cartesian morphisms (for the simple fibration on $\mathbb{B}$) to pullback squares in $\mathbb{B}$ (*i.e.* Cartesian morphisms for the codomain functor on $\mathbb{B}$).

3.4.2. Assume $\begin{smallmatrix} \mathbb{E} \\ \downarrow p \\ \mathbb{B} \end{smallmatrix}$ is a fibred CCC with equality.
   (i) Show that the Frobenius property for equality holds automatically.

(ii) By definition, each contraction functor $\delta(I, J)^*$ has a left adjoint $\coprod_\delta$. Show that it also has a right adjoint $\prod_\delta$, given by

$$\big(\mathbb{E}_{I \times J} \ni X\big) \mapsto \big(\mathrm{Eq}_{I,J}(1) \Rightarrow \pi^*(X) \in \mathbb{E}_{(I \times J) \times J}\big).$$

[Notice that equality $=$ is left adjoint $\coprod_\delta$ at 1 and inequality $\neq$ is right adjoint $\prod_\delta$ at 0.]

3.4.3.   Verify that the Beck-Chevalley condition for the fibration $\begin{smallmatrix} \mathcal{L}(\Sigma, \mathcal{H}) \\ \downarrow \\ \mathcal{C}\ell(\Sigma) \end{smallmatrix}$ in Example 3.4.4 regulates the proper distribution of substitution over equations. Describe the isomorphism $\cong$ which was used in proving the Frobenius property.

3.4.4.   Describe how the canonical natural transformations in Definitions 3.4.1 and 3.4.7 are obtained.

3.4.5.   Consider the projections $\pi' \circ \pi, \pi' : (I \times J) \times J \rightrightarrows J$ in the base category of a fibration with equality. Show that

$$\mathrm{Eq}(\pi' \circ \pi, \pi') \cong \mathrm{Eq}(1).$$

3.4.6.   Let $\begin{smallmatrix} \mathbb{E} \\ \downarrow p \\ \mathbb{B} \end{smallmatrix} \overset{(K, L)}{\longrightarrow} \begin{smallmatrix} \mathbb{E}' \\ \downarrow p' \\ \mathbb{B}' \end{smallmatrix}$ be a morphism between fibrations p and p' with fibred terminal object and equality.

(i) Assume (K, L) preserves the terminal object and equality. Verify that for parallel arrows u, v in $\mathbb{B}$ the canonical vertical map

$$\mathrm{Eq}'(Ku, Kv) \longrightarrow L(\mathrm{Eq}(u, v)) \qquad\qquad (*)$$

is an isomorphism

(ii) Assume that p and p' also have fibred finite products and that (K, L) preserves all of these. Assume additionally that Frobenius holds both for p and for p'. Show that if the maps $(*)$ in (i) are isomorphisms (for all parallel u, v), then (K, L) preserves equality.

[*Hint.* Use the previous exercise.]

3.4.7.   The point of this exercise is to check the details of Lawvere's proof (from [193]) of Proposition 3.4.6 for a fibration $\begin{smallmatrix} \mathbb{E} \\ \downarrow p \\ \mathbb{B} \end{smallmatrix}$ with coproducts $\coprod_u$, satisfying Frobenius. By Lemma 3.4.3 this fibration then has equality satisfying Frobenius. Check that

(i) $\coprod_{w \times \mathrm{id}}(\pi^*(X_1) \times \pi'^*(X_2)) \cong \pi^* \coprod_w(X_1) \times \pi'^*(X_2)$.

(ii) $\coprod_{w_1 \times w_2}(\pi^*(X_1) \times \pi'^*(X_2)) \cong \pi^* \coprod_{w_1}(X_1) \times \pi'^* \coprod_{w_2}((X_2)$.

(iii) $\coprod_{\delta(I, J \times K)}(1) \cong \coprod_{\delta(I, J)}(1) \times \coprod_{\delta(I, K)}(1)$, using that there is a pullback square

$$I \times (J \times K) \longrightarrow (I \times J) \times (I \times K)$$

$$\delta \downarrow \qquad\qquad\qquad \downarrow \delta$$

$$((I \times (J \times K)) \times (J \times K) \longrightarrow ((I \times J) \times J) \times ((I \times K) \times K)$$

in which the horizontal arrows are the obvious maps.

(iv) And finally, that $\mathrm{Eq}(\langle u_1, u_2 \rangle, \langle v_1, v_2 \rangle) \cong \mathrm{Eq}(u_1, v_1) \wedge \mathrm{Eq}(u_2, v_2)$.

3.5 Fibrations for equational logic

In this section we give meaning to equations in fibrations with equality, as described in the previous section. This fibred approach has as main advantages that it is very general and flexible and that it scales up smoothly to other logics. We start with the definition of validity of a (conditional) equation in a fibration. Then we show how different fibrations on the same base category can capture different notions of equality for arrows in this base category. This is what we mean by the flexibility of the fibred approach: using different logics to reason about one (base) category can be done by putting different fibrations on this same base category.

3.5.1. Definition. An **Eq-fibration** is a fibration which

(i) is a fibred preorder (*i.e.* all its fibre categories are preorders);

(ii) has fibred finite products $(\top, \wedge)$ and finite products $(1, \times)$ in its base category;

(iii) has equality Eq satisfying Frobenius.

We impose the restriction to fibred preorders in (i) because we limit our attention in this chapter to models of logics, interpreting provability and not proofs (like in type theories).

3.5.2. Examples. (i) An important example of an Eq-fibration is the syntactically constructed fibration $\begin{smallmatrix} \mathcal{L}(\Sigma, \mathcal{H}) \\ \downarrow \\ \mathcal{C}\!\ell(\Sigma) \end{smallmatrix}$ associated with an equational specification $(\Sigma, \mathcal{H})$, see Example 3.4.4 (iv). It will be called the **classifying Eq-fibration** of $(\Sigma, \mathcal{H})$.

(ii) For each category $\mathbb{B}$ with finite limits, the fibration $\begin{smallmatrix} \mathrm{Sub}(\mathbb{B}) \\ \downarrow \\ \mathbb{B} \end{smallmatrix}$ of subobjects of $\mathbb{B}$ is an Eq-fibration, see Example 3.4.4 (ii).

(iii) Let X be a poset (or a preorder) with finite meets and a bottom element. The family fibration $\begin{smallmatrix} \mathrm{Fam}(X) \\ \downarrow \\ \mathbf{Sets} \end{smallmatrix}$ is then an Eq-fibration, see Example 3.4.4 (iii).

In the beginning of Section 3.3 we briefly described what it means for a conditional equation to hold in a category with finite limits. Below we show that validity of equations can be described more generally in Eq-fibrations, with the special case of subobject fibrations capturing this earlier mentioned situation.

3.5.3. Definition (Validity in Eq-fibrations). Consider a situation

$$
\begin{array}{ccc}
 & & \mathbb{E} \\
 & & \downarrow p \\
\mathcal{C}\!\ell(\Sigma) & \xrightarrow{\ \ \mathcal{M}\ \ } & \mathbb{B}
\end{array}
$$

where p is an Eq-fibration and $\mathcal{M}$ is a model of the signature Σ in the base category $\mathbb{B}$. We say that a Σ-equation

$$
\Gamma \mid M_1 =_{\sigma_1} M_1', \ldots, M_n =_{\sigma_n} M_n' \vdash N =_\tau N'
$$

holds in $\mathcal{M}$ or **is validated by** $\mathcal{M}$ with respect to p if

$$
\mathrm{Eq}\big(\mathcal{M}(M_1), \mathcal{M}(M_1')\big) \wedge \cdots \wedge \mathrm{Eq}\big(\mathcal{M}(M_n), \mathcal{M}(M_n')\big) \leq \mathrm{Eq}\big(\mathcal{M}(N), \mathcal{M}(N')\big)
$$

in the preorder fibre category above the interpretation $\mathcal{M}(\Gamma)$ of the type context Γ in $\mathbb{B}$. Often we simply say that such an equation holds in $\mathcal{M}$ without reference to the fibration p if the latter is understood from the context.

Thus it becomes clear that a fibred category $\begin{smallmatrix}\mathbb{E}\\\downarrow\\\mathbb{B}\end{smallmatrix}$ on $\mathbb{B}$ provides us with a logic to reason about what happens in $\mathbb{B}$. This shows how fibred (preorder) categories play a rôle in logic. We will expand on this point shortly, but first, we notice that for a model $\mathcal{M} : \mathcal{C}\!\ell(\Sigma) \to \mathbb{B}$ in a category $\mathbb{B}$ with finite limits one has a situation

$$
\begin{array}{ccc}
 & & \mathrm{Sub}(\mathbb{B}) \\
 & & \downarrow \\
\mathcal{C}\!\ell(\Sigma) & \xrightarrow{\ \ \mathcal{M}\ \ } & \mathbb{B}
\end{array}
$$

in which an equation holds in $\mathcal{M}$ as defined above with respect to the subobject fibration if and only if it holds in $\mathcal{M}$ as described in the beginning of Section 3.3 (after Example 3.3.1). Thus we can conclude that the previous fibred definition does not lead to ambiguity and that its notion of validation of equations extends the earlier one for ordinary categories.

In Example 3.4.4 (iv) we saw how different equational specifications $(\Sigma, \mathcal{H})$ and $(\Sigma, \mathcal{H}')$ give rise to different fibrations $\begin{smallmatrix}\mathcal{L}(\Sigma,\mathcal{H})\\\downarrow\\\mathcal{C}\!\ell(\Sigma)\end{smallmatrix}$ and $\begin{smallmatrix}\mathcal{L}(\Sigma,\mathcal{H}')\\\downarrow\\\mathcal{C}\!\ell(\Sigma)\end{smallmatrix}$ to reason about the same base category. Next we give two mathematical examples of this phenomenon: we show how different notions of equality—for continuous functions between dcpos, and for relations (as morphisms) between sets—can be captured by different fibrations on the base category **Dcpo** of dcpos and continuous functions, and on the base category **REL** of sets and relations. These different notions of equality really require different fibrations because equality is determined by its defining adjunction (Eq ⊣ δ^*), and is thus directly linked to the fibrations reindexing operation $(-)^*$.

Recall from 3.4.2 that two parallel arrows u, v in the base category of an Eq-fibration are internally equal if an inequality $\top \leq \mathrm{Eq}(u, v)$ holds over their domain. External equality simply means $u = v$.

3.5.4. Extended example. The category of directed complete partial orders (dcpos) and (Scott-) continuous (*i.e.* directed suprema $\bigvee^{\uparrow}$-preserving) functions will be written as **Dcpo**. The singleton dcpo forms a terminal object, and the Cartesian product of the underlying sets of two dcpos, with componentwise order, yields the product in **Dcpo**. A subset $A \subseteq X$ of a dcpo X is called **admissible** if it is closed under directed suprema: for each directed $a \subseteq X$ with $a \subseteq A$ one has $\bigvee^{\uparrow} a \in A$. A category ASub(**Dcpo**) is formed with such admissible subsets as objects. We consider these as certain predicates on dcpos. A morphism $(A \subseteq X) \to (B \subseteq Y)$ in ASub(**Dcpo**) is a continuous function $f: X \to Y$ with the property that $x \in A$ implies $f(x) \in B$, for all $x \in X$. This means that there is a commuting diagram (in **Sets** or in **Dcpo**)

$$
\begin{array}{ccc}
A & \dashrightarrow & B \\
\big\uparrow & & \big\uparrow \\
\big\downarrow & & \big\downarrow \\
X & \xrightarrow{\ f\ } & Y
\end{array}
$$

There is an obvious forgetful functor $\begin{array}{c} \text{ASub}(\mathbf{Dcpo}) \\ \downarrow \\ \mathbf{Dcpo} \end{array}$, namely $(A \subseteq X) \mapsto X$ sending a predicate to its carrier (type). It is a split fibration, with reindexing $B \subseteq Y$ along $f: X \to Y$ given by

$$f^*(B) = \{x \in X \mid f(x) \in B\}.$$

In particular we have for a diagonal $\delta = \delta(X, Y) = \langle \mathrm{id}, \pi' \rangle: X \times Y \to (X \times Y) \times Y$ and for an admissible subset $B \subseteq (X \times Y) \times Y$ that

$$\delta^*(B) = \{(x, y) \mid (x, y, y) \in B\} \subseteq X \times Y.$$

A left adjoint Eq to this δ^* is then defined by

$$\mathrm{Eq}(A) = \{(x, y, y') \mid y = y' \text{ and } (x, y) \in A\} \subseteq (X \times Y) \times Y.$$

Notice that this is an admissible subset again. Hence the usual definitions for sets work in this case as well. For parallel arrows $f, g: X \rightrightarrows Y$ in **Dcpo** the corresponding equality on X is

$$\mathrm{Eq}(f,g) = \{x \in X \mid f(x) = g(x)\}.$$

Since the terminal object over $X \in \mathbf{Dcpo}$ is $(X \subseteq X)$ we get

$$
\begin{aligned}
f, g \text{ are internally equal } &\Leftrightarrow\ X \subseteq \mathrm{Eq}(f,g) \\
&\Leftrightarrow\ \forall x \in X.\, f(x) = g(x) \\
&\Leftrightarrow\ f = g \colon X \to Y \\
&\Leftrightarrow\ f, g \text{ are externally equal.}
\end{aligned}
$$

We now put a different logic on $\mathbf{Dcpo}$ by taking different subsets as predicates. Call a subset $A \subseteq X$ **down closed** if $y \leq x$ and $x \in A$ implies $y \in A$. These down closed subsets are organised in a category $\mathrm{DSub}(\mathbf{Dcpo})$ as before: a morphism $(A \subseteq X) \to (B \subseteq Y)$ is a continuous function $f \colon X \to Y$ with $f(x) \in B$ for all $x \in A$. Again we get a split fibration $\begin{smallmatrix}\mathrm{DSub}(\mathbf{Dcpo})\\ \downarrow\\ \mathbf{Dcpo}\end{smallmatrix}$ by $(A \subseteq X) \mapsto X$, with reindexing as above. The earlier definition of equality does not yield a down closed subset, so we now define for $A \subseteq X \times Y$,

$$\mathrm{Eq}(A) = \{(x,y,y') \mid \exists z \in Y.\, y \leq z \text{ and } y' \leq z \text{ and } (x,z) \in A\}.$$

We then get bijective correspondences

$$
\frac{\mathrm{Eq}(A) \subseteq B \quad \text{over } (X \times Y) \times Y}{A \subseteq \delta^*(B) \quad \text{over } X \times Y}
$$

as follows.

- Assuming $\mathrm{Eq}(A) \subseteq B$ we have for $(x,y) \in A$ that $(x,y,y) \in \mathrm{Eq}(A) \subseteq B$, so that $(x,y) \in \delta^*(B)$.
- And assuming $A \subseteq \delta^*(B)$, we get for $(x,y,y') \in \mathrm{Eq}(A)$, say with $y, y' \leq z$ where $(x,z) \in A$, that $(x,z,z) \in B$. Since B is down closed and $(x,y,y') \leq (x,z,z)$ we have $(x,y,y') \in B$.

For $f, g \colon X \rightrightarrows Y$ in $\mathbf{Dcpo}$ we now get

$$\mathrm{Eq}(f,g) = \{x \in X \mid \exists z \in Y.\, f(x) \leq z \text{ and } g(x) \leq z\}$$

so that

$$
\begin{aligned}
f, g \text{ are internally equal } &\Leftrightarrow\ X \subseteq \mathrm{Eq}(f,g) \\
&\Leftrightarrow\ \forall x \in X.\, \exists z \in Y.\, f(x) \leq z \text{ and } g(x) \leq z.
\end{aligned}
$$

This fibration $\begin{smallmatrix}\mathrm{DSub}(\mathbf{Dcpo})\\ \downarrow\\ \mathbf{Dcpo}\end{smallmatrix}$ thus captures a different logic to reason about $\mathbf{Dcpo}$: in the logic incorporated by this fibration two morphisms $f, g \colon X \rightrightarrows Y$ are equal if and only $f(x), g(x)$ have an upper bound in Y, for each $x \in X$.

We describe a similar phenomenon for relations.

3.5.5. Extended example. We write **REL** for the category of sets and relations. Objects are sets I, and morphisms $I \to J$ are relations $R \subseteq I \times J$. Often one uses the notation $R : I \nrightarrow J$ to indicate that R is a relation from I to J. The identity $I \nrightarrow I$ is then the diagonal relation (or equality) on I, and the composite of $R : I \nrightarrow J$ and $S : J \nrightarrow K$ is the relational composite:

$$S \circ R = \{(i, k) \mid \exists j \in J.\, R(i, j) \text{ and } S(j, k)\} \subseteq I \times K.$$

A relation $R : I \nrightarrow J$ can be understood as a **multifunction** $I \to J$. That is, as a function $I \to PJ$, given by $i \mapsto R_i = \{j \in J \mid R(i, j)\}$, which may have many outputs. Under this view one considers the category **REL** as the Kleisli category of the powerset monad P on **Sets**.

The terminal object in **REL** is the empty set $\emptyset$, and the Cartesian product of sets I, J is the disjoint union $I + J$, with graphs of the coprojections $\kappa : I \to I + J$ and $\kappa' : J \to I + J$ as projections:

$$\pi = \{(z, i) \mid z = \kappa i\} \qquad \text{and} \qquad \pi' = \{(z, j) \mid z = \kappa' j\}.$$

The tuple of two maps $R : K \nrightarrow I$ and $S : K \nrightarrow J$ is the relation,

$$\langle R, S \rangle = \{(i, z) \mid (z = \kappa i \text{ and } R(i, j)) \text{ or } (z = \kappa' i \text{ and } S(i, j))\}.$$

Thus **REL** has finite products. Actually, it also has finite coproducts, given by these same formulas $\emptyset$ and $I + J$ (on objects).

There may be different notions of equality for multifunctions $R, S : I \rightrightarrows PJ$. For example, there is the extensional view that such multifunctions are equal if and only if they yield the same output sets for each input. But one may also consider two multifunctions R, S as equal if for each input i the output sets R_i and S_i have elements in common (*i.e.* are not disjoint).

The point of this example is to show that these different notions of equality live in different fibrations on top of the category **REL** of sets and relations. These give us different ways to reason about relations. The two fibrations incorporating the two abovementioned notions of equality, will be written as $\begin{smallmatrix} \textbf{PredREL} \\ \downarrow \\ \textbf{REL} \end{smallmatrix}$ and $\begin{smallmatrix} \textbf{EPredREL} \\ \downarrow \\ \textbf{REL} \end{smallmatrix}$, where the latter fibration gives the extensional view that relations $R, S : I \rightrightarrows J$ are equal (in the logic of the fibration) if and only if the subsets $R, S \subseteq I \times J$ are equal.

We start with the 'non-extensional' example. The total category **PredREL** is a category of relations with predicates. It has

objects	pairs (I, X) where I is a set and $X \subseteq PI$ is a set of subsets of I.
morphisms	$(X \subseteq PI) \to (Y \subseteq PJ)$ are relations $R \subseteq I \times J$ from I to J satisfying: for each non-empty $a \in X$, there is a non-empty $b \subseteq \bigcup_{i \in a} R_i$ with $b \in Y$.

Identities and composites in **PredREL** are as in the category **REL** of rela-tions. This gives us a forgetful functor $\begin{smallmatrix} \mathbf{PredREL} \\ \downarrow \\ \mathbf{REL} \end{smallmatrix}$ by $(I, X) \mapsto I$. In the fibre over $I \in \mathbf{REL}$ we define a preordering:

$$X \dashrightarrow Y \iff \begin{cases} \text{for each non-empty } a \in X, \\ \text{there is a non-empty } b \subseteq a \text{ with } b \in Y. \end{cases}$$

In this preorder the predicate $PI \subseteq PI$ is top element $\top$.

The functor $\begin{smallmatrix} \mathbf{PredREL} \\ \downarrow \\ \mathbf{REL} \end{smallmatrix}$ is a fibration, since for a relation $R : I \nrightarrow J$ and a predicate $Y \subseteq PJ$ on J, we can substitute along R by

$$R^*(Y) = \{ a \subseteq I \mid \exists b \subseteq \bigcup_{i \in a} R_i. \, b \neq \emptyset \text{ and } b \in Y \}.$$

In particular, for the diagonal $\delta : J \nrightarrow J + J$ we get

$$\delta^*(Y) = \{ a \subseteq J \mid \exists b \subseteq \kappa(a) \cup \kappa'(a). \, b \neq \emptyset \text{ and } b \in Y \}.$$

where $\kappa(a) = \{ \kappa j \mid j \in a \}$. For a predicate $X \subseteq PJ$ on J, we define an equality predicate $\mathrm{Eq}(X) \subseteq P(J + J)$ as

$$\mathrm{Eq}(X) = \{ c \subseteq J + J \mid \kappa^{-1}(c) \cup \kappa'^{-1}(c) \in X \text{ and }$$
$$\forall j \in J. \, \kappa j \in c \iff \kappa' j \in c \}.$$

Then we get a bijective correspondence,

$$\frac{\mathrm{Eq}(X) \dashrightarrow Y \quad \text{over } J + J}{X \dashrightarrow \delta^*(Y) \quad \text{over } J}$$

which is given as follows.

- Assume $\mathrm{Eq}(X) \to Y$, and let $a \in X$ be non-empty. Then $\kappa(a) \cup \kappa'(a) \in \mathrm{Eq}(X)$, so there is a non-empty $b \subseteq \kappa(a) \cup \kappa'(a)$ with $b \in Y$. But then $a \in \delta^*(Y)$.
- In the reverse direction, assume $X \to \delta^*(Y)$, and let $c \in \mathrm{Eq}(X)$ be non-empty. Then $\kappa^{-1}(c) \cup \kappa'^{-1}(c) \in X$ is non-empty, so there is a non-empty $a \subseteq \kappa^{-1}(c) \cup \kappa'^{-1}(c)$ with $a \in \delta^*(Y)$. The latter yields a non-empty $b \subseteq \kappa(a) \cup \kappa'(a)$ with $b \in Y$. But then $b \subseteq c$, since for $z \in b$, either $z = \kappa j$ or $z = \kappa' j$ with $j \in a \subseteq \kappa^{-1}(c) \cup \kappa'^{-1}(c)$; in both cases we get $z \in c$ since $\forall j \in J. \, \kappa j \in c \iff \kappa' j \in c$.

Notice that—for reasons of simplicity—we define equality without parameters.

Now we are in a position to compute the equality predicate $\text{Eq}(R, S)$ for two parallel maps $R, S: I \rightrightarrows J$ in the base category **REL** as

$\text{Eq}(R, S)$

$\quad = \langle R, S \rangle^* \text{Eq}(\top)$

$\quad = \langle R, S \rangle^* \{ b \subseteq J + J \mid \forall j \in J. \kappa j \in b \Leftrightarrow \kappa' j \in b \}$

$\quad = \{ a \subseteq I \mid \exists b \subseteq \bigcup_{i \in a} \langle R, S \rangle_i . b \neq \emptyset \text{ and } \forall j \in J. \kappa j \in b \Leftrightarrow \kappa' j \in b \}.$

where $\top = (PI \subseteq PI)$ is the top element over I. Our claim is then the following. Two maps $R, S: I \rightrightarrows J$ are equal in the logic of the fibration $\begin{smallmatrix} \textbf{PredREL} \\ \downarrow \\ \textbf{REL} \end{smallmatrix}$, that is, there is a map $\top \leq \text{Eq}(R, S)$ over I, if and only if for each $i \in I$ there is a $j \in J$ for which $R(i, j)$ and $S(i, j)$ both hold. The latter can also be expressed as: $R_i \cap S_i \neq \emptyset$, for each $i \in I$. This is the second view of equality of multifunctions mentioned above (which is thus captured by the fibration).

To support the claim, assume $\top \leq \text{Eq}(R, S)$ over I. Then for each $i \in I$, we have $\{i\} \in 1 = PY$, so there is a non-empty $a \subseteq \{i\}$ with $a \in \text{Eq}(R, S)$. Thus a must be $\{i\}$, which yields that there is a non-empty $b \subseteq \langle R, S \rangle_i$ with $b \in \text{Eq}(1)$. Let $z \in b$; if $z = \kappa j$, then also $\kappa' j \in b$ and vice-versa, so we may assume a pair $\{ \kappa j, \kappa' j \} \subseteq \langle R, S \rangle_i$. This yields both $R(i, j)$ and $S(i, j)$.

In the reverse direction, if for each $i \in I$ there is a $j \in J$ with $R(i, j)$ and $S(i, j)$, then for each non-empty $a \in PI$, say with $i \in a$, we can find a $j \in J$ with $R(i, j)$ and $S(i, j)$. Then $\{i\} \subseteq \text{Eq}(R, S)$, since $b = \{ \kappa j, \kappa' j \} \in \langle R, S \rangle_i$ is non-empty and is in $\text{Eq}(\top)$. This means $\top \leq \text{Eq}(R, S)$.

We turn to the second fibration $\begin{smallmatrix} \textbf{EPredREL} \\ \downarrow \\ \textbf{REL} \end{smallmatrix}$, which gives us a logic incorporating the 'extensional' equality. We shall be a bit more sketchy and leave details to the reader. The total category **EPredREL** has

objects   pairs (I, X) where I is a set and $X \subseteq PI$.

morphisms $(X \subseteq PI) \to (Y \subseteq PJ)$ are relations $R \subseteq I \times J$ such that for each $a \in X$ we have $\bigcup_{i \in a} R_i \in Y$.

Identities and composites are inherited from **REL**, so that we get a forgetful functor $\begin{smallmatrix} \textbf{EPredREL} \\ \downarrow \\ \textbf{REL} \end{smallmatrix}$. In the fibre over I we define $X \dashrightarrow Y$ if and only if $X \subseteq Y$. Reindexing of $Y \subseteq PJ$ along $R: I \twoheadrightarrow J$ is given by $R^*(Y) = \{ a \subseteq I \mid \bigcup_{i \in a} R_i \}$. In particular $\delta^*(Y) = \{ a \mid \kappa(a) \cup \kappa'(a) \in Y \}$. Equality $\text{Eq}(X)$ is defined as before. We then get for $R, S: I \rightrightarrows J$ that

$$\text{Eq}(R, S) = \{ a \subseteq I \mid \forall j \in J. (\exists i \in a. R(i, j)) \Leftrightarrow (\exists i \in a. S(i, j)) \}.$$

The maps R, S are equal in the logic of this second fibration on **REL** if and only if $\top = PI \subseteq \text{Eq}(R, S)$ if and only if for each $i \in I$ one has $\forall j \in J. R(i, j) \Leftrightarrow S(i, j)$, if and only if R and S are 'extensionally equal'. This concludes the example.

In Example 3.3.1 we saw what it means to have validity of the defining equations of a group in a category. At this stage we recognise this as external equality. We can also describe internal equality of these equations in a fibration. This will give us the notion of an "internal group in a fibration". It is a fibration with in its base category an object carrying the operations of a group (multiplication, unit, inverse) which internally satisfy the group equations.

We conclude with the following two lemmas which together yield familiar soundness and completeness results for equational logic in Eq-fibrations.

3.5.6. Lemma (Soundness). *Let* $(\Sigma, \mathcal{H})$ *be an equational specification. In case a model* $\mathcal{M} \colon \mathcal{C\!l}(\Sigma) \to \mathbb{B}$ *validates all equations in* $\mathcal{H}$ *(with respect to some Eq-fibration with base* $\mathbb{B}$ *), then it validates all equations in the theory of* $(\Sigma, \mathcal{H})$.

Proof. By Lemma 3.4.5, which gives the appropriate combinators for the soundness of the rules specific to equational logic. As the reader may verify, all the context rules from Section 3.1 are sound, so we are done. $\qquad \square$

We should be more explicit about what is going on in this proof with respect to substitution: syntactic substitution $[L/z]$ in a proposition $M =_\sigma M'$ is interpreted by categorical substitution (or reindexing) in a fibration. This is because

$$\mathrm{Eq}(u \circ w, v \circ w) = w^* \mathrm{Eq}(u, v)$$

as already mentioned in 3.4.2. Notice that composition $u \circ w, v \circ w$ on the left hand side is substitution in terms, whereas w^* on the right hand side is substitution in propositions. Syntactically this equation is

$$\bigl(M[L/z] =_\sigma M'[L/z] \bigr) \equiv \bigl(M =_\sigma M' \bigr)[L/z].$$

The weakening and contraction rules are handled as special cases of substitution, see Example 3.1.1.

3.5.7. Lemma (Completeness). *For an equational specification* $(\Sigma, \mathcal{H})$, *consider the situation:*

$$\mathcal{L}(\Sigma, \mathcal{H})$$
$$\downarrow$$
$$\mathcal{C\!l}(\Sigma) \xrightarrow{\quad \mathrm{id} \quad} \mathcal{C\!l}(\Sigma)$$

This generic model of Σ *validates precisely the equations in the theory of* $(\Sigma, \mathcal{H})$. $\qquad \square$

As a result, an equation is derivable from an equational specification $(\Sigma, \mathcal{H})$ if and only if it holds in all (fibred) models of $(\Sigma, \mathcal{H})$.

Exercises

3.5.1. Check that internal equality in the fibrations $\begin{smallmatrix} \mathrm{DSub}(\mathbf{Dcpo}) \\ \downarrow \\ \mathbf{Dcpo} \end{smallmatrix}$ and $\begin{smallmatrix} \mathbf{PredREL} \\ \downarrow \\ \mathbf{REL} \end{smallmatrix}$ is not transitive. Conclude that the Frobenius property does not hold. [*Hint.* Inspect the construction of the transitivity combinator in the proof of Lemma 3.4.5.]

3.5.2. Prove that in the fibration $\begin{smallmatrix} \mathrm{DSub}(\mathbf{Dcpo}) \\ \downarrow \\ \mathbf{Dcpo} \end{smallmatrix}$ internal and external equality coincide on $Y \in \mathbf{Dcpo}$ if and only if the order on Y is discrete.

3.5.3. Let $\mathbf{Sp}$ be the category of topological spaces and continuous functions.

(i) Define a poset fibration $\begin{smallmatrix} \mathrm{ClSub}(\mathbf{Sp}) \\ \downarrow \\ \mathbf{Sp} \end{smallmatrix}$ of closed subsets ($A \subseteq X$) over topological spaces X.

(ii) Show that a contraction functor δ^* associated with a diagonal $\delta \colon X \times Y \to X \times Y \times Y$ in $\mathbf{Sp}$ has a left adjoint Eq, given on a closed subset $A \subseteq X \times Y$ by

$$\mathrm{Eq}(A) = \overline{\{(x, y, y') \in X \times Y \times Y \mid (x, y) \in A \text{ and } y = y'\}}.$$

(iii) Prove that equality on $Y \in \mathbf{Sp}$ is very strong (*i.e.* that internal equality and external equality on Y coincide) in this fibration of closed subsets if and only if Y is a Hausdorff space.

3.6 Fibred functorial semantics

In this section we start by describing appropriate morphisms between Eq-fibrations preserving the relevant structure. These allow us to describe functorial models of an equational specification in a fibration as morphisms of Eq-fibrations with the classifying fibration of the specification as domain. We also associate an equational specification with an Eq-fibration so that an adjoint correspondence between morphisms of Eq-fibrations and morphisms of equational specifications can be established (see Proposition 3.6.5).

In the last part of this section we show how every Eq-fibration gives rise to a (quotient) Eq-fibration in which internal and external equality are forced to be equal. This quotient enjoys a universal property, which is described in terms of morphisms of Eq-fibrations.

3.6.1. Definition. A morphism (or map) of Eq-fibrations $p \to q$ is a morphism of fibrations $p \to q$ which preserves the structure of Eq-fibrations: it preserves finite products in the base category, finite products in the fibre categories and equality Eq.

In this way we obtain a subcategory $\mathbf{EqFib} \hookrightarrow \mathbf{Fib}$ of Eq-fibrations. We may see $\mathbf{EqFib}$ as a 2-category, by letting the inclusion be full and faithful on 2-cells. Thus, 2-cells between maps of Eq-fibrations are the same as 2-cells between these maps as maps of fibrations.

If we have a morphism of Eq-fibrations in a situation

$$
\begin{array}{ccc}
\mathbb{E} & \xrightarrow{\quad H \quad} & \mathbb{D} \\
{\scriptstyle p}\downarrow & & \downarrow{\scriptstyle q} \\
\mathbb{B} & \xrightarrow{\quad K \quad} & \mathbb{A}
\end{array}
$$

then it is obvious that $K \colon \mathbb{B} \to \mathbb{A}$ preserves external equality: $u = v$ in $\mathbb{B}$ implies $Ku = Kv$ in $\mathbb{A}$. But K also preserves internal equality, since

$$
\begin{aligned}
u, v \text{ are internally equal} \;\;\Rightarrow\;\; & \top \leq \mathrm{Eq}(u, v) \\
\Rightarrow\;\; & \top \cong H(\top) \leq H(\mathrm{Eq}(u, v)) \cong \mathrm{Eq}(Ku, Kv) \\
& \text{(see Exercise 3.4.6)} \\
\Rightarrow\;\; & Ku, Kv \text{ are internally equal.}
\end{aligned}
$$

As an example, every finite limit preserving functor $\mathbb{B} \to \mathbb{A}$ between categories $\mathbb{B}, \mathbb{A}$ with finite limits induces a map of Eq-fibrations $\begin{smallmatrix} \mathrm{Sub}(\mathbb{B}) \\ \downarrow \\ \mathbb{B} \end{smallmatrix} \to \begin{smallmatrix} \mathrm{Sub}(\mathbb{A}) \\ \downarrow \\ \mathbb{A} \end{smallmatrix}$ between the corresponding subobject fibrations. This map between fibrations is described in Lemma 1.7.5.

3.6.2. Definition. Let $(\Sigma, \mathcal{H})$ be an equational specification and $\begin{smallmatrix} \mathbb{E} \\ \downarrow{\scriptstyle p} \\ \mathbb{B} \end{smallmatrix}$ an Eq-fibration. A **model of** $(\Sigma, \mathcal{H})$ **in** p is a morphism of Eq-fibrations:

$$
\left(\begin{smallmatrix} \mathcal{L}(\Sigma, \mathcal{H}) \\ \downarrow \\ \mathcal{C}\ell(\Sigma) \end{smallmatrix} \right) \longrightarrow \left(\begin{smallmatrix} \mathbb{E} \\ \downarrow{\scriptstyle p} \\ \mathbb{B} \end{smallmatrix} \right)
$$

This fibred functorial definition of model will be justified by the following result.

3.6.3. Theorem. *Let* $(\Sigma, \mathcal{H})$ *be an equational specification and let* $\begin{smallmatrix} \mathbb{E} \\ \downarrow{\scriptstyle p} \\ \mathbb{B} \end{smallmatrix}$ *be an Eq-fibration. Every model* $\mathcal{M}$ *in* $\mathbb{B}$

$$
\begin{array}{ccc}
 & & \mathbb{E} \\
 & & \downarrow{\scriptstyle p} \\
\mathcal{C}\ell(\Sigma) & \xrightarrow{\quad \mathcal{M} \quad} & \mathbb{B}
\end{array}
$$

validates the equations in $\mathcal{H}$ *(with respect to p) if and only if it extends to a (up-to-isomorphism unique) morphism of Eq-fibrations:*

$$
\begin{array}{ccc}
\mathcal{L}(\Sigma, \mathcal{H}) & \xrightarrow{\;\;\mathcal{M}'\;\;} & \mathbb{E} \\
\downarrow & & \downarrow{\scriptstyle p} \\
\mathcal{C}\!\ell(\Sigma) & \xrightarrow[\;\;\mathcal{M}\;\;]{} & \mathbb{B}
\end{array}
$$

A model of equations can thus be identified functorially as a morphism of Eq-fibrations. This extends functorial semantics from ordinary categories (see Sections 2.2 and 3.3) to fibred categories.

Proof. Suppose $\mathcal{M}$ validates the equations in $\mathcal{H}$ with respect to p. We can then define a functor $\mathcal{M}' \colon \mathcal{L}(\Sigma, \mathcal{H}) \to \mathbb{E}$ by

$$
(\Gamma \mid M_1 =_{\sigma_1} M_1', \ldots, M_n =_{\sigma_n} M_n') \mapsto
$$
$$
\mathrm{Eq}(\mathcal{M}(M_1), \mathcal{M}(M_1')) \wedge \cdots \wedge \mathrm{Eq}(\mathcal{M}(M_n), \mathcal{M}(M_n')).
$$

By soundness (Lemma 3.5.6) $\mathcal{M}'$ extends to a functor. The resulting pair $(\mathcal{M}, \mathcal{M}')$ preserves equality by Exercise 3.4.6.

Conversely, given the above extension $\mathcal{M}'$, for terms $\Gamma \vdash N, N' \colon \sigma$ we have

$$
\mathcal{M}'(\Gamma \mid N =_\sigma N') = \mathcal{M}'(\mathrm{Eq}(N, N')) \cong \mathrm{Eq}(\mathcal{M}(N), \mathcal{M}(N')).
$$

This shows that $\mathcal{M}'$ is determined up-to-isomorphism by $\mathcal{M}$. Further, because $\mathcal{M}'$ is a functor preserving fibred finite products, we obtain

$$
\begin{aligned}
& \Gamma \mid \vec{N} =_{\vec{\sigma}} \vec{N}' \vdash M =_\tau M' \quad \text{is derivable (from } \mathcal{H}) \\
\Rightarrow\quad & \mathrm{Eq}(N_1, N_1') \wedge \cdots \wedge \mathrm{Eq}(N_m, N_m') \\
& \qquad \leq \mathrm{Eq}(M, M') \quad \text{in } \mathcal{L}(\Sigma, \mathcal{H}) \\
\Rightarrow\quad & \mathcal{M}'(\mathrm{Eq}(N_1, N_1')) \wedge \cdots \wedge \mathcal{M}'(\mathrm{Eq}(N_m, N_m')) \\
& \qquad \leq \mathcal{M}'(\mathrm{Eq}(M, M')) \quad \text{in } \mathbb{E} \\
\Rightarrow\quad & \mathrm{Eq}(\mathcal{M}(N_1), \mathcal{M}(N_1')) \wedge \cdots \wedge \mathrm{Eq}(\mathcal{M}(N_m), \mathcal{M}(N_m')) \\
& \qquad \leq \mathrm{Eq}(\mathcal{M}(M), \mathcal{M}(M')) \\
\Rightarrow\quad & \Gamma \mid \vec{N} =_{\vec{\sigma}} \vec{N}' \vdash M =_\tau M' \quad \text{holds in } p.
\end{aligned}
$$

Hence $\mathcal{M}$ validates the theory of the equational specification $(\Sigma, \mathcal{H})$, and thus certainly its subset $\mathcal{H}$ of axioms. $\qquad\square$

In Section 3.3 we have defined for an (ordinary) category with finite products an associated theory of non-conditional equations. Similarly, for an Eq-fibration we will define a theory of conditional equations. Then, a model as above in Definition 3.6.2, can alternatively be described as a morphism of equational specifications using this induced theory.

3.6.4. Definition. Let $\begin{smallmatrix} \mathbb{E} \\ \downarrow p \\ \mathbb{B} \end{smallmatrix}$ be an Eq-fibration. The underlying signature $\mathrm{Sign}(\mathbb{B})$ of $\mathbb{B}$ comes naturally equipped with a set of equations $\mathcal{H}(p)$, namely:

$$\Gamma \mid M_1 =_{\sigma_1} M_1', \ldots, M_n =_{\sigma_n} M_n' \vdash N =_\tau N'$$

is in $\mathcal{H}(p)$, if and only if

$$\mathrm{Eq}(\varepsilon M_1, \varepsilon M_1') \wedge \cdots \wedge \mathrm{Eq}(\varepsilon M_n, \varepsilon M_n') \leq \mathrm{Eq}(\varepsilon N, \varepsilon N')$$

holds in the fibre over $\varepsilon\Gamma$—where ε is the model $\mathcal{C}\ell(\mathrm{Sign}(\mathbb{B})) \to \mathbb{B}$ of the signature of $\mathbb{B}$ in $\mathbb{B}$ itself. Thus $\mathcal{H}(p)$ contains all **Sign**($\mathbb{B}$)-equations which hold in

$$
\begin{array}{ccc}
 & & \mathbb{E} \\
 & & \downarrow p \\
\mathcal{C}\ell(\mathbf{Sign}(\mathbb{B})) & \xrightarrow{\quad \varepsilon \quad} & \mathbb{B}
\end{array}
$$

3.6.5. Proposition. *Let* $(\Sigma, \mathcal{H})$ *be an equational signature and* $\begin{smallmatrix} \mathbb{E} \\ \downarrow p \\ \mathbb{B} \end{smallmatrix}$ *an Eq-fibration. There is a bijective correspondence (up-to-isomorphism)*

$$
\frac{\left(\begin{smallmatrix} \mathcal{L}(\Sigma,\mathcal{H}) \\ \downarrow \\ \mathcal{C}\ell(\Sigma) \end{smallmatrix} \right) \xrightarrow{\;(\mathcal{M},\mathcal{N})\;} \left(\begin{smallmatrix} \mathbb{E} \\ \downarrow p \\ \mathbb{B} \end{smallmatrix} \right) \quad in \ \mathbf{EqFib}}{(\Sigma, \mathcal{H}) \xrightarrow{\quad\phi\quad} (\mathrm{Sign}(\mathbb{B}), \mathcal{H}(p)) \quad in \ \mathbf{EqSpec}}
$$

which makes the classifying fibration $\begin{smallmatrix} \mathcal{L}(\Sigma,\mathcal{H}) \\ \downarrow \\ \mathcal{C}\ell(\Sigma) \end{smallmatrix}$ *the free Eq-fibration generated by* $(\Sigma, \mathcal{H})$.

Proof. Remember from Theorem 2.2.5 the bijective correspondence

$$
\frac{\mathcal{C}\ell(\Sigma) \xrightarrow{\;\mathcal{M}\;} \mathbb{B} \quad in \ \mathbf{FPCat}}{\Sigma \xrightarrow{\quad\phi\quad} \mathrm{Sign}(\mathbb{B}) \quad in \ \mathbf{Sign}}
$$

It is then easy to see that $\mathcal{M}$ validates the equations in $\mathcal{H}$ with respect to p if and only if ϕ extends to a morphism of equational signatures $(\Sigma, \mathcal{H}) \to (\mathrm{Sign}(\mathbb{B}), \mathcal{H}(p))$. This because $\mathcal{M}$ and ϕ are related via ε: for a term N one has $\mathcal{M}(N) = \varepsilon(\phi N)$. $\qquad\square$

The above result gives rise to a model

$$
\left(
\begin{array}{c}
\mathcal{L}(\mathrm{Sign}(\mathbb{B}), \mathcal{H}(p)) \\
\downarrow \\
\mathcal{C}\ell(\mathrm{Sign}(\mathbb{B}))
\end{array}
\right)
\longrightarrow
\left(
\begin{array}{c}
\mathbb{E} \\
\downarrow p \\
\mathbb{B}
\end{array}
\right)
$$

of the theory of an Eq-fibration p in p itself. One can ask whether it is in general an equivalence, *i.e.* whether our equational logic is rich enough to reconstruct an Eq-fibration from its signature—as in the case of non-conditional equations for categories with finite products, see Theorem 3.3.8. The answer here is *no*, because an Eq-fibration may have many more 'predicates' (objects in the total category), than just equations $\mathrm{Eq}(u, v)$—which are the only propositions that we have in equational logic. In the next chapter on (first order) predicate logic we describe how these extra predicates can be incorporated in logic.

3.6.6. Remark. The notion of morphism between Eq-fibrations introduced in Definition 3.6.1 is in a sense the obvious one. But there is a reasonable alternative. One may wish to consider the functors between the base categories upto internal equality: call two morphisms of Eq-fibrations $(K, H), (K', H'): p \rightrightarrows q$ between Eq-fibrations $\begin{array}{c} \mathbb{E} \\ \downarrow p \\ \mathbb{B} \end{array}$ and $\begin{array}{c} \mathbb{D} \\ \downarrow q \\ \mathbb{A} \end{array}$ equivalent if

- $H = H': \mathbb{E} \to \mathbb{D}$ and on objects $K = K': \mathrm{Obj}\,\mathbb{B} \to \mathrm{Obj}\,\mathbb{A}$;
- Ku and $K'u$ are internally equal in q, for each morphism u in $\mathbb{B}$.

Equivalence classes of such morphisms then yield an alternative notion of map between Eq-fibrations. Its usefulness may be illustrated via the following two very simple algebraic specifications.

- Σ_1 has one type Ω, one function symbol $a: () \longrightarrow \Omega$ and no equations; so the set $\mathcal{H}_1$ of equations in this specification is empty.
- Σ_2 also has one type Ω, but two function symbols $b: () \longrightarrow \Omega$, $c: () \longrightarrow \Omega$ with a singleton set of equations $\mathcal{H}_2$ containing $\emptyset \mid \emptyset \vdash b =_\Omega c$.

One would expect these specifications to be (logically) equivalent (in an informal sense). Certainly, the signature Σ_2 has two function symbols, but they are required to be (internally) equal in the logic of $(\Sigma_2, \mathcal{H}_2)$. The classifying categories $\mathcal{C}\ell(\Sigma_1)$ and $\mathcal{C}\ell(\Sigma_2)$ of the signatures—without the equations—are not equivalent, simply because Σ_2 has more function symbols. Hence the classifying Eq-fibrations $\begin{array}{c} \mathcal{L}(\Sigma_1, \mathcal{H}_1) \\ \downarrow \\ \mathcal{C}\ell(\Sigma_1) \end{array}$ and $\begin{array}{c} \mathcal{L}(\Sigma_2, \mathcal{H}_2) \\ \downarrow \\ \mathcal{C}\ell(\Sigma_2) \end{array}$ are not equivalent with the notion of morphism in Definition 3.6.1. But we do have an equivalence of Eq-fibrations if we use the adapted notion of morphism that we just described, since it takes internal equality into account.

We have seen that putting an Eq-fibration $\begin{smallmatrix}\mathbb{E}\\\downarrow p\\\mathbb{B}\end{smallmatrix}$ on a base category $\mathbb{B}$ allows us to consider certain parallel morphisms in $\mathbb{B}$ as (internally) equal. This gives the possibility to identify these morphisms in $\mathbb{B}$ in a quotient category. Doing so actually leads to a quotient fibration $p \twoheadrightarrow p/\mathrm{Eq}$ in which internal and external equality are forced to coincide. We shall use the following notation for $p \twoheadrightarrow p/\mathrm{Eq}$.

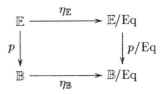

3.6.7. Definition. For an Eq-fibration $\begin{smallmatrix}\mathbb{E}\\\downarrow p\\\mathbb{B}\end{smallmatrix}$ we define two categories $\mathbb{B}/\mathrm{Eq}$ and $\mathbb{E}/\mathrm{Eq}$ as follows.

$\mathbb{B}/\mathrm{Eq}$	**objects**	$I \in \mathbb{B}$.
	morphisms	$[u]: I \to J$ are equivalence classes $[u]$ of morphisms $u: I \to J$ in $\mathbb{B}$, where $u, u': I \rightrightarrows J$ in $\mathbb{B}$ are equivalent if they are internally equal, *i.e.* if $\top \le \mathrm{Eq}(u, u')$ holds in the fibre $\mathbb{E}_I$.
$\mathbb{E}/\mathrm{Eq}$	**objects**	$X \in \mathbb{E}$.
	morphisms	$[f]: X \to Y$ are equivalence classes of maps $f: X \to Y$ in $\mathbb{E}$, with $f, f': X \rightrightarrows Y$ equivalent if pf, pf' are equivalent in $\mathbb{B}$.

These categories $\mathbb{B}/\mathrm{Eq}$ and $\mathbb{E}/\mathrm{Eq}$ are quotients of $\mathbb{B}$ and $\mathbb{E}$ via obvious functors $\eta_{\mathbb{B}}: \mathbb{B} \twoheadrightarrow \mathbb{B}/\mathrm{Eq}$ and $\eta_{\mathbb{E}}: \mathbb{E} \twoheadrightarrow \mathbb{E}/\mathrm{Eq}$.

Finally, the functor $p/\mathrm{Eq}: \mathbb{E}/\mathrm{Eq} \to \mathbb{B}/\mathrm{Eq}$ is defined by $X \mapsto pX$ and $([f]: X \to Y) \mapsto ([pf]: pX \to pY)$.

3.6.8. Proposition. (i) *The functor* $\begin{smallmatrix}\mathbb{E}/\mathrm{Eq}\\\downarrow p/\mathrm{Eq}\\\mathbb{B}/\mathrm{Eq}\end{smallmatrix}$ *introduced above is an Eq-fibration in which internal and external equality coincide.*

(ii) *The pair of functors* $\eta = (\eta_{\mathbb{B}}, \eta_{\mathbb{E}})$ *forms a morphism of Eq-fibrations* $\eta: p \twoheadrightarrow p/\mathrm{Eq}$, *which is universal in the following sense: every map of Eq-fibrations* $p \to q$ *to an Eq-fibration q in which internal and external equality* *coincide factors via a unique map of Eq-fibrations $p/\mathrm{Eq} \dashrightarrow q$ as:*

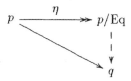

Before we give the proof, we recall that Eq-fibrations—like all preorder fibrations—are faithful, as functors, see Exercise 1.3.11. And in these preorders we have vertical isomorphisms $u^*(X) \cong v^*(X)$ for internally equal parallel maps u, v, since by the substitution combinator from Lemma 3.4.5 we get:

$$u^*(X) \cong u^*(X) \wedge \top \cong u^*(X) \wedge \mathrm{Eq}(u, v) \leq v^*(X).$$

The inequality $u^*(X) \geq v^*(X)$ is obtained by symmetry.

Proof. (i) We first show that p/Eq is a fibration. For an object $Y \in \mathrm{Obj}\,(\mathbb{E}/\mathrm{Eq}) = \mathrm{Obj}\,(\mathbb{E})$ and a morphism $[u]\colon I \to pY$ in $\mathbb{B}/\mathrm{Eq}$, we choose a representative $u\colon I \to pY$ in $\mathbb{B}$, and a Cartesian lifting $\overline{u}(Y)\colon u^*(Y) \to Y$ of u in $\mathbb{E}$. It gives a morphism $[\overline{u}(Y)]\colon u^*(Y) \to Y$ in $\mathbb{E}/\mathrm{Eq}$ over $[u]\colon I \to pY$ in $\mathbb{B}/\mathrm{Eq}$. We claim that it is a p/Eq-Cartesian lifting: for a map $[f]\colon X \to Y$ in $\mathbb{E}/\mathrm{Eq}$ with $[f] = [u] \circ [v]$ in $\mathbb{B}/\mathrm{Eq}$, so that $\top \leq \mathrm{Eq}(pf, u \circ v)$, we obtain a mediating map $X \to u^*(Y)$ in $\mathbb{E}$ as composite:

$$X \leq (pf)^*(Y) \cong (u \circ v)^*(Y) \cong v^*u^*(Y) \longrightarrow u^*(Y)$$

It yields the required mediating map in $\mathbb{E}/\mathrm{Eq}$.

We notice that the fibre category of p/Eq over $I \in \mathbb{B}/\mathrm{Eq}$ is the same as the fibre category of p over $I \in \mathbb{B}$. Hence p/Eq has fibred finite products $(\top, \wedge)$ and equality Eq as in p. By construction:

> $[u], [v]\colon I \rightrightarrows J$ in $\mathbb{B}/\mathrm{Eq}$ are internally equal in p/Eq
>
> $\Leftrightarrow \top \leq \mathrm{Eq}(u, v)$
>
> $\Leftrightarrow [u] = [v]$
>
> $\Leftrightarrow [u], [v]$ are externally equal.

(ii) Assume $\begin{smallmatrix} \mathbb{D} \\ \downarrow q \\ \mathbb{A} \end{smallmatrix}$ is an Eq-fibration in which internal and external equality coincide, and $(K\colon \mathbb{B} \to \mathbb{A}, H\colon \mathbb{E} \to \mathbb{D})$ is a morphism of Eq-fibrations $p \to q$.

We define two functors $\overline{K}, \overline{H}$ in

by

$$\overline{K} = \left\{ \begin{array}{c} I \mapsto KI \\ (I \xrightarrow{[u]} J) \mapsto (I \xrightarrow{Ku} J) \end{array} \right. \qquad \overline{H} = \left\{ \begin{array}{c} X \mapsto HI \\ (X \xrightarrow{[f]} Y) \mapsto (HX \xrightarrow{Hf} HY) \end{array} \right.$$

These functors $\overline{K}, \overline{H}$ are well-defined, since

$$\begin{aligned} u \sim u' \text{ in } \mathbb{B} &\Rightarrow \top \leq \mathrm{Eq}(u, u') \text{ in } \mathbb{E} \\ &\Rightarrow \top \leq \mathrm{Eq}(Ku, Ku') \text{ in } \mathbb{D} \\ &\Rightarrow Ku = Ku'. \end{aligned}$$

The last implication holds because internal and external equality coincide in q. Similarly:

$$\begin{aligned} f \sim f' \text{ in } \mathbb{E} &\Rightarrow \top \leq \mathrm{Eq}(pf, pf') \text{ in } \mathbb{E} \\ &\Rightarrow \top \leq \mathrm{Eq}(Kpf, Kpf') = \mathrm{Eq}(qHf, qHf') \text{ in } \mathbb{D} \\ &\Rightarrow qHf = qHf' \\ &\Rightarrow Hf = Hf' \qquad \text{(see Exercise 1.3.11).} \end{aligned}$$

Since p/Eq inherits its Eq-fibration structure from p, this pair $(\overline{K}, \overline{H})$ forms a morphism of Eq-fibrations. □

Exercises

3.6.1.   Show that for an Eq-fibration p, the result $K^*(p)$ of change-of-base along a finite product preserving functor K is again an Eq-fibration and that the morphism $K^*(p) \to p$ involved forms a morphism of Eq-fibrations.

3.6.2.   Let $\begin{smallmatrix} \mathbb{E} \\ \downarrow p \\ \mathbb{B} \end{smallmatrix}$ be an Eq-fibration, and let $\mathrm{Eq}(\mathbb{E}) \hookrightarrow \mathbb{E}$ be the full subcategory of those $X \in \mathbb{E}$ for which there is a vertical isomorphism $X \cong \mathrm{Eq}(u, v)$, for certain maps $u, v \colon pX \rightrightarrows \bullet$ in $\mathbb{B}$. Prove that $\begin{smallmatrix} \mathrm{Eq}(\mathbb{E}) \\ \downarrow \\ \mathbb{B} \end{smallmatrix}$ is also an Eq-fibration, and that the inclusion is a morphism of Eq-fibrations.

[Remember Proposition 3.4.6 and Exercise 3.4.5.]

3.6.3.   Consider the category $\mathbb{B}/\mathrm{Eq}$ in Definition 3.6.7 and show in detail that

(i)   its composition can be defined from composition in $\mathbb{B}$ via representatives;

(ii)   it has finite products.

Chapter 4

First order predicate logic

Equational logic, as studied in the previous chapter, is not very expressive. It allows us to formulate statements like $x : \mathbb{N} \mid x + 2 =_{\mathbb{N}} 5 \vdash x =_{\mathbb{N}} 3$, but not much more. In the present chapter we will study (first order) predicate logic (over simple type theory), in which we can formulate more interesting statements like $x, y : \mathbb{Q} \mid x <_{\mathbb{Q}} y \vdash \exists z : \mathbb{Q}. \, x <_{\mathbb{Q}} z \land z <_{\mathbb{Q}} y$. This requires more general atomic propositions $x <_{\mathbb{Q}} y$ than just equations $x =_{\mathbb{Q}} y$. And further, it requires additional logical operations like $\land, \exists$.

In this chapter we consider *first order* predicate logic where one can quantify over types σ, in propositions of the form $\forall x : \sigma. \, \varphi$ and $\exists x : \sigma. \, \varphi$. In the next chapter we study higher order predicate logic in which one can additionally quantify over propositions and predicates, like in $\forall \alpha : \mathsf{Prop}. \, \varphi$ and $\exists \alpha : \mathsf{Prop}. \, \varphi$. What we consider here is *simple* predicate logic (SPL),indexSSimple!– predicate logic or predicate logic over *simple type theory*, in contrast to dependent predicate logic (over dependent type theory) or polymorphic predicate logic (over polymorphic type theory), see Sections 8.6 and 11.1. This means that the types in our simple predicate logic are types from simple type theory, which do not contain (term or type) variables: they are built up from constants, using type constructors like $+, \times, \to$ as studied in Chapter 2. In standard texts on mathematical logic it is common to consider only single-typed (or single-sorted, in more traditional terminology) predicate logic with only one type, but in computer science, many-typed logic is more natural.

The categorical models that we shall use to describe predicate logics are certain kinds of preordered fibrations. The preorderedness makes these fibrations so-called proof-irrelevance models in which provability, and not proof, is captured: $\varphi \vdash \psi$ in predicate logic means that *there is* a proof of ψ which assumes φ. This makes the turnstile $\vdash$ a preorder relation. In contrast, formal (type theoretic) systems with explicitly proof-terms $x : \varphi \vdash P : \psi$—providing an actual proof of ψ, assuming a proof $x : \varphi$ of φ—lead to non-preordered models with proof-terms as arrows $\varphi \rightarrow \psi$.

The operations in predicate logic are described categorically via adjunctions in these fibrations for predicate logics. Existential $\exists$ and universal $\forall$ quantification form left and right adjoints to weakening functors, equality $=_{\sigma}$ forms left adjoints to contraction functors, subset types $\{x : \sigma \mid \varphi\}$ form a right adjoint to a truth predicate functor, and quotient types σ / R form a left adjoint to an equality relation functor. Equivalently, one can describe subset types by a right adjoint to this equality relation functor. The adjunctions for $\exists, \forall, =_{\sigma}$ are between fibre categories, whereas the adjunctions for $\{x : \sigma \mid \varphi\}$ and σ / R are between the total and base category of a fibration. The introductory Section 0.2 gives a brief presentation of these adjunctions for the familiar logic of predicates on sets.

Lawvere may be seen as the first to use fibred (or indexed) categories in logic, for example in [193]. Some of the details involved are elaborated in [305]. Fibred categories for predicate logic are used subsequently for example in [62, 209, 210, 336]. Since the 1970s much of categorical logic has been done in direct contact with topos theory. As a result, logic is often described in terms of subobject fibrations, see for example [211] and [85]. Here we use general fibred categories for predicate logic, and subobject fibrations occur as special instances. The advantages of this more general approach are that it provides

- more flexibility: a base category $\mathbb{B}$ may carry different logics, and not just its subobject logic, see Examples 3.5.4 and 3.5.5 where we have two different logics on the category $\mathbb{B} = \mathbf{Rel}$ of relations given by two different fibrations.
- natural, unified presentations of examples as they come from realisability, frames (complete Heyting algebras), Kripke models, or cylindric algebras, see Section 4.2.
- a framework in which all the logical operations can be studied separately. In subobject fibrations much structure comes for granted, like equality, unique existence $\exists!$ or subset types, see Section 4.9.
- a presentation which scales up from logic to type theory in a direct manner.

This chapter starts with appropriate signatures for predicate logic, containing not only typed function symbols $F : \sigma_1, \ldots, \sigma_n \longrightarrow \sigma_{n+1}$ but also typed predicate symbols $P : \sigma_1, \ldots, \sigma_n$. With these one can form besides equations $M =_{\sigma} M'$ also other atomic propositions $P(M_1, \ldots, M_n)$, for example for reasoning with inequalities $M \leq_{\sigma} M'$. Next, in Section 4.2 we describe fibrations for first order logic—and for the subsystems of what is called regular and coherent logic. The main novelty is that the quantifiers $\exists$ and $\forall$ are left and right adjoints to weakening functors. A series of examples of such fibrations is included. Especially we shall elaborate classifying fibrations (or term models) built up from syntax. Predicate logic is rich enough to reconstruct the fibration we use from the classifying fibration of its own signature. The logic thus associated with a fibration will be called its **internal language**, or **internal logic**. This language facilitates dealing with fibrations of predicate logic, because it allows one to replace categorical calculation by logical reasoning. The internal language will be described and used in Section 4.3. Then, in Sections 4.4 and 4.5 we concentrate on subobject fibrations. Among the fibred categories used to model predicate logics, subobject fibrations are very special (for example because of their rôle in topos theory) and will therefore they be investigated separately. The subsequent three sections will be about subset types and quotient types. With these fundamental mathematical constructions we can form new types of the form $\{x \colon \sigma \,|\, \varphi\}$, where σ is a type and φ a proposition, and of the form σ/R, where R is a binary relation on σ. Both subset types and quotient types can be described categorically by adjoints. We conclude this chapter with a characterisation of subobject fibrations. They are fibrations for predicate logic in which one has: very strong equality, full subset types, and unique choice $\exists!$.

4.1 Signatures, connectives and quantifiers

Up-to-now we have studied (typed) terms and equations between them in equational logic. These equations are the only kind of propositions that we have seen so far. Our next step is also to allow predicates as atomic propositions and form derived propositions using logical connectives, like implication $\supset$ or existential quantification $\exists$. In this section we describe the syntactic aspects of these extensions. It will involve signatures which not only have function symbols but also predicate symbols, and specifications, which are signatures together with a collection of axioms. After these preliminaries on how to form the atomic propositions, we describe the (standard) rules of first order predicate logic. In the end we reformulate the rules for typed equality $=_\sigma$, universal $\forall x \colon \sigma. \, (-)$ and existential $\exists x \colon \sigma. \, (-)$ quantification as 'mate' rules. These essentially exhibit these logical operations as adjoints.

In equational logic one only has equations

$$M =_\sigma M'$$

for terms M and M' of the same type σ, as (atomic) propositions. In this chapter we will also allow atomic propositions

$$P(M_1, \ldots, M_n)$$

where $P: \sigma_1, \ldots, \sigma_n$ is a **predicate symbol** and $M_1: \sigma_1, \ldots, M_n: \sigma_n$ are appropriately typed terms. To allow such predicate symbols, we have to extend our notion of signature.

We say that a **signature with predicates** is a pair (Σ, Π) where Σ is a many-typed signature and Π is a function $|\Sigma|^\star \to \mathbf{Sets}$, which yields for each sequence $\sigma_1, \ldots, \sigma_n$ of types a set $\Pi(\sigma_1, \ldots, \sigma_n)$ of predicate symbols of this type. We shall write

$$P: \sigma_1, \ldots, \sigma_n \qquad \text{for} \qquad P \in \Pi(\sigma_1, \ldots, \sigma_n).$$

A morphism $\phi: (\Sigma, \Pi) \to (\Sigma', \Pi')$ of such signatures with predicates consists of three mappings, sending types to types, function symbols to function symbols, and predicate symbols to predicate symbols, in such a way that arities are preserved. Following Convention 1.6.2, we shall use the symbol ϕ for all three mappings. This yields the following requirements.

$$F: \sigma_1, \ldots, \sigma_n \longrightarrow \sigma_{n+1} \; \Rightarrow \; \phi(F): \phi(\sigma_1), \ldots, \phi(\sigma_n) \longrightarrow \phi(\sigma_{n+1})$$
$$P: \sigma_1, \ldots, \sigma_n \; \Rightarrow \; \phi(P): \phi(\sigma_1), \ldots, \phi(\sigma_n).$$

As a result, we get a category, which is written as **SignPred**. Using change-of-base, it can be obtained simply as follows (see also Definition 1.6.1).

4.1.1. Definition. The category **SignPred** of **signatures with predicates** arises in the following change-of-base situation.

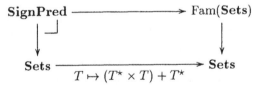

(In the first section of the next chapter on higher order logic we use many-typed signatures containing a distinguished type Prop of propositions. In such signatures there is no need for predicate symbols $P: \sigma_1, \ldots, \sigma_n$, since they can be described as function symbols $P: \sigma_1, \ldots, \sigma_n \longrightarrow$ Prop.)

One can view a signature with predicates as a first order specification—without axioms yet; these will be added in Definition 4.1.3 below.

4.1.2. Example. In a signature containing a type N of natural numbers together with a type NList of finite lists of these, one may have function and predicate symbols

$$\text{insert: N} \longrightarrow \text{NList}, \qquad \text{IsEmpty: NList}, \qquad \text{Occurs?: N, NList}$$

where the latter predicate tells of a natural number n and a list ℓ whether or not n occurs in ℓ. One expects as an axiom

$$n\colon \mathsf{N} \mid \emptyset \vdash \mathsf{Occurs?}\,(n, \mathsf{insert}\,(n)).$$

Let (Σ, Π) be a signature with predicates. The associated **atomic propositions** have the form

$$M =_\sigma M' \qquad \text{and} \qquad P(M_1, \ldots, M_n)$$

where M, M' are of the same type σ, and $P\colon \sigma_1, \ldots, \sigma_n$ is a predicate symbol with appropriately typed terms $M_i\colon \sigma_i$. A bit formally, using a new syntactic category (or universe) Prop, we can write these as formation rules:

atomic equation proposition

$$\frac{\Gamma \vdash M\colon \sigma \qquad \Gamma \vdash M'\colon \sigma}{\Gamma \vdash (M =_\sigma M')\colon \mathsf{Prop}}$$

atomic predicate proposition

$$\frac{\Gamma \vdash M_1\colon \sigma_1 \quad \cdots \quad \Gamma \vdash M_n\colon \sigma_n}{\Gamma \vdash P(M_1, \ldots, M_n)\colon \mathsf{Prop}} \;(\text{for } P\colon \sigma_1, \ldots, \sigma_n)$$

Substitution over these atomic propositions takes the form

$$(M =_\sigma M')[N/x] \;\equiv\; (M[N/x]) =_\sigma (M'[N/x])$$
$$P(M_1, \ldots, M_n)[N/x] \;\equiv\; P(M_1[N/x], \ldots, M_n[N/x]).$$

The following **connectives** or **logical operations** may be used in first order logic to construct new propositions.

$\bot$	*falsum*, absurdity, the universally false proposition;
$\top$	truth, the universally true proposition;
$\neg\varphi$	negation: not φ;
$\varphi \wedge \psi$	conjunction: φ and ψ;
$\varphi \vee \psi$	disjunction: φ or ψ;
$\varphi \supset \psi$	implication: if φ then ψ;
$\forall x\colon \sigma.\,\varphi$	universal quantification over type σ: for all x in σ, φ;
$\exists x\colon \sigma.\,\varphi$	existential quantification over type σ: for some x in σ, φ.

In the last two cases the term variable x becomes bound in the quantified propositions $\forall x\colon \sigma.\,\varphi$ and $\exists x\colon \sigma.\,\varphi$. Formally, one can write all of these as formation rules. For example,

$$\frac{\Gamma \vdash \varphi\colon \mathsf{Prop} \qquad \Gamma \vdash \psi\colon \mathsf{Prop}}{\Gamma \vdash \varphi \wedge \psi\colon \mathsf{Prop}} \qquad\qquad \frac{\Gamma, x\colon \sigma \vdash \varphi\colon \mathsf{Prop}}{\Gamma \vdash \exists x\colon \sigma.\,\varphi\colon \mathsf{Prop}}$$

but that is a bit cumbersome, really. Then, given propositions

$$\Gamma \vdash \varphi_1 : \mathsf{Prop}, \quad \cdots \quad \Gamma \vdash \varphi_n : \mathsf{Prop}, \qquad \Gamma \vdash \psi : \mathsf{Prop}$$

we can form a sequent,

$$\Gamma \mid \varphi_1, \ldots, \varphi_n \vdash \psi$$

which is read as: assuming term variable declarations $x : \sigma$ in Γ, then the proposition ψ follows as conclusion from propositions $\varphi_1, \ldots, \varphi_n$. The latter sequence $\varphi_1, \ldots, \varphi_n$ will be called the proposition context; we often abbreviate it as Θ, Ξ, like in the previous section. Recall that Γ is called the type context.

For completeness, we list the natural deduction rules of predicate logic in Figure 4.1 (apart from the context rules, as already described in section 3.1). These rules have all of the assumptions explicit at every stage, in type and proposition contexts Γ and Θ.

Recall that a sequent $\Gamma \mid \Theta \vdash \varphi$ is **derivable** if there is a derivation tree regulated by these rules, with $\Gamma \mid \Theta \vdash \varphi$ as conclusion. The sequents at the top of this tree may be axioms. We sometimes write

$$\blacktriangleright \Gamma \mid \Theta \vdash \varphi$$

to express that the sequent $\Gamma \mid \Theta \vdash \varphi$ is derivable. For example, one has

$$\blacktriangleright \Gamma \mid \varphi, (\varphi \wedge \psi) \supset \chi \vdash \psi \supset \chi$$

for propositions $\Gamma \vdash \varphi, \psi, \chi : \mathsf{Prop}$ in type context Γ, as shown by the following derivation.

$$
\cfrac{
\cfrac{
\Gamma \mid (\varphi \wedge \psi) \supset \chi \vdash (\varphi \wedge \psi) \supset \chi
}{
\Gamma \mid \varphi, (\varphi \wedge \psi) \supset \chi, \psi \vdash (\varphi \wedge \psi) \supset \chi
}
\qquad
\cfrac{
\cfrac{
\cfrac{\Gamma \mid \varphi \vdash \varphi}{\Gamma \mid \varphi, \psi \vdash \varphi}
\qquad
\cfrac{\Gamma \mid \psi \vdash \psi}{\Gamma \mid \varphi, \psi \vdash \psi}
}{
\Gamma \mid \varphi, \psi \vdash \varphi \wedge \psi
}
}{
\Gamma \mid \varphi, (\varphi \wedge \psi) \supset \chi, \psi \vdash \varphi \wedge \psi
}
}{
\cfrac{
\Gamma \mid \varphi, (\varphi \wedge \psi) \supset \chi, \psi \vdash \chi
}{
\Gamma \mid \varphi, (\varphi \wedge \psi) \supset \chi \vdash \psi \supset \chi
}
}
$$

The reader may notice that negation ($\neg$) does not occur among these rules. The reason is that negation is defined as

$$\neg \varphi \stackrel{\text{def}}{=} \varphi \supset \bot.$$

Classical logic is then obtained by adding the rule

reductio ad absurdum

$$\cfrac{\Gamma \mid \Theta, \neg \varphi \vdash \bot}{\Gamma \mid \Theta \vdash \varphi}$$

$$\frac{}{\Gamma \mid \Theta \vdash \top}$$

$$\frac{}{\Gamma \mid \Theta, \bot \vdash \psi}$$

$$\frac{\Gamma \mid \Theta \vdash \varphi \quad \Gamma \mid \Theta \vdash \psi}{\Gamma \mid \Theta \vdash \varphi \wedge \psi}$$

$$\frac{\Gamma \mid \Theta \vdash \varphi \wedge \psi}{\Gamma \mid \Theta \vdash \varphi}$$

$$\frac{\Gamma \mid \Theta \vdash \varphi \wedge \psi}{\Gamma \mid \Theta \vdash \psi}$$

$$\frac{\Gamma \mid \Theta \vdash \varphi}{\Gamma \mid \Theta \vdash \varphi \vee \psi}$$

$$\frac{\Gamma \mid \Theta \vdash \psi}{\Gamma \mid \Theta \vdash \varphi \vee \psi}$$

$$\frac{\Gamma \mid \Theta, \varphi \vdash \chi \quad \Gamma \mid \Theta, \psi \vdash \chi}{\Gamma, \mid \Theta, \varphi \vee \psi \vdash \chi}$$

$$\frac{\Gamma \mid \Theta, \varphi \vdash \psi}{\Gamma \mid \Theta \vdash \varphi \supset \psi}$$

$$\frac{\Gamma \mid \Theta \vdash \varphi \supset \psi \quad \Gamma \mid \Theta \vdash \varphi}{\Gamma \mid \Theta \vdash \psi}$$

$$\frac{\Gamma, x{:}\sigma \mid \Theta \vdash \psi}{\Gamma \mid \Theta \vdash \forall x{:}\sigma. \psi}$$
$$(x \text{ not free in } \Theta)$$

$$\frac{\Gamma \vdash M{:}\sigma \quad \Gamma \mid \Theta \vdash \forall x{:}\sigma. \psi}{\Gamma \mid \Theta \vdash \psi[M/x]}$$

$$\frac{\Gamma \vdash M{:}\sigma \quad \Gamma \mid \Theta \vdash \psi[M/x]}{\Gamma \mid \Theta \vdash \exists x{:}\sigma. \psi}$$

$$\frac{\Gamma \mid \Theta \vdash \exists x{:}\sigma. \psi \quad \Gamma, x{:}\sigma \mid \Xi, \psi \vdash \chi}{\Gamma \mid \Theta, \Xi \vdash \chi}$$
$$(x \text{ not free in } \Xi, \chi)$$

$$\frac{\Gamma \vdash M = M'{:}\sigma}{\Gamma \mid \Theta \vdash M =_\sigma M'}$$

$$\frac{\Gamma \mid \Theta \vdash M =_\sigma M' \quad \Gamma \mid \Theta \vdash M' =_\sigma M''}{\Gamma \mid \Theta \vdash M =_\sigma M''}$$

$$\frac{\Gamma \mid \Theta \vdash M =_\sigma M'}{\Gamma \mid \Theta \vdash M' =_\sigma M}$$

$$\frac{\Gamma \mid \Theta \vdash M =_\sigma M' \quad \Gamma \mid \Theta \vdash \psi[M/x]}{\Gamma \mid \Theta \vdash \psi[M'/x]}$$
(this rule will be called **replacement**)

Fig. 4.1. Rules for (many-typed) first order predicate logic

This rule says that if it is absurd to assume that φ is false, then φ must be true. It is an indirect, non-constructive principle of reasoning. One can show that it is equivalent to the excluded middle $\varphi \vee \neg \varphi$ axiom, also called *tertium non datur*, see Exercise 4.1.3. This rule will not be assumed, unless stated explicitly.

As another abbreviation, we shall use

$$\varphi \mathbb{I} \psi \stackrel{\text{def}}{=} (\varphi \supset \psi) \wedge (\psi \supset \varphi)$$

for logical equivalence.

The following rule in this list,

$$\frac{\Gamma \vdash M = M' : \sigma}{\Gamma \mid \Theta \vdash M =_\sigma M'}$$

deserves some special attention. It tells us first of all that convertible terms $(M = M' : \sigma)$ in the underlying type theory give rise to derivable equality propositions $(M =_\sigma M')$ in the logic. Thus propositional equality includes conversion, or in different terminology, internal equality includes external equality. The converse may also be required as a rule, but that is not done here. Because conversion in type theory is reflexive, this rule tells us in particular that logical equality $=_\sigma$ is reflexive. (Symmetry and transitivity of $=_\sigma$ are given by explicit rules.) And in case one considers an elementary term calculus without basic conversions (*e.g.* because there are no type constructors like $\rightarrow$, $\times$ or $+$), then one may still consider $M = M : \sigma$ as part of a trivial conversion relation, guaranteeing the presence of a reflexivity rule.

Since term variables $x : \sigma$ may occur in propositions φ, the question arises how propositions $\varphi[M/x]$ and $\varphi[M'/x]$ for convertible terms $M = M' : \sigma$ are related. It turns out that they are equivalent—*i.e.* that $\varphi[M/x] \mathbb{I} \varphi[M'/x]$ is derivable: the conversion $M = M' : \sigma$ leads to a proposition $M =_\sigma M'$ by the above rule, which can be used to derive $\varphi[M/x]$ from $\varphi[M'/x]$ by the replacement rule, and thus to derive $\varphi[M/x] \supset \varphi[M'/x]$. The reverse implication is obtained similarly.

Substitution over these propositions is done in the familiar way, *i.e.*,

$$\top[L/z] \equiv \top$$
$$(\varphi \wedge \psi)[L/z] \equiv (\varphi[L/z]) \wedge (\psi[L/z])$$
$$\bot[L/z] \equiv \bot$$
$$(\varphi \vee \psi)[L/z] \equiv (\varphi[L/z]) \vee (\psi[L/z])$$
$$(\varphi \supset \psi)[L/z] \equiv (\varphi[L/z]) \supset (\psi[L/z])$$
$$(\forall x : \sigma. \psi)[L/z] \equiv \forall x : \sigma. (\psi[L/z])$$
$$(\exists x : \sigma. \psi)[L/z] \equiv \exists x : \sigma. (\psi[L/z])$$

where in the latter two cases it is assumed that the variable x is different from z and does not occur free in L. By renaming of bound variables, this can always be assured.

In what we call (full) first order logic all of the above connectives and quantifiers can be used. We mention two interesting subsystems explicitly:

| **regular logic** | only has | $=, \wedge, \top, \exists$ |
| **coherent logic** | only has | $=, \wedge, \top, \vee, \bot, \exists.$ |

The expressions 'regular' and 'coherent' will also be used for propositions in these logics (which may contain only the above symbols as connectives). We note that *classical* coherent logic—with a negation operation $\varphi \mapsto \neg\varphi$ behaving as in the *reductio ad absurdum* rule—is the same as classical (full) first order logic, since in classical coherent logic $\supset$ and $\forall$ are definable. So a restriction to coherent logic is only meaningful in a constructive setting.

4.1.3. Definition. (i) A **first order specification** is a triple $(\Sigma, \Pi, \mathcal{A})$ where (Σ, Π) is a signature with predicates and $\mathcal{A}$ is a collection of **axioms**; the latter are sequents in the language of (Σ, Π).

(ii) A **regular** (or **coherent**) specification has regular (or coherent) propositions as axioms.

4.1.4. Definition. (i) A first order specification $(\Sigma, \Pi, \mathcal{A})$ is a **first order theory** if the collection $\mathcal{A}$ of axioms is closed under derivability. Every such specification evidently determines a theory $Th(\Sigma, \Pi, \mathcal{A})$ which can be obtained by closing $\mathcal{A}$ under derivability.

(ii) A **morphism of first order specifications** $(\Sigma, \Pi, \mathcal{A}) \to (\Sigma', \Pi', \mathcal{A}')$ is a morphism $\phi \colon (\Sigma, \Pi) \to (\Sigma', \Pi')$ of signatures with predicates such that for each sequent

$$\Gamma \mid \Theta \vdash \chi$$

in $\mathcal{A}$, one has that the sequent obtained by ϕ-translation

$$\phi(\Gamma) \mid \phi(\Theta) \vdash \phi(\chi)$$

is in $Th(\Sigma', \Pi', \mathcal{A}')$. This yields a category **FoSpec**.

Similar definitions can be given for regular and coherent signatures.

4.1.5. Remark. Earlier we have been careful in distinguishing equality in internal (propositional, in the logic of a fibration) and external (in the base category) form. Axioms, as described in the previous definition, can only capture internal equations. If we wish to have external equations in our logic, then we have to add these explicitly as an additional set of (algebraic) equations, like in Definition 3.2.5 (i). In general, we shall not do so, except when reconstructing a fibration from its internal logic, see Section 4.3 (notably in

Definition 4.3.5). A morphism preserving the structure of such extended specifications is a morphism as in (ii) above, which is additionally a morphism of algebraic specifications, as in Definition 3.2.6.

In the remainder of this section, we shall reformulate the rules for $=, \exists$ and $\forall$ in Figure 4.1 in order to make them more amenable to a categorical description in the next section. First we take a closer look at the last four rules on equality in Figure 4.1. We show that replacement rule from equational logic, denoted as (EL-R) for convenience, and the replacement rule from predicate logic, written as (PL-R), are of equal strength. This shows that there is no omission in the list of rules in Figure 4.1.

4.1.6. Lemma. *The* replacement *rule from equational logic (see Section 3.2),*

$$\frac{\Gamma \mid \Theta \vdash M =_\sigma M' \qquad \Gamma, x \colon \sigma \vdash N \colon \tau}{\Gamma \mid \Theta \vdash N[M/x] =_\tau N[M'/x]} \text{ (EL-R)}$$

is a consequence of the replacement *rule from predicate logic*

$$\frac{\Gamma \mid \Theta \vdash M =_\sigma M' \qquad \Gamma \mid \Theta \vdash \varphi[M/x]}{\Gamma \mid \Theta \vdash \varphi[M'/x]} \text{ (PL-R)}$$

as given in Figure 4.1. In the reverse direction, the rule (PL-R) *restricted to equations follows from the rule* EL-R. *Here we assume reflexivity, symmetry and transitivity of equality in the background.*

Proof. Assume the rule (PL-R) and let φ be the equation $N[M/x] =_\tau N$. Notice that the variable x occurs free in φ, only in N on the right hand side. By reflexivity one has

$$\Gamma \mid \Theta \vdash \varphi[M/x]$$

and thus by (EL-R)

$$\Gamma \mid \Theta \vdash \varphi[M'/x]$$

i.e.

$$\Gamma \mid \Theta \vdash N[M/x] =_\tau N[M'/x].$$

In the reverse direction, assume (EL-R). Let φ be an equation $N =_\tau N'$. Then, the assumption

$$\Gamma \mid \Theta \vdash N[M/x] =_\tau N'[M/x]$$

together with the two conclusions of (EL-R), applied to N and to N',

$$\Gamma \mid \Theta \vdash N[M/x] =_\tau N[M'/x] \quad \text{and} \quad \Gamma \mid \Theta \vdash N'[M/x] =_\tau N'[M'/x]$$

yield by symmetry and transitivity the required result:

$$\Gamma \mid \Theta \vdash N[M'/x] =_\tau N'[M'/x]. \qquad \square$$

We continue with an adaptation of Lemma 3.2.3 to first order logic. It enables us to use Lawvere's description of equality as left adjoints to contraction functors also in predicate logic; it will be used in the next section.

4.1.7. Lemma. *The last four rules in Figure 4.1 about equality are of the same strength as the (double) rule*

Lawvere equality

$$\frac{\Gamma, x\colon \sigma \mid \Theta \vdash \varphi[x/y]}{\Gamma, x\colon \sigma, y\colon \sigma \mid \Theta, x =_\sigma y \vdash \varphi} \ \text{(Eq-mate)}$$

involving a proposition φ *in type context* $\Gamma, x\colon \sigma, y\colon \sigma$.

The extended "Frobenius" version of this result (involving a proposition context of the form $\Theta[x/y]$ instead of Θ above the lines), as in Lemma 3.2.4 is left to the reader in Exercise 4.1.6.

Proof. Let us split this rule of Lawvere's in (Eq-TD), for top-down, and (Eq-BU), for bottom-up. Assuming this rule, we obtain reflexivity, symmetry and transitivity as in the proof of Lemma 3.2.3. The replacement rule is obtained as follows. For a proposition φ with free variables declared in $\Gamma, x\colon \sigma$, put $\varphi' = \varphi[y/x]$. Then

$$\frac{\Gamma \mid \Theta \vdash M =_\sigma M' \quad \dfrac{\Gamma \mid \Theta, \varphi[M/x], M =_\sigma M' \vdash \varphi[M'/x]}{\Gamma \mid \Theta, M =_\sigma M' \vdash \varphi[M'/x]} \text{(cut)}}{\Gamma \mid \Theta \vdash \varphi[M'/x]} \text{(cut)}$$

where the right premise comes from

$$\frac{\dfrac{\Gamma, x\colon \sigma \mid \Theta, \varphi \vdash \varphi'[x/y]}{\Gamma, x\colon \sigma, y\colon \sigma \mid \Theta, \varphi, x =_\sigma y \vdash \varphi'} \text{(Eq-TD)}}{\Gamma \mid \Theta, \varphi[M/x], M =_\sigma M' \vdash \varphi[M'/x]} \text{(subst)}$$

and on the left $\Gamma \mid \Theta \vdash \varphi[M/x]$.

In the reverse direction, assuming these last four equality rules from Figure 4.1, one obtains (Eq-BU) simply by substituting x for y and using reflexivity (as in the proof of Lemma 3.2.3). We shall derive (Eq-TD) from replacement.

$$\frac{\dfrac{}{\Gamma, x\colon \sigma, y\colon \sigma \mid \Theta, x =_\sigma y \vdash x =_\sigma y} \quad \dfrac{\Gamma, x\colon \sigma \mid \Theta \vdash \varphi[x/y]}{\Gamma, x\colon \sigma, y\colon \sigma \mid \Theta, x =_\sigma y \vdash \varphi[x/y]}}{\Gamma, x\colon \sigma, y\colon \sigma \mid \Theta, x =_\sigma y \vdash \varphi[y/y]} \qquad \square$$

The next result paves the way for a categorical characterisation of existential and universal quantification in terms of left and right adjoints to weakening functors adding a dummy assumption to the type context, see Example 3.1.1.

4.1.8. Lemma. *The rules for existential $\exists$ and universal $\forall$ quantification in Figure 4.1 can equivalently be described as the following two (double) rules.*

$$\frac{\Gamma \mid \Theta, \exists x{:}\sigma.\, \psi \vdash \varphi}{\Gamma, x{:}\sigma \mid \Theta, \psi \vdash \varphi} \;(\exists\text{-mate}) \qquad\qquad \frac{\Gamma \mid \Theta, \varphi \vdash \forall x{:}\sigma.\, \psi}{\Gamma, x{:}\sigma \mid \Theta, \varphi \vdash \psi} \;(\forall\text{-mate})$$

These rules express that $\exists$ is left adjoint and $\forall$ is right adjoint to weakening $(\Gamma \vdash \chi{:}\, \mathsf{Prop}) \mapsto (\Gamma, x{:}\sigma \vdash \chi{:}\, \mathsf{Prop})$.

Proof. We shall do the case of $\exists$, since $\forall$ is much simpler. The rules in the proposition follow from the rules for $\exists$ in Figure 4.1, since

$$\frac{\dfrac{\Gamma, x{:}\sigma \vdash x{:}\sigma \qquad \Gamma, x{:}\sigma \mid \psi \vdash \psi[x/x]}{\Gamma, x{:}\sigma \mid \psi \vdash \exists x{:}\sigma.\, \psi} \qquad \Gamma \mid \Theta, \exists x{:}\sigma.\, \psi \vdash \varphi}{\dfrac{}{}}$$

$$\Gamma, x{:}\sigma \mid \Theta, \exists x{:}\sigma.\, \psi \vdash \varphi$$

$$\Gamma, x{:}\sigma \mid \Theta, \psi \vdash \varphi$$

and

$$\frac{\Gamma \mid \exists x{:}\sigma.\, \psi \vdash \exists x{:}\sigma.\, \psi \qquad \Gamma, x{:}\sigma \mid \Theta, \psi \vdash \varphi}{\Gamma \mid \Theta, \exists x{:}\sigma.\, \psi \vdash \varphi}$$

Conversely, assuming the above ($\exists$-mate) rule, we can derive the two rules for $\exists$ in Figure 4.1.

$$\frac{\Gamma \vdash M{:}\sigma \qquad \dfrac{\dfrac{\Gamma \mid \Theta, \exists x{:}\sigma.\, \psi \vdash \exists x{:}\sigma.\, \psi}{\Gamma, x{:}\sigma \mid \Theta, \psi \vdash \exists x{:}\sigma.\, \psi}\,(\exists\text{-mate})}{\Gamma \mid \Theta, \psi[M/x] \vdash \exists x{:}\sigma.\, \psi}\,(\text{subst})}{\Gamma \mid \Theta \vdash \exists x{:}\sigma.\, \psi}$$

and

$$\frac{\Gamma \mid \Theta \vdash \exists x{:}\sigma.\, \psi \qquad \dfrac{\Gamma, x{:}\sigma \mid \Xi, \psi \vdash \chi}{\Gamma \mid \Xi, \exists x{:}\sigma.\, \psi \vdash \chi}\,(\exists\text{-mate})}{\Gamma \mid \Theta, \Xi, \psi \vdash \chi} \qquad\qquad \square$$

We close this section by examining a subtle point in many-typed predicate logic which is related to "empty types". We recall from Section 3.1 that the rule

strengthening

$$\frac{\Gamma, x{:}\sigma \mid \varphi_1, \ldots, \varphi_n \vdash \psi}{\Gamma \mid \varphi_1, \ldots, \varphi_n \vdash \psi} \;(\text{if } x \text{ not free in } \varphi_1, \ldots, \varphi_n, \psi)$$

will *not* be assumed. It may fail in models where the interpretation of σ is empty, see Exercise 3.1.3. This rule can be used without harm in single-typed logic because there, one has the common requirement that the interpretation of the single type is non-empty. This point often leads to confusion. For example in [102, 11.8], one finds the following reasoning against the rule *Modus Ponens* (or, $\supset$-elimination): for a variable $x\colon\sigma$ both

$$(x =_\sigma x) \supset (\exists x\colon\sigma.\, x =_\sigma x) \qquad \text{and} \qquad x =_\sigma x$$

hold if the interpretation of σ is empty; but then

$$\exists x\colon\sigma.\, x =_\sigma x$$

does not hold. If we recast this line of thought in our notation with explicit type contexts of term variables, it becomes clear that there is an illegal use of the above strengthening rule involved, and that there is nothing wrong with implication.

$$\frac{\dfrac{\overline{x\colon\sigma \vdash x\colon\sigma} \qquad \overline{x\colon\sigma \mid x =_\sigma x \vdash x =_\sigma x}}{\dfrac{x\colon\sigma \mid x =_\sigma x \vdash \exists x\colon\sigma.\, x =_\sigma x}{x\colon\sigma \mid \emptyset \vdash (x =_\sigma x) \supset (\exists x\colon\sigma.\, x =_\sigma x)}} \qquad \dfrac{\overline{x\colon\sigma \vdash x\colon\sigma}}{x\colon\sigma \mid \emptyset \vdash x =_\sigma x}\,(\text{refl})}{\dfrac{x\colon\sigma \mid \emptyset \vdash \exists x\colon\sigma.\, x =_\sigma x}{\emptyset \mid \emptyset \vdash \exists x\colon\sigma.\, x =_\sigma x}\,(\text{strengthening!})}$$

This point is also stressed in [186, top of p. 131].

Exercises

4.1.1.   Consider the rules for the connectives $\wedge$ of conjunction and $\vee$ of disjunction. Show that $\wedge$ and $\vee$ distribute over each other, *i.e.* that the following two equivalences are derivable.

$$\Gamma \mid \emptyset \vdash \varphi \wedge (\psi \vee \chi) \mathbin{\supset\subset} (\varphi \wedge \psi) \vee (\varphi \wedge \chi)$$
$$\Gamma \mid \emptyset \vdash \varphi \vee (\psi \wedge \chi) \mathbin{\supset\subset} (\varphi \vee \psi) \wedge (\varphi \vee \chi).$$

See also Exercise 2.6.2.

4.1.2.   In the same vain, assume x does not occur free in φ; derive

(i) $\Gamma \mid \emptyset \vdash \exists x{:}\,\sigma.\,(\varphi \wedge \psi) \mathbin{\supset\subset} (\varphi \wedge (\exists x{:}\,\sigma.\,\psi))$;

(ii) $\Gamma \mid \emptyset \vdash \forall x{:}\,\sigma.\,(\psi \supset \varphi) \mathbin{\supset\subset} ((\exists x{:}\,\sigma.\,\psi) \supset \varphi)$.

[Related to (ii) is Exercise 1.9.7.]

4.1.3.   Show that the *reductio ad absurdum* rule is equivalent to the excluded middle (or *tertium non datur*) rule:

$$\frac{\Gamma \vdash \varphi{:}\,\mathsf{Prop}}{\Gamma \mid \emptyset \vdash \varphi \vee \neg\varphi}$$

4.1.4.   Assume a proposition $x{:}\,\sigma \vdash \varphi{:}\,\mathsf{Prop}$. Is it possible to derive

$$\emptyset \mid \emptyset \vdash (\forall x{:}\,\sigma.\,\varphi) \supset (\exists x{:}\,\sigma.\,\varphi) \ ?$$

4.1.5.   Prove Lemma 4.1.8 for $\forall$.

4.1.6.   Give a strengthened, "Frobenius" version of the rule in Lemma 4.1.7 in which the variable y is also allowed to occur in the proposition context Θ (as in Lemma 3.2.4).

4.1.7.   A ring is called **local** if it has a unique maximal ideal. Show that a ring R is local if and only if it satisfies the (coherent!) proposition

$$x{:}\,R \mid \emptyset \vdash ((\exists y{:}\,R.\,x \cdot y = 1) \vee (\exists y{:}\,R.\,(1 - x) \cdot y = 1)).$$

[This is the standard example of a notion definable in coherent logic.]

4.2 Fibrations for first order predicate logic

In the previous section the syntax of first order predicate logic was given—and of the subsystems of regular logic (with $=, \wedge, \top, \exists$) and coherent logic (with $=, \wedge, \top, \exists, \vee, \bot$). In this section we shall define appropriate fibrations to captures such logics categorically. Further, we shall describe several standard examples of such fibred categories. These include the topological model by Tarski, the realisability model by Kleene, and so-called Kripke models. Among these, the 'realisability' fibration $\begin{smallmatrix} \mathrm{UFam}(P\!N) \\ \downarrow \\ \mathbf{Sets} \end{smallmatrix}$ incorporating Kleene's realisability interpretation of constructive logic will play an important rôle in the sequel (in the construction of the effective topos). Also subobject fibrations form important examples, but since they are rather special, they will be investigated separately in the later Sections 4.4, 4.5 and 4.9.

The categorical structure used for the connectives and quantifiers can be read off almost immediately from their rules—by keeping in mind the syntactic fibrations $\begin{smallmatrix} \mathcal{L}(\Sigma,\Pi,\mathcal{A}) \\ \downarrow \\ \mathcal{C}\!\ell(\Sigma) \end{smallmatrix}$ from Section 3.1 (built on top of a signature with predicates (Σ,Π) with set of axioms $\mathcal{A}$): the connectives $\top, \wedge, \bot, \vee, \supset$ correspond to fibred finite products $(\top, \wedge)$, coproducts $(\bot, \vee)$ and exponents $\supset$. Equality $=$ is described by fibred equality as in Section 3.4 (that is, by left adjoints to contraction functors δ^*, using Lemma 4.1.7), and the quantifiers $\exists, \forall$ are described by simple coproducts and products (*i.e.* by left and right adjoints to weakening functors π^* as in Lemma 4.1.8) from Section 1.9. We thus come to the following definitions.

4.2.1. Definition. (i) A **regular fibration** is an Eq-fibration with simple coproducts satisfying Frobenius. That is, $\begin{smallmatrix} \mathbb{E} \\ \downarrow p \\ \mathbb{B} \end{smallmatrix}$ is a regular fibration if

- p is a fibred preorder with finite products in its base category $\mathbb{B}$;
- p has fibred finite products (for $\top, \wedge$);
- p has fibred equality ($\mathrm{Eq}_{I,J} \dashv \delta(I,J)^*$) satisfying Frobenius (for $=$);
- p has simple coproducts ($\coprod_{(I,J)} \dashv \pi^*_{I,J}$) satisfying Frobenius (for $\exists$).

(ii) A **coherent fibration** is a regular fibration which has fibred finite coproducts $(\bot, \vee)$ which are fibrewise distributive, *i.e.* $X \wedge (Y \vee Z) \cong (X \wedge Y) \vee (X \wedge Z)$ in each fibre. Thus each fibre is a (preorder) distributive lattice.

(iii) A **first order fibration** is a coherent fibration which is a fibred CCC and has simple products $\prod_{(I,J)}$.

We recall from Definition 3.5.1 that Eq-fibrations are preordered. Hence also regular, coherent and first order fibrations have preordered fibre categories. For a non-preordered version of a regular fibration, see [256]. There are obvious "split" versions of the above notions of regular / coherent / first order fibration, in which all of the relevant structure is split.

The rest of this section will be devoted to examples of the above kind of fibred categories. Details of interpreting predicate logics in such fibrations may be found in the next section, but it may be useful to have in mind when reading the examples below that objects I of the base category are to be thought of as type contexts (or as types), and objects X of the total category above I as predicates in context I. Validity of this predicate X corresponds to the presence of an inequality $\top(I) \leq X$ (over I), where $\top(I)$ is the terminal object in the fibre over I. This view will be formalised in the 'internal language' of such a fibration towards the end of the next section.

4.2.2. Syntactic examples. Let us fix a signature with predicates (Σ, Π), as introduced in (or before) Definition 4.1.1.

(i) Consider (Σ, Π) with regular logic and assume a collection $\mathcal{A}$ of (regular) axioms. One can construct a (preorder) classifying fibration $\begin{smallmatrix} \mathcal{L}(\Sigma,\Pi,\mathcal{A}) \\ \downarrow \\ \mathcal{Cl}(\Sigma) \end{smallmatrix}$ as in Section 3.1. The objects of the total category $\mathcal{L}(\Sigma, \Pi, \mathcal{A})$ are type-plus-proposition contexts $(\Gamma \mid \varphi_1, \ldots, \varphi_n)$. In the sequel we usually assume finite conjunctions $\top, \wedge$ in our logic and so we may conveniently assume this sequence of φ's in $(\Gamma \mid \varphi_1, \ldots, \varphi_n)$ to be of length one. A morphism $(\Gamma \vdash \varphi: \mathsf{Prop}) \to (\Delta \vdash \psi: \mathsf{Prop})$ is then a context morphism $\vec{M}: \Gamma \to \Delta$ for which one can derive $\Gamma \mid \varphi \vdash \psi(\vec{M})$, using the sequents in $\mathcal{A}$ as axioms.

It is easy to see that the rules for $\top, \wedge$ induce fibred finite products for $\begin{smallmatrix} \mathcal{L}(\Sigma,\Pi,\mathcal{A}) \\ \downarrow \\ \mathcal{Cl}(\Sigma) \end{smallmatrix}$. As before, in Example 3.4.4 (iv), this fibration has equality satisfying Frobenius by the rules for $=$ (as formulated suitably in Lemma 4.1.7). Similarly, for simple coproducts we have to show that for contexts $\Gamma, \Gamma' \in \mathcal{Cl}(\Sigma)$, the reindexing functor π^* induced by the projection $\pi: (\Gamma, \Gamma') \to \Gamma$ has a left adjoint. For convenience we assume Γ' to be $x: \sigma$ of length one. This weakening functor π^* then sends

$$\Gamma \vdash \varphi: \mathsf{Prop} \qquad \text{to} \qquad \Gamma, x: \sigma \vdash \varphi: \mathsf{Prop}$$

by adding an extra hypothesis. Its left adjoint sends

$$\Gamma, x: \sigma \vdash \psi: \mathsf{Prop} \qquad \text{to} \qquad \Gamma \vdash \exists x: \sigma. \, \psi: \mathsf{Prop}.$$

The adjunction $\exists x: \sigma. \, (-) \dashv \pi^*$ requires a bijective correspondence

$$\frac{\Gamma \mid \exists x: \sigma. \, \psi \vdash \varphi}{\Gamma, x: \sigma \mid \psi \vdash \varphi} \ (\exists\text{-mate})$$

which follows from the reformulation of the $\exists$-rules in Lemma 4.1.8. By Exercise 4.1.2 (i) these coproducts satisfy Frobenius. In the general case where $\Gamma' = (x_1: \sigma_1, \ldots, x_n: \sigma_n)$ need not be of length one, a left adjoint to π^* associated with the projection $\pi: (\Gamma, \Gamma') \to \Gamma$ sends a proposition $\Gamma, \Gamma' \vdash \psi: \mathsf{Prop}$ to $\Gamma \vdash \exists x_1: \sigma_1. \cdots \exists x_n: \sigma_n. \, \psi: \mathsf{Prop}$. We conclude that the rules of regular logic make $\begin{smallmatrix} \mathcal{L}(\Sigma,\Pi,\mathcal{A}) \\ \downarrow \\ \mathcal{Cl}(\Sigma) \end{smallmatrix}$ into a regular fibration.

(ii) If one further adds finite disjunctions $(\bot, \vee)$ to the logic, then the fibration $\begin{smallmatrix} \mathcal{L}(\Sigma,\Pi,\mathcal{A}) \\ \downarrow \\ \mathcal{Cl}(\Sigma) \end{smallmatrix}$ has fibred finite coproducts. These are distributive over conjunctions by Exercise 4.1.1. Thus coherent logic leads to coherent classifying fibrations.

(iii) It will probably not come as a surprise anymore that full first order logic (obtained by adding $\supset$ and $\forall$) makes the syntactic fibration a first order fibration. Implication yields fibred exponents and universal quantification induces *right* adjoints to the weakening functors π^* mentioned in (i), which send

$$\Gamma, x: \sigma \vdash \psi: \mathsf{Prop} \qquad \text{to} \qquad \Gamma \vdash \forall x: \sigma.\, \psi: \mathsf{Prop}.$$

The adjunction $\pi^* \dashv \forall x: \sigma.\, (-)$ involves the bijective correspondence

$$\frac{\Gamma \mid \varphi \vdash \forall x: \sigma.\, \psi}{\Gamma, x: \sigma \mid \varphi \vdash \psi} \; (\forall\text{-mate})$$

which follows from the reformulation of the $\forall$-rules in Lemma 4.1.8.

4.2.3. Set theoretic example. Let (Σ, Π) be a signature with predicates. A **(set theoretic) model** of (Σ, Π)—or, a (Σ, Π)-**algebra**—consists of

(a) a collection $(A_\sigma)_{\sigma \in |\Sigma|}$ of 'carrier' sets, indexed by the underlying set $|\Sigma|$ of types of the signature;

(b) for each function symbol $F: \sigma_1, \ldots, \sigma_n \longrightarrow \sigma_{n+1}$ in Σ a function

$$A_{\sigma_1} \times \cdots \times A_{\sigma_n} \xrightarrow{\quad [\![F]\!] \quad} A_{\sigma_{n+1}}$$

(c) for each predicate symbol $P: \sigma_1, \ldots, \sigma_n$ in Π a subset

$$[\![P]\!] \longhookrightarrow A_{\sigma_1} \times \cdots \times A_{\sigma_n}.$$

Note that (a)+(b) constitute a Σ-algebra as described in Section 1.6.

For such an algebra $\langle (A_\sigma), [\![-]\!], [\![-]\!] \rangle$ we construct a first order fibration. Let $\mathbb{A}$ be the base category with

objects   sequences $(\sigma_1, \ldots, \sigma_n)$ of types $\sigma_i \in |\Sigma|$.

morphisms $(\sigma_1, \ldots, \sigma_n) \to (\tau_1, \ldots, \tau_m)$ are m-tuples $(f_1, \ldots, f_m)$ of functions $f_i: A_{\sigma_1} \times \cdots \times A_{\sigma_n} \to A_{\tau_i}$.

We leave it to the reader to verify that $\mathbb{A}$ is a category with finite products.

An indexed category $\mathbb{A}^{\mathrm{op}} \to \mathbf{Cat}$ is obtained by

$$(\sigma_1, \ldots, \sigma_n) \mapsto \text{the power poset } \langle P(A_{\sigma_1} \times \cdots \times A_{\sigma_n}), \subseteq \rangle$$

$$f = (f_1, \ldots, f_m) \mapsto \text{the functor } f^* \text{ sending}$$

$$Y \mapsto \{\vec{x} \mid (f_1(\vec{x}), \ldots, f_m(\vec{x})) \in Y\}.$$

Applying the Grothendieck construction to this indexed category yields a split fibration over $\mathbb{A}$. The total category of this fibration has as objects pairs consisting of a sequence $(\sigma_1, \ldots, \sigma_n)$ of types together with a predicate $X \subseteq A_{\sigma_1} \times \cdots \times A_{\sigma_n}$ on the associated product of carriers. And morphisms

$\langle(\sigma_1,\dots,\sigma_n),X\rangle \to \langle(\tau_1,\dots,\tau_m),Y\rangle$ consist of an m-tuple of $(f_1,\dots,f_m)$ of functions $f_i\colon A_{\sigma_1}\times\cdots\times A_{\sigma_n}\to A_{\tau_i}$ satisfying $(f_1(\vec{x}),\dots,f_m(\vec{x}))\in Y$ for all $\vec{x}\in X$.

We claim that this is a first order fibration. The fibre categories $P(A_{\sigma_1}\times\cdots\times A_{\sigma_n})$ are Boolean algebras. Hence we have $\top,\wedge,\bot,\vee$ and $\supset$ (and also *reductio ad absurdum* as in classical logic). Quantification along the projection $\pi\colon(\sigma_1,\dots,\sigma_n,\sigma_{n+1})\to(\sigma_1,\dots,\sigma_n)$ is given by

$$X \xmapsto{\;\coprod\;} \{\vec{x}\in A_{\sigma_1}\times\cdots\times A_{\sigma_n}\mid \text{for some } y\in A_{\sigma_{n+1}},\ (\vec{x},y)\in X\}$$

$$X \xmapsto{\;\prod\;} \{\vec{x}\in A_{\sigma_1}\times\cdots\times A_{\sigma_n}\mid \text{for all } y\in A_{\sigma_{n+1}},\ (\vec{x},y)\in X\}.$$

And equality along $\delta\colon(\sigma_1,\dots,\sigma_n,\sigma_{n+1})\to(\sigma_1,\dots,\sigma_n,\sigma_{n+1},\sigma_{n+1})$ is

$$X \mapsto \{(\vec{x},y,z)\mid (\vec{x},y)\in X \text{ and } y=z\}.$$

4.2.4. Kripke model example. For a signature with predicates (Σ,Π) there is a category $\mathbf{Alg}(\Sigma,\Pi)$ of (Σ,Π)-algebras (as in the previous example). Morphisms $H\colon\langle(A_\sigma),[\![-]\!],[\![-]\!]\rangle\to\langle(B_\sigma),[\![-]\!],[\![-]\!]\rangle$ in $\mathbf{Alg}(\Sigma,\Pi)$ are collections of functions $H=(H_\sigma\colon A_\sigma\to B_\sigma)_{\sigma\in|\Sigma|}$ between the carrier sets which commute with the interpretations of function symbols $F\colon\sigma_1,\dots,\sigma_n\longrightarrow\sigma_{n+1}$ and predicate symbols $P\colon\sigma_n,\dots,\sigma_n$, as in:

A **Kripke model** for (Σ,Π) consists of an index poset $\mathbb{I}=\langle\mathbb{I},\le\rangle$ together with a functor

$$\mathbb{I}\xrightarrow{\;\mathcal{K}\;}\mathbf{Alg}(\Sigma,\Pi)$$

It involves for each element $i\in\mathbb{I}$ a (Σ,Π)-algebra

$$\mathcal{K}(i)=\langle(\mathcal{K}(i)_\sigma),[\![-]\!](i),[\![-]\!](i)\rangle$$

and for each pair $i, j \in \mathbb{I}$ with $i \leq j$ a morphism $\mathcal{K}(ij)$ of (Σ, Π)-algebras. The latter consists of a collection of $\mathcal{K}(ij) = (\mathcal{K}(ij)_\sigma \colon \mathcal{K}(i)_\sigma \to \mathcal{K}(j)_\sigma)$ of functions commuting with the interpretations of function and predicate symbols, like the H_σ's above. One thinks of the elements of $\mathbb{I}$ as stages in (branching) time and of the algebra $\mathcal{K}(i)$ as the state of knowledge at stage $i \in \mathbb{I}$.

In order to construct a first order fibration from such a $\mathcal{K}$ we need some notation: for a sequence $(\sigma_1, \ldots, \sigma_n)$ of types (from Σ), there is a functor $\mathcal{K}(\sigma_1, \ldots, \sigma_n) \colon \mathbb{I} \to \mathbf{Sets}$ given by

$$\begin{cases} i \mapsto \mathcal{K}(i)_{\sigma_1} \times \cdots \times \mathcal{K}(i)_{\sigma_n} \\ i \leq j \mapsto \mathcal{K}(ij)_{\sigma_1} \times \cdots \times \mathcal{K}(ij)_{\sigma_n}. \end{cases}$$

Application of the morphism part of this functor will be abbreviated as follows. For $i, j \in \mathbb{I}$ with $i \leq j$ and $\vec{x} \in \mathcal{K}(\sigma_1, \ldots, \sigma_n)(i)$ we write

$$(\vec{x})^j \quad \text{for} \quad (\mathcal{K}(ij)_{\sigma_1}(x_1), \ldots, \mathcal{K}(ij)_{\sigma_n}(x_n)) \in \mathcal{K}(\sigma_1, \ldots, \sigma_n)(j).$$

A collection of subsets $\left(X_i \subseteq \mathcal{K}(\sigma_1, \ldots, \sigma_n)(i) \right)_{i \in \mathbb{I}}$ is called **monotone** if

$$i \leq j \text{ and } \vec{x} \in X_i \quad \text{implies} \quad (\vec{x})^j \in X_j.$$

Notice that for a predicate symbol $P \colon \sigma_1, \ldots, \sigma_n$ in Π, the interpretations of P in the $\mathcal{K}(i)$'s,

$$[\![P]\!](i) \subseteq \mathcal{K}(i)_{\sigma_1} \times \cdots \times \mathcal{K}(i)_{\sigma_n} = \mathcal{K}(\sigma_1, \ldots, \sigma_n)(i)$$

form such a monotone collection, because the $\mathcal{K}(ij)$'s, for $i \leq j$, are morphisms of (Σ, Π)-algebras.

In [182] Kripke showed how to interpret intuitionistic (or constructive) predicate logic in such an $\mathbb{I}$-indexed collection $\mathcal{K} \colon \mathbb{I} \to \mathbf{Alg}(\Sigma, \Pi)$ of models of classical first order logic. A proposition $\Gamma \vdash \varphi \colon \mathsf{Prop}$—say with type context $\Gamma = x_1 \colon \sigma_1, \ldots, x_n \colon \sigma_n$—is interpreted as a monotone collection of subsets

$$[\![\Gamma \vdash \varphi \colon \mathsf{Prop}]\!](i) \subseteq \mathcal{K}(\sigma_1, \ldots, \sigma_n)(i) \qquad \text{for } i \in \mathbb{I}.$$

The main clauses of Kripke's interpretation are:

$$[\![\Gamma \vdash \varphi \vee \psi \colon \mathsf{Prop}]\!](i) = [\![\Gamma \vdash \varphi \colon \mathsf{Prop}]\!](i) \cup [\![\Gamma \vdash \psi \colon \mathsf{Prop}]\!](i)$$

$$[\![\Gamma \vdash \varphi \wedge \psi \colon \mathsf{Prop}]\!](i) = [\![\Gamma \vdash \varphi \colon \mathsf{Prop}]\!](i) \cap [\![\Gamma \vdash \psi \colon \mathsf{Prop}]\!](i)$$

$$[\![\Gamma \vdash \varphi \supset \psi \colon \mathsf{Prop}]\!](i) = \{ \vec{x} \mid \text{for all } j \geq i, (\vec{x})^j \in [\![\Gamma \vdash \varphi \colon \mathsf{Prop}]\!](j)$$
$$\text{implies } (\vec{x})^j \in [\![\Gamma \vdash \psi \colon \mathsf{Prop}]\!](j) \}$$

$$[\![\Gamma \vdash \exists y \colon \sigma. \, \psi \colon \mathsf{Prop}]\!](i) = \{ \vec{x} \mid \text{for some } y \in \mathcal{K}(i)_\sigma,$$
$$(\vec{x}, y) \in [\![\Gamma, y \colon \sigma \vdash \psi \colon \mathsf{Prop}]\!](i) \}$$

$$[\![\Gamma \vdash \forall y \colon \sigma. \, \psi \colon \mathsf{Prop}]\!](i) = \{ \vec{x} \mid \text{for all } j \geq i \text{ and } y \in \mathcal{K}(j)_\sigma,$$
$$((\vec{x})^j, y) \in [\![\Gamma, y \colon \sigma \vdash \psi \colon \mathsf{Prop}]\!](j) \}.$$

Notice that knowledge whether a proposition involving $\supset$ or $\forall$ holds at stage i, involves knowledge about future stages $j \geq i$. Indeed, these connectives have far-reaching consequences.

We shall construct a first order fibration (from $\mathcal{K}$) in which these clauses hold. As base category $\mathbb{B}$ we take

objects   sequences of types $(\sigma_1, \ldots, \sigma_n)$.

morphisms $(\sigma_1, \ldots, \sigma_n) \to (\tau_1, \ldots, \tau_m)$ are natural transformations α between the corresponding functors, as in:

$$
\begin{array}{c}
\mathcal{K}(\sigma_1, \ldots, \sigma_n) \\
\mathbb{I} \underset{\mathcal{K}(\tau_1, \ldots, \tau_m)}{\overset{}{\rightleftharpoons}} \Downarrow \alpha \quad \mathbf{Sets}
\end{array}
$$

It is easy to verify that $\mathbb{B}$ is a category with finite products. For each sequence $(\sigma_1, \ldots, \sigma_n) \in \mathbb{B}$ there is a poset fibre category with

objects   monotone collections $\big(X_i \subseteq \mathcal{K}(\sigma_1, \ldots, \sigma_n)(i)\big)_{i \in \mathbb{I}}$.

morphisms $(X_i)_{i \in \mathbb{I}} \to (Y_i)_{i \in \mathbb{I}}$ exist if and only if $X_i \subseteq Y_i$ for each $i \in \mathbb{I}$.

Every morphism $\alpha \colon (\sigma_1, \ldots, \sigma_n) \to (\tau_1, \ldots, \tau_m)$ in $\mathbb{B}$ determines a reindexing functor α^* between the fibres, by pointwise inverse image:

$$
\big(X_i \subseteq \mathcal{K}(\tau_1, \ldots, \tau_m)(i)\big)_{i \in \mathbb{I}} \mapsto \big(\{\vec{y} \in \mathcal{K}(\sigma_1, \ldots, \sigma_n)(i) \mid \alpha_i(\vec{y}) \in X_i\}\big)_{i \in \mathbb{I}}.
$$

By naturality of α, this new collection is monotone again. Our claim is that the result is a first order fibration.

Each fibre category—say over $(\sigma_1, \ldots, \sigma_n)$—is a Heyting algebra with structure

$$
\begin{aligned}
\bot(i) &= \emptyset \\
\top(i) &= \mathcal{K}(\sigma_1, \ldots, \sigma_n)(i) \\
(X \vee Y)(i) &= X(i) \cup Y(i) \\
(X \wedge Y)(i) &= X(i) \cap Y(i) \\
(X \supset Y)(i) &= \{\vec{x} \in \mathcal{K}(\sigma_1, \ldots, \sigma_n)(i) \mid \text{for all } j \geq i, \\
&\qquad (\vec{x})^j \in X_j \text{ implies } (\vec{x})^j \in Y_j\}.
\end{aligned}
$$

We check that $X \supset Y$ is an exponent, *i.e.* that there is a bijective correspondence

$$
\frac{(Z \wedge X)(i) \subseteq Y(i) \quad \text{for all } i \in \mathbb{I}}{Z(i) \subseteq (X \supset Y)(i) \quad \text{for all } i \in \mathbb{I}}
$$

For the implication downwards, assume $\vec{x} \in Z(i)$ and for some $j \geq i$, $(\vec{x})^j \in X(j)$. Then by monotony of Z, $(\vec{x})^j \in Z(j)$ and thus $(\vec{x})^j \in Z(j) \cap X(j) \subseteq$

$Y(j)$. Hence $\vec{x} \in (X \supset Y)(i)$. Upwards, assume $\vec{x} \in Z(i) \cap X(i)$. Then $\vec{x} \in Z(i)$ and thus $\vec{x} \in (X \supset Y)(i)$. Since $i \geq i$ and $(\vec{x})^i = \vec{x} \in X(i)$ one obtains $\vec{x} \in Y(i)$ as required.

Quantification along a projection $\pi \colon (\sigma_1, \ldots, \sigma_n, \sigma_{n+1}) \to (\sigma_1, \ldots, \sigma_n)$ takes the following form.

$$\coprod(Y)(i) = \{\vec{x} \in \mathcal{K}(\sigma_1, \ldots, \sigma_n)(i) \mid$$
$$\text{for some } y \in \mathcal{K}(i)_{\sigma_{n+1}}, \ (\vec{x}, y) \in Y(i)\}$$
$$\prod(Y)(i) = \{\vec{x} \in \mathcal{K}(\sigma_1, \ldots, \sigma_n)(i) \mid$$
$$\text{for all } j \geq i \text{ and } y \in \mathcal{K}(j)_{\sigma_{n+1}}, \ ((\vec{x})^j, y) \in Y(j)\}.$$

Equality is left as an exercise below.

4.2.5. Order theoretic examples. Let A be a **frame**, *i.e.* a poset with finite meets and infinite joins such that these joins distribute over meets: $(\bigvee_i a_i) \wedge b = \bigvee_i (a_i \wedge b)$. A frame is sometimes called a **complete Heyting algebra** or a **locale**. The prime examples of such a structure are posets $\langle \mathcal{O}(X), \subseteq \rangle$ of open subsets of a topological space X. For such a frame A, the family fibration $\begin{smallmatrix} \mathrm{Fam}(A) \\ \downarrow \\ \mathbf{Sets} \end{smallmatrix}$ is a coherent fibration: the finite meets and joins in A induce fibred finite products and coproducts; arbitrary joins $\bigvee$ in A induce simple coproducts, which satisfy Frobenius by the above distribution, see Exercise 1.9.4. Finally, the bottom element leads to equality satisfying Frobenius, see Example 3.4.4 (iii).

Finite meets and infinite joins form the essential structure of frames: morphisms of frames preserve these by definition. But in a frame one can *define* infinite meets by

$$\bigwedge_i a_i = \bigvee \{b \mid b \text{ is lower bound of } (a_i)\} = \bigvee \{b \mid \forall i.\, b \leq a_i\}$$

and implication by

$$a \supset b = \bigvee \{c \mid a \wedge c \leq b\}.$$

Thus $\begin{smallmatrix} \mathrm{Fam}(A) \\ \downarrow \\ \mathbf{Sets} \end{smallmatrix}$ is actually a first order fibration.

The special case where A is the frame $\mathcal{O}(X)$ of opens of a topological space X captures Tarski's [328] interpretation of (constructive) first order logic in the opens of a topological space X, formulated in the 1930s. We mention the main points. For opens $(U_{(i,j)})_{(i,j) \in I \times J}$ one has

$$\text{coproduct:} \left(\bigcup_{j \in J} U_{(i,j)} \right)_{i \in I} \quad \text{and product:} \quad \left(\mathrm{Int}(\bigcap_{j \in J} U_{(i,j)}) \right)_{i \in I}$$

where $\text{Int}(-)$ is the interior operation. (If $J = \emptyset$ we take this intersection to be X.)

For I-indexed collections $(U_i)_{i \in I}$ and $(V_i)_{i \in I}$ one has as implication over I, the I-indexed implication in the Heyting algebra $\mathcal{O}(X)$:

$$(U_i)_{i \in I} \supset (V_i)_{i \in I} = \big(\text{Int}((X - U_i) \cup V_i)\big)_{i \in I}.$$

For more information, see [335, Chapter 9] or [281].

4.2.6. Realisability example. Under constructive reading, a proof of a proposition consists of a method of establishing it. In 1945 Kleene [178] gave the so-called **realisability interpretation** of constructive logic (see also [333, 23, 335]), in which such a proof is understood as a code of a partial recursive function. Kleene introduced a relation $nr\varphi$, to be read as '$n \in \mathbb{N}$ realises proposition φ'. That is, n is a code for a partial recursive function $m \mapsto n \cdot m$ which is a method for establishing φ. Kleene stipulated,

$$nr(\varphi \wedge \psi) \;\Leftrightarrow\; n \text{ is (recursive) pair } \langle n_1, n_2 \rangle \text{ with } n_1 r \varphi \text{ and } n_2 r \psi$$

$$nr(\varphi \vee \psi) \;\Leftrightarrow\; n \text{ is (recursive) pair } \langle n_1, n_2 \rangle \text{ with } \begin{cases} n_2 r \varphi & \text{if } n_1 = 0 \\ n_2 r \psi & \text{if } n_1 = 1 \end{cases}$$

$$nr(\varphi \supset \psi) \;\Leftrightarrow\; \text{for each } m \text{ with } mr\varphi \text{ one has } (n \cdot m) r \psi.$$

See *e.g.* [335, 23] for more information on realisability. In order to deal with first order logic, we shall describe a set-indexed version of this interpretation. We write $P\mathbb{N}$ for the powerset of $\mathbb{N}$. For an arbitrary set I, consider the set of functions $(P\mathbb{N})^I$ from I to $P\mathbb{N}$. Elements of $(P\mathbb{N})^I$ are also called **non-standard predicates** on I. Such a predicate $X \in (P\mathbb{N})^I$ is called **valid** if

$$\left(\bigcap_{i \in I} X(i) \right) \neq \emptyset.$$

Thus X is valid if there is a single natural number which is member of all X_i's.

In line with Kleene's stipulations, put for $X, Y \in (P\mathbb{N})^I$,

$$\begin{aligned}
(X \wedge Y)(i) &= \{\langle n, m \rangle \mid n \in X(i) \text{ and } m \in Y(i)\} \\
(X \vee Y)(i) &= \{\langle 0, n \rangle \mid n \in X(i)\} \cup \{\langle 1, m \rangle \mid m \in Y(i)\} \\
(X \supset Y)(i) &= \{n \mid \text{for each } m \in X(i), n \cdot m \in Y(i)\}.
\end{aligned}$$

The latter gives rise to a preorder on $(P\mathbb{N})^I$ by

$$X \leq Y \;\Leftrightarrow\; \left(\bigcap_{i \in I} X(i) \supset Y(i) \right) \neq \emptyset.$$

Notice that this ordering is not pointwise but uniform: a single code must be member of every $X(i) \supset Y(i)$. For more information on this ordering, see Exercise 4.2.5.

There is a bottom element $\perp = \lambda i \in I. \emptyset$ and a top element $\top = \lambda i \in I. \mathbb{N}$ in $(P\mathbb{N})^I$. In this way $\langle P\mathbb{N}, \leq \rangle$ becomes a Heyting pre-algebra (or a preorder bicartesian closed category).

The quantifiers are described as follows. For a predicate $X \in (P\mathbb{N})^{I \times J}$ one takes

$$\coprod_{(I,J)}(X)(i) = \bigcup_{j \in J} X(i,j), \qquad \prod_{(I,J)}(X)(i) = \bigcap_{j \in J} X(i,j)$$

where we understand the latter intersection to be $\mathbb{N}$ in case J is empty.

The assignment $I \mapsto (P\mathbb{N})^I$ extends to a functor (or split indexed category) **Sets**$^{\mathrm{op}} \to$ **Cat** with reindexing by pre-composition. The fibration resulting from this indexed category (via the Grothendieck construction) will be written as $\begin{smallmatrix} \mathrm{UFam}(P\mathbb{N}) \\ \downarrow \\ \mathbf{Sets} \end{smallmatrix}$ in which the letter 'U' in 'UFam' emphasises the uniform character of vertical maps. This will be called the **realisability fibration**. In Example 5.3.4 in the next chapter it will be shown how, more generally, one can construct a similar fibration from a 'partial combinatory algebra' (like the Kleene structure $\langle \mathbb{N}, \cdot \rangle$ used here).

We claim that $\begin{smallmatrix} \mathrm{UFam}(P\mathbb{N}) \\ \downarrow \\ \mathbf{Sets} \end{smallmatrix}$ is a first order fibration. In fact all the relevant structure, except equality, has already been described above. Equality is given as follows. For $X \in (P\mathbb{N})^{I \times J}$ put,

$$\mathrm{Eq}(X)(i,j,j') = \begin{cases} X(i,j) & \text{if } j = j' \\ \emptyset & \text{else.} \end{cases}$$

4.2.7. Recursive enumerability example. Recall that a relation $X \subseteq \mathbb{N}^n$ is recursively enumerable ('r.e.') if and only if there is a partial recursive function $f: \mathbb{N}^n \to \mathbb{N}$ such that $\vec{x} \in X \Leftrightarrow f(\vec{x}) \downarrow$ (*i.e.* $f(\vec{x})$ is defined precisely on $\vec{x} \in X$); also that such r.e. relations on $\mathbb{N}^n$ are closed under intersection $\cap$ and union $\cup$. Hence r.e. relations on $\mathbb{N}^n$ ordered by inclusion form a distributive lattice (with bottom $\emptyset$ and top $\mathbb{N}^n$). Further, for an r.e. relation $Y \subseteq \mathbb{N}^{n+1}$ the set

$$\{ \vec{x} \in \mathbb{N}^n \mid \text{for some } y \in \mathbb{N}, (\vec{x}, y) \in Y \}$$

is r.e. again. All this suggests there is a coherent fibration (with $\top, \wedge, \perp, \vee$ and $\exists, =$) involved.

We first form a base category **PR** of partial recursive functions: objects are $n \in \mathbb{N}$ and morphisms $m \to n$ are n-tuples $(f_1, \ldots, f_n)$ of partial recursive functions $f_i: \mathbb{N}^m \to \mathbb{N}$. Composition is done in the obvious way. One has that $0 \in \mathbf{PR}$ is terminal and that $n + m$ is the product of n and m.

The next step is to define an indexed category $\mathbf{PR}^{\mathrm{op}} \to \mathbf{Cat}$. We assign to n the poset of r.e. relations on $\mathbb{N}^n$, ordered by inclusion. Reindexing along $(f_1, \ldots, f_n) \colon m \to n$ in $\mathbf{PR}$ is done by substitution:

$$X \subseteq \mathbb{N}^n \mapsto \{\vec{y} \in \mathbb{N}^m \mid (f_1(\vec{y}), \ldots, f_n(\vec{y})) \in X\} \subseteq \mathbb{N}^m.$$

Obviously, the relation on the right hand side is r.e. again. The resulting fibration will be written as $\begin{smallmatrix}\mathbf{RE}\\\downarrow\\\mathbf{PR}\end{smallmatrix}$. The above ingredients yield fibred finite products and coproducts and simple coproducts. Equality is given as follows. For $X \subseteq \mathbb{N}^{n+m}$ put,

$$\mathrm{Eq}(X) = \{(\vec{x}, \vec{y}, \vec{z}) \in \mathbb{N}^{n+m+m} \mid (\vec{x}, \vec{y}) \in X \text{ and } \vec{y} = \vec{z}\}.$$

Since the base category $\mathbf{PR}$ of this fibration $\begin{smallmatrix}\mathbf{RE}\\\downarrow\\\mathbf{PR}\end{smallmatrix}$ is an 'algebraic theory' (this means that the objects are of the form 1^n for $n \in \mathbb{N}$) one obtains a structure in which one can interpret single-typed coherent logic: the object n in the base category stands for the type context in which n term variables (of this single type) are declared.

We close this section by sketching how so-called cylindric algebras give rise to (single-typed, classical) first order fibrations. These cylindric algebras have been introduced by Tarski (see [121]) as algebraisations of predicate logic. They essentially consist of a Boolean algebra with distinguished operations (the c_n and $d_{n,m}$ below) for existential quantification and equality. This Boolean algebra is to be thought of as the collection of all propositions (with free term variables). What is lacking in this approach is the presence of separate structures for each type context (which is such a prominent aspect of the indexed/fibred approach). We will briefly discuss a way to construct a first order fibration from a cylindric algebra.

4.2.8. Cylindric algebra example. A **cylindric algebra** ("of dimension ω") consists of a Boolean algebra $A = \langle A, \bot, \top, \wedge, \vee \rangle$ together with cylindrification operations $c_n \colon A \to A$ and diagonal elements $d_{n,m} \in A$, for $n, m \in \mathbb{N}$, satisfying the following seven postulates.

    (i) $c_n \bot = \bot$;

    (ii) $x \leq c_n x$;

    (iii) $c_n(x \wedge c_n y) = c_n x \wedge c_n y$;

    (iv) $c_n c_m x = c_m c_n x$;

    (v) $d_{n,n} = \top$;

    (vi) $d_{n,k} = c_m(d_{n,m} \wedge d_{m,k})$, for $n \neq m, k$;

    (vii) $c_n(d_{n,m} \wedge x) \wedge c_n(d_{n,m} \wedge \neg x) = \bot$, for $n \neq m$.

The intuition that one should keep in mind is that $c_n x$ is the proposition $\exists v_n. x(v_n)$ and that $d_{n,m}$ is the proposition $v_n = v_m$, assuming we have a countable collection (v_n) of term variables (corresponding to dimension ω).

An arbitrary set U gives rise to a cylindric algebra $P(U^{\mathbb{N}})$ consisting of subsets α of functions $\varphi \colon \mathbb{N} \to U$, with obvious Boolean algebra structure, and with cylindrification and diagonalisation operations:

$$c_n(\alpha) = \bigcup_{x \in U} \{\varphi[x/n] \mid \varphi \in \alpha\}, \qquad d_{n,m} = \{\varphi \in U^{\mathbb{N}} \mid \varphi(n) = \varphi(m)\}$$

where $\varphi[x/n]$ is the function which is x on n, and $\varphi(m)$ on $m \neq n$. See [121] for more information.

In order to turn a cylindric algebra A as above into a fibration, we first need a base category $\mathbb{B}$. We take

objects   natural numbers $n \in \mathbb{N}$. For such an object $n \in \mathbb{B}$ we write $[n]$ for the finite set $\{0, 1, \ldots, n-1\}$ of numbers below n (so that $[0] = \emptyset$).

morphisms $n \to m$ are functions $[m] \to [n]$. Identities and composites are as for functions, except that the order is reversed.

This category $\mathbb{B}$ has finite products: $0 \in \mathbb{B}$ is terminal object, and $n + m \in \mathbb{B}$ is the Cartesian product of $n, m \in \mathbb{B}$: the projections $\pi \colon n + m \to n$ and $\pi' \colon n + m \to m$ are given on $i \in [n] = \{0, \ldots, n-1\}$ and $j \in [m] = \{0, \ldots, m-1\}$ by $\pi(i) = i$ and $\pi'(j) = n + j$; and the pairing of $f \colon k \to n$ and $g \colon k \to m$ as a function $\langle f, g \rangle \colon [n + m] \to [k]$ is $\langle f, g \rangle(i)$ is given by: $f(i)$ if $i < n$, and $g(i - n)$ otherwise. Notice that a diagonal $\delta \colon n + m \to (n + m) + m$ in $\mathbb{B}$, as a function $\delta \colon [n + 2m] \to [n + m]$ is defined as: $\delta(i)$ is i if $i < n + m$, and $i - m$ otherwise.

The next step is put an appropriate indexed category $\mathcal{A} \colon \mathbb{B}^{\mathrm{op}} \to \mathbf{Cat}$ on $\mathbb{B}$ with Boolean algebras as fibres. To this end we identify sub-Boolean algebras $\mathcal{A}(n) \hookrightarrow A$, meant as fibres, as follows.

$$\mathcal{A}(n) = \{x \in A \mid \forall m \geq n.\, c_m x = x\}.$$

(We thus only consider the "finitary" part of the cylindric algebra A.) It follows from the above postulates that the Boolean algebra structure from A restricts to $\mathcal{A}(n)$. The main difficulty is to construct for a morphism $f \colon n \to m$ in $\mathbb{B}$ a substitution functor $f^* \colon \mathcal{A}(m) \to \mathcal{A}(n)$. One can do this by first defining $f^* \colon A \to A$ on A, and by subsequently checking that f^* restricts appropriately. In the theory of cylindric algebras there are substitution operators $s_\ell^k \colon A \to A$, for $k, \ell \in \mathbb{N}$, defined as

$$s_\ell^k x = \begin{cases} x & \text{if } k = \ell \\ c_k(d_{k,\ell} \wedge x) & \text{otherwise} \end{cases}$$

It may be understood as "substitution of variable v_ℓ for v_k". These functions s_ℓ^k will be used to describe categorical substitution f^*, but they cannot be used directly, because we need simultaneous substitution, in which unintended overwrites should be avoided. The standard trick to avoid such clashes is to use a "parking area": let $k = \max\{n, m\}$; the area above k can safely be used for parking, so that for $i < m$

$$s^i_{f(i)}x = x[f(i)/i] = x[k+i/i][f(i)/k+i] = s^{k+i}_{f(i)}s^i_{k+i}x.$$

In this way we can define for $x \in A$

$$f^*(x) = s^k_{f(0)} \cdots s^{k+m-1}_{f(m-1)} s^0_k \cdots s^{m-1}_{k+m-1} x.$$

(Any $k \geq \max\{n, m\}$ yields the same outcome.) The verifications that f^* restricts to $A(m) \to A(n)$ and preserves the Boolean algebra structure, and additionally, that the assignment $f \mapsto f^*$ preserves identities and composition, are quite involved. We are especially interested in the cases where f is a projection $\pi: n + m \to n$ or a diagonal $\delta: n + m \to n + 2m$. In that case one can calculate that

$$\pi^*(x) = x \qquad \text{and} \qquad \delta^*(x) = s^{n+m}_n \cdots s^{n+2m-1}_{n+m-1} x.$$

Left adjoints $\coprod_{(n,m)}$ and $\mathrm{Eq}_{(n,m)}$ to these π^* and δ^* are obtained as:

$$\coprod_{(n,m)}(x) = c_n \cdots c_{n+m-1} x$$
$$\mathrm{Eq}_{(n,m)}(x) = x \wedge d_{n,n+m} \wedge \cdots \wedge d_{n+m-1,n+2m-1}.$$

(A right adjoint to π^* is then induced because we are in a Boolean situation.) We check the adjunction correspondence $\coprod_{(n,m)}(x) \leq y \Leftrightarrow x \leq \pi^*(y)$, for $y \in \mathcal{A}(n)$ and $z \in \mathcal{A}(n + 2m)$: if $\coprod_{(n,m)}(x) \leq y$, then $x \leq c_n \cdots c_{n+m-1}x = \coprod_{(n,m)}(x) \leq y = \pi^*(y)$. For the converse, if $x \leq \pi^*(y) = y$, then $\coprod_{(n,m)}(x) = c_n \cdots c_{n+m-1}x \leq c_n \cdots c_{n+m-1}y = y$, since $y \in \mathcal{A}(n)$, so $c_{n+i}y = y$.

Exercises

4.2.1. Check that the coproducts in Example 4.2.2 satisfy Beck-Chevalley.

4.2.2. In the same Example 4.2.2 check that the weakening functor π^* associated with a (general) projection $\pi: (\Gamma, \Gamma') \to \Gamma$, with Γ' not necessarily of length one, has both a left and a right adjoint.

4.2.3. Consider the fibration constructed in Example 4.2.4 from a Kripke model $\mathcal{K}: \mathbb{I} \to \mathbf{Alg}(\Sigma, \Pi)$.
  (i) Prove that $\coprod(Y)$ and $\prod(Y)$ are monotone collections (for Y monotone).
  (ii) Establish the bijective correspondence

$$\frac{\pi^*(X)(i) \subseteq Y(i) \quad \text{for all } i \in \mathbb{I}}{X(i) \subseteq \prod(Y)(i) \quad \text{for all } i \in \mathbb{I}}$$

which produces the adjunction required for simple products.

(iii) Define equality.

(iv) Check that each (Σ, Π)-algebra forms a Kripke model $1 \to \mathbf{Alg}(\Sigma, \Pi)$ and that the fibration associated with this Kripke model (as in Example 4.2.4) is the same as the fibration associated with the algebra (as in Example 4.2.3).

4.2.4.   Consider a fibration with simple coproducts satisfying Frobenius. Prove that for objects X over $I \times J$ and Y over $I \times K$ there is a vertical isomorphism

$$\coprod_{(I, J \times K)} \big((\mathrm{id} \times \pi^*)(X) \times (\mathrm{id} \times \pi')^*(Y) \big) \cong \coprod_{(I,J)}(X) \times \coprod_{(I,K)}(Y)$$

over I. Explain the logical meaning of this isomorphism.

4.2.5.   This exercise shows that the order in the fibres $(P\mathbb{N})^I$ of the realisability fibration $\begin{smallmatrix} \mathrm{UFam}(P\mathbb{N}) \\ \downarrow \\ \mathbf{Sets} \end{smallmatrix}$ is not the pointwise order. We consider $I = 2 = \{0, 1\}$. Fix an arbitrary subset $A \subseteq \mathbb{N}$, and consider the following two predicates $X, Y : 2 \rightrightarrows P\mathbb{N}$.

$$X(n) = \begin{cases} \mathbb{N} \backslash A & \text{if } n = 0 \\ A & \text{if } n = 1, \end{cases} \qquad Y(n) = \begin{cases} \{0\} & \text{if } n = 0 \\ \{1\} & \text{if } n = 1. \end{cases}$$

(i)   Check that $X(0) \leq Y(0)$ and $X(1) \leq Y(1)$, so that X is pointwise less than Y.

(ii)   Prove that if $X \leq Y$, say via $e \in (X(0) \supset Y(0)) \cap (X(1) \supset Y(1))$, then e yields a decision code for A. But $A \subseteq \mathbb{N}$ is arbitrary, so we may take A to be the halting set.

4.2.6.   Check that for a proposition φ in predicate logic, the result $\varphi[v_\ell / v_k]$ of substituting a term variable v_ℓ for a different variable v_k is logically equivalent to the proposition $\exists v_k . (v_k = v_\ell \wedge \varphi)$. This a logical justification of the definition of the substitution operation s_ℓ^k in Example 4.2.8.

4.2.7.   In which of the examples in this section does one have classical logic? That is, $\neg\neg X \leq X$, where negation $\neg(-)$ is $(-) \supset \bot$, so that each fibre is a Boolean algebra.

4.2.8.   A coherent (or first order) fibration $\begin{smallmatrix} \mathbb{E} \\ \downarrow p \\ \mathbb{B} \end{smallmatrix}$ will be called **Boolean** if for each $X \in \mathbb{E}$ above $I \in \mathbb{B}$ there is a **complement** X' above I with vertical isomorphisms $X \wedge X' \cong \bot$ and $X \vee X' \cong \top$. In that case, each fibre is a Boolean pre-algebra, see *e.g.* Example 4.2.3.

(i)   Show that such a complement X' is unique up-to-isomorphism.

Suppose now p is Boolean and choose for each $X \in \mathbb{E}$ such a complement and write it as $\neg X$.

(ii)   Show that $\neg$ forms a functor $\mathbb{E}_I^{\mathrm{op}} \to \mathbb{E}_I$, which commutes with substitution.

(iii)   Prove that a Boolean coherent fibration is already a (Boolean) first order fibration.

4.3 Functorial interpretation and internal language

In the previous section we introduced appropriate fibrations for predicate logic and listed a series of examples. Here we turn to the (functorial) interpretation of predicate logics (as described in Section 4.1) in such fibrations. It leads to the concept of the internal language, which gives us a convenient means to reason directly in such fibrations, as will be shown in several examples.

We start with validity in a fibration, as first described in Definition 3.5.3 for equations. Here it will be extended to arbitrary predicates. Let therefore (Σ, Π) be a signature with predicates. Consider a situation

$$
\begin{array}{ccc}
 & & \mathbb{E} \\
 & & \downarrow p \\
\mathcal{Cl}(\Sigma) & \xrightarrow{\quad \mathcal{M} \quad} & \mathbb{B}
\end{array}
$$

where p is a preorder fibration with finite products both fibrewise and in its base category, and where $\mathcal{M}$ is a (functorial) model of Σ in $\mathbb{B}$. A **model of** (Σ, Π) **in** $\begin{smallmatrix}\mathbb{E}\\\downarrow p\\\mathbb{B}\end{smallmatrix}$ consists of such a model $\mathcal{M}$ of Σ in the base category $\mathbb{B}$ of p together with for each predicate symbol $P\colon \sigma_1, \ldots, \sigma_n$ in Π, a predicate object

$$
\mathcal{N}(P) \in \mathbb{E} \quad \text{above} \quad \mathcal{M}(x_1\colon \sigma_1, \ldots, x_n\colon \sigma_n) \cong \mathcal{M}(\sigma_1) \times \cdots \times \mathcal{M}(\sigma_n) \in \mathbb{B}.
$$

Such a model of (Σ, Π) in p can be identified with a morphism in the category **Fib** of fibrations,

$$
\begin{array}{ccc}
\mathcal{L}(\Sigma, \Pi) & \xrightarrow{\quad \mathcal{N} \quad} & \mathbb{E} \\
\downarrow & & \downarrow p \\
\mathcal{Cl}(\Sigma) & \xrightarrow[\quad \mathcal{M} \quad]{} & \mathbb{B}
\end{array}
$$

where $\begin{smallmatrix}\mathcal{L}(\Sigma,\Pi)\\\downarrow\\\mathcal{Cl}(\Sigma)\end{smallmatrix}$ is the syntactically constructed classifying fibration from Section 3.1, which has only predicates from Π as (basic) propositions. This is because the only rules that can be used in this restricted logic are the context rules from Section 3.1 (which are unaffected by the presence of the atomic predicates in Π). And these context rules are sound, as we already saw in Lemma 3.5.6. Like in Example 4.2.2 we shall use finite conjunctions $(\top, \wedge)$ and logical contexts of length one in this classifying fibration.

Assume now that p is a regular fibration. Then one can extend the above interpretation $\mathcal{N}$ to propositions with $=, \top, \wedge, \exists$. In a straightforward way one puts

$$\mathcal{N}(\Gamma \vdash M =_\sigma M' \colon \mathsf{Prop}) \;=\; \mathrm{Eq}(\mathcal{M}(M), \mathcal{M}(M'))$$
$$\text{(where Eq is as in 3.4.2)}$$
$$\mathcal{N}(\Gamma \vdash \top \colon \mathsf{Prop}) \;=\; \top$$
$$\mathcal{N}(\Gamma \vdash \varphi \wedge \psi \colon \mathsf{Prop}) \;=\; \mathcal{N}(\Gamma \vdash \varphi \colon \mathsf{Prop}) \wedge \mathcal{N}(\Gamma \vdash \psi \colon \mathsf{Prop})$$
$$\mathcal{N}(\Gamma \vdash \exists x \colon \sigma.\, \varphi \colon \mathsf{Prop}) \;=\; \coprod_{(\mathcal{M}(\Gamma), \mathcal{M}(\sigma))} \big(\mathcal{N}(\Gamma, x \colon \sigma \vdash \varphi \colon \mathsf{Prop}) \big)$$

where $\coprod_{(I,J)}$ is the coproduct functor $\mathbb{E}_{I \times J} \to \mathbb{E}_I$, left adjoint to the weakening functor $\pi^*_{I,J}$.

We say that **a sequent** $\Gamma \mid \varphi \vdash \psi$ **holds** or is **valid** in the above model $(\mathcal{M}, \mathcal{N})$ if

$$\mathcal{N}(\Gamma \vdash \varphi \colon \mathsf{Prop}) \leq \mathcal{N}(\Gamma \vdash \psi \colon \mathsf{Prop})$$

in the fibre above $\mathcal{M}(\Gamma)$. And that **a predicate** $\Gamma \vdash \varphi \colon \mathsf{Prop}$ **holds** or is **valid** in case the sequent $\Gamma \mid \top \vdash \varphi$ holds (*i.e.* in case $\top \leq \mathcal{N}(\Gamma \vdash \varphi \colon \mathsf{Prop})$ over $\mathcal{M}(\Gamma)$).

As we have seen in Lemma 4.1.8, the rules for $\exists$ can be cast in 'mate' form:

$$\frac{\Gamma \mid \Theta, \exists x \colon \sigma.\, \varphi \vdash \psi}{\Gamma, x \colon \sigma \mid \Theta, \varphi \vdash \psi}$$

We shall write χ for the finite conjunction of the propositions in Θ. Soundness of this (double) rule then follows from the adjointness $(\coprod \dashv \pi^*)$ with indices $\mathcal{M}(\Gamma)$ and $\mathcal{M}(\sigma)$, together with the Frobenius condition:

$\Gamma \mid \chi, \exists x \colon \sigma.\, \varphi \vdash \psi$ is valid

$\Leftrightarrow \; \mathcal{N}(\Gamma \vdash \chi \wedge (\exists x \colon \sigma.\, \varphi) \colon \mathsf{Prop}) \leq \mathcal{N}(\Gamma \vdash \psi \colon \mathsf{Prop})$

$\Leftrightarrow \; \mathcal{N}(\Gamma \vdash \chi \colon \mathsf{Prop}) \wedge \coprod \mathcal{N}(\Gamma, x \colon \sigma \vdash \varphi \colon \mathsf{Prop}) \leq \mathcal{N}(\Gamma \vdash \psi \colon \mathsf{Prop})$

$\Leftrightarrow \; \coprod \big(\pi^* \mathcal{N}(\Gamma \vdash \chi \colon \mathsf{Prop}) \wedge \mathcal{N}(\Gamma, x \colon \sigma \vdash \varphi \colon \mathsf{Prop}) \big) \leq \mathcal{N}(\Gamma \vdash \psi \colon \mathsf{Prop})$

$\Leftrightarrow \; \mathcal{N}(\Gamma, x \colon \sigma \vdash \chi \colon \mathsf{Prop}) \wedge \mathcal{N}(\Gamma, x \colon \sigma \vdash \varphi \colon \mathsf{Prop}) \leq \pi^* \mathcal{N}(\Gamma \vdash \psi \colon \mathsf{Prop})$

$\Leftrightarrow \; \mathcal{N}(\Gamma, x \colon \sigma \vdash \chi \wedge \varphi \colon \mathsf{Prop}) \leq \mathcal{N}(\Gamma, x \colon \sigma \vdash \psi \colon \mathsf{Prop})$

$\Leftrightarrow \; \Gamma, x \colon \sigma \mid \chi \wedge \varphi \vdash \psi$ is valid.

The fact that the rules for $\exists$ hold in such a model $\begin{smallmatrix} \mathbb{E} \\ \downarrow p \\ \mathbb{B} \end{smallmatrix}$ can alternatively be expressed by: the morphism $(\mathcal{M}, \mathcal{N})$ of fibrations (in the previous diagram) preserves simple coproducts $\coprod$, where the coproducts in the classifying fibration arise as in Example 4.2.2.

Thus, given a set $\mathcal{A}$ of regular axioms for (Σ, Π), we can say that the above model satisfies $\mathcal{A}$ if all sequents in $\mathcal{A}$ are validated. In that case one obtains

a morphism of fibrations

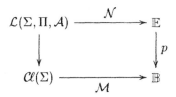

where the classifying fibration on the left captures the logic with axioms from $\mathcal{A}$. The total category $\mathcal{L}(\Sigma, \Pi, \mathcal{A})$ contains the propositions $\Gamma \vdash \varphi \colon \mathsf{Prop}$ that can be formed from equations $M =_\sigma M'$ and from atomic predicates from Π. The fibred preorder structure $(\Gamma \vdash \varphi \colon \mathsf{Prop}) \leq (\Gamma \vdash \psi \colon \mathsf{Prop})$ in $\mathcal{L}(\Sigma, \Pi, \mathcal{A})$ over $\Gamma \in \mathcal{C}\ell(\Sigma)$ is given by derivability of the sequence $\Gamma \mid \varphi \vdash \psi$ from the axioms in $\mathcal{A}$.

In case this fibration p is coherent (*i.e.* additionally has distributive fibred coproducts), then one can interpret finite disjunctions as

$$\mathcal{N}(\Gamma \vdash \bot \colon \mathsf{Prop}) \; = \; \bot$$
$$\mathcal{N}(\Gamma \vdash \varphi \vee \psi \colon \mathsf{Prop}) \; = \; \mathcal{N}(\Gamma \vdash \varphi \colon \mathsf{Prop}) \vee \mathcal{N}(\Gamma \vdash \psi \colon \mathsf{Prop}).$$

And if p is a first order fibration, then one can interpret the remaining logical operations of implication and universal quantification as

$$\mathcal{N}(\Gamma \vdash \varphi \supset \psi \colon \mathsf{Prop}) \; = \; \mathcal{N}(\Gamma \vdash \varphi \colon \mathsf{Prop}) \supset \mathcal{N}(\Gamma \vdash \psi \colon \mathsf{Prop})$$
$$\mathcal{N}(\Gamma \vdash \forall x \colon \sigma. \, \psi \colon \mathsf{Prop}) \; = \; \textstyle\prod_{(\mathcal{M}(\Gamma), \mathcal{M}(\sigma))} \big(\mathcal{N}(\Gamma, x \colon \sigma \vdash \psi \colon \mathsf{Prop}) \big).$$

Validity of the rules involved is left as an exercise below.

We see that the main aspect of Lawvere's functorial semantics can be used also for the interpretation of predicate logic: namely that interpretation is preservation of the relevant structure.

We proceed along (by now) fairly predictable lines: firstly we say formally what a morphism of regular / coherent / first order fibrations is; this enables us to say what a (functorial) model of a predicate logic is. Secondly, we describe the signature with predicates (plus the axioms) associated with a regular / coherent / first order fibration; then a bijective correspondence between models and morphisms of specifications can be given.

4.3.1. Definition. A morphism of regular fibrations is a morphism of Eq-fibrations which preserves simple coproducts $\coprod$. **A morphism of coherent fibrations** additionally preserves fibred finite coproducts $(\bot, \vee)$ and a **morphism of first order fibrations** is a morphism of coherent fibrations which also preserves fibred exponents $\supset$ and simple products $\prod$.

The appropriate 2-cells are as for (Eq-)fibrations, see Definition 3.6.1 (ii).

4.3.2. Definition. A model of a regular / coherent / first order spec-ification $(\Sigma, \Pi, \mathcal{A})$ consists of a regular / coherent / first order fibration $\begin{smallmatrix}\mathbb{E}\\\downarrow p\\\mathbb{B}\end{smallmatrix}$ together with a morphism

$$\begin{pmatrix} \mathcal{L}(\Sigma,\Pi,\mathcal{A}) \\ \downarrow \\ \mathcal{C}\ell(\Sigma) \end{pmatrix} \longrightarrow \begin{pmatrix} \mathbb{E} \\ \downarrow p \\ \mathbb{B} \end{pmatrix}$$

of regular / coherent / first order fibrations.

Recall from Remark 4.1.5 that a specification in predicate logic may be ex-tended with external equations, so that it becomes a four-tuple $(\Sigma, \Pi, \mathcal{A}_i, \mathcal{A}_e)$, where $\mathcal{A}_i$ is the set of internal axioms (as used so far), and $\mathcal{A}_e$ are the addi-tional external equations. In this extended case one should describe a functo-rial model as a structure preserving morphism

$$\begin{pmatrix} \mathcal{L}(\Sigma,\Pi,\mathcal{A}_i) \\ \downarrow \\ \mathcal{C}\ell(\Sigma,\mathcal{A}_e) \end{pmatrix} \longrightarrow \begin{pmatrix} \mathbb{E} \\ \downarrow \\ \mathbb{B} \end{pmatrix}$$

involving a quotient base category $\mathcal{C}\ell(\Sigma) \twoheadrightarrow \mathcal{C}\ell(\Sigma, \mathcal{A}_e)$ incorporating the addi-tional identifications.

The following two lemmas form the basis for soundness and completeness results.

4.3.3. Lemma (Soundness). *Let $(\Sigma, \Pi, \mathcal{A})$ be a regular / coherent / first or-der signature. Every (Σ, Π)-sequent which is derivable from $\mathcal{A}$ in regular / coherent / first order logic, holds in a model of $(\Sigma, \Pi, \mathcal{A})$.*   □

4.3.4. Lemma (Completeness). *Let $(\Sigma, \Pi, \mathcal{A})$ be a regular / coherent / first order signature. A (Σ, Π)-sequent is derivable from $\mathcal{A}$ in regular / coherent / first order logic if and only if it holds in the* **generic model**:

$$\begin{pmatrix} \mathcal{L}(\Sigma,\Pi,\mathcal{A}) \\ \downarrow \\ \mathcal{C}\ell(\Sigma) \end{pmatrix} \overset{=}{\longrightarrow} \begin{pmatrix} \mathcal{L}(\Sigma,\Pi,\mathcal{A}) \\ \downarrow \\ \mathcal{C}\ell(\Sigma) \end{pmatrix}$$ □

4.3.5. Definition. (i) Let $\begin{smallmatrix}\mathbb{E}\\\downarrow p\\\mathbb{B}\end{smallmatrix}$ be a (regular) fibration. The many-typed signature $\mathrm{Sign}(\mathbb{B})$—containing objects $I \in \mathbb{B}$ as types and morphisms $u\colon I_1 \times \cdots \times I_n \to J$ in $\mathbb{B}$ as function symbols—can be extended to a signature with predicates $(\mathrm{Sign}(\mathbb{B}), \Pi(p))$ where

$$X\colon I_1, \ldots, I_n \text{ is in } \Pi(p) \qquad \text{if and only if} \qquad X \in \mathbb{E}_{I_1 \times \cdots \times I_n}.$$

There is an obvious functorial model of the signature with predicates of the fibration p in p itself:

$$
\begin{array}{ccc}
\mathcal{L}(\mathrm{Sign}(\mathbb{B},\Pi(p))) & \xrightarrow{\;\;\varepsilon'\;\;} & \mathbb{E} \\
\downarrow & & \downarrow p \\
\mathcal{C}\ell(\mathrm{Sign}(\mathbb{B})) & \xrightarrow{\;\;\varepsilon\;\;} & \mathbb{B}
\end{array}
$$

(ii) The collection $\mathcal{A}(p)$ of axioms of p contains the $(\mathrm{Sign}(\mathbb{B}),\Pi(p))$-sequents of the form

$$i\colon I \mid X(i) \vdash Y(i) \qquad \text{where } X,Y \in \mathbb{E} \text{ satisfy} \qquad X \leq Y \text{ in } \mathbb{E}_I.$$

The model in (i) can be extended to a morphism of fibrations

$$
\begin{array}{ccc}
\mathcal{L}(\mathrm{Sign}(\mathbb{B},\Pi(p),\mathcal{A}(p))) & \xrightarrow{\;\;\varepsilon'\;\;} & \mathbb{E} \\
\downarrow & & \downarrow p \\
\mathcal{C}\ell(\mathrm{Sign}(\mathbb{B}),\mathcal{A}(\mathbb{B})) & \xrightarrow{\;\;\varepsilon\;\;} & \mathbb{B}
\end{array}
$$

by interpreting the specification of p in p itself—where $\mathcal{A}(\mathbb{B})$ is the collection of (external) equations which hold in the base category $\mathbb{B}$, as described in Definition 3.3.6. A fibration thus gives rise to an extended specification as in Remark 4.1.5.

4.3.6. Theorem. *Let $(\Sigma,\Pi,\mathcal{A}_i,\mathcal{A}_e)$ be a regular signature and $\begin{smallmatrix}\mathbb{E}\\\downarrow p\\\mathbb{B}\end{smallmatrix}$ a regular fibration. There is a bijective correspondence between*

$$\text{regular models}\ \left(\begin{array}{c}\mathcal{L}(\Sigma,\Pi,\mathcal{A}_i)\\\downarrow\\\mathcal{C}\ell(\Sigma,\mathcal{A}_e)\end{array}\right) \longrightarrow \left(\begin{array}{c}\mathbb{E}\\\downarrow p\\\mathbb{B}\end{array}\right)$$

$$\overline{\phantom{\text{maps of regular specifications}}}$$

maps of regular specifications $(\Sigma,\Pi,\mathcal{A}_i,\mathcal{A}_e) \longrightarrow (\mathrm{Sign}(\mathbb{B}),\Pi(p),\mathcal{A}(p),\mathcal{A}(\mathbb{B}))$

The 'counit' regular model in (ii) in the previous definition is an equivalence. Similar results hold for coherent and first order fibrations.

Proof. The correspondence follows from a by now standard argument. In order to obtain the equivalence, we have to define functors θ and θ' in:

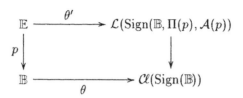

The functor θ' maps a predicate $X \in \mathbb{E}_I$ to the predicate $i\colon I \vdash X(i)\colon \mathsf{Prop}$ in the internal language of p. It is a functor since a map $X \leq Y$ in $\mathbb{E}_I$ yields an axiom $i\colon I \mid X(i) \vdash Y(i)$ in $\mathcal{A}(p)$.

The functor θ maps $I \in \mathbb{B}$ to the context $x\colon I$. It is easy to see that $\varepsilon \circ \theta = \mathrm{id}$. Using the operations from Example 3.3.7 one obtains that there are maps $(\theta \circ \varepsilon) \rightleftarrows \mathrm{id}$, the composites of which are equal to identities in the classifying category $\mathcal{C}\ell(\Sigma, \mathcal{A})$. Here we crucially need the equations from $\mathbb{B}$ as external equations in the logic. $\qquad\square$

Internal language

The starting point in the remainder of this section is our last Theorem 4.3.6. It tells us that a regular / coherent / first order fibration can be reconstructed from its specification, *i.e.* from its signature with predicates plus its axioms. Therefore we can conveniently use the logical language associated with this specification in order to reason in such fibred categories. Below, we present several examples of this approach, but many more examples occur in the course of this book, where the internal language of a (preorder) fibration will be used frequently.

For a fibration $\begin{smallmatrix} \mathbb{E} \\ \downarrow{p} \\ \mathbb{B} \end{smallmatrix}$ we shall call the predicate logic built on top of its signature with predicates $(\mathrm{Sign}(\mathbb{B}), \Pi(p))$ the **internal language** of p. And the **internal logic** of p is the logic which starts from the specification $(\mathbf{Sign}(\mathbb{B}), \Pi(p), \mathcal{A}(p))$ of p. This logic incorporates everything that holds in p (via its axioms).

In this internal language, an object $I \in \mathbb{B}$ is a type and an object $X \in \mathbb{E}$ above $I \in \mathbb{B}$ is a proposition in context $i\colon I$, *i.e.* a predicate on I. Therefore we often write such an X as

$$i\colon I \vdash X\colon \mathsf{Prop} \quad \text{or as} \quad i\colon I \vdash X(i)\colon \mathsf{Prop} \quad \text{or as} \quad i\colon I \vdash X_i\colon \mathsf{Prop}.$$

In the latter two cases we have made the dependence on I explicit in $X(i)$ and X_i. This is convenient notation. Also for example, when $X \in \mathbb{E}_{I \times J}$ is a predicate on a product type, we can write this as

$$i: I, j: J \vdash X_{(i,j)}: \mathsf{Prop}$$

And if we have $X, Y \in \mathbb{E}_I$, then,

$$i: I \mid X_i \vdash Y_i \quad \text{is derivable in the internal logic of } p$$
$$\text{if and only if} \quad X \leq Y \text{ over } I.$$

The result of reindexing a predicate $Y = (j: J \vdash Y_j: \mathsf{Prop}) \in \mathbb{E}_J$ in p over J along a morphism (or term) $u: I \to J$ in the basis will be written as

$$u^*(Y) = (i: I \vdash Y_{u(i)}: \mathsf{Prop}).$$

As a special case we could write weakening of $X \in \mathbb{E}_I$ by adding a dummy assumption $j: J$ as $i: I, j: J \vdash X_{\pi(i,j)}: \mathsf{Prop}$. This is would be different from ordinary predicate logic, where weakening is not an explicit operation—but see the explicit notation of [256], as mentioned in Example 3.1.1 (ii).

We will use $=_I, \top, \wedge, \bot, \vee, \supset, \exists, \forall$ with obvious meaning in the underlying fibration. The internal language (or logic) has the advantage that it is easy to manipulate, in contrast to categorical calculations, which are often more complicated. This will be illustrated in the next series of examples.

4.3.7. Examples (Quantification). (i) Let $\begin{smallmatrix} \mathbb{E} \\ \downarrow p \\ \mathbb{B} \end{smallmatrix}$ be a regular fibration. By definition, each weakening functor π^* induced by a Cartesian projection π then has a left adjoint. We shall show in the internal language of p that in fact *each* functor u^* has a left adjoint $\coprod_u$. Later in Section 9.1 we shall see that this makes p an 'opfibration'.

Assume an arbitrary map $u: I \to J$ in $\mathbb{B}$; the functor $\coprod_u$ is defined as

$$X = (i: I \vdash X_i: \mathsf{Prop}) \longmapsto (j: J \vdash \exists i: I. (u(i) =_J j \wedge X_i): \mathsf{Prop}).$$

The adjunction $(\coprod_u \dashv u^*)$ follows from the following derivation.

$$\frac{\dfrac{j: J \mid \exists i: I. (u(i) = j \wedge X_i) \vdash Y_j}{i: I, j: J \mid u(i) = j, X_i \vdash Y_j}}{i: I \mid X_i \vdash Y_{u(i)}}$$

Notice that the Beck-Chevalley condition need not hold: it is an external condition involving pullbacks in the base category. These are not required to exist and—in case they happen to be there—they need not be expressible in the internal language. See Exercise 4.9.2 for some more relevant details.

(ii) Assume now that $\begin{smallmatrix} \mathbb{E} \\ \downarrow p \\ \mathbb{B} \end{smallmatrix}$ is a first order fibration. One can now show that each reindexing functor has a right adjoint as well. For $u: I \to J$ in $\mathbb{B}$ define

$\prod_u(X)$ by

$$X = \big(i\colon I \vdash X_i\colon \mathsf{Prop}\big) \longmapsto \big(j\colon J \mid \forall i\colon I.\,(u(i) =_J j \supset X_i)\colon \mathsf{Prop}\big).$$

Then

$$\frac{\dfrac{\dfrac{j\colon J \mid Y_j \vdash \forall i\colon I.\,(u(i) = j \supset X_i)}{i\colon I,j\colon J \mid Y_j \vdash u(i) = j \supset X_i}}{i\colon I,j\colon J \mid Y_j, u(i) = j \vdash X_i}}{i\colon I \mid Y_{u(i)} \vdash X_i}$$

Notice that the formulas for $\coprod_u$ and $\prod_u$ are the familiar set-theoretic ones, as used for example in Lemma 1.9.5.

Next we show how one can conveniently describe in the internal language a category of relations associated with a regular fibration.

4.3.8. Example (Relations). Recall from Example 3.5.5 that the category **Rel** of sets and relations has sets as objects and relations $R \subseteq I \times J$ as morphisms $I \to J$. One usually writes $R\colon I \nrightarrow J$ for $R \subseteq I \times J$ in this setting. Identity morphisms and composition in **Rel** can be expressed using the connectives $=, \wedge, \exists$ of regular logic, see the beginning of Example 3.5.5. This leads us to the following construction.

For a regular fibration $\begin{smallmatrix} \mathbb{E} \\ \downarrow p \\ \mathbb{B} \end{smallmatrix}$ let $\mathrm{Rel}(p)$ be the category with

objects $\quad I \in \mathbb{B}$.

morphisms $\quad I \nrightarrow J$ are equivalence classes of objects $R \in \mathbb{E}$ above $I \times J$.

The equivalence relation is the one induced by the preorder of entailment in the fibres (*i.e.* equivalence in the internal logic). Often we shall write $R\colon I \nrightarrow J$ in the internal language as a predicate

$$i\colon I, j\colon J \vdash R(i,j)\colon \mathsf{Prop}.$$

The identity $I \nrightarrow I$ is then given by (internal) equality on I:

$$i\colon I, i'\colon I \vdash i =_I i'\colon \mathsf{Prop}.$$

And composition of $R\colon I \nrightarrow J$ and $S\colon J \nrightarrow K$ by the 'composite' relation:

$$i\colon I, k\colon K \vdash \exists j\colon J.\,R(i,j) \wedge S(j,k)\colon \mathsf{Prop}.$$

One easily checks that $\mathrm{Rel}(p)$ is a category; for associativity of composition of

$$I \xrightarrow{\;R\;} J \xrightarrow{\;S\;} K \xrightarrow{\;T\;} L$$

one can reason informally:

$$((T \circ S) \circ R)(i, \ell) \iff \exists j \colon J. \, R(i, j) \wedge \exists k \colon K. \big[S(j, k) \wedge T(k, \ell) \big]$$
$$\iff \exists k \colon K. \big[\exists j \colon J. \, R(i, j) \wedge S(j, k) \big] \wedge T(k, \ell)$$
$$\iff (T \circ (S \circ R))(i, \ell).$$

One can similarly describe the subcategory $\mathrm{FRel}(p) \hookrightarrow \mathrm{Rel}(p)$ of **functional relations**. This category $\mathrm{FRel}(p)$ has objects I of the base category as objects. A morphism $I \to J$ in $\mathrm{FRel}(p)$ is a morphism $R \colon I \nrightarrow J$ in $\mathrm{Rel}(p)$ which is internally **single-valued** and **total**, as expressed by the following two sequents.

$$i \colon I, j \colon J, j' \colon J \mid R(i, j), R(i, j') \vdash j =_J j', \qquad i \colon I \mid \emptyset \vdash \exists j \colon J. \, R(i, j).$$

It is easy to check that identities and composition from $\mathrm{Rel}(p)$ can be used in $\mathrm{FRel}(p)$, so that we get an inclusion functor $\mathrm{FRel}(p) \hookrightarrow \mathrm{Rel}(p)$.

The expressiveness of first order predicate logic allows us to formulate various mathematical notions in a very general situation where one has a fibred category $\begin{smallmatrix} \mathbb{E} \\ \downarrow \\ \mathbb{B} \end{smallmatrix}$ which allows us to reason about $\mathbb{B}$ in the logic of this fibration. As an easy example we mention the following.

4.3.9. Definition. Consider a regular fibration $\begin{smallmatrix} \mathbb{E} \\ \downarrow p \\ \mathbb{B} \end{smallmatrix}$. A morphism $u \colon I \to J$ in the base category $\mathbb{B}$ is called **internally injective** if the following sequent in the internal language of p holds in p.

$$i \colon I, i' \colon I \mid u(i) =_J u(i') \vdash i =_I i'.$$

Similarly, u is **internally surjective** if

$$j \colon J \mid \emptyset \vdash \exists i \colon I. \, j =_J u(i)$$

holds.

Notice that 'internal injectivity' and 'internal surjectivity' are relative notions in the sense that they are not intrinsic to the base category, but depend on the fibration that one puts on top of the base category (to get a certain logic).

It is not hard to see that if internal and external equality coincide in a fibration, then 'internally injective' means 'monic in the base category'. This is more subtle with internal surjectivity, due to the occurrence of the existential quantifier $\exists$. Consider for example a family fibration $\begin{smallmatrix} \mathrm{Fam}(A) \\ \downarrow \\ \mathbf{Sets} \end{smallmatrix}$ for a frame A. For a set I and an I-indexed collection $X = (X_i)_{i \in I}$ of objects $X_i \in A$ over I, the proposition $\exists i \colon I. \, X$ holds in this fibration if and only if the join $\bigvee_{i \in I} X_i$ is the top element $\top$ in A. But this need not mean "external existence", *i.e.* that there is an actual element X_{i_0} in this collection X for which $X_{i_0} = \top$. We can conclude that internal existence ($\exists i : I. X_i$ holds) need not imply external existence (X_{i_0} holds for some specific $i_0 : 1 \rightarrow I$).

Conversely, external existence trivially implies internal existence. We return to this delicate matter of existence at the end of Section 4.5 in connection with the Axiom of Choice, especially in Exercises 4.5.4 and 4.5.5.

Exercises

4.3.1. (i) Prove the soundness of the 'traditional' rules for $\exists$ as in Figure 4.1 in a regular fibration.

      (ii) Verify that the rules for implication $\supset$ and universal quantification $\forall$ are sound with respect to the interpretation described above.

4.3.2. Give a purely categorical proof of the inequality

$$\coprod\nolimits_{(I,K)} (\mathrm{id} \times u)^*(X) \leq \coprod\nolimits_{(I,J)} X$$

for $u : K \rightarrow J$ and X over $I \times J$. That is, of the entailment

$$i : I \mid \exists k : K. X(i, u(k)) \vdash \exists j : J. X(i, j).$$

4.3.3. Show that the *reductio ad absurdum* rule of classical logic (see Section 4.1) is sound in a Boolean coherent fibration, as defined in Exercise 4.2.8.

4.3.4. Investigate internal injectivity / surjectivity in the fibrations

      (i) $\begin{array}{c} \mathrm{DSub}(\mathbf{Dcpo}) \\ \downarrow \\ \mathbf{Dcpo} \end{array}$ of down closed subsets on dcpos, in Example 3.5.4;

      (ii) $\begin{array}{c} \mathbf{PredRel} \\ \downarrow \\ \mathbf{Rel} \end{array}$ in Example 3.5.5.

4.3.5. Consider an Eq-fibration on a distributive base category. Prove that the coproduct coprojections κ, κ' are *internally injective*, see Proposition 2.3.4 (ii).

4.3.6. In an Eq-fibration $\begin{array}{c} \mathbb{E} \\ \downarrow p \\ \mathbb{B} \end{array}$ there are left adjoints $\coprod_{\delta(I)} = \mathrm{Eq}_I$ to contraction functors $\delta(I)^*$, where $\delta(I) = \langle \mathrm{id}, \mathrm{id} \rangle : I \rightarrow I \times I$ is the (unparametrised) diagonal on I. Prove that the implication

$$u \text{ is mono in } \mathbb{B} \;\Rightarrow\; u \text{ is internally injective}$$

holds if and only if these $\coprod_{\delta(I)}$'s satisfy the Beck-Chevalley condition with respect to pullback squares of the form

$$
\begin{array}{ccc}
I & \xrightarrow{\;u\;} & J \\
{\scriptstyle \delta(I)}\downarrow & \lrcorner & \downarrow{\scriptstyle \delta(J)} \\
I \times I & \xrightarrow{u \times u} & J \times J
\end{array}
$$

4.3.7. Give a purely categorical argument to show that a reindexing functor of a regular fibration has a left adjoint, as in Example 4.3.7 (i).

4.3.8. For a regular fibration $\begin{smallmatrix} \mathbb{E} \\ \downarrow P \\ \mathbb{B} \end{smallmatrix}$, describe a sequence of functors

$$\mathbb{B} \longrightarrow \mathrm{FRel}(p) \longrightarrow \mathrm{Rel}(p)$$

mapping a morphism $I \to J$ in $\mathbb{B}$ to its graph relation $I \relbar\joinrel\twoheadrightarrow J$.

4.4 Subobject fibrations I: regular categories

This section is entirely devoted to examples of regular fibrations which arise from subobject fibrations $\begin{smallmatrix} \mathrm{Sub}(\mathbb{B}) \\ \downarrow \\ \mathbb{B} \end{smallmatrix}$. We shall find conditions on a category $\mathbb{B}$ ensuring that subobjects in $\mathbb{B}$ form such a regular fibration. A category will then be called a regular. In the next section we concentrate on the case where the subobject fibration is a coherent or first order fibration. All the material in this (and the next) section is standard (early sources are [17] and [284]), but usually it is not presented in terms of fibrations. For a slightly different approach to regular categories, in which not all finite limits are assumed to exist, see [36, II, Chapter 2].

Subobject fibrations have received much attention because they incorporate the logic of toposes, see Section 5.4 later on. These fibrations are in fact rather special. For example, they always support very strong equality and full subsets (or comprehension). Later, in Section 4.9 we will give a precise characterisation of subobject fibrations in terms of logical structure. Part of this structure is given by the following result, which can be interpreted as saying that in subobject fibrations one always has unique choice ($\exists!$), see Proposition 4.9.2.

4.4.1. Observation. Each subobject fibration $\begin{smallmatrix} \mathrm{Sub}(\mathbb{B}) \\ \downarrow \\ \mathbb{B} \end{smallmatrix}$ admits quantification along monos: if $m: X \rightarrowtail I$ is a mono, then composition with m forms a left adjoint $\coprod_m$ to reindexing m^* along m. Moreover, these $\coprod$'s satisfy a Beck-Chevalley condition: for every pullback square

$$\begin{array}{ccc} Y & \overset{n}{\rightarrowtail} & J \\ v\downarrow & \lrcorner & \downarrow u \\ X & \underset{m}{\rightarrowtail} & I \end{array}$$

the canonical natural transformation $\coprod_n v^* \Rightarrow u^* \coprod_m$ is an isomorphism. Further, these coproducts satisfy the Frobenius property: there is a (canonical) isomorphism $\coprod_m (m^*(n) \wedge k) \xrightarrow{\cong} n \wedge \coprod_m (k)$. Both Beck-Chevalley and Frobenius follow from the Pullback Lemma (see Exercise 1.1.5).

Later in Section 10.5, we develop tools to give an alternative formulation of this result; then we shall say that the 'comprehension category $\mathrm{Sub}(\mathbb{B}) \hookrightarrow \mathbb{B}^{\rightarrow}$ has (very strong) coproducts'.

The following definitions are standard.

4.4.2. Definition. (i) A category has **images** if every morphism has an **image factorisation**; that is, every morphism $u: I \to J$ factors as

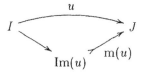

where $m(u)$ is the least mono through which u factors: for an arbitrary factorisation $I \to K \rightarrowtail J$ of u, there is a necessarily unique map $\mathrm{Im}(u) \dashrightarrow K$ making the diagram below commute.

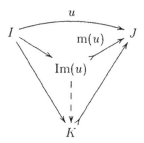

(ii) A category with images has **stable images** if its images are stable under pullback: if the diagram on the left below is a pullback, then so is the one on the right,

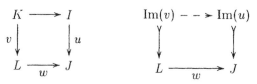

The map $\mathrm{Im}(v) \dashrightarrow \mathrm{Im}(u)$ is uniquely determined by the universal property of the image factorisation of v, since v factors through $w^*(m(u))$.

(iii) A **regular category** is a category which has finite limits and stable images.

4.4.3. Example. In the category of sets the image of a function $u \colon I \to J$ exists and is given by the subset

$$\{j \in J \mid \exists i \in I.\, u(i) = j\} \ \overset{\displaystyle \hookrightarrow}{} \ J$$

It may be clear that this is the least mono (injection) through which $u \colon I \to J$ factors.

Note that there is an existential quantifier involved. It is made explicit in the next result. It shows that regular categories can be characterised in various ways. Of most interest to us is the equivalence of (i) and (v) below.

4.4.4. Theorem. *Let $\mathbb{B}$ be a category with finite limits. The following points are equivalent.*

(i) *The category $\mathbb{B}$ is regular.*

(ii) *The inclusion functor $\mathrm{Sub}(\mathbb{B}) \hookrightarrow \mathbb{B}^{\to}$ (obtained by choosing representatives) has a fibred left adjoint.*

(iii) *The subobject fibration $\begin{smallmatrix} \mathrm{Sub}(\mathbb{B}) \\ \downarrow \\ \mathbb{B} \end{smallmatrix}$ has coproducts $\coprod_u \dashv u^*$ satisfying Frobenius.*

(iv) *The fibration $\begin{smallmatrix} \mathrm{Sub}(\mathbb{B}) \\ \downarrow \\ \mathbb{B} \end{smallmatrix}$ has simple coproducts $\coprod_{(I,J)} \dashv \pi^*_{I,J}$ satisfying Frobenius.*

(v) *The subobject fibration $\begin{smallmatrix} \mathrm{Sub}(\mathbb{B}) \\ \downarrow \\ \mathbb{B} \end{smallmatrix}$ is regular.*

We recall that for $\mathbb{B} = \mathbf{Sets}$, the subobject fibration $\begin{smallmatrix} \mathrm{Sub}(\mathbb{B}) \\ \downarrow \\ \mathbb{B} \end{smallmatrix}$ is the fibration $\begin{smallmatrix} \mathbf{Pred} \\ \downarrow \\ \mathbf{Sets} \end{smallmatrix}$ of predicates over sets as described in Section 0.2 in the Introduction.

Proof. The equivalence (iv) $\Leftrightarrow$ (v) is obvious, because the fibration $\begin{smallmatrix} \mathrm{Sub}(\mathbb{B}) \\ \downarrow \\ \mathbb{B} \end{smallmatrix}$ already has fibred finite products and equality satisfying Frobenius. Further, (iii) $\Rightarrow$ (iv) is immediate. We shall do (i) $\Leftrightarrow$ (iii) and (iv) $\Rightarrow$ (iii) and leave the equivalence of (ii) to the other points as an exercise.

(i) $\Rightarrow$ (iii). For a morphism $u \colon I \to J$ in $\mathbb{B}$ and a mono $m \colon X \rightarrowtail I$ one defines a coproduct by

$$\coprod_u(m) = \Big(\mathrm{Im}(u \circ m) \rightarrowtail J\Big)$$

Notice that $\coprod_u(m)$ is simply $u \circ m$ if u is a mono (as in Observation 4.4.1). There is then a bijective correspondence

$$\frac{\coprod_u(m) \leq n}{m \leq u^*(n)}$$

establishing an adjunction $(\coprod_u \dashv u^*)$, as follows.

- if $m \leq u^*(n)$, then $u \circ m$ factors through n and thus $\coprod_u(m)$, being the least mono for which this holds, satisfies $\coprod_u(m) \leq n$.
- if $\coprod_u(m) \leq n$, then, using that $u \circ m$ factors—by definition of image—through $\coprod_u(m)$, one obtains $m \leq u^*(\coprod_u(m))$; hence $m \leq u^*(\coprod_u(m)) \leq u^*(n)$.

The stability of the image factorisation ensures that the Beck-Chevalley condition holds. For Frobenius, consider for $u\colon I \to J$, $m \in \mathrm{Sub}(I)$ and $n \in \mathrm{Sub}(J)$ the following pullback squares.

$$
\begin{array}{ccccc}
X' & \rightarrowtail & Y' & \xrightarrow{\ u'\ } & Y \\
\downarrow & \lrcorner & \downarrow u^*(n) & \lrcorner & \downarrow n \\
X & \xrightarrow{\ m\ } & I & \xrightarrow{\ u\ } & J
\end{array}
$$

Then

$$
\begin{aligned}
n \wedge \coprod_u(m) &= \coprod_n n^*(\coprod_u(m)) && \text{by definition of } \wedge \\
&\cong \coprod_n \coprod_{u'}(u^*(n))^*(m) && \text{by Beck-Chevalley} \\
&\cong \coprod_u \coprod_{u^*(n)}(u^*(n))^*(m) \\
&= \coprod_u(u^*(n) \wedge m).
\end{aligned}
$$

(iii) $\Rightarrow$ (i). Given a morphism $u\colon I \to J$, define the image of u as:

$$
m(u) = \Big(\mathrm{Im}(u) \xrightarrow{\ \coprod_u(\top) \,=\, \coprod_u(\mathrm{id}_I)\ } J\Big)
$$

Using the unit η in the diagram:

$$
\begin{array}{ccc}
I \xrightarrow{\ \eta\ } K & \xrightarrow{\hspace{2cm}} & \mathrm{Im}(u) \\
\diagdown \quad \downarrow u^*(\coprod_u(\mathrm{id}_I)) & & \downarrow m(u) = \coprod_u(\mathrm{id}_I) \\
\mathrm{id}_I \searrow \quad I & \xrightarrow{\ u\ } & J
\end{array}
$$

one obtains that u factors through $m(u)$. If also $u = n \circ f$ with n monic, then $\mathrm{id}_I \leq u^*(n)$ and thus, by transposition, $m(u) = \coprod_u(\mathrm{id}_I) \leq n$. Stability follows from the Beck-Chevalley condition:

$$
w^*(m(u)) = w^*(\coprod_u(\top)) \cong \coprod_{w^*(u)}(\top) = m(w^*(u)).
$$

(iv) $\Rightarrow$ (iii). Coproducts along an arbitrary map $u\colon I \to J$ are obtained by writing $u = \pi \circ \langle u, \mathrm{id}\rangle\colon I \rightarrowtail J \times I \to J$ as composite of a mono and a Cartesian projection. Since we have coproducts along projections (by assumptions) and along monos (by Observation 4.4.1) we are done by composition of adjoints. Beck-Chevalley follows from a similar argument.   □

Image factorisation in regular categories gives rise to a class of maps which are called 'covers'. The best way to think about these is as surjections, see explicitly in Lemma 4.4.7 below.

4.4.5. Definition. In a regular category, a morphism $u: I \to J$ is called a **cover** if its monic part $m(u): \text{Im}(u) \rightarrowtail J$ is an isomorphism. One often writes $u: I \twoheadrightarrow J$ to indicate that u is a cover.

In Example 4.4.3, the covers in **Sets** are precisely the surjective functions. The next lemma lists a series of results about these covers; it includes four alternative characterisations: in (i), (iii), (vii) and (viii).

4.4.6. Lemma. *In a regular category the following holds.*

   (i) *A morphism u is a cover if and only if u is **extremal**: for each factorisation $u = m \circ u'$ one has: m is a mono implies that m is an isomorphism.*

   (ii) *A monic cover is an isomorphism.*

   (iii) *A morphism u is a cover if and only if the map $\coprod_u(\top) \to \top$ is an isomorphism, where $\coprod_u$ is the induced left adjoint to u^*, see (iii) in Theorem 4.4.4.*

   (iv) *Every isomorphism is a cover. Covers are closed under composition: if u, v are (composable) covers, then $v \circ u$ is a cover. Also, if $v \circ u$ and u are covers, then v is a cover.*

   (v) *Covers are stable under pullback.*

   (vi) *Every map factorises as a cover followed by a mono.*

   (vii) *A morphism u is a cover if and only if u is **orthogonal** to all monos. The latter means that in a commuting square*

there is a unique diagonal as indicated, making everything in sight commute.

   (viii) *A morphism u is cover if and only if u is regular epimorphism.*

From (vii) one can conclude that the factorisation in (vi) is essentially unique (in an obvious sense). This yields that the collections (Monos) and (Covers) form a **factorisation system** (see [18]) in a regular category.

Most of the results in this lemma are easy to prove, except the implication ($\Rightarrow$) in (viii) which tells that covers are regular epis (*i.e.* epis which occur as coequalisers). This is a folklore result. The proof that we present is essentially as in [169, Theorem 1.52].

Proof. (i) The implication ($\Leftarrow$) is obvious by definition of 'cover'. In the reverse direction, assume a cover $u: I \twoheadrightarrow J$ is written as $u = m \circ u'$, where m is a mono. Then the image $m(u)$ of u must satisfy $m(u) \leq m$. But since $m(u)$ is an isomorphism by assumption, we get that m is an isomorphism as well.

(ii) Write $u = u \circ \mathrm{id}$ and apply ($\Rightarrow$) in (i).

(iii) Notice that for a morphism u one has

$$\coprod_u(\top) = \coprod_u(\mathrm{id}) = m(u \circ \mathrm{id}) = m(u).$$

Hence u is a cover if and only if $m(u): \mathrm{Im}(u) \to \mathrm{id}$ is an isomorphism (in the slice category), *i.e.* if and only if $\coprod_u(\top) \to \top$ is an isomorphism.

(iv) If u is an isomorphism, then one can take as monic part $m(u) = u$. And if u, v are covers, then so is v:

$$\coprod_{v \circ u}(\top) \cong \coprod_v(\coprod_u(\top)) \cong \coprod_v(\top) \cong \top$$

so we are done by (iii). Similarly, if $v \circ u$ and u are covers, then

$$\top \cong \coprod_{v \circ u}(\top) \cong \coprod_v(\coprod_u(\top)) \cong \coprod_v(\top).$$

(v) Consider a pullback square

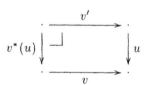

Then

$$\begin{aligned}
\coprod_{v^*(u)}(\top) &\cong \coprod_{v^*(u)}(v'^*(\top)) \\
&\cong v^*(\coprod_u(\top)) \qquad \text{by Beck-Chevalley} \\
&\cong v^*(\top) \\
&\cong \top.
\end{aligned}$$

Hence $v^*(u)$ is a cover again by (iii).

(vi) Every morphism $u: I \to J$ can be written as $I \xrightarrow{u'} \mathrm{Im}(u) \xrightarrow{m(u)} J$. We show that u' is a cover using (i). Assume $u' = n \circ u''$, where n is a mono. Then $u = (m(u) \circ n) \circ u''$ and thus $m(u) \leq m(u) \circ n$, since $m(u)$ is the least mono through which u factorises. But then n must be an isomorphism.

(vii) First, assume that u is cover in a commuting square

$$
\begin{array}{ccc}
I & \xrightarrow{u} & J \\
{\scriptstyle f}\downarrow & & \downarrow{\scriptstyle g} \\
Z & \underset{m}{\rightarrowtail} & K
\end{array}
$$

Then u factors through the mono $g^*(m)$. Hence by (i), $g^*(m)$ is an isomorphism. This yields the required diagonal $J \dashrightarrow Z$.

Conversely, if $u: I \to J$ is orthogonal to all monos, and u can be written as $u = n \circ f$ with n a mono. Then we get a commuting square

$$
\begin{array}{ccc}
I & \xrightarrow{u} & J \\
{\scriptstyle f}\downarrow & & \| \\
Y & \underset{n}{\rightarrowtail} & J
\end{array}
$$

This yields a fill-in $s: J \dashrightarrow Y$ with $s \circ u = f$ and $n \circ s = \mathrm{id}$. So n is a split mono and thus an isomorphism.

(viii) It is easy to see that regular epis are covers: suppose $u: I \twoheadrightarrow J$ is coequaliser of $f, g: K \rightrightarrows I$ and can be factorised as $u = m \circ u'$ where m is a mono. Then $u' \circ f = u' \circ g$ so there is a unique n with $n \circ u = u'$. Hence $m \circ n \circ u = m \circ u' = u$, which yields $m \circ n = \mathrm{id}$ by the fact that u is epi. Thus m is an isomorphism and u is a cover by (i).

For the converse we first prove that a cover $u: I \dashrightarrow J$ is an epi: suppose $f, g: J \rightrightarrows K$ are given with $f \circ u = g \circ u$. Then u factorises through the (monic) equaliser of f and g. Hence this equaliser must be an isomorphism by (i). Thus $f = g$.

We come to the proof that such a cover $u: I \dashrightarrow J$ is a *regular* epi. We form the kernel pair $\pi_1, \pi_2: R \rightrightarrows I$, by taking the pullback of u against itself, and intend to show that u is the coequaliser of this pair π_1, π_2. Assume therefore that $v: I \to K$ also satisfies $v \circ \pi_1 = v \circ \pi_2$. We factorise the tuple $\langle u, v \rangle: I \to J \times K$ as

$$
\left(I \xrightarrow{\langle u, v \rangle} I \times K \right) = \left(I \xrightarrow{w} W \rightarrowtail^{m} J \times K \right)
$$

and intend to show that $\pi \circ m: W \to J$ is an isomorphism; then we are done since it yields that $\alpha = \pi' \circ m \circ (\pi \circ m)^{-1}: J \to K$ is the (unique) required mediating map with $\alpha \circ u = v$ (since $(\pi \circ m)^{-1} \circ u = w$).

Firstly, $\pi \circ m$ is a cover since both w and $(\pi \circ m) \circ w = u$ are covers, using (iv). In order to see that $\pi \circ m$ is monic, assume $f, g: Z \rightrightarrows W$ are given with

$\pi \circ m \circ f = \pi \circ m \circ g$. Form the pullback square

$$
\begin{array}{ccc}
Z' & \xrightarrow{\quad h \quad} & Z \\
{\scriptstyle \langle f', g' \rangle} \downarrow \;\;\lrcorner & & \downarrow {\scriptstyle \langle f, g \rangle} \\
I \times I & \xrightarrow[w \times w]{} & W \times W
\end{array}
$$

Both $\mathrm{id} \times w$ and $w \times \mathrm{id}$ can be obtained from w by pullback along a Cartesian projection. Hence they are covers by (v) and thus $w \times w = (\mathrm{id} \times w) \circ (w \times \mathrm{id})$ as well. But then also h is a cover and in particular an epimorphism. One has

$$
\begin{aligned}
u \circ f' &= \pi \circ \langle u, v \rangle \circ f' \\
&= \pi \circ m \circ w \circ f' \\
&= \pi \circ m \circ f \circ h \\
&= \pi \circ m \circ g \circ h \qquad \text{by assumption about } f, g \\
&= u \circ g'.
\end{aligned}
$$

Hence there is a unique $k \colon Z' \to R$ with $\pi_1 \circ k = f'$ and $\pi_2 \circ k = g'$. Then

$$
\begin{aligned}
(\pi' \circ m \circ f) \circ h &= \pi' \circ m \circ \pi \circ w \times w \circ \langle f', g' \rangle \\
&= \pi' \circ \langle u, v \rangle \circ f' \\
&= v \circ \pi_1 \circ k \\
&= v \circ \pi_2 \circ k \qquad \text{by assumption about } v \\
&= \pi' \circ \langle u, v \rangle \circ g' \\
&= (\pi' \circ m \circ g) \circ h.
\end{aligned}
$$

But then $\pi' \circ m \circ f = \pi' \circ m \circ g$, since h is an epi. Hence $m \circ f = m \circ g$ and thus $f = g$, since m is a mono. By (ii) we conclude that $\pi \circ m$ is isomorphism and so we are done. $\qquad \square$

The attention in this section has been focussed on monos and covers in regular categories. It is therefore appropriate to close with the following result, which tells that monos and covers are the internal injections and surjections in subobject fibrations (see Definition 4.3.9). Some more information on monos and covers is given in the exercises.

4.4.7. Lemma. *With respect to a subobject fibration* $\begin{smallmatrix} \mathrm{Sub}(\mathbb{B}) \\ \downarrow \\ \mathbb{B} \end{smallmatrix}$ *of a regular category* $\mathbb{B}$, *a morphism in* $\mathbb{B}$ *is internally injective if and only if it is monic in* $\mathbb{B}$, *and it is internally surjective if and only if it is a cover in* $\mathbb{B}$.

Proof. Consider a morphism $u \colon I \to J$ in $\mathbb{B}$. The first part of the statement that u is internally injective if and only if it is a mono in $\mathbb{B}$ follows from the fact that internal and external equality coincide in subobject fibrations. Explicitly, the proposition stating that u is internally injective amounts to

$$\text{Eq}(u \circ \pi, u \circ \pi') \leq \text{Eq}(\pi, \pi') \quad \text{over } I \times I.$$

Since equality in subobject fibrations is given by equalisers (see Examples 3.4.4 (i) and (ii)), this reduces to the statement that the equaliser of $u \circ \pi$ and $u \circ \pi'$ factors through the diagonal $\delta = \langle \text{id}, \text{id} \rangle \colon I \rightarrowtail I \times I$. One easily verifies that the latter holds if and only if u is a mono in $\mathbb{B}$.

In the same vain, the statement 'u is internally surjective' unravels to

$$\text{id}_J \leq \text{Image of} \left(\bullet \overset{\text{Eq}(\pi,\, u\, \circ\, \pi')}{\rightarrowtail} J \times I \overset{\pi}{\longrightarrow} J \right).$$

But since the equaliser $\text{Eq}(\pi, u \circ \pi')$ of π and $u \circ \pi'$ is $\langle u, \text{id} \rangle \colon I \rightarrowtail J \times I$, this amounts to

$$\text{id}_J \leq \text{Image of } u,$$

which says that u is a cover. $\square$

Exercises

4.4.1.   Show that images in the category of sets (as described in Example 4.4.3) are stable.

4.4.2.   A **split epi** is morphism u which has a section (*i.e.* for which there is an s with $u \circ s = \text{id}$).

(i)   Check that a split epi is an epi.

(ii)   Show that in a regular category, a split epi is a cover.

4.4.3.   Show that for a subobject fibration, the internal description of $\coprod_u$ in Example 4.3.7 (i) coincides with the description in the proof of Theorem 4.4.4 in terms of images.

4.4.4.   In a regular category, consider maps

$$K \overset{e}{\twoheadrightarrow} I \overset{u}{\longrightarrow} J \overset{m}{\rightarrowtail} L$$

and show that there are the following equalities of subobjects.

(i) $\text{Im}(u) \rightarrowtail J$ is $\text{Im}(u \circ e) \rightarrowtail J$;

(ii) $\text{Im}(u) \rightarrowtail J \overset{m}{\rightarrowtail} L$ is $\text{Im}(m \circ u) \rightarrowtail L$.

4.4.5.   Let $I \overset{e}{\twoheadrightarrow} I' \overset{m}{\rightarrowtail} J$ be the factorisation of $u \colon I \to J$ in a regular category. Form the kernel pairs

$$K \underset{\pi_1}{\overset{\pi_0}{\rightrightarrows}} I \overset{u}{\longrightarrow} J \quad \text{and} \quad L \underset{\lambda_1}{\overset{\lambda_0}{\rightrightarrows}} I \overset{e}{\twoheadrightarrow} I'$$

and show that the tuples $\langle \pi_0, \pi_1 \rangle$ and $\langle \lambda_0, \lambda_1 \rangle$ are the same, as subobjects of $I \times I$.

4.4.6.   Let I be an object in a regular category. Show that $I \twoheadrightarrow 1$ (*i.e.* the unique map $I \to 1$ is a cover) if and only if $I \times I \rightrightarrows I \to 1$ is a coequaliser diagram.

4.4.7.   Let $\mathbb{B}$ be a regular category with an object $I \in \mathbb{B}$. Show that the following statements are equivalent.

(i) $I \twoheadrightarrow 1$;

(ii) The functor $I^*: \mathbb{B} \to \mathbb{B}/I$ reflects isomorphisms;

(iii) The functor $I^*: \mathbb{B} \to \mathbb{B}/I$ is **conservative** (*i.e.* reflects isomorphisms and is faithful).

4.4.8.   Let $\mathbb{B}$ be a category with finite limits and coequalisers. Show that $\mathbb{B}$ is regular if and only if regular epimorphisms in $\mathbb{B}$ are stable under pullback. (Sometimes one finds (for example in [18]) this latter formulation as definition of regular category—in presence of coequalisers.)

   [*Hint.* For a map $I \to J$ consider its kernel pair $R \rightrightarrows I$ and their coequaliser $I \twoheadrightarrow J'$. One gets a mono $J' \rightarrowtail J$ which is the image of u.]

4.4.9.   Establish that the category **Sp** of topological spaces and continuous functions is *not* regular.

   [*Hint.* Use the previous exercise, or see [36, II, Counterexample 2.4.5].]

4.4.10.   In Example 4.3.8 we have associated with a regular fibration $\begin{smallmatrix} \mathbb{E} \\ \downarrow{\scriptstyle p} \\ \mathbb{B} \end{smallmatrix}$ two categories $\mathrm{Rel}(p)$ and $\mathrm{FRel}(p)$ of types and (functional) relations in p. Here we define a slightly different category $\mathrm{FRelP}(p)$ of predicates and functional relations. It has

objects   objects $X \in \mathbb{E}$.

morphisms $X \nrightarrow Y$, say with X over I and Y over J, are (equivalence classes of) relations $R \in \mathbb{E}_{I \times J}$ which satisfy

$$i: I, j: J \mid R(i, j) \vdash X_i \wedge Y_i,$$
$$i: I, j: J, j': J \mid R(i, j), R(i, j') \vdash j =_J j',$$
$$i: I \mid X_i \vdash \exists j: J. R(i, j).$$

Verify that $\mathrm{FRelP}(p)$ is a regular category. This is like the construction of a regular category from a regular theory, like for example in [211, Chapter 8, 2].

4.5 Subobject fibrations II: coherent categories and logoses

We continue our investigation of subobject fibrations. In particular we investigate when a subobject fibration $\begin{smallmatrix} \mathrm{Sub}(\mathbb{B}) \\ \downarrow \\ \mathbb{B} \end{smallmatrix}$ is a coherent fibration (*i.e.* has fibred distributive coproducts $\bot, \vee$) and when it is a first order fibration (*i.e.* additionally has implication $\supset$ and universal quantification $\forall$). In the first case we call $\mathbb{B}$ a coherent category, and in the second case a logos. In [85] a coherent category is called a pre-logos, and in [211] it is called a logical category.

4.5.1. Definition. **A coherent category** is a regular category with

- binary joins $\vee$ in each subobject poset $\mathrm{Sub}(I)$, which are preserved by pull-back functors $u^*\colon \mathrm{Sub}(J) \to \mathrm{Sub}(I)$;
- a strict initial object 0.

Recall that strictness means that each arrow $X \to 0$ is an isomorphism.

The way in which these joins $\vee$ are usually obtained is as follows.

4.5.2. Lemma. *In a regular category with universal coproducts $+$, the sub-object fibre $\mathrm{Sub}(I)$ over I has joins $\vee$ of subobjects $X \rightarrowtail I$ and $Y \rightarrowtail I$, by taking the image of the cotuple, as in*

These $\vee$'s are stable under pullback. □

4.5.3. Theorem. *A regular category $\mathbb{B}$ is coherent if and only if its subobject fibration $\begin{smallmatrix} \mathrm{Sub}(\mathbb{B}) \\ \downarrow \\ \mathbb{B} \end{smallmatrix}$ is coherent.*

Proof. Assume $\mathbb{B}$ is a coherent category. We have to show that the subobject fibration $\begin{smallmatrix} \mathrm{Sub}(\mathbb{B}) \\ \downarrow \\ \mathbb{B} \end{smallmatrix}$ has a fibred initial object and that its joins $\vee$ are distributive. We begin with the latter: one has

$$
\begin{aligned}
n \wedge (m_1 \vee m_2) &\cong \coprod_n n^*(m_1 \vee m_2) && \text{by definition of } \wedge \\
&\cong \coprod_n n^*(m_1) \vee n^*(m_2) \\
&\cong \coprod_n n^*(m_1) \vee \coprod_n n^*(m_2) && \text{since } \coprod_n \text{ is left adjoint} \\
&\cong (n \wedge m_1) \vee (n \wedge m_2).
\end{aligned}
$$

Further, for $I \in \mathbb{B}$, let $\bot_I$ be the unique vertical map $0 \rightarrowtail I$; it is a mono, since for maps $f, g\colon K \rightrightarrows 0$, both f and g are isomorphism with $f^{-1} = g^{-1}$ by initiality. Thus $f = g$. For each mono $m\colon X \rightarrowtail I$ one obviously has $\bot_I \leq m$ in the poset $\mathrm{Sub}(I)$ of subobjects on I. And for $u\colon I \to J$, in the pullback

diagram,

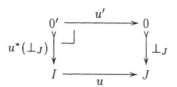

the map u' is an isomorphism by strictness. By initiality one gets that the composite $0 \overset{\cong}{\Rightarrow} 0' \to I$ is $\perp_I$, so that $u^*(\perp_J) \cong \perp_I$ over I. Thus the subobject fibration on $\mathbb{B}$ has a fibred initial object.

In the other direction, we follow the argument in [85, 1.61] to show that if $\overset{\text{Sub}(\mathbb{B})}{\underset{\mathbb{B}}{\downarrow}}$ is a coherent fibration, then $\mathbb{B}$ has a strict initial object (and is thus a coherent category). Let 0 be the domain of the bottom element $\perp_1 : 0 \rightarrowtail 1$ in Sub(1). For an object $I \in \mathbb{B}$, consider the pullback diagram,

$$
\begin{array}{ccc}
0' & \overset{h}{\longrightarrow} & 0 \\
{\scriptstyle \perp_I}\downarrow & \lrcorner & \downarrow{\scriptstyle \perp_1} \\
I & \underset{!_I}{\longrightarrow} & 1
\end{array}
$$

Assume there is an arrow $f: I \to 0$; we show that f is necessarily an isomorphism. The above pullback yields a map $f': I \to 0'$ with $\perp_I \circ f' = \mathrm{id}_I = \top_I$. But then $\perp_I \cong \top_I$ in Sub(I). Applying this same argument to $f \circ \pi: I \times I \to I \to 0$ yields $\perp_{I \times I} \cong \delta_I \cong \top_{I \times I}$, where δ_I is the diagonal $\langle \mathrm{id}, \mathrm{id} \rangle: I \rightarrowtail I \times I$. Hence δ_I is an isomorphism and so $\pi = \pi': I \times I \to I$. The unique arrow $!_I: I \to 1$ is then monic and so an object of Sub(1). Hence $\perp_1 \leq !_I$ which yields an inverse for $f: I \to 0$.

In particular, h in the above diagram is an isomorphism. We obtain a map $0 \to 0' \to I$. It is the only map $0 \to I$, since given two such maps, their equaliser has codomain 0 and is thus an isomorphism. □

Next we look at first order subobject fibrations.

4.5.4. Definition. A **logos** is a coherent category for which each pullback functor $u^*: \mathrm{Sub}(J) \to \mathrm{Sub}(I)$ has a right adjoint $\prod_u$.

4.5.5. Theorem. *A category $\mathbb{B}$ with finite limits is a logos if and only if its subobject fibration $\overset{\text{Sub}(\mathbb{B})}{\underset{\mathbb{B}}{\downarrow}}$ is a first order fibration.*

Proof. Assume $\mathbb{B}$ is a logos. We first notice that the subobject fibration $\overset{\text{Sub}(\mathbb{B})}{\underset{\mathbb{B}}{\downarrow}}$ has products: there are adjunctions $(u^* \dashv \prod_u)$ and Beck-Chevalley holds for these products, because it already holds for coproducts $\coprod_u$ (see Lemma 1.9.7). In particular, this fibration has simple products. It also has fibred exponents: for $m_1, m_2 \in \mathrm{Sub}(I)$ put,

$$m_1 \supset m_2 \stackrel{\text{def}}{=} \textstyle\prod_{m_1} m_1^*(m_2).$$

Then, for a subobject $m_1 \colon X \rightarrowtail I$ with domain X,

$$\mathrm{Sub}(I)\big(n, \, m_1 \supset m_2\big) \;\cong\; \mathrm{Sub}(X)\big(m_1^*(n), \, m_1^*(m_2)\big)$$

$$\cong\; \mathrm{Sub}(I)\big(\textstyle\coprod_{m_1} m_1^*(n), \, m_2\big)$$

$$\cong\; \mathrm{Sub}(I)\big(m_1 \wedge n, \, m_2\big).$$

The latter by definition of $\wedge$. Exponents are preserved under reindexing by Beck-Chevalley for $\prod$.

The reverse implication follows from the construction of $\prod_u$ in a first order fibration, as described in Example 4.3.7 (ii). $\square$

We conclude this section with examples of logoses, involving in particular the categories of sets and of PERs. Logic in ω-**Sets** will be described later in Section 5.3 (especially in Proposition 5.3.9) in terms of its *regular* subobjects. There one also finds a description of regular subobjects in the category **PER**, giving rise to classical logic.

4.5.6. Example. (i) The category **Sets** of sets and functions is a logos. We have already seen that it is a regular category. Its posets of subobjects $\mathrm{Sub}(I)$, occurring as fibre categories of the fibration $\begin{smallmatrix}\mathrm{Sub}(\mathbf{Sets})\\\downarrow\\\mathbf{Sets}\end{smallmatrix} = \begin{smallmatrix}\mathbf{Pred}\\\downarrow\\\mathbf{Sets}\end{smallmatrix}$, can be identified with the powersets $\langle PI, \subseteq\rangle$. These posets are Boolean algebras, so we certainly have distributive joins (namely $\emptyset$ and $\cup$) making **Sets** into a coherent category. Further, there are products $\prod \colon \mathrm{Sub}(I \times J) \to \mathrm{Sub}(I)$ by

$$(X \subseteq I \times J) \mapsto (\{i \mid \forall j \in J.\, (i,j) \in X\} \subseteq I).$$

In the next chapter we shall see that, more generally, every topos is a logos.

(ii) The category **PER** of partial equivalence relations is also a logos (as shown in [143]). Recall from Section 1.2 that **PER** has finite limits. It is not hard to see that a morphism $f \colon R \to S$ is a mono in **PER** if and only if the function $f \colon \mathbb{N}/R \to \mathbb{N}/S$ between the underlying quotient sets is injective.

For a morphism $f \colon R \to S$ we define the image as the PER,

$$\mathrm{Im}(f) = \{(n, n') \in |R| \times |R| \mid f([n]_R) = f([n']_R)\},$$

together with the monomorphism,

$$\mathrm{Im}(f) \overset{\mathrm{m}(f)}{\rightarrowtail} S, \qquad\qquad [n]_{\mathrm{Im}(f)} \;\longmapsto\; f([n]_R).$$

This is indeed a mono, because the underlying function is injective by construction:

$$m(f)([n']_{\mathrm{Im}(f)}) = m(f)([n']_{\mathrm{Im}(f)}) \Leftrightarrow f([n]_R) = f([n']_R)$$
$$\Leftrightarrow [n]_{\mathrm{Im}(f)} = [n']_{\mathrm{Im}(f)}.$$

There is then a morphism $f' \colon R \to \mathrm{Im}(f)$ by $[n]_R \mapsto [n]_{\mathrm{Im}(f)}$. It obviously satisfies $m(f) \circ f' = f$. It is not hard to see that this image $\mathrm{Im}(f) \rightarrowtail S$ is appropriately minimal, and stable under pullback. Thus we get that **PER** is a regular category. A characterisation of covers in **PER** is given in Exercise 4.5.2 below.

One can conclude that **PER** is a coherent category from the fact that it has universal finite coproducts $(0, +)$, using Lemma 4.5.2. The initial object 0 is the empty PER $\emptyset \subseteq \mathbb{N} \times \mathbb{N}$, which has quotient set $\mathbb{N}/\emptyset = \emptyset$. And the coproduct of PERs R, S is

$$R + S = \{(\langle 0, n\rangle, \langle 0, m\rangle) \mid n R m\} \cup \{(\langle 1, n\rangle, \langle 1, m\rangle) \mid n S m\}.$$

Finally, **PER** is a logos, because it is locally Cartesian closed, see Exercise 1.2.7. The product functors $\prod_f \colon \mathbf{PER}/R \to \mathbf{PER}/S$ between slices restrict to product functors $\prod_f \colon \mathrm{Sub}(R) \to \mathrm{Sub}(S)$, because right adjoints preserve monos.

(iii) In Example 4.2.5 we have seen how a frame (or complete Heyting algebra) A gives rise to a first order fibration $\begin{smallmatrix} \mathrm{Fam}(A) \\ \downarrow \\ \mathbf{Sets} \end{smallmatrix}$. It turns out that first order logic is also present at a different level: the total category $\mathrm{Fam}(A)$ of this fibration is itself a logos. We sketch the main points.

A terminal object in $\mathrm{Fam}(A)$ is the family $(\top)_{*\in 1}$ consisting of the top element $\top \in A$ over a singleton (terminal) set $1 = \{*\}$. The pullback of morphisms $u \colon (x_i)_{i\in I} \to (z_k)_{k\in K}$ and $v \colon (y_j)_{j\in J} \to (z_k)_{k\in K}$ consist of the family $(x_i \wedge y_j)_{(i,j)\in I\times_K J}$ over the pullback $I \times_K J$ in **Sets** of $u \colon I \to K$ and $v \colon J \to K$, with obvious pullback projections. A morphism $u \colon (x_i)_{i\in I} \to (y_j)_{j\in J}$ is monic in $\mathrm{Fam}(A)$ if and only if the underlying function $u \colon I \to J$ is injective. A subobject of a family $(x_i)_{i\in I}$ may thus be identified with a "subfamily" $(x'_i)_{i\in V}$ for $V \subseteq I$ with $x'_i \le x_i$, for all $i \in V$.

For an arbitrary such map u in $\mathrm{Fam}(A)$, we can first factorise $u \colon I \to J$ in **Sets** as $u' \colon I \twoheadrightarrow J' = \{j \in J \mid \exists i \in I. u(i) = j\}$ followed by $m(u) \colon J' \rightarrowtail J$. This factorisation can be lifted to $\mathrm{Fam}(A)$ as:

$$(x_i)_{i\in I} \xrightarrow{\ u'\ } \left(\bigvee_{i\in u^{-1}(j)} x_i \right)_{j\in J'} \xrightarrow{\ m(u)\ } (y_j)_{j\in J}$$

It is easy to verify that this factorisation in $\mathrm{Fam}(A)$ is appropriately minimal and stable under pullback. This shows that $\mathrm{Fam}(A)$ is a regular category. It is coherent since the empty family is a strict initial object, and since for an arbitrary family $(x_i)_{i \in I} \in \mathrm{Fam}(A)$, the join of two "subfamilies" $(y_i)_{i \in U \subseteq I}$ and $(z_i)_{i \in V \subseteq I}$ is a family over $U \cup V \subseteq I$ given by:

$$((y_i) \vee (z_i))(i) = \begin{cases} y_i & \text{if } i \in U \backslash V \\ z_i & \text{if } i \in V \backslash U \\ y_i \vee z_i & \text{if } i \in U \cap V. \end{cases}$$

These joins are distributive and stable. Finally, we have to produce a right adjoint $\prod_u$ to pullback $u^{\#}$ along a morphism $u \colon (x_i)_{i \in I} \to (y_j)_{j \in J}$. Notice that

$$u^{\#}\big((w_j)_{j \in V \subseteq J}\big) = (w_{u(i)})_{i \in u^{\bullet}(V) \subseteq I}.$$

The required right adjoint is then given by:

$$\prod_u ((y_i)_{i \in U \subseteq I}) = \left(\bigwedge_{i \in u^{-1}(j)} y_i \right)_{j \in \forall_u(U)}$$

where $\forall_u(U) = \{j \in J \mid \forall i \in I. \, u(i) = j \Rightarrow i \in U\}$.

The second example above, showing that subobjects in **PER** form a first order fibration, leads to the following associated result, showing that also *regular* subobjects in **PER** from a first order fibration—but with classical logic. It is based on a (folklore) correspondence between regular subobjects in **PER** and subsets of quotients.

4.5.7. Proposition. (i) *For a PER R there is a bijective (order preserving) correspondence between*

  (a) subsets $A \subseteq \mathbb{N}/R$ of the quotient of R;

  (b) subsets $B \subseteq |R|$ of the domain of R which are **saturated**: *if $n \in B$ and nRn' then $n' \in B$;*

  (c) regular subobjects $R' \rightarrowtail R$ of R.

  (ii) *The fibration* $\begin{array}{c} \mathrm{RegSub}(\mathbf{PER}) \\ \downarrow \\ \mathbf{PER} \end{array}$ *is a first order fibration with classical logic.*

Proof. (i) The equivalence (a) $\Leftrightarrow$ (b) is easy, so we concentrate on (b) $\Leftrightarrow$ (c). For two parallel morphisms $f, g \colon R \rightrightarrows S$ in **PER**, one can describe their equaliser by restricting attention to the saturated subset $B \subseteq |R|$ given by $B = \{n \in |R| \mid f([n]_R) = g([n]_R)\}$. And conversely, given a saturated subset $B \subseteq |R|$, define a PER S by

$$S = \{(\langle i, n \rangle, \langle j, n' \rangle) \mid i, j \in \{0, 1\} \text{ and } nRn' \text{ and } n \in B\}$$

$$\cup \{\langle 0, n \rangle, \langle 0, n' \rangle) \mid nRn' \text{ and } n \notin B\} \cup \{\langle 1, n \rangle, \langle 1, n' \rangle) \mid nRn' \text{ and } n \notin B\}$$

Then there are morphisms $f, g: R \rightrightarrows S$ defined by

$$f([n]_R) = [\langle 0, n \rangle]_S \quad \text{and} \quad g([n]_R) = [\langle 1, n \rangle]_S.$$

It is not hard to see that the set $B \subseteq |R|$ can be recovered from the equaliser subobject $R' \rightarrowtail R$ resulting from these f, g.

(ii) Using (a) in (i), the set-theoretic operations of (classical) first order logic can be used for this fibration of regular subobjects. □

Exercises

4.5.1. Prove in a regular category with universal coproducts: if $X \rightarrowtail I$ and $Y \rightarrowtail I$ are disjoint (*i.e.* $X \wedge Y = \bot$), then $X \vee Y = X + Y$.

4.5.2. (From [143]) Show that a map $f: R \to S$ in **PER** is a cover if and only if

$$\exists e \in \mathbb{N}. \forall n \in |S|. f([e \cdot n]_R) = [n]_S.$$

[Notice that such an e need not give us a morphism $S \to R$, since we do not know that $nSm \Rightarrow e \cdot nRe \cdot m$.]

4.5.3. The following combined formulation of $\supset$ and $\forall$ comes from [211]. Show that in a category with finite limits, one has implication in subobject posets and right adjoints $\prod_u$ to pullback functor u^* if and only if: for each $u: I \to J$ and for each pair of subobjects $m: X \rightarrowtail I$ and $n: Y \rightarrowtail I$ there is a largest subobject $k: Z \rightarrowtail J$ with $u^*(k) \wedge m \leq n$.

4.5.4. One can say that the **Axiom of Choice** (AC) holds in a regular category if and only if every cover $c: I \twoheadrightarrow J$ splits (*i.e.* has a section $s: J \to I$ with $c \circ s = \text{id}$).

   (i) Check that this formulation in **Sets** is equivalent to (one of) the usual formulations of (AC).

   (ii) Verify that (AC) holds in a regular category if and only if the covers are precisely the split epis (see Exercise 4.4.2).

   (iii) Prove that if (AC) holds, then internal and external existence coincide in the subobject fibration. This means that for a subobject $X \rightarrowtail I \times J$, the proposition

$$i: I \vdash \exists j: J. X(i, j): \mathsf{Prop}$$

   holds if and only if there is a map $s: I \to J$ such that

$$i: I \vdash X(i, s(i)): \mathsf{Prop}$$

   holds.

4.5.5. In a first order fibration with exponents in its base category, we say that the **internal Axiom of Choice** (IAC) holds if the following proposition holds.

$$f: J^I \mid \text{``}f \text{ is internally surjective''} \vdash \text{``}f \text{ has a section''}$$

   (i) Describe explicitly the predicates "f is internally surjective" and "f has a section".

(ii) Consider a logos $\mathbb{B}$ with exponents. Show that (IAC) holds (in the associated subobject fibration) if and only if for each object K the exponent functor $(-)^K \colon \mathbb{B} \to \mathbb{B}$ preserves covers.
[A proof making use of Kripke-Joyal semantics can be found in [188, Chapter VI].]

4.6 Subset types

Most of the structure of fibrations that we considered so far was structure in fibres (like $\wedge, \vee$) or between fibres (like $\coprod, \prod$). In the next three sections we shall study subset types and quotient types. These are new in the sense that they involve structure between a total category and a base category of a fibration, given by adjoints.

In this section we will give the syntax and categorical semantics of subset types (or also called subsets). This involves the operation which maps a proposition $(x \colon \sigma \vdash \varphi \colon \mathsf{Prop})$ to a type $\{x \colon \sigma \,|\, \varphi\} \colon \mathsf{Type}$. The intended meaning of the latter is the subtype of σ consisting of those terms $M \colon \sigma$ for which $\varphi[M/x]$ holds. The categorical description that we give below, will turn out to be a special case of a general form of comprehension (see Section 10.4).

Subset types involve a new type formation rule, namely:

Each proposition $(x \colon \sigma \vdash \varphi \colon \mathsf{Prop})$ gives rise to a type $\{x \colon \sigma \,|\, \varphi\}$.

Formally we write this as a rule,

formation

$$\frac{x \colon \sigma \vdash \varphi \colon \mathsf{Prop}}{\vdash \{x \colon \sigma \,|\, \varphi\} \colon \mathsf{Type}}$$

It comes with introduction and elimination rules for terms of this newly formed type:

introduction

$$\frac{x \colon \sigma \vdash \varphi \colon \mathsf{Prop} \qquad \Gamma \vdash M \colon \sigma \qquad \Gamma \,|\, \emptyset \vdash \varphi[M/x]}{\Gamma \vdash \mathsf{i}(M) \colon \{x \colon \sigma \,|\, \varphi\}}$$

elimination

$$\frac{\Gamma \vdash N \colon \{x \colon \sigma \,|\, \varphi\}}{\Gamma \vdash \mathsf{o}(N) \colon \sigma} \quad \text{with} \quad \frac{\Gamma, x \colon \sigma \,|\, \Theta, \varphi \vdash \psi}{\Gamma, y \colon \{x \colon \sigma \,|\, \varphi\} \,|\, \Theta[\mathsf{o}(y)/x] \vdash \psi[\mathsf{o}(y)/x]}$$

The associated conversions are

$$\mathsf{o}(\mathsf{i}(M)) = M \qquad \text{and} \qquad \mathsf{i}(\mathsf{o}(N)) = N.$$

The letters 'i' and 'o' stand for 'in' and 'out'. It is more appropriate to write $i_\varphi(M)$ and $o_\varphi(N)$ with the proposition φ explicit as a subscript, but we often find this notation too cumbersome. In mathematical practice these i's and o's are usually omitted altogether.

We say we have **full subset types** if we also have the converse of the last rule:

full subset types

$$\frac{\Gamma, y \colon \{x \colon \sigma \mid \varphi\} \mid \Theta[\mathsf{o}(y)/x] \vdash \psi[\mathsf{o}(y)/x]}{\Gamma, x \colon \sigma \mid \Theta, \varphi \vdash \psi}$$

This is a useful additional rule. Consider for example two propositions $x \colon \sigma \vdash \varphi, \psi \colon \mathsf{Prop}$. With this rule we can conclude that $\{x \colon \sigma \mid \varphi\}$ is included in $\{x \colon \sigma \mid \psi\}$ if and only if φ implies ψ. In one way this obvious; we give a derivation of the other way, to indicate where fullness is used:

$$\frac{\dfrac{y \colon \{x \colon \sigma \mid \varphi\} \vdash i_\psi(\mathsf{o}_\varphi(y)) \colon \{x \colon \sigma \mid \psi\}}{y \colon \{x \colon \sigma \mid \varphi\} \mid \emptyset \vdash \psi[\mathsf{o}_\varphi(y)/x]}}{x \colon \sigma \mid \varphi \vdash \psi} \text{ (full subset types)}$$

As this example suggests, fullness of subset types corresponds to fullness of an associated functor. This will be made explicit in Definition 4.6.1 below.

Notice that in the above type formation rule, we have a subset type $\{x \colon \sigma \mid \varphi\}$ in which x is the *only* variable which may occur in φ. We could have stated a more general formation rule with type context Γ,

$$\frac{\Gamma, x \colon \sigma \vdash \varphi \colon \mathsf{Prop}}{\Gamma \vdash \{x \colon \sigma \mid \varphi\} \colon \mathsf{Type}}$$

But that leads to *type dependency*: one gets a type $\{x \colon \sigma \mid \varphi\}$ which may contain variables $y \colon \tau$ of types τ declared in Γ. In the present chapter we only consider "simple" predicate logic (SPL) over simple type theory, in which we wish to exclude this type dependency. We postpone such subset types with contexts to what will be called "dependent" predicate logic (DPL) in Section 11.1. But we would like to stress here that the extended formation rule is quite natural, for example in forming the subset type

$$\frac{n \colon \mathsf{N}, m \colon \mathsf{N} \vdash m \leq n \colon \mathsf{Prop}}{n \colon \mathsf{N} \vdash \{m \colon \mathsf{N} \mid m \leq n\} \colon \mathsf{Type}}$$

of natural numbers less than n. This is clearly a type in which a term variable n occurs. One could say this more strongly: the restricted formation rule without type context is the more unnatural version.

Categorically, a logic is described by a preorder fibration $\begin{smallmatrix}\mathbb{E}\\\downarrow p\\\mathbb{B}\end{smallmatrix}$ where we standardly assume that the base category $\mathbb{B}$ has finite products and that the fibration p has fibred finite products. Objects $I \in \mathbb{B}$ are seen as types and objects $X \in \mathbb{E}$ as propositions. One thus expects that subset types involve a functor $\{-\}: \mathbb{E} \to \mathbb{B}$ which maps a proposition $Y = (j: J \mid Y_j) \in \mathbb{E}$ to a type $\{j: J \mid Y_j\} \in \mathbb{B}$. One further expects there to be a monic 'projection' morphism

$$\{j: J \mid Y_j\} \xrightarrow{\quad \pi_Y \quad} J$$

making $\{j: J \mid Y_j\}$ a subtype of J. Our use of the word 'projection' here comes from the more general treatment of comprehension in Section 10.4. Sometimes we call the object (or type) $\{Y\} = \{j: J \mid Y_j\}$ the **extent** of Y.

The natural requirement is that an element $k: J$ is in $\{j: J \mid Y_j\}$ if and only if Y_k holds. In arrow-theoretic language:

a morphism $u: I \to J$ factors through $\pi_Y: \{Y\} \rightarrowtail J$

if and only if   (∗)

the proposition $(i: I \vdash Y_{u(i)}: \mathsf{Prop})$ holds (*i.e.* $\top \le u^*(Y)$).

All this structure comes about by the requirement that the functor $\{-\}: \mathbb{E} \to \mathbb{B}$ is right adjoint to the terminal object functor $\top: \mathbb{B} \to \mathbb{E}$. This is (a preorder version of) what is called a **D-category** in [74, 75]. It is a simplification of a structure used by Lawvere to capture comprehension, see Exercise 4.6.7. Later in Section 10.4 these notions will be studied more systematically under the name 'comprehension category with unit'.

4.6.1. Definition. A preorder fibration $\begin{smallmatrix}\mathbb{E}\\\downarrow p\\\mathbb{B}\end{smallmatrix}$ with terminal object functor $\top: \mathbb{B} \to \mathbb{E}$ is said to have **subsets** (or **subset types**) if this functor $\top$ has a right adjoint.

We usually write $\{-\}: \mathbb{E} \to \mathbb{B}$ for such a right adjoint. For $X \in \mathbb{E}$, the counit $\varepsilon_X: \top\{X\} \to X$ induces a morphism $p(\varepsilon_X): \{X\} \to pX$ in $\mathbb{B}$. We write $\pi_X = p(\varepsilon_X)$ for this map and call it a (**subset**) **projection**.

The assignment $X \mapsto \pi_X$ extends to a (faithful) functor $\mathbb{E} \to \mathbb{B}^{\to}$. We say that the fibration p has **full subset types** if this functor $\pi_{(-)}: \mathbb{E} \to \mathbb{B}^{\to}$ is full (and faithful).

Notice that having subset types is a property of a fibration, because it is expressed by an adjunction. Later, in Theorem 4.8.3 we shall see an equivalent description of subset types in terms of a right adjoint to an equality functor.

The next lemma gives several useful results involving subset types. In particular, in (ii) it is shown that the earlier expected property (∗) of subset types is captured by the above definition.

4.6.2. Lemma. *Let* $\begin{smallmatrix} \mathbb{E} \\ \downarrow p \\ \mathbb{B} \end{smallmatrix}$ *be a preorder fibration with subset types as described above.*

(i) *Each projection morphism* $\pi_X \colon \{X\} \to pX$ *in* $\mathbb{B}$ *is monic.*

(ii) *For each map* $u \colon I \to J$ *in* $\mathbb{B}$ *and object* $Y \in \mathbb{E}$ *over* J, *there is a bijective correspondence*

$$\frac{\top \leq u^*(Y) \quad \text{over } I}{u \dashrightarrow \pi_Y \quad \text{in } \mathbb{B}/J}$$

This says that $u \colon I \to J$ *factors through* π_Y *if and only if* $\top \leq u^*(Y)$ *as in* $(*)$ *above.*

(iii) *The assignment* $X \mapsto \pi_X$ *extends to a functor* $\mathcal{P} \colon \mathbb{E} \to \mathbb{B}^{\to}$ *which maps Cartesian morphisms to pullback squares. This functor restricts to* $\mathbb{E} \to$ Sub$(\mathbb{B})$ *by (i).*

(iv) *The functor* $\mathcal{P}$ *in (iii) preserves all fibred limits.*

Proof. (i) Suppose that parallel maps $v, w \colon K \rightrightarrows \{X\}$ are given with $\pi_X \circ v = \pi_X \circ w$. The transposes $v^\wedge, w^\wedge \colon \top \rightrightarrows X$ then satisfy

$$p(v^\wedge) = p(\varepsilon_X \circ \top(v)) = \pi_X \circ v = \pi_X \circ w = p(w^\wedge).$$

But then $v^\wedge = w^\wedge$, because we have a fibred preorder (see Exercise 1.3.11). Hence $v = w$.

(ii) For a vertical map $f \colon \top \to u^*(Y)$ over I, one obtains a map $\overline{u}(Y) \circ f \colon \top \to Y$ in $\mathbb{E}$ over u and, by transposition, a map:

$$\widehat{f} = \{\overline{u}(Y) \circ f\} \circ \eta_I \colon I \longrightarrow \{Y\} \quad \text{in } \mathbb{B}.$$

This $\widehat{f}$ is a map $u \to \pi_Y$ in the slice category $\mathbb{B}/J$, since

$$\begin{aligned}
\pi_Y \circ \widehat{f} &= p(\varepsilon_Y \circ \top\{\overline{u}(Y) \circ f\} \circ \top \eta_I) \\
&= p(\overline{u}(Y) \circ f) \circ p(\varepsilon_{\top(I)} \circ \top \eta_I) \\
&= u.
\end{aligned}$$

Conversely, given $v \colon I \to \{Y\}$ satisfying $\pi_Y \circ v = u$, then by transposition one obtains a map $\top \to Y$ over u, by an argument as in (i). Thus one gets a vertical morphism $\widehat{v} \colon \top \to u^*(Y)$ over I. These operations $f \mapsto \widehat{f}$ and $v \mapsto \widehat{v}$ are each others inverses.

(iii) For a morphism $f: X \to Y$ in $\mathbb{E}$, there is a commuting diagram in $\mathbb{B}$,

$$
\begin{array}{ccc}
\{X\} & \xrightarrow{\ \{f\}\ } & \{Y\} \\
{\scriptstyle \pi_X}\downarrow & & \downarrow{\scriptstyle \pi_Y} \\
pX & \xrightarrow[\ pf\ ]{} & pY
\end{array}
$$

since

$$
\pi_Y \circ \{f\} = p(\varepsilon_Y \circ \mathsf{T}\{f\}) = p(f \circ \varepsilon_X) = pf \circ \pi_X.
$$

In case f is Cartesian in $\mathbb{E}$, this diagram is a pullback in $\mathbb{B}$: if maps $u: I \to pX$ and $v: I \to \{Y\}$ are given with $pf \circ u = \pi_Y \circ v$, then v is a morphism $(pf \circ u) \to \pi_Y$ in $\mathbb{B}/pY$. Hence one obtains by (ii) a morphism in

$$
\begin{aligned}
\mathbb{E}_I\big(\mathsf{T},\, (pf \circ u)^*(Y)\big) &\cong \mathbb{E}_I\big(\mathsf{T},\, u^*(pf)^*(Y)\big) \\
&\cong \mathbb{E}_I\big(\mathsf{T},\, u^*(X)\big) \qquad \text{because } f \text{ is Cartesian} \\
&\cong \mathbb{B}/pX\big(u,\, \pi_X\big).
\end{aligned}
$$

This resulting map in $\mathbb{B}/pX(u, \pi_X)$ is the required mediating map.

(iv) We write $\mathcal{P}$ for the functor $X \mapsto \pi_X$, and shall show that $\mathcal{P}$ preserves fibred finite products (which is of most interest at this stage).

Since T is a full and faithful functor, the unit $\eta_I: I \to \{\mathsf{T}I\}$ is an isomorphism. Thus $\pi_{(\mathsf{T}I)} \cong \mathrm{id}_I$ in $\mathbb{B}/I$, which shows that $\mathcal{P}$ preserves fibred terminal objects.

For X, Y over J, we have for an arbitrary map $u: I \to J$ in $\mathbb{B}$,

$$
\begin{aligned}
\mathbb{B}/J\big(u,\, \mathcal{P}(X \times Y)\big) &\cong \mathbb{E}_I\big(\mathsf{T},\, u^*(X \times Y)\big) \\
&= \mathbb{E}_I\big(\mathsf{T},\, u^*(X) \times u^*(Y)\big) \\
&= \mathbb{E}_I\big(\mathsf{T},\, u^*(X)\big) \times \mathbb{E}_I\big(\mathsf{T},\, u^*(Y)\big) \\
&\cong \mathbb{B}/pX\big(u,\, \mathcal{P}X\big) \times \mathbb{B}/pX\big(u,\, \mathcal{P}Y\big). \qquad \square
\end{aligned}
$$

It is now easy to see that having (full) subset types in a fibration (as in Definition 4.6.1) gives us validity of the rules of (full) subset types as described in the beginning of this section: for a term $u: I \to J$ and a proposition $Y = (j: J \vdash Y_j: \mathsf{Prop})$ above J, a morphism $\mathsf{T} \leq u^*(Y)$ over I induces by (ii) a map

$$
I \xrightarrow{\ \mathsf{i}(u)\ } \{Y\} \qquad \text{with} \qquad \pi_Y \circ \mathsf{i}(u) = u.
$$

This gives validity of the introduction rule. As to elimination, for a term $v: I \to \{Y\}$ we put $o(v) = \pi_Y \circ v: I \to J$. Further, assume we have an entailment,

$$i: I, j: J \mid X(i,j) \wedge Y(j) \vdash Z(i,j),$$

that is, an inequality above $I \times J$,

$$X \wedge Y' \leq Z \qquad \text{where} \qquad Y' = \pi'^*(Y).$$

Then we have to show

$$i: I, k: \{j: J \mid Y_j\} \mid X(i, o(k)) \vdash Z(i, o(k)),$$

which translates into

$$(\mathrm{id} \times \pi_Y)^*(X) \leq (\mathrm{id} \times \pi_Y)^*(Z).$$

This is equivalent to

$$\pi_{Y'}^*(X) \leq \pi_{Y'}^*(Z) \quad \text{above } \{Y'\},$$

Since both diagrams below are pullbacks.

$$
\begin{array}{ccc}
I \times \{Y\} & \xrightarrow{\;\pi'\;} & \{Y\} \\
{\scriptstyle \mathrm{id} \times \pi_Y}\big\downarrow\;\lrcorner & & \big\downarrow{\scriptstyle \pi_Y} \\
I \times J & \xrightarrow[\;\pi'\;]{} & J
\end{array}
\qquad
\begin{array}{ccc}
\{\pi'^*(Y)\} = \{Y'\} & \xrightarrow{\;\{\overline{\pi'(Y)}\}\;} & \{Y\} \\
{\scriptstyle \pi_{Y'}}\big\downarrow\;\lrcorner & & \big\downarrow{\scriptstyle \pi_Y} \\
I \times J & \xrightarrow[\;\pi'\;]{} & J
\end{array}
$$

The latter in equality $\pi_{Y'}^*(X) \leq \pi_{Y'}^*(Z)$ follows from $X \wedge Y' \leq Z$ by applying $\pi_{Y'}^*$ to $X \leq Z$ and using $\top \leq \pi_{Y'}^*(Y')$, which is the vertical part of the counit $\varepsilon_{Y'}: \top = \top\{Y'\} \to Y'$.

If we additionally assume that the fibration has *full* subset types, then the corresponding full subset rule is valid. Therefore we have to establish the converse

$$\pi_{Y'}^*(X) \leq \pi_{Y'}^*(Z) \Rightarrow X \wedge Y' \leq Z$$

of what we just proved. This is done as follows.

$\pi_{Y'}^*(X) \leq \pi_{Y'}^*(Z)$

$\Rightarrow$ there is a (unique) map $\pi_{Y'}^*(X) \dashrightarrow Z$ over $\pi_{Y'}$

$\Rightarrow$ there is a (unique) map $\dashrightarrow$ in

$$
\begin{array}{ccccc}
\{X\} & \longleftarrow & \{\pi_{Y'}^*(X)\} & \dashrightarrow & \{Z\} \\
\Big\downarrow{\scriptstyle \pi_X} & & \Big\downarrow & & \Big\downarrow{\scriptstyle \pi_Z} \\
I \times J & \xleftarrow{\;\pi_{Y'}\;} & \{Y'\} & \xrightarrow{\;\pi_{Y'}\;} & I \times J
\end{array}
$$

$\Rightarrow$ $\pi_{(X \wedge Y')} \cong \pi_X \wedge \pi_{Y'} \le \pi_Z$ $\quad$ with $\cong$ from (iv) in Lemma 4.6.2

$\Rightarrow$ $X \wedge Y' \le Z$ $\quad$ because the projection functor $\mathbb{E} \to \mathbb{B}^{\to}$ is full.

Notice that the square on the left is a pullback because the projection functor maps Cartesian morphism to pullback squares (as in (iii) in the lemma).

4.6.3. Examples. (i) Every subobject fibration $\begin{smallmatrix} \mathrm{Sub}(\mathbb{B}) \\ \downarrow \\ \mathbb{B} \end{smallmatrix}$ has full subset types. The associated functor $\{-\} \colon \mathrm{Sub}(\mathbb{B}) \to \mathbb{B}$ takes a representation $(m \colon X \rightarrowtail J)$ of a subobject to its domain $X \in \mathbb{B}$. There is an obvious correspondence

$$
\left(\begin{smallmatrix} I \\ \downarrow{\scriptstyle \mathrm{id}} \\ I \end{smallmatrix} \right) \longrightarrow \left(\begin{smallmatrix} X \\ \downarrow{\scriptstyle m} \\ J \end{smallmatrix} \right) \quad \text{in } \mathrm{Sub}(\mathbb{B})
$$
$$
\overline{\overline{\qquad\qquad\qquad\qquad\qquad\qquad\qquad}}
$$
$$
I \longrightarrow X - \{m\} \quad \text{in } \mathbb{B}
$$

establishing that $\{-\}$ is right adjoint to the terminal object functor $I \mapsto \mathrm{id}_I$. The resulting projection functor $\mathrm{Sub}(\mathbb{B}) \to \mathbb{B}^{\to}$ sends a subobject to a representative. It is then a full and faithful (fibred) functor. Hence subobject fibrations always have full subset types.

(ii) For each poset X with top element $\top$, the family fibration $\begin{smallmatrix} \mathrm{Fam}(X) \\ \downarrow \\ \mathbf{Sets} \end{smallmatrix}$ comes equipped with a subset functor given by

$$
(x_j)_{j \in J} \mapsto \{j \in J \mid x_j = \top\}.
$$

It singles out the indices of elements that 'are true'. In general, this does not lead to a full functor $\mathrm{Fam}(X) \to \mathbf{Sets}^{\to}$.

(iii) Consider a predicate logic with (full) subset types, built on top of a specification $(\Sigma, \Pi, \mathcal{A})$. The associated classifying fibration $\begin{smallmatrix} \mathcal{L}(\Sigma, \Pi, \mathcal{A}) \\ \downarrow \\ \mathcal{Cl}(\Sigma) \end{smallmatrix}$ then has (full) subset types in a categorical sense. One defines a functor $\{-\} \colon \mathcal{L}(\Sigma, \Pi, \mathcal{A}) \to \mathcal{Cl}(\Sigma)$ by $(x \colon \sigma \vdash \varphi \colon \mathsf{Prop}) \mapsto (\{x \colon \sigma \mid \varphi\} \colon \mathsf{Type})$. The required adjunction $\top \dashv \{-\}$ boils down to a correspondence between terms M

and N in:

$$(y: \tau \vdash \top: \mathsf{Prop}) \xrightarrow{\quad M \quad} (x: \sigma \vdash \varphi: \mathsf{Prop})$$
$$\overline{\quad\quad\quad\quad\quad\quad\quad\quad\quad\quad\quad\quad}$$
$$\tau \xrightarrow[N]{\quad\quad} \{x: \sigma \,|\, \varphi\}$$

I.e. between M and N in:

$$y: \tau \vdash M(y): \sigma \quad \text{with} \quad y: \tau \,|\, \top \vdash \varphi[M/x]$$
$$\overline{\quad\quad\quad\quad\quad\quad\quad\quad\quad\quad\quad\quad}$$
$$y: \tau \vdash N: \{x: \sigma \,|\, \varphi\}$$

This correspondence is given by

$$M \mapsto \mathsf{i}(M) \quad \text{and} \quad N \mapsto \mathsf{o}(N).$$

(To make this work, we must have equivalence classes (under conversion) of terms as morphisms in the base category $\mathcal{Cl}(\Sigma)$. Also we are assuming finite products of types here, so that we may restrict ourselves to singleton type contexts—which may be identified with types—as object of this base category.)

One gets a functor $\mathcal{L}(\Sigma, \mathcal{A}) \to \mathcal{Cl}(\Sigma)^{\to}$ which maps a proposition $(x: \sigma \vdash \varphi: \mathsf{Prop})$ to the term $\mathsf{o}(z): (z: \{x: \sigma \,|\, \varphi\}) \to (x: \sigma)$. It is full if and only if it is fibrewise full. The latter means that for a term $z: \{x: \sigma \,|\, \varphi\} \vdash M: \{y: \tau \,|\, \psi\}$ with $\mathsf{o}_\psi(M) = \mathsf{o}_\varphi(z)$, we have an entailment $x: \sigma \,|\, \varphi \vdash \psi$ (which is a morphism over $(x: \sigma)$. Since $M = \mathsf{i}_\psi(\mathsf{o}_\varphi(z))$, this follows from an argument as in the beginning of this section, using the full subset rule.

(iv) Next we describe an example of subset types involving 'metric predicates'. It is an adaptation of a construction in [177] (which is based on [194]). For a metric space X we conveniently write X for the underlying set and $X(-, -)$ for the metric involved. That is, for the function $X(-, -): X \times X \to [0, \infty]$ satisfying for $x, y, z \in X$,

$$X(x, y) = 0 \Leftrightarrow x = y,$$
$$X(x, y) = X(y, x), \quad\quad X(x, y) + X(y, z) \leq X(x, z).$$

For convenience we have included ∞ in the range $[0, \infty]$ of the distance function; one can also take $[0, 1] \subseteq \mathbb{R}$ as range.

A function $f: X \to Y$ between the underlying sets of two such metric spaces in called **non-expansive** if $Y(f(x), f(x')) \leq X(x, x')$ for all $x, x' \in X$. We write **MS** for the resulting category of metric spaces and non-expansive functions.

A **metric predicate** on a metric space X is a non-expansive map $\varphi: X \to [0, \infty]$ where $[0, \infty]$ has the obvious metric. One can show that these metric predicates on X form a metric space with distance between $\varphi, \psi: X \rightrightarrows [0, \infty]$

given by

$$\sup_{x \in X} |\varphi(x) - \psi(x)|.$$

They can be ordered by

$$\varphi \sqsubseteq \psi \iff \forall x \in X. \, \psi(x) \leq \varphi(x).$$

Note the inversion. This yields a poset $MP(X)$. One thinks of such a metric predicate φ as absolutely true in x if $\varphi(x) = 0$ and as almost true in x if $\varphi(x)$ is very small. Thus $\varphi \sqsubseteq \psi$ if and only if ψ is everywhere more true than φ.

The assignment $X \mapsto MP(X)$ extends to a split indexed category $\mathbf{MS}^{op} \to \mathbf{Cat}$ by composition; hence to a split fibration, which will be written as $\begin{smallmatrix} \mathbf{MP} \\ \downarrow \\ \mathbf{MS} \end{smallmatrix}$. It has a terminal object functor $\top : \mathbf{MS} \to \mathbf{MP}$ which sends a metric space X to the top metric predicate

$$\top_X = \lambda x \in X. \, 0$$

in the poset $MP(X)$ of metric predicates on X. We also have a subset functor $\{-\} : \mathbf{MP} \to \mathbf{MS}$ by

$$\left(X \xrightarrow{\varphi} [0, \infty] \right) \mapsto \{x \in X \mid \varphi(x) = 0\}, \quad \text{with metric as in } X.$$

It singles out the points where the predicate is absolutely true. The adjunction $(\top \dashv \{-\})$ is then easily established.

One does not get full subset types.

4.6.4. Remark. Subset types are often used in implicit, hidden form. For example, one often conveniently writes

$$(i > 0) \land \psi(i - 1)$$

where i ranges over natural numbers $\mathbb{N}$, and ψ is a predicate on $\mathbb{N}$. Formally, $i : \mathbb{N} \vdash \psi(i) : \mathsf{Prop}$. The proposition $\psi(i - 1)$ only makes sense if $i > 0$ holds, so we cannot interpret $(i > 0) \land \psi(i - 1)$ as a predicate on natural numbers. In fact, $\psi'(i) = \psi(i - 1)$ is a predicate on the extent $\{x : \mathbb{N} \mid x > 0\}$ of the predicate $\varphi(i) = (i > 0)$. Then we can correctly write the above conjunction as:

$$\varphi(i) \land \coprod_{\pi_\varphi} (\psi')(i)$$
$$= (i > 0) \land \exists j : \{x : \mathbb{N} \mid x > 0\}. \left(\mathsf{o}(j) =_\mathbb{N} i \land \psi(\mathsf{o}(j) - 1) \right)$$

so that it becomes a predicate on the natural numbers. Of course, this is rather cumbersome, especially because it is clear that we should take i(i) as instantiation of j.

Notice, by the way, that by using Frobenius we obtain: $\varphi \land \coprod_{\pi_\varphi} (\psi') \cong \coprod_{\pi_\varphi} (\pi_\varphi^*(\varphi) \land \psi') \cong \coprod_{\pi_\varphi} (\psi')$. In [262] a new connective φ andalso ψ' (with its own rules) is introduced for $\varphi \land \coprod_{\pi_\varphi} (\psi')$.

We conclude this section with an example of how subset types can be used to get a factorisation of maps in base categories of regular fibrations. This gives a more abstract description of the factorisation that we have seen for regular subobject fibrations in Section 4.4.

4.6.5. Example. Let $\begin{smallmatrix}\mathbb{E}\\\downarrow p\\\mathbb{B}\end{smallmatrix}$ be a regular fibration with subset types. An arbitrary map $u: I \to J$ in the base category $\mathbb{B}$ can then be written as composite

$$\left(I \xrightarrow{\ u\ } J \right) = \left(I \xrightarrow{\ u'\ } \{ \textstyle\coprod_u(\top) \} \xrightarrowtail{\ \pi\ } J \right)$$

where $\coprod_u$ is the left adjoint to the reindexing functor u^* associated with u, as in Example 4.3.7 (i). And the morphism u' is then obtained from the unit of the adjunction $(\coprod_u \dashv u^*)$ at the terminal object $\top \in \mathbb{E}_I$ over I,

$$\top \leq u^* \textstyle\coprod_u(\top),$$

which yields a map $u': u \to \pi_{\coprod_u(\top)}$ in the slice $\mathbb{B}/J$, by Lemma 4.6.2 (ii).

Further, this factorisation has the following universal property: for each object $X \in \mathbb{E}_J$ and morphism $v: I \to \{X\}$ in $\mathbb{B}$ with $u = \pi_X \circ v$, there is a unique map $f: \coprod_u(\top) \dashrightarrow X$ in $\mathbb{E}_J$ with

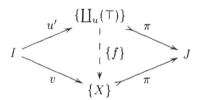

This map f is obtained as follows. By the correspondence in Lemma 4.6.2 (ii), the map v, considered as a morphism $u \to \pi_X$ in $\mathbb{B}/J$, gives rise to a vertical map $\top \to u^*(X)$ over I, and thus by transposition to the required $f: \coprod_u(\top) \to X$ over J. Because this f is vertical one gets $\pi_X \circ \{f\} = \pi_{\coprod_u(\top)}$. Hence $\{f\} \circ u' = v$ holds because $\pi_X: \{X\} \to J$ is a monomorphism.

This is precisely the universal property of the image factorisation in Definition 4.4.2 (i), when considered in the associated subobject fibration.

Exercises

4.6.1.   Describe the resulting projection functor $\mathrm{Fam}(X) \to \mathbf{Sets}^{\to}$ in Example 4.6.3 (ii) and show that in general it need not be full. Do the same in Example 4.6.3 (iv).

4.6.2.   (i)   Check that the category **MS** in Example 4.6.3 (iv) has finite limits.

(ii) Show that the fibration $\begin{smallmatrix} \mathbf{MP} \\ \downarrow \\ \mathbf{MS} \end{smallmatrix}$ has simple products and coproducts.

4.6.3. Consider the regular family fibration $\begin{smallmatrix} \mathrm{Fam}(X) \\ \downarrow \\ \mathbf{Sets} \end{smallmatrix}$ associated with a (non-trivial) frame X. Prove that the factorisation of a function $u: I \to J$ as in Example 4.6.5 is the usual factorisation of u as a surjection followed by an injection.

4.6.4. Show that a projection $\{X\} \rightarrowtail I$ is an isomorphism if and only if $\top \leq X$.

4.6.5. Prove that the 'monic part' $\{\coprod_u(\top)\} \rightarrowtail J$ of $u: I \to J$ in Example 4.6.5 is an isomorphism if and only if u is internally surjective.

4.6.6. Show that a fibration with equality and subset types has equalisers "in the internal logic": for parallel maps $u, v: I \rightrightarrows J$ in the base category we have a diagram

$$\{\mathrm{Eq}(u,v)\} \overset{\pi}{\rightarrowtail} I \overset{u}{\underset{v}{\rightrightarrows}} J \qquad \text{with } \top \leq \mathrm{Eq}(u \circ \pi, v \circ \pi).$$

And for each $w: K \to I$ with $\top \leq \mathrm{Eq}(u \circ w, v \circ w)$ there is a unique $\overline{w}: K \to \{\mathrm{Eq}(u,v)\}$ with $\pi \circ \overline{w} = w$.

Conclude that if internal and external equality coincide, then the base category has (ordinary) equalisers.

4.6.7. Let $\begin{smallmatrix} \mathbb{E} \\ \downarrow p \\ \mathbb{B} \end{smallmatrix}$ be a regular fibration with subset types.

(i) Extend the operation $u \mapsto \coprod_u(\top)$ to a functor $\mathcal{S}: \mathbb{B}^\to \to \mathbb{E}$.

(ii) Show that the projection functor $X \mapsto \pi_X$ is right adjoint to this functor $\mathcal{S}$.

[Lawvere [193] originally described comprehension (or subset types) by requiring such a right adjoint to $\mathcal{S}$; the approach above with a right adjoint to a terminal object functor is somewhat simpler.]

4.6.8. Let $\begin{smallmatrix} \mathbb{E} \\ \downarrow p \\ \mathbb{B} \end{smallmatrix}$ be a fibration with subset types and let $\mathcal{P}: \mathbb{E} \to \mathbb{B}^\to$ be the induced projection functor.

(i) Show that $\mathcal{P}$ preserves any kind of fibred limit as defined in Exercise 1.8.8.

(ii) Suppose that p has products $\prod$; prove that $\mathcal{P}$ preserves these.

4.7 Quotient types

In the previous section we have presented subset types via a right adjoint to a truth predicate functor. In an almost dual fashion we shall now present quotient types via a left adjoint to an equality relation functor. It shows again the rôle played by adjunctions in capturing the essentials of the structures used in logic and mathematics. We split the material on quotients in two parts: in this section we describe the syntax and use of quotient types in (simple) predicate logic. And in the next section we present the categorical description of quotients, involving an appropriate adjunction. In higher order logic quotient types become more powerful and behave better; this will be shown later in Section 5.1. For more information on quotient types, see [132, 135, 133, 21].

We start with the syntax of quotient types (also called quotients, for short). We assume we are in a predicate logic over simple type theory, with at least propositional (or internal) equality $M =_\sigma M'$: Prop, for terms M, M' of the same type σ (as in Section 3.2). The following rule tells us how to obtain a quotient type.

formation

$$\frac{x\colon \sigma, y\colon \sigma \vdash R(x, y)\colon \mathsf{Prop}}{\vdash \sigma/R\colon \mathsf{Type}}$$

Thus, given a type σ with a (binary) relation R on σ, we can form the quotient type σ/R. Notice that we do not require that R is an equivalence relation. Set theoretically, one can think of σ/R as the quotient by the equivalence relation generated by R. This can be made more precise in higher order logic, see Lemma 5.1.8 (but see also Exercise 4.7.3 below). Associated with the formation rule, we have introduction and elimination rules for quotient types.

introduction

$$\frac{\Gamma \vdash M\colon \sigma}{\Gamma \vdash [M]_R\colon \sigma/R} \quad \text{with} \quad \frac{\Gamma \vdash M\colon \sigma \qquad \Gamma \vdash M'\colon \sigma}{\Gamma \mid R(M, M') \vdash [M]_R =_{\sigma/R} [M']_R}$$

This yields the equivalence class $[M]_R$ associated with an inhabitant M of σ. Often we write $[M]$ for $[M]_R$ if the relation R is understood. The associated equality rule tells that if terms are related by R, then their classes are equal. We thus get the "canonical" map $[-]_R\colon \sigma \to \sigma/R$.

elimination

$$\frac{\Gamma, x\colon \sigma \vdash N\colon \tau \qquad \Gamma, x\colon \sigma, y\colon \sigma \mid R(x, y) \vdash N(x) =_\tau N(y)}{\Gamma, a\colon \sigma/R \vdash \mathsf{pick}\ x\ \mathsf{from}\ a\ \mathsf{in}\ N(x)\colon \tau}$$

The intuition is as follows: by assumption, the term $N(x)$ is constant on equivalence classes of R. Hence we may define a new term pick x from a in $N(x)$, which, given a class $a\colon \sigma/R$, picks an element x from the class a, and uses it in $N(x)$. The outcome does not depend on which x we pick. Notice that the variable x thus becomes bound in the elimination term pick x from a in $N(x)$. By α-conversion, this term is then the same as pick y from a in $N(y)$.

The associated conversions are

$$(\beta) \qquad \mathsf{pick}\ x\ \mathsf{from}\ [M]_R\ \mathsf{in}\ N\ =\ N[M/x]$$
$$(\eta) \qquad \mathsf{pick}\ x\ \mathsf{from}\ Q\ \mathsf{in}\ N[[x]_R/a]\ =\ N[Q/a].$$

In the (η)-conversion it is assumed—as usual—that the variable x does not occur free in N. In the calculations below, (η) turns out to be very useful, especially in 'expansion' form: from right to left. An alternative formulation of (η) involving a commutation rule is presented in Exercise 4.7.1.

For completeness we should also mention the behaviour of the new terms under substitution:

$$[M]_R[P/z] \equiv [M[P/z]]_R$$
$$(\mathsf{pick}\ x\ \mathsf{from}\ Q\ \mathsf{in}\ N)[P/z] \equiv \mathsf{pick}\ x\ \mathsf{from}\ Q[P/z]\ \mathsf{in}\ N[P/z].$$

The latter if x does not occur free in P. And also the compatibility rules:

$$M = M' \ \Rightarrow\ [M]_R = [M']_R$$
$$N = N'\ \text{and}\ Q = Q'\ \Rightarrow\ \mathsf{pick}\ x\ \mathsf{from}\ Q\ \mathsf{in}\ N = \mathsf{pick}\ x\ \mathsf{from}\ Q'\ \mathsf{in}\ N'.$$

where in the latter case it is implicitly understood that both N and N' are constant on equivalence classes. We recall that in these rules the equality symbol $=$ without subscript refers to conversion, whereas $=_\tau$ with subscript refers to propositional equality (of type τ).

In the special case where the relation R that we started from is an equivalence relation (provable in the logic), then we can require an additional rule, which is a converse of the equation in the introduction rule. This extra rule can be described categorically by the requirement that a certain functor associated with quotients is full (as will be explained in the next section). Therefore, it makes sense to speak of full quotients, in case this additional rule is added (in analogy with full subset types in the previous section). In category theory one usually calls these quotients effective.

effective or full quotients

$$\frac{\Gamma \vdash M\colon\sigma \qquad\qquad \Gamma \vdash M'\colon\sigma}{\Gamma \mid [M]_R =_{\sigma/R} [M']_R \vdash R(M, M')}\ (R\ \text{is an equivalence relation})$$

Thus, effectiveness says that inhabitants of σ which have the same R-classes must be related by R.

In the above description of quotients we have restricted the relation $R = R(x, y)$ on σ in such a way that it contains only the variables $x, y\colon\sigma$. If we drop this restriction, we get a formation rule

$$\frac{\Gamma, x\colon\sigma, y\colon\sigma \vdash R(x, y)\colon\mathsf{Prop}}{\Gamma \vdash \sigma/R\colon\mathsf{Type}}$$

involving a context Γ of term variables. This leads to *type dependency*: the newly formed quotient type σ/R may contain term variables z in R declared in Γ. A typical example is the group $\mathbb{Z}_n$ of integers modulo n, obtained as quotient type $\mathbb{Z}/n\mathbb{Z}$, for $n: \mathbb{N}$.

This is very much like what we have seen for subset types in the previous section. The natural setting in which to use subset and quotient types is what we shall later call "dependent predicate logic" in Section 11.1. But for the moment we restrict ourselves to quotient types without type context Γ in the formation rule, so that we remain within simple predicate logic.

Propositional equality $=_\sigma$ is essential in formulating the above rules for quotient types. But the presence of these quotients also has an effect on equality, as the following result (from [133, 3.2.7]) shows.

4.7.1. Lemma. *In the presence of quotient types, propositional equality on function types is extensional: one can derive*

$$f: \sigma \to \tau, g: \sigma \to \tau \mid \forall x: \sigma.\, fx =_\tau gx \vdash f =_{\sigma \to \tau} g.$$

(The categorical counterpart of this result states that quotients satisfy a "Frobenius property" (as in Exercise 4.7.6) if and only if the equality functor Eq preserves exponents, see Section 9.2.)

Proof. Consider the following relation $\sim$ on the arrow type $\sigma \to \tau$,

$$f: \sigma \to \tau, g: \sigma \to \tau \vdash f \sim g \overset{\text{def}}{=} \forall x: \sigma.\, fx =_\tau gx : \mathsf{Prop}.$$

Form the associated quotient type $\sigma \Rightarrow \tau \overset{\text{def}}{=} (\sigma \to \tau)/\sim$, with canonical map $[-]: (\sigma \to \tau) \to (\sigma \Rightarrow \tau)$. There is a term P in the reverse direction, obtained via

$$\frac{x: \sigma, f: \sigma \to \tau \vdash fx: \tau \qquad x: \sigma, f: \sigma \to \tau, g: \sigma \to \tau \mid f \sim g \vdash fx =_\tau gx}{x: \sigma, a: (\sigma \Rightarrow \tau) \vdash \mathsf{pick}\ f\ \text{from}\ a\ \text{in}\ fx: \tau}$$

$$a: (\sigma \Rightarrow \tau) \vdash P(a) \overset{\text{def}}{=} \lambda x: \sigma.\,\mathsf{pick}\ f\ \text{from}\ a\ \text{in}\ fx: \sigma \to \tau$$

Obviously, for $f: \sigma \to \tau$,

$$P([f]) = \lambda x: \sigma.\, fx = f,$$

by first using (β) for quotients, and then (η) for $\to$. Thus if $f \sim g$, then $[f] =_{\sigma \Rightarrow \tau} [g]$, and so $f =_{\sigma \to \tau} g$. This completes the proof. Notice by the way, that one also has that $[P(a)] = a$, so that we have an isomorphism of types $(\sigma \to \tau) \cong (\sigma \Rightarrow \tau)$. $\qquad\square$

4.7.2. Notation. Assume we have a relation R on a type σ, and a relation S on a type τ. Then we conveniently write

$$\mathsf{pick}\ x, y\ \text{from}\ a, b\ \text{in}\ N(x, y)$$

for
$$\text{pick } x \text{ from } a \text{ in (pick } y \text{ from } b \text{ in } N(x,y))$$

whenever the latter expression makes sense. This is the case when we can derive the following equations.

$$\Gamma, x{:}\,\sigma, y, y'{:}\,\tau \mid S(y,y') \vdash N(x,y) =_\rho N(x,y')$$
$$\Gamma, x, x'{:}\,\sigma, y{:}\,\tau \mid R(x,x') \vdash N(x,y) =_\rho N(x',y).$$

Via the first of these equations we can form the term pick y from b in $N(x,y)$. By substituting x' for x we also get pick y from b in $N(x',y)$. We now obtain the required multiple pick term via the following derivation.

$$\dfrac{\dfrac{\Gamma, x, x'{:}\,\sigma, y{:}\,\tau \mid R(x,x') \vdash N(x,y) =_\rho N(x',y)}{\begin{array}{l}\Gamma, x, x'{:}\,\sigma, b{:}\,\tau/S \mid R(x,x') \vdash \\[2pt] \qquad \text{pick } y \text{ from } b \text{ in } N(x,y) =_\rho \text{pick } y \text{ from } b \text{ in } N(x',y)\end{array}}}{\Gamma, a{:}\,\sigma/R, b{:}\,\tau/S \vdash \text{pick } x \text{ from } a \text{ in (pick } y \text{ from } b \text{ in } N(x,y))\mathpunct{:}\rho}$$

The first step follows from Exercise 4.7.5.

The remainder of this section is devoted to an elementary example of the use of quotients in (simple) predicate logic. It involves the construction of the integers from the natural numbers, as a free Abelian group.

4.7.3. Example. Recall that the set of integers $\mathbb{Z}$ can be constructed from the natural numbers $\mathbb{N}$ by considering a pair of natural numbers (n,m) as representation for the integer $m-n$. Then one identifies two pairs (n_1, m_1) and (n_2, m_2) of naturals if $m_1 - n_1 = m_2 - n_2$. Or equivalently, if $n_1 + m_2 = n_2 + m_1$. Thus one introduces $\mathbb{Z}$ as a quotient of $\mathbb{N} \times \mathbb{N}$. One can then define addition $+{:}\,\mathbb{Z} \times \mathbb{Z} \to \mathbb{Z}$, zero $0 \in \mathbb{Z}$ and minus $-{:}\,\mathbb{Z} \to \mathbb{Z}$ via representatives. For example, one takes for $a \in \mathbb{Z}$,

$$-a \stackrel{\text{def}}{=} [m,n] \qquad \text{if} \qquad a = [n,m].$$

This construction of $\mathbb{Z}$ form $\mathbb{N}$ can be described in a slightly more abstract way as the formation of the free Abelian group on a commutative monoid via a quotient. Indeed, $(\mathbb{Z}, 0, +, -(\cdot))$ is the free Abelian group on $(\mathbb{N}, 0, +)$.

In our predicate logic over simple type theory we now assume that we have a commutative monoid $(\mathsf{N}, 0, +)$, consisting of a type $\mathsf{N}{:}\,\mathsf{Type}$ with constants $0, +$ in

$$\vdash 0{:}\,\mathsf{N} \qquad \text{and} \qquad x{:}\,\mathsf{N}, y{:}\,\mathsf{N} \vdash x + y{:}\,\mathsf{N}$$

satisfying the commutative monoid equations

$$0 + x =_{\mathsf{N}} x, \qquad x + y =_{\mathsf{N}} y + x, \qquad x + (y + z) =_{\mathsf{N}} (x + y) + z,$$

for $x, y, z \colon \mathsf{N}$. We think of these as (internal) equalities which come with the data type $(\mathsf{N}, +, 0)$. One may read N as natural numbers, but all we need are these commutative monoid equations.

We then consider the relation $\sim$ on $\mathsf{N} \times \mathsf{N}$,

$$u \colon \mathsf{N} \times \mathsf{N}, v \colon \mathsf{N} \times \mathsf{N} \vdash u \sim v \stackrel{\text{def}}{=} (\pi u + \pi' v =_{\mathsf{N}} \pi' u + \pi v) \colon \mathsf{Prop}$$

which corresponds to the identification of pairs (n_1, m_1), (n_2, m_2) via $n_1 + m_2 = n_2 + m_1$ above. We write

$$\mathsf{Z} = (\mathsf{N} \times \mathsf{N})/\!\!\sim \quad \text{and} \quad [x, y] \text{ for } [\langle x, y \rangle] \quad \text{in} \quad \mathsf{N} \times \mathsf{N} \xrightarrow{\;[-]\;} \mathsf{Z}.$$

Of course we think of Z as the type of integers.

The next step is to provide Z with an Abelian group structure $\widehat{0}$, $\widehat{+}$ and inverse $-$. This is done, as in the set-theoretic construction, via representatives. And the syntax we have allows us to reason conveniently with these representatives inside pick ... terms.

The neutral element is easily obtained as

$$\widehat{0} \stackrel{\text{def}}{=} [0, 0] \colon \mathsf{Z}.$$

The inverse operation $-(\cdot)$ is

$$-a \stackrel{\text{def}}{=} \text{pick } w \text{ from } a \text{ in } [\pi' w, \pi w] \colon \mathsf{Z},$$

which is very much like the set-theoretic minus $-(\cdot)$ mentioned above. Notice that this term is well-defined because from $u \sim v$ one derives $\langle \pi' u, \pi u \rangle \sim \langle \pi' v, \pi v \rangle$.

Finally, addition $\widehat{+}$ on Z is then

$$a \,\widehat{+}\, b \stackrel{\text{def}}{=} \text{pick } u, v \text{ from } a, b \text{ in } [\pi u + \pi v, \pi' u + \pi' v]$$
$$= \text{pick } u \text{ from } a \text{ in } (\text{pick } v \text{ from } b \text{ in } [\pi u + \pi v, \pi' u + \pi' v]).$$

This operation is well-defined, since we can derive

$$u_1, u_2 \colon \mathsf{N} \times \mathsf{N}, v \colon \mathsf{N} \times \mathsf{N} \mid u_1 \sim u_2 \vdash$$
$$\langle \pi u_1 + \pi v, \pi' u_1 + \pi' v \rangle \sim \langle \pi u_2 + \pi v, \pi' u_2 + \pi' v \rangle$$
$$u \colon \mathsf{N} \times \mathsf{N}, v_1, v_2 \colon \mathsf{N} \times \mathsf{N} \mid v_1 \sim v_2 \vdash$$
$$\langle \pi u + \pi v_1, \pi' u + \pi' v_1 \rangle \sim \langle \pi u + \pi v_2, \pi' u + \pi' v_2 \rangle.$$

Then $\hat{0}$ is neutral element, since we can compute:

$$
\begin{aligned}
a \,\hat{+}\, \hat{0} &= \text{pick } u \text{ from } a \text{ in (pick } v \text{ from } [0,0] \text{ in } [\pi u + \pi v, \pi' u + \pi' v]) \\
&= \text{pick } u \text{ from } a \text{ in } [\pi u + 0, \pi' u + 0] \\
&= \text{pick } u \text{ from } a \text{ in } [\pi u, \pi' u] \\
&= \text{pick } u \text{ from } a \text{ in } [u] \\
&= a.
\end{aligned}
$$

We leave it to the reader to verify that $(\mathsf{Z}, \hat{+}, \hat{0}, -)$ is an Abelian group. Therefore one needs the conversions in Exercises 4.7.1 and 4.7.4 below. We do show that Z has the appropriate universal property making it the *free* Abelian group on N. First, we have an extension map $c: \mathsf{N} \to \mathsf{Z}$ by $c(x) = [0, x]$. This is a monoid homomorphism, since by definition $c(0) = [0,0] = \hat{0}$, and

$$
\begin{aligned}
c(x) \,\hat{+}\, c(y) &= \text{pick } u, v \text{ from } [0,x], [0,y] \text{ in } [\pi u + \pi v, \pi' u + \pi' v] \\
&= [0 + 0, x + y] \\
&= c(x + y).
\end{aligned}
$$

Further, if we are given an arbitrary Abelian group $(G, \bullet, 1, (\cdot)^{-1})$ together with a monoid homomorphism $M: \mathsf{N} \to G$, then there is a unique homomorphism $\widehat{M}: \mathsf{Z} \to G$ with $\widehat{M} \circ c = M$ in,

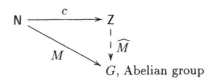

$$
G, \text{ Abelian group}
$$

Therefore, write

$$
N(u) = M(\pi' u) \bullet M(\pi u)^{-1} : G, \quad \text{for } u : \mathsf{N} \times \mathsf{N}
$$

To see that the term

$$
\widehat{M}(a) \stackrel{\text{def}}{=} \text{pick } u \text{ from } a \text{ in } N(u) : G, \quad \text{for } a : \mathsf{Z}
$$

is well-defined we need to derive

$$
u, v: \mathsf{N} \times \mathsf{N} \mid u \sim v \vdash N(u) =_G N(v)
$$

But this follows because G is an Abelian group: if $u \sim v$, then by definition $\pi u + \pi' v =_{\mathsf{N}} \pi' u + \pi v$. Hence $M(\pi u) \bullet M(\pi' v) =_G M(\pi u + \pi' v) =_G M(\pi' u + \pi v) =_G M(\pi' u) \bullet M(\pi v)$, and so $N(u) =_G M(\pi' u) \bullet M(\pi u)^{-1} =_G M(\pi' v) \bullet$

$M(\pi v)^{-1} =_G N(v)$. Then indeed,

$$
\begin{aligned}
(\widehat{M} \circ c)(x) &= \quad \text{pick } u \text{ from } [0, x] \text{ in } N(u) \\
&= \quad N(\langle 0, x \rangle) \\
&= \quad M(x) \bullet M(0)^{-1} \\
&=_G \quad M(x) \bullet 1^{-1} \\
&=_G \quad M(x).
\end{aligned}
$$

We leave it to the reader to verify that $\widehat{M}$ is a homomorphism. And if another term (-homomorphism) $P: Z \to G$ satisfies $P(c(x)) =_G M(x)$, then $N(u) = M(\pi' u) \bullet M(\pi u)^{-1} =_G P(c(\pi' u)) \bullet P(c(\pi u))^{-1} =_G P([0, \pi' u]) \bullet P(-[0, \pi u]) =_G P([0, \pi' u] \widehat{+} [\pi u, 0]) =_G P([\pi u, \pi' u]) =_G P([u])$. Hence

$$\widehat{M}(a) = \text{pick } u \text{ from } a \text{ in } N(u) = \text{pick } u \text{ from } a \text{ in } P([u]) = P(a).$$

This concludes the example.

The first two of the exercises below give conversions which are quite useful in computations with quotient types.

Exercises

4.7.1. Prove that in the presence of (β)-conversion for quotients, the (η)-conversion is equivalent to the combination of

$$
\begin{aligned}
P[(\text{pick } x \text{ from } Q \text{ in } N)/z] &= \text{pick } x \text{ from } Q \text{ in } P[N/z] \\
\text{pick } x \text{ from } Q \text{ in } [x]_R &= Q.
\end{aligned}
$$

The first of these is a 'commutation' conversion, and is comparable to the conversion in Lemma 2.3.3 for coproduct types $+$.

4.7.2. Prove that the term

$$x: \sigma \vdash [x]_{=_\sigma} : \sigma / =_\sigma$$

is invertible.

4.7.3. (i) Let R, S be two relations on the same type σ. Show how an entailment $x, x': \sigma \mid R(x, x') \vdash S(x, x')$ gives rise to a term $a: \sigma/R \vdash M(a): \sigma/S$.

(ii) For a relations R on σ, define the reflexive and symmetric closure of R as two relations on σ given (respectively) by

$$R^{\mathrm{r}}(x, x') \stackrel{\mathrm{def}}{=} R(x, x') \vee (x =_\sigma x'), \qquad R^{\mathrm{s}}(x, x') \stackrel{\mathrm{def}}{=} R(x, x') \vee R(x', x).$$

Show that taking $S = R^{\mathrm{r}}$ and $S = R^{\mathrm{s}}$ in (i) leads in both cases to invertible terms.

4.7.4. Prove the following derived conversions.

(i) In case a term $\Gamma \vdash N: \tau$ that we apply elimination to, does not contain a variable x of type σ, then we get in context $\Gamma, a: \sigma/R$ a conversion,

$$\text{pick } x \text{ from } a \text{ in } N = N.$$

(ii) And in case we have two variables $\Gamma, x, y\colon \sigma \vdash N(x, y)\colon \tau$ and equalities

$$\Gamma, x\colon \sigma, y, y'\colon \sigma \mid R(y, y') \vdash N(x, y) =_\tau N(x, y')$$
$$\Gamma, x, x'\colon \sigma, y\colon \sigma \mid R(x, x') \vdash N(x, y) =_\tau N(x', y)$$

then in context $\Gamma, a\colon \sigma/R$ we have a conversion,

pick x, y from a, a in $N(x, y)$ = pick x from a in $N(x, x)$.

4.7.5.   Derive the following replacement rule for internal equality $=_\tau$.

$$\frac{\Gamma, x\colon \sigma \mid \Theta \vdash N =_\tau N'}{\Gamma, a\colon \sigma/R \mid \Theta \vdash (\text{pick } x \text{ from } a \text{ in } N) =_\tau (\text{pick } x \text{ from } a \text{ in } N')} \quad (x \text{ not in } \Theta)$$

where both N and N' are assumed to be constant on equivalence classes. It is used to justify the multiple pick's in Notation 4.7.2.

4.7.6.   Let types σ, ρ and a relation $x, y\colon \sigma \vdash R(x, y)\colon \mathsf{Prop}$ be given. Form a new relation $\rho^*(R)$ on $\rho \times \sigma$ by

$$u\colon \rho \times \sigma, v\colon \rho \times \sigma \vdash \rho^*(R)(u, v) \stackrel{\text{def}}{=} (\pi u =_\rho \pi v) \wedge R(\pi' u, \pi' v)\colon \mathsf{Prop}.$$

Prove that the canonical map

$$(\rho \times \sigma)/\rho^*(R) \longrightarrow \rho \times (\sigma/R)$$

given by

$$a\colon (\rho \times \sigma)/\rho^*(R) \vdash P(a) \stackrel{\text{def}}{=} \text{pick } u \text{ from } a \text{ in } \langle \pi u, [\pi' u]_R \rangle \colon \rho \times (\sigma/R)$$

is invertible.

[This is shows that a "Frobenius distributivity" for quotient types is inherent in the syntax that we use (with explicitly contexts Γ in the elimination rule). It is like for other 'colimits' such as $+$ and $\exists$.]

4.7.7.   Consider a predicate logic with a commutative monoid N of natural numbers as in Example 4.7.3, and with Z as the Abelian group of integers constructed from N, as in the example.

(i)   Give a formal description of the construction of the rationals Q as quotient of $\mathsf{Z} \times \mathsf{N}$, where the pair (n, m) represents the rational $\frac{n}{m+1}$.

(ii) Assume now that one also has exponent types $\to$ and subset types. Try to formalise the construction of the Cauchy reals (see for Example [335, Chapter V]).

4.8 Quotient types, categorically

In this section we describe quotient types (in simple predicate logic) in categorical terms. These quotients, like subsets, involve an adjunction between a base category and a total category of a fibration. But where subsets involve a right adjoint to a truth predicate functor, quotients involve a left adjoint to an equality (relation) functor. Interestingly, it turns out that subsets can also be described in terms of a right adjoint to this equality functor.

We recall from Definition 3.5.1 that an Eq-fibration $\begin{smallmatrix}\mathbb{E}\\\downarrow p\\\mathbb{B}\end{smallmatrix}$ is a fibred pre-order which has fibred finite products and finite products in its base category, and also has equality satisfying the Frobenius property. Below we describe quotients only for such preordered fibrations, but the main definition 4.8.1 applies to non-preordered fibrations as well. We shall write Eq_I for the left adjoint of the diagonal $\delta(I) = \langle \mathrm{id}, \mathrm{id}\rangle \colon I \to I \times I$ in $\mathbb{B}$.

For such a fibration $\begin{smallmatrix}\mathbb{E}\\\downarrow p\\\mathbb{B}\end{smallmatrix}$ we form the fibration $\begin{smallmatrix}\mathrm{Rel}(\mathbb{E})\\\downarrow\\\mathbb{B}\end{smallmatrix}$ of binary predicates (or relations) in p by the following change-of-base situation.

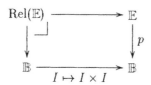

A fibre $\mathrm{Rel}(\mathbb{E})_I$ is then the same as the fibre $\mathbb{E}_{I \times I}$ of relations on $I \in \mathbb{B}$. Note however, that in the notation $\mathrm{Rel}(\mathbb{E})$ the dependence on the fibration p is left implicit.

There is then an "equality relation" functor

$$\mathbb{B} \xrightarrow{\ \mathrm{Eq}\ } \mathrm{Rel}(\mathbb{E}) \qquad \text{by} \qquad I \longmapsto \mathrm{Eq}(I) = \mathrm{Eq}_I(\top),$$

where $\top = \top(I)$ is the terminal object in the fibre $\mathbb{E}_I$. A morphism $u \colon I \to J$ in $\mathbb{B}$ is mapped to the composite

$$\mathrm{Eq}(I) = \mathrm{Eq}_I(\top(I)) \longrightarrow (u \times u)^*(\mathrm{Eq}_J(\top(J))) \longrightarrow \mathrm{Eq}_J(\top(J)) = \mathrm{Eq}(J)$$

where the first part of this map is obtained by transposing the following composite across the adjunction $\mathrm{Eq}_I \dashv \delta(I)^*$.

$$\top(I) \cong u^*(\top(J)) \xrightarrow{\ u^*(\eta_J)\ } u^*\delta(J)^*\mathrm{Eq}_J(\top(J)) \cong \delta(I)^*(u \times u)^*\mathrm{Eq}_J(\top(J))$$

It may be clear that the functor Eq is a section of the fibration of relations.

For a morphism $u \colon I \to J$ we write

$$\mathrm{Ker}(u) \stackrel{\mathrm{def}}{=} (u \times u)^*(\mathrm{Eq}(J)) \in \mathrm{Rel}(\mathbb{E})_{I \times I}$$

for the kernel relation $u(i) = u(i')$ on $I \times I$. This operation $u \mapsto \mathrm{Ker}(u)$ can be extended to a functor $\mathbb{B}^{\rightarrow} \to \mathrm{Rel}(\mathbb{E})$ commuting with the domain functor (or fibration) dom: $\mathbb{B}^{\rightarrow} \to \mathbb{B}$, see Exercise 4.8.8.

We can now state our main notion (in this section).

4.8.1. Definition. Let $\begin{smallmatrix}\mathbb{E}\\\downarrow p\\\mathbb{B}\end{smallmatrix}$ be a fibration as above. We say that p **has quotients** or **quotient types** if the equality functor Eq: $\mathbb{B} \to \mathrm{Rel}(\mathbb{E})$ has a left adjoint.

This left adjoint maps a binary relation $R \in \mathrm{Rel}(\mathbb{E})_I = \mathbb{E}_{I \times I}$ to the quotient object $I/R \in \mathbb{B}$. The unit η_R is a map $R \to \mathrm{Eq}(I/R)$ in $\mathrm{Rel}(\mathbb{E})$. Its underlying map in $\mathbb{B}$ will be written as $c_R \colon I \to I/R$. It is the "canonical quotient map" associated with the quotient.

The next result is the analogue for quotients of Lemma 4.6.2 for subset types.

4.8.2. Lemma. *Consider a fibration* $\begin{smallmatrix}\mathbb{E}\\\downarrow p\\\mathbb{B}\end{smallmatrix}$ *with quotients, as above.*

  (i) *The canonical maps* $c_R \colon I \to I/R$ *are epis in the base category.*

  (ii) *For each morphism* $u \colon I \to J$ *in* $\mathbb{B}$ *and for each relation* $R \in \mathbb{E}_{I \times I}$ *on* I, *there is a bijective correspondence*

$$\frac{R \leq \mathrm{Ker}(u) \quad in\ \mathrm{Rel}(\mathbb{E})_I}{c_R \dashrightarrow u \quad in\ I \backslash \mathbb{B}}$$

where $I \backslash \mathbb{B}$ *is the 'opslice' category of maps with domain* I *and commuting triangles.*

  (iii) *The assignment* $R \mapsto c_R$ *extends to a "canonical quotient map" functor* $\mathcal{C}$ *in*

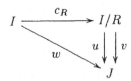

which maps 'opcartesian' morphisms in $\mathrm{Rel}(\mathbb{E})$ *to pushout squares in* $\mathbb{B}$.

(An 'opcartesian' map is for an 'opfibration' what a Cartesian map is for a fibration, as we shall see in Section 9.1. In this situation a morphism $f \colon R \to S$ in $\mathrm{Rel}(\mathbb{E})$ over $u \colon I \to J$ is opcartesian if and only if $S \leq \coprod_{u \times u}(R)$.)

Proof. (i) Consider a situation

where $u \circ c_R = v \circ c_R = w$. The transposes $u^\vee = \mathrm{Eq}(u) \circ \eta_R$ and $v^\vee = \mathrm{Eq}(v) \circ \eta_R$ are maps $R \rightrightarrows \mathrm{Eq}(J)$ which are both above $u \circ c_R = v \circ c_R$. But then $u^\vee = v^\vee$, because $\begin{smallmatrix} \mathrm{Rel}(\mathbb{E}) \\ \downarrow \\ \mathbb{B} \end{smallmatrix}$, like $\begin{smallmatrix} \mathbb{E} \\ \downarrow \\ \mathbb{B} \end{smallmatrix}$, is a preordered fibration (see Exercise 1.3.11). Hence $u = v$.

(ii) Assume we have an inequality $R \leq \mathrm{Ker}(u) = (u \times u)^*(\mathrm{Eq}(J))$ over I. There is then a unique map $f \colon R \to \mathrm{Eq}(J)$ in $\mathrm{Rel}(\mathbb{E})$ over $u \colon I \to J$. By transposing it we get a morphism $f^\wedge \colon I/R \to J$ in $\mathbb{B}$, satisfying $f^\wedge \circ c_R = u$. Conversely, assume we have a morphism $v \colon I/R \to J$ in $\mathbb{B}$ with $v \circ c_R = u$. The transpose $v^\vee \colon R \to \mathrm{Eq}(J)$ is then above u, by an argument as in (i). This yields the required inequality $R \leq \mathrm{Ker}(u) = (u \times u)^*(\mathrm{Eq}(J))$ over I.

(iii) For a morphism $f \colon R \to S$ in $\mathrm{Rel}(\mathbb{E})$ over $u \colon I \to J$ we have to find a map $I/R \dashrightarrow J/S$ in a commuting square,

$$
\begin{array}{ccc}
I/R & \dashrightarrow & J/S \\
{\scriptstyle c_R} \uparrow & & \uparrow {\scriptstyle c_S} \\
I & \xrightarrow{\;\;u\;\;} & J
\end{array}
$$

This requires a map $c_R \dashrightarrow (c_S \circ u)$ in the opslice $I \backslash \mathbb{B}$. By combining the inequality $R \leq (u \times u)^*(S)$ with $S \leq (c_S \times c_S)^*(\mathrm{Eq}(J/S))$ we obtain the following inequality.

$$
\begin{aligned}
R &\leq (u \times u)^*(S) \\
&\leq (u \times u)^*(c_S \times c_S)^*(\mathrm{Eq}(J/S)) \\
&\cong ((c_S \circ u) \times (c_S \circ u))^*(\mathrm{Eq}(J/S)) \\
&= \mathrm{Ker}(c_S \circ u).
\end{aligned}
$$

Then we get the required map by (ii).

If our map $f \colon R \to S$ is opcartesian over u—i.e. if $S \leq \coprod_{u \times u}(R)$—then the above square becomes a pushout in $\mathbb{B}$: assume maps $v \colon I/R \to K$ and $w \colon J \to K$ in $\mathbb{B}$ with $v \circ c_R = w \circ u$. Then v is a morphism $c_R \to (w \circ u)$ in $I \backslash \mathbb{B}$. This yields $R \leq \mathrm{Ker}(w \circ u)$, by (ii). Now

$$
\begin{aligned}
S &\leq \coprod_{u \times u}(R) & \text{because } f \text{ is opcartesian} \\
&\leq \coprod_{u \times u}(u \times u)^*(w \times w)^*(\mathrm{Eq}(K)) & \text{because } R \leq \mathrm{Ker}(w \circ u) \\
&\leq (w \times w)^*(\mathrm{Eq}(K)) & \text{by } \coprod_{u \times u} \dashv (u \times u)^* \\
&= \mathrm{Ker}(w).
\end{aligned}
$$

Hence we get the required mediating map $c_S \dashrightarrow w$ in $J \backslash \mathbb{B}$ by (ii). $\qquad\square$

Notice that the canonical maps $\{X\} \rightarrowtail I$ for subset types are monos, whereas the canonical maps $I \twoheadrightarrow I/R$ for quotient types are epis. But there is a deeper duality between subset types and quotient types, as we will show next. Recall that we have introduced subset types via a right adjoint to truth, and quotient types via a left adjoint to equality. It turns out that subset types can equivalently be described by a right adjoint to equality.

4.8.3. Theorem. *Let* $\begin{smallmatrix} \mathbb{E} \\ \downarrow p \\ \mathbb{B} \end{smallmatrix}$ *be an Eq-fibration. The induced equality functor* Eq: $\mathbb{B} \to \mathrm{Rel}(\mathbb{E})$ *then has a right adjoint if and only if p admits subset types.*

This result, and its proof below, also hold for non-preordered fibrations.

Proof. Assume the fibration p has subset types, via a right adjoint $\{-\}: \mathbb{E} \to \mathbb{B}$ to the truth predicate functor $\top$. For a relation $R \in \mathbb{E}_{I \times I}$ on I we have the following (natural) isomorphisms.

$$
\begin{aligned}
\mathrm{Rel}(\mathbb{E})\big(\mathrm{Eq}(J),\ R \big) &\cong \coprod_{u:J \to I} \mathbb{E}_{J \times J}\big(\mathrm{Eq}(J),\ (u \times u)^*(R) \big), \text{ see Lemma 1.4.10} \\
&\cong \coprod_{u:J \to I} \mathbb{E}_J\big(\top(J),\ \delta^*(u \times u)^*(R) \big) \\
&\cong \coprod_{u:J \to I} \mathbb{E}_J\big(\top(J),\ u^*\delta^*(R) \big) \\
&\cong \mathbb{E}\big(\top(J),\ \delta^*(R) \big) \\
&\cong \mathbb{B}\big(J,\ \{\delta^*(R)\} \big).
\end{aligned}
$$

Hence $R \mapsto \{\delta^*(R)\}$ is right adjoint to Eq: $\mathbb{B} \to \mathrm{Rel}(\mathbb{E})$.

Conversely, assume that the equality functor Eq has a right adjoint $K: \mathrm{Rel}(\mathbb{E}) \to \mathbb{B}$. For an object $X \in \mathbb{E}$ over $I \in \mathbb{B}$, put $\{X\} = K(\pi^*(X))$, where π is the first projection $I \times I \to I$. Then $X \mapsto \{X\}$ is right adjoint to truth $\top$:

$$
\begin{aligned}
\mathbb{E}\big(\top(J),\ X \big) &\cong \coprod_{u:J \to I} \mathbb{E}_J\big(\top(J),\ u^*(X) \big), \quad \text{by Lemma 1.4.10} \\
&\cong \coprod_{u:J \to I} \mathbb{E}_J\big(\top(J),\ u^*\delta^*\pi^*(X) \big) \\
&\cong \coprod_{u:J \to I} \mathbb{E}_J\big(\top(J),\ \delta^*(u \times u)^*\pi^*(X) \big) \\
&\cong \coprod_{u:J \to I} \mathbb{E}_{J \times J}\big(\mathrm{Eq}(J),\ (u \times u)^*\pi^*(X) \big) \\
&\cong \mathrm{Rel}(\mathbb{E})\big(\mathrm{Eq}(J),\ \pi^*(X) \big) \\
&\cong \mathbb{B}\big(J,\ K(\pi^*(X)) \big) \\
&= \mathbb{B}\big(J,\ \{X\} \big). \qquad\qquad \square
\end{aligned}
$$

We consider the following two additional requirements for quotients in a fibration.

4.8.4. Definition. Let $\begin{smallmatrix}\mathbb{E}\\\downarrow p\\\mathbb{B}\end{smallmatrix}$ be a fibration with quotients as above.

(i) We say that the quotients satisfy the **Frobenius property** in case the following holds. If for a relation R on I and an object $J \in \mathbb{B}$ we form the relation

$$J^*(R) \stackrel{\text{def}}{=} (\pi \times \pi)^*(\text{Eq}(J)) \times (\pi' \times \pi')^*(R) \quad \text{on } J \times I$$

then the canonical map

$$(J \times I)/J^*(R) \longrightarrow J \times (I/R)$$

is an isomorphism.

(ii) And if p is a preorder fibration, then we say that quotients are **effective** or **full** if for each equivalence relation R on I (in the logic of the fibration p), the unit map $\eta_R: R \to \text{Eq}(I/R)$ is Cartesian over $c_R: I \twoheadrightarrow I/R$ in the fibration $\begin{smallmatrix}\text{Rel}(\mathbb{E})\\\downarrow\\\mathbb{B}\end{smallmatrix}$ of relations.

The canonical map in (i) is obtained by transposing the following composite

$$J^*(R) = (\pi \times \pi)^*(\text{Eq}(J)) \times (\pi' \times \pi')^*(R)$$

$$\Big\downarrow \text{id} \times (\pi' \times \pi')^*(\eta)$$

$$(\pi \times \pi)^*(\text{Eq}(J)) \times (\pi' \times \pi')^*(\text{Eq}(I/R)) \cong \text{Eq}(J \times (I/R))$$

accross the quotient-adjunction. The latter isomorphism comes from the fact that Eq is a right adjoint and must thus preserve products. In the total category $\text{Rel}(\mathbb{E})$ these products are given by the formula on the left of $\cong$, as we shall see in more detail in Section 9.2.

We briefly discuss the interpretation of the quotient type syntax from the previous section in a fibration $\begin{smallmatrix}\mathbb{E}\\\downarrow p\\\mathbb{B}\end{smallmatrix}$ with quotients satisfying the Frobenius property. The latter is used—as always—to get an appropriate elimination rule with contexts.

A relation R on a type $I \in \mathbb{B}$ is an object $R \in \mathbb{E}_{I \times I} = \text{Rel}(\mathbb{E})_I$. We can form the associated quotient type $I/R \in \mathbb{B}$ with its canonical map $c_R = [-]_R: I \twoheadrightarrow I/R$ satisfying $R \leq (c_R \times c_R)^*(\text{Eq}(I/R))$. This gives us for each $i: I$ a class $[i]_R: I/R$, together with an entailment $i, i': I \mid R(i, i') \vdash [i]_R =_{I/R} [i']_R$. This yields validity of the formation and introduction rules.

For the elimination rule, assume we have a term,

$$j: J, i: I \vdash u(j, i): K \qquad \text{as a map in } \mathbb{B} \qquad J \times I \stackrel{u}{\longrightarrow} K,$$

which is constant on elements related by R:

$$j: J, i, i': I \mid R(i, i') \vdash u(j, i) =_K u(j, i').$$

The latter yields an entailment

$$j, j': J, i, i': I \mid J^*(R)((j, i), (j', i')) \vdash u(j, i) =_K u(j', i'),$$

since $J^*(R)((j, i), (j', i')) = (j =_J j') \wedge R(i, i')$. We thus get a map $J^*(R) \to$ Eq(K) in Rel$(\mathbb{E})$ over u. Transposition across the quotient adjunction yields a map $(J \times I)/J^*(R) \to K$, and so by Frobenius we get our required map

$$J \times (I/R) \xrightarrow{\ \cong\ } (J \times I)/J^*(R) \longrightarrow K$$

which may be read as

$$j: J, a: I/R \vdash \text{pick } i \text{ from } a \text{ in } u(j, i): K.$$

We leave validity of the quotient conversions as exercise to the reader.

What are traditionally called 'effective' quotients in category theory may also be called 'full' quotients, because of the following result, and because of the analogy with 'full' subset types.

4.8.5. Proposition. *Let* $\begin{smallmatrix} \mathbb{E} \\ \downarrow p \\ \mathbb{B} \end{smallmatrix}$ *be a (preorder) fibration with quotients. We write* ERel$(\mathbb{E}) \hookrightarrow$ Rel$(\mathbb{E})$ *for the full subcategory of equivalence relations (in the logic of p). The quotients in p are then effective (or full) if and only if the "canonical quotient map" functor* $\mathcal{C}$: ERel$(\mathbb{E}) \to \mathbb{B}^{\to}$ *is full (and faithful).*

Proof. Assume quotients in p are effective, and consider a commuting square in $\mathbb{B}$ of the form:

$$
\begin{array}{ccc}
I/R & \xrightarrow{\quad v \quad} & J/S \\
\mathcal{C}(R) = c_R \uparrow & & \uparrow c_S = \mathcal{C}(S) \\
I & \xrightarrow[\quad u \quad]{} & J
\end{array}
$$

where R, S are equivalence relations. We must show $R \leq (u \times u)^*(S)$ to get fullness of $\mathcal{C}$. This is done as follows.

$$
\begin{aligned}
R &\cong (c_R \times c_R)^* \text{Eq}(I/R) & \text{by effectiveness} \\
&\leq (c_R \times c_R)^* (v \times v)^* \text{Eq}(J/S) & \\
&\cong (u \times u)^* (c_S \times c_S)^* \text{Eq}(J/S) & \\
&\cong (u \times u)^*(S) & \text{by effectiveness again.}
\end{aligned}
$$

Conversely, we need to show that a unit map $\eta_S: S \to \text{Eq}(J/S)$ is Cartesian in Rel$(\mathbb{E})$, for $S \in \mathbb{E}_{J \times J}$ an equivalence relation. That is, we need to show that $\mathrm{Ker}(c_S) = (c_S \times c_S)^* \mathrm{Eq}(J/S) \leq S$. Since $\mathrm{Ker}(c_S)$ is (also) an equivalence relation, it suffices by fullness of the functor $\mathcal{C}$ to produce a map $c_{\mathrm{Ker}(c_S)} \to c_S$ in $J \backslash \mathbb{B}$. But this follows from $\mathrm{Ker}(c_S) \leq \mathrm{Ker}(c_S)$, using Lemma 4.8.2 (ii). $\square$

We continue this section with several examples of fibrations with quotients, starting with subobject fibrations.

4.8.6. Proposition. *Consider a category $\mathbb{B}$ with finite limits.*

(i) *If $\mathbb{B}$ has coequalisers, then the subobject fibration* $\begin{smallmatrix} \mathrm{Sub}(\mathbb{B}) \\ \downarrow \\ \mathbb{B} \end{smallmatrix}$ *on $\mathbb{B}$ has quotients. These are effective if and only if each equivalence relation $R \rightarrowtail I \times I$ in $\mathbb{B}$ is* **effective**, *i.e. is a kernel pair $R \rightrightarrows I$ of some map $I \to J$ in $\mathbb{B}$.*

(ii) *In case $\mathbb{B}$ is a regular category the converse of (i) also holds: $\mathbb{B}$ has coequalisers if and only if* $\begin{smallmatrix} \mathrm{Sub}(\mathbb{B}) \\ \downarrow \\ \mathbb{B} \end{smallmatrix}$ *has quotients.*

(iii) *And in the situation of (ii), the coequalisers in $\mathbb{B}$ are preserved by functors $J \times (-) \colon \mathbb{B} \to \mathbb{B}$ if and only if the Frobenius property holds for the quotients in* $\begin{smallmatrix} \mathrm{Sub}(\mathbb{B}) \\ \downarrow \\ \mathbb{B} \end{smallmatrix}$.

A regular category with coequalisers, as in (ii), is often called an **exact** category, see *e.g.* [36, II, 2.6].

Proof. (i) Assume $\mathbb{B}$ has coequalisers. For a relation $\langle r_0, r_1 \rangle \colon R \rightarrowtail I \times I$ on I, we find a quotient object I/R by forming the coequaliser

$$R \underset{r_1}{\overset{r_0}{\rightrightarrows}} I \xrightarrow{c_R} I/R$$

This assignment $R \mapsto I/R$ yields a left adjoint to the equality functor since there is a bijective correspondence between morphisms u and v in:

$$
\begin{array}{ccc}
R & \dashrightarrow & J \\
{\scriptstyle \langle r_0, r_1 \rangle} \downarrow & & \downarrow {\scriptstyle \delta} \\
I \times I & \xrightarrow{u \times u} & J \times J \\
\hline
I/R & \xrightarrow{\quad v \quad} & J
\end{array}
$$

In case R is an equivalence relation, then its quotient is effective—according to Definition 4.8.4 (ii)—if and only if there is a pullback diagram

$$
\begin{array}{ccc}
R & \longrightarrow & I/R \\
{\scriptstyle \langle r_0, r_1 \rangle} \downarrow \quad \lrcorner & & \downarrow {\scriptstyle \delta} \\
I \times I & \xrightarrow{\ c_R \times c_R\ } & (I/R) \times (I/R)
\end{array}
$$

This diagram expresses that R is the kernel of its own coequaliser c_R. But that is equivalent so saying that R is the coequaliser of some map $I \to J$ in $\mathbb{B}$.

(ii) If $\mathbb{B}$ is a regular category then quotients for the subobject fibration on $\mathbb{B}$ induce coequalisers in $\mathbb{B}$. Given a parallel pair of maps $u, v: K \rightrightarrows I$ in $\mathbb{B}$, we first factorise

$$
\Big(K \xrightarrow{\ \langle u, v \rangle\ } I \times I \Big) = \Big(K \xrightarrow{\ e\ } R \xrightarrowtail{\ \langle r_0, r_1 \rangle\ } I \times I \Big)
$$

and then take the quotient of the relation $R \rightarrowtail I \times I$,

$$
I \xrightarrow{\quad c_R \quad} I/R
$$

The unit map $\eta_R: R \to \mathrm{Eq}(I/R)$ consists of a square

$$
\begin{array}{ccc}
R & \longrightarrow & I/R \\
{\scriptstyle \langle r_0, r_1 \rangle} \downarrow & & \downarrow {\scriptstyle \delta} \\
I \times I & \xrightarrow{\ c_R \times c_R\ } & (I/R) \times (I/R)
\end{array}
$$

This gives us $c_R \circ r_0 = c_R \circ r_1$, and thus $c_R \circ u = c_R \circ v$. If also $w: I \to J$ satisfies $w \circ u = w \circ v$, then, because e is an epi, we get $w \circ r_0 = w \circ r_1$. The latter tells us that we have a map $R \to \mathrm{Eq}(J)$ in the category of relations over $w: I \to J$. By transposition we then get the required mediating map $I/R \to J$. This shows that c_R is the coequaliser of u, v in $\mathbb{B}$.

(iii) The main point is that for a relation $\langle r_0, r_1 \rangle: R \rightarrowtail I \times I$ on I and an object $J \in \mathbb{B}$ the relation $J^*(R)$ on $J \times I$ in Definition 4.8.4 (i) is the subobject

$$
J \times R \xrightarrowtail{\ \langle J \times r_0, J \times r_1 \rangle\ } (J \times I) \times (J \times I)
$$

Thus, assuming that $\mathbb{B}$ has coequalisers that are preserved by functors $J \times (-)$, we obtain: if $c_R: I \to I/R$ is the quotient of R—i.e. the coequaliser of $r_0, r_1: R \rightrightarrows I$—then $J \times c_R$ is the coequaliser of $J \times r_0, J \times r_1: J \times R \rightrightarrows J \times I$. Thus we get $(J \times I)/J^*(R) \xrightarrow{\cong} J \times (I/R)$.

Conversely, if the quotients in the subobject fibration satisfy Frobenius, then coequalisers in $\mathbb{B}$ are preserved by functors $J \times (-)$. This is because the above factorisation of the tuple $\langle u, v \rangle$ in (ii) yields a factorisation of $\langle J \times u, J \times v \rangle$,

$$J \times K \xrightarrow{\quad J \times e \quad} J \times R \, \rightarrowtail \xrightarrow{\quad \langle J \times r_0, J \times r_1 \rangle \quad} (J \times I) \times (J \times I)$$

In this diagram $J \times e$ is still a cover because covers are stable under pullback, and the relation $J \times R$ is $J^*(R)$. This shows that as coequaliser of $J \times u$ and $J \times v$ one can take the quotient $J \times c_R \colon J \times I \twoheadrightarrow J \times (I/R)$ of $J^*(R)$. Thus $J \times (-)$ preserves coequalisers. $\qquad\qquad\square$

In the next series of examples it will be shown that family fibrations (for a poset) always have quotients, that the classifying fibration of a predicate logic with quotient types has quotients in the categorical sense, and that a fibration of "admissible" subsets of complete lattices also has quotients. The latter order theoretic construction follows [174, Chapter I].

4.8.7. Examples. (i) Recall from Example 3.4.4 (iii) that if X is a poset with bottom $\bot$ and top $\top$ elements, then the family fibration $\begin{smallmatrix} \mathrm{Fam}(X) \\ \downarrow \\ \mathbf{Sets} \end{smallmatrix}$ has equality. For a family $x = (x_{(i,j)})_{(i,j)\in I \times J}$ over $I \times J$, it is given by

$$\mathrm{Eq}(x)_{(i,j,j')} = \begin{cases} x_{(i,j)} & \text{if } j = j' \\ \bot & \text{otherwise.} \end{cases}$$

Then for a function $u \colon I \to J$, the kernel relation $\mathrm{Ker}(u)$ on I is given by

$$\mathrm{Ker}(u)_{(i,i')} = \begin{cases} \top & \text{if } u(i) = u(i') \\ \bot & \text{otherwise.} \end{cases}$$

For a relation $r = (r_{(i,i')})_{i,i'\in I}$ on I in the family fibration on X, consider the set theoretic relation $R = \{(i,i') \mid r_{(i,i')} \neq \bot\} \subseteq I \times I$. Let $\overline{R} \subseteq I \times I$ be the least equivalence relation containing R. Then we get a quotient $I/r \stackrel{\mathrm{def}}{=} I/\overline{R}$ in **Sets**, which serves as quotient in the fibred sense. It comes with canonical map $c_r = [-] \colon I \to I/\overline{R}$. The adjunction boils down to $r \leq \mathrm{Ker}(u) \Leftrightarrow c_r$ factors through u, as in Lemma 4.8.2 (ii).

(ii) We can form a classifying fibration of a (simple) predicate logic with quotient types using the fibration $\begin{smallmatrix} \mathcal{L}(\Sigma,\Pi,\mathcal{A}) \\ \downarrow \\ \mathcal{C}\ell(\Sigma) \end{smallmatrix}$ associated with the logic on the signature with predicates (Σ, Π) (plus axioms $\mathcal{A}$) as described in Section 3.1. We can then form the category $\mathrm{Rel}(\mathcal{L}(\Sigma, \Pi, \mathcal{A}))$ of relations in this logic via the change of base situation preceding Definition 4.8.1. This category has relations $(x, x' \colon \sigma \vdash R(x, x') \colon \mathsf{Prop})$ as objects. And a morphism $(x, x' \colon \sigma \vdash$

$R(x, x')$: Prop) $\rightarrow$ $(y, y' : \tau \vdash S(y, y')$: Prop) in $\mathrm{Rel}(\mathcal{L}(\Sigma, \Pi, \mathcal{A}))$ is a morphism $M : \sigma \rightarrow \tau$ in the base category $\mathcal{Cl}(\Sigma)$ for which one can derive

$$x, x' : \sigma \mid R(x, x') \vdash S(M(x), M(x')).$$

The equality relation functor Eq: $\mathcal{Cl}(\Sigma) \rightarrow \mathrm{Rel}(\mathcal{L}(\Sigma, \Pi, \mathcal{A}))$ is then given by the assignment $\tau \mapsto (y, y' : \tau \vdash y =_\tau y'$: Prop).

Quotient types as described in the previous section determine a left adjoint to this functor Eq. It maps a relation $(x, x' : \sigma \vdash R(x, x')$: Type) to the quotient object σ/R in the base category $\mathcal{Cl}(\Sigma)$. The adjunction involves a bijective correspondence between (equivalence classes of) terms M and N in:

$$(x, x' : \sigma \vdash R(x, x') : \mathsf{Prop}) \xrightarrow{\quad M \quad} (y, y' : \tau \vdash y =_\tau y' : \mathsf{Prop})$$
$$\overline{\qquad\qquad\qquad\qquad \sigma/R \xrightarrow[\ N\ ]{} \tau \qquad\qquad\qquad\qquad}$$

That is, between terms M and N in:

$$x : \sigma \vdash M : \tau \quad \text{with} \ \left(x, x' : \sigma \mid R(x, x') \vdash M(x) =_\tau M(x') \right.$$
$$\overline{\qquad\qquad\qquad a : \sigma/R \vdash N : \tau \qquad\qquad\qquad}$$

This correspondence is precisely given by

$$M(x) \mapsto \mathsf{pick} \ x \ \mathsf{from} \ a \ \mathsf{in} \ M(x) \qquad \text{and} \qquad N(a) \mapsto N[[x]_R/a].$$

The (β)- and (η)-conversions precisely state that these operations are each others inverses. And Exercise 4.7.6 tells that the Frobenius property automatically holds. Thus the quotient types in the logic induce quotients for the fibration associated with the logic.

(iii) Let **CL** be the category of complete lattices (posets with joins of all subsets) and with functions preserving all these joins between them. It is well-known that requiring the existence of joins of all subsets is equivalent to requiring the existence of meets of all supsets. A morphism $f : X \rightarrow Y$ in **CL** always has a right adjoint $f_* : Y \rightarrow X$ (between poset categories), given as $f_*(y) = \bigvee \{x \in X \mid f(x) \leq y\}$. It is easy to see that $(g \circ f)_* = f_* \circ g_*$ and that $(f_*)_* = f$. (Using these right adjoints one can show that **CL** is a self-dual category.) The category **CL** has finite products in the obvious manner: one uses finite products of the underlying sets, with componentwise ordering.

Let us a call a subset $A \subseteq X$ of a complete lattice X **admissible** if A is closed under (all) joins in X. Such subsets can be organised in a fibration
$$\begin{array}{c} \mathrm{ASub}(\mathbf{CL}) \\ \downarrow \\ \mathbf{CL} \end{array}$$
in which the total category $\mathrm{ASub}(\mathbf{CL})$ has such admissible subsets $(A \subseteq X)$ as objects. A morphism $(A \subseteq X) \rightarrow (B \subseteq Y)$ in $\mathrm{ASub}(\mathbf{CL})$ is then a morphism $f : X \rightarrow Y$ in **CL** between the underlying carrier sets which satisfies:

$x \in A \Rightarrow f(x) \in B$, for all $x \in X$. This fibration has a terminal object functor $\top \colon \mathbf{CL} \to \mathrm{ASub}(\mathbf{CL})$ sending a complete lattice X to the admissible subset $(X \subseteq X)$. It has (full) subsets, via a functor $\{-\} \colon \mathrm{ASub}(\mathbf{CL}) \to \mathbf{CL}$ which maps an admissible subset $(A \subseteq X)$ to A, considered as a complete lattice in itself. Our aim is to show that this fibration also has quotients.

We therefore first consider the fibration $\begin{array}{c} \mathrm{ARel}(\mathbf{CL}) \\ \downarrow \\ \mathbf{CL} \end{array}$ of admissible relations, obtained by change-of-base along $X \mapsto X \times X$ (as described in the beginning of this section). There is an equality functor $\mathrm{Eq} \colon \mathbf{CL} \to \mathrm{ARel}(\mathbf{CL})$, mapping a complete lattice Y to the admissible subset $(\{(y,y) \mid y \in Y\} \subseteq Y \times Y)$. So far, these constructions are all straightforward. The quotient adjoint (to equality) is less standard. It maps an admissible subset $(R \subseteq X \times X)$ to the complete lattice

$$X/R = \{x \in X \mid \forall (y, y') \in R.\, y \le x \text{ iff } y' \le x\}.$$

It is easy to see that X/R, with order as on X, is closed under all meets. Therefore it is a complete lattice. The inclusion function $i \colon X/R \to X$ has a left adjoint $c_R \colon X \to X/R$, given by

$$c_R(x) = \bigwedge \{z \in X/R \mid x \le z\}.$$

Since c_R is a left adjoint, it preserve joins and is a morphism $X \to X/R$ in $\mathbf{CL}$ (with $(c_R)_* = i$). Obviously, $R(x, x') \Rightarrow c_R(x) = c_R(x')$, so that c_R is a map of relations $R \to \mathrm{Eq}(X/R)$. The quotient adjunction requires a bijective correspondence between morphisms f and g in:

$$\frac{(R \subseteq X \times X) \xrightarrow{\ f\ } \mathrm{Eq}(Y) \quad \text{in } \mathrm{ARel}(\mathbf{CL})}{X/R \xrightarrow{\ g\ } Y \quad \text{in } \mathbf{CL}}$$

For $f \colon (R \subseteq X \times X) \to \mathrm{Eq}(Y)$ one takes $\overline{f} = f \circ i \colon X/R \to Y$. And conversely, given $g \colon X/R \to Y$, one takes $\overline{g} = g \circ c_R \colon (R \subseteq X \times X) \to \mathrm{Eq}(Y)$. Then it is easy to see that $\overline{\overline{g}} = g \circ c_R \circ i = g$, because $c_R \circ i = \mathrm{id}$. But showing that $\overline{\overline{f}} = f \circ i \circ c_R = f$ is harder. First we notice that $f_*(y)$ is in X/R, for $y \in Y$. Indeed, for a pair $(x, x') \in R$ we have $f(x) = f(x')$, and thus $x \le f_*(y) \Leftrightarrow f(x) \le y \Leftrightarrow f(x') \le y \Leftrightarrow x' \le f_*(y)$. But this means that $i(c_R(f_*(y))) = f_*(y)$. Then we are done, since

$$f = (f_*)_* = (i \circ c_R \circ f_*)_* = (f_*)_* \circ (c_R)_* \circ i_* = f \circ i \circ c_R.$$

This completes the example.

In Example 4.6.5 we have seen how subset types give rise to a certain factorisation of maps in the base category. One also gets a factorisation from quotients, as we will show next. In higher order logic this factorisation has a slightly different universal property, see Example 5.1.9 (i) and Exercise 5.1.6.

4.8.8. Example. Assume $\begin{smallmatrix} \mathbb{E} \\ \downarrow p \\ \mathbb{B} \end{smallmatrix}$ is a fibration with quotients. For a morphism $u: I \to J$ in the base category $\mathbb{B}$ we can form the kernel relation $\text{Ker}(u) = (u \times u)^*(\text{Eq}(J))$ on I, and its quotient $I/\text{Ker}(u) \in \mathbb{B}$. It gives a factorisation,

$$\left(I \xrightarrow{\quad u \quad} J \right) = \left(I \xrightarrow{\quad c_{\text{Ker}(u)} \quad} I/\text{Ker}(u) \ -\ \overset{u'}{-}\ \to\ J \right)$$

where the map $u': c_{\text{Ker}(u)} \dashrightarrow u$ in the opslice category $I\backslash\mathbb{B}$ comes by Lemma 4.8.2 (ii) from the inequality $\text{Ker}(u) \leq \text{Ker}(u)$.

This factorisation is universal in the following sense. Given an arbitrary relation $R \in \mathbb{E}_{I \times I}$ on I and a morphism $v: I/R \to J$ in $\mathbb{B}$ with $v \circ c_R = u$, there is a unique map of relations $f: R \dashrightarrow \text{Ker}(u)$ over I such that

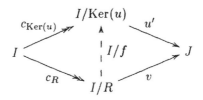

This mediating map arises as follows. The morphism $v: c_R \to u$ in the opslice $I\backslash\mathbb{B}$ gives rise to an inequality $f: R \leq \text{Ker}(u)$ over I. By applying the quotient functor we get a morphism $I/f: I/R \to I/\text{Ker}(u)$ in $\mathbb{B}$ which commutes with the quotient maps. Finally, $u' \circ I/f = v$ holds because c_R is an epi.

Exercises

4.8.1. Assume a fibration with quotients. Prove that for a relation R on I, the canonical map $c_R: I \twoheadrightarrow I/R$ is an isomorphism if and only if $R \leq \text{Eq}(I)$ over I. (A special case is $I \xrightarrow{\cong} I/\text{Eq}(I)$, see Exercise 4.7.2.)

4.8.2. Prove that the 'epic part' $c_{\text{Ker}(u)}: I \twoheadrightarrow I/\text{Ker}(u)$ of $u: I \to J$ in Example 4.8.8 is an isomorphism if and only if u is internally injective.

4.8.3. Check that quotients in a predicate logic are effective if and only if the quotients in the associated fibration—as in Example 4.8.7 (ii)—are effective in the categorical sense. Describe fullness of the canonical map functor in Definition 4.8.5 type theoretically.

        [*Hint.* Remember Exercise 3.1.1.]

4.8.4. Show that a relation $R \rightarrowtail I \times I$ is the kernel pair $R \rightrightarrows I$ of its own coequaliser if and only if it is the kernel of some map $I \to J$.

4.8.5. Let $\begin{smallmatrix}\mathbb{E}\\\downarrow p\\\mathbb{B}\end{smallmatrix}$ be a regular fibration with quotients. For parallel maps $u, v \colon K \rightrightarrows$ I in $\mathbb{B}$, form the relation $R = \coprod_{\langle u,v\rangle}(\top) \in \mathbb{E}_{I\times I}$ and its quotient

$$K \underset{v}{\overset{u}{\rightrightarrows}} I \xrightarrow{\ c_R\ } I/R$$

Show that this forms a coequaliser diagram in the internal logic: one has $\top \leq \mathrm{Eq}(c_R \circ u, c_R \circ v)$, and if $w \colon I \to J$ satisfies $\top \leq \mathrm{Eq}(w \circ u, w \circ v)$, then there is a unique map $\overline{w} \colon I/R \dashrightarrow J$ with $\overline{w} \circ c_R = w$.

Notice that Proposition 4.8.6 (ii) is a special case of this construction—which is dual to the one for subsets in Exercise 4.6.6. And also that co-equalisers in **Sets** are obtained in this way.

4.8.6. Let $\mathrm{RRel}(\mathbb{E}) \hookrightarrow \mathrm{Rel}(\mathbb{E})$ be the full subcategory of reflexive relations in an Eq-fibration $\begin{smallmatrix}\mathbb{E}\\\downarrow p\\\mathbb{B}\end{smallmatrix}$, where $R \in \mathbb{E}_{I\times I}$ is reflexive if and only if $\top \leq \delta(I)^*(R)$, if and only if $\mathrm{Eq}(I) \leq R$.

(i) Show that the composite $\mathrm{RRel}(\mathbb{E}) \hookrightarrow \mathrm{Rel}(\mathbb{E}) \to \mathbb{B}$ is a fibration.

(ii) Prove that $I \mapsto \mathrm{Eq}(J)$ yields a functor $\mathbb{B} \to \mathrm{RRel}(\mathbb{E})$ which is left adjoint to the fibration $\mathrm{RRel}(\mathbb{E}) \to \mathbb{B}$.

[Thus, equality on I is the **least reflexive relation** on I.]

(iii) Check that $S \mapsto S \vee \mathrm{Eq}(I)$ for $S \in \mathbb{E}_{J\times J}$ yields a fibred left adjoint to the inclusion $\mathrm{RRel}(\mathbb{E}) \hookrightarrow \mathrm{Rel}(\mathbb{E})$.

4.8.7. Notice that the restriction to preorder fibrations in Definition 4.8.1 is unnecessary, and that the definition of quotients applies to arbitrary fibrations with equality. In particular it applies to codomain fibrations. Prove that a category $\mathbb{B}$ with finite limits has coequalisers if and only if its codomain fibration $\begin{smallmatrix}\mathbb{B}^{\to}\\\downarrow\\\mathbb{B}\end{smallmatrix}$ has quotients.

[Note that the category $\mathrm{Rel}(\mathbb{B}^{\to})$ in this situation is the category $\mathbb{B}^{\rightrightarrows}$ of parallel arrows in $\mathbb{B}$.]

4.8.8. Let $\begin{smallmatrix}\mathbb{E}\\\downarrow p\\\mathbb{B}\end{smallmatrix}$ be a fibration with equality. Describe the kernel operation $\mathrm{Ker}(-)$ as a functor in a commuting diagram,

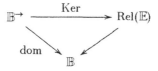

And prove that p has quotients if and only if this functor Ker has a left adjoint, with vertical unit and counit. This could be used as a definition of quotients, dual to Lawvere's definition of subset types as described in Exercise 4.6.7.

4.9 A logical characterisation of subobject fibrations

In this chapter on (first order, simple) predicate logic we have seen various fibrations capturing various systems of predicate logic. Among these fibrations, subobject fibrations have received special attention in Sections 4.4 and 4.5. They will play an important rôle in later chapters, notably in topos theory. In this final section we ask ourselves: when is a fibration (equivalent to) a subobject fibration? Such a fibration should certainly have the logical operations that come for free in subobject fibrations, namely full subtypes and so-called very strong equality. Recall that this means that internal and external equality coincide, see Notation 3.4.2. There is a third logical operation that is available in subobject fibrations, namely unique choice $\exists!$. And the combination of these three: full subset types, very strong equality and unique choice, characterise subobject fibrations, as will be shown in the present section.

We start with unique choice.

4.9.1. Definition. Let $\begin{smallmatrix} \mathbb{E} \\ \downarrow p \\ \mathbb{B} \end{smallmatrix}$ be an Eq-fibration with subset types.

(i) A relation $R \in \mathbb{E}_{I \times J}$ is called **single-valued** if it satisfies

$$i: I, j, j': J \mid R(i,j) \wedge R(i,j') \vdash j =_J j'.$$

Or, more categorically, if above $I \times (J \times J)$ there is an inequality

$$(\mathrm{id} \times \pi)^*(R) \wedge (\mathrm{id} \times \pi')^*(R) \leq \pi'^*(\mathrm{Eq}(J)).$$

(ii) The fibration p has **unique choice** $\exists!$ if for each single-valued relation $R \in \mathbb{E}_{I \times J}$, the coproduct $\coprod_{(I,J)}(R) \in \mathbb{E}_I$ exists, and the canonical map $\dashrightarrow$ in the following diagram

$$
\begin{array}{ccc}
\{R\} & \dashrightarrow & \{\coprod_{(I,J)}(R)\} \\
\pi_R \downarrow & & \downarrow \pi_{\coprod_{(I,J)}(R)} \\
I \times J & \xrightarrow{\ \ \pi\ \ } & I
\end{array}
$$

is an isomorphism.

The canonical map $\pi \circ \pi_R \to \pi_{\coprod_{(I,J)}(R)}$ in the slice category $\mathbb{B}/I$ comes by Lemma 4.6.2 (ii) from applying the reindexing functor π_R^* to the unit map $\eta: R \to \pi^* \coprod_{(I,J)}(R)$. This yields $\top \leq (\pi \circ \pi_R)^* \coprod_{(I,J)}(R)$.

The idea behind this definition is that if for $i \in I$ there is a unique $j \in J$ with $R(i,j)$, then the canonical (projection) map

$$\{(i,j) \mid R(i,j)\} \xrightarrow{\hspace{3cm}} \{i \mid \exists j.\, R(i,j)\}$$

is an isomorphism.

4.9.2. Proposition. *Subobject fibrations have unique choice.*

Proof. Let $\mathbb{B}$ be a category with finite limits, and let $\langle r_0, r_1 \rangle : R \rightarrowtail I \times I$ be a relation on I which is single-valued. The latter means $(\text{id} \times \pi)^*(R) \wedge (\text{id} \times \pi')^*(R)$ factors through $\pi'^*(\delta(J)) = \text{id} \times \delta$, in a situation,

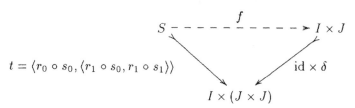

$$t = \langle r_0 \circ s_0, \langle r_1 \circ s_0, r_1 \circ s_1 \rangle \rangle$$

where S is obtained in the pullback diagram,

$$
\begin{array}{ccc}
S & \xrightarrow{\ s_1\ } & R \\
{\scriptstyle s_0}\downarrow & \lrcorner & \downarrow{\scriptstyle r_0} \\
R & \xrightarrow[\ r_0\ ]{} & I
\end{array}
$$

We will show that the map $r_0 : R \to I$ is a mono in $\mathbb{B}$. Assume therefore parallel maps $u, v : K \rightrightarrows R$ with $r_0 \circ u = r_0 \circ v$. There is then a unique map $w : K \to S$ with $s_0 \circ w = u$ and $s_1 \circ w = v$. But then $r_1 \circ u = r_1 \circ v$, as witnessed by the following computation.

$$
\begin{aligned}
r_1 \circ u &= r_1 \circ s_0 \circ w \\
&= \pi \circ \pi' \circ t \circ w \\
&= \pi \circ \pi' \circ \text{id} \times \delta \circ f \circ w \\
&= \pi' \circ \pi' \circ \text{id} \times \delta \circ f \circ w \\
&= \pi' \circ \pi' \circ t \circ w \\
&= r_1 \circ s_1 \circ w \\
&= r_1 \circ v.
\end{aligned}
$$

Now we have an equation $\langle r_0, r_1 \rangle \circ u = \langle r_0, r_1 \rangle \circ v$, so that we may conclude $u = v$.

We can thus take the coproduct in the usual way by composition: $\coprod_{(I,J)}(R) = \pi \circ \langle r_0, r_1 \rangle = r_0 : R \rightarrowtail I$, so that the unique map $\dashrightarrow$ in Definition 4.9.1 (ii) is the identity. $\qquad \square$

In the formulation of unique choice we have made use of subset types. In a similar manner we can express very strong equality in a fibration via subset types. Recall that equality is very strong if external equality $u = v: I \to J$ and internal equality $\top \leq \mathrm{Eq}(u, v)$ coincide—for two parallel maps in the base category.

4.9.3. Proposition. *Let* $\begin{smallmatrix}\mathbb{E}\\\downarrow P\\\mathbb{B}\end{smallmatrix}$ *be an Eq-fibration with subset types. This fibration has very strong equality if and only if for each object* $I \in \mathbb{B}$ *the canonical morphism* κ *in the triangle*

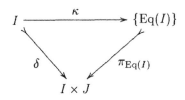

is an isomorphism.

(This morphism κ *is obtained by Lemma 4.6.2 (ii) from the unit map* $\top \leq \delta(I)^* \mathrm{Eq}(I) = \delta(I)^* \mathrm{Eq}_I(\top)$.*)*

This result may be read as: equality is very strong if and only if diagonals occur as (subset) projections.

Proof. Assume that the above map $\kappa: I \to \{\mathrm{Eq}(I)\}$ is an isomorphism in $\mathbb{B}$. Then for parallel morphisms $u, v: K \rightrightarrows I$ there are equivalences:

> u, v are internally equal
>
> $\Leftrightarrow \top \leq \mathrm{Eq}(u, v) = \langle u, v \rangle^* \mathrm{Eq}(I)$
>
> $\Leftrightarrow \langle u, v \rangle$ factors through $\pi_{\mathrm{Eq}(I)}$ by Lemma 4.6.2 (ii)
>
> $\Leftrightarrow \langle u, v \rangle$ factors through $\delta(I)$ by the isomorphism κ
>
> $\Leftrightarrow u = v$
>
> $\Leftrightarrow u, v$ are externally equal.

Conversely, assume that internal and external equality coincide. As candidate for the required inverse for κ we have $\pi \circ \pi_{\mathrm{Eq}(I)}: \{\mathrm{Eq}(I)\} \to I$, since

$$(\pi \circ \pi_{\mathrm{Eq}(I)}) \circ \kappa = \pi \circ \delta(I) = \mathrm{id}.$$

Further, we have above $\{\mathrm{Eq}(I)\}$,

$$\begin{aligned} \top &\leq \pi^*_{\mathrm{Eq}(I)} \mathrm{Eq}(I) && \text{by Lemma 4.6.2 (ii)}\\ &\cong \pi^*_{\mathrm{Eq}(I)} \mathrm{Eq}(\pi, \pi') && \text{see Exercise 3.4.5}\\ &\cong \mathrm{Eq}(\pi \circ \pi_{\mathrm{Eq}(I)}, \pi' \circ \pi_{\mathrm{Eq}(I)}) && \text{see Notation 3.4.2.} \end{aligned}$$

This tells us that the maps $\pi \circ \pi_{\mathrm{Eq}(I)}$ and $\pi' \circ \pi_{\mathrm{Eq}(I)}$ are internally equal.

Hence they are also externally equal, by assumption. But then,

$$\pi_{\mathrm{Eq}(I)} \circ \kappa \circ (\pi \circ \pi_{\mathrm{Eq}(I)}) = \delta(I) \circ \pi \circ \pi_{\mathrm{Eq}(I)}$$
$$= \langle \pi \circ \pi_{\mathrm{Eq}(I)}, \pi \circ \pi_{\mathrm{Eq}(I)} \rangle$$
$$= \langle \pi \circ \pi_{\mathrm{Eq}(I)}, \pi' \circ \pi_{\mathrm{Eq}(I)} \rangle$$
$$= \pi_{\mathrm{Eq}(I)},$$

so that we can conclude $\kappa \circ (\pi \circ \pi_{\mathrm{Eq}(I)}) = \mathrm{id}$, since the subset projection $\pi_{\mathrm{Eq}(I)}$ is a mono. Hence κ is an isomorphism. $\square$

Given this characterisation, it is immediate that subobject fibrations have very strong equality, because their equality predicate $\mathrm{Eq}(I)$ is simply the diagonal on I.

We now come to the main result in this section.

4.9.4. Theorem. *Let* $\begin{smallmatrix} \mathbb{E} \\ \downarrow p \\ \mathbb{B} \end{smallmatrix}$ *be an Eq-fibration. This fibration is (equivalent to) the subobject fibration on its base category* $\mathbb{B}$ *if and only if*

- *equality in p is very strong;*
- *p has full subset types;*
- *p has unique choice.*

Proof. It may be clear that a subobject fibration satisfies the above three properties: it has very strong equality as we just noted, it has full subset types by Example 4.6.3 (i), and unique choice by Proposition 4.9.3.

Conversely, let $\begin{smallmatrix} \mathbb{E} \\ \downarrow p \\ \mathbb{B} \end{smallmatrix}$ be an Eq-fibration satisfying the above three properties. We first note that by Exercise 4.6.6—using that equality is very strong—the base category $\mathbb{B}$ has finite limits, so that it makes sense to talk about the subobject fibration on $\mathbb{B}$. Full subset types give us a full and faithful fibred functor

$$\mathbb{E} \xrightarrow{\ \pi_{(-)}\ } \mathrm{Sub}(\mathbb{B})$$

with p and the projection forming a commuting triangle over $\mathbb{B}$.

We show that it is a fibred equivalence. We can define in the reverse direction a functor $\mathcal{S} \colon \mathrm{Sub}(\mathbb{B}) \to \mathbb{E}$ by

$$\left(J \xrightarrowtail{\ m\ } I \right) \xmapsto{\ \mathcal{S}\ } \coprod_{(I,J)}(G_m) \in \mathbb{E}_I,$$

where G_m is the 'graph relation' of m:

$$G_m = \mathrm{Eq}(\pi, m \circ \pi') = \langle \pi, m \circ \pi' \rangle^* \mathrm{Eq}(I) \in \mathbb{E}_{I \times J}.$$

This relation is single-valued because m is a mono and equality is very strong:

$$G_m(i,j) \wedge G_m(i,j') \;\Rightarrow\; i = m(j) \wedge i = m(j')$$
$$\Rightarrow\; j = j'.$$

Hence the coproduct $\mathcal{S}(m) = \coprod_{(I,J)}(G_m) \in \mathbb{E}_I$ exists by unique choice. Its subset projection is the original mono m, since there are isomorphisms in $\mathbb{B}/I$,

$$
\begin{aligned}
\pi_{\mathcal{S}(m)} \;&\cong\; \pi \circ \pi_{G_m} && \text{by definition of unique choice} \\
&\cong\; \pi \circ \langle \pi, m \circ \pi' \rangle^*(\pi_{\mathrm{Eq}(I)}) && \text{because } \pi_{(-)} \text{ is a fibred functor} \\
&\cong\; \pi \circ \langle \pi, m \circ \pi' \rangle^*(\delta(I)) && \text{since equality is very strong} \\
&\cong\; \pi \circ \langle m, \mathrm{id} \rangle && \text{because of the pullback square,}
\end{aligned}
$$

$$
\begin{array}{ccc}
J & \xrightarrow{\quad m \quad} & I \\
{\scriptstyle \langle m, \mathrm{id} \rangle} \downarrow \quad \lrcorner & & \downarrow {\scriptstyle \delta(I)} \\
I \times J & \xrightarrow[\langle \pi, m \circ \pi' \rangle]{} & I \times I
\end{array}
$$

$$= m.$$

We thus get $\pi_{(-)} \circ \mathcal{S} \cong \mathrm{id}$. But then also $\mathcal{S} \circ \pi_{(-)} \cong \mathrm{id}$, since $\pi_{(-)}$ is a full and faithful functor. $\square$

In similar fashion we can characterise regular subobject fibrations.

4.9.5. Theorem. *An Eq-fibration $\begin{smallmatrix} \mathbb{E} \\ \downarrow p \\ \mathbb{B} \end{smallmatrix}$ is (equivalent to) the regular subobject fibration on its base category $\mathbb{B}$ if and only if*

- *equality in p is very strong;*
- *p has full subset types;*
- *every predicate is an equation: for every $X \in \mathbb{E}_I$ there are maps $u, v \colon I \rightrightarrows J$ in $\mathbb{B}$ with $X \cong \mathrm{Eq}(u,v)$.*

Proof. We concentrate on the (if)-part of the statement. As in Exercise 4.6.6, the base category $\mathbb{B}$ has finite limits. And from the way equalisers are constructed in $\mathbb{B}$, we conclude that each projection $\pi_X \colon \{X\} \rightarrowtail I$ (for $X \in \mathbb{E}_I$) is a regular mono, using that X is an equation. We construct a functor $\mathcal{R} \colon \mathrm{RegSub}(\mathbb{B}) \to \mathbb{E}$ as follows. Let $m \colon K \rightarrowtail I$ be equaliser of $u, v \colon I \rightrightarrows J$. Put then $\mathcal{R}(m) = \mathrm{Eq}(u,v) \in \mathbb{E}_I$. Then $\pi_{\mathcal{R}(m)} \cong m$. But also $\mathcal{R}(\pi_X) \cong X$, because X is an equation. $\square$

Exercises

4.9.1.   Let $\begin{smallmatrix} \mathbb{E} \\ \downarrow p \\ \mathbb{B} \end{smallmatrix}$ be an Eq-fibration with subset types. Say that one has **unique choice on** $J \in \mathbb{B}$ if for every single-valued relation $R \in \mathbb{E}_{I \times J}$ from I to

J, one has unique choice as in Definition 4.9.1 (ii). And say that **equality on J is very strong** if one has a canonical isomorphism $\delta(J) \overset{\cong}{\Rightarrow} \pi_{\mathrm{Eq}(J)}$ like in Proposition 4.9.3. Prove that unique choice on J implies very strong equality on J.

4.9.2.   Let $\begin{smallmatrix} \mathbb{E} \\ \downarrow p \\ \mathbb{B} \end{smallmatrix}$ be an Eq-fibration with very strong equality and full subset types.

  (i) Express the induced pullbacks in $\mathbb{B}$ in the internal language of the fibration, see Exercise 4.6.6.

  (ii) Assume now that p is also regular, *i.e.* additionally has simple coproducts $\coprod_{(I,J)}$. Prove that the induced coproduct functors $\coprod_u$ from Example 4.3.7 (i) satisfy the Beck-Chevalley condition.

     [*Hint.* The usual set theoretic argument may be carried out internally.]

  (iii) Prove also that if p is a first order fibration then the induced products $\prod_u$ also satisfy the Beck-Chevalley condition.

Chapter 5

Higher order predicate logic

Moving from equational logic to first order predicate logic leads to a clear increase of expressive power. But certain concepts cannot be expressed in first order predicate logic because they require "higher order" quantification over subsets (or predicates). A typical example in algebra is the concept of a Noetherian ring: it is a ring R in which every ideal $I \subseteq R$ has a finite basis (*i.e.* is finitely generated). This cannot be expressed in first order predicate logic, because it requires higher order quantification. By the latter we mean quantification over propositions (inhabitants of Prop) and over predicates (inhabitants of $\sigma \to$ Prop, where σ is a type). In contrast, in first order predicate logic one only quantifies over inhabitants of types. So the easiest way to introduce higher order quantification is to make Prop a type, *i.e.* to introduce a 'higher order' axiom ⊢ Prop: Type. This approach will be followed. Propositions $x_1 : \sigma_1, \ldots, x_n : \sigma_n \vdash \varphi$: Prop are then terms of type Prop: Type. Quantification $\exists, \forall$ over types can take the particular form $\forall \alpha$: Prop. φ and $\exists \alpha$: Prop. φ of quantification over propositions, since Prop is a type. This forms the essential aspect of higher order logic.

The resulting formal system will be referred to as higher order simple predicate logic, or higher order logic for short. The qualification 'simple' refers to the fact that the underlying type theory is simple (like in the previous chapter), and not polymorphic or dependent. Tool support for higher order logic is provided by the HOL system [104] (and also by a special configuration of the ISABELLE system [250]). The PVS system [242, 241] is a tool for *dependent* higher order predicate logic, see Section 11.1. These tools are used for machine-assisted verifications in higher order logic.

311

This chapter contains the syntax of higher order logic in its first section. The second section is on generic objects. These are the categorical counterparts of the earlier mentioned distinguished type Prop, which relates predicates $x\colon \sigma \vdash \varphi\colon$ Prop on a type σ and "classifying" terms $\sigma \to$ Prop. For split fibrations this correspondence can be described in a straightforward manner, but for arbitrary fibrations there are some complications to be investigated. This will involve a version of the Yoneda Lemma which is suitable for fibred categories. The third section gives the appropriate fibred structures to capture higher order logic. Examples include realisability triposes, which generalise the realisability fibration from the previous chapter, and the regular subobject fibration over ω-sets (but not over PERs). In the same section we first encounter the notion of a topos: it is a category for which its subobject fibration is such a 'higher order fibration'. This is a distinctly logical definition. The remainder of this chapter will be devoted to the (standard) theory of these toposes. In Section 5.4 we present the ordinary 'elementary' definition of a topos, and show that it is equivalent to the 'logical' one. Further, we describe nuclei (or Lawvere-Tierney topologies) in toposes. Such a nucleus **j** gives rise to an associated higher order fibration of **j**-closed subobjects. Also, for a nucleus one can define separated objects and sheaves in a topos. Especially the double negation nucleus $\neg\neg$ is of logical importance. Its categories of separated objects and of sheaves come with classical logic (via their regular and ordinary subobjects).

The expositions on toposes form a preparation for the special example of the 'effective topos' **Eff** in the next chapter.

5.1 Higher order signatures

We start our description of higher order predicate logic (over simple type theory) by identifying an appropriate notion of signature for such logic—like we did for equational logic and for first order predicate logic. For higher order predicate logic, signatures are actually simpler than for first order predicate logic: higher order signatures will contain a distinguished type Prop, making it no longer necessary to describe function symbols and predicate symbols separately: predicate symbols can be identified with function symbols with codomain Prop.

In a logical setting, we shall always write Prop for this distinguished type and we view inhabitants of Prop as propositions. A variable of type Prop is therefore a proposition variable, for which we shall use letters $\alpha, \beta, \gamma, \ldots$ from the beginning of the Greek alphabet. These proposition variables may occur in propositions—because, in general, variables inhabiting types may occur in propositions.

A higher order signature Σ consists first of all of an underlying set $T = |\Sigma|$ of atomic types containing a special type **Prop**. Thus, $|\Sigma|$ can be understood as a pointed set. Further, Σ contains function symbols $F: \sigma_1, \ldots, \sigma_n \longrightarrow \sigma_{n+1}$ as in ordinary signatures. A morphism $\phi: \Sigma \to \Sigma'$ of higher order signatures consists of a function $\phi: |\Sigma| \to |\Sigma'|$ between the underlying sets of atomic types with $\phi(\mathsf{Prop}) = \mathsf{Prop}$ (making ϕ a morphism in the category **Sets**$_\bullet$ of pointed sets), and of a collection of functions (also written as ϕ) mapping function symbols $F: \sigma_1, \ldots, \sigma_n \longrightarrow \sigma_{n+1}$ in Σ to function symbols $\phi(F): \phi(\sigma_1), \ldots, \phi(\sigma_n), \longrightarrow \phi(\sigma_{n+1})$ in Σ'. There are no explicit predicate symbols in a higher order signature, like in a signature with predicates (see Definition 4.1.1). Instead, function symbols $F: \sigma_1, \ldots, \sigma_n \longrightarrow \mathsf{Prop}$ are understood as predicate symbols.

We recapitulate in concise fibred terminology.

5.1.1. Definition. The category **HoSign** of **higher order signatures** is defined in the following change-of-base situation,

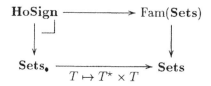

where **Sets**$_\bullet$ is the category of pointed sets (see Exercise 1.2.3 for the definition). Implicitly, in the base functor **Sets**$_\bullet \to$ **Sets** there is a coercion turning a pointed set into an ordinary set.

A higher order signature forms the basis for a logic, much like in the previous two chapters. What is new is that terms $\Gamma \vdash \varphi: \mathsf{Prop}$ will be taken as propositions. They can be built up inductively from atomic propositions $M =_\sigma M'$ and $P(M_1, \ldots, M_n)$, where P is a function (or predicate) symbol $\sigma_1, \ldots, \sigma_n \longrightarrow \mathsf{Prop}$. Thus, in higher order logic, propositions are not some external entities but live as terms inside the type theory. Notice that **Prop** is itself a (atomic) type. Explicitly, via an axiom:

$$\vdash \mathsf{Prop}: \mathsf{Type}$$

In what we call higher order logic (on top of a higher order signature Σ) we shall use finite product and exponent types (as in the calculus $\lambda 1_\times(\Sigma)$ in Section 2.3), and connectives and quantifiers as in first order predicate logic. Thus we have in particular constants $\top, \bot: \mathsf{Prop}$ for true and false.

Since Prop is now a type, we may quantify over it as in

$$\forall \alpha \colon \mathsf{Prop}.\, (\alpha \supset \alpha)$$

or in

$$\forall \alpha \colon \mathsf{Prop}.\, \exists P \colon \alpha \to \mathsf{Prop}.\, \exists x \colon \alpha.\, Px.$$

This gives typical propositions in a higher order logic.

For propositions $\varphi_1, \ldots, \varphi_n, \psi$ in context Γ we continue to use sequents $\Gamma \mid \varphi_1, \ldots, \varphi_n \vdash \psi$ as regulated by the rules for first order predicate logic in Figure 4.1. But there is a crucial difference: in higher order logic, propositions are special terms (with type Prop), which leads to the question of how logical equivalence $\supset\!\subset$ (for propositions) is related to (internal) equality $=_{\mathsf{Prop}}$ on the type Prop of propositions (for terms of type Prop). The following rule equates them (for predicates).

extensionality (of entailment)

$$\frac{\Gamma \vdash P, Q \colon \sigma \to \mathsf{Prop} \qquad \Gamma, x \colon \sigma \mid \Theta, Px \vdash Qx \qquad \Gamma, x \colon \sigma \mid \Theta, Qx \vdash Px}{\Gamma \mid \Theta \vdash P =_{\sigma \to \mathsf{Prop}} Q}$$

See also [91, Definition 2.2.9]. We shall not standardly assume this rule in higher order logic, and shall mention explicitly when it is used.

5.1.2. Example. An important consequence of this extensionality of entailment rule is that a proposition $\alpha \colon \mathsf{Prop}$ is derivable if and only if the equality $(\alpha =_{\mathsf{Prop}} \top)$ is derivable. In one direction this is easy via Lawvere's equality rule (see Lemma 3.2.3):

$$\frac{\emptyset \mid \emptyset \vdash \alpha[\top/\alpha]}{\alpha \colon \mathsf{Prop} \mid \alpha =_{\mathsf{Prop}} \top \vdash \alpha}$$

For the converse one uses the above extensionality of entailment rule with $\sigma = 1$, so that $1 \to \mathsf{Prop} \cong \mathsf{Prop}$. We then get

$$\frac{\alpha \colon \mathsf{Prop} \vdash \alpha, \top \colon \mathsf{Prop} \qquad \alpha \colon \mathsf{Prop} \mid \alpha, \alpha \vdash \top \qquad \alpha \colon \mathsf{Prop} \mid \alpha, \top \vdash \alpha}{\alpha \colon \mathsf{Prop} \mid \alpha \vdash \alpha =_{\mathsf{Prop}} \top}$$

Let us summarise.

5.1.3. Definition. Let Σ be a higher order signature. It gives rise to a **higher order logic** with

- for types: finite product 1, $\times$ and exponent types $\to$;
- for propositions: finite conjunctions $\top$, $\wedge$ and disjunctions $\bot$, $\vee$, equality $=_\sigma$, for $\sigma \colon \mathsf{Type}$, and existential and universal quantification $\exists x \colon \sigma.\, (-), \forall x \colon \sigma.\, (-)$

over types σ (including the important special case were σ is Prop), satisfying the rules in Figure 4.1.

We will say that this higher order logic on Σ has **extensional entailment** if it includes the above "extensionality of entailment" rule.

A **higher order specification** consists of a higher order signature Σ together with a collection $\mathcal{A}$ of sequents in the higher order logic associated with Σ; these are taken as axioms. Such a higher order specification determines a higher order theory, by closing $\mathcal{A}$ under derivability.

Notice that we do not standardly include subset types $\{x\!:\!\sigma \,|\, \varphi\}$ or quotient types σ/R in higher order logic. Neither the requirement that equality is very strong. All this may be added separately. Recall that we call a logic extensional if its equality is very strong (*i.e.* if its internal and external equality coincide); this is not related to the above extensionality of entailement rule.

5.1.4. Example. Higher order logic as formulated above contains considerable redundancy. For example, one can define

$$\bot \;=\; \forall \alpha\!:\!\mathsf{Prop}.\,\alpha$$
$$\top \;=\; \bot \supset \bot$$
$$\varphi \vee \psi \;=\; \forall \alpha\!:\!\mathsf{Prop}.\,(\varphi \supset \alpha) \supset ((\psi \supset \alpha) \supset \alpha)$$
$$\varphi \wedge \psi \;=\; \forall \alpha\!:\!\mathsf{Prop}.\,(\varphi \supset (\psi \supset \alpha)) \supset \alpha$$
$$\exists x\!:\!\sigma.\,\varphi \;=\; \forall \alpha\!:\!\mathsf{Prop}.\,(\forall x\!:\!\sigma.\,(\varphi \supset \alpha)) \supset \alpha.$$

And for terms $M, N\!:\!\sigma$,

$$(M =_\sigma N) \;=\; \forall P\!:\!\sigma \to \mathsf{Prop}.\,PM \supset PN$$

The latter definition yields what is commonly called **Leibniz equality**; it says that terms are equal if they have the same properties. Thus implication $\supset$ and universal quantification $\forall$ are the essential connectives.

5.1.5. Lemma. *The above definitions yield connectives that satisfy the rules in Figure 4.1.*

Proof. We shall do $\exists$ and $=_\sigma$ and leave the remaining connectives as an exercise below. The introduction rule for $\exists$ is obtained as follows.

$$
\cfrac{
\cfrac{
\Gamma \vdash M\!:\!\sigma \quad \Gamma, \alpha\!:\!\mathsf{Prop} \mid \Theta, \forall x\!:\!\sigma.\,(\varphi \supset \alpha) \vdash \forall x\!:\!\sigma.\,(\varphi \supset \alpha)
}{
\cfrac{
\Gamma, \alpha\!:\!\mathsf{Prop} \mid \Theta, \forall x\!:\!\sigma.\,(\varphi \supset \alpha) \vdash \varphi[M/x] \supset \alpha \qquad \Gamma \mid \Theta \vdash \varphi[M/x]
}{
\cfrac{
\Gamma, \alpha\!:\!\mathsf{Prop} \mid \Theta, \forall x\!:\!\sigma.\,(\varphi \supset \alpha) \vdash \alpha
}{
\Gamma, \alpha\!:\!\mathsf{Prop} \mid \Theta \vdash (\forall x\!:\!\sigma.\,(\varphi \supset \alpha)) \supset \alpha
}
}
}
}{
\Gamma \mid \Theta \vdash \forall \alpha\!:\!\mathsf{Prop}.\,(\forall x\!:\!\sigma.\,(\varphi \supset \alpha)) \supset \alpha \;=\; \exists x\!:\!\sigma.\,\varphi
}
$$

and the elimination rule (with x not in Θ, ψ) as

$$\frac{\Gamma \vdash \psi: \mathsf{Prop} \qquad \Gamma \mid \Theta \vdash \exists x: \sigma.\, \varphi \qquad \dfrac{\Gamma, x: \sigma \mid \Theta, \varphi \vdash \psi}{\dfrac{\Gamma, x: \sigma \mid \Theta \vdash \varphi \supset \psi}{\Gamma \mid \Theta \vdash \forall x: \sigma.\, (\varphi \supset \psi)}}}{\Gamma \mid \Theta \vdash (\forall x: \sigma.\, (\varphi \supset \psi)) \supset \psi \qquad\qquad}$$

$$\frac{}{\Gamma \mid \Theta \vdash \psi}$$

We turn to Leibniz equality. The reflexivity, transitivity and replacement rules for equality are easily established. For symmetry, assume $(M =_\sigma N) = \forall P: \sigma \to \mathsf{Prop}.\, PM \supset PN$. In order to get $(N =_\sigma M)$, assume $P: \sigma \to \mathsf{Prop}$ with PN. Take

$$P' = \lambda x: \sigma.\, Px \supset PM: \sigma \to \mathsf{Prop}.$$

Then, instantiating the assumption $M =_\sigma N$ with P' yields $P'M \supset P'N$. Since $P'M$ we get $P'N = (PN \supset PM)$, and PM follows, as required. $\qquad\square$

Power types

In higher order logic we can write

$$P\sigma \overset{\text{def}}{=} \sigma \to \mathsf{Prop}: \mathsf{Type}$$

for the σ-powerset type. We can think of terms in $P\sigma$ either as predicates on σ, or as subsets of σ. Such a type allows us to quantify over predicates, as in $\forall a: P\sigma.\, a =_{P\sigma} a$. It comes equipped with a typed membership relation $\in_\sigma$ described by

$$x: \sigma, a: P\sigma \vdash x \in_\sigma a \overset{\text{def}}{=} a \cdot x: \mathsf{Prop}.$$

There is then the familiar (typed) inclusion relation $\subseteq_\sigma$ on $P\sigma$, as:

$$a: P\sigma, b: P\sigma \vdash a \subseteq_\sigma b \overset{\text{def}}{=} \forall x: \sigma.\, (x \in_\sigma a) \supset (x \in_\sigma b): \mathsf{Prop}.$$

For a proposition $x: \sigma, y: \tau \vdash \varphi(x, y): \mathsf{Prop}$ it makes sense to write

$$x: \sigma \vdash \{y \in \tau \mid \varphi(x, y)\} \overset{\text{def}}{=} \lambda y: \tau.\, \varphi(x, y): P\tau.$$

So that we get a subset *term*. Note that this is different from subset *types* as described in Section 4.6, since there, $\{y: \tau \mid \varphi(x, y)\}$ was a *type*. Notice the difference in notation between the term $\{x \in \sigma \mid \psi\}$ and the type $\{x: \sigma \mid \psi\}$. By construction, the terms $\varphi(x, y)$ and $z \in_\tau \{y \in \tau \mid \varphi(x, y)\}$ are (β)-convertible, so that we may replace one by the other. In particular, they are logically equivalent.

We mention some elementary results concerning these constructs.

5.1.6. Lemma. *Assume we are in higher order logic with extensional entailment.*

(i) *Write for* $x\colon\sigma$,

$$\{x\}_\sigma = \lambda z\colon\sigma.\,(x =_\sigma z)\colon P\sigma,$$

for the singleton predicate associated with x. *Then,*

$$x\colon\sigma, y\colon\sigma \mid \{x\}_\sigma =_{P\sigma} \{y\}_\sigma \vdash x =_\sigma y.$$

(ii) *The inclusion relation* $\subseteq_\sigma$ *is (internally) a partial order on the power-type* $P\sigma$:

$$a\colon P\sigma, b\colon P\sigma \mid a \subseteq_\sigma b, b \subseteq_\sigma a \vdash a =_{P\sigma} b.$$

Proof. (i) If $\{x\}_\sigma =_{P\sigma} \{y\}_\sigma$, then we have equalities of propositions

$$\top =_{\mathsf{Prop}} (x =_\sigma x) =_{\mathsf{Prop}} \{x\}_\sigma \cdot x =_{\mathsf{Prop}} \{y\}_\sigma \cdot x =_{\mathsf{Prop}} (y =_\sigma x).$$

Hence we get $x =_\sigma y$, by Example 5.1.2.

(ii) Assume $a \subseteq_\sigma b$ and $b \subseteq_\sigma a$. Then $a, b\colon P\sigma = \sigma \to \mathsf{Prop}$ satisfy the premises of the extensionality of entailment rule, so that $a =_{P\sigma} b$. □

The singleton map $\{-\}_\sigma\colon \sigma \to P\sigma$ described in (i) will play an important rôle in the rest of this chapter. Above one sees that it is internally injective. This result thus holds in all models of higher order logic with extensional entailment.

Quotient types in higher order logic

What we call higher order logic does not include quotient types. But of course one can additionally require these quotient types. It turns out that within higher order predicate logic, quotient types behave much better than within first order predicate logic. For example, we have the following result from [133, Proposition 5.1.10].

5.1.7. Lemma. *In higher order logic with extensional entailment, quotients are automatically effective: for an equivalence relation* R *on a type* σ, *one can derive:*

$$x\colon\sigma, y\colon\sigma \mid [x]_R =_{\sigma/R} [y]_R \vdash R(x, y).$$

Proof. If we have a relation $x\colon\sigma, y\colon\sigma \vdash R(x, y)\colon\mathsf{Prop}$ which is provably an equivalence relation, then by transitivity and symmetry, we can form the pick-term

$$\frac{x\colon\sigma, y\colon\sigma \vdash R(x, y)\colon\mathsf{Prop} \qquad x\colon\sigma, y\colon\sigma, z\colon\sigma \mid R(y, z) \vdash R(x, y) =_{\mathsf{Prop}} R(x, z)}{x\colon\sigma, a\colon\sigma/R \vdash \mathsf{pick}\ w\ \mathsf{from}\ a\ \mathsf{in}\ R(x, w)\colon\mathsf{Prop}}$$

Hence by using reflexivity, we get,

$$x\colon\sigma, y\colon\sigma \mid [x]_R =_{\sigma/R} [y]_R \vdash \top =_{\mathsf{Prop}} R(x, x)$$
$$= \quad \text{pick } w \text{ from } [x]_R \text{ in } R(x, w)$$
$$=_{\mathsf{Prop}} \quad \text{pick } w \text{ from } [y]_R \text{ in } R(x, w)$$
$$= \quad R(x, y). \qquad \square$$

When we first introduced quotient types in Section 4.7, we explained that the quotient σ/R by an arbitrary relation R should be understood as the quotient by the equivalence relation $\overline{R}$ generated by R. This can be made precise in higher order logic.

5.1.8. Lemma. *For an arbitrary relation $x\colon\sigma, y\colon\sigma \vdash R(x, y)\colon\mathsf{Prop}$ one can form in higher order logic the least equivalence relation $\overline{R}$ containing R as*

$$x\colon\sigma, y\colon\sigma \vdash \overline{R}(x, y) \stackrel{\text{def}}{=} \forall S\colon \sigma \times \sigma \to \mathsf{Prop}.$$
$$(\mathrm{Equiv}(S) \wedge \mathrm{Incl}(R, S)) \supset S(x, y) : \mathsf{Prop}.$$

In this expression we use the abbreviations,

$$\mathrm{Equiv}(S) \stackrel{\text{def}}{=} \forall x\colon\sigma.\, S(x, x) \wedge \forall x, y\colon\sigma.\, S(x, y) \supset S(y, x)$$
$$\wedge\; \forall x, y, z\colon\sigma.\, S(x, y) \wedge S(y, z) \supset S(x, z)$$
$$\mathrm{Incl}(R, S) \stackrel{\text{def}}{=} \forall x, y\colon\sigma.\, R(x, y) \supset S(x, y).$$

The relation R then yields the same quotient type as the equivalence relation $\overline{R}$ that it generates, in the sense that there is an isomorphism of types,

$$\sigma/R \cong \sigma/\overline{R}.$$

Proof. The isomorphism is given by the two terms

$$a\colon\sigma/R \vdash P(a) \stackrel{\text{def}}{=} \text{pick } x \text{ from } a \text{ in } [x]_{\overline{R}}\colon \sigma/\overline{R}$$
$$b\colon\sigma/\overline{R} \vdash Q(b) \stackrel{\text{def}}{=} \text{pick } y \text{ from } b \text{ in } [y]_R\colon \sigma/R,$$

where Q is well-defined because $\overline{R}(x, y)$ implies $[x]_R =_{\sigma/R} [y]_R$, since the latter is an equivalence relation containing R. Then for $b\colon\sigma/\overline{R}$,

$P[Q(b)/a]$
$= \text{pick } y \text{ from } b \text{ in } P[[y]_R/a] \qquad \text{by commutation from Exercise 4.7.1}$
$= \text{pick } y \text{ from } b \text{ in } (\text{pick } x \text{ from } [y]_R \text{ in } [x]_{\overline{R}})$
$= \text{pick } y \text{ from } b \text{ in } [y]_{\overline{R}}$
$= b.$

In a similar manner one obtains a conversion $Q[P(a)/b] = a$. $\qquad \square$

We conclude this section with two examples of the use of quotient types in higher order logic with extensional entailment. The first example involves the standard factorisation of terms as a surjection followed by an injection. And the second example describes the Abelian quotient of an arbitrary group.

5.1.9. Examples. Assume we have quotient types in higher order logic with extensional entailment.

(i) We first notice that for a relation R on σ, the canonical map $[-]: \sigma \to \sigma/R$ is always surjective (in the logic). Consider therefore the proposition,

$$a: \sigma/R \vdash \varphi(a) \stackrel{\text{def}}{=} \exists x: \sigma. \, a =_{\sigma/R} [x] : \text{Prop}.$$

Obviously, $y: \sigma \mid \emptyset \vdash \varphi([y]) =_{\text{Prop}} \top$, and thus for $a: \sigma/R$,

$$\varphi(a) \stackrel{(\eta)}{=} \text{pick } y \text{ from } a \text{ in } \varphi([y]) =_{\text{Prop}} \text{pick } y \text{ from } a \text{ in } \top = \top.$$

Thus $\exists x: \sigma. \, a =_{\sigma/R} [x]$ holds for $a: \sigma/R$.

This result can be used to factor an arbitrary term $x: \sigma \vdash M(x): \tau$ as a surjection followed by an injection:

$$\left(\sigma \xrightarrow{\;\;M\;\;} \tau \right) = \left(\sigma \xrightarrow{\;\;[-]\;\;\;} \sigma/K \xrightarrow{\;\;\overline{M}\;\;} \tau \right)$$

In this diagram, K is the kernel relation,

$$x: \sigma, y: \sigma \vdash K(x, y) \stackrel{\text{def}}{=} (M(x) =_\tau M(y)) : \text{Prop}$$

and $\overline{M}(a) = \text{pick } x \text{ from } a \text{ in } M(x)$ for $a: \sigma/K$. Then obviously $\overline{M}([x]) = M(x)$. Moreover, this term $\overline{M}$ is (internally) injective: one can derive

$$a: \sigma/K, b: \sigma/K \mid \overline{M}(a) =_\tau \overline{M}(b) \vdash a =_{\sigma/K} b.$$

as follows.

$$
\begin{aligned}
\overline{M}(a) =_\tau \overline{M}(b) &= \text{pick } x, y \text{ from } a, b \text{ in } M([x]) =_\tau M([y]) \\
&= \text{pick } x, y \text{ from } a, b \text{ in } M(x) =_\tau M(y) \\
&= \text{pick } x, y \text{ from } a, b \text{ in } K(x, y) \\
&= \text{pick } x, y \text{ from } a, b \text{ in } [x] =_{\sigma/K} [y] \\
&= a =_{\sigma/K} b.
\end{aligned}
$$

This factorisation is the one from Example 4.8.8. Its universal property is described in Exercise 5.1.6 below. This factorisation is also familiar from topos theory: it is almost literally as in the proof of [169, Theorem 1.52] (describing the factorisation of an arbitrary map in a topos as an epi followed by a mono).

(ii) Let G: Type be a type which (internally) carries a group structure $(0, +, -(\cdot))$. Consider the following relation $\sim$ on G,

$$u, v : G \vdash u \sim v \stackrel{\mathrm{def}}{=} \forall R : G \times G \to \mathsf{Prop}. \big(\mathrm{Equiv}(R) \wedge \mathrm{Cong}(R)$$
$$\wedge\, \forall x, y : G.\, R(x+y, y+x)\big) \supset R(u,v) : \mathsf{Prop}$$

where the predicates $\mathrm{Equiv}(R)$ and $\mathrm{Cong}(R)$ express that R is an equivalence relation, and is a congruence. The latter is described by

$$\mathrm{Cong}(R) \stackrel{\mathrm{def}}{=} \forall x_1, x_2, y_1, y_2 : G.\, R(x_1, x_2) \wedge R(y_1, y_2) \supset R(x_1 - y_1, x_2 - y_2).$$

It is then easy to see that if we have $\mathrm{Equiv}(R) \wedge \mathrm{Cong}(R)$ then $R(0,0)$, and $R(x,y) \supset R(-x, -y)$. Also, with some elementary reasoning one obtains that $\mathrm{Equiv}(\sim) \wedge \mathrm{Cong}(\sim)$, where $\sim$ is the relation on G defined above.

We now put $\dot{G} \stackrel{\mathrm{def}}{=} G/\!\sim$, with canonical map $x : G \vdash [x] : \dot{G}$. Then we can define for $a, b : \dot{G}$,

$$\dot{0} \stackrel{\mathrm{def}}{=} [0]$$
$$a \mathbin{\dot{+}} b \stackrel{\mathrm{def}}{=} \mathsf{pick}\ u, v\ \mathsf{from}\ a, b\ \mathsf{in}\ [u + v]$$
$$\dot{-}\, a \stackrel{\mathrm{def}}{=} \mathsf{pick}\ u\ \mathsf{from}\ a\ \mathsf{in}\ [-u]$$

so that we get group operations on $\dot{G}$ via representatives. This yields an Abelian group structure, as may be verified by the interested reader. The canonical map $[-] : G \to \dot{G}$ is a universal group homomorphism: for any homomorphism $M : G \to H$ into an Abelian group H, we get a unique homomorphism $\overline{M}$ in

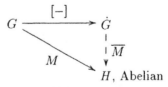

One puts $\overline{M}(a) \stackrel{\mathrm{def}}{=} \mathsf{pick}\ u\ \mathsf{from}\ a\ \mathsf{in}\ M(u)$. This is well-defined because one can form the kernel relation $x, y : G \vdash K(x, y) = (M(x) =_H M(y)) : \mathsf{Prop}$ and show that it is a congruence and an equivalence relation. It also satisfies $K(x+y, y+x)$, since $M(x+y) =_H M(x) \bullet M(y) =_H M(y) \bullet M(x) =_H M(y+x)$, because the group operation $\bullet$ of H is commutative. Thus if $u \sim v$, then $K(u, v)$ and so $M(u) =_H M(v)$.

Exercises

5.1.1.   Use the extensionality of entailment rule to derive in higher order logic,

$$\alpha: \mathsf{Prop}, \beta: \mathsf{Prop} \mid \alpha \supset \beta, \beta \supset \alpha \vdash \alpha =_{\mathsf{Prop}} \beta.$$

5.1.2.   Check that the connectives $\bot, \vee, \top, \wedge$ as defined in Example 5.1.4 satisfy the rules in Figure 4.1.

5.1.3.   For a proposition $x: \sigma, y: \tau \vdash \varphi(x, y): \mathsf{Prop}$ in higher order logic with extensional entailment, show that $\forall y: \tau. \varphi(x, y)$ is logically equivalent to the equation $\{y \in \tau \mid \varphi(x, y)\} =_{P\tau} \{y \in \tau \mid \top\}$.

5.1.4.   Prove that in higher order logic with quotient types there are conversions,

pick x from a in $\top$ $=$ $\top$

pick x from a in $(\varphi \wedge \psi)$ $=$ (pick x from a in φ) $\wedge$ (pick x from a in ψ).

Conclude that in higher order logic with extensional entailment the pick-operation preserves entailment $\vdash$.

5.1.5.   Define in higher order logic an order $\leq$ on Prop by

$$\alpha: \mathsf{Prop}, \beta: \mathsf{Prop} \vdash \alpha \leq \beta \stackrel{\mathrm{def}}{=} \alpha \supset \beta: \mathsf{Prop},$$

and show that $(\mathsf{Prop}, \leq)$ is (internally) a Heyting pre-algebra.

5.1.6.   Consider the factorisation $M = \overline{M} \circ [-]_K: \sigma \to \tau$ from Example 5.1.9 (i), and show that it is universal in the following sense. If we can write $M = Q \circ P$, where $Q: \rho \to \tau$ is internally injective, then there is a unique term (up-to-conversion) $\overline{P}: \sigma/K \to \rho$ with conversions $\overline{P} \circ [-]_K = P$ and $Q \circ \overline{P} = \overline{M}$.

5.1.7.   Let R be a relation on σ and S a reflexive relation on σ/R. Prove that in higher order logic with extensionality of entailment and with quotient types, the quotient $\sigma/R/S$ is isomorphic to the type σ/T, where T is the relation $T(x, x') = S([x]_R, [x']_R)$ on σ.

5.2 Generic objects

Higher order signatures as described in the previous section involve a special atomic type Prop, which is such that predicates on σ correspond to "characteristic" terms $\sigma \to \mathsf{Prop}$. Categorically, such a correspondence is described in terms of so-called 'generic objects'. These can be defined easily for split fibrations, but for arbitrary fibrations there are some complications. In order to describe these matters properly, we need a fibred Yoneda lemma. But we shall start with the easy case of split fibrations.

5.2.1. Definition. A split fibration $\begin{smallmatrix}\mathbb{E}\\\downarrow p\\\mathbb{B}\end{smallmatrix}$ has a **split generic object** if there is an object $\Omega \in \mathbb{B}$ together with a collection of isomorphisms

$$\mathbb{B}(I, \Omega) \xrightarrow[\cong]{\theta_I} \mathrm{Obj}\,\mathbb{E}_I$$

natural in I; that is, $\theta_J(u \circ v) = v^*(\theta_I(u))$ for $v : J \to I$.

It may be clear that the above $\Omega \in \mathbb{B}$ plays the rôle of Prop, and that θ identifies terms $I \to \Omega$ with predicates $X \in \mathbb{E}_I$ on I. The following result gives a slightly different formulation of the same notion.

5.2.2. Lemma. *A split fibration* $\begin{smallmatrix}\mathbb{E}\\\downarrow p\\\mathbb{B}\end{smallmatrix}$ *has a split generic object if and only if there is an object* $T \in \mathbb{E}$ *with the property that*

$$\forall X \in \mathbb{E}.\ \exists! u \colon pX \to pT.\ u^*(T) = X.$$

Proof. Assume $\begin{smallmatrix}\mathbb{E}\\\downarrow p\\\mathbb{B}\end{smallmatrix}$ has a split generic object (Ω, θ) as described in the above definition. Take $T = \theta_\Omega(\mathrm{id}_\Omega) \in \mathbb{E}_\Omega$. Then for $X \in \mathbb{E}_I$ we have that $\theta_I^{-1}(X) \colon I \to \Omega = pT$ satisfies

$$\theta_I^{-1}(X)^*(T) = \theta_I^{-1}(X)^*(\theta_\Omega(\mathrm{id}_\Omega)) = \theta_I(\mathrm{id}_\Omega \circ \theta_I^{-1}(X)) = X.$$

And it is easy to see that $\theta_I^{-1}(X)$ is unique in satisfying this property: if $X = u^*(T) = \theta_I(u)$, then $u = \theta_I^{-1}(X)$.

In the reverse direction, assume $T \in \mathbb{E}$ as in the lemma, and write $\Omega = pT \in \mathbb{B}$. For $I \in \mathbb{B}$ and $u \colon I \to \Omega$, let $\theta_I(u) = u^*(T)$. It is then clearly a bijection. And $\theta_I(u \circ v) = (u \circ v)^*(T) = v^*(u^*(T)) = v^*(\theta_I(u))$, for $v : J \to I$. □

5.2.3. Examples. (i) Let $\mathbb{C}$ be a category with a small collection $\Omega = \mathrm{Obj}\,\mathbb{C}$ of objects. Then $\Omega \in \mathbf{Sets}$ forms a split generic object for the family fibration $\begin{smallmatrix}\mathrm{Fam}(\mathbb{C})\\\downarrow\\\mathbf{Sets}\end{smallmatrix}$. The set of functions $I \to \Omega$ is actually equal to the collection of objects of the fibre $\mathrm{Fam}(\mathbb{C})_I$ over I.

In [252] a (split) fibration is called 'globally small' if it has a (split) generic object. This family example provides a justification for this terminology. The smallness aspect will become more apparent in Proposition 5.2.7 below. Later, in section 9.5 a fibration is called 'locally small' if its fibred homsets are small (in a suitable sense).

(ii) A special case of (i) is $\mathbb{C} = \mathbf{2} = \{\bot, \top\}$ with $\bot \leq \top$. The family fibration $\begin{smallmatrix}\mathrm{Fam}(\mathbf{2})\\\downarrow\\\mathbf{Sets}\end{smallmatrix}$ is then isomorphic to the subobject fibration $\begin{smallmatrix}\mathrm{Sub}(\mathbf{Sets})\\\downarrow\\\mathbf{Sets}\end{smallmatrix}$ on $\mathbf{Sets}$, see Exercise 1.7.3. The generic object is the set $\{\bot, \top\}$ in $\mathbf{Sets}$, together with the isomorphism between subsets of I and 'characteristic' functions $I \to \{\bot, \top\}$.

(iii) Consider a split classifying fibration $\begin{smallmatrix} \mathcal{L}(\Sigma,\Pi,\mathcal{A}) \\ \downarrow \\ \mathcal{Cl}(\Sigma) \end{smallmatrix}$ constructed syntactically (see Section 3.1) from a higher order specification $(\Sigma, \Pi, \mathcal{A})$. Such a fibration also has a split generic object, namely the type $\mathsf{Prop} \in \mathcal{Cl}(\Sigma)$. For an object (or type) $\sigma \in \mathcal{Cl}(\Sigma)$, the morphisms $M \colon \sigma \to \mathsf{Prop}$ in $\mathcal{Cl}(\Sigma)$ are by definition the terms $x \colon \sigma \vdash M \colon \mathsf{Prop}$, and thus the propositions in context σ, *i.e.* the objects over $\sigma \in \mathcal{Cl}(\Sigma)$.

(iv) Let $\mathbb{E}$ be a distributive category. Write $\Omega = 1 + 1 \in \mathbb{E}$. Then Ω forms a boolean algebra, see Exercise 2.6.1. Each homset $\mathbb{E}(I, \Omega)$ is then partially ordered by $\varphi \leq \psi \Leftrightarrow \varphi \wedge \psi = \varphi$. Hence the assignment $I \mapsto \mathbb{E}(I, \Omega)$ yields an indexed category $\mathbb{E}^{\mathrm{op}} \to \mathbf{Cat}$. The resulting split fibration has Ω as split generic object by construction.

These generic objects can be described on a more abstract level in terms of a fibred Yoneda lemma. This result—and also the subsequent corollary—may be found in [27]. We recall from Exercise 1.10.2 that the Grothendieck construction applied to the representable functor $\mathbb{E}(-, I) \colon \mathbb{E}^{\mathrm{op}} \to \mathbf{Cat}$ yields the domain fibration $\begin{smallmatrix} \mathbb{E}/I \\ \downarrow \mathrm{dom}_I \\ \mathbb{E} \end{smallmatrix}$. These fibrations dom_I play the rôle of representable objects in fibred category theory.

5.2.4. Lemma (Fibred Yoneda). (i) *For a cloven fibration* $\begin{smallmatrix} \mathbb{E} \\ \downarrow p \\ \mathbb{B} \end{smallmatrix}$ *and an object* $I \in \mathbb{B}$ *there is an equivalence of categories*

$$\mathbb{E}_I \simeq \mathrm{Hom}\bigl(\mathrm{dom}_I, \, p\bigr)$$

between the fibre category over I *and the hom-category of fibred functors* $\mathbb{B}/I \to \mathbb{E}$ *over* $\mathbb{B}$ *and vertical natural transformations between them.*

The equivalence is natural in I *in the sense that for each morphism* $w \colon I \to J$ *in* $\mathbb{B}$, *the diagram*

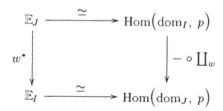

commutes up-to-unique-isomorphism.

(ii) *In case* p *is a* split *fibration, the equivalence in (i) is an isomorphism and the naturality diagram commutes on-the-nose.*

Proof. (i) Each object $X \in \mathbb{E}_I$ gives rise to a functor $F_X \colon \mathbb{B}/I \to \mathbb{E}$ by $u \mapsto u^*(X)$ on objects, and on morphisms by,

$$
\left(
\begin{array}{c}
J \xrightarrow{\ \phi\ } K \\
u \searrow \quad \swarrow v \\
I
\end{array}
\right)
\longmapsto
\left(
\begin{array}{c}
u^*(X) \dashrightarrow^{F_X(\phi)} v^*(Y) \\
\searrow \quad \swarrow \\
X
\end{array}
\right)
$$

where $F_X(\phi)$ is the unique arrow in $\mathbb{E}$ above ϕ satisfying $\overline{v}(X) \circ F_X(\phi) = \overline{u}(X)$.

In the reverse direction, every fibred functor $G \colon \mathbb{B}/I \to \mathbb{E}$ over $\mathbb{B}$ gives an object $G(\mathrm{id}_I) \in \mathbb{E}_I$. These operations $X \mapsto F_X$ and $G \mapsto G(\mathrm{id}_I)$ constitute an equivalence, since

$$
\begin{aligned}
F_X(\mathrm{id}_I) &= \mathrm{id}_I^*(X) \\
&\cong X.
\end{aligned}
$$

$$
\begin{aligned}
F_{G(\mathrm{id}_I)}(u) &= u^* G(\mathrm{id}_I) \\
&\cong G(u^*(\mathrm{id}_I)) \quad \text{since } G \text{ is a fibred functor} \\
&= G(u).
\end{aligned}
$$

Naturality in I holds, because for $w \colon I \to J$, one has

$$
\begin{aligned}
F_{w^*(X)}(u) &= u^* w^*(X) \\
&\cong (w \circ u)^*(X) \\
&= F_X(w \circ u) \\
&= F_X(\textstyle\coprod_w(u)).
\end{aligned}
$$

(ii) Obvious, since in the split case, all the above isomorphisms are identities. $\qquad\square$

5.2.5. Corollary. *Every fibration is equivalent to a split fibration.*

Proof. For a fibration $\begin{smallmatrix}\mathbb{E}\\\downarrow p\\\mathbb{B}\end{smallmatrix}$, define a split indexed category on $\mathbb{B}$ by

$$
I \mapsto \mathrm{Hom}\big(\mathrm{dom}_I, p\big) \qquad \text{and} \qquad \big(I \xrightarrow{\ w\ } J\big) \mapsto \big(- \circ \textstyle\coprod_w\big).
$$

The resulting split fibration (obtained by the Grothendieck construction) is by the previous lemma equivalent to p. $\qquad\square$

This result may be used to transform fibred models of certain type theories into equivalent split models, see *e.g.* [134].

5.2.6. Definition. A fibration $\begin{smallmatrix} \mathbb{E} \\ \downarrow p \\ \mathbb{B} \end{smallmatrix}$ is **representable** if it is equivalent to a domain fibration $\begin{smallmatrix} \mathbb{B}/\Omega \\ \downarrow \mathrm{dom}_\Omega \\ \mathbb{B} \end{smallmatrix}$ for some object $\Omega \in \mathbb{B}$.

The fibred Yoneda lemma tells us that if a (cloven) fibration $\begin{smallmatrix} \mathbb{E} \\ \downarrow p \\ \mathbb{B} \end{smallmatrix}$ is representable, say with an equivalence $p \simeq \mathrm{dom}_\Omega$ for $\Omega \in \mathbb{B}$, then there is an object $T \in \mathbb{E}$ above Ω yielding a functor $\mathbb{B}/\Omega \xrightarrow{\simeq} \mathbb{E}$ by $u \mapsto u^*(T)$. Further, this means that in the reverse direction there is a fibred functor $H: \mathbb{E} \to \mathbb{B}/\Omega$ together with vertical natural isomorphisms $\varphi: \mathrm{id} \xRightarrow{\cong} H(-)^*(T)$ and $\psi: \mathrm{id} \xRightarrow{\cong} H((-)^*(T))$. But since the fibration dom_Ω has discrete fibre categories ψ must be the identity and thus $H(u^*(T)) = u$ in $\mathbb{B}/\Omega$.

For a *split* fibration $\begin{smallmatrix} \mathbb{E} \\ \downarrow p \\ \mathbb{B} \end{smallmatrix}$ there is a **split fibration of objects** $\begin{smallmatrix} \mathrm{Split}(\mathbb{E}) \\ \downarrow \|p\| \\ \mathbb{B} \end{smallmatrix}$ where $\mathrm{Split}(\mathbb{E})$ is the subcategory of $\mathbb{E}$ with all objects from $\mathbb{E}$, but with Cartesian maps coming from the splitting only, see also Exercise 1.8.9. Next we will show how a split generic object for p exists if and only if this fibration of objects $\|p\|$ is representable.

5.2.7. Proposition. *A split fibration* $\begin{smallmatrix} \mathbb{E} \\ \downarrow p \\ \mathbb{B} \end{smallmatrix}$ *has a split generic object if and only if the associated fibration of objects* $\begin{smallmatrix} \mathrm{Split}(\mathbb{E}) \\ \downarrow \|p\| \\ \mathbb{B} \end{smallmatrix}$ *is representable.*

Proof. Assume that the split fibration of objects $\begin{smallmatrix} \mathrm{Split}(\mathbb{E}) \\ \downarrow \|p\| \\ \mathbb{B} \end{smallmatrix}$ is representable, say via $H: \mathrm{Split}(\mathbb{E}) \to \mathbb{B}/\Omega$ together with isomorphisms $\varphi_X: X \to HX^*(T)$ as described above. Then $\varphi_X = \mathrm{id}$, since $\|p\|$ has discrete fibre categories. Thus we obtain isomorphisms $\mathrm{Obj}\,\mathbb{E}_I = \mathrm{Split}(\mathbb{E})_I \cong (\mathbb{B}/\Omega)_I = \mathbb{B}(I, \Omega)$, which commute (on-the-nose) with reindexing.

Conversely, given Ω with the isomorphisms θ_I as in Definition 5.2.1. The object $T = \theta_\Omega(\mathrm{id}_\Omega) \in \mathbb{E}_\Omega$ induces a fibred functor $\mathbb{B}/\Omega \to \mathrm{Split}(\mathbb{E})$ by $u \mapsto u^*(T) = \theta_I(u)$, which is an obvious isomorphism. This shows that $\|p\|$ is representable. □

Next we turn to generic objects for non-split fibrations. This is a subtle matter: an equality like in Lemma 5.2.2 has to be replaced by an isomorphism. There are several alternatives.

5.2.8. Definition. Consider a fibration $\begin{smallmatrix} \mathbb{E} \\ \downarrow p \\ \mathbb{B} \end{smallmatrix}$ and an object T in the total category $\mathbb{E}$. We call T a

(i) **weak generic object** if

$$\forall X \in \mathbb{E}.\, \exists f: X \to T.\, f \text{ is Cartesian,}$$

or, equivalently,

$$\forall X \in \mathbb{E}. \; \exists u: pX \to pT. \; \exists f: u^*(T) \to X. \; f \text{ is a vertical isomorphism.}$$

(ii) **generic object** if

$$\forall X \in \mathbb{E}. \; \exists! u: pX \to pT. \; \exists f: X \to T. \; f \text{ is Cartesian over } u$$

or, equivalently,

$$\forall X \in \mathbb{E}. \; \exists! u: pX \to pT. \; \exists f: u^*(T) \to X. \; f \text{ is a vertical isomorphism.}$$

(iii) and a **strong generic object** if

$$\forall X \in \mathbb{E}. \; \exists! f: X \to T. \; f \text{ is Cartesian}$$

or, equivalently,

$$\forall X \in \mathbb{E}. \; \exists! u: pX \to pT. \; \exists! f: u^*(T) \to X. \; f \text{ is a vertical isomorphism.}$$

5.2.9. Lemma. *Generic and strong generic objects are determined up-to-isomorphism (but not the weak ones).*

In a preorder fibration, there is no difference between a generic object and a strong generic object.

Proof. Exercise. □

For these generic objects we seek a reformulation of the above notions in terms of representable fibrations. This will be achieved for (ordinary) generic objects, so that they form the most natural notion among the above three options (weak, ordinary, and strong).

Recall (*e.g.* from Exercise 1.1.4) that for every fibration $\begin{smallmatrix} \mathbb{E} \\ \downarrow p \\ \mathbb{B} \end{smallmatrix}$ there is a **fibration of objects** $\begin{smallmatrix} \mathrm{Cart}(\mathbb{E}) \\ \downarrow |p| \\ \mathbb{B} \end{smallmatrix}$ where $\mathrm{Cart}(\mathbb{E})$ is the subcategory of $\mathbb{E}$ with all objects but Cartesian morphisms only.

5.2.10. Proposition. *A fibration* $\begin{smallmatrix} \mathbb{E} \\ \downarrow p \\ \mathbb{B} \end{smallmatrix}$ *has a generic object if and only if the associated fibration of objects* $\begin{smallmatrix} \mathrm{Cart}(\mathbb{E}) \\ \downarrow |p| \\ \mathbb{B} \end{smallmatrix}$ *is representable.*

Proof. Assume p has a generic object $T \in \mathbb{E}$, say with $\Omega = pT \in \mathbb{B}$, satisfying the description in Definition 5.2.8 (ii). We intend to show that the fibration $\begin{smallmatrix} \mathrm{Cart}(\mathbb{E}) \\ \downarrow |p| \\ \mathbb{B} \end{smallmatrix}$ is equivalent to dom_Ω (and thus representable). One defines a functor $H: \mathrm{Cart}(\mathbb{E}) \to \mathbb{B}/\Omega$ by mapping X to the unique arrow $u_X: pX \to \Omega$ with

$u^*(T) \cong X$, vertically. For a Cartesian morphism $f: X \to Y$ we get $u_Y \circ pf = u_X$, by uniqueness, since

$$(u_Y \circ pf)^*(T) \cong (pf)^* u_Y^*(T) \cong (pf)^*(Y) \cong X,$$

the latter because f is Cartesian. By definition we have $(-)^*(T) \circ H \cong$ id. We get $H(u^*(T)) = u$ by uniqueness, since by definition $H(u^*(T))^*(T) \cong u^*(T)$.

Conversely, assume the fibration $|p|$ is representable, say via $H: \mathrm{Cart}(\mathbb{E}) \to \mathbb{B}/\Omega$ with isomorphisms $\varphi_X: X \to HX^*(T)$ natural in $X \in \mathbb{E}$, where $HX: pX \to \Omega$. For any $u: pX \to \Omega$ which also satisfies $u^*(T) \cong X$ vertically, we get a vertical isomorphism $H(u^*(T)) \cong HX$ in $\mathbb{B}/\Omega$, and thus $u = H(u^*(T)) = HX$. □

5.2.11. Examples. (i) Consider the (non-split) fibration $\begin{array}{c}\mathbf{Mono(Sets)} \\ \downarrow \\ \mathbf{Sets}\end{array}$ of monos (injections) in **Sets**. The inclusion

$$T = \left(1 = \{\top\} \subseteq \{\bot, \top\} = 2\right)$$

is a (strong) generic object for this fibration: for every injection $m: X \rightarrowtail I$ there is a unique map $\chi_m: I \to 2$ for which there is a pullback square,

This map χ_m is then determined by $\chi_m(i) = \top \Leftrightarrow \exists x \in X.\, m(x) = i$.

(ii) A weak generic object often arises in the following situation. Let $\mathbb{B}$ be a category with finite limits and $a: A \to B$ be an arbitrary morphism. Write $\mathcal{D}$ for the collection of morphism of the form $u^*(a)$ which are obtained from a by pullback along some u. We write $\mathcal{D}^\to$ for the full subcategory of $\mathbb{B}^\to$ with objects in $\mathcal{D}$. Then the codomain functor $\begin{array}{c}\mathcal{D}^\to \\ \downarrow \\ \mathbb{B}\end{array}$ is a fibration with a as weak generic object.

5.2.12. Remark. Suppose $\begin{array}{c}\mathbb{E} \\ \downarrow p \\ \mathbb{B}\end{array}$ is a fibration with a generic object, say given by $T \in \mathbb{E}_\Omega$. Fibred structure for p then induces structure on Ω which captures the fibred structure on objects. For example, if p has fibred Cartesian products $\times$, then one obtains a map $\&: \Omega \times \Omega \to \Omega$ such that for parallel maps $u, v: I \rightrightarrows \Omega$ there is an isomorphism:

$$u^*(T) \times v^*(T) \cong (\& \circ \langle u, v \rangle)^*(T).$$

Thus the map $\&$ describes the object part of fibred Cartesian products—since every object X is isomorphic to $u^*(T)$ for a unique u.

This map $\&$ comes about as follows. The object part of the Cartesian product functor on $\mathbb{E}$ works also on $\mathrm{Cart}(\mathbb{E})$ and hence—by Proposition 5.2.10—on $\mathbb{B}/\Omega$. It leads to a natural transformation with components

$$\mathbb{B}\big(I,\,\Omega\times\Omega\big)\longrightarrow\mathbb{B}\big(I,\,\Omega\big)$$

and thus by the Yoneda lemma to a map $\&\colon\Omega\times\Omega\to\Omega$.

In a similar way fibred exponents yield a map $\Rightarrow\colon\Omega\times\Omega\to\Omega$. And if $\mathbb{B}$ is Cartesian closed, simple coproducts and products lead to collections of maps (for every $I\in\mathbb{B}$),

$$\Omega^I\xrightarrow{\ \exists_I\ }\Omega\qquad\text{and}\qquad\Omega^I\xrightarrow{\ \forall_I\ }\Omega$$

A similar phenomenon occurs for split fibrations with split generic objects.

We conclude this section with morphisms and generic objects.

5.2.13. Definition. Let $\left(\begin{smallmatrix}\mathbb{E}\\ \downarrow p\\ \mathbb{B}\end{smallmatrix}\right)\xrightarrow{(K,L)}\left(\begin{smallmatrix}\mathbb{E}'\\ \downarrow p'\\ \mathbb{B}'\end{smallmatrix}\right)$ be a morphism between fibrations p and p', each with a (weak, strong) generic objects, say $T\in\mathbb{E}_\Omega$ and $T'\in\mathbb{E}'_{\Omega'}$. We say that (K,L) **preserves these generic objects** if the induced Cartesian map $LT\to T'$ is an isomorphism.

A bit stronger, (K,L) preserves these generic objects **on-the-nose** if this map $LT\to T'$ is an identity.

Preservation on-the-nose is most appropriate for split generic objects.

5.2.14. Lemma. *Suppose* $\left(\begin{smallmatrix}\mathbb{E}\\ \downarrow p\\ \mathbb{B}\end{smallmatrix}\right)\xrightarrow{(K,L)}\left(\begin{smallmatrix}\mathbb{E}'\\ \downarrow p'\\ \mathbb{B}'\end{smallmatrix}\right)$ *is a morphism of split fibrations with split generic objects. If (K,L) preserves these on-the-nose, then the following diagram commutes.*

$$
\begin{array}{ccc}
\mathrm{Obj}\,\mathbb{E}_I & \xrightarrow{\ \ L\ \ } & \mathrm{Obj}\,\mathbb{E}'_{KI}\\[4pt]
{\scriptstyle\cong}\big\uparrow & & \big\uparrow{\scriptstyle\cong}\\[4pt]
\mathbb{B}(I,\Omega) & \xrightarrow{\ \ K\ \ } & \mathbb{B}'(KI,\Omega')
\end{array}
$$

Proof. Exercise. $\square$

Exercises

5.2.1. (From [81, Section 2] Define a PER $\Sigma = \{(n, n') \mid n \cdot 0 \downarrow \Leftrightarrow n' \cdot 0 \downarrow\}$.

    (i) Prove that for a PER R, there is a bijective correspondence between maps $R \to \Sigma$ in **PER** (or in ω-**Sets**) and subsets $A \subseteq |R|$ which are saturated(*i.e.* which satisfy: $n \in A$ and nRn' imply $n' \in A$) and for which there is a r.e. subset $B \subseteq \mathbb{N}$ with $A = |R| \cap B$. These subsets are called "natural subobjects" of R.

    (ii) Define a fibration $\begin{array}{c} \text{NatSub}(\mathbf{PER}) \\ \downarrow \\ \mathbf{PER} \end{array}$ of natural subobjects, with split generic object using $\Sigma \in \mathbf{PER}$.

5.2.2. Recall the natural numbers object $\mathsf{N} = (\text{Eq}(\mathbb{N}) \subseteq \mathbb{N} \times \mathbb{N})$ in **PER** from Exercise 1.2.10.

    (i) Conclude from the previous exercise that maps $\Sigma \to \mathsf{N}$ in **PER** can be identified with r.e. subsets of $\mathbb{N}$.

    (ii) Check that maps $R \to 2\,(= 1 + 1)$ can be identified with recursive subsets of $\mathbb{N}$.

5.2.3. Consider a fibration $\begin{array}{c} \mathbb{E} \\ \downarrow p \\ \mathbb{B} \end{array}$ and a functor $F \colon \mathbb{A} \to \mathbb{B}$ with a right adjoint, and the resulting fibration $F^*(p)$ obtained by change-of-base along F.

    (i) Assume first that p is split and has a split generic object. Show that $F^*(p)$ also has a split generic object.

    (ii) Assume next that p has a generic object, and show that $F^*(p)$ also has a generic object.

5.2.4. Let $\begin{array}{c} \mathbb{E} \\ \downarrow p \\ \mathbb{B} \end{array}$ be a split fibration on a base category with Cartesian products.

    (i) Show that for an object $I \in \mathbb{B}$, the (split) exponent fibration $\text{dom}_I \Rightarrow p$ from Exercise 1.10.6 is isomorphic to the fibration obtained from p by change-of-base along $I \times (-) \colon \mathbb{B} \to \mathbb{B}$.

    (ii) Conclude that p has split simple products/coproducts if and only if each diagonal functor $p \to (\text{dom}_I \Rightarrow p)$ has a split fibred right/left adjoint.

5.2.5. Recall the category **MS** of metric spaces and non-expansive functions from Example 4.6.3 (iv), see also Exercise 4.6.2. A subset $A \subseteq X$ of a metric space X is closedf each limit point of A is contained in A.

    (i) Check that these closed subsets are stable under pullback. Organise them in a (poset) fibration over **MS**.

    (ii) Show that for a closed subset $A \subseteq X$ there is a characteristic metric predicate $\chi_A \colon X \to [0, \infty]$ forming a pullback diagram,

[*Hint.* Define $\chi_A(x) = \inf\{X(x,y) \mid y \in A\}$.]

(iii) Conclude that the fibration $\begin{smallmatrix} \text{ClSub(MS)} \\ \downarrow \\ \textbf{MS} \end{smallmatrix}$ of closed subsets has a weak generic object.

(iv) Use this to show that the regular subobjects (*i.e.* those subobjects which have an equaliser as underlying mono) in **MS** are precisely the closed subsets.

5.2.6.   Prove Lemma 5.2.14.

5.2.7.   Show how the maps $\exists_I$ and $\forall_I$ come about in Remark 5.2.12.

5.2.8.   Consider a (split) fibration $\begin{smallmatrix} \mathbb{E} \\ \downarrow{\scriptstyle P} \\ \mathbb{B} \end{smallmatrix}$ with a (split) generic object Ω on a Cartesian closed base category $\mathbb{B}$. Prove that for a morphism $u\colon I \to J$ in $\mathbb{B}$ the following diagram commutes.

$$
\begin{array}{ccccc}
\mathbb{E}_{K \times J} & \longrightarrow & \mathbb{B}(K \times J,\, \Omega) & \overset{\cong}{\longrightarrow} & \mathbb{B}(K,\, \Omega^J) \\
{\scriptstyle (\mathrm{id} \times u)^{*}}\big\downarrow & & & & \big\downarrow{\scriptstyle \Omega^u \,\circ\, -} \\
\mathbb{E}_{K \times I} & \longrightarrow & \mathbb{B}(K \times I,\, \Omega) & \overset{\cong}{\longrightarrow} & \mathbb{B}(K,\, \Omega^I)
\end{array}
$$

5.3 Fibrations for higher order logic

In this section we define appropriate 'higher order' fibrations as models of higher order logic. Several examples are given as instances of a general "tripos" construction. But most importantly, a topos is defined as a category $\mathbb{B}$ for which its subobject fibration $\begin{smallmatrix} \text{Sub}(\mathbb{B}) \\ \downarrow \\ \mathbb{B} \end{smallmatrix}$ is such a higher order fibration. This will turn out to be a powerful notion. It can be defined in various other and more elementary ways (as will be shown in the next two sections), but the approach via higher order fibrations is appropriate from a purely logical perspective. Towards the end of this section we also describe the higher order fibrations resulting from *regular* subobjects in the categories of ω-sets and of PERs.

Definition 5.1.3 in the first section of this chapter describes higher order logic. The aspects which are not captured in first order fibrations (as described in the previous chapter) are the presence of a type Prop of propositions and of exponent types.

5.3.1. Definition. A **higher order fibration** is a first order fibration with

- a generic object;
- a Cartesian closed base category.

Such a higher order fibration will be called **split** if the fibration is split and all of its fibred structure (including the generic object) is split.

5.3.2. Examples. For a frame (or, a complete Heyting algebra) X, the family fibration $\begin{smallmatrix} \mathrm{Fam}(X) \\ \downarrow \\ \mathbf{Sets} \end{smallmatrix}$ is a split higher order fibration. It is a first order fibration as described in Example 4.2.5 and has the underlying set $X \in \mathbf{Sets}$ as split generic object by Examples 5.2.3 (i) and (ii). Obviously, the base category **Sets** is Cartesian closed.

In a similar way, the realisability fibration $\begin{smallmatrix} \mathrm{UFam}(P\mathbb{N}) \\ \downarrow \\ \mathbf{Sets} \end{smallmatrix}$ from Example 4.2.6 is a split higher order fibration. Its split generic object is the set $P\mathbb{N} \in \mathbf{Sets}$.

We need not say much about the interpretation of higher order logic in higher order fibrations, since we have already seen how to interpret simply typed λ-calculus in Cartesian closed categories, and predicate logic in (preorder) fibrations. But there is something to say about the extensionality of entailment rule

$$\frac{\Gamma \vdash P, Q : \sigma \to \mathsf{Prop} \qquad \Gamma, x : \sigma \mid \Theta, Px \vdash Qx \qquad \Gamma, x : \sigma \mid \Theta, Qx \vdash Px}{\Gamma \mid \Theta \vdash P =_{\sigma \to \mathsf{Prop}} Q}$$

since it may fail. In a family fibration $\begin{smallmatrix} \mathrm{Fam}(X) \\ \downarrow \\ \mathbf{Sets} \end{smallmatrix}$ of a frame X, the assumptions of this rule applied to predicates $P, Q : J \rightrightarrows X^I$ in **Sets** express that

$$P(j)(i) \le Q(j)(i) \qquad \text{and} \qquad Q(j)(i) \le P(j)(i),$$

for all $j \in J$ and $i \in I$—where $\le$ is the order on X. Hence we may conclude that $P = Q$, as required (since internal and external equality coincides).

In the realisability fibration $\begin{smallmatrix} \mathrm{UFam}(P\mathbb{N}) \\ \downarrow \\ \mathbf{Sets} \end{smallmatrix}$ the same assumptions for $P, Q : J \rightrightarrows (P\mathbb{N})^I$ yield that

$$\left(\bigcap_{(j,i) \in J \times I} P(j)(i) \supset Q(j)(i) \right) \neq \emptyset, \qquad \left(\bigcap_{(j,i) \in J \times I} Q(j)(i) \supset P(j)(i) \right) \neq \emptyset.$$

This means that there are realisers inhabiting $P(j)(i) \supset Q(j)(i)$ and $Q(j)(i) \supset P(j)(i)$, for all j, i. But this is not enough to conclude $P = Q$: take for example $P = \lambda j \in J. \lambda i \in I. \{0\}$ and $Q = \lambda j \in J. \lambda i \in I. \{1\}$. In this realisability example the truth of a proposition $\varphi \subseteq P\mathbb{N}$ means $\varphi \neq \emptyset$, which is not the same as $\varphi = \top$, since $\top = P\mathbb{N}$. Hence, this realisability fibration is *not* a model of higher order logic with extensional entailment.

These examples both form instances of what is called a tripos in [145, 267]. Mostly, these triposes are considered with **Sets** as base category. We recall that in a higher order fibration we require 'simple' quantification $\coprod, \prod$ along Cartesian projections. By the constructions in Examples 4.3.7 (i) and (ii) we then get quantification $\coprod_u, \prod_u$ along arbitrary maps u in the base category, but Beck-Chevalley need not hold for these. This Beck-Chevalley condition is required explicitly for triposes, although it is not needed to model higher order simple predicate logic (but it does lead to a model of higher order *dependent* predicate logic, as in Proposition 11.2.2 (ii)).

5.3.3. Definition (See [145, 267]). A **tripos** is a higher order fibration $\begin{smallmatrix} \mathbb{E} \\ \downarrow \\ \textbf{Sets} \end{smallmatrix}$ over **Sets** for which the induced products $\prod_u$ and coproducts $\coprod_u$ along an arbitrary function u satisfy the Beck-Chevalley condition.

(By Lemma 1.9.7 it suffices that Beck-Chevalley holds either for products or for coproducts.)

These triposes are mostly used as an intermediate step in the construction of certain toposes (see Section 6.1). But [267] is a study of "tripos theory" on its own.

5.3.4. Example (Triposes built from partial combinatory algebras). A partial combinatory algebra (PCA) consists of a set A together with a partial application function $\cdot : A \times A \rightharpoonup A$ and two elements $\boldsymbol{K}, \boldsymbol{S} \in A$ such that

$$\boldsymbol{K}x\downarrow, \quad \boldsymbol{S}x\downarrow, \quad \boldsymbol{S}xy\downarrow \qquad \text{and} \qquad \boldsymbol{K}xy \simeq x, \quad \boldsymbol{S}xyz \simeq xz(yz),$$

where $P\downarrow$ means that P is defined and where Kleene equality $P \simeq Q$ means that P is defined if and only if Q is defined, and in that case they are equal. As above, we often omit the application dot $\cdot$. The element $\boldsymbol{I} = \boldsymbol{SKK} \in A$ satisfies $I \cdot a = a$, for all $a \in A$. Examples of PCAs include the natural numbers $\mathbb{N}$ with Kleene application $\cdot$ and all models of the untyped λ-calculus, see [32] for more information.

For such a PCA $(A, \cdot)$ one can prove combinatory completeness: for every polynomial term $M(x_1, \ldots, x_n)$ built from variables $x_1, \ldots, x_n$, constants $\underline{c}$ for $c \in A$, and application $\cdot$, there is an element $a \in A$ such that for all elements $b_1, \ldots, b_n \in A$,

$$ab_1 \cdots b_n \simeq [\![M]\!](b_1, \ldots, b_n),$$

where $[\![M]\!]$ is the function $A^n \rightharpoonup A$ obtained by interpreting the polynomial M. One uses Schönfinkels abstraction rules:

$$\begin{aligned} \lambda x. x &= \boldsymbol{I} = \boldsymbol{SKK} \\ \lambda x. M &= \boldsymbol{K}M \qquad \qquad \text{if } x \text{ is not free in } M \\ \lambda x. MN &= \boldsymbol{S}(\lambda x. M)(\lambda x. N). \end{aligned}$$

Then one takes $a = \lambda x_1 \cdots x_n. M$ to get combinatory completeness.

In this way one can define pairing in PCAs as in the untyped λ-calculus:

$$\langle a, b \rangle \stackrel{\text{def}}{=} \lambda z.\, zab = S(SI(Ka))(Kb)$$

with projections

$$\pi c \stackrel{\text{def}}{=} cK \qquad \text{and} \qquad \pi' c \stackrel{\text{def}}{=} c(KI).$$

Then $\pi \langle a, b \rangle = a$ and $\pi' \langle a, b \rangle = b$.

In [145] it is shown how each such PCA A gives rise to a tripos $\begin{array}{c} \text{UFam}(PA) \\ \downarrow \\ \textbf{Sets} \end{array}$.
As predicates on a set I one takes functions $\varphi: I \to PA$. These are pre-ordered
by the relation $\vdash$, given as:

$$\varphi \vdash \psi \;\Leftrightarrow\; \left(\bigcap_{i \in I} \varphi(i) \supset \psi(i) \right) \neq \emptyset$$

where for subsets $X, Y \subseteq A$,

$$X \supset Y = \{ f \in A \mid \forall a \in X.\, f \cdot a \!\downarrow\ \text{and}\ f \cdot a \in Y \}.$$

Notice the uniformity: for $\varphi \vdash \psi$ to hold, there must be a single "realiser"
$a \in A$ with $a \in \varphi(i) \supset \psi(i)$ for all $i \in I$.

There are the usual propositional connectives for these predicates on I:

$$
\begin{aligned}
\top_I &= \lambda i \in I.\, A \\
\bot_I &= \lambda i \in I.\, \emptyset \\
\varphi \wedge \psi &= \lambda i \in I.\, \{ \langle a, b \rangle \mid a \in \varphi(i) \text{ and } b \in \psi(i) \} \\
\varphi \vee \psi &= \lambda i \in I.\, \{ \langle K, a \rangle \mid a \in \varphi(i) \} \cup \{ \langle KI, b \rangle \mid b \in \psi(i) \} \\
\varphi \supset \psi &= \lambda i \in I.\, \varphi(i) \supset \psi(i).
\end{aligned}
$$

We show that $\vee$ is join and leave the other cases as exercises. We have $\varphi \vdash$
$\varphi \vee \psi$, since

$$\lambda x.\, \langle x, K \rangle \in \bigcap_{i \in I} \varphi(i) \supset (\varphi \vee \psi)(i)$$

and similarly $\psi \vdash \varphi \vee \psi$. Next suppose we have $\varphi \vdash \chi$ and $\psi \vdash \chi$, say via
realisers

$$f \in \bigcap_{i \in I} \varphi(i) \supset \chi(i) \qquad \text{and} \qquad g \in \bigcap_{i \in I} \psi(i) \supset \chi(i).$$

Then

$$h = \lambda z.\, \langle f, g \rangle (\pi z)(\pi' z) \in \bigcap_{i \in I} (\varphi \vee \psi)(i) \supset \chi(i).$$

Indeed, if $\langle K, a \rangle \in (\varphi \vee \psi)(i)$, with $a \in \varphi(i)$, then

$$h \langle K, a \rangle = \langle f, g \rangle K a = \pi \langle f, g \rangle a = f a \in \chi(i)$$

and similarly $h\langle \mathbf{KI}, b \rangle = gb \in \chi(i)$.

For a predicate $\varphi \colon I \times J \to PA$, we define an equality predicate $\mathrm{Eq}(\varphi)$ by

$$\mathrm{Eq}(\varphi) = \lambda(i, j, j') \in (I \times J \times J). \begin{cases} \varphi(i, j) & \text{if } j = j' \\ \emptyset & \text{otherwise.} \end{cases}$$

Then it can be shown that $\mathrm{Eq}(\varphi) \vdash \psi \iff \varphi \vdash \delta(I, J)^*(\psi)$. This yields equality. Finally, for a function $u \colon I \to J$ in **Sets** and a predicate $\varphi \colon I \to PA$, we put

$$\prod_u(\varphi) = \lambda j \in J. \bigcap_{i \in I} (u(i) =_J j) \supset \varphi(i)$$

$$\coprod_u(\varphi) = \lambda j \in J. \bigcup_{i \in I} (u(i) =_J j) \wedge \varphi(i)$$

where

$$(u(i) =_J j) = \mathrm{Eq}(u \circ \pi, \pi')(i, j) = \begin{cases} A & \text{if } u(i) = j \\ \emptyset & \text{else} \end{cases}$$

(In case $I = \emptyset$, the above intersection over I equals A.) Then one easily checks that $\psi \vdash \prod_u(\varphi) \iff (\psi \circ u) \vdash \varphi$ and $\coprod_u(\varphi) \vdash \psi \iff \varphi \vdash (\psi \circ u)$. Beck-Chevalley holds for these products and coproducts.

It may be clear that if we apply this construction to the PCA $(\mathbb{N}, \cdot)$ with $\cdot$ for Kleene application, then we get the realisability tripos $\begin{smallmatrix} \mathrm{UFam}(P\mathbb{N}) \\ \downarrow \\ \mathbf{Sets} \end{smallmatrix}$ which was first introduced in Example 4.2.6. But the construction also yields 'realisability triposes' starting from models of the untyped λ-calculus, like $(D_\infty, \cdot)$ or $(P\omega, \cdot)$; the latter (and especially the resulting topos) is investigated in [261] within 'synthetic domain theory'.

There are variations on the above construction: in [147] the starting point is a 'right-absorptive C-PCA' which serves as a bases for a tripos using modified realisability. The resulting topos (as in Section 6.1) is used to give generic proofs of strong normalisation for various typed λ-calculi. In [239, Chapter IV] a tripos is constructed which captures another version of realisability, namely Lifschitz' realisability—and the resulting topos is studied.

Next we turn to an important class of examples of higher order fibrations.

5.3.5. Definition. A **topos** is a category $\mathbb{B}$ with finite limits such that its subobject fibration $\begin{smallmatrix} \mathrm{Sub}(\mathbb{B}) \\ \downarrow \\ \mathbb{B} \end{smallmatrix}$ is a (split) higher order fibration.

The subobject fibration of a topos thus has a generic object. In this situation with a poset fibration it does not matter whether we call this generic object 'split' or not. The same applies for the rest of the higher order structure.

Such a generic object corresponds to a 'subobject classifier', which gives a correspondence between subobjects and characteristic maps, as in Example 5.2.3 (ii). This will be made explicit in the next result. It forms the core of a more elementary description of toposes in the next section.

5.3.6. Lemma. *A subobject fibration* $\begin{smallmatrix} \mathrm{Sub}(\mathbb{B}) \\ \downarrow \\ \mathbb{B} \end{smallmatrix}$ *has a (split) generic object if and only if* $\mathbb{B}$ *has a* **subobject classifier***. The latter is a (monic) map* **true**$: 1 \to \Omega$ *such for each mono* $m\colon X \rightarrowtail I$ *there is a unique 'character- istic' or 'classifying' morphism* $\mathrm{char}(m)\colon I \to \Omega$ *forming a pullback diagram,*

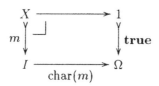

Proof. By Lemma 5.2.2 the fibration $\begin{smallmatrix} \mathrm{Sub}(\mathbb{B}) \\ \downarrow \\ \mathbb{B} \end{smallmatrix}$ has a split generic object if and only if there is a subobject **true**$: \Omega_0 \rightarrowtail \Omega$ such that for each subobject $m\colon X \rightarrowtail I$ there is a unique map $\mathrm{char}(m)\colon I \to \Omega$ with $\mathrm{char}(m)^*(\mathbf{true}) = m$, as subobjects. The latter means that there is a pullback diagram,

$$\begin{array}{ccc} X & \longrightarrow & \Omega_0 \\ {\scriptstyle m}\downarrow & \lrcorner & \downarrow {\scriptstyle \mathbf{true}} \\ I & \underset{\mathrm{char}(m)}{\longrightarrow} & \Omega \end{array}$$

Thus if $\mathbb{B}$ has a subobject classifier **true**$: 1 \to \Omega$ as in the lemma, then the subobject fibration obviously has a split generic object. The converse holds if we can show that for the above mono **true**$: \Omega_0 \rightarrowtail \Omega$ the object Ω_0 is terminal. This will be done: for each object $I \in \mathbb{B}$, the identity mono $I \rightarrowtail I$ yields a unique map $f = \mathrm{char}(\mathrm{id})\colon I \to \Omega$ and a pullback diagram as on the left below. Thus we have at least one map $f'\colon I \to \Omega_0$. If also $g\colon I \to \Omega_0$, then we get a pullback as on the right, since **true** is a mono.

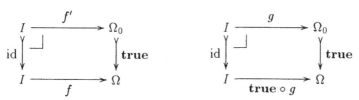

But then by uniqueness $f = \mathbf{true} \circ g$. Hence $g = f'$, since $\mathbf{true} \circ g = f = \mathbf{true} \circ f'$ and **true** is monic. $\qquad \square$

The above notion of subobject classifier was first formulated by Lawvere and Tierney in 1969 in their axiomatisation of set theory and sheaf theory. Here we treat a subobject classifier as an instance of a generic object. More about toposes may be found in the next few sections. At this stage we only mention that the category of sets is a topos. The subobject classifier is the generic object $1 \to 2$ described in Example 5.2.3 (ii).

In toposes the extensionality of entailment rule from Section 5.1 comes for free.

5.3.7. Lemma. *The subobject fibration of a topos is a model of higher order logic with extensional entailment.*

Proof. Assume $\mathbb{B}$ is a topos, and let $f, g \colon J \rightrightarrows \Omega^I$ be predicates satisfying the assumptions of the extensionality of entailment rule. This means that

$$(\text{ev} \circ f \times \text{id})^*(\textbf{true}) = (\text{ev} \circ g \times \text{id})^*(\textbf{true}),$$

as subobjects of $J \times I$. But then, by uniqueness, one gets $\text{ev} \circ f \times \text{id} = \text{ev} \circ g \times \text{id}$, and thus $f = g$. $\qquad \square$

We mention two further examples of a higher order fibration, involving the fibration of regular subobjects (see Exercise 1.3.6) in the categories of ω-sets and of PERs. It is left to the reader to check that a split fibration $\begin{smallmatrix} \text{RegSub}(\mathbb{B}) \\ \downarrow \\ \mathbb{B} \end{smallmatrix}$ of regular subobjects in a category $\mathbb{B}$ with finite limits has a (split) generic object if and only if the category $\mathbb{B}$ has a **regular subobject classifier**: a regular mono $\textbf{true}\colon 1 \rightarrowtail \Omega$ such that for any regular mono $m \colon I' \rightarrowtail I$ there is a unique classifying map $I \to \Omega$ which yields m as pullback of $\textbf{true}$.

5.3.8. Lemma. *Consider the category ω-Sets of ω-sets described in Section 1.2. Recall that it comes with a left adjoint $\nabla \colon \textbf{Sets} \to \omega\text{-}\textbf{Sets}$ to the forgetful functor $(I, E) \mapsto I$.*

(i) *Regular subobjects of an object $(I, E) \in \omega\text{-}\textbf{Sets}$ correspond to subsets $X \subseteq I$, with existence predicate inherited from (I, E).*

(ii) *The fibration $\begin{smallmatrix} \text{RegSub}(\omega\text{-}\textbf{Sets}) \\ \downarrow \\ \omega\text{-}\textbf{Sets} \end{smallmatrix}$ of regular subobjects in ω-Sets has $\nabla 2 \in \omega\text{-}\textbf{Sets}$ as (split) generic object—where $2 = \{\bot, \top\}$.*

Proof. (i) Given an object $(I, E) \in \omega\text{-}\textbf{Sets}$ and a subset $X \subseteq I$ of its carrier set, consider its characteristic function $I \to 2$ and the function $I \to 2$ which is constantly $\top \in 2$. These form a pair of parallel maps $(I, E) \rightrightarrows \nabla 2$ in ω-Sets, the equaliser of which is given by the inclusion $(X, E \restriction X) \rightarrowtail (I, E)$.

Conversely, if $m \colon (X, E_X) \rightarrowtail (I, E_I)$ is equaliser of $f, g \colon (I, E_I) \rightrightarrows (J, E_J)$, then $X' = \{i \in I \mid f(i) = g(i)\} \subseteq I$ comes with an inclusion $(X', E_I \restriction X') \rightarrowtail (I, E_I)$ which equalises f, g. Therefore it must be isomorphic to m.

(ii) For $(I, E) \in \omega\text{-}\mathbf{Sets}$, there are isomorphisms between the sets of

(a) regular subobjects of (I, E)
(b) subsets $X \subseteq I$
(c) functions $I \to 2$ in **Sets**
(d) morphisms $(I, E) \to \nabla 2$ in $\omega\text{-}\mathbf{Sets}$

Thus $\nabla 2 \in \omega\text{-}\mathbf{Sets}$ is a split generic object for the fibration of regular subobjects. $\qquad\square$

5.3.9. Proposition. *The regular subobjects in $\omega\text{-}\mathbf{Sets}$ give rise to a split higher order fibration* $\begin{array}{c} \text{RegSub}(\omega\text{-}\mathbf{Sets}) \\ \downarrow \\ \omega\text{-}\mathbf{Sets} \end{array}$ *. Its logic is classical.*

Proof. The generic object comes from the previous lemma. Fibred finite conjunctions and disjunctions are given by finite intersections and unions. The exponent $X \Rightarrow Y$ of $X, Y \subseteq I$ for $(I, E) \in \omega\text{-}\mathbf{Sets}$ is given by $X \Rightarrow Y = (I - X) \cup Y$. Thus the negation $\neg X$ of X is its complement $(I - X)$. Quantification along a projection $\pi\colon (I, E) \times (J, E) \to (I, E)$ in $\omega\text{-}\mathbf{Sets}$ are also given by the set theoretic formulas:

$$\text{product:} \quad (X \subseteq I \times J) \mapsto \{i \in I \mid \forall j \in J. (i, j) \in X\}$$
$$\text{coproduct:} \quad (X \subseteq I \times J) \mapsto \{i \in I \mid \exists j \in J. (i, j) \in X\}. \qquad\square$$

This will turn out to be an instance of a more general result: the regular subobject fibration of a category of separated objects in a topos is a higher order fibration with classical logic, see Corollary 5.7.12. This general result applies, since $\omega\text{-}\mathbf{Sets}$ will turn out to be the category of regular objects in the effective topos **Eff**, see Section 6.2.

The situation for regular subobjects in the category **PER** is different.

5.3.10. Proposition. *Regular subobjects in the category **PER** form a first order fibration, but not a higher order fibration.*

Proof. The first order structure of the fibration $\begin{array}{c} \text{RegSub}(\mathbf{PER}) \\ \downarrow \\ \mathbf{PER} \end{array}$ is described in Proposition 4.5.7. Here we show that it does not have a generic object, following an argument due to Streicher. Suppose, towards a contradiction, that $\Omega \in \mathbf{PER}$ is a generic object. Then for $R \in \mathbf{PER}$ there should be isomorphisms

$$\mathbf{PER}(R, \Omega) \cong \text{RegSub}(R)$$
$$\cong P(\mathbb{N}/R) \quad \text{by Proposition 4.5.7 (i).}$$

The homset $\mathbf{PER}(R, \Omega)$, like any homset in **PER**, is countable. But the powerset $P(\mathbb{N}/R)$ can be uncountable, for example if R is the natural numbers object $N = (\text{Eq}(\mathbb{N}) \subseteq \mathbb{N} \times \mathbb{N})$ with quotient $\mathbb{N}/N \cong \mathbb{N}$. $\qquad\square$

We conclude by noting that these fibrations of regular subobjects in ω-**Sets** and in **PER** arise in the following change-of-base situations.

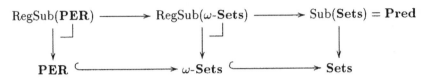

Exercises

5.3.1.   Verify that $\mathbb{N}$ with Kleene application $\cdot$ is a PCA.

5.3.2.   Check some more details in the realisability tripos construction in Example 5.3.4, especially,

     (i)   that the connectives $\top, \bot, \wedge, \supset$ in the fibre have the required properties;

     (ii)   that $\prod_u, \coprod_u$ are right and left adjoint to substitution u^*;

     (iii)   that Beck-Chevalley holds for these products and coproducts.

5.3.3.   Show that the category of finite sets is a topos.

5.3.4.   Let $\begin{smallmatrix}\mathbb{E}\\\downarrow p\\\mathbb{B}\end{smallmatrix}$ be a higher order fibration, say with generic object $T \in \mathbb{E}_\Omega$. For parallel morphisms $u, v : I \rightrightarrows \Omega$ put $u \leq v$ if and only if $u^*(T) \leq v^*(T)$ in the fibre over I.

     (i)   Show that each homset $\mathbb{B}(I, \Omega)$ is a Heyting pre-algebra.
         [*Hint.* Use Remark 5.2.12.]

     (ii)   Show that Ω is internally complete and cocomplete in the following sense. For each pair of objects $I, J \in \mathbb{B}$, the functor (between preorders)

$$\mathbb{B}\big(I, \, \Omega \big) \xrightarrow{\quad - \circ \pi \quad} \mathbb{B}\big(I \times J, \, \Omega \big)$$

     has both a right and a left adjoint.

     (iii)   Assume that the equaliser exists of $\wedge, \pi : \Omega \times \Omega \rightrightarrows \Omega$ and write it as $\leq \rightarrowtail \Omega \times \Omega$—where $\wedge$ is the induced conjunction map on Ω as in Remark 5.2.12. Prove that for $u, v : I \rightrightarrows \Omega$ one has $u \leq v$ as above if and only if $\langle u, v \rangle : I \rightarrow \Omega \times \Omega$ factors through $\leq \rightarrowtail \Omega \times \Omega$.

5.3.5.   Prove that a category $\mathbb{B}$ with finite limits has a regular subobject classifier if and only if its regular subobject fibration $\begin{smallmatrix}\mathrm{RegSub}(\mathbb{B})\\\downarrow\\\mathbb{B}\end{smallmatrix}$ has a (split) generic object.

5.4 Elementary toposes

In the previous section we introduced toposes as categories whose subobject fibrations are higher order fibrations. This gives a distinctly logical description of toposes. It turns out that there are more elementary formulations of this notion. The first alternative formulation will be given below. We will show that it is equivalent to the previous definition. This involves some basic (logical) constructions in toposes. Two other alternatives will be discussed in the next section

There is much more to say about toposes than the few logical aspects that we touch upon below, and in the next four sections. Here, we merely collect some useful facts for the readers convenience, mainly as a preparation for the effective topos **Eff**, to be introduced in the next chapter. Not all details are given; more information may be found in the extensive literature, see *e.g.* [188, 169, 18, 24, 218] and the references given there.

5.4.1. Definition. (i) An (**elementary**) **topos** is a category $\mathbb{B}$ which has

- finite limits;
- exponents (so that $\mathbb{B}$ is Cartesian closed);
- a subobject classifier **true**: $1 \rightarrowtail \Omega$. Thus for each mono $m: X \rightarrowtail I$, there is a unique characteristic map $\text{char}(m): I \to \Omega$ with $m \cong \text{char}(m)^*(\textbf{true})$, as in,

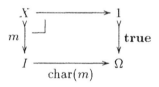

(ii) A **logical morphism** between two toposes $\mathbb{B}, \mathbb{B}'$ is a functor $F: \mathbb{B} \to \mathbb{B}'$ which preserves finite limits, exponents and the subobject classifier. The latter means that the canonical map $F\Omega \to \Omega'$ is an isomorphism in

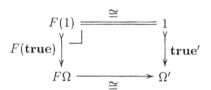

We can immediately see (by Lemma 5.3.6) that if a subobject fibration $\underset{\mathbb{B}}{\overset{\text{Sub}(\mathbb{B})}{\downarrow}}$ is a higher order fibration—so that $\mathbb{B}$ is a topos as defined in the previous section—then $\mathbb{B}$ is an elementary topos. Our aim in this section is to prove that the converse also holds, *i.e.* that the elementary description coincides with the logical description.

5.4.2. Example. As we have already seen in the previous section, the category **Sets** is a topos with subobject classifier

$$1 \xrightarrow{\quad \mathbf{true} = \lambda x.\, 1 \quad} \{0,1\} = 2$$

More generally, for each locally small category $\mathbb{C}$, the category $\widehat{\mathbb{C}} = \mathbf{Sets}^{\mathbb{C}^{\mathrm{op}}}$ of presheaves $\mathbb{C}^{\mathrm{op}} \to \mathbf{Sets}$ and natural transformations between them, is a topos. Finite limits are computed pointwise as in **Sets**. The exponents and subobject classifier are obtained via the Yoneda Lemma, as will be sketched.

For presheaves $F, G\colon \mathbb{C}^{\mathrm{op}} \rightrightarrows \mathbf{Sets}$, the exponent $F \Rightarrow G\colon \mathbb{C}^{\mathrm{op}} \to \mathbf{Sets}$ should satisfy

$$
\begin{aligned}
(F \Rightarrow G)(X) &\cong \widehat{\mathbb{C}}\big(\mathbb{C}(-, X),\; F \Rightarrow G\big) \quad \text{by Yoneda} \\
&\cong \widehat{\mathbb{C}}\big(\mathbb{C}(-, X) \times F,\; G\big) \quad \text{because } F \Rightarrow G \text{ is exponent.}
\end{aligned}
$$

Therefore, one simply defines,

$$(F \Rightarrow G)(X) \stackrel{\mathrm{def}}{=} \widehat{\mathbb{C}}\big(\mathbb{C}(-, X) \times F,\; G\big).$$

The verification that this indeed yields exponents in $\widehat{\mathbb{C}}$ is a bit involved, but in essence straightforward.

A subobject $S \rightarrowtail \mathbb{C}(-, X)$ of a representable presheaf $\mathbb{C}(-, X)$ can be identified with a **sieve on** $X \in \mathbb{C}$. That is, with a set S of arrows with codomain X (*i.e.* $S \subseteq \mathrm{Obj}\,\mathbb{C}/X$) which is "down closed":

$$
\left.
\begin{array}{c}
\xrightarrow{\;g,\;f\;} \text{in } \mathbb{C} \\[-2pt]
f \in S
\end{array}
\right\} \;\Rightarrow\; f \circ g \in S
$$

Thus an appropriate presheaf $\Omega\colon \mathbb{C}^{\mathrm{op}} \to \mathbf{Sets}$ should satisfy

$$
\begin{aligned}
\Omega(X) &\cong \widehat{\mathbb{C}}\big(\mathbb{C}(-, X),\; \Omega\big) &&\text{by Yoneda} \\
&\cong \mathrm{Sub}\big(\mathbb{C}(-, X)\big) &&\text{because } \Omega \text{ classifies subobjects} \\
&\cong \{S \mid S \text{ is a sieve on } X\}.
\end{aligned}
$$

Hence one simply puts

$$\Omega(X) \stackrel{\mathrm{def}}{=} \{S \mid S \text{ is a sieve on } X\}.$$

And for a morphism $f\colon X \to Y$ in $\mathbb{C}$ there is a map $\Omega(f)\colon \Omega(X) \to \Omega(Y)$ defined by

$$(T, \text{ sieve on } Y) \mapsto \{g\colon Y \to X \mid f \circ g \in T\}.$$

The generic subobject $\mathbf{true}\colon 1 \to \Omega$ is then given by maximal sieves:

$$\mathbf{true}_X(*) = {\downarrow} X = \{f \in \mathrm{Arr}\,\mathbb{C} \mid \mathrm{cod}(f) = X\}.$$

We leave it as an exercise to verify all remaining details.

5.4.3. Notation. In a topos, we write PI for the **power object** Ω^I. It comes equipped with a **membership predicate** $\in_I \rightarrowtail PI \times I$; it is the subobject corresponding to the evaluation map ev: $PI \times I \rightarrow \Omega$ as $\mathrm{ev}^*(\mathbf{true})$. For maps $x: J \rightarrow I$ and $a: J \rightarrow PI$ we can then write

$$x \in_I a \iff \langle a, x \rangle \text{ factors through } \in_I \rightarrowtail PI \times I.$$

Also there is a **singleton map** $\{\ \}: I \rightarrow PI$, obtained in the following way. The diagonal morphism $\delta(I) = \langle \mathrm{id}, \mathrm{id} \rangle: I \rightarrowtail I \times I$ on I has a characteristic map $\mathrm{char}(\delta(I)): I \times I \rightarrow \Omega$. The exponential transpose of the latter is the singleton map:

$$\{\ \} \overset{\mathrm{def}}{=} \left(I \xrightarrow{\quad \Lambda(\mathrm{char}(\delta(I))) \quad} \Omega^I = PI \right)$$

Next, consider this singleton map in the mono $\langle \{\ \}, \mathrm{id} \rangle: I \rightarrowtail PI \times I$, and its characteristic map $PI \times I \rightarrow \Omega$. Exponentiation yields a morphism $s: PI \rightarrow PI$. Informally, $s(a) = \{x \mid \{x\} = a\}$. We form the lift object $\bot I$ via the equaliser:

$$\bot I \rightarrowtail PI \underset{\mathrm{id}}{\overset{s}{\rightrightarrows}} PI$$

In **Sets**, the power exponent 2^I is the ordinary powerset PI of I. And $\bot I$ is the **lift** of I: the pointed set $\bot I = \{\emptyset\} \cup \{\{i\} \mid i \in I\} \subseteq PI$ obtained from I by adding a base point. Partial functions $J \rightharpoonup I$ between sets correspond to total functions $J \rightarrow \bot I$. This will be generalised to arbitrary toposes. But first we need an elementary result.

5.4.4. Lemma. (i) *The singleton map* $\{\ \}: I \rightarrow PI$ *is monic.*

(ii) *The singleton map* $\{\ \}$ *factors through* $\bot I \rightarrowtail PI$, *i.e. it restricts to a map* $\{\ \}: I \rightarrow \bot I$, *which is a mono again by* (i).

As a result of this lemma, x is the only element of $\{x\} = \{\ \} \circ x$, see Exercise 5.4.3 below.

Proof. (i) Assume parallel maps $u, v: J \rightrightarrows I$ with $\{\ \} \circ u = \{\ \} \circ v$. Then one gets $\mathrm{char}(\delta(I)) \circ u \times \mathrm{id} = \mathrm{char}(\delta(X)) \circ v \times \mathrm{id} = w$, say. Consider the corresponding subobject $w^*(\mathbf{true})$ in,

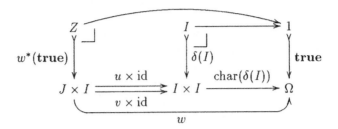

Both $\langle u, \mathrm{id}\rangle\colon I \rightarrowtail J \times I$ and $\langle v, \mathrm{id}\rangle\colon I \rightarrowtail J \times I$ are obtained by pullback of **true** along w—since

$$\langle u, \mathrm{id}\rangle = (u \times \mathrm{id})^*(\delta(I)) \qquad \text{and} \qquad \langle v, \mathrm{id}\rangle = (v \times \mathrm{id})^*(\delta(I)).$$

Hence $\langle u, \mathrm{id}\rangle = \langle v, \mathrm{id}\rangle$, as subobjects of $J \times I$, and so $u = v$.

(ii) We have to show that $s \circ \{\,\} = \{\,\}$, where $s\colon PI \to PI$ is as introduced in Notation 5.4.3. We compute:

$$s \circ \{\,\} = \Lambda(\mathrm{char}(\langle\{\,\}, \mathrm{id}\rangle) \circ \{\,\} \times \mathrm{id}) \stackrel{(*)}{=} \Lambda(\mathrm{char}(\delta(I)) = \{\,\}$$

where the equality $(*)$ comes from the fact that the left square $(**)$ below is a pullback by (i).

$$
\begin{array}{ccccc}
I & \!=\!=\!=\!=\!=\! & I & \longrightarrow & 1 \\[2pt]
{\scriptstyle \delta(I)}\big\downarrow & (**) & {\scriptstyle \langle\{\,\}, \mathrm{id}\rangle}\big\downarrow & & \big\downarrow{\scriptstyle \textbf{true}} \\[2pt]
I \times I & \underset{\{\,\} \times \mathrm{id}}{\rightarrowtail} & PI \times I & \underset{\mathrm{char}\langle\{\,\}, \mathrm{id}\rangle}{\longrightarrow} & \Omega
\end{array}
\qquad \square
$$

Categorically, a partial map $I \rightharpoonup J$ is (an equivalence class of) a span

$$I \xleftarrowtail{\;\;m\;\;} X \xrightarrow{\;\;u\;\;} J$$

It tells that u is defined on a subset X of I. Two such spans $I \xleftarrowtail{m} X \xrightarrow{u} J$ and $I \xleftarrowtail{n} Y \xrightarrow{v} J$ are equivalent if there is a necessarily unique isomorphism $\varphi\colon X \xrightarrow{\cong} Y$ with $n \circ \varphi = m$ and $v \circ \varphi = u$. As for subobjects, one usually does not distinguish notationally between such a span and its equivalence class.

5.4.5. Proposition. *The singleton map* $\{\,\}\colon J \rightarrowtail \bot J$ *is a* **partial map classifier**: *for each partial map* $I \xleftarrowtail{m} X \xrightarrow{u} J$, *there is a unique morphism*

$v: I \to \bot J$ *forming a pullback square,*

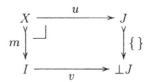

Proof. Given $I \xleftarrow{m} X \xrightarrow{u} J$, consider the 'graph' mono $\langle m, u \rangle: X \rightarrowtail I \times J$, its characteristic map $I \times J \to \Omega$, and the resulting exponential transpose $I \to \Omega^J = PJ$. The latter factors through $\bot J \rightarrowtail PJ$. $\square$

5.4.6. Corollary. *The assignment* $I \mapsto \bot I$ *is functorial, and the singleton maps* $\{\,\}: I \to \bot I$ *are components of a natural transformation* id $\Rightarrow \bot$.

Proof. For a map $u: I \to J$, there is a partial map $\bot I \xleftarrow{\{\,\}_I} I \xrightarrow{u} J$ and thus a unique morphism $\bot u: \bot I \to \bot J$, in a pullback square

$$
\begin{array}{ccc}
I & \xrightarrow{\ \ u\ \ } & J \\
{\scriptstyle \{\,\}_I} \downarrow & \quad\lrcorner & \downarrow {\scriptstyle \{\,\}_J} \\
\bot I & \xrightarrow[\bot u]{} & \bot J
\end{array}
$$
$\square$

Such classification of partial maps is an important first step in the axiomatisation of domain theory, see *e.g.* [144, 81, 259], and also [296].

Next we are going to show that every topos is locally Cartesian closed (*i.e.* that all of its slice categories are Cartesian closed). In **Sets**, the exponent in the slice **Sets**/I of two I-indexed families $(X_i)_{i \in I}$ and $(Y_i)_{i \in I}$ is the pointwise exponent (function space) $(X_i \Rightarrow Y_i)_{i \in I}$. It can alternatively be described in terms of suitable partial maps $f: X \rightharpoonup Y$, namely those f with for all $x \in X_i$, $f(x)$ is defined and $f(x) \in Y_i$. This will be used below.

5.4.7. Proposition. *A topos is a locally Cartesian closed category (LCCC). A logical morphism preserves the LCCC-structure.*

Proof. A topos $\mathbb{B}$ has finite limits by definition, so that each slice category $\mathbb{B}/I$ has finite limits. We only have to show that $\mathbb{B}/I$ is Cartesian closed. Assume therefore families $\left(\begin{smallmatrix} X \\ \downarrow \varphi \\ I \end{smallmatrix} \right)$ and $\left(\begin{smallmatrix} Y \\ \downarrow \psi \\ I \end{smallmatrix} \right)$. We then have a partial map $I \times X \rightharpoonup I$, namely

$$
I \times X \xleftarrow{\ \langle \varphi, \mathrm{id} \rangle\ } X \xrightarrow{\ \varphi\ } I
$$

Let it be classified by $\widehat{\varphi} \colon I \times X \to \bot I$. We define the exponent family $\varphi \Rightarrow \psi$ to be

$$\begin{array}{ccc}
W & \xrightarrow{\ \ w\ \ } & (\bot Y)^X \\
{\scriptstyle \varphi \Rightarrow \psi} \downarrow \ \lrcorner & & \downarrow {\scriptstyle (\bot \psi)^X \,=\, \Lambda(\bot \psi \,\circ\, \mathrm{ev})} \\
I & \xrightarrow[\ \Lambda(\widehat{\varphi})\ ]{} & (\bot I)^X
\end{array}$$

For an arbitrary family $\left(\begin{smallmatrix} Z \\ \downarrow X \\ I \end{smallmatrix} \right)$, we have to establish a bijective correspondence between maps $f \colon \chi \times \varphi \to \psi$ and $g \colon \chi \to (\varphi \Rightarrow \psi)$ in $\mathbb{B}/I$. It arises as follows.

- Given $f \colon Z \times_I X \to Y$, consider the partial map $Z \times X \rightharpoonup Y$,

$$Z \times X \longleftarrowtail Z \times_I X \xrightarrow{\ \ f\ \ } Y$$

It induces a map $\widehat{f} \colon Z \times X \to \bot Y$, and hence $\Lambda(\widehat{f}) \colon Z \to (\bot Y)^X$. The latter, together with $\chi \colon Z \to I$ yields a mediating map $Z \dashrightarrow W$ with respect to the above pullback. It is the map we want.
- Conversely, given $g \colon Z \to W$, one obtains the appropriate map $Z \times_I X \dashrightarrow Y$ in,

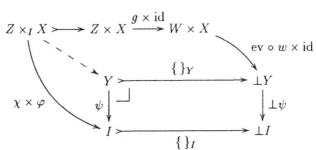

These exponents in the slices are preserved by logical morphisms, because they are defined in terms of the topos structure as in Definition 5.4.1. $\qquad \square$

This result has important consequences for the codomain and subobject fibrations of a topos.

5.4.8. Corollary. *If $\mathbb{B}$ is a topos, then its codomain fibration $\begin{smallmatrix} \mathbb{B}^{\to} \\ \downarrow \\ \mathbb{B} \end{smallmatrix}$ is fibrewise a topos: each fibre $\mathbb{B}/I$ is a topos and reindexing functors u^* are logical morphisms (they preserve the topos structure).*

Proof. Each slice has finite limits and exponents; the latter by the previous result. And if **true**: $1 \rightarrowtail \Omega$ is subobject classifier in $\mathbb{B}$, then the map $I^*(\textbf{true}) = \text{id} \times \textbf{true}: I \times 1 \to I \times \Omega$ is a morphism between families,

$$1 = \begin{pmatrix} I \times 1 \\ \downarrow\pi \\ I \end{pmatrix} \xrightarrow{\;I^*(\textbf{true})\;} \begin{pmatrix} I \times \Omega \\ \downarrow\pi \\ I \end{pmatrix} = I^*(\Omega)$$

which is a subobject classifier in $\mathbb{B}/I$. The proof uses that a map between families in $\mathbb{B}/I$ is a mono in $\mathbb{B}/I$ if and only if it is a mono in $\mathbb{B}$.

Obviously, pullback functors preserve all this structure. $\qquad\square$

5.4.9. Corollary. *If $\mathbb{B}$ is a topos, then its subobject fibration* $\genfrac{}{}{0pt}{}{\text{Sub}(\mathbb{B})}{\underset{\mathbb{B}}{\downarrow}}$ *is a higher order fibration.*

A category $\mathbb{B}$ is thus a topos as in Definition 5.4.1 in this section if and only if $\mathbb{B}$ is a topos as defined in the previous section.

Proof. Because a topos $\mathbb{B}$ is locally Cartesian closed, each pullback functor $u^*: \mathbb{B}/J \to \mathbb{B}/I$ has a right adjoint $\prod_u$, by Proposition 1.9.8 (iii). These functors $\prod_u$ restrict to functors $\prod_u: \text{Sub}(I) \to \text{Sub}(J)$, because right adjoints preserves monos. With these products we can define implication $\supset$ as in the proof of Theorem 4.5.5. Thus, in the subobject fibration of a topos, we already have $\Omega, \forall, \supset, \wedge, \top$ and $=$. The latter three always exist in subobject fibrations. The missing logical operations $\bot, \vee, \exists$ are then definable, using $\Omega, \forall$ and $\supset$, as in Example 5.1.4. Thus we have a higher order subobject fibration. $\qquad\square$

Exercises

5.4.1. Use characteristic maps to show that each mono in a topos is a regular mono. Conclude that a map in a topos which is both a mono and an epi is an isomorphism. Categories with this property are sometimes called **balanced**.

5.4.2. Check that the constructions described in Example 5.4.2 indeed yield a topos of presheaves $\textbf{Sets}^{\mathbb{C}^{\text{op}}}$. Describe the subobject classifier **true**: $1 \to \Omega$ for $\mathbb{C}$ is (a) a monoid, (b) $2 = (\cdot \to \cdot)$, (c) $\mathbb{N} = (\cdot \to \cdot \to \cdot \to \cdots)$, and (d) an arbitrary poset,

5.4.3. Show that for 'generalised elements' $x, y: J \rightrightarrows I$ in a topos, one has

$$x \in_I \{y\} \iff x = y$$

where $\{y\} = \{\,\} \circ y: J \to PI$.

5.4.4. Show that
   (i) $\{\,\}: 1 \rightarrowtail \bot 1$ is **true**: $1 \rightarrowtail \Omega$;
   (ii) $\bot I \rightarrowtail PI$ is a split mono.

5.4.5. (From [169, 1.45]) For an object I in a topos, consider I as family over
        1, and form the product $\prod_{\mathbf{true}}(I)$ over Ω along $\mathbf{true}\colon 1 \to \Omega$. Notice that
        $\begin{pmatrix} I \\ \downarrow \\ 1 \end{pmatrix}$ can be obtained as pullback $\mathbf{true}^* \begin{pmatrix} I \\ \downarrow \\ \Omega \end{pmatrix}$, where $I \to \Omega$ is $\mathbf{true} \circ\ !$.
        There is thus a unit map $I \to \prod_{\mathbf{true}}(I)$. Prove that it gives an alternative
        description of the partial map classifier $\{\,\}\colon I \to \bot I$.

5.4.6. In the subobject fibration of a topos there are implication $\supset$ operations in
        the fibres. It induces a map $\supset\colon \Omega \times \Omega \to \Omega$, like in Remark 5.2.12. Prove
        that this map $\supset$ is the classifying map of the order $\leq\ \rightarrowtail\ \Omega \times \Omega$, obtained
        as equaliser of $\wedge, \pi\colon \Omega \times \Omega \rightrightarrows \Omega$.

5.4.7. Verify that a logical morphism preserves images. More generally, that a log-
        ical morphism between toposes yields a morphism preserving the structure
        of the corresponding higher order subobject fibrations.

5.5 Colimits, powerobjects and well-poweredness in a topos

In this section we mention some further results on toposes, which are of less
importance for the main line of this book. They involve two more alternative
formulations of the notion of topos: one involving powerobjects $PI = \Omega^I$,
and one involving well-poweredness of the associated codomain fibration (in
a fibred sense). Also we show that every topos has finite colimits. The proof
involves some special properties of subobject fibrations.

We start with powerobjects.

5.5.1. Theorem. *A category $\mathbb{B}$ is a topos if and only if it has both*

- *finite limits;*
- **power objects**: *for each object I there is a power object PI together with
  a "membership" relation $\in_I\ \rightarrowtail\ PI \times I$ which is universal in the following
  sense: for each relation $R\ \rightarrowtail\ J \times I$ there is a unique "relation classifier"
  $r\colon J \to PI$ forming a pullback square,*

$$
\begin{array}{ccc}
R & \longrightarrow & \in_I \\
\downarrow & \lrcorner & \downarrow \\
J \times I & \xrightarrow{\ r \times \mathrm{id}\ } & PI \times I
\end{array}
$$

One thinks of $r\colon J \to PI$ as $j \mapsto \{i \in I \mid R(j,i)\}$.

Proof. If $\mathbb{B}$ is a topos, then one takes $PI = \Omega^I$ and $\in_I$ as classifier of
evaluation, as in Notation 5.4.3. Every relation $R\ \rightarrowtail\ J \times I$, as a mono,
has a classifying map $\mathrm{char}(R)\colon J \times I \to \Omega$ and thus we obtain a map

$r = \Lambda(\mathrm{char}(R)): J \to \Omega^I = PI$ by abstraction. The outer rectangle below is a pullback:

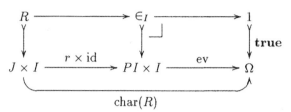

so that we can conclude from the Pullback Lemma that the rectangle on the left is a pullback.

In the reverse direction, if one has powerobjects, then the relation $\in_1 : 1 \rightarrowtail P1 \times 1 \cong P1$ is a subobject classifier. Further, exponents J^I can be constructed as suitable subobjects $J^I \rightarrowtail P(I \times J)$ of relations which are both single-valued and total. For the details, see *e.g.* [188, IV, 2]. $\qquad\square$

This result gives the most economical formulation of 'topos'; it is due to Kock. It is remarkable that the above two requirements suffice to give us all of the structure of higher order logic.

We also like to mention that taking powerobjects is functorial (and yields a monad, see Exercise 5.5.2 below).

5.5.2. Proposition. *For a topos* $\mathbb{B}$, *the assignment* $I \mapsto PI$ *extends to a functor* $\mathbb{B} \to \mathbb{B}$. *The singleton maps* $\{\ \}_I : I \to PI$ *are components of a natural transformation* $\mathrm{id}_{\mathbb{B}} \Rightarrow P$.

Proof. For a morphism $u: I \to J$, consider the image $\coprod_{\mathrm{id} \times u}(\in_I) \rightarrowtail PI \times J$ of the composite

$$\in_I \rightarrowtail PI \times I \xrightarrow{\ \mathrm{id} \times u\ } PI \times J$$

By the previous result, there is a unique classifying map $P(u): PI \to PJ$ in a pullback square

$$
\begin{array}{ccc}
\coprod_{\mathrm{id} \times u}(\in_I) & \longrightarrow & \in_J \\
\downarrow & & \downarrow \\
PI \times J & \xrightarrow[P(u) \times \mathrm{id}]{} & PJ \times J
\end{array}
$$

We get $P(u) \circ \{\ \}_I = \{\ \}_J \circ u: I \to PJ$, because both maps classify the same relation, namely the graph $\langle \text{id}, u \rangle: I \rightarrowtail I \times J$ of u, in:

$$
\begin{array}{ccccc}
I & \xrightarrow{\;\;u\;\;} & J & \longrightarrow & \in_J \\
{\scriptstyle \langle \text{id}, u \rangle}\downarrow & \lrcorner & {\scriptstyle \delta(I)}\downarrow & \lrcorner & \downarrow \\
I \times J & \xrightarrow[u \times \text{id}]{} & J \times J & \xrightarrow[\{\}_J \times \text{id}]{} & PJ \times J
\end{array}
$$

and

$$
\begin{array}{ccccc}
I & \longrightarrow & \coprod_{\text{id} \times u}(\in_I) & \longrightarrow & \in_J \\
{\scriptstyle \langle \text{id}, u \rangle}\downarrow & \lrcorner & \downarrow & \lrcorner & \downarrow \\
I \times J & \xrightarrow[\{\}_I \times \text{id}]{} & PI \times J & \xrightarrow[P(u) \times \text{id}]{} & PJ \times J
\end{array}
$$

where in the latter case the square on the left is a pullback by Beck-Chevalley:

$$
\begin{aligned}
(\{\}_I \times \text{id})^* \coprod\nolimits_{\text{id} \times u}(\in_I) &\cong \coprod\nolimits_{\text{id} \times u}((\{\}_I \times \text{id})^*(\in_I)) \\
&\cong \coprod\nolimits_{\text{id} \times u}(\delta(I)) \\
&\cong \langle \text{id}, u \rangle. \qquad \qquad \square
\end{aligned}
$$

We turn to finite colimits in a topos. Remarkably, they come for free. To see this we need the following auxiliary result.

5.5.3. Lemma. *Since a topos is a coherent category, it has a strict initial object* 0, *see Theorem 4.5.3. Write* **false**$: 1 \to \Omega$ *for the classifying map obtained in*

$$
\begin{array}{ccc}
0 & \longrightarrow & 1 \\
\downarrow & \lrcorner & \downarrow {\scriptstyle \text{true}} \\
1 & \xrightarrow[\text{false}]{} & \Omega
\end{array}
$$

For an arbitrary object I, *put* $\emptyset_I = \Lambda(\textbf{false} \circ \pi): 1 \rightarrowtail PI$. *Then* $\emptyset_I$ *and the singleton map* $\{\}_I: I \rightarrowtail PI$ *are disjoint: there is a pullback square*

$$
\begin{array}{ccc}
0 & \rightarrowtail & 1 \\
\downarrow & \lrcorner & \downarrow {\scriptstyle \emptyset_I} \\
I & \xrightarrow[\{\}_I]{} & PI
\end{array}
$$

Proof. Exercise. $\square$

5.5.4. Proposition. *Each topos has finite colimits; they are preserved by pullback functors (i.e. they are* **universal***).*

Moreover, these colimits are preserved by logical morphisms between toposes.

Proof. We already know that a topos has an initial object 0. For objects I, J consider the subobjects

$$\langle \{ \}_I, \emptyset_J \rangle \colon I \rightarrowtail PI \times PJ \qquad \text{and} \qquad \langle \emptyset_I, \{ \}_J \rangle \colon I \rightarrowtail PI \times PJ.$$

By the previous lemma, these are disjoint. Hence their join $I \vee J \rightarrowtail PI \times PJ$ is $I + J \rightarrowtail PI \times PJ$ by Exercise 4.5.1.

(More informally, one constructs the coproduct $I + J$ as the set

$$\{(a, b) \in PI \times PJ \mid (a \text{ is a singleton and } b \text{ is empty})$$

$$\text{or } (a \text{ is empty and } b \text{ is a singleton})\}.$$

See also [186, II,5, Exercise 2].)

In order to show that a topos has coequalisers, we use Proposition 4.8.6 (ii), and construct for a relation $\langle r_0, r_1 \rangle \colon R \rightarrowtail I \times I$ a quotient object I/R. Let $\langle \bar{r}_0, \bar{r}_1 \rangle \colon \overline{R} \rightarrowtail I \times I$ be the least equivalence relation containing R; it may be obtained as in Lemma 5.1.8. Write $r = \Lambda(\mathrm{char}(\overline{R})) \colon I \to PI$ for the relation classifier of $\overline{R}$, and factor this map as

$$I \xrightarrow{\;c\;} I/R \overset{m}{\rightarrowtail} PI$$

We must show that maps $I/R \to J$ are in bijective correspondence with maps of relations $R \to \mathrm{Eq}(J)$, *i.e.* with maps $u \colon I \to J$ satisfying $u \circ r_0 = u \circ r_1$.

For $i, i', j \in I$ with $R(i, i')$ one has a logical equivalence

$$\overline{R}(i, j) \supset\!\subset \overline{R}(i', j)$$

because $\overline{R}$ is symmetric and transitive. More categorically, one has an equality of subobjects,

$$(\bar{r}_0 \times \mathrm{id})^*(\overline{R}) = (\bar{r}_1 \times \mathrm{id})^*(\overline{R}) \quad \text{over } \overline{R} \times I.$$

As a result, $\mathrm{char}(\overline{R}) \circ \bar{r}_0 \times \mathrm{id} = \mathrm{char}(\overline{R}) \circ \bar{r}_1 \times \mathrm{id}$, and so $r \circ \bar{r}_0 = r \circ \bar{r}_1$. This gives us $c \circ \bar{r}_0 = c \circ \bar{r}_1$, because $r = m \circ c$ and m is a mono. We also get $c \circ r_0 = c \circ r_1$, since $R \leq \overline{R}$. Hence a morphism $v \colon I/R \to J$ gives rise to a morphism $u = v \circ c \colon I \to J$ with $u \circ r_0 = u \circ r_1$.

And if we have a morphism $u \colon I \to J$ with $u \circ r_0 = u \circ r_1$, then $R \leq (u \times u)^*(\delta(J)) = \mathrm{Ker}(u)$. Since this kernel $\mathrm{Ker}(u)$ is an equivalence relation containing R, we get $\overline{R} \leq \mathrm{Ker}(u)$. The required mediating map $I/R \to J$ is now obtained from the fact that covers are orthogonal to monos, see Lemma 4.4.6 (vii), in a diagram:

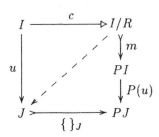

The outer rectangle commutes, as may be concluded from the following computation.

$$
\begin{aligned}
(P(u) \circ r)(i) &= P(u)(\{i' \in I \mid \overline{R}(i, i')\}) \\
&= \{u(i') \mid \overline{R}(i, i')\} \\
&= \{u(i)\} \qquad\qquad \text{by reflexivity of } \overline{R} \\
&= (\{\,\}_J \circ u)(i).
\end{aligned}
$$

Colimits in a topos are preserved by pullback functors u^*, because each u^* has a right adjoint $\prod_u$. And since these colimits are described in terms of the logical structure of a topos, they are preserved by logical morphisms between toposes.   □

In this proof we rely on logical tools. There is a more categorical argument due to Paré: by using Beck's Theorem one can show that for a topos $\mathbb{B}$, the opposite category $\mathbb{B}^{\mathrm{op}}$ is monadic over $\mathbb{B}$. Thus $\mathbb{B}^{\mathrm{op}}$ inherits limits from $\mathbb{B}$, *i.e.* $\mathbb{B}$ inherits colimits. Details may be found in [188, 169, 18]. This proof has the advantage that it directly applies to non-finite colimits as well; they exist in a topos as soon as the corresponding limits exist.

Notice that since colimits are stable under pullback, epimorphisms are preserved by pullback functors (since the fact that a map is an epi can be expressed in a pushout diagram). One can further show that coproducts are disjoint, but the argument is non-trivial, see *e.g.* [188, IV,6, Corollary 5].

5.5.5. Corollary. *The epis in a topos are precisely the covers (i.e. the regular epimorphisms).*

Proof. Images can be constructed as in Exercise 4.4.8 and in Example 5.1.9 (i). Explicitly, given a map $u : I \to J$ one forms the coequaliser

$I \twoheadrightarrow J'$ of u's kernel pair $I \times_J I \rightrightarrows I$, as in,

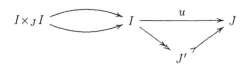

Then $J' \to J$ is internally injective, as proved in Example 5.1.9 (i). In a subobject fibration this means that it is monic. In case u is an epi itself, then so is $J' \rightarrowtail J$. Hence the latter is an isomorphism (see Exercise 5.4.1). Thus every epi is a regular epi and hence a cover, see Lemma 4.4.6 (viii). $\qquad \Box$

We turn to another characterisation of toposes, in terms of well-powered-ness. An (ordinary) category $\mathbb{C}$ is called **well-powered** if for each object $X \in \mathbb{C}$ the collection of subobjects of X is a small set (as opposed to a proper class). In a fibred definition of this concept the reference to small sets is eliminated and replaced by a reference to objects of a base category of a fibration.

For a fibration $\begin{smallmatrix} \mathbb{E} \\ \downarrow p \\ \mathbb{B} \end{smallmatrix}$ we say that a map in $\mathbb{E}$ is **vertically monic** if it is vertical, say in the fibre over I, and is a mono in this fibre category $\mathbb{E}_I$. We say that substitution functors **preserve monos** if for each morphism $u \colon I \to J$ in $\mathbb{B}$ and vertical mono $m \colon X' \rightarrowtail X$ over J, one has that $u^*(m) \colon u^*(X') \to u^*(X)$ is vertically monic over I. In case substitution functors preserve fibred pull-backs or have left adjoints, then they preserve monos. A **vertical subobject** is a subobject in a fibre category which comes from a vertical mono. Below we shall write $\mathrm{VSub}_I(X)$ for the collection of vertical subobjects of an object X over I.

5.5.6. Definition. A fibration $\begin{smallmatrix} \mathbb{E} \\ \downarrow p \\ \mathbb{B} \end{smallmatrix}$ is said to be **well-powered** if both

- substitution functors preserve monos;
- for each $X \in \mathbb{E}$, say over $I \in \mathbb{B}$, the functor

$$(\mathbb{B}/I)^{\mathrm{op}} \longrightarrow \mathbf{Sets} \qquad \text{given by} \qquad \left(J \xrightarrow{u} I \right) \longmapsto \mathrm{VSub}_J(u^*(X))$$

is representable.

The latter means that for each $X \in \mathbb{E}$ there is a map $\mathcal{S}X \colon \underline{\mathrm{Sub}}(X) \to I$ in $\mathbb{B}$ together with a vertical mono $s \colon X' \rightarrowtail \mathcal{S}X^*(X)$ which is universal: for each $u \colon J \to I$ with a vertical mono $m \colon Y \rightarrowtail u^*(X)$ over J, there is a unique map

$v: J \to \underline{\mathrm{Sub}}(X)$ in $\mathbb{B}$ with $\mathcal{S}X \circ v = u$ such that there is a commuting diagram

$$
\begin{array}{ccc}
Y & \dashrightarrow & X' \\
m \downarrow & & \downarrow s \\
u^*(X) & \dashrightarrow & \mathcal{S}X^*(X)
\end{array}
$$

where $u^*(X) \dashrightarrow \mathcal{S}X^*(X)$ is the unique Cartesian arrow over v with $u^*(X) \to \mathcal{S}X^*(X) \to X$ is $u^*(X) \to X$, so that $Y \dashrightarrow X'$ is uniquely determined. All this says is that there are isomorphisms

$$
\mathbb{B}/I\big(u,\, \mathcal{S}X\big) \cong \mathrm{VSub}_J(u^*(X)), \quad \text{natural in } u: J \to I.
$$

5.5.7. Theorem (See also [246, 4.2.1]). *A category $\mathbb{B}$ with finite limits is a topos if and only if its codomain fibration $\begin{array}{c}\mathbb{B}^{\to}\\ \downarrow\\ \mathbb{B}\end{array}$ is well-powered.*

Proof. If the codomain fibration on $\mathbb{B}$ is well-powered, then for each object $I \in \mathbb{B}$ we have can view I has a constant family over the terminal object $1 \in \mathbb{B}$. We take as powerobject

$$
P(I) = \mathrm{dom}\, \mathcal{S}\left(\begin{array}{c} I \\ \downarrow \\ 1 \end{array}\right).
$$

The isomorphism characterising well-poweredness gives us

$$
\mathbb{B}\big(J,\, P(I)\big) \cong \mathbb{B}/1\left(\left(\begin{array}{c} J \\ \downarrow \\ 1 \end{array}\right),\, \mathcal{S}\left(\begin{array}{c} I \\ \downarrow \\ 1 \end{array}\right)\right) \cong \mathrm{VSub}_J(J^*(I)) \cong \mathrm{Sub}(J \times I).
$$

This shows that $P(I)$ indeed behaves like a powerobject.

Conversely, assume $\mathbb{B}$ is a topos. Since every slice category $\mathbb{B}/I$ is a topos, it carries a power object functor $P/I: \mathbb{B}/I \to \mathbb{B}/I$. For a family $\left(\begin{array}{c} X \\ \downarrow \varphi \\ I \end{array}\right)$ and a map $u: J \to I$ we then get isomorphisms

$$
\mathbb{B}/I\big(u,\, P/I(\varphi)\big) \cong \mathrm{Sub}(u \times_I \varphi) \cong \mathrm{VSub}_J(u^*(\varphi)). \qquad \square
$$

Exercises

5.5.1. Describe an 'undefined' element $\perp_I: 1 \to \perp I$. And prove that $\perp 0 \cong 1$.

5.5.2. (i) Check that the assignment $u \mapsto P(u)$ in Proposition 5.5.2 preserves identities and composites.

   (ii) Extend P to a monad on $\mathbb{B}$ with singleton maps $\{\,\}$ as unit.
   [It can then be shown that algebras of this monad are the (internally) complete lattices in $\mathbb{B}$, see [169, after Proposition 5.36].]

(iii) Extend also $\bot$ to a monad and show that the maps $\bot I \rightarrowtail PI$ form a morphism of monads $\bot \rightarrowtail P$.

[Algebras for this $\bot$ monad are investigated in [179].]

(iv) Define the **non-empty power object** $P^+I \rightarrowtail PI$ by the following pullback diagram,

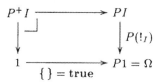

and show that P^+ also forms a submonad of P.

5.5.3. Let $\mathbb{B}$ be a topos. For every object $I \in \mathbb{B}$ we have a slice topos $\mathbb{B}/I$, and thus a power object functor $P/I : \mathbb{B}/I \to \mathbb{B}/I$.

(i) Show that these fibrewise functors combine into one single fibred power object functor $\mathbb{B}^\to \to \mathbb{B}^\to$.

(ii) Check that in the case where $\mathbb{B} = \mathbf{Sets}$, this fibred functor is given by

$$(X_i)_{i \in I} \mapsto (PX_i)_{i \in I}.$$

5.5.4. Prove in purely categorical terms that the rectangle in the proof of Proposition 5.5.4 commutes.

5.5.5. Consider an ordinary category $\mathbb{C}$, recall Exercise 1.2.2, and prove that

(i) substitution functors of the family fibration $\begin{smallmatrix} \mathrm{Fam}(\mathbb{C}) \\ \downarrow \\ \mathbf{Sets} \end{smallmatrix}$ always preserve vertical monos;

(ii) the category $\mathbb{C}$ is well-powered if and only if its family fibration $\begin{smallmatrix} \mathrm{Fam}(\mathbb{C}) \\ \downarrow \\ \mathbf{Sets} \end{smallmatrix}$ is well-powered.

5.6 Nuclei in a topos

In this section we describe nuclei (also called Lawvere-Tierney topologies) in a topos and study the associated closed and dense subobjects in some detail. We show in particular that the fibration of closed subobjects is a higher order fibration, just like the fibration of ordinary subobjects in the underlying topos. Thus we can do higher order logic with closed subobjects. For the special case of the double negation nucleus $\neg\neg$, the logic of this fibration of $\neg\neg$-closed subobjects is classical: the entailment $\neg\neg\varphi \vdash \varphi$ is (forced to be) valid.

We start with the standard definition of a nucleus in a topos as a morphism $j : \Omega \to \Omega$ satisfying some special properties. An alternative, more logical, approach is possible in which a nucleus is introduced as an operation $\varphi \mapsto \overline{\varphi}$

on propositions, see Exercise 5.6.6 (and also [339]). This operation may also be studied as an operation of modal logic, see [103].

The conjunction operation $\wedge\colon \Omega \times \Omega \to \Omega$ in the next definition arises as in Remark 5.2.12.

5.6.1. Definition. In a topos, a **nucleus** (or **Lawvere-Tierney topology**) is a map $j\colon \Omega \to \Omega$ making the following three diagrams commute.

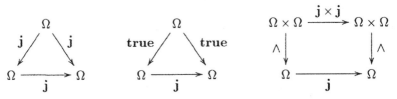

5.6.2. Example (Double negation). The example of a nucleus that we will be most interested in is the double negation nucleus $\neg\neg\colon \Omega \to \Omega$ in a topos. The negation map $\neg\colon \Omega \to \Omega$ is the unique map giving rise to a pullback square,

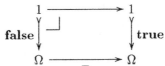

see Exercise 5.6.1—where **false**: $1 \to \Omega$ is the characteristic map of $0 \rightarrowtail 1$.

Since Ω is an (internal) Heyting algebra (see Exercise 5.1.5), we have,

$$\neg\varphi = \neg\neg\neg\varphi, \qquad \neg\neg\textbf{true} = \textbf{true}, \qquad \neg\neg(\varphi \wedge \psi) = \neg\neg\varphi \wedge \neg\neg\psi$$

which yields that $\neg\neg$ is a nucleus. The difficult case is to deduce $\neg\neg(\varphi \wedge \psi)$ from $\neg\neg\varphi$ and $\neg\neg\psi$. We shall give a derivation in propositional logic.

$$
\cfrac{
  \cfrac{
    \cfrac{
      \cfrac{
        \cfrac{\varphi \vdash \varphi \qquad \psi \vdash \psi}{\varphi, \psi \vdash \varphi \wedge \psi} \qquad \neg(\varphi \wedge \psi) \vdash \neg(\varphi \wedge \psi)
      }{
        \cfrac{\varphi, \psi, \neg(\varphi \wedge \psi) \vdash \bot}{\psi, \neg(\varphi \wedge \psi) \vdash \neg\varphi} \qquad \neg\neg\varphi \vdash \neg\neg\varphi
      }
    }{
      \cfrac{\psi, \neg\neg\varphi, \neg(\varphi \wedge \psi) \vdash \bot}{\neg\neg\varphi, \neg(\varphi \wedge \psi) \vdash \neg\psi} \qquad \neg\neg\psi \vdash \neg\neg\psi
    }
  }{
    \neg\neg\varphi, \neg\neg\psi, \neg(\varphi \wedge \psi) \vdash \bot
  }
}{
  \neg\neg\varphi, \neg\neg\psi \vdash \neg\neg(\varphi \wedge \psi)
}
$$

5.6.3. Example (Grothendieck topologies). A nucleus **j** on a presheaf topos $\widehat{\mathbb{C}} = \mathbf{Sets}^{\mathbb{C}^{\mathrm{op}}}$ corresponds to a **Grothendieck topology** on $\mathbb{C}$. The latter consists of a mapping $\mathcal{J}$ that assigns to every object $X \in \mathbb{C}$ a collection $\mathcal{J}(X)$ of sieves on X, such that the following three conditions are satisfied.

Identity. The maximal sieve $\downarrow X = \{f \mid \mathrm{cod}(f) = X\}$ is in $\mathcal{J}(X)$.

Stability. If a sieve S is in $\mathcal{J}(X)$ and $f: Y \to X$ is an arbitrary map, then the sieve $f^*(S) = \{g: Z \to Y \mid f \circ g \in S\}$ is in $\mathcal{J}(Y)$.

Transitivity. If $S \in \mathcal{J}(X)$, then also $R \in \mathcal{J}(X)$, for a sieve R with $f^*(R) \in \mathcal{J}(Y)$ for each $f: Y \to X$ in S.

Such a pair $(\mathbb{C}, \mathcal{J})$ is also called a **site**, in case $\mathbb{C}$ is a small category. The elements of $\mathcal{J}(X)$ are **covers** or **covering families**. Details of the correspondence between **j** and $\mathcal{J}$ may be found in [188].

Often one describes such a Grothendieck topology via a basis, *i.e.* via collections $\mathcal{K}(X)$ of families of maps with codomain X, such that the induced collections of sieves

$$\mathcal{J}(X) = \Big(\downarrow R = \{f \circ g \mid f \in R\} \Big)_{R \in \mathcal{K}(X)}$$

form a Grothendieck topology. One says that $\mathcal{J}$ is generated by $\mathcal{K}$. The families $R \in \mathcal{K}(X)$ are also called covers of X.

As particular examples of sites we mention the following.

(i) Each frame A, considered as a poset category, carries the **sup topology**, with as covers of $x \in A$ the down sets $S \subseteq \downarrow x$ with $\bigvee S = x$. The frame distributivity ensures that we get a topology.

This applies in particular to the case where A is the frame $\mathcal{O}(X)$ of opens of a topological space X. In terms of bases, we get that a collection $S = (U_i)_{i \in I}$ of opens U_i covers $U \in \mathcal{O}(X)$ if $\bigcup_{i \in I} U_i = U$.

(ii) For a regular category $\mathbb{C}$ there is what is called the **regular epi topology**, given by the following basis. The covers of an object $X \in \mathbb{C}$ are singleton sets $\{Y \twoheadrightarrow X\}$ of covers (*i.e.* regular epis, see Lemma 4.4.6) with codomain X. The associated sieves are sets with elements of the form $\to\twoheadrightarrow$.

5.6.4. Notation. For a subobject $m: X \rightarrowtail I$ one writes $\overline{m}: \overline{X} \rightarrowtail I$ for the **closure** obtained as pullback

$$
\begin{array}{ccc}
\overline{X} & \longrightarrow & 1 \\
{\scriptstyle \overline{m}} \downarrow \;\; \lrcorner & & \downarrow {\scriptstyle \mathbf{true}} \\
I & \xrightarrow[\mathrm{char}(m)]{} \Omega \xrightarrow{\;\mathbf{j}\;} & \Omega
\end{array}
$$

A subobject $X \rightarrowtail I$ is called **closed** if $\overline{X} = X$ and **dense** if $\overline{X} = I$. We write

$\mathrm{ClSub_j}(I) \hookrightarrow \mathrm{Sub}(I)$ and $\mathrm{ClSub_j}(\mathbb{B}) \hookrightarrow \mathrm{Sub}(\mathbb{B})$ for the full subcategories of closed subobjects (on $I \in \mathbb{B}$ and in $\mathbb{B}$).

5.6.5. Lemma. (i) *Closure commutes with pullbacks: if the diagram on the left below is a pullback, then so is the one on the right.*

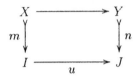

Briefly, $u^*(\overline{n}) = \overline{u^*(n)}$. *Especially, if n is closed or dense, then so is $u^*(n)$.*
(ii) *For a subobjects $X \rightarrowtail I$ and $Y \rightarrowtail I$ one has,*

$$X \leq \overline{X}, \qquad \overline{\overline{X}} = \overline{X}, \qquad \overline{X} \wedge \overline{Y} = \overline{X \wedge Y}, \qquad X \leq Y \Rightarrow \overline{X} \leq \overline{Y}.$$

Proof. (i) Because

$$
\begin{aligned}
\mathrm{char}\big(u^*(\overline{n})\big) &= \mathrm{char}(\overline{n}) \circ u \\
&= \mathbf{j} \circ \mathrm{char}(n) \circ u \\
&= \mathbf{j} \circ \mathrm{char}\big(u^*(n)\big) \\
&= \mathrm{char}\big(\overline{u^*(n)}\big).
\end{aligned}
$$

(ii) Easy, using the diagrams in Definition 5.6.1. □

The following equaliser will be important.

$$\Omega_j \rightarrowtail \Omega \underset{\mathbf{j}}{\overset{\mathrm{id}}{\rightrightarrows}} \Omega$$

As a result, $X \overset{m}{\rightarrowtail} I$ is closed if and only if $\mathrm{char}(m): I \to \Omega$ factors through $\Omega_j \rightarrowtail \Omega$. In particular, we can write **true**: $1 \rightarrowtail \Omega_j$.

For an object I, write $\overline{\delta(I)}: \overline{I} \rightarrowtail I \times I$ for the closure of the diagonal $\delta(I) = \langle \mathrm{id}, \mathrm{id} \rangle: I \rightarrowtail I \times I$. It yields a map $I \times I \to \Omega_j$, and hence by exponential transpose the **j**-singleton map

$$I \xrightarrow{\ \{\}_j\ } \Omega_j^I$$

5.6.6. Proposition. *Let* $\mathbf{j}: \Omega \to \Omega$ *be a nucleus in a topos* $\mathbb{B}$. *Since closed subobjects are closed under pullback, we get a split fibration* $\begin{smallmatrix} \mathrm{ClSub_j}(\mathbb{B}) \\ \downarrow \\ \mathbb{B} \end{smallmatrix}$ *of closed* *subobjects. It is a higher order fibration with extensional entailment, in which:*

- $\top, \wedge, \supset$ *and* $\prod$ *are as for ordinary subobjects.*
- $\perp_j = \overline{\perp}$, $X \vee_j Y = \overline{X \vee Y}$, $\coprod_j(X) = \overline{\coprod(X)}$ *and* $\mathrm{Eq}_j(X) = \overline{\mathrm{Eq}(X)}$, *and thus* $\neg_j X = X \supset \overline{\perp}$.
- **true**: $1 \rightarrowtail \Omega_j$ *is a (split) generic object.*

Hence closure $\overline{(-)}$ *defines a fibred functor* $\mathrm{Sub}(\mathbb{B}) \to \mathrm{ClSub}_j(\mathbb{B})$ *over* $\mathbb{B}$ *which preserves all this structure, except the generic object.*

Notice by the way that these fibrations of closed subobjects have full subset types.

Proof. By (ii) in the previous lemma, closed subobjects are closed under finite meets. And if $X \rightarrowtail I$, $Y \rightarrowtail I$ are closed, then so is $X \supset Y$:

$$\overline{(X \supset Y)} \leq (X \supset Y) \Leftrightarrow \overline{(X \supset Y)} \wedge X \leq Y$$
$$\Leftrightarrow \overline{(X \supset Y) \wedge X} \leq Y \quad \text{since } X \text{ is closed}$$
$$\Leftrightarrow (X \supset Y) \wedge X \leq Y \quad \text{since } Y \text{ is closed.}$$

The latter obviously holds since it is evaluation (or the counit). Similarly for a closed subobject $X \rightarrowtail I \times J$,

$$\overline{\prod(X)} \leq \prod(X) \Leftrightarrow \pi^* \overline{\prod(X)} \leq X$$
$$\Leftrightarrow \overline{\pi^* \prod(X)} \leq X \quad \text{by Lemma 5.6.5 (i)}$$
$$\Leftrightarrow \pi^* \prod(X) \leq X \quad \text{since } X \text{ is closed.}$$

And the latter holds (it is the counit again).

As for coproducts, one obviously has $\overline{\perp} \leq X$, for closed X. And for closed subobjects X, Y, Z in the same fibre,

$$X \leq Y \text{ and } X \leq Z \Leftrightarrow X \vee Y \leq Z$$
$$\Leftrightarrow \overline{X \vee Y} = X \vee_j Y \leq Z.$$

Similarly for simple coproducts $\coprod$ and equality Eq. Finally, **true**: $1 \rightarrowtail \Omega_j$ is split generic object because the characteristic maps of closed subobjects factor through Ω_j. And extensionality of entailment follows like in toposes, using this generic object as in Lemma 5.3.7. □

This result allows us to do higher order logic with closed subobjects. For example, a map $u: I \to J$ in the base category of a fibration $\begin{smallmatrix} \mathrm{ClSub}_j(\mathbb{B}) \\ \downarrow \\ \mathbb{B} \end{smallmatrix}$ of closed subobjects is internally injective (with respect to this fibration) in case one has in the internal language,

$$i: I, i': I \mid u(i) =_J u(i') \vdash i =_I i'$$

see Definition 4.3.9. Spelled out in categorical language, this means that

$$(u \times u)^* \left(\overline{\delta(J)} \right) \leq \overline{\delta(I)}$$

or equivalently,

$$\mathrm{Ker}(u) = (u \times u)^* \left(\delta(J) \right) \leq \overline{\delta(I)}$$

But this means that $\overline{\delta(I)} \rightrightarrows I$ is the kernel pair of $u \colon I \to J$. Equivalently, that the inclusion map $\delta(I) \rightarrowtail \mathrm{Ker}(u)$ is dense. A map u with this property is sometimes called **almost monic**. And an arbitrary map u is called **almost epic** if it is internally surjective in the fibration of closed subobjects. This means that in u's epi-mono factorisation the mono-part is dense. Finally, a map is called **bidense** if it is both internally injective and surjective in this logic, see [169, Definition 3.41].

We finish with the following two results about such fibrations of closed subobjects.

5.6.7. Lemma. *If* $X \overset{m}{\rightarrowtail} I$ *is a dense mono in a topos* $\mathbb{B}$ *with nucleus* **j**, *then the associated reindexing functor* m^* *for the fibration* $\begin{smallmatrix} \mathrm{ClSub_j}(\mathbb{B}) \\ \downarrow \\ \mathbb{B} \end{smallmatrix}$ *of closed subobjects is an isomorphism:*

$$\mathrm{ClSub_j}(I) \xrightarrow[\cong]{\quad m^* \quad} \mathrm{ClSub_j}(X)$$

Proof. The inverse of m^* maps a closed subobject $Y \overset{n}{\rightarrowtail} X$ to $\overline{m \circ n}$. Indeed,

$$\begin{aligned} m^*(\overline{m \circ n}) &= \overline{m^*(m \circ n)} \\ &= \overline{n} \qquad \text{since } m \text{ is a mono} \\ &= n. \end{aligned}$$

And, the other way round, for a closed $Z \overset{k}{\rightarrowtail} I$,

$$\begin{aligned} \overline{m \circ m^*(k)} &= \overline{m \wedge k} \\ &= \overline{m} \wedge \overline{k} \\ &= \overline{k} \qquad \text{since } m \text{ is dense} \\ &= k. \end{aligned} \qquad \square$$

5.6.8. Lemma. *For* $\mathbf{j} = \neg\neg$ *the double negation nucleus, the logic of the fibration* $\begin{smallmatrix} \mathrm{ClSub_j}(\mathbb{B}) \\ \downarrow \\ \mathbb{B} \end{smallmatrix}$ *of closed subobjects is classical: one has* $\neg\neg X = X$.

Proof. Since

$$\neg_j\neg_j X = (X \supset \overline{\bot}) \supset \overline{\bot} \quad \text{see the description of } \neg_j \text{ in Proposition 5.6.6}$$
$$= (X \supset \bot) \supset \bot \quad \text{since } \overline{\bot} = \neg\neg\bot = \bot \text{ for this nucleus}$$
$$= \neg\neg X$$
$$= X \qquad \text{because } X \text{ is } \neg\neg\text{-closed.} \qquad \square$$

Exercises

5.6.1. Recall that negation $\neg\varphi$ is defined in predicate logic as $\varphi \supset \bot$. Show that $\neg\colon \Omega \to \Omega$ as defined in Example 5.6.2 coincides with $(-) \supset \bot\colon \Omega \to \Omega$, where $(-) \supset \bot = \, \supset \circ \langle \text{id}, \bot \circ \, !\rangle\colon \Omega \to \Omega$ is obtained from the induced structure $\supset\colon \Omega \times \Omega \to \Omega$ and $\bot\colon 1 \to \Omega$ as in Remark 5.2.12.

5.6.2. Consider the following commuting square of monos,

and prove

(i) k is dense $\Rightarrow$ $\overline{n} = \overline{m}$;

(ii) n is closed and $\overline{m} = \overline{n} \, (= n)$ $\Rightarrow$ k is dense;

(iii) k, n are dense $\Rightarrow$ $m = n \circ k$ is dense.

[*Hint.* Write the triangle as a pullback.]

5.6.3. Let $j\colon \Omega \to \Omega$ be a nucleus in a topos $\mathbb{B}$.

(i) Show that for each object $I \in \mathbb{B}$, the map $I^*(j)\colon I^*(\Omega) \to I^*(\Omega)$ is a nucleus in the slice topos $\mathbb{B}/I$.

(ii) Check that a mono $\begin{pmatrix} X' \\ \downarrow \\ I \end{pmatrix} \overset{m}{\rightarrowtail} \begin{pmatrix} X \\ \downarrow \\ I \end{pmatrix}$ is closed / dense in $\mathbb{B}/I$ if and only if $X' \overset{m}{\rightarrowtail} X$ is closed / dense in $\mathbb{B}$.

5.6.4. Consider in a topos with a nucleus an arbitrary map $u\colon I \to J$ with an arbitrary subobject $X \rightarrowtail I$ on its domain. Prove that

$$\overline{\coprod_u(\overline{m})} = \overline{\coprod_u(m)}$$

as (closed) subobjects of J.

5.6.5. Say a map $u\colon I \to J$ has **dense image** if its image $\text{Im}(u) \rightarrowtail J$ is dense. Show that $\mathcal{M} = $ (closed monos) and $\mathcal{E} = $ (maps with dense image) form a factorisation system on a topos (see *e.g.* [18] for the definition).

5.6.6. Let $\begin{smallmatrix} \mathbb{E} \\ \downarrow p \\ \mathbb{B} \end{smallmatrix}$ be a preorder fibration with fibred finite products $(\top, \wedge)$. Define a **nucleus** on p to be a fibred "closure" monad $T\colon \mathbb{E} \to \mathbb{E}$ which preserves fibred finite products.

(i)   Show that for a topos $\mathbb{B}$, a nucleus $\mathbf{j}: \Omega \to \Omega$ as in Definition 5.6.1, corresponds to a nucleus $T = \overline{(-)}$ on the associated subobject fibration.

(ii) For a frame A, a nucleus on A is a map $j: A \to A$ satisfying $x \leq j(x)$, $j(j(x)) \leq j(x)$, and $j(x \wedge y) = j(x) \wedge j(y)$, see [170, II,2]. Show that there is a correspondence between nuclei on A and nuclei on the corresponding (regular) family fibration $\begin{array}{c} \mathrm{Fam}(A) \\ \downarrow \\ \mathbf{Sets} \end{array}$.

(iii) Let T be a nucleus on a regular fibration $\begin{array}{c} \mathbb{E} \\ \downarrow p \\ \mathbb{B} \end{array}$. Show that the fibred category p^T of algebras (see Exercise 1.7.9; the fibred category of "closed" predicates) is again a regular fibration. Also that the map $p \to p^T$ preserves this structure.

5.7 Separated objects and sheaves in a topos

In this section we present some basic results about separated objects and sheaves in a topos with a nucleus, and describe (fibred) sheafification and separated reflection. We will later use these constructions in the special case of the effective topos **Eff**, to be introduced in the next chapter: the categories of sets and of ω-sets can be characterised as the categories of sheaves and of separated objects in **Eff**, for the double negation nucleus $\neg\neg$.

5.7.1. Definition. Consider a topos $\mathbb{B}$ with nucleus $\mathbf{j}$. An **extension** of a partial map $I \overset{m}{\rightarrowtail} X \overset{u}{\to} J$ is a morphism $v: I \to J$ with $v \circ m = u$. We call (m, u) a **dense partial map** in case the mono m is dense.

An object $J \in \mathbb{B}$ is called a **separated object** if each dense partial map $I \rightharpoonup J$ has at most one extension $I \to J$. And J is a **sheaf** if there exists precisely one extension (again, for each such dense $I \rightharpoonup J$).

We write $\mathrm{Sep}_{\mathbf{j}}(\mathbb{B})$ and $\mathrm{Sh}_{\mathbf{j}}(\mathbb{B})$ for the full categories of separated objects and of sheaves. There are then obvious inclusion functors $\mathrm{Sh}_{\mathbf{j}}(\mathbb{B}) \hookrightarrow \mathrm{Sep}_{\mathbf{j}}(\mathbb{B}) \hookrightarrow \mathbb{B}$.

In a diagram, J is a separated object / sheaf if there is at most / precisely one dashed arrow:

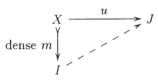

Put differently, for dense $X \overset{m}{\rightarrowtail} I$, the function "pre-compose with m"

$$\mathbb{B}(I, J) \xrightarrow{\quad - \circ m \quad} \mathbb{B}(X, J)$$

is injective / bijective if and only if J is a separated object / sheaf.

The above notion of sheaf is an abstraction of the notion of sheaf on a site. As we mentioned in Example 5.6.3, a nucleus on a topos $\widehat{\mathbb{C}} = \textbf{Sets}^{\mathbb{C}^{\text{op}}}$ of presheaves corresponds to a Grothendieck topology J on $\mathbb{C}$. Expressed in terms of J, a presheaf $P \colon \mathbb{C}^{\text{op}} \to \textbf{Sets}$ is a sheaf if and only if the following holds. Assume that S covers $X \in \mathbb{C}$ and that elements $\alpha_f \in P(Y)$, for $(f \colon Y \to X) \in S$, form a "matching" family: for $f \colon Y \to X$ in S and arbitrary $g \colon Z \to Y$ one has $\alpha_{(f \circ g)} = P(g)(\alpha_f)$. Then there is a unique element $a \in P(X)$ such that $\alpha_f = P(f)(a)$, for all $f \in S$.

It is useful to notice the following.

5.7.2. Lemma. (i) *If J is a separated object and $Y \rightarrowtail J$ is a mono, then Y is separated.*

(ii) *If J is a sheaf then a mono $Y \rightarrowtail J$ is closed if and only if Y is a sheaf.*

Proof. The first point is obvious, so we only do the second. Assume J is a sheaf and $n \colon Y \rightarrowtail J$ is closed, and consider a dense partial map $I \overset{m}{\rightarrowtail} X \overset{u}{\to} Y$. Then we get a unique extension v in

$$
\begin{array}{ccc}
X & \overset{u}{\longrightarrow} & Y \\
{\scriptstyle \text{dense } m}\big\downarrow & & \big\downarrow{\scriptstyle n} \\
I & \underset{v}{\dashrightarrow} & J
\end{array}
$$

using that J is a sheaf. But then $m \leq v^*(n)$ and so

$$ \text{id} = \overline{m} \leq \overline{v^*(n)} = v^*(\overline{n}) = v^*(n), $$

which shows that v factors through n. This yields the required extension $I \to Y$. It is unique by (i). Hence Y is a sheaf.

Conversely, if Y is a sheaf, consider the closure $\overline{n}$ of $n \colon Y \rightarrowtail J$ on the left,

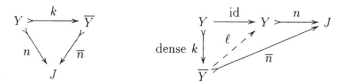

We get ℓ as indicated on the right, with $\ell \circ k = \text{id}$, because Y is a sheaf. But then, since J is also a sheaf, we get $n \circ \ell = \overline{n}$. Hence $\overline{n} \leq n$, so n is closed. $\square$

5.7.3. Lemma. *Consider a nucleus $\mathbf{j}$ in a topos $\mathbb{B}$.*

(i) *An object $J \in \mathbb{B}$ is separated if and only if the diagonal $\delta(J) = \langle \text{id}, \text{id} \rangle \colon J \rightarrowtail J \times J$ on J is closed. The latter means that internal and ex-*

ternal equality on J coincide in the fibration of closed subobjects, i.e. that equality on J is very strong in this fibration.

(ii) *The categories* $\mathrm{Sep_j}(\mathbb{B})$ *and* $\mathrm{Sh_j}(\mathbb{B})$ *are closed under finite limits in* $\mathbb{B}$.

(iii) *If* $J \in \mathbb{B}$ *is a separated object / sheaf then so is each exponent* $J^I = (I \Rightarrow J) \in \mathbb{B}$.

(iv) $\Omega_{\mathbf{j}}$ *is a sheaf. Hence each object* $P_{\mathbf{j}}(I) = \Omega_{\mathbf{j}}^I$ *is also a sheaf.*

(v) *The* $\mathbf{j}$-*singleton map* $\{\ \}_{\mathbf{j}} \colon I \to P_{\mathbf{j}}(I) = \Omega_{\mathbf{j}}^I$ *is internally injective in the fibration of* $\mathbf{j}$-*closed subobjects (or, almost monic): its kernel pair is the closure of the diagonal* $\delta(I)$.

Proof. (i) If J is separated, then $\pi \circ \overline{\delta(J)} = \pi' \circ \overline{\delta(J)}$. Hence $\overline{\delta(J)}$ factors through the equaliser $\delta(J)$ of $\pi, \pi' \colon J \times J \rightrightarrows J$. Thus $\delta(J)$ is closed.

Conversely, given a dense partial map $I \overset{m}{\leftarrowtail} X \overset{u}{\to} J$ with two extensions $v, w \colon I \rightrightarrows J$, then $m \leq \langle v, w \rangle^*(\delta(J))$, so that

$$\mathrm{id}_I = \overline{m} \leq \overline{\langle v, w \rangle^*(\delta(J))} = \langle v, w \rangle^*(\overline{\delta(J)}) = \langle v, w \rangle^*(\delta(J)).$$

Hence $\langle v, w \rangle$ factors through $\delta(J)$, and thus $v = w$.

(ii) + (iii) Left as exercises (or see *e.g.* [188, V,2 Lemma 1]).

(iv) We must show that for dense $m \colon X \rightarrowtail I$, the "pre-compose with m" function $- \circ m \colon \mathbb{B}(I, \Omega_{\mathbf{j}}) \to \mathbb{B}(X, \Omega_{\mathbf{j}})$ is an isomorphism. But since **true**: $1 \to \Omega_{\mathbf{j}}$ is split generic object for the fibration of closed subobjects we can describe this operation $- \circ m$ also as composite,

$$\mathbb{B}\big(I, \Omega_{\mathbf{j}}\big) \cong \mathrm{ClSub_j}(I) \xrightarrow[\cong]{\ m^*\ } \mathrm{ClSub_j}(X) \cong \mathbb{B}\big(X, \Omega_{\mathbf{j}}\big)$$

in which m^* is an isomorphism by Lemma 5.6.7.

(v) The map $\{\ \}_{\mathbf{j}} \colon I \to P_{\mathbf{j}}(I)$ is the singleton map as defined in Lemma 5.1.6 (ii), for the higher order fibration of closed subobjects. In the same lemma it is shown that this map is internally injective. The argument may be carried out in (the internal language of) the fibration of $\mathbf{j}$-closed subobjects. $\square$

Ordinary singleton maps $\{\ \} \colon I \to P(I) = \Omega^I$ are internally injective for the fibration of ordinary subobjects, and $\mathbf{j}$-singleton maps $\{\ \}_{\mathbf{j}} \colon I \to P_{\mathbf{j}}(I) = \Omega_{\mathbf{j}}^I$ are internally injective for the fibration of $\mathbf{j}$-closed subobjects. We need to know that $\mathbf{j}$-singleton maps form a natural transformation, like ordinary singleton maps, see Proposition 5.5.2. The proof goes analogously.

5.7.4. Lemma. *Given a nucleus* $\mathbf{j}$ *in a topos* $\mathbb{B}$, *the assignment* $I \mapsto P_{\mathbf{j}}(I) = \Omega_{\mathbf{j}}^I$ *extends to a functor* $\mathbb{B} \to \mathbb{B}$, *and the* $\mathbf{j}$-*singleton maps* $\{\ \}_{\mathbf{j}}$ *form a natural transformation* $\mathrm{id}_{\mathbb{B}} \Rightarrow P_{\mathbf{j}}$. $\square$

The **j**-singleton maps can be used to characterise separated objects and sheaves.

5.7.5. Lemma. *Consider a topos* $\mathbb{B}$ *with nucleus* **j** *and an arbitrary object* $I \in \mathbb{B}$. *Then*

(i) *I is a separated object if and only if the* **j**-*singleton map* $\{\ \}_{\mathbf{j}}\colon I \to P_{\mathbf{j}}(I)$ *on I is a mono in* $\mathbb{B}$;

(ii) *I is a sheaf if and only if* $\{\ \}_{\mathbf{j}}\colon I \to P_{\mathbf{j}}(I)$ *is a closed mono.*

Proof. (i) The (if)-part follows from the first and last point of the previous lemma: if I is a separated object then the diagonal $\delta(I)$ on I is closed, so that the kernel of $\{\ \}_{\mathbf{j}}$ is contained in $\overline{\delta(I)} = \delta(I)$. This makes $\{\ \}_{\mathbf{j}}$ a mono. The (only if)-part follows directly from Lemma 5.7.2 (i).

(ii) By Lemma 5.7.2 (ii), using (i) and the fact that $P_{\mathbf{j}}(I)$ is a sheaf. $\square$

This result suggests how to obtain a separated object or sheaf from an arbitrary object, simply by taking the monic or closed monic parts of the corresponding **j**-singleton map $\{\ \}_{\mathbf{j}}$. This will lead to left adjoints to the corresponding inclusion functors.

5.7.6. Definition. In a topos $\mathbb{B}$ with nucleus **j**, write for an object $I \in \mathbb{B}$,

$$\left(I \xrightarrow{\ \{\ \}_{\mathbf{j}}\ } P_{\mathbf{j}}(I) \right) = \left(I \xrightarrow{\ e_I\ } \mathbf{s}(I) \xrightarrow{\ m_I\ } P_{\mathbf{j}}(I) \right)$$

for the epi-mono factorisation of the **j**-singleton map $\{\ \}_{\mathbf{j}}$. And write

$$\mathbf{a}(I) \stackrel{\text{def}}{=} \overline{\mathbf{s}(I)}$$

for the closure of $\mathbf{s}(I) \rightarrowtail P_{\mathbf{j}}(I)$.

We notice that $\mathbf{s}(I)$ is the image $\{a\colon P_{\mathbf{j}}(I) \mid \exists i\colon I.\, a =_{P_{\mathbf{j}}(I)} \{i\}_{\mathbf{j}}\}$ of the singleton map $\{\ \}_{\mathbf{j}}$ in the fibration $\begin{smallmatrix} \mathrm{Sub}(\mathbb{B}) \\ \downarrow \\ \mathbb{B} \end{smallmatrix}$ of ordinary subobjects in $\mathbb{B}$, and that $\mathbf{a}(I)$ is the image $\{a\colon P_{\mathbf{j}}(I) \mid \exists i\colon I.\, a =_{P_{\mathbf{j}}(I)} \{i\}_{\mathbf{j}}\}$ of the **j**-singleton map $\{\ \}_{\mathbf{j}}$ in the fibration $\begin{smallmatrix} \mathrm{ClSub}_{\mathbf{j}}(\mathbb{B}) \\ \downarrow \\ \mathbb{B} \end{smallmatrix}$ of **j**-closed subobjects in $\mathbb{B}$—where we use that $P_{\mathbf{j}}(I)$ is separated. It is almost immediate—using Lemma 5.7.4—that the assignments $I \mapsto \mathbf{s}(I)$ and $I \mapsto \mathbf{a}(I)$ are functorial, using the universal properties of epi-mono factorisations.

5.7.7. Theorem. *The assignment* $I \mapsto \mathbf{s}(I)$ *is left adjoint to the inclusion* $\mathrm{Sep}_{\mathbf{j}}(\mathbb{B}) \hookrightarrow \mathbb{B}$. *And* $I \mapsto \mathbf{a}(I)$ *is left adjoint to* $\mathrm{Sh}_{\mathbf{j}}(\mathbb{B}) \hookrightarrow \mathbb{B}$.

The functor $\mathbf{a}(-)$ is called **sheafification**. And $\mathbf{s}(-)$ is **separated reflection**. A proof of this result using internal languages is described in [337].

Proof. We first consider separated reflection. For a map $u\colon I \to J$ in $\mathbb{B}$ with a separated object J as codomain we have to produce a unique map $v\colon \mathbf{s}(I) \to J$

with $v \circ e_I = u$. But this follows from functoriality of $\mathbf{s}$ and the fact that $\{\ \}_{\mathbf{j}}: J \to P_{\mathbf{j}}(J)$ is a mono by Lemma 5.7.5 (i)—so that its epi-part e_J is an isomorphism.

The argument for sheafification is similar. $\qquad\qquad\qquad\square$

Later in this section we shall make use of the following two standards facts about sheafification. A proof of the first result occurs in almost any text on topos theory. Exercise 5.7.5 below elaborates on a proof of the second result.

5.7.8. Lemma. *The sheafification functor* $\mathbf{a}$ *preserves finite limits.* $\qquad\square$

5.7.9. Lemma. *A morphism f is bidense (i.e. both internally injective and surjective in the logic of closed subobjects) if and only if $\mathbf{a}(f)$ is an isomorphism.* $\qquad\square$

5.7.10. Remarks. (i) The above adjunction $\mathrm{Sh}_{\mathbf{j}}(\mathbb{B}) \leftrightarrows \mathbb{B}$ forms an example of a **geometric morphism**. This is a second notion of morphism between toposes, the first one being 'logical morphism', see Definition 5.4.1. In general, a geometric morphism $F: \mathbb{A} \to \mathbb{B}$ between toposes $\mathbb{A}, \mathbb{B}$ consists of a pair of adjoint functors

$$
F = \left(
\begin{array}{c}
\xrightarrow{\quad F^* \quad} \\
\mathbb{A} \ \underset{F_*}{\overset{\perp}{\rightleftarrows}} \ \mathbb{B}
\end{array}
\right)
\qquad \text{with } F^* \text{ finite limit preserving.}
$$

One calls F^* the 'inverse image' and F_* is 'direct image' part of F. These geometric morphisms play a more important rôle in topos theory than logical morphisms. They satisfy a factorisation property and are used, for example, in functorial semantics for geometric logic, see *e.g.* [188, Chapters VII and X].

(ii) If $\mathbb{B}$ is a topos with nucleus $\mathbf{j}$, then (by Exercise 5.6.3) each slice topos $\mathbb{B}/I$ has a nucleus $I^*(\mathbf{j})$. Hence one can define what families $\left(\begin{array}{c} X \\ \downarrow \\ I \end{array} \right)$ of $I^*(\mathbf{j})$-separated objects and families of $I^*(\mathbf{j})$-sheaves are. This is to be understood fibrewise: each X_i is a $\mathbf{j}$-separated object or a $\mathbf{j}$-sheaf. In this way one gets fibrations $\begin{array}{c} \mathrm{FSep}_{\mathbf{j}}(\mathbb{B}) \\ \downarrow \\ \mathbb{B} \end{array}$ and $\begin{array}{c} \mathrm{FSh}_{\mathbf{j}}(\mathbb{B}) \\ \downarrow \\ \mathbb{B} \end{array}$ of such families.

Also, one can define separated reflection and sheafification in $\mathbb{B}/I$. This gives rise to fibred functors $\mathbf{Fs}: \mathbb{B}^{\to} \to \mathrm{FSep}_{\mathbf{j}}(\mathbb{B})$ and $\mathbf{Fa}: \mathbb{B}^{\to} \to \mathrm{FSh}_{\mathbf{j}}(\mathbb{B})$, which are fibred left adjoints to the respective inclusions. Everything is fibred because separated reflection $\mathbf{s}$ and sheafification $\mathbf{a}$ are defined by constructions that are preserved by pullback functors. We shall return to this point towards the end of this section.

We are now in a position to give a more refined version of Lemma 5.7.2.

5.7.11. Lemma. *Consider a mono $X \rightarrowtail I$ in a topos $\mathbb{B}$ with nucleus j.*
  (i) *If I is a sheaf, then*

$$X \rightarrowtail I \text{ is closed} \iff X \rightarrowtail I \text{ is a mono in } \mathrm{Sh}_j(\mathbb{B}).$$

  (ii) *And if I is a separated object, then*

$$X \rightarrowtail I \text{ is closed} \iff X \rightarrowtail I \text{ is a regular mono in } \mathrm{Sep}_j(\mathbb{B}).$$

Thus (i) and (ii) say that there are change-of-base situations,

$$
\begin{array}{ccccc}
\mathrm{Sub}(\mathrm{Sh}_j(\mathbb{B})) & \longrightarrow & \mathrm{RegSub}(\mathrm{Sep}_j(\mathbb{B})) & \longrightarrow & \mathrm{ClSub}_j(\mathbb{B}) \\
\downarrow\lrcorner & & \downarrow\lrcorner & & \downarrow \\
\mathrm{Sh}_j(\mathbb{B}) & \hookrightarrow & \mathrm{Sep}_j(\mathbb{B}) & \hookrightarrow & \mathbb{B}
\end{array}
$$

Proof. (i) The implication ($\Rightarrow$) follows from Lemma 5.7.2 (ii). Conversely, by the reflection $\mathrm{Sh}_j(\mathbb{B}) \leftrightarrows \mathbb{B}$, a mono in $\mathrm{Sh}_j(\mathbb{B})$ is also a mono in $\mathbb{B}$. It is then closed because I is a sheaf, see Lemma 5.7.2 (ii) again.

(ii) If $m: X \rightarrowtail I$ is closed, then X is separated by Lemma 5.7.2 (i). It is an equaliser in $\mathrm{Sep}_j(\mathbb{B})$, namely of $\mathrm{char}(m), \mathbf{true} \circ \,!: I \rightrightarrows \Omega_j$. In the reverse direction, assume $m: X \rightarrowtail I$ is an equaliser in $\mathrm{Sep}_j(\mathbb{B})$, say of $u, v: I \rightrightarrows K$. Since $X \rightarrowtail \overline{X}$ is dense and K is separated, the two maps $u \circ \overline{m}, v \circ \overline{m}: \overline{X} \rightrightarrows K$ must be equal. But then $\overline{m} \leq m$, since m is an equaliser, and so m is closed. $\qquad\square$

5.7.12. Corollary. *Both the fibrations* $\begin{array}{c} \mathrm{Sub}(\mathrm{Sh}_j(\mathbb{B})) \\ \downarrow \\ \mathrm{Sh}_j(\mathbb{B}) \end{array}$ *and* $\begin{array}{c} \mathrm{RegSub}(\mathrm{Sep}_j(\mathbb{B})) \\ \downarrow \\ \mathrm{Sep}_j(\mathbb{B}) \end{array}$ *of subobjects in a category of j-sheaves, and of regular subobjects in a category of j-separated objects are higher order fibrations. In particular, $\mathrm{Sh}_j(\mathbb{B})$ is a topos.*

And in case j is the double negation nucleus $\neg\neg$, both these fibrations are models of classical logic.

Proof. By Proposition 5.6.6 and Lemma 5.6.8. $\qquad\square$

Thus, by forming the category $\mathrm{Sh}_j(\mathbb{B})$ of j-sheaves in a topos $\mathbb{B}$, we get a new topos. It can be characterised in a universal way: for example, every geometric morphism $F: \mathbb{A} \to \mathbb{B}$ whose inverse image part F^* sends bidense maps to isomorphisms, factors through $\mathrm{Sh}_j(\mathbb{B}) \hookrightarrow \mathbb{B}$ (see [169, Theorem 3.47]; similar such universal properties are described there). The passage from $\mathbb{B}$ to $\mathrm{Sh}_j(\mathbb{B})$ can thus be understood as forcing "j-isomorphisms" (*i.e.* bidense maps) to be actual isomorphisms. An important special case of this $\mathrm{Sh}_j(-)$ construction is the ' "Grothendieck" topos $\mathrm{Sh}(\mathbb{C}, \mathcal{J}) \hookrightarrow \mathbf{Sets}^{\mathbb{C}^{\mathrm{op}}}$ of sheaves on a site $(\mathbb{C}, \mathcal{J})$. It plays a central role in [188].

We conclude this section with an explicit description of *fibred* sheafification, as mentioned in Remark 5.7.10 (ii). This requires the following auxiliary result.

5.7.13. Lemma. *Consider in a topos with a nucleus two morphisms* $m \colon X \to X'$ *and* $u \colon X \to J$ *with common domain, where* m *is bidense and* J *is a sheaf. Then there is a unique morphism* $v \colon X' \to J$ *with* $v \circ m = u$.

Proof. Simply take $v = (\eta_J)^{-1} \circ \mathbf{a}(u) \circ \mathbf{a}(m)^{-1} \circ \eta_{X'}$, using Lemma 5.7.9 (where η is the unit of the sheafification adjunction). $\qquad\square$

This lemma applies in particular when m is itself a unit component η_I. The next result occurs (without proof) in [150] (before Lemma 6.5). It gives an explicit description of fibred sheafification.

5.7.14. Proposition. *Consider a topos* $\mathbb{B}$ *with a nucleus* $\mathbf{j}$, *and for an arbitrary object* $I \in \mathbb{B}$ *the slice topos* $\mathbb{B}/I$ *with the induced nucleus* $I^*(\mathbf{j})$, *see Exercise 5.6.3. Sheafification* $\mathbf{Fs}$ *in this slice category can be described as the mapping which sends a family* $\left(\begin{smallmatrix} X \\ \downarrow \varphi \\ I \end{smallmatrix} \right)$ *to the family* $\left(\begin{smallmatrix} X' \\ \downarrow \varphi' \\ I \end{smallmatrix} \right)$ *obtained in the following pullback diagram.*

$$
\begin{array}{ccc}
X' & \xrightarrow{\;\;\theta\;\;} & \mathbf{a}(X) \\
{\scriptstyle \varphi'} \downarrow & \lrcorner & \downarrow {\scriptstyle \mathbf{a}(\varphi)} \\
I & \xrightarrow[\;\;\eta_I\;\;]{} & \mathbf{a}(I)
\end{array}
$$

Proof. It is not hard to see that the family φ' is a sheaf in $\mathbb{B}/I$: consider a dense mono $\left(\begin{smallmatrix} Z \\ \downarrow \chi \\ I \end{smallmatrix} \right) \xrightarrow{\;m\;} \left(\begin{smallmatrix} Y \\ \downarrow \psi \\ I \end{smallmatrix} \right)$ and a morphism $\left(\begin{smallmatrix} Z \\ \downarrow \chi \\ I \end{smallmatrix} \right) \xrightarrow{\;f\;} \left(\begin{smallmatrix} X' \\ \downarrow \varphi' \\ I \end{smallmatrix} \right)$. Because m is a dense mono $Z \rightarrowtail Y$ in $\mathbb{B}$, we get a unique $g \colon Y \to \mathbf{a}(X)$ with $g \circ m = \theta \circ f$. Then $\mathbf{a}(\varphi) \circ g = \eta_I \circ \psi$ because $\mathbf{a}(I)$ is a sheaf. The required map $Y \to X'$ is then obtained as mediating map for the pullback.

What remains to show is that the family φ' is universal. Assume therefore a morphism $\left(\begin{smallmatrix} X \\ \downarrow \varphi \\ I \end{smallmatrix} \right) \xrightarrow{\;f\;} \left(\begin{smallmatrix} Y \\ \downarrow \psi \\ I \end{smallmatrix} \right)$ to a sheaf ψ in $\mathbb{B}/I$. Let us write $\eta_\varphi \colon X \to X'$ for the unique map—obtained from the above pullback—with $\varphi' \circ \eta_\varphi = \varphi$ and $\theta \circ \eta_\varphi = \eta_X$. We claim that $\mathbf{a}(\eta_\varphi)$ is an isomorphism. This follows from the fact that sheafification $\mathbf{a}$ preserves pullbacks (as stated in Lemma 5.7.8): applying $\mathbf{a}$ to the pullback $\varphi' \to \mathbf{a}(\varphi)$ in the proposition yields a new pullback. As a result, $\mathbf{a}(\theta)$ is an isomorphism, since $\mathbf{a}(\eta_I)$ is an isomorphism. But then $\mathbf{a}(\eta_\varphi) = \mathbf{a}(\theta)^{-1} \circ \mathbf{a}(\eta_X)$ must be an isomorphism as well. Hence η_φ is bidense in $\mathbb{B}$, by Lemma 5.7.9, and also in $\mathbb{B}/I$—since epi-mono factorisations and closures in $\mathbb{B}/I$ are the same as in $\mathbb{B}$. Therefore Lemma 5.7.13 applies (in $\mathbb{B}/I$); it yields the required map $\varphi' \to \psi$. $\qquad\qquad\square$

Exercises

5.7.1. Show that an object I in a topos with a nucleus is separated if and only if the unit $\eta_I \colon I \to \mathbf{a}(I)$ of the adjunction $\mathrm{Sh}_{\mathbf{j}}(\mathbb{B}) \leftrightarrows \mathbb{B}$ is a mono.

5.7.2. Show that the inclusion $\mathrm{Sh}_{\mathbf{j}}(\mathbb{B}) \hookrightarrow \mathrm{Sep}_{\mathbf{j}}(\mathbb{B})$ also has a left adjoint.

5.7.3. Prove that if X is a sheaf and I is a separated object, then any mono $X \rightarrowtail I$ is automatically closed.

5.7.4. Show that J is a sheaf if and only if $\perp J$ is a sheaf and $\{\,\} \colon J \rightarrowtail \perp J$ is closed.

5.7.5. Consider a topos with a nucleus $\mathbf{j}$. The aim of this exercise is to get a proof of Lemma 5.7.9 (following [169, 3.42 and 3.43]).

(i) Prove that if a mono $m \colon X \rightarrowtail X'$ is dense, then $\mathbf{a}(m) \colon \mathbf{a}(X) \rightarrowtail \mathbf{a}(X')$ is an isomorphism.

(ii) Prove also the converse of (i): if $\mathbf{a}(m)$ is an isomorphism for a mono m, then m is dense.

[*Hint.* Notice first that if $\mathbf{a}(m)$ is an isomorphism, then for any map $u \colon X \to J$ to a sheaf J there is a unique $v \colon X' \to J$ with $v \circ m = u$, like in Lemma 5.7.13. Apply this to the case $J = \Omega_{\mathbf{j}}$. It yields a unique map $v \colon X' \to \Omega_{\mathbf{j}}$ with $g \circ m = \mathbf{true} \circ !_X$. This v then classifies both the identity $X' \rightarrowtail X'$ and m's closure $\overline{m} \colon \overline{X} \rightarrowtail X'$.]

(iii) Conclude from (i) and (ii) that if $u \colon I \to J$ is internally injective/surjective (*i.e.* almost monic/epic) if and only if $\mathbf{a}(u) \colon \mathbf{a}(I) \to \mathbf{a}(J)$ is monic/epic. And also that u is bidense if and only if $\mathbf{a}(u)$ is an isomorphism—as stated in Lemma 5.7.9.

5.7.6. Let $\mathbb{B}$ be a topos with nucleus $\mathbf{j}$.

(i) Assume that $\left(\begin{smallmatrix} X \\ \downarrow \\ I \end{smallmatrix} \right)$ is an $I^*(\mathbf{j})$-separated object. Show that for each map $u \colon J \to I$ the pullback $u^* \left(\begin{smallmatrix} X \\ \downarrow \\ I \end{smallmatrix} \right)$ in $\mathbb{B}/J$ is $J^*(\mathbf{j})$-separated. Prove the same for sheaves.

(ii) Prove also that if $\left(\begin{smallmatrix} Y \\ \downarrow \\ J \end{smallmatrix} \right)$ is a $J^*(\mathbf{j})$-separated object/sheaf, then the product $\prod_u \left(\begin{smallmatrix} Y \\ \downarrow \\ J \end{smallmatrix} \right)$ is an $I^*(\mathbf{j})$-separated object/sheaf, for $u \colon J \to I$.

5.7.7. Assume a topos $\mathbb{B}$ with a nucleus. Use the characterisation of sheafification on slices $\mathbb{B}/I$ from Proposition 5.7.14 to show that sheafification on families leads to a *fibred* functor $\mathbf{Fs}$ on $\mathbb{B}^{\rightarrow}$ (with respect to the codomain fibration).

5.8 A logical description of separated objects and sheaves

We conclude this chapter with a logical description of separated objects and sheaves in terms very strong equality and unique choice. Recall from Exercise 4.9.1 that for a fibration $\begin{smallmatrix} \mathbb{E} \\ \downarrow \\ \mathbb{B} \end{smallmatrix}$ with equality and subset types we say that **equality on** $J \in \mathbb{B}$ **is very strong** if internal and external equality on J coincide, or—in terms of subset types—if the canonical map $J \to \{\mathrm{Eq}(J)\}$ is an isomorphism. And **unique choice on** $J \in \mathbb{B}$ holds if for each single-valued relation $R \in \mathbb{E}_{I \times J}$ the canonical map $\{R\} \to \{\coprod(R)\}$ is an isomorphism. The main result below is that with respect to the fibration $\begin{smallmatrix} \mathrm{ClSub}_{\mathbf{j}}(\mathbb{B}) \\ \downarrow \\ \mathbb{B} \end{smallmatrix}$ of **j**-closed subobjects, (a) an object $J \in \mathbb{B}$ is separated if and only if equality on J is very strong, and (b) $J \in \mathbb{B}$ is a sheaf if and only if unique choice holds on J. It gives us a purely logical characterisation of separated objects and sheaves. It may be used in more general situations like in Exercise 5.6.6 where one has a notion of nucleus suitably generalised to fibrations.

Recall from Section 4.9 that what characterises subobject fibrations is: full subset types, very strong equality, and unique choice. The first of these points is present in fibrations of **j**-closed subobjects; the second point is by (a) above obtained by restricting to separated objects, and the third one comes by (b) by a further restriction to sheaves. Thus, by restriction to sheaves, the fibration of closed subobjects becomes a subobject fibration (on these sheaves). And since we already know (from Proposition 5.6.6) that such a fibration of closed subobjects is a higher order fibration, we obtain an alternative road to the result that a category of sheaves in topos is itself a topos, see Lemma 5.7.11.

To obtain the main result, we prepare the grounds in the lemma below. As already stated, we consider the logic of closed subobjects with respect to some nucleus **j** on a topos $\mathbb{B}$. That is, we work in the internal language of the higher order fibration $\begin{smallmatrix} \mathrm{ClSub}_{\mathbf{j}}(\mathbb{B}) \\ \downarrow \\ \mathbb{B} \end{smallmatrix}$. A functional relation $R : I \rightarrowtail J$ herein is a predicate $i : I, j : J \vdash R(i,j) : \mathsf{Prop}$ satisfying:

(R is single valued) $i : I, j : J, j' : J \mid R(i,j), R(i,j') \vdash j =_J j'$

(R is total) $i : I \mid \top \vdash \exists j : J. R(i,j).$

See Example 4.3.8. In $\begin{smallmatrix} \mathrm{ClSub}_{\mathbf{j}}(\mathbb{B}) \\ \downarrow \\ \mathbb{B} \end{smallmatrix}$ this means that the relation $\langle r_0, r_1 \rangle : R \rightarrowtail I \times J$ is a closed subobject satisfying

- The tuple $\langle r_1 \circ \pi_0, r_1 \circ \pi_1 \rangle : R' \rightarrowtail J \times J$ factors through $\overline{\delta(J)}$, where R' is obtained as kernel (equaliser):

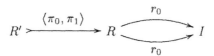

$$R' \;\rightarrowtail\; \xrightarrow{\langle \pi_0, \pi_1 \rangle}\; R \; \underset{r_0}{\overset{r_0}{\rightrightarrows}} \; I$$

- the image of $r_0: R \to I$ is dense.

See the description of equality and existence in Proposition 5.6.6. It turns out that there is a close connection between these functional relations in $\begin{smallmatrix} \mathrm{ClSub}_j(\mathbb{B}) \\ \downarrow \\ \mathbb{B} \end{smallmatrix}$ and dense partial maps in $\mathbb{B}$. This correspondence is the basis for the alternative logical description of separated objects and sheaves below.

5.8.1. Lemma. *Consider the logic of closed subobjects, as above.*

(i) *If $I \xleftarrow{m} X \xrightarrow{u} J$ is a dense partial map, then the closure of its ordinary graph $\langle m, u \rangle: X \rightarrowtail I \times J$ is a functional relation $I \nrightarrow J$.*

(ii) *Consider a map $v: I \to J$ and a dense partial map $(m, u): I \to J$. If v is an extension of (m, u), then the closures $\overline{\langle \mathrm{id}, v \rangle}$ and $\overline{\langle m, u \rangle}$ of the (ordinary) graphs are equal. And if J is separated, the converse also holds.*

(iii) *If $R: I \nrightarrow J$ is a single-valued relation $\langle r_0, r_1 \rangle: R \rightarrowtail I \times J$, then*

(a) *$r_0: R \to I$ is a mono if J is separated;*

(b) *r_0 is closed, if J is a sheaf.*

Proof. (i) Notice that the graph $\langle m, u \rangle$ is described in the internal language of the fibration $\begin{smallmatrix} \mathrm{Sub}(\mathbb{B}) \\ \downarrow \\ \mathbb{B} \end{smallmatrix}$ of ordinary subobjects as the proposition

$$i: I, j: J \vdash \exists x: X.\, m(x) =_I i \wedge u(x) =_J j: \mathsf{Prop}.$$

Thus the closure $\overline{\langle m, u \rangle}$ is described by the same expression, call it $G(i, j)$, but this time in the internal language of the fibration $\begin{smallmatrix} \mathrm{ClSub}_j(\mathbb{B}) \\ \downarrow \\ \mathbb{B} \end{smallmatrix}$ of closed subobjects. We reason informally in this language to show that this graph G is functional.

- If $G(i, j)$ and $G(i, j')$, say with x, x' such that $m(x) =_I i \wedge u(x) =_J j$ and $m(x') =_I i \wedge u(x') =_J j'$, then $m(x) =_I i =_I m(x')$. Hence $x =_X x'$ since m is internally injective, by Exercise 4.3.6. So $j =_J u(x) =_J u(x') =_J j'$.
- For $i: I$ we have $\exists x: X.\, m(x) =_I i$ since m is dense. Take such an x and put $j = u(x)$. Then $G(i, j)$.

(ii) If v extends (m, u), then we get $\overline{\langle m, u \rangle} = \overline{\langle \mathrm{id}, v \rangle}$ by applying Exercise 5.6.2 (i) to the triangle of monos

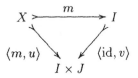

For the converse, assume J is separated and $\overline{\langle \mathrm{id}, v \rangle} = \overline{\langle m, u \rangle}$. Write $\overline{\langle m, u \rangle} = \langle r_0, r_1 \rangle \colon R \rightarrowtail I \times J$. There are then dense monos $n \colon X \rightarrowtail R$ and $k \colon I \rightarrowtail R$, so we can form their pullback, as in,

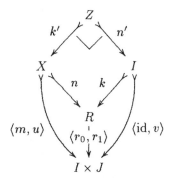

Since dense maps are closed under pullback, this k' is dense. Thus we can conclude that $v \circ m = u$—and thus that v extends (m, u)—from the fact that J is separated, and the calculation,

$$
\begin{aligned}
v \circ m \circ k' &= v \circ r_0 \circ n \circ k' \\
&= v \circ r_0 \circ k \circ n' \\
&= v \circ n' \\
&= r_1 \circ k \circ n' \\
&= r_1 \circ n \circ k' \\
&= u \circ k'.
\end{aligned}
$$

(iii) For a single-valued relation $\langle r_0, r_1 \rangle \colon R \rightarrowtail I \times J$, where J is separated, we first establish that r_0 is a mono. Assume therefore $u, v \colon K \rightrightarrows R$ with $r_0 \circ u = r_0 \circ v$. Then there is a (unique) $w \colon K \to R'$ with $\pi_0 \circ w = u$ and $\pi_1 \circ w = v$, where $\pi_0, \pi_1 \colon R' \rightrightarrows R$ is the kernel pair of $r_0 \colon R \to I$, as described before the lemma. We get that $\langle r_1 \circ u, r_1 \circ v \rangle$ factors through $\langle r_1 \circ \pi_0, r_1 \circ \pi_1 \rangle$, and hence through $\overline{\delta(J)} = \delta(J)$. Then $r_1 \circ u = r_1 \circ v$ and so $\langle r_0, r_1 \rangle \circ u = \langle r_0, r_1 \rangle \circ v$. Hence $u = v$ so that we may conclude: r_0 is a mono.

Next we assume that J is a sheaf, and show that the closure $\overline{r_0}\colon \overline{R} \rightarrowtail I$ of $r_0\colon R \rightarrowtail I$ is r_0. By pulling the closed mono $\mathbf{true}\colon 1 \rightarrowtail \Omega_{\mathbf{j}}$ back along the evaluation map $\mathrm{ev}\colon \Omega_{\mathbf{j}}^J \times J \to \Omega_{\mathbf{j}}$, we get a closed mono $\in_J \rightarrowtail \Omega_{\mathbf{j}}^J \times J$. Hence the object $\in_J$ is a sheaf. This yields a map v in

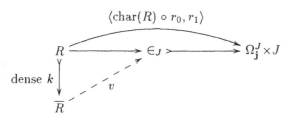

where we use that the closed relation $R \rightarrowtail I \times J$ has a relation classifier $\mathrm{char}(R)\colon I \to \Omega_{\mathbf{j}}^J$ as in the square below. We write w for the composite $\overline{R} \dashrightarrow \in_J \rightarrowtail \Omega_{\mathbf{j}}^J \times J$ in the diagram. Then $\pi \circ w = \mathrm{char}(R) \circ \overline{r_0}\colon \overline{R} \to \Omega_{\mathbf{j}}^J$, because $\Omega_{\mathbf{j}}^J$ is separated. We then get a map

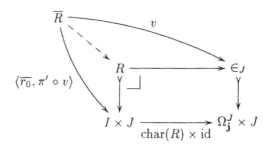

which shows that $\overline{r_0} \leq r_0$, and so that r_0 is closed. $\qquad\square$

5.8.2. Theorem. *Let $\mathbf{j}$ be a nucleus in a topos $\mathbb{B}$.*

(i) *An object $J \in \mathbb{B}$ is separated if and only if equality on J is very strong in the fibration* $\begin{array}{c}\mathrm{ClSub}_{\mathbf{j}}(\mathbb{B})\\\downarrow\\\mathbb{B}\end{array}$ *of $\mathbf{j}$-closed subobjects in $\mathbb{B}$.*

(ii) *And $J \in \mathbb{B}$ is a sheaf if and only if unique choice holds on J, in this same fibration of closed subobjects.*

Proof. (i) By Proposition 5.6.6, internal equality on J is the closed subobject $\mathrm{Eq}(J) = \overline{\delta(J)}$ on $J \times J$. It coincides with the diagonal $\delta(J)$ on J if and only if J is separated, by Lemma 5.7.3 (i).

(ii) Assume J is a sheaf, and $\langle r_0, r_1 \rangle\colon R \rightarrowtail I \times J$ is a single-valued relation in the fibration of closed subobjects. Then

$$
\begin{aligned}
\coprod_{\mathbf{j}}(R) \;=\; & \overline{\coprod(R)} & & \text{by Proposition 5.6.6} \\
=\; & \overline{\mathrm{Im}(\pi \circ \langle r_0, r_1 \rangle)} & & \text{see Theorem 4.4.4} \\
=\; & \overline{r_0} & & \text{because } r_0 \text{ is a mono,} \\
& & & \text{see Lemma 5.8.1 (iii) (a)} \\
=\; & r_0 & & \text{since } r_0 \text{ is closed, by (b).}
\end{aligned}
$$

Hence the canonical map $\{R\} \to \{\coprod_{\mathbf{j}}(R)\}$ is the identity, which is certainly an isomorphism.

Conversely, assume unique choice holds on J. From Exercise 4.9.1 we know that equality on J is then very strong, so that J is a separated object. Let $I \xleftarrow{m} X \xrightarrow{u} J$ be a dense partial map. Then $R = \overline{\langle m, u \rangle}$ is a functional relation $I \twoheadrightarrow J$, by Lemma 5.8.1 (i). The coproduct $\coprod_{\mathbf{j}}(R) = \exists j\colon J.\,R(i,j)$ is $\top$ because R is total. Hence we get two isomorphisms in the diagram

$$
\begin{array}{ccc}
R & \xrightarrow{\;\cong\;} & \{\coprod_{\mathbf{j}}(R)\} \\
{\scriptstyle \overline{\langle m,u\rangle}}\big\downarrow & & \big\downarrow{\scriptstyle \cong} \\
I \times J & \xrightarrow[\;\pi\;]{} & I
\end{array}
$$

so that there is a $v\colon I \to J$ with

$$
\begin{array}{ccc}
I & \xrightarrow{\;\cong\;} & R \\
{\scriptstyle \langle \mathrm{id}, v\rangle}\searrow & & \swarrow{\scriptstyle \overline{\langle m,u\rangle}} \\
& I \times J &
\end{array}
$$

But then v extends (m, u) by Lemma 5.8.1 (ii).   □

Exercises

5.8.1.   Prove that in a topos $\mathbb{B}$ with nucleus $\mathbf{j}$,
   (i)   an object $J \in \mathbb{B}$ is separated if and only if each functional relation $R\colon I \twoheadrightarrow J$ in $\begin{smallmatrix} \mathrm{ClSub}_{\mathbf{j}}(\mathbb{B}) \\ \downarrow \\ \mathbb{B} \end{smallmatrix}$ is the graph (in this fibration) of at most one map $I \to J$.
   (ii) And that J is a sheaf if and only if there is a unique such map, for each functional relation R.

Chapter 6

The effective topos

This chapter concentrates on one particular topos, namely the effective topos **Eff**. It can be seen as a topos in which the ordinary set theoretic world is combined with the recursion theoretic world. For example, there is a full and faithful functor **Sets** → **Eff**. But also the endomorphisms $N \to N$ on the natural numbers object N in **Eff** can be identified with the total recursive functions $\mathbb{N} \to \mathbb{N}$.

We shall be mostly interested in this topos as a universe for modelling various type theories. Therefore our view and description of **Eff** is rather limited in scope. For example, we only sketch in the last section of this chapter how one can do mathematics inside **Eff**, and suggest that this is "recursive mathematics". For a more elaborate account we refer to the last part of Hyland's original paper [142] on the effective topos. Another interesting aspect of **Eff**, namely the combination of higher types and effectivity, see [296] is ignored. For the rôle of **Eff** in the analysis of (higher order) Kleene realisability, we refer to [239, 240]. And Turing degrees within **Eff** may be found in [258]. Here we simply use **Eff** as a "forum" or "universe" in which we can discuss sets, ω-sets and PERs. Especially, the (internal) category of PERs inside ω-**Sets** and **Eff**—complete in the first case, and nearly so in the second—will interest us. In this chapter we shall describe families of PERs and of ω-sets over **Eff** in a concrete fashion. These will be used later to model type theories.

Our presentation of the effective topos is the "logical one", based on the higher order predicate logic of the realisability fibration $\begin{smallmatrix} \mathrm{UFam}(P\mathbb{N}) \\ \downarrow \\ \mathbf{Sets} \end{smallmatrix}$, as used by Hyland. There are alternative ways to introduce **Eff**, namely as completion with colimits of certain categories. For example, in [41] **Eff** is obtained by adding quotients to the category ω-**Sets** of ω-sets (or 'assemblies', as they are called there). And in [292] **Eff** results from a two-step completion of **Sets**, by first adding recursively indexed coproducts and then quotients of (pseudo) equivalence relations.

The material in this chapter is entirely standard, except the indexing of ω-sets by objects of **Eff** via the *split* fibration $\begin{smallmatrix} \mathrm{UFam}(\omega\text{-}\mathbf{Sets}) \\ \downarrow \\ \mathbf{Eff} \end{smallmatrix}$ instead of via the *non-split* fibration $\begin{smallmatrix} \mathrm{FSep}(\mathbf{Eff}) \\ \downarrow \\ \mathbf{Eff} \end{smallmatrix}$ of $(\neg\neg\text{-})$separated families in Section 6.3. Later, in Section 11.7, the relation between these fibrations will be described: the last one is the so-called "stack completion" of the former.

6.1 Constructing a topos from a higher order fibration

In Example 4.3.8 we have associated with a regular fibration p the categories $\mathrm{Rel}(p)$ of types (objects of the base category) with ordinary relations and $\mathrm{FRel}(p)$ of types with functional relations. In this section we shall describe a similar construction, which yields for a higher order fibration p an associated topos $\mathrm{Set}(p)$. Objects of $\mathrm{Set}(p)$ are types I with an (abstract) equality relation $\approx$; morphisms are suitable functional relations (in the logic of p) between the types. This construction includes the topos of sets with a Heyting-valued equality of Fourman & Scott [80] and the effective topos **Eff** of Hyland [142]. The latter is of most concern to us and will be further investigated in the next three sections of this chapter.

The construction of the topos $\mathrm{Set}(p)$ will be described in purely logical terms; that is, in the internal language of the fibration p. As such, it may be found in [145], except that there, one starts from a 'tripos' instead from the slightly more general notion of 'higher order fibration' that we use, see Example 5.3.4. A more detailed investigation may be found in [267].

Although we are essentially only interested in the special case where the fibration p is the realisability fibration $\begin{smallmatrix} \mathrm{UFam}(P\mathbb{N}) \\ \downarrow \\ \mathbf{Sets} \end{smallmatrix}$ from Section 4.2, we do present the construction of the topos at a more general level. We do so, because all the time we reason in the internal language, and nothing particular of this realisability fibration is used.

6.1.1. Definition. Let $\begin{smallmatrix} \mathbb{E} \\ \downarrow p \\ \mathbb{B} \end{smallmatrix}$ be a regular fibration. Write $\mathrm{Set}(p)$ for the category with

    objects   pairs $(I, \approx_I)$ where $I \in \mathbb{B}$ is an object of the base category and $\approx_I \in \mathbb{E}_{I \times I}$ is an 'equality' predicate on I. The latter is required to be symmetric and transitive in the logic of p. This means that validity in p is required of:

$$i_1, i_2 : I \mid i_1 \approx_I i_2 \vdash i_2 \approx_I i_1$$
$$i_1, i_2, i_3 : I \mid i_1 \approx_I i_2, i_2 \approx_I i_3 \vdash i_1 \approx_I i_3.$$

objects $(I, \approx_I) \to (J, \approx_J)$ are equivalence classes of relations $F \in \mathbb{E}_{I \times J}$ from I to J, which are

- extensional:

$$i_1, i_2 : I, j_1, j_2 : J \mid i_1 \approx_I i_2, j_1 \approx_J j_2, F(i_1, j_1) \vdash F(i_2, j_2)$$

- strict:

$$i : I, j : J \mid F(i, j) \vdash (i \approx_I i) \wedge (j \approx_J j)$$

- single-valued:

$$i : I, j_1, j_2 : J \mid F(i, j_1), F(i, j_2) \vdash j_1 \approx_J j_2$$

- total:

$$i : I \mid i \approx_I i \vdash \exists j : J. F(i, j).$$

The equivalence relation on these relations F is logical equivalence (in the internal language) as described by isomorphisms in the fibre. For convenience, we usually write representatives F instead of equivalence classes $[F]$.

Sometimes we also omit the subscript and write $\approx$ for $\approx_I$. And we write $|i_1 \approx_I i_2|$ for $i_1 \approx_I i_2$ Notice that the abstract equality $\approx_I$ is not required to be reflexive. We write $E_I(i)$, or $E(i)$, for $|i \approx_I i|$. Thus $E_I(-)$ is a unary "existence" predicate on I, defined categorically by $E_I(-) = \langle \mathrm{id}, \mathrm{id} \rangle^*(\approx_I) \in \mathbb{E}_I$.

The identity morphism on an object $(I \approx_I)$ of $\mathrm{Set}(p)$ is the (equivalence class of the) relation $\approx_I$ itself:

$$i_1, i_2 : I \vdash i_1 \approx_I i_2 : \mathsf{Prop}.$$

And composition of $(I, \approx_I) \xrightarrow{F} (J, \approx_J) \xrightarrow{G} (K, \approx_K)$ in $\mathrm{Set}(p)$ is the composite relation $G \circ F$, given as:

$$i : I, k : K \vdash \exists j : J. F(i, j) \wedge G(j, k) : \mathsf{Prop}.$$

Some elementary verifications in the internal language demonstrate that these identities and composites are again extensional, strict, single-valued and total.

Notice that all the logical machinery from p that we need in order to define $\mathrm{Set}(p)$ is $\exists, =, \wedge, \top$, as in a regular logic.

The following are the main examples.

6.1.2. Examples. (i) If Ω is a complete Heyting algebra (*i.e.* a frame), then the above Set$(-)$ construction applied to the regular family fibration $\begin{smallmatrix} \mathrm{Fam}(\Omega) \\ \downarrow \\ \mathbf{Sets} \end{smallmatrix}$ of set-indexed families of elements of Ω described in Example 4.2.5, yields the category Ω-**set** of Heyting valued sets as introduced in [80]. Objects of Ω-**set** are sets I together with a Ω-valued equality predicate $\approx_I : I \times I \to \Omega$ satisfying (in Ω)

$$|i_1 \approx_I i_2| \leq |i_2 \approx_I i_1|$$
$$|i_1 \approx_I i_2| \wedge |i_2 \approx_I i_3| \leq |i_1 \approx_I i_3|$$

for all elements $i_1, i_2, i_3 \in I$. Morphisms $(I, \approx_I) \to (J, \approx_J)$ in Ω-set are Ω-valued functions $F : I \times J \to \Omega$ which are extensional, strict, single-valued and total. The latter means for example, that for each $i \in I$,

$$E(i) = |i \approx_I i| \leq \bigvee_{j \in J} F(i, j).$$

(This category Ω-**set** should not be confused with the category ω-**Sets** of ω-sets (I, E) with $E : I \to P\mathbb{N}$ from Section 1.2.)

(ii) The same construction applied to the realisability fibration $\begin{smallmatrix} \mathrm{UFam}(P\mathbb{N}) \\ \downarrow \\ \mathbf{Sets} \end{smallmatrix}$ from Example 4.2.6 produces the **effective topos Eff** from [142]. Objects of **Eff** are sets I together with a $P\mathbb{N}$-valued equality predicate $\approx_I : I \times I \to P\mathbb{N}$ satisfying

$$\bigcap_{i_1, i_2 \in I} \left(|i_1 \approx_I i_2| \supset |i_2 \approx_I i_1| \right) \neq \emptyset$$
$$\bigcap_{i_1, i_2, i_3 \in I} \left(|i_1 \approx_I i_2| \wedge |i_2 \approx_I i_3| \supset |i_1 \approx_I i_3| \right) \neq \emptyset$$

where $\wedge$ and $\supset$ are the operations $P\mathbb{N} \times P\mathbb{N} \rightrightarrows P\mathbb{N}$ described in Example 4.2.6.

A morphism $(I, \approx_I) \to (J, \approx_J)$ in **Eff** is then an equivalence class of a relation $F : I \times J \to P\mathbb{N}$ which is extensional, strict, single-valued and total. Explicitly, this means that there are realisers

$$n_1 \in \bigcap_{i_1, i_2 \in I, j_1, j_2 \in J} \left(|i_1 \approx_I i_2| \wedge |j_1 \approx_J j_2| \wedge F(i_1, j_1) \supset F(i_2, j_2) \right)$$
$$n_2 \in \bigcap_{i \in I, j \in J} \left(F(i, j) \supset E_I(i) \wedge E_J(j) \right)$$

$$n_3 \in \bigcap_{i \in I, j_1, j_2 \in J} \left(F(i, j_1) \wedge F(i, j_2) \supset |j_1 \approx_J j_2| \right)$$

$$n_4 \in \bigcap_{i \in I} \left(E(i) \supset \bigcup_{j \in J} F(i, j) \right).$$

Thus, for example, for $i \in I$ and $m \in E(i)$ one has that $n_4 \cdot m$ is an element of $F(i, j)$, for some $j \in J$.

Similar realisability examples arise by applying the $\mathrm{Set}(-)$ construction to triposes, as constructed in Example 5.3.4, starting from a partial combinatory algebra. This yields many more such examples, see for instance [239, 261, 264].

(iii) If we start from a topos $\mathbb{B}$, we have a higher order fibration $\begin{smallmatrix} \mathrm{Sub}(\mathbb{B}) \\ \downarrow \\ \mathbb{B} \end{smallmatrix}$ of subobjects. Also here we can apply the $\mathrm{Set}(-)$ construction, which will yield a new topos.

Our aim in this section is to show that these categories of the form $\mathrm{Set}(p)$ are toposes, in case p is a higher order fibration. The following is the first step.

6.1.3. Proposition. *For* $\begin{smallmatrix} \mathbb{E} \\ \downarrow p \\ \mathbb{B} \end{smallmatrix}$ *a higher order fibration, the category* $\mathrm{Set}(p)$ *has finite limits and is Cartesian closed.*

Proof. As terminal object in $\mathrm{Set}(p)$ one takes the terminal object $1 \in \mathbb{B}$ with the truth predicate

$$x: 1, x': 1 \vdash |x \approx_1 x'| \overset{\mathrm{def}}{=} \top: \mathsf{Prop}$$

as equality. We shall write $1 = (1, \approx_1) \in \mathrm{Set}(p)$ for this object. A morphism $F: (I, \approx) \to 1$ in $\mathrm{Set}(p)$ is (an equivalence class of) a predicate $F \in \mathbb{E}_{I \times 1} \cong \mathbb{E}_I$ satisfying

$$i_1, i_2: I \mid F(i_1), i_1 \approx i_2 \vdash F(i_2)$$
$$i: I \mid F(i) \vdash E(i)$$
$$i: I \mid E(i) \vdash \exists x: 1.\, F(i)$$

so that $F(i)$ is logically equivalent to $E(i)$. There is thus a unique such map $(I, \approx) \to 1$.

The Cartesian product of $(I, \approx_I)$ and $(J, \approx_J)$ is the object $I \times J \in \mathbb{B}$ together with equality predicate

$$z, w: I \times J \vdash |\pi z \approx_I \pi w| \wedge |\pi' z \approx_J \pi' w|: \mathsf{Prop}.$$

The projection maps $(I, \approx) \leftarrow (I \times J, \approx) \to (J, \approx)$ are then given by the predicates

$$\begin{cases} z: I \times J, i: I \vdash |\pi z \approx_I i| \wedge E_J(\pi' z): \mathsf{Prop} \\ z: I \times J, j: J \vdash |\pi' z \approx_J j| \wedge E_I(\pi z): \mathsf{Prop}. \end{cases}$$

And tupleing of two maps $F\colon (K, \approx) \to (I, \approx)$ and $G\colon (K, \approx) \to (J, \approx)$ involves the predicate

$$k\colon K, z\colon I \times J \vdash F(k, \pi z) \wedge G(k, \pi' z)\colon \mathsf{Prop}.$$

For parallel maps F, G, an equaliser,

$$(I, \cong) \xrightarrow{\ \ \approx\ \ } (I, \approx) \underset{G}{\overset{F}{\rightrightarrows}} (J, \approx)$$

is obtained by taking as new equality predicate $\cong$ on I,

$$i_1, i_2\colon I \vdash |i_1 \cong i_2| \overset{\mathrm{def}}{=} |i_1 \approx i_2| \wedge \exists j\colon J.\, F(i_1, j) \wedge G(i_2, j)\colon \mathsf{Prop}.$$

This predicate $\cong$ is also the equaliser map $\cong\colon (I, \cong) \rightarrowtail (I, \approx)$ of F, G.

Finally, in order to form the exponent of objects $(I, \approx), (J, \approx) \in \mathrm{Set}(p)$, we take $P(I \times J) = \Omega^{(I \times J)}$ as underlying object with existence predicate

$$f\colon P(I \times J) \vdash E(f) \overset{\mathrm{def}}{=} \text{``}f \text{ is extensional and strict}$$
$$\text{and single-valued and total''}\colon \mathsf{Prop}.$$

That is,

$$E(f) \overset{\mathrm{def}}{=} \forall i_1, i_2\colon I.\, \forall j_1, j_2\colon J.\, |i_1 \approx_I i_2| \wedge |j_1 \approx_J j_2| \wedge f(i_1, j_1) \supset f(i_2, j_2)$$
$$\wedge \ \forall i\colon I.\, \forall j\colon J.\, f(i, j) \supset E_I(i) \wedge E_J(j)$$
$$\wedge \ \forall i\colon I.\, \forall j_1, j_2\colon J.\, f(i, j_1) \wedge f(i, j_2) \supset |j_1 \approx_J j_2|$$
$$\wedge \ \forall i\colon I.\, E_I(i) \supset \exists j\colon J.\, f(i, j).$$

The equality relation on the object $P(I \times J)$ underlying the exponent $(I, \approx) \Rightarrow (J, \approx)$ is then

$$f, g\colon P(I \times J) \vdash |f \approx g| \overset{\mathrm{def}}{=} E(f) \wedge E(g) \wedge \forall i\colon I.\, \forall j\colon J.\, f(i, j) \mathrel{\rlap{\supset}{\subset}} g(i, j)\colon \mathsf{Prop}.$$

The evaluation map $\mathrm{Ev}\colon ((I, \approx) \Rightarrow (J, \approx)) \times (I, \approx) \to (J, \approx)$ is given by

$$f\colon P(I \times J), i\colon I, j\colon J \vdash \mathrm{Ev}(f, i, j) \overset{\mathrm{def}}{=} f(i, j) \wedge E(f)\colon \mathsf{Prop}.$$

And for a morphism $H\colon (K, \approx) \times (I, \approx) \to (J, \approx)$, we get an abstraction map $\Lambda(H)\colon (K, \approx) \to (I, \approx) \Rightarrow (J, \approx)$ by

$$k\colon K, f\colon I \times J \vdash \Lambda(H)(k, f) \overset{\mathrm{def}}{=} E(k) \wedge E(f) \wedge$$
$$\forall i \in I.\, \forall j \in J.\, H(f, i, j) \mathrel{\rlap{\supset}{\subset}} f(i, j)\colon \mathsf{Prop}. \qquad \square$$

As a first step towards understanding (the logic of) subobjects in $\mathrm{Set}(p)$, there is the following quite useful result.

6.1.4. Lemma. *A morphism* $F: (I, \approx) \to (J, \approx)$ *in* $\mathrm{Set}(p)$ *is a monomorphism if and only if the following entailment holds in* p:

$$i_1, i_2: I, j: J \mid F(i_1, j), F(i_2, j) \vdash i_1 \approx_I i_2.$$

Proof. Assume validity of the above statement, and assume that two morphisms $G, H: (K, \approx) \rightrightarrows (I, \approx)$ in $\mathrm{Set}(p)$ are given with $F \circ G = F \circ H$, *i.e.* with

$$k: K, j: J \mid \emptyset \vdash \big[\exists i: I.\, F(i, j) \wedge G(k, i)\big] \rightleftharpoons \big[\exists i: I.\, F(i, j) \wedge H(k, i)\big].$$

We need to show that $G(k, i)$ implies $H(k, i)$. Assume therefore $G(k, i)$; then $E_I(i)$, so $F(i, j)$ for some $j: J$. We then have $\exists i: I.\, F(i, j) \wedge G(k, i)$, so $F(i', j) \wedge H(k, i')$ for some $i': I$. But then $F(i, j) \wedge F(i', j)$, and so $i \approx_I i'$ by the assumption. Hence $H(k, i)$.

Conversely, the pullback of F against itself is the object $(I \times I, \cong)$ where

$$z, w: I \times I \vdash |z \cong w| \overset{\mathrm{def}}{=} |z \approx_{I \times I} w| \wedge \exists j: J.\, F(\pi z, j) \wedge F(\pi' z, j): \mathsf{Prop}.$$

In case F is a mono in $\mathrm{Set}(p)$, this predicate must be equivalent to $|\pi z \approx_I \pi w| \wedge |\pi' z \approx_I \pi' w| \wedge |\pi z \approx \pi' z|$, see Exercise 6.1.2. We then get the statement as in the lemma. $\qquad\square$

In order to get a better handle on subobjects in $\mathrm{Set}(p)$, one uses so-called strict predicates.

6.1.5. Definition. Let $\begin{smallmatrix} \mathbb{E} \\ \downarrow p \\ \mathbb{B} \end{smallmatrix}$ be a higher order fibration.

(i) For an object $(I, \approx) \in \mathrm{Set}(p)$, a **strict predicate** on $(I, \approx)$ is a predicate $A \in \mathbb{E}_I$ which satisfies in p

$$i_1, i_2: I \mid A(i_1), i_1 \approx_I i_2 \vdash A(i_2) \qquad \text{and} \qquad i: I \mid A(i) \vdash E_I(i).$$

(ii) We form a category $\mathrm{SPred}(p)$ of $\rightleftharpoons$-equivalence classes of strict predicates, by stipulating that a morphism from a strict predicate A on $(I, \approx)$ to a strict predicate B on $(J, \approx)$ consists of a map $F: (I, \approx) \to (J, \approx)$ in $\mathrm{Set}(p)$ for which we have in p

$$i: I \mid A(i) \vdash \exists j: J.\, F(i, j) \wedge B(j).$$

This category $\mathrm{SPred}(p)$ comes with an obvious forgetful functor $\mathrm{SPred}(p) \to \mathrm{Set}(p)$. As usual, we do not distinguish notationally between a strict predicate and its equivalence class.

6.1.6. Proposition. *Assume* $\begin{smallmatrix} \mathbb{E} \\ \downarrow p \\ \mathbb{B} \end{smallmatrix}$ *is a higher order fibration.*

(i) *The above functor* $\begin{smallmatrix} \mathrm{SPred}(p) \\ \downarrow \\ \mathrm{Set}(p) \end{smallmatrix}$ *is a poset fibration. The order in the fibre over* $(I, \approx)$ *is the order inherited from* p's *fibre over* I: *for strict predicates*

A, B on $(I, \approx)$ one has

$$A \leq B \text{ in } \mathrm{SPred}(p) \text{ over } (I, \approx) \quad \Leftrightarrow \quad i: I \mid A(i) \vdash B(i) \text{ in } p.$$

(This may look confusing, since $\leq$ on the left is a partial order, whereas $\vdash$ on the right is a preorder; but we should have written the equivalence classes of A, B on the left.)

(ii) *Strict predicates on $(I, \approx)$ correspond to subobjects of $(I, \approx)$ in $\mathrm{Set}(p)$ in the sense that there is an isomorphism of fibred categories,*

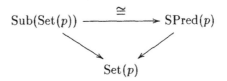

(iii) *The fibration $\begin{smallmatrix} \mathrm{SPred}(p) \\ \downarrow \\ \mathrm{Set}(p) \end{smallmatrix}$ of strict predicates is a higher order fibration. If, for the time being, we mark its connectives with a tilde $\tilde{\ }$, then expressed in terms of the connectives of p (which are written in ordinary fashion), we have*

- *propositional connectives in the fibre over $(I, \approx)$ are $\tilde{\bot} = \bot$, $\tilde{\vee} = \vee$, $\tilde{\top} = E_I$, $\tilde{\wedge} = \wedge$ and $A \tilde{\supset} B = E_I \wedge (A \supset B)$.*
- *For a strict predicate A over $(I, \approx) \times (J, \approx)$,*

$$\tilde{\exists} j{:} J. A(i, j) = \exists j{:} J. E(j) \wedge A(i, j) = \exists j{:} J. A(i, j)$$
$$\tilde{\forall} j{:} J. A(i, j) = E(i) \wedge \forall j{:} J. E(j) \supset A(i, j)$$
$$\widetilde{\mathrm{Eq}}(A)(i, j, j') = A(i, j) \wedge |j \approx_J j'|.$$

- *The object $\Omega \in \mathbb{B}$ in the basis of p with logical equivalence $\supset\!\subset$ as equality $\approx_\Omega$, carries a (split) generic object, namely the strict predicate*

$$\alpha{:} \, \mathsf{Prop} \vdash \mathbf{true}(\alpha) \stackrel{\mathrm{def}}{=} |\alpha \approx_\Omega \top|{:} \mathsf{Prop}.$$

Thus, in the logic of strict predicates in p—or subobjects of $\mathrm{Set}(p)$—the operations $\top, \supset, \forall$ and Eq are only slightly different from those of p.

Proof. (i) For a morphism $F{:} (I, \approx) \to (J, \approx)$ in $\mathrm{Set}(p)$ and a strict predicate B on $(J, \approx)$, one gets a strict predicate on $(I, \approx)$ by

$$i{:} I \vdash F^*(B)(i) \stackrel{\mathrm{def}}{=} \exists j{:} J. F(i, j) \wedge B(j){:} \mathsf{Prop}.$$

For strict predicates A, B on $(I, \approx)$ one has $A \leq B$ over $(I, \approx)$ if and only if

$$i{:} I \mid A(i) \vdash \exists i'{:} I. |i \approx_I i'| \wedge B(i').$$

But the predicate on the right of the turnstile $\vdash$ is clearly equivalent to $B(i)$.

(ii) For a strict predicate A on $(I, \approx)$ one forms a new object $(I, \approx_A)$ by

$$i_1, i_2 \colon I \vdash |i_1 \approx_A i_2| \stackrel{\text{def}}{=} A(i_1) \wedge |i_1 \approx_I i_2| \colon \mathsf{Prop}$$

It gives rise to an obvious mono in $\mathrm{Set}(p)$,

$$\approx_A \colon (I, \approx_A) \rightarrowtail (I, \approx).$$

And conversely, given a mono $M \colon (X, \approx_X) \rightarrowtail (I, \approx)$ one forms a strict predicate A_M on $(I, \approx)$ by

$$i \colon I \vdash A_M(i) \stackrel{\text{def}}{=} \exists x \colon X.\, M(x, i) \colon \mathsf{Prop}.$$

Then, starting from a strict predicate A, we get

$$
\begin{aligned}
A_{\approx_A}(i) &= \exists i' \colon I.\, |i' \approx_A i| \\
&= \exists i' \colon I.\, A(i') \wedge |i' \approx_I i| \\
&\cong A(i).
\end{aligned}
$$

And in order to show that M and $\approx_{A_M}$ give rise to the same subobject of $(I, \approx)$ we define maps G, H in

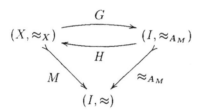

by $G(x, i) = M(x, i) = H(i, x)$. It is easy to see that G and H form an isomorphism between M and $\approx_{A_M}$.

For a morphism $F \colon (I, \approx) \to (J, \approx)$ from A to B in $\mathrm{SPred}(p)$ one gets a commuting square

$$
\begin{array}{ccc}
(I, \approx_A) & \xrightarrow{\ \ F'\ \ } & (J, \approx_B) \\
{\scriptstyle \approx_A}\Big\downarrow & & \Big\downarrow{\scriptstyle \approx_B} \\
(I, \approx) & \xrightarrow{\ \ F\ \ } & (J, \approx)
\end{array}
$$

by putting

$$i \colon I, j \colon J \vdash F'(i, j) \stackrel{\text{def}}{=} F(i, j) \wedge A(i) \wedge B(j) \colon \mathsf{Prop}.$$

Remaining details are left to the reader.

(iii) It is easily verified that the propositional connectives are as described above. The existential and universal quantification have this form since weakening along the projection $\pi\colon (I,\approx) \times (J,\approx) \to (I,\approx)$ is given by

$$\begin{aligned}
\pi^*(B)(i,j) &= \exists i'\colon I.\, \pi(i,i',j) \wedge B(i') \\
&= \exists i'\colon I.\, |i \approx i'| \wedge E(j) \wedge B(i') \\
&\cong E(j) \wedge B(i).
\end{aligned}$$

For equality, consider the parametrised diagonal

$$(I,\approx) \times (J,\approx) \xrightarrow{\ \ \delta = \mathrm{id} \times \pi'\ \ } ((I,\approx) \times (J,\approx)) \times (J,\approx)$$

Then for a strict predicate B one has

$$\delta^*(B)(i,j) = B(i,j,j),$$

so that putting $\widetilde{\mathrm{Eq}}(i,j,j') = A(i,j) \wedge |j \approx_J j'|$ yields the required adjunction $\widetilde{\mathrm{Eq}} \dashv \delta^*$.

Finally, for each strict predicate A on $(I,\approx)$ we get a characteristic map $\mathrm{char}(A)\colon (I,\approx) \to (\Omega,\approx)$ in $\mathrm{Set}(p)$ by

$$i\colon I, \alpha\colon \mathsf{Prop} \vdash \mathrm{char}(A)(i,\alpha) \stackrel{\mathrm{def}}{=} E(i) \wedge |A(i) \approx_\Omega \alpha|\colon \mathsf{Prop}.$$

Then

$$\begin{aligned}
\mathrm{char}(A)^*(\mathbf{true})(i) &= \exists \alpha\colon \mathsf{Prop}.\, \mathrm{char}(A)(i,\alpha) \wedge \mathbf{true}(\alpha) \\
&= \exists \alpha\colon \mathsf{Prop}.\, |A(i) \approx_\Omega \alpha| \wedge E(i) \wedge |\alpha \approx_\Omega \top| \\
&\cong E(i) \wedge |A(i) \approx_\Omega \top| \\
&\cong A(i).
\end{aligned}$$

We still have to show that $\mathrm{char}(A)$ is unique with this property. So assume $F\colon (I,\approx) \to (\Omega,\approx)$ also satisfies $F^*(\mathbf{true}) \cong A$, then

$$A(i) \cong F^*(\mathbf{true})(i) = \exists \alpha\colon \mathsf{Prop}.\, F(i,\alpha) \wedge |\alpha \approx_\Omega \top| \cong F(i,\top),$$

so that

$$\mathrm{char}(A)(i,\alpha) \cong E(i) \wedge |A(i) \approx_\Omega \alpha| \cong E(i) \wedge |F(i,\top) \approx_\Omega \alpha| \stackrel{(*)}{\cong} F(i,\alpha)$$

where the last isomorphism $(*)$ arises as follows. Given $F(i,\alpha)$, one has $E(i)$ by definition; also one has $F(i,\top) \approx_\Omega \alpha$, since: from $F(i,\top)$ we get $\alpha \approx_\Omega \top$ by single-valuedness; this yields α. Conversely, given α, we get $\alpha \approx_\Omega \top$, and thus $F(i,\top)$ from $F(i,\alpha)$. For the reverse of $(*)$, assume $E(i) \wedge |F(i,\top) \approx_\Omega \alpha|$. From $E(i)$ we get $F(i,\beta)$ for some β. But then, as we have just shown, $F(i,\top) \approx_\Omega \beta$. This yields $\alpha \approx_\Omega F(i,\top) \approx_\Omega \beta$, and thus $F(i,\alpha)$. $\qquad\square$

6.1.7. Corollary. *If p is a higher order fibration, then $\mathrm{Set}(p)$ is a topos.*

Proof. The isomorphism of fibred categories in the previous proposition makes the subobject fibration $\begin{smallmatrix} \mathrm{Sub}(\mathrm{Set}(p)) \\ \downarrow \\ \mathrm{Set}(p) \end{smallmatrix}$ a higher order fibration, and it hence makes $\mathrm{Set}(p)$ a topos, see Corollary 5.4.9. $\square$

6.1.8. Examples. (i) If Ω is a complete Heyting algebra, then we get Fourman and Scott's category Ω-**set** as a result of applying the $\mathrm{Set}(-)$ construction to the higher order family fibration $\begin{smallmatrix} \mathrm{Fam}(\Omega) \\ \downarrow \\ \mathbf{Sets} \end{smallmatrix}$. The object Ω with ordinary equality forms a subobject classifier with truth map **true**: $1 \to \Omega$ given by

$$\mathbf{true}(x, \alpha) = \begin{cases} \top & \text{if } \alpha = \top \\ \bot & \text{otherwise.} \end{cases}$$

It can be shown that this category Ω-**set** is equivalent to the category of sheaves on Ω—with "sup" topology as in Example 5.6.3 (i), see [80] or [36, III, 2.8 and 2.9] for details. This result is due to Higgs.

(ii) Our main example is Hyland's **effective topos Eff** arising from applying the above construction to the realisability fibration $\begin{smallmatrix} \mathrm{UFam}(P\mathbb{N}) \\ \downarrow \\ \mathbf{Sets} \end{smallmatrix}$. Its subobject classifier **true**: $1 \to \Omega$ has codomain $\Omega = P\mathbb{N}$ with equality given by

$$|\alpha \approx_\Omega \beta| = \alpha \mathbin{\supset\kern-0.6em\subset} \beta = (\alpha \supset \beta) \wedge (\beta \supset \alpha)$$

where $\alpha, \beta \subseteq \mathbb{N}$. Thus α and β are identified if and only if there are codes $n, m \in \mathbb{N}$ with

$$\forall k \in \alpha.\, n \cdot k \in \beta \qquad \text{and} \qquad \forall k \in \beta.\, m \cdot k \in \alpha.$$

This effective topos **Eff**—or als called realisability topos—will be further investigated in the remainder of this chapter.

We conclude this section with a characterisation of epis and quotients.

6.1.9. Lemma. *A morphism $F \colon (I, \approx) \to (J, \approx)$ is an epimorphism in $\mathrm{Set}(p)$ if and only if in the logic of the fibration p it is the case that*

$$j \colon J \mid E(j) \vdash \exists i \colon I.\, F(i, j).$$

Proof. Since $\mathrm{Set}(p)$ is a topos, F is an epi if and only if it is a cover, and the latter if and only if it is an internal epi in the logic of the subobject fibration of $\mathrm{Set}(p)$, see Lemma 4.4.7. If we use the equivalence between subobjects and strict predicates from Proposition 6.1.6 (ii), and express 'internal surjectivity' in terms of strict predicates, we get $j \colon J \mid E(j) \vdash \exists i \colon I.\, F(i, j)$, see also Exercise 6.1.3. $\square$

6.1.10. Lemma. *Let $(I, \approx)$ be an object of* Set(p). *Call a strict relation $R = (i, i': I \vdash R(i, i'): \mathsf{Prop})$* **almost an equivalence relation on** $(I, \approx)$ *if R is symmetric and transitive and satisfies $i: I \mid E(i) \vdash R(i, i)$.*

In that case R is obviously an equality predicate on I, and it forms an epi $R: (I, \approx) \twoheadrightarrow (I, R)$ in Set(p).

Moreover, every quotient $(I, \approx) \twoheadrightarrow (J, \approx)$ is of this form $(I, \approx) \twoheadrightarrow (I, R)$ for such an almost equivalence relation R.

Proof. The first part of the lemma is easy. For the last part, assume an epi $F: (I, \approx) \twoheadrightarrow (J, \approx)$. We get a strict predicate $R(i, i') \stackrel{\text{def}}{=} \exists j: J. \, F(i, j) \wedge F(i', j)$, which is almost an equivalence relation on $(I, \approx)$. The logical characterisation of F as epi in the previous lemma yields that $(I, \approx) \twoheadrightarrow (J, \approx)$ is the same quotient as $(I, \approx) \twoheadrightarrow (I, R)$. $\qquad\square$

Exercises

6.1.1.　In order to check that two representing elements F, G of maps $(I, \approx) \rightrightarrows (J, \approx)$ in Set(p) are equivalent, show that it suffices to have $F \leq G$ (in the fibre over $I \times J$).

6.1.2.　Prove that an object $(I, \approx) \in$ Set(p) is isomorphic to the object $(I \times I, \cong)$ with equality

$$z, w: I \times I \vdash |z \cong w| = |\pi z \approx_I \pi w| \wedge |\pi' z \approx_I \pi' w| \wedge |\pi z \approx \pi' z|: \mathsf{Prop}.$$

6.1.3.　Let $F: (I, \approx) \rightarrow (J, \approx)$ be a morphism in Set(p). Check that its graph $\langle \mathrm{id}, F \rangle: (I, \approx) \rightarrowtail (I, \approx) \times (J, \approx)$, considered as a strict predicate, is simply $i: I, j: J \vdash F(i, j): \mathsf{Prop}$.

6.1.4.　For $F: (I, \approx) \rightarrow (J, \approx)$ in Set(p), prove that the resulting substitution functor $F^*: \mathrm{SPred}(p)_{(J, \approx)} \rightarrow \mathrm{SPred}(p)_{(I, \approx)}$ has both a left and a right adjoint given on $B = (j: J \vdash B(j): \mathsf{Prop})$ by

$$i: I \vdash \exists_F(B)(i) \stackrel{\text{def}}{=} \exists j: J. \, F(i, j) \wedge B(j): \mathsf{Prop}$$
$$i: I \vdash \forall_F(B)(i) \stackrel{\text{def}}{=} E(i) \wedge \forall j: J. \, F(i, j) \supset B(j): \mathsf{Prop}.$$

6.1.5.　Describe singleton maps $\{\,\}: (I, \approx) \rightarrow P(I, \approx)$ in Set(p).

6.1.6.　Let $\begin{smallmatrix} \mathbb{E} \\ \downarrow p \\ \mathbb{B} \end{smallmatrix}$ be a higher order fibration. Define a functor Eq: $\mathbb{B} \rightarrow$ Set(p) by equipping an object $I \in \mathbb{B}$ with internal equality $(I, \mathrm{Eq}(I))$. Show that one can recover p from Set(p) via the change-of-base situation,

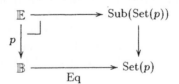

6.2 The effective topos and its subcategories of sets, ω-sets, and PERs

We now start the investigation of the effective topos **Eff**, arising as a special case of the construction described in the previous section. We focus on how the categories of sets, of ω-sets and of PERs are related to **Eff** (following [142]). In particular, we will describe how the double negation nucleus ¬¬ on **Eff** has **Sets** as its category of sheaves and ω-**Sets** as its category of separated objects. In the next section we shall deal with families of ω-sets and of PERs indexed by objects of **Eff**.

We start by identifying global elements in **Eff**.

6.2.1. Lemma. (i) *For an object* $(I, \approx) \in$ **Eff**, *the set* $\Gamma(I, \approx)$ *of global elements* $1 \to (I, \approx)$ *can be described as the quotient set*

$$\Gamma(I, \approx) = \{i \in I \mid E(i) \neq \emptyset\}/\sim \qquad where \qquad i \sim i' \Leftrightarrow |i \approx i'| \neq \emptyset.$$

(ii) *For a map* $F: (I, \approx) \to (J, \approx)$ *in* **Eff** *we get a function* $\Gamma F: \Gamma(I, \approx) \to \Gamma(J, \approx)$ *by*

$$[i] \mapsto \{j \in J \mid F(i, j) \neq \emptyset\}.$$

This describes the global elements (or sections) functor $\Gamma:$ **Eff** $\to$ **Sets**.

Notice that $\sim$ is a partial equivalence relation on I, and so an equivalence relation on the subset of I for which $i \sim i$ (that is $E(i) \neq \emptyset$), see Exercise 1.2.5.

Proof. (i) A morphism $G: 1 \to (I, \approx)$ in **Eff** is a function $G: I \to P\mathbb{N}$ with $G(i_0) \neq \emptyset$ for some $i_0 \in I$. Then $E(i_0) \neq \emptyset$ and we have an equivalence,

$$i: I \mid \top \vdash G(i) \sqsupset\!\sqsubset |i \approx i_0|,$$

so that G may be identified with the class $[i_0]$. Conversely, given such a class, we get a (unique) map $1 \to (I, \approx)$ in **Eff** by this logical equivalence $\sqsupset\!\sqsubset$.

(ii) Assume an equivalence class $[i] \in \Gamma(I, \approx)$; then $E(i) \neq \emptyset$, and so $F(i, j_0) \neq \emptyset$ for some $j_0 \in J$. Hence $E(j_0) \neq \emptyset$ and we have an equivalence $F(i, j) \sqsupset\!\sqsubset |j \approx j_0|$ like above. $\qquad\square$

6.2.2. Proposition. *There is a functor* $\nabla:$ **Sets** $\to$ **Eff** *which maps a set* J *to the object* $(J, =)$ *with*

$$|j = j'| \stackrel{\text{def}}{=} \bigcup\{\mathbb{N} \mid j = j'\} = \begin{cases} \mathbb{N} & if \ j = j' \\ \emptyset & otherwise. \end{cases}$$

This functor ∇ *is full and faithful and has the global sections functor* $\Gamma:$ **Eff** $\to$ **Sets** *as left adjoint.*

Because Γ preserves finite limits, we have a geometric morphism $\mathbf{Sets} \to \mathbf{Eff}$. It is what is called an "inclusion of toposes", since the direct image part ∇ is full and faithful.

Proof. For a function $f: J \to K$ one gets a morphism $\nabla f: \nabla J \to \nabla K$ in $\mathbf{Eff}$ by the predicate

$$(\nabla f)(j, k) = \begin{cases} \mathbb{N} & \text{if } f(j) = k \\ \emptyset & \text{otherwise.} \end{cases}$$

We establish a bijective correspondence

$$\frac{\Gamma(I, \approx) \xrightarrow{\ f\ } J \quad \text{in } \mathbf{Sets}}{(I, \approx) \xrightarrow[\ G\ ]{} \nabla J \quad \text{in } \mathbf{Eff}}$$

in the following manner. Assuming a function $f: \Gamma(I, \approx) \to J$, there is a predicate $\overline{f}: I \times J \to P\mathbb{N}$ given by

$$\overline{f}(i, j) = \begin{cases} E(i) & \text{if } E(i) \neq \emptyset \text{ and } f([i]) = j \\ \emptyset & \text{otherwise} \end{cases}$$

which yields a map $(I, \approx) \to \nabla J$ in $\mathbf{Eff}$. And conversely, given a map G as above, one gets a function $\Gamma(I, \approx) \to J$ by the following recipe. For a class $[i] \in \Gamma(I, \approx)$ one has $E(i) \neq \emptyset$, and so $G(i, j) \neq \emptyset$ for some $j \in J$. But this j is unique, since the equality of ∇J is actual equality on J.

Notice that the counit $\varepsilon: \Gamma \nabla J \to J$ is thus given by $[j] = \{j\} \mapsto j$. Hence it is an isomorphism and, as a result, $\nabla: \mathbf{Sets} \to \mathbf{Eff}$ is full and faithful. For future reference, the unit $\eta: (I, \approx) \to \nabla \Gamma(I, \approx)$ is given by the predicate

$$\mathrm{Eta}(i, [i']) = \begin{cases} E(i) & \text{if } [i] = [i'], \text{ i.e. if } |i \approx i'| \neq \emptyset \\ \emptyset & \text{otherwise.} \end{cases} \qquad \square$$

If we replace $\mathbf{Sets}$ by ω-$\mathbf{Sets}$ then we have a similar result.

6.2.3. Proposition. *There is an inclusion functor $\mathcal{I}: \omega\text{-}\mathbf{Sets} \to \mathbf{Eff}$ which maps an ω-set (J, E) to the object $(J, =_E)$ with*

$$|j =_E j'| = \begin{cases} E(j) & \text{if } j = j' \\ \emptyset & \text{otherwise.} \end{cases}$$

This functor $\mathcal{I}$ is full and faithful and has a left adjoint $\mathbf{s}$, which maps an object $(I, \approx) \in \mathbf{Eff}$ to the set $\Gamma(I, \approx)$ of global elements of $(I, \approx)$ with existence predicate

$$E([i]) = \bigcup_{i' \in [i]} E(i').$$

The notation **s** for this left adjoint is in accordance with the notation **s** for separated reflection in Section 5.7, because that is what this functor will turn out to be.

Proof. We seek correspondences

$$\frac{\mathbf{s}(I,\approx) \xrightarrow{\ f\ } (J,E) \quad \text{in } \omega\text{-Sets}}{(I,\approx) \xrightarrow[\ G\ ] \mathcal{I}(J,E) = (J,=_E) \quad \text{in } \mathbf{Eff}}$$

They are essentially as in the proof of the previous proposition, except that we have to be more careful about codes.

- Starting from a morphism $f\colon \mathbf{s}(I,\approx) \to (J,E_J)$ in ω-**Sets**, we have by definition a function $f\colon \Gamma(I,\approx) \to J$ for which there is a code e satisfying: for each $m \in E([i]) = \bigcup_{i' \in [i]} E_I(i)$ we have $e \cdot m \in E_J(f([i]))$. We get a predicate $\overline{f}$ as in the previous proof. The code e is then used for the validity of strictness and single-valuedness:

$$i\colon I, j\colon J \mid \overline{f}(i,j) \ \vdash\ E(i) \wedge E(j)$$
$$i\colon I, j, j'\colon J \mid \overline{f}(i,j), \overline{f}(i,j') \ \vdash\ |j =_J j'|.$$

- Conversely, for a map $G\colon (I,\approx) \to (J,=_E)$ in **Eff** we may assume codes

$$n_1 \ \in\ \bigcap_{i \in I} \Big(E(i) \supset \bigcup_{j \in J} G(i,j) \Big)$$
$$n_2 \ \in\ \bigcap_{(i,j) \in I \times J} \Big(G(i,j) \supset E(i) \wedge E(j) \Big)$$

For each $[i] \in \Gamma(I,\approx)$ there is a unique $j \in J$ with $G(i,j) \neq \emptyset$. A code for the resulting function $\Gamma(I,\approx) \to J$ is obtained as follows: if $m \in E([i]) = \bigcup_{i' \in [i]} E_I(i)$, say $m \in E_I(i')$ for i' with $|i = i'| \neq \emptyset$; then the unique $j \in J$ with $G(i,j) \neq \emptyset$ also satisfies $G(i',j) \neq \emptyset$. Using the above codes n_1, n_2 we get $n_1 \cdot m \in G(i',j)$, and so the second component of $n_2 \cdot (n_1 \cdot m)$ is in $E_J(j)$.

Again, the counit of the adjunction is an isomorphism, so that the functor ω-**Sets** $\to$ **Eff** is full and faithful. $\qquad\square$

It is easy to see that the functor $\nabla\colon \mathbf{Sets} \to \mathbf{Eff}$ factors through the embedding ω-**Sets** $\to$ **Eff** via the functor **Sets** $\to \omega$-**Sets** from Section 1.2. If we also involve the inclusion $\mathbf{PER} \to \omega\text{-}\mathbf{Sets}$, then we get a summarising diagram

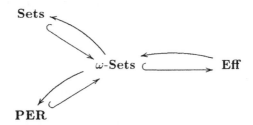

in which all arrows $\hookrightarrow$ from left to right are full and faithful functors, and the arrows in opposite direction are their left adjoints. Thus, these adjoints $\leftrightarrows$ are reflections. In this diagram, the categories $\mathbf{Sets}$ and $\mathbf{Eff}$ are toposes, but $\omega\text{-}\mathbf{Sets}$ is not a topos, see Exercise 6.2.5 below.

The images of these functors $\hookrightarrow$ in $\mathbf{Eff}$ can also be described intrinsically. Therefore we need the following notions.

6.2.4. Definition. (i) An object $(I, \approx) \in \mathbf{Eff}$ will be called **canonically separated** if both

$$|i \approx_I i'| \neq \emptyset \;\Rightarrow\; i = i' \qquad \text{and} \qquad E_I(i) \neq \emptyset$$

for all $i, i' \in I$. Or equivalently, if

$$|i \approx_I i'| \neq \emptyset \;\Leftrightarrow\; i = i',$$

so that $\approx_I$ is completely determined by its existence map $E_I \colon I \to P\mathbb{N}$.

(ii) And $(I, \approx)$ is **canonically a sheaf** if both

$$|i \approx_I i'| \neq \emptyset \;\Rightarrow\; i = i' \qquad \text{and} \qquad \bigcap_{i \in I} E(i) \neq \emptyset.$$

Such a canonical sheaf $(I, \approx)$ is thus certainly canonically separated.

(iii) Finally, $(I, \approx) \in \mathbf{Eff}$ is called **modest** (or a **modest set**, or also an **effective object**) if it is canonically separated and satisfies

$$E_I(i) \cap E_I(i') \neq \emptyset \;\Rightarrow\; i = i'.$$

Often we say that an object is canonically separated, a sheaf or modest if it is isomorphic to an object which is canonically separated, a sheaf or modest.

6.2.5. Proposition. (i) *The category $\omega\text{-}\mathbf{Sets}$ is equivalent to the full subcategory of $\mathbf{Eff}$ on the canonically separated objects.*

(ii) *And $\mathbf{Sets}$ is equivalent to the full subcategory of canonical sheaves.*

(iii) *Finally, $\mathbf{PER}$ is equivalent to the full subcategory on the modest sets.*

Proof. The canonically separated objects $(I, \approx_I)$ are determined by their existence predicates $E_I \colon I \to P\mathbb{N}$, which satisfy $E_I(i) \neq \emptyset$. These correspond to ω-sets. And the $E_I \colon I \to P\mathbb{N}$ with disjoint images (*i.e.* the modest sets) correspond to PERs, see Exercise 1.2.9. And if $\bigcap_{i \in I} E_I(i) \neq \emptyset$, then we may as well assume $E_I(i) = \mathbb{N}$. But such objects correspond to sets. □

The double negation nucleus in **Eff**

In the remainder of this section, the notions of 'closed', 'dense', 'separated object' and 'sheaf' will refer to the double negation nucleus $\neg\neg$ on the effective topos **Eff**. We first notice that for $\alpha \in \Omega = P\mathbb{N}$ one has

$$\neg\alpha = (\alpha \supset \bot) = \{n \mid \forall m \in \alpha.\, n \cdot m \in \emptyset\} = \begin{cases} \mathbb{N} & \text{if } \alpha = \emptyset \\ \emptyset & \text{otherwise.} \end{cases}$$

And thus

$$\neg\neg\alpha = \{n \mid \forall m \in \neg\alpha.\, n \cdot m \in \emptyset\} = \begin{cases} \mathbb{N} & \text{if } \alpha \neq \emptyset \\ \emptyset & \text{otherwise.} \end{cases}$$

Thus $\neg\neg\alpha$ simply tells whether α is empty or not; it forgets about all the realisers (elements of α).

A subobject of $(I, \approx)$ in **Eff** can by Proposition 6.1.6 (ii) be identified with a strict predicate $A \colon I \to P\mathbb{N}$ on $I \in$ **Sets** in the realisability fibration $\begin{smallmatrix} \mathrm{UFam}(P\mathbb{N}) \\ \downarrow \\ \mathbf{Sets} \end{smallmatrix}$. Its double negation is given by the predicate

$$\begin{aligned} \neg\neg A(i) &= E(i) \wedge ((E(i) \wedge (A(i) \supset \bot)) \supset \bot) \\ &\cong \begin{cases} E(i) & \text{if } A(i) \neq \emptyset \\ \emptyset & \text{otherwise.} \end{cases} \end{aligned}$$

Such a predicate A is thus **closed** if

$$\left(\bigcap_{i \in I,\, A(i) \neq \emptyset} E(i) \supset A(i) \right) \neq \emptyset.$$

And A is **dense** if for each $i \in I$

$$E(i) \neq \emptyset \;\Rightarrow\; A(i) \neq \emptyset.$$

The latter is of course different from 'A **holds**', which is

$$\left(\bigcap_{i \in I} E(i) \supset A(i) \right) \neq \emptyset.$$

(This validity requires a uniform realiser.)

We start with two basic lemmas.

6.2.6. Lemma. *Closed subobjects of* $(I, \approx) \in$ **Eff** *can be identified with ordinary subsets of* $\Gamma(I, \approx) \in$ **Sets***. More formally, there is a change-of-base situation,*

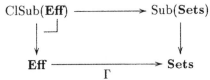

Proof. Closed strict predicates A on $(I, \approx)$ are turned into subsets by

$$A \mapsto \{[i] \in \Gamma(I, \approx) \mid A(i) \neq \emptyset\}.$$

And conversely, given a subset $S \subseteq \Gamma(I, \approx)$ one gets a closed strict predicate

$$B(i) = \begin{cases} E(i) & \text{if } E(i) \neq \emptyset \text{ and } [i] \in S \\ \emptyset & \text{otherwise.} \end{cases} \qquad \square$$

6.2.7. Lemma. (i) *Consider a canonical subobject* $(I, \approx_A) \rightarrowtail (I, \approx)$ *in* **Eff** *given by a strict predicate A on $(I, \approx)$. Then*

$$(I, \approx) \text{ is canonically separated/modest}$$
$$\Rightarrow \quad (I, \approx_A) \text{ is canonically separated/modest.}$$

(ii) *If R is almost an equivalence relation on $(I, \approx)$ which is closed, then (I, R) in the quotient $(I, \approx) \twoheadrightarrow (I, R)$ is canonically separated. And it is modest if $(I, \approx)$ is. Vice-versa, if (I, R) is canonically separated, then R is closed.*

Proof. (i) Consider the subset $I' = \{i \in I \mid A(i) \neq \emptyset\}$ of I, with equality $|i \approx_{I'} i'| = A(i) \wedge |i \approx_I i'|$, which was written as $\approx_A$ earlier. Then $(I, \approx_A)$ is equal (as a subobject) to $(I', \approx_{I'})$. Hence it is canonically separated/modest if $(I, \approx)$ is.

(ii) Assume $R \colon I \times I \to P\mathbb{N}$ is a closed and almost an equivalence relation on $(I, \approx)$. Closedness means that $R(i, i')$ is $E(i) \wedge E(i')$ if $R(i, i') \neq \emptyset$ and $\emptyset$ otherwise. One can then show that (I, R) is equal (as a quotient) to $(\Gamma(I, \approx), \widehat{R})$ where

$$\widehat{R}([i_1], [i_2]) = \bigcup \{R(i_1', i_2') \mid i_1' \in [i_1] \text{ and } i_2' \in [i_2]\}.$$

And if $(I, \approx)$ is modest, then $\Gamma(I, \approx) \cong I$, so that $\widehat{E}(i_1) \cap \widehat{E}(i_2) \neq \emptyset$ implies $E(i_1) \cap E(i_2) \neq \emptyset$ and thus $i_1 = i_2$. Finally, if (I, R) is canonically separated, then, if $R(i, i') \neq \emptyset$ we get $i = i'$. Hence closedness follows from reflexivity $i \colon I \mid E(i) \vdash R(i, i)$. $\qquad \square$

Our next aim is to show that the $(\neg\neg\text{-})$sheaves in **Eff** are sets and that the $(\neg\neg\text{-})$separated objects are ω-sets. The first can be established by a three-line topos-theoretic argument using the geometric inclusion **Sets** $\hookrightarrow$ **Eff**, see [142]: there must be a nucleus **j** on **Eff** with **Sets** $\simeq$ Sh$_{\mathbf{j}}$(**Eff**); since ∇: **Sets** $\rightarrow$ **Eff** preserves the initial object, one gets $\mathbf{j} \leq \neg\neg$. But since **Sets** is Boolean, one must also have $\mathbf{j} \geq \neg\neg$.

Because we have not seen all the topos theory used in this argument, we give a direct proof, based on the theory in Sections 5.7 and 5.8.

6.2.8. Theorem. *The category of sheaves of the double negation nucleus* $\neg\neg$ *on the effective topos* **Eff** *is equivalent to* **Sets**. *And the category of separated objects is equivalent to* ω-**Sets**.

Proof. We use the 'logical' characterisations of separated objects and sheaves in Theorem 5.8.2. For separated objects, to prove the result, it suffices to show that

$$(J, \approx) \text{ if canonically separated} \;\Leftrightarrow\; \bigcap_{j, j' \in J} \big(\neg\neg |j \approx_J j'| \supset |j \approx_J j'| \big) \neq \emptyset.$$

where the right hand side expresses that internal and external equality on $(J, \approx)$ coincide (in the fibration of closed subobjects).

For the implication ($\Rightarrow$) assume

$$m \in \neg\neg |j \approx_J j'| = \begin{cases} E(j) \wedge E(j') & \text{if } |j \approx_J j'| \neq \emptyset \\ \emptyset & \text{otherwise.} \end{cases}$$

Then $m \in E(j) \wedge E(j')$ and $|j \approx_J j'| \neq \emptyset$. The latter yields $j = j'$, because $(J, \approx)$ is canonically separated. Hence the first projection $\mathbf{p}m$ is in $E(j) = |j \approx j| = |j \approx j'|$.

For the implication ($\Leftarrow$), assume e is a code in the non-empty set on the right. We show that the unit $\eta \colon (J, \approx) \rightarrow \nabla\Gamma(J, \approx)$ is a mono. Since $\nabla\Gamma(J, \approx)$ is separated, Lemma 6.2.7 (i) yields that $(J, \approx)$ is isomorphic to a canonically separated object, as required. By Lemma 6.1.4, the unit η is monic if and only if we have validity of the following entailment.

$$j_1, j_2 \colon J, [j_3] \colon \Gamma(J, \approx) \mid \text{Eta}(j_1, [j_3]), \text{Eta}(j_2, [j_3]) \vdash |j_1 \approx j_2|.$$

(See the end of the proof of Proposition 6.2.2 where the predicate Eta is described explicitly.) If we have elements $m_1 \in \text{Eta}(j_1, [j_3])$ and $m_2 \in \text{Eta}(j_2, [j_3])$, then $m_1 \in E(j_1)$ and $|j_1 \approx j_3| \neq \emptyset$ and also $m_2 \in E(j_2)$ and $|j_2 \approx j_3| \neq \emptyset$. Then $|j_1 \approx j_2| \neq \emptyset$, so that the pair $\langle m_1, m_2 \rangle$ is in $\neg\neg |j_1 \approx j_2|$, and so we may conclude that $e \cdot \langle m_1, m_2 \rangle$ is in $|j_1 \approx j_2|$.

Now we come to sheaves. Let $(J, \approx)$ be canonically a sheaf. In order to prove that $(J, \approx)$ is a sheaf, we use Theorem 5.8.2 (ii). (An alternative is in Exercise 6.2.6 below.) We thus have to show that unique choice holds on $(J, \approx)$ in the fibration of closed subobjects. Following Definition 4.9.1, let

$R: I \times J \to P\mathbb{N}$ be a closed single-valued strict relation on $(I, \approx) \times (J, \approx)$. We have to show that the canonical map F is an isomorphism in

$$
\begin{array}{ccc}
(I \times J, \approx_R) & \overset{F}{\dashrightarrow} & (I, \cong) \\
\downarrow & & \downarrow \\
(I, \approx) \times (J, \approx) & \underset{\pi}{\longrightarrow} & (I, \approx)
\end{array}
$$

where $\cong$ is the equality of the coproduct $\overline{\exists_\pi(R)} = \neg\neg \bigcup_{j \in J} R(-, j)$. The canonical map F is given by

$$F(i, j, i') = |i \approx i'| \wedge R(i, j) = G(i, i', j).$$

We use Lemmas 6.1.4 and 6.1.9 to show that F is both a mono and an epi, and thus an isomorphism in **Eff**. Since the relation R is single-valued, F is a mono. In order to show that F is an epi, we need to find a realiser in

$$\bigcap_{i \in I} |i \cong i| \supset \bigcup_{j \in J} R(i, j),$$

i.e. in

$$\bigcap \left\{ E(i) \supset \bigcup_{j \in J} R(i, j) \;\middle|\; i \in I \text{ with } \bigcup_{j \in J} R(i, j) \neq \emptyset \right\}.$$

We use codes $k \in \bigcap_{j \in J} E(j)$, and $e \in \bigcap_{(i,j) \in I \times J} \neg\neg R(i, j) \supset R(i, j)$. Assuming $m \in E(i)$ where $\bigcup_{j \in J} R(i, j) \neq \emptyset$, we get $R(i, j) \neq \emptyset$ for some $j \in J$. Then $\langle \mathbf{p}m, k \rangle \in E(i) \wedge E(j)$, so that $e \cdot \langle \mathbf{p}m, k \rangle \in R(i, j)$. And this for all $i \in I$, as required.

Conversely, if $(J, \approx)$ is a sheaf, then we already know that the unit map $\eta: (J, \approx) \to \nabla\Gamma(J, \approx)$ is a mono. It is easily seen to be dense. But it is closed by Exercise 5.7.3 since $(J, \approx)$ is a sheaf, so this unit is an isomorphism. $\square$

Exercises

6.2.1.   Verify that the global sections functor $\Gamma: \mathbf{Eff} \to \mathbf{Sets}$ satisfies

$$\Gamma\left(1 \overset{\mathbf{true}}{\longrightarrow} \Omega\right) = \left(1 \overset{1}{\longrightarrow} \{0, 1\}\right).$$

6.2.2.   Give direct proofs that the functors $\mathbf{Sets} \to \mathbf{Eff}$ and $\omega\text{-}\mathbf{Sets} \to \mathbf{Eff}$ are full and faithful (without using the adjunctions like in Propositions 6.2.2 and 6.2.3). Check also that $\Gamma: \mathbf{Eff} \to \mathbf{Sets}$ preserves finite limits.

6.2.3.   Prove that the separated reflection functor s: $\mathbf{Eff} \to \omega\text{-}\mathbf{Sets}$ preserves finite products. (It does not preserve finite limits.)

6.2.4.   Show that an object $(I, \approx) \in \mathbf{Eff}$ is a modest set if and only if it is canoni-
cally separated and satisfies: $|i \approx_I i'| = E_I(i) \cap E_I(i')$ for each pair $i, i' \in I$.

6.2.5.   The point of this exercise is to show that the category ω-**Sets** of ω-sets is
not a topos. We use the ω-set $(\mathbb{N}, E)$ with $E(n) = \{n\}$, which is a natural
numbers object in ω-**Sets**, see Exercise 1.2.10.

(i)   Let $f: (I, E) \to (J, E)$ be a morphism in ω-**Sets**. Show that f is monic
in ω-**Sets** if and only if f is injective (between the underlying sets I, J).
And similarly that f is epic in ω-**Sets** if and only if f is surjective.

(ii)   Let $A \subseteq \mathbb{N}$ be an arbitrary subset. We define an ω-set $(\mathbb{N}, E_A)$ by

$$E_A(n) = \begin{cases} \{\langle 1, n \rangle\} & \text{if } n \in A \\ \{\langle 0, n \rangle\} & \text{otherwise.} \end{cases}$$

There is then an obvious morphism of ω-sets $i: (\mathbb{N}, E_A) \to (\mathbb{N}, E)$,
which is tracked by a code for the second projection $\langle m, n \rangle \mapsto n$. By (i)
this map i is both a mono and an epi. If ω-**Sets** were a topos, then i
should be an isomorphism, see Exercise 5.4.1. Verify that a code for an
inverse of i yields a decision code for A. But A is an arbitrary set...

6.2.6.   Here is an alternative proof that objects of the form ∇X are $(\neg\neg\text{-})$sheaves
in **Eff**; it involves less hacking with realisers than above. First show that
for a dense mono $(I, \approx_A) \rightarrowtail (I, \approx)$, given by a strict predicate $A: I \to P\mathbb{N}$
on $(I, \approx)$ one gets $\Gamma(I, \approx_A) = \Gamma(I, \approx)$ in **Sets**. Deduce now that any dense
partial map $(I, \approx) \leftarrowtail (I, \approx_A) \to \nabla X$ has a (unique) extension $(I, \approx) \to$
∇X, using the reflection $(\Gamma \dashv \nabla)$.

6.2.7.   Notice that the fact that the regular subobjects in ω-**Sets** yield a higher
order fibration with classical logic (as stated in Proposition 5.3.9) is a consequence of Theorem 6.2.8 and Corollary 5.7.12.

6.2.8.   Prove the fact, used in the proof of Theorem 6.2.8, that if $(J, \approx)$ is separated,
then the mono $\eta: (J, \approx) \rightarrowtail \nabla\Gamma(J, \approx)$ is dense.

6.3 Families of PERs and ω-sets over the effective topos

Later on we shall be using the effective topos **Eff** as a universe for models
of type theories, in which types are interpreted as PERs (in polymorphic
type theories, see Chapter 8), and kinds as ω-sets (in polymorphic/dependent
dependent type theories, see Chapter 11). In order to do so we need to consider
PERs and ω-sets suitably indexed by objects of **Eff**. This is the subject of the
present section. We shall define split fibrations $\begin{smallmatrix} \mathrm{UFam}(\mathbf{PER}) \\ \downarrow \\ \mathbf{Eff} \end{smallmatrix}$ and $\begin{smallmatrix} \mathrm{UFam}(\omega\text{-}\mathbf{Sets}) \\ \downarrow \\ \mathbf{Eff} \end{smallmatrix}$,
which are very much like the fibrations $\begin{smallmatrix} \mathrm{UFam}(\mathbf{PER}) \\ \downarrow \\ \omega\text{-}\mathbf{Sets} \end{smallmatrix}$ and $\begin{smallmatrix} \mathrm{UFam}(\omega\text{-}\mathbf{Sets}) \\ \downarrow \\ \omega\text{-}\mathbf{Sets} \end{smallmatrix}$ of ω-set-
indexed-PERs and ω-set-indexed-ω-sets, that we introduced in Section 1.2.
Although we use each of the names UFam($\mathbf{PER}$) and UFam(ω-**Sets**) twice
(over ω-**Sets** and over **Eff**), we refer in each case to two different categories: the

total categories UFam($\mathbf{PER}$) and UFam(ω-$\mathbf{Sets}$) will be different, whether we consider them over ω-$\mathbf{Sets}$ or over $\mathbf{Eff}$. But as long as we present them together with their base categories, confusion is not likely to occur.

(In the next chapter we shall see that PERs form an "internal category" in ω-$\mathbf{Sets}$ and also in $\mathbf{Eff}$, and that the fibrations $\begin{smallmatrix} \text{UFam}(\mathbf{PER}) \\ \downarrow \\ \omega\text{-}\mathbf{Sets} \end{smallmatrix}$ and $\begin{smallmatrix} \text{UFam}(\mathbf{PER}) \\ \downarrow \\ \mathbf{Eff} \end{smallmatrix}$ are the "externalisations" of these internal categories.)

We start with PERs indexed by objects of $\mathbf{Eff}$. This is as in [143].

6.3.1. Definition. For an object $(I, \approx) \in \mathbf{Eff}$, consider the following "fibre" category.

> **objects** $\Gamma(I, \approx)$-indexed families $(R_{[i]})_{[i] \in \Gamma(I, \approx)}$ of PERs.
>
> **morphisms** $f : (R_{[i]}) \to (S_{[i]})$ are $\Gamma(I, \approx)$-indexed collections $f = (f_{[i]})_{[i] \in \Gamma(I, \approx)}$ of maps $f_{[i]} : R_{[i]} \to S_{[i]}$ in $\mathbf{PER}$ which are **uniformly** (or, **effectively**) **tracked**: for some code $e \in \mathbb{N}$,
>
> $$\forall i \in I. \, \forall m \in E(i). \, e \cdot m \text{ tracks } f_{[i]} \text{ in } \mathbf{PER}.$$

A morphism $F : (J, \approx) \to (I, \approx)$ in $\mathbf{Eff}$ induces a reindexing functor F^* by

$$(R_{[i]}) \mapsto (R_{\Gamma(F)([j])}) \qquad \text{and} \qquad (f_{[i]}) \mapsto (f_{\Gamma(F)([j])}).$$

Application of the Grothendieck construction yields a split fibration, which will be written as $\begin{smallmatrix} \text{UFam}(\mathbf{PER}) \\ \downarrow \\ \mathbf{Eff} \end{smallmatrix}$.

This indexing of PERs by objects $(I, \approx) \in \mathbf{Eff}$ is clearly similar to the indexing of PERs by objects $(I, E) \in \omega$-$\mathbf{Sets}$, as in Definition 1.4.8. This will be made precise in the proposition below. But first we introduce a similar indexing of ω-sets over $\mathbf{Eff}$ via global sections. This yields a fibration $\begin{smallmatrix} \text{UFam}(\omega\text{-}\mathbf{Sets}) \\ \downarrow \\ \mathbf{Eff} \end{smallmatrix}$ where the total category UFam(ω-$\mathbf{Sets}$) has, in the fibre over $(I, \approx) \in \mathbf{Eff}$:

> **objects** $\Gamma(I, \approx)$-indexed families $(X_{[i]})$ of ω-sets
>
> **morphisms** $(X_{[i]}) \to (Y_{[i]})$ are $\Gamma(I, \approx)$-indexed collections $(f_{[i]})$ of maps $f_{[i]} : X_{[i]} \to Y_{[i]}$ in ω-$\mathbf{Sets}$ which are **uniformly tracked**: for some code $e \in \mathbb{N}$ one has $\forall i \in I. \, \forall m \in E(i). \, e \cdot m$ tracks $f_{[i]}$ in ω-$\mathbf{Sets}$.

Reindexing F^* is above, for PERs.

In the next result we relate indexing over $\mathbf{Eff}$ and indexing over ω-$\mathbf{Sets}$, both of PERs and of ω-sets

6.3.2. Proposition. *Indexing of PERs and of ω-sets over* **Eff** *and over ω-Sets can be related in the following manner.*

(i) *There are change-of-base situations,*

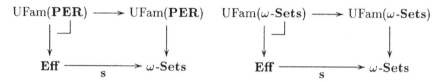

where s: **Eff** → ω-**Sets** *is the separated reflection functor, as introduced in Proposition 6.2.3.*

(ii) *Similarly, we have change-of-base situations,*

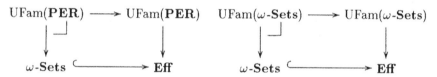

(iii) *The reflection* **PER** ⇆ ω-**Sets** *lifts to a fibred reflection over* **Eff**, *in a diagram:*

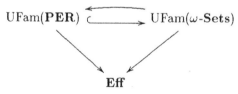

Proof. (i) By definition of indexing over **Eff**.

(ii) Because if $(I, \approx) \in$ **Eff** comes from an ω-set (I, E), we have $\Gamma(I, \approx) = I$, so indexing over $(I, \approx) \in$ **Eff** takes the form of indexing over $(I, E) \in$ ω-**Sets**.

(iii) By a pointwise construction.   □

In the previous section we saw that ω-sets can be identified within **Eff** as the (¬¬-) separated objects. So an alternative approach to ω-sets indexed by objects $(I, \approx) \in$ **Eff** would be to consider families $\begin{pmatrix} (X, \approx) \\ \downarrow \\ (I, \approx) \end{pmatrix}$ in **Eff** which are separated in the slice topos **Eff**/$(I, \approx)$. We recall from Exercises 5.7.1 and Proposition 5.7.14 that such a family $\begin{pmatrix} (X, \approx) \\ \downarrow \varphi \\ (I, \approx) \end{pmatrix}$ in **Eff**/$(I, \approx)$ is separated if and only if the mediating map η_φ is a mono in:

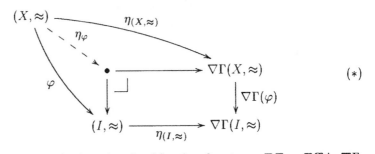

$$(*)$$

The underlying reason is that the sheafification functor $\mathbf{a} \colon \mathbf{Eff} \to \mathbf{Eff}$ is $\nabla\Gamma$, and that sheafification in the slice $\mathbf{Eff}/(I, \approx)$ of $\begin{pmatrix} (X,\approx) \\ \downarrow \\ (I,\approx) \end{pmatrix}$ is $\begin{pmatrix} \bullet \\ \downarrow \\ (I,\approx) \end{pmatrix}$, see Proposition 5.7.14.

The full subcategory of $\mathbf{Eff}^{\to}$ of $(\neg\neg\text{-})$separated families is then written as FSep($\mathbf{Eff}$). The codomain functor is a fibration $\begin{smallmatrix} \text{FSep}(\mathbf{Eff}) \\ \downarrow \\ \mathbf{Eff} \end{smallmatrix}$.

We first show that, when we restrict ourselves to ω-sets as indices, we indeed get families of ω-sets in this way.

6.3.3. Proposition. *If $(I, \approx) \in \mathbf{Eff}$ is a separated object, then a family $\begin{pmatrix} (X,\approx) \\ \downarrow\varphi \\ (I,\approx) \end{pmatrix}$ is separated if and only if the object $(X, \approx)$ is separated. This means that there is a change-of-base situation,*

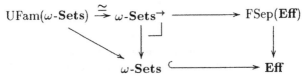

Proof. If $(I, \approx)$ is separated, then the unit $\eta_{(I,\approx)} \colon (I, \approx) \to \nabla\Gamma(I, \approx)$ in the above diagram $(*)$ is a mono. Hence the map $\bullet \to \nabla\Gamma(X, \approx)$ as well, because it is obtained by pullback. Thus we have: the family $\begin{pmatrix} (X,\approx) \\ \downarrow\varphi \\ (I,\approx) \end{pmatrix}$ is separated $\Leftrightarrow$ the map η_φ is a mono $\Leftrightarrow$ the unit $\eta_{(X,\approx)}$ is a mono $\Leftrightarrow$ $(X, \approx)$ is separated. $\qquad\square$

The relation between the fibrations $\begin{smallmatrix} \text{UFam}(\omega\text{-}\mathbf{Sets}) \\ \downarrow \\ \mathbf{Eff} \end{smallmatrix}$ and $\begin{smallmatrix} \text{FSep}(\mathbf{Eff}) \\ \downarrow \\ \mathbf{Eff} \end{smallmatrix}$ is rather subtle, and involves the manner in which the indexed ω-sets are given to us.

The full story appears in Section 11.7 where it will be shown that the latter fibration is the 'stack completion' of the former. At this stage we merely note the following result, which will be crucial in modelling type dependency with indexed ω-sets.

6.3.4. Proposition. *There is a full and faithful fibred functor over* **Eff**

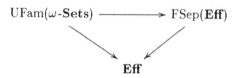

As a result, we have full and faithful fibred functors (over **Eff**),

$$\text{UFam}(\textbf{PER}) \longrightarrow \text{UFam}(\omega\text{-}\textbf{Sets}) \longrightarrow \text{FSep}(\textbf{Eff}) \longrightarrow \textbf{Eff}^{\rightarrow}$$

Proof. For a family $X = (X_{[i]})_{[i] \in \Gamma(I, \approx)}$ of ω-sets in UFam(ω-**Sets**) over $(I, \approx) \in \textbf{Eff}$, form the set

$$\{X\} = \{(i, x) \mid i \in I \text{ with } E_I(i) \neq \emptyset, \text{ and } x \in X_{[i]}\}$$

with equality

$$|(i, x) \cong (i', x')| = \begin{cases} |i \approx i'| \wedge E(x) & \text{if } [i] = [i'] \text{ and } x = x' \\ \emptyset & \text{otherwise.} \end{cases}$$

Then $E(i, x) = E(i) \wedge E(x) \neq \emptyset$ for all $(i, x) \in \{X\}$. Further, the set of global sections of this object is the disjoint union

$$\Gamma(\{X\}, \cong) = \coprod_{[i] \in \Gamma(I, \approx)} X_{[i]}.$$

We can define a projection map $\pi_X : (\{X\}, \cong) \to (I, \approx)$ in **Eff** by

$$\pi_X((i, x), i') = |i \approx_I i'| \wedge E(x).$$

·We claim this yields a separated family over $(I, \approx)$. Consider therefore the following pullback diagram.

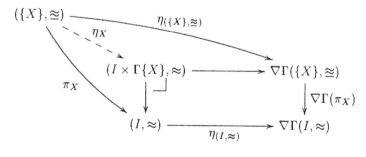

The map η_X herein is described by

$$\eta_X\left((i_1, x_1), (i_2, (i_3, x_2))\right) = \begin{cases} |i_1 \approx_I i_2| \wedge E(x_1) & \text{if } [i_1] = [i_3] \text{ and } x_1 = x_2 \\ \emptyset & \text{otherwise} \end{cases}$$

which is a mono by Lemma 6.1.4. Hence the family $\begin{pmatrix} (\{X\}, \approx) \\ \downarrow \pi_X \\ (I, \approx) \end{pmatrix}$ is separated.

For a morphism $(F, f) \colon (X_{[i]}) \to (Y_{[j]})$ in UFam(ω-**Sets**), where $F \colon (I, \approx) \to (J, \approx)$ is a map in **Eff** between the underlying index sets and f is a (uniformly tracked) family of maps $f_{[i]} \colon X_{[i]} \to Y_{\Gamma(F)([i])}$ in ω-**Sets**, we get a morphism $\{F, f\} \colon \{X\} \to \{Y\}$ in **Eff** by

$$\{F, f\}((i, x), (j, y)) = \begin{cases} F(i, j) \wedge E(x) & \text{if } f_{[i]}(x) = y \\ \emptyset & \text{otherwise} \end{cases}$$

such that $\pi_Y \circ \{F, f\} = F \circ \pi_X$. We leave it to the reader to verify that this yields a full and faithful fibred functor.   □

Exercises

6.3.1.   Show that the following diagram commutes, combining the functors from Propositions 6.3.3 and 6.3.4.

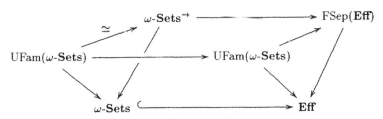

6.4 Natural numbers in the effective topos and some associated principles

In the final section of this chapter, we outline some aspects of the effective topos as a mathematical universe. In particular, we mention the natural numbers object N in **Eff**. It appears that **Eff** is the world of 'recursive mathematics': maps $N \to N$ in **Eff** can be identified with total recursive functions. This may even be internalised to give a stronger statement. We mention some of the principles that hold in **Eff**, like Markov's Principle, the Axiom of Countable Choice and Troelstra's Uniformity Principle. And in the end we investigate an alternative description of PERs (or modest sets) as subquotients of N. This description also applies to families of PERs (over separated objects).

The material in this section comes from [142]. More information may be found there, and also in, for example, [297, 262, 218].

6.4.1. Proposition. *The object $N = (\mathbb{N}, E)$ with $E(n) = \{n\}$ and $|n \approx m| = E(n) \cap E(m)$ is natural numbers object in* **Eff**. *It is a modest set, by definition.*

Proof. The zero map $0: 1 \to N$ in **Eff** comes from the zero element $0 \in \mathbb{N} = \Gamma(\mathbb{N}, E)$ in **Sets** as $|\cdot = 0|: 1 \to (\mathbb{N}, E)$, or as predicate

$$0(n) = \begin{cases} \{0\} & \text{if } n = 0 \\ \emptyset & \text{otherwise.} \end{cases}$$

And the successor morphism $S: N \to N$ is the predicate

$$S(n, m) = \begin{cases} \{m + 1\} & \text{if } n = m + 1 \\ \emptyset & \text{otherwise.} \end{cases}$$

Consider a diagram $1 \xrightarrow{[i_0]} (I, \approx) \xrightarrow{F} (I, \approx)$, where $i_0 \in I$ is some element with $E(i_0) \neq \emptyset$ (see Lemma 6.2.1). Write $F^{(n)} = F \circ \cdots \circ F$ (n times), so that $F^{(0)}(i_1, i_2) = |i_1 \approx i_2|$ and $F^{(n+1)}(i_1, i_2) = \exists i_3: I. F^{(n)}(i_1, i_3) \wedge F(i_3, i_2)$, and define a morphism $G: (\mathbb{N}, E) \to (I, \approx)$ by

$$G(n, i) = E(n) \wedge F^{(n)}(i_0, i).$$

It is easy to see that G is then a mediating map: $G \circ 0 = [i_0]$ and $G \circ S = F \circ G$. Uniqueness of G is a bit more involved. Assume H also satisfies $H \circ 0 = [i_0]$ and $H \circ S = F \circ H$, then we may assume codes

$$a \in \bigcap_i G(0, i) \supset H(0, i) \qquad c \in \bigcap_{n,i} G(n + 1, i) \supset \left(\bigcup_{i'} G(n, i') \wedge F(i', i) \right)$$

$$b \in \bigcap_{n,i} G(n, i) \supset E(n) \qquad d \in \bigcap_{n,i} \left(\bigcup_{i'} H(n, i') \wedge F(i', i) \right) \supset H(n + 1, i).$$

The following serves as motivation. Assume that we already have what we have to produce, namely a realiser e in $E(n) \supset (G(n, i) \supset H(n, i))$, for all n, i. Then $e \cdot n$ is in $G(n, i) \supset H(n, i)$, and this $e \cdot n$ can be used to construct a code in $G(n + 1, i) \supset H(n + 1, i)$. If $m \in G(n + 1, i)$, then $c \cdot m = \langle m_1, m_2 \rangle$ where $m_1 \in G(n, i')$ and $m_2 \in F(i', i)$ for some $i' \in I$. But then $(e \cdot n) \cdot m_1 \in H(n, i')$. So if we write $m_3 = \langle (e \cdot n) \cdot m_1, m_2 \rangle$, then

$$m_3 \in H(n, i') \wedge F(i', i) \subseteq \bigcup_{i'} H(n, i') \wedge F(i', i),$$

so that $d \cdot m_3 \in H(n + 1, i)$. And this for every $i \in I$.

This tells us how to obtain such a code e by primitive recursion on n, as

$$e \cdot 0 = a \qquad \text{and} \qquad e \cdot (n+1) = \Lambda m.\, d \cdot \langle (e \cdot n) \cdot \mathbf{p}(c \cdot m),\, \mathbf{p}'(c \cdot m) \rangle.$$

Finally, if we put

$$e' = \Lambda k.\, (e \cdot (b \cdot k)) \cdot k$$

then $e' \in \bigcap_{n,i} G(n,i) \supset H(n,i)$, making G, H equal maps $(\mathbb{N}, E) \to (I, \approx)$. $\square$

In Brouwer's intuitionism, the real numbers $\mathbb{R}$ are understood in such a way that all functions $\mathbb{R} \to \mathbb{R}$ are continuous, see [335, 4.12]. Classically this is not true of course, but there are toposes in which this does hold, see [188, VI, 9]. In a similar manner, all functions $N \to N$ in the effective topos **Eff** are recursive. This suggests that in **Eff** we are in the world of "recursive mathematics", where all functions are computable.

In this way, toposes provide a rich supply of universes for various kinds of mathematics.

6.4.2. Theorem. *Morphisms* $F \colon (\mathbb{N}, E) \to (\mathbb{N}, E)$ *in* **Eff** *can be identified with (total) recursive functions* $f \colon \mathbb{N} \to \mathbb{N}$.

Proof. Since the natural numbers object $N = (\mathbb{N}, E)$ is separated, and the embedding ω-**Sets** $\to$ **Eff** is full and faithful, we have that a morphism $F \colon (\mathbb{N}, E) \to (\mathbb{N}, E)$ in **Eff** corresponds uniquely to a morphism $f \colon (\mathbb{N}, E) \to (\mathbb{N}, E)$ in ω-**Sets**, where $E(n) = \{n\}$, see Exercise 1.2.10. But the latter has a code $e \in \mathbb{N}$ satisfying $f(n) = e \cdot n$. Thus f is a total recursive function. Translating this back to F in **Eff**, we can write F as

$$F(n, m) = \begin{cases} \{n\} & \text{if } m = e \cdot n \\ \emptyset & \text{otherwise.} \end{cases} \qquad \square$$

We would like to have an internal version of this result telling that for each $f \colon N^N$ there is an $e \colon N$ with $f(n) = e \cdot n$ in the internal language of **Eff**. Then one can properly say that Church's Thesis holds. But therefore we need to know what Kleene application $e \cdot n$ is in **Eff**. It turns out to be the same as in **Sets**, but this requires the following result about the exponent object N^N in **Eff**, and Markov's Principle.

6.4.3. Lemma. *The exponent* N^N *in* **Eff** *may be described as the canonically separated object* (TR, E) *where*

$$\mathrm{TR} = \{\textit{total recursive functions } f \colon \mathbb{N} \to \mathbb{N}\}$$
$$E(f) = \{e \in \mathbb{N} \mid e \textit{ is a code for } f\}.$$

Proof. Essentially this is because separated objects form an exponential ideal. Thus the exponent N^N in **Eff** is the same as in ω-**Sets**. $\square$

In an arbitrary topos $\mathbb{B}$ we have the object $2 = 1 + 1$ with cotuple [**false, true**]: $2 \to \Omega$. This map is a mono, by Exercise 4.5.1, since **false, true** are by definition disjoint. So one can identify the subobject $2 \rightarrowtail \Omega$ as the set $\{\alpha: \Omega \mid \alpha \vee \neg\alpha\}$ of **decidable** propositions. It is not hard to show that a subobject $X \rightarrowtail I$ is decidable (*i.e.* $X \vee \neg X = I$) if and only if its characteristic map $I \to \Omega$ factors through $2 \rightarrowtail \Omega$. By exponentiation we get that

$$2^I = \{\beta: \Omega^I \mid \forall i: I. \beta(i) \vee \neg\beta(i)\} \rightarrowtail \Omega^I$$

is the object of decidable predicates.

6.4.4. Proposition. *In **Eff** Markov's Principle holds: one has*

$$\forall \beta: 2^N. \neg\neg(\exists n: N. \beta(n)) \supset \exists n: N. \beta(n).$$

Recall that in constructive logic, the premise $\neg\neg(\exists n: N. \beta(n))$ is logically equivalent to $\neg\forall n: N. \neg\beta(n)$. Markov's Principle says that if for a decidable predicate β on N we know that it is impossible that β fails for all natural numbers, then β must hold for some number n. This n can for example be obtained by testing $\beta(0), \beta(1), \beta(2), \ldots, \beta(n)$ until a candidate n is found. But there is of course no bound for this search. Therefore, the status of Markov's Principle within constructive mathematics is not uncontroversial. But since this search for the candidate n can be done via minimalisation in a computable way, Markov's Principle holds in **Eff**.

Proof. It is convenient to identify 2^N with $\{f: N^N \mid \forall n: N. f(n) = 0 \vee f(n) = 1\}$, so that we can use Lemma 6.4.3. For a total recursive function $f \in \mathrm{TR}$, let

$$A_f = \exists n: N. (E(n) \wedge |f(n) \approx 1|) \subseteq \mathbb{N}.$$

We have to produce an inhabitant of

$$\bigcap_{f \in \mathrm{TR}} E(f) \supset (\neg\neg A_f \supset A_f).$$

Recall from Section 6.2 that $\neg\neg A_f$ is either $\emptyset$ of $\mathbb{N}$, so a realiser for it will not contain any useful information. For $f \in \mathrm{TR}$ and $e \in E(f)$, we know that e is a code for f. If $m \in \neg\neg A_f$, then there is an $n \in \mathbb{N}$ with $f(n) = 1$. Thus minimalisation $\mu n. (e \cdot n = 1)$ is defined and gives us a least such n. Therefore, $e' = \Lambda m. \mu n. (e \cdot n = 1)$ takes $\neg\neg A_f$ to A_f.

In the other case when $\neg\neg A_f = \emptyset$, then e' vacuously takes $\neg\neg A_f$ to A_f. Hence $\Lambda e. \Lambda m. \mu n. (e \cdot n = 1)$ is a realiser for Markov's Principle. $\square$

6.4.5. Example. Recall Kleene's Normal Form Theorem (*e.g.* from [236, Theorem II.1.2] or [66, 5, Corollary 1.4]): the basic predicates of recursion theory,

$$e \cdot n \uparrow, \qquad e \cdot n \downarrow, \qquad e \cdot n = m$$

are definable in first order arithmetic via the T-predicate and output function U, as

$$e \cdot n \uparrow \iff \neg \exists x : N. T(e, n, x)$$
$$e \cdot n \downarrow \iff \exists x : N. T(e, n, x)$$
$$e \cdot n = m \iff \exists x : N. T(e, n, x) \wedge U(x) = m$$

The predicates T and $U(-) = (-)$ are primitive recursive. Hence they are decidable, and so $\neg\neg$-closed. But then also the above predicates $e \cdot n \uparrow$, $e \cdot n \downarrow$ and $e \cdot n = m$ are $\neg\neg$-closed, by Markov's Principle. Hence their interpretation is as in **Sets**. (They are 'almost negative formulas', see [142] for a complete account.)

We can formulate and prove **Church's Thesis** inside **Eff**:

$$\forall f : N^N . \exists e : N. \forall n : N. f(n) = e \cdot n.$$

Again we use Lemma 6.4.3. We have to produce a realiser in

$$\bigcap_{f \in \mathrm{TR}} E(f) \supset \left(\bigcup_{e \in \mathbb{N}} E(e) \wedge \bigcap_{n \in \mathbb{N}} E(n) \supset |f(n) \approx e \cdot n| \right)$$

But one can simply take $\Lambda e. \langle e, e \rangle$.

Markov's principle is one of the corner stones of the Russian school of constructive mathematics, founded by Markov.

We turn to the Uniformity Principle.

6.4.6. Definition. A type (or object) U is said to be **uniform** (with respect to N) if

$$\forall \alpha : \Omega^{U \times N} . (\forall u : U. \exists n : N. \alpha(u, n)) \supset (\exists n : N. \forall u : U. \alpha(u, n)).$$

6.4.7. Proposition. *In* **Eff** *the powerobject* $PN = \Omega^N$ *is uniform.* $\square$

We refer to [142] for the proof. This principle can be understood as follows: PN is a very amorphous collection since predicates on N can be described in very many ways. Thus if we have assigned a natural number n to each predicate $u : PN$, then the only conceivable way of so doing would be to pick the *same* n for every u. See also [335, 4.9]. Uniformity may be used to prove the existence of products of PERs over PERs in **Eff**, see [297], so that one can interpret second order products in polymorphic type theory.

PERs as subquotients of N

In the remainder of this section we will use the natural numbers object $N = (\mathbb{N}, E)$ in **Eff** to give an alternative description of (families of) PERs. Recall from Exercise 1.2.5 that PERs may be described as equivalence relations on subsets of $\mathbb{N}$, *i.e.* as quotients of subsets of $\mathbb{N}$. The latter are also called 'subquotients' of $\mathbb{N}$. This alternative description can also be given inside **Eff**.

6.4.8. Proposition. *The category* **PER** *of partial equivalence relations is equivalent to the full subcategory of* **Eff** *on the separated 'subquotients'* (X, E) *of* $N = (\mathbb{N}, E)$. *That is to those separated* (X, E) *occurring in a diagram*

$$(X, E) \longleftarrow\!\!\!\longleftarrow (Y, \approx) \rightarrowtail\!\!\!\rightarrow (\mathbb{N}, E)$$

(One may equivalently require the mono to be closed.)

Recall from Lemma 6.2.7 (ii) that the quotient $(Y, \approx) \twoheadrightarrow (X, E)$ with (X, E) separated may be described via a closed, almost equivalence relation R on $(Y, \approx)$.

Proof. Given a PER $R \subseteq \mathbb{N} \times \mathbb{N}$, we obtain such a subquotient,

$$(\mathbb{N}/R, \in) \xleftarrow{\ P\ } (|R|, \approx) \xrightarrow{\ M\ } (\mathbb{N}, E)$$

where the (closed) mono M is obtained from the inclusion $|R| = \{n \mid nRn\} \subseteq \mathbb{N}$. Thus $M(n, m) = \{n\} \cap \{m\}$. And the epi $P \colon (|R|, \approx) \twoheadrightarrow (\mathbb{N}/R, \in)$ is then given by $P(n, [m]) = \{n\} \cap [m]$.

And if we have a subquotient

$$(X, E) \xleftarrow{\ P\ } (Y, \approx) \xrightarrow{\ M\ } (\mathbb{N}, E)$$

then (X, E) is modest (and hence comes from a PER): if $n \in E(x_1) \cap E(x_2)$ for $x_1, x_2 \in X$, then, using appropriate codes a, b, c, d we get:

$$
\begin{aligned}
a \cdot n &\in P(y_1, x_1) \cap P(y_2, x_2) && \text{for some } y_1, y_2 \in Y \\
b \cdot (a \cdot n) &\in E(y_1) \cap E(y_2) \\
c \cdot (b \cdot (a \cdot n)) &\in M(y_1, m_1) \cap M(y_2, m_2) && \text{for certain } m_1, m_2 \in \mathbb{N} \\
d \cdot (c \cdot (b \cdot (a \cdot n))) &\in E(m_1) \cap E(m_2).
\end{aligned}
$$

But since $N = (\mathbb{N}, E)$ is modest, the latter implies $m_1 = m_2$. Reasoning backwards we get $|y_1 \approx y_2|$ since M is a mono, $|x_1 \approx x_2|$ and thus $x_1 = x_2$, since (X, E) is separated. $\qquad\square$

This result has an extension to families of PERs, provided we restrict ourselves to a separated index object.

6.4.9. Proposition. *For a separated index object* $(I, E) \in$ **Eff***, the fibre category* $\mathrm{UFam}(\mathbf{PER})_{(I,E)}$ *of* (I, E)*-indexed families of PERs is equivalent to the full subcategory of the slice category* **Eff**$/(I, E)$ *on the separated subquotients of the constant family* $(I, E)^*(N)$.

Proof. First notice that $\Gamma(I, E) = I$. So assume an I-indexed family of PERs $R = (R_i)_{i \in I}$ and form the diagram

$$(\{R\}, \approxeq) \xleftarrow{\quad P \quad} (\|R\|, \approx) \xrightarrowtail{\quad M \quad} (I, E) \times (\mathbb{N}, E)$$

$$\pi_R \searrow \quad \downarrow \quad \swarrow \pi$$

$$(I, E)$$

where M is a (closed) mono determined by the subset $\|R\| = \{(i, n) \mid i \in I \text{ and } n R_i n\} \subseteq I \times \mathbb{N} = \Gamma((I, E) \times (\mathbb{N}, \approx))$ and $(\{R\}, \approxeq)$ is as defined in the proof of Proposition 6.3.4, resulting from the inclusion UFam($\mathbf{PER}$) $\hookrightarrow$ UFam(ω-**Sets**) $\hookrightarrow$ **Eff**$^{\rightarrow}$. The morphism $P : (\|R\|, \approx) \twoheadrightarrow (\{R\}, \approxeq)$ is given by

$$P((i, n), (i', [n'])) = |i \approx i'| \cap (\{n\} \cap [n']).$$

It is an epi by Lemma 6.1.9.

And if we have such a diagram

$$(X, \approx) \xleftarrow{\quad P \quad} (Y, \approx) \xrightarrowtail{\quad M \quad} (I, E) \times (\mathbb{N}, E)$$

$$\varphi \searrow \quad \downarrow \quad \swarrow \pi$$

$$(I, E)$$

where P is an epi, M is a mono and φ is a separated family, then $(X, \approx)$ must be separated by Proposition 6.3.3 so that we can identify $\begin{pmatrix} (X, \approx) \\ \downarrow \varphi \\ (I, E) \end{pmatrix}$ with an (I, E)-indexed family $(X_i, E_i)_{i \in I}$ in $\begin{smallmatrix} \text{UFam}(\omega\text{-}\mathbf{Sets}) \\ \downarrow \\ \omega\text{-}\mathbf{Sets} \end{smallmatrix}$. The same argument as in the previous proof, applied fibrewise, yields that each (X_i, E_i) is modest. $\quad\square$

Exercises

6.4.1. Prove that a decidable predicate $(\varphi \vee \neg\varphi)$ is $\neg\neg$-closed $(\neg\neg\varphi \supset \varphi)$.

6.4.2. Consider for a subset $A \subseteq \mathbb{N}$, the canonical subobject $(\mathbb{N}, E_{\widehat{A}}) \rightarrowtail (\mathbb{N}, E)$ in **Eff** arising from $\widehat{A} : \mathbb{N} \to P\mathbb{N}$ as a strict predicate:

$$\widehat{A}(n) = \begin{cases} \{\langle n, 1 \rangle\} & \text{if } n \in A \\ \{\langle n, 0 \rangle\} & \text{if } n \notin A. \end{cases}$$

Show that A is decidable in recursion theoretic sense if and only if $\widehat{A}$—i.e. $(\mathbb{N}, E_{\widehat{A}}) \rightarrowtail (\mathbb{N}, E)$—is decidable in topos theoretic sense.

6.4.3. (i) Prove that if U is uniform and $U \twoheadrightarrow V$, then V is uniform.

(ii) Let $(U, \approx) \in \mathbf{Eff}$ be such that $\bigcap_{u \in U} E(u) \neq \emptyset$. Prove that $(U, \approx)$ is uniform.

(iii) Conclude that every quotient of a sheaf is uniform in $\mathbf{Eff}$. It can be shown that every uniform object is in fact a quotient of a sheaf.

6.4.4.   Prove that the following **Axiom of Countable Choice** holds in $\mathbf{Eff}$:

$$\forall \alpha : \Omega^{N \times N}. \, (\forall n : N. \, \exists m : N. \, \alpha(n, m)) \supset (\exists f : N^N. \, \forall n : N. \, \alpha(n, f(n))).$$

Chapter 7

Internal category theory

So far, indexing has been described mainly in terms of fibred categories, but in Section 1.10 also (briefly) in terms of indexed categories. In this chapter a third formalism for indexing will be presented, namely internal categories. These are categories described internally in another "ambient" category, using the diagrammatic language of category theory to express the familiar constituents of a category. An internal category is thus like an internal group: it is obtained by interpreting the defining requirements of a category in some more general universe than the category **Sets**. One can also describe functors and natural transformations internally. And one can say when an internal category is, for example, Cartesian closed.

Below we describe the basic (standard) ingredients of internal category theory. Our emphasis is on the relation between internal and fibred categories. It turns out that with an internal category one canonically associates a (split) fibration, which is called the externalisation of the internal category. Many concepts in internal category theory can also be expressed externally using fibred category theory. One of the more important results is that if a fibration is complete (in a fibred sense: it has fibred finite limits and products ($u^* \dashv \prod_u$) satisfying Beck-Chevalley), then every "internal diagram" has a limit. Such a diagram can be understood as a functor from an internal category to the fibration. We thus have a fibred version of a familiar result in ordinary category theory: if a category has equalisers and arbitrary products, then every small diagram has a limit.

Partial equivalence relations (PERs) form an example of an internal category in the category of ω-sets, and also in the effective topos. Externalisation gives the (already defined) fibrations of PERs indexed by ω-sets, and of PERs indexed by objects in the effective topos.

The rôle of internal categories in the semantics of (higher order) type theories was first emphasised by Moggi and Hyland [143]. It will be investigated in the next chapter (and also in Chapter 11).

7.1 Definition and examples of internal categories

As we have seen in Section 3.3 the concept of a group (and of other algebraic structures) can be expressed in diagrammatic language, and can thus be described in arbitrary categories (with finite products). The idea is that a group in a category—*i.e.* an 'internal group'—is an object G equipped with morphisms

$$ G \times G \xrightarrow{\ m\ } G \qquad 1 \xrightarrow{\ e\ } G \qquad G \xrightarrow{\ i\ } G $$

for multiplication m, neutral element e and inverse i. These are required to make some diagrams commute, expressing the fact that m is associative, e is a neutral element for m and that i provides an inverse for m, see Example 3.3.1. Clearly a group in the category **Sets** of sets and functions is just a group in the original sense. But an internal group in the category **Sp** of topological spaces and continuous functions is what is called a topological group. And an internal group in the category **PoSets** may be called an ordered group.

An obvious question is whether one can also describe categories diagrammatically. This involves describing one category inside another. It gives us extra generality, comparable to the generality provided by the description of a group as an internal group in a category, as in the examples just mentioned. For the description of internal categories we shall need more than just finite products in the 'ambient' category: we also need equalisers to form objects of composable arrows.

7.1.1. Definition. Let $\mathbb{B}$ be a category with finite limits. An **internal category C** in $\mathbb{B}$ consists of the following data. First there are two objects $C_0, C_1 \in \mathbb{B}$ which should be understood as the object C_0 of objects of **C** and the object C_1 of arrows of **C**. These come equipped with morphisms in $\mathbb{B}$

$$ C_1 \underset{\partial_1}{\overset{\partial_0}{\rightrightarrows}} C_1 \qquad \text{and} \qquad C_0 \xrightarrow{\ i\ } C_1 $$

representing the operations of domain ∂_0, codomain ∂_1 and identity (on an object) in $\mathbf{C}$. These should make the following diagram commute.

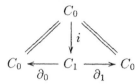

Further, a composition morphism m is required. Therefore, we need the following two pullback diagrams.

$$C_2 = C_1 \times_{C_0} C_1 \xrightarrow{\ \pi_1\ } C_1 \qquad\qquad C_3 = C_1 \times_{C_0} C_1 \times_{C_0} C_1 \longrightarrow C_1$$

$$\pi_0 \downarrow \qquad\quad \downarrow \partial_0 \qquad\qquad\qquad\qquad\qquad \downarrow \qquad\quad \downarrow \partial_0$$

$$C_1 \xrightarrow{\ \partial_1\ } C_0 \qquad\qquad\qquad\qquad\qquad C_2 \xrightarrow{\ \partial_1 \circ \pi_1\ } C_0$$

They yield the objects C_2 and C_3 of composable tuples and triples of arrows in $\mathbf{C}$. The composition morphism of $\mathbf{C}$ is then a morphism $m\colon C_2 \to C_1$ satisfying

$$C_1 \xleftarrow{\ \pi_0\ } C_2 \xrightarrow{\ \pi_1\ } C_1$$

$$\partial_0 \downarrow \qquad\qquad \downarrow m \qquad\qquad \downarrow \partial_1$$

$$C_0 \xleftarrow{\ \partial_0\ } C_1 \xrightarrow{\ \partial_1\ } C_0$$

to get the domain and codomain of the composites right. Further, there are the familiar categorical equations,

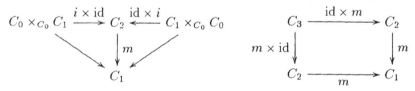

An internal category $\mathbf{C}$ is thus a 6-tuple $\langle C_0, C_1, \partial_0, \partial_1, i, m \rangle$ as described. Usually we refer to $\mathbf{C}$ by writing the graph $\mathbf{C} = (C_1 \rightrightarrows C_0)$ of domain and codomain maps only. We often call the category $\mathbb{B}$ in which $\mathbf{C}$ lives the ambient category of $\mathbf{C}$.

In the internal language of the ambient category $\mathbb{B}$—*i.e.* in the internal language of the subobject fibration $\begin{smallmatrix} \mathrm{Sub}(\mathbb{B}) \\ \downarrow \\ \mathbb{B} \end{smallmatrix}$ on $\mathbb{B}$—one can express commutativity of these diagrams via the familiar equations describing ordinary categories. Therefore we make use of the fact that subobject fibrations always have very strong equality and full subset types. For instance,

$$C_2 = \{\langle f, g \rangle : C_1 \times C_1 \mid \partial_1(f) = \partial_0(g)\}$$
$$C_3 = \{\langle \langle f, g \rangle, h \rangle : C_2 \times C_1 \mid \partial_1(g) = \partial_0(h)\}.$$

Commutativity of one of the above diagrams may then be expresses equationally as

$$\langle f, g \rangle : C_2 \mid \emptyset \vdash \partial_0(m(f,g)) =_{C_0} \partial_0(f).$$

Or, equivalently, using that we have *full* subset types, as

$$f : C_1, g : C_1 \mid \partial_1(f) =_{C_0} \partial_0(g) \vdash \partial_0(m(f,g)) =_{C_0} \partial_0(f).$$

Similarly we have an equation,

$$x : C_0, f : C_1 \mid x =_{C_0} \partial_0(f) \vdash m(i(x), f) =_{C_1} f.$$

Thus, in the internal language of the ambient category we can reason with internal categories as if they were ordinary categories. Indeed, an internal category $\mathbf{C}$ in $\mathbb{B}$ is an ordinary category within the world of $\mathbb{B}$, just like an internal group in $\mathbb{B}$ is an ordinary group seen from the perspective of $\mathbb{B}$.

Notice, by the way, that in order to formulate what an internal category is, we only need the pullbacks C_2 and C_3 of composable tuples and triples. Hence the requirement that the base category has *all* pullbacks is really too strong. But in order to use the internal language of $\mathbb{B}$ we need more pullbacks to express substitution.

Notice also that (variables for) internal categories are written in boldface: $\mathbf{C}$, instead of $\mathbb{C}$ for ordinary categories. Sometimes one calls an internal category a small category (in some base category). This terminology comes from the first of the following examples.

7.1.2. Examples. (i) An ordinary category $\mathbb{C}$ is by definition small if its collections of objects and of morphism are sets (as opposed to proper classes). Thus, $\mathbb{C}$ is small if and only if $\mathbb{C}$ is an internal category in **Sets**, with

$$C_0 = \mathrm{Obj}\,\mathbb{C} \qquad \text{and} \qquad C_1 = \coprod_{X, Y \in \mathbb{C}} \mathbb{C}(X, Y)$$

and with obvious domain and codomain maps $C_1 \rightrightarrows C_0$ from the disjoint union C_1 of all homsets to C_0.

(ii) Every object I in a category $\mathbb{B}$ forms a (discrete) internal category $\underline{I}$ with I both as object of objects and as object of morphisms. The domain and codomain maps $I \rightrightarrows I$ are identities, and so are the maps for identities and composition.

In general, an internal category **C** is called **discrete** if its morphism of identities $i: C_0 \to C_1$ is an isomorphism.

(iii) Here are some finite internal categories. A terminal object 1 yields a discrete internal category $\underline{1}$ with one object and one arrow. It is terminal among internal categories (in a fixed base category), as will become clear in the next section, once internal functors have been introduced.

An internal category $\underline{2}$ with two objects and only two (identity) arrows is the discrete category $\underline{2} = 1 + 1$.

An internal category $(\cdot \to \cdot)$ can be constructed with $1 + 1$ as object of objects and $1 + 1 + 1$ as object of morphisms. Domain and codomain maps $(1 + 1 + 1) \rightrightarrows (1 + 1)$ are obtained by making appropriate case distinctions. Similarly one gets an internal category $(\cdot \rightrightarrows \cdot)$ with $1 + 1$ as object of objects and $1 + 1 + 1 + 1$ as object as morphisms.

(To make these examples work, we have to assume that the coproducts $+$ are disjoint and universal.)

These are first, in a sense degenerate, examples of internal categories. Next we describe how PERs form an internal category in ω-**Sets** and also in **Eff**.

7.1.3. Example (Internal categories of PERs). Recall the full subcategory **PER** $\hookrightarrow \omega$-**Sets** of partial equivalence relations from Section 1.2. This category **PER** is small, so it is internal in **Sets**.

More interestingly, PERs also form an internal category in ω-**Sets** with as object of objects

$$\mathbf{PER}_0 = \nabla\mathrm{PER}$$

where ∇ is the inclusion functor **Sets** $\to \omega$-**Sets** described in Section 1.2. Thus $\mathbf{PER}_0$ is the set PER of PERs with trivial existence predicate $E(R) = \mathbb{N}$. As object of morphisms $\mathbf{PER}_1$ one uses a disjoint union as underlying set in

$$\mathbf{PER}_1 = \left(\coprod_{R, S \in \mathrm{PER}} \mathbb{N}/(R \Rightarrow S), \ E \right)$$

where the sets $\mathbb{N}/(R \Rightarrow S)$ are quotients by the exponent PERs $R \Rightarrow S = \{(n, n') \mid \forall m, m'. mRm' \Rightarrow n \cdot mSn' \cdot m'\}$, and where E is the existence predicate on the disjoint union given by

$$E(R, S, [n]_{R \Rightarrow S}) = \{m \in \mathbb{N} \mid m(R \Rightarrow S)n\} = [n]_{R \Rightarrow S}.$$

There are then obvious projection morphisms $\mathbf{PER}_1 \rightrightarrows \mathbf{PER}_0$ forming domain and codomain maps. And the identity morphism $\mathbf{PER}_0 \to \mathbf{PER}_1$ maps a per R to $(R, R, [\Lambda n. n])$. The description of the internal composition map is left to the reader. Some more details may be found in [143, 8] or [199, Section 8].

The inclusion functor ω-**Sets** $\hookrightarrow$ **Eff** is right adjoint, and thus preserves finite limits. This means that the diagram $(\textbf{PER}_1 \rightrightarrows \textbf{PER}_0)$ is also an internal category in **Eff**.

We close this section with two general constructions that yield internal categories, starting from a single map.

7.1.4. Examples. Let $a\colon A \to B$ be an arbitrary, but fixed morphism in a category $\mathbb{B}$ with finite limits. We describe how to construct two internal categories in $\mathbb{B}$ from a. In the first one, the object of objects will be the domain A and in the second case it will be the codomain B^1.

(i) Form the kernel pair of $a\colon A \to B$ by pullback of a against itself. This yields two morphisms ∂_0, ∂_1 in

$$
\begin{array}{ccc}
A \times_B A & \xrightarrow{\ \partial_1\ } & A \\
{\scriptstyle \partial_0}\big\downarrow \ \lrcorner & & \big\downarrow {\scriptstyle a} \\
A & \xrightarrow{\ a\ } & B
\end{array}
\qquad \text{and also} \qquad
\begin{array}{ccc}
A \times_B A \times_B A & \xrightarrow{\ \pi_1\ } & A \times_B A \\
{\scriptstyle \pi_0}\big\downarrow \ \lrcorner & & \big\downarrow {\scriptstyle \partial_0} \\
A \times_B A & \xrightarrow{\ \partial_1\ } & A
\end{array}
$$

There is then a (unique) diagonal $i\colon A \to A \times_B A$ with $\partial_0 \circ i = \mathrm{id} = \partial_1 \circ i$. Also, there is a unique mediating $m\colon A \times_B A \times_B A \to A \times_B A$ with $\partial_0 \circ m = \partial_0 \circ \pi_0$ and $\partial_1 \circ m = \partial_1 \circ \pi_1$. In one single graph:

$$
A \times_B A \times_B A \rightrightarrows\!\!\!\rightarrow A \times_B A \leftrightarrows A \xrightarrow{\ a\ } B
$$

It is not hard to verify that one gets an internal category this way. Informally, the objects are elements $x \in A$ and a morphism $x \to y$ exists if and only if $a(x) = a(y)$. The identity i maps $x \in A$ to the pair (x, x) and the composition of (x, y, z) is simply (x, z).

What one gets is an internal groupoid, which plays an important rôle in descent theory, see *e.g.* [168] for a recent reference with pointers to the literature. If one continues forming pullbacks $A \times_B \cdots \times_B A$ one obtains a simplicial object in $\mathbb{B}$, see [169, Remark 2.13].

(ii) For the second construction we need an exponent in a slice category of $\mathbb{B}$, and so we now assume that $\mathbb{B}$ is locally Cartesian closed. This construction may be found in [169, Example 2.38] or [268, 3.2]. The resulting internal category is called the **full internal category** associated with $a\colon A \to B$ and is written as $\mathrm{Full}_{\mathbb{B}}(a)$, or just $\mathrm{Full}(a)$ if the base category $\mathbb{B}$ is clear from the context. The reason for this terminology may become clear from

---

[1] An early source for these constructions is Lawvere's "Perugia Lecture Notes" (1972/1973).

Example 7.3.4 (ii). Later this construction will be described for an arbitrary "locally small" fibration, see Theorem 9.5.5.

We start the construction of the internal category $\text{Full}(a)$ by forming the families $\pi^*(a), \pi'^*(a)$ in the slice category $\mathbb{B}/(B \times B)$ by pullback, and writing $\langle \partial_0, \partial_1 \rangle$ for the exponent $\pi^*(a) \Rightarrow \pi'^*(a)$ in $\mathbb{B}/(B \times B)$, say with domain B_1. This yields a pair of parallel maps $(B_1 \rightrightarrows B)$ with B as object of objects. Informally, the morphisms between objects $x, y \in B$ are all maps $a^{-1}(x) \to a^{-1}(y)$ in $\mathbb{B}$ between the fibres of $a \colon A \to B$ over x and y.

In order to describe internal identity and composition maps we need the following correspondence $(*)$, between maps in appropriate slice categories:

$$\frac{\langle u, v \rangle \longrightarrow \langle \partial_0, \partial_1 \rangle}{u^*(a) \longrightarrow v^*(a)} \quad (*)$$

It arises from

$$\frac{\coprod_{\langle u,v \rangle} (\text{id}) = \langle u, v \rangle \longrightarrow \langle \partial_0, \partial_1 \rangle = \pi^*(a) \Rightarrow \pi'^*(a)}{\dfrac{\text{id} \longrightarrow \langle u, v \rangle^*(\pi^*(a) \Rightarrow \pi'^*(a)) \cong u^*(a) \Rightarrow v^*(a)}{u^*(a) \longrightarrow v^*(a)}}$$

The morphism of identities $i \colon \langle \text{id}, \text{id} \rangle \to \langle \partial_0, \partial_1 \rangle$ then arises by applying this correspondence $(*)$ to the identity map $\text{id}^*(a) \to \text{id}^*(a)$. And the composition morphism $m \colon \langle \partial_0 \circ \pi_0, \partial_1 \circ \pi_1 \rangle \to \langle \partial_0, \partial_1 \rangle$, with π_0, π_1 as in Definition 7.1.1, is obtained as follows. By applying $(*)$ downwards to the identity $\langle \partial_0, \partial_1 \rangle \to \langle \partial_0, \partial_1 \rangle$ we get a morpism $\partial_0^*(a) \to \partial_1^*(a)$. We can apply both π_0^* and π_1^* and then compose, as in:

$$(\partial_0 \circ \pi_0)^*(a) \longrightarrow (\partial_0 \circ \pi_1)^*(a) \cong (\partial_1 \circ \pi_0)^*(a) \longrightarrow (\partial_1 \circ \pi_1)^*(a)$$

This yields, by $(*)$ upwards, the required internal composition map.

In [87] this last construction plays a special rôle in a categorical characterization of the definition of a type theory within a logical framework (see also Section 10.2) as an internal category in an ambient category corresponding to the framework. This internal category arises from a family $\begin{pmatrix} U \\ \downarrow \\ \text{Type} \end{pmatrix}$ where U is the universe of the type theory, elements of which name types of the framework.

Exercises

7.1.1.   Prove formally using the rules for full subset types that one can derive (i) from (ii) and vice-versa.

(i) $\alpha: C_2 \mid \emptyset \vdash \partial_0(m(\alpha)) =_{C_0} \partial_0(\pi o(\alpha))$;

(ii) $f: C_1, g: C_1 \mid \partial_1(f) =_{C_0} \partial_0(g) \vdash \partial_0(m(i\langle f, g\rangle)) =_{C_0} \partial_0(f)$,

where $o(-)$ and $i(-)$ the "out" and "in" operation associated with subset types, as in Section 4.6.

7.1.2. Describe in diagrammatic language when an internal category is an internal groupoid (*i.e.* a category in which every morphism is an isomorphism). Show that each internal group G yields a groupoid internal category ($G \rightrightarrows 1$).

7.1.3. Describe the internal composition morphism for the internal category **PER** in ω-**Sets** in Example 7.1.3.

7.1.4. Consider a preorder in a category: a relation $R \rightarrowtail A \times A$ which is reflexive and transitive (see Section 1.3). Show that it forms a (preorder) internal category in which the pair $\langle \partial_0, \partial_1 \rangle$ is a mono. (In fact, this mono-condition defines internal preorders.)

7.1.5. Prove that the category $\text{Full}_{\mathbb{B}}(a)$ in Example 7.1.4 (ii) is an internal preorder in $\mathbb{B}$ if and only if the map $a: A \to B$ is a mono in $\mathbb{B}$.

7.1.6. The following example of an internal category is in the overlap of the constructions in Example 7.1.4 (i) and (ii). Let $\mathbb{B}$ be a category with finite products, and let A be an object in $\mathbb{B}$.

(i) Describe the internal category resulting from applying construction in Example 7.1.4 (i) to the unique map $A \to 1$ from A to the terminal object $1 \in \mathbb{B}$.

(ii) Do the same for the construction in (ii) starting from the identity $A \to A$. Notice that one gets the same internal category.

[This particular internal category has a universal property, see Exercise 7.2.6 (ii).]

7.1.7. Give a definition of an internal 2-category.

7.1.8. Notice that an internal category in an ambient category $\mathbb{B}$ consists of several maps in $\mathbb{B}$ satisfying certain equations in the logic of the subobject fibration $\begin{smallmatrix} \text{Sub}(\mathbb{B}) \\ \downarrow \\ \mathbb{B} \end{smallmatrix}$ on $\mathbb{B}$. See if the latter equational aspect can be generalised to a (preorder) fibration $\begin{smallmatrix} \mathbb{E} \\ \downarrow \\ \mathbb{B} \end{smallmatrix}$, by using the internal equality of the fibration, so that one gets a notion of "category with respect to a fibration". What logical structure should the fibration have, so that one can express such a notion?

7.2 Internal functors and natural transformations

Just like categories can be described inside an ambient (or base) category, also functors and natural transformations can be described internally. The descriptions are the ones familiar for ordinary categories. They can be expressed either in diagrammatic language, or in the internal language of the subobject fibration on the base category $\mathbb{B}$. In this way we get a (2-)category $\mathbf{cat}(\mathbb{B})$ of internal categories and internal functors (and internal natural transformations), in a fixed base category $\mathbb{B}$. It allows us in particular to say what an internal adjunction is. And in terms of these adjunctions we can define familiar structure, like products $\times$, in internal categories. This will be the subject of the present section. In the subsequent section we shall see that these internal notions correspond to fibred notions for the "externalisation" of the internal category; the latter is a fibration which is (canonically) associated with the internal category.

7.2.1. Definition. Assume two internal categories $\mathbf{C} = (C_1 \rightrightarrows C_0)$ and $\mathbf{D} = (D_1 \rightrightarrows D_0)$ in an ambient category $\mathbb{B}$.

(i) An **internal functor** $F: \mathbf{C} \to \mathbf{D}$ is given by two morphisms

$$C_0 \xrightarrow{\;F_0\;} D_0 \quad \text{and} \quad C_1 \xrightarrow{\;F_1\;} D_1$$

(in $\mathbb{B}$) mapping objects and arrows of $\mathbf{C}$ to objects and arrows of $\mathbf{D}$, in such a way that domains, codomains, identities and compositions are preserved. That is, such that the following four diagrams commute.

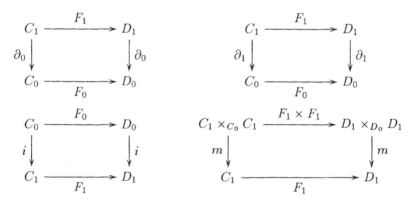

It is easy to see that one obtains a category in this way, with composition in the obvious way. We shall write $\mathbf{cat}(\mathbb{B})$ for the resulting category of internal categories in $\mathbb{B}$ and internal functors between them.

(ii) An **internal natural transformation** α between two internal functors $F, G: \mathbf{C} \rightrightarrows \mathbf{D}$ consists of a single map $\alpha: C_0 \to D_1$ yielding for each object in $\mathbf{C}$ a morphism in $\mathbf{D}$ such that the following diagrams commute.

$$
\begin{array}{ccc}
 & C_0 & \\
F_0 \swarrow & \downarrow \alpha & \searrow G_0 \\
D_0 \xleftarrow{\ \partial_0\ } & D_1 & \xrightarrow{\ \partial_1\ } D_0
\end{array}
\qquad\qquad
\begin{array}{ccc}
 & C_1 & \\
\langle \alpha \circ \partial_0, G_1 \rangle \swarrow & & \searrow \langle F_1, \alpha \circ \partial_1 \rangle \\
D_1 \times_{D_0} D_1 \xrightarrow{\ m\ } & D_1 & \xleftarrow{\ m\ } D_1 \times_{D_0} D_1
\end{array}
$$

In the internal language of (the subobject fibration on the) base or ambient category $\mathbb{B}$ we can express commutativity of, for example, the last diagram in (ii), as

$$ f \colon C_1 \mid \emptyset \vdash m(\alpha(\partial_0(f)), G_1(f)) =_{C_1} m(F_1(f), \alpha(\partial_1(f))). $$

Or, in more readable form, as

$$ f \colon C_1, x \colon C_0, y \colon C_0 \mid x =_{C_0} \partial_0(f), y =_{C_0} \partial_1(f) \vdash G_1(f) \circ \alpha_x =_{C_1} \alpha_y \circ F_1(f). $$

7.2.2. Proposition. *Assume $\mathbb{B}$ is a category with finite limits. Then the category $\mathbf{cat}(\mathbb{B})$ of internal categories in $\mathbb{B}$ also has finite limits. And if $\mathbb{B}$ is additionally Cartesian closed, then so is $\mathbf{cat}(\mathbb{B})$.*

Proof. One can either prove this purely categorically, or by making use of the internal language of $\mathbb{B}$. We shall sketch the first approach for finite limits of internal categories, and the second one for exponents.

The discrete internal category 1 on the terminal object $1 \in \mathbb{B}$ is terminal in $\mathbf{cat}(\mathbb{B})$. And the Cartesian product of

$$
\mathbf{C} = \left(C_1 \ \underset{\partial_1}{\overset{\partial_0}{\rightrightarrows}} \ C_0 \right)
\qquad \text{and} \qquad
\mathbf{D} = \left(D_1 \ \underset{\partial_1}{\overset{\partial_0}{\rightrightarrows}} \ D_0 \right)
$$

is

$$
\mathbf{C} \times \mathbf{D} = \left(C_1 \times D_1 \ \underset{\partial_1 \times \partial_1}{\overset{\partial_0 \times \partial_0}{\rightrightarrows}} \ C_0 \times D_0 \right)
$$

with similar componentwise maps for composition and identities.

The equaliser $\mathbf{Eq}(F, G)$ of two internal functors $F, G \colon \mathbf{C} \rightrightarrows \mathbf{D}$ is obtained from equalisers in $\mathbb{B}$,

$$
\mathbf{Eq}(F, G)_0 \rightarrowtail C_0 \ \underset{G_0}{\overset{F_0}{\rightrightarrows}} \ D_0
\qquad \text{and} \qquad
\mathbf{Eq}(F, G)_1 \rightarrowtail C_1 \ \underset{G_1}{\overset{F_1}{\rightrightarrows}} \ D_1
$$

Using the universal properties of these equalisers we get domain and codomain maps $\big(\mathbf{Eq}(F, G)_1 \rightrightarrows \mathbf{Eq}(F, G)_0\big)$, and also maps for identities and composition.

In case the base category $\mathbb{B}$ is Cartesian closed, then we may use $\forall$ in the logic of subobjects in $\mathbb{B}$, by Corollary 1.9.9. For the **internal functor category** $\mathbf{D}^{\mathbf{C}} = \big((\mathbf{D}^{\mathbf{C}})_1 \rightrightarrows (\mathbf{D}^{\mathbf{C}})_0\big)$ we define

$$(\mathbf{D}^{\mathbf{C}})_0 = \{(F_0, F_1)\colon D_0^{C_0} \times D_1^{C_1} \mid \text{``}F_0, F_1 \text{ form an internal functor } \mathbf{C} \to \mathbf{D}\text{''}\},$$

where the latter predicate may be written out as

$$\begin{aligned}
&\text{``}F_0, F_1 \text{ form an internal functor } \mathbf{C} \to \mathbf{D}\text{''}\\
&\Leftrightarrow \forall f\colon C_1.\, \partial_0(F_1(f)) =_{C_0} F_0(\partial_0(f)) \wedge \partial_1(F_1(f)) =_{C_0} F_0(\partial_1(f))\\
&\qquad \wedge \forall x\colon C_0.\, F_1(i(x)) =_{C_1} i(F_0(x))\\
&\qquad \wedge \forall\langle f, g\rangle\colon C_2.\, F_1(m(f, g)) =_{C_1} m(F_1(f), F_1(g)).
\end{aligned}$$

Similarly we take

$$(\mathbf{D}^{\mathbf{C}})_1 = \{(F, G, \alpha)\colon (\mathbf{D}^{\mathbf{C}})_0 \times (\mathbf{D}^{\mathbf{C}})_0 \times \mathbf{D}_1^{\mathbf{C}_0} \mid \text{``}\alpha \text{ is an internal}$$
$$\text{natural transformation from } F \text{ to } G\text{''}\}. \qquad \square$$

For $\mathbf{cat}(\mathbb{B})$ to be a 2-category, we need to know about the interaction between internal functors and internal natural transformations. Consider therefore the diagram of internal functors

$$\mathbf{C}' \xrightarrow{\ H\ } \mathbf{C} \underset{G}{\overset{F}{\rightrightarrows}}{}^{\Downarrow \alpha} \mathbf{D} \xrightarrow{\ K\ } \mathbf{D}'$$

with an internal natural transformation α as indicated. One gets two new internal natural transformations:

$$C_0' \xrightarrow{\ H_0\ } C_0 \xrightarrow{\ \alpha\ } D_1 \qquad \text{yields} \qquad \mathbf{C}' \underset{GH}{\overset{FH}{\rightrightarrows}}{}^{\Downarrow \alpha H} \mathbf{D}$$

And similarly:

$$C_0 \xrightarrow{\ \alpha\ } D_1 \xrightarrow{\ H_1\ } D_1' \qquad \text{yields} \qquad \mathbf{C} \underset{KG}{\overset{KF}{\rightrightarrows}}{}^{\Downarrow K\alpha} \mathbf{D}'$$

All this permits us to say what an **internal adjunction** is: it is given by two internal functors

together with two internal natural transformations

and

satisfying the familiar triangular identities:

$$G\varepsilon \circ \eta G = \mathrm{id} \qquad \text{and} \qquad \varepsilon F \circ F\eta = \mathrm{id}.$$

Of course, this is just and adjunction $(F \dashv G)$ in the 2-category $\mathbf{cat}(\mathbb{B})$.

By combining these internal adjunctions with the finite products that we have for internal categories, we can describe structure like products or equalisers inside internal categories. Let $\mathbf{C}$ be an internal category in a base category $\mathbb{B}$. One says that $\mathbf{C}$ has an **internal terminal object** if the unique internal functor $!\colon \mathbf{C} \to \underline{1}$ from $\mathbf{C}$ to the terminal internal category $\underline{1}$ in $\mathbb{B}$, has an internal right adjoint, say $\mathrm{t}\colon \underline{1} \to \mathbf{C}$. The internal counit natural transformation $(! \circ \mathrm{t}) \Rightarrow \mathrm{id}_{\underline{1}}$ is then the identity. The unit $\eta\colon \mathrm{id}_{\mathbf{C}} \Rightarrow (\mathrm{t} \circ !)$ consists of a map $\eta\colon C_0 \to C_1$ with (among other things),

$$C_0 \xrightarrow{\ \partial_0 \circ \eta = \mathrm{id}\ } C_0 \qquad \text{and} \qquad C_0 \xrightarrow{\ \partial_1 \circ \eta = \mathrm{t} \circ !\ } C_0$$

This η thus maps an internal object X in $\mathbf{C}$ to an arrow $X \to \mathrm{t}$ in $\mathbf{C}$. It is of course the unique internal map from X to t. This uniqueness follows from naturality and the triangular identities. But it may also be expressed in more elementary terms, see Exercise 7.2.4 below.

In a similar way, the usual external definition of Cartesian products $\times$ can be internalised readily. One says that $\mathbf{C}$ has **internal Cartesian products** if the internal diagonal functor $\Delta\colon \mathbf{C} \to \mathbf{C} \times \mathbf{C}$ has an internal right adjoint.

Equalisers are a bit more involved. We first have to construct the internal category $\mathbf{C}^{\rightrightarrows}$ of parallel arrows in $\mathbf{C}$, together with an internal diagonal functor $\Delta\colon \mathbf{C} \to \mathbf{C}^{\rightrightarrows}$ sending

$$X \mapsto \left(X \underset{\mathrm{id}}{\overset{\mathrm{id}}{\rightrightarrows}} X \right).$$

We then say that $\mathbf{C}$ has **internal equalisers** if this diagonal $\Delta\colon \mathbf{C} \to \mathbf{C}^{\rightrightarrows}$ has an internal right adjoint. One can form $\mathbf{C}^{\rightrightarrows}$ as the category of internal functors from $(\cdot \rightrightarrows \cdot)$ to $\mathbf{C}$—where $(\cdot \rightrightarrows \cdot)$ is as in Example 7.1.2 (iii). But one can also build $\mathbf{C}^{\rightrightarrows}$ using the internal language of $\mathbb{B}$. For example, the object of objects $\mathbf{C}_0^{\rightrightarrows}$ is obtained as equaliser,

$$\mathbf{C}_0^{\rightrightarrows} \rightarrowtail C_1 \times C_1 \underset{\langle \partial_0 \circ \pi', \partial_1 \circ \pi \rangle}{\overset{\partial_0 \times \partial_1}{\rightrightarrows}} C_0 \times C_0$$

i.e. as interpretation

$$\mathbf{C}_0^{\rightrightarrows} = \{(f, g)\colon C_1 \times C_1 \mid \partial_0(f) =_{C_0} \partial_0(g) \wedge \partial_1(f) =_{C_0} \partial_1(g)\}.$$

For internal exponents we use the approach of Exercise 1.8.9. Therefore assume $\mathbf{C}$ has internal Cartesian products $\times$. Write $|C_0| = \underline{C_0}$ for the discrete internal category with objects as in $\mathbf{C}$. There is an obvious internal inclusion functor $D\colon |C_0| \to \mathbf{C}$. Thus we can define an extended internal product functor

$$|C_0| \times \mathbf{C} \xrightarrow{\;\mathsf{prod} = \langle \pi, \times \circ (D \times \mathrm{id}) \rangle\;} |C_0| \times \mathbf{C}$$

and say that $\mathbf{C}$ has **internal exponents** if this functor has an internal right adjoint exp. This exp then yields a functor $\mathbf{C}^{\mathrm{op}} \times \mathbf{C} \to \mathbf{C}$ as usual. Here, the internal category $\mathbf{C}^{\mathrm{op}}$ is obtained from $\mathbf{C}$ by interchanging the domain and codomain maps, and by adapting the map for composition accordingly. An internal Cartesian closed category is then, as one expects, an internal category with internal terminal object, internal Cartesian products, and internal exponents.

Finally, one can say that $\mathbf{C}$ has **internal simple products** (or **coproducts**) if for each object $I \in \mathbb{B}$, the diagonal functor $\mathbf{C} \to \mathbf{C}^{\underline{I}}$ has an internal right (or left) adjoint—where $\mathbf{C}^{\underline{I}}$ is the internal functor category from the discrete internal category $\underline{I}$ to $\mathbf{C}$.

In the next section we shall see how this internal structure may also be described "externally" via the "externalisation" of an internal category.

Earlier in this—and in the previous—section we used the internal language of the ambient category to describe internal categories, functors and natural transformations. We should warn that this does not extend smoothly to internal structure: for an internal category $\mathbf{C}$ in $\mathbb{B}$, having an internal terminal as above is not the same as validity of the statement:

$$\exists t\colon C_0.\, \forall x\colon C_0.\, \exists ! f\colon C_1.\, \partial_0(f) =_{C_0} x \wedge \partial_1(f) =_{C_0} t$$

(in the internal language of the subobject fibration on $\mathbb{B}$) since internal existence is not the same as external existence, see the last few paragraphs of Section 4.3. Similarly, internal Cartesian products via explicit internal functors as above is stronger than the validity of the statement describing the existence of a pair $x \leftarrow x \times y \rightarrow y$ of projection maps, for all $x, y: C_0$. Unless of course, the Axiom of Choice holds in the ambient category $\mathbb{B}$. But recall that also for ordinary categories there is a difference between, say, Cartesian products as given functorially by an adjunction, and Cartesian products as given by the existence of universal diagrams $X \leftarrow X \times Y \rightarrow Y$ for each pair of objects X, Y. To get from these diagrams to an adjunction one has to *choose* projection maps for each such pair X, Y. There one uses the Axiom of Choice in the meta-language, which is relatively harmless. In internal category theory however, this meta-language is the internal language of the ambient category, in which the Axiom of Choice may fail.

Exercises

7.2.1.   Write out in the internal language the predicate "α is an internal natural transformation from F to G" in the proof of Proposition 7.2.2. Give a purely categorical description of the object $(\mathbf{D}^{\mathbf{C}})_0$—*e.g.* as intersection of four equalisers.

7.2.2.   Verify that αH and $K\alpha$ (as described after the proof of Proposition 7.2.2) are internal natural transformations.

7.2.3.   Verify that the internal category $\mathbf{PER} = (\mathbf{PER}_1 \rightrightarrows \mathbf{PER}_0)$ in ω-**Sets** (and also in **Eff**) has Cartesian products $\times$, via an appropriate internal adjunction.

7.2.4.   Show that an internal category $\mathbf{C}$ has an internal terminal object (as defined above) if and only if there are maps

$$1 \xrightarrow{\ t\ } C_0 \quad \text{and} \quad C_0 \xrightarrow{\ \eta\ } C_1$$

with

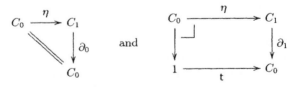

7.2.5.   Let $\mathbf{PreOrd}(\mathbb{B}) \hookrightarrow \mathbf{cat}(\mathbb{B})$ be the full subcategory of internal preorders in $\mathbb{B}$, *i.e.* of internal categories $(C_1 \rightrightarrows C_0)$ for which the tuple $\langle \partial_0, \partial_1 \rangle: C_1 \rightarrow C_0 \times C_0$ is a mono. Prove that this inclusion functor $\mathbf{PreOrd}(\mathbb{B}) \hookrightarrow \mathbf{cat}(\mathbb{B})$ has a left adjoint, in case $\mathbb{B}$ is a regular category.

7.2.6. Consider the "underlying object" functor $U: \mathbf{cat}(\mathbb{B}) \to \mathbb{B}$ given by $(C_1 \rightrightarrows C_0) \mapsto C_0$.

(i) Show that the assignment $I \mapsto$ (the discrete internal category $\underline{I}$ on I) extends to a functor $\mathbb{B} \to \mathbf{cat}(\mathbb{B})$, which is left adjoint to U.

(ii) Prove that U also has a right adjoint, which maps $I \in \mathbb{B}$ to the "indiscrete" internal category $(I \times I \rightrightarrows I)$ with Cartesian projections as domain and codomain maps (see Exercise 7.1.6).

(iii) And prove that if $\mathbb{B}$ has (reflexive) coequalisers, then the discrete category functor $I \mapsto \underline{I}$ has a left adjoint Π_0, which maps $\mathbf{C} \in \mathbf{cat}(\mathbb{B})$ to the codomain of the coequaliser

$$C_1 \overset{\partial_0}{\underset{\partial_1}{\rightrightarrows}} C_0 \longrightarrow \Pi_0(\mathbf{C})$$

7.3 Externalisation

Just like indexed categories in Section 1.10 correspond to certain fibrations (namely to cloven ones), internal categories also correspond to certain fibrations, namely to so-called *small* fibrations. This correspondence really is a tautology by the way we define small fibrations below, as coming from an internal category. But there is an alternative characterisation of small fibrations in terms of so-called "local smallness" and generic objects, see Corollary 9.5.6 later on.

We shall start by describing how an internal category gives rise to a (split) fibration. The construction is known as *externalisation*.

7.3.1. Definition (Externalisation). Let $\mathbf{C} = (C_1 \rightrightarrows C_0)$ be an internal category in $\mathbb{B}$. For each object $I \in \mathbb{B}$ form a category $\mathbf{C}^I$ with

objects   maps $X: I \to C_0$ in $\mathbb{B}$. We think of these as I-indexed collections $(X_i)_{i \in I}$ of objects of $X_i \in \mathbf{C}$.

morphisms $(I \overset{X}{\to} C_0) \longrightarrow (I \overset{Y}{\to} C_0)$ are maps $f: I \to C_1$ with

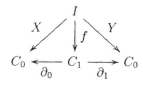

This morphism $f: X \to Y$ can be seen as an I-indexed family $f = (f_i: X_i \to Y_i)_{i \in I}$ of maps in $\mathbf{C}$.

The identity in $\mathbf{C}^I$ on $X: I \to C_0$ is the composite map

$$I \xrightarrow{\ X\ } C_0 \xrightarrow{\ i\ } C_1$$

where i is the map of internal identities in $\mathbf{C}$. And composition of morphisms $X \xrightarrow{f} Y \xrightarrow{g} Z$ in $\mathbf{C}^I$ is

$$I \xrightarrow{\ \langle f, g \rangle\ } C_1 \times_{C_0} C_1 \xrightarrow{\ m\ } C_1$$

In a next step we notice that the assignment

$$I \mapsto \mathbf{C}^I$$

extends to a split indexed category $\mathbb{B}^{\mathrm{op}} \to \mathbf{Cat}$. A morphism $u: I \to J$ in $\mathbb{B}$ yields a functor $u^* = - \circ u: \mathbf{C}^J \to \mathbf{C}^I$ by composition. Hence we get a split fibration on $\mathbb{B}$; it is written as

$$\begin{array}{c} \mathrm{Fam}_\mathbb{B}(\mathbf{C}) \\ \downarrow p_\mathbf{C} \\ \mathbb{B} \end{array} \quad \text{or as} \quad \begin{array}{c} \mathrm{Fam}_\mathbb{B}(\mathbf{C}) \\ \downarrow \\ \mathbb{B} \end{array} \quad \text{or simply as} \quad \begin{array}{c} \mathrm{Fam}(\mathbf{C}) \\ \downarrow \\ \mathbb{B} \end{array}$$

The total category $\mathrm{Fam}_\mathbb{B}(\mathbf{C})$ thus has maps $(I \xrightarrow{X} C_0)$ in $\mathbb{B}$ as objects. And a morphism $(I \xrightarrow{X} C_0) \longrightarrow (J \xrightarrow{Y} C_0)$ in $\mathrm{Fam}_\mathbb{B}(\mathbf{C})$ is a pair of maps $u: I \to J$ and $f: I \to C_1$ in $\mathbb{B}$ with

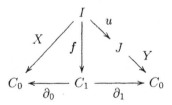

We explicitly describe the splitting of $\begin{array}{c} \mathrm{Fam}(\mathbf{C}) \\ \downarrow \\ \mathbb{B} \end{array}$. For a map $u: I \to J$ in $\mathbb{B}$ and an object $Y: J \to C_0$ in $\mathrm{Fam}(\mathbf{C})$ over J, one gets an object $Y \circ u: I \to C_0$ over I, together with a splitting map in $\mathrm{Fam}(\mathbf{C})$

$$\left(I \xrightarrow{\ Y \circ u\ } C_0 \right) \xrightarrow{\quad (u, i \circ Y \circ u)\quad} \left(J \xrightarrow{\ Y\ } C_0 \right)$$

The following holds by construction, and is worth mentioning explicitly.

7.3.2. Lemma. *The externalisation* $\begin{array}{c} \mathrm{Fam}(\mathbf{C}) \\ \downarrow \\ \mathbb{B} \end{array}$ *of an internal category* $\mathbf{C} = (C_1 \rightrightarrows C_0)$ *in* $\mathbb{B}$ *is a split fibration with a split generic object* $(C_0 \xrightarrow{\mathrm{id}} C_0) \in \mathrm{Fam}(\mathbf{C})$ *above* $C_0 \in \mathbb{B}$. $\qquad\square$

As mentioned, an ordinary category is small if and only if it is an internal category in the standard universe of sets. This terminology is extended to fibred categories: smallness can now be defined with respect to an arbitrary base category.

7.3.3. Definition. A fibration is called **small** if it is equivalent to the externalisation of an internal category in its base category.

7.3.4. Examples. (i) A small category $\mathbb{C}$ is, as we have seen, an internal category in **Sets**. Its externalisation is the familiar family fibration $\begin{smallmatrix}\mathrm{Fam}(\mathbb{C})\\\downarrow\\\mathbf{Sets}\end{smallmatrix}$. Thus using the Fam notation above is justified. But note that we write $\mathrm{Fam}_{\mathbb{B}}(\mathbf{C})$ or $\mathrm{Fam}(\mathbf{C})$ with boldface **C**, in case **C** is an internal category in $\mathbb{B}$.

(ii) Consider the externalisation of a full internal category $\mathrm{Full}(a)$ associated with a morphism $a\colon A \to B$ in a category $\mathbb{B}$, as described in Example 7.1.4 (ii). We claim that morphisms

$$\left(I \xrightarrow{\;X\;} B\right) \xrightarrow{\hspace{3cm}} \left(J \xrightarrow{\;Y\;} B\right)$$

in the total category $\mathrm{Fam}(\mathrm{Full}(a))$ can be identified with morphisms

$$X^*(a) \xrightarrow{\hspace{3cm}} Y^*(a)$$

in $\mathbb{B}^{\to}$. Indeed, using the correspondence $(*)$ in Example 7.1.4 (ii) we get:

$$\frac{\dfrac{\dfrac{X \longrightarrow Y \quad \text{in } \mathrm{Fam}\big(\mathrm{Full}(a)\big) \text{ over } u}{\langle X, Y \circ u\rangle \longrightarrow \langle \partial_0, \partial_1\rangle \quad \text{in } \mathbb{B}/B \times B}}{X^*(a) \longrightarrow u^*Y^*(a) \quad \text{in } \mathbb{B}/I}}{X^*(a) \longrightarrow Y^*(a) \quad \text{in } \mathbb{B}^{\to} \text{ over } u} \quad (*)$$

In this way we obtain a full and faithful fibred functor $\mathrm{Fam}\big(\mathrm{Full}(a)\big) \to \mathbb{B}^{\to}$. In particular, the fibre over $I \in \mathbb{B}$ of the externalisation $\mathrm{Fam}\big(\mathrm{Full}(a)\big)$ may be identified with the category with

objects   maps $(X\colon I \to B)$.

morphisms $f\colon X \to Y$ are morphisms in $\mathbb{B}/I$:

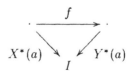

With composition and identities as in $\mathbb{B}/I$. This greatly simplifies matters.

The situation in this last example deserves an explicit name.

7.3.5. Definition. An internal category $\mathbf{C}$ in $\mathbb{B}$ for which there is a full and faithful fibred functor $\mathrm{Fam}(\mathbf{C}) \to \mathbb{B}^{\to}$ (over $\mathbb{B}$) will be called a **full internal (sub)category** in $\mathbb{B}$.

In a locally Cartesian closed base category, these full internal categories are the ones of the form $\mathrm{Full}(a)$ for some morphism a. This is the content of the next result.

7.3.6. Proposition. *Let $\mathbf{C} = (C_1 \rightrightarrows C_0)$ be a full internal category in a locally Cartesian closed category $\mathbb{B}$, say via a full and faithful fibred functor $\mathcal{P}: \mathrm{Fam}(\mathbf{C}) \to \mathbb{B}^{\to}$. Then $\mathbf{C}$ is isomorphic to an internal category of the form $\mathrm{Full}(a)$, namely for $a = \mathcal{P}(C_0 \xrightarrow{\mathrm{id}} C_0) \in \mathbb{B}/C_0$.*

Proof. By Yoneda: for a family $\langle X, Y \rangle: I \to C_0 \times C_0$ we have:

$$\mathbb{B}/C_0 \times C_0 \big(\langle X, Y \rangle, \ \pi^*(a) \Rightarrow \pi'^*(a) \big)$$
$$\cong \ \mathbb{B}/C_0 \times C_0 \big(\langle X, Y \rangle \times \pi^*(a), \ \pi'^*(a) \big)$$
$$\cong \ \mathbb{B}/C_0 \times C_0 \big(\coprod_{\langle X, Y \rangle} \langle X, Y \rangle^* \pi^*(a), \ \pi'^*(a) \big) \quad \text{by definition of } \times$$
$$\cong \ \mathbb{B}/I \big(\langle X, Y \rangle^* \pi^*(a), \ \langle X, Y \rangle^* \pi'^*(a) \big) \qquad \text{since } \coprod_{\langle X, Y \rangle} \dashv \langle X, Y \rangle^*$$
$$\cong \ \mathbb{B}/I \big(X^*(\mathcal{P}(\mathrm{id})), \ Y^*(\mathcal{P}(\mathrm{id})) \big)$$
$$\cong \ \mathbb{B}/I \big(\mathcal{P}(X), \ \mathcal{P}(Y) \big) \qquad \qquad \mathcal{P} \text{ is a fibred functor}$$
$$\cong \ \mathrm{Fam}(\mathbf{C})_I \big(X, \ Y \big) \qquad \qquad \mathcal{P} \text{ is full and faithful}$$
$$\cong \ \mathbb{B}/C_0 \times C_0 \big(\langle X, Y \rangle, \ \langle \partial_0, \partial_1 \rangle \big).$$

Hence $\langle \partial_0, \partial_1 \rangle \cong \pi^*(a) \Rightarrow \pi'^*(a)$, as families over $C_0 \times C_0$. $\qquad \square$

PERs in ω-**Sets** and in **Eff** form such a full internal category. In fact, the externalisations are familiar fibrations that we introduced earlier.

7.3.7. Proposition. (i) *The internal category* $\mathbf{PER} = (\mathbf{PER}_1 \rightrightarrows \mathbf{PER}_0)$ *in ω-**Sets** has as externalisation the fibration* $\begin{smallmatrix} \mathrm{UFam}(\mathbf{PER}) \\ \downarrow \\ \omega\text{-}\mathbf{Sets} \end{smallmatrix}$ *of PERs over ω-sets from Definition 1.4.8. It is a full internal category in ω-**Sets** via the composite of the full and faithful fibred functors*

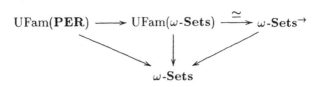

obtained from Example 1.8.7 (ii) and Proposition 1.4.7.

(ii) *And* **PER** *in* **Eff** *has as externalisation the fibration* $\begin{smallmatrix} \text{UFam}(\mathbf{PER}) \\ \downarrow \\ \mathbf{Eff} \end{smallmatrix}$ *from Definition 6.3.1. It is also a full internal category in* **Eff** *via the composite*

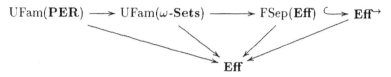

from Propositions 6.3.2 and 6.3.4.   □

There is more to say about externalisation; it is functorial in the following sense.

7.3.8. Proposition (Externalisation, continued). *Fix a base category* $\mathbb{B}$. *The assignment,*

$$\mathbf{C} \mapsto \begin{pmatrix} \text{Fam}(\mathbf{C}) \\ \downarrow \\ \mathbb{B} \end{pmatrix} \quad \text{extends to a (2-)functor} \quad \mathbf{cat}(\mathbb{B}) \longrightarrow \mathbf{Fib_{split}}(\mathbb{B})$$

which is "locally full and faithful", i.e. full and faithful both on 1-cells and on 2-cells. Moreover, it preserves finite products, and also exponents (where exponents of split fibrations are as described in Exercise 1.10.6).

Proof. Most of this is straightforward formal manipulation. For example, an internal functor $F = \langle F_0, F_1 \rangle \colon \mathbf{C} \to \mathbf{D}$ yields a (split) fibred functor $\text{Fam}(\mathbf{C}) \to \text{Fam}(\mathbf{D})$, call it $\text{Fam}(F)$, by

$$\left(I \xrightarrow{\ X\ } C_0 \right) \longmapsto \left(I \xrightarrow{\ F_0 \circ X\ } D_0 \right), \quad \left(I \xrightarrow{\ f\ } C_1 \right) \longmapsto \left(I \xrightarrow{\ F_1 \circ f\ } D_1 \right)$$

And an internal natural transformation $\alpha \colon F \Rightarrow G$ between internal functors $F, G \colon \mathbf{C} \rightrightarrows \mathbf{D}$ yields a vertical natural transformation $\text{Fam}(F) \Rightarrow \text{Fam}(G)$ between the induced fibred functors, with components

$$\left(I \xrightarrow{\ X\ } C_0 \right) \longmapsto \left(I \xrightarrow{\ \alpha \circ X\ } D_1 \right)$$

Every split fibred functor $H \colon \text{Fam}(\mathbf{C}) \to \text{Fam}(\mathbf{D})$ arises in this a way: such an H gives rise to a pair of maps

$$H_0 = H \left(C_0 \xrightarrow{\ \text{id}\ } C_0 \right) \colon C_0 \longrightarrow D_0, \quad H_1 = H \left(C_1 \xrightarrow{\ \text{id}\ } C_1 \right) \colon C_1 \longrightarrow D_1$$

forming an internal functor $\mathbf{C} \to \mathbf{D}$, whose externalisation is H again. And these H_0, H_1 are unique in doing so.

Similarly, every vertical natural transformation $\alpha: \mathrm{Fam}(F) \Rightarrow \mathrm{Fam}(G)$ comes from a unique internal natural transformation. Notice that for each object $X: I \to C_0$ in $\mathrm{Fam}(\mathbf{C})$ one can describe α_X as a map $I \to D_1$ forming a morphism in $\mathrm{Fam}(\mathbf{D})$,

$$\left(I \xrightarrow{F_0 \circ X} D_0 \right) \xrightarrow{\quad \alpha_X \quad} \left(I \xrightarrow{G_0 \circ X} D_0 \right)$$

Hence by taking the object X to be $\mathrm{id}: C_0 \to C_0$ we get a (unique) map $\alpha = \alpha_{\mathrm{id}_{C_0}}: C_0 \to D_1$ whose externalisation is the natural transformation α. Therefore we must show that $\alpha_X = \alpha \circ X$. But this follows from naturality of α, if we consider the morphism

$$\left(I \xrightarrow{\ X\ } C_0 \right) \xrightarrow{\quad (X, i) \quad} \left(C_0 \xrightarrow{\ \mathrm{id}\ } C_0 \right)$$

in $\mathrm{Fam}(\mathbf{C})$.

Finally, showing that externalisation preserves finite products and exponents is a matter of writing out definitions. $\qquad\square$

This technical result has some important consequences.

7.3.9. Corollary. *An internal category $\mathbf{C}$ in $\mathbb{B}$ is internally Cartesian closed if and only if its externalisation* $\begin{smallmatrix} \mathrm{Fam}(\mathbf{C}) \\ \downarrow \\ \mathbb{B} \end{smallmatrix}$ *is a split Cartesian closed fibration.*

Proof. Because externalisation preserves and reflects 1- and 2-cells, and preserves finite products of internal categories. We shall illustrate the details for Cartesian products $\times$.

$\mathbf{C}$ has internal Cartesian products

$$\Leftrightarrow \begin{cases} \text{there is a 1-cell} \\[4pt] \qquad \mathbf{C} \times \mathbf{C} \xrightarrow{\ \mathrm{prod}\ } \mathbf{C} \\[4pt] \text{in } \mathbf{cat}(\mathbb{B}) \text{ with 2-cells} \\[4pt] \qquad \mathrm{id} \overset{\eta}{\Longrightarrow} \mathrm{prod}\Delta \quad \text{and} \quad \Delta\mathrm{prod} \overset{\varepsilon}{\Longrightarrow} \mathrm{id} \\[4pt] \text{in } \mathbf{cat}(\mathbb{B}) \text{ satisfying the triangular identities} \end{cases}$$

$$\Leftrightarrow \begin{cases} \text{there is a 1-cell} \\ \left(\begin{array}{c} \text{Fam}(\mathbf{C}\times\mathbf{C}) \\ \downarrow \\ \mathbb{B} \end{array}\right) \cong \left(\begin{array}{c} \text{Fam}(\mathbf{C}) \\ \downarrow \\ \mathbb{B} \end{array}\right) \times \left(\begin{array}{c} \text{Fam}(\mathbf{C}) \\ \downarrow \\ \mathbb{B} \end{array}\right) \xrightarrow{\text{prod}} \left(\begin{array}{c} \text{Fam}(\mathbf{C}) \\ \downarrow \\ \mathbb{B} \end{array}\right) \\ \text{in } \mathbf{Fib}_{\mathbf{split}}(\mathbb{B}) \text{ with 2-cells} \\ \text{id} \xRightarrow{\eta} \text{prod}\Delta \quad \text{and} \quad \Delta\text{prod} \xRightarrow{\varepsilon} \text{id} \\ \text{in } \mathbf{Fib}_{\mathbf{split}}(\mathbb{B}) \text{ satisfying the triangular identities} \end{cases}$$

$$\Leftrightarrow \left(\begin{array}{c} \text{Fam}(\mathbf{C}) \\ \downarrow \\ \mathbb{B} \end{array}\right) \text{ has split Cartesian products.}$$

$\square$

There is also the following result.

7.3.10. Corollary. *An internal category* $\mathbf{C}$ *in a Cartesian closed category* $\mathbb{B}$ *has internal simple (co)products if and only if its externalisation* $\begin{array}{c} \text{Fam}(\mathbf{C}) \\ \downarrow \\ \mathbb{B} \end{array}$ *has split simple (co)products.*

Proof. One can reason as before, using Exercises 7.3.3 and 5.2.4:

$\mathbf{C}$ has internal simple products

$$\Leftrightarrow \begin{cases} \text{for each object } I \in \mathbb{B}, \text{ the internal diagonal functor} \\ \mathbf{C} \longrightarrow \mathbf{C}^{\underline{I}} \\ \text{has an internal right adjoint} \end{cases}$$

$$\Leftrightarrow \begin{cases} \text{for each } I \in \mathbb{B}, \text{ the fibred diagonal functor} \\ \left(\begin{array}{c} \text{Fam}(\mathbf{C}) \\ \downarrow \\ \mathbb{B} \end{array}\right) \longrightarrow \left(\begin{array}{c} \text{Fam}(\mathbf{C}^{\underline{I}}) \\ \downarrow \\ \mathbb{B} \end{array}\right) \cong \left(\begin{array}{c} \mathbb{B}/I \\ \downarrow \\ \mathbb{B} \end{array}\right) \Rightarrow \left(\begin{array}{c} \text{Fam}(\mathbf{C}) \\ \downarrow \\ \mathbb{B} \end{array}\right) \\ \text{has a split fibred right adjoint} \end{cases}$$

$$\Leftrightarrow \left(\begin{array}{c} \text{Fam}(\mathbf{C}) \\ \downarrow \\ \mathbb{B} \end{array}\right) \text{ has split simple products.}$$

But one can also simply unravel the definitions. Briefly, internal right adjoints $\Pi^I : \mathbf{C}^{\underline{I}} \to \mathbf{C}$ give right adjoints $\text{Fam}(\mathbf{C})_{J \times I} \to \text{Fam}(\mathbf{C})_J$ to weakening π^* by

$$\left(J \times I \xrightarrow{X} C_0\right) \longmapsto \left(J \xrightarrow{\Lambda(X)} C_0^I \xrightarrow{\Pi_0^I} C_0\right)$$

And conversely, right adjoints $\text{Fam}(\mathbf{C})_{J \times I} \to \text{Fam}(\mathbf{C})_J$ yield maps $C_i^I \to C_i$ for $i = 0, 1$, by considering for $J = C_i^I$ their action at evaluation $C_i^I \times I \to C_i$.

$\square$

7.3.11. Example. We can now conclude that the internal category **PER** in ω-**Sets** is internally Cartesian closed and has all internal simple products and

coproducts over ω-sets (see Example 1.8.3 (v) and Lemma 1.9.6). But **PER**
in ω-**Sets** also has internal equalisers, by a similar argument, so it has all
internal finite limits.

Similarly, **PER** is internally Cartesian closed in **Eff**, via the (first) change-
of-base situation in Proposition 6.3.2 (i) and the fact that separated reflection
preserves finite products (see Exercise 6.2.3).

Thus, internal and external structure are closely related. These matters will
be further investigated in the next section. If there is a choice, we prefer to
describe structure externally using fibred terminology, for two reasons.

- Doing internal category theory diagrammatically is rather cumbersome.
- Using the internal language makes it more convenient, but has its limi-
  tations, due to the difference between internal and external existence, as
  mentioned at the end of the previous section. And it is this external existence that we want in modelling logics and type theories.

We close this section by noting that under certain size restrictions, a split
fibration can also be internalised in the topos of presheaves on its base category.

7.3.12. Proposition (Internalisation). *Let* $\begin{smallmatrix} \mathbb{E} \\ \downarrow p \\ \mathbb{B} \end{smallmatrix}$ *be a split fibration which is
'fibrewise small' (i.e. all its fibre categories are small), on a base category*
$\mathbb{B}$ *which is locally small. Then there is an internal category* **P** *in the topos*
$\widehat{\mathbb{B}} = \mathbf{Sets}^{\mathbb{B}^{\mathrm{op}}}$ *of presheaves of* $\mathbb{B}$, *and a change-of-base situation,*

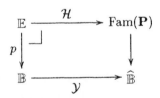

(where $\mathcal{Y}: \mathbb{B} \to \widehat{\mathbb{B}}$ *is the Yoneda embedding* $I \mapsto \mathbb{B}(-, I)$*) in which both* $\mathcal{Y}$ *and*
$\mathcal{H}$ *are full and faithful functors.*

Thus we can reconstruct the fibration p *with base category* $\mathbb{B}$ *from the associated internal category* **P** *in* $\widehat{\mathbb{B}}$.

Proof. We have to define suitable presheaves $P_0, P_1 : \mathbb{B}^{\mathrm{op}} \rightrightarrows \mathbf{Sets}$ forming
objects of objects and of arrows of $\mathbf{P} = (P_1 \rightrightarrows P_0)$ in $\widehat{\mathbb{B}}$. Firstly, P_0 sends

$$I \mapsto \mathrm{Obj}\,\mathbb{E}_I$$

and secondly, P_1 is given by disjoint unions of maps in the fibres of p, as in

$$I \mapsto \coprod_{X,Y \in \mathbb{E}_I} \mathbb{E}_I(X, Y).$$

In order to get a commuting diagram, the functor $\mathcal{H}$ has to send an object $X \in \mathbb{E}$, say above $I \in \mathbb{B}$, to a natural transformation

$$\mathcal{Y}(I) = \mathbb{B}(-, I) \Longrightarrow P_0$$

Its component at K is defined as

$$\left(K \xrightarrow{\; w \;} I \right) \longmapsto w^*(X)$$

For a morphism $f: X \to Y$ in $\mathbb{E}$, say above $u: I \to J$ in $\mathbb{B}$, a suitable natural transformation

$$\mathcal{Y}(I) = \mathbb{B}(-, I) \Longrightarrow P_1$$

is required. Here one first determines the vertical part $f': X \to u^*(Y)$ of f (with $\overline{u}(Y) \circ f' = f$). Then one can take for $\mathcal{H}f$ at K

$$\left(K \xrightarrow{\; w \;} I \right) \longmapsto \langle w^*(X), w^* u^*(Y), w^*(f') \rangle \qquad \square$$

Exercises

7.3.1. Show that the subobject fibration $\begin{smallmatrix} \text{Sub}(\mathbf{Sets}) \\ \downarrow \\ \mathbf{Sets} \end{smallmatrix}$ is the externalisation of $(\cdot \to \cdot)$ in **Sets**, *i.e.* of the poset category $\{0, 1\}$ (with $0 \leq 1$). More generally, show that for a topos $\mathbb{B}$ the subobject fibration $\begin{smallmatrix} \text{Sub}(\mathbb{B}) \\ \downarrow \\ \mathbb{B} \end{smallmatrix}$ is the externalisation of the internal poset $(\Omega, \leq)$ as in Exercises 5.3.4.

7.3.2. Elaborate out the details of the construction $\text{Full}_{\mathbb{B}}(a)$ in Example 7.1.4 (ii) in case $\mathbb{B} = \mathbf{Sets}$, and describe the resulting full and faithful fibred functor from the externalisation of this (small) category to $\mathbf{Sets}^{\to}$.

7.3.3. Show that for an object $I \in \mathbb{B}$ the externalisation $\begin{smallmatrix} \text{Fam}(\underline{I}) \\ \downarrow \\ \mathbb{B} \end{smallmatrix}$ of the discrete internal category $\underline{I} \in \mathbf{cat}(\mathbb{B})$ on I is the domain fibration $\begin{smallmatrix} \mathbb{B}/I \\ \downarrow \\ \mathbb{B} \end{smallmatrix}$. Conclude that $\mathbf{C} \in \mathbf{cat}(\mathbb{B})$ has internal exponents if and only if its externalisation has split fibred exponents (like in the proof of Corollary 7.3.9, using the exp functor mentioned towards the end of the previous section).

7.3.4. Consider a full internal category $\text{Full}(a)$ built on top of a morphism $a: A \to B$ in a locally Cartesian closed category $\mathbb{B}$, as in Example 7.1.4 (ii). Show that it has internal Cartesian products and exponents which are preserved by the associated functor $\text{Fam}\big(\text{Full}(a)\big) \to \mathbb{B}^{\to}$ if and only if there are maps prod, exp: $B \times B \rightrightarrows B$ for which one has pullback diagrams:

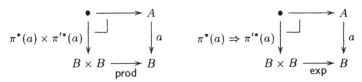

where $\times$ and $\Rightarrow$ are Cartesian product and exponent in the slice over $B \times B$. [These elementary formulations occur in [268]. There one can also find a similar formulation for simple products (using a single pullback square as above).]

7.3.5.   Definition 7.3.5 says that $\mathbf{C} = (C_1 \rightrightarrows C_0)$ is *full* internal category in $\mathbb{B}$ if there is a fibred full and faithful functor $\mathrm{Fam}(\mathbf{C}) \to \mathbb{B}^{\to}$. A good candidate for such a functor is the (internal) global sections functor Γ. Assume $\mathbf{C}$ has an internal terminal object $t\colon 1 \to C_0$. Define for $X\colon I \to C_0$ in $\mathrm{Fam}(\mathbf{C})$ a family $\Gamma(X)$ over I as pullback:

$$
\begin{array}{ccccc}
\mathbf{C}(t, X) & \longrightarrow & \mathbf{C}(t, -) & \longrightarrow & C_1 \\
\Gamma(X)\Big\downarrow \quad \llcorner & & \Big\downarrow \quad \llcorner & & \Big\downarrow \langle \partial_0, \partial_1 \rangle \\
I & \xrightarrow{\quad X \quad} & C_0 & \xrightarrow{\langle t \,\circ\, !,\, \mathrm{id}\rangle} & C_0 \times C_0
\end{array}
$$

(i)   Describe $\Gamma\colon \mathrm{Fam}(\mathbb{C}) \to \mathbf{Sets}^{\to}$ for $\mathbb{C}$ a small category with terminal object $t \in \mathbb{C}$.

(ii)   Show that $X \mapsto \Gamma(X)$ yields a fibred functor $\mathrm{Fam}(\mathbf{C}) \to \mathbb{B}^{\to}$.

(iii)   Check that the functors $\mathrm{UFam}(\mathbf{PER}) \to \omega\text{-}\mathbf{Sets}^{\to}$ and $\mathrm{UFam}(\mathbf{PER}) \to \mathbf{Eff}^{\to}$ in Proposition 7.3.7 are such global sections functors.

[This definition of Γ occurs explicitly in [143, 0.1]. Its "external" version as in (i) will be investigated later in Section 10.4, see especially Example 10.4.8 (iii) and Exercise 10.4.4.]

7.4 Internal diagrams and completeness

The aim in this section is to give a fibred version of a familiar result: if a category $\mathbb{A}$ has equalisers and arbitrary products then each diagram (functor) $\mathbb{C} \to \mathbb{A}$ from a small category $\mathbb{C}$, has a limit in $\mathbb{A}$. Here we replace $\mathbb{A}$ by a fibred category and the small category $\mathbb{C}$ by a internal category in the basis. Then the result also holds in a fibred setting. This is part of the folklore of fibred category theory, see also [36, II, 8.5]. The only ingredient of this result which still has to be explained is the notion of a functor from an internal category to a fibred category. These are called "internal diagrams", and will occupy us first. (They also play an important rôle in descent theory, see *e.g.* [88] for a survey.)

There are several alternative formulations available for these internal diagrams, see Remark 7.4.2 below. The most satisfying formulation from a fibred perspective is mentioned there as (i). But we choose to start with a different formulation, because this one is more common in the literature, and because it will be used in computing limits (in the proof of Lemma 7.4.8).

7.4.1. Definition. Let $\begin{smallmatrix} \mathbb{E} \\ \downarrow p \\ \mathbb{B} \end{smallmatrix}$ be a fibration and $\mathbf{C}$ an internal category in $\mathbb{B}$. An **internal diagram of type C** in p is a pair (U, μ) where $U \in \mathbb{E}_{C_0}$ is an object of the total category above the object C_0 of objects of $\mathbf{C}$, and μ is a vertical morphism $\partial_0^*(U) \to \partial_1^*(U)$ in $\mathbb{E}$, called the **action** of the internal diagram, satisfying:

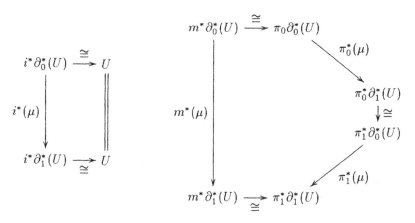

This definition may be found, for instance, in [189, Section 5]. It is in fact most familiar for the case where the above fibration p is a codomain fibration. An internal diagram can then be described as a family $\begin{pmatrix} U \\ \downarrow \varphi \\ C_0 \end{pmatrix}$ together with an action morphism μ in a commuting diagram

$$
\begin{array}{ccc}
U \times_{C_0} C_1 & \xrightarrow{\ \mu\ } & U \\
{\scriptstyle \partial_0^*(\varphi)} \big\downarrow & & \big\downarrow {\scriptstyle \varphi} \\
C_1 & \xrightarrow[\ \partial_1\ ]{} & C_0
\end{array}
$$

satisfying the equations

$$U \xrightarrow{\langle \mathrm{id},\, i \circ \varphi \rangle} U \times_{C_0} C_1 \quad C_2 \times_{C_0} U = C_1 \times_{C_0} C_1 \times_{C_0} U \xrightarrow{\mathrm{id} \times \mu} C_1 \times_{C_0} U$$

$$\downarrow \mu \qquad\qquad m \times \mathrm{id} \downarrow \qquad\qquad\qquad \downarrow \mu$$

$$U \qquad\qquad\qquad C_1 \times_0 U \xrightarrow{\qquad \mu \qquad} U$$

see *e.g.* [169, Definition 2.14], [36, I, Definition 8.21] (where these internal diagrams in codomain fibrations are called internal base-valued functors) or [188, V,7] (where they are called category actions). In the even more familiar case of sets, an internal diagram in the codomain fibration $\begin{smallmatrix} \mathbf{Sets}^{\rightarrow} \\ \downarrow \\ \mathbf{Sets} \end{smallmatrix}$ can be identified with a presheaf $\mathbf{C} \to \mathbf{Sets}$. Indeed the family $\left(\begin{smallmatrix} U \\ \downarrow \varphi \\ C_0 \end{smallmatrix} \right)$ yields a functor $\mathbf{C} \to \mathbf{Sets}$ by $X \mapsto U_X = \varphi^{-1}(X)$ and $(f \colon X \to Y) \mapsto \lambda u \in U_X . \mu(u, f)$. And conversely, such a functor $D \colon \mathbf{C} \to \mathbf{Sets}$ gives an internal diagram with family $\coprod_{X \in C_0} D(X) \xrightarrow{\pi} C_0$ and action $(X, a, f) \mapsto (Y, D(f)(a))$, for $f \colon X \to Y$ in $\mathbf{C}$.

7.4.2. Remark. There are alternative descriptions of internal diagrams. We briefly mention three of these below, without going into all the details.

(i) First of all, an internal diagram of type $\mathbf{C}$ in $\begin{smallmatrix} \mathbb{E} \\ \downarrow p \\ \mathbb{B} \end{smallmatrix}$ is nothing but a fibred functor $\mathrm{Fam}(\mathbf{C}) \to \mathbb{E}$ from the externalisation of $\mathbf{C}$ to p. Indeed, given an internal diagram (U, μ) as defined above, one obtains a functor $\mathrm{Fam}(\mathbf{C}) \to \mathbb{E}$ by

$$\left(I \xrightarrow{\ X\ } C_0 \right) \longmapsto X^*(U)$$

And conversely, given a fibred functor $F \colon \mathrm{Fam}(\mathbf{C}) \to \mathbb{E}$ one takes as carrier object $U = F(\mathrm{id}_{C_0})$ in $\mathbb{E}_{C_0}$. Further, the identity on C_1 is a morphism $\partial_0 \to \partial_1$ in $\mathrm{Fam}(\mathbf{C})$. And $F(\partial_i) = F(\mathrm{id}_{C_0} \circ \partial_i) \cong \partial_i^*(F(\mathrm{id}_{C_0})) = \partial_i^*(U)$ since F is a fibred functor. Therefore we get an action morphism μ as composite

$$\partial_0^*(U) \cong F(\partial_0) \xrightarrow{\ F(\mathrm{id}_{C_1})\ } F(\partial_1) \cong \partial_1^*(U).$$

It is not hard to verify that these passages from internal diagrams to fibred functors and vice-versa, are each others inverses.

These fibred functors $\mathrm{Fam}(\mathbf{C}) \to \mathbb{E}$ are sometimes easier to handle than the internal diagrams (U, μ). For example, with this fibred functor formulation, it is almost immediate that internal functors $G \colon \mathbf{D} \to \mathbf{C}$ take internal diagrams of type $\mathbf{C}$ to internal diagrams of type $\mathbf{D}$, simply by pre-composition with the external functor $\mathrm{Fam}(G)$. And by post-composition one transforms internal diagrams in one fibration into internal diagrams in another fibration. Also, this correspondence extends to appropriate morphisms, see Exercise 7.4.3 below.

As an example, we see that an internal category of the form $\mathrm{Full}_{\mathbb{B}}(a)$ comes equipped with a canonical internal diagram, namely the functor $\mathrm{Full}_{\mathbb{B}}(a) \to \mathbb{B}^{\to}$ described in Example 7.1.4 (ii).

(ii) There is however a second alternative description of internal diagrams, which is closer to the original description in Defintion 7.4.1. The correspondence occurs for codomain fibrations in [169, Proposition 2.21] and in [188, V, 7, Theorem 2]. It is mentioned in full generality in [30], but is independently due to Beck.

For a fibration $\begin{smallmatrix} \mathbb{E} \\ \downarrow p \\ \mathbb{B} \end{smallmatrix}$ with coproducts $\coprod_u$ and an internal category $\mathbf{C}$ in $\mathbb{B}$, consider the functor

$$\mathbb{E}_{C_0} \xrightarrow{\quad T = \coprod_{\partial_1} \partial_0^* \quad} \mathbb{E}_{C_0}$$

It is a monad, and its algebras $T(U) \to U$ are precisely the internal diagrams $\partial_0^*(U) \to \partial_1^*(U)$.

We only sketch how to get a unit and multiplication for this monad, and leave further details to the meticulous reader. A unit $\mathrm{id}_{\mathbb{E}_{C_0}} \Rightarrow T$ is obtained as composite

$$\mathrm{id}_{\mathbb{E}_{C_0}} \cong \coprod_{\partial_1} \coprod_i i^* \partial_0^* \xrightarrow{\coprod_{\partial_1} \varepsilon \partial_0^*} \coprod_{\partial_1} \partial_0^* = T$$

and a multiplication $T^2 \Rightarrow T$ as composite,

$$
\begin{aligned}
T^2 = \coprod_{\partial_1} \partial_0^* \coprod_{\partial_1} \partial_0^* &\overset{(\mathrm{BC})}{\cong} \coprod_{\partial_1} \coprod_{\pi_1} \pi_0^* \partial_0^* \\
&\cong \coprod_{\partial_1 \circ \pi_1} (\partial_0 \circ \pi_0)^* \\
&\cong \coprod_{\partial_1 \circ m} (\partial_0 \circ m)^* \\
&\cong \coprod_{\partial_1} \coprod_m m^* \partial_0^* \xrightarrow{\coprod_{\partial_1} \varepsilon \partial_0^*} \coprod_{\partial_1} \partial_0^* = T.
\end{aligned}
$$

(Where 'BC' stands for Beck-Chevalley.)

(iii) If the fibration additionally has products $(u^* \dashv \prod_u)$ then the above monad $T = \coprod_{\partial_1} \partial_0^*$ has a right adjoint $\prod_{\partial_0} \partial_1^*$. By the Eilenberg-Moore Theorem (see *e.g.* [188, V, 8, Theorems 1 and 2] or [36, II, Proposition 4.4.6]) this right adjoint is a comonad and its category of co-algebras is isomorphic to the category of T-algebras. Hence internal diagrams can also be described as co-algebras of $\prod_{\partial_0} \partial_1^*$.

What we are most interested in are internal diagrams that are parametrised by an object I of the base category.

7.4.3. Definition. Consider an internal category $\mathbf{C}$ in the base category of a fibration $\begin{smallmatrix}\mathbb{E}\\\downarrow p\\\mathbb{B}\end{smallmatrix}$. Let I be an object in $\mathbb{B}$. An *I-parametrised internal diagram* of type $\mathbf{C}$ in p is an internal diagram of type $\underline{I} \times \mathbf{C}$ in p, where $\underline{I}$ is the discrete internal category associated with I, see Example 7.1.2 (ii).

More concretely, it is given by an object $U \in \mathbb{E}_{I \times C_0}$ together with a vertical action morphism $\mu\colon (\mathrm{id} \times \partial_0)^*(U) \to (\mathrm{id} \times \partial_1)^*(U)$ over $I \times C_1$, making some diagrams commute (as in Definition 7.4.1).

7.4.4. Lemma. *Let $\begin{smallmatrix}\mathbb{E}\\\downarrow p\\\mathbb{B}\end{smallmatrix}$ be a fibration and $\mathbf{C}$ be an internal category in $\mathbb{B}$, with externalization $\begin{smallmatrix}\mathrm{Fam}(\mathbf{C})\\\downarrow p_{\mathbf{C}}\\\mathbb{B}\end{smallmatrix}$. For an object $I \in \mathbb{B}$ the following are essentially the same.*

(i) *I-parametrised internal diagrams (of type $\mathbf{C}$ in p);*

(ii) *fibred functors F in*

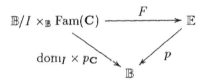

(iii) *objects of the fibre over I of the 'exponent' fibration $p_{\mathbf{C}} \Rightarrow p$, described in Exercise 1.10.6.*

Proof. The correspondence between (i) and (ii) is obtained essentially as in Remark 7.4.2 (i): given an I-parametrised internal diagram (U, μ) as above one gets a fibred functor $\mathbb{B}/I \times_{\mathbb{B}} \mathrm{Fam}(\mathbf{C}) \to \mathbb{E}$ by

$$\left(\begin{smallmatrix} & I' & \\ v\swarrow & & \searrow v' \\ I & & C_0 \end{smallmatrix}\right) \mapsto \langle v, v'\rangle^*(U) \in \mathbb{E}_{I'}.$$

And conversely, given a fibred functor F as above, one takes

$$U = F\left(\begin{smallmatrix} & I \times C_0 & \\ \pi\swarrow & & \searrow \pi' \\ I & & C_0 \end{smallmatrix}\right) \in \mathbb{E}_{I \times C_0}.$$

In order to define the action μ, we notice that for $i = 0, 1$, there is are isomorphisms:

$$F\left(\begin{array}{c} I \times C_1 \\ \pi \swarrow \searrow \partial_i \circ \pi' \\ I \qquad C_0 \end{array}\right) = F\left((\mathrm{id} \times \partial_i)^* \left(\begin{array}{c} I \times C_0 \\ \pi \swarrow \searrow \pi' \\ I \qquad C_0 \end{array}\right)\right) \cong (\mathrm{id} \times \partial_i)^*(U)$$

And also that we have a morphism in $\mathbb{B}/I \times_{\mathbb{B}} \mathrm{Fam}(\mathbf{C})$,

$$\left(\begin{array}{c} I \times C_1 \\ \pi \swarrow \searrow \partial_0 \circ \pi' \\ I \qquad C_0 \end{array}\right) \xrightarrow{(\mathrm{id}_{I \times C_1}, \pi')} \left(\begin{array}{c} I \times C_1 \\ \pi \swarrow \searrow \partial_1 \circ \pi' \\ I \qquad C_0 \end{array}\right).$$

Hence, by applying F and using the appropriate isomorphisms, we obtain a vertical action morphism $\mu \colon (\mathrm{id} \times \partial_0)^*(U) \to (\mathrm{id} \times \partial_1)^*(U)$ over $I \times C_1$.

The correspondence between (ii) and (iii) is immediate from the description of the exponent fibration in Exercise 1.10.6. □

In a next step, these parametrised internal diagrams can be organised in a fibred category.

7.4.5. Definition. Consider a fibration $\begin{smallmatrix} \mathbb{E} \\ \downarrow{\scriptstyle p} \\ \mathbb{B} \end{smallmatrix}$ with an internal category $\mathbf{C}$ in $\mathbb{B}$. We form a category $\mathbb{E}^{\mathbf{C}}$ with

objects triples (I, U, μ) where (U, μ) is an I-parametrised internal diagram of type $\mathbf{C}$ in p.

morphisms $(I, U, \mu) \to (J, V, \nu)$ are tuples (w, f) where $w \colon I \to J$ is a morphism in $\mathbb{B}$ and $f \colon U \to V$ is a morphism in $\mathbb{E}$ above $w \times \mathrm{id} \colon I \times C_0 \to J \times C_0$ for which there is a commuting diagram,

$$\begin{array}{ccc} (\mathrm{id} \times \partial_0)^*(U) & \dashrightarrow & (\mathrm{id} \times \partial_0)^*(V) \\ {\scriptstyle\mu}\downarrow & & \downarrow{\scriptstyle\nu} \\ (\mathrm{id} \times \partial_1)^*(U) & \dashrightarrow & (\mathrm{id} \times \partial_1)^*(V) \end{array}$$

in which the dashed arrows are the unique ones above $w \times \mathrm{id} \colon I \times C_1 \to J \times C_1$, induced by $f \colon U \to V$.

Notice that in the notation $\mathbb{E}^{\mathbf{C}}$ the rôle of the fibration p is left implicit. In the literature one sometimes finds for an internal category $\mathbf{C}$ in $\mathbb{B}$ the notation $\mathbb{B}^{\mathbf{C}}$ for the fibre over 1 of the category of diagrams $(\mathbb{B}^{\to})^{\mathbf{C}}$ in the codomain fibration on $\mathbb{B}$.

7.4.6. Lemma. *Let* $\overset{\mathbb{E}}{\underset{\mathbb{B}}{\downarrow}}{}^{p}$ *and* **C** *be as in the previous definition.*

(i) *The category* $\mathbb{E}^{\mathbf{C}}$ *of internal diagrams is fibred over* $\mathbb{B}$ *via the projection* $(I, U, \mu) \mapsto I$.

(ii) *There is a fibred diagonal functor* $\Delta \colon \mathbb{E} \to \mathbb{E}^{\mathbf{C}}$ *which maps an object* $X \in \mathbb{E}$, *say over* $I \in \mathbb{B}$, *to the object* $\pi^*(X) \in \mathbb{E}_{I \times C_0}$ *together with the action*

$$(\mathrm{id} \times \partial_0)^* \pi^*(X) \overset{\cong}{\longrightarrow} \pi^*(X) \overset{\cong}{\longrightarrow} (\mathrm{id} \times \partial_1)^*(X).$$

Proof. (i) For a morphism $w \colon I \to J$ in $\mathbb{B}$ and an object (J, V, ν) in $\mathbb{E}^{\mathbf{C}}$ above J, take for U the reindexed object $(w \times \mathrm{id})^*(V) \in \mathbb{E}_{I \times C_0}$, and for μ the following composite over $I \times C_1$.

$$
\begin{array}{c}
(\mathrm{id} \times \partial_0)^*(U) \\[4pt]
\cong \downarrow \\[4pt]
(w \times \mathrm{id})^*(\mathrm{id} \times \partial_0)^*(V) \xrightarrow{\quad (w \times \mathrm{id})^*(\nu) \quad} (w \times \mathrm{id})^*(\mathrm{id} \times \partial_1)^*(V) \\[4pt]
\downarrow \cong \\[4pt]
(\mathrm{id} \times \partial_1)^*(U).
\end{array}
$$

(ii) Easy.   □

We are finally in a position to say what limits with respect to internal categories are. We shall do this in two stages (as in Section 1.9, see especially Theorem 1.9.10) by first describing 'simple' limits, and taking ordinary limits to be simple limits relativised to all slices of the base category.

7.4.7. Definition. Consider an internal category **C** in the base category of a fibration $\overset{\mathbb{E}}{\underset{\mathbb{B}}{\downarrow}}{}^{p}$.

(i) We say that this fibration p has **simple limits of type C** if the fibred diagonal functor $\Delta \colon \mathbb{E} \to \mathbb{E}^{\mathbf{C}}$ has a fibred right adjoint.

(ii) And we say that p has **all small limits** in case for each object $I \in \mathbb{B}$ and for every internal category **C** in $\mathbb{B}/I$, the localisation fibration $I^*(p)$ in

$$
\begin{array}{ccc}
\mathbb{B}/I \times_{\mathbb{B}} \mathbb{E} & \longrightarrow & \mathbb{E} \\
I^*(p) \downarrow \;\;\; \llcorner & & \downarrow p \\
\mathbb{B}/I & \xrightarrow[\mathrm{dom}_I]{} & \mathbb{B}
\end{array}
$$

has simple limits of type **C**.

Note that for split fibrations one can also have split (simple) limits. These involve split adjunctions in the above definition. Here one uses that for a split fibration, the fibration of diagrams is also split.

The main technical work is in the following result.

7.4.8. Lemma. *Assume* $\begin{smallmatrix} \mathbb{E} \\ \downarrow p \\ \mathbb{B} \end{smallmatrix}$ *is a fibration with fibred equalisers and simple products* $\prod_{(I,J)}$ *(i.e. products along Cartesian projections* $\pi\colon I \times J \to I$*). Then p has simple limits of type* $\mathbf{C}$*, for every internal category* $\mathbf{C}$ *in* $\mathbb{B}$.

Proof. For an I-parametrised diagram consisting of an object $U \in \mathbb{E}_{I \times C_0}$ with action $\mu\colon (\mathrm{id} \times \partial_0)^*(U) \to (\mathrm{id} \times \partial_1)^*(U)$, we have to construct an appropriate limit object $L = \varprojlim(I, U, \mu) \in \mathbb{E}_I$. Therefore consider the following two maps (for $i = 0, 1$).

$$
\begin{array}{c}
\pi_{I,C_1}^* \prod_{(I,C_0)}(U) \\
\Big\downarrow \cong \\
(\mathrm{id} \times \partial_i)^* \pi_{I,C_0}^* \prod_{(I,C_0)}(U) \xrightarrow{\ (\mathrm{id} \times \partial_i)^*(\varepsilon)\ } (\mathrm{id} \times \partial_i)^*(U)
\end{array}
$$

They give rise to two (vertical) maps

$$
\pi_{I,C_1}^* \prod_{(I,C_0)}(U) \rightrightarrows (\mathrm{id} \times \partial_1)^*(U),
$$

the first one by taking $i = 0$ and composing with the action μ, and the second one directly for $i = 1$. By transposing these two, we get maps

$$
\prod_{(I,C_0)}(U) \rightrightarrows \prod_{(I,C_1)}(\mathrm{id} \times \partial_1)^*(U)
$$

and so we can construct $L \rightarrowtail \prod_{(I,C_0)}(U)$ as the equaliser of these (in the fibre over I). This yields an appropriate limit: for $X \in \mathbb{E}_I$ transposition yields a correspondence

$$
\frac{X \longrightarrow \prod_{(I,C_0)}(U) \ \text{equalising} \ \prod_{(I,C_0)}(U) \rightrightarrows \prod_{(I,C_1)}(\mathrm{id} \times \partial_1)^*(U)}{\pi_{I,C_0}^*(X) \longrightarrow U \ \text{forming a map of diagrams} \ \Delta(X) \to (U, \mu)}
$$

$\square$

It may be interesting to note that the construction in this proof is essentially the one used for ordinary categories, see [187, Diagram (1) on p. 109].

Finally, the main result can now be obtained without much difficulty.

7.4.9. Theorem. *A fibration with products* $\prod_u$ *and fibred equalisers has all small limits.*

Proof. For a fibration p as in the theorem, each fibration $I^*(p)$ has simple products, by Theorem 1.9.10. Hence we are done by the previous lemma. □

We close this section by noting that for *small* fibrations p a converse of this theorem can be established. Indeed, assume a fibration $p = p_{\mathbf{D}}$ arising as externalization $\begin{smallmatrix} \text{Fam}(\mathbf{D}) \\ \downarrow \\ \mathbb{B} \end{smallmatrix}$ of an internal category $\mathbf{D}$ in $\mathbb{B}$, which has all small limits. In particular, it has simple limits, so each diagonal functor

$$\left(\begin{smallmatrix} \text{Fam}(\mathbf{D}) \\ \downarrow \\ \mathbb{B} \end{smallmatrix} \right) \xrightarrow{\quad \Delta \quad} \left(\begin{smallmatrix} \text{Fam}(\mathbf{D})^{\mathbf{C}} \\ \downarrow \\ \mathbb{B} \end{smallmatrix} \right)$$

has a (split) fibred right adjoint. Using Lemma 7.4.4 this means that each fibred diagonal functor

$$p_{\mathbf{D}} \xrightarrow{\quad \Delta \quad} (p_{\mathbf{C}} \Rightarrow p_{\mathbf{D}})$$

has a (split) fibred right adjoint. But this corresponds to the statement that each internal diagonal functor

$$\mathbf{D} \xrightarrow{\quad \Delta \quad} \mathbf{D}^{\mathbf{C}}$$

has an internal right adjoint. This yields that $\mathbf{D}$ has

$$\begin{cases} \text{an internal terminal object:} & \text{take } \mathbf{C} = \underline{0} \\ \text{internal binary products:} & \text{take } \mathbf{C} = \underline{2} = (\cdot \ \cdot) \\ \text{internal equalisers:} & \text{take } \mathbf{C} = (\cdot \rightrightarrows \cdot) \end{cases}$$

In order to show that the fibration $\begin{smallmatrix} \text{Fam}(\mathbf{D}) \\ \downarrow p_{\mathbf{D}} \\ \mathbb{B} \end{smallmatrix}$ also has products $\prod_u$, it suffices by Theorem 1.9.10 to show that each fibration

$$I^*(p_{\mathbf{D}}) = \left(\begin{smallmatrix} \text{Fam}(I^*(\mathbf{D})) \\ \downarrow \\ \mathbb{B}/I \end{smallmatrix} \right)$$

has simple products. Therefore it suffices to show that for each object (or family) $u \in \mathbb{B}/I$ the diagonal functor

$$I^*(\mathbf{D}) \xrightarrow{\quad \Delta \quad} (I^*(\mathbf{D}))^{\underline{u}}$$

has an internal right adjoint—where $\underline{u}$ is the discrete category associated with the object u in $\mathbb{B}/I$. But this holds because $\begin{smallmatrix} \text{Fam}(\mathbf{D}) \\ \downarrow \\ \mathbb{B} \end{smallmatrix}$ has all small limits. Thus we have the following result.

7.4.10. Theorem. *A small fibration* $\begin{smallmatrix}\text{Fam}(\mathbf{D})\\\downarrow\\\mathbb{B}\end{smallmatrix}$ *has all small limits if and only if it is a complete fibration.* □

There is a standard result (due to Freyd, see Fact 8.3.3 later on) saying that there are no (ordinary) categories which are both small and complete. In contrast, *fibred* categories which are both small and complete do exist: (the externalisation of) **PER** in ω-**Sets** is an example, see Lemma 1.9.6 and Example 1.8.3 (v). It will provide a model for various typed calculi later on.

For a small complete category **C** in a universe $\mathbb{B}$ there are various pleasant properties which are lacking for ordinary categories. For example, completeness automatically yields cocompleteness, see [143, 255], which in **Sets** only holds for posets. In what is called "synthetic domain theory" one tries to exploit such properties to obtain a smooth theory of domains within universes other than **Sets** (especially toposes), see for example [144, 331, 260]. In these more general universes one treats domains and continuous functions simply as sets and ordinary functions (following ideas of D. Scott). This should make it easier to describe models of term (or program) languages with various kinds of fixed points.

Exercises

7.4.1. Describe the fibred functor $\text{Fam}(\mathbf{C}) \to \mathbf{Sets}^{\to}$ corresponding to a presheaf $\mathbf{C} \to \mathbf{Sets}$, considered as an internal diagram in $\begin{smallmatrix}\mathbf{Sets}^{\to}\\\downarrow\\\mathbf{Sets}\end{smallmatrix}$.

7.4.2. Consider an internal category **C** in a base category $\mathbb{B}$ with finite limits. Describe an internal diagram of type **C** in the simple fibration $\begin{smallmatrix}s(\mathbb{B})\\\downarrow\\\mathbb{B}\end{smallmatrix}$ on $\mathbb{B}$, and compare it with an internal diagram in the codomain fibration $\begin{smallmatrix}\mathbb{B}^{\to}\\\downarrow\\\mathbb{B}\end{smallmatrix}$ on $\mathbb{B}$. Check that if **C** is an internal monoid (*i.e.* if C_0 is terminal object), then there is no difference between diagrams of type **C** in the simple fibration and in the codomain fibration.

7.4.3. Consider internal diagrams (U, μ) and (V, ν) of type **C** in a fibration $\begin{smallmatrix}\mathbb{E}\\\downarrow p\\\mathbb{B}\end{smallmatrix}$ as described in Definition 7.4.1. A morphism $f: (U, \mu) \to (V, \beta)$ of such internal diagrams (as defined in [189]) is a vertical map $f: U \to V$ making the following diagram commute.

$$
\begin{array}{ccc}
\partial_0^*(U) & \xrightarrow{\ \partial_0^*(f)\ } & \partial_0^*(V) \\
{\scriptstyle \mu}\downarrow & & \downarrow{\scriptstyle \nu} \\
\partial_1^*(U) & \xrightarrow[\ \partial_1^*(f)\ ]{} & \partial_1^*(V)
\end{array}
$$

Notice that such morphisms of internal diagrams are in fact morphisms in the fibre over the terminal object $1 \in \mathbb{B}$ of the category $\mathbb{E}^{\mathbf{C}}$ of internal diagrams as above.

Show that the resulting category of internal diagrams is equivalent to the category of fibred functors $\mathrm{Fam}(\mathbf{C}) \to \mathbb{E}$ and vertical natural transformations between them.

7.4.4. Show that a fibration p has simple limits of type $\mathbf{C}$ if and only if the diagonal functor $\Delta: p \to (p_{\mathbf{C}} \Rightarrow p)$ has a fibred right adjoint. This diagonal maps an object $X \in \mathbb{E}$, say over $I \in \mathbb{B}$, to the functor $\Delta(X): \mathbb{B}/I \times_{\mathbb{B}} \mathrm{Fam}(\mathbf{C}) \to \mathbb{E}$ given by

$$
\begin{pmatrix}
& I' & \\
& {}^{v} \swarrow \quad \searrow {}^{v'} & \\
I & & C_0
\end{pmatrix}
\mapsto v^*(X) \in \mathbb{E}_{I'}.
$$

7.4.5. Check that if $\begin{smallmatrix} \mathbb{E} \\ \downarrow \\ \mathbb{B} \end{smallmatrix}$ has fibred (finite) limits, then so has a fibration $\begin{smallmatrix} \mathbb{E}^{\mathbf{C}} \\ \downarrow \\ \mathbb{B} \end{smallmatrix}$ of internal diagrams.

7.4.6. Let $\mathbf{C}$ be an internal category in $\mathbb{B}$.

(i) Prove that the family $\begin{pmatrix} C_1 \\ \downarrow \langle \partial_0, \partial_1 \rangle \\ C_0 \times C_0 \end{pmatrix}$ carries a "Hom" action for a diagram of type $\mathbf{C}^{\mathrm{op}} \times \mathbf{C}$ in $\begin{smallmatrix} \mathbb{B}^{\to} \\ \downarrow \\ \mathbb{B} \end{smallmatrix}$.

(ii) Show that the resulting fibred "Yoneda" functor

$$
\mathrm{Fam}(\mathbf{C}) \longrightarrow (\mathbb{B}^{\to})^{\mathbf{C}^{\mathrm{op}}}
$$

is full and faithful.

Chapter 8

Polymorphic type theory

Types in simple type theory (STT) are built up from atomic types using type constructors like $\to, \times, 1$ or $+, 0$, as described in Chapter 2. In polymorphic type theory (PTT) one may also use type variables $\alpha, \beta, \gamma \ldots$ to build types. This is the main innovation in PTT; it gives rise to an extra level of indexing: not only by term variables (as in STT) but also by type variables. This forms the topic of the present chapter.

We distinguish three versions of polymorphic type theory, called first order PTT $\lambda\to$, second order PTT $\lambda 2$, and higher order PTT $\lambda\omega$. We informally describe the differences: in **first order polymorphic type theory** $\lambda\to$ there is an identity function

$$\lambda x : \alpha. \, x : \alpha \to \alpha$$

where α is a type variable. It yields by substituting a specific type σ for α, the identity function

$$\lambda x : \sigma. \, x : \sigma \to \sigma$$

on σ. In **second order polymorphic type theory** (denoted by $\lambda 2$) one may abstract type variables, as in:

$$I = \lambda \alpha : \mathsf{Type}. \, \lambda x : \alpha. \, x : \Pi \alpha : \mathsf{Type}. \, (\alpha \to \alpha).$$

We then get the identity on a type σ by application (and β-reduction):

$$I\sigma = \lambda x : \sigma. \, x : \sigma \to \sigma.$$

Similarly, one can have a polymorphic conditional term,

$$\mathsf{if} : \Pi \alpha : \mathsf{Type}. \, (\mathsf{bool} \times \alpha \times \alpha) \to \alpha$$

441

which can be instantiated to a specific type σ by application if σ. Notice that such polymorphic product $\Pi\alpha\colon \mathsf{Type}.\,\sigma(\alpha)$ is impredicative: it involves quantification over all types, and in particular over the product type itself. This impredicativity makes second (and higher) order PTT very powerful, but it introduces various (semantical) complications.

There is one further system, namely **higher order polymorphic type theory** $\lambda\omega$ in which one can form finite products and exponents of 'kinds' like Type (and quantify over all of these). Then one can form terms like

$$\lambda\alpha\colon \mathsf{Type} \to \mathsf{Type}.\,\lambda\beta\colon \mathsf{Type}.\,\alpha\,\beta \to \beta \colon (\mathsf{Type} \to \mathsf{Type}) \to (\mathsf{Type} \to \mathsf{Type}).$$

Such polymorphic type theories were first introduced by Girard in [95] for proof theoretic purposes, and independently by Reynolds in [285] with motivation stemming from computer science. In computing it is not practical to have, for example, a sorting algorithm for natural numbers and also one for strings *etc.*, but instead, one would like to have an algorithm which is parametric, in the sense of Strachey: it should work uniformly for an arbitrary type with an arbitrary (linear) order on it. One turns such a parametric algorithm into a specific one (which works on natural numbers or strings) by suitably instantiating it—by substitution or application, as for the identity $\sigma \to \sigma$ above, see [44, 228] for more details. This parametricity puts a certain uniformity restriction on the indexing by type variables α in terms $\big(M(\alpha)\colon \sigma(\alpha)\big)_{\alpha\colon \mathsf{Type}}$.

The functional programming language ML (see [224, 223]) is loosely based on the first order polymorphic lambda calculus $\lambda\to$ (and so is the type theory of the proof assistant ISABELLE [250]). But there are some subtle differences, as will be explained in Section 8.2. There is also a higher order functional programming language QUEST, see [43] and the references there for more information, which is based on $\lambda\omega$.

The names $\lambda\to$ 'lambda-arrow', $\lambda2$ 'lambda-two' and $\lambda\omega$ 'lambda-omega' come from Barendregt [14]. However, the calculi as described there are not precisely as we use them: here they are described on top of a so-called 'polymorphic signature'. Further, we standardly consider them with finite product types (and kinds for $\lambda\omega$) and with sums Σ. Equality however, is an additional feature. In contrast, only the 'minimal' systems (with only $\to$ and Π) are considered in [14].

Our treatment of the syntax of polymorphic type theory will concentrate on the essentials, and, for example, Curch-Rosser (CR) and Strong Normalisation (SN) properties will not form part of it. Actually, CR is straightforward, but SN is non-trivial for (minimal versions of) second and higher order polymorphic type theory. CN was first proved by Girard in [94], using so-called *candidats de reducibilité* (also called saturated sets, see [300]). Girard's proof can be simplified by erasing types, see [325, 225]. It is presented via a (modified) realisability interpretation in [147].

The fibred categories needed to model polymorphic type theories are essentially as for (higher order) predicate logics, except that the fibres need not be preordered (as in Chapters 3, 4, 5). This extra structure is used to accommodate for terms inhabiting types, or, under a propositions-as-types reading, for proof-objects inhabiting propositions. These proof-objects give rise to nontrivial morphisms in the fibres, by considering proofs rather than provability. In the language of internal categories one can say that for PTT we do not use internal preorders (as for predicate logics), but proper internal categories.

The chapter starts with the syntactic aspects of polymorphic type theories: in the first section we describe how specific calculi can be defined on top of a polymorphic signature—giving atomic *types* and *kinds* with function symbols—and we establish their relation to predicate logics, via propositions-as-types. The subsequent section is about actual use of especially the second order polymorphic calculus in encoding inductively and co-inductively defined types, and data-types with encapsulation. The semantic study starts in the third Section 8.3 with a naive set theoretic approach. It turns out to work only for first order polymorphic type theory, but it is illuminating as a starting point, because it gives a clear picture of the double forms of indexing in PTT—via term variables and via type variables. Also, it will be used to explain the famous negative result of Reynolds [287, 288], stating that there are no set-theoretic models of (impredicative) polymorphic products $\Pi\alpha\colon \mathsf{Type}.\,\sigma(\alpha)$. We analyse this result along the lines of Pitts [269], and see that in higher order logic there cannot be an embedding $\mathsf{Prop} \rightarrowtail \sigma$ of propositions into a type σ of a polymorphic calculus. This rules out set-theoretic models—except trivial ones, since in a set theoretic model Prop is $\{0,1\}$. But we shall see that for PERs R in ω-**Sets** and in **Eff** there are indeed no such inclusions $\mathsf{Prop} \rightarrowtail R$—for $\mathsf{Prop} = \nabla 2$ in ω-**Sets** and $\mathsf{Prop} = P\mathbb{N}$ in **Eff**. In the fourth section we present the general definitions of fibred categories corresponding to the three versions $\lambda\!\to$, $\lambda 2$ and $\lambda\omega$ of PTT. Attention is devoted in particular to examples involving PERs indexed over sets, over ω-sets, and over objects of the effective topos **Eff**. But also to a PER model which is relationally parametric (in the sense of [286]). In the fifth section we concentrate on two constructions to turn fibred categories for PTT into internal categories for PTT. And in the last and sixth section we describe how one can have "logic over PTT" in a way similar to how predicate logic is a "logic over STT". We will use such a logic in a categorical description of relational parametricity (as in [204, 293]). This relational parametricity was originally introduced by Reynolds in an attempt to circumvent the abovementioned problems with set theoretic models of second order PTT. The hope was that restriction to a certain class of "parametric" functions would allow the existence of set theoretic models—just like there are no set theoretic models $X \cong X^X$ of the untyped λ-calculus (for cardinality reasons), but restriction to continuous functions does yield examples of solutions X. This plan was later abandoned by Reynolds. But relational parametricity survived as a notion because it turned out to be important as a criterion for "good models", in which various syntactically definable operations satisfy the appropriate universal properties, and in which polymorphic maps are automatically "natural", see [340, 11, 273, 118] (or Exercises 8.4.5 and 8.4.6 below). But see also [237] for other applications of parametricity.

The literature on (the semantics of) polymorphic type theory is extensive. We mention [38, 226] for set theoretic notions of model, and [307, 268, 56, 37] for category theoretic (indexed and internal) notions. An easy going introduction to the indexed models is [61]. Translations between the set theoretic and the indexed notions may be found in [220, 155], and between the indexed and internal notions in [268, 8]. References about parametricity in models (besides the above ones) may be found in Remark 8.6.4 (v).

8.1 Syntax

In simple type theory (STT) we used an infinite set of *term* variables

$$\mathrm{Var} = \{v_1, v_2, v_3, \ldots\}.$$

In polymorphic type theory (PTT) we will additionally use an infinite set of *type* variables

$$\mathrm{TypeVar} = \{\alpha_1, \alpha_2, \alpha_3, \ldots\}.$$

Just like we sometimes used $x, y, z, \ldots$ for term variables, we shall be using $\alpha, \beta, \gamma, \ldots$ for type variables.

In STT there are types σ: Type and terms M: σ inhabiting such types, but in PTT the situation is more complicated: there are *types* σ: Type and terms M: σ inhabiting types, but also what are often called *kinds* A: Kind and terms M: A inhabiting these kinds. This gives a picture with two simple type theories (with Type and with Kind), but on different levels. However, these levels are connected, since the types σ: Type occur as terms of the distinguished kind Type: Kind. This is like in higher order logic where propositions φ: Prop are terms of the type Prop: Type. Indeed, there is a close correspondence between these polymorphic type theories and higher order predicate logics. It will be described as a propositions-as-types analogy below, in which inhabitants of types will appear as proofs of the corresponding propositions. In the type theories $\lambda \rightarrow$ and $\lambda 2$ the only (atomic) kind is Type, but in $\lambda \omega$ one may have more kinds (and form finite products and exponents of these). Thus $\lambda \rightarrow$ and $\lambda 2$ are 'single-kinded', whereas $\lambda \omega$ is 'many-kinded'.

As in simple type theory, a specific polymorphic calculus is built on top of a suitable signature—called "polymorphic" signature in this context—describing the atomic ingredients of the calculus. Although these polymorphic signatures are the appropriate starting point for a completely formal presentation, we postpone their definition and introduce the rules of PTT first. It is better, we think, first to get acquainted with the systems, since these polymorphic signatures are somewhat complicated (two-level) structures.

First we are going to set up a calculus of kinds, types and terms. We shall write

$$\Xi = (\alpha_1 : A_1, \ldots, \alpha_n : A_n) \qquad \text{and} \qquad \Gamma = (x_1 : \sigma_1, \ldots, x_m : \sigma_m)$$

for **kind context** Ξ and **type context** Γ. A well-formed term $M : \tau$ with free type variables $\alpha_1 : A_1, \ldots, \alpha_n : A_n$ and free term variables $x_1 : \sigma_1, \ldots, x_m : \sigma_m$ takes the form

$$\alpha_1 : A_1, \ldots, \alpha_n : A_n \mid x_1 : \sigma_1, \ldots, x_m : \sigma_m \vdash M : \tau$$

where the sign '|' works as a separator between the kind context and the type context—just as in predicate logic, where it separates the type context and the proposition context there, see Section 3.1. In such a sequent it is assumed that the σ_i and τ are well-formed types in kind context $\alpha_1 : A_1, \ldots, \alpha_n : A_n$, which we shall write in explicit form as

$$\alpha_1 : A_1, \ldots, \alpha_n : A_n \vdash \sigma_i : \mathsf{Type}.$$

We sometimes write $\sigma(\vec{\alpha})$ instead of σ in order to see the free type variables $\vec{\alpha}$ in a type σ explicit. Similarly, we also write $M(\vec{\alpha}, \vec{x})$ to make both the type and term variables $\vec{\alpha}$ and $\vec{x}$ explicit in a term M.

Here are two examples of terms:

$$\alpha : \mathsf{Type} \mid f : \alpha \to \alpha, x : \alpha \vdash f(f(x)) : \alpha \qquad\qquad \beta : \mathsf{Type} \mid x : \beta \vdash \mathbf{I}\beta x : \beta,$$

where in the latter case $\mathbf{I}$ is the polymorphic identity $\lambda\alpha : \mathsf{Type}.\, \lambda x : \alpha.\, x : \Pi\alpha : \mathsf{Type}.\, (\alpha \to \alpha)$.

Substitution of terms inhabiting kinds for variables of appropriate kinds will provide indexed categorical structure. This will be described in more detail later, and at this stage we only suggest how this works. Suppose for a kind context $\Xi = (\alpha_1 : A_1, \ldots, \alpha_n : A_n)$ we have terms inhabiting the kinds $B_1, \ldots, B_m$ in Ξ, say

$$\Xi \vdash \sigma_1 : B_1, \qquad \cdots \qquad \Xi \vdash \sigma_m : B_m.$$

Then we can transfer types and terms in kind context $(\beta_1 : B_1, \ldots, \beta_m : B_m)$ to types and terms in context Ξ by substituting $\sigma_1, \ldots, \sigma_m$ for $\beta_1, \ldots, \beta_m$. This is done by

$$\tau(\vec{\beta}) \mapsto \tau[\vec{\sigma}/\vec{\beta}] \qquad \text{and} \qquad M(\vec{\beta}, \vec{x}) \mapsto M[\vec{\sigma}/\vec{\beta}, \vec{x}].$$

We have prepared the grounds so that we can describe the rules of the three polymorphic type theories $\lambda{\rightarrow}$, $\lambda 2$ and $\lambda\omega$. For the first two systems, $\lambda{\rightarrow}$ and $\lambda 2$, we require Type to be the sole kind, so that there are no rules for kind formation except the axiom

$$\vdash \mathsf{Type}\colon \mathsf{Kind}.$$

This is the type theoretic analogue of the higher order axiom $\vdash \mathsf{Prop}\colon \mathsf{Type}$ in higher order predicate logic. Categorically, the PTT axiom $\vdash \mathsf{Type}\colon\mathsf{Kind}$ will be captured by a generic object, which gives a correspondence between types $\alpha\colon A \vdash \sigma\colon\mathsf{Type}$ over $A\colon\mathsf{Kind}$ and "classifying" maps $\sigma\colon A \rightarrow \mathsf{Type}$ between kinds.

First order polymorphic type theory $\lambda{\rightarrow}$

In the system $\lambda{\rightarrow}$ of first order PTT we use finite product types $(1, \times)$ and exponent types $\rightarrow$ in every kind context $\Xi = (\alpha_1\colon\mathsf{Type}, \ldots, \alpha_n\colon\mathsf{Type})$. The rules for these type constructors are

$$\frac{}{\Xi \vdash 1\colon\mathsf{Type}} \qquad \frac{\Xi \vdash \sigma\colon\mathsf{Type} \quad \Xi \vdash \tau\colon\mathsf{Type}}{\Xi \vdash \sigma \times \tau\colon\mathsf{Type}} \qquad \frac{\Xi \vdash \sigma\colon\mathsf{Type} \quad \Xi \vdash \tau\colon\mathsf{Type}}{\Xi \vdash \sigma \rightarrow \tau\colon\mathsf{Type}}$$

plus the rules for finite tuples and projections, and abstraction and application terms. These are as in STT, see Section 2.3, except that the extra kind contexts are written. For example, the rules for abstraction and application are (essentially) the STT-rules:

$$\frac{\Xi \mid \Gamma, x\colon\sigma \vdash M\colon\tau}{\Xi \mid \Gamma \vdash \lambda x\colon\sigma.\, M\colon\sigma \rightarrow \tau} \qquad \frac{\Xi \mid \Gamma \vdash M\colon\sigma \rightarrow \tau \quad \Xi \mid \Gamma \vdash N\colon\sigma}{\Xi \mid \Gamma \vdash MN\colon\tau}$$

with conversions as in STT. Basically, $\lambda{\rightarrow}$ is $\lambda 1_\times$ with type variables.

Second order polymorphic type theory $\lambda 2$

Our second system $\lambda 2$ of PTT has new type constructors: products Π and sums Σ for forming (second order) product types $\Pi\alpha\colon\mathsf{Type}.\,\sigma$ and sum types $\Sigma\alpha\colon\mathsf{Type}.\,\sigma$, which bind the type variable $\alpha\colon\mathsf{Type}$ in σ. These products and sums thus have formation rules:

$$\frac{\Xi, \alpha\colon\mathsf{Type} \vdash \sigma\colon\mathsf{Type}}{\Xi \vdash \Pi\alpha\colon\mathsf{Type}.\,\sigma\colon\mathsf{Type}} \qquad \frac{\Xi, \alpha\colon\mathsf{Type} \vdash \sigma\colon\mathsf{Type}}{\Xi \vdash \Sigma\alpha\colon\mathsf{Type}.\,\sigma\colon\mathsf{Type}}$$

Associated with these new types there are rules for introducing new terms: they allow us to abstract over types via polymorphic functions $\lambda\alpha\colon\mathsf{Type}.\,M$, and to form polymorphic tuples $\langle\tau, M\rangle$ where τ is a type and M is a term of type $\sigma[\tau/\alpha]$. The use of these new terms is illustrated in the next section. Here we merely present the rules. The introduction rules for Π and Σ are:

$$\frac{\Xi, \alpha\colon \mathsf{Type} \mid \Gamma \vdash M\colon \sigma}{\Xi \mid \Gamma \vdash \lambda\alpha\colon \mathsf{Type}.\, M\colon \Pi\alpha\colon \mathsf{Type}.\, \sigma} \quad (\alpha \text{ not in } \Gamma)$$

$$\frac{\Xi \vdash \tau\colon \mathsf{Type} \qquad \Xi \mid \Gamma \vdash M\colon \sigma[\tau/\alpha]}{\Xi \mid \Gamma \vdash \langle \tau, M\rangle\colon \Sigma\alpha\colon \mathsf{Type}.\, \sigma}$$

An the elimination rules are:

$$\frac{\Xi \mid \Gamma \vdash M\colon \Pi\alpha\colon \mathsf{Type}.\, \sigma \qquad \Xi \vdash \tau\colon \mathsf{Type}}{\Xi \mid \Gamma \vdash M\tau\colon \sigma[\tau/\alpha]}$$

$$\frac{\Xi \vdash \rho\colon \mathsf{Type} \qquad \Xi, \alpha\colon \mathsf{Type} \mid \Gamma, x\colon \sigma \vdash N\colon \rho}{\Xi \mid \Gamma, z\colon \Sigma\alpha\colon \mathsf{Type}.\, \sigma \vdash \mathsf{unpack}\ z\ \mathsf{as}\ \langle \alpha, x\rangle\ \mathsf{in}\ N\colon \rho} \quad (\alpha \text{ not in } \Gamma, \rho)$$

In the term unpack z as $\langle \alpha, x\rangle$ in N in the latter rule, the type and term variable α and x in N become bound. They are linked, as a tuple, to z in N. Possible alternatives for this "unpack" notation use "let" and "where" as in:

$$\mathsf{let}\ \langle \alpha, x\rangle := z\ \mathsf{in}\ N, \qquad N\,\mathsf{where}\ \langle \alpha, x\rangle := z$$

The associated conversions are

$$(\lambda\alpha\colon \mathsf{Type}.\, M)\,\tau \;=\; M[\tau/\alpha] \qquad (\beta)$$

$$\lambda\alpha\colon \mathsf{Type}.\, M\alpha \;=\; M \qquad (\eta)$$

$$\mathsf{unpack}\ \langle \tau, M\rangle\ \mathsf{as}\ \langle \alpha, x\rangle\ \mathsf{in}\ N \;=\; N[\tau/\alpha, M/x] \qquad (\beta)$$

$$\mathsf{unpack}\ M\ \mathsf{as}\ \langle \alpha, x\rangle\ \mathsf{in}\ N[\langle \alpha, x\rangle/z] \;=\; N[M/z] \qquad (\eta)$$

To be more precise, with explicit contexts, types and restrictions, these conversions read as follows.

$$\frac{\Xi, \alpha\colon \mathsf{Type} \mid \Gamma \vdash M\colon \sigma \qquad \Xi \vdash \tau\colon \mathsf{Type}}{\Xi \mid \Gamma \vdash (\lambda\alpha\colon \mathsf{Type}.\, M)\tau = M[\tau/\alpha]\colon \sigma[\tau/\alpha]} \quad (\alpha \text{ not in } \Gamma)$$

$$\frac{\Xi \mid \Gamma \vdash M\colon \Pi\alpha\colon \mathsf{Type}.\, \sigma}{\Xi \mid \Gamma \vdash \lambda\alpha\colon \mathsf{Type}.\, M\alpha = M\colon \Pi\alpha\colon \mathsf{Type}.\, \sigma}$$

$$\frac{\Xi \vdash \tau\colon \mathsf{Type} \quad \Xi \mid \Gamma \vdash M\colon \sigma[\tau/\alpha] \quad \Xi, \alpha\colon \mathsf{Type} \mid \Gamma, x\colon \sigma \vdash N\colon \rho}{\Xi \mid \Gamma \vdash (\mathsf{unpack}\ \langle \tau, M\rangle\ \mathsf{as}\ \langle \alpha, x\rangle\ \mathsf{in}\ N) = N[\tau/\alpha, M/x]\colon \rho} \quad (\alpha \notin \Gamma, \rho)$$

$$\frac{\Xi \mid \Gamma \vdash M\colon \Sigma\alpha\colon \mathsf{Type}.\, \sigma \qquad \Xi \mid \Gamma, z\colon \Sigma\alpha\colon \mathsf{Type}.\, \sigma \vdash N\colon \rho}{\Xi \mid \Gamma \vdash (\mathsf{unpack}\ M\ \mathsf{as}\ \langle \alpha, x\rangle\ \mathsf{in}\ N[\langle \alpha, x\rangle/z]) = N[M/z]\colon \rho}$$

Higher order polymorphic type theory $\lambda\omega$

In our final system $\lambda\omega$ the requirement that Type is the only kind is dropped: in $\lambda\omega$ one can have more atomic kinds A: Kind than just Type. Additionally, there are rules for forming finite products and exponents of kinds. Thus one can form 1: Kind, $A \times B$: Kind and $A \to B$: Kind, for A, B: Kind, regulated by the STT-rules for $(1, \times, \to)$ as in Section 2.3. The calculus $\lambda\omega$ has all features of $\lambda2$ with (higher order) polymorphic products and sums $\Pi\alpha\colon A.\,\sigma$ and $\Sigma\alpha\colon A.\,\sigma$ over all kinds A: Kind (and not just over Type: Kind). Since this gives us conversions for inhabitants of kinds A: Kind, we have in particular conversions for types σ: Type. This calls for a rule which tells that type-inhabitation is stable under conversion:

conversion

$$\frac{\Gamma \vdash M\colon \sigma \qquad \Gamma \vdash \sigma = \tau\colon \mathsf{Type}}{\Gamma \vdash M\colon \tau}$$

This rule forms part of higher order polymorphic type theory $\lambda\omega$.

We summarise the type and kind constructors in the following table.

PTT	kinds	types
$\lambda\to$	Type	$1,\ \sigma \times \tau,\ \sigma \to \tau$
$\lambda2$	Type	$1,\ \sigma \times \tau,\ \sigma \to \tau,$ $\Pi\alpha\colon \mathsf{Type}.\,\sigma,\ \Sigma\alpha\colon \mathsf{Type}.\,\sigma$
$\lambda\omega$	Type, $1,\ A \times B,\ A \to B$	$1,\ \sigma \times \tau,\ \sigma \to \tau,$ $\Pi\alpha\colon A.\,\sigma,\ \Sigma\alpha\colon A.\,\sigma$

As we mentioned in the beginning, the proper starting point for a specific polymorphic calculus is a polymorphic signature, containing basic kinds and types together with function symbols for these. Such a polymorphic signature consists of two connected levels of ordinary signatures. It involves a combination of the higher order signatures used in higher order logic, and the ordinary signatures used in simple type theory.

Here is an example of what we may wish to specify in a polymorphic signature for a first or second order calculus. Remember that there are no kinds other than Type (for first and second order), so there are only function symbols in a kind signature:

$$\mathsf{List}\colon \mathsf{Type} \longrightarrow \mathsf{Type}, \qquad\qquad \mathsf{Tree}\colon \mathsf{Type}, \mathsf{Type} \longrightarrow \mathsf{Type}.$$

These may come with function symbols in type signatures:

$$\text{nil:}\,() \longrightarrow \text{List}(\alpha), \qquad\qquad \text{cons:}\,\alpha, \text{List}(\alpha) \longrightarrow \text{List}(\alpha).$$

involving a type variable α: Type. And for trees one may wish to have function symbols

$$\text{nil:}\,() \longrightarrow \text{Tree}(\alpha, \beta), \qquad \text{node:}\,\alpha, \text{Tree}(\alpha, \beta), \text{Tree}(\alpha, \beta), \beta \longrightarrow \text{Tree}(\alpha, \beta).$$

involving two type variables α, β: Type. In general, one may have such a signature of function symbols between types for every sequence $\vec{\alpha}$ of type variables, *i.e.* for every kind context. Here is the general notion to capture such structures.

8.1.1. Definition. (i) A **polymorphic signature** consists of

(1) a higher order signature Σ; we call the elements of the underlying set $K = |\Sigma|$ **(atomic) kinds** and write Type for the base point in $|\Sigma|$, see Definition 5.1.1.

(2) a collection of signatures $(\Sigma_a)_{a \in K^*}$ where for each sequence of kinds $a = \langle A_1, \ldots, A_n \rangle \in K^*$, the underlying set $|\Sigma_a|$ of types of the signature Σ_a is the set of Σ-terms $\alpha_1 \colon A_1, \ldots, \alpha_n \colon A_n \vdash \sigma \colon$ Type that can be built with the kind signature from (1). Elements of this set $|\Sigma_a|$ will therefore be called **types** with free type variables $\alpha_1 \colon A_1, \ldots, \alpha_n \colon A_n$.

In such a polymorphic signature $\langle \Sigma, (\Sigma_a) \rangle$ we shall refer to Σ as the **kind signature** and to Σ_a as the **type signature over** $a \in |\Sigma|^*$.

(ii) For the first and second order polymorphic calculi $\lambda{\to}$ and $\lambda 2$ one restricts oneself to polymorphic signatures $\langle \Sigma, (\Sigma_a) \rangle$ where the kind signature Σ is single-typed—or better, single-kinded; that is, the underlying set $|\Sigma|$ is {Type}. The type signatures are then of the form $(\Sigma_n)_{n \in \mathbb{N}}$ (as in the above example).

A specific polymorphic calculus may now be written as $\lambda \Diamond (\mathcal{S})$ where $\Diamond$ is $\to, 2$ or ω, and $\mathcal{S} = \langle \Sigma, (\Sigma_a) \rangle$ is an appropriate polymorphic signature.

Equality in polymorphic type theory

One can also extend these polymorphic λ-calculi with **equality types** $\text{Eq}_A(\sigma, \tau) \colon$ Type for $\sigma, \tau \colon A$ where $A \colon$ Kind. Inhabitation of such a type $\text{Eq}_A(\sigma, \tau)$ is intended to mean that σ and τ are equal terms of kind A. This is like internal equality in equational logic or in predicate logic. Such equality types only really make sense for the higher order calculus $\lambda\omega$, because only $\lambda\omega$ involves a non-trivial subcalculus of terms for kinds. However, equality types can also be added to the calculi $\lambda{\to}$ and $\lambda 2$—in which case the only kind is Type. We shall write $\lambda{\to}=$, $\lambda2=$ and $\lambda\omega=$ for the calculi $\lambda{\to}$, $\lambda2$ and $\lambda\omega$ extended with the following rules for equality types.

$$\frac{\Xi \vdash \sigma\colon A \qquad \Xi \vdash \tau\colon A}{\Xi \vdash \operatorname{Eq}_A(\sigma, \tau)\colon \mathsf{Type}} \qquad\qquad \frac{\Xi \vdash \sigma\colon A}{\Xi \mid \emptyset \vdash r_A(\sigma)\colon \operatorname{Eq}_A(\sigma, \sigma)}$$

$$\frac{\Xi, \alpha\colon A, \beta\colon A \vdash \rho\colon \mathsf{Type} \qquad \Xi, \alpha\colon A \mid \Gamma[\alpha/\beta] \vdash N\colon \rho[\alpha/\beta]}{\Xi, \alpha\colon A, \beta\colon A \mid \Gamma, z\colon \operatorname{Eq}_A(\alpha, \beta) \vdash N \text{ with } \beta = \alpha \text{ via } z\colon \rho}$$

with (β)- and (η)-conversions:

$$\frac{\Xi, \alpha\colon A \mid \Gamma[\alpha/\beta] \vdash N\colon \rho[\alpha/\beta]}{\Xi, \alpha\colon A \mid \Gamma[\alpha/\beta] \vdash (N \text{ with } \alpha = \alpha \text{ via } r_A(\alpha)) = N\colon \rho[\alpha/\beta]}$$

$$\frac{\Xi, \alpha\colon A, \beta\colon A \vdash \rho\colon \mathsf{Type} \qquad \Xi, \alpha\colon A, \beta\colon A \mid \Gamma, z\colon \operatorname{Eq}(\alpha, \beta) \vdash L\colon \rho}{\Xi, \alpha\colon A, \beta\colon A \mid \Gamma, z\colon \operatorname{Eq}(\alpha, \beta) \vdash (L[\alpha/\beta, r_A(\alpha)/z]) \text{ with } \beta = \alpha \text{ via } z) = L\colon \rho}$$

In the above introduction rule there is a proof-term $r = r_A(\sigma)$ for reflexivity of equality on A, for $\sigma\colon A$. Often we omit the A and σ when they clear from the context. In the next lemma we show that similar proof-terms for symmetry, transitivity and replacement are definable.

8.1.2. Lemma. *There are proof-terms*

$$\alpha, \beta\colon A \mid x\colon \operatorname{Eq}_A(\alpha, \beta) \;\vdash\; s(x)\colon \operatorname{Eq}_A(\beta, \alpha)$$
$$\alpha, \beta, \gamma\colon A \mid x\colon \operatorname{Eq}_A(\alpha, \beta), y\colon \operatorname{Eq}_A(\beta, \gamma) \;\vdash\; t(x, y)\colon \operatorname{Eq}_A(\alpha, \gamma)$$
$$\Xi \mid \Gamma, x\colon \operatorname{Eq}_A(\sigma, \tau), y\colon \rho[\sigma/\alpha] \;\vdash\; \operatorname{rep}(x, y)\colon \rho[\tau/\alpha]$$

yielding combinators for symmetry, transitivity and replacement for the polymorphic equality type Eq. *Moreover, types indexed by equal type variables are equal: there is a proof-term*

$$\alpha, \beta\colon A \mid x\colon \operatorname{Eq}_A(\alpha, \beta) \vdash i(x)\colon \operatorname{Eq}_{\mathsf{Type}}(\rho[\alpha/\gamma], \rho[\beta/\gamma]).$$

Proof. A proof-term for symmetry is obtained in:

$$\frac{\alpha, \beta\colon A \vdash \operatorname{Eq}_A(\alpha, \beta)\colon \mathsf{Type} \qquad \alpha\colon A \mid \emptyset \vdash r_A(\alpha)\colon \operatorname{Eq}_A(\beta, \alpha)[\alpha/\beta]}{\alpha\colon A, \beta\colon A \mid x\colon \operatorname{Eq}_A(\alpha, \beta) \vdash r_A(\alpha) \text{ with } \beta = \alpha \text{ via } z\colon \operatorname{Eq}_A(\beta, \alpha)}$$

And for transitivity in:

$$\frac{\alpha, \beta, \gamma\colon A \vdash \operatorname{Eq}_A(\alpha, \gamma)\colon \mathsf{Type} \qquad \alpha, \beta\colon A \mid x\colon \operatorname{Eq}_A(\alpha, \beta) \vdash x\colon \operatorname{Eq}_A(\alpha, \gamma)[\beta/\gamma]}{\alpha, \beta, \gamma\colon A \mid x\colon \operatorname{Eq}_A(\alpha, \beta), y\colon \operatorname{Eq}_A(\beta, \gamma) \vdash x \text{ with } \gamma = \beta \text{ via } y\colon \operatorname{Eq}_A(\alpha, \gamma)}$$

For replacement, consider the following instantiation of the elimination rule.

$$\frac{\Xi, \alpha, \beta\colon A \vdash \rho[\beta/\alpha]\colon \mathsf{Type} \qquad \Xi, \alpha\colon A \mid x\colon \rho \vdash x\colon \rho[\beta/\alpha][\alpha/\beta]}{\Xi, \alpha, \beta\colon A \mid x\colon \rho, y\colon \mathrm{Eq}_A(\alpha, \beta) \vdash x \text{ with } \beta = \alpha \text{ via } y\colon \rho[\beta/\alpha]}$$

By substituting $[\sigma/\alpha, \tau/\beta]$ we obtain a term $\mathsf{rep}(x, y) = x$ with $\tau = \sigma$ via $y\colon \rho[\tau/\alpha]$ as required. Finally, to see that equal index variables yield equal indexed types, consider for $\Xi, \gamma\colon A \vdash \rho\colon \mathsf{Type}$ the following instance of the equality elimination rule.

$$\frac{\Xi, \alpha\colon A \mid \emptyset \vdash r_{\mathsf{Type}}(\rho[\beta/\gamma])\colon \mathrm{Eq}_{\mathsf{Type}}(\rho[\alpha/\gamma], \rho[\beta/\gamma])[\alpha/\beta]}{\begin{array}{l}\Xi, \alpha\colon A, \beta\colon A \mid x\colon \mathrm{Eq}_A(\alpha, \beta) \\ \qquad \vdash r_{\mathsf{Type}}(\rho[\beta/\gamma]) \text{ with } \beta = \alpha \text{ via } x\colon \mathrm{Eq}_{\mathsf{Type}}(\rho[\alpha/\gamma], \rho[\beta/\gamma])\end{array}} \qquad \square$$

Propositions as types

In the beginning of the section we spoke of a formal similarity between predicate logic (as in Chapter 4) and polymorphic type theory. This takes the form of a propositions-as-types correspondence, like in Section 2.3 between propositional logic and simple type theory:

predicate logic		polymorphic type theory
propositions	as	types
types	as	kinds

Indeed, one can view a type σ in predicate logic as a kind $\widehat{\sigma}$ in polymorphic type theory and a proposition φ as a type $\widehat{\varphi}$. Under this translation the propositional connectives $\top, \wedge, \supset$ become the type constructors $1, \times, \rightarrow$, and the quantifiers $\forall x\colon \sigma.\, \varphi, \exists x\colon \sigma.\, \varphi$ become the polymorphic product and sum $\Pi\alpha\colon \widehat{\sigma}.\, \widehat{\varphi}.\, \Sigma\alpha\colon \widehat{\sigma}.\, \widehat{\varphi}$. At the level of kinds we take $\widehat{\mathsf{Prop}} = \mathsf{Type}$, $\widehat{\mathsf{Type}} = \mathsf{Kind}$, and the type constructors $1, \times, \rightarrow$ in predicate logic become the kind constructors $1, \times, \rightarrow$ in polymorphic type theory. Then one can prove that

$$x_1\colon \sigma_1, \ldots, x_n\colon \sigma_n \mid \varphi_1, \ldots, \varphi_m \vdash \psi$$

is derivable in higher order predicate logic

if and only if $\qquad (*)$

there is a term $\alpha_1\colon \widehat{\sigma}_1, \ldots \alpha_n\colon \widehat{\sigma}_n \mid z_1\colon \widehat{\varphi}_1, \ldots, z_m\colon \widehat{\varphi}_m \vdash M\colon \widehat{\psi}$ in $\lambda\omega{=}$

This establishes a typical propositions-as-types relation between provability in logic and inhabitation in type theory.

For example, consider the proposition

$$\left(\exists x\colon\sigma.\,\varphi\wedge\psi\right)\supset\left(\exists x\colon\sigma.\,\varphi\right)\wedge\left(\exists x\colon\sigma.\,\psi\right)$$

together with its translation into type theory:

$$\left(\Sigma x\colon\widehat{\sigma}.\,\widehat{\varphi}\times\widehat{\psi}\right)\rightarrow\left(\Sigma x\colon\widehat{\sigma}.\,\widehat{\varphi}\right)\times\left(\Sigma x\colon\widehat{\sigma}.\,\widehat{\psi}\right).$$

Obviously, the proposition is derivable: take an $x\colon\sigma$ with $\varphi\wedge\psi$. Then this same x can be used to establish φ, and also to establish ψ. This proof outline is recognisable in the following term, inhabiting the translated proposition:

$$\lambda z\colon(\Sigma x\colon\widehat{\sigma}.\,\widehat{\varphi}\wedge\widehat{\psi}).\,\mathsf{unpack}\ z\ \mathsf{as}\ \langle x,y\rangle\ \mathsf{in}\ \langle\langle x,\pi y\rangle,\langle x,\pi'y\rangle\rangle.$$

A few remarks are in order about the details of the propositions-as-types correspondence $(*)$.

(1) One can prove the correspondence by annotating the rules for predicate logic in Figure 4.1 with appropriate proof-terms. For example, the deduction step in logic

$$\frac{x_1\colon\sigma_1,\ldots,x_n\colon\sigma_n,y\colon\tau\mid\varphi_1,\ldots,\varphi_m\vdash\psi}{x_1\colon\sigma_1,\ldots,x_n\colon\sigma_n\mid\varphi_1,\ldots,\varphi_m\vdash\forall y\colon\tau.\,\psi}\ (y\ \text{not in}\ \vec{\varphi})$$

becomes in type theory

$$\frac{\alpha_1\colon\widehat{\sigma}_1,\ldots\alpha_n\colon\widehat{\sigma}_n,\beta\colon\widehat{\tau}\mid z_1\colon\widehat{\varphi}_1,\ldots,z_m\colon\widehat{\varphi}_m\vdash M\colon\widehat{\psi}}{\alpha_1\colon\widehat{\sigma}_1,\ldots\alpha_n\colon\widehat{\sigma}_n\mid z_1\colon\widehat{\varphi}_1,\ldots,z_m\colon\widehat{\varphi}_m\mid\lambda\beta\colon\widehat{\tau}.\,M\colon\Pi\beta\colon\widehat{\tau}.\,\widehat{\psi}=\widehat{\forall y\colon\tau.\,\psi}}$$

Since we have not standardly included rules for finite coproduct types $(+,0)$ in our polymorphic calculi, we have to restrict ourselves to the fragment of higher order predicate logic without disjunctions. Of course, such finite coproduct types can be added in PTT as well, following the description in Section 2.3.

Lemma 8.1.2 gives proof-terms for the equality rules of predicate logic in Figure 4.1. In PTT we do not have the extensionality of entailment rule (see Section 5.1). So the above propositions-as-types correspondence involves higher order predicate logic without extensionality of entailment.

(2) A higher order predicate logic is built on top of a higher order signature plus a number of axioms. Due to the way that we have defined polymorphic signatures, these axioms can be translated into a polymorphic signature only if they involve solely predicates and no connectives. For each such an axiom

$$\vec{x}\colon\vec{\sigma}\mid P_1(\vec{x}),\ldots,P_m(\vec{x})\vdash P_{m+1}(\vec{x})$$

one postulates a function symbol

$$F\colon\widehat{P}_1,\ldots,\widehat{P}_m\longrightarrow\widehat{P}_{m+1}$$

in the type signature over $\langle\widehat{\sigma}_1,\ldots,\widehat{\sigma}_n\rangle$. This function symbol serves as 'atomic proof term' for the axiom.

In order to accommodate for all axioms—and not just these restricted ones—we should adapt the notion of polymorphic signature in such a way that it can also have function symbols between composite kinds and types involving $\rightarrow, \times, 1$ and $\Pi, \Sigma, \rightarrow, \times, 1, \mathsf{Eq}$.

(3) There is a border case of this propositions-as-types correspondence which should be mentioned separately. Propositional logic can be seen as a degenerated form of predicate logic in which all predicates are closed (*i.e.* do not contain free term variables and are thus merely propositions). This means that higher order logic without atomic predicate symbols (and without equations) is higher order propositional logic. If one further restricts oneself in logic to second order quantification only (*i.e.* of the form $\forall \alpha \colon \mathsf{Prop}. \varphi$) then one can get a correspondence between derivability in this logic and inhabitation in "pure" second order polymorphic calculus $\lambda 2(\emptyset)$ on the empty signature. This correspondence occurs in [14].

Exercises

8.1.1. Show that in $\lambda 2$ one can assign the type $\Pi \alpha \colon \mathsf{Type}. (\Pi \beta \colon \mathsf{Type}. \beta) \rightarrow \alpha$ to the self-application term $\lambda x. xx$.

8.1.2. Write down how substitution, both in type variables and in term variables, distributes properly over the type constructors Π, Σ, Eq.

8.1.3. Prove the 'mate' versions of the rules for Π and Σ. That is, assume α is not in Γ, ρ below and establish bijective correspondences between terms M and N in

$$\frac{\Xi, \alpha \colon A \mid \Gamma \vdash M \colon \sigma}{\Xi \mid \Gamma \vdash N \colon \Pi \alpha \colon A. \sigma} \qquad \frac{\Xi, \alpha \colon A \mid \Gamma, x \colon \sigma \vdash M \colon \rho}{\Xi \mid \Gamma, z \colon \Sigma \alpha \colon A. \sigma \vdash N \colon \rho}$$

This shows that $\Pi \alpha \colon A. -$ and $\Sigma \alpha \colon A. -$ are right and left adjoints to the weakening functor which adds an extra variable $\alpha \colon A$ to $\Xi \vdash \rho \colon \mathsf{Type}$.

8.1.4. Formulate and derive a similar mate version for equality.

8.1.5. (i) In $\lambda 2$ (and in $\lambda \omega$) a sum type Σ_d is definable in terms of Π and $\rightarrow$: for $\Xi, \alpha \colon \mathsf{Type} \vdash \sigma \colon \mathsf{Type}$, put

$$\Sigma_d \alpha \colon \mathsf{Type}. \sigma \stackrel{\mathrm{def}}{=} \Pi \beta \colon \mathsf{Type}. (\Pi \alpha \colon \mathsf{Type}. \sigma \rightarrow \beta) \rightarrow \beta$$

Establish that there are introduction and elimination rules for Σ_d as for Σ, but that only the (β)-conversion rule holds.

(ii) Show that likewise finite product and coproduct types are definable:

$$1_d \stackrel{\mathrm{def}}{=} \Pi \alpha \colon \mathsf{Type}. (\alpha \rightarrow \alpha)$$

$$0_d \stackrel{\mathrm{def}}{=} \Pi \alpha \colon \mathsf{Type}. \alpha$$

$$\sigma \times_d \tau \stackrel{\mathrm{def}}{=} \Pi \alpha \colon \mathsf{Type}. (\sigma \rightarrow \tau \rightarrow \alpha) \rightarrow \alpha$$

$$\sigma +_d \tau \stackrel{\mathrm{def}}{=} \Pi \alpha \colon \mathsf{Type}. (\sigma \rightarrow \alpha) \rightarrow (\tau \rightarrow \alpha) \rightarrow \alpha$$

[These definitions are the type theoretic versions of the ones we already saw in higher order logic, see Example 5.1.5. If one is interested in (β)-conversions only, these translations show that it is sufficient to restrict oneself to versions of $\lambda 2$ and $\lambda \omega$ with only $\to$ and Π, as is done for example in [14]. But the (η)-conversion does not hold, since the type $\Sigma_d \alpha$: Type. σ contains "junk": *e.g.* for $\sigma \equiv \alpha$ it contains the term $\lambda \beta$: Type. λf: $\Pi \alpha$: Type. $(\alpha \to \beta)$. $f(\beta \to \beta)(\lambda x: \beta. x)$, which is not (convertible to) a pair.]

8.1.6. Along the same lines, describe the rules for Leibniz equality in $\lambda \omega$, see Example 5.1.5. (Here one needs the conversion rule of $\lambda \omega$. And the "non-Frobenius" version of equality—with type contexts Γ instead of $\Gamma[\alpha/\beta]$ in the elimination rule—is obtained. But using arrow types one gets the Frobenius version, as above.)

8.1.7. Show that the (η)-conversion unpack M as $\langle \alpha, x \rangle$ in $N[\langle \alpha, x \rangle/z] = N[M/z]$ for the polymorphic sum Σ is equivalent to the combination of the following two conversions, called (commutation) and (η').

$$L[(\text{unpack } M \text{ as } \langle \alpha, x \rangle \text{ in } N)/z] \;=\; \text{unpack } M \text{ as } \langle \alpha, x \rangle \text{ in } L[N/z]$$
$$\text{unpack } \langle \alpha, x \rangle \text{ as } \langle \alpha, x \rangle \text{ in } M \;=\; M.$$

[Non-extensional polymorphic sums (as described by 'semi-adjunctions', see [155]) satisfy the commutation conversion, but not (η'). The above definable sums Σ_d in Exercise 8.1.5 are even weaker: they do not satisfy (commutation).]

8.1.8. Consider the proof-terms in Lemma 8.1.2. Show that there are conversions:
$$\alpha: A, \beta: A \mid z: \text{Eq}_A(\alpha, \beta) \vdash t(z, r_A(\beta)) = z: \text{Eq}(\alpha, \beta)$$
$$\alpha: A, \beta: A \mid z: \text{Eq}_A(\alpha, \beta) \vdash t(r_A(\alpha), z) = z: \text{Eq}(\alpha, \beta).$$

8.1.9. Prove the propositions-as-types correspondence relating provability in higher order logic and inhabitation in higher order polymorphic λ-calculus $\lambda \omega =$ (under the restrictions as mentioned in (1), (2) above).

8.1.10. In Section 2.3 it was shown how (β)- and (η)-conversions correspond under the propositions-as-types reading to certain identifications on derivations.
(i) Write down similar identifications involving polymorphic product Π and sum Σ.
(ii) Do the same for equality Eq.

8.1.11. Define an appropriate notion of 'morphism of polymorphic signatures'.

8.2 Use of polymorphic type theory

In this section we briefly discuss three aspects of the use of polymorphic type theory, namely:

- the polymorphic type system of the functional programming language ML;

- encoding of inductively and co-inductively defined types in second order polymorphic type theory $\lambda 2$;
- encoding of abstract types and of classes (as in object-oriented programming) in $\lambda 2$, by encapsulation via sum types Σ.

ML-style polymorphism

As already mentioned briefly in the introduction to this chapter, the type system of the functional programming language ML is loosely based on first order polymorphic type theory $\lambda \rightarrow$, see [223, 224, 251, 108, 228]. But there are certain differences between the type systems.

(1) ML is a calculus which is implicitly—and not explicitly—typed (like the calculi in this book). In the terminology of [14], ML has typing *à la* Curry, whereas here we consider typing *à la* Church. This means that ML-terms are essentially untyped terms, to which one can assign types 'on the outside'. These types however, are not part of the syntax of terms. This difference will not be emphasised here.

(2) In ML one distinguishes *types* and *type schemes*. The types are our $\lambda \rightarrow$-types (if we forget about inductively defined types for a moment), and the type schemes are of the form

$$\Pi \vec{\alpha} : \mathsf{Type}. \, \sigma$$

where σ is a type and $\vec{\alpha}$ is a sequence of type variables. Hence one does have Π-quantifiers inside these type schemes, but only on the outside. There is an instantiation rule as in $\lambda 2$: if a term M is of type scheme $\Pi \alpha : \mathsf{Type}. \, \sigma$, then M has type $\sigma[\tau / \alpha]$ for every type τ. Notice that one can instantiate with types, but not with type schemes, because that would lead to quantifiers inside. This is called "ML-style polymorphism".

The crucial difference in the way that types and type schemes are used is that term variables of types may be bound by λ whereas variables of type schemes may be bound by let only. We consider two examples. First, assume a polymorphic term $\mathsf{len} : \Pi \alpha : \mathsf{Type}. \, (\mathsf{list}(\alpha) \rightarrow \mathsf{N})$ which gives the length of a list. Then one can derive the type

$$\mathsf{list}(\beta) \rightarrow \mathsf{list}(\gamma) \rightarrow \mathsf{N}$$

for the (untyped) term

$$\lambda x. \, \lambda y. \, \mathsf{if} \, \mathsf{len}(x) \leq 100 \, \mathsf{then} \, \mathsf{len}(y) \, \mathsf{else} \, 100$$

Notice that one needs to instantiate the type of len twice for this type assignment: once with β and once with γ. The above mechanism with Π's on the outside takes care of this. And the abstractions λx and λy are over types: in the first case, x has type $\mathsf{list}(\beta)$ and in the second case, y is of type $\mathsf{list}(\gamma)$. But note that if we wish to λ-abstract len in this term as in

$$\lambda f. \lambda x. \lambda y. \text{ if } (fx) \leq 100 \text{ then } (fy) \text{ else } 100,$$

then we cannot use a Π-type for f. However, for this term we can derive the (less general) type

$$(\alpha \to \mathsf{N}) \to \alpha \to \alpha \to \mathsf{N}.$$

Or, by abstracting α,

$$\Pi\alpha\colon \mathsf{Type}. \, (\alpha \to \mathsf{N}) \to \alpha \to \alpha \to \mathsf{N}.$$

Secondly, one makes essential use of type schemes in typing the term

$$\mathsf{let} \, f = \lambda x. \, x \text{ in } ff.$$

One can type f with the type scheme $\Pi\alpha\colon \mathsf{Type}. \, (\alpha \to \alpha)$. The first f in the self-application ff can then be instantiated with $[(\sigma \to \sigma)/\alpha]$ and the second one with $[\sigma/\alpha]$.

This use of let is characteristic of what is sometimes called "ML-style" or "let" polymorphism. It is quite successful because there are algorithms which produce appropriate ML-types for untyped terms, see *e.g.* [222, 68]. A semantic study of ML (in realisability toposes) can be found in [265, 198].

Encoding of inductively and co-inductively defined data types

The standard encoding of the natural numbers $\mathbb{N}$ in the untyped λ-calculus uses the so-called Church numerals, see [13]. These encoded numerals are the terms $c_n = \lambda f x. \, f^{(n)} x$ for $n \in \mathbb{N}$. In $\lambda 2$ all these terms can be typed with type

$$\mathsf{Nat} = \Pi\alpha\colon \mathsf{Type}. \, (\alpha \to \alpha) \to \alpha \to \alpha.$$

It turns out that Nat comes with zero and successor terms

$$\mathsf{zero} \; = \; \lambda\alpha\colon \mathsf{Type}. \, \lambda f\colon \alpha \to \alpha. \, \lambda x\colon \alpha. \, x : \mathsf{Nat}$$

$$\mathsf{succ} \; = \; \lambda z\colon \mathsf{Nat}. \, \lambda\alpha\colon \mathsf{Type}. \, \lambda f\colon \alpha \to \alpha. \, \lambda x\colon \alpha. \, f(z \, \alpha \, f \, x) : \mathsf{Nat} \to \mathsf{Nat}$$

which form a "weak" natural numbers object (NNO): for terms $P\colon \sigma$ and $Q\colon \sigma \to \sigma$ there is a (not necessarily unique) mediating term $M\colon \mathsf{Nat} \to \sigma$ with $M\mathsf{zero} = P$ and $M \circ \mathsf{succ} = \mathsf{succ} \circ P$, *i.e.* $\lambda z\colon \mathsf{Nat}. \, M(\mathsf{succ}\, z) = \lambda z\colon \mathsf{Nat}. \, P(M \, z)$. Just take $M = \lambda z\colon \mathsf{Nat}. \, z \, \sigma \, Q \, P$. This weak NNO property is the essence of the encoding of the natural numbers.

Below we show how to encode more general inductive (and co-inductive) types in $\lambda 2$. Such types are given by appropriate signatures, introduced in Definition 2.3.7 as 'Hagino signatures': they consist of a single type $\sigma(X)$ involving a distinguished type variable X, built with finite product and coproduct types, together with either a constructor $\mathsf{constr}\colon \sigma(X) \to X$ or a destructor destr: $X \to \sigma(X)$. Their models are described in terms of initial algebras (for constructors) and terminal co-algebras (for destructors), see Section 2.6. Here it will be shown that such models always exist in second order polymorphic type theory $\lambda 2$, but that they are only "weakly" initial and "weakly" terminal. This means that given any other (co-)algebra there is always a mediating map, but it need not be unique. We shall describe the basics of this encoding, but we do not consider advanced topics like iteration of data types (as in CHARITY [53] and also [124]).

We first have to extend polymorphic type theory with finite coproduct types $(+, 0)$. Let us write $\lambda{\to}_+$, $\lambda 2_+$, $\lambda \omega_+$ for the calculi $\lambda{\to}$, $\lambda 2$, $\lambda \omega$ extended with coproducts $(0, +)$ in each kind context, *i.e.* with formation rules

$$\frac{}{\vdash 0: \mathsf{Type}} \qquad \frac{\Xi \vdash \sigma: \mathsf{Type} \qquad \Xi \vdash \tau: \mathsf{Type}}{\Xi \vdash \sigma + \tau: \mathsf{Type}}$$

and with introduction, elimination and conversion rules as in Section 2.3—but extended with kind contexts. In second and higher order type theory one can also use the encoded weak finite coproducts described in Exercise 8.1.5.

A Hagino type $\sigma(X)$ as above occurs in a polymorphic calculus as a type

$$\Xi, X: \mathsf{Type} \vdash \sigma: \mathsf{Type}$$

where the kind context Ξ contains the type variables $\vec{\alpha}$ in σ which are different from X. We assume that such a Hagino type is formed from constants and type variables using finite product and coproduct types. A constructor is then a term

$$\Xi, X: \mathsf{Type} \mid x: \sigma \vdash \mathsf{constr}: X \quad \text{or} \quad \Xi \mid \emptyset \vdash \mathsf{constr}: \Pi X: \mathsf{Type}. (\sigma \to X).$$

Dually, a destructor is a term

$$\Xi, X: \mathsf{Type} \mid x: X \vdash \mathsf{destr}: \sigma \quad \text{or} \quad \Xi \mid \emptyset \vdash \mathsf{destr}: \Pi X: \mathsf{Type}. (X \to \sigma).$$

Before we can describe algebras and co-algebras in $\lambda 2$ we have to explain where the underlying functors come from. They arise by substitution in types.

8.2.1. Lemma. *For each kind context Ξ (in $\lambda 2$ or $\lambda \omega$), let $\mathbb{C}(\Xi)$ be the category of types and terms in context Ξ: object of $\mathbb{C}(\Xi)$ are types $\Xi \vdash \tau: \mathsf{Type}$ and, since we have exponent types around, morphisms $\tau \to \rho$ may be described as equivalence classes $[M]$ of terms $\Xi \mid \emptyset \vdash M: \tau \to \rho$. Each Hagino type $\Xi, X: \mathsf{Type} \vdash \sigma: \mathsf{Type}$ then induces an endofunctor $\sigma[-/X]: \mathbb{C}(\Xi) \to \mathbb{C}(\Xi)$ by*

$$\big(\Xi \vdash \tau: \mathsf{Type}\big) \longmapsto \big(\Xi \vdash \sigma[\tau/X]: \mathsf{Type}\big).$$

The proof proceeds by induction on the structure of the type σ. At this stage we are mostly interested in the cases where σ is built with finite products $\times$

and coproducts $+$ only. The result can be extended easily to include $\to, \Pi, \Sigma$ as type formers, see Exercise 8.2.5 below, but some care with positive and negative occurrences is needed for arrow types.

Proof. We need to describe the action of $\sigma[-/X]$ on morphisms. Therefore we distinguish the following cases.

- If $\sigma \equiv X$, then $\sigma[-/X]$ is simply the identity functor.
- If $\sigma \equiv \alpha$ occurring in Ξ, then $\sigma[-/X]$ is the functor which is constantly α.
- Similarly $0[-/X]$ is constantly 0 and $1[-/X]$ is constantly 1.
- Suppose $\sigma \equiv \sigma_1 \times \sigma_2$ and $[M]$ is a morphism $\tau \to \rho$ in $\mathbb{C}(\Xi)$. By induction hypothesis we have morphisms $\sigma_1[M/X] \colon \sigma_1[\tau/X] \to \sigma_1[\rho/X]$ and $\sigma_2[M/X] \colon \sigma_2[\tau/X] \to \sigma_2[\rho/X]$. We need to define a morphism

$$(\sigma_1 \times \sigma_2)[\tau/X] \equiv \sigma_1[\tau/X] \times \sigma_2[\tau/X] \longrightarrow \sigma_1[\rho/X] \times \sigma_2[\rho/X] \equiv (\sigma_1 \times \sigma_2)[\rho/X]$$

  We take:

$$\lambda z \colon \sigma_1[\tau/X] \times \sigma_2[\tau/X]. \, \langle \sigma_1[M/X] \cdot (\pi z), \, \sigma_2[M/X] \cdot (\pi' z) \rangle.$$

- Similarly, for $\sigma \equiv \sigma_1 + \sigma_2$ and $[M] \colon \tau \to \rho$ we take

$$(\sigma_1 + \sigma_2)[M/X] = \lambda z \colon \sigma_1[\tau/X] + \sigma_2[\tau/X].$$
$$\mathsf{unpack} \; z \; \mathsf{as} \; [\kappa x \; \mathsf{in} \; \kappa(\sigma_1[M/X] \cdot x), \kappa'y \; \mathsf{in} \; \kappa'(\sigma_2[M/X] \cdot y)]. \qquad \square$$

The category $\mathbb{C}(\Xi)$ that we use in this lemma will turn out to be the fibre category above kind context Ξ in the fibration associated with this polymorphic calculus, see Example 8.4.2 later on in this chapter.

The basis for the next result is [196, 35]. There it is shown how certain many-typed signatures can be expressed in $\lambda 2$ in a (weakly) free way. This result is reformulated in [346] in terms of existence of weakly initial algebras of "polynomial" functors $\sigma[-/X]$. The dual version involving co-algebras appears there as well; it is independently due to Hasegawa. A more general approach involving "expressible" functors may be found in [288].

8.2.2. Theorem. *Let $\Xi, X \colon \mathsf{Type} \vdash \sigma \colon \mathsf{Type}$ be a Hagino type. In second order polymorphic type theory, the induced functor $\sigma[-/X]$ has both a weakly initial algebra and a weakly terminal co-algebra.*

Proof. A weakly initial algebra $\mathsf{constr} \colon \sigma[\tau_0/X] \to \tau_0$ is obtained as follows. Put

$$\tau_0 \;=\; \Pi X \colon \mathsf{Type}.\, (\sigma \to X) \to X$$
$$\mathsf{constr} \;=\; \lambda x \colon \sigma[\tau_0/X].\, \lambda X \colon \mathsf{Type}.\, \lambda y \colon \sigma \to X.\; y\,(\sigma[(\lambda z \colon \tau_0.\, zXy)/X]\, x)$$

where in the second case we use the action of the functor $\sigma[-/X]$ on morphisms, as defined in the previous lemma. This constructor is well-typed, since the term $\lambda z \colon \tau_0 . zXy$ is of type $\tau_0 \to X$, so that $\sigma[(\lambda z \colon \tau_0 . zXy)/X]$ is of type $\sigma[\tau_0/X] \to \sigma$.

If we have an arbitrary algebra $\sigma[\tau/X] \to \tau$, given by a term $M \colon \sigma[\tau/X] \to \tau$, then we obtain a mediating morphism $\widehat{M} = \lambda x \colon \tau_0 . x\tau M \colon \tau_0 \to \tau$. It is an algebra morphism, since

$$
\begin{aligned}
\widehat{M} \circ \mathsf{constr} &= \lambda x \colon \sigma[\tau_0/X].\ \mathsf{constr}\ x\ \tau\ M \\
&= \lambda x \colon \sigma[\tau_0/X].\ M\left(\sigma[(\lambda z \colon \tau_0 . z\tau M)/X]\,x\right) \\
&= \lambda x \colon \sigma[\tau_0/X].\ M\left(\sigma[\widehat{M}/X]\,x\right) \\
&= M \circ \sigma[\widehat{M}/X].
\end{aligned}
$$

Similarly, there is a weakly terminal co-algebra $\mathsf{destr} \colon \tau_1 \to \sigma[\tau_1/X]$, where

$$
\tau_1 = \Sigma X \colon \mathsf{Type}.\, X \times (X \to \sigma).
$$

Before we define the associated term destr, we introduce for an arbitrary term $N \colon \tau \to \sigma[\tau/X]$ a term $\widehat{N} \colon \tau \to \tau_1$ by

$$
\widehat{N} = \lambda x \colon \tau.\, \langle \tau, \langle x, N \rangle \rangle.
$$

This will actually be the mediating map. The construction $\widehat{\ \ }$ is also used in the following definition of the weakly terminal co-algebra $\mathsf{destr} \colon \tau_1 \to \sigma[\tau_1/X]$.

$$
\mathsf{destr} = \lambda y \colon \tau_1.\, \mathsf{unpack}\ y\ \mathsf{as}\ \langle X, z \rangle\ \mathsf{in}\ \sigma[\widehat{\pi' z}/X]((\pi' z)(\pi z)).
$$

It forms a morphism of co-algebras:

$$
\begin{aligned}
\mathsf{destr} \circ \widehat{N} &= \lambda x \colon \tau.\, \mathsf{destr}(\widehat{N}\, x) \\
&= \lambda x \colon \tau.\, \mathsf{unpack}\ \langle \tau, \langle x, N \rangle \rangle\ \mathsf{as}\ \langle X, z \rangle\ \mathsf{in}\ \sigma[\widehat{\pi' z}/X]((\pi' z)(\pi z)) \\
&= \lambda x \colon \tau.\, \sigma[\widehat{N}/X](N\, x) \\
&= \sigma[\widehat{N}/X] \circ N. \hspace{3cm} \square
\end{aligned}
$$

We notice that the encoding of weakly initial algebras only uses the type constructors Π and $\to$. But also the encoding of weakly terminal co-algebras can be done entirely in terms of Π and $\to$, since the sum Σ in the definition $\tau_1 = \Sigma X \colon \mathsf{Type}.\, X \times (X \to \sigma)$ can be replaced by the definable sum Σ_{d} from Exercise 8.1.5. This yields a different formulation for τ_1 as:

$$
\tau_1 = \Pi\alpha \colon \mathsf{Type}.\, (\Pi X \colon \mathsf{Type}.\, (X \to \sigma) \to X \to \alpha) \to \alpha.
$$

Let us apply this Theorem 8.2.2 to our earlier example of the Church numerals, with type $\mathsf{Nat} = \Pi\alpha \colon \mathsf{Type}.\, (\alpha \to \alpha) \to \alpha \to \alpha$. Categorically one sees a natural numbers object as an initial algebra of the functor $X \mapsto 1 + X$. Taking the corresponding type $\sigma(X) = 1 + X$ yields, according to the definition of τ_0 in the proof,

$$\Pi X \colon \mathsf{Type}. \, ((1 + X) \to X) \to X)$$

Or equivalently, the type of the Church numerals:

$$\Pi X \colon \mathsf{Type}. \, (X \to X) \to X \to X$$

using the isomorphisms,

$$(1 + X) \to X \cong (1 \to X) \times (X \to X) \cong X \times (X \to X) \cong (X \to X) \times X$$

and Currying. Hence the general approach of the theorem gives an *a posteriori* justification for the type of the Church numerals.

We consider another, co-algebraic, example of this encoding. For a type $\mathsf{stream}(\alpha)$ of streams (or infinite sequences of elements) of type α we seek a pair of destructors $\langle \mathsf{head}, \mathsf{tail} \rangle \colon \mathsf{stream}(\alpha) \to \alpha \times \mathsf{stream}(\alpha)$ as (weakly) terminal co-algebra for the type $\sigma(X) = \alpha \times X$. The definition of τ_1 in the above proof yields

$$\mathsf{stream}(\alpha) \; = \; \Sigma X \colon \mathsf{Type}. \, X \times (X \to \alpha \times X).$$

The associated head and tail operations are given by

$$\mathsf{head} \; = \; \lambda y \colon \mathsf{stream}(\alpha). \, \mathsf{unpack} \; y \; \mathsf{as} \; \langle \alpha, x \rangle \; \mathsf{in} \; \pi\big((\pi' x)(\pi x)\big)$$
$$\mathsf{tail} \; = \; \lambda y \colon \mathsf{stream}(\alpha). \, \mathsf{unpack} \; y \; \mathsf{as} \; \langle \alpha, x \rangle \; \mathsf{in} \; \pi'\big((\pi' x)(\pi x)\big).$$

For terms $P \colon \sigma \to \alpha$ and $Q \colon \sigma \to \sigma$ we then get a mediating term $M \colon \sigma \to \mathsf{stream}(\alpha)$ by

$$M \; = \; \lambda x \colon \sigma. \, \langle x, \lambda y \colon \sigma. \, \langle P \, y, Q \, y \rangle \rangle. \; .$$

It is easy to see that $\mathsf{head} \circ M = P$ and $\mathsf{tail} \circ M = M \circ Q$. An application of this encoding of streams to a formulation of the sieve of Erastosthenes occurs in [195].

In [118, 273, 11] it is shown how under certain "parametricity" conditions the weakly initial algebras and weakly terminal co-algebras in Theorem 8.2.2 become truly initial and terminal (see also Exercise 8.4.5). This also works if there are enough equalisers around, see [288]. It is desirable to have the additional uniqueness of true initiality / terminality because it yields reasoning properties for these (co-)inductively defined data types. Weak initiality / terminality only gives definition principles, see [167].

Encapsulation via sum types

Let us consider a signature given by a finite number of atomic types $\alpha_1, \ldots, \alpha_n$ and a finite number of function symbols $F \colon \sigma_1 \longrightarrow \tau_1, \ldots, F_m \colon \sigma_m \longrightarrow \tau_m,$

where the σ's and τ's are built from the α's using finite products and coproducts (say). The $n + m$-tuple

$$\langle \alpha_1, \ldots, \alpha_n, F_1, \ldots, F_m \rangle$$

can then be seen as a term of the (second order) polymorphic sum type

$$\Sigma \alpha_1 : \mathsf{Type}. \cdots \Sigma \alpha_n : \mathsf{Type}. (\sigma_1 \to \tau_1) \times \cdots \times (\sigma_m \to \tau_m).$$

This type thus captures the structure of the signature that we started from. An arbitrary term $\langle \rho_1, \ldots, \rho_n, M_1, \ldots, M_m \rangle$ inhabiting this type can be seen as an instantiation (or model) of the signature.

To be more specific, the signature of monoids contains a unit function symbol $e: 1 \longrightarrow \alpha$ and a multiplication $m: \alpha \times \alpha \longrightarrow \alpha$. It gives rise to a type

$$\mu = \Sigma \alpha : \mathsf{Type}. \alpha \times ((\alpha \times \alpha) \to \alpha).$$

As a specific instantiation we can take the triple

$$\langle \sigma \to \sigma, \lambda x : \sigma. x, \lambda y : (\sigma \to \sigma) \times (\sigma \to \sigma). \lambda x : \sigma. \pi'y((\pi y)x) \rangle$$

describing the monoid of terms $\sigma \to \sigma$ for an arbitrary type σ. Notice that at present we have no means for expressing equalities like $m(x, e) = x$. Therefore one needs a logic over polymorphic type theory, just like the logic of equations and predicates in Chapters 3 and 4 is a logic over simply type theory. Elements of such a logic will be described in Section 8.6.

We thus see that a signature (a collection of types and function symbols) can be represented as a single sum type, in which the atomic types α_i are hidden. In this setting one often speaks of an **abstract type** instead of a signature, following [227]. Abstractness refers to a presentation without reference to any particular implementation. The following syntax is used.

$$\mathsf{abstype}\ \alpha\ \mathsf{with}\ x_1 : \sigma_1, \ldots, x_n : \sigma_n\ \mathsf{is}\ M\ \mathsf{in}\ N : \rho \qquad (*)$$

involving types and terms

$$\Xi, \alpha : \mathsf{Type} \vdash \sigma_i : \mathsf{Type}$$
$$\Xi \mid \Gamma \vdash M : \Sigma \alpha : \mathsf{Type}. (\sigma_1 \times \cdots \times \sigma_n)$$
$$\Xi \mid \Gamma, x_1 : \sigma_1, \ldots, x_n : \sigma_n \vdash N : \rho$$

with the restriction that the type variable α is not free in Γ, ρ.

In the syntax that we use for Σ-elimination, the abstype-term $(*)$ would be written as

$$\mathsf{unpack}\ M\ \mathsf{as}\ \langle \alpha, \langle x_1, \ldots, x_n \rangle \rangle\ \mathsf{in}\ N.$$

The idea is thus that M is a specific representation of the abstract types $\Sigma \alpha : \mathsf{Type}. (\sigma_1 \times \cdots \times \sigma_n)$ which is bound to the free variables $\alpha, x_1, \ldots, x_n$ in

N. As an example consider the type $\mu = \Sigma\alpha\colon \mathsf{Type}.\,\alpha \times ((\alpha \times \alpha) \to \alpha)$ we used earlier for the monoid function symbols. The operation which takes a monoid and replaces its multiplication $(x, y) \mapsto x \cdot y$ by $(x, y) \mapsto x^2 \cdot y^2$ can be described as the following term of type $\mu \to \mu$.

$\lambda z\colon \mu.\,\mathsf{abstype}\ \alpha\ \mathsf{with}\ e\colon \alpha, m\colon \alpha \times \alpha \to \alpha$

$\quad$ is z in $\langle \alpha, e, \lambda z\colon \alpha \times \alpha.\,m(m(\pi z)(\pi z))(m(\pi' z)(\pi' z)) \rangle$.

There is a related encoding of classes of objects in object-oriented programming using sum types Σ, see *e.g.* [266, 138, 1] (and the collection [109]). If one understands methods and attributes (object-oriented operations) as an "embedded" part of objects, then it is natural to encode an object as consisting of three parts: its (hidden) state space X, its current state as an element of X, and its operations as a co-algebra $X \to \sigma(X)$, where σ describes the interface of the object. These data are collected into the type

$$\Sigma X\colon \mathsf{Type}.\,X \times (X \to \sigma(X))$$

of the class of "objects with interface σ". Here, like for abstract types above, the state space X is encapsulated via type variable binding in sums.

Interestingly, objects with interface $\sigma(X)$ may be interpreted as inhabitants of the carrier τ_1 of the (weakly) terminal co-algebra of the associated functor $\sigma[-/X]$ as in the proof of Theorem 8.2.2—provided X occurs positively. This connection between co-algebras and object-orientation is further elaborated in [283, 138, 162, 164].

Exercises

8.2.1. Define addition and multiplication terms of type $\mathsf{Nat} \to \mathsf{Nat} \to \mathsf{Nat}$ for the Church numerals with type $\mathsf{Nat} = \Pi\alpha\colon \mathsf{Type}.\,(\alpha \to \alpha) \to \alpha \to \alpha$, via suitable mediating algebra maps $\mathsf{Add}(x)$, $\mathsf{Mult}(x)\colon \mathsf{Nat} \rightrightarrows \mathsf{Nat}$, using weak initiality.

8.2.2. Give the encoding in $\lambda 2$ of
   (i) Booleans (with constructors $\mathsf{true}\colon 1 \to \mathsf{bool}$ and $\mathsf{false}\colon 1 \to \mathsf{bool}$);
   (ii) finite lists of type α;
   (iii) binary trees of type α of finite depth;
   (iv) binary trees of type α of infinite depth;
   (v) finitely branching trees of type α which are of infinite depth.
   [For the last three examples, see *e.g.* [124].]

8.2.3. Consider the formulation of the carrier of the weakly terminal co-algebra τ_1 in terms of Π and $\to$,

$$\tau_1 = \Pi\alpha\colon \mathsf{Type}.\,(\Pi X\colon \mathsf{Type}.\,(X \to \sigma) \to X \to \alpha) \to \alpha.$$

as mentioned after the proof of Theorem 8.2.2. Define for this formulation an associated destructor destr: $\tau_1 \rightarrow \sigma[\tau_1/X]$, and show that it is weakly terminal.

8.2.4. Define the product of (the signatures of) two monoids as a term $\mu \times \mu \rightarrow \mu$ using the above abstype notation.

8.2.5. One defines the sets $\mathrm{PV}(\sigma)$ and $\mathrm{NV}(\sigma)$ of **positively** and **negatively** occurring variables in a type σ as follows.

$$
\begin{aligned}
\mathrm{PV}(\alpha) &= \{\alpha\} & \mathrm{NV}(\alpha) &= \{\alpha\} \\
\mathrm{PV}(\sigma \times \tau) &= \mathrm{PV}(\sigma) \cup \mathrm{PV}(\tau) & \mathrm{NV}(\sigma \times \tau) &= \mathrm{NV}(\sigma) \cup \mathrm{NV}(\tau) \\
\mathrm{PV}(\sigma + \tau) &= \mathrm{PV}(\sigma) \cup \mathrm{PV}(\tau) & \mathrm{NV}(\sigma + \tau) &= \mathrm{NV}(\sigma) \cup \mathrm{NV}(\tau) \\
\mathrm{PV}(\sigma \rightarrow \tau) &= \mathrm{NV}(\sigma) \cup \mathrm{PV}(\tau) & \mathrm{NV}(\sigma \rightarrow \tau) &= \mathrm{PV}(\sigma) \cup \mathrm{NV}(\tau) \\
\mathrm{PV}(\Pi\alpha\colon\mathsf{Type}.\,\sigma) &= \mathrm{PV}(\sigma) - \{\alpha\} & \mathrm{NV}(\Pi\alpha\colon\mathsf{Type}.\,\sigma) &= \mathrm{NV}(\sigma) - \{\alpha\} \\
\mathrm{PV}(\Sigma\alpha\colon\mathsf{Type}.\,\sigma) &= \mathrm{PV}(\sigma) - \{\alpha\} & \mathrm{NV}(\Sigma\alpha\colon\mathsf{Type}.\,\sigma) &= \mathrm{NV}(\sigma) - \{\alpha\}
\end{aligned}
$$

Extend Lemma 8.2.1 in the following way. Let $\Xi, X\colon \mathsf{Type} \vdash \sigma\colon \mathsf{Type}$ be a type with a free type variable X. If X occurs only positively in σ, then $\sigma[-/X]$ extends to a covariant functor $\mathbb{C}(\Xi) \rightarrow \mathbb{C}(\Xi)$; and it extends to a contravariant functor $\mathbb{C}(\Xi)^{\mathrm{op}} \rightarrow \mathbb{C}(\Xi)$ in case X occurs only negatively.

8.3 Naive set theoretic semantics

As we have seen in Chapter 2 on simple type theory, set theoretic semantics of types and terms involves interpreting a type τ as a set $[\![\tau]\!]$ and a closed term M of type τ as an element $[\![M]\!] \in [\![\tau]\!]$. If M is not closed, say with free term variables $x_1\colon\sigma_1, \ldots, x_n\colon\sigma_n$, then it gets interpreted as a function $[\![M]\!]\colon [\![\sigma_1]\!] \times \cdots \times [\![\sigma_n]\!] \rightarrow [\![\tau]\!]$.

In polymorphic type theory there are type variables α, which introduce an extra level of indexing. In this section we write out the corresponding (naive) set theoretic semantics. This works fine for first order polymorphic type theory $\lambda\rightarrow$. The type theories $\lambda 2$ and $\lambda\omega$ however, involve (among other things) second and higher order polymorphic products Π. As shown by Reynolds in [287], see also [288], these cannot be interpreted in this naive set theoretic way. So one may ask: why consider this set theoretic semantics? First of all, the negative result in itself is significant. And in order to explain it, we have to describe the set theoretic interpretation to some extent. The result shows that we have to look for other categories than **Sets** in order to model second and higher order polymorphism. Secondly, there are indeed other universes— like the category ω-**Sets** of ω-sets, or the effective topos **Eff**—in which these polymorphic type theories can be interpreted. Moreover, this can essentially be done in a naive way, *i.e.* using the set theoretic formulations, but applied to the full internal category of PERs in ω-**Sets** and in **Eff**.

We begin by considering what a set theoretic model of a polymorphic signature $\langle \Sigma, (\Sigma_a) \rangle$ should be. A model of a higher order signature in predicate logic is a model of that signature in such a way that the type **Prop** of propositions gets interpreted as the set (of sets) $2 = \{0, 1\} = \{\emptyset, \{\emptyset\}\}$, where $0 = \emptyset$ stands for 'false' and $1 = \{\emptyset\}$ for 'true'. In higher order (predicate) logic it is enough to have this true/false distinction, but in polymorphic type theory there is more structure (given by terms in types, or, by proofs in propositions, using the propositions-as-types analogy). Thus we let the interpretation of the kind **Type** be an arbitrary set of sets $\mathcal{U}$, so that functions between the sets in $\mathcal{U}$ can serve as interpretations of terms.

A set theoretic model of a polymorphic signature $\langle \Sigma, (\Sigma_a) \rangle$ consists of

- a set theoretic model of the kind signature Σ, where the interpretation $[\![\text{Type}]\!]$ of **Type** is a set of sets $\mathcal{U}$. Each type

$$\alpha_1 : A_1, \ldots, \alpha_n : A_n \vdash \sigma : \text{Type}$$

i.e. each type in the type context over $\langle A_1, \ldots, A_n \rangle$, is then interpreted as a function

$$[\![A_1]\!] \times \cdots \times [\![A_n]\!] \xrightarrow{\;[\![\sigma]\!]\;} \mathcal{U}$$

It thus forms a collection of sets $[\![\sigma]\!](\vec{a}) \in \mathcal{U}$ for $\vec{a} \in [\![A_1]\!] \times \cdots \times [\![A_n]\!]$.
- for each function symbol $F : \sigma_1, \ldots, \sigma_m \longrightarrow \sigma_{m+1}$ in the type signature $\Sigma_{\langle A_1, \ldots, A_n \rangle}$ over kind context $\langle A_1, \ldots, A_n \rangle$, a collection of functions

$$[\![\sigma_1]\!](\vec{a}) \times \cdots \times [\![\sigma_m]\!](\vec{a}) \xrightarrow{\;[\![F]\!](\vec{a})\;} [\![\sigma_{m+1}]\!](\vec{a})$$

for $\vec{a} \in [\![A_1]\!] \times \cdots \times [\![A_n]\!]$. Put a bit differently, F is interpreted as a function $[\![F]\!]$ in the big 'dependent' product

$$[\![F]\!] \in \prod_{\vec{a} \in [\![A_1]\!] \times \cdots \times [\![A_n]\!]} \left([\![\sigma_1]\!](\vec{a}) \times \cdots \times [\![\sigma_m]\!](\vec{a}) \Rightarrow [\![\sigma_{m+1}]\!](\vec{a}) \right),$$

where $\Rightarrow$ is function space.

8.3.1. Remarks. (i) For convenience we may assume that the set $\mathcal{U}$ is closed under finite products of sets, see Exercise 8.3.1 below. This allows us to restrict ourselves for the interpretation of terms to sets of the form

$$\prod_{a \in [\![A]\!]} \left([\![\sigma]\!](a) \Rightarrow [\![\tau]\!](a) \right)$$

which makes things more manageable. One sees how the 'double indexing' by type variables $a \in [\![A]\!]$ and by term variables $x \in [\![\sigma]\!](a)$ takes place.

(ii) Let us provide some categorical clarity about what we are doing: we consider the full subcategory $\mathcal{U} \hookrightarrow \mathbf{Sets}$ with $\mathcal{U}$ as set of objects. This $\underline{\mathcal{U}}$ is a small category, and hence an internal category in $\mathbf{Sets}$. A type $\alpha\colon A \vdash \sigma(\alpha)\colon \mathsf{Type}$ gets interpreted as a map $[\![\sigma]\!]\colon [\![A]\!] \to \mathcal{U}$, *i.e.* as an object in the fibre over $[\![A]\!]$ in the externalisation of this internal category $\underline{\mathcal{U}}$. And a function symbol $F\colon \sigma(\alpha) \longrightarrow \tau(\alpha)$ becomes a morphism $[\![\sigma]\!] \to [\![\tau]\!]$ in this fibre over $[\![A]\!]$.

(One can also describe $\underline{\mathcal{U}}$ as coming from a family in $\mathbf{Sets}$, see Exercise 8.3.5 (iv) below.)

(iii) We will assign meanings $[\![\sigma]\!]$ and $[\![M]\!]$ to all types $\Xi \vdash \sigma\colon \mathsf{Type}$ and terms $\Xi \mid \Gamma \vdash M\colon \sigma$ (and not just to the atomic ones in the signature). In writing $[\![\sigma]\!]$ and $[\![M]\!]$ we ignore these kind and type contexts Ξ and Γ. It would be better to carry them along, but that is notationally rather cumbersome (see Exercise 8.3.3).

We continue with the naive set theoretic semantics of the polymorphic calculi $\lambda{\to}$, $\lambda2$ and $\lambda\omega$, on top of a model of a polymorphic signature as above.

(1) For $\lambda{\to}$ one assumes that the set $\mathcal{U}$ is closed under exponents (function spaces), *i.e.* that it satisfies

$$X, Y \in \mathcal{U} \;\Rightarrow\; Y^X \in \mathcal{U}.$$

By an argument similar to the one in Exercise 8.3.1 one may extend $\mathcal{U}$ to a set which is closed in such a way.

The exponent type $\sigma \to \tau$ of types σ, τ interpreted as functions

$$[\![A]\!] \xRightarrow[{[\![\tau]\!]}]{{[\![\sigma]\!]}} \mathcal{U}$$

is then given by pointwise function spaces:

$$[\![A]\!] \xrightarrow{[\![\sigma \to \tau]\!] \overset{\text{def}}{=} \lambda a.\, [\![\tau]\!](a)^{[\![\sigma]\!](a)}} \mathcal{U}$$

Abstraction is done as follows. Assume we have a term $\alpha\colon A \mid x\colon\rho, y\colon\sigma \vdash M\colon\tau$, which is interpreted as $[\![M]\!] \in \prod_{a\in[\![A]\!]} \big([\![\rho]\!](a) \times [\![\sigma]\!](a) \Rightarrow [\![\tau]\!](a)\big)$, then we get

$$[\![\lambda y\colon\sigma.\, M]\!] \overset{\text{def}}{=} \lambda a.\, \lambda x.\, \lambda y.\, [\![M]\!](a)(x, y)$$
$$\in \prod_{a\in[\![A]\!]} \big([\![\rho]\!](a) \Rightarrow [\![\sigma \to \tau]\!](a)\big).$$

For application assume terms $\alpha\colon A \mid x\colon\rho \vdash M\colon\sigma \to \tau$ and $\alpha\colon A \mid x\colon\rho \vdash N\colon\sigma$. These are interpreted as dependent functions $[\![M]\!] \in \prod_{a\in[\![A]\!]}([\![\rho]\!](a) \Rightarrow [\![\sigma \to \tau]\!](a))$ and $[\![N]\!] \in \prod_{a\in[\![A]\!]}([\![\rho]\!](a) \Rightarrow [\![\sigma]\!](a))$, then one takes

$$[\![MN]\!] \stackrel{\text{def}}{=} \lambda a.\, \lambda x.\, [\![M]\!](a)(x)([\![N]\!](a)(x))$$
$$\in \prod_{a\in[\![A]\!]} \big([\![\rho]\!](a) \Rightarrow [\![\tau]\!](a)\big).$$

(2) For $\lambda 2$ one would naively assume that $\mathcal{U}$ is closed under exponents and also under dependent products over itself, in order to interpret quantification of the from $\Pi\alpha\colon\mathsf{Type}.\,\sigma$ for $\sigma\colon\mathsf{Type}$. That is, one additionally assumes that

$$F \in \mathcal{U}^{\mathcal{U}} \;\Rightarrow\; \Pi F \in \mathcal{U}$$

where ΠF is the set

$$\Pi F = \{f\colon\mathcal{U} \to \textstyle\bigcup_{X\in\mathcal{U}} F(X) \mid \forall X \in \mathcal{U}.\, f(X) \in F(X)\}.$$

Below it will be shown that this assumption cannot be true, but for the time being, we simply continue.

For a type $\alpha\colon A, \beta\colon\mathsf{Type} \vdash \sigma\colon\mathsf{Type}$, interpreted by

$$[\![A]\!] \times \mathcal{U} \xrightarrow{\;\;[\![\sigma]\!]\;\;} \mathcal{U}$$

one puts

$$[\![A]\!] \xrightarrow[\displaystyle \lambda a.\, \Pi(\lambda b.\, [\![\sigma]\!](a,b))]{\displaystyle [\![\Pi\beta\colon\mathsf{Type}.\,\sigma]\!] \stackrel{\text{def}}{=}} \mathcal{U}$$

Second order abstraction goes as follows. Take a term $\alpha\colon A, \beta\colon\mathsf{Type} \mid x\colon\sigma \vdash M\colon\tau$, where β does not occur in σ. This M is interpreted as a function $[\![M]\!] \in \prod_{(a,b)\in[\![A]\!]\times\mathcal{U}}([\![\sigma]\!](a) \Rightarrow [\![\tau]\!](a,b))$, so that we can put

$$[\![\lambda\beta\colon\mathsf{Type}.\,M]\!] \stackrel{\text{def}}{=} \lambda a.\, \lambda x.\, \lambda b.\, [\![M]\!](a,b)(x)$$
$$\in \prod_{a\in[\![A]\!]} \big([\![\sigma]\!](a) \Rightarrow [\![\Pi\beta\colon\mathsf{Type}.\,\tau]\!](a)\big).$$

Notice in particular that by the restriction in the Π-introduction rule on the occurrence of β, the interpretation of $[\![\sigma]\!](a) \in \mathcal{U}$ does not depend on $b \in \mathcal{U}$. For application of terms to types, assume $\alpha\colon A \mid x\colon\sigma \vdash N\colon\Pi\beta\colon\mathsf{Type}.\,\tau$ and $\alpha\colon A \vdash \rho\colon\mathsf{Type}$, interpreted as $[\![N]\!] \in \prod_{a\in[\![A]\!]}([\![\sigma]\!](a) \Rightarrow [\![\Pi\beta\colon\mathsf{Type}.\,\tau]\!](a))$

and $[\![\rho]\!] : [\![A]\!] \to \mathcal{U}$. Then one takes

$$[\![N\rho]\!] \overset{\text{def}}{=} \lambda a.\, \lambda x.\, [\![N]\!](a)(x)([\![\rho]\!](a))$$
$$\in \prod_{a \in [\![A]\!]} \Big([\![\sigma]\!](a) \Rightarrow [\![\tau[\rho/\beta]]\!](a) \Big).$$

The set theoretic interpretation of sums Σ is left as an exercise below.

(3) For $\lambda\omega$ one naively assumes that $\mathcal{U}$ is closed under all dependent products (and not just over itself):

$$F \in \mathcal{U}^I \;\Rightarrow\; \Pi_I(F) \in \mathcal{U}$$

for any set I. The dependent product Π_I is given by

$$\Pi_I(F) = \{ f : I \to \textstyle\bigcup_{i \in I} F(i) \mid \forall i \in I.\, f(i) \in F(i) \}.$$

Thus ΠF in (2) is $\Pi_{\mathcal{U}}(F)$. One can then extend the quantification to all kinds: for $\alpha : A, \beta : B \vdash \sigma : \mathsf{Type}$, one puts

$$[\![A]\!] \xrightarrow{\;\dfrac{[\![\Pi\beta : B.\, \sigma]\!] \overset{\text{def}}{=}}{\lambda a.\, \Pi_{[\![B]\!]}(\lambda b.\, [\![\sigma]\!](a, b))}\;} \mathcal{U}$$

The rest is as before.

In the remainder of this section we show (and analyse) the impossibility of having a set of sets $\mathcal{U}$ closed under exponents and products as in (2) above. Notice that we have assumed these exponents and products in $\mathcal{U}$ to be the same as in the ambient category **Sets**. The argument we use is very much like in the proof of the following famous result.

8.3.2. Fact (Freyd [82, Chapter 3, Exercise D]). There are no small complete categories except preorders.

Proof. Let $\mathbb{C}$ be a category with a small collection C_0 of objects and small collections of hom-sets. Form the set C_1 as disjoint union of arrows in $\mathbb{C}$:

$$C_1 = \coprod_{X \in \mathbb{C}} \coprod_{Y \in \mathbb{C}} \mathbb{C}(X, Y).$$

Assume we have two different parallel arrows $f \neq g : X \rightrightarrows Y$ in $\mathbb{C}$. Every subset $A \subseteq C_1$ gives rise to an arrow $h(A) : X \to \prod_{C_1}(Y)$ with component $h(A)_\alpha$ for $\alpha \in C_1$ equal to f if $\alpha \in A$, and equal to g else. The assignment $A \mapsto h(A)$ is clearly injective. This gives a composite of injections $P(C_1) = 2^{C_1} \hookrightarrow \mathbb{C}(X, \prod_{C_1}(Y)) \hookrightarrow C_1$, which is impossible, since a powerset $P(C_1)$ cannot be embedded into C_1. $\qquad\square$

Roughly the same argument as used in this proof can be applied to the set theoretic models that we consider above.

8.3.3. Fact (After Reynolds [287]). There are no sets of sets $\mathcal{U}$ closed under exponents and dependent product over itself—as in (2) above—except trivial ones (for which every set $X \in \mathcal{U}$ has at most one element).

Proof. Assume towards a contradiction that some set $X \in \mathcal{U}$ has two different elements $x, y \in X$. Form the sets

$$D = \Pi_{Z \in \mathcal{U}} X^Z \in \mathcal{U} \qquad \text{and} \qquad \mathcal{V} = \coprod_{Z \in \mathcal{U}} Z = \{(Z, z) \mid Z \in \mathcal{U} \text{ and } z \in Z\}.$$

There is an obvious injection $D \hookrightarrow \mathcal{V}$, namely $f \mapsto (\Pi_{Z \in \mathcal{U}} X^Z, f)$. And for every subset $A \subseteq \mathcal{V}$ there is an element in $h(A) \in D$, with components $h(A)(Z)(z)$ equal to $x \in X$ if $(Z, z) \in A$ and to $y \in X$ else. It is clear that $A \mapsto h(A)$ is injective, so we have a sequence of injections $P(\mathcal{V}) = 2^{\mathcal{V}} \hookrightarrow D \hookrightarrow \mathcal{V}$. Again this is impossible. $\qquad \square$

It is useful to analyse the logical aspects of this negative result. The contradictions above come from the impossibility of having an injection $2^A \hookrightarrow A$ in (classical) set theory. But in a topos 2^A is the object of *decidable* subobjects of A, which may very well be embedded into A. The corresponding negative result in a topos is the following. We explicitly formulate it in higher order logic, so that it applies to other models of higher order logic than toposes. In particular we wish to use it for the (classical) higher order logic of regular subobjects in ω-**Sets**.

8.3.4. Proposition. *An injection $P\sigma \rightarrowtail \sigma$ cannot exist in higher order logic—where $P\sigma$ is the powertype of σ: Type defined as exponent type $P\sigma = \sigma \rightarrow \mathsf{Prop}$.*

Proof. We reason informally using an argument due to van Oosten. Assume $m: P\sigma \rightarrowtail \sigma$, and form terms $a: P\sigma$ and $e: \sigma$ by

$$a = \{x: \sigma \mid \forall b: P\sigma.\, x \in_\sigma b \supset x \neq_\sigma m(b)\} \qquad \text{and} \qquad e = m(a).$$

It is then clear that $e \notin_\sigma a$. But according to the definition of a we do get $e \in_\sigma a$. Indeed, for every $b: P\sigma$ with $e \in_\sigma b$, the equation $e =_\sigma m(b)$ does not hold: by injectivity of m it leads to $a =_{P\sigma} b$ and so to $e \in_\sigma a$. Hence e is both in a and not in a. $\qquad \square$

8.3.5. Fact (Higher order logic version of Fact 8.3.3). For a set of sets $\mathcal{U}$ which is closed under exponents and dependent products over itself, there is no set $X \in \mathcal{U}$ with an injection $\mathsf{Prop} \rightarrowtail X$ in higher order logic with extensional entailment.

It may be clear that the previous negative result (Fact 8.3.3) is a special case, since in classical set theory (*i.e.* in the higher order subobject fibration on **Sets**) Prop is $2 = \{0, 1\}$: if there cannot be an injection $\{0, 1\} \rightarrowtail X$, then X cannot have more than one element. The use of a set of sets $\mathcal{U}$ here is rather informal. A more precise version may be obtained by doing Exercise 8.3.6. This more general Fact 8.3.5 is a special case of the main theorem in [269]. The proof used there is more complicated than the one below, and uses expressible functors (as in [288]).

Proof. We can give virtually the same proof as before (with the same D and $\mathcal{V}$) starting from $m: \mathsf{Prop} \rightarrowtail X \in \mathcal{U}$. One now defines $P(\mathcal{V}) \rightarrowtail D$ by

$$(A \subseteq \mathcal{V}) \mapsto \lambda Z \in \mathcal{U}. \lambda z \in Z. m((Z, z) \in A).$$

This yields an injection by extensionality. Hence we get $P(\mathcal{V}) \hookrightarrow D \hookrightarrow \mathcal{V}$, which is impossible by Proposition 8.3.4.　　　　　　　　　　　　　□

8.3.6. Remarks. (i) As we shall see in the next section there are indeed models of higher order logic containing a "set" which is suitably closed under exponents and products (as in the ambient category) to allow interpretation of second and higher order polymorphic type theory. These examples involve the internal category of PERs in ω-**Sets** and in **Eff**. It is instructive to see why there is no injection $\mathsf{Prop} \rightarrowtail R$ for R a PER in these cases.

(a) In the classical higher order predicate logic of regular subobjects in ω-**Sets** (see Proposition 5.3.9) Prop is $\nabla 2$. As one easily checks, every map $\nabla 2 \to (\mathbb{N}/R, \in)$ for R a PER is constant: if e realises such a map f, then since $0 \in E_{\nabla 2}(0) \cap E_{\nabla 2}(1)$, we get $e \cdot 0 \in f(0) \cap f(1)$. Hence $f(0) = f(1)$ and f is constant. Thus there is certainly no mono $\nabla 2 \to (\mathbb{N}/R, \in)$.

(b) By essentially the same argument one shows that every map $\mathsf{Prop} = (P\mathbb{N}, \mathfrak{X}) \to (\mathbb{N}/R, \in)$ in **Eff** is constant. This is left to the reader.

These considerations show that Fact 8.3.5 does not obstruct suitable completeness of PERs in ω-**Sets** and in **Eff**. What it also suggests is to investigate the collection of those objects X for which all maps $\mathsf{Prop} \to X$ are constant, as a possible candidate for a suitably complete set of sets $\mathcal{U}$. Such objects X are called 'orthogonal to Prop'. This aspect of orthogonality of PER models was much emphasised by Freyd (see also [41, 150]). It will be further investigated in Section 11.7.

(ii) It has been argued (see notably Pitts [268]) that the impossibility of having a suitably complete model of polymorphism in **Sets** is due to the classical logic that we have for sets. The above arguments show that it is not the classical nature of the logic (since ω-**Sets** with regular subobjects also has classical logic, see Proposition 5.3.9) that is to blame, but rather the nature of the type **Prop** of propositions (as described in Fact 8.3.5) which obstructs the existence of such models. Shortly after [268] Pitts expressed this adapted analysis in [269].

In the end one may think that our approach is too naive in the sense that we have required products $\Pi F \in \mathcal{U}$ for *all* $F \in \mathcal{U}^{\mathcal{U}}$, whereas for the interpretation of $\lambda 2$ one only needs products of certain $F \in \mathcal{U}^{\mathcal{U}}$, namely of those F that come from the interpretation of types. The original result of Reynolds in [287] is that this situation already cannot occur. The above Fact 8.3.3 is a diluted version of Reynolds' result. Also in the sense that it involves products with (η).

Exercises

8.3.1. Let $\mathcal{U}$ be a set of sets. Put
$$\begin{aligned}\mathcal{U}_0 &= \mathcal{U} \cup \{1\}, \quad \text{where } 1 = \{\emptyset\} \\ \mathcal{U}_{n+1} &= \mathcal{U}_n \cup \{X \times Y \mid X, Y \in \mathcal{U}_n\} \\ \mathcal{U}_\infty &= \bigcup_{n \in \mathbb{N}} \mathcal{U}_n.\end{aligned}$$
Show that $\mathcal{U}_\infty$ is the least set containing $\mathcal{U}$ which is closed under finite products.

8.3.2. Check the validity of the conversions for $\to$ and Π in (1) and (2) above.

8.3.3. Rewrite the above interpretations as $[\![\, \Xi \vdash \sigma : \mathsf{Type} \,]\!]$ and $[\![\, \Xi \mid \Gamma \vdash M : \sigma \,]\!]$ with explicit contexts (for a few crucial cases). Describe the interpretation of the weakening and contraction rules using this notation.

8.3.4. Let $\mathcal{U}$ be a set of sets closed under dependent sums:
$$F \in \mathcal{U}^I \;\Rightarrow\; \Sigma_I(F) \in \mathcal{U}$$
where $\Sigma_I(F) = \{\langle i, x\rangle \mid i \in I \text{ and } x \in F(i)\}$. Interpret the Σ-types of $\lambda\omega$.

8.3.5. Let $\mathcal{U}$ be a set of sets closed under finite products. One can consider $\mathcal{U}$ has a full subcategory $\underline{\mathcal{U}} \hookrightarrow \mathbf{Sets}$. Notice that $\underline{\mathcal{U}}$ is a small category, and is thus internal in **Sets**.
   (i)   Show that in assumption (1) for the interpretation of $\lambda\to$, $\underline{\mathcal{U}}$ is assumed to be Cartesian closed, with exponents as in **Sets**.
   (ii) Show that the dependent product Π_I and the dependent sum Σ_I form (internal) right and left adjoints to the diagonal functor $\mathcal{U} \to \mathcal{U}^I$.
   (iii) Verify that in the special case where $\mathcal{U} = 2 = \{\emptyset, \{\emptyset\}\}$ (which is used for logic), $\mathcal{U}$ is Cartesian closed and has right and left adjoints Π_I, Σ_I.
   (iv) Let $\mathcal{V} = \coprod_{X \in \mathcal{U}} X = \{(X, x) \mid X \in \mathcal{U} \text{ and } x \in X\}$ with obvious projection $\left(\begin{smallmatrix} \mathcal{V} \\ \downarrow \\ \mathcal{U} \end{smallmatrix}\right)$. Describe the full internal category resulting from this family as in Example 7.1.4 (ii) and notice that this gives another way to describe $\underline{\mathcal{U}}$.

8.3.6. Consider in higher order logic an internal category $\mathbf{C} = (C_1 \rightrightarrows C_0)$—given by types $C_0, C_1 : \mathsf{Type}$ with suitable terms—which is internally Cartesian closed and has an internal right adjoint to the diagonal $\mathbf{C} \to \mathbf{C}^{C_0}$. Show that there cannot be an object $X: C_0$ with an injection $\mathsf{Prop} \rightarrowtail \mathbf{C}(1, X) = \{f: C_1 \mid \partial_0(f) =_{C_0} 1 \wedge \partial_1(f) =_{C_0} X\}$.

8.4 Fibrations for polymorphic type theory

The double indexing in polymorphic type theory—by type variables and by term variables, or, by kind contexts Ξ and by type contexts Γ—asks for a fibred description. After all, indexing is what fibred categories are all about. The aim in this section is to define appropriate fibred categories for polymorphic type theories, and to provide several examples. Prominent among these are models of PERs, suitably indexed over **Sets**, over ω-**Sets** and over **Eff**.

As we already mentioned, the set of sets $\mathcal{U}$ used in the previous section corresponds to a full internal subcategory in **Sets**, and hence by externalisation to a small fibration over **Sets**. The next definition captures the essential fibred aspects used so far.

8.4.1. Definition. A **polymorphic fibration** is a fibration with a generic object, with fibred finite products and with finite products in its base category. It will be called **split** whenever all this structure is split.

Such a polymorphic fibration captures the structure of contexts in polymorphic type theory, just like categories with finite products capture the structure of contexts in simple type theory. One thinks of the objects I of the base category as kinds, and of objects X of the fibre over I as types in kind context I, which we can write as $X = (i: I \vdash X_i: \mathsf{Type})$. The generic object involves an object Type in the base category and a correspondence between types $i: I \vdash X_i: \mathsf{Type}$ over I and classifying morphisms of kinds $I \to \mathsf{Type}$ in the basis.

(As an aside, we mention that a fibration with fibred finite products and finite products in its base category is an object with finite products in the 2-category **Fib** of fibrations over arbitrary bases. See Exercise 1.8.10 for details. Hence, fibred categories with finite products capture the context structure of polymorphic type theory, just like ordinary categories with finite products capture the context structure of simple type theory.)

8.4.2. Example. Let $\langle \Sigma, (\Sigma_a) \rangle$ be a polymorphic signature. It forms the starting point for a syntactically constructed (split) polymorphic fibration $\begin{smallmatrix} \mathcal{C}\ell(\Sigma,(\Sigma_a)) \\ \downarrow \\ \mathcal{C}\ell(\Sigma) \end{smallmatrix}$. The base category $\mathcal{C}\ell(\Sigma)$ is the classifying category of the (higher order) signature Σ, as described in Chapter 2. Objects are kind contexts $\Xi = \alpha_1: A_1, \ldots, \alpha_n: A_n$. Morphisms $\Xi' \to \Xi$ are sequences of terms $(M_1, \ldots, M_n)$

with $\Xi' \vdash M_i\colon A_i$. For a kind A we often write $A \in \mathcal{Cl}(\Sigma)$ for the singleton context $(\alpha\colon A) \in \mathcal{Cl}(\Sigma)$. Especially we have $\mathsf{Type} \in \mathcal{Cl}(\Sigma)$. Products in $\mathcal{Cl}(\Sigma)$ are given by concatenation of kind contexts.

There is a split indexed category on $\mathcal{Cl}(\Sigma)$; the functor $\mathcal{Cl}(\Sigma)^{\mathrm{op}} \to \mathbf{Cat}$ assigns to a kind context Ξ, the **category of types and terms in context** Ξ:

objects   types $\Xi \vdash \sigma\colon \mathsf{Type}$ in context Ξ.

morphisms $\sigma \to \tau$ are terms $N = N(x)$ with $\Xi \mid x\colon\sigma \vdash N(x)\colon\tau$.

The morphism part of this functor $\mathcal{Cl}(\Sigma)^{\mathrm{op}} \to \mathbf{Cat}$ is described by substitution in type variables, as described in the beginning of Section 8.1. The Grothendieck construction then yields a split fibration $\begin{smallmatrix}\mathcal{Cl}(\Sigma,(\Sigma_a))\\ \downarrow\\ \mathcal{Cl}(\Sigma)\end{smallmatrix}$. It has $\mathsf{Type} \in \mathcal{Cl}(\Sigma)$ as split generic object—since objects over Ξ are morphisms $\Xi \to \mathsf{Type}$ in the base, by construction. The fibration has (split) fibred finite products because we assume finite products of types (in all polymorphic calculi).

(We have used terms for morphisms, but what we really mean in this example are equivalence classes (under conversion) of terms. But explicitly writing these classes is cumbersome.)

An arbitrary split polymorphic fibration $\begin{smallmatrix}\mathbb{E}\\ \downarrow p\\ \mathbb{B}\end{smallmatrix}$ can serve as a categorical model of a polymorphic signature $\langle \Sigma, (\Sigma_a) \rangle$ in the following manner.

- Such a model is given first of all by a model of Σ in $\mathbb{B}$, which interprets Type as the generic object $\Omega \in \mathbb{B}$. We conveniently describe this model as a finite product preserving functor $\mathcal{M}\colon \mathcal{Cl}(\Sigma) \to \mathbb{B}$ with $\mathcal{M}(\mathsf{Type}) \cong \Omega$. Each type $\Xi \vdash \sigma\colon \mathsf{Type}$ then gets interpreted as a morphism $\mathcal{M}(\Xi) \to \Omega$. Hence it corresponds to an object X_σ above $\mathcal{M}(\Xi)$.
- Secondly such a model of $\langle \Sigma, (\Sigma_a) \rangle$ involves for each function symbol $F\colon \sigma_1, \ldots, \sigma_m \longrightarrow \sigma_{m+1}$ in a type context $\Sigma_{\langle A_1, \ldots, A_n \rangle}$ over $\langle A_1, \ldots, A_n \rangle$, a morphism $X_{\sigma_1} \times \cdots \times X_{\sigma_m} \to X_{\sigma_{m+1}}$ in the fibre over $\mathcal{M}(\alpha_1\colon A_1, \ldots, \alpha_n\colon A_n)$.

It is not hard to see that such a model corresponds to a morphism of fibrations

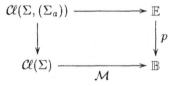

which preserves the structure of split polymorphic fibrations. This gives another instance of functorial semantics—as used earlier for simple type theory and logic—where interpretations are structure preserving functors.

By putting some further structure on polymorphic fibrations we get appropriate fibred categories for first, second and higher order polymorphic calculi $\lambda\to$, $\lambda 2$, $\lambda\omega$. This structure is as in the fibrations used for predicates logics, but is now used in a situation with proper (non-preordered) fibre categories. In $\lambda\to$ there are exponent types $\sigma \to \tau$, which are modelled as fibred exponents. In $\lambda 2$ one additionally has polymorphic products and sums $\Pi\alpha\colon \mathsf{Type}.\,\sigma$ and $\Sigma\alpha\colon \mathsf{Type}.\,\sigma$ over Type. These are modelled categorically by quantification along Cartesian projections $\pi\colon I \times \Omega \to I$, where Ω interprets Type. This was called "simple Ω-products/coproducts" in Section 2.4 for the CT-structure with Ω as sole type. In $\lambda\omega$ there is quantification over all kinds (and not just over Type), which is modelled by quantification along all Cartesian projections $\pi\colon I \times J \to I$. This was called "simple products/coproducts" in Section 1.9. Further, the exponent kinds require the base category to be Cartesian closed. All this will be summarised in (point (i) in) the following definition. It is by now quite standard; early work in this direction was done by Seely [307], Lamarche [185] and Pitts [268].

8.4.3. Definition. (i) A polymorphic fibration with Ω in the base as generic object, will be called

    (a) a $\lambda\to$**-fibration** if it has fibred exponents;

    (b) a $\lambda 2$**-fibration** if it has fibred exponents and also simple Ω-products and coproducts;

    (c) a $\lambda\omega$**-fibration** if it has fibred exponents, simple products and coproducts, and exponents in its base category.

    (ii) A $\lambda\to$= / $\lambda 2$= / $\lambda\omega$=-fibration is a $\lambda\to$ / $\lambda 2$ / $\lambda\omega$-fibration with (simple) equality. (The Frobenius condition holds automatically in the presence of fibred exponents, see Lemma 1.9.12 (i).)

8.4.4. Definition. Let $\Diamond$ be $\to, 2, \omega$ or $\to=, 2=, \omega=$.

    (i) A $\lambda\Diamond$-fibration will be called **split** if all of its structure is split. In particular, its underlying polymorphic fibration is then split. A $\lambda\Diamond$-fibration is called **small** if the fibration is small.

    (ii) A morphism of (split) $\lambda\Diamond$-fibrations is a morphism of the underlying (split) fibrations which preserves all of the structure (on-the-nose).

The functorial interpretation as described after Example 8.4.2 for polymorphic fibrations extends in a rather straightforward to $\lambda\to$-, $\lambda 2$- and $\lambda\omega$-fibrations. The interpretation of $\Pi\alpha\colon A.\,\sigma$ and $\Sigma\alpha\colon A.\,\sigma$ is much like the interpretation of $\forall x\colon \sigma.\,\varphi$ and $\exists x\colon \sigma.\,\varphi$ from predicate logic in Section 4.3, except that the extra structure of terms has to be taken into account. (See the 'mate' correspondences in Lemma 4.1.8 and Exercise 8.1.3.)

The interpretation of equality is also like in equational logic and in predicate logic. For objects I, J in the base category, the contraction functor δ^* induced by the diagonal $\delta = \langle \mathrm{id}, \pi' \rangle \colon I \times J \to (I \times J) \times J$ has a left adjoint $\mathrm{Eq}_{(I,J)}$, see Definition 3.4.1. We then have an equality type $\mathrm{Eq}_J = \mathrm{Eq}_{(I,J)}(1)$ over $(I \times J) \times J$; it may be written as

$$i \colon I, j \colon J, j' \colon J \vdash \mathrm{Eq}_J(j, j') \colon \mathsf{Type}$$

where i is a parameter. The unit of the equality adjunction yields the reflexivity term

$$i \colon I, j \colon J \mid \emptyset \vdash r_J(j) \colon \mathrm{Eq}_J(j, j) = \delta^*(\mathrm{Eq}_J)(j).$$

For elimination, assume we have a term

$$i \colon I, j \colon J \mid x \colon X_{(i,j,j)} \vdash M \colon Y_{(i,j,j)}.$$

This is a map $\delta^*(X) \to \delta^*(Y)$ over $I \times J$. By transposition and Frobenius we get a composite

$$X \times \mathrm{Eq}_J = X \times \mathrm{Eq}_{(I,J)}(1) \cong \mathrm{Eq}_{(I,J)}(\delta^*(X) \times 1) \cong \mathrm{Eq}_{(I,J)}(\delta^*(X)) \longrightarrow Y$$

which is the required elimination term in

$$i \colon I, j \colon J, j' \colon J \mid x \colon X_{(i,j,j')}, y \colon \mathrm{Eq}_J(j, j') \vdash M \text{ with } j' = j \text{ via } y \colon Y_{(i,j,j')}.$$

In the remainder of this section we describe various examples of polymorphic fibrations. Later, at the end of Section 11.3 there is an example of a model with only exponent types $\to$ plus polymorphic products and sums Π, Σ, but without Cartesian product types $1, \times$. The categorical description of fewer type formers requires more sophisticated fibred category theory—in the form of extra levels of indexing, like in Section 2.4 where we used simple fibrations (instead of ordinary categories) to describe models of simple type theory with exponents but without Cartesian products.

PER-examples

Next we describe three examples of $\lambda\omega{=}$-fibrations where types are interpreted as PERs, indexed over **Sets**, over ω-**Sets** and over **Eff**. The three fibrations consist of "uniform families of PERs" and can be organised in change-of-base situations:

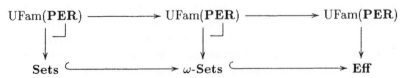

In this diagram we have used the name UFam(**PER**) for three *different* categories. One could write them as UFam$_{\text{Sets}}$(**PER**), UFam$_{\omega\text{-Sets}}$(**PER**) and UFam$_{\text{Eff}}$(**PER**), but that involves rather heavy notation. And confusion is not likely as long as we use these total categories together with their base category, as in $\begin{smallmatrix} \text{UFam}(\mathbf{PER}) \\ \downarrow \\ \omega\text{-}\mathbf{Sets} \end{smallmatrix}$, instead of $\begin{smallmatrix} \text{UFam}_{\omega\text{-}\mathbf{Sets}}(\mathbf{PER}) \\ \downarrow \\ \omega\text{-}\mathbf{Sets} \end{smallmatrix}$

Recall from Proposition 7.3.7 that the two fibrations of PERs over ω-**Sets** and over **Eff** are small. But the fibration of PERs over **Sets** is not small, as will be shown in Example 9.5.3 later on.

An efficient way to establish that these three fibrations of PERs are $\lambda\omega$=-fibrations is first to show that the fibration $\begin{smallmatrix} \text{UFam}(\mathbf{PER}) \\ \downarrow \\ \mathbf{Eff} \end{smallmatrix}$ on the right is such a $\lambda\omega$=-fibration and then to appeal to these change-of-base situations. But computations over ω-**Sets** are easier than over **Eff** so we prefer to start with the fibration of PERs over ω-sets in the middle. Then we can still appeal to change-of-base since the fibration of PERs over **Eff** can be obtained via change-of-base along the separated reflection functor, see Proposition 6.3.2 (i).

8.4.5. Proposition (Moggi and Hyland [143]). *The fibration* $\begin{smallmatrix} \text{UFam}(\mathbf{PER}) \\ \downarrow \\ \omega\text{-}\mathbf{Sets} \end{smallmatrix}$ *of PERs over ω-sets is a small split $\lambda\omega$=-fibration.*

Proof. Since the fibration is obtained by externalisation it is by definition small and split, and has a split generic object. It has a split fibred CCC-structure, inherited pointwise from **PER**. It has simple products, coproducts and equality by Lemma 1.9.6. We give the formulas explicitly: the product $\prod$ along a projection $\pi\colon (I, E) \times (J, E) \to (I, E)$ in ω-**Sets** is given on a family $R = (R_{i,j})_{i \in I, j \in J}$ of PERs over $(I, E) \times (J, E)$ by

$$\textstyle\prod(R)_i = \{(n, n') \mid \forall j \in J.\, \forall m, m' \in E(j).\, n \cdot m R_{i,j} n' \cdot m'\}.$$

And the coproduct $\coprod$ along this same projection is:

$$\textstyle\coprod(R)_i = \mathbf{r}\big(\bigcup_{j \in J} \mathbb{N}/R_{i,j}, E \big)$$

where $E(j, [n]) = E(j) \wedge [n] = \{\langle k, m\rangle \mid k \in E(j) \text{ and } m \in [n]\}$, and where $\mathbf{r}\colon \omega\text{-}\mathbf{Sets} \to \mathbf{PER}$ is the left adjoint to the inclusion $\mathbf{PER} \hookrightarrow \omega\text{-}\mathbf{Sets}$. Equality along a diagonal $\delta\colon (I, E) \times (J, E) \to ((I, E) \times J, E)) \times (J, E)$ on a family $R = (R_{i,j})$ over $(I, E) \times (J, E)$ is

$$\text{Eq}(R)_{i,j,j'} = \begin{cases} R_{i,j} & \text{if } j = j' \\ \emptyset & \text{else.} \end{cases} \qquad \square$$

We turn to families of PERs over sets. We define the fibration $\begin{smallmatrix} \text{UFam}(\mathbf{PER}) \\ \downarrow \\ \mathbf{Sets} \end{smallmatrix}$ of these as the one obtained by change-of-base from the fibration of PERs over ω-sets along the inclusion $\nabla\colon \mathbf{Sets} \hookrightarrow \omega\text{-}\mathbf{Sets}$. The fibre category over $I \in \mathbf{Sets}$ thus consists of

objects I-indexed families $R = (R_i)_{i \in I}$ of PERs R_i.

morphisms $R \to S$ are families $f = (f_i)_{i \in I}$ of morphisms of PERs $f_i\colon R_i \to S_i$ which have a common realiser: there is a single code $e \in \mathbb{N}$ such that e tracks every f_i.

8.4.6. Corollary (After [94]). *The fibration* $\begin{array}{c}\text{UFam}(\mathbf{PER})\\\downarrow\\\mathbf{Sets}\end{array}$ *of PERs over sets is a split $\lambda\omega{=}$-fibration.*

Proof. This follows from the change-of-base situation

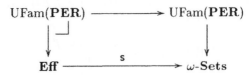

since the inclusion $\mathbf{Sets} \hookrightarrow \omega\text{-}\mathbf{Sets}$ preserves finite products. Explicitly, for a family $R = (R_{i,j})_{i \in I, j \in J}$ over $I \times J$, the product and coproduct along $\pi\colon I \times J \to I$ are, respectively,

$$\textstyle\prod(R)_i = \bigcap_{j \in J} R_{i,j} \qquad \text{and} \qquad \coprod(R)_i = \bigvee_{j \in J} R_{i,j},$$

where the intersection $\bigcap$ is the maximal PER $\mathbb{N} \times \mathbb{N}$ in case J is empty, and where $\bigvee$ is the join in the poset $(\mathrm{PER}, \subseteq)$ of PERs. $\qquad\qquad\square$

We shall see later in Example 9.5.3 that this fibration of PERs over **Sets** is not small. Hence the non-existence results in the previous section do not apply in this case.

Our last PER-example involves indexing over the effective topos **Eff**.

8.4.7. Corollary. *The fibration* $\begin{array}{c}\text{UFam}(\mathbf{PER})\\\downarrow\\\mathbf{Eff}\end{array}$ *of PERs over **Eff** is a small split $\lambda\omega{=}$-fibration.*

Proof. This follows from the change-of-base situation

$$\begin{array}{ccc}\text{UFam}(\mathbf{PER}) & \longrightarrow & \text{UFam}(\mathbf{PER})\\ \downarrow{\lrcorner} & & \downarrow\\ \mathbf{Eff} & \xrightarrow{\;\;\mathsf{s}\;\;} & \omega\text{-}\mathbf{Sets}\end{array}$$

from Proposition 6.3.2 (i), because the separated reflection functor $\mathsf{s}\colon \mathbf{Eff} \to \omega\text{-}\mathbf{Sets}$ preserves finite products (see Exercise 6.2.3). $\qquad\qquad\square$

These PER-examples may be called "parametric in the sense of Strachey", since a term inhabiting a polymorphic product $\Pi\alpha\colon \mathsf{Type}.\,\sigma(\alpha)$ has a single underlying code—which may be seen as an untyped program—that computes each instantiation of the type variable α. The term thus behaves uniformly for each type.

Next we describe an alternative PER-model, which is parametric in the sense of Reynolds.

A relationally parametric PER-example

Below we present a different PER-model of second order polymorphic type theory $\lambda 2$. It is different from the earlier ones (over **Sets**, ω-**Sets** and **Eff**) in the sense that it may be called relationally parametric in the sense of Reynolds [286]. What we present is a categorical version of the concrete interpretation described in [11]. Noteworthy is that the two properties known as "identity extension" (for types) and "abstraction" (for terms) which are explicitly proved for the interpretation of [11] are incorporated here as as conditions in the categorical model. Towards the end of the next section we give a further analysis of the parametricity in this model, along the lines of [204, 293].

Recall from Proposition 4.5.7 that a regular subobject of a PER R (in the category of PERs) may be identified with a subset $A \subseteq \mathbb{N}/R$ of R's quotient set. A (regular) relation on a pair R, S of PERs may then be described as a subset of $\mathbb{N}/R \times \mathbb{N}/S$. Especially, there is an equality relation $\mathrm{Eq}(R) \subseteq \mathbb{N}/R \times \mathbb{N}/R$ given as $\mathrm{Eq}(R) = \{([n]_R, [n]_R) \mid n \in |R|\}$. The main idea in the model of [11] is to interpret a type as a pair of maps, one mapping PERs to PERs (as in the earlier PER-models) and one mapping relations to relations, in an appropriate manner. This is formalised in the following definition, giving the intended base category of a $\lambda 2$-fibration of parametric families of PERs.

8.4.8. Definition. Let **PPER** be the category with natural numbers $n \in \mathbb{N}$ as objects. We first define morphisms $f\colon n \to 1$ with codomain 1. These are pairs $f = \langle f^p, f^r \rangle$ of functions

$$\mathsf{PER}^n \xrightarrow{\quad f^p \quad} \mathsf{PER}$$

and

$$f^r \in \prod_{\vec{R}, \vec{S} \in \mathsf{PER}^n} \left[\prod_{1 \le i \le n} P(\mathbb{N}/R_i \times \mathbb{N}/S_i) \right] \Rightarrow P\big(\mathbb{N}/f^p(\vec{R}) \times \mathbb{N}/f^p(\vec{S})\big)$$

satisfying the **identity extension** condition: for each n-tuple of PERs $\vec{R} \in$

PER^n one has

$$\text{Eq}\left(f^p(\vec{R})\right) = f^r_{\vec{R},\vec{R}}\left(\overrightarrow{\text{Eq}(R_i)}\right)$$

as subsets of $\mathbb{N}/f^p(\vec{R}) \times \mathbb{N}/f^p(\vec{R})$. We use the superscripts $(-)^p$ and $(-)^r$ for the components acting on *per*s and on *r*elations.

More generally, morphisms $n \to m$ in **PPER** are m-tuples $(f_1, \ldots, f_m)$ of morphisms $f_i: n \to 1$.

It is not hard to see that we get a category in this way. For $1 \leq i \leq n$ there are projection morphisms $\text{proj}(i) = \langle \text{proj}(i)^p, \text{proj}(i)^r \rangle: n \to 1$ by

$$\text{proj}(i)^p(\vec{R}) = R_i \qquad \text{and} \qquad \text{proj}(i)^r_{\vec{R},\vec{S}}(\vec{A}) = A_i.$$

These clearly satisfy "identity extension". The identity morphism $n \to n$ is then the n-tuple $(\text{proj}(1), \ldots, \text{proj}(n))$. Composition of $(f_1, \ldots, f_m): n \to m$ and $(g_1, \ldots, g_k): m \to k$ is the k-tuple $(h_1, \ldots, h_k): n \to k$, where

$$h_i^p(\vec{R}) = g_i^p\big(f_1^p(\vec{R}), \ldots, f_m^p(\vec{R})\big)$$
$$\big(h_i^r\big)_{\vec{R},\vec{S}}(\vec{A}) = (g_i^r)_{\vec{U},\vec{V}}\big((f_1^r)_{\vec{R},\vec{S}}(\vec{A}), \ldots, (f_m^r)_{\vec{R},\vec{S}}(\vec{A})\big),$$
$$\text{where } U_i = f_i^p(\vec{R}) \text{ and } V_i = f_i^p(\vec{S}).$$

By construction, this category **PPER** has finite products: $0 \in$ **PPER** is terminal object, and $n + m$ is the Cartesian product of $n, m \in$ **PPER**. In a next step we define a fibration over **PPER** with maps $n \to 1$ as objects over n. In this way the object $1 \in$ **PPER** becomes generic object of the fibration.

8.4.9. Definition. The split fibration $\begin{smallmatrix} \text{PFam}(\textbf{PER}) \\ \downarrow \\ \textbf{PPER} \end{smallmatrix}$ of parametric families of PERs is defined as arising from applying the Grothendieck construction to the indexed category **PPER**$^{\text{op}} \to$ **Cat** which has as fibre category of $n \in$ **PPER**:

objects	maps $f = \langle f^p, f^r \rangle: n \to 1$ in **PPER**.
morphisms	$\alpha: f \to g$ are families of morphisms

$$\alpha = \left(f^p(\vec{R}) \xrightarrow{\ \ \alpha_{\vec{R}}\ \ } g^p(\vec{R}) \right)_{\vec{R} \in \text{PER}^n}$$

between PERs, which:

- are tracked by a single code: some $e \in \mathbb{N}$ tracks every $\alpha_{\vec{R}}$ in **PER**:

$$\forall \vec{R} \in \text{PER}^n. \forall a \in |f^p(\vec{R})|. \alpha_{\vec{R}}([a]_{f^p(\vec{R})}) = [e \cdot a]_{g^p(\vec{R})}.$$

- satisfy the **abstraction** condition: for n-tuples $\vec{R}, \vec{S} \in$ PER^n of PERs and relations $A_i \subseteq \mathbb{N}/R_i \times \mathbb{N}/S_i$ on them, if

$$\langle [a], [b] \rangle \in f^r_{\vec{R}, \vec{S}}(\vec{A}) \subseteq \mathbb{N}/f^p(\vec{R}) \times \mathbb{N}/f^p(\vec{S})$$

then

$$\langle \alpha_{\vec{R}}([a]), \alpha_{\vec{S}}([b]) \rangle = \langle [e \cdot a], [e \cdot b] \rangle \in g^r_{\vec{R}, \vec{S}}(\vec{A})$$
$$\subseteq \mathbb{N}/g^p(\vec{R}) \times \mathbb{N}/g^p(\vec{S}).$$

(Where we have omitted subscripts of equivalence classes $[-]$ to increase the readability.)

Reindexing is done simply by composition.

8.4.10. Proposition (After [11]). *The above fibration* $\begin{array}{c} \text{PFam}(\mathbf{PER}) \\ \downarrow \\ \mathbf{PPER} \end{array}$ *of parametric families of PERs is a split $\lambda 2$-fibration.*

Proof. By construction, the fibration $\begin{array}{c} \text{PFam}(\mathbf{PER}) \\ \downarrow \\ \mathbf{PPER} \end{array}$ has a split generic object $1 \in \mathbf{PPER}$. We show that it is a split fibred CCC, and that it has split simple 1-products. In the fibre over $n \in \mathbf{PPER}$ there is a terminal object $1_n : n \to 1$ by $1^p_n(\vec{R}) = \mathbb{N} \times \mathbb{N}$, which is pointwise the terminal PER. This function 1^p_n on PERs determines its associated function 1^r_n on relations by the identity extension condition. The Cartesian product of objects $f, g : n \rightrightarrows 1$ is $f \times g : n \to 1$ with

$$\begin{aligned}(f \times g)^p(\vec{R}) &= f^p(\vec{R}) \times g^p(\vec{R}) \\ &= \{\langle a, b \rangle \mid \langle \mathbf{p}a, \mathbf{p}b \rangle \in f^p(\vec{R}) \text{ and } \langle \mathbf{p}'a, \mathbf{p}'b \rangle \in g^p(\vec{R})\},\end{aligned}$$

which is the pointwise product of PERs; and with action on relations:

$$\begin{aligned}(f \times g)^r_{\vec{R}, \vec{S}}(\vec{A}) = \{([\langle a, b \rangle], [\langle a', b' \rangle]) \mid ([a], [a']) &\in f^r_{\vec{R}, \vec{S}}(\vec{A}) \\ \text{and } ([b], [b']) &\in g^r_{\vec{R}, \vec{S}}(\vec{A})\}.\end{aligned}$$

The exponent of $f, g : n \rightrightarrows 1$ is $f \Rightarrow g : n \to 1$ with

$$\begin{aligned}(f \Rightarrow g)^p(\vec{R}) &= f^p(\vec{R}) \Rightarrow g^p(\vec{R}) \\ &= \{\langle a, a' \rangle \mid \forall \langle b, b' \rangle \in f^p(\vec{R}). \langle a \cdot b, a' \cdot b' \rangle \in g^p(\vec{R})\}.\end{aligned}$$

This is the pointwise exponent of PERs. On relations we take

$$\begin{aligned}(f \Rightarrow g)^r_{\vec{R}, \vec{S}}(\vec{A}) = \{([a], [a']) \mid \forall ([b], [b']) &\in f^r_{\vec{R}, \vec{S}}(\vec{A}). \\ ([a \cdot b], [a' \cdot b']) &\in g^r_{\vec{R}, \vec{S}}(\vec{A})\}.\end{aligned}$$

We leave it to the reader to verify that these products $f \times g$ and exponents $f \Rightarrow g$ satisfy "identity extension", and that the associated (pointwise) projection, tupleing, evaluation and abstraction operations satisfy "abstraction".

For products $\prod$ we define for an object $f\colon n + 1 \to 1$ over $n + 1$ an object $\prod(f)\colon n \to 1$ over n by

$$
\begin{aligned}
\prod(f)^p(\vec{R}) \;=\; & \{\langle a, a' \rangle \mid \forall U, V \in \mathrm{PER}.\, \forall B \subseteq \mathbb{N}/U \times \mathbb{N}/V. \\
& a \in |f^p(\vec{R}, U)| \text{ and } a' \in |f^p(\vec{R}, V)| \text{ and} \\
& ([a]_{f^p(\vec{R},U)}, [a']_{f^p(\vec{R},V)}) \in f^r_{(\vec{R},U),(\vec{R},V)}(\overrightarrow{\mathrm{Eq}(R_i)}, B)\} \\
\prod(f)^r_{\vec{R},\vec{S}}(\vec{A}) \;=\; & \{([a]_{\prod(f)^p(\vec{R})}, [a']_{\prod(f)^p(\vec{S})} \mid \\
& \forall U, V \in \mathrm{PER}.\, \forall B \subseteq \mathbb{N}/U \times \mathbb{N}/V. \\
& ([a]_{f^p(\vec{R},U)}, [a']_{f^p(\vec{S},V)}) \in f^r_{(\vec{R},U),(\vec{S},V)}(\vec{A}, B)\}.
\end{aligned}
$$

We need to prove "identity extension" for this product $\prod(f)$, *i.e.* we need to show that for $\vec{R} \in \mathrm{PER}^n$

$$
\{([a], [a]) \mid a \in |\prod(f)^p(\vec{R})|\} = \prod(f)^r_{\vec{R},\vec{R}}(\overrightarrow{\mathrm{Eq}(R_i)}).
$$

($\subseteq$) If $a \in |\prod(f)^p(\vec{R})|$, then for each relation $B \subseteq \mathbb{N}/U \times \mathbb{N}/V$ on PERs U, V we have $([a], [a]) \in f^r_{(\vec{R},U),(\vec{R},V)}(\overrightarrow{\mathrm{Eq}(R_i)}, B)$ by definition of $\prod(f)^p(\vec{R})$.

($\supseteq$) If the pair of classes $([a], [a'])$ is in the relation on the right hand side, then, as special case, for each $U \in \mathrm{PER}$ we have

$$
([a], [a']) \in f^r_{(\vec{R},U),(\vec{R},U)}(\overrightarrow{\mathrm{Eq}(R_i)}, \mathrm{Eq}(U)) = \mathrm{Eq}(f^p(\vec{R}, U)),
$$

the latter by identity extension for f. Thus $[a] = [a']$, and the pair $([a], [a'])$ of equal classes is in the equality relation on the left hand side.

The adjoint correspondences between maps $g \to \prod(f)$ over n and maps $\pi^*(g) \to f$ over $n + 1$ are as usual.   □

Relational parametricity in the sense of Reynolds [286] means the following. Consider a term $M\colon \Pi\alpha\colon \mathrm{Type}.\, \sigma(\alpha)$ inhabiting a polymorphic product type. A reasonable condition to expect of such an M is that it acts uniformly on types in the sense that it maps related types to related types. Specifically, if we have a relation $r \subset \tau \times \rho$ between types τ and ρ, then the terms $M\tau\colon \sigma[\tau/\alpha]$ and $M\rho\colon \sigma[\rho/\alpha]$ should be in the relation $\sigma[r/\alpha] \subset \sigma[\tau/\alpha] \times \sigma[\rho/\alpha]$. The latter relation $\sigma[r/\alpha]$ is defined by induction on the structure of σ, using appropriate operations on relations (discussed as operations on "logical relations" on a category $\mathrm{Rel}(\mathbb{E})$ in Section 9.2, see Example 9.2.5 (ii)).

But this is precisely what happens in the fibration $\begin{smallmatrix} \text{PFam}(\mathbf{PER}) \\ \downarrow \\ \mathbf{PPER} \end{smallmatrix}$ of parametric families of PERs, since a type's interpretation carries this action on relations along: if $\sigma(\alpha)$ is an object $f = \langle f^p, f^r \rangle : 1 \to 1$ over 1, then for a term $[n] \in \mathbb{N}/\prod(f)^p$ inhabiting the polymorphic product $\prod(f) : 0 \to 1$, we have by definition of $\prod(f)$: for types $U, V \in \mathbf{PER}$ with a relation $B \subseteq \mathbb{N}/U \times \mathbb{N}/V$, instantiation at U and at V,

$$[n]_{f^p(U)} \in \mathbb{N}/f^p(U) \qquad \text{and} \qquad [n]_{f^p(V)} \in \mathbb{N}/f^p(V)$$

yields two terms which are related by $f^r_{U,V}(B) \subseteq \mathbb{N}/f^p(U) \times \mathbb{N}/f^p(V)$, *i.e.* which satisfy $([n]_{f^p(U)}, [n]_{f^p(V)}) \in f^r_{U,V}(B)$. The latter is the type's action on the relation B. Hence Reynolds' parametricity holds by construction in this model of parametric families of PERs.

Towards the end of the next section we give a more formal description of the parametricity in this model. Consequences of parametricity may be found in Exercises 8.4.5 and 8.4.6.

Other examples

8.4.11. Preorder models. Any (split) higher order fibration is a (split) $\lambda\omega=$-fibration. In this way one gets preorder models (where all fibre categories are preorders) in which all structure of terms $M : \sigma$ in types σ is destroyed. In a propositions-as-types view these are 'proof-irrelevant' or 'truth-value' models. In particular, examples include,

    (1) the subobject fibration of a topos;

    (2) the fibration of regular subobjects of ω-**Sets**.

    (3) the family fibration $\begin{smallmatrix} \text{Fam}(\Omega) \\ \downarrow \\ \mathbf{Sets} \end{smallmatrix}$ for Ω a frame (*i.e.* a complete Heyting algebra).

    (4) the realisability fibration $\begin{smallmatrix} \text{UFam}(P\mathbb{N}) \\ \downarrow \\ \mathbf{Sets} \end{smallmatrix}$. In fact, any realisability fibration $\begin{smallmatrix} \text{UFam}(A) \\ \downarrow \\ \mathbf{Sets} \end{smallmatrix}$ coming from a PCA A, as in Example 5.3.4.

8.4.12. Term models. Let $\langle \Sigma, (\Sigma_a) \rangle$ be a polymorphic signature. The polymorphic fibration $\begin{smallmatrix} \mathcal{C}\ell(\Sigma, (\Sigma_a)) \\ \downarrow \\ \mathcal{C}\ell(\Sigma) \end{smallmatrix}$ described in Example 8.4.2 can be upgraded to a $\lambda{\to} / \lambda 2 / \lambda\omega$ fibration by incorporating the structure of these type theories—and taking appropriate equivalence classes of terms as morphisms. For the categorical products $\prod$ and coproducts $\coprod$ one uses the mate formulations of $\forall$ and $\exists$ in Exercise 8.1.3. We shall describe equality in some detail. The left adjoint to the diagonal $\delta : A \times B \to (A \times B) \times B$ in $\mathcal{C}\ell(\Sigma)$ is given by

$$\big(\alpha : A, \beta : B \vdash \sigma : \mathsf{Type}\big) \mapsto \big(\alpha : A, \beta : B, \beta' : B \vdash \sigma \times \mathsf{Eq}_B(\beta, \beta') : \mathsf{Type}\big).$$

The required adjunction involves a correspondence between (equivalence classes of) terms M and N in

$$\frac{\alpha\colon A, \beta\colon B, \beta'\colon B \mid x\colon\sigma, y\colon \mathrm{Eq}_B(\beta, \beta') \vdash M\colon\tau}{\alpha\colon A, \beta\colon B \mid x\colon\sigma \vdash N\colon\tau[\beta/\beta']} \quad \text{(Eq-mate)}$$

It is given by

$$M \mapsto M[\beta/\beta', \mathsf{r}_B(\beta)/y] \qquad \text{and} \qquad N \mapsto N \text{ with } \beta' = \beta \text{ via } y.$$

A model of one of these type theories in a specific fibration can then conveniently be described as a morphism of suitable fibrations with this syntactically constructed "classifying fibration" as domain and the specific fibration as codomain. This functorial semantics is as described in more detail in earlier chapters for simple type theory, equational logic and (higher order) predicate logic.

8.4.13. Domain theoretic models. There are models of the second order λ-calculus $\lambda 2$ (with at least polymorphic products) which use Scott domains (see [57] or [108, 61]) or coherence spaces (see [96] or [98]) or algebraic complete lattices (see [56] or [61]). The constructions are somewhat similar, in the sense that they rely on embedding-projections (in order to circumvent problems with contravariance of exponent types). We shall describe these embedding-projections in Section 10.6 for a model of dependent type theory. In [277] a model of $\lambda 2$ is obtained as a solution of a big domain equation capturing BMM-models as described in set theoretic terms in [38].

Towards the end of Section 10.6 there is also the fibration $\begin{smallmatrix}\mathbf{Fam(Clos)}\\\downarrow\\\mathbf{Clos}\end{smallmatrix}$ of closures indexed by closures (as originally described in [302]). It is a small $\lambda\omega$-fibration (see also Exercise 10.6.3 there).

Exercises

8.4.1. Be very naive (as in the previous section) and assume a set of sets $\mathcal{U}$ closed under products and sums Π_I, Σ_I, with $\underline{\mathcal{U}} \hookrightarrow \mathbf{Sets}$ as associated full subcategory. Show that $\begin{smallmatrix}\mathrm{Fam}(\underline{\mathcal{U}})\\\downarrow\\\mathbf{Sets}\end{smallmatrix}$ is a $\lambda\omega$-fibration. What does it take to interpret equality types as well?

8.4.2. (i) Give the precise interpretation for Π and Σ in a $\lambda\omega$-fibration and establish the validity of the associated conversions.

(ii) Do the same for equality in a $\lambda\omega$=-fibration (using the interpretation as described after Definition 8.4.4).

(iii) Check that the interpretation for Π in the $\lambda\omega$-fibration in the previous exercise, coincides with the set theoretic interpretation given in the last section.

8.4.3. Since **PER** is a small category, it is internal in **Sets**. Check that the fibra-
tion $\begin{smallmatrix} \text{UFam}(\textbf{PER}) \\ \downarrow \\ \omega\textbf{-Sets} \end{smallmatrix}$ of families of PERs over sets is *not* the externalisation of this internal category.

8.4.4. Define two parallel maps $u, v: I \rightrightarrows J$ in the base category of a (not necessarily preordered) fibration with equality Eq to be **internally equal** if
there is a vertical map $1(I) \to \text{Eq}(u, v)$ from the terminal object $1(I)$ over
I to the object $\text{Eq}(u, v)$ as in Notation 3.4.2. Check that in the above three
PER-models internal and external equality coincide.

8.4.5. (i) Consider the fibration $\begin{smallmatrix} \text{PFam}(\textbf{PER}) \\ \downarrow \\ \textbf{PPER} \end{smallmatrix}$ of parametric families of PERs. The
definable coproducts $+_d$ from Exercise 8.1.5, is given in this situation
on objects $f, g: n \rightrightarrows 1$ over n as

$$(f +_d g)^P(\vec{R})$$
$$= \{ \langle a, a' \rangle \mid \forall U, V \in \textbf{PER}. \forall B \subseteq \mathbb{N}/U \times \mathbb{N}/V.$$
$$a \in |(f^P(\vec{R}) \Rightarrow U) \times (g^P(\vec{R}) \Rightarrow U) \Rightarrow U|$$
$$\text{and } a' \in |(f^P(\vec{R}) \Rightarrow V) \times (g^P(\vec{R}) \Rightarrow V) \Rightarrow V|$$
$$\text{and } ([a], [a']) \in (\text{Eq}(f^P(\vec{R})) \Rightarrow B) \times (\text{Eq}(g^P(\vec{R})) \Rightarrow B) \Rightarrow B \}.$$

Prove that $+_d$ is the ordinary (fibred) coproduct using parametricity.

[*Hint.* Use graph relations of appropriate maps, and prove first that
$[\kappa, \kappa'] = \text{id}: f +_d g \to f +_d g$.]

(ii) Show also that the definable product $\times_d$ from Exercise 8.1.5 is the
ordinary product in this fibration.

[In parametric models the second order definable operations in Exercise 8.1.5 have the appropriate (non-weak) universal properties; similarly,
the weakly initial algebras and weakly terminal co-algebras from Theorem 8.2.2 become properly initial/terminal, see [118, 273, 11]. In the ordinary PER-model of uniform families the weakly initial algebras are shown
to be initial in [149].]

8.4.6. Let $\Xi, X: \textsf{Type} \vdash \sigma(X): \textsf{Type}$ be a type built up with finite products and
coproducts, as in Lemma 8.2.1, and consider its interpretation $[\![\sigma(X)]\!]: n +$
$1 \to 1$ as object over $n + 1$ in the fibration $\begin{smallmatrix} \text{PFam}(\textbf{PER}) \\ \downarrow \\ \textbf{PPER} \end{smallmatrix}$ of parametric
families of PERs.

(i) Describe the corresponding endofunctor between fibres

$$[\![\sigma[-/X]\!]\!]: \text{PFam}(\textbf{PER})_n \longrightarrow \text{PFam}(\textbf{PER})_n$$

following Lemma 8.2.1.

(ii) For a morphism $\gamma: g \to h$ in the fibre $\text{PFam}(\textbf{PER})_n$ define the graph
relation $\mathcal{G}(\gamma) \subset g \times h$ as

$$\mathcal{G}(\gamma)_{\vec{R}} = \{ ([a], \gamma_{\vec{R}}([a])) \mid a \in |g^P(\vec{R})| \} \subseteq \mathbb{N}/g^P(\vec{R}) \times \mathbb{N}/h^P(\vec{R}).$$

Prove by induction on the structure of σ that

$$\mathcal{G}\Big(\llbracket \sigma[\gamma/X]\rrbracket\Big)_{\vec{R}} = \llbracket \sigma(X)\rrbracket^r_{(\vec{R},g^p(\vec{R})),(\vec{R},h^p(\vec{R}))}\Big(\overrightarrow{\mathrm{Eq}(R_i)},\mathcal{G}(\gamma)_{\vec{R}}\Big)$$

(iii) Assume now two types $\Xi, X: \mathsf{Type} \vdash \sigma_1(X),\sigma_2(X):\mathsf{Type}$ as above. Show that every inhabitant β of $\llbracket \Pi X:\mathsf{Type}.\,\sigma_1(X) \to \sigma_2(X)\rrbracket$ satisfies the following naturality condition: for each $\gamma: g \to h$ the following diagram commutes

$$
\begin{array}{ccc}
\sigma_1[g/X] & \xrightarrow{\quad\beta_g\quad} & \sigma_2[g/X] \\[2pt]
{\scriptstyle\sigma_1[\gamma/X]}\big\downarrow & & \big\downarrow{\scriptstyle\sigma_2[\gamma/X]} \\[2pt]
\sigma_1[h/X] & \xrightarrow[\quad\beta_h\quad]{} & \sigma_2[h/X]
\end{array}
$$

when interpreted in the fibration of parametric families of PERs. [Such naturality properties stemming from parametricity are described in [340]. More generally, "dinaturality" conditions (which arise from negative occurrences of X) may be found in [11].]

8.4.7. Let $\mathbb{C}$ be a CCC. We form the category $\mathrm{NP}(\mathbb{C})$—where 'N' stands for negative and 'P' for positive—as follows. Objects are natural numbers $n \in \mathbb{N}$. Morphisms $(F_1,\ldots,F_m):n \to m$ are given by functors $F_i: (\mathbb{C}^{\mathrm{op}})^n \times \mathbb{C}^n \to \mathbb{C}$. Especially, we have projections $\mathsf{proj}_i: n \to 1$ described by $(\vec{X},\vec{Y}) \mapsto Y_i$ and $(\vec{f},\vec{g}) \mapsto g_i$. Given $F: n \to 1$, i.e. $F: (\mathbb{C}^{\mathrm{op}})^n \times \mathbb{C}^n \to \mathbb{C}$, we write $F^{\mathrm{tw}}: (\mathbb{C}^{\mathrm{op}})^n \times \mathbb{C}^n \to \mathbb{C}^{\mathrm{op}}$ for the 'twisted' version of F obtained as the composite of

$$(\mathbb{C}^{\mathrm{op}})^n \times \mathbb{C}^n \cong (\mathbb{C}^{\mathrm{op}\,\mathrm{op}})^n \times (\mathbb{C}^{\mathrm{op}})^n \cong ((\mathbb{C}^{\mathrm{op}})^n \times \mathbb{C}^n)^{\mathrm{op}} \xrightarrow{\ F^{\mathrm{op}}\ } \mathbb{C}^{\mathrm{op}}$$

Now one can define composition in $\mathrm{NP}(\mathbb{C})$ by $(G_1,\ldots,G_k) \circ (F_1,\ldots,F_m) = (H_1,\ldots,H_k)$, where $H_i = G_i \circ (F_1^{\mathrm{tw}},\ldots,F_m^{\mathrm{tw}},F_1,\ldots,F_m)$. Notice that $\mathsf{id}_n = (\mathsf{proj}_1,\ldots,\mathsf{proj}_n)$. In this way one obtains a category $\mathrm{NP}(\mathbb{C})$. It has finite products: 0 is terminal and $n+m$ is the products of n and m. For arrows $F,G: n \to 1$, we put

$$F \times G \;=\; \mathsf{prod} \circ (F,G): (\mathbb{C}^{\mathrm{op}})^n \times \mathbb{C}^n \longrightarrow \mathbb{C} \times \mathbb{C} \longrightarrow \mathbb{C}$$
$$F \Rightarrow G \;=\; \mathsf{exp} \circ (F^{\mathrm{tw}},G): (\mathbb{C}^{\mathrm{op}})^n \times \mathbb{C}^n \longrightarrow \mathbb{C}^{\mathrm{op}} \times \mathbb{C} \longrightarrow \mathbb{C}.$$

Next we define an indexed category $\Psi: \mathrm{NP}(\mathbb{C})^{\mathrm{op}} \to \mathbf{Cat}$ by giving the fibre categories $\Psi(n)$ morphisms $F: n \to 1$ in $\mathrm{NP}(\mathbb{C})$ as objects. Morphisms $F \to G$ in $\Psi(n)$ are families $\sigma = (\sigma_{\vec{X}})_{\vec{X}\in\mathbb{C}^n}$ of arrows $\sigma_{\vec{X}}: F(\vec{X},\vec{X}) \to G(\vec{X},\vec{X})$ in $\mathbb{C}$. There is no dinaturality requirement for such families. Show that applying the Grothendieck construction to $\Psi: \mathrm{NP}(\mathbb{C})^{\mathrm{op}} \to \mathbf{Cat}$ yields a (split) $\lambda\to$-category.

[This construction comes from [154, Section 3.3]; it is based on [11]. Also in [154] one finds how a restriction on the above F's and σ's yields the free

$\lambda\!\to\!$-fibration generated by $\mathbb{C}$. And in [99] it is shown that the interpretations of terms in this model are dinatural transformations.]

8.5 Small polymorphic fibrations

We recall that a polymorphic $\lambda\!\to\!$-, $\lambda 2$- or $\lambda\omega$-fibration is called *small* if it arises via externalisation from an internal category in its base category. In this section we shall investigate how a polymorphic fibration can be turned into a small one. We describe two ways of doing so. The first internalisation is based on the "standard construction" as described in Proposition 7.3.12: a fibration $\overset{\mathbb{E}}{\underset{\mathbb{B}}{\downarrow}}$ (which is split and satisfies suitable size conditions) yields an internal category in the topos of presheaves $\widehat{\mathbb{B}} = \mathbf{Sets}^{\mathbb{B}^{\mathrm{op}}}$. This standard construction is applied to polymorphic fibrations (actually indexed categories) in [8]. The second construction is from [268]. It yields for a polymorphic fibration $\overset{\mathbb{E}}{\underset{\mathbb{B}}{\downarrow}}$ a *full* internal category in the total category $\mathbb{E}$. Moreover, if we move this internal category via the Yoneda embedding to the topos of presheaves $\widehat{\mathbb{E}} = \mathbf{Sets}^{\mathbb{E}^{\mathrm{op}}}$ then we again have a full internal category, but this time in a topos. Below we shall show that this second step (the move from $\mathbb{E}$ to $\widehat{\mathbb{E}}$) is in fact an application of the standard construction. What is emphasised in [268] is that his construction turns a polymorphic fibration into a small one in a suitably rich ambient category (a topos) where the induced internal category has its Cartesian closed structure and products as in this ambient category. This facilitates reasoning about such structures.

We shall assume in this section that all polymorphic fibrations (with all their structure) are split. Further, we shall write Cartesian closed structure in fibres of a fibration with symbols $\top, \wedge, \supset$ for finite products and exponents. This is not to suggest that we have preordered fibres, but in order to be able to use different notation for the induced structure (see below) on the total category of the fibration. Such structure in total categories is studied systematically in Section 9.2.

We start with the standard construction for internalisation (as in Proposition 7.3.12). It is repeated in (i) below. The second additional point comes from [8].

8.5.1. Theorem. *Let* $\overset{\mathbb{E}}{\underset{\mathbb{B}}{\downarrow}}p$ *be a split fibration where the base category* $\mathbb{B}$ *is locally small and all fibre categories* $\mathbb{E}_I$ *are small.*

(i) [Internalisation as in Proposition 7.3.12] *There is an internal category*

P *in* $\widehat{\mathbb{B}} = \mathbf{Sets}^{\mathbb{B}^{op}}$ *with a change-of-base situation:*

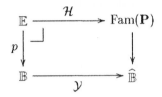

where $\mathcal{Y}$ *(for Yoneda) and* $\mathcal{H}$ *are full and faithful functors.*

(ii) *If* $\begin{smallmatrix} \mathbb{E} \\ \downarrow p \\ \mathbb{B} \end{smallmatrix}$ *is a* $\lambda{\to}$- *or* $\lambda 2$-*fibration, then so is the externalisation* $\begin{smallmatrix} \mathrm{Fam}(\mathbf{P}) \\ \downarrow \\ \mathbb{B} \end{smallmatrix}$, *and the morphism of fibrations* $(\mathcal{Y}, \mathcal{H})$ *in (i) preserves this structure.*

(The abovementioned size conditions can be ignored if one is willing to replace **Sets** by a suitably larger category of classes.)

Proof. (i) The presheaves P_0 and P_1 of objects and of morphisms of the internal category $\mathbf{P} = (P_1 \rightrightarrows P_0)$ in $\widehat{\mathbb{B}}$ are, on objects, respectively,

$$I \mapsto \mathrm{Obj}\,\mathbb{E}_I \qquad \text{and} \qquad I \mapsto \coprod_{X, Y \in \mathbb{E}_I} \mathbb{E}_I \big(X, Y \big).$$

See the proof of Proposition 7.3.12 for further details.

(ii) The fibred Cartesian closed structure of $\begin{smallmatrix} \mathrm{Fam}(\mathbf{P}) \\ \downarrow \\ \mathbb{B} \end{smallmatrix}$ is defined pointwise using the structure in the fibres of p. For example, in the fibre over $F \in \widehat{\mathbb{B}}$ the product of objects $\alpha, \beta \colon F \Rightarrow P_0$ is given by $(\alpha \times \beta)_I(a) = \alpha_I(a) \times \beta_I(a)$, where $I \in \mathbb{B}$ and $a \in F(I)$.

Next we notice that the split generic object $\Omega \in \mathbb{B}$ of p—with isomorphisms $\mathbb{B}(I, \Omega) \cong \mathrm{Obj}\,\mathbb{E}_I$—makes the presheaf P_0 of objects representable. Hence we can identify the exponent presheaf $P_0^{P_0}$ at $I \in \mathbb{B}$ as

$$\begin{aligned} P_0^{P_0}(I) &\cong \widehat{B}\big(\mathcal{Y}(I) \times \mathcal{Y}(\Omega),\, \mathcal{Y}(\Omega)\big) & \text{see exponents in Example 5.4.2} \\ &\cong \widehat{B}\big(\mathcal{Y}(I \times \Omega),\, \mathcal{Y}(\Omega)\big) & \text{because } \mathcal{Y} \text{ preserves products} \\ &\cong \mathbb{B}\big(I \times \Omega,\, \Omega\big) & \text{since } \mathcal{Y} \text{ is full and faithful} \\ &\cong \mathrm{Obj}\,\mathbb{E}_{I \times \Omega} & \text{because } \Omega \text{ is split generic object.} \end{aligned}$$

Hence the object parts of the required internal product and coproduct functors $\prod_0, \coprod_0 \colon P_0^{P_0} \rightrightarrows P_0$ can be described via functions $\mathrm{Obj}\,\mathbb{E}_{I \times \Omega} \rightrightarrows \mathrm{Obj}\,\mathbb{E}_I$ sending $X \mapsto \prod_I(X)$ and $X \mapsto \coprod_I(X)$, where $\prod_I, \coprod_I$ are the right and left adjoint to the weakening functor π^* induced by $\pi \colon I \times \Omega \to \Omega$. $\qquad \square$

Notice that we explicitly use representability of the object presheaf in order to obtain polymorphic products and coproducts. The argument extends to arbitrary representable presheaves (see Exercise 8.5.1 below), but it is not clear how to obtain quantification with respect to non-representable presheaves, so that the result can be extended to $\lambda\omega$-fibrations.

We turn to the internalisation of [268]. We shall present it in different form, using "simple fibrations over a fibration". These generalise "simple fibrations over categories" as in Section 1.3, see Exercise 8.5.3. But first, we need the following result, which shows how fibred ("local") finite products and exponents induce ("global") finite products and exponents in total categories. A more complete account of the interaction between local and global structure occurs in Section 9.2 in the next chapter, and in [125, 127].

8.5.2. Lemma. (i) *Let* $\begin{smallmatrix} \mathbb{E} \\ \downarrow p \\ \mathbb{B} \end{smallmatrix}$ *be a fibration with finite products* $(1, \times)$ *in its base category, and fibred finite products* $(\top, \wedge)$ *in its fibre categories. The total category* $\mathbb{E}$ *then has finite products:*

- *the terminal object* $\top \in \mathbb{E}_1$ *in the fibre over the terminal object* $1 \in \mathbb{B}$ *is terminal in* $\mathbb{E}$.
- *the Cartesian product in* $\mathbb{E}$ *of* $X \in \mathbb{E}_I$ *and* $Y \in \mathbb{E}_J$ *is the object* $X \times Y = \pi^*(X) \wedge \pi'^*(Y)$ *in the fibre over* $I \times J \in \mathbb{B}$.

(ii) *If* p *additionally has fibred exponents* $\supset$ *and simple products* $\prod_{(I,J)}$, *then the category* $\mathbb{E}$ *is Cartesian closed. The exponent of* $X \in \mathbb{E}_I$ *and* $Y \in \mathbb{E}_J$ *is the object*

$$ X \Rightarrow Y = \prod_{(I \Rightarrow J, I)} \left(\pi^*(X) \supset \mathrm{ev}^*(Y) \right) $$

in the fibre over $I \Rightarrow J \in \mathbb{B}$, *where* π *and* ev *in this expression are the projection and evaluation maps* $(I \Rightarrow J) \leftarrow (I \Rightarrow J) \times I \rightarrow J$.

Proof. (i) Reasoning in the total category $\mathbb{E}$ will be reduced to reasoning in fibre categories by Lemma 1.4.10. For $Z \in \mathbb{E}$ over $K \in \mathbb{B}$ one has in $\mathbb{E}$,

$$
\begin{aligned}
\mathbb{E}\left(Z, X \times Y \right) &\cong \coprod_{u:K \rightarrow I \times J} \mathbb{E}_K\left(Z, u^*(X \times Y) \right) \\
&\cong \coprod_{u:K \rightarrow I \times J} \mathbb{E}_K\left(Z, (\pi \circ u)^*(X) \wedge (\pi' \circ u)^*(Y) \right) \\
&\cong \coprod_{v:K \rightarrow I} \coprod_{w:K \rightarrow J} \mathbb{E}_K\left(Z, v^*(X) \wedge w^*(Y) \right) \\
&\cong \coprod_{v:K \rightarrow I} \mathbb{E}_K\left(Z, v^*(X) \right) \times \coprod_{w:K \rightarrow J} \mathbb{E}_K\left(Z, w^*(Y) \right) \\
&\cong \mathbb{E}\left(Z, X \right) \times \mathbb{E}\left(Z, Y \right).
\end{aligned}
$$

(ii) We similarly compute, again for $Z \in \mathbb{E}$ above $K \in \mathbb{B}$,

$$
\begin{aligned}
&\mathbb{E}\big(Z,\ X \Rightarrow Y \big) \\
&\cong \coprod_{u:K \to (I \Rightarrow J)} \mathbb{E}_K \left(Z,\ u^* \textstyle\prod_{(I \Rightarrow J, I)} (\pi'^*(X) \supset \mathrm{ev}^*(Y)) \right) \\
&\cong \coprod_{u:K \to (I \Rightarrow J)} \mathbb{E}_K \left(Z,\ \textstyle\prod_{(K,I)} (u \times \mathrm{id})^* (\pi'^*(X) \supset \mathrm{ev}^*(Y)) \right) \\
&\cong \coprod_{u:K \to (I \Rightarrow J)} \mathbb{E}_K \left(Z,\ \textstyle\prod_{(K,I)} (\pi'^*(X) \supset (\mathrm{ev} \circ u \times \mathrm{id})^*(Y)) \right) \\
&\cong \coprod_{v:K \times I \to J} \mathbb{E}_{K \times I} \left(\pi^*(Z),\ \pi'^*(X) \supset v^*(Y) \right) \\
&\cong \coprod_{v:K \times I \to J} \mathbb{E}_{K \times I} \left(\pi^*(Z) \wedge \pi'^*(X),\ v^*(Y) \right) \\
&\cong \mathbb{E}\big(Z \times X,\ Y \big). \hspace{4cm} \square
\end{aligned}
$$

Notice that, by construction, the functor $p \colon \mathbb{E} \to \mathbb{B}$ strictly preserves these finite products and exponents in $\mathbb{E}$.

8.5.3. Definition. Let $\begin{smallmatrix}\mathbb{E}\\\downarrow p\\\mathbb{B}\end{smallmatrix}$ be a fibration with fibred Cartesian products $\wedge$. We write $s_p(\mathbb{E})$ for the category with

objects   pairs $X, X' \in \mathbb{E}$ in the same fibre.

morphisms $(X, X') \to (Y, Y')$ are pairs of morphisms $f \colon X \to Y$ and $f' \colon X \wedge X' \to Y'$ in $\mathbb{E}$ over the same map in $\mathbb{B}$, *i.e.* with $p(f) = p(f')$.

There is then a projection functor $s_p(\mathbb{E}) \to \mathbb{E}$, namely $(X, X') \mapsto X$, which will be written as s_p. It will turn out to be a fibration, which we call the **simple fibration on** p.

The identity morphism on $(X, X') \in s_p(\mathbb{E})$ is the pair (id, π') consisting of the identity $X \to X$ together with the (horizontal) projection $X \wedge X' \to X'$. The composite of

$$
(X, X') \xrightarrow{\ (f, f')\ } (Y, Y') \xrightarrow{\ (g, g')\ } (Z, Z')
$$

is given by the pair of maps

$$
X \xrightarrow{\ f\ } Y \xrightarrow{\ g\ } Z, \qquad X \wedge X' \xrightarrow{\ \langle f \circ \pi, f' \rangle\ } Y \wedge Y' \xrightarrow{\ g'\ } Z
$$

where the map $\langle f \circ \pi, f' \rangle$ arises as follows. Write $u = p(f) = p(f')$ in $\mathbb{B}$ and consider the vertical parts $X \wedge X' \to u^*(Y)$ of $f \circ \pi$ and $X \wedge X' \to u^*(Y')$ of

f'. Tupleing these in the fibre yields a map $X \wedge X' \to u^*(Y) \wedge u^*(Y')$. And using that substitution preserves products $\wedge$, we get $\langle f \circ \pi, f' \rangle$ as composite

$$
\begin{array}{c}
X \wedge X' \\
\downarrow \\
u^*(Y) \wedge u^*(Y') \\
\cong \downarrow \\
u^*(Y \wedge Y') \longrightarrow Y \wedge Y'
\end{array}
$$

It is by construction the unique map over u with $\pi \circ \langle f \circ \pi, f' \rangle = f \circ \pi$ and $\pi' \circ \langle f \circ \pi, f' \rangle = f'$.

Notice that applying the definition to the trivial fibration $\begin{smallmatrix} \mathbb{B} \\ \downarrow \\ 1 \end{smallmatrix}$ yields the simple fibration $\begin{smallmatrix} s(\mathbb{B}) \\ \downarrow \\ \mathbb{B} \end{smallmatrix}$ on $\mathbb{B}$ from Section 1.3, for $\mathbb{B}$ with Cartesian products. We thus have a (fibred) generalisation of our earlier "simple" construction for (ordinary) categories.

8.5.4. Lemma. *Let* $\begin{smallmatrix} \mathbb{E} \\ \downarrow p \\ \mathbb{B} \end{smallmatrix}$ *be a fibration with fibred finite products* $(\top, \wedge)$.

(i) *The above functor* $\begin{smallmatrix} s_p(\mathbb{E}) \\ \downarrow s_p \\ \mathbb{E} \end{smallmatrix}$ *is a fibration with fibred finite products. It is split (with split products) if p is split (with split products).*

(ii) *The original fibration p can be recovered from the simple fibration s_p via change-of-base along the terminal object functor $\top$ as in:*

$$
\begin{array}{ccc}
\mathbb{E} & \longrightarrow & s_p(\mathbb{E}) \\
p \downarrow & \lrcorner & \downarrow s_p \\
\mathbb{B} & \xrightarrow{\top} & \mathbb{E}
\end{array}
$$

(iii) $\begin{smallmatrix} s_p(\mathbb{E}) \\ \downarrow \\ \mathbb{E} \end{smallmatrix}$ *is a (split) fibred CCC if and only if* $\begin{smallmatrix} \mathbb{E} \\ \downarrow \\ \mathbb{B} \end{smallmatrix}$ *is a (split) fibred CCC.*

(iv) *There is a full and faithful fibred functor $s_p(\mathbb{E}) \to \mathbb{E}^{\to}$ given by*

$$
(X, X') \mapsto \left(\begin{smallmatrix} X \wedge X' \\ \downarrow \pi \\ X \end{smallmatrix} \right).
$$

It preserves the above fibred CCC-structure.

Proof. (i) Reindexing along a morphism $f \colon X \to Y$ in $\mathbb{E}$ above $u \colon I \to J$ in $\mathbb{B}$ takes the form

$$
(Y, Y') \mapsto (X, u^*(Y'))
$$

with as Cartesian lifting $(X, u^*(Y')) \to (Y, Y')$ the pair $f: X \to Y$ and $\overline{u}(Y') \circ \pi': X \wedge u^*(Y') \to u^*(Y') \to Y'$.

In the fibre of $s_p(\mathbb{E})$ over $X \in \mathbb{E}$ one has as terminal object and Cartesian product:

$$(X, \top(pX)) \quad \text{and} \quad (X, X') \wedge (X, X'') = (X, X' \wedge X'').$$

(ii) For $I \in \mathbb{B}$ the objects in $s_p(\mathbb{E})$ over $\top(I) \in \mathbb{E}$ are the objects in $\mathbb{E}$ over I. And for $u: I \to J$ in $\mathbb{B}$ the morphisms in $s_p(\mathbb{E})$ over $\top(u)$ are the maps $\top(I) \wedge X \to Y$ in $\mathbb{E}$ over u, which can be identified with the morphisms $X \to Y$ in $\mathbb{E}$ over u.

(iii) Given fibred exponents $\supset$ in p, one defines fibred exponents in $s_p(\mathbb{E})$ by

$$(X, X') \supset (X, X'') = (X, X' \supset X'').$$

The converse is easy by the change-of-base situation in (ii).

(iv) Easy. $\square$

8.5.5. Theorem. (i) *If* $\begin{smallmatrix} \mathbb{E} \\ \downarrow p \\ \mathbb{B} \end{smallmatrix}$ *is a split* $\lambda{\to}$-*fibration, then the simple fibration* $\begin{smallmatrix} s_p(\mathbb{E}) \\ \downarrow s_p \\ \mathbb{E} \end{smallmatrix}$ *on* p *is a full small* $\lambda{\to}$-*fibration: there is a full internal category* **C** *in* $\mathbb{E}$ *whose externalisation is this simple fibration.*

(ii) *And* $\begin{smallmatrix} s_p(\mathbb{E}) \\ \downarrow \\ \mathbb{E} \end{smallmatrix}$ *is a split* $\lambda 2$- *(or* $\lambda\omega$-*) fibration if and only if* $\begin{smallmatrix} \mathbb{E} \\ \downarrow \\ \mathbb{B} \end{smallmatrix}$ *is a split* $\lambda 2$- *(or* $\lambda\omega$-*) fibration.*

Proof. (i) Let $\Omega \in \mathbb{B}$ be split generic object for p with natural isomorphisms $\theta_I: \mathbb{B}(I, \Omega) \xrightarrow{\cong} \mathrm{Obj}\,\mathbb{E}_I$. Write $T = \theta_\Omega(\mathrm{id}) \in \mathbb{E}_\Omega$ and $C_0 = \top\Omega \in \mathbb{E}_\Omega$ for the terminal object in the fibre above Ω. We are going to form the full internal category $\mathrm{Full}(a)$ in $\mathbb{E}$ that comes from the (vertical) map in $\mathbb{E}$

$$a = \begin{pmatrix} T \\ \downarrow! \\ \top\Omega \end{pmatrix}$$

as in Example 7.1.4 (ii). Therefore we need the following two observations.

(a) For each morphism $f: X \to \top\Omega$ in $\mathbb{E}$, say over $u: I \to \Omega$ in $\mathbb{B}$, we have a pullback square in $\mathbb{E}$,

$$
\begin{array}{ccc}
X \wedge u^*(T) & \xrightarrow{\overline{u}(T)\,\circ\,\pi'} & T \\
\pi \downarrow \quad \lrcorner & & \downarrow\, ! = a \\
X & \xrightarrow{\quad f \quad} & \top\Omega
\end{array}
$$

This is not hard to check.

(b) The exponent $\pi^*(a) \Rightarrow \pi'^*(a)$ in the slice category $\mathbb{E}/T\Omega \times T\Omega$ exists, namely as the (vertical) projection

$$\left(\begin{array}{c} (T\Omega \times T\Omega) \wedge (\pi^*(T) \supset \pi'^*(T)) \\ \downarrow \pi \\ T\Omega \times T\Omega \end{array} \right)$$

(It is used as pair $\langle \partial_0, \partial_1 \rangle : C_1 \to C_0 \times C_0$ in the Full$(-)$-construction in Example 7.1.4 (ii).) Indeed, for a family $\langle f, g \rangle : X \to T\Omega \times T\Omega$ in $\mathbb{E}$ we have

$$\mathbb{E}/T\Omega \times T\Omega \big(\langle f, g \rangle \times \pi^*(a), \ \pi'^*(a) \big)$$

$$\cong \ \mathbb{E}/X \big(\langle f, g \rangle^* \pi^*(a), \ \langle f, g \rangle^* \pi'^*(a) \big)$$

$$\qquad \text{since } \langle f, g \rangle \times \pi^*(a) \cong \coprod_{\langle f, g \rangle} \langle f, g \rangle^* \pi^*(a)$$

$$\cong \ \mathbb{E}/X \left(\begin{array}{cc} X \wedge (pf)^*(T) & X \wedge (pg)^*(T) \\ \downarrow \pi & \downarrow \pi \\ X & X \end{array} \right) \qquad \text{by (a)}$$

$$\cong \ \mathbb{E}_{pX} \big(X \wedge (pf)^*(T), \ (pg)^*(T) \big)$$

$$\cong \ \mathbb{E}_{pX} \big(X, \ \langle pf, pg \rangle^* (\pi^*(T) \supset \pi'^*(T)) \big)$$

$$\cong \ \mathbb{E}_{\langle pf, pg \rangle} \big(X, \ \pi^*(T) \supset \pi'^*(T) \big)$$

$$\cong \ \mathbb{E}/T\Omega \times T\Omega \left(\langle f, g \rangle, \begin{array}{c} (T\Omega \times T\Omega) \wedge (\pi^*(T) \supset \pi'^*(T)) \\ \downarrow \pi \\ T\Omega \times T\Omega \end{array} \right).$$

The latter projection thus behaves as exponent in the slice.

Hence we have enough structure to form the full internal category Full(a). The object of its externalisation Fam$\big(\text{Full}(a)\big)$ over $X \in \mathbb{E}$ are the maps $X \to C_0 = T\Omega$ in $\mathbb{E}$. Because the terminal object functor T is right adjoint to p, these objects correspond to maps $pX \to \Omega$ in $\mathbb{B}$, and hence to objects in $\mathbb{E}_{pX} = s_p(\mathbb{E})_X$ because Ω is generic object. Similarly, morphisms of Fam$\big(\text{Full}(a)\big)$ over X correspond to morphisms of $s_p(\mathbb{E})$ over X.

(ii) Products $\prod$ and coproducts $\coprod$ for the simple fibration along the projections $X \times Y \to X$ in $\mathbb{E}$ are obtained via the products $\prod$ and coproducts $\coprod$ along the projections $pX \times pY \to pX$ in $\mathbb{B}$ and via products and exponents in the fibres:

$$\prod_{(X,Y)}(X \times Y, Z) = (X, \textstyle\prod_{(pX, pY)} (\pi'^*(Y) \supset Z))$$
$$\coprod_{(X,Y)}(X \times Y, Z) = (X, \textstyle\coprod_{(pX, pY)} (\pi'^*(Y) \wedge Z)).$$

Then

$$s_p(\mathbb{E})_{X \times Y}\left(\pi^*(X,W),\ (X \times Y, Z)\right)$$
$$= \mathbb{E}_{pX \times pY}\left((X \times Y) \wedge \pi^*(W),\ Z\right)$$
$$\cong \mathbb{E}_{pX \times pY}\left(\pi^*(X) \wedge \pi'^*(Y) \wedge \pi^*(W),\ Z\right)$$
$$\cong \mathbb{E}_{pX \times pY}\left(\pi^*(X \wedge W),\ \pi'^*(Y) \supset Z\right)$$
$$\cong \mathbb{E}_{pX}\left(X \wedge W,\ \textstyle\prod_{(pX,pY)}\left(\pi'^*(Y) \supset Z\right)\right)$$
$$= s_p(\mathbb{E})_X\left((X,W),\ \textstyle\prod_{(X,Y)}(X \times Y, Z)\right).$$

And similarly for $\coprod_{(X,Y)}$. Exponents in $\mathbb{E}$ exist by Lemma 8.5.2 (ii). $\qquad\square$

The full internal category $\mathbf{C}$ in this theorem has its fibred CCC-structure and products $\prod$ as in its ambient category $\mathbb{E}$. The next step is to form a full internal category in a richer ambient category, namely in the topos $\widehat{\mathbb{E}} = \mathbf{Sets}^{\mathbb{E}^{\mathrm{op}}}$ of presheaves on the total category $\mathbb{E}$ of our fibration. In principle there are two constructions: apply the standard construction from Theorem 8.5.1 to the simple fibration $\begin{smallmatrix} s_p(\mathbb{E}) \\ \downarrow \\ \mathbb{E} \end{smallmatrix}$, or apply the Full$(-)$ construction in $\widehat{\mathbb{E}}$ to the image $\mathcal{Y}(a)$ of the Yoneda functor $\mathcal{Y}$ applied the map $a = \begin{pmatrix} T \\ \downarrow \\ T\Omega \end{pmatrix}$ which gives the earlier full internal category $\mathbf{C}$ in $\mathbb{E}$. It turns out that these two constructions coincide. The outcome is the main result of [268]. There, only the second constructions is described, and it is not explicit that there is an application of the standard construction involved via the mediating rôle played by the simple fibration in the previous theorem.

In the proof below we need the following technicality.

8.5.6. Lemma. *Yoneda functors $\mathcal{Y} \colon \mathbb{C} \to \mathbf{Sets}^{\mathbb{C}^{\mathrm{op}}}$ preserve finite limits and exponents in slice categories.* $\qquad\square$

Preservation of finite limits is standard (and easy). For the proof that also exponents in slices are preserved we refer to [268].

8.5.7. Theorem (Pitts [268]). *Every split $\lambda\!\to\!$- (or $\lambda 2$-) fibration $\begin{smallmatrix} \mathbb{E} \\ \downarrow p \\ \mathbb{B} \end{smallmatrix}$ gives rise to a full internal category $\mathbf{P}$ in the topos of presheaves $\widehat{\mathbb{E}} = \mathbf{Sets}^{\mathbb{E}^{\mathrm{op}}}$, whose externalisation is again a $\lambda\!\to\!$- (or $\lambda 2$-) fibration with its structure as in the ambient category $\widehat{\mathbb{E}}$. Moreover, there is a change-of-base situation*

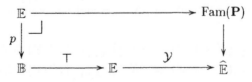

so that the fibration p can be recovered from the internal category $\mathbf{P}$ in $\widehat{\mathbb{E}}$.

(There are also some size restrictions in this theorem, analogous to those in Theorem 8.5.1.)

Proof. We show that the full internal category $\mathbf{D} = (D_1 \rightrightarrows D_0)$ in $\widehat{\mathbb{E}}$ arising from the family

$$\mathcal{Y}(a) = \begin{pmatrix} \mathcal{Y}(T) \\ \downarrow{!} \\ \mathcal{Y}(T\Omega) \end{pmatrix}$$

is the same as the outcome $\mathbf{P} = (P_1 \rightrightarrows P_0)$ of applying the standard construction from Theorem 8.5.1 to the simple fibration s_p. For $X \in \mathbb{E}$ we have, according to the definition of the object P_0 of objects of $\mathbf{P}$ in the proof of Theorem 8.5.1 (i):

$$
\begin{aligned}
P_0(X) &= \mathrm{Obj}\, s(\mathbb{E})_X \\
&= \mathrm{Obj}\, \mathbb{E}_{pX} \\
&\cong \mathbb{B}\big(pX,\, \Omega\big) \\
&\cong \mathbb{E}\big(X,\, T\Omega\big) \\
&\cong \mathcal{Y}(T\Omega)(X) \\
&= D_0(X).
\end{aligned}
$$

And

$$
\begin{aligned}
P_1(X) &= \coprod_{Y,Z \in s(\mathbb{E})_X} s(\mathbb{E})_X\big(Y,\, Z\big) \\
&\cong \coprod_{u,v:pX \rightrightarrows \Omega} \mathbb{E}_{pX}\big(X \wedge u^*(T),\, v^*(T)\big) \\
&\cong \coprod_{u,v:pX \rightrightarrows \Omega} \mathbb{E}_{\langle u,v\rangle}\big(X,\, \pi^*(T) \supset \pi'^*(T)\big) \\
&\cong \mathbb{E}\big(X,\, (T\Omega \times T\Omega) \wedge (\pi^*(T) \supset \pi'^*(T))\big) \\
&= \mathcal{Y}\big(\text{domain of } \pi^*(a) \Rightarrow \pi'^*(a)\big)(X) \\
&\cong \big(\text{domain of } \pi^*\mathcal{Y}(a) \Rightarrow \pi'^*\mathcal{Y}(a)\big)(X) \quad \text{by Lemma 8.5.6} \\
&= D_1(X). \qquad\qquad\qquad\qquad\qquad\qquad\qquad\qquad \square
\end{aligned}
$$

Recall that in a polymorphic fibration we require the presence of a generic object, but such an object need not come from an internal category—uniform families of PERs over **Sets** in Corollary 8.4.6 form an example where this is not the case. Having a small polymorphic fibration gives something extra, and this 'something' will be characterised later as: for a $\lambda{\to}$-fibration p there are equivalences

$$p \text{ is small} \overset{9.5.6}{\iff} p \text{ is "locally small"}$$

$$\overset{10.4.11}{\iff} p \text{ is a "comprehension category with unit"}$$

The latter structures are used for dependent type theory. Thus, in such *small* polymorphic fibrations there is more structure than strictly needed for modelling polymorphic type theory.

Exercises

8.5.1. Consider the internalisation $\mathbf{P}$ in $\widehat{\mathbb{B}}$ of a fibration $\begin{smallmatrix} \mathbb{E} \\ \downarrow p \\ \mathbb{B} \end{smallmatrix}$ as in Theorem 8.5.1 (i). Let $R \subseteq \widehat{\mathbb{B}}$ be the collection of representable presheaves $\mathbb{B}(-, I)$. Extend the description of (co)products $\prod, \coprod$ in the proof of Theorem 8.5.1 (ii) in the following way: p has simple (co)products $\Leftrightarrow$ the externalisation of $\mathbf{P}$ has simple (co)products with respect to the CT-structure $(\mathbb{B}, R)$. (What the latter means is in Definition 2.4.3.)

8.5.2. Prove that the induced product functor $\times : \mathbb{E} \times \mathbb{E} \to \mathbb{E}$ in Lemma 8.5.2 (i) is a fibred functor (over $\times : \mathbb{B} \times \mathbb{B} \to \mathbb{B}$).

8.5.3. (i) Let $\mathbb{B}$ be a category with Cartesian products $\times$. Show that the functor $(X, X') \mapsto (X, X \times X')$ forms a comonad $\mathbb{B} \times \mathbb{B} \to \mathbb{B} \times \mathbb{B}$, and that its Kleisli category is the total category $s(\mathbb{B})$ of the simple fibration $\begin{smallmatrix} s(\mathbb{B}) \\ \downarrow \\ \mathbb{B} \end{smallmatrix}$ on $\mathbb{B}$.

(ii) Let $\begin{smallmatrix} \mathbb{E} \\ \downarrow p \\ \mathbb{B} \end{smallmatrix}$ be a fibration with fibred Cartesian products $\wedge$. Prove that the assignment $(X, X') \mapsto (X, X \wedge X')$ now yields a fibred comonad $p \times p \to p \times p$ (over $\mathbb{B}$), and that its (fibred) Kleisli category (see Exercise 1.7.9) is the total category $s_p(\mathbb{E})$ of the simple fibration $\begin{smallmatrix} s_p(\mathbb{E}) \\ \downarrow \\ \mathbb{E} \end{smallmatrix}$ on p.

8.5.4. Consider the full internal category $\mathbf{C}$ from Theorem 8.5.5 in the total category $\mathbb{E}$ of a $\lambda \to$-fibration $\begin{smallmatrix} \mathbb{E} \\ \downarrow p \\ \mathbb{B} \end{smallmatrix}$.

(i) Show that the object C_2 of composable tuples of morphisms of $\mathbf{C}$ can be described as

$$C_2 = (\mathsf{T}\Omega \times \mathsf{T}\Omega \times \mathsf{T}\Omega) \wedge (\pi^*(T) \supset \pi'^*(T)) \wedge (\pi'^*(T) \supset \pi''^*(T)).$$

(ii) Give a similar description of the object C_3 of composable triples in $\mathbf{C}$.

8.5.5. Prove that the functor $s_p(\mathbb{E}) \to \mathbb{E}^{\to}$ from Lemma 8.5.4 (iv) also preserves simple products $\prod$.

8.6 Logic over polymorphic type theory

In the three chapters 3, 4 and 5 on equational logic, and on first and higher order predicate logic, we have considered logics in which one can reason about terms in *simple* type theory (STT). Semantically this involved putting a (pre-order) fibration $\begin{smallmatrix} \mathbb{E} \\ \downarrow \\ \mathbb{B} \end{smallmatrix}$ on the base category $\mathbb{B}$ incorporating the simple type theory. An obvious next step at this stage is to consider logics over *polymorphic* type theory (PTT). Semantically, this will again involve putting a (preorder) fibration on top of a model of PTT, but it turns out that there are several ways of doing so, depending on whether one wishes to reason about terms inhabiting kinds A: Kind, or about terms inhabiting types σ: Type (or even about both at the same time). In this section we briefly discuss some aspects of "logic over PTT". Such a logic may be called **polymorphic predicate logic** (PPL), in contrast to simple predicate logic SPL, as studied so far. The approach in this section is sketchy because

- PPL is not so very well-known in the literature, but there are exceptions, like [273, 326, 120];
- some of the relevant techniques—like fibrations over a fibration, and general forms of quantification—will be developed only in the next chapter.

Towards the end of this section we will describe how PPL may be used to give an abstract formulation (following [204, 293]) of what makes a $\lambda2$-fibration relationally parametric.

Logic of kinds

To start, consider a polymorphic $\lambda\rightarrow\!\!\!-$, $\lambda2$- or $\lambda\omega$-fibration $\begin{smallmatrix} \mathbb{E} \\ \downarrow \\ \mathbb{B} \end{smallmatrix}$. The objects of the base category $\mathbb{B}$ are considered as kinds A: Kind, and the objects of the total category $\mathbb{E}$ over A as types σ: Type in context A, written formally as $\alpha: A \vdash \sigma(\alpha)$: Type.

We first investigate what it means to put a logic on $\mathbb{B}$, say via a (preorder) fibration $\begin{smallmatrix} \mathbb{D} \\ \downarrow \\ \mathbb{B} \end{smallmatrix}$. This would add a new syntactic category Prop, with inhabitants $\alpha: A \vdash \varphi(\alpha)$: Prop depending on kinds A: Kind. These propositions are objects in $\mathbb{D}$ over A: Kind in $\mathbb{B}$. They allow us to reason about inhabitants of kinds, and especially, since Type: Kind, about types.

For example, one may have a logic of subtyping propositions (like in [263, 161]) in this way, with propositions

$$\alpha: \text{Type}, \beta: \text{Type} \vdash \alpha <: \beta: \text{Prop}$$

as objects of the category $\mathbb{D}$. With these, one can consider entailments between

propositions, like

$$\alpha, \alpha', \beta, \beta' \colon \mathsf{Type} \mid \alpha <: \alpha', \beta <: \beta' \vdash \alpha' \to \beta <: \alpha \to \beta'$$

expressing the usual contra- and co-variance of exponent types with respect
to the subtyping $<:$ relation. Semantically, such entailments $\vdash$ are to be con-
sidered as morphisms $\leq$ in a fibre of $\begin{smallmatrix}\mathbb{D}\\\downarrow\\\mathbb{B}\end{smallmatrix}$, see Exercise 8.6.1 below for a PER
model of such a logic of subtyping.

To mention a concrete example, consider the fibrations over ω-**Sets**

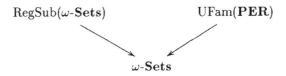

The (higher order) fibration of regular subobject in ω-**Sets** on the left hand
side gives us a powerful (classical) logic to reason about ω-sets, as kinds for
the $\lambda\omega$-fibration $\begin{smallmatrix}\text{UFam}(\mathbf{PER})\\\downarrow\\\omega\text{-}\mathbf{Sets}\end{smallmatrix}$ of ω-set-indexed families of PERs on the right
hand side. In this situation we have a *logic of kinds*.

Logic of types

We return to the general situation with a polymorphic fibration $\begin{smallmatrix}\mathbb{E}\\\downarrow\\\mathbb{B}\end{smallmatrix}$ and
consider—as alternative—what it means to put a (preorder) fibration $\begin{smallmatrix}\mathbb{D}\\\downarrow\\\mathbb{E}\end{smallmatrix}$ on
top of the total category $\mathbb{E}$ of types. Again we may see this as adding a new
syntactic category Prop of propositions, but this time inhabitants $\varphi \colon \mathsf{Prop}$ depend both on (variables in) kinds and on (variables in) types, as in

$$\alpha_1 \colon A_1, \ldots, \alpha_n \colon A_n \mid x_1 \colon \sigma_1, \ldots, x_m \colon \sigma_m \vdash \varphi(\vec{\alpha}, \vec{x}) \colon \mathsf{Prop}.$$

These propositions should be considered as objects of $\mathbb{D}$. And we may have
entailments

$$\alpha_1 \colon A_1, \ldots, \alpha_n \colon A_n \mid x_1 \colon \sigma_1, \ldots, x_m \colon \sigma_m \mid \varphi_1, \ldots, \varphi_k \vdash \psi$$

in which we have to deal with three separate contexts: for kinds, types and
propositions, corresponding to objects in the three categories $\mathbb{B} \leftarrow \mathbb{E} \leftarrow \mathbb{D}$
that we have. These entailments are morphisms $\leq$ in a fibre of $\mathbb{D}$ (over $\mathbb{E}$).

For such a *logic of types* one expects the usual propositional connectives
$\bot, \vee, \top, \wedge, \supset$. Interestingly, one can have quantification $\exists x \colon \sigma. \varphi$ and $\forall x \colon \sigma. \varphi$
over types, but also quantification $\exists \alpha \colon A. \varphi$ and $\forall \alpha \colon A. \varphi$ over kinds—with
the restriction that in first and second order PTT $A = \mathsf{Type}$ is the only

kind. For the latter form of quantification over kinds one has to impose the restriction that the variable α cannot occur free in the types on which φ depends. Formally in a formation rule:

$$\frac{\alpha_1:A_1,\ldots,\alpha_n:A_n,\alpha:A \mid x_1:\sigma_1,\ldots,x_m:\sigma_m \vdash \varphi:\mathsf{Prop}}{\alpha_1:A_1,\ldots,\alpha_n:A_n \mid x_1:\sigma_1,\ldots,x_m:\sigma_m \vdash \forall\alpha:A.\,\varphi:\mathsf{Prop}}\quad(\alpha \text{ not free in } \vec{\sigma})$$

Let us see what these logical operations mean in fibrations $\overset{\mathbb{D}}{\underset{\mathbb{E}}{\downarrow}}$ and $\overset{\mathbb{E}}{\underset{\mathbb{B}}{\downarrow}}$ of propositions-over-types, and of types-over-kinds. To illustrate this, we first syntactically construct a term model example of such a fibration. Let $\mathbb{K}$ the category with kinds $A:\mathsf{Kind}$ as objects, and with terms $\alpha:A \vdash M(\alpha):B$ as morphisms $A \to B$. Next, let $\mathbb{T}$ over $\mathbb{K}$ be the category with types $\alpha:A \vdash \sigma(\alpha):\mathsf{Type}$ as objects (over A), and with terms $\alpha:A \mid x:\sigma(\alpha) \vdash N(\alpha,x):\tau(\alpha)$ as morphisms $\sigma \to \tau$ over A. Finally, let $\mathbb{P}$ over $\mathbb{T}$ be the category of propositions $\alpha:A \mid x:\sigma(\alpha) \vdash \varphi(\alpha,x):\mathsf{Prop}$ as objects, and with entailments $\alpha:A \mid x:\sigma(\alpha) \mid \varphi(\alpha,x) \vdash \psi(\alpha,x)$ as morphisms $\varphi \to \psi$ (or $\varphi \leq \psi$) over σ. Then we have forgetful functors $\overset{\mathbb{P}}{\underset{\mathbb{T}}{\downarrow}}$ and $\overset{\mathbb{T}}{\underset{\mathbb{K}}{\downarrow}}$ which are split fibrations. For convenience we assume finite products $(1,\times)$ for kinds and for types. In this situation:

- The propositional connectives $\bot,\vee,\top,\wedge,\supset$ correspond to fibred preorder BiCCC structure for the fibration $\overset{\mathbb{P}}{\underset{\mathbb{T}}{\downarrow}}$ of propositions over types. Applying these connectives to propositions does not change the contexts of types and kinds in which the propositions live.
- Quantification $\exists y:\tau.\,\varphi$ and $\forall y:\tau.\,\varphi$ over types changes the type context by binding the variable $y:\tau$, but it leaves the kind context $\alpha:A$ unchanged. These quantifiers $\exists y:\tau.(-)$ and $\forall y:\tau.(-)$ are thus left and right adjoints to the weakening functors induced by the Cartesian projection morphism

$$\left(\alpha:A \vdash \sigma \times \tau:\mathsf{Type}\right) \longrightarrow \left(\alpha:A \vdash \sigma:\mathsf{Type}\right)$$

between types, given by the (vertical) projection term

$$\alpha:A \mid z:\sigma \times \tau \vdash \pi z:\sigma \qquad \text{or by} \qquad \alpha:A \mid x:\sigma,y:\tau \vdash x:\sigma.$$

Explicitly, we then have "mate" correspondences

$$\frac{\alpha:A \mid x:\sigma,y:\tau \mid \psi(\alpha,x) \vdash \varphi(\alpha,x,y)}{\alpha:A \mid x:\sigma \mid \psi(\alpha,x) \vdash \forall y:\tau.\,\varphi(\alpha,x,y)}$$

$$\frac{\alpha:A \mid x:\sigma,y:\tau \mid \varphi(\alpha,x,y) \vdash \psi(\alpha,x)}{\alpha:A \mid x:\sigma \mid \exists y:\tau.\,\varphi(\alpha,x,y) \vdash \psi(\alpha,x)}$$

These correspondences are as in (simple) predicate logic over STT, except that they involve an extra level of indexing given by variables $\alpha\colon A$ in kinds. Categorically, this quantification over types in polymorphic predicate logic (PPL) is given by simple coproducts and products, as in first order predicate logic, but the relevant adjunctions are vertical with respect to the fibration $\begin{smallmatrix}\mathbb{T}\\\downarrow\\\mathbb{K}\end{smallmatrix}$ of types over kinds, since the kind contexts are not affected. This will be made precise in Section 9.4.

- Quantification $\exists\beta\colon B.\,\varphi$ and $\forall\beta\colon B.\,\varphi$ over kinds $B\colon\mathsf{Kind}$ is more subtle. The first thought is probably that it means that the composite fibration $\begin{smallmatrix}\mathbb{P}\\\downarrow\\\mathbb{K}\end{smallmatrix}$ of propositions over kinds has simple coproducts and products. This would give a functor

$$\bigl(\alpha\colon A,\beta\colon B \mid x\colon\sigma(\alpha,\beta)\vdash\varphi(\alpha,\beta,x)\colon\mathsf{Prop}\bigr)$$
$$\xmapsto{\exists}\bigl(\alpha\colon A \mid x\colon\sigma(\alpha,\beta)\vdash\exists\beta\colon B.\,\varphi(\alpha,\beta,x)\colon\mathsf{Prop}\bigr)$$

which ignores the abovementioned restriction on the occurrence in type contexts of the variable β that becomes bound.

What really happens is that the projection morphism in $\mathbb{K}$ between kinds

$$A\times B\xrightarrow{\ \pi\ }A\qquad\text{described as}\qquad\alpha\colon A,\beta\colon B\vdash\alpha\colon A$$

lifts to a Cartesian morphism $\overline{\pi}$ in $\mathbb{T}$, namely

$$\pi^*\bigl(\alpha\colon A\vdash\sigma(\alpha)\colon\mathsf{Type}\bigr)\xrightarrow{\ \overline{\pi}\ }\bigl(\alpha\colon A\vdash\sigma(\alpha)\colon\mathsf{Type}\bigr)$$
$$\|$$
$$\bigl(\alpha\colon A,\beta\colon B\vdash\sigma(\alpha)\colon\mathsf{Type}\bigr)$$

$$\bigl(\alpha\colon A,\beta\colon B\bigr)\xrightarrow{\hspace{3cm}\pi\hspace{3cm}}\bigl(\alpha\colon A\bigr)$$

Then $\exists\beta\colon B.\,(-)$ and $\forall\beta\colon B.\,(-)$ are left and right adjoint to the weakening functor $\overline{\pi}^*$ induced in $\begin{smallmatrix}\mathbb{P}\\\downarrow\\\mathbb{T}\end{smallmatrix}$ by this lifted projection functor $\overline{\pi}$. This functor $\overline{\pi}^*$ maps

$$\bigl(\alpha\colon A \mid x\colon\sigma(\alpha)\vdash\psi(\alpha,x)\colon\mathsf{Prop}\bigr)\mapsto\bigl(\alpha\colon A,\beta\colon B \mid x\colon\sigma(\alpha)\vdash\psi(\alpha,x)\colon\mathsf{Prop}\bigr)$$

by adding a dummy assumption $\beta\colon B$. Quantification $\exists\beta\colon B.\,(-)$ and $\forall\beta\colon B.\,(-)$ over kinds is then characterised by the "mate" rules

$$\frac{\alpha\colon A,\beta\colon B \mid x\colon\sigma(\alpha)\mid\varphi(\alpha,\beta,x)\vdash\psi(\alpha,x)}{\alpha\colon A \mid x\colon\sigma(\alpha)\mid\exists\beta\colon B.\,\varphi(\alpha,\beta,x)\vdash\psi(\alpha,x)}$$

$$\frac{\alpha\colon A, \beta\colon B \mid x\colon \sigma(\alpha) \mid \psi(\alpha, x) \vdash \varphi(\alpha, \beta, x)}{\alpha\colon A \mid x\colon \sigma(\alpha) \mid \psi(\alpha, x) \vdash \forall \beta\colon B.\, \varphi(\alpha, \beta, x)}$$

In this case we have "lifted simple" coproducts and products. A precise categorical description will follow in Section 9.3, involving an appropriate associated Beck-Chevalley condition (which regulates the proper distribution of substitution over $\exists \beta\colon B.\, \varphi$ and $\forall \beta\colon B.\, \varphi$).

Of course, one can do a further step and generalise these preorder fibrations $\begin{smallmatrix} \mathbb{D} \\ \downarrow \\ \mathbb{E} \end{smallmatrix}$ and $\begin{smallmatrix} \mathbb{P} \\ \downarrow \\ \mathbb{T} \end{smallmatrix}$ of propositions over types to proper, non-preordered fibrations. This naturally to type theories like λHOL [91, 154], or $\lambda\omega_L$ [276].

8.6.1. Example. Recall from Proposition 4.5.7 that one can do classical logic with PERs, using regular subobjects as predicates. And also that such a regular subobject of a PER R can be identified with a subset $A \subseteq \mathbb{N}/R$ of its quotient set. We organise these regular subobjects of (individual) PERs into families of subobjects of families of PERs over ω-sets so that we get a model of a "logic over types" in PTT. Therefore we form a fibration

UFamRegSub(**PER**)
↓
UFam(**PER**)

with as fibre over an (I, E)-indexed collection $(R_i)_{i \in I}$ of PERs R_i:

objects   families of subsets $\left(A_i \subseteq \mathbb{N}/R_i \right)_{i \in I}$.

morphisms $\left(A_i \subseteq \mathbb{N}/R_i \right)_{i \in I} \longrightarrow \left(B_i \subseteq \mathbb{N}/R_i \right)_{i \in I}$ exist if and only if for each $i \in I$ there is an inclusion $A_i \subseteq B_i$. This fibre category is thus a poset.

For a morphism $(u, f)\colon (R_i)_{i \in I} \to (S_j)_{j \in J}$ in UFam(**PER**) between families of PERs—*i.e.* for a morphism of ω-sets $u\colon (I, E) \to (J, E)$ and a uniformly tracked family $f = (f_i\colon R_i \to S_{u(i)})_{i \in I}$ of morphisms of PERs—we get a substitution functor $(u, f)^*$ which maps

$$\left(B_j \subseteq \mathbb{N}/S_j \right)_{j \in J} \longmapsto \left(\{[n]_{R_i} \mid f_i([n]_{R_i}) \in B_{u(i)}\} \subseteq \mathbb{N}/R_i \right)_{i \in I}.$$

We now have two fibrations

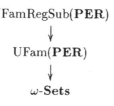

UFamRegSub(**PER**)
↓
UFam(**PER**)
↓
ω-**Sets**

incorporating the fibration $\begin{array}{c}\text{RegSub}(\mathbf{PER})\\\downarrow\\\mathbf{PER}\end{array}$, see Exercise 8.6.2. Each fibre of

$\begin{array}{c}\text{UFamRegSub}(\mathbf{PER})\\\downarrow\\\text{UFam}(\mathbf{PER})\end{array}$ is a Boolean algebra, so that we have classical propositional logic in this situation. Moreover, this fibration has simple products,
along the vertical projections $R \times S \to R$ over (I, E)—consisting of (uniformly tracked) families of PER-projections $\pi = (\pi_i \colon R_i \times S_i \to R_i)_{i \in I}$. For
$\left(B_i \subseteq \mathbb{N}/R_i \times \mathbb{N}/S_i\right)_{i \in I}$ we use set-theoretic quantification to get products
and coproducts along these projections:

$$\forall_S(B)_i = \{[n]_{R_i} \mid \forall m \in |S_i|. ([n]_{R_i}, [m]_{S_i}) \in B_i\} \subseteq \mathbb{N}/R_i$$
$$\exists_S(B)_i = \{[n]_{R_i} \mid \exists m \in |S_i|. ([n]_{R_i}, [m]_{S_i}) \in B_i\} \subseteq \mathbb{N}/R_i.$$

In a similar way we have "lifted simple" products and coproducts: for a
projection map $\pi \colon (I, E) \times (J, E) \to (I, E)$ in the base category ω-**Sets** of
kinds, we have a lifting $\overline{\pi}$ in UFam(**PER**) at $R = (R_i)_{i \in I}$ over (I, E), which
in its turn acts on predicates on R as

$$\overline{\pi}^*\left(\left(A_i \subseteq \mathbb{N}/R_i\right)_{i \in I}\right) = \left(A_i \subseteq \mathbb{N}/R_i\right)_{i \in I, j \in J} \xrightarrow{\hspace{3cm}} \left(A_i \subseteq \mathbb{N}/R_i\right)_{i \in I}$$

$$\pi^*\left(\left(R_i\right)_{i \in I}\right) = \left(R_i\right)_{i \in I, j \in J} \xrightarrow{\overline{\pi} = (\pi, \text{id})} \left(R_i\right)_{i \in I}$$

$$(I, E) \times (J, E) \xrightarrow{\hspace{1.5cm} \pi \hspace{1.5cm}} (I, E)$$

It acts as weakening of kinds on predicates. Products and coproducts (of
propositions over kinds) along this map $\overline{\pi}$ are again essentially set-theoretic:
for a predicate $\left(B_{i,j} \subseteq \mathbb{N}/R_i\right)_{i \in I, j \in J}$ one takes

$$\forall_{(J,E)}(B) = \bigcap_{j \in J} B_{i,j} \subseteq \mathbb{N}/R_i \quad \text{and} \quad \exists_{(J,E)}(B) = \bigcup_{j \in J} B_{i,j} \subseteq \mathbb{N}/R_i$$

(where the intersection $\bigcap$ is $\mathbb{N}/R_i$ if $J = \emptyset$).

In Section 8.4 we have seen the $\lambda2$-fibration $\begin{array}{c}\text{PFam}(\mathbf{PER})\\\downarrow\\\mathbf{PPER}\end{array}$ of parametric families of PERs, and we have argued informally that it is relationally parametric
in the sense of Reynolds. We now investigate whether we can express this
parametricity more formally in a suitable logic over this fibration. This line
of thought is followed in [273] where a logical system is formulated to express
parametricity. In essence, the key requirement there is that equality should
preserve polymorphic products. In [204, 293] there is a slightly more general

requirement involving a reflexive graphs $\bullet \rightrightarrows \bullet$ preserving appropriate structure, in which the arrow $\bullet \longleftarrow \bullet$ need not be the equality functor (as in [273]).
We are going to explicitly construct such a reflexive graph for the fibration of
parametric families of PERs.

In fact we shall be slightly more general than [204, 293] in the sense that
we formulate parametricity for $\lambda2$-fibrations with respect to a "logic of types"
over polymorphic type theory. In the approach below, this logic is a parameter,
whereas in [204, 293] the standard logic of subobjects in a category is taken.
We shall be using the logic of *regular* subobjects of PERs, as in the above
example.

8.6.2. Definition (After [204, 293]). Let $\begin{smallmatrix}\mathbb{E}\\\downarrow p\\\mathbb{B}\end{smallmatrix}$ be a $\lambda2$-fibration, provided
with a logic over types, given by a preorder fibration $\begin{smallmatrix}\mathbb{D}\\\downarrow q\\\mathbb{E}\end{smallmatrix}$. We say that p
is a **relationally parametric $\lambda2$-fibration** if there is another $\lambda2$-fibration
$\begin{smallmatrix}\mathbb{F}\\\downarrow r\\\mathbb{C}\end{smallmatrix}$ and a "reflexive graph" of $\lambda2$-fibrations:

$$\left(\begin{smallmatrix}\mathbb{F}\\\downarrow r\\\mathbb{C}\end{smallmatrix}\right) \overset{\mathcal{D}_0}{\underset{\mathcal{D}_1}{\overset{\longrightarrow}{\underset{\longrightarrow}{\longleftarrow \mathcal{I} \longrightarrow}}}} \left(\begin{smallmatrix}\mathbb{E}\\\downarrow p\\\mathbb{B}\end{smallmatrix}\right) \qquad \text{where } \mathcal{D}_0 \circ \mathcal{I} = \text{id} = \mathcal{D}_1 \circ \mathcal{I}$$

in such a way that that the fibre category $\mathbb{F}_1$ over the terminal object in $\mathbb{C}$ is
the category of relations in the preorder fibration $\begin{smallmatrix}\mathbb{D}\\\downarrow\\\mathbb{E}\end{smallmatrix}$ on the fibre category $\mathbb{E}_1$
over the terminal object in $\mathbb{B}$. Formally, this requirement is expressed by the
change-of-base situation:

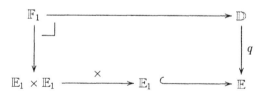

8.6.3. Proposition. *The $\lambda2$-fibration* $\begin{smallmatrix}\text{PFam}(\mathbf{PER})\\\downarrow\\\mathbf{PPER}\end{smallmatrix}$ *of parametric families of
PERs from Proposition 8.4.10 is relationally parametric (as formulated in the
previous definition).*

Proof. We construct fibrations $\begin{smallmatrix}\text{PFamRegSub}(\mathbf{PER})\\\downarrow\\\text{PFam}(\mathbf{PER})\end{smallmatrix}$ and $\begin{smallmatrix}\text{RFam}(\mathbf{PER})\\\downarrow\\\mathbf{RPER}\end{smallmatrix}$ giving us
a logic over parametric families of PERs, and a $\lambda2$-fibration in a reflexive

graph of $\lambda 2$-fibrations:

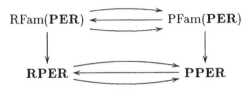

The "logical" fibration $\begin{array}{c}\text{PFamRegSub}(\mathbf{PER})\\ \downarrow\\ \text{PFam}(\mathbf{PER})\end{array}$ has in its total category:

objects   pairs $\langle A, f : n \to 1 \rangle$ where $f = (f^p, f^r)$ is a morphism in $\mathbf{PPER}$—*i.e.* an object of PFam($\mathbf{PER}$) over $n \in \mathbf{PPER}$— and A is a family of predicates on f of the form:

$$A = \big(A_{\vec{R}} \subseteq \mathbb{N}/f^p(\vec{R}) \big)_{\vec{R} \in \mathbf{PER}^n}.$$

morphisms $\langle A, f : n \to 1 \rangle \to \langle B, g : m \to 1 \rangle$ are morphisms $(\vec{h}, \alpha) : f \to g$ in UFam($\mathbf{PER}$) satisfying for $\vec{R} \in \mathbf{PER}^n$

$$[a] \in A_{\vec{R}} \subseteq \mathbb{N}/f^p(\vec{R}) \;\Rightarrow$$
$$\alpha_{\vec{R}}([a]) \in B_{\vec{h}(\vec{R})} \subseteq \mathbb{N}/(g \circ \vec{h})(\vec{R}).$$

The fibre category PFam($\mathbf{PER}$)$_1$ over the terminal object $1 \in \mathbf{PPER}$ is the category $\mathbf{PER}$ of PERs. Thus the fibre category RFam($\mathbf{PER}$)$_1$ over the terminal object $1 \in \mathbf{RPER}$ must be the category RegRel($\mathbf{PER}$) of regular relations $A \subseteq \mathbb{N}/R \times \mathbb{N}/R'$, obtained in the change-of-base situation:

$$
\begin{array}{ccccc}
\text{RegRel}(\mathbf{PER}) & \longrightarrow & \text{RegSub}(\mathbf{PER}) & \longrightarrow & \text{PFamRegSub}(\mathbf{PER}) \\
\downarrow\;\lrcorner & & \downarrow\;\lrcorner & & \downarrow \\
\mathbf{PER} \times \mathbf{PER} & \xrightarrow{\;\times\;} & \mathbf{PER} & \lhook\joinrel\longrightarrow & \text{PFam}(\mathbf{PER})
\end{array}
$$

as required in Definition 8.6.2. Since the fibration $\begin{array}{c}\text{RFam}(\mathbf{PER})\\ \downarrow\\ \mathbf{RPER}\end{array}$ must be a $\lambda 2$-fibration we know that the morphisms $1 \to \Omega$ to the generic object Ω in $\mathbf{RPER}$ must be regular relations on PERs. We thus define a category $\mathbf{RPER}$ with mappings between relations as morphisms.

objects   pairs of maps $\big(n \xrightarrow{\vec{f}} p \xleftarrow{\vec{f'}} n' \big)$ in $\mathbf{PPER}$ with common codomain $p \in \mathbf{PPER}$.

morphisms $(n \xrightarrow{\vec{f}} p \xleftarrow{\vec{f'}} n') \longrightarrow (m \xrightarrow{\vec{g}} q \xleftarrow{\vec{g'}} m')$ are triples $(\vec{h}, \vec{h'}, \vec{\varphi})$ of sequences where

$$\vec{h} \colon n \longrightarrow m \qquad \text{and} \qquad \vec{h'} \colon n' \longrightarrow m'$$

are morphisms in **PPER**, and where $\vec{\varphi} = (\varphi_1, \ldots, \varphi_q)$ is a sequence of functions φ_j in the function space

$$\prod_{\vec{R} \in \mathbf{PER}^n, \vec{S} \in \mathbf{PER}^{n'}} \left[\prod_{1 \leq i \leq p} P\big(\mathbb{N}/f_i^p(\vec{R}) \times \mathbb{N}/f_i'^p(\vec{S})\big) \right]$$
$$\Longrightarrow P\big(\mathbb{N}/(g_j \circ \vec{h})(\vec{R}) \times \mathbb{N}/(g_j' \circ \vec{h'})(\vec{S})\big).$$

We leave it to the reader that this category has finite products: $(0 \to 0 \leftarrow 0)$ is terminal object, and the product of $(n \to p \leftarrow n')$ and $(m \to q \leftarrow m')$ is $(n + m \to p + q \leftarrow n' + m')$. As generic object we shall take the pair of identities $(1 \to 1 \leftarrow 1)$.

There are obvious projection functors **PPER** $\longleftarrow$ **RPER** $\longrightarrow$ **PPER**, namely $n \leftarrow\!\mid (n \xrightarrow{\vec{f}} p \xleftarrow{\vec{f'}} n') \mapsto n'$. But there is also a functor **PPER** $\longrightarrow$ **RPER**, namely

$$\begin{cases} n \mapsto (n \xrightarrow{\mathrm{id}} n \xleftarrow{\mathrm{id}} n) \\ [\vec{h} \colon n \to m] \mapsto [(\vec{h}, \vec{h}, \vec{h}^r) \colon (n \xrightarrow{\mathrm{id}} n \xleftarrow{\mathrm{id}} n) \longrightarrow (m \xrightarrow{\mathrm{id}} m \xleftarrow{\mathrm{id}} m)] \end{cases}$$

where for $j \leq n$

$$h_j^r \in \prod_{\vec{R}, \vec{S} \in \mathbf{PER}^n} \left[\prod_{1 \leq i \leq n} P(\mathbb{N}/R_i \times \mathbb{N}/S_i) \right] \Rightarrow P\big(\mathbb{N}/h_j^p(\vec{R}) \times \mathbb{N}/h_j^p(\vec{S})\big)$$

as in Definition 8.4.8. We thus already have a reflexive graph **RPER** $\underset{\longrightarrow}{\overset{\longrightarrow}{\rightleftarrows}}$ **PPER** between base categories. It is easy to see that these functors preserve finite products (and also the generic object).

We turn to the fibration $\begin{array}{c} \mathrm{RFam}(\mathbf{PER}) \\ \downarrow \\ \mathbf{RPER} \end{array}$. The objects over $(n \xrightarrow{\vec{f}} p \xleftarrow{\vec{f'}} n')$ are the morphisms $(n \xrightarrow{\vec{f}} p \xleftarrow{\vec{f'}} n') \longrightarrow (1 \xrightarrow{\mathrm{id}} 1 \xleftarrow{\mathrm{id}} 1)$ in **PPER**. These consist of triples $(h \colon n \to 1, h' \colon n' \to 1, \varphi)$, where for sequences $\vec{R} \in \mathbf{PER}^n$, $\vec{S} \in \mathbf{PER}^{n'}$ of PERs and $A_i \subseteq \mathbb{N}/f_i^p(\vec{R}) \times \mathbb{N}/f_i'^p(\vec{S})$ of relations, we have a relation $\varphi_{\vec{R}, \vec{S}}(\vec{A}) \subseteq \mathbb{N}/h^p(\vec{R}) \times \mathbb{N}/h'^p(\vec{S})$. Morphisms $(h, h', \varphi) \to (k, k', \psi)$ in this fibre are pairs of maps

$$\alpha \colon h \longrightarrow k \text{ over } n \qquad \text{and} \qquad \alpha' \colon h' \longrightarrow k' \text{ over } n'$$

in UFam(**PER**) for which we have for all $\vec{R} \in \mathbf{PER}^n$, $\vec{S} \in \mathbf{PER}^{n'}$ and $A_i \subseteq \mathbb{N}/f^p(\vec{R}) \times \mathbb{N}/f_i'^p(\vec{S})$

$$([a],[b]) \in \varphi_{\vec{R},\vec{S}}(\vec{A}) \subseteq \mathbb{N}/h^p(\vec{R}) \times \mathbb{N}/h'^p(\vec{S}) \Rightarrow$$
$$(\alpha_{\vec{R}}([a]), \alpha'_{\vec{S}}([b])) \in \psi_{\vec{R},\vec{S}}(\vec{A}) \subseteq \mathbb{N}/k^p(\vec{R}) \times \mathbb{N}/k'^p(\vec{S}).$$

Substitution along a morphism in **RPER** is done by composition. It is not hard to see that these fibre categories are Cartesian closed (by seeing the objects as "logical relations", essentially using the constructions for the fibred CCC-structure in the proof of Proposition 8.4.10), and that substitution functors preserve this CCC-structure.

Products $\prod$ are most interesting. Consider an object (h, h', φ) over

$$\left(n \xrightarrow{\;\vec{f}\;} p \xleftarrow{\;\vec{f'}\;} n'\right) \times \left(1 \xrightarrow{\;\mathrm{id}\;} 1 \xleftarrow{\;\mathrm{id}\;} 1\right) = \left(n+1 \xrightarrow{\;\vec{f},\mathrm{id}\;} p+1 \xleftarrow{\;\vec{f'},\mathrm{id}\;} n'+1\right)$$

so that $\varphi_{(\vec{R},U),(\vec{S},V)}(\vec{A}, B) \subseteq \mathbb{N}/h^p(\vec{R}, U) \times \mathbb{N}/h'^p(\vec{S}, V)$ is a relation for $\vec{R}, U \in \mathbf{PER}^{n+1}$, $\vec{S}, V \in \mathbf{PER}^{n'+1}$ and $A_i \subseteq \mathbb{N}/f_i^p(\vec{R}, U) \times \mathbb{N}/f_i'^p(\vec{S}, V)$, $B \subseteq \mathbb{N}/U \times \mathbb{N}/V$. We have to produce an object $\forall(h, h', \varphi)$ over $\left(n \xrightarrow{\;\vec{f}\;} p \xleftarrow{\;\vec{f'}\;} n'\right)$. We define

$$\forall(h, h', \varphi) = (\textstyle\prod(h) \colon n \to 1, \prod(h') \colon n' \to 1, \forall(\varphi))$$

where $\prod(h)$, $\prod(h')$ are as defined in the proof of Proposition 8.4.10, and $\forall(\varphi)$ is

$$\forall(\varphi)_{\vec{R},\vec{S}}(\vec{A}) = \{([a]_{\prod(h)^p(\vec{R})}, [a']_{\prod(h')^p(\vec{S})}) \mid$$
$$\forall U, V \in \mathbf{PER}. \forall B \subseteq \mathbb{N}/U \times \mathbb{N}/V.$$
$$([a]_{h^p(\vec{R},U)}, [a']_{h'^p(\vec{S},V)}) \in \varphi_{(\vec{R},U),(\vec{S},V)}(\vec{A}, B)\}.$$

We leave it to the interested reader to check that this makes $\begin{smallmatrix} \mathrm{RFam}(\mathbf{PER}) \\ \downarrow \\ \mathbf{RPER} \end{smallmatrix}$ a (split) $\lambda2$-fibration. It remains to show that we have a reflexive graph $\mathrm{RFam}(\mathbf{PER}) \rightrightarrows \mathrm{PFam}(\mathbf{PER})$ over $\mathbf{RPER} \rightrightarrows \mathbf{PPER}$ preserving the $\lambda2$-structure. There are obvious functors $\mathrm{PFam}(\mathbf{PER}) \longleftarrow \mathrm{RFam}(\mathbf{PER}) \longrightarrow \mathrm{PFam}(\mathbf{PER})$ namely $h \leftarrowtail (h, h', \varphi) \mapsto h$. And there is also a functor $\mathrm{PFam}(\mathbf{PER}) \longrightarrow \mathrm{RFam}(\mathbf{PER})$ which maps

$$\begin{cases} [f \colon n \to 1] \mapsto \left[(f, f, f^r) \colon (n \xrightarrow{\;\mathrm{id}\;} n \xleftarrow{\;\mathrm{id}\;} n) \longrightarrow (1 \to 1 \leftarrow 1)\right] \\ \qquad \alpha \mapsto (\alpha, \alpha). \end{cases}$$

This functor is well-defined by the abstraction condition for morphisms in $\mathrm{PFam}(\mathbf{PER})$. It is not hard to check that it preserves the $\lambda2$-structure as described in the proof of Proposition 8.4.10. $\qquad\square$

8.6.4. Remarks. (i) The $\lambda2$-fibration $\begin{array}{c}\text{RFam}(\textbf{PER})\\\downarrow\\\textbf{RPER}\end{array}$ of relations on PERs described in the above proof looks quite complicated, and possibly even *ad hoc*. But it follows in essence from a general construction in [293]. This construction applies to (suitably complete) small $\lambda2$-fibrations, and yields an associated parametric $\lambda2$-fibration.

(ii) The structure that makes $\begin{array}{c}\text{RFam}(\textbf{PER})\\\downarrow\\\textbf{RPER}\end{array}$ a $\lambda2$-fibration may be called the "logical relations $\lambda2$-structure". It is explicitly described in logical terms in [340, 273].

(iii) Earlier in this section we have put a fibration on top of the base category of a polymorphic fibration to get a "logic on kinds", and we have put a fibration on top of the total category of a polymorphic fibration to obtain a "logic on types". The latter will turn out to be a fibration in the 2-category **Fib**($\mathbb{B}$) of fibrations on the (fixed) base category $\mathbb{B}$ of the polymorphic fibration. What happens in the previous proof is that we have a logic of relations both on kinds and on types. Categorically it involves putting a fibration on top of (the product with itself of the) polymorphic fibration $\begin{array}{c}\text{PFam}(\textbf{PER})\\\downarrow\\\textbf{PPER}\end{array}$ in the 2-category **Fib** of fibrations over arbitrary bases, as in:

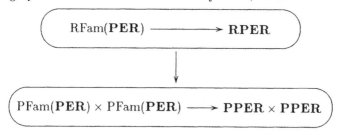

Details about fibrations in 2-categories **Fib**($\mathbb{B}$) and **Fib** of fibrations will be given in Section 9.4.

(iv) It is not clear in what sense the functor PFam(**PER**) $\rightarrow$ RFam(**PER**) used in the proof may be seen as an equality functor (as used in [273]) within the logic given by the structure in (iii).

(v) Essentially, all that we have done is describe one relationally parametric (PER-) model of second order PTT (in fibred form). We do not claim to have covered a substantial part of the theory on parametricity, and refer to the literature [286, 340, 11, 149, 117, 204, 84, 273, 2, 118, 291, 26, 326] for more information and further references.

Exercises

8.6.1.　(From [161]) Let $\mathbf{PER}_{<:}$ be the category of "indexed PERs and inclusions".
It has

objects　　　　triples (I, R, R') where $I = (I, E)$ is an ω-set and R, R'
are I-indexed PERs. Hence we may describe such an
object as a pair of parallel maps $R, R' : (I, E) \rightrightarrows \nabla\mathrm{PER}$
in ω-**Sets**. (Thus $\nabla\mathrm{PER} \times \nabla\mathrm{PER}$ is split generic object.)

morphisms　　$(I, R, R') \rightarrow (J, S, S')$ are morphisms $u : (I, E) \rightarrow (J, E)$
of ω-sets for which one has for all $i \in I$

$$R_i \subseteq R'_i \;\Rightarrow\; S_{u(i)} \subseteq S'_{u(i)}.$$

(i)　Check that the functor $\mathbf{PER}_{<:} \rightarrow \omega\text{-}\mathbf{Sets}$ given by $(I, R, R') \mapsto I$ is a
fibration, with posets as fibre categories.

(ii)　Show that there is a full and faithful fibred functor $\mathbf{PER}_{<:} \rightarrow$
$\mathrm{RegSub}(\omega\text{-}\mathbf{Sets})$ over ω-**Sets**. Hence this category of PERs with subtypings gives us a "sublogic" of the logic of regular subobjects of ω-sets.
In combination with the fibration $\begin{array}{c}\mathrm{UFam}(\mathbf{PER})\\\downarrow\\\omega\text{-}\mathbf{Sets}\end{array}$ we get a logic of kinds,
as above.

(iii)　Prove that the fibration $\begin{array}{c}\mathbf{PER}_{<:}\\\downarrow\\\omega\text{-}\mathbf{Sets}\end{array}$ of subtypings has fibred finite meets.

8.6.2.　Check that there are change-of-base situations:

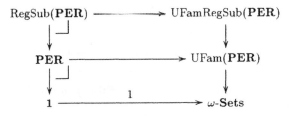

8.6.3.　Check that the $\forall$'s and $\exists$'s (over types and over kinds) in Example 8.6.1
are right and left adjoints to appropriate weakening functors.

8.6.4.　Investigate equality $\mathrm{Eq}_B(\beta, \beta') : \mathrm{Prop}$ for inhabitants $\beta, \beta' : B$ of $B : \mathrm{Kind}$ in
the (term model of the) logic over types in PTT, in terms of left adjoints
to lifted diagonals $\bar{\delta}$. See also what this means in the model of families of
regular subobjects of PERs in Example 8.6.1.

8.6.5.　Check that the morphism of split fibrations

$$\left(\begin{array}{c}\mathrm{PFam}(\mathbf{PER})\\\downarrow\\\mathbf{PPER}\end{array}\right) \longrightarrow \left(\begin{array}{c}\mathrm{RFam}(\mathbf{PER})\\\downarrow\\\mathbf{RPER}\end{array}\right)$$

in the proof of Proposition 8.6.3 preserves polymorphic products $\prod$.

8.6.6.   Describe inside ω-**Sets** an internal category **RegSub(PER)** of regular subobjects in the internal category **PER** in ω-**Sets**. Define an "internal fibration" **RegSub(PER)** $\rightarrow$ **PER** in ω-**Sets** and show that the situation

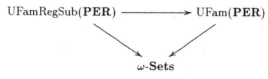

arises by externalisation.

Chapter 9

Advanced fibred category theory

This is the second chapter in this book—after Chapter 1—in which fibred category theory is studied on its own, and not in relation to a specific logic or type theory. This chapter collects some miscellaneous topics in fibred category theory, which are of a more advanced nature. Most of these will re-appear in the subsequent last two chapters on (first and higher order) dependent type theory. The notions and results that will be of greatest importance in the sequel are in Section 9.3 on quantification.

We start this chapter with opfibrations, which are suitable duals of fibrations; they involve an "initial lifting" property, as opposed to a "terminal lifting" property defining fibrations. These are then used in the second section to describe categorical structure in total categories of fibrations. This structure is often used for so-called "logical" predicates and relations. It gives rise to a categorical description of induction and co-induction principles associated with inductively and co-inductively defined data types. The third section gives a general notion of quantification along a certain class of maps in a base category, presented either via a "weakening and contraction comonad", or equivalently, via a "comprehension category". This general notion of quantification encompasses all forms that we have seen so far. The fourth section 9.4 deals with another generalisation: a fibration involves objects in a total category which are indexed by objects in a base category. Logically, propositions in a total category are indexed by types, to reason about the type theory of the base category. One can go a step further and investigate multiple levels of indexing. We already saw examples in Section 8.6 where we had propositions to reason about types, which in turn were indexed by kinds in polymorphic type theory. Such double levels of indexing are captured categorically by having one fibration being fibred over another fibration. This can basically happen in two ways, depending on whether one keeps the base category fixed or not. In the last two Sections 9.5 and 9.6 we describe Bénabou's notions of 'locally small' and 'definable' fibration. They involve representation in the base category of: homsets in the total category (for locally small fibrations), and best approximations of an object in the total category, with respect to a certain property (for definable fibrations). One of the main results is that small fibrations (coming as externalisations of internal categories) can be characterised as fibrations which are both locally small and globally small (where the latter means that the fibration has a generic object). And this will imply that "definable subfibrations" of a small fibration are again small fibrations.

9.1 Opfibrations and fibred spans

We recall that what determines a fibration is a "terminal lifting" property for morphisms in the base category. There is a dual notion of opfibration for which one has an "initial lifting" property instead: a functor $p: \mathbb{E} \to \mathbb{B}$ is an opfibration if p, considered as functor $p: \mathbb{E}^{\mathrm{op}} \to \mathbb{B}^{\mathrm{op}}$ between opposite categories, is a fibration. That is, if each morphism $pX \to I$ in $\mathbb{B}$ has a lifting $X \to \bullet$ in $\mathbb{E}$, which is initial in a suitable sense. One then says that $\mathbb{E}$ is **opfibred** over $\mathbb{B}$. Sometimes opfibrations are called cofibrations.

In this section these opfibrations will be investigated. Most examples of opfibrations are straightforward dualisations of examples of fibrations. In fibrations one can choose reindexing functors u^*, and similarly in opfibrations one can choose opreindexing functors $u_!$. One can think of u^* as restriction, and of $u_!$ as extension, see the examples after the proof of Proposition 9.1.4 below. If a fibration is also an opfibration then there are adjunctions $u_! \dashv u^*$. Lemma 9.1.2 below states that these adjunctions characterise such fibrations which are also opfibrations (called bifibrations). We further describe categories which are fibred over one category and opfibred over another. Such structures will be called fibred spans; they give rise to interesting examples. However, they form a side topic.

The following is a direct dualisation of Definition 1.1.3, which introduces fibrations.

9.1.1. Definition. Let $p: \mathbb{E} \to \mathbb{B}$ be a functor.

(i) A morphism $f: X \to Y$ in $\mathbb{E}$ over $u = pf: I \to J$ in $\mathbb{B}$ is called **op-cartesian** over u if it is Cartesian over u for $p: \mathbb{E}^{\mathrm{op}} \to \mathbb{B}^{\mathrm{op}}$. That is, if each morphism $g: X \to Z$ in $\mathbb{E}$ with $pg = v \circ u$ in $\mathbb{B}$, uniquely determines a map

$h: Y \to Z$ over v with $h \circ f = g$, as in:

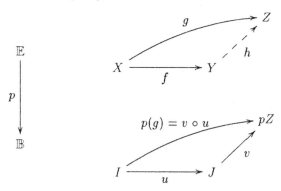

(ii) The functor $p: \mathbb{E} \to \mathbb{B}$ is an **opfibration** if $p: \mathbb{E}^{\mathrm{op}} \to \mathbb{B}^{\mathrm{op}}$ is a fibration. Equivalently, if above each morphism $pX \to J$ in $\mathbb{B}$ there is an opcartesian map $X \to Y$ in $\mathbb{E}$.

An opfibration $p: \mathbb{E} \to \mathbb{B}$ will be called **cloven** or **split** whenever $p: \mathbb{E}^{\mathrm{op}} \to \mathbb{B}^{\mathrm{op}}$ is a cloven or split fibration.

(iii) A **bifibration** is a functor which is at the same time a fibration and an opfibration.

Earlier we wrote $u^*(X) \to X$ for a Cartesian lifting of a morphism $u: I \to pX$. Similarly, we shall write $X \to u_!(X)$ or $X \to \coprod_u (X)$ for an opcartesian lifting of $u: pX \to J$ in a base category. The assignment $X \mapsto u_!(X)$ extends to a functor, which may be called **opreindexing**, **extension** or **sum** functor— because it will turn out to be a left adjoint to substitution u^*, see Lemma 9.1.2 below.

(The notation u_* (or $\prod_u$) is often used for a *right* adjoint to a functor u^*.)

We present two easy examples of opfibrations; they are obvious dualisations of the type theoretic 'codomain' and 'simple' fibrations to 'domain' and 'simple' opfibrations. First, the domain functor $\mathrm{dom}: \mathbb{B}^{\to} \to \mathbb{B}$ is an opfibration if and only if the category $\mathbb{B}$ has pushouts. Pushouts in $\mathbb{B}$ are precisely the opcartesian morphism in $\mathbb{B}^{\to}$. The fibre above $I \in \mathbb{B}$ is the **opslice** $I \backslash \mathbb{B}$.

Secondly, for a category $\mathbb{B}$ with binary coproducts $+$ there is a **simple opfibration** $\begin{smallmatrix} \mathrm{so}(\mathbb{B}) \\ \downarrow \\ \mathbb{B} \end{smallmatrix}$ where the total category $\mathrm{so}(\mathbb{B})$ has

objects   pairs (I, X) of objects in $\mathbb{B}$.

morphisms $(I, X) \to (J, Y)$ are pairs of morphisms $u: I \to J$ and $f: X \to J + Y$ in $\mathbb{B}$.

The first projection $\mathrm{so}(\mathbb{B}) \to \mathbb{B}$ is then a split opfibration. One obtains an opcartesian lifting starting from $(I, X) \in \mathrm{so}(\mathbb{B})$ over the domain of $u: I \to J$

in $\mathbb{B}$:

$$(I, X) \;-\;-\;\overset{?}{-}\;-\;-\;\blacktriangleright\; ??$$

$$I \xrightarrow{\quad u \quad} J$$

by taking $?? = (J, X)$ and $? = (u, \kappa')$, where $\kappa' \colon X \to J + X$ in the second coprojection. The fibre over $I \in \mathbb{B}$ will be called the **simple opslice category** and will be written as $I \backslash\!\backslash \mathbb{B}$.

The following result is quite useful.

9.1.2. Lemma. *A fibration is a bifibration (i.e. additionally is an opfibration) if and only if each reindexing functor u^* has a left adjoint (written as $u_!$ or $\coprod_u$).*

Notice that the Beck-Chevalley condition is not required for these $\coprod_u$'s in the lemma. But the result implies that every fibration with coproducts (*i.e.* with adjunctions $\coprod_u \dashv u^*$ satisfying Beck-Chevalley) is a bifibration. Codomain fibrations are thus bifibrations (see Proposition 1.9.8 (i)).

Proof. Let $\begin{smallmatrix} \mathbb{E} \\ \downarrow p \\ \mathbb{B} \end{smallmatrix}$ be a fibration. For a morphism $u \colon I \to J$ in $\mathbb{B}$ and objects $X \in \mathbb{E}_I$ and $Y \in \mathbb{E}_J$ consider the chain:

$$\mathbb{E}_J\big(\coprod\nolimits_u(X),\, Y\big) \overset{*}{\cong} \mathbb{E}_I\big(X,\, u^*(Y)\big) \cong \mathbb{E}_u\big(X,\, Y\big) \overset{**}{\cong} \mathbb{E}_J\big(\coprod\nolimits_u(X),\, Y\big)$$

where the isomorphism $\cong$ in the middle comes from the fact that p is a fibration. Then:

$$\begin{aligned}
\text{left adjoints } \coprod\nolimits_u \text{ exist} \;&\Leftrightarrow\; \text{isomorphisms } \overset{*}{\cong} \text{ exist} \\
&\Leftrightarrow\; \text{isomorphisms } \overset{**}{\cong} \text{ exist} \\
&\Leftrightarrow\; p \text{ is an opfibration.} \qquad \square
\end{aligned}$$

The next result is similar to Lemma 1.9.5 for family fibrations over sets.

9.1.3. Lemma. *Let $\mathbb{C}$ be a fixed category. The assignment*

$$\mathbb{A} \mapsto (\text{the functor category } \mathbb{C}^{\mathbb{A}})$$

extends to a functor $\mathbf{Cat}^{\mathrm{op}} \to \mathbf{Cat}$. The resulting split fibration will be written as $\begin{smallmatrix} \mathrm{Fam}(\mathbb{C}) \\ \downarrow \\ \mathbf{Cat} \end{smallmatrix}$. Then

$$\mathbb{C} \text{ is cocomplete} \;\Leftrightarrow\; \begin{smallmatrix} \mathrm{Fam}(\mathbb{C}) \\ \downarrow \\ \mathbf{Cat} \end{smallmatrix} \text{ is an opfibration.}$$

Proof. For a functor $U: \mathbb{A} \to \mathbb{B}$ the reindexing functor $U^*: \mathbb{C}^{\mathbb{B}} \to \mathbb{C}^{\mathbb{A}}$ is given by pre-composition with U, that is, by $U^* = (-) \circ U$. A left adjoint is thus given by left Kan extension. If $\mathbb{C}$ is cocomplete, then one can use the pointwise formula, see *e.g.* [187].

Conversely, if $\begin{smallmatrix} \mathrm{Fam}(\mathbb{C}) \\ \downarrow \\ \mathbf{Cat} \end{smallmatrix}$ is an opfibration, then, by the previous lemma, every functor $F \in \mathbb{C}^{\mathbb{A}}$ has a colimit $\coprod_{\mathbb{A}}(F)$ in the fibre over the terminal category 1—which is isomorphic to $\mathbb{C}$. $\qquad \square$

There is a similar result (due to Lawvere) which is of relevance in the semantics of parametrised specifications. This will be explained after the proof.

9.1.4. Proposition. *Let $\mathbb{B}$ be a cocomplete category with finite products, such that functors $I \times (-): \mathbb{B} \to \mathbb{B}$ preserve colimits. There is an indexed category* $\mathbf{AlgSpec}^{\mathrm{op}} \to \mathbf{Cat}$ *given by*

$$(\Sigma, \mathcal{A}) \mapsto \Big(\text{the category } \mathbf{FPCat}\big(\mathcal{C}\ell(\Sigma, \mathcal{A}), \mathbb{B}\big) \text{ of models of } (\Sigma, \mathcal{A}) \text{ in } \mathbb{B} \Big).$$

The resulting fibration $\begin{smallmatrix} \mathbf{Model}(\mathbb{B}) \\ \downarrow \\ \mathbf{AlgSpec} \end{smallmatrix}$ *is a bifibration.*

Recall from Section 3.3 that these categories of (functorial) models of algebraic specifications contain finite product preserving functors as objects, with natural transformations between them.

Proof. For a morphism $\phi: (\Sigma, \mathcal{A}) \to (\Sigma', \mathcal{A}')$ of algebraic specifications and a model $\mathcal{M}: \mathcal{C}\ell(\Sigma, \mathcal{A}) \to \mathbb{B}$ one obtains a functor $\phi_!(\mathcal{M}): \mathcal{C}\ell(\Sigma', \mathcal{A}') \to \mathbb{B}$ with the required universal property by pointwise left Kan extension. Some detailed computations—using that the functors $I \times (-): \mathbb{B} \to \mathbb{B}$ preserve colimits—prove that this functor $\phi_!(\mathcal{M})$ preserves finite products. $\qquad \square$

This result gives for a morphism $\phi: (\Sigma, \mathcal{A}) \to (\Sigma', \mathcal{A}')$ of algebraic specifications and models $\mathcal{M}: \mathcal{C}\ell(\Sigma, \mathcal{A}) \to \mathbb{B}$ and $\mathcal{N}: \mathcal{C}\ell(\Sigma', \mathcal{A}') \to \mathbb{B}$ of $(\Sigma, \mathcal{A})$ and $(\Sigma', \mathcal{A}')$ in $\mathbb{B}$, a bijective correspondence between natural transformations

$$\frac{\mathcal{M} \implies \phi^*(\mathcal{N}) = \mathcal{N} \circ \mathcal{C}\ell(\phi)}{\phi_!(\mathcal{M}) \implies \mathcal{N}}$$

If $\phi: (\Sigma, \mathcal{A}) \to (\Sigma', \mathcal{A}')$ is an inclusion then $\phi^*(\mathcal{N})$ is **restriction** and $\phi_!(\mathcal{M})$ is **extension**: $\phi_!(\mathcal{M})$ yields an interpretation of all the extra types and function symbols in $(\Sigma', \mathcal{A}')$. Moreover, the above correspondence tells us that it is the best possible extension. For more information, see [152] (and [160]) and the references mentioned there.

In [344] there is a similar bifibration of labelled transition systems over pointed sets of labels: reindexing u^* is restriction along a relabelling map, and opreindexing $u_!$ is extension.

There are many more examples of bifibrations. The classical example in algebra is given by modules over rings. For a homomorphism of rings $f: R \to S$ there is a reindexing functor $f^*: \mathrm{Mod}_S \to \mathrm{Mod}_R$ from modules over S to modules over R (much like in Exercise 1.1.11 for vector spaces). It has a left adjoint $f_!: \mathrm{Mod}_R \to \mathrm{Mod}_S$, called "extension of the base", see *e.g.* [190, XVI, 4] or [36, II, 4.7]. It plays an important rôle in descent theory for modules.

Fibred spans

The notion to be introduced next is due to Bénabou. It can be understood as follows. Indexed categories $\mathbb{B}^{\mathrm{op}} \to \mathbf{Cat}$ are categorical generalisations of presheaves $\mathbb{B}^{\mathrm{op}} \to \mathbf{Sets}$. Similarly, one can generalise profunctors $\mathbb{A}^{\mathrm{op}} \times \mathbb{B} \to \mathbf{Sets}$ to $\mathbb{A}^{\mathrm{op}} \times \mathbb{B} \to \mathbf{Cat}$. And the fibred versions of the latter are described as what we call fibred spans. This correspondence is made explicit in Proposition 9.1.8.

9.1.5. Definition. (i) A **fibred span** consists of a diagram

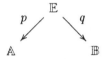

where p is a fibration and q is an opfibration and these fibred structures are compatible in the following way.

(a) Every morphism $I \to pX$ in $\mathbb{A}$ has a p-Cartesian lifting $\bullet \to X$ in $\mathbb{E}$ which is q-vertical. Also, every $qY \to J$ in $\mathbb{B}$ has an opcartesian lifting $Y \to \bullet$ in $\mathbb{E}$ which is p-vertical.

(b) Consider a commuting diagram in $\mathbb{E}$

$$
\begin{array}{ccc}
U & \xrightarrow{\;\;k\;\;} & X \\
{\scriptstyle g}\downarrow & & \downarrow{\scriptstyle f} \\
V & \xrightarrow{\;\;h\;\;} & Y
\end{array}
$$

where f, g are q-vertical and h, k are p-vertical. Then

$$
\left.\begin{array}{r}
h, k \text{ are } p\text{-Cartesian} \\
f \text{ is } q\text{-opcartesian}
\end{array}\right\} \Rightarrow g \text{ is } q\text{-opcartesian.}
$$

Or, equivalently,

$$
\left.\begin{array}{r}
f, g \text{ are } q\text{-opcartesian} \\
k \text{ is } p\text{-Cartesian}
\end{array}\right\} \Rightarrow h \text{ is } p\text{-Cartesian.}
$$

(ii) A fibred span will be called **split** if both the fibration p and the opfibration q are split, and the compatibility conditions (a)+(b) hold for the morphisms given by the splitting and opsplitting.

It is easy to see that the notion of fibred span comprises both fibrations, namely as fibred spans $\overset{\mathbb{E}}{\underset{\mathbb{B}\quad 1}{\swarrow\,\searrow}}$ and opfibrations, as fibred spans $\overset{\mathbb{E}}{\underset{1\quad \mathbb{B}}{\swarrow\,\searrow}}$. For this reason, Street [317] calls these fibred spans fibrations. What we call fibrations and opfibrations then appear as special cases.

The next result gives rise to many more examples of fibred spans. One half of it appeared as Exercise 1.4.6.

9.1.6. Lemma. *Consider two functors* $\mathbb{A} \xrightarrow{F} \mathbb{C} \xleftarrow{G} \mathbb{B}$ *with common codomain* $\mathbb{C}$. *The comma category* $(F \downarrow G)$ *together with the associated projection functors to* $\mathbb{A}$ *and* $\mathbb{B}$ *forms a split fibred span*

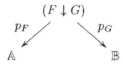

Proof. Let $\varphi \colon FX \to GY$ be an object of $(F \downarrow G)$. It is sent to:

$$\left(FX \xrightarrow{\varphi} GY \right)$$
$$p_F \nearrow \qquad \nwarrow p_G$$
$$X \qquad\qquad Y$$

For morphisms $u \colon I \to X$ in $\mathbb{A}$ and $v \colon Y \to J$ in $\mathbb{B}$ there are Cartesian liftings (on the left) and opcartesian liftings (on the right):

$$
\begin{array}{ccc}
 & F(u) & \\
FI & \longrightarrow & FX \\
u^*(\varphi) = {\Big\downarrow} & & {\Big\downarrow} \varphi \\
\varphi \circ F(u) & & \\
GY & \longrightarrow & GY \\
 & G(\mathrm{id}) &
\end{array}
\qquad
\begin{array}{ccc}
 & F(\mathrm{id}) & \\
FX & \longrightarrow & FX \\
\varphi {\Big\downarrow} & & {\Big\downarrow} \; v_!(\varphi) = \\
 & & G(v) \circ \varphi \\
GY & \longrightarrow & GJ \\
 & G(v) &
\end{array}
$$

$$
\begin{array}{ccc}
I & \xrightarrow{\;\;u\;\;} & X
\end{array}
\qquad\qquad
\begin{array}{ccc}
Y & \xrightarrow{\;\;v\;\;} & J
\end{array}
\qquad\qquad \square
$$

Fibred spans often arise in situations where one has a category $\mathbb{E}$ in which morphisms $X \to Y$ consist of two maps $X \rightrightarrows Y$ in some other category $\mathbb{B}$.

One then obtains fibred spans of the form $\overset{\mathbb{E}}{\underset{\mathbb{B} \quad \mathbb{B}^{\mathrm{op}}}{\swarrow \quad \searrow}}$. This is quite common for models of linear logic, see the first example below, and Exercise 9.1.6.

9.1.7. Examples. (i) Let K be an arbitrary set. Consider the identity functor Id: **Sets** $\to$ **Sets** and the exponent functor $(-) \Rightarrow K$: **Sets**^{op} $\to$ **Sets**. The resulting comma category $(\mathrm{Id} \downarrow (-) \Rightarrow K)$ is the category Game$_K$ of Lafont and Streicher [184]. It is thus fibred over **Sets** and opfibred over **Sets**^{op} in the fibred span

(ii) Vickers [338] defines a category of what he calls **topological systems**, which form a common generalisation of topological spaces and locales. We write **Loc** = **Frm**^{op} for the category which has locales (or frames, or complete Heyting algebras) as objects. A morphism $f: A \to B$ in **Loc** is a morphism of frames $f: B \to A$ in the reverse direction. It preserves arbitrary joins and finite meets.

An object of this category **TS** of topological systems is a triple $(A, X, \models)$, where A is a locale, X a set and $\models \subseteq X \times A$ a relation satisfying

$$x \models \bigvee_{i \in I} a_i \quad \Leftrightarrow \quad x \models a_i \text{ for some } i \in I$$
$$x \models a_1 \wedge \cdots \wedge a_n \quad \Leftrightarrow \quad x \models a_i \text{ for all } i.$$

Equivalently, $\models$ is a morphism of locales $\mathcal{P}X \to A$.

A morphism $(A, X, \models) \to (B, Y, \models)$ is a pair (f, g) where $f: A \to B$ is a morphism of locales and $g: X \to Y$ is an ordinary function, satisfying

$$x \models f(b) \quad \Leftrightarrow \quad g(x) \models b.$$

The resulting category **TS** of topological systems can thus be understood as the comma category obtained from the powerset functor $\mathcal{P}$: **Sets** $\to$ **Loc** and the identity **Loc** $\to$ **Loc**. Thus we have a fibred span

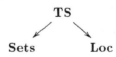

In Section 1.10 we have seen a "Grothendieck" correspondence between cloven fibrations on $\mathbb{B}$ and pseudo functors $\mathbb{B}^{\mathrm{op}} \to$ **Cat**. This correspondence extends to fibred spans $\overset{\bullet}{\underset{\mathbb{A} \quad \mathbb{B}}{\swarrow \quad \searrow}}$ and suitable pseudo functors $\mathbb{A}^{\mathrm{op}} \times \mathbb{B} \to$ **Cat**. The latter can be understood as "**Cat**-valued distributors". In order not to complicate matters, we shall restrict ourselves to the split case.

9.1.8. Proposition (See [316]). *Every split fibred span* $\begin{smallmatrix} & E & \\ \swarrow & & \searrow \\ A & & B \end{smallmatrix}$ *determines a functor* $A^{op} \times B \to \mathbf{Cat}$. *Conversely there is a generalised Grothendieck construction which yields for every such functor a split fibred span of the form* $\begin{smallmatrix} & \bullet & \\ \swarrow & & \searrow \\ A & & B \end{smallmatrix}$. *These constructions are mutually inverse.*

Proof. Given a split fibred span $\begin{smallmatrix} & E & \\ p\swarrow & & \searrow q \\ A & & B \end{smallmatrix}$ one defines a functor $\Phi \colon A^{op} \times B \to \mathbf{Cat}$ as follows. For $I \in A$ and $I' \in B$, take $\Phi(I, I')$ to be the category with

objects $X \in E$ with $pX = I$ and $qX = I'$.

morphisms $f \colon X \to Y$ are maps $f \colon X \to Y$ in E with $pf = \mathrm{id}$ and $qf = \mathrm{id}$.

For a morphism $(u, u') \colon (I, I') \to (J, J')$ in $A^{op} \times B$ one obtains a functor $\Phi(u, u') \colon \Phi(I, I') \to \Phi(J, J')$ by

$$\begin{cases} X \mapsto u^*(u'_!(X)) = u'_!(u^*(X)) \\ f \mapsto u^*(u'_!(f)) = u'_!(u^*(f)). \end{cases}$$

In the reverse direction, given such a $\Psi \colon A^{op} \times B \to \mathbf{Cat}$, let $\int \Psi$ be the category with

objects (I, I', X) where $X \in \Psi(I, I')$.

morphisms $(I, I', X) \to (J, J', Y)$ are triples $u \colon I \to J$ in A, $u' \colon I' \to J'$ in B and $f \colon \Psi(\mathrm{id}_I, u')(X) \to \Psi(u, \mathrm{id}_{J'})(Y)$ in $\Psi(I, J')$.

There are obvious projection functors $\begin{smallmatrix} & \int\Psi & \\ \swarrow & & \searrow \\ A & & B \end{smallmatrix}$ forming a fibred span. $\square$

The above result finds application in the (functorial) semantics of logics and type theories. Reindexing along a morphism of signatures (or specifications) in syntax works contravariantly on models, whereas reindexing along a morphism of models works covariantly. It is precisely this aspect that Goguen and Burstall [152] seek to capture in the notion of institution. Here we shall describe these phenomena via fibred spans. Algebraic specifications form again the paradigmatic example: there is a 'canonical' model functor $\mathbf{AlgSpec}^{op} \times \mathbf{FPCat} \to \mathbf{Cat}$ given by

$$\langle (\Sigma, \mathcal{A}), B \rangle \longmapsto \mathbf{FPCat}\big(\mathcal{C}\ell(\Sigma, \mathcal{A}), B\big).$$

A morphism (ϕ, F) in $\mathbf{AlgSpec}^{op} \times \mathbf{FPCat}$ is sent to the functor

$$\mathcal{M} \longmapsto F \circ \mathcal{M} \circ \mathcal{C}\ell(\phi).$$

The total category of the resulting fibred span is the category of categorical models described earlier in Exercise 2.2.4, in case $\mathcal{A} = \emptyset$—that is, if the specification consists only of a signature Σ without axioms.

Exercises

9.1.1.   (i) Establish a Grothendieck correspondence between functors $\mathbb{B} \to \mathbf{Cat}$
         and split opfibrations on $\mathbb{B}$.
         (ii) Prove that the composite of two opfibrations is an opfibration again.
9.1.2.   What are the opcartesian morphism for a codomain functor?
9.1.3.   Consider a category $\mathbb{B}$ with finite coproducts $(0, +)$.
         (i) Show that the simple opslice $0 \backslash\backslash \mathbb{B}$ over the initial object $0 \in \mathbb{B}$ is
              isomorphic to $\mathbb{B}$.
         (ii) Describe the opreindexing functor $I_! : \mathbb{B} \cong 0 \backslash\backslash \mathbb{B} \to I \backslash\backslash \mathbb{B}$ resulting from
              $0 \to I$.
         (iii) Define a full and faithful functor $I \backslash\backslash \mathbb{B} \to I \backslash \mathbb{B}$ from the simple to the
              ordinary opslice category in a commuting diagram:

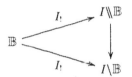

9.1.4.   Note that if $\begin{smallmatrix} & \mathbb{E} & \\ \swarrow & & \searrow \\ \mathbb{A} & & \mathbb{B} \end{smallmatrix}$ is a fibred span, then so is $\begin{smallmatrix} & \mathbb{E} & \\ \swarrow & & \searrow \\ \mathbb{B}^{\mathrm{op}} & & \mathbb{A}^{\mathrm{op}} \end{smallmatrix}$.

9.1.5.   (i) Verify that the fibred span $\begin{smallmatrix} & (F \downarrow G) & \\ \swarrow & & \searrow \\ \mathbb{A} & & \mathbb{B} \end{smallmatrix}$ in Lemma 9.1.6 can be obtained
              as instance of the generalised Grothendieck construction in Proposi-
              tion 9.1.8, applied to a functor $\mathbb{A}^{\mathrm{op}} \times \mathbb{B} \to \mathbf{Cat}$ with discrete categories
              as fibres.
         (ii) Check that the Grothendieck constructions for fibrations and for op-
              fibrations are also special cases of this generalised Grothendieck con-
              struction.
9.1.6.   Describe the dialectica category $\mathbf{G}\mathbb{C}$ from [244] in a fibred span $\begin{smallmatrix} & \mathbf{G}\mathbb{C} & \\ \swarrow & & \searrow \\ \mathbb{C} & & \mathbb{C}^{\mathrm{op}} \end{smallmatrix}$.

9.2 Logical predicates and relations

Consider a coherent predicate logic over a (simple) type theory with finite
products $(1, \times)$ and coproducts $(0, +)$ and exponents $\to$ for types. Propo-
sitions are written as $(x : \sigma \vdash \varphi : \mathsf{Prop})$, where $x : \sigma$ is the only possible free
variable in the proposition φ. Thus one can view φ as a predicate on σ. A
category of such predicates can be formed by stipulating that a morphism
$(x : \sigma \vdash \varphi : \mathsf{Prop}) \to (y : \tau \vdash \psi : \mathsf{Prop})$ is given by the equivalence class (under
conversion) of a term M with

$$x : \sigma \vdash M : \tau \qquad \text{and} \qquad x : \sigma \mid \varphi \vdash \psi[M/y].$$

This gives us a category of predicates; it is essentially the total category of a logic as described in Section 3.1. A reasonable question to ask is how to obtain finite products and coproducts and exponents in such a (total) category of predicates. Finite products are easy; they are given by the formulas,

$$1 = (x{:}1 \vdash \top{:}\mathsf{Prop})$$
$$(x{:}\sigma \vdash \varphi{:}\mathsf{Prop}) \times (y{:}\tau \vdash \psi{:}\mathsf{Prop})$$
$$= (z{:}\sigma \times \tau \vdash \varphi[\pi z/x] \wedge \psi[\pi' z/y]{:}\mathsf{Prop}).$$

These finite products of predicates thus sit over the finite products of their underlying types. Finite coproducts are slightly more complicated:

$$0 = (x{:}0 \vdash \bot{:}\mathsf{Prop})$$
$$(x{:}\sigma \vdash \varphi{:}\mathsf{Prop}) + (y{:}\tau \vdash \psi{:}\mathsf{Prop})$$
$$= (z{:}\sigma + \tau \vdash (\exists x{:}\sigma.\, z = \kappa x \wedge \varphi) \vee (\exists y{:}\tau.\, z = \kappa' y \wedge \psi){:}\mathsf{Prop}).$$

And exponents are given by

$$(x{:}\sigma \vdash \varphi{:}\mathsf{Prop}) \Rightarrow (y{:}\tau \mid \psi{:}\mathsf{Prop})$$
$$= (f{:}\sigma \to \tau \vdash \forall x{:}\sigma.\, (\varphi \supset \psi[(fx)/y]){:}\mathsf{Prop}).$$

Predicates with this BiCCC-structure are often referred to as "logical predicates". Similar formulas apply for relations, and in this context one talks about "logical relations". They can be used in the analysis of properties of type theories (like strong normalisation), see *e.g.* [311, 228] and the references there for more information. In this section we describe the categorical aspects of the above formulas: they can be described as products, coproducts and exponents in total categories of fibrations. Parts of these descriptions already occurred in Section 8.5.

The issue in this section is the interaction between local structure (in fibre categories) and total structure (in total categories), and the use in logic of the global structure. The first result states that if a base category has a certain type of limit, then these limits exist fibrewise if and only if they exist in the total category (and are preserved by the functor). The same holds for colimits if one works with a fibration that is additionally an opfibration (*i.e.* with a bifibration). We discuss these (folklore) results in some detail. A further analysis of this interaction between local and global structure may be found in [125–127] (along the lines of Exercise 1.8.11). The global structure is used towards the end of this section in a uniform description of the (logical) induction principles associated with (co-)inductively defined data types, following [128, 130]. See also [124].

Although the abovementioned motivation comes from logic, there is nothing in this section that holds only for preorder fibrations. But we shall use logical notation like $\top, \wedge, \bot, \vee, \supset$ for products, coproducts and exponents in fibre categories—as in Section 8.5. This allows us to distinguish this fibre-wise structure from such structure in total categories. But it is not meant to suggest that fibres are preorders.

The next result holds for arbitrary limits, but for simplicity we only consider finite products. The general case is left as Exercises 9.2.3 and 9.2.4.

9.2.1. Proposition. *Consider a fibration* $\begin{smallmatrix} \mathbb{E} \\ \downarrow p \\ \mathbb{B} \end{smallmatrix}$ *with finite products in its base category* $\mathbb{B}$. *Then* p *has finite products in* **Fib**$(\mathbb{B})$ *if and only if* p *has finite products in* **Fib**. *That is,* p *has fibred finite products (i.e. finite products in each fibre category, preserved by reindexing) if and only if the total category* $\mathbb{E}$ *has finite products via fibred functors, and the functor* p *strictly preserves these products.*

Explicitly, given fibred finite products $(\top, \wedge)$ *one can define in* $\mathbb{E}$:

$$1 = \top \in \mathbb{E}_1 \qquad and \qquad X \times Y = \pi^*(X) \wedge \pi'^*(Y) \in \mathbb{E}_{I \times J}$$

where $X \in \mathbb{E}_I$ *and* $Y \in \mathbb{E}_J$. *Conversely, given finite products* $(1, \times)$ *in* $\mathbb{E}$ *one takes in the fibre over* I,

$$\top = {!}_I^*(1) \in \mathbb{E}_I \qquad and \qquad X \wedge Y = \delta_I^*(X \times Y) \in \mathbb{E}_I,$$

where ${!}_I \colon I \dashrightarrow 1$ *and* $\delta_I = \langle \mathrm{id}, \mathrm{id} \rangle \colon I \to I \times I$.

Proof. Lemma 8.5.2 (i) states already that $1 \in \mathbb{E}$ and $\times$ as defined above are finite products in $\mathbb{E}$. And p preserves them by construction. For the fibration p to have finite products in **Fib** we need to check that the induced product functor $\times \colon \mathbb{E} \times \mathbb{E} \to \mathbb{E}$ over $\times \colon \mathbb{B} \times \mathbb{B} \to \mathbb{B}$ is fibred. This follows easily:

$$\begin{aligned}
(u \times v)^*(X \times Y) &\cong (u \times v)^* \pi^*(X) \wedge (u \times v)^* \pi'^*(Y) \\
&\cong \pi^* u^*(X) \wedge \pi'^* v^*(Y) \\
&= u^*(X) \times v^*(Y).
\end{aligned}$$

Conversely we show that $X \wedge Y = \delta_I^*(X \times Y) \in \mathbb{E}_I$ is the binary product of X and Y in the fibre over I:

$$\mathbb{E}_I\big(Z, X \wedge Y\big) = \mathbb{E}_\delta\big(Z, X \times Y\big) \cong \mathbb{E}_I\big(Z, X\big) \times \mathbb{E}_I\big(Z, Y\big).$$

The latter because p strictly preserves the binary products. These products $\wedge$ in the fibres are preserved by reindexing functors: for $u \colon J \to I$,

$$\begin{aligned}
u^*(X \wedge Y) &= u^* \delta^*(X \times Y) \\
&\cong \delta^*(u \times u)^*(X \times Y) \\
&\cong \delta^*(u^*(X) \times u^*(Y)) \quad \text{because } \times \text{ is a fibred functor} \\
&= u^*(X) \wedge u^*(Y). \qquad\qquad\qquad\qquad\qquad\qquad\quad \square
\end{aligned}$$

For coproducts the situation is slightly more complicated, and some additional assumptions are needed to get a smooth correspondences between fibrewise and global structure as for products. We shall use opreindexing, as conveniently described via adjunctions $\coprod_u \dashv u^*$ in Lemma 9.1.2. Part of the next result occurs as [105, Corollary 4.3]. As before, we only do the finite case.

9.2.2. Proposition. *Let* $\begin{smallmatrix} \mathbb{E} \\ \downarrow p \\ \mathbb{B} \end{smallmatrix}$ *be a bifibration with finite coproducts in its base category* $\mathbb{B}$.

(i) *Every fibre category has finite coproducts* $(\bot, \vee)$ *if and only if the total category* $\mathbb{E}$ *has finite coproducts* $(0, +)$ *and* p *strictly preserves these.*

The formulas are the following. Given $(\bot, \vee)$, *one takes in* $\mathbb{E}$:

$$0 = \bot \in \mathbb{E}_0 \qquad and \qquad X + Y = \coprod_\kappa(X) \vee \coprod_{\kappa'}(Y) \in \mathbb{E}_{I+J}$$

where $X \in \mathbb{E}_I$ *and* $Y \in \mathbb{E}_J$. *Conversely, given* $(0, +)$ *one defines over* I:

$$\bot = \coprod_{!_I}(0) \qquad and \qquad X \vee Y = \coprod_{\nabla_I}(X + Y),$$

where $!_I: 0 \dashrightarrow I$ *and* $\nabla_I = [\mathrm{id}, \mathrm{id}]: I + I \to I$.

(ii) *Under the additional assumptions that the category* $\mathbb{B}$ *is extensive (i.e. that its coproducts are disjoint and universal, see Section 1.5), and that the coproducts* $\coprod_u$ *satisfy Beck-Chevalley, one obtains the following strengthening:* p *has finite coproducts in* **Fib**$(\mathbb{B})$ *if and only if* p *has finite coproducts in* **Fib***.*

An alternative formulation of the coproduct $+$ in the total category $\mathbb{E}$ in terms of products $\prod$ instead of coproducts $\coprod$ occurs in Exercise 9.5.9.

Proof. (i) We do the binary case only. Consider $X + Y \in \mathbb{E}_{I+J}$ as defined above. For $Z \in \mathbb{E}$ above K, we have:

$$\begin{aligned}
\mathbb{E}(X + Y, Z) &\cong \coprod_{u:I+J \to K} \mathbb{E}_{I+J}\Big(\coprod_u(X + Y), Z\Big) \\
&\cong \coprod_{u:I+J \to K} \mathbb{E}_{I+J}\Big(\coprod_u \coprod_\kappa(X) \vee \coprod_u \coprod_{\kappa'}(Y), Z\Big) \\
&\cong \coprod_{u:I+J \to K} \mathbb{E}_{I+J}\Big(\coprod_{u \circ \kappa}(X) \vee \coprod_{u \circ \kappa'}(Y), Z\Big) \\
&\cong \coprod_{v:I \to K \ w:J \to K} \mathbb{E}_I\Big(\coprod_v(X), Z\Big) \times \mathbb{E}_I\Big(\coprod_w(Y), Z\Big) \\
&\cong \coprod_{v:I \to K} \mathbb{E}_I\Big(\coprod_v(X), Z\Big) \times \coprod_{w:J \to K} \mathbb{E}_I\Big(\coprod_w(Y), Z\Big) \\
&\cong \mathbb{E}(X, Z) \times \mathbb{E}(Y, Z).
\end{aligned}$$

Conversely, we simply have,

$$\mathbb{E}_I\Big(X \vee Y, Z\Big) \cong \mathbb{E}_{\nabla_I}\Big(X + Y, Z\Big) \cong \mathbb{E}_I\Big(X, Z\Big) \times \mathbb{E}_I\Big(Y, Z\Big)$$

because p preserves coproducts.

(ii) In one direction we have to show that $+: \mathbb{E} \times \mathbb{E} \to \mathbb{E}$ is fibred over $+: \mathbb{B} \times \mathbb{B} \to \mathbb{B}$. This follows from the isomorphisms:

$$
\begin{aligned}
(u + v)^*(X + Y) &\cong (u + v)^* \textstyle\coprod_\kappa (X) \vee (u + v)^* \textstyle\coprod_{\kappa'} (Y) \\
&\cong \textstyle\coprod_\kappa u^*(X) \vee \textstyle\coprod_{\kappa'} v^*(Y) \ \text{ by Beck-Chevalley,} \\
&\qquad\qquad\qquad\qquad\qquad\qquad\text{using Exercise 1.5.7} \\
&= u^*(X) + v^*(Y).
\end{aligned}
$$

And in the other direction, assuming that this global functor $+$ is fibred, we have to show that the induced coproducts $\vee$ in the fibres are preserved under reindexing:

$$
\begin{aligned}
u^*(X \vee Y) &= u^* \textstyle\coprod_\nabla (X + Y) \\
&\cong \textstyle\coprod_\nabla (u + u)^*(X + Y) \ \text{ by Beck-Chevalley} \\
&\simeq \textstyle\coprod_\nabla (u^*(X) + u^*(Y)) \ \text{ because } + \text{ is a fibred functor} \\
&= u^*(X) \vee u^*(Y). \qquad\qquad\qquad\qquad\qquad\qquad\square
\end{aligned}
$$

In a next step also local and global distributivity can be related. Therefore we use the additional assumption that Frobenius holds.

9.2.3. Proposition. *Let $\begin{smallmatrix} \mathbb{E} \\ \downarrow p \\ \mathbb{B} \end{smallmatrix}$ be a fibration with coproducts $\coprod_u$ satisfying Beck-Chevalley and Frobenius. Assume $\mathbb{B}$ is a distributive category. Then, p is a distributive fibred category if and only if the total category $\mathbb{E}$ is a distributive category and p strictly preserves this structure.*

Proof. First we assume that all fibre categories are distributive. We consider arbitrary objects $Z \in \mathbb{E}_K$, $X \in \mathbb{E}_I$ and $Y \in \mathbb{E}_J$, and write φ for the inverse of the canonical map $[\mathrm{id} \times \kappa, \mathrm{id} \times \kappa']: (K \times I) + (K \times J) \to K \times (I + J)$ in $\mathbb{B}$. We can prove distributivity in $\mathbb{E}$ as follows.

$$
\begin{aligned}
Z \times (X + Y) &= \pi^*(Z) \wedge \pi'^*(\textstyle\coprod_\kappa (X) \vee \textstyle\coprod_{\kappa'} (Y)) \\
&\cong \pi^*(Z) \wedge (\textstyle\coprod_{\mathrm{id} \times \kappa} (\pi'^*(X)) \vee \textstyle\coprod_{\mathrm{id} \times \kappa'} (\pi'^*(Y))) \\
&\qquad\qquad \text{by Beck-Chevalley} \\
&\cong (\pi^*(Z) \wedge \textstyle\coprod_{\mathrm{id} \times \kappa} (\pi'^*(X))) \vee (\pi^*(Z) \wedge \textstyle\coprod_{\mathrm{id} \times \kappa'} (\pi'^*(Y))) \\
&\qquad\qquad \text{by distributivity} \\
&\cong \textstyle\coprod_{\mathrm{id} \times \kappa} (\pi^*(Z) \wedge \pi'^*(X)) \vee \textstyle\coprod_{\mathrm{id} \times \kappa'} (\pi^*(Z) \wedge \pi'^*(Y)) \\
&\qquad\qquad \text{by Frobenius}
\end{aligned}
$$

$$\cong \coprod_\varphi \coprod_{\mathrm{id} \times \kappa} (Z \times X) \vee \coprod_\varphi \coprod_{\mathrm{id} \times \kappa'} (Z \times Y)$$
$$\text{because } \varphi \text{ is an isomorphism}$$
$$\cong \coprod_\kappa (Z \times X) \vee \coprod_{\kappa'} (Z \times Y)$$
$$= (Z \times X) + (Z \times Y).$$

In the reverse direction, assuming distributivity in $\mathbb{E}$, distributivity in the fibres follows from the following computation.

$$
\begin{aligned}
Z \wedge (X \vee Y) &= Z \wedge \coprod_\nabla (X + Y)\\
&\cong \coprod_\nabla (\nabla^*(Z) \wedge (X + Y)) \quad \text{by Frobenius}\\
&= \coprod_\nabla \delta^*(\nabla^*(Z) \times (X + Y))\\
&\cong \coprod_\nabla \delta^* \varphi^*(\nabla^*(Z) \times X + \nabla^*(Z) \times Y) \quad \text{by distributivity}\\
&\cong \coprod_\nabla \delta^* \varphi^*(\nabla \times \mathrm{id} + \nabla \times \mathrm{id})^*(Z \times X + Z \times Y)\\
&\cong \coprod_\nabla (\delta + \delta)^*(Z \times X + Z \times Y)\\
&\cong \coprod_\nabla (\delta^*(Z \times X) + \delta^*(Z \times Y))\\
&= (Z \wedge X) \vee (Z \wedge Y). \qquad\qquad \square
\end{aligned}
$$

We notice that the proof requires only coproducts ($\coprod_\kappa \dashv \kappa^*$) along coprojections, satisfying Beck-Chevalley and Frobenius. So the assumptions in the proposition can actually be weakened.

Finally, we consider the relation between exponents in fibre categories and in total categories.

9.2.4. Proposition. *Let $\begin{smallmatrix}\mathbb{E}\\\downarrow p\\\mathbb{B}\end{smallmatrix}$ be a fibration with a Cartesian closed base category $\mathbb{B}$ and with simple products $\pi^* \dashv \prod_\pi$. If the fibre categories are Cartesian closed, then so is the total category and this global CCC-structure is strictly preserved by p. For $X \in \mathbb{E}_I$ and $Y \in \mathbb{E}_J$ one defines*

$$X \Rightarrow Y = \prod_\pi (\pi'^*(X) \supset \mathrm{ev}^*(Y)) \in \mathbb{E}_{I \Rightarrow J}.$$

If there are additionally right adjoints $\delta^* \dashv \prod_\delta$ to contraction functors, then the converse also holds: exponents $\supset$ in fibres can be obtained from exponents in $\mathbb{E}$ via the formula

$$X \supset Y = \Lambda(\mathrm{id})^*(X \Rightarrow \prod_\delta(Y))$$

where $\Lambda(\mathrm{id}): I \to (I \Rightarrow (I \times I))$.

Proof. The first part of the statement is Lemma 8.5.2 (ii). For the second part we note that in the fibre over $I \in \mathbb{B}$ we have

$$
\begin{aligned}
\mathbb{E}_I\big(Z \wedge X,\, Y\big) &= \mathbb{E}_I\big(\delta^*(Z \times X),\, Y\big) \\
&\cong \mathbb{E}_{I \times I}\big(Z \times X,\, \textstyle\prod_\delta(Y)\big) \\
&\cong \mathbb{E}_{\Lambda(\mathrm{id})}\big(Z,\, X \Rightarrow \textstyle\prod_\delta(Y)\big) \\
&\cong \mathbb{E}_I\big(Z,\, \Lambda(\mathrm{id})^*(X \Rightarrow \textstyle\prod_\delta(Y))\big). \qquad\qquad \square
\end{aligned}
$$

We discuss some examples and consequences of these results.

9.2.5. Examples. (i) If $\begin{smallmatrix}\mathbb{E}\\\downarrow p\\\mathbb{B}\end{smallmatrix}$ is a bifibration with fibred finite products and coproducts over a base category $\mathbb{B}$ with finite products and coproducts, then the total category $\mathbb{E}$ has finite products and coproducts, and p preserves them. This result is a consequence of Propositions 9.2.1 and 9.2.2 above. It applies in particular when p is the classifying fibration of a coherent predicate logic with $\top, \wedge, \bot, \vee, =$ and $\exists$ as described in the beginning of this section. This is a bifibration by Example 4.3.7 (i) together with Lemma 9.1.2. The resulting finite products and coproducts in the total category of predicates as described in Propositions 9.2.1 and 9.2.2 are the "logical predicate" products and coproducts as in the beginning of the section.

We mention two similar applications. If $\mathbb{B}$ is a coherent category with distributive coproducts, then its category of subobjects $\mathrm{Sub}(\mathbb{B})$ is a distributive category. And if $\mathbb{B}$ is Cartesian closed, then the associated scone $\mathrm{Sc}(\mathbb{B})$ is also Cartesian closed, and the functor (or fibration) $\mathrm{Sc}(\mathbb{B}) \to \mathbb{B}$ preserves the CCC structure, see Example 1.5.2 (ii) and Exercise 1.5.4.

(ii) Suppose $\begin{smallmatrix}\mathbb{E}\\\downarrow p\\\mathbb{B}\end{smallmatrix}$ is a fibration as before, and form the fibration $\begin{smallmatrix}\mathrm{Rel}(\mathbb{E})\\\downarrow\\\mathbb{B}\end{smallmatrix}$ of binary relations in p via the change-of-base situation

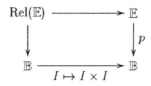

like in Section 4.8. It is easy to see that $\begin{smallmatrix}\mathrm{Rel}(\mathbb{E})\\\downarrow\\\mathbb{B}\end{smallmatrix}$ is also a bifibration with fibred finite products and coproducts, so that we may conclude that the total category $\mathrm{Rel}(\mathbb{E})$ of relations has finite products and coproducts, and that they are preserved by $\mathrm{Rel}(\mathbb{E}) \to \mathbb{B}$. If we apply this to a classifying fibration of a coherent logic, we get, for example, as global product of relations:

$$(x, x' \colon \sigma \vdash R(x, x') \colon \mathsf{Prop}) \times (y, y' \colon \tau \mid S(y, y') \colon \mathsf{Prop})$$
$$= (z, z' \colon \sigma \times \tau \vdash R(\pi z, \pi z') \wedge S(\pi' z, \pi' z') \colon \mathsf{Prop})$$

This is the "logical relations" product, see *e.g.* [125, 228].

For a fibration $\begin{smallmatrix} \mathbb{E} \\ \downarrow p \\ \mathbb{B} \end{smallmatrix}$ as above, we can define an equality functor $\mathrm{Eq} \colon \mathbb{B} \to$ $\mathrm{Rel}(\mathbb{E})$ by $I \mapsto \coprod_{\delta_I}(\top I)$—essentially as in the beginning of Section 4.8. For a morphism $u \colon I \to J$ in $\mathbb{B}$ we obtain $\mathrm{Eq}(u) \colon \mathrm{Eq}(I) \to \mathrm{Eq}(J)$ using that $\top I \to \mathrm{Eq}(I)$ is opcartesian. Notice that the statement "$\mathrm{Eq} \colon \mathbb{B} \to \mathrm{Rel}(\mathbb{E})$ preserves products $\times$" means that equality on a product is componentwise equality. This preservation of products is proved explicitly in Proposition 3.4.6).

If our fibration p is a fibred CCC on a CCC $\mathbb{B}$ and has simple products, then the same holds for $\begin{smallmatrix} \mathrm{Rel}(\mathbb{E}) \\ \downarrow \\ \mathbb{B} \end{smallmatrix}$. Hence the global category $\mathrm{Rel}(\mathbb{E})$ of relations is Cartesian closed. The exponent of relations R on I and S on J is given by the formula

$$R \Rightarrow S = \textstyle\prod_{\pi} \alpha^*\big((\pi' \times \pi')^*(R) \supset (\mathrm{ev} \times \mathrm{ev})^*(S)\big)$$

where α is the isomorphism

$$((I \Rightarrow J) \times (I \Rightarrow J)) \times (I \times I) \xrightarrow{\;\cong\;} ((I \Rightarrow J) \times I) \times ((I \Rightarrow J) \times I)$$

In the special case when R is the equality relation $\mathrm{Eq}(I)$ on I one can simplify the right hand side to

$$\mathrm{Eq}(I) \Rightarrow S = \textstyle\prod_{\pi} \langle \mathrm{ev} \circ \pi \times \mathrm{id}, \mathrm{ev} \circ \pi' \times \mathrm{id} \rangle^*(S).$$

This says informally that maps $f, g \colon I \rightrightarrows J$ are in the relation $\mathrm{Eq}(I) \Rightarrow S$ if and only if for all $i \in I$ the pair $f(i), g(i)$ is in the relation S. Thus we can say that equality on maps in $\mathbb{B}$ is **pointwise** if the equality functor $\mathrm{Eq} \colon \mathbb{B} \to \mathrm{Rel}(\mathbb{E})$ preserves exponents.

If the fibration p admits quotients—*i.e.* if the equality functor Eq has a left adjoint (see Definition 4.8.1)—then one can prove: equality on maps in p is pointwise if and only if p's quotients satisfy Frobenius. The argument is standard, like in the proof of Lemma 1.9.12.

(iii) Still in the same situation, we assume now that the terminal object functor $\top \colon \mathbb{B} \to \mathbb{E}$ preserves coproducts $+$. This happens for example when $\top$ has a right adjoint, giving comprehension in the logic of p. Explicitly, this preservation means that for objects $I, J \in \mathbb{B}$ the canonical map

$$\textstyle\coprod_{\kappa}(\top I) \vee \coprod_{\kappa'}(\top J) \longrightarrow \top(I + J)$$

is an isomorphism. Logically, this amounts to validity of the axiom (scheme):

$$z: \sigma + \tau \mid \top \vdash (\exists x: \sigma. \, z =_{\sigma+\tau} \kappa x) \vee (\exists y: \tau. \, z =_{\sigma+\tau} \kappa' y).$$

It is not hard to see that this is equivalent to the following rule: for a term $\Gamma \vdash M: \sigma + \tau$ and propositions $\Gamma \vdash \varphi, \psi: \mathsf{Prop}$

$$\frac{\Gamma, x: \sigma \mid \varphi, M =_{\sigma+\tau} \kappa x \vdash \psi \qquad \Gamma, y: \tau \mid \varphi, M =_{\sigma+\tau} \kappa' y \vdash \psi}{\Gamma \mid \varphi \vdash \psi}$$

This rule is more convenient than the axiom, and has the advantage that it does not involve existential quantification $\exists$ so that it can be formulated within (conditional) equational logic. It gives a fibred formulation of **universality** for coproducts in a base category, see Exercise 9.2.9 below.

(Disjointness of coproducts $+$ simply means that the coprojections are internally injective: $x, x': \sigma \mid \kappa x =_{\sigma+\tau} \kappa x' \vdash x =_{\sigma} x'$, and similarly for κ', together with: $x: \sigma, y: \tau \mid \kappa x =_{\sigma+\tau} \kappa' y \vdash \varphi$, expressing that anything (or equivalently: *falsum* $\perp$) follows from equality of $\kappa x, \kappa' y$.)

(iv) Finally, some concrete examples: the fibration $\begin{smallmatrix} \mathbf{Sign} \\ \downarrow \\ \mathbf{Sets} \end{smallmatrix}$ of signatures over sets is obtained by change-of-base from $\begin{smallmatrix} \mathbf{Fam(Sets)} \\ \downarrow \\ \mathbf{Sets} \end{smallmatrix}$ see Definition 1.6.1. It thus has all fibred colimits. Hence the total category **Sign** is cocomplete. The same argument occurs in [327, Example 1]. Actually, Proposition 9.2.2 only gives finite coproducts, but, as mentioned, the result also holds for arbitrary colimits, see Exercise 9.2.4 (ii) below. This cocompleteness of the category of signatures is instrumental in the theory of specifications. It lets us define instantiation and put together parametrised signatures via pushouts and other colimits, see [327, 152] for details.

Winskel [344] defines a bifibration $\begin{smallmatrix} \mathbf{T} \\ \downarrow \\ \mathbf{Sets}_{\bullet} \end{smallmatrix}$ of labelled transition systems over pointed sets. Local products and coproducts of transition systems are defined over a particular pointed set of labels. And in terms of these, also global products and coproducts of transition systems, following the constructions in this section.

Similar constructions appear in the theory of deliverables, see [40, 217].

Reasoning principles via global structure

We describe how the induction (and co-induction) principles associated with data types given as initial algebras (and terminal co-algebras) of Hagino signatures can be described in a uniform and concise way using global structure in total categories of fibrations. This follows [125, 128, 130]. We concentrate on induction, and mention co-induction in Exercise 9.2.10 below (but see also [270, 299, 79]). The approach that we describe here leads to "mixed induction/co-induction" proof principles in [130], and to proof principles for iterated data types in [124]. Here we concentrate on the basic ideas: we first give a formulation of the induction principle in a fibred setting, and then show that in logics with comprehension this principle holds automatically.

Let $\mathbb{B}$ be a category with finite products $(1, \times)$ and coproducts $(0, +)$. We think of $\mathbb{B}$ as a model of a simple type theory. Assume $T: \mathbb{B} \to \mathbb{B}$ is a polynomial functor built up from constant functors $X \mapsto C$, the identity functor $X \mapsto X$, and constructors $\times, +$. We can think of T as the functor which is canonically associated with a type $\sigma(X)$ in a Hagino signature, see Definition 2.6.4. We assume we have a fibration $\begin{smallmatrix} \mathbb{E} \\ \downarrow p \\ \mathbb{B} \end{smallmatrix}$ giving us a logic to reason about $\mathbb{B}$. We further assume this p to be a bifibration with fibred finite products and coproducts. The total category $\mathbb{E}$ of predicates then has finite products and coproducts. This enables us to **lift** the functor $T: \mathbb{B} \to \mathbb{B}$ to a functor $\mathrm{Pred}(T): \mathbb{E} \to \mathbb{E}$ in a commuting diagram

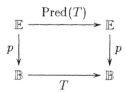

by induction on the structure of T. We replace the identity functor, the product $\times$ and the coproduct $+$ on $\mathbb{B}$ occurring in T by the identity functor, product and coproduct on $\mathbb{E}$. And we replace the constants $C \in \mathbb{B}$ occurring in T by the constant $\top(C)$ in $\mathrm{Pred}(T)$—where $\top(C)$ is the terminal object in the fibre over C. (In terms of Definition 2.6.4: we replace the model $A: S \to \mathbb{B}$ of the constants in $\mathbb{B}$ by the model $\top \circ A: S \to \mathbb{E}$.)

This lifting of T to $\mathrm{Pred}(T)$ is an important step. It brings us from the world of data types (in the form of algebras and co-algebras of the functor T) to the world of logic about such data types (in the form of algebras and co-algebras of the lifted functor $\mathrm{Pred}(T)$). Essentially, an algebra $f: \mathrm{Pred}(T)(X) \to X$ of this lifted functor $\mathrm{Pred}(T)$ consists of an underlying T-algebra $u = pf: T(pX) = p(\mathrm{Pred}(T)(X)) \to pX$, together with a proof that the predicate X is closed under the operations of the T-algebra u. Dually, a co-algebra $g: Y \to \mathrm{Pred}(T)(Y)$ consists of a T-co-algebra $v = pg: pY \to T(pY)$ together with a proof that Y is closed under the operations (or transitions) of v. Such predicates Y are called invariants in [164]: once they hold in a state x, they continue to hold no matter which of the operations in the co-algebra v are applied to x. Here we concentrate on $\mathrm{Pred}(T)$-algebras. They incorporate the assumptions of induction arguments for data types whose form is determined by the functor T.

First we need a further assumption, namely that the terminal object functor $\top: \mathbb{B} \to \mathbb{E}$ preserves finite coproducts—*e.g.* because it has a right adjoint, describing comprehension in the logic of p—then one gets a canonical (vertical) isomorphism

$$\top \circ T \overset{\cong}{\Longrightarrow} \mathrm{Pred}(T) \circ \top$$

by induction on the structure of T. Here one uses that $\top$ preserves finite products because it is right adjoint to p (see Lemma 1.8.8).

9.2.6. Definition. Consider a bifibration $\begin{smallmatrix} \mathbb{E} \\ \downarrow p \\ \mathbb{B} \end{smallmatrix}$ with a polynomial functor $T: \mathbb{B} \to \mathbb{B}$ as above. An initial algebra $a: T(K) \overset{\cong}{\Rightarrow} K$ of this functor is called **inductive** if the resulting isomorphism in $\mathbb{E}$

$$\mathrm{Pred}(T)(\top K) \overset{\cong}{\longrightarrow} \top T(K) \xrightarrow[\cong]{\top a} \top K$$

is initial algebra of the lifted functor $\mathrm{Pred}(T): \mathbb{E} \to \mathbb{E}$. Inductive initial algebras in a fibration p come equipped with an associated induction proof principle in the logic of p.

.

We give an illustration of this notion of inductive initial algebra. Consider the standard fibration $\begin{smallmatrix} \mathrm{Sub}(\mathbf{Sets}) \\ \downarrow \\ \mathbf{Sets} \end{smallmatrix}$ of predicates (subsets) over sets capturing the classical logic of sets, with polynomial functor $T(X) = 1 + X$. This functor $T: \mathbf{Sets} \to \mathbf{Sets}$ has the cotuple $[0, \mathsf{S}]: 1 + \mathbb{N} \overset{\cong}{\Rightarrow} \mathbb{N}$ of zero and successor on the natural numbers $\mathbb{N}$ as initial algebra. The associated lifted functor $\mathrm{Pred}(T): \mathrm{Sub}(\mathbf{Sets}) \to \mathrm{Sub}(\mathbf{Sets})$ send a predicate $(Y \subseteq J)$ to the predicate $(1 + Y \subseteq 1 + J)$. An algebra of this functor thus consists of a map $(1 + Y \subseteq J) \to (Y \subseteq J)$ in $\mathrm{Sub}(\mathbf{Sets})$. It consists of a pair of maps

$$\begin{cases} j: 1 \longrightarrow J \\ u: J \longrightarrow J \end{cases} \quad \text{satisfying} \quad \begin{cases} j \in Y \\ y \in Y \Rightarrow u(y) \in Y. \end{cases}$$

Clearly the truth predicate $\top \mathbb{N} = (\mathbb{N} \subseteq \mathbb{N})$ carries the algebra $[0, \mathsf{S}]: (1 + \mathbb{N} \subseteq 1 + \mathbb{N}) \to (\mathbb{N} \subseteq \mathbb{N})$. Inductiveness of this initial T-algebra $[0, \mathsf{S}]: 1 + \mathbb{N} \overset{\cong}{\Rightarrow} \mathbb{N}$ means that this $\mathrm{Pred}(T)$-algebra $[0, \mathsf{S}]: (1 + \mathbb{N} \subseteq 1 + \mathbb{N}) \to (\mathbb{N} \subseteq \mathbb{N})$ is again initial. When we spell this out we get: for a $\mathrm{Pred}(T)$-algebra $[j, u]: (1 + Y \subseteq$

$J) \rightarrow (Y \subseteq J)$ as above there is a unique map of $\mathrm{Pred}(T)$-algebras:

$$
\begin{array}{ccc}
(1 + \mathbb{N} \subseteq 1 + \mathbb{N}) & \xrightarrow{\ \mathrm{id} + v\ } & (1 + Y \subseteq 1 + J) \\
{\scriptstyle [0,\,S]} \downarrow & & \downarrow {\scriptstyle [j,\,u]} \\
(\mathbb{N} \subseteq \mathbb{N}) & \dashrightarrow[v]{\ \ } & (Y \subseteq J)
\end{array}
$$

This means that the unique mediating map $v: \mathbb{N} \dashrightarrow J$ with $v \circ 0 = j$ and $v \circ S = u \circ v$ is a map of predicates $(\mathbb{N} \subseteq \mathbb{N}) \rightarrow (Y \subseteq J)$. Hence for all $n \in \mathbb{N}$ we have $v(n) \in Y$. This is the conclusion of the induction principle. It is obtained by initiality of the algebra $\mathrm{Pred}(T)(T\mathbb{N}) \overset{\cong}{\Rightarrow} T\mathbb{N}$ on the truth predicate $T\mathbb{N}$ on $\mathbb{N}$.

(We have given the "unary" induction principle for predicates; there is also an equivalent "binary" version for (congruence) relations, see Exercise 9.2.11 (iii).)

The following result shows that in a fibration with comprehension (giving us a logic with subset types) every initial algebra is automatically inductive. Hence one can use induction as a reasoning principle in such a logic. We give a concrete proof below, but a more abstract proof using a "transfer of adjunction" lemma occurs in [128, 130].

9.2.7. Proposition. *Let* $\begin{smallmatrix} \mathbb{E} \\ \downarrow{\scriptstyle P} \\ \mathbb{B} \end{smallmatrix}$ *be a bifibration with*

- *finite products and coproducts in its base category* $\mathbb{B}$;
- *fibred finite products and coproducts;*
- *comprehension, given by a right adjoint* $\{-\}: \mathbb{E} \rightarrow \mathbb{B}$ *to the terminal object functor* $\top: \mathbb{B} \rightarrow \mathbb{E}$.

An initial algebra $a: T(K) \overset{\cong}{\Rightarrow} K$ *of a polynomial functor* $T: \mathbb{B} \rightarrow \mathbb{B}$ *is then automatically inductive.*

Proof. Write φ for the natural isomorphism $\mathrm{Pred}(T) \circ \top \overset{\cong}{\Rightarrow} T \circ \top$. We have to show initiality of the $\mathrm{Pred}(T)$-algebra $\top a \circ \varphi_K: \mathrm{Pred}(T)(TK) \overset{\cong}{\Rightarrow} \top T(K) \overset{\cong}{\Rightarrow} \top K$. Assume a $\mathrm{Pred}(T)$-algebra $f: \mathrm{Pred}(T)(X) \rightarrow X$ in $\mathbb{E}$, say over $b: T(I) \rightarrow I$ in $\mathbb{B}$. Let $u: K \dashrightarrow I$ in $\mathbb{B}$ be the unique mediating T-algebra map in $\mathbb{B}$ with $b \circ T(u) = u \circ a$. Write

$$
g = \left(TT\{X\} \xrightarrow[\cong]{\varphi^{-1}_{\{X\}}} \mathrm{Pred}(T)(T\{X\}) \xrightarrow{\mathrm{Pred}(T)(\varepsilon_X)} \mathrm{Pred}(T)(X) \xrightarrow{f} X \right)
$$

It gives us a transpose $g^{\vee} = \{g\} \circ \eta_{T(\{X\})}: T(\{X\}) \rightarrow \{X\}$, which is an algebra on the extent $\{X\}$ of X. An easy calculation shows that the (comprehension) projection $\pi_X \colon \{X\} \to I$ is a homomorphism of algebras from $(g^\vee \colon T(\{X\}) \to \{X\})$ to $(b \colon T(I) \to I)$. Hence we also get a unique mediating algebra map $v \colon K \dashrightarrow \{X\}$ in:

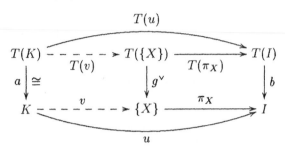

with $\pi_X \circ v = u$ by initiality. Transposing this map $v \colon K \to \{X\}$ in $\mathbb{B}$ back to $\mathbb{E}$ as $v^\wedge = \varepsilon_X \circ Tv \colon TK \to X$ gives us the required (unique) mediating map of $\operatorname{Pred}(T)$-algebras:

$$
\begin{aligned}
v^\wedge \circ Ta \circ \varphi_K &= \varepsilon_X \circ T(v \circ a) \circ \varphi_K \\
&= \varepsilon_X \circ T(g^\vee \circ T(v)) \circ \varphi_K \\
&= g \circ TT(v) \circ \varphi_K \\
&= f \circ \operatorname{Pred}(T)(\varepsilon_X) \circ \operatorname{Pred}(T)(Tv) \circ \varphi_K^{-1} \circ \varphi_K \\
&= f \circ \operatorname{Pred}(v^\wedge). \qquad\qquad \square
\end{aligned}
$$

The above approach to the logic of (co-)inductive data types is essentially syntax-driven: the types of the operations of a data type determine a functor T, and thereby its lifting $\operatorname{Pred}(T)$, in terms of which the proof principles are formulated. This approach—and its extension to more complicated (co-)data types—has been implemented in a (front-end) tool for formal reasoning about such data types, see [123, 122].

Exercises

9.2.1. Describe the bicartesian closed structure in the following total categories (of obvious fibrations) according to the formulas given in this section.
   (i)   The category of subsets Sub(**Sets**).
   (ii) The category of set-indexed sets Fam(**Sets**).

9.2.2. Give the "logical relations" coproduct in the term model (of coherent predicate logic).

9.2.3. Consider a fibration $\begin{smallmatrix}\mathbb{E}\\ \downarrow{\scriptstyle p}\\ \mathbb{B}\end{smallmatrix}$ with equalisers in its base category. Show that $\mathbb{B}$ has fibred equalisers if and only if $\mathbb{E}$ has equalisers and p strictly preserves them.

9.2.4.　(i)　Extend Proposition 9.2.1 to arbitrary limits in the following way. Let $\mathbb{I}$ be some index category and assume the base category $\mathbb{B}$ of a fibration $\begin{smallmatrix}\mathbb{E}\\\downarrow p\\\mathbb{B}\end{smallmatrix}$ has limits of shape $\mathbb{I}$. Show that p has fibred limits of shape $\mathbb{I}$ (as defined in Exercise 1.8.8) if and only if the total category $\mathbb{E}$ has limits of shape $\mathbb{I}$ and p strictly preserves them.
　　　　[*Hint.* Use Exercise 1.8.11 (ii).]
　　(ii)　Do the same for Proposition 9.2.2 (for a bifibration).

9.2.5.　Conclude from Proposition 9.2.3 that if a category $\mathbb{B}$ has pullbacks and disjoint and universal coproducts $(0, +)$, then its category of arrows $\mathbb{B}^{\rightarrow}$ is distributive.

9.2.6.　Prove that the exponent $\mathrm{Eq}(I) \Rightarrow S$ in a category of relations is isomorphic to $\prod_\pi \langle \mathrm{ev} \circ \pi \times \mathrm{id}, \mathrm{ev} \circ \pi' \times \mathrm{id}\rangle^*(S)$, as claimed above.

9.2.7.　Prove that for a topos $\mathbb{B}$
　　(i)　the comprehension functor $\{-\} \colon \mathrm{Sub}(\mathbb{B}) \to \mathbb{B}$—which is right adjoint to truth and gives subset types—preserves finite coproducts;
　　(ii)　the quotient functor $Q \colon \mathrm{ERel}(\mathbb{B}) \to \mathbb{B}$ which takes the quotient of equivalence relations preserves finite products.
　　　　[The fibration $\begin{smallmatrix}\mathrm{ERel}(\mathbb{B})\\\downarrow\\\mathbb{B}\end{smallmatrix}$ is described at the end of Section 1.3. The second point may also be proved type theoretically using Exercise 5.1.7.]

9.2.8.　Assume that $\begin{smallmatrix}\mathbb{E}\\\downarrow p\\\mathbb{B}\end{smallmatrix}$ is a fibred CCC with simple products which also has coproducts $\coprod_u \dashv u^*$ satisfying Beck-Chevalley (and Frobenius by Lemma 1.9.12 (i)). Prove that the global exponent $\Rightarrow$ from Proposition 9.2.4 forms a fibred functor in the situation:

$$
\begin{array}{ccc}
\mathbb{E}^{\mathrm{op}} \times \mathbb{E} & \xrightarrow{\;\;\Rightarrow\;\;} & \mathbb{E} \\
{\scriptstyle p \times p}\downarrow & & \downarrow{\scriptstyle p} \\
\mathbb{B}^{\mathrm{op}} \times \mathbb{B} & \xrightarrow[\;\;\Rightarrow\;\;]{} & \mathbb{B}
\end{array}
$$

That is: $(u \Rightarrow v)^*(X \Rightarrow Y) \cong \coprod_u(X) \Rightarrow v^*(Y)$.

9.2.9.　Let $\mathbb{B}$ be a category with finite limits and disjoint coproducts $+$. Show that these coproducts $+$ are universal if and only if the type theoretic rule

$$
\frac{\Gamma, x \colon \sigma \mid \varphi, M =_{\sigma+\tau} \kappa x \vdash \psi \qquad \Gamma, y \colon \tau \mid \varphi, M =_{\sigma+\tau} \kappa' y \vdash \psi}{\Gamma \mid \varphi \vdash \psi}
$$

is valid in the subobject fibration $\begin{smallmatrix}\mathrm{Sub}(\mathbb{B})\\\downarrow\\\mathbb{B}\end{smallmatrix}$ on $\mathbb{B}$. Conclude that in a coherent category $\mathbb{B}$ with disjoint coproducts $+$, one gets universality for free.

9.2.10.　(From [130]) We consider a fibration $\begin{smallmatrix}\mathbb{E}\\\downarrow p\\\mathbb{B}\end{smallmatrix}$ as in Example 9.2.5 (ii).
　　(i)　Prove that if the terminal object functor $\top \colon \mathbb{B} \to \mathbb{E}$ preserves finite coproducts, then so does the equality functor $\mathrm{Eq} \colon \mathbb{B} \to \mathrm{Rel}(\mathbb{E})$.

We now assume that this equality functor Eq preserves finite coproducts and products; the latter *e.g.* because it has a right adjoint giving quotients in the logic of the fibration. Since $\mathrm{Rel}(\mathbb{E})$ has finite products and coproducts we can lift a polynomial functor $T \colon \mathbb{B} \to \mathbb{B}$ to $\mathrm{Rel}(T) \colon \mathrm{Rel}(\mathbb{E}) \to \mathrm{Rel}(\mathbb{E})$ by induction on T, whereby we replace a constant $C \in \mathbb{B}$ occurring in T by the constant $\mathrm{Eq}(C) \in \mathrm{Rel}(\mathbb{E})$. By construction we have $\mathrm{Rel}(T) \circ \mathrm{Eq} \cong \mathrm{Eq} \circ T$. A co-algebra of this lifted functor $\mathrm{Rel}(T)$ can be identified with bisimulation relation, forming an assumption of a co-induction argument.

Define a terminal co-algebra $c \colon K \stackrel{\cong}{\Rightarrow} T(K)$ to be **co-inductive** if the resulting map

$$\mathrm{Eq}(K) \xrightarrow[\cong]{\mathrm{Eq}(c)} \mathrm{Eq}(TK) \xrightarrow{\cong} \mathrm{Rel}(T)(\mathrm{Eq}(K))$$

is terminal $\mathrm{Rel}(T)$-co-algebra. This means that the terminal co-algebra comes with a co-induction proof principle in this fibration.

(ii) Investigate what this means for the fibration $\begin{array}{c} \mathrm{Sub}(\mathbf{Sets}) \\ \downarrow \\ \mathbf{Sets} \end{array}$ and the "stream" functor $T(-) = C \times (-)$.

(iii) Prove that if p has quotients (*i.e.* if Eq has a left adjoint), then every terminal co-algebra is automatically co-inductive. This gives the dual of Proposition 9.2.7.

9.2.11. (From [130]) Let $\begin{array}{c} \mathbb{E} \\ \downarrow p \\ \mathbb{B} \end{array}$ be a fibration with fibred finite products and coproducts on a distributive base category $\mathbb{B}$. Assume that p is a bifibration with coproducts $\coprod_u \dashv u^*$ satisfying Beck-Chevalley and Frobenius, and that truth $\top \colon \mathbb{B} \to \mathbb{E}$ preserves finite coproducts and equality $\mathrm{Eq} \colon \mathbb{B} \to \mathrm{Rel}(\mathbb{E})$ preserves finite products. (For convenience one may assume that p is a fibred preorder.)

First relate the "Pred" and "Rel" liftings of a polynomial functor $T \colon \mathbb{B} \to \mathbb{B}$ in the following way.

(i) Prove, by induction on T, that there are vertical (natural) isomorphisms

$$\delta^* \mathrm{Rel}(T)(R) \cong \mathrm{Pred}(T)(\delta^*(R)).$$

(ii) Prove similarly that

$$\coprod_\delta \mathrm{Pred}(T)(X) \cong \mathrm{Rel}(T)(\coprod_\delta(X)).$$

These isomorphisms actually determine a pseudo map of adjunctions $(\delta_I \dashv \coprod_{\delta_I}) \to (\delta_{T(I)} \dashv \coprod_{\delta_{T(I)}})$, see Exercise 1.8.7.

(iii) Prove that for an initial T-algebra $a \colon T(K) \stackrel{\cong}{\Rightarrow} K$ the following three points are equivalent:

 • a is inductive (*i.e.* the canonical map $\mathrm{Pred}(T)(\top K) \stackrel{\cong}{\Rightarrow} \top K$ is initial $\mathrm{Pred}(T)$-algebra);

- the canonical map $\mathrm{Rel}(T)(\mathrm{Eq}(K)) \overset{\ast}{\to} \mathrm{Eq}(K)$ is initial $\mathrm{Rel}(T)$-algebra;
- every congruence on a is reflexive, *i.e.* for every $\mathrm{Rel}(T)$-algebra $R \in \mathbb{E}_{K \times K}$ on a we have $\mathrm{Eq}(K) \leq R$.

[The latter formulation gives the so-called binary induction principle, which is formulated in terms of congruence relations being reflexive. See also [299].]

9.2.12. Let $\begin{smallmatrix}\mathbb{E}\\\downarrow p\\\mathbb{B}\end{smallmatrix}$ be a fibration with fibred pullbacks and pullbacks in its base category $\mathbb{B}$. By Proposition 9.2.1 and Exercise 9.2.3 we know that the total category $\mathbb{E}$ also has pullbacks and that p strictly preserves them.

(i) Describe these pullbacks in $\mathbb{E}$ in detail.

Assume now that p has coproducts, via adjunctions $(\coprod_u \dashv u^*)$ satisfying Beck-Chevalley. Following Bénabou, these coproducts are called **disjoint** if for each map $u: I \to J$ in $\mathbb{B}$ and opcartesian morphism $\underline{u}(X): X \to \coprod_u(X)$ in $\mathbb{E}$ over u, the diagonal $\delta: X \to X \times_{\coprod_u(X)} X$ of the pullback of $\underline{u}(X)$ against itself, is opcartesian again (over the diagonal in $\mathbb{B}$). And these coproducts of p are called **universal** if for each pullback square in $\mathbb{E}$ of the form

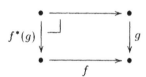

one has: g is opcartesian implies $f^*(g)$ is opcartesian.

(ii) Check that a category $\mathbb{C}$ has set-indexed disjoint / universal coproducts if and only if its family fibration $\begin{smallmatrix}\mathrm{Fam}(\mathbb{C})\\\downarrow\\\mathbf{Sets}\end{smallmatrix}$ has disjoint / universal coproducts (see Lemma 1.9.5).

(iii) Verify that for a category $\mathbb{B}$ with finite limits, the codomain fibration $\begin{smallmatrix}\mathbb{B}^{\to}\\\downarrow\\\mathbb{B}\end{smallmatrix}$ always has universal and disjoint coproducts.

(iv) Show that it suffices for universality to restrict the requirement to *vertical* f, as above.

9.2.13. Consider a fibration $\begin{smallmatrix}\mathbb{E}\\\downarrow p\\\mathbb{B}\end{smallmatrix}$ with disjoint and universal coproducts $\coprod_u$ as in the previous exercise. In the following we establish a fibred version of Proposition 1.5.4, involving change-of-base of a codomain fibration along a copower functor.

(i) Prove that if maps $f, g: Y \rightrightarrows X$ in $\mathbb{E}$ above the same map in $\mathbb{B}$ satisfy $\underline{u}(X) \circ f = \underline{u}(X) \circ g: Y \to \coprod_u(X)$, then $f = g$.

(ii) Consider vertical maps f, g in a commuting square in $\mathbb{E}$ of the form:

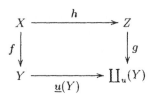

Show that this square is a pullback in $\mathbb{E}$ if and only if the canonical map $\coprod_u(X) \to Z$ is an isomorphism. This shows that for $u: I \to J$ the canonical functor $\mathbb{E}_I/Y \to \mathbb{E}_J/\coprod_u(Y)$ is an equivalence.

(iii) Prove the converse of (ii): if for each map $u: I \to J$ in $\mathbb{B}$ and object $Y \in \mathbb{E}_I$ the canonical functor $\mathbb{E}_I/Y \to \mathbb{E}_J/\coprod_u(Y)$ is an equivalence, then coproducts $\coprod_u$ are disjoint and universal.
[A fibration with this property may be called **extensive**.]

(iv) Consider for a map $u: I \to J$ in $\mathbb{B}$ and for objects $X \in \mathbb{E}_J$ and $Y \in \mathbb{E}_I$ the following diagram in $\mathbb{F}$.

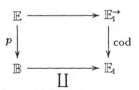

(The Cartesian products are in the fibres, and the two projections are vertical.) Prove that this diagram is a pullback, and conclude from (ii) that the Frobenius property automatically holds.

(v) Define for each $I \in \mathbb{B}$ an object $\coprod(I)$ in the fibre $\mathbb{E}_1$ over the terminal object $1 \in \mathbb{B}$ by

$$\coprod(I) = \coprod_{!_I}(\top I)$$

where $\top: \mathbb{B} \to \mathbb{E}$ is the terminal object functor. Extend $\coprod$ to a functor $\mathbb{B} \to \mathbb{E}_1$ and show that it preserves finite limits.

(vi) Check that the whole fibration p can be recovered from its fibre category $\mathbb{E}_1$ over $1 \in \mathbb{B}$ via the following change-of-base situation.

$$
\begin{array}{ccc}
\mathbb{E} & \longrightarrow & \vec{\mathbb{E}_1} \\
p \downarrow & & \downarrow \text{cod} \\
\mathbb{B} & \underset{\coprod}{\longrightarrow} & \mathbb{E}_1
\end{array}
$$

9.3 Quantification

Earlier we have seen various forms of quantification: along Cartesian projection morphisms $\pi \colon I \times J \to I$ in Definition 1.9.1, along arbitrary morphisms $u \colon I \to J$ in Definition 1.9.4, along Cartesian diagonal morphisms $\delta(I, J) = \langle \mathrm{id}, \pi' \rangle \colon I \times J \to (I \times J) \times J$ in Definition 3.4.1 and also along the Cartesian projection maps $\pi \colon I \times X \to I$ induced by a CT-structure, see Definition 2.4.3, and along monomorphisms $m \colon X \rightarrowtail I$ in Observation 4.4.1. In this section we shall introduce a general description of quantification which captures all of the above examples (and many more). This will enable us to establish some results for all these forms of quantification at once. The general theme is that products and coproducts $\prod, \coprod$ are right and left adjoints to weakening functors, and that equality Eq is left adjoint to contraction functors. (With fibred exponents one then gets right adjoints to contraction functors for free, see Exercise 3.4.2.)

Weakening introduces an extra dummy variable $x \colon \sigma$, and contraction replaces two variables $x, y \colon \sigma$ of the same type by a single one (by substituting $[x/y]$). Indeed, categorically, weakening and contraction are special cases of substitution, see explicitly in Example 3.1.1. These operations of weakening and contraction carry the structure of a comonad (W, ε, δ), where the counit $\varepsilon_X \colon WX \to X$ corresponds to weakening, and the comultiplication (also called diagonal) $\delta_X \colon WX \to W^2 X$ to contraction. Intuitively, the comonad equations $\delta \circ \varepsilon W = \mathrm{id}$ and $\delta \circ W\varepsilon = \mathrm{id}$ correspond to the fact that first weakening and then contracting is the identity, in two forms:

$$(\Gamma, x \colon \sigma \vdash M \colon \tau) \;\mapsto\; (\Gamma, x \colon \sigma, y \colon \sigma \vdash M \colon \tau) \;\mapsto\; (\Gamma, x \colon \sigma \vdash M[x/y] \colon \tau)$$
$$(\Gamma, x \colon \sigma \vdash M \colon \tau) \;\mapsto\; (\Gamma, y \colon \sigma, x \colon \sigma \vdash M \colon \tau) \;\mapsto\; (\Gamma, y \colon \sigma \vdash M[y/x] \colon \tau)$$

For the second one we need α-conversion (change of variables).

Below we shall describe quantification $\prod, \coprod$, Eq with respect to an abstract "weakening and contraction comonad" (W, ε, δ). Since weakening and contraction are operations which—by their nature—change the context, this comonad will be on the total category of a fibration. Product and coproduct are then right and left adjoints to the weakening functors induced by ε and equality is left adjoint to the contraction functor induced by δ. This sums up the basic framework.

It turns out that a "weakening and contraction comonad" on the total category of a fibration corresponds to certain structure in the base category, in terms of which we can also describe quantification. Part of this correspondence was noted by Hermida. The latter structure will be called a "comprehension category" (following [154, 157]). It is somewhat more elementary than a weakening and contraction comonad. It will be fundamental for type dependency in the next chapter. Therefore, in the sequel we shall mostly be using these comprehension categories—instead of weakening and contraction comonads.

But we start with comonads.

9.3.1. Definition. Let $\downarrow p$ be a fibration. A **weakening and contraction comonad** on p consists of a comonad (W, ε, δ) on the total category $\mathbb{E}$, satisfying:
(1) Each counit component $\varepsilon_X \colon WX \to X$ is Cartesian.
(2) For each Cartesian morphism $f \colon X \to Y$ in $\mathbb{E}$, the naturality square,

$$
\begin{array}{ccc}
WX & \xrightarrow{\ \ Wf\ \ } & WY \\
{\scriptstyle \varepsilon_X}\downarrow & & \downarrow{\scriptstyle \varepsilon_Y} \\
X & \xrightarrow{\ \ f\ \ } & Y
\end{array}
$$

is a pullback in the total category $\mathbb{E}$.

The first of these conditions says that weakening is a special case of substitution (see also Example 3.1.1), and the second one is a stability condition. As a consequence of this second condition the functor W preserves Cartesianness. But note that W is not a fibred functor $p \to p$ since we do not necessarily have $p \circ W = p$. And note also that the counit ε and comultiplication δ need not be vertical.

If we apply the second condition with $f = \varepsilon_X$ we get a pullback square

$$
\begin{array}{ccc}
W^2X & \xrightarrow{\ \ W(\varepsilon_X)\ \ } & WX \\
{\scriptstyle \varepsilon_{WX}}\downarrow\ \ \lrcorner & & \downarrow{\scriptstyle \varepsilon_X} \\
WX & \xrightarrow{\ \ \varepsilon_X\ \ } & X
\end{array}
$$

This shows that the diagonal $\delta_X \colon WX \to W^2X$ is completely determined as the unique mediating map for the cone $WX \xleftarrow{\ \mathrm{id}\ } WX \xrightarrow{\ \mathrm{id}\ } WX$, with respect to this pullback. It is not hard to see that these δ_X's are also Cartesian. In fact, if we only have a pair (W, ε) satisfying (1) and (2) above and define δ_X in this way we get a natural transformation satisfying the comonad equations. Hence the above definition contains some redundancy. But we like to keep it as it stands in order to stress that weakening and contraction form a comonad.

Two standard examples are as follows. For a category $\mathbb{B}$ with finite products there is a weakening and contraction comonad on the associated simple fibration $\overset{s(\mathbb{B})}{\underset{\mathbb{B}}{\downarrow}}$ with functor $W\colon s(\mathbb{B}) \to s(\mathbb{B})$ given by

$$(I, X) \mapsto (I \times X, X)$$

and (Cartesian) counit

$$(I \times X, X) \xrightarrow{\;\varepsilon_{(I,X)} = (\pi, \pi')\;} (I, X)$$

The (induced) comultiplication is

$$(I \times X, X) \xrightarrow{\;\delta_{(I,X)} = (\langle \mathrm{id}, \pi' \rangle, \pi')\;} ((I \times X) \times X, X)$$

For a category $\mathbb{B}$ with finite limits there is a weakening and contraction comonad on the associated codomain fibration $\overset{\mathbb{B}^{\to}}{\underset{\mathbb{B}}{\downarrow}}$. The functor $\mathbb{B}^{\to} \to \mathbb{B}^{\to}$ maps

$$\begin{pmatrix} X \\ \downarrow \\ I \end{pmatrix} \mapsto \begin{pmatrix} X \times_I X \\ \downarrow \\ X \end{pmatrix}.$$

We turn to quantification with respect to a weakening and contraction comonad.

9.3.2. Definition. Let $W = (W, \varepsilon, \delta)$ be a weakening and contraction comonad on $\overset{\mathbb{E}}{\underset{\mathbb{B}}{\downarrow}}p$ and let $\overset{\mathbb{D}}{\underset{\mathbb{B}}{\downarrow}}q$ be another fibration with base $\mathbb{B}$.

(i) This fibration q is said to have W-**products** if for each object $X \in \mathbb{E}$ there is a right adjoint

$$p(\varepsilon_X)^* \dashv \textstyle\prod_X$$

to the "weakening functor" $p(\varepsilon_X)^* \colon \mathbb{D}_{pX} \to \mathbb{D}_{pWX}$, together with a Beck-Chevalley condition: for each Cartesian map $f\colon X \to Y$ in $\mathbb{E}$, the canonical natural transformation

$$\begin{array}{ccc}
\mathbb{D}_{pWX} & \xrightarrow{\;\;\prod_X\;\;} & \mathbb{D}_{pX} \\[4pt]
{\scriptstyle (pWf)^*}\Big\uparrow & \diagdown\kern-0.5em\diagup & \Big\uparrow{\scriptstyle (pf)^*} \\[4pt]
\mathbb{D}_{pWY} & \xrightarrow[\;\;\prod_Y\;\;]{} & \mathbb{D}_{pY}
\end{array}$$

is an isomorphism.

(ii) Similarly, q has W-**coproducts** if there are left adjoints $\coprod_X \dashv p(\varepsilon_X)^*$, and the canonical natural transformation $\coprod_X (pWf)^* \Rightarrow (pf)^* \coprod_Y$ is an isomorphism, for each Cartesian map $f\colon X \to Y$ in $\mathbb{E}$.

(iii) Finally, q has W-**equality** if there are left adjoints $\mathrm{Eq}_X \dashv p(\delta_X)^*$ to the "contraction functors" $p(\delta_X)^*$, and for each Cartesian map $f: X \to Y$ in $\mathbb{E}$, the canonical natural transformation

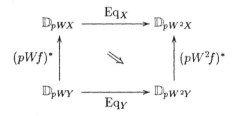

is an isomorphism.

9.3.3. Definition. If the fibration q above has fibred Cartesian products $\times$, then we say that q has W-**coproducts satisfying Frobenius** if it has W-coproducts $\coprod_X$ such that the canonical maps

$$\coprod_X (p(\varepsilon_X)^*(Z) \times Y) \longrightarrow Z \times \coprod_X(Y)$$

are isomorphisms.

Similarly, for W-**equality satisfying Frobenius**, one requires the canonical maps

$$\mathrm{Eq}_X (p(\delta_X)^*(Z) \times Y) \longrightarrow Z \times \mathrm{Eq}_X(Y)$$

to be isomorphisms.

We can thus define quantification with respect to a weakening and contraction comonad in terms of the induced structure $p(\varepsilon_X)$ and $p(\delta_X)$ in the base category. It is not hard to see that quantification with respect to the weakening and contraction comonads described before Definition 9.3.2 gives simple and ordinary quantification. These weakening and contraction comonads are somewhat unpractical to work with, since what one is really interested in are these induced maps $p(\varepsilon_X): pWX \to pX$ and $p(\delta_X): pWX \to pW^2X$ in the base category. It turns out that such a comonad can be recovered from this induced structure, if it is suitably described as a "comprehension category". This structure is much easier to use. Moreover, it will be of fundamental importance in the categorical description of type dependency in the next chapter. Therefore we shall from now on mostly use these comprehension categories.

(The aspect of 'comprehension' will be explained in the next chapter, see especially Corollary 10.4.6.)

9.3.4. Theorem. *Weakening and contraction comonads on a fibration* $\begin{smallmatrix}\mathbb{E}\\\downarrow p\\\mathbb{B}\end{smallmatrix}$
are in one-one-correspondence with functors $\mathcal{P}\colon \mathbb{E} \to \mathbb{B}^{\to}$ *satisfying*

(1) $\mathrm{cod} \circ \mathcal{P} = p\colon \mathbb{E} \to \mathbb{B};$

(2) for each Cartesian map f in $\mathbb{E}$, the induced square $\mathcal{P}(f)$ in $\mathbb{B}$ is a pullback.

Such a functor $\mathcal{P}$ will be called a **comprehension category** *(on p). We shall often write $\{-\} = \mathrm{dom} \circ \mathcal{P}\colon \mathbb{E} \to \mathbb{B}$. Thus $\mathcal{P}$ is a natural transformation $\{-\} \Rightarrow p$.*

Proof. Assume a comprehension category $\mathcal{P}\colon \mathbb{E} \to \mathbb{B}^{\to}$ on $\begin{smallmatrix}\mathbb{E}\\\downarrow p\\\mathbb{B}\end{smallmatrix}$. Define a functor $W\colon \mathbb{E} \to \mathbb{E}$ by

$$W(X) = \mathcal{P}X^*(X)$$

with (Cartesian) counit

$$\varepsilon_X = \left(WX \xrightarrow{\ \overline{\mathcal{P}X}(X)\ } X \right)$$

A morphism $f\colon X \to Y$ in $\mathbb{E}$ is mapped to the unique map $Wf\colon \mathcal{P}X^*(X) \to \mathcal{P}Y^*(Y)$ above $\{f\}$ with $\overline{\mathcal{P}Y}(Y) \circ Wf = f \circ \overline{\mathcal{P}X}(X)$. This makes ε a natural transformation.

Conversely, if a weakening and contraction comonad (W, ε, δ) is given, we get a functor $\mathcal{P}\colon \mathbb{E} \to \mathbb{B}^{\to}$ by

$$X \mapsto \left(\begin{matrix} pWX \\ \downarrow p(\varepsilon_X) \\ pX \end{matrix} \right) \qquad \text{and} \qquad f \mapsto (pf, pWf).$$

Then $cod \circ \mathcal{P} = p$ by construction. Condition (2) in the theorem holds if and only if condition (2) in Definition 9.3.1 holds, since for a Cartesian map $f\colon X \to Y$ in $\mathbb{E}$ the diagram on the left below is a pullback in $\mathbb{E}$ if and only if the diagram on the right is a pullback in $\mathbb{B}$, by Exercise 1.4.4.

$$
\begin{array}{ccc}
WX & \xrightarrow{\ Wf\ } & WY \\
{\scriptstyle \varepsilon_X}\downarrow & & \downarrow{\scriptstyle \varepsilon_Y} \\
X & \xrightarrow[\ f\ ]{} & Y
\end{array}
\qquad \text{above} \qquad
\begin{array}{ccc}
\{X\} & \xrightarrow{\ \{f\}\ } & \{Y\} \\
{\scriptstyle \mathcal{P}X}\downarrow & & \downarrow{\scriptstyle \mathcal{P}Y} \\
pX & \xrightarrow[\ pf\ ]{} & pY
\end{array}
$$

Finally, if we start from a comprehension category $\mathcal{P}$, turn it into a comonad, with $\varepsilon_X = \overline{\mathcal{P}X}(X)$, and then into a comprehension category again, we get back the original one: $p(\varepsilon_X) = p(\overline{\mathcal{P}X}(X)) = \mathcal{P}X$. And if we start with a weakening and contraction comonad, we get back a comonad which is (vertically) isomorphic to the original one.   □

By this result we have a notion of quantification with respect to an arbitrary comprehension category. This is the form which will be used mostly. Therefore, we make it explicit. An alternative formulation in terms of fibred adjunctions occurs in Exercise 9.3.8.

9.3.5. Definition. Consider a comprehension category $\mathcal{P} \colon \mathbb{E} \to \mathbb{B}^{\to}$ and a fibration $\begin{smallmatrix} \mathbb{D} \\ \downarrow q \\ \mathbb{B} \end{smallmatrix}$ in a situation

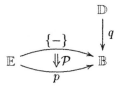

We say that q has $\mathcal{P}$-products / -coproducts / -equality if q has products / coproducts / equality with respect to the weakening and contraction comonad associated with $\mathcal{P}$. Explicitly, this means the following.

(i) The fibration q has $\mathcal{P}$-products (resp. coproducts) if there is for each object $X \in \mathbb{E}$ an adjunction

$$\mathcal{P}X^* \dashv \textstyle\prod_X \qquad (\text{resp. } \textstyle\coprod_X \dashv \mathcal{P}X^*(X))$$

plus a Beck-Chevalley condition: for each Cartesian map $f \colon X \to Y$ in $\mathbb{E}$, the canonical natural transformation

$$(pf)^* \textstyle\prod_Y \Longrightarrow \textstyle\prod_X \{f\}^* \qquad (\text{resp. } \textstyle\coprod_X \{f\}^* \Longrightarrow (pf)^* \textstyle\coprod_Y)$$

is an isomorphism.

(ii) And q has $\mathcal{P}$-equality if for each $X \in \mathbb{E}$ there is an adjunction

$$\mathrm{Eq}_X \dashv \delta_X^*$$

where δ_X is the unique mediating diagonal $\langle \mathrm{id}, \mathrm{id} \rangle \colon \{X\} \to \{\mathcal{P}X^*(X)\}$ in

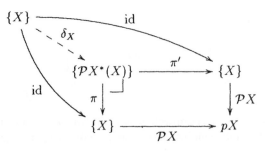

where $\pi = \mathcal{P}(\mathcal{P}X^*(X))$ and $\pi' = \{\overline{\mathcal{P}X}(X)\}$ are the pullback projections. Additionally there is a Beck-Chevalley requirement: for each Cartesian map $f: X \to Y$ in $\mathbb{E}$, the canonical natural transformation

$$\text{Eq}_X \{f\}^* \Longrightarrow \{f'\}^* \text{Eq}_Y$$

should be an isomorphism—where f' is the unique morphism in $\mathbb{E}$ over $\{f\}$ in,

$$
\begin{array}{ccc}
\mathcal{P}X^*(X) & \xdashrightarrow{\quad f' \quad} & \mathcal{P}Y^*(Y) \\
\downarrow & & \downarrow \\
X & \xrightarrow{\quad f \quad} & Y
\end{array}
$$

This lemma provides us with various examples of quantification. We consider the most important ones explicitly.

9.3.6. Examples. (i) The weakening and contraction comonad on a simple fibration $\begin{smallmatrix} s(\mathbb{B}) \\ \downarrow \\ \mathbb{B} \end{smallmatrix}$ described earlier in this section corresponds to a comprehension category $s(\mathbb{B}) \to \mathbb{B}^\to$ mapping

$$(I, X) \mapsto \left(\begin{array}{c} I \times X \\ \downarrow \pi \\ I \end{array} \right)$$

(as described in Exercise 1.3.1). It is now much easier to see that simple products and coproducts (along Cartesian projections) is quantification with respect to this "simple" comprehension category $s(\mathbb{B}) \to \mathbb{B}^\to$.

The diagonals $\delta(I, J)$ of this simple comprehension category are the morphisms

$$I \times J \xrightarrow{\quad \delta(I, J) = \langle \text{id}, \pi' \rangle \quad} (I \times J) \times J$$

That is, the Cartesian diagonals $\delta(I, J)$ as used in Definition 3.4.1 for defining simple equality.

And if we have a CT-structure $(\mathbb{B}, T)$ where $T \subseteq \text{Obj } \mathbb{B}$ then we have a similar comprehension category $s(T) \to \mathbb{B}^\to$ on the simple fibration $\begin{smallmatrix} s(T) \\ \downarrow \\ \mathbb{B} \end{smallmatrix}$ mapping a pair $I \in \mathbb{B}$ and $X \in T$ to the Cartesian projection $I \times X \to I$. Products and coproducts with respect to this comprehension category are what we have called "simple T-products" and "simple T-coproducts" in Definition 2.4.3.

(ii) Next we consider the earlier weakening and contraction comonad on the codomain fibration $\begin{smallmatrix} \mathbb{B}^\to \\ \downarrow \\ \mathbb{B} \end{smallmatrix}$ of a category $\mathbb{B}$ with finite limits. The corresponding comprehension category $\mathbb{B}^{\rightarrow} \to \mathbb{B}^{\rightarrow}$ is simply the identity functor. Products and coproducts with respect to this comprehension category are products and coproducts along all morphisms in $\mathbb{B}$, as in Definition 1.9.4. The diagonal on a family $\left(\begin{smallmatrix} X \\ \downarrow \\ I \end{smallmatrix} \right)$ is the mediating map $X \to X \times_I X$.

(iii) The inclusion $\mathrm{Sub}(\mathbb{B}) \to \mathbb{B}^{\rightarrow}$ for a category $\mathbb{B}$ with finite limits is a comprehension category on the subobject fibration on $\mathbb{B}$. Products and coproducts with respect to this comprehension category $\mathrm{Sub}(\mathbb{B}) \to \mathbb{B}^{\rightarrow}$ involves products $\prod_m$ and coproducts $\coprod_m$ along all monomorphisms m in the base category $\mathbb{B}$.

In due course, we shall see many more examples of quantification with respect to a comprehension category. In the remainder of this section we describe some basic notions and results associated with the general form of quantification with respect to a comprehension category. We start by making explicit what preservation of products, coproducts and equality means in this general setting.

9.3.7. Definition. Let $\begin{smallmatrix} \mathbb{D} \\ \downarrow q \\ \mathbb{B} \end{smallmatrix} \xrightarrow{\ H\ } \begin{smallmatrix} \mathbb{F} \\ \downarrow r \\ \mathbb{B} \end{smallmatrix}$ be a fibred functor (over $\mathbb{B}$). Suppose that the fibrations q and r have products (resp. coproducts or equality) with respect to some comprehension category $\mathcal{P} \colon \mathbb{E} \to \mathbb{B}^{\rightarrow}$. We say that H **preserves** $\mathcal{P}$-products (resp. coproducts or equality) if for each appropriate pair of objects $X \in \mathbb{E}$, $A \in \mathbb{D}$, the canonical map

$$H \prod_X (A) \longrightarrow \prod_X (HA)$$

(resp. $\coprod_X (HA) \longrightarrow H \coprod_X (A)$ or $\mathrm{Eq}_X(HA) \longrightarrow H \mathrm{Eq}_X(A)$)

is an isomorphism.

There is the following situation, which is as for ordinary categories. The proof is easy and left to the reader.

9.3.8. Lemma. *Fibred right adjoints preserve products $\prod$ (with respect to some comprehension category). Fibred left adjoints preserve coproducts $\coprod$ and equality* Eq. $\qquad\qquad\square$

A standard categorical result is that a reflection $\mathbb{A} \leftrightarrows \mathbb{B}$ induces products and coproducts (like $\times$ and $+$) in $\mathbb{A}$ if they exist in $\mathbb{B}$. (By a 'reflection' we mean an adjunction $(F \dashv G)$ with an isomorphism $FG \overset{\cong}{\Rightarrow} \mathrm{id}$ as counit; or equivalently, with a full and faithful functor G as right adjoint.) In Exercise 9.3.9 below one finds the fibred analogue applying to fibred products like $\times$ and plus $+$. Here we mention the version applying to products $\prod$, coproducts $\coprod$ and equality Eq with respect to a comprehension category.

9.3.9. Lemma. *Consider a fibred reflection $(F \dashv G)$ and a comprehension category $\mathcal{P}$ in the following situation.*

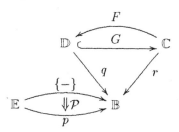

Then, if the fibration r has products / coproducts / equality with respect to the comprehension category $\mathcal{P}: \mathbb{E} \to \mathbb{B}^{\to}$, then so has the fibration q. Moreover, the right adjoint G preserves the induced products.

Proof. The case of coproducts is easy: for objects $X \in \mathbb{E}$ and $A \in \mathbb{D}$ above $\{X\}$, define

$$\exists_X(A) = F \coprod{}_X (GA)$$

where $\coprod_X$ is the assumed left adjoint to $\mathcal{P}X^* : \mathbb{C}_{pX} \to \mathbb{C}_{\{X\}}$. Then we get coproduct correspondences:

$$
\begin{array}{c}
\exists_X(A) = F \coprod{}_X (GA) \longrightarrow B \\
\hline\hline
\coprod{}_X (GA) \longrightarrow GB \\
\hline\hline
GA \longrightarrow \mathcal{P}X^*(GB) \cong G(\mathcal{P}X^*(B)) \\
\hline\hline
A \longrightarrow \mathcal{P}X^*(B)
\end{array}
$$

the latter, because G is a full and faithful functor. Thus we have adjunctions $(\exists_X \dashv \mathcal{P}X^*)$. Equality is transferred to q in the same manner.

For products, we define, as for coproducts,

$$\forall_X(A) = F \prod{}_X (GA).$$

We first show that the unit

$$\prod{}_X(GA) \longrightarrow GF \prod{}_X(GA) = G\forall_X(A)$$

is an isomorphism. Its inverse is obtained by transposing the composite

$$\mathcal{P}X^*(GF \prod{}_X(GA))) \xrightarrow{\cong} GF\mathcal{P}X^* \prod{}_X(GA)) \xrightarrow{GF(\varepsilon)} GFGA \cong GA$$

Then there are product correspondences:

$$\frac{\dfrac{\dfrac{B \longrightarrow \forall_X(A)}{GB \longrightarrow G\forall_X(A) \cong \prod_X(GA)}}{G(\mathcal{P}X^*(B)) \cong \mathcal{P}X^*(GB) \longrightarrow GA}}{\mathcal{P}X^*(B) \longrightarrow A}$$

$\square$

9.3.10. Lemma. *Let* $\begin{smallmatrix} \mathbb{E} \\ \downarrow{p} \\ \mathbb{B} \end{smallmatrix}$ *and* $\begin{smallmatrix} \mathbb{D} \\ \downarrow{q} \\ \mathbb{B} \end{smallmatrix}$ *be two fibrations on the same basis* $\mathbb{B}$
(with q cloven). Form the fibration $\begin{smallmatrix} \mathbb{D}\times_{\mathbb{B}}\mathbb{E} \\ \downarrow{q^*(p)} \\ \mathbb{D} \end{smallmatrix}$ *by change-of-base. A com-*
prehension category $\mathcal{P}:\mathbb{E} \to \mathbb{B}^{\to}$ *on p can then be lifted to a comprehension*
category $q^*(\mathcal{P})$ *on* $q^*(p)$, *in a diagram:*

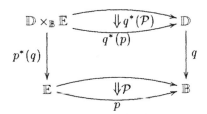

by assigning to $A \in \mathbb{D}$ *and* $X \in \mathbb{E}$ *above the same object in* $\mathbb{B}$ *the map*

$$q^*(\mathcal{P})(A,X) = \left(\mathcal{P}X^*(A) \xrightarrow{\ \overline{\mathcal{P}X}(A)\ } A\right)$$

in $\mathbb{D}$ *above* $\mathcal{P}X$.

Proof. For a morphism $(f,g) \colon (A,X) \to (B,Y)$ in $\mathbb{D}\times_{\mathbb{B}}\mathbb{E}$ we get a commuting square in $\mathbb{D}$

$$q^*(\mathcal{P})(f,g) = \left(\begin{array}{ccc} \mathcal{P}X^*(A) & \xdashrightarrow{\ f'\ } & \mathcal{P}Y^*(B) \\ \downarrow & & \downarrow \\ A & \xrightarrow{\ \ f\ \ } & B \end{array}\right) \quad \text{above} \quad \mathcal{P}(g)$$

where f' is the unique map above $\{g\}$ making the square commute. This yields a comprehension category since if the map (f,g) is Cartesian in $\mathbb{D} \times_{\mathbb{B}} \mathbb{E}$, then g is Cartesian in $\mathbb{E}$, so that the underlying square $\mathcal{P}(g)$ in $\mathbb{B}$ is a pullback. Hence the above square $q^*(\mathcal{P})(f,g)$ is a pullback in $\mathbb{D}$ by Exercise 1.4.4. $\square$

In the "logic of types" in polymorphic predicate logic (over polymorphic type theory) in Section 8.6 we have been somewhat vague about the precise categorical formulation of quantification $\forall \alpha\colon A.\,\varphi$ and $\exists \alpha\colon A.\,\varphi$ of propositions over kinds. In the categorical set-up we used one fibration $\begin{smallmatrix}\mathbb{E}\\\downarrow p\\\mathbb{D}\end{smallmatrix}$ over types over kinds, and another (preorder) fibration $\begin{smallmatrix}\mathbb{D}\\\downarrow q\\\mathbb{E}\end{smallmatrix}$ of propositions over types. We suggested that this form of quantification involved quantification along the liftings in $\mathbb{E}$ of the Cartesian projections in $\mathbb{B}$. At this stage we have the technical means to be more precise: what one needs is that the fibration $\begin{smallmatrix}\mathbb{D}\\\downarrow\\\mathbb{E}\end{smallmatrix}$ has products and coproducts with respect to the simple comprehension category $\mathcal{S}_{\mathbb{B}} = (\mathrm{s}(\mathbb{B}) \to \mathbb{B}^{\to})$ lifted along p. That is, q should have $p^*(\mathcal{S}_{\mathbb{B}})$-products and coproducts.

Exercises

9.3.1. (i) Show that a weakening and contraction comonad on the total category $\mathbb{E}$ of a fibration, restricts to a comonad on the subcategory $\mathrm{Cart}(\mathbb{E}) \hookrightarrow \mathbb{E}$ with Cartesian maps only.

(ii) Assume we have a functor W on the total category of a fibration together with a natural transformation $\varepsilon\colon W \Rightarrow \mathrm{id}$ satisfying conditions (1) and (2) in Definition 9.3.1. Prove that there is a unique natural transformation $\delta\colon W \Rightarrow W^2$ making (W, ε, δ) a (weakening and contraction) comonad.

9.3.2. Let $\begin{smallmatrix}\mathbb{D}\\\downarrow q\\\mathbb{B}\end{smallmatrix}$ be a fibred CCC which has coproducts or equality with respect to a comprehension category $\mathcal{P}\colon \mathbb{E} \to \mathbb{B}^{\to}$. Show that the Frobenius property is automatically satisfied.

9.3.3. Give explicit descriptions of all canonical maps in this section: in Definitions 9.3.2, 9.3.3 and 9.3.7.

9.3.4. Prove Lemma 9.3.8.

9.3.5. Investigate what the Beck-Chevalley condition says about substitution in products $\forall \alpha\colon A.\,\varphi$ and coproducts $\exists \alpha\colon A.\,\varphi$ of propositions over kinds in the term model $\begin{smallmatrix}\mathbb{P}\\\downarrow\\\mathbb{T}\end{smallmatrix}$ over $\begin{smallmatrix}\mathbb{T}\\\downarrow\\\mathbb{K}\end{smallmatrix}$ (as described in Section 8.6), assuming that these $\forall$ and $\exists$ are described categorically *wrt.* a lifted simple comprehension category, as mentioned after the proof of Lemma 9.3.10.

9.3.6. Let $\begin{smallmatrix}\mathbb{D}\\\downarrow q\\\mathbb{B}\end{smallmatrix}$ be a fibration and $\mathcal{P}\colon \mathbb{E} \to \mathbb{B}^{\to}$ a comprehension category on p. Show that q has $\mathcal{P}$-coproducts if and only if both:

- for each $X \in \mathbb{E}$ and $A \in \mathbb{D}$ above $\{X\}$, there is an opcartesian map $A \to \coprod_X(A)$ above $\mathcal{P}X$;

- for each Cartesian map $f: X \to Y$ in $\mathbb{E}$ and each commuting diagram

$$
\begin{array}{ccc}
A & \xrightarrow{\;g\;} & B \\
{\scriptstyle r}\downarrow & & \downarrow{\scriptstyle s} \\
A' & \xrightarrow[\;h\;]{} & B'
\end{array}
$$

in $\mathbb{D}$, above the pullback square $\mathcal{P}(f)$ in $\mathbb{B}$, one has:

$$
\left.\begin{array}{l}
s \text{ is opcartesian over } \mathcal{P}Y \\
g \text{ is Cartesian over } \{f\} \\
h \text{ is Cartesian over } pf
\end{array}\right\} \Rightarrow r \text{ is opcartesian over } \mathcal{P}X.
$$

Give a similar reformulation for equality in terms of opcartesian maps.

9.3.7.   Generalise Exercise 3.4.2 (ii) to arbitrary comprehension categories: show that if a fibred CCC q has $\mathcal{P}$-equality then each $\mathcal{P}$-contraction functor (also) has a right adjoint in q.

9.3.8.   Consider a comprehension category $\mathcal{P}: \mathbb{E} \to \mathbb{B}^{\to}$ and a fibration 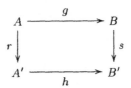 as in Definition 9.3.5. The natural transformation $\mathcal{P}: \{-\} \Rightarrow p$ gives by Lemma 1.7.10 rise to a fibred functor $\langle \mathcal{P} \rangle: p^*(q) \to \{-\}^*(q)$. It turns out that quantification with respect to $\mathcal{P}$ can be described in terms of adjoints to $\langle \mathcal{P} \rangle$, as in [74, Definition 7].

   (i)   Prove that the fibration q has products (resp. coproducts) with respect to $\mathcal{P}$ if and only if $\langle \mathcal{P} \rangle$ has a fibred right (resp. left) adjoint.

   (ii)   Formulate and prove a similar result for equality.

9.3.9.   Let $G: \mathbb{A} \to \mathbb{B}$ be a full and faithful functor, which has a left adjoint F (so that we have a reflection). Recall the following basic results.

   (a)   If $\mathbb{B}$ has coproducts $+$, then so has $\mathbb{A}$, given by $X \vee Y = F(GX + GY)$.

   (b)   If $\mathbb{B}$ has products $\times$, then so has $\mathbb{A}$—given by $X \wedge Y = F(GX \times GY)$—and G preserves them.

   [The proof of (a) is straightforward, but (b) is slightly more complicated; one first shows that the unit $\eta: GX \times GY \to G(X \wedge Y)$ is an isomorphism. Its inverse is $\langle G(\varepsilon_X) \circ GF(\pi), G(\varepsilon_Y) \circ GF(\pi') \rangle$.]

   (i)   Extend the above results to other limits.

   (ii)   Extend it also to fibred reflections: assume fibrations $\begin{smallmatrix}\mathbb{E}\\[-2pt]\downarrow p\\[-2pt]\mathbb{B}\end{smallmatrix}$ and $\begin{smallmatrix}\mathbb{D}\\[-2pt]\downarrow q\\[-2pt]\mathbb{B}\end{smallmatrix}$ and a full and faithful fibred functor $G: \mathbb{E} \to \mathbb{D}$, which has a fibred left adjoint F. Show that if q has fibred (co)products (like $\times$ and $+$), then so has p. And that G preserves the products.

9.4 Category theory over a fibration

Up to now we have mostly been studying a single level of indexing, given by one category being fibred over another. Typical are situations with a logic fibred over a simple type theory—as in logics of equations or of predicates—or with a calculus of types fibred over a calculus of kinds—as in polymorphic calculi. But in order to study a "logic of types" over a polymorphic calculus, one has propositions fibred over types and types fibred over kinds, as described in Section 8.6. Such double levels of indexing will be investigated systematically in this section. Technically, this will involve fibrations, not in the 2-category **Cat** of categories, but in the 2-categories **Fib**($\mathbb{B}$) and **Fib** of fibrations over a fixed basis $\mathbb{B}$, and over arbitrary bases (see Section 1.7). We shall use the 2-categorical formulation of the notion of fibration from Street [315, 317], based on Chevalley's result in Exercise 1.4.8. This gives a suitably abstract reformulation of when a functor $p: \mathbb{E} \to \mathbb{B}$ is a cloven fibration, in terms of the existence of a right adjoint to a canonical functor $(\mathbb{E} \downarrow \mathbb{E}) \to (\mathbb{B} \downarrow p)$ between comma categories. Since adjunctions can be formulated in arbitrary 2-categories, we only need suitable comma-objects.

Recall, *e.g.* from [36, 1.6.1–1.6.3], that the universal property of a **comma object** $(f \downarrow g)$ of two 1-cells $A \xrightarrow{f} C \xleftarrow{g} B$ in a 2-category can be formulated as follows. There are projection maps $A \xleftarrow{\text{fst}} (f \downarrow g) \xrightarrow{\text{snd}} B$ together with a 2-cell

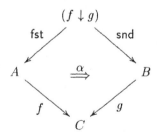

which are universal: for any object D with 1-cells $A \xleftarrow{a} D \xrightarrow{b} B$ and a 2-cell $\gamma: (f \circ a) \Rightarrow (g \circ b)$ there is a unique 1-cell $d: D \dashrightarrow (f \downarrow g)$ with

$$\text{fst} \circ d = a, \qquad \text{snd} \circ d = b, \qquad \alpha d = \gamma.$$

A special case is the **arrow object** on an object C; it is written as

$$C^{\to} = (C \downarrow C) = (\text{id}_C \downarrow \text{id}_C).$$

It is characterised by the correspondence between

$$\text{1-cells} \quad A \longrightarrow C^{\rightarrow}$$
$$\overline{}$$
$$\text{2-cells} \quad A \underset{\Downarrow}{\overset{}{\rightrightarrows}} C$$

Thus, the arrow category $C^{\rightarrow}$ on a category C is the arrow-object in the 2-category **Cat** of categories. And the comma category $(F \downarrow G)$ of functors $A \xrightarrow{F} C \xleftarrow{G} B$ is the comma object in **Cat**. Recall that this comma category $(F \downarrow G)$ has

objects   triples (X, φ, Y) where $X \in A, Y \in B$ and $\varphi \colon FX \to GY$
              in C.

morphisms $(X, \varphi, Y) \to (X', \varphi', Y')$ are pairs of maps $f \colon X \to X'$ in A
              and $g \colon Y \to Y'$ in B with $\varphi' \circ Ff = Gg \circ \varphi$.

9.4.1. Definition (Following [315, 317]). Consider a 1-cell $p \colon E \to B$ in a 2-category, and assume that the comma objects $E^{\rightarrow} = (E \downarrow E) = (\mathrm{id}_E \downarrow \mathrm{id}_E)$ and $(B \downarrow p) = (\mathrm{id}_B \downarrow p)$ exist. There is then a unique mediating 1-cell $\widehat{p} \colon E^{\rightarrow} \dashrightarrow (B \downarrow p)$, in the following situation.

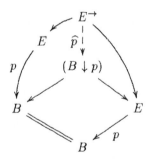

(For the obvious 2-cell $p \circ \mathsf{fst} \Rightarrow \mathsf{snd} \circ p \colon E^{\rightarrow} \rightrightarrows B$.)

    We say that p is a **(cloven) fibration** in this 2-category if the induced map $\widehat{p}$ has a right-adjoint-right-inverse; that is, if $\widehat{p}$ has a right adjoint $\widetilde{p}$ such that the counit of the adjunction $(\widehat{p} \dashv \widetilde{p})$ is the identity 2-cell.

    For an ordinary functor $p \colon E \to B$ the induced functor $\widehat{p} \colon E^{\rightarrow} \to (B \downarrow p)$ sends $(f \colon X \to Y) \mapsto (pX, f, X)$. And p is a fibration in **Cat** (according to this definition) if and only if p is a cloven fibration, see Exercise 1.4.8.

    We split this section in two, by first considering fibrations in a 2-category **Fib**(B) and then in **Fib**.

Fibrations over fibrations with a fixed base category

We first observe the following.

9.4.2. Lemma. *The 2-category* **Fib**$(\mathbb{B})$ *of fibrations over a fixed base category* $\mathbb{B}$ *has comma objects: for two maps of fibrations* $p \xrightarrow{F} r \xleftarrow{G} q$ *in a situation*

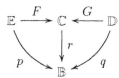

The total category of their comma fibration $\begin{smallmatrix} V(F\downarrow G) \\ \downarrow \\ \mathbb{B} \end{smallmatrix}$ *is the full subcategory*

$$V(F \downarrow G) \lhook\joinrel\longrightarrow (F \downarrow G)$$

of the ordinary comma category, on the vertical *maps* $\varphi \colon FX \to GY$ *in* $\mathbb{C}$*. It is fibred over* $\mathbb{B}$ *via the functor* $(\varphi \colon FX \to GY) \mapsto pX = qY$.

In particular, for a fibration $\begin{smallmatrix} \mathbb{E} \\ \downarrow p \\ \mathbb{B} \end{smallmatrix}$ *the arrow object* $p^{\to}$ *in* **Fib**$(\mathbb{B})$ *exists. We write it as* $\begin{smallmatrix} V(\mathbb{E}) \\ \downarrow \\ \mathbb{B} \end{smallmatrix}$ *where* $V(\mathbb{E}) \hookrightarrow \mathbb{E}^{\to}$ *is the full subcategory on the vertical maps. Sometimes we write* $V_p(\mathbb{E})$ *instead of* $V(\mathbb{E})$*, in order to have the dependence on* p *explicit.*

Proof. The category $V(F \downarrow G)$ is fibred over $\mathbb{B}$ since for an object $(\varphi \colon FX \to GY) \in V(F \downarrow G)$ and a map $u \colon I \to pX$ in $\mathbb{B}$ we have a lifting given by combining the liftings $\overline{u}(X) \colon u^*(X) \to X$ in $\mathbb{E}$ and $\overline{u}(Y) \colon u^*(Y) \to Y$ in $\mathbb{D}$ in a diagram:

$$
\begin{array}{ccc}
F(u^*(X)) & \xrightarrow{\;F(\overline{u}(X))\;} & FX \\[2pt]
{\scriptstyle u^*(\varphi)}\downarrow & & \downarrow{\scriptstyle \varphi} \\[2pt]
G(u^*(Y)) & \xrightarrow{\;G(\overline{u}(Y))\;} & GY
\end{array}
$$

$$
I \xrightarrow{\;\;u\;\;} pX
$$

There is then an obvious vertical natural transformation

$$V(F \downarrow G) \; \rightrightarrows^{\Downarrow} \; \mathbb{C}$$

with the required universal property.   □

We can now characterise fibrations in a 2-category $\mathbf{Fib}(\mathbb{B})$ of fibrations over a fixed base category $\mathbb{B}$.

9.4.3. Proposition. *Consider two fibrations* $\begin{smallmatrix}\mathbb{D}\\\downarrow q\\\mathbb{B}\end{smallmatrix}$ *and* $\begin{smallmatrix}\mathbb{E}\\\downarrow p\\\mathbb{B}\end{smallmatrix}$ *over* $\mathbb{B}$, *and a morphism* $r: q \to p$ *of fibrations, as in:*

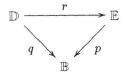

The following statements are then equivalent:

(i) $r: q \to p$ *is a fibration in* $\mathbf{Fib}(\mathbb{B})$, *i.e. the induced fibred functor* $\widehat{r}$ *in*

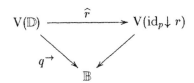

has a fibred right-adjoint-right-inverse (in $\mathbf{Fib}(\mathbb{B})$*);*

(ii) *r is itself a cloven fibration;*

(iii) *r is "fibrewise a cloven fibration": for each object* $I \in \mathbb{B}$ *the functor*

$$q^{-1}(I) = \mathbb{D}_I \xrightarrow{\quad r_I \quad} \mathbb{E}_I = p^{-1}(I)$$

obtained by restricting r to the fibres over I, *is a cloven fibration, and for each morphism* $u: J \to I$ *in* $\mathbb{B}$ *and p-reindexing functor* $u^*: \mathbb{E}_I \to \mathbb{E}_J$ *there is a q-reindexing functor* $u^{\#}: \mathbb{D}_I \to \mathbb{D}_J$ *forming a morphism of fibrations (in* $\mathbf{Fib}$*):*

$$
\begin{array}{ccc}
\mathbb{D}_I & \xrightarrow{\quad u^{\#} \quad} & \mathbb{D}_J \\
{\scriptstyle r_I}\downarrow & & \downarrow{\scriptstyle r_J} \\
\mathbb{E}_I & \xrightarrow[\quad u^* \quad]{} & \mathbb{E}_J
\end{array}
$$

If one of these conditions holds, then one calls r a **fibration over** *p.*

Proof. (i) $\Rightarrow$ (ii) Assume the functor $\widehat{r}$ has a fibred right adjoint $\widetilde{r}$, and that the counit $\widehat{r}\widetilde{r} \Rightarrow \mathrm{id}$ is the identity. For an object $A \in \mathbb{D}$ and a map $f: X \to rA$ in $\mathbb{E}$, we first write f as a vertical map $g: X \to r(u^{\#}(A))$ followed by the

Cartesian map $r(\overline{u}(A))$ as below. This g is then an object $(X, g, u^{\#}(A)) \in V(\text{id}_p \downarrow r)$, so we can apply the right adjoint $\tilde{r}$. This yields a vertical map $\tilde{r}(g): A' \to u^{\#}(A)$, as in:

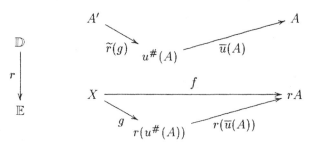

It is not hard to see that the resulting composite $A' \to A$ is an r-Cartesian lifting of f.

(ii) $\Rightarrow$ (iii) The fact that all the r_I's are fibrations is already stated in Lemma 1.5.5 (ii). One obtains a morphism of fibrations as in the square in (iii) above, since for each reindexing functor $u^*: \mathbb{E}_I \to \mathbb{E}_J$ one can choose $u^{\#}$ as follows: for $A \in \mathbb{D}_I$ with $r_I(A) = X \in \mathbb{E}_I$, take $u^{\#}(A)$ to be the domain of the r-lifting of $\overline{u}(X): u^*(X) \to X$, like in the proof of Lemma 1.5.5 (i).

(iii) $\Rightarrow$ (i) One defines a right adjoint $\tilde{r}$ as follows. For a vertical map $\varphi: X \to rA$, say over $I \in \mathbb{B}$, take $\tilde{r}(\varphi)$ to be the r_I-Cartesian lifting $\overline{\varphi}(A): \varphi^{\bullet}(A) \to A$ in $\mathbb{D}$. Commutativity of the diagram in (iii) ensures that $\tilde{r}$ is a fibred functor. $\qquad\square$

We mention some examples of fibrations over a fibration (in **Fib**($\mathbb{B}$)). Recall from Definition 8.5.3 (and Lemma 8.5.4) the **simple fibration over a fibration**: for a fibration $\begin{smallmatrix} \mathbb{E} \\ \downarrow p \\ \mathbb{B} \end{smallmatrix}$ with fibred finite products, one obtains a fibration $\begin{smallmatrix} s_p(\mathbb{E}) \\ \downarrow s_p \\ \mathbb{E} \end{smallmatrix}$ on top of p. Notice that for each object $I \in \mathbb{B}$, the fibrewise fibration $\begin{smallmatrix} s_p(\mathbb{E})_I \\ \downarrow \\ \mathbb{E}_I \end{smallmatrix}$ is the (ordinary) simple fibration $\begin{smallmatrix} s(\mathbb{E}_I) \\ \downarrow \\ \mathbb{E}_I \end{smallmatrix}$ on the fibre category $\mathbb{E}_I$.

If $\begin{smallmatrix} \mathbb{E} \\ \downarrow p \\ \mathbb{B} \end{smallmatrix}$ is a fibration with fibred finite limits, then the codomain functor $V(\mathbb{E}) \to \mathbb{E}$ is also a fibration over p. It will be called the **codomain fibration over** p (in **Fib**($\mathbb{B}$)). The fibre fibration $\begin{smallmatrix} V(\mathbb{E})_I \\ \downarrow \\ \mathbb{E}_I \end{smallmatrix}$ is the codomain fibration $\begin{smallmatrix} \mathbb{E}_I^{\to} \\ \downarrow \\ \mathbb{E}_I \end{smallmatrix}$ on the fibre category $\mathbb{E}_I$. See Exercise 9.4.2 below for some more details.

These two examples generalise the simple and codomain fibrations over categories to simple and codomain fibrations over fibrations: one recovers the ordinary simple and codomain fibrations "fibrewise".

The "logic over types" in polymorphic predicate logic (PPL) as described in the beginning of Section 8.6 is described by a preorder fibration (giving us a logic) over a polymorphic fibration. In this same section we have described a term model example of a fibration $\begin{smallmatrix}\mathbb{P}\\\downarrow\\\mathbb{T}\end{smallmatrix}$ of propositions over types, which is a fibration over a fibration $\begin{smallmatrix}\mathbb{T}\\\downarrow\\\mathbb{K}\end{smallmatrix}$ of types over kinds.

As another example of such a fibration for PPL we mentioned the logic of regular subobjects of families of PERs in $\begin{smallmatrix}\text{UFamRegSub}(\mathbf{PER})\\\downarrow\\\text{UFam}(\mathbf{PER})\end{smallmatrix}$ over families of PERs in $\begin{smallmatrix}\text{UFam}(\mathbf{PER})\\\downarrow\\\omega\text{-Sets}\end{smallmatrix}$, as defined in Example 8.6.1.

Along the lines of the previous example, we can form for a category $\mathbb{B}$ with finite limits, the fibration $\begin{smallmatrix}\text{FamSub}(\mathbb{B})\\\downarrow\\\mathbb{B}^{\rightarrow}\end{smallmatrix}$ giving us the logic of subobjects on families in $\mathbb{B}$ via the change-of-base situation:

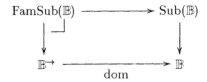

Such fibrations will be used later in a "logic over dependent type theory", see Section 11.2.

Fibrations over fibrations over arbitrary base categories

As a first step towards the description of fibrations in the 2-category **Fib** we start by identifying comma objects in **Fib**.

9.4.4. Lemma. *The 2-category* **Fib** *of fibrations over arbitrary base categories has comma objects. For two maps of fibrations* $q \xrightarrow{\langle F,H\rangle} r \xleftarrow{\langle G,K\rangle} p$ *in a diagram*

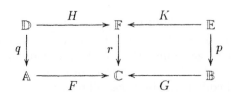

there is a comma fibration in **Fib**,

$$\Big(\langle F, H\rangle \downarrow \langle G, K\rangle\Big) = \begin{pmatrix} (H \downarrow K) \\ \downarrow \\ (F \downarrow G) \end{pmatrix}$$

where $(H \downarrow K)$ *and* $(F \downarrow G)$ *are the ordinary comma categories in* **Cat**, *and where the functor* $(H \downarrow K) \to (F \downarrow G)$ *sends*

$$\Big(HA \xrightarrow{\varphi} KX\Big) \mapsto \Big(FqA = rHA \xrightarrow{r(\varphi)} rKX = GpX\Big)$$

In particular, for a fibration $\begin{smallmatrix} \mathbb{E} \\ \downarrow p \\ \mathbb{B} \end{smallmatrix}$ *the arrow object* $p^{\to}$ *exists in* **Fib**, *and can be identified as* $\begin{smallmatrix} \mathbb{E}^{\to} \\ \downarrow \\ \mathbb{B}^{\to} \end{smallmatrix}$

Notice that we have overloaded the notation $p^{\to}$ by using it both for the arrow fibration $\begin{smallmatrix} V(\mathbb{E}) \\ \downarrow \\ \mathbb{B} \end{smallmatrix}$ in **Fib**($\mathbb{B}$) and for the arrow fibration $\begin{smallmatrix} \mathbb{E}^{\to} \\ \downarrow \\ \mathbb{B}^{\to} \end{smallmatrix}$ in **Fib**. Whenever confusion is likely, we shall mention explicitly in which 2-category the arrow construction $(-)^{\to}$ lives.

Proof. We only show that the functor $\begin{smallmatrix} (H\downarrow K) \\ \downarrow \\ (F\downarrow G) \end{smallmatrix}$ is a fibration, since it is easy to check that it has the appropriate universal property (as comma object in **Fib**). Consider an object $(HA \xrightarrow{\varphi} KX)$ in the total category $(H \downarrow K)$, and a morphism $(u, v)\colon \alpha \to r(\varphi)$ in the base category $(F \downarrow G)$ in a diagram:

$$\begin{array}{ccc} F(I) & \xrightarrow{\ F(u)\ } & F(qA) = r(HA) \\ {\scriptstyle \alpha}\downarrow & & \downarrow{\scriptstyle r(\varphi)} \\ G(J) & \xrightarrow[\ G(v)\ ]{} & G(pX) = r(KX) \end{array}$$

We first take Cartesian maps $\overline{u}(A)\colon u^*(A) \to A$ in $\mathbb{D}$ over $u\colon I \to qA$ in $\mathbb{A}$, and $\overline{v}(X)\colon v^*(X) \to X$ in $\mathbb{E}$ over $v\colon J \to pX$ in $\mathbb{B}$. Then we let $(u, v)^*(\varphi)$ be the unique mediating map $H(u^*(A)) \dashrightarrow K(v^*(X))$ over α with $K(\overline{v}(X)) \circ \langle u, v\rangle^*(\varphi) = \varphi \circ H(\overline{u}(A))$. This gives a Cartesian morphism $(u, v)^*(\varphi) \to \varphi$ in $(H \downarrow K)$ over (u, v) in $(F \downarrow G)$. $\qquad \square$

9.4.5. Proposition. *Consider two fibrations* $\begin{smallmatrix} \mathbb{D} \\ \downarrow q \\ \mathbb{A} \end{smallmatrix}$ *and* $\begin{smallmatrix} \mathbb{E} \\ \downarrow p \\ \mathbb{B} \end{smallmatrix}$ *and a morphism* $\langle K, H \rangle \colon q \to p$ *of fibrations:*

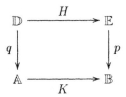

The following are then equivalent:

    (i) $\langle K, H \rangle \colon q \to p$ *is a* **fibration (over a fibration)** *in* **Fib**, i.e. *the induced fibred functor* $\widehat{\langle K, H \rangle} = \langle \widehat{K}, \widehat{H} \rangle$ *in*

$$
\begin{array}{ccc}
\mathbb{D}^{\to} & \xrightarrow{\ \widehat{H}\ } & (\mathbb{E} \downarrow H) \\
{\scriptstyle q^{\to}} \downarrow & & \downarrow {\scriptstyle (p \downarrow \langle K, H \rangle)} \\
\mathbb{A}^{\to} & \xrightarrow[\ \widehat{K}\ ]{} & (\mathbb{B} \downarrow K)
\end{array}
$$

has a fibred right-adjoint-right-inverse (in **Fib***);*

    (ii) *both H and K are cloven fibrations, and* $\langle p, q \rangle$ *is a map of fibrations* $H \to K$ *which strictly preserves the cleavage.*

Proof. Because: H and K are cloven fibrations if and only if the induced functors $\widehat{H}$ and $\widehat{K}$ both have a right-adjoint-right inverse, say $\widetilde{H}$ and $\widetilde{K}$ respectively. And: commutation of the diagram $q^{\to} \circ \widetilde{H} = \widetilde{K} \circ (p \downarrow \langle K, H \rangle)$ means that the cleavage is preserved. Finally, $\widetilde{H}$ is automatically a fibred functor by Exercise 9.4.4.   □

We can also describe codomain and simple fibrations in **Fib**. Let $\begin{smallmatrix} \mathbb{E} \\ \downarrow p \\ \mathbb{B} \end{smallmatrix}$ be a fibration with pullbacks in **Fib**. This means that the base category $\mathbb{B}$ has (chosen) pullbacks and that the total category $\mathbb{E}$ has (chosen) pullbacks, which are strictly preserved by the functor p. (Equivalent to the existence of pullbacks in $\mathbb{E}$ is the existence of fibred pullbacks in p, see Proposition 9.2.1.) We then have two codomain fibrations $\begin{smallmatrix} \mathbb{E}^{\to} \\ \downarrow \\ \mathbb{E} \end{smallmatrix}$ and $\begin{smallmatrix} \mathbb{B}^{\to} \\ \downarrow \\ \mathbb{B} \end{smallmatrix}$ on $\mathbb{E}$ and on $\mathbb{B}$ forming a fibration $\begin{smallmatrix}\mathbb{E}^{\rightarrow}\\\downarrow\\\mathbb{B}^{\rightarrow}\end{smallmatrix}$ over $\begin{smallmatrix}\mathbb{E}\\\downarrow\\\mathbb{B}\end{smallmatrix}$ in:

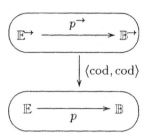

This describes the **codomain fibration** $\begin{smallmatrix}p^{\rightarrow}\\\downarrow\\p\end{smallmatrix}$ on p in **Fib**.

There is also a simple fibration in **Fib** $\begin{smallmatrix}\mathbb{E}\\\downarrow p\\\mathbb{B}\end{smallmatrix}$ with Cartesian products in **Fib** (see [129]). The latter means that $\mathbb{B}$ and $\mathbb{E}$ have Cartesian products and that p strictly preserves them. These Cartesian products in $\mathbb{E}$ may be described via fibred Cartesian products $\wedge$ in p—like in Proposition 9.2.1. We form a fibration $\begin{smallmatrix}S(\mathbb{E})\\\downarrow\\s(\mathbb{B})\end{smallmatrix}$ over p, by stipulating that the new total category $S(\mathbb{E})$ has

objects   triples (X, I, X') where $X, X' \in \mathbb{E}$ and $I \in \mathbb{B}$ satisfy $p(X') = pX \times I$.

morphisms $(X, I, X') \rightarrow (Y, J, Y')$ are triples of maps $f: X \rightarrow Y$ in $\mathbb{E}$, $u: pX \times I \rightarrow J$ in $\mathbb{B}$ and $f': \pi^*(X) \wedge X' \rightarrow Y'$ in $\mathbb{E}$ over $\langle pf \circ \pi, u \rangle : pX \times I \rightarrow pY \times J$.

One gets a commuting diagram

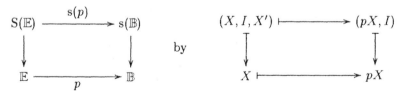

giving us a fibration $\begin{smallmatrix}S(\mathbb{E})\\\downarrow\\s(\mathbb{B})\end{smallmatrix}$ over $\begin{smallmatrix}\mathbb{E}\\\downarrow\\\mathbb{B}\end{smallmatrix}$. It is the simple fibration $\begin{smallmatrix}s(p)\\\downarrow\\p\end{smallmatrix}$ on p in **Fib**.

For an object $X \in \mathbb{E}$ over $I \in \mathbb{B}$, the "fibre fibration" $\begin{smallmatrix}S(\mathbb{E})_X\\\downarrow\\s(\mathbb{B})_I\end{smallmatrix}$ is used in [129] as the fibration resulting from p by adjoining an indeterminate of a "predicate" X to p (in **Fib**).

Here is a different example. For a category $\mathbb{B}$ with finite limits, let $\text{Del}(\mathbb{B})$ be the category of **deliverables** in $\mathbb{B}$. It has

objects   pairs of subobjects $X \rightarrowtail I$ and $U \rightarrowtail I \times A$.

morphisms $(X \rightarrowtail I, U \rightarrowtail I \times A) \longrightarrow (Y \rightarrowtail J, V \rightarrowtail J \times B)$ are pairs of maps $u\colon I \to J$ and $v\colon I \times A \to B$ for which there are (necessarily unique) dashed maps:

$$
\begin{array}{ccc}
X & - - - - \to & Y \\
\downarrow & & \downarrow \\
I & \xrightarrow{\;\;u\;\;} & J
\end{array}
\qquad \text{and} \qquad
\begin{array}{ccc}
U & - - - - - - - \to & V \\
\downarrow & & \downarrow \\
I \times A & \xrightarrow{\;\;\langle u \circ \pi, v\rangle\;\;} & J \times B
\end{array}
$$

Following the terminology of [217] one calls u a "first order deliverable" and v a "second order deliverable". These u and v can be seen as programs, and the dashed arrows indicate that they satisfy certain specifications—given as predicates. Such deliverables are used in a combined development of a program together with a proof that it satisfies a specification, see [217].

One now gets a fibration $\genfrac{}{}{0pt}{}{\text{Del}(\mathbb{B})}{\downarrow\ s(\mathbb{B})}$ over the subobject fibration $\genfrac{}{}{0pt}{}{\text{Sub}(\mathbb{B})}{\downarrow\ \mathbb{B}}$ in:

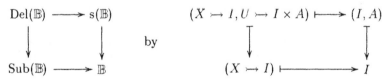

$$
\begin{array}{ccc}
\text{Del}(\mathbb{B}) & \longrightarrow & s(\mathbb{B}) \\
\downarrow & & \downarrow \\
\text{Sub}(\mathbb{B}) & \longrightarrow & \mathbb{B}
\end{array}
\qquad \text{by} \qquad
\begin{array}{ccc}
(X \rightarrowtail I, U \rightarrowtail I \times A) & \longmapsto & (I, A) \\
\downarrow & & \downarrow \\
(X \rightarrowtail I) & \longmapsto & I
\end{array}
$$

We mention one last example: in Remark 8.6.4 (iv) we have a fibration $\genfrac{}{}{0pt}{}{\text{RFam}(\mathbf{PER})}{\downarrow\ \mathbf{RPER}}$ over $\genfrac{}{}{0pt}{}{\text{PFam}(\mathbf{PER})^2}{\downarrow\ \mathbf{PPER}^2}$. It gives the fibration of relations over the (product with itself of) fibration of parametric families of PERs.

Doubtlessly, many more examples of fibrations in **Fib** may be found in the literature. It is not clear at this stage, precisely what kind of logics one can describe with these fibrations over a fibration in **Fib**. Logics over polymorphic type theory, like in [273, 326], to reason with parametricity are likely candidates. The last of the above examples suggests such a link.

Exercises

9.4.1.   Consider the fibred functors F, G in Lemma 9.4.2. Prove for a morphism
$(f, g) \colon (X, \varphi, Y) \to (X', \varphi', Y')$ that

$$(f, g) \text{ is Cartesian in } V(F \downarrow G) \;\Leftrightarrow\; \begin{cases} f \colon X \to X' \text{ is Cartesian in } \mathbb{E} \\ \text{and} \\ g \colon Y \to Y' \text{ is Cartesian in } \mathbb{D}. \end{cases}$$

9.4.2.   Let $\begin{smallmatrix} \mathbb{E} \\ \downarrow p \\ \mathbb{B} \end{smallmatrix}$ be a fibration with fibred finite limits, and consider its arrow
fibration $\begin{smallmatrix} V(\mathbb{E}) \\ \downarrow \\ \mathbb{B} \end{smallmatrix}$ in $\mathbf{Fib}(\mathbb{B})$.
   (i)   Prove that the codomain functor $V(\mathbb{E}) \to \mathbb{E}$ is a fibration. Show that
         it has fibred finite limits again.
   (ii) Check that there is a change-of-base situation

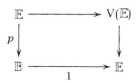

         where $1 \colon \mathbb{B} \to \mathbb{E}$ is the terminal object functor.
   (iii) Notice that for a category $\mathbb{A}$ with finite limits, the ordinary codomain
         fibration $\begin{smallmatrix} \mathbb{A}^{\to} \\ \downarrow \\ \mathbb{A} \end{smallmatrix}$ is the codomain fibration $\begin{smallmatrix} V(\mathbb{A}) \\ \downarrow \\ \mathbb{A} \end{smallmatrix}$ over $\begin{smallmatrix} \mathbb{A} \\ \downarrow \\ 1 \end{smallmatrix}$.
   (iv) Describe the arrow fibration over an arrow fibration.

9.4.3.   Let $\begin{smallmatrix} \mathbb{E} \\ \downarrow p \\ \mathbb{B} \end{smallmatrix}$ be a fibration with fibred finite products. Call a subset $T \subseteq \mathrm{Obj}\,\mathbb{E}$
closed under substitution if for every Cartesian map $f \colon X \to Y$ in $\mathbb{E}$ with
$Y \in T$ also $X \in T$. Such a collection T forms a **fibred CT-structure**.
Define an associated (generalised) simple fibration $\begin{smallmatrix} s_p(T) \\ \downarrow \\ \mathbb{E} \end{smallmatrix}$ over p.

9.4.4.   Consider adjunctions $\langle F \dashv G, \eta, \varepsilon \rangle$ and $\langle F' \dashv G', \eta', \varepsilon' \rangle$ in a situation

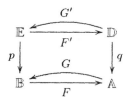

where $q \circ F' = F \circ p$, $p \circ G' = G \circ q$ and η', ε' sit over η, ε. Prove that the
pair $\langle G, G' \rangle$ is then automatically a morphism of fibrations $q \to p$ (*i.e.* that
G' is a fibred functor).
[Notice that Exercise 1.8.5 is a special case.]

9.4.5. (i) Define composition in the category $S(\mathbb{E})$ in the construction of the simple fibration in **Fib**, and show that $S(\mathbb{E})$ is fibred over $s(\mathbb{B})$ and over $\mathbb{E}$, in such a way that liftings are preserved appropriately.

(ii) Check also that the fibration $\begin{array}{c} \mathrm{Del}(\mathbb{B}) \\ \downarrow \\ s(\mathbb{B}) \end{array}$ of deliverables is fibred over the subobject fibration $\begin{array}{c} \mathrm{Sub}(\mathbb{B}) \\ \downarrow \\ \mathbb{B} \end{array}$.

9.5 Locally small fibrations

Recall that an ordinary category $\mathbb{C}$ is called locally small if for each pair of objects $X, Y \in \mathbb{C}$ the collection $\mathbb{C}(X, Y)$ of morphisms $X \to Y$ in $\mathbb{C}$ is a *set*, as opposed to a proper class. That is, each collection $\mathbb{C}(X, Y)$ is an object of the category **Sets** of sets and functions. From a fibred perspective, this dependence of the notion of local smallness on the universe of sets is an unnatural restriction. In fibred category theory one looks for a more general formulation which applies to arbitrary universes (or base categories). This is the same generalisation that led us in Section 1.9 from set-indexed products and coproducts to products and coproducts with respect to (objects and arrows in) an arbitrary universe (given as base category of a fibration). In this section we shall investigate the fibred version of the notion of local smallness. It will turn out to have a close connection to comprehension.

In a fibration $\begin{array}{c} \mathbb{E} \\ \downarrow p \\ \mathbb{B} \end{array}$, the base category $\mathbb{B}$ provides a universe for the total category $\mathbb{E}$. Local smallness for fibrations will involve a representation of homsets in $\mathbb{E}$ as objects of $\mathbb{B}$. The situation in ordinary category theory is captured as a special case via the family fibration $\begin{array}{c} \mathrm{Fam}(\mathbb{C}) \\ \downarrow \\ \mathbf{Sets} \end{array}$. Notice that, if $\mathbb{C}$ is locally small, then for every pair of objects $X = (X_i)_{i \in I}$ and $Y = (Y_i)_{i \in I}$ in the total category $\mathrm{Fam}(\mathbb{C})$ over a set I we can form the disjoint union of all homsets $\mathbb{C}(X_i, Y_i)$. It comes equipped with a projection function π_0 to I:

$$\underline{\mathrm{Hom}}_I(X, Y) \stackrel{\mathrm{def}}{=} \left(\coprod_{i \in I} \mathbb{C}(X_i, Y_i) \right) \xrightarrow{\ \ \pi_0\ \ } I$$

It forms a morphism in the base category **Sets**. There is then an obvious vertical map in $\mathrm{Fam}(\mathbb{C})$ over $\underline{\mathrm{Hom}}_I(X, Y)$, namely

$$\pi_0^*(X) \xrightarrow{\ \ \pi_1\ \ } \pi_0^*(Y) \qquad \text{given by} \qquad (\pi_1)_{(i,f)} = \left(X_i \xrightarrow{\ f_i\ } Y_i \right)$$

This pair (π_0, π_1) satisfies a universal property: for any function $u \colon J \to I$

with a morphism $f : u^*(X) \to u^*(Y)$ in $\mathrm{Fam}(\mathbb{C})$, there is a unique function $v : J \dashrightarrow \underline{\mathrm{Hom}}_I(X, Y)$ with $\pi_0 \circ v = u$ and $v^*(\pi_1) = f$. Clearly this function v sends an element $j \in J$ to the pair $\langle u(j), f_j \rangle \in \coprod_{i \in I} \mathbb{C}(X_i, Y_i)$.

We have described the homsets of $\mathbb{C}$ using a language that makes sense for any fibred category. This formulation will be used in the definition of local smallness for fibrations below, in the form of a representability condition. The definition comes from Bénabou [27, 29] (see also [36, II, 8.6]), just like almost all of the results in this section.

The definition below makes use of a particular cleavage in a fibration, but it does not depend on the cleavage. Lemma 9.5.4 gives an intrinsic (cleavage-free) reformulation. It is however less intuitive. Exercise 9.5.2 below contains another alternative formulation.

9.5.1. Definition. A fibration $\begin{smallmatrix} \mathbb{E} \\ \downarrow p \\ \mathbb{B} \end{smallmatrix}$ is called **locally small** if for each pair of objects $X, Y \in \mathbb{E}$ in the same fibre, say over $I \in \mathbb{B}$, the functor from $(\mathbb{B}/I)^{\mathrm{op}}$ to **Sets**—or to some suitably larger universe than **Sets**—given by

$$\left(J \xrightarrow{\;u\;} I \right) \longmapsto \mathbb{E}_J\left(u^*(X),\ u^*(Y) \right)$$

is representable.

In that case we write

$$\underline{\mathrm{Hom}}_I(X, Y) \xrightarrow{\quad \pi_0 \quad} I$$

for the representing arrow in $\mathbb{B}$, which comes equipped with a vertical morphism "of arrows" in $\mathbb{E}$ over $\underline{\mathrm{Hom}}_I(X, Y)$, written as

$$\pi_0^*(X) \xrightarrow{\quad \pi_1 \quad} \pi_0^*(Y)$$

It is such that for each map $u : J \to I$ in $\mathbb{B}$ together with a vertical morphism $f : u^*(X) \to u^*(Y)$ in $\mathbb{E}$ over $J \in \mathbb{B}$ there is a unique map $v : J \dashrightarrow \underline{\mathrm{Hom}}_I(X, Y)$ in $\mathbb{B}$ making the following two diagrams commute:

$$
\begin{array}{ccc}
J \xrightarrow{\quad v \quad} \underline{\mathrm{Hom}}_I(X, Y) & \qquad &
\begin{array}{ccc}
u^*(X) & \dashrightarrow & \pi_0^*(X) \\
{\scriptstyle f}\downarrow & & \downarrow{\scriptstyle \pi_1} \\
u^*(Y) & \dashrightarrow & \pi_0^*(Y)
\end{array}
\end{array}
$$

where the dashed arrows are the (unique, Cartesian) mediating maps over v.

Sometimes the subscript I in $\underline{\mathrm{Hom}}_I(X, Y)$ is omitted if it is clear from the context in which fibre X, Y live. The intuition is that the fibre over $i \in I$ of the family $\begin{pmatrix} \underline{\mathrm{Hom}}_I(X,Y) \\ \downarrow \pi_0 \\ I \end{pmatrix}$ in $\mathbb{B}/I$ is the homset of vertical maps $X_i \to Y_i$ in $\mathbb{E}$.

We consider some examples. We have already seen that if a category $\mathbb{C}$ is locally small in the ordinary sense, then the associated family fibration $\begin{smallmatrix} \mathrm{Fam}(\mathbb{C}) \\ \downarrow \\ \mathbf{Sets} \end{smallmatrix}$ is locally small in the fibred sense. The converse is also true, since one can consider objects X, Y of $\mathbb{C}$ as objects of $\mathrm{Fam}(\mathbb{C})$ over a one-element (terminal) set 1. The resulting set $\underline{\mathrm{Hom}}_1(X, Y)$ is then the homset $\mathbb{C}(X, Y)$, since elements $1 \to \underline{\mathrm{Hom}}_1(X, Y)$ correspond to maps $X \to Y$ in $\mathbb{C}$.

The externalisation $\begin{smallmatrix} \mathrm{Fam}(\mathbf{C}) \\ \downarrow \\ \mathbb{B} \end{smallmatrix}$ of an internal (small) category $\mathbf{C}$ in an ambient category $\mathbb{B}$ yields a similar example of a locally small fibration. For objects $X, Y: I \rightrightarrows C_0$ in the fibre of $\mathrm{Fam}(\mathbf{C})$ over I, one forms a representing family $\begin{pmatrix} \underline{\mathrm{Hom}}(X,Y) \\ \downarrow \pi_0 \\ I \end{pmatrix}$ via the pullback:

$$
\begin{array}{ccc}
\underline{\mathrm{Hom}}(X,Y) & \xrightarrow{\;\;\pi_1\;\;} & C_1 \\[4pt]
{\scriptstyle \pi_0}\Big\downarrow \quad \lrcorner & & \Big\downarrow {\scriptstyle \langle \partial_0, \partial_1 \rangle} \\[4pt]
I & \xrightarrow[\;\langle X, Y \rangle\;]{} & C_0 \times C_0
\end{array}
$$

For each $u: J \to I$ in $\mathbb{B}$ with vertical $f: u^*(X) \to u^*(Y)$ in $\mathrm{Fam}(\mathbf{C})$ over J, we have f as a morphism $f: J \to C_1$ in $\mathbb{B}$ satisfying $\langle \partial_0, \partial_1 \rangle \circ f = \langle X, Y \rangle \circ u$. This yields the required unique map $v: J \dashrightarrow \underline{\mathrm{Hom}}(X, Y)$ as mediating map for the above pullback.

Finally, for an arbitrary category $\mathbb{B}$ with finite limits, the associated codomain fibration $\begin{smallmatrix} \mathbb{B}^{\to} \\ \downarrow \\ \mathbb{B} \end{smallmatrix}$ is locally small if and only if the category $\mathbb{B}$ is locally Cartesianlosed: for a morphism $u: J \to I$ in $\mathbb{B}$ and for families $\varphi, \psi \in \mathbb{B}/I$ over I, we have isomorphisms

$$
\mathbb{B}/I\big(u^*(\varphi),\, u^*(\psi)\big) \cong \mathbb{B}/I\big(\textstyle\coprod_u u^*(\varphi),\, \psi\big) \cong \mathbb{B}/I\big(u \times \varphi,\, \psi\big).
$$

Hence the left hand side has a representing object if and only if the right hand side has one. That is, the codomain fibration is locally small if and only if all slices are Cartesian closed.

In order to produce a non-example, we use the following easy result. It gives a relation between local smallness and comprehension (or subset types), as described in Section 4.6. The proof is not hard, but is postponed until Section 10.4 where we shall have more to say about comprehension (see especially Proposition 10.4.10).

9.5.2. Lemma. *Let* $\overset{\mathbb{E}}{\underset{\mathbb{B}}{\downarrow}}p$ *be a fibred CCC. Then p is locally small if and only the fibred terminal object functor* $1: \mathbb{B} \to \mathbb{E}$ *has a right adjoint. This right adjoint is then written as* $\{-\}$ *since it provides the fibration with comprehension (also called subset types, if p is a preorder fibration).* □

9.5.3. Example. In Example 7.1.3 we saw that the fibration $\overset{\text{UFam}(\mathbf{PER})}{\underset{\omega\text{-Sets}}{\downarrow}}$ of ω-set-indexed PERs is small; hence it is locally small. But the fibration $\overset{\text{UFam}(\mathbf{PER})}{\underset{\text{Sets}}{\downarrow}}$ of (ordinary) set-indexed PERs, introduced as an example of a $\lambda\omega$-fibration in Corollary 8.4.6, is *not* small. We will show here that this fibration is not locally small, and hence certainly not small. We assume towards a contradiction that the fibration is locally small, and thus—by the previous lemma—has a right adjoint $\{-\}$ to the terminal object functor $1: \mathbf{Sets} \to \text{UFam}(\mathbf{PER})$.

Call a PER S non-empty if its domain $|S| = \{n \in \mathbb{N} \mid nSn\}$ is non-empty, and write $\text{PER}_{\neq \emptyset} \hookrightarrow \text{PER}$ for the set of non-empty PERs. Consider the family of PERs $X = (R)_{R \in \text{PER}_{\neq \emptyset}}$ in UFam($\mathbf{PER}$), and the resulting set $\{X\}$ with projection $\pi_X: \{X\} \to \text{PER}_{\neq \emptyset}$. For every non-empty PER S we have S as a map $S: 1 \to \text{PER}_{\neq \emptyset}$ in **Sets**, together with a morphism $f(S): 1(1) \to S^*(X) = S$ over 1 in UFam($\mathbf{PER}$). Namely, $f(S) = [n_S]_S$, where $n_S \in |S|$ is a chosen inhabitant of the domain of S. By the adjunction, $f(S)$ gives a map $v(S): 1 \to \{X\}$ with $\pi_X \circ v(S) = S$. Collecting these $v(S)$'s together yields a function $v: \text{PER}_{\neq \emptyset} \to \{X\}$, with $\pi_X \circ v = \text{id}$. Transposing v gives a vertical map $f: 1(\text{PER}_{\neq \emptyset}) \to X$ in UFam($\mathbf{PER}$) over **Sets**. It must have—as any other morphism in this category—a realiser, say e, which works for all indices. Thus, for every non-empty PER S, $f_S = [e]_S$ where $e \in |S|$. This leads to the absurd conclusion that there is an element e which is in the domain of every non-empty PER S.

Next we give an intrinsic alternative formulation of local smallness.

9.5.4. Lemma. *A fibration* $\overset{\mathbb{E}}{\underset{\mathbb{B}}{\downarrow}}p$ *is locally small if and only if for each pair of objects* $X, Y \in \mathbb{E}$ *in the same fibre, we can find a span* $X \overset{f}{\leftarrow} A \overset{g}{\to} Y$ *in* $\mathbb{E}$ *with f Cartesian over* $p(g)$, *which is universal in the following sense. For each span* $X \overset{h}{\leftarrow} B \overset{k}{\to} Y$ *with h Cartesian over* $p(k)$, *there is a unique map*

$\varphi\colon B \dashrightarrow A$ *in* $\mathbb{E}$ *making the following diagram commute.*

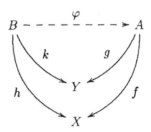

(Such a mediating map φ is Cartesian, since f and $f \circ \varphi = h$ are Cartesian.)

Proof. Suppose $\begin{smallmatrix}\mathbb{E}\\\downarrow p\\\mathbb{B}\end{smallmatrix}$ is a locally small fibration. For $X, Y \in \mathbb{E}_I$, the required universal span $X \leftarrow \pi_0^*(X) \rightarrow Y$ is:

$$
\begin{array}{ccc}
X & \xleftarrow{\;\;\overline{\pi_0}(X)\;\;} & \pi_0^*(X) \\
 & & \downarrow{\scriptstyle \pi_1} \\
 & & \pi_0^*(Y) \xrightarrow[\overline{\pi_0}(Y)]{} Y
\end{array}
$$

Conversely, assume for $X, Y \in \mathbb{E}_I$ there is a universal span $X \xleftarrow{f} A \xrightarrow{g} Y$ as in the lemma. Write $\pi_0\colon \underline{\mathrm{Hom}}_I(X, Y) \rightarrow I$ for $p(f) = p(g)$ in $\mathbb{B}$ and π_1 for the vertical part of $\pi_0^*(X) \xrightarrow{\cong} A \longrightarrow Y$. This pair π_0 and π_1 is universal as described in Definition 9.5.1. $\qquad\square$

Earlier in Example 7.1.4 (ii) we have seen how a full internal category Full(a) can be constructed from a morphism $a\colon A \rightarrow B$ in a locally Cartesian closed category $\mathbb{B}$. Below we shall describe a generalisation of this construction—due to Penon [257]—which may be performed in a locally small fibration. This gives rise to an important corollary, stating that a fibration is small if and only if it is locally small and has a generic object.

9.5.5. Theorem. *Let* $\begin{smallmatrix}\mathbb{E}\\\downarrow p\\\mathbb{B}\end{smallmatrix}$ *be a locally small fibration. Then every object* $U \in \mathbb{E}$ *induces an internal category* Full(U) *in the base category* $\mathbb{B}$ *(provided there are enough pullbacks in $\mathbb{B}$ to say what this means). This internal category*

Full(U) *is "full in p": it comes equipped with a full and faithful fibred functor:*

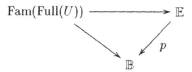

Proof. For $U \in \mathbb{E}$, write $U_0 = pU \in \mathbb{B}$ for the intended object of objects of Full(U). Let us write $\langle \partial_0, \partial_1 \rangle \colon U_1 \to U_0 \times U_0$ for the representing arrow of the two objects $\pi^*(U)$ and $\pi'^*(U)$ in the fibre above $U_0 \times U_0$. It comes equipped with an arrow (the π_1), which we now write as $\mu \colon \partial_0^*(U) \to \partial_1^*(U)$. By construction we have for $X, Y \colon I \rightrightarrows U_0$ natural isomorphisms

$$\mathbb{B}/(U_0 \times U_0)\big(\langle X, Y \rangle, \langle \partial_0, \partial_1 \rangle\big) \xrightarrow[\cong]{\Phi} \mathbb{E}_I\big(X^*(U), Y^*(U)\big)$$

Internal identities and composition in Full(U) are borrowed from p via this isomorphism Φ (like for the earlier "Full($-$)" construction in Example 7.1.4 (ii) for locally Cartesian closed categories): the morphism of internal identities $i \colon \langle \text{id}, \text{id} \rangle \to \langle \partial_0, \partial_1 \rangle$ in $\mathbb{B}/(U_0 \times U_0)$ is obtained by applying Φ^{-1} to the identity map $\text{id}^*(U) \to \text{id}^*(U)$. For internal composition, consider the pullback of composable maps:

$$
\begin{array}{ccc}
U_2 = U_1 \times_{U_0} U_1 & \xrightarrow{\ \xi_1\ } & U_1 \\
{\scriptstyle \xi_0} \downarrow \quad \lrcorner & & \downarrow {\scriptstyle \partial_0} \\
U_1 & \xrightarrow[\ \partial_1\ ]{} & U_0
\end{array}
$$

where we have written ξ_0, ξ_1 for the pullback projections. What we need is a composition map $m \colon \langle \partial_0 \circ \xi_0, \partial_1 \circ \xi_1 \rangle \to \langle \partial_0, \partial_1 \rangle$ in $\mathbb{B}/(U_0 \times U_0)$. It is constructed as follows. By applying both ξ_0^* and ξ_1^* to $\mu \colon \partial_0^*(U) \to \partial_1^*(U)$, we get a composite:

$$
\begin{array}{l}
(\partial_0 \circ \xi_0)^*(U) \xrightarrow{\ \cong\ } \xi_0^* \partial_0^*(U) \\[4pt]
\qquad {\scriptstyle \xi_0^*(\mu)} \downarrow \\[4pt]
\qquad \xi_0^* \partial_1^*(U) \xrightarrow{\ \cong\ } \xi_1^* \partial_0^*(U) \\[4pt]
\qquad\qquad\qquad {\scriptstyle \xi_1^*(\mu)} \downarrow \\[4pt]
\qquad\qquad\qquad \xi_1^* \partial_1^*(U) \xrightarrow{\ \cong\ } (\partial_1 \circ \xi_1)^*(U)
\end{array}
$$

The composition map $m: U_2 \to U_1$ is then obtained by applying Φ^{-1} to this composite map $(\partial_0 \circ \xi_0)^*(U) \longrightarrow (\partial_1 \circ \xi_1)^*(U)$.

The morphism $\mu: \partial_0^*(U) \to \partial_1^*(U)$ is by construction the action of an internal diagram of type $\mathrm{Full}(U)$ in p. It corresponds, following Remark 7.4.2 (i), to a fibred functor $\mathrm{Fam}(\mathrm{Full}(U)) \to \mathbb{E}$, given by

$$\left(I \xrightarrow{\ X\ } U_0 \right) \longmapsto X^*(U)$$

This yields a full and faithful functor: we may restrict ourselves to a fibre (see Exercise 1.7.2), and there we have:

$$\mathrm{Fam}(\mathrm{Full}(U))_I \left(I \xrightarrow{X} U_0, \ I \xrightarrow{Y} U_0 \right) \ \cong \ \mathbb{B}/(U_0 \times U_0)\big(\langle X, Y \rangle, \ \langle \partial_0, \partial_1 \rangle \big)$$
$$\cong \ \mathbb{E}_I \big(X^*(U), \ Y^*(U) \big). \qquad \Box$$

It is not hard to see that the construction of the full internal category $\mathrm{Full}(a)$ starting from a morphism a in a locally Cartesian closed category $\mathbb{B}$ in Example 7.1.4 (ii), and of the associated full and faithful fibred functor $\mathrm{Fam}(\mathrm{Full}(a)) \to \mathbb{B}^{\to}$ in Example 7.3.4 (ii), are special cases of the construction in the above proof, when applied to a as an object in the total category $\mathbb{B}^{\to}$ of the (locally small) codomain fibration on $\mathbb{B}$.

Recall that an ordinary category is small if and only if it has a small collection of objects and also a small collection of morphisms. The above theorem yields a similar description for fibred categories.

9.5.6. Corollary. *A fibration is small if and only if it is locally small and has a generic object.*

Proof. We have already seen that a small fibration is locally small and has a generic object, so we concentrate on the (if)-part. Let a locally small fibration $\begin{smallmatrix} \mathbb{E} \\ \downarrow p \\ \mathbb{B} \end{smallmatrix}$ have $T \in \mathbb{E}_\Omega$ as generic object. By the previous result we can form an internal category $\mathrm{Full}(T)$ in $\mathbb{B}$, for which there is a full and faithful functor

$$\mathrm{Fam}(\mathrm{Full}(T)) \longrightarrow \mathbb{E} \qquad \text{sending} \qquad \left(I \xrightarrow{\ u\ } \Omega \right) \longmapsto u^*(T)$$

This functor is an equivalence because each object $X \in \mathbb{E}$ is of the form $u^*(T)$ for a unique morphism $u: pX \to \Omega$, since T is generic object. $\qquad \Box$

This corollary comes from [27]. In somewhat different formulation, it also occurs in [246, II, Theorem 3.11.1] for indexed categories.

We close this section with a fibred version of a familiar homset description of ordinary products and coproducts: if a category $\mathbb{C}$ has Cartesian products $\times$, then there are the (natural) isomorphisms in **Sets** between homsets:

$$\mathbb{C}\big(Z, \ X \times Y \big) \cong \mathbb{C}\big(Z, \ X \big) \times \mathbb{C}\big(Z, \ Y \big).$$

Similar isomorphisms exist for coproducts $+$, and for arbitrary limits and colimits. The next result gives a fibred analogue, for locally small fibrations.

9.5.7. Lemma. *Let* $\begin{smallmatrix} \mathbb{E} \\ \downarrow p \\ \mathbb{B} \end{smallmatrix}$ *be a locally small fibrations with products* $\prod_u$ *and coproducts* $\coprod_u$ *on a locally Cartesian closed base category* $\mathbb{B}$*. Then, for a map* $u \colon I \to J$*, there are isomorphisms in* $\mathbb{B}/J$*:*

$$\left(\begin{smallmatrix} \underline{\mathrm{Hom}}(X,\prod_u(Y)) \\ \downarrow \\ J \end{smallmatrix} \right) \cong \prod_u \left(\begin{smallmatrix} \underline{\mathrm{Hom}}(u^*(X),Y) \\ \downarrow \\ I \end{smallmatrix} \right)$$

$$\left(\begin{smallmatrix} \underline{\mathrm{Hom}}(\coprod_u(Y),X) \\ \downarrow \\ J \end{smallmatrix} \right) \cong \prod_u \left(\begin{smallmatrix} \underline{\mathrm{Hom}}(Y,u^*(X)) \\ \downarrow \\ I \end{smallmatrix} \right).$$

Notice that $\prod_u, \coprod_u$ *on the left hand side of the isomorphisms* $\cong$ *are the product and coproduct in* p*, and* $\prod_u$ *on the right hand side is the product of the locally Cartesian closed category* $\mathbb{B}$ *(see Proposition 1.9.8).*

Proof. We shall do the first one. For a family $\left(\begin{smallmatrix} K \\ \downarrow \psi \\ J \end{smallmatrix} \right)$ over J in $\mathbb{B}$, consider the following pullback square.

$$
\begin{array}{ccc}
L & \xrightarrow{\;\;u'\;\;} & K \\
{\scriptstyle \psi' = u^*(\psi)}\downarrow & \lrcorner & \downarrow{\scriptstyle \psi} \\
I & \xrightarrow[\;\;u\;\;]{} & J
\end{array}
$$

We get the required isomorphism by Yoneda:

$$\mathbb{B}/J \left(\left(\begin{smallmatrix} K \\ \downarrow \psi \\ J \end{smallmatrix} \right), \left(\begin{smallmatrix} \underline{\mathrm{Hom}}(X,\prod_u(Y)) \\ \downarrow \\ J \end{smallmatrix} \right) \right)$$

$$\cong \mathbb{E}_K \left(\psi^*(X),\ \psi^*(\textstyle\prod_u(Y)) \right) \qquad\qquad \text{by local smallness}$$

$$\cong \mathbb{E}_K \left(\psi^*(X),\ \textstyle\prod_{u'} \psi'^*(Y) \right) \qquad\qquad \text{by Beck-Chevalley}$$

$$\cong \mathbb{E}_L \left(u'^*\psi^*(X),\ \psi'^*(Y) \right)$$

$$\cong \mathbb{E}_L \left(\psi'^*u^*(X),\ \psi'^*(Y) \right)$$

$$\cong \mathbb{B}/I \left(\left(\begin{smallmatrix} L \\ \downarrow \psi' \\ I \end{smallmatrix} \right), \left(\begin{smallmatrix} \underline{\mathrm{Hom}}(u^*(X),Y) \\ \downarrow \\ I \end{smallmatrix} \right) \right) \qquad \text{by local smallness}$$

$$\cong \mathbb{B}/J \left(\left(\begin{smallmatrix} K \\ \downarrow \psi \\ J \end{smallmatrix} \right), \prod_u \left(\begin{smallmatrix} \underline{\mathrm{Hom}}(u^*(X),Y) \\ \downarrow \\ I \end{smallmatrix} \right) \right) \qquad \text{because } \psi' = u^*(\psi). \quad \square$$

Exercises

9.5.1. Let $\begin{smallmatrix} \mathbb{E} \\ \downarrow p \\ \mathbb{B} \end{smallmatrix}$ be a locally small fibration. Prove that for an epimorphism $u\colon J \twoheadrightarrow I$ in $\mathbb{B}$ the reindexing functor $u^*\colon \mathbb{E}_I \to \mathbb{E}_J$ is faithful.

9.5.2. Let $\begin{smallmatrix} \mathbb{E} \\ \downarrow p \\ \mathbb{B} \end{smallmatrix}$ be a cloven fibration with finite limits in its base category. Show that p is locally small if and only if for each pair of objects $X, Y \in \mathbb{E}$—not necessarily in the same fibre!—the functor $\mathbb{B}/(pX \times pY)^{\mathrm{op}} \to \mathbf{Sets}$ given by

$$\left(I \xrightarrow{\ u\ } pX \times pY \right) \;\longmapsto\; \mathbb{E}_I\Big((\pi \circ u)^*(X),\, (\pi' \circ u)^*(Y) \Big)$$

is representable.
[This formulation occurs in [169, A2].]

9.5.3. Consider Theorem 9.5.5 for a family fibration $\begin{smallmatrix} \mathrm{Fam}(\mathbb{C}) \\ \downarrow \\ \mathbf{Sets} \end{smallmatrix}$ of a locally small fibration $\mathbb{C}$. Show that for a family $U = (X_\alpha)_{\alpha \in A} \in \mathrm{Fam}(\mathbb{C})$ of objects $X_\alpha \in \mathbb{C}$ one gets a small category $\mathrm{Full}(U)$ with A as set of objects, and arrows $\alpha \to \beta$ given by morphisms $X_\alpha \to X_\beta$ in $\mathbb{C}$.

9.5.4. Let $(\mathbb{B}, T)$ be a $\lambda 1$-category. Show that the associated simple fibration $\begin{smallmatrix} s(T) \\ \downarrow \\ \mathbb{B} \end{smallmatrix}$ is locally small.

9.5.5. Let $\begin{smallmatrix} \mathbb{E} \\ \downarrow p \\ \mathbb{B} \end{smallmatrix}$ be a fibration and $F\colon \mathbb{A} \to \mathbb{B}$ be a functor with a right adjoint, where $\mathbb{A}$ is a category with pullbacks. Consider the fibration $F^*(p)$ obtained by change-of-base along F.
   (i)   Prove that if p is locally small, then so is $F^*(p)$.
   (ii) Conclude that if p is small then $F^*(p)$ is also small.
         [*Hint.* Remember Exercise 5.2.3.]

9.5.6. Let $\begin{smallmatrix} \mathbb{E} \\ \downarrow p \\ \mathbb{B} \end{smallmatrix}$ be a locally small fibration.
   (i)   Show that the assignment $(X, Y) \mapsto \left(\begin{smallmatrix} \underline{\mathrm{Hom}}(X,Y) \\ \downarrow \\ pX \end{smallmatrix} \right)$ —for X, Y in the same fibre—extends to a functor $\mathcal{H}$ (for Hom) in

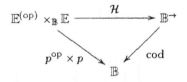

which maps Cartesian morphisms to pullback squares. This gives a comprehension category $\mathcal{H}$ on the product fibration $p^{\mathrm{op}} \times p$.
   (ii) Define the associated exponential transpose $\mathcal{Y}\colon p \to \mathrm{cod}^{p^{\mathrm{op}}}$ of $\mathcal{H}\colon p^{\mathrm{op}} \times p \to \mathrm{cod}$, see Exercise 1.10.6, and show that $\mathcal{Y}$ is a full and faithful

functor.

[This generalises Exercise 7.4.6.]

9.5.7. Let $\begin{smallmatrix} \mathbb{E} \\ \downarrow p \\ \mathbb{B} \end{smallmatrix}$ be a locally small fibration on a base category $\mathbb{B}$ with finite limits. Show that each fibre category $\mathbb{E}_I$ is enriched over the slice category $\mathbb{B}/I$, with respect to its Cartesian structure.

9.5.8. (Bénabou) Consider a locally small fibration $\begin{smallmatrix} \mathbb{E} \\ \downarrow p \\ \mathbb{B} \end{smallmatrix}$. The aim is to obtain (canonical) equivalences

$$\mathbb{E}_0 \xrightarrow{\ \simeq\ } 1 \qquad \text{and} \qquad \mathbb{E}_{I+J} \xrightarrow{\ \simeq\ } \mathbb{E}_I \times \mathbb{E}_J \qquad (*)$$

(i) Show that every object in $\mathbb{E}$ over an initial object in $\mathbb{B}$ is initial in $\mathbb{E}$. Conclude that if $\mathbb{B}$ has an initial object 0, then for all $X, Y \in \mathbb{E}_0$ one gets $X \cong Y$ in $\mathbb{E}_0$. And that if $\mathbb{E}$ has at least one object, then $\mathbb{E}_0 \simeq 1$.

Assume now that the base category $\mathbb{B}$ has binary coproducts $+$.

(ii) Prove that the functor

$$\mathbb{E}_{I+J} \xrightarrow{\ \langle \kappa^*, \kappa'^* \rangle\ } \mathbb{E}_I \times \mathbb{E}_J$$

is full and faithfull.

(iii) Show that this functor is an equivalence if one additionally assumes that p has coproducts $\coprod$ and fibred coproducts $\vee$, and that the coproducts $+$ in $\mathbb{B}$ are disjoint.

9.5.9. Assume a locally small fibration with fibred finite products and coproducts $\wedge, \vee$ and with products and coproducts $\prod, \coprod$. Assume additionally that the base category has disjoint coproducts $+$. Show that the coproduct $+$ in the total category

$$X + Y = \coprod_\kappa(X) \vee \coprod_{\kappa'}(Y)$$

as described in Proposition 9.2.2,, can alternatively be described in terms of products as:

$$X + Y = \prod_{\kappa'}(X) \wedge \prod_{\kappa'}(Y).$$

9.5.10. Consider a locally small fibration with fibred coproducts $\vee$ and coproducts $\coprod_u$ over a distributive base category. Prove that for an object X over $I \times (J + K)$ there is an isomorphism

$$\coprod_{\pi_{I,J+K}}(X) \cong \left(\coprod_{\pi_{I,J}} (\mathrm{id} \times \kappa)^*(X)) \right) \vee \left(\coprod_{\pi_{I,K}} (\mathrm{id} \times \kappa')^*(X)) \right).$$

Explain the logical significance of this isomorphism.

9.5.11. (Bénabou) Assume a fibration $\begin{smallmatrix} \mathbb{E} \\ \downarrow p \\ \mathbb{B} \end{smallmatrix}$ with products $\prod_u$. The point of this exercise is to show that if one has equivalences $(*)$ as in Exercise 9.5.8, then p automatically has fibred finite products $(1, \times)$. [This shows that under suitable additional assumptions in Definition 1.9.11 of completeness for fibrations, it is enough to require products $\prod_u$ and fibred *equalisers*, instead of products $\prod_u$ and fibred *finite limits*.]

(i) Let $* \in \mathbb{E}_0$; show that the objects $1(I) = \prod_{!_I}(*) \in \mathbb{E}_I$ provide the fibration p with fibred terminal objects.

(ii) For $X, Y \in \mathbb{E}_I$, let $Z \in \mathbb{E}_{I+I}$ be such that $\kappa^*(Z) \cong X$ and $\kappa'^*(Z) \cong Y$. Show that $\prod_{\nabla_I}(Z)$ is Cartesian product of X, Y in $\mathbb{E}_I$, where $\nabla_I = [\mathrm{id}, \mathrm{id}]: I + I \to I$ is the codiagonal. (For preservation under reindexing one has to assume that the coproducts $+$ in $\mathbb{B}$ are disjoint and universal.)

9.5.12. (See [88]) Call a fibration $\begin{smallmatrix} \mathbb{E} \\ \downarrow{\scriptstyle p} \\ \mathbb{B} \end{smallmatrix}$ a **geometric fibration** if

(a) the base category $\mathbb{B}$ is a topos, and the fibration p is "fibrewise a topos": each fibre is a topos and each reindexing functor is logical;

(b) the fibration p has coproducts $\coprod_u$ which are disjoint and universal (see Exercises 9.2.12 and 9.2.13);

(c) the fibration p is locally small, or equivalently, by Lemma 9.5.2, p has comprehension, via a right adjoint $\{-\}: \mathbb{E} \to \mathbb{B}$ to the terminal object functor 1.

Prove that geometric fibrations $\begin{smallmatrix} \mathbb{E} \\ \downarrow \\ \mathbb{B} \end{smallmatrix}$ on a topos $\mathbb{B}$ correspond to geometric morphisms $F: \mathbb{A} \to \mathbb{B}$ with codomain $\mathbb{B}$.

[*Hint.* Given $\begin{smallmatrix} \mathbb{E} \\ \downarrow \\ \mathbb{B} \end{smallmatrix}$, consider the coproduct functor $\coprod: \mathbb{B} \to \mathbb{E}_1$ from Exercise 9.2.13 as inverse image part. And given $F: \mathbb{A} \to \mathbb{B}$, define a fibration $\begin{smallmatrix} \mathbb{A}/F \\ \downarrow \\ \mathbb{B} \end{smallmatrix}$ by pulling the codomain fibration on $\mathbb{A}$ back along $F^*: \mathbb{B} \to \mathbb{A}$.]

Prove also that the geometric morphism $F: \mathbb{A} \to \mathbb{B}$ is an inclusion of toposes (*i.e.* the direct image part $F_*: \mathbb{A} \to \mathbb{B}$ is full and faithful) if and only if the fibration $\begin{smallmatrix} \mathbb{A}/F \\ \downarrow \\ \mathbb{B} \end{smallmatrix}$ has full comprehension (*i.e.* the induced functor $\mathbb{A}/F \to \mathbb{B}^{\to}$ is full and faithful).

9.6 Definability

Subset types $\{i: I \mid X_i\}$ (with X a predicate on I) in the logic of a preorder fibration are described in Section 4.6 via a functor $\{-\}$ from the total category to the base category of the fibration, which is right adjoint to the truth (or terminal object) functor $\top$. The idea is that $\{-\}$ singles out those instantiations $i: I$ for which X_i holds. There is a notion of "definability" for fibrations which gives more general means for singling out certain instantiations of objects in a total category and representing them in the base category. This allows us to consider in an arbitrary base category (and not just in **Sets**) the (universal) subobject $I' \rightarrowtail I$ of those indices $i: I'$ for which X_i satisfies some property P

on the objects of the total category. The notion of definability and the associated results in this section are due to Bénabou, see *e.g.* [29] or [36, II, 8.7]. Type theoretic use of definability can be found in [263], where "powerkinds" are described as objects in a base category of kinds, associated with a collection of inclusion maps between types in a total category of a polymorphic fibration.

We start with a formulation of definability which is intrinsic, and we later give an alternative formulation involving a cleavage.

9.6.1. Definition. Let $\begin{smallmatrix}\mathbb{E}\\\downarrow p\\\mathbb{B}\end{smallmatrix}$ be a fibration and $P \subseteq \mathrm{Obj}\,\mathbb{E}$ be a collection of objects (or predicates) in the total category.

(i) We call P **closed under substitution** if for each Cartesian map $Y \to X$ with its codomain X in P, also the domain Y is in P. This is a minimal condition to make P a sensible collection of predicates. Notice that such a collection P is closed under isomorphism: if $X \in P$ and $X \cong Y$, then $Y \in P$.

(ii) The collection P is **definable** if it is closed under substitution and satisfies: for each object $X \in \mathbb{E}$ there is an object $X' \in P$ and a Cartesian morphism $i_X \colon X' \to X$ which is universal in the following sense. For each Cartesian map $f \colon Y \to X$ with its domain Y in P, there is a unique (necessarily Cartesian) map $f' \colon Y \dashrightarrow X'$ with $i_X \circ f' = f$.

The idea is to think of $X' \in P$ in this definition as the best approximation of $X \in \mathbb{E}$ by an element of P, as suggested in the following picture.

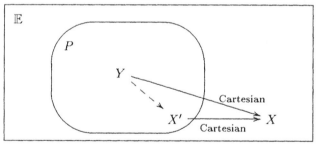

Informally, if $X = (X_i)_{i \in I}$, then one may think of X' as the family $(X_i)_{i \in \{j \in I \,|\, X_j \in P\}}$ obtained by suitably restricting X, see Lemma 9.6.4 below. In more abstract form, this is made explicit in the next reformulation of definability.

9.6.2. Lemma. *Let* $\begin{smallmatrix}\mathbb{E}\\\downarrow p\\\mathbb{B}\end{smallmatrix}$ *be a cloven fibration with a collection of objects* $P \subseteq \mathrm{Obj}\,\mathbb{E}$ *which is closed under substitution. Then P is definable if and only if for each object $X \in \mathbb{E}$ above $I \in \mathbb{B}$ the functor $\mathbb{B}^{\mathrm{op}} \to \mathbf{Sets}$ given by*

$$J \mapsto \{u \colon J \to I \mid u^*(X) \in P\}$$

is representable.

Explicitly, this representability means that there is a representing morphism in $\mathbb{B}$

$$\{X \in P\} \xrightarrow{\ \theta_X\ } I, \quad with \quad \theta_X^*(X) \in P$$

such that each $u: J \to I$ *with* $u^*(X) \in P$ *factors in a unique way through* θ_X. *(This map* θ_X *must then be a monomorphism.)*

Proof. Assume the collection P is definable as described in the above definition. Choose for each object $X \in \mathbb{E}$ a universal Cartesian morphism $i_X: X' \to X$ and write $\theta_X: \{X \in P\} \to pX$ for $p(i_X)$ in $\mathbb{B}$. There is then a vertical isomorphism $\theta_X^*(X) \cong X' \in P$. For a morphism $u: J \to pX$ with $u^*(X) \in P$ there is a Cartesian morphism $\overline{u}(X): u^*(X) \to X$ in $\mathbb{E}$ whose domain is in P. Because i_X is the best approximation of X in P, there is a unique map $f: u^*(X) \dashrightarrow X'$ with $i_X \circ f = \overline{u}(X)$. Then $pf: J \to \{X \in P\}$ satisfies $\theta_X \circ pf = p(i_X \circ f) = p(\overline{u}(X)) = u$. If there is another map $w: J \to \{X \in P\}$ satisfying $\theta_X \circ w = u$, then we get a unique morphism $g: u^*(X) \to X'$ over w with $i_X \circ g = \overline{u}(X)$, because i_X is Cartesian. But then $f = g$, by uniqueness, and thus $w = pg = pf$.

Conversely, assume that representing morphisms θ_X as in the lemma exist. Write $X' = \theta_X^*(X) \in P$ and $i_X = \overline{\theta_X}(X): \theta_X^*(X) \to X$ for the associated Cartesian lifting. For a Cartesian morphism $f: Y \to X$ with $Y \in P$ we get a unique map $v: pY \to \{X \in P\}$ with $\theta_X \circ v = pf$. But then there must be a (unique) mediating $f': Y \to X'$ over v with $i_X \circ f' = f$. And if also $g: Y \to X'$ satisfies $i_X \circ g = f$, then $pg: pY \to \{X \in P\}$ satisfies $\theta_X \circ pg = pf$. Hence $pg = v = pf'$, and thus $g = f'$. $\square$

The following two results describe what definability means in familiar situations.

9.6.3. Lemma. *Let* $\begin{smallmatrix}\mathbb{E}\\\downarrow p\\\mathbb{B}\end{smallmatrix}$ *be a preorder fibration with a terminal object (or truth) functor* $\top: \mathbb{B} \to \mathbb{E}$. *We write* Truth $\subseteq$ Obj $\mathbb{E}$ *for the collection of predicates which are true, i.e.* Truth $= \{X \in \mathbb{E} \mid \top \leq X \text{ vertically}\}$. *Then* Truth *is definable if and only if the fibration* p *has subset types (that is, if and only if the terminal object functor* $\top$ *has a right adjoint* $\{-\}: \mathbb{E} \to \mathbb{B}$).

Proof. Assume Truth $= \{X \in \mathbb{E} \mid \top \leq X \text{ vertically}\}$ is definable. Then there is a right adjoint $X \mapsto \{X \in \text{Truth}\}$ to $\top: \mathbb{B} \to \mathbb{E}$ since

$$\begin{aligned}
\mathbb{E}\big(\top(J), X\big) &\cong \{u: J \to pX \mid \top(J) \leq u^*(X) \text{ over } J\} \\
&= \{u: J \to pX \mid u^*(X) \in \text{Truth}\} \\
&\cong \mathbb{B}\big(J, \{X \in \text{Truth}\}\big).
\end{aligned}$$

Conversely, if such a right adjoint $\{-\}$ to $\top$ exists, then the associated subset projections $\pi_X \colon \{X\} \rightarrowtail pX$ form representing arrows for the functors in the previous lemma: for a map $u \colon J \to pX$ in $\mathbb{B}$ we get by Lemma 4.6.2 (ii):

$$u^*(X) \in \text{Truth} \;\Leftrightarrow\; \top \le u^*(X) \text{ over } J \;\Leftrightarrow\; u \dashrightarrow \pi_X \text{ in } \mathbb{B}/pX. \qquad \square$$

Next we show that for family fibrations (which incorporate the world of ordinary categories), definable collections correspond simply to subsets of objects.

9.6.4. Lemma. *There is bijective correspondence between definable collec-*
$$\begin{array}{c} \text{Fam}(\mathbb{C}) \\ \downarrow \\ \text{Sets} \end{array}$$
tions for a family fibration $\quad$ and collections of objects in $\mathbb{C}$.

Proof. Let $P \subseteq \text{Obj Fam}(\mathbb{C})$ be a definable collection. Because P is closed under substitution, if a family $Y = (Y_i)_{i \in I}$ is in P, then each of the indexed objects $Y_j = j^*(Y)$ is in P, for $j \in I$. But the converse is also true: if each $Y_j = j^*(Y) \in P$, then $Y \in P$. To see this, consider the best approximation $Y' \to Y$ with $Y' \in P$. Since $Y' \to Y$ is Cartesian, we may write $Y' = (Y_{v(k)})_{k \in K}$ for a function $v \colon K \to I$. Since for each $j \in I$ there is a Cartesian map $Y_j \to Y$ over $j \colon 1 \to I$ with $Y_j \in P$ there must be by definability of P a unique map $w(j) \colon 1 \to K$ with $v(w(j)) = j$. These $w(j)$'s combine into a function $w \colon I \to K$, which is inverse of v. Hence $Y \cong Y'$, and so $Y \in P$.

Thus we consider the collection $P_1 = \{X \in \mathbb{C} \mid X \in P \text{ in the fibre over } 1\}$ of objects in $\mathbb{C}$. Then one easily shows that an arbitrary family $X = (X_i)_{i \in I}$ is represented by the inclusion

$$\{i \in I \mid X_i \in P_1\} \overset{\theta_X}{\lhook\joinrel\longrightarrow} I$$

Clearly, $\theta_X^*(X)$ is in P, because for each $j \in \{i \in I \mid X_i \in P\}$ we have $\theta_X^*(X)_j = X_j \in P$ by construction. And indeed, for a function $u \colon J \to I$, if $u^*(X) = (X_{u(j)})_{j \in J}$ is in P, then each object $X_{u(j)}$ is in P_1, so that u factors trough $\theta_X \colon \{i \in I \mid X_i \in P_1\} \hookrightarrow I$.

Conversely, given a collection $Q \subseteq \text{Obj}\,\mathbb{C}$, let

$$\overline{Q} = \{(Y_i)_{i \in I} \in \text{Fam}(\mathbb{C}) \mid \forall i \in I.\, Y_i \in Q\}$$

be the closure of Q under substitution. Then $\overline{Q}$ is definable, since for an arbitrary collection $X = (X_i)_{i \in I}$, take the subset $I' = \{i \in I \mid X_i \in Q\} \hookrightarrow I$ and the subfamily $X' = (X_i)_{i \in I'}$. Then $X' \to X$ is Cartesian over the inclusion $I' \hookrightarrow I$, and is the best approximation of X in $\overline{Q}$.

As we have seen in the beginning of the proof, for a definable collection $P \subseteq \text{Obj Fam}(\mathbb{C})$ we have $\overline{(P_1)} = \{(Y_i)_{i \in I} \mid \forall i \in I.\, Y_i \in P_1\} = P$. And for a collection $Q \subseteq \text{Obj}\,\mathbb{C}$ we have $(\overline{Q})_1 = Q$. $\qquad \square$

Consider for example the family fibration $\begin{smallmatrix}\text{Fam}(\mathbf{Sets})\\\downarrow\\\mathbf{Sets}\end{smallmatrix}$ of set-indexed sets. Let Fin $\subseteq$ Obj Fam($\mathbf{Sets}$) be the collection of those families $(X_i)_{i\in I}$ for which each set X_i is finite. This collection is definable, since it comes from the subcollection of Obj $\mathbf{Sets}$ consisting of finite sets: given an arbitrary family $(X_i)_{i\in I}$, there is an inclusion

$$\{i \in I \mid X_i \text{ is finite}\} \lhook\joinrel\longrightarrow I$$

serving as representing function.

Next consider the family fibration $\begin{smallmatrix}\text{Fam}(\mathbf{FinSets})\\\downarrow\\\mathbf{Sets}\end{smallmatrix}$ of set-indexed collections of finite sets. We look at bounded families of these. More precisely, families which have a common bound: let

$$\text{CB} = \{(X_i)_{i\in I} \in \text{Fam}(\mathbf{FinSets}) \mid \exists m \in \mathbb{N}.\, \forall i \in I.$$
$$X_i \text{ has less than } m \text{ elements}\}.$$

We claim that CB is *not* definable—but it is closed under substitution. Take as counterexample the family $A = \big([n]\big)_{n\in\mathbb{N}}$ of finite sets $[n] = \{0, 1, \ldots, n-1\}$, which is clearly not an element of the collection CB. If CB is definable, then there is a subset $\{A \in \text{CB}\} \hookrightarrow \mathbb{N}$ such that each function $u\colon J \to \mathbb{N}$ satisfies: $u^*(A) = \big([u(j)]\big)_{j\in J}$ is in CB if and only if u factors through $\{A \in \text{CB}\} \hookrightarrow \mathbb{N}$. For each $n \in \mathbb{N}$, considered as map $n\colon 1 \to \mathbb{N}$, we have $n^*(A) = [n] \in \text{CB}$. Hence $n \in \{A \in \text{CB}\}$. This shows that the identity function $\text{id}\colon \mathbb{N} \to \mathbb{N}$ factors through $\{A \in \text{CB}\} \hookrightarrow \mathbb{N}$. But we do not have $\text{id}^*(A) = A \in \text{CB}$. Thus CB is not definable.

Here is a another illustration, involving PERs. It is adapted from [262].

9.6.5. Example. A partial equivalence relation $R \subseteq \mathbb{N} \times \mathbb{N}$ will be called **decidable** if there is a "decision code" $e \in \mathbb{N}$ such that for all $n, m \in \mathbb{N}$

$$e \cdot \langle n, m \rangle = \begin{cases} 1 & \text{if } nRm \\ 0 & \text{otherwise.} \end{cases}$$

We write DPER for the set of decidable PERs. We will show that the notion of decidability is definable in the fibration $\begin{smallmatrix}\text{UFam}(\mathbf{PER})\\\downarrow\\\omega\text{-}\mathbf{Sets}\end{smallmatrix}$ of ω-set-indexed PERs but not in the fibration $\begin{smallmatrix}\text{UFam}(\mathbf{PER})\\\downarrow\\\mathbf{Sets}\end{smallmatrix}$ of set-indexed PERs. We begin with the latter.

Call a set-indexed family $(R_i)_{i\in I}$ of PERs decidable if there is a single decision code which works for each R_i. Consider the collection of decidable families of PERs in the fibration $\begin{smallmatrix}\text{UFam}(\mathbf{PER})\\\downarrow\\\mathbf{Sets}\end{smallmatrix}$. This collection is not definable. Otherwise, there is a for each family $R = (R_i)_{i\in I}$ a subset $I' \hookrightarrow I$ such that every function $u: J \to I$ satisfies: $u^*(R) = (R_{u(j)})_{j \in J}$ is a decidable family if and only if u factors through $I' \hookrightarrow I$. Consider in particular the family $D = (S)_{S \in \mathrm{DPER}}$, say represented by an inclusion $\{D\} \hookrightarrow \mathrm{PER}$. Then there is an inclusion $\mathrm{DPER} \hookrightarrow \{D\}$, since for each $R \in \mathrm{DPER}$, the family (over 1) $R = R^*(D)$, consisting only of R, is itself a decidable family, which is obtained by reindexing D along $R: 1 \to \mathrm{PER}$ in Sets. Thus, $D = (S)_{S \in \mathrm{DPER}}$ is a decidable family. But clearly no single decision code can work for all decidable PERs.

For ω-set-indexed families of PERs we can do better. Call such a family $R = (R_i)_{i \in (I,E)}$ of PERs over an ω-set (I, E) decidable if there is a "uniform" code $e \in \mathbb{N}$ such that

$$\forall i \in I. \forall n \in E(i).\, e \cdot n \text{ is a decision code for } R_i.$$

The crucial difference with the earlier indexing over sets is that the decision code for the PER R_i now depends on (a code of) the index $i \in I$. This enables us to show that the set of decidable families in the fibration $\begin{smallmatrix} \mathrm{UFam}(\mathbf{PER}) \\ \downarrow \\ \omega\text{-}\mathbf{Sets} \end{smallmatrix}$ is definable: for an arbitrary family of PERs $(R_i)_{i \in (I,E)}$ consider the subset

$$I' = \{i \in I \mid R_i \text{ is a decidable PER}\} \xrightarrow{\;\;\theta_R\;\;} I$$

with existence predicate

$$E'(i) = \{\langle n, m \rangle \mid n \in E(i) \text{ and } m \text{ is a decision code for } R_i\}$$

A code for the first projection tracks θ_R. And $\theta_R^*(R) = (R_i)_{i \in (I',E')}$ is a decidable family, since for each $i \in I'$ and $k \in E'(i)$ the second projection $\mathbf{p}'k$ is by construction a decision code for R_i.

For a morphism $u: (J, E) \to (I, E)$ in ω-**Sets**, say tracked by d, such that $u^*(R) = (R_{u(j)})_{j \in (J,E)}$ is a decidable family, say with decision code e, we have that u factors as $u: (J, E) \to (I', E')$ through θ_R, tracked by the code $\Lambda x. \langle d \cdot x, e \cdot x \rangle$.

The notion of definability can be formulated in much greater generality. Therefore we need the exponent fibration $p^{\mathbb{C}}$ of a fibration $\begin{smallmatrix} \mathbb{E} \\ \downarrow p \\ \mathbb{B} \end{smallmatrix}$ with an arbitrary category $\mathbb{C}$, as described in Exercise 1.8.8. Its objects in the fibre over I are functors $\mathbb{C} \to \mathbb{E}_I$. Reindexing along $u: J \to I$ in $\mathbb{B}$ is done by post-composition with $u^*: \mathbb{E}_I \to \mathbb{E}_J$.

9.6.6. Definition. Let $\begin{smallmatrix} \mathbb{E} \\ \downarrow p \\ \mathbb{B} \end{smallmatrix}$ be a fibration and $\mathbb{C}$ an arbitrary category. A collection of functors $\mathbb{C} \to \mathbb{E}$ (factoring through some fibre) is **definable** if it is definable with respect to the exponent fibration $p^{\mathbb{C}}$.

(This involves the requirement that the collection must be closed under substitution with respect to this fibration $p^{\mathbb{C}}$.)

This definition generalises the earlier one, since a collection $P \subseteq \mathrm{Obj}\,\mathbb{E}$ of objects of the total category is definable as in Definition 9.6.1 if and only if, considered as a collection of functors $(X : \mathbf{1} \to \mathbb{E})_{X \in P}$ from the terminal category $\mathbf{1}$ to $\mathbb{E}$, it is definable as above in Definition 9.6.6.

Now we can also talk about definability of collections of vertical morphisms: a collection $V \subseteq \mathrm{Arr}\,\mathbb{E}$ of vertical maps of a fibration $\begin{smallmatrix} \mathbb{E} \\ \downarrow p \\ \mathbb{B} \end{smallmatrix}$ is definable if it is definable when considered as a collection of functors $(\cdot \to \cdot) \to \mathbb{E}$. Explicitly, this means that for each vertical map $f : X' \to X$ above $I \in \mathbb{B}$, there is a mono $\{f \in V\} \rightarrowtail I$ in $\mathbb{B}$ such that for each morphism $u : J \to I$ one has $u^*(f) \in V$ if and only if u factors through $\{f \in V\} \rightarrowtail I$.

(Notice that this is not really an extension since it is the same as definability with respect to the arrow fibration $p^{\to} = \begin{smallmatrix} V(\mathbb{E}) \\ \downarrow \\ \mathbb{B} \end{smallmatrix}$ in $\mathbf{Fib}(\mathbb{B})$.)

As a special case, consider the collection

$$\mathrm{vIso} = \{f \mid f \text{ is vertical and Cartesian}\}$$
$$= \{f \mid f \text{ is a vertical isomorphism}\} \subseteq \mathrm{Obj}\,V(\mathbb{E}).$$

One says that **isomorphisms** are definable if this collection vIso of vertical isomorphisms is definable.

Similarly, one can say that **equality** of vertical maps is definable if the collection of functors $(\cdot \rightrightarrows \cdot) \to \mathbb{E}$ given by

$$\{F : (\cdot \overset{a}{\underset{b}{\rightrightarrows}} \cdot) \to \mathbb{E}_I \mid I \in \mathbb{B} \text{ and } Fa = Fb\}$$

is definable. This means that for each parallel pair $f, f' : X \rightrightarrows Y$ of maps in the fibre over $I \in \mathbb{B}$, there is a mono $\{f = f'\} \rightarrowtail I$ such that each $u : J \to I$ satisfies: $u^*(f) = u^*(f')$ if and only if u factors through $\{f = f'\} \rightarrowtail I$.

Sometimes the expression "definable subfibration" is used. What this means is the following. First, a **subfibration** of a fibration $\begin{smallmatrix} \mathbb{E} \\ \downarrow p \\ \mathbb{B} \end{smallmatrix}$ consists of a subcategory $\mathbb{D} \hookrightarrow \mathbb{E}$ with the property that for each $X \in \mathbb{D}$ and Cartesian morphism $f : Y \to X$ in $\mathbb{E}$ one has that f is already a map in $\mathbb{D}$. This implies that the inclusion $\mathbb{D} \hookrightarrow \mathbb{E}$ is a fibred functor. If this inclusion is full, one calls the subfibration **full**. Such a subfibration $\mathbb{D}$ is **definable** if both its collections of objects and of vertical morphisms are definable. For a full subfibration this simply means definability of objects.

We list some results (due to Bénabou) related to definability of isomorphisms.

9.6.7. Lemma. *Let* $\overset{\mathbb{E}}{\underset{\mathbb{B}}{\downarrow}}p$ *be a locally small fibration with finite limits in its base category. Then isomorphisms are definable in p.*

Proof. Recall from Exercise 9.5.7 that each fibre category $\mathbb{E}_I$ is enriched over the slice $\mathbb{B}/I$. For objects $X, Y \in \mathbb{E}_I$ we can combine composition maps c into a morphism in $\mathbb{B}$

$$\underline{\mathrm{Hom}}(X,Y) \times_I \underline{\mathrm{Hom}}(Y,X) \xrightarrow{\langle c,c \rangle} \underline{\mathrm{Hom}}(X,X) \times_I \underline{\mathrm{Hom}}(Y,Y)$$

There is also a morphism

$$\underline{\mathrm{Hom}}(X,Y) \times_I \underline{\mathrm{Hom}}(Y,X) \xrightarrow{\;!\;} 1 \xrightarrow{\langle i,i \rangle} \underline{\mathrm{Hom}}(X,X) \times_I \underline{\mathrm{Hom}}(Y,Y)$$

obtained from identity maps. Taking the equaliser in $\mathbb{B}$ yields a mono

$$\underline{\mathrm{vIso}}(X,Y) \xrightarrowtail{\;e\;} \underline{\mathrm{Hom}}(X,Y) \times_I \underline{\mathrm{Hom}}(Y,X)$$

The construction is such that for each $u\colon J \to I$ with an isomorphism $u^*(X) \overset{\cong}{\Rightarrow} u^*(Y)$ there is a unique morphism $v\colon J \dashrightarrow \underline{\mathrm{vIso}}(X,Y)$ with

$$u = \left(J \xrightarrow{\;v\;} \underline{\mathrm{vIso}}(X,Y) \xrightarrow{\;e\;} \underline{\mathrm{Hom}}(X,Y) \times_I \underline{\mathrm{Hom}}(Y,X) \longrightarrow I \right)$$

For a morphism $f\colon X \to Y$ over I, let $\hat{f}\colon I \to \underline{\mathrm{Hom}}(X,Y)$ be the corresponding section of the canonical projection $\pi_0\colon \underline{\mathrm{Hom}}(X,Y) \to I$. The required representing map θ_f is obtained by pullback in

$$\begin{array}{ccc}
\{f \in \mathrm{vIso}\} & \xrightarrow{\;\;f'\;\;} & \underline{\mathrm{vIso}}(X,Y) \\
{\scriptstyle \theta_f} \downarrow \;\;\lrcorner & & \downarrow {\scriptstyle \pi \circ e} \\
I & \xrightarrow[\;\;\hat{f}\;\;]{} & \underline{\mathrm{Hom}}(X,Y)
\end{array}$$

Then $\theta_f^*(f)\colon \theta_f^*(X) \to \theta_f^*(Y)$ is an isomorphism, with inverse resulting from the map

$$\{f \in \mathrm{vIso}\} \xrightarrow{\;\pi' \circ e \circ f'\;} \underline{\mathrm{Hom}}(Y,X)$$

with θ_f and I below.

Finally, if we have a morphism $u: J \to I$ in $\mathbb{B}$ for which $u^*(f): u^*(X) \to u^*(Y)$ is an isomorphism, then we get a morphism $v: J \to \underline{\mathrm{vIso}}(X, Y)$ as described above. This map, together with $u: J \to I$ satisfies $\pi \circ e \circ v = \hat{f} \circ u$, so that we get our required mediating map $J \dashrightarrow \{f \in \mathrm{vIso}\}$ using the pullback. $\qquad \square$

Definability of isomorphisms is important because it implies definability of some other classes.

9.6.8. Lemma. *Let* $\begin{smallmatrix} \mathbb{E} \\ \downarrow p \\ \mathbb{B} \end{smallmatrix}$ *be a fibration with fibred finite limits and definable isomorphisms. Then*

(i) *Equality is definable.*

(ii) *The collection* $\{X \in \mathbb{E} \mid X \text{ is terminal object in its fibre}\}$ *is definable.*

(iii) *Among the vertical diagrams* $X \leftarrow Z \to Y$ *the ones where* Z *is product of* X *and* Y *in the fibre, are definable.*

(iv) *Monomorphisms are definable.*

Proof. (i) For parallel maps $f, f': X \rightrightarrows Y$ in the fibre over $I \in \mathbb{B}$, let $e: X' \rightarrowtail X$ be their equaliser in $\mathbb{E}_I$. The morphism $\theta_e: \{e \in \mathrm{vIso}\} \rightarrowtail I$ then represents equality of f, f': a map $u: J \to I$ satisfies $u^*(f) = u^*(f')$ if and only if $u^*(e)$ is an isomorphism, and the latter if and only if u factors through θ_e.

(ii) For $X \in \mathbb{E}$ over I, consider the morphism $\theta_X: \{(!_X: X \to 1I) \in \mathrm{vIso}\} \rightarrowtail I$. It is the required representing arrow: a map $u: J \to I$ satisfies $u^*(X)$ is terminal over J if and only if $!_{u^*(X)}$ is an isomorphism if and only if u factors through θ_X.

(iii) Similarly, for a vertical diagram $X \leftarrow Z \to Y$ consider the representing morphism for the induced tuple $Z \to X \times Y$.

(iv) One uses that a vertical $f: Y \to X$ is monic if and only if the diagram

$$
\begin{array}{ccc}
Y & \xrightarrow{\;\mathrm{id}\;} & Y \\
{\scriptstyle \mathrm{id}} \downarrow & & \downarrow {\scriptstyle f} \\
Y & \xrightarrow[\;f\;]{} & X
\end{array}
$$

is a pullback. Thus one considers the representing arrow for the induced diagonal $Y \to X \times_Y X$. $\qquad \square$

Along the same lines we get that families of separated objects or sheaves are definable in a topos $\mathbb{B}$ with nucleus $\mathbf{j}$. Recall therefore from Remark 5.7.10 (ii) the description of the full subfibrations $\mathrm{FSep}_{\mathbf{j}}(\mathbb{B}) \hookrightarrow \mathbb{B}^{\to}$ and $\mathrm{FSh}_{\mathbf{j}}(\mathbb{B}) \hookrightarrow \mathbb{B}^{\to}$, with respective fibred left adjoints $\mathbf{Fs}$ and $\mathbf{Fa}$.

9.6.9. Proposition. *Let $\mathbb{B}$ be a topos with nucleus $\mathbf{j}$. Then the full subfibrations* $\mathrm{FSep}_{\mathbf{j}}(\mathbb{B}) \hookrightarrow \mathbb{B}^{\rightarrow}$ *and* $\mathrm{FSh}_{\mathbf{j}}(\mathbb{B}) \hookrightarrow \mathbb{B}^{\rightarrow}$ *of the codomain fibration, consisting of families of separated objects and of families of sheaves, are both definable.*

Proof. For a family $\left(\begin{smallmatrix} X \\ \downarrow \varphi \\ I \end{smallmatrix} \right)$ consider the vertical unit $\eta_{\varphi} \colon \varphi \to \mathbf{Fs}(\varphi)$ as a map in the slice category $\mathbb{B}/I$. Since the codomain fibration of a topos is locally small (because a topos is locally Cartesian closed) isomorphisms are definable. Hence we take, as before, a representing morphism $\{\eta_{\varphi} \in \mathrm{vIso}\} \rightarrowtail I$. Then, for a map $u \colon J \to I$:

$$u^*(\varphi) \text{ is separated in } \mathbb{B}/J \;\Leftrightarrow\; \eta_{u^*(\varphi)} \text{ is an isomorphism}$$
$$\Leftrightarrow\; u^*(\eta_{\varphi}) \text{ is an isomorphism}$$
$$\Leftrightarrow\; u \text{ factors through } \{\eta_{\varphi} \in \mathrm{vIso}\} \rightarrowtail I.$$

For sheaves one similarly uses the unit $\varphi \to \mathbf{Fa}(\varphi)$. $\qquad\qquad\square$

We close this section by showing that the properties of being globally, respectively locally small, are inherited by subfibrations with definable collections of objects respectively arrows. Global smallness means that there is a generic object. This implies that definable subfibrations of small fibrations are also small.

9.6.10. Theorem. *Let* $\begin{smallmatrix} \mathbb{E} \\ \downarrow p \\ \mathbb{B} \end{smallmatrix}$ *be a fibration.*

(i) *Assume p is globally small, i.e. has a generic object. Consider a definable collection of objects, given as a full subcategory $\mathbb{D} \hookrightarrow \mathbb{E}$. Then the induced subfibration $\mathbb{D} \hookrightarrow \mathbb{E} \to \mathbb{B}$ also has a generic object.*

(ii) *Assume now that p is locally small; then a definable collection of vertical maps, given as a subcategory $\mathbb{D} \hookrightarrow \mathbb{E}$ gives rise to a subfibration $\mathbb{D} \hookrightarrow \mathbb{E} \to \mathbb{B}$ which is locally small again.*

Proof. (i) Let $T \in \mathbb{E}$ over $\Omega \in \mathbb{B}$ be the generic object of p. This means that for every $X \in \mathbb{E}$ there is a unique $u \colon pX \to \Omega$ with $u^*(T) \cong X$, vertically. Consider the representing arrow $\theta \colon \{T \in \mathbb{D}\} \rightarrowtail \Omega$, and write $T' = \theta^*(T) \in \mathbb{D}$ over $\{T \in \mathbb{D}\}$. This is the generic object for the subfibration $\begin{smallmatrix} \mathbb{D} \\ \downarrow \\ \mathbb{B} \end{smallmatrix}$. For an object $Y \in \mathbb{D}$ there is a unique arrow $v \colon pY \to \Omega$ with $v^*(T) \cong Y \in \mathbb{D}$. Hence v factors through θ, say as $v = \theta \circ w$. Then $w^*(T') \cong Y$, and $w \colon pY \to \{T \in \mathbb{D}\}$ is the only map with this property.

(ii) If p is locally small, then we have for each pair of objects $X, Y \in \mathbb{D}_I$ in the same fibre, a map $\pi_0 \colon \underline{\mathrm{Hom}}(X, Y) \to I$ in $\mathbb{B}$ together with a vertical morphism $\pi_1 \colon \pi_0^*(X) \to \pi_0^*(Y)$ in $\mathbb{E}$, satisfying a universal property as in

Lemma 9.5.4. Since vertical maps in $\mathbb{D}$ are definable, we get an appropriate representing morphism $\theta: \{\pi_1 \in V(\mathbb{D})\} \rightarrowtail \underline{\mathrm{Hom}}(X, Y)$, such that $\theta^*(\pi_1)$ is the best approximation of π_1 in $V(\mathbb{D})$. Put $\lambda_0 = \pi_0 \circ \theta: \{\pi_1 \in V(\mathbb{D})\} \to I$, and let $\lambda_1: \lambda_0^*(X) \dashrightarrow \lambda_0^*(Y)$ be the unique vertical map (isomorphic to $\theta^*(\pi_1)$) in

$$
\begin{array}{ccc}
\lambda_0^*(X) & \longrightarrow & \pi_0^*(X) \\
\lambda_1 \big\downarrow & & \big\downarrow \pi_1 \\
\lambda_0^*(Y) & \longrightarrow & \pi_0^*(Y)
\end{array}
$$

where the horizontal arrows are the unique Cartesian ones over $\theta: \{\pi_1 \in V(\mathbb{D})\} \rightarrowtail \underline{\mathrm{Hom}}(X, Y)$. This pair (λ_0, λ_1) satisfies the appropriate universal property in $\mathbb{D}$, making $\mathbb{D} \to \mathbb{B}$ locally small. $\qquad\square$

9.6.11. Corollary. *If* $\begin{smallmatrix}\mathbb{E}\\\downarrow P\\\mathbb{B}\end{smallmatrix}$ *is a small fibration, and* $\mathbb{D} \hookrightarrow \mathbb{E}$ *forms a definable subfibration (i.e. both the objects and vertical arrows of $\mathbb{D}$ are definable), then* $\begin{smallmatrix}\mathbb{D}\\\downarrow\\\mathbb{B}\end{smallmatrix}$ *is also a small fibration.*

Proof. Because a fibration is small if and only if it is both globally and locally small, see Corollary 9.5.6. $\qquad\square$

This result shows that the notion of 'definable subfibration' satisfies a reasonable criterion for the concept of a 'part' of a fibration; namely that a part of a small fibration should itself be small.

Exercises

9.6.1.   Recall from Lemma 4.4.6 (i) that a cover $c: I \twoheadrightarrow J$ is a morphism which factors through a mono $J' \rightarrowtail J$ only if this mono is an isomorphism. Similarly, a family of morphisms $(c_\alpha: I_\alpha \to J)_{\alpha \in A}$ is a **collective cover** if every mono $J' \rightarrowtail J$ through which each c_α factors is an isomorphism.

       (i)   Assume a collective cover $(c_\alpha: I_\alpha \to J)_{\alpha \in A}$. Prove that for a definable collection P and an object X above J

$$
X \in P \;\Leftrightarrow\; \forall \alpha \in A.\, u_\alpha^*(X) \in P.
$$

       (ii) Show that Lemma 9.6.4 is a consequence of this observation.

9.6.2.   Prove that definable subfibrations of a family fibration $\begin{smallmatrix}\mathrm{Fam}(\mathbb{C})\\\downarrow\\\mathbf{Sets}\end{smallmatrix}$ correspond to subcategories of $\mathbb{C}$.

9.6.3.   Let $\mathbb{E}$ be a category with pullbacks, which is fibred over a category $\mathbb{B}$. Prove that if $P_1, P_2 \subseteq \mathrm{Obj}\,\mathbb{E}$ are definable collections, then their intersection $P_1 \cap P_2 \subseteq \mathrm{Obj}\,\mathbb{E}$ is also definable.

9.6.4. Call a family of PERs $R = (R_i)_{i \in (I,E)}$ over an ω-set (I, E) **bounded** if there is a code $e \in \mathbb{N}$ such that

$$\forall i \in I. \forall n \in E(i). \forall m \in |R_i|. m \leq e \cdot n.$$

Show that the collection of such bounded families is definable in the fibration $\begin{array}{c} \text{UFam}(\mathbf{PER}) \\ \downarrow \\ \omega\text{-}\mathbf{Sets} \end{array}$ of PERs over ω-sets.

9.6.5. Prove that subfunctors of a presheaf $G: \mathbb{C}^{\mathrm{op}} \to \mathbf{Sets}$ correspond to full subfibrations $\mathbb{D} \hookrightarrow \int G$ of the Grothendieck completion $\begin{array}{c} \int G \\ \downarrow \\ \mathbb{C} \end{array}$ of G. (A subfunctor of G may be identified with a natural transformation $\alpha: G \Rightarrow \Omega$, where $\Omega: \mathbb{C}^{\mathrm{op}} \to \mathbf{Sets}$ is the subobject classifier in the topos $\mathbf{Sets}^{\mathbb{C}^{\mathrm{op}}}$ of presheaves from Example 5.4.2.) Describe in terms of a subfunctor when the associated full subfibration is definable.

9.6.6. Let $\begin{array}{c} \mathbb{E} \\ \downarrow p \\ \mathbb{B} \end{array}$ be a fibration with a definable collection $P \subseteq \mathrm{Obj}\,\mathbb{E}$. Show that the assignment $X \mapsto \begin{pmatrix} \{X \in P\} \\ \downarrow \\ pX \end{pmatrix}$ extends to a functor $\mathrm{Cart}(\mathbb{E}) \to \mathbb{B}^{\to}$, and that all squares in its image are pullbacks in $\mathbb{B}$.

Chapter 10

First order dependent type theory

In simple type theory (STT), types σ: Type are built from atomic types (constants) using type constructors like $\to$, $\times$ or $+$. The distinguishing aspect of polymorphic type theory (PTT) is that one may additionally have type variables α: Type occurring in types and terms. This introduces an extra level of indexing, described syntactically by an extra context. In this chapter we study another variation on STT. In dependent type theory (DTT), a term variable x: σ may occur in another type $\tau(x)$: Type. Typical examples are the types

$$n: \mathsf{N} \vdash \mathsf{Nat}(n): \mathsf{Type} \qquad \text{and} \qquad n: \mathsf{N} \vdash \mathsf{NatList}(n): \mathsf{Type}$$

of natural numbers from 1 to n, and of lists of natural numbers of length n. Clearly, they contain a term variable n: N. Notice that such types do not exist in simple or polymorphic type theory. We thus have examples of "types depending on types". This is like "sets depending on sets", for example in an I-indexed collection $X = (X_i)_{i \in I}$, written formally as

$$i: I \vdash X_i: \mathbf{Set} \qquad \text{where} \qquad \vdash I: \mathbf{Set}.$$

Indeed, set-indexed-sets form obvious models of dependent type theory—as formalised by the family fibration $\begin{smallmatrix} \mathbf{Fam(Sets)} \\ \downarrow \\ \mathbf{Sets} \end{smallmatrix}$

Dependent types are widely used in mathematical practice. For example, as an n-fold Cartesian product X^n, where n: N is a parameter. Or in algebra as a set (or type) of $n \times m$ matrices, say with entries from the reals. The following is a typical example in computer science. In the description of hardware, bit vectors play an important rôle. They are finite sequences of bits (which can be represented as booleans $\mathsf{true}, \mathsf{false}$). In the description of digital

581

systems one usually deals with types of bit vectors of a specific length, for example, as types of input or output signals. This leads to dependent types $\mathsf{bvec}(n) = \mathsf{bool}^n : \mathsf{Type}$, depending on $n : \mathsf{N}$. See also [114] for more such examples in hardware. These dependent types are so common that often one is not explicitly aware of using them. They are very convenient in expressing various results and arguments, see also Section 10.2 below.

Dependent types were first studied systematically in the AUTOMATH project in the late 1960s at Eindhoven University in the research group headed by de Bruijn, see [231] for an overview. The aim of the project was to formalise mathematical arguments and to have them checked by a computer. For example, Landau's entire *Grundlagen* book was checked in this way (which brought forward a few minor bugs). But since this project was not so well publicised and the AUTOMATH notation was somewhat confusing and unstable, it did not get the attention right from the beginning that it deserved.

Later in the 1970s, Martin-Löf proposed comparable calculi of dependent types, see *e.g.* [213–215]. His aim was not so much mechanical checking of mathematical arguments, but more the formulation of a foundational language for constructive mathematics. The original calculus contained a type of all types ($\mathsf{Type} : \mathsf{Type}$). As Girard showed in his thesis [94], this leads to inconsistency; the result is known as Girard's paradox, see Exercise 11.5.3 in the next chapter. Subsequently, the system was adapted, see *e.g.* [215, 232]. Nowadays one often uses the name "Martin-Löf type theory" for a variety of (first order) dependent type theories. Although these theories were originally developed for foundational reasons, they have also been used as a basis for proof tools (like NUPRL [78], ALF [207], VERITAS [113] and also PVS [242, 241],) which are employed for the verification of mathematical theories and of software and hardware systems in computer science.

In the next chapter we consider two kinds of extensions of DTT: logical extensions, leading to dependently typed predicate logic to reason about terms and types in DTT, and type theoretic extensions, combining type dependency and polymorphism (forming the basis of proof tools like LEGO or COQ). There we study (higher order) dependent type theories with two syntactic categories (or universes) like Type and Prop, or Kind and Type. In this chapter we restrict ourselves to only a single universe Type.

Categorically, a dependent type $n : \mathsf{N} \vdash \mathsf{NatList}(n) : \mathsf{Type}$, as described in the beginning of this section, is understood as a family of types, indexed by term variables $n : \mathsf{N}$. Hence we could write $\mathsf{NatList} = \left(\mathsf{NatList}(n)\right)_{n : \mathsf{N}}$, using a notation as for set-indexed-sets. More categorically, such a family of types can be described as a morphism $\left(\begin{smallmatrix}\mathsf{NatList}\\\downarrow\\\mathsf{N}\end{smallmatrix}\right)$, such that $\mathsf{NatList}(n)$ appears as fibre over $n : \mathsf{N}$. This leads to the view that dependent types are objects of slice categories—in the example of a slice $\mathbb{C}/N$, for some category $\mathbb{C}$. Or, a bit more formally, dependent types appear as objects of the total category $\mathbb{C}^{\rightarrow}$ of a codomain fibration $\begin{smallmatrix} \mathbb{C}^{\rightarrow} \\ \downarrow \\ \mathbb{C} \end{smallmatrix}$. Along these lines a categorical description of dependent type theory was first given in terms of locally Cartesian closed categories by Seely [306]. The categorical products $\prod$ and coproducts $\coprod$ of the associated codomain fibrations correspond to the dependent products Π and Σ of dependent type theory.

This all works well, but codomain fibrations $\begin{smallmatrix} \mathbb{C}^{\rightarrow} \\ \downarrow \\ \mathbb{C} \end{smallmatrix}$ suffer from the same defect that subobject fibrations $\begin{smallmatrix} \mathrm{Sub}(\mathbb{C}) \\ \downarrow \\ \mathbb{C} \end{smallmatrix}$ do: there are a number of features built-in that we would like to study in isolation. As we saw in Chapter 4, subobject fibrations always have subset types, unique existence and (very strong) equality. Similarly in codomain fibrations, one always has dependent (strong) coproducts and (strong) equality. Therefore we shall be working with certain generalisations of codomain fibrations. The obvious approach (as initiated in [329]) is to consider not all morphisms in as families of types, but only a restricted subclass of them. These are then called "display maps". They form a subfibration of the codomain fibration. In terms of these display maps, all the syntactic features can be described separately, see Section 10.3 below. We shall pursue this approach, but use "comprehension categories", instead of "display map categories". These comprehension categories are fibrations with certain additional structure. They have been used for a general notion of quantification in Section 9.3. They have certain technical advantages over display map categories, but there is no essential difference.

(Many other categorical notions have been proposed to capture type dependency: contextual categories [45, 319], categories with attributes [45] (see also [230, 271, 72]), D-categories [74, 62], higher level indexed categories [234], comprehensive fibrations [252]. There are no great differences between these notions, see [157] for a comparison between some of them.)

In this chapter we introduce calculi with dependent types and describe them categorically. The first two sections will be devoted to the (essentials of the) syntax and use of such calculi. Subsequently in the third section we start the categorical investigations. We first organise the syntactic material of an arbitrary dependently typed calculus in a term model. This brings forward the importance of a distinguished class of morphisms, called "display maps" or "(dependent) projections". In the term model they will be the context projections $\pi\colon (\Gamma, x\colon \sigma) \to \Gamma$ for a type $\Gamma \vdash \sigma\colon \mathsf{Type}$ in context Γ (containing the free term variables in σ). Appropriate categorical descriptions of these projections will be given in Section 10.4, first in terms of display map categories, and then in terms of comprehension categories. Fibrations with a right adjoint to the terminal object functor (like in Section 4.6 describing subset types for preorder fibrations) will be called fibrations with comprehension. They give rise to important examples of comprehension categories. In Section 10.5 we identify the notion of a "closed" comprehension category (CCompC). It describes models of DTT with unit type 1, dependent product Π and strong dependent sum Σ. The last two Sections 10.5 and 10.6 contain many examples of these CCompCs.

10.1 A calculus of dependent types

Our first aim is to describe the syntax of a calculus of dependent types, see also [215, 334, 232, 137]. This is a subtle matter, since terms may occur in types. Thus, types cannot be introduced separately, and a big simultaneous recursion is required. The basic (new) type forming operations are

$\Pi x\colon \sigma.\, \tau(x)$ the dependent product of $\tau(x)$ where x ranges over σ

$\Sigma x\colon \sigma.\, \tau(x)$ the dependent sum of $\tau(x)$ where x ranges over σ

$\mathrm{Eq}_\sigma(x, x')$ the type of σ-equality for x, x' ranging over σ.

Notice that term variables x, x' occur explicitly in the latter equality type $\mathrm{Eq}_\sigma(x, x')$. The intuition is that this type is inhabited if and only if $x, x'\colon \sigma$ are equal. Sometimes, these equality types are called identity types. We recall that the dependent product and sum of a family $(X_i)_{i \in I}$ of sets are given by

$$\Pi i \in I.\, X_i \;=\; \{f\colon I \to \bigcup_{i \in I} X_i \mid \forall i \in I.\, f(i) \in X_i\}$$
$$\Sigma i \in I.\, X_i \;=\; \{\langle i, x \rangle \mid i \in I \text{ and } x \in X_i\}.$$

This gives an intuition for these dependent product and sum: $\Pi x\colon \sigma.\, \tau(x)$ is the collection of functions f such that for each $a\colon \sigma$ one has $fa\colon \tau[a/x]$. And $\Sigma x\colon \sigma.\, \tau(x)$ is the set of pairs $\langle a, b \rangle$ with $a\colon \sigma$ and $b\colon \tau[a/x]$. Notice the substitution $[a/x]$ in type τ, which is typical for DTT. As we shall see formally in Example 10.1.2 and Exercise 10.1.1 below, dependent products Π generalise exponents $\to$ and dependent sums Σ generalise Cartesian products $\times$. For elements $x, x' \in I$ one may think of the associated equality type as

$$\mathrm{Eq}_I(x, x') = \begin{cases} \{*\} & \text{if } x = x' \\ \emptyset & \text{otherwise.} \end{cases}$$

The calculus we are about to set up has contexts of variable declarations

$$\Gamma = x_1\colon \sigma_1, \ldots, x_n\colon \sigma_n$$

satisfying the condition that each type σ_{i+1} is well-formed in the preceding context $x_1: \sigma_1, \ldots, x_i: \sigma_i$, which we write as:

$$x_1: \sigma_1, \ldots, x_i: \sigma_i \vdash \sigma_{i+1}: \mathsf{Type}.$$

As a consequence, each term variable y which is free in σ_{i+1} must already have been declared in the part of Γ preceding σ_{i+1}: it must be one of $x_1, \ldots, x_i$. So in particular, σ_1 is a closed type; it does not contain any term variables. Thus a well-formed context is

$$n: \mathsf{N}, \ell: \mathsf{NatList}(n)$$

but

$$n: \mathsf{N}, z: \mathsf{Matrix}(n, m)$$

is not well-formed (since m is not declared).

Sequents have one of the following four forms.

(1) $\Gamma \vdash \sigma: \mathsf{Type}$ (2) $\Gamma \vdash M: \sigma$

(3) $\Gamma \vdash M = N: \sigma$ (4) $\Gamma \vdash \sigma = \tau: \mathsf{Type}.$

The first (1) of these is type formation as we have seen in other calculi. Likewise for inhabitation in (2) and equality (conversion) of terms in (3). Equality of types in (4) arises because terms may occur in types and so conversions may take place inside types. The main rule associated with equality of types is the following

conversion

$$\frac{\Gamma \vdash M: \sigma \qquad \Gamma \vdash \sigma = \tau: \mathsf{Type}}{\Gamma \vdash M: \tau}$$

We shall not pay much attention to this equality of types, because we take the categorically motivated extensional view that types are equal if they are inhabited by the same terms. Instead we concentrate on rules for sequents of the form (1), (2) and (3). We let $\mathcal{J}$ stand for an arbitrary expression which may occur on the right of a turnstile $\vdash$.

First there are the following five basic rules of DTT.

projection

$$\frac{\Gamma \vdash \sigma: \mathsf{Type}}{\Gamma, x: \sigma \vdash x: \sigma}$$

substitution

$$\frac{\Gamma \vdash M: \sigma \qquad \Gamma, x: \sigma, \Delta \vdash \mathcal{J}}{\Gamma, \Delta[M/x] \vdash \mathcal{J}[M/x]}$$

contraction

$$\frac{\Gamma, x: \sigma, y: \sigma, \Delta \vdash \mathcal{J}}{\Gamma, x: \sigma, \Delta[x/y] \vdash \mathcal{J}[x/y]}$$

weakening

$$\frac{\Gamma \vdash \sigma: \mathsf{Type} \qquad \Gamma \vdash \mathcal{J}}{\Gamma, x: \sigma \vdash \mathcal{J}}$$

exchange

$$\frac{\Gamma, x{:}\,\sigma, y{:}\,\tau, \Delta \vdash \mathcal{J}}{\Gamma, y{:}\,\tau, x{:}\,\sigma, \Delta \vdash \mathcal{J}} \text{ (if } x \text{ not free in } \tau)$$

For convenience we assume a singleton (or unit) type (as in Section 2.3), with rules

$$\frac{}{\vdash 1{:}\,\mathsf{Type}} \qquad \frac{}{\vdash \langle \rangle{:}\,1} \qquad \frac{\Gamma \vdash M{:}\,1}{\Gamma \vdash M = \langle \rangle{:}\,1}$$

Binary product types $\sigma \times \tau$ come for free if one has sum types Σ, see Example 10.1.2. And finite coproduct types $(0, +)$ will be considered separately at the end of this section.

Then there are the formation rules for dependent product Π, sum Σ and equality Eq.

$$\frac{\Gamma, x{:}\,\sigma \vdash \tau{:}\,\mathsf{Type}}{\Gamma \vdash \Pi x{:}\,\sigma.\,\tau{:}\,\mathsf{Type}} \qquad \frac{\Gamma, x{:}\,\sigma \vdash \tau{:}\,\mathsf{Type}}{\Gamma \vdash \Sigma x{:}\,\sigma.\,\tau{:}\,\mathsf{Type}} \quad .$$

$$\frac{\Gamma \vdash \sigma{:}\,\mathsf{Type}}{\Gamma, x{:}\,\sigma, x'{:}\,\sigma \vdash \mathrm{Eq}_\sigma(x, x'){:}\,\mathsf{Type}}$$

These type constructors change the context, and will be described categorically via adjunctions between fibre categories. The variable $x{:}\,\sigma$ becomes bound in $\Pi x{:}\,\sigma.\,\tau$ and $\Sigma x{:}\,\sigma.\,\tau$. Hence substitution in these types takes the following form.

$$(\Pi x{:}\,\sigma.\,\tau)[L/z] \;\equiv\; \Pi x{:}\,\sigma[L/z].\,\tau[L/z]$$
$$(\Sigma x{:}\,\sigma.\,\tau)[L/z] \;\equiv\; \Sigma x{:}\,\sigma[L/z].\,\tau[L/z]$$
$$\mathrm{Eq}_\sigma(x, x')[L/z] \;\equiv\; \mathrm{Eq}_{\sigma[L/z]}(x[L/z], x'[L/z]).$$

where in the first two cases it is assumed that the variable z is different from the (bound) variable x; otherwise the substitution has no effect. Also, x should not be free in L.

There are associated introduction and elimination rules:

$$\frac{\Gamma, x{:}\,\sigma \vdash M{:}\,\tau}{\Gamma \vdash \lambda x{:}\,\sigma.\,M{:}\,\Pi x{:}\,\sigma.\,\tau} \qquad \frac{\Gamma \vdash M{:}\,\Pi x{:}\,\sigma.\,\tau \quad \Gamma \vdash N{:}\,\sigma}{\Gamma \vdash MN{:}\,\tau[N/x]}$$

$$\frac{\Gamma \vdash \sigma{:}\,\mathsf{Type} \quad \Gamma, x{:}\,\sigma \vdash \tau{:}\,\mathsf{Type}}{\Gamma, x{:}\,\sigma, y{:}\,\tau \vdash \langle x, y \rangle{:}\,\Sigma x{:}\,\sigma.\,\tau}$$

$$\frac{\Gamma \vdash \rho{:}\,\mathsf{Type} \quad \Gamma, x{:}\,\sigma, y{:}\,\tau \vdash Q{:}\,\rho}{\Gamma, z{:}\,\Sigma x{:}\,\sigma.\,\tau \vdash (\text{unpack } z \text{ as } \langle x, y \rangle \text{ in } Q){:}\,\rho} \text{ (weak)}$$

$$\frac{\Gamma \vdash \sigma \colon \mathsf{Type}}{\Gamma, x \colon \sigma \vdash \mathsf{r}_\sigma(x) \colon \mathrm{Eq}_\sigma(x, x)}$$

$$\frac{\Gamma, x \colon \sigma, x' \colon \sigma, \Delta \vdash \rho \colon \mathsf{Type} \qquad \Gamma, x \colon \sigma, \Delta[x/x'] \vdash Q \colon \rho[x/x']}{\Gamma, x \colon \sigma, x' \colon \sigma, z \colon \mathrm{Eq}_\sigma(x, x'), \Delta \vdash (Q \text{ with } x' = x \text{ via } z) \colon \rho} \text{ (weak)}$$

These rules involve abstraction and application for dependent products Π, pairing (or packing) and unpacking for dependent sum Σ and a reflexivity combinator $\mathsf{r} = \mathsf{r}_\sigma(x)$ for equality, together with an associated elimination rule. The variables x, y are bound in the sum elimination term unpack z as $\langle x, y \rangle$ in Q. Substituting in these terms is done in the obvious way. The elimination rules for sums and equality are labelled "weak", because also "strong" versions exist; they will be described later in this section.

Notice the explicit occurrence of a parameter context Δ in the equality elimination rule. It is sometimes forgotten, but plays a crucial rôle. (Categorically, it involves a Frobenius property.) In presence of product types Π this Δ may be omitted, since the rules with Δ's are derivable. For sums it comes for free, even without Π's, see Exercise 10.1.8 (i). In the following such parameter contexts are used explicitly.

10.1.1. Example. If we assume terms

$$\Gamma \vdash M \colon \sigma, \qquad \Gamma \vdash M' \colon \sigma \qquad \text{with} \qquad \Gamma \vdash P \colon \mathrm{Eq}_\sigma(M, M')$$

and a type

$$\Gamma, x \colon \sigma \vdash \rho \colon \mathsf{Type} \qquad \text{with inhabitant} \qquad \Gamma \vdash Q \colon \rho[M/x]$$

then we can also find an inhabitant of the type $\rho[M'/x]$ in context Γ. This is replacement, see also Lemma 8.1.2. One obtains such an inhabitant from the rules

$$\frac{\Gamma, x \colon \sigma \vdash \rho \colon \mathsf{Type}}{\Gamma, x \colon \sigma, y \colon \rho[x/x'] \vdash y \colon \rho[x'/x][x/x']}$$
$$\Gamma, x \colon \sigma, x' \colon \sigma, z \colon \mathrm{Eq}_\sigma(x, x'), y \colon \rho \vdash (y \text{ with } x' = x \text{ via } z) \colon \rho[x'/x]$$

Then by substituting $[M/x, M'/x', P/z]$ we get,

$$\Gamma, y \colon \rho[M/x] \vdash y \text{ with } M' = M \text{ via } P \colon \rho[M'/x]$$

and so we are done by further substituting $[Q/y]$. This yields the required inhabitant

$$\Gamma \vdash Q \text{ with } M' = M \text{ via } P \colon \rho[M'/x],$$

and concludes the example.

The conversions associated with products, sums and equality are presented below. The (β)-conversions are the ones mentioned first, followed by the associated (η)-conversions.

$$(\lambda x{:}\,\sigma.\,M)N \;=\; M[N/x]$$
$$\lambda x{:}\,\sigma.\,Mx \;=\; M$$
$$\text{unpack } \langle M, N\rangle \text{ as } \langle x, y\rangle \text{ in } Q \;=\; Q[M/x, N/y]$$
$$\text{unpack } P \text{ as } \langle x, y\rangle \text{ in } Q[\langle x, y\rangle/z] \;=\; Q[P/z]$$
$$Q \text{ with } x = x \text{ via } \mathsf{r} \;=\; Q$$
$$Q[x/x', \mathsf{r}/z] \text{ with } x' = x \text{ via } z \;=\; Q.$$

With the usual proviso that in the (η)-conversion for products the variable x is not allowed to occur free in M.

10.1.2. Example. For types $\Gamma \vdash \sigma, \tau\colon \mathsf{Type}$ in the same context, with a term variable $x\colon \sigma$ which does not occur in τ, we write

$$\sigma \times \tau \;\stackrel{\mathrm{def}}{=}\; \Sigma x{:}\,\sigma.\,\tau \qquad \text{and} \qquad \sigma \to \tau \;\stackrel{\mathrm{def}}{=}\; \Pi x{:}\,\sigma.\,\tau.$$

(in which τ is used as a type in context $\Gamma, x\colon \sigma$ via weakening). It is then almost immediate that $\to$ satisfies all the rules which are required for exponent types, see Section 2.3. For $\times$ we have to do some work. For a term $P\colon \sigma \times \tau$ write

$$\pi P \;\stackrel{\mathrm{def}}{=}\; \text{unpack } P \text{ as } \langle x, y\rangle \text{ in } x \quad \text{and} \quad \pi'P \;\stackrel{\mathrm{def}}{=}\; \text{unpack } P \text{ as } \langle x, y\rangle \text{ in } y.$$

Notice that the second projection $\pi'P$ can be defined because the variable x does not occur in τ. It is easy to see that with these definitions one obtains (β)-conversions $\pi\langle M, N\rangle = M$ and $\pi'\langle M, N\rangle = N$, but also the (η)-conversion holds:

$$
\begin{aligned}
\langle \pi P, \pi'P\rangle \;&=\; \langle \pi z, \pi'z\rangle[P/z] \\
&\stackrel{\eta}{=}\; \text{unpack } P \text{ as } \langle x, y\rangle \text{ in } \langle \pi z, \pi'z\rangle[\langle x, y\rangle/z] \\
&=\; \text{unpack } P \text{ as } \langle x, y\rangle \text{ in } \langle \pi\langle x, y\rangle, \pi'\langle x, y\rangle\rangle \\
&\stackrel{\beta}{=}\; \text{unpack } P \text{ as } \langle x, y\rangle \text{ in } \langle x, y\rangle \\
&\stackrel{\eta}{=}\; P.
\end{aligned}
$$

These dependent product Π and sum Σ thus generalise the exponent $\to$ and binary product $\times$. Since we assume a unit type 1, we get the type constructors of $\lambda 1_\times$ calculi (and of Cartesian closed categories), in every context Γ.

Strong sums and strong equality

As already mentioned, the above elimination rules for sums and equality are the so-called weak ones. The strong versions are obtained by allowing the type ρ with respect to which one eliminates, to depend on an extra variable—of type $\Sigma x\colon\sigma.\,\tau$ or $\mathrm{Eq}_\sigma(x,x')$. This leads to the following strong versions of the elimination rules.

$$\frac{\Gamma, z\colon\Sigma x\colon\sigma.\,\tau \vdash \rho\colon\mathsf{Type} \qquad \Gamma, x\colon\sigma, y\colon\tau \vdash Q\colon\rho[\langle x,y\rangle/z]}{\Gamma, z\colon\Sigma x\colon\sigma.\,\tau \vdash (\mathsf{unpack}\ z\ \mathsf{as}\ \langle x,y\rangle\ \mathsf{in}\ Q)\colon\rho} \quad \text{(strong)}$$

$$\frac{\Gamma, x\colon\sigma, x'\colon\sigma, z\colon\mathrm{Eq}_\sigma(x,x') \vdash \rho\colon\mathsf{Type} \qquad \Gamma, x\colon\sigma \vdash Q\colon\rho[x/x',\mathsf{r}/z]}{\Gamma, x\colon\sigma, x'\colon\sigma, z\colon\mathrm{Eq}_\sigma(x,x') \vdash (Q\ \mathsf{with}\ x' = x\ \mathsf{via}\ z)\colon\rho} \quad \text{(strong)}$$

The conversions remain the same. Notice that there is no parameter context anymore in the strong elimination rule for equality. The rule with such a context is derivable, using the result below. It gives alternative formulations of these strong rules, which are probably more familiar. But the above formulations with additional free variable (the z in ρ) are more appropriate since they scale up to calculi with more universes than just Type, see Section 11.4 in the next chapter.

The first point below comes from [215], and the second one from [322].

10.1.3. Proposition. *The strong elimination rules for sum and equality can equivalently be formulated in the following way.*

(i)

$$\frac{\Gamma \vdash P\colon\Sigma x\colon\sigma.\,\tau}{\Gamma \vdash \pi P\colon\sigma} \qquad\qquad \frac{\Gamma \vdash P\colon\Sigma x\colon\sigma.\,\tau}{\Gamma \vdash \pi'P\colon\tau[\pi P/x]}$$

with conversions $\pi\langle M,N\rangle = M$, $\pi'\langle M,N\rangle = N$ *and* $\langle\pi P,\pi'P\rangle = P$.

(ii)

$$\frac{\Gamma \vdash P\colon\mathrm{Eq}_\sigma(M,M')}{\Gamma \vdash M = M'\colon\sigma} \qquad\qquad \frac{\Gamma \vdash P\colon\mathrm{Eq}_\sigma(M,M')}{\Gamma \vdash P = \mathsf{r}\colon\mathrm{Eq}_\sigma(M,M')}$$

Proof. (i) Assuming the above strong sum elimination rule, we get the projection rules as in Example 10.1.2 for binary products $\times$; the approach can be extended to sums, since the second Σ-projection can be defined with the strong elimination rule:

$$\frac{\Gamma, z\colon\Sigma x\colon\sigma.\,\tau \vdash \tau[\pi z/x]\colon\mathsf{Type} \qquad \Gamma, x\colon\sigma, y\colon\tau \vdash y\colon\tau[\pi z/x][\langle x,y\rangle/z]}{\Gamma, z\colon\Sigma x\colon\sigma.\,\tau \vdash \pi'z \overset{\mathrm{def}}{=} \mathsf{unpack}\ z\ \mathsf{as}\ \langle x,y\rangle\ \mathsf{in}\ y\colon\tau[\pi z/x]}$$

In the reverse direction, one simply puts

$$\mathsf{unpack}\ P\ \mathsf{as}\ \langle x,y\rangle\ \mathsf{in}\ Q \overset{\mathrm{def}}{=} Q[\pi P/x,\ \pi'P/y].$$

(ii) Assume the strong equality elimination rule as formulated above. For variables $x, x' : \sigma$ and $z : \mathrm{Eq}_\sigma(x, x')$ we have

$$
\begin{aligned}
x \; &\overset{\eta}{=} \; x[x/x', \mathsf{r}/z] \text{ with } x' = x \text{ via } z \\
&= \; x'[x/x', \mathsf{r}/z] \text{ with } x' = x \text{ via } z \\
&\overset{\eta}{=} \; x'.
\end{aligned}
$$

and so by substituting $[M/x, M'/x', P/z]$ we get the conversion $M = M' : \sigma$. Likewise, we have

$$
\begin{aligned}
z \; &\overset{\eta}{=} \; z[x/x', \mathsf{r}/z] \text{ with } x' = x \text{ via } z \\
&= \; \mathsf{r}[x/x', \mathsf{r}/z] \text{ with } x' = x \text{ via } z \\
&\overset{\eta}{=} \; \mathsf{r}.
\end{aligned}
$$

So by substituting $[M/x, M'/x', P/z]$, we obtain $P = \mathsf{r} : \mathrm{Eq}_\sigma(M, M')$ as required.

In the reverse direction, for a term $\Gamma, x : \sigma \vdash Q : \rho[x/x', \mathsf{r}/z]$ we can put

$$
\Gamma, x : \sigma, x' : \sigma, z : \mathrm{Eq}_\sigma(x, x') \vdash (Q \text{ with } x' = x \text{ via } z) \overset{\mathrm{def}}{=} Q : \rho
$$

since if $z : \mathrm{Eq}_\sigma(x, x')$, then $x = x' : \sigma$ and $z = \mathsf{r} : \mathrm{Eq}_\sigma(x, x')$ by the assumption. The conversions associated for 'with = via' hold since

$$
\begin{aligned}
Q \text{ with } x = x \text{ via } \mathsf{r} \; &= \; (Q \text{ with } x' = x \text{ via } z)[x/x', \mathsf{r}/z] \\
&= \; Q \text{ with } x' = x \text{ via } z \\
&\qquad \text{by the conversions } x = x' \text{ and } z = \mathsf{r} \\
&= \; Q.
\end{aligned}
$$

And similarly

$$
\begin{aligned}
Q[x/x', \mathsf{r}/z] \text{ with } x' = x \text{ via } z \; &= \; Q \text{ with } x' = x \text{ via } z \\
&= \; Q. \qquad\qquad \square
\end{aligned}
$$

Equality types in dependent type theory form a non-trivial subject, because there are many possible (combinations of) rules. We have chosen to present the above "weak" and "strong" versions, with both (β)- and (η)-conversions since these come out most naturally from a categorical perspective. The two strong equality elimination rules obtained in the previous proposition have some disadvantages from a type theoretic perspective, because they make conversion between terms depend on inhabitation of types. This leads to undecidability of conversion, see [133, 3.2.2]. DTT with strong equality is often called extensional DTT. Since the (η)-conversion for equality types is crucial in proving the above result, it is often dropped. See for example [232, Chapter 8] where "extensional equality" is what we call "strong equality", and "intensional equality" is "strong equality without the (η)-conversion". See also [139, 133, 136] for more information and results, in particular about "setoids", which yield a model with extensional equality inside dependent type theory with intensional equality.

The strong equality types (with (η)-conversion) do have the following syntactic advantage. Consider a type $\Gamma \vdash \sigma$: Type, and another type $\Gamma, x{:}\sigma \vdash \tau$: Type parametrised by σ. If we have inhabitation of a strong equality type $\mathrm{Eq}_\sigma(M, M')$ for terms $M, M'{:}\sigma$, then we may conclude (by the above proposition) that the types $\tau[M/x]$ and $\tau[M'/x]$ are equal. This is problematic otherwise, see [322, 133] for detailed investigations.

10.1.4. Convention. Unless stated otherwise, Σ and Eq will refer to the strong versions of sum and equality. These are most natural (and unproblematic) from a semantic perspective, see *e.g.* Theorem 10.5.10 (i), where they are linked to equalisers.

Weak and strong coproduct types +

The rules for finite coproduct types $(0, +)$ as given in Section 2.3 extend in a straightforward way from simple type theory to dependent type theory. They give us finite coproducts in every context Γ. We shall call these **weak coproducts**. One can also define **strong coproducts** in DTT, analogous to strong sum and equality. The difference between weak and strong coproducts $+$ lies in their elimination rules. In the weak case there is a direct extension of the rule used in simple type theory: for types $\Gamma \vdash \sigma, \tau$: Type in the same context, one can form the coproduct type $\Gamma \vdash \sigma + \tau$: Type in this context, with first and second coprojection terms $\Gamma, x{:}\sigma \vdash \kappa x{:}\sigma + \tau$ and $\Gamma, y{:}\tau \vdash \kappa'y{:}\sigma + \tau$. The weak elimination rule takes the form

$$\frac{\Gamma \vdash \rho\text{: Type} \qquad \Gamma, x{:}\sigma \vdash Q{:}\rho \qquad \Gamma, y{:}\tau \vdash R{:}\rho}{\Gamma, z{:}\sigma + \tau \vdash \mathsf{unpack}\ z\ \mathsf{as}\ [\kappa x\ \mathsf{in}\ Q, \kappa'y\ \mathsf{in}\ R]{:}\rho}\ \text{(weak)}$$

An extended rule with a parameter context is derivable from this one, see Exercise 10.1.8 (ii) below. In the strong $+$-elimination rule one allows this type ρ to depend on a variable $z{:}\sigma + \tau$. Additionally we put in parameter contexts, but they may be omitted in the presence of Π-types.

$$\frac{\Gamma, z{:}\sigma + \tau, \Delta \vdash \rho\text{: Type}}{} $$
$$\frac{\Gamma, x{:}\sigma, \Delta[\kappa x/z] \vdash Q{:}\rho[\kappa x/z] \qquad \Gamma, y{:}\tau, \Delta[\kappa'y/z] \vdash R{:}\rho[\kappa'y/z]}{\Gamma, z{:}\sigma + \tau, \Delta \vdash \mathsf{unpack}\ z\ \mathsf{as}\ [\kappa x\ \mathsf{in}\ Q, \kappa'y\ \mathsf{in}\ R]{:}\rho}\ \text{(strong)}$$

These strong coproducts are called **disjoint union** types in [215] (but the parameter context Δ is omitted there). The strong coproducts are more natural in DTT than the weak ones. They give rise to appropriate distribution isomorphisms, see Exercise 10.1.7 below.

We close this subsection on coproducts + with the following observation. We have formulated the elimination rules for coproduct + and sum Σ to that we can form new *terms* by "unpacking". In DTT one can in principle also formulate similar rules to form new *types* by unpacking, like in

$$\frac{\Gamma, x\colon\sigma \vdash \mu\colon \mathsf{Type} \qquad \Gamma, y\colon \tau \vdash \nu\colon \mathsf{Type}}{\Gamma, z\colon\sigma + \tau \vdash \mathsf{unpack}\ z\ \mathsf{as}\ [\kappa x\ \mathsf{in}\ \mu, \kappa' y\ \mathsf{in}\ \nu]\colon \mathsf{Type}}$$

Together with appropriate conversions for this newly formed type. A similar rule may be formulated for (weak) sums. We shall not investigate the effect of such type definition rules.

The diligent reader may have noticed that we did not start this section with "dependent signatures", just like we began descriptions of simple and polymorphic calculi with appropriate signatures. The reason is that signatures for the latter two calculi have elementary algebraic descriptions. In dependent type theory one cannot separate types and terms, and the whole calculus has to be defined in one big recursion. Hence a signature has to be introduced at the same time as the entire calculus, for example by describing some constants in context, of the form

$$\Gamma \vdash C\colon \mathsf{Type} \qquad \text{and} \qquad \Delta \vdash F\colon C[\vec{M}/\vec{x}]$$

where $\vec{x}$ are the variables declared in Γ and $\vec{M}$ is an appropriately typed sequence of terms.

We will gloss over these (non-trivial) matters. The usual way to be precise here is to first introduce pre-contexts, pre-types and pre-terms by induction, and then to single out from these the well-formed contexts, types and terms using rules as above, see [45, 46, 319, 137, 271] for more details. Dependent signatures introduce appropriate pre-types and pre-terms in this approach.

Church-Rosser and strong normalisation proofs for dependent type theory may be found in [231, 115].

Exercises

10.1.1. Consider the set-theoretic sum and product $\Sigma i\colon I.\,X_i$ and $\Pi i\colon I.\,X_i$ as described in the beginning of this section.
   (i) Show that if the X_i do not depend on $i \in I$, i.e. $X_i = X$ for some set X, then
   $$\Sigma i\colon I.\,X_i \cong I \times X \qquad \text{and} \qquad \Pi i\colon I.\,X_i \cong X^I.$$
   (ii) And if I is finite, say $I = \{1, \ldots, n\}$, then
   $$\Sigma i\colon I.\,X_i \cong X_1 + \cdots + X_n \qquad \text{and} \qquad \Pi i\colon I.\,X_i \cong X_1 \times \cdots \times X_n.$$

10.1.2. Construct proof-terms for symmetry and transitivity of equality types.

10.1.3. Formulate explicitly how substitution distributes over the various term forming constructs in this section.

10.1.4. Derive the "commutation conversion" for (weak) sum types:

$$R[(\text{unpack } P \text{ as } \langle x, y \rangle \text{ in } Q)/z] = \text{unpack } P \text{ as } \langle x, y \rangle \text{ in } R[Q/z].$$

10.1.5. Show that the definition unpack P as $\langle x, y \rangle$ in $Q \overset{\text{def}}{=} Q[\pi P/x, \pi'P/y]$ in the proof of Proposition 10.1.3 (i) satisfies the required conversions.

10.1.6. Call two types $\Gamma \vdash \sigma, \tau$: Type in the same context isomorphic, and write this as $\Gamma \vdash \sigma \cong \tau$, if there are terms $\Gamma, x: \sigma \vdash P: \tau$ and $\Gamma, y: \tau \vdash Q: \sigma$ with conversions $\Gamma, x: \sigma \vdash Q[P/y] = x: \sigma$ and $\Gamma, y: \tau \vdash P[Q/x] = y: \tau$. Use weak sums to prove for types $\Gamma \vdash \sigma$: Type, $\Gamma, x: \sigma \vdash \tau$: Type and $\Gamma, z: \Sigma x: \sigma. \tau \vdash \rho$: Type that there are isomorphisms:

(i) $\Gamma \vdash \Sigma z: (\Sigma x: \sigma. \tau). \rho \cong \Sigma x: \sigma. \Sigma y: \tau. \rho[\langle x, y \rangle / z]$

(ii) $\Pi z: (\Sigma x: \sigma. \tau). \rho \cong \Pi x: \sigma. \Pi y: \tau. \rho[\langle x, y \rangle / z].$

And for $\Gamma \vdash \mu$: Type

(iii) $\Gamma \vdash (\Pi x: \sigma. \tau \rightarrow \mu) \cong (\Sigma x: \sigma. \tau) \rightarrow \mu.$

10.1.7. (See also [73]) Prove similarly for types $\Gamma \vdash \sigma, \tau$: Type and $\Gamma, z: \sigma + \tau \vdash \rho$: Type that

(i) $\Gamma \vdash \Sigma z: \sigma + \tau. \rho \cong \Sigma x: \sigma. \rho[(\kappa x)/z] + \Sigma y: \tau. \rho[(\kappa'y)/z]$

(ii) $\Gamma \vdash \Pi z: \sigma + \tau. \rho \cong \Pi x: \sigma. \rho[(\kappa x)/z] \times \Pi y: \tau. \rho[(\kappa'y)/z]$

where $+$ is strong coproduct and Σ is weak sum. (The parameter context in the strong coproduct elimination rule is needed.)

10.1.8. (i) Prove that the following extended weak sum elimination rule with parameter context Δ is derivable from the rule given above (without parameter).

$$\frac{\Gamma, \Delta \vdash \rho: \text{Type} \qquad \Gamma, x: \sigma, y: \tau, \Delta \vdash Q: \rho}{\Gamma, z: \Sigma x: \sigma. \tau, \Delta \vdash (\text{unpack } z \text{ as } \langle x, y \rangle \text{ in } Q): \rho} \text{ (weak)}$$

[Categorically this will mean that the Frobenius property comes for free in a full comprehension category with coproducts, see Exercise 10.5.4 (ii).]

(ii) Prove similarly that the following extended weak coproduct elimination rule is derivable.

$$\frac{\Gamma, \Delta \vdash \rho: \text{Type} \qquad \Gamma, x: \sigma, \Delta \vdash Q: \rho \qquad \Gamma, y: \tau, \Delta \vdash R: \rho}{\Gamma, z: \sigma + \tau, \Delta \vdash \text{unpack } z \text{ as } [\kappa x \text{ in } Q, \kappa'y \text{ in } R]: \rho}$$

[See Exercise 10.4.8.]

10.1.9. Assume a unit type and strong sum types.

(i) Show that there is a bijective correspondence between types

$$x_1: \sigma, \ldots, x_n: \sigma_n \vdash \tau: \text{Type}$$

in contexts of length n, and types

$$z: (\Sigma x_1: \sigma. \cdots \Sigma x_{n-1}: \sigma_{n-1}. \sigma_n) \vdash \rho: \text{Type}$$

in context of length 1, where $\Sigma x_1 \colon \sigma. \ \cdots \Sigma x_{n-1} \colon \sigma_{n-1} . \sigma_n$ is 1 in case $n = 0$.

(ii) Give a similar correspondence between terms

$$x_1 \colon \sigma, \ldots, x_n \colon \sigma_n \vdash M \colon \tau$$

and

$$z \colon (\Sigma x_1 \colon \sigma. \ \cdots \Sigma x_{n-1} \colon \sigma_{n-1} . \sigma_n) \vdash N \colon \rho.$$

(iii) Conclude that in the presence of a unit type and strong sum types, we may (for convenience) assume that contexts have length one.

[Hence 1 and strong Σ play the same rôle in dependent type theory as 1 and $\times$ in simple type theory, see Exercise 2.3.3.]

10.2 Use of dependent types

This section contains some illustrations of the use of dependent type theory. Especially it contains expositions of the propositions-as-types correspondence: once *à la* Howard and once *à la* de Bruijn.

Our first example is intended to show the expressiveness and usefulness of a calculus with dependent types. It is taken from [114], and mentioned also in [165]. It involves the representation of a type date: Type whose inhabitants describe a particular year, month and day. On first thought one may view date as a Cartesian product $N \times N \times N$ involving the type N of natural numbers, where the first component represents the year, the second the month and the third the day. This can obviously be done in a far more precise way, since the number of months in a year does not exceed 12 and the number of days in a month does not exceed 31. So a second try is $N \times Nat(12) \times Nat(31)$, where for $n \colon N$, $Nat(n)$ is the type of natural numbers from 1 to n. This is already much better. But not every month has 31 days; even worse, the length of the month of February depends on the year. So the best representation gives dates as "dependent tuples" with their type given by sums:

$$\text{date} = \Sigma y \colon N. \ \Sigma m \colon Nat(12). \ Nat(\text{length of month } m \text{ in year } y)$$

where the term 'length of month m in year y' is defined by cases in an obvious way. A typical term of type date is $\langle 1993, \langle 7, 15 \rangle \rangle$. With this type date, correctness of a date representation becomes a well-typedness issue.

This example illustrates an important point: using dependent types may lead to a precise and concise typing discipline. This is convenient in various applications.

Propositions-as-types

The general idea of a propositions-as-types correspondence is that propositions in some logical system can be seen as (or translated into) types in a type theory, in such a way that derivable propositions give rise to inhabited types. Sometimes, one also has that if the translated proposition is inhabited, then the original proposition is derivable. This latter property may be seen as a form of completeness, see *e.g.* [91] for details.

We have already seen that connectives $\top, \wedge, \supset$ in propositional logic correspond to type formers $1, \times, \rightarrow$ in simple type theory. Universal and existential quantifiers $\forall, \exists$ in predicate logic can be translated as polymorphic product and sum $\prod, \coprod$ in polymorphic type theory, see Section 8.1: the types in predicate logic become kinds, and the propositions become types under this translation. It turns out that one can translate predicate logic also into dependent type theory whereby both types and propositions in predicate logic become types in dependent type theory. Hence inside DTT the distinction between (proper) types and propositions is blurred. This has some advantages because it yields a formal system in which one can do both type theory (in the proper sense) and logic. This gives for example the possibility to develop a program together with a proof that it satisfies a specification (described by some logical formula) within the same formalism, see [232]. But blurring the distinction between types and propositions also creates confusion. It is for this reason that the type theories of the proof tools LEGO and COQ (which are based on higher order dependent type theory) make a careful distinction between propositions and types.

Once we agree to see types within DTT sometimes as proper types and sometimes as propositions, we may view an inhabitant $M : \tau$ of a type τ as a proof-term, or as a proof (of τ). The term calculus in dependent type theory yields that

> a proof $P : \Pi x : \sigma. \tau$ in a product is a function $P = \lambda x : \sigma. Px$ which gives for each element $M : \sigma$ of the domain type σ of quantification, a proof $PM : \tau[M/x]$ showing that the proposition τ is true for this element M.

And

> A proof $P : \Sigma x : \sigma. \tau$ in a (strong) sum is a pair $P = \langle \pi P, \pi' P \rangle$ consisting of an element $\pi P : \sigma$ of the domain type of quantification together with a proof $\pi' P : \tau[\pi P/x]$ showing that the proposition τ is true for this element πP.

Here one sees σ as a proper type and τ as a proposition. Since $\rightarrow$ and $\times$ are special instances of Π and Σ, see Example 10.1.2, we get as special cases:

> A proof $P : \sigma \rightarrow \tau$ in an exponent is a function that transforms each proof

$M : \sigma$ of the proposition σ into a proof $PM : \tau$ of the proposition τ.

And

A proof $P : \sigma \times \tau$ in a Cartesian product is a pair consisting of a proof $\pi P : \sigma$ of the proposition σ together with a proof $\pi' P : \tau$ of the proposition τ.

In these last two cases one sees both σ and τ as propositions. What we have here may be understood as the so-called **Brouwer-Heyting-Kolmogorov** interpretation of the connectives of constructive logic, see the introduction to [335]. The proof-terms that one uses correspond to constructive reasonings. For example, consider the proposition

$$\exists x : \sigma. \, (\varphi \wedge \psi) \supset (\exists x : \sigma. \, \varphi) \wedge (\exists x : \sigma. \, \psi).$$

How does one see that it holds? Assume we have an $x : \sigma$ for which both φ and ψ hold. Then we certainly have an $x : \sigma$ for which φ holds and also there is an $x : \sigma$ for which ψ holds. Just take this same x twice.

In the notation of dependent type theory, one writes the proposition as

$$\Sigma x : \sigma. \, (\varphi \times \psi) \to (\Sigma x : \sigma. \, \varphi) \times (\Sigma x : \sigma. \, \psi).$$

The inhabitant of this type corresponding to the proof that we just sketched is the term

$$\lambda z : \Sigma x : \sigma. \, (\varphi \times \psi). \, \langle \langle \pi z, \pi(\pi' z) \rangle, \langle \pi z, \pi'(\pi' z) \rangle \rangle.$$

One sees that a proof $z : \Sigma x : \sigma. \, (\varphi \times \psi)$ is decomposed into an element $\pi z : \sigma$ for which we have a proof $\pi' z = \langle \pi(\pi' z), \pi'(\pi' z) \rangle$ showing that $\varphi \times \psi$ holds of πz. These components are put together again to form proofs of $\Sigma x : \sigma. \, \varphi$ and of $\Sigma x : \sigma. \, \psi$.

(Notice that we can get a different proof-term if we take Σ as weak sum, namely

$$\lambda z : \Sigma x : \sigma. \, (\varphi \times \psi). \, \mathsf{unpack} \, z \, \mathsf{as} \, \langle x, w \rangle \, \mathsf{in} \, \langle \langle x, \pi w \rangle, \langle x, \pi' w \rangle \rangle.$$

Essentially this is the proof-term that arose when we translated the above proposition into polymorphic type theory in Section 8.1—where we only have weak sums.)

We do a more complicated example (taken from [215]). The Axiom of Choice can be formulated in predicate logic and can thus be translated into dependent type theory. It yields the type

$$\Pi x : \sigma. \, \Sigma y : \tau. \, \varphi(x, y) \to \Sigma f : (\Pi x : \sigma. \, \tau). \, \Pi x : \sigma. \, \varphi(x, f \, x).$$

We are going to prove the Axiom of Choice in DTT by deriving a term **ac** which inhabits this type. Here we use the strong sum Σ. We shall reason informally: assume a variable

$$z: \Pi x{:}\, \sigma.\, \Sigma y{:}\, \tau.\, \varphi(x, y).$$

Then for each $x{:}\, \sigma$ we get

$$zx{:}\, \Sigma y{:}\, \tau.\, \varphi(x, y)$$

so that we have first and second projections

$$\pi(zx){:}\, \sigma \qquad \text{and} \qquad \pi'(zx){:}\, \varphi(x, \pi(zx)).$$

Thus if we put $f = \lambda x.\, \pi(zx)$, then we get

$$\pi'(zx){:}\, \varphi(x, fx).$$

Hence we take

$$\mathsf{ac} = \lambda z.\, \langle \lambda x.\, \pi(zx),\, \lambda x.\, \pi'(zx) \rangle$$

where we have omitted the types of the abstracted variables in order to increase readability.

Our next example is taken from [323], again within the "types-and-propositions-as-types" perspective. In predicate logic one formalises inferences like the following: 'All men are mortal; Socrates is a man; hence Socrates is mortal.' But not everything can be formulated in predicate logic. A famous elusive phrase is the so-called *donkey sentence* (due to Geach):

> Every man who owns a donkey beats it.

The problem in formalising this sentence lies in 'it' referring back to the donkey. One might wish to try

$$\forall d{:}\, \mathsf{Donkey}.\, \forall m{:}\, \mathsf{Man}.\, \mathsf{Owns}(m, d) \supset \mathsf{Beats}(m, d).$$

But the quantification here is over all donkeys instead of over men owning a donkey. This problem can be solved in dependent type theory in the following elegant way.

$$\Pi m{:}\, \mathsf{Man}.\, \Pi x{:}\, (\Sigma d{:}\, \mathsf{Donkey}.\, \mathsf{Owns}(m, d)).\, \mathsf{Beats}(m, \pi x).$$

Or, alternatively, as:

$$\Pi x{:}\, (\Sigma m{:}\, \mathsf{Man}.\, \Sigma d{:}\, \mathsf{Donkey}.\, \mathsf{Owns}(m, d)).\, \mathsf{Beats}(\pi x, \pi(\pi'x)).$$

What is the (subtle) difference between these formalisations? For more information and a discussion, see [324]. But see also Exercise 10.2.4 below.

Dependent type theory as a logical framework

In [91] a distinction is made between the propositions-as-types perspective *à la Howard* and *à la de Bruijn*. The first is what we have considered so far—above, but also in simple and polymorphic type theory. The latter *à la de Bruijn* works differently: types are not used as propositions directly, but as a means to formulate a logic. In this way one gets what is called a logical framework: a system in which various logics and type theories can be formulated. This was first done systematically in the AUTOMATH project, and later further developed, for example, in the Edinburgh Logical Framework ELF, see [115] for the basics and [89] for further details, and also in Martin-Löf's Logical Framework, see [232, Part IV]. Logical frameworks form the basis for proof tools like ISABELLE [250] and ALF [207].

The way dependent type theory can be used as a logical framework will be explained by way of example. One starts by postulating a constant

$$\vdash \Omega : \mathsf{Type}$$

which one thinks of as representing propositions; it comes with a 'lifting operation' $T(-)$

$$\alpha : \Omega \vdash T(\alpha) : \mathsf{Type}$$

which maps a proposition to the type of its proofs. One can then describe implication as a constant,

$$\alpha : \Omega, \beta : \Omega \vdash \alpha \supset \beta : \Omega \qquad \text{or as} \qquad \vdash \supset : \Omega \to \Omega \to \Omega$$

together with two constants for introduction and elimination:

$$\alpha : \Omega, \beta : \Omega \ \vdash \ \mathsf{impl\_intro} : (T(\alpha) \to T(\beta)) \to T(\alpha \supset \beta)$$
$$\alpha : \Omega, \beta : \Omega \ \vdash \ \mathsf{impl\_elim} : T(\alpha \supset \beta) \to T(\alpha) \to T(\beta).$$

These constants encode the introduction and elimination rules of implication. The first impl_intro transforms a function f from proofs of α to proofs of β, into a proof impl_intro f of $\alpha \supset \beta$. And if $g : T(\alpha \supset \beta)$ is a proof of $\alpha \supset \beta$ and $x : T(\alpha)$ is a proof of α, then impl_elim $g\,x$ yields a proof of β. It is *Modus Ponens*.

In order to do (single-typed) predicate logic we assume a constant

$$\vdash D : \mathsf{Type}$$

which serves as domain over which the term variables of the logic range. Of course, there may be more such constants, giving rise to many-typed predicate logic. Notice that the variables which are used for the logic that we are encoding, are the variables of the framework (*i.e.* of the dependent type theory in which we are working). This is convenient in implementations: the mechanism for handling variables can be described once and for all for the framework (see *e.g.* [250]).

Universal quantification is now described via a formation constant

$$\vdash \forall_D : (D \to \Omega) \to \Omega$$

with introduction and elimination constants:

$$\alpha\colon D \to \Omega \;\;\vdash\;\; \mathsf{all\_intro}\colon (\Pi x\colon D.\,T(\alpha\,x)) \to T(\forall_D(\alpha))$$
$$\alpha\colon D \to \Omega \;\;\vdash\;\; \mathsf{all\_elim}\colon T(\forall_D(\alpha)) \to (\Pi x\colon D.\,T(\alpha\,x)).$$

Notice that in this approach one does have a clear distinction between propositions (inhabitants of Ω) and types (in the usual sense).

Similarly one describes other connectives. Also for negation one can postulate a constant

$$\alpha\colon \Omega \vdash \mathsf{double\_neg}\colon T(\neg\neg\alpha) \to T(\alpha)$$

where $\neg\alpha = \alpha \supset \bot$ and $\bot\colon\Omega$ is a constant for *falsum*. This gives us classical logic. As an example, we can now construct a proof-term

$$\alpha\colon \Omega,\, \beta\colon\Omega,\, z\colon T(\neg\beta \supset \neg\alpha) \vdash L(z)\colon T(\alpha \supset \beta).$$

Therefore, we will need double negation, since the corresponding entailment $\neg\beta \supset \neg\alpha \vdash \alpha \supset \beta$ does not hold constructively. We assume variables $x\colon T(\alpha)$ and $y\colon T(\neg\beta)$. Then we can form a term

$$M(x,y,z) = \mathsf{impl\_elim}\,(\mathsf{impl\_elim}\,z\,y)\,x\colon T(\bot)$$

and also

$$N(x,z) = \mathsf{impl\_intro}\,(\lambda y\colon T(\neg\beta).\,M(x,y,z))\colon T(\neg\neg\beta).$$

Hence we get our required proof-term

$$L(z) = \mathsf{impl\_intro}\,(\lambda x\colon T(\alpha).\,\mathsf{double\_neg}\,N(x,z))\colon T(\alpha \supset \beta).$$

A categorical analysis of such logical framework encodings in terms of internal categories is presented in [87]. There, a framework λ_{TT} is presented in which every universe $U\colon \mathsf{Type}$ of an encoded system comes with a "lifting" operation $t_U\colon U \to \mathsf{Type}$. The associated context projection morphism $(x\colon U, y\colon t_U(x)) \to (x\colon U)$ in the category of contexts of the framework (as will be explained in the next section) then gives rise to an internal category as in Example 7.1.4 (ii). It captures the (U-part of) the system that is encoded as an internal category in the ambient category of the framework.

Encapsulation and specification in dependent type theory

In Section 8.2 we discussed encapsulation in polymorphic type theory via sum types of the form:

$$\Sigma\alpha_1\colon \mathsf{Type}.\,\cdots\Sigma\alpha_n\colon\mathsf{Type}.\,(\sigma_1 \to \tau_1) \times \cdots \times (\sigma_m \to \tau_m).$$

Such types capture certain operations with types $\sigma_i \to \tau_i$, containing type variables $\vec{\alpha}$. In dependent type theory one can also use sum types for encapsulation: this typically takes the form:

$$\Sigma f_1 : \sigma_1 \to \tau_1. \cdots \Sigma f_m : \sigma_m \to \tau_m. \, \mathsf{Axiom}(f_1, \ldots, f_m)$$

where the f_i are operations (programs) $\sigma_i \to \tau_i$ satisfying a certain axiom (or specification) $\mathsf{Axiom}(\vec{f}) \colon \mathsf{Type}$. Inhabitants of $\mathsf{Axiom}(\vec{f})$ are seen as proofs of the axiom). For example, for a given type $\sigma \colon \mathsf{Type}$, the type of monoid structures on σ is:

$$\mathsf{Mon}(\sigma) = \Sigma m \colon \sigma \to \sigma \to \sigma. \, \Sigma e \colon \sigma.$$
$$\big(\Pi x \colon \sigma. \, \mathsf{Eq}_\sigma(mxe, x) \times \mathsf{Eq}_\sigma(mex, x) \big)$$
$$\times \big(\Pi x, y, z \colon \sigma. \, \mathsf{Eq}_\sigma(mx(myz), m(mxy)z) \big).$$

An inhabitant $a \colon \mathsf{Mon}(\sigma)$ can be identified with a tuple $\langle m, \langle e, \langle p, q \rangle \rangle \rangle$ where p proves that $e \colon \sigma$ is a neutral element for $m \colon \sigma \to \sigma \to \sigma$, and q proves associativity of m.

Higher order dependent type theories (as will be discussed in the next chapter) enable us to combine the encapsulation of PTT with the encapsulation of DTT, so that, for example, we can collect all monoids as:

$$\Sigma \alpha \colon \mathsf{Type}. \, \mathsf{Mon}(\alpha)$$

(This will be an inhabitant of Kind.) Also, the higher order structure can be used for modularisation and structured specification, see [201, 202], and also [205, 232, 217]. This use of sum types in dependent type theory also forms the basis for program extraction (via sum projections), see [247, 249] for more information and further references.

Exercises

10.2.1. We have constructed a proof term for the Axiom of Choice, namely

$$\vdash \mathsf{ac} \colon (\Pi x \colon \sigma. \, \Sigma y \colon \tau. \, \varphi(x, y)) \to (\Sigma f \colon \Pi x \colon \sigma. \, \tau. \, \Pi x \colon \sigma. \, \varphi(x, fx))$$

Show that in the reverse direction there is a 'canonical' term

$$\vdash \mathsf{ca} \colon (\Sigma f \colon \Pi x \colon \sigma. \, \tau. \, \Pi x \colon \sigma. \, \varphi(x, fx)) \to (\Pi x \colon \sigma. \, \Sigma y \colon \tau. \, \varphi(x, y))$$

which is the inverse of ac, *i.e.* which satisfies $\mathsf{ca}(\mathsf{ac}\, z) = z$ and $\mathsf{ac}(\mathsf{ca}\, w) = w$ (as in Exercise 10.1.6). This gives us complete distributivity, see Exercise 1.9.9.

10.2.2. Give the proof of $\neg\beta \supset \neg\alpha \vdash \alpha \supset \beta$ in ordinary (classical) logic corresponding to the proof-term L described above.

10.2.3. (i) Formulate appropriate formation, introduction and elimination constants for conjunction $\wedge$ and for existential quantification $\exists_D$ (using dependent type theory as a logical framework).

(ii) Give proof terms inhabiting the types

$$T(\forall_D(\alpha)) \to T(\neg\exists_D(\neg\alpha)) \quad \text{and} \quad T(\neg\exists_D(\neg\alpha)) \to T(\forall_D(\alpha))$$

where we have written $\neg\alpha$ for (the formally correct) $\lambda x \colon D. \, \neg(\alpha\, x)$.

10.2.4. Show that the two types given above for the donkey sentence are isomorphic (in the sense of Exercise 10.1.6).

10.2.5. Define a type $\mathsf{Gr}(\sigma)$ of group structures on a type σ—in analogy with the type $\mathsf{Mon}(\sigma)$ of monoid structures, as described above. Describe a "forget-ful" term $\mathsf{Gr}(\sigma) \to \mathsf{Mon}(\sigma)$.

10.3 A term model

This section starts the categorical investigation of type dependency. It does not yet present an abstract categorical notion of what constitutes a model of type dependency, but it contains a rather detailed investigation of the term model construction. It will bring forward the importance of a distinguished class of morphisms, called "display maps" (after Taylor [329, 330]) or "(dependent) projections", in a category of contexts. In the next section we capture this situation abstractly by the notion of "display map category", which forms a particular instance of a "comprehension category".

In previous chapters we saw how one could construct certain categories of contexts from the syntax of various logics, but also from the syntax of simply or polymorphically typed calculi. These term model constructions are interesting since they show us some of the underlying categorical structure. Term models can also be formed for dependently typed calculi, but things are slightly more complicated. First of all, since terms can occur in types, one may have conversions between types. But then one may also have conversions between contexts (componentwise). In a term model, one should therefore not consider contexts Γ as objects, but equivalences $[\Gamma]$ of contexts (under conversion). We shall leave these square braces $[-]$ implicit for contexts, types and terms, and consider syntax up-to-conversion. This keeps notation manageable. Also, we shall consider types and terms in contexts up to α-conversion. This allows us to assume that whenever we are dealing with two (or more) contexts, their sets of variables are disjoint.

We now consider some arbitrary, but fixed, dependently typed calculus and form a category of contexts $\mathbb{C}$ from this syntactic material. This category $\mathbb{C}$ has

objects   (equivalence classes of) contexts Γ

morphisms $\Gamma \to \Delta$, say with Δ of the form $(x_1 \colon \sigma_1, \ldots, x_n \colon \sigma_n)$ where the variables $x_1, \ldots, x_{i-1}$ may occur free in the type σ_i, are n-tuples $(M_1, \ldots, M_n)$ of (equivalence classes of) terms M_i typed as:

$$\Gamma \vdash M_i : \sigma_i[M_1/x_1, \ldots, M_{i-1}/x_{i-1}].$$

Such a sequence of terms $\vec{M} : \Gamma \to \Delta$ will often be called a **context morphism**.

Notice the explicit substitutions in types. This is typical for DTT, as opposed to STT and PTT. These substitutions are performed simultaneously.

For a context $\Gamma = (x_1 : \sigma_1, \ldots, x_n : \sigma_n)$, the identity map $\Gamma \to \Gamma$ is the n-tuple of variables $(x_1, \ldots, x_n)$. The composite of morphisms

$$\Gamma \xrightarrow{\;\;(M_1, \ldots, M_n)\;\;} \Delta \xrightarrow{\;\;(N_1, \ldots, N_m)\;\;} \Theta$$

say with contexts and types

$$\Delta = (x_1 : \sigma_1, \ldots, x_n : \sigma_n) \quad \text{and} \quad \Gamma \vdash M_i : \sigma_i[M_1/x_1, \ldots, M_{i-1}/x_{i-1}]$$
$$\Theta = (y_1 : \tau_1, \ldots, y_m : \tau_m) \quad \text{and} \quad \Delta \vdash N_j : \tau_i[N_1/y_1, \ldots, N_{j-1}/y_{j-1}],$$

is the m-tuple $(L_1, \ldots, L_m) : \Gamma \to \Theta$ with components

$$L_j = N_j[\vec{M}/\vec{x}] = N_j[M_1/x_1, \ldots, M_n/x_n].$$

These are well-typed:

$$\Gamma \vdash L_j : \tau_j[N_1/y_1, \ldots, N_{j-1}/y_{j-1}][\vec{M}/\vec{x}] = \tau_j[L_1/y_1, \ldots, L_{j-1}/y_{j-1}].$$

By using appropriate substitution lemmas one can establish associativity of composition. This is a non-trivial matter. Already in STT it required considerable work to get associativity, see the exercises of Section 2.1. Here we choose to gloss over these details, and we refer to [332, 271] for more information.

Remember that categories of contexts in STT and PTT always have finite products. The empty context is a terminal object, and concatenation of contexts yields binary products. These are the basic operations for contexts, corresponding to simple categorical structure. An important question is what structure is induced in the above category $\mathbb{C}$ by these operations on contexts in DTT. It is easy to see that the empty context again yields a terminal object. But concatenation of contexts in DTT does not yield Cartesian products, but rather "dependent sums". Consider the case of a context $(x : \sigma, y : \tau)$ of two types, where crucially x may occur in τ. A context morphism $\Gamma \to (x : \sigma, y : \tau)$ does not correspond to two morphisms $\Gamma \to (x : \sigma)$ and $\Gamma \to (y : \tau)$, but to two morphisms $M : \Gamma \to (x : \sigma)$ and $N : \Gamma \to (y : \tau[M/x])$. This dependent pairing property can be described via the existence of certain pullbacks in the category $\mathbb{C}$. More precisely, pullbacks along certain "projection" maps in $\mathbb{C}$, of the form

$$(\Gamma, z{:}\rho) \xrightarrow{\ \pi\ } \Gamma$$

Explicitly, for Γ of the form $(x_1{:}\sigma_1, \ldots, x_n{:}\sigma_n)$ this map $\pi{:}(\Gamma, z{:}\rho) \to \Gamma$ is the n-tuple $(x_1, \ldots, x_n)$ of variables in Γ. Such maps will be called **display maps**, or simply **projections**. They are "dependent" projections, because all the variables $x_1, \ldots, x_n$ declared in Γ may occur free in the type ρ that is projected away.

It turns out that the collection of these display maps in $\mathbb{C}$ is closed under pullback (see the lemma below). This gives us the abovementioned correspondence between context morphisms $(M, N){:}\Gamma \to (x{:}\sigma, y{:}\tau)$ and pairs of morphisms $M{:}\Gamma \to (x{:}\sigma)$ and $N{:}\Gamma \to (y{:}\tau[M/x])$, in a situation:

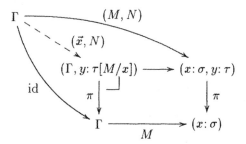

The rôle of Cartesian products in simple and polymorphic type theory is taken over by (certain) pullbacks in dependent type theory.

10.3.1. Lemma. *The above display maps in the category $\mathbb{C}$ of contexts are stable under pullback in the following sense. For an arbitrary context morphism $\vec{M}{:}\Gamma \to \Delta$ and a display map $\pi{:}(\Delta, y{:}\tau) \to \Delta$ on Δ, there is a display map on Γ in a pullback square of the following form.*

$$
\begin{array}{ccc}
(\Gamma, z{:}\tau[\vec{M}/\vec{x}]) & \longrightarrow & (\Delta, y{:}\tau) \\
{\scriptstyle\pi}\big\downarrow \ \ \llcorner & & \big\downarrow{\scriptstyle\pi} \\
\Gamma & \xrightarrow[\ \vec{M}\ ]{} & \Delta
\end{array}
\qquad (*)
$$

Proof. Assume Δ is of the form $(x_1{:}\sigma_1, \ldots, x_n{:}\sigma_n)$ so that the terms M_i are typed as:

$$\Gamma \vdash M_i{:}\sigma_i[M_1/x_1, \ldots, M_{i-1}/x_{i-1}].$$

Then we have as top morphism $(\Gamma, z{:}\tau[\vec{M}/\vec{x}]) \to (\Delta, y{:}\tau)$ in the above diagram the sequence of terms

$$(\Gamma, z: \tau[\vec{M}/\vec{x}]) \xrightarrow{\ (M_1, \ldots, M_n, z)\ } (\Delta, y: \tau)$$

This certainly makes the diagram commute. If we have another context Θ and context morphisms

$$\Theta \xrightarrow{\ (N_1, \ldots, N_m)\ } \Gamma \qquad \Theta \xrightarrow{\ (L_1, \ldots, L_n, L_{n+1})\ } (\Delta, y: \tau)$$

with $\vec{M} \circ \vec{N} = \pi \circ \vec{L}$ then for each $i \leq n$ there is a conversion

$$\Theta \vdash M_i[\vec{N}/\vec{x}] = L_i: \sigma_i[L_1/x_1, \ldots, L_{i-1}/x_{i-1}].$$

Thus we get as unique mediating map $\Theta \to (\Gamma, z: \tau[\vec{M}/\vec{x}])$, the $m + 1$-tuple

$$(N_1, \ldots, N_m, L_{n+1}). \qquad \Box$$

Notice that for this result does not require any of the type constructors $1, \Pi$, weak Σ or strong Σ of dependent type theory—as introduced in Section 10.1. The lemma describes some basic categorical structure induced by context concatenation (actually extension with a single type only) and substitution. The type constructors correspond to certain additional categorical properties of the above class of display maps in $\mathbb{C}$. This will be described next.

But first we need some notation. We write $\mathcal{D}$ for the collection of display maps $\pi: (\Gamma, x: \sigma) \to \Gamma$ in $\mathbb{C}$ induced by types $\Gamma \vdash \sigma: \mathsf{Type}$ in context. Often we shall write these maps simply as $(\Gamma, x: \sigma) \to \Gamma$, leaving the projection π implicit. By the previous lemma, these maps in $\mathcal{D}$ form a split fibration over the category $\mathbb{C}$ of contexts. This (codomain) fibration will be written as $\begin{smallmatrix} \mathcal{D}^{\to} \\ \downarrow \\ \mathbb{C} \end{smallmatrix}$, where $\mathcal{D}^{\to} \hookrightarrow \mathbb{C}^{\to}$ is the full subcategory with display maps as objects. This inclusion actually forms a "comprehension category" (see the description in Theorem 9.3.4).

Notice that substitution π^* along a display map $\pi: (\Gamma, x: \sigma) \to \Gamma$ in this fibration is **weakening**, since it moves a type $\Gamma \vdash \tau: \mathsf{Type}$ to a bigger context $\Gamma, x: \sigma \vdash \tau: \mathsf{Type}$ in the pullback square

$$
\begin{array}{ccc}
(\Gamma, x: \sigma, y: \tau) & \longrightarrow & (\Gamma, y: \tau) \\
\pi \downarrow \quad\ \lrcorner & & \downarrow \pi \\
(\Gamma, x: \sigma) & \xrightarrow{\ \pi\ } & \Gamma
\end{array}
$$

The next two propositions describe the correspondence between type theoretical and categorical structure (for unit and dependent product and sums).

10.3.2. Proposition. *A unit type* $1 \colon \mathsf{Type}$ *in our calculus corresponds to a terminal object functor* $1 \colon \mathbb{C} \to \mathcal{D}^{\to}$ *for the associated fibration* $\begin{smallmatrix} \mathcal{D}^{\to} \\ \downarrow \\ \mathbb{C} \end{smallmatrix}$ *of display maps. The domain functor* $\mathcal{D}^{\to} \to \mathbb{C}$ *is then right adjoint to* 1.

Recall that a right adjoint to a terminal object functor was used in Section 4.6 to describe subset types $\{-\}$ for a preorder fibration. It plays a similar rôle of "comprehension" here, as formalised by the notion of comprehension category with unit in the next section.

Proof. Assume a unit type $\vdash 1 \colon \mathsf{Type}$ with (sole) inhabitant $\vdash \langle\,\rangle \colon 1$. Define a functor $1 \colon \mathbb{C} \to \mathcal{D}^{\to}$ by

$$\Gamma \mapsto \begin{pmatrix} \Gamma, z{:}1 \\ \downarrow \\ \Gamma \end{pmatrix}.$$

Then for an arbitrary display map $\begin{pmatrix} \Gamma, x{:}\sigma \\ \downarrow \\ \Gamma \end{pmatrix}$ on Γ there is precisely one morphism $\begin{pmatrix} \Gamma, x{:}\sigma \\ \downarrow \\ \Gamma \end{pmatrix} \to \begin{pmatrix} \Gamma, z{:}1 \\ \downarrow \\ \Gamma \end{pmatrix}$ over Γ, namely $(\vec{v}, \langle\,\rangle)$ where $\vec{v}$ is the sequence of variables declared in Γ.

Further, for display maps $\begin{pmatrix} \Delta, y{:}\tau \\ \downarrow \\ \Delta \end{pmatrix}$ there is a bijective correspondence

$$\frac{\begin{pmatrix} \Gamma, z{:}1 \\ \downarrow \\ \Gamma \end{pmatrix} \longrightarrow \begin{pmatrix} \Delta, y{:}\tau \\ \downarrow \\ \Delta \end{pmatrix} \quad \text{in } \mathcal{D}^{\to}}{\Gamma \longrightarrow (\Delta, y{:}\tau) \quad \text{in } \mathbb{C}}$$

which establishes that the domain functor $\mathrm{dom} \colon \mathcal{D}^{\to} \to \mathbb{C}$ mapping a display map $\begin{pmatrix} \Delta, y{:}\tau \\ \downarrow \\ \Delta \end{pmatrix}$ to its domain context $(\Delta, y{:}\tau)$ is right adjoint to the terminal object functor.

Conversely, assume that the fibration $\begin{smallmatrix} \mathcal{D}^{\to} \\ \downarrow \\ \mathbb{C} \end{smallmatrix}$ comes with a (split) terminal object functor $1 \colon \mathbb{C} \to \mathcal{D}^{\to}$. Write $\vdash 1 \colon \mathsf{Type}$ for the domain of the functor 1 applied to the empty context $\emptyset$. Take an arbitrary closed inhabited type, say $\vdash \sigma_0 \colon \mathsf{Type}$ with inhabitant $\vdash M_0 \colon \sigma_0$, consider the unique morphism $\begin{pmatrix} x{:}\sigma_0 \\ \downarrow \\ \emptyset \end{pmatrix} \overset{P}{\longrightarrow} \begin{pmatrix} z{:}1 \\ \downarrow \\ \emptyset \end{pmatrix}$ in the fibre over the empty context $\emptyset$, and write $\langle\,\rangle = P[M_0/x] \colon 1$. This is the sole inhabitant of $1 \colon \mathsf{Type}$, since if $\Gamma \vdash N \colon 1$, then we get two morphisms over Γ

$$\begin{pmatrix} \Gamma,x{:}\sigma_0 \\ \downarrow \\ \Gamma \end{pmatrix} \xrightarrow[\;(\vec{v},N)\;]{\;(\vec{v},P)\;} \begin{pmatrix} \Gamma,z{:}1 \\ \downarrow \\ \Gamma \end{pmatrix} = !_\Gamma^* \begin{pmatrix} z{:}1 \\ \downarrow \\ \emptyset \end{pmatrix} = !_\Gamma^*(1\emptyset) = 1\Gamma$$

where both N and P are considered as terms in context $(\Gamma, x{:}\sigma_0)$ via weakening, and where $\vec{v}$ is the sequence of variables in Γ. These maps are equal, because 1Γ is terminal, and so we get a conversion $\Gamma, x{:}\sigma_0 \vdash P = N{:}1$. Because x does not occur in N we obtain:

$$\Gamma \vdash \langle\rangle = P[M_0/x] = N[M_0/x] = N{:}1. \qquad \square$$

(In the last part of the proof we have used the assumption that our dependently typed calculus is non-trivial in the sense that there is at least one inhabited closed type.)

10.3.3. Proposition. (i) *Dependent products Π in the calculus correspond to the fibration* $\begin{smallmatrix} \mathcal{D}^\rightarrow \\ \downarrow \\ \mathbb{C} \end{smallmatrix}$ *of display maps having products $\prod$ along display maps* $\pi{:}(\Gamma, x{:}\sigma) \to \Gamma$ *in* $\mathbb{C}$.

(This means: right adjoints $\prod$ to weakening functors π^* together with a Beck-Chevalley condition: for a pullback square of the form $(*)$ in Lemma 10.3.1 the induced canonical natural transformation is an identity—in this split case.)

(ii) *Weak dependent sums Σ correspond to left adjoints $\coprod$ along display maps satisfying Beck-Chevalley.*

(iii) *And strong dependent sums Σ correspond to weak dependent sums for which the canonical map κ in the diagram*

$$\begin{array}{ccc}
(\Gamma, x{:}\sigma, y{:}\tau) & \xrightarrow{\;\kappa\;} & (\Gamma, z{:}\Sigma x{:}\sigma.\,\tau) \\
{\scriptstyle \pi}\downarrow & & \downarrow{\scriptstyle \pi} \\
(\Gamma, x{:}\sigma) & \xrightarrow[\;\pi\;]{} & \Gamma
\end{array}$$

is an isomorphism. This map κ has the dependent tuple term $\Gamma, x{:}\sigma, y{:}\tau \vdash \langle x, y\rangle{:}\Sigma x{:}\sigma.\,\tau$ as last component.

Notice that the products and coproducts in (i) and (ii) are products and coproducts in the fibration $\begin{smallmatrix} \mathcal{D}^\rightarrow \\ \downarrow \\ \mathbb{C} \end{smallmatrix}$ of display maps, with respect to its own comprehension category $\mathcal{D}^\rightarrow \hookrightarrow \mathbb{C}^\rightarrow$, as described in Definition 9.3.5. This more abstract formulation will be used later on. Interestingly, the dependent product $\Pi x\colon \sigma.\,(-)$ and sum $\Sigma x\colon \sigma.\,(-)$, being right and left adjoint to weakening π^*, fit in the same pattern as the polymorphic product $\Pi\alpha\colon A.\,(-)$ and sum $\Sigma\alpha\colon A.\,(-)$ in Chapter 8, and as universal $\forall x\colon \sigma.\,(-)$ and existential $\exists x\colon \sigma.\,(-)$ quantification in Chapter 4.

Proof. (i) Assume dependent products Π and define for a type $\Gamma \vdash \sigma\colon \mathsf{Type}$ quantification along a display map $\pi\colon (\Gamma, x\colon \sigma) \to \Gamma$ by

$$
\begin{pmatrix} \Gamma,x:\sigma,y:\tau \\ \downarrow \\ \Gamma,x:\sigma \end{pmatrix}
\overset{\Pi}{\longmapsto}
\begin{pmatrix} \Gamma,z:\Pi x:\sigma.\,\tau \\ \downarrow \\ \Gamma \end{pmatrix}.
$$

Then one gets an adjoint correspondence between maps over Γ and over $(\Gamma, x\colon \sigma)$ in

$$
\dfrac{\begin{pmatrix} \Gamma,w:\rho \\ \downarrow \\ \Gamma \end{pmatrix} \xrightarrow{(\vec{v},\,M)} \begin{pmatrix} \Gamma,z:\Pi x:\sigma.\,\tau \\ \downarrow \\ \Gamma \end{pmatrix} = \Pi \begin{pmatrix} \Gamma,x:\sigma,y:\tau \\ \downarrow \\ \Gamma,x:\sigma \end{pmatrix}}{\pi^*\begin{pmatrix} \Gamma,w:\rho \\ \downarrow \\ \Gamma \end{pmatrix} = \begin{pmatrix} \Gamma,x:\sigma,w:\rho \\ \downarrow \\ \Gamma,x:\sigma \end{pmatrix} \xrightarrow{(\vec{v},\,x,\,N)} \begin{pmatrix} \Gamma,x:\sigma,y:\tau \\ \downarrow \\ \Gamma,x:\sigma \end{pmatrix}}
$$

where π^* is the "weakening" pullback functor induced by the projection $\pi\colon (\Gamma, x\colon \sigma) \to \Gamma$ along which we quantify, and where $\vec{v}$ is the sequence of variables declared in Γ. The bijection is obtained by sending a term $\Gamma, w\colon \rho \vdash M\colon \Pi x\colon \sigma.\,\tau$ to $\Gamma, w\colon \rho, x\colon \sigma \vdash \overline{M} \overset{\text{def}}{=} M x\colon \tau$; and $\Gamma, x\colon \sigma, w\colon \rho \vdash N\colon \tau$ with x not occurring free in ρ (by weakening) to $\Gamma, w\colon \rho \vdash \overline{N} \overset{\text{def}}{=} \lambda x\colon \sigma.\,N\colon \Pi x\colon \sigma.\,\tau$. The equations $\overline{\overline{M}} = M$ and $\overline{\overline{N}} = N$ are the (β)- and (η)-conversions for Π. Beck-Chevalley follows from the appropriate distribution of substitution over Π's.

In the reverse direction, assume that (categorical) products $\prod$ along display maps exist, satisfying Beck-Chevalley. For a type $\Gamma, x\colon \sigma \vdash \tau\colon \mathsf{Type}$, write $\Pi x\colon \sigma.\,\tau$ for the type in context Γ for which

$$
\begin{pmatrix} \Gamma,z:\Pi x:\sigma.\,\tau \\ \downarrow \\ \Gamma \end{pmatrix}
= \prod \begin{pmatrix} \Gamma,x:\sigma,y:\tau \\ \downarrow \\ \Gamma,x:\sigma \end{pmatrix}.
$$

where $\prod$ is quantification along $\pi\colon (\Gamma, x\colon \sigma) \to \Gamma$. The adjunction $(\pi^* \dashv \prod)$ gives a bijective correspondence between terms $\Gamma, w\colon \rho \vdash M\colon \Pi x\colon \sigma.\,\tau$ and $\Gamma, w\colon \rho, x\colon \sigma \vdash N\colon \tau$. It yields abstraction and application operations for Π in the calculus, with an extra assumption $w\colon \rho$. One obtains the rules product rules from Section 10.1 by using for ρ any inhabited closed type.

(ii) Similarly. One obtaines a coproduct functor from sum types by

$$\begin{pmatrix} \Gamma,x{:}\sigma,y{:}\tau \\ \downarrow \\ \Gamma,x{:}\sigma \end{pmatrix} \overset{\coprod}{\longmapsto} \begin{pmatrix} \Gamma,z{:}\Sigma x{:}\sigma.\,\tau \\ \downarrow \\ \Gamma \end{pmatrix}$$

which captures quantification along $\pi\colon (\Gamma, x{:}\sigma) \to \Gamma$ in a correspondence

$$\coprod\begin{pmatrix} \Gamma,x{:}\sigma,y{:}\tau \\ \downarrow \\ \Gamma,x{:}\sigma \end{pmatrix} = \begin{pmatrix} \Gamma,z{:}\Sigma x{:}\sigma.\,\tau \\ \downarrow \\ \Gamma \end{pmatrix} \xrightarrow{(\vec{v},\,M)} \begin{pmatrix} \Gamma,w{:}\rho \\ \downarrow \\ \Gamma \end{pmatrix}$$

$$\overline{\begin{pmatrix} \Gamma,x{:}\sigma,y{:}\tau \\ \downarrow \\ \Gamma,x{:}\sigma \end{pmatrix} \xrightarrow{(\vec{v},\,x,\,N)} \begin{pmatrix} \Gamma,x{:}\sigma,w{:}\rho \\ \downarrow \\ \Gamma,x{:}\sigma \end{pmatrix} = \pi^{*}\begin{pmatrix} \Gamma,w{:}\rho \\ \downarrow \\ \Gamma \end{pmatrix}}$$

between terms $\Gamma, z\colon \Sigma x\colon \sigma.\,\tau \vdash M\colon \rho$ and $\Gamma, x\colon \sigma, y\colon \tau \vdash N\colon \rho$. The transpose operations correspond to "packing" via tupling $\langle -, - \rangle$ and "unpacking".

(iii) An inverse morphism $(\Gamma, z\colon \Sigma x\colon \sigma.\,\tau) \to (\Gamma, x\colon \sigma, y\colon \tau)$ to the pairing morphism $\kappa = (\vec{v}, \langle x, y \rangle)$ must have two terms

$$\Gamma, z\colon \Sigma x\colon \sigma.\,\tau \vdash \mathsf{fst}(z)\colon \sigma \qquad \text{and} \qquad \Gamma, z\colon \Sigma x\colon \sigma.\,\tau \vdash \mathsf{snd}(z)\colon \tau[\mathsf{fst}(z)/x].$$

as last two components. The fact that fst and snd form an inverse to κ determines equations

$$\mathsf{fst}(\langle x, y \rangle) = x, \qquad \mathsf{snd}(\langle x, y \rangle) = y, \qquad \langle \mathsf{fst}(z), \mathsf{snd}(z) \rangle = z.$$

Hence, if κ has an inverse then we have first and second projections for Σ, making this sum strong. The converse is trivial. □

As we have seen, the projections $\pi\colon (\Gamma, x\colon \sigma) \to \Gamma$ along which we quantify in dependent type theory are not Cartesian projections (like in predicate logic and in polymorphic type theory) but are more general "dependent" projections. In Exercise 10.3.3 below we mention how equality (types) in DTT correspond (in the term model) to left adjoints to contractions functors δ^{*} induced by "dependent" diagonals $\delta\colon (\Gamma, x\colon \sigma) \to (\Gamma, x\colon \sigma, x'\colon \sigma)$. This follows the abstract pattern of equality with respect to a comprehension category (see Definition 9.3.5), which also underlies equality (and quantification) in predicate logic and in polymorphic type theory, via "simple" comprehension categories. This fundamental rôle of comprehension categories will be further investigated in the next two sections.

Exercises

10.3.1.   Establish an isomorphism in the category of contexts $\pi\colon (\Gamma, z\colon 1) \overset{\cong}{\to} \Gamma$, where $\vdash 1\colon \mathsf{Type}$ is a unit type. As a result, the singleton context $(z\colon 1)$ is isomorphic to the terminal, empty context $\emptyset$.

10.3.2. Describe explicitly so-called "mate" rules (see *e.g.* Lemma 4.1.8 and Exercise 8.1.3) for dependent products Π and sums Σ, as used implicitly in the proofs of Proposition 10.3.3 (i) and (ii).

10.3.3. For a type $\Gamma \vdash \sigma : \mathsf{Type}$ write $\delta : (\Gamma, x : \sigma) \to (\Gamma, x : \sigma, x' : \sigma)$ for the context morphism in the category $\mathbb{C}$ of contexts given by the sequence of terms $(\vec{y}, x, x)$, where $\vec{y}$ is the sequence of variables declared in context Γ.

(i) Verify that δ is the unique mediating map for the pullback of the display map $\pi : (\Gamma, x : \sigma) \to \Gamma$ against itself. Hence this is a diagonal as associated with the comprehension category $\mathcal{D}^{\to} \hookrightarrow \mathbb{C}^{\to}$ in Definition 9.3.5 (ii).

(ii) Check that the associated pullback functor δ^* performs contraction.

Assume now that we have finite product types $(1, \times)$ in our calculus—the latter for example because we have dependent sums Σ.

(iii) Show that *weak* equality types corresponds to left adjoints to these contraction functors δ^* plus Beck-Cevalley, as formulated in Definition 9.3.5 (ii).

(iv) Prove that *strong* equality types correspond to left adjoints as in (iii), satisfying the additional property that the canonical map κ in

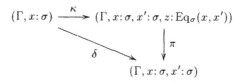

is an isomorphism.

(Note the similarity with the formulations of 'strong equality' in Propositions 10.3.3 (iii) and 4.9.3.)

10.4 Display maps and comprehension categories

In the previous section we have defined a (term model) category of contexts in dependent type theory, and we have seen how a distinguished class of "projection" or "display" maps $\pi : (\Gamma, x : \sigma) \to \Gamma$ in this category allowed us to describe the type constructors $1, \Pi, \Sigma, \mathsf{Eq}$ categorically. In this section we shall describe the notion of a "display map category" following [329, 330, 148], which captures this situation abstractly. In a next generalisation step, we shall describe the relevant context structure in terms of comprehension categories— which were used earlier in Section 9.3 to give a general notion of quantification. Display map categories form instances of such comprehension categories. We close this section with the notion of "comprehension category with unit"; essentially it is a fibration with subset types—as given by a right adjoint to the terminal object functor for preordered fibrations in Section 4.6—except that the preorderedness restriction is lifted.

10.4.1. Definition. A **display map category** (or a **category with display maps**) is a pair $(\mathbb{B}, \mathcal{D})$ where $\mathbb{B}$ is a category with a terminal object 1, and $\mathcal{D} \subseteq \mathrm{Arr}\,\mathbb{B}$ is a collection of morphisms in $\mathbb{B}$, called **display maps** or **projections** which is closed under pullback: for each family $\begin{pmatrix} X \\ \downarrow \varphi \\ I \end{pmatrix} \in \mathcal{D}$ and each morphism $u: J \to I$ in $\mathbb{B}$, a pullback square

$$
\begin{array}{ccc}
X' & \xrightarrow{\;\;u'\;\;} & X \\
{\scriptstyle u^*(\varphi)}\big\downarrow & \raisebox{1ex}{$\lrcorner$} & \big\downarrow{\scriptstyle \varphi} \\
J & \xrightarrow[\;\;u\;\;]{} & I
\end{array}
$$

exists in $\mathbb{B}$, and for every such pullback square one has that $u^*(\varphi) \in \mathcal{D}$.

We write $\mathcal{D}^{\to}$ for the full subcategory of the arrow category $\mathbb{B}^{\to}$ with display maps as objects, and $\begin{smallmatrix} \mathcal{D}^{\to} \\ \downarrow \\ \mathbb{B} \end{smallmatrix}$ for the associated codomain fibration. The fibre over $I \in \mathbb{B}$ will be written as $\mathcal{D}/I$ using a slice notation.

For such a display map category $(\mathbb{B}, \mathcal{D})$ we formulate the following additional conditions.

(**unit**) All isomorphisms in $\mathbb{B}$ are in $\mathcal{D}$.

(**product**) For each display map $\varphi: X \to I$, the pullback functor $\varphi^*: \mathcal{D}/I \to \mathcal{D}/X$ has a right adjoint $\prod_\varphi$ and a Beck-Chevalley condition holds: for a pullback square as above, the canonical natural transformation $u^* \prod_\varphi \Rightarrow \prod_{u^*(\varphi)} u'^*$ is an isomorphism.

(**weak sum**) For each display map φ, the pullback functor φ^* has a left adjoint $\coprod_\varphi$ plus Beck-Chevalley.

(**strong sum**) Display maps are closed under composition. This implies (weak sum).

(**weak equality**) For each display map $\varphi: X \to I$, write $\delta(\varphi) = \langle \mathrm{id}, \mathrm{id} \rangle: X \to X \times_I X$ for the mediating diagonal of the pullback of φ against itself. The requirement is then that each $\delta(\varphi)^*: \mathcal{D}/(X \times_I X) \to \mathcal{D}/X$ has a left adjoint Eq_φ, plus a Beck-Chevalley condition (see Exercise 10.4.2 below for the details).

(**strong equality**) Each such diagonal $\delta(\varphi)$ is a display map. The condition (weak equality) can then be proved.

A display map category is called a **relatively Cartesian closed category** (RCCC, for short) if it satisfies (unit), (product) and (strong sum).

Notice that the class $\mathcal{D}$ in this definition is required to be closed under *all* pullbacks—and not under particular chosen ones. As a result, the class $\mathcal{D}$ is closed under (vertical) isomorphisms. One could have relaxed the condition so that $\mathcal{D}$ is required to be closed under certain chosen pullbacks only, but these pullbacks should then be such that they compose, in order to get appropriate substitution properties—and a fibration. But requiring particular pullbacks to compose is not so natural, and such matters are better handled by using fibrations from the start, as is done with comprehension categories. Notice by the way that in the term model $\mathbb{C}$ in the previous section we do have distinguished pullbacks, which compose, see Lemma 10.3.1. In the comprehension category version of the term model (see below) this happens because we have a split fibration, and this splitting induces the particular pullbacks of projections.

By closing the class of context projection maps $\left(\begin{array}{c} \Gamma, x{:}\sigma \\ \downarrow \\ \Gamma \end{array} \right)$ under vertical isomorphisms we do get a (term model) display map category. The rôle of the above conditions on display map categories can be seen in Proposition 10.3.3 and Exercise 10.3.3. Following [329, 330], display map categories with such conditions have been used in many places, see *e.g.* [148, 272, 320, 185].

We notice that the strong versions of dependent sum and equality have much easier formulations than the weak ones. This will be different for comprehension categories, see Definitions 10.5.1 and 10.5.2 in the next section.

Finally, we note the following similarity. The general categorical description of simple type theories was given in Section 2.4 in terms of CT-structures. The latter are pairs $(\mathbb{B}, T)$ where $\mathbb{B}$ is a category (of contexts) with finite products and $T \subseteq \mathrm{Obj}\,\mathbb{B}$ is a collection of types, considered as singleton contexts. For dependently typed calculi, we do not use a subcollection $T \subseteq \mathrm{Obj}\,\mathbb{B}$ of objects, but a subcollection $\mathcal{D} \subseteq \mathrm{Arr}\,\mathbb{B} = \mathrm{Obj}\,(\mathbb{B}^{\rightarrow})$ of morphisms. More formally, we do not use subfibrations of simple fibrations, but subfibrations of codomain fibrations. Hence CT-structures and display map categories form similar generalisations of these type theoretic (simple and codomain) fibrations.

(Display map categories are clearly more general than CT-structures, since every CT-structure $(\mathbb{B}, T)$ gives rise to an associated collection of display maps given by constant families $I^*(X) = \left(\begin{array}{c} I \times X \\ \downarrow \pi \\ I \end{array} \right)$ for $I \in \mathbb{B}$ and $X \in T$.)

Comprehension categories

The above notion of display map category is not entirely satisfactory, because it does not give a central position to types, but instead to the projections *induced by* types. Therefore, we shall be using certain presentations of display map categories, called 'comprehension categories' in [155, 157]. These were used earlier in Section 9.3 to give a general form of quantification along certain "projection maps". These projection maps will play the rôle of the display maps above. This double rôle of comprehension categories—providing both domains of quantification and context structure for type dependency—will turn out to be very convenient (see *e.g.* in Definition 10.5.1 in the next section).

We consider again the syntactically constructed category $\mathbb{C}$ of contexts from the previous section for explaining the intuition behind the use of comprehension categories for modelling type dependency. We already have a category $\mathbb{C}$ of contexts, see the beginning of Section 10.3. But one can also form a category $\mathbb{T}$ of types, as follows.

objects   types $\Gamma \vdash \sigma : \mathsf{Type}$ in context.

morphisms $(\Gamma \vdash \sigma : \mathsf{Type}) \to (\Delta \vdash \tau : \mathsf{Type})$ are pairs $(\vec{M}, N)$ with $\vec{M} : \Gamma \to \Delta$ a context morphism in $\mathbb{C}$ and N a term

$$\Gamma, x : \sigma \vdash N : \tau[\vec{M}/\vec{v}]$$

where $\vec{v}$ is the sequence of variables declared in Δ.

The obvious projection functor $\begin{smallmatrix} \mathbb{T} \\ \downarrow \\ \mathbb{C} \end{smallmatrix}$ mapping a type $\Gamma \vdash \sigma : \mathsf{Type}$ to its underlying context Γ is then a split fibration. Cartesian (split) morphisms are of the form $(\vec{M}, x)$ where x is a variable. This fibration actually comes about by applying the Grothendieck construction to the functor $\mathbb{C}^{\mathrm{op}} \to \mathbf{Cat}$ which assigns to a context Γ the category $\mathbb{T}_\Gamma$ of types and terms in context Γ; it has types $\Gamma \vdash \sigma : \mathsf{Type}$ as objects and terms $\Gamma, x : \sigma \vdash N : \tau$ as morphisms $(\Gamma \vdash \sigma : \mathsf{Type}) \to (\Gamma \vdash \tau : \mathsf{Type})$.

The projection morphisms $\pi : (\Gamma, x : \sigma) \to \Gamma$ in the category $\mathbb{C}$ of contexts now arise in the following way. There is a functor $\mathcal{P}$ from the category $\mathbb{T}$ of types to the arrow category $\mathbb{C}^\to$, which maps a type $\Gamma \vdash \sigma : \mathsf{Type}$ to its corresponding projection morphisms $\pi : (\Gamma, x : \sigma) \to \Gamma$ in $\mathbb{C}$. Here one sees how the types are taken as primitive, and the associated projections as induced by a functor acting on types. A morphism $(\vec{M}, N) : (\Gamma \vdash \sigma : \mathsf{Type}) \to (\Delta \vdash \tau : \mathsf{Type})$ in the category $\mathbb{T}$ of types is sent by $\mathcal{P}$ to the following commuting square in the category $\mathbb{C}$ of contexts.

$$
\begin{array}{ccc}
(\Gamma, x : \sigma) & \xrightarrow{\ (M_1 \circ \pi, \ldots, M_n \circ \pi, N)\ } & (\Delta, y : \tau) \\
{\scriptstyle \pi} \downarrow & & \downarrow {\scriptstyle \pi} \\
\Gamma & \xrightarrow[\ (M_1, \ldots, M_n)\ ]{} & \Delta
\end{array}
$$

The functor $\mathcal{P}\colon \mathbb{T} \to \mathbb{C}^{\to}$ that we construct in this way satisfies the following four properties.

(1) The composite $\mathrm{cod} \circ \mathcal{P}\colon \mathbb{T} \to \mathbb{C}$ of $\mathcal{P}$ with the codomain functor $\mathrm{cod}\colon \mathbb{C}^{\to} \to \mathbb{C}$ is a fibration, namely the fibration $\begin{smallmatrix} \mathbb{T} \\ \downarrow \\ \mathbb{C} \end{smallmatrix}$ of dependent types over their contexts.

(2) The functor $\mathcal{P}$ sends Cartesian morphisms in $\mathbb{T}$ to pullback squares in $\mathbb{C}$. This is how the pullback squares in Lemma 10.3.1 arise.

(3) The functor $\mathcal{P}$ is full and faithful.

(4) The base category $\mathbb{C}$ has a terminal object (namely the empty context).

Points (1) and (4) are obvious. As for (2), a Cartesian morphism $(\vec{M}, x)$ in $\mathbb{T}$ is sent to a pullback square $(*)$ as described in Lemma 10.3.1. And (3) is left as an exercise below.

We conclude that this functor $\mathcal{P}\colon \mathbb{T} \to \mathbb{C}^{\to}$ gives a presentation of the class of display maps $\pi\colon (\Gamma, x\colon \sigma) \to \Gamma$ in $\mathbb{C}$, that we described directly in the previous section. The pullbacks in Lemma 10.3.1 arise by applying $\mathcal{P}$ to split morphisms in $\mathbb{T}$. We can now see that these pullbacks compose (in $\mathbb{C}^{\to}$) precisely because split morphisms compose in $\mathbb{T}$. This structure in $\mathbb{C}$ is thus induced by the structure in $\mathbb{T}$ via the functor $\mathcal{P}$. Since $\mathcal{P}$ is a full and faithful functor, everything that we want for these display maps (like the various conditions in Definition 10.4.1) can be described for the fibration $\begin{smallmatrix} \mathbb{T} \\ \downarrow \\ \mathbb{C} \end{smallmatrix}$. For convenience we repeat (from Theorem 9.3.4) the two conditions defining a comprehension category.

10.4.2. Definition. A **comprehension category** is a functor of the form $\mathcal{P}\colon \mathbb{E} \to \mathbb{B}^{\to}$ for which

(i) the composite $\mathrm{cod} \circ \mathcal{P}\colon \mathbb{E} \to \mathbb{B}^{\to} \to \mathbb{B}$ is a fibration;

(ii) if f is a Cartesian morphism in $\mathbb{E}$ with respect to this fibration $\begin{smallmatrix} \mathbb{E} \\ \downarrow \\ \mathbb{B} \end{smallmatrix}$, then $\mathcal{P}(f)$ is a pullback square in $\mathbb{B}$.

Such a comprehension category $\mathcal{P}$ will be called **full** if $\mathcal{P}$ is a full and faithful functor $\mathbb{E} \to \mathbb{B}^{\to}$. And it is called **split** (or **cloven**) whenever the fibration $\mathrm{cod} \circ \mathcal{P}\colon \mathbb{E} \to \mathbb{B}$ is split (or cloven).

This definition captures the above points (1) and (2). Points (3) and (4) are not part of the definition, since they are not relevant for the use of comprehension categories in quantification. In models of type theories however, full and faithfulness and a terminal object in the basis will always be there.

We thus see that the above constructions yield a "term model" full comprehension category $\mathbb{T} \to \mathbb{C}^{\to}$. And for every display map category $(\mathbb{B}, \mathcal{D})$, where $\mathcal{D}$ is a class of maps in $\mathbb{B}$ closed under pullback, the inclusion $\mathcal{D}^{\to} \hookrightarrow \mathbb{B}^{\to}$ is an instance of a full comprehension category.

We recall from Section 9.3 some important examples of (full) comprehension categories.

(i) The identity functor $\mathbb{B}^{\to} \to \mathbb{B}^{\to}$ for a category $\mathbb{B}$ with pullbacks forms the identity comprehension category. This example arises from the display map category $(\mathbb{B}, \mathcal{D})$ where $\mathcal{D}$ contains all morphisms from $\mathbb{B}$.

(ii) The simple comprehension category $s(\mathbb{B}) \to \mathbb{B}^{\to}$ for a category $\mathbb{B}$ with Cartesian products $\times$. It maps an object $(I, X) \in s(\mathbb{B})$ to the Cartesian projection $\begin{pmatrix} I \times X \\ \downarrow \\ I \end{pmatrix}$. In this example there is no real type dependency.

(iii) The subobject comprehension category $\mathrm{Sub}(\mathbb{B}) \to \mathbb{B}^{\to}$ for a category $\mathbb{B}$ with pullbacks. It takes a subobject to a representing arrow in $\mathbb{B}$.

(iv) Recall from Definition 7.3.5 that a *full* internal category $\mathbf{C}$ in a base category $\mathbb{B}$ has a full and faithful fibred functor $\mathrm{Fam}(\mathbf{C}) \to \mathbb{B}^{\to}$ over $\mathbb{B}$. Hence $\mathbf{C}$ is full if and only if its externalisation is part of a full comprehension category.

But there are many more examples, see *e.g.* Exercise 9.5.6.

10.4.3. Notation and terminology. Let $\mathcal{P} \colon \mathbb{E} \to \mathbb{B}^{\to}$ be a comprehension category, and write $p = \mathrm{cod} \circ \mathcal{P} \colon \mathbb{E} \to \mathbb{B}$ for the fibration involved. For an object $X \in \mathbb{E}$ in the total category, the corresponding morphism $\mathcal{P}X$ in $\mathbb{B}$ will be called a **projection** or a **display map**. We therefore often write π_X for $\mathcal{P}X$ when this functor $\mathcal{P}$ is understood from the context. An induced reindexing functor $\pi_X^* = \mathcal{P}X^*$ will be called a **weakening functor**.

We write $\{-\} = \mathrm{dom} \circ \mathcal{P} \colon \mathbb{E} \to \mathbb{B}$ for the second functor $\mathbb{E} \to \mathbb{B}$ induced by $\mathcal{P}$—again whenever $\mathcal{P}$ is understood from the context. Thus for a morphism $f \colon X \to Y$ in $\mathbb{E}$ we get in $\mathbb{B}$ a commuting diagram written as

$$
\begin{array}{ccc}
\{X\} & \xrightarrow{\ \{f\}\ } & \{Y\} \\
{\scriptstyle \pi_X} \downarrow & & \downarrow {\scriptstyle \pi_Y} \\
pX & \xrightarrow[\ pf\ ]{} & pY
\end{array}
$$

This square is a pullback in case f is Cartesian. Besides weakening functors π_X^*, there are also **contraction** functors δ_X^* induced by the diagonal $\delta_X \colon \{X\} \to \{\pi_X^*(X)\}$ obtained in the pullback of π_X against itself (as special case of the lemma below).

Often we use the corresponding small Roman letters $p = \mathrm{cod} \circ \mathcal{P}$ for the fibration involved. Thus we write q for the fibration $\mathrm{cod} \circ \mathcal{Q}$ when $\mathcal{Q}$ is a comprehension category.

A section of a projection π_X—that is a morphism $s: pX \to \{X\}$ with $\pi_X \circ s = \mathrm{id}$—may be called a **term (of type X)**.

It is worthwhile to formulate the following abstract version of Lemma 10.3.1 explicitly.

10.4.4. Lemma. *Let $\mathcal{P}: \mathbb{E} \to \mathbb{B}^{\to}$ be a comprehension category. Its projections are stable under pullback in the following sense. For each object $X \in \mathbb{E}$ and morphism $u: I \to pX = (\mathrm{cod} \circ \mathcal{P})(X)$ in $\mathbb{B}$ there is a projection with codomain I and a pullback of the form:*

$$
\begin{array}{ccc}
\{u^*(X)\} & \xrightarrow{\ \{\overline{u}(X)\}\ } & \{X\} \\
{\scriptstyle \pi_{u^*(X)}}\big\downarrow & \raisebox{1ex}{$\lrcorner$} & \big\downarrow{\scriptstyle \pi_X} \\
I & \xrightarrow[\ u\ ]{} & pX
\end{array}
$$

Proof. Since $p = \mathrm{cod} \circ \mathcal{P}: \mathbb{E} \to \mathbb{B}$ is a fibration, we can choose a Cartesian morphism $\overline{u}(X): u^*(X) \to X$ over u. It is mapped by $\mathcal{P}$ to the above pullback square in $\mathbb{B}$. □

This result allows us to describe substitution in terms. Consider a morphism $u: J \to I$ in the basis $\mathbb{B}$ of a comprehension category $\mathcal{P}: \mathbb{E} \to \mathbb{B}^{\to}$, and a term $s: I \to \{X\}$ of type $X \in \mathbb{E}_I$. Then we can define a term $u^{\#}(s)$ of type $u^*(X) \in \mathbb{E}_J$ as the unique mediating map $u^{\#}(s): J \dashrightarrow \{u^*(X)\}$ for the above pullback, with $\pi_{u^*(X)} \circ u^{\#}(s) = \mathrm{id}_J$ and $\{\overline{u}(X)\} \circ u^{\#}(s) = s \circ u$.

As a result of the lemma, projections of a comprehension category thus give rise to a display map category.

10.4.5. Corollary. *For a comprehension category $\mathcal{P}: \mathbb{E} \to \mathbb{B}^{\to}$, write $[\mathcal{P}] \subseteq \mathrm{Arr}\,\mathbb{B}$ for the image of $\mathcal{P}$ closed under vertical isomorphisms. That is, $[\mathcal{P}]$ is smallest class containing all projections $\mathcal{P}X$ and satisfying: $\varphi \cong \psi$ in $\mathbb{B}/I$ and $\varphi \in [\mathcal{P}]$ implies $\psi \in [\mathcal{P}]$. Then $(\mathbb{B}, [\mathcal{P}])$ is a display map category.* □

And the pullback in the above lemma enables a form of dependent tupleing.

10.4.6. Corollary. *For a comprehension category $\mathcal{P}: \mathbb{E} \to \mathbb{B}^{\to}$ we have an isomorphism*

$$
\mathbb{B}\big(I, \{X\}\big) \cong \coprod_{u: I \to pX} \{s \mid s \text{ is a term of type } u^*(X)\}. \qquad \square
$$

In type theoretic formulation, the functor $\{-\} = \mathrm{dom} \circ \mathcal{P}$ of a comprehension category $\mathcal{P}$ sends a type $\Gamma \vdash \sigma: \mathsf{Type}$ to the corresponding extended context $(\Gamma, x: \sigma)$. It is precisely this operation which is difficult in dependent type theory, since one cannot use Cartesian products $\Gamma \times \sigma$ as in simple of polymorphic type theory (where one has no type dependency). A form of disjoint union $\coprod_{\vec{v} \in \Gamma} . \sigma(\vec{v})$ is needed in DTT—as in the above corollary.

Comprehension categories with unit

The following definition gives rise to an important class of examples of comprehension categories.

10.4.7. Definition (See [74, 62]). A fibration $\begin{smallmatrix} \mathbb{E} \\ \downarrow p \\ \mathbb{B} \end{smallmatrix}$ with a terminal object functor $1 \colon \mathbb{B} \to \mathbb{E}$ is said to admit **comprehension** if this functor 1 has a right adjoint, which we commonly write as $\{-\} \colon \mathbb{E} \to \mathbb{B}$. We then have adjunctions

$$p \dashv 1 \dashv \{-\}.$$

In this situation we get a functor $\mathbb{E} \to \mathbb{B}^{\to}$ by $X \mapsto p(\varepsilon_X)$, where ε_X is the counit $1\{X\} \to X$ of the adjunction $(1 \dashv \{-\})$ at X. This functor is actually a comprehension category, see below. In such a situation, we shall call this functor a **comprehension category with unit**. And we shall say that p admits **full comprehension** if this induced comprehension category is full (*i.e.* if $\mathbb{E} \to \mathbb{B}^{\to}$ is a full and faithful functor).

A "fibration with subset types" as introduced in Definition 4.6.1 is thus simply a preorder fibration with comprehension. Many of the properties which hold for such fibrations with subset types also hold for fibrations with comprehension. But there is one important exception: the fact that the induced projections $\pi_X \colon \{X\} \to pX$ are monos for fibrations with subset types crucially depends on preorderedness, see the proof of Lemma 4.6.2 (i).

We check that the functor $\mathbb{E} \to \mathbb{B}^{\to}$ as in the definition is a comprehension category indeed. Therefore we need to show that for a Cartesian morphism $f \colon X \to Y$ in $\mathbb{E}$, the naturality square

$$
\begin{array}{ccc}
\{X\} & \xrightarrow{\{f\}} & \{Y\} \\
{\scriptstyle \pi_X = p(\varepsilon_X)} \downarrow & & \downarrow {\scriptstyle \pi_Y = p(\varepsilon_Y)} \\
pX & \xrightarrow[pf]{} & pY
\end{array}
$$

is a pullback in the base category $\mathbb{B}$. If morphisms $u \colon I \to pX$ and $v \colon I \to \{Y\}$ are given with $pf \circ u = \pi_Y \circ v$, then the transpose $v^\wedge = \varepsilon_Y \circ 1v \colon 1I \to Y$ satisfies

$$p(v^\wedge) = p(\varepsilon_Y) \circ p1(v) = \pi_Y \circ v = pf \circ u.$$

Since f is Cartesian, we get a unique map $g: 1I \to X$ in $\mathbb{E}$ over u with $f \circ g = v^\wedge$. But then, taking the transpose $g^\vee = \{g\} \circ \eta_I : I \to \{X\}$, yields the required mediating morphism:

$$\pi_X \circ g^\vee = \pi_X \circ \{g\} \circ \eta_I = p(g) \circ \pi_{1I} \circ \eta_I = u \circ p(\varepsilon_{1I}) \circ p1(\eta_I) = u$$

and

$$\{f\} \circ g^\vee = \{f \circ g\} \circ \eta_I = \{v^\wedge\} \circ \eta_I = v^{\wedge\vee} = v.$$

Uniqueness is left to the reader.

We recall that a terminal object functor 1 is full and faithful, because the counit $p1 \Rightarrow \mathrm{id}$ of the adjunction $(p \dashv 1)$ is the identity, see Lemma 1.8.8. Therefore, the unit components $I \to \{1I\}$ of the adjunction $(1 \dashv \{-\})$ are isomorphisms.

Notice that being a comprehension category with unit is a *property* of a fibration; namely that it has a fibred terminal object and that the resulting (terminal object) functor has a right adjoint. In contrast, an arbitrary comprehension category $\mathcal{P}: \mathbb{E} \to \mathbb{B}^\to$ provides the underlying fibration $p = \mathrm{cod} \circ \mathcal{P}$ with certain *structure*. Therefore we may say that a fibration 'is' a comprehension category with unit.

This description of comprehension via a right adjoint to the terminal object functor is a simplification of the description originally proposed by Lawvere [193], see also Exercise 4.6.7. A slightly different notion of comprehension is proposed by Pavlović [252]. See [157] for a comparison of these notions.

10.4.8. Examples. (i) For a category $\mathbb{B}$ with finite products $(\times, 1)$, the simple comprehension category $s(\mathbb{B}) \to \mathbb{B}^\to$ is a comprehension category with unit. The underlying simple fibration $\begin{smallmatrix} s(\mathbb{B}) \\ \downarrow \\ \mathbb{B} \end{smallmatrix}$ has a terminal object functor $\mathbb{B} \to s(\mathbb{B})$, namely $I \mapsto (I, 1)$. This functor has a right adjoint, given by $(I, X) \mapsto I \times X$. The resulting projection $\pi_{(I,X)}$ (as in the above definition) is the Cartesian projection $\pi: I \times X \to I$.

(ii) For a category $\mathbb{B}$ with pullbacks, the identity functor $\mathbb{B}^\to \to \mathbb{B}^\to$ is a comprehension category with unit, since there are the following two basic adjunctions

$$\mathrm{cod} \left(\dashv \Big\uparrow \dashv \right) \mathrm{dom}$$

with $\mathbb{B}^\to$ above and $\mathbb{B}$ below,

where the up-going arrow in the middle is the terminal object functor $I \mapsto \mathrm{id}_I$ for the underlying codomain fibration.

(iii) Let $\mathbb{C}$ be a category with terminal object $1 \in \mathbb{C}$, and small hom-sets $\mathbb{C}(1, X)$. The family fibration $\begin{smallmatrix} \mathsf{Fam}(\mathbb{C}) \\ \downarrow \\ \mathsf{Sets} \end{smallmatrix}$ then has a terminal object functor $1: \mathbf{Sets} \to \mathsf{Fam}(\mathbb{C})$ by $J \mapsto (1)_{j \in J}$. And 1 has a right adjoint by disjoint union:

$$(X_i)_{i \in I} \mapsto \coprod_{i \in I} \mathbb{C}(1, X_i) = \{(i, x) \mid i \in I \text{ and } x: 1 \to X_i\}.$$

This will be called the **family comprehension category** $\mathsf{Fam}(\mathbb{C}) \to \mathbf{Sets}^{\to}$. It is rarely full, see Exercise 10.4.4 (ii). Notice that the special case where $\mathbb{C} = \mathbf{Sets}$ yields the equivalence $\mathsf{Fam}(\mathbf{Sets}) \overset{\simeq}{\to} \mathbf{Sets}^{\to}$ from Proposition 1.2.2.

(iv) For a calculus of dependent types with a unit type $1: \mathsf{Type}$ we form the term model fibration $\begin{smallmatrix} \mathbb{T} \\ \downarrow \\ \mathbb{C} \end{smallmatrix}$ of dependent types $\Gamma \vdash \sigma: \mathsf{Type}$ over their contexts Γ, as described earlier in this section. This fibration then has a fibred terminal object $\Gamma \vdash 1: \mathsf{Type}$ in the fibre over context Γ. The associated functor $1: \mathbb{C} \to \mathbb{T}$ has a right adjoint, which maps a type $\Gamma \vdash \sigma: \mathsf{Type}$ to the extended context $(\Gamma, x: \sigma)$. We have an adjoint correspondence

$$\frac{(\Gamma \vdash 1: \mathsf{Type}) \longrightarrow (\Delta \vdash \tau: \mathsf{Type}) \quad \text{in } \mathbb{T}}{\Gamma \longrightarrow (\Delta, y: \tau) \quad \text{in } \mathbb{C}}$$

amounting to an (obvious) correspondence between

$$\frac{\text{context morphisms } \vec{M}: \Gamma \longrightarrow \Delta \text{ with a term } \Gamma, x: 1 \vdash N: \tau[\vec{M}/\vec{v}]}{\text{context morphisms } (\vec{P}, Q): \Gamma \longrightarrow (\Delta, y: \tau)}$$

The functor $\mathbb{T} \to \mathbb{C}^{\to}$ induced by this adjunction maps a type $\Gamma \vdash \sigma: \mathsf{Type}$ to the associated projection $\pi: (\Gamma, x: \sigma) \to \Gamma$. (But these projections can of course be described directly, without using the unit type.)

(v) As already mentioned, every fibration with subset types is a comprehension category with unit. Notice that the latter is full if and only if the fibration has full subset types (as defined in Section 4.6). In particular, every subobject fibration $\begin{smallmatrix} \mathsf{Sub}(\mathbb{B}) \\ \downarrow \\ \mathbb{B} \end{smallmatrix}$ forms a full comprehension category with unit $\mathsf{Sub}(\mathbb{B}) \to \mathbb{B}^{\to}$.

It is not the case that a comprehension category with a fibred terminal object for its underlying fibration is automatically a comprehension category with unit. As counter example, consider the family fibration $\begin{smallmatrix} \mathsf{Fam}(\mathbf{Sets}_{\bullet}) \\ \downarrow \\ \mathbf{Sets} \end{smallmatrix}$ of set-indexed-pointed sets. This fibration has a terminal object functor $1: \mathbf{Sets} \to \mathsf{Fam}(\mathbf{Sets}_{\bullet})$ since the category $\mathbf{Sets}_{\bullet}$ has a terminal object $\{\bullet\}$—which, at the same time, is initial object. The fibration also carries a comprehension category structure $\mathrm{Fam}(\mathbf{Sets}_\bullet) \to \mathbf{Sets}^\to$, namely

$$(X_i)_{i \in I} \mapsto \left(\begin{array}{c} \coprod_{i \in I} X_i \\ \downarrow \\ I \end{array} \right)$$

In this situation we do not have that the functor $(X_i)_{i \in I} \mapsto \coprod_{i \in I} X_i$ is right adjoint to the terminal object functor, since morphisms $(1)_{j \in J} \to (X_i)_{i \in I}$ in $\mathrm{Fam}(\mathbf{Sets}_\bullet)$ correspond simply to functions $J \to I$, and not to functions $J \to \coprod_{i \in I} X_i$.

(But the fibration $\begin{array}{c} \mathrm{Fam}(\mathbf{Sets}_\bullet) \\ \downarrow \\ \mathbf{Sets} \end{array}$ has a fibred monoidal structure given point-wise by the monoidal structure on pointed sets $\mathbf{Sets}_\bullet$ (with "smash" product as tensor and $2 = \{0,1\}$ as neutral element). The comprehension functor $(X_i)_{i \in I} \mapsto \coprod_{i \in I} X_i$ is a right adjoint, namely to the neutral element functor $2 \colon \mathbf{Sets} \to \mathrm{Fam}(\mathbf{Sets}_\bullet)$.)

The inclusion $\mathbf{Fib} \hookrightarrow \mathbf{Cat}^\to$ forms a comprehension category on the fibration of fibrations over their base categories (see Lemma 1.7.2). The latter does admit comprehension, given by a right adjoint $\left(\begin{array}{c} \mathbb{E} \\ \downarrow \\ \mathbb{B} \end{array} \right) \mapsto \mathrm{Cart}(\mathbb{E})$ to the terminal object functor, but this gives a different comprehension structure. A similar fibration $\begin{array}{c} \mathrm{Fam}(\mathbf{Cat}) \\ \downarrow \\ \mathbf{Cat} \end{array}$ of $\mathbf{Cat}$-valued presheaves is shown not to admit comprehension in [106]. (But $\mathbf{Sets}$-valued presheaves do admit comprehension, see Exercise 10.5.1.)

Next we reformulate the main points of Lemma 4.6.2, adapted to the present setting. The proofs are as in Section 4.6.

10.4.9. Lemma. *Let* $\begin{array}{c} \mathbb{E} \\ \downarrow{\scriptstyle p} \\ \mathbb{B} \end{array}$ *be a fibration with comprehension.*

(i) *Fix an object* $J \in \mathbb{B}$; *for maps* $u \colon I \to J$ *in* $\mathbb{B}$ *and objects* $Y \in \mathbb{E}$ *over* J, *there is an isomorphism*

$$\mathbb{E}_I\left(1I,\, u^*(Y)\right) \cong \mathbb{B}/J\left(u,\, \pi_Y\right).$$

Moreover, these isomorphisms are natural in u *and* Y *—when both sides are seen as functors* $(\mathbb{B}/J)^{\mathrm{op}} \times \mathbb{E}_J \rightrightarrows \mathbf{Sets}.$)

(ii) *The induced functor* $\mathbb{E} \to \mathbb{B}^\to$ *preserves all fibred limits. It also preserves products* $\prod$.

As a result of (i), terms of type $X \in \mathbb{E}_I$, which were defined as sections of the projection $\pi_X \colon \{X\} \to I$ in Notation 10.4.3 above, can be identified in a fibration admitting comprehension with the global sections $1(I) \to X$ in the fibre over I. This description of terms as maps in fibres is often more convenient.

Proof. We only prove the last statement about preservation of products $\prod$, say with respect to a comprehension category $Q\colon \mathbb{D} \to \mathbb{B}^{\to}$. For $A \in \mathbb{D}$, we have a projection QA in $\mathbb{B}$, say with domain and codomain $QA\colon J \to K$. For a morphism $u\colon I \to K$ we can form a pullback square

$$
\begin{array}{ccc}
L = \mathrm{dom}Q(u^*(A)) & \xrightarrow{\ \ u' = QA^*(u)\ \ } & \mathrm{dom}QA = J \\[2pt]
{\scriptstyle Q(u^*(A))}\Big\downarrow \ \lrcorner & & \Big\downarrow {\scriptstyle QA} \\[6pt]
I & \xrightarrow[\quad u \quad]{} & \mathrm{cod}QA = K
\end{array}
$$

The assumed right adjoint $\prod_A$ to QA^* in $\mathbb{E}$ satisfies, by (i):

$$
\begin{aligned}
\mathbb{B}/K\big(u,\ \pi_{\prod_A(X)}\big)
&\cong \mathbb{E}_I\big(1I,\ u^*(\textstyle\prod_A(X))\big) \\
&\cong \mathbb{E}_I\big(1I,\ \textstyle\prod_{u^*(A)} u'^*(X)\big) \qquad \text{by Beck-Chevalley} \\
&\cong \mathbb{E}_L\big(Q(u^*(A))^*(1I),\ u'^*(X)\big) \\
&\cong \mathbb{E}_L\big(1L,\ u'^*(X)\big) \\
&\cong \mathbb{B}/J\big(QA^*(u),\ \pi_X\big).
\end{aligned}
$$

Hence the projection $\pi_{\prod_A(X)}$ in $\mathbb{B}/K$ behaves like a product $\prod_{QA}(\pi_X)$ in the arrow fibration on $\mathbb{B}$. $\qquad\square$

Finally we have the following alternative "representability" description of comprehension categories with unit.

10.4.10. Proposition. *Let* $\begin{smallmatrix}\mathbb{E}\\ \downarrow p\\ \mathbb{B}\end{smallmatrix}$ *be a fibration with fibred terminal object. Then p admits comprehension if and only if fibred global sections are representable; that is, if for each $X \in \mathbb{E}$, say above $I \in \mathbb{B}$, the functor*

$$
(\mathbb{B}/I)^{\mathrm{op}} \longrightarrow \mathbf{Sets} \qquad \text{given by} \qquad \big(J \xrightarrow{\ u\ } I\big) \longmapsto \mathbb{E}_J\big(1J,\ u^*(X)\big)
$$

is representable.

Proof. (only if) For $X \in \mathbb{E}_I$, the projection $\pi_X\colon \{X\} \to I$ is a representing arrow, since for $u\colon J \to I$ one has $\mathbb{B}/I(u, \pi_X) \cong \mathbb{E}_J(1J, u^*(X))$, by (i) in the previous result.

(if) Choose for each object $X \in \mathbb{E}$ a representing arrow π_X and write $\{X\}$ for its domain. Then

$$
\begin{aligned}
\mathbb{E}\big(1J,\, X\big) &\cong \coprod_{u:J \to I} \mathbb{E}_J\big(1J,\, u^*(X)\big) \quad \text{by Lemma 1.4.10} \\
&\cong \coprod_{u:J \to I} \mathbb{B}/I\big(u,\, \pi_X\big) \\
&\cong \mathbb{B}\big(J,\, \{X\}\big).
\end{aligned}
$$

So the terminal object functor $1 \colon \mathbb{B} \to \mathbb{E}$ has a right adjoint $\{-\}$. □

This result provides a link with local smallness for fibrations (which was already announced as Lemma 9.5.2).

10.4.11. Corollary. *Let* $\begin{smallmatrix}\mathbb{E}\\ \downarrow p\\ \mathbb{B}\end{smallmatrix}$ *be a fibred CCC. Then p is a comprehension category with unit if and only if p is locally small.*

Proof. Since each fibre is a CCC, vertical morphisms $X \to Y$ correspond to global sections $1 \to (X \Rightarrow Y)$. So the former are representable if and only if the latter are. □

Exercises

10.4.1. Let $(\mathbb{B}, \mathcal{D})$ be a display map category. Prove that if there is a family $\left(\begin{smallmatrix}X\\ \downarrow\\ 1\end{smallmatrix}\right) \in \mathcal{D}$ which is inhabited (*i.e.* for which there is a map $1 \to X$), then a terminal object functor $\mathbb{B} \to \mathcal{D}^{\to}$ is automatically left adjoint to the domain functor $\mathcal{D}^{\to} \to \mathbb{B}$.

10.4.2. Consider a display map category $(\mathbb{B}, \mathcal{D})$ which satisfies the (strong equality) condition in Definition 10.4.1 and in which the collection $\mathcal{D}$ is closed under Cartesian products (in slices $\mathbb{B}/I$). Show that the associated fibration admits equality with respect to the comprehension category $\mathcal{D}^{\to} \hookrightarrow \mathbb{B}^{\to}$, as described in Definition 9.3.5. Give an explicit formulation of the Beck-Chevalley condition (from this definition) for equality in display map categories.

10.4.3. Verify that the term model comprehension category $\mathcal{P} \colon \mathbb{T} \to \mathbb{C}^{\to}$ is full.

10.4.4. (i) Show in detail that one gets a "family" comprehension category with unit $\mathrm{Fam}(\mathbb{C}) \to \mathbf{Sets}^{\to}$ in Example 10.4.8 (iii).

(ii) Prove that $\begin{smallmatrix}\mathrm{Fam}(\mathbb{C})\\ \downarrow\\ \mathbf{Sets}\end{smallmatrix}$ is a full comprehension category if and only if the global sections functor $\mathbb{C}(1, -) \colon \mathbb{C} \to \mathbf{Sets}$ is full and faithful.

10.4.5. Verify that the functor $\Gamma \colon \mathrm{Fam}(\mathbf{C}) \to \mathbb{B}^{\to}$ in Exercise 7.3.5 forms a comprehension category with unit. (It gives an internal version of the family example described above.)

10.4.6. A **category with families** according to [72] (see also [137]) consists of a
category $\mathbb{C}$ with a terminal object together with

(a) a functor $F = (\mathrm{Ty}, \mathrm{Te}): \mathbb{C}^{\mathrm{op}} \to \mathrm{Fam}(\mathbf{Sets})$; for an object $I \in \mathbb{C}$ we
write $F(I) = \big(\mathrm{Te}(I, \sigma)\big)_{\sigma \in \mathrm{Ty}(I)}$, and for a morphism $u: J \to I$ in $\mathbb{C}$
we write $F(u)$ as consisting of functions $\mathrm{Ty}(u): \mathrm{Ty}(I) \to \mathrm{Ty}(J)$ and
$\mathrm{Te}(u): \mathrm{Te}(I, \sigma) \to \mathrm{Te}(J, \mathrm{Ty}(u)(\sigma))$, for each $\sigma \in \mathrm{Ty}(I)$;

(b) for each $I \in \mathbb{C}$ and $\sigma \in \mathrm{Ty}(I)$ a **comprehension**: an object
$I \cdot \sigma \in \mathbb{C}$ together with a morphism $p(\sigma): I \cdot \sigma \to I$ and an element
$v_\sigma \in \mathrm{Te}(I \cdot \sigma, \mathrm{Ty}(p(\sigma))(\sigma))$ such that: for each $u: J \to I$ in $\mathbb{C}$ and
$t \in \mathrm{Te}(J, \mathrm{Ty}(u)(\sigma))$ there is a unique morphism $\langle u, t\rangle: J \to I \cdot \sigma$ with
$p(\sigma) \circ \langle u, t\rangle = u$ and $\mathrm{Te}(\langle u, t\rangle)(v_\sigma) = t$.

Define for such a functor F a category $\mathbb{F}$ with

> **objects**   pairs (I, σ) with $I \in \mathbb{C}$ and $\sigma \in \mathrm{Ty}(I)$.
>
> **morphisms** $(I, \sigma) \to (J, \tau)$ are pairs (u, t) where $u: I \to J$ is a
> morphism in $\mathbb{C}$ and t is an element of $\mathrm{Te}(I \cdot \sigma, \mathrm{Ty}(u \circ p(\sigma))(\tau))$.

Define a split full comprehension category $\mathbb{F} \to \mathbb{C}^{\to}$.

10.4.7. Let $\begin{smallmatrix}\mathbb{E}\\\downarrow p\\\mathbb{B}\end{smallmatrix}$ be a fibration with comprehension. We write η and ε for the unit
and counit of the adjunction $(1 \dashv \{-\})$. Recall that the projection π_X is
defined as $p(\varepsilon_X)$.

(i) Show that the projection $\pi_{1I}: \{1I\} \to I$ is the inverse of the unit component η_I.

(ii) For $X \in \mathbb{E}$ above I, prove that

$$\{!_X\} = \eta_I \circ \pi_X: \{X\} \longrightarrow \{1I\}$$

where $!_X$ is the unique map $X \to 1I$ over I.

(iii) Prove that if $!_X: X \to 1I$ is a mono in $\mathbb{E}$, then the projection
$\pi_X: \{X\} \to I$ is a mono in $\mathbb{B}$.

10.4.8. Let $\begin{smallmatrix}\mathbb{E}\\\downarrow p\\\mathbb{B}\end{smallmatrix}$ be a fibration with fibred finite products and full comprehension.

(i) Prove that for $X, Y, Z \in \mathbb{E}$ in the same fibre, say over $I \in \mathbb{B}$, there is
a bijective correspondence

$$\frac{X \times Y \longrightarrow Z \quad \text{over } I}{\pi_X^*(Y) \longrightarrow \pi_X^*(Z) \quad \text{over } \{X\}}$$

natural in Y, Z.

(ii) Conclude that in such a situation, fibred colimits are automatically
distributive, *i.e.* preserved by functors $X \times (-)$.

10.4.9. Consider two contexts $\Gamma = (x_1: \sigma_1, \ldots, x_n: \sigma_n)$ and $\Delta = (y_1: \tau_1, \ldots, y_m: \tau_m)$
in DTT together with a context morphism $s = (M_1, \ldots, M_m): \Gamma \to \Delta$. It
yields a substitution functor $s^*: \mathbb{T}_\Delta \to \mathbb{T}_\Gamma$ in the reverse direction. Assume

weak sum and weak equality types. Then one can define for a type $\Gamma \vdash \rho$: Type

$$\textstyle\coprod_s(\rho) \;=\; \Sigma x_1\colon \sigma_1.\,\dots\,\Sigma x_n\colon \sigma_n.$$
$$\mathrm{Eq}_{\tau_1}(y_1, M_1) \times \cdots \times \mathrm{Eq}_{\tau_m}(y_m, M_m) \times \rho$$

which yields a type in context Δ.

(i) Prove that $\coprod_s$ is left adjoint to s^*.

(ii) Assume dependent products Π. Show that one can also define a right adjoint to s^* by

$$\textstyle\prod_s(\rho) \;=\; \Pi x_1\colon \sigma_1.\,\dots\,\Pi x_n\colon \sigma_n.$$
$$(\mathrm{Eq}_{\tau_1}(y_1, M_1) \times \cdots \times \mathrm{Eq}_{\tau_m}(y_m, M_m)) \to \rho.$$

10.4.10. Let $\mathcal{P}\colon \mathbb{E} \to \mathbb{B}^{\to}$ be a comprehension category. Write $\mathbb{E}^{\heartsuit}$ for the full image of $\mathcal{P}$, *i.e.* for the category with

objects $\quad X \in \mathbb{E}.$

morphisms $\quad X \to Y$ in $\mathbb{E}^{\heartsuit}$ are morphisms $\mathcal{P}X \to \mathcal{P}Y$ in $\mathbb{B}^{\to}$.

Define a *full* comprehension category $\mathcal{P}^{\heartsuit}\colon \mathbb{E}^{\heartsuit} \to \mathbb{B}^{\to}$. (It is the full completion of $\mathcal{P}$ and called the 'heart of $\mathcal{P}$' by Ehrhard.)

10.4.11. Let $\begin{smallmatrix}\mathbb{E}\\ \downarrow p\\ \mathbb{B}\end{smallmatrix}$ be a fibration with (full) comprehension and let $F\colon \mathbb{A} \to \mathbb{B}$ be a functor with a right adjoint, where $\mathbb{A}$ is a category with pullbacks. Prove that the fibration $F^*(p)$, obtained by change-of-base along F, also admits (full) comprehension.

[See also Exercise 9.5.5.]

10.5 Closed comprehension categories

From a semantical perspective, the most natural combination of type constructors in dependent type theory is: unit type 1, dependent product Π, and strong sum Σ. In the present section we introduce a categorical structure which combines these operations, and call it a "closed comprehension category". (Such a notion has also been identified in terms of display maps as a "relatively Cartesian closed category", (RCCC) see [329, 148], or Definition 10.4.1.) There are many instances of these closed comprehension categories, some of which will be presented in this section. The next section is devoted to two domain theoretic examples. There are many more examples, *e.g.* in [148] (based on "lim theories", forming categorical generalisations of domain theoretic examples), in [185] (generalisations Girard's qualitative domains), in [132, 133] (with so-called setoids), and in [137] (a presheaf model for proving a conservativity result).

What we have not defined yet is what it means for a comprehension category to have (dependent) products and coproducts. We take this to mean: the underlying fibration has products and coproducts with respect to the (projections of the) comprehension category itself. (Recall from Definition 9.3.5 that one can define quantification in a fibration with respect to a comprehension category.) Here we make convenient use of the double rôle of comprehension categories: in quantification and in modelling type dependency.

10.5.1. Definition. Let $\mathcal{P}: \mathbb{E} \to \mathbb{B}^{\to}$ be a comprehension category. We say that $\mathcal{P}$ admits **products** if its underlying fibration $\begin{smallmatrix} \mathbb{E} \\ \downarrow p \\ \mathbb{B} \end{smallmatrix}$ —where $p = \text{cod} \circ \mathcal{P}$—admits products with respect to the comprehension category $\mathcal{P}: \mathbb{E} \to \mathbb{B}^{\to}$; that is, if p has products $\pi_X^* \dashv \prod_X$ along $\mathcal{P}$'s projections $\pi_X = \mathcal{P}X$ (plus a Beck-Chevalley condition), see Definition 9.3.5.

Similarly we say that $\mathcal{P}$ admits **weak coproducts** if the fibration $\begin{smallmatrix} \mathbb{E} \\ \downarrow \\ \mathbb{B} \end{smallmatrix}$ has coproducts with respect to $\mathcal{P}$; this involves adjunctions $\coprod_X \dashv \pi_X^*$ plus Beck-Chevalley.

And $\mathcal{P}$ admits **weak equality** if $\begin{smallmatrix} \mathbb{E} \\ \downarrow \\ \mathbb{B} \end{smallmatrix}$ has equality with respect to $\mathcal{P}$. This involves left adjoints $\text{Eq}_X \dashv \delta_X^*$ along its own diagonals, again with Beck-Chevalley.

Next we should say what *strong* coproduct and equality are.

10.5.2. Definition. Let $\mathcal{P}: \mathbb{E} \to \mathbb{B}^{\to}$ be a comprehension category.

(i) We say that $\mathcal{P}$ admits **strong coproducts** if it has coproducts as above in such a way that the canonical maps κ are isomorphisms in diagrams of the form:

$$
\begin{array}{ccc}
\{Y\} & \xrightarrow[\cong]{\kappa} & \{\coprod_X(Y)\} \\
\pi \downarrow & & \downarrow \pi \\
\{X\} & \xrightarrow[\pi]{} & pX
\end{array}
$$

(ii) Similarly, $\mathcal{P}$ admits **strong equality** if there are canonical isomorphisms:

$$
\begin{array}{ccc}
\{Y\} & \xrightarrow[\cong]{\kappa} & \{\text{Eq}_X(Y)\} \\
\pi \downarrow & & \downarrow \pi \\
\{X\} & \xrightarrow[\delta]{} & \{\pi_X^*(X)\}
\end{array}
$$

The canonical maps $\{Y\} \to \{\coprod_X(Y)\}$ and $\{Y\} \to \{\text{Eq}_X(Y)\}$ in this definition arise by applying the functor $\{-\} = \text{dom} \circ \mathcal{P}$ to the (opcartesian) composites

$$
\begin{array}{ccc}
Y & & Y \\
\eta\downarrow & & \eta\downarrow \\
\pi_X^* \coprod_X(Y) \longrightarrow \coprod_X(Y) & \quad & \delta_X^* \mathrm{Eq}_X(Y) \longrightarrow \mathrm{Eq}_X(Y)
\end{array}
$$

Finally we come to the main notion of this section.

10.5.3. Definition. A **closed comprehension category** (CCompC) is a full comprehension category with unit, which admits products and strong coproducts, and which has a terminal object in its base category. It will be called split if all of its fibred structure is split.

The easiest model of type-indexed-types in dependent type theory uses set-indexed-sets, as formalised by the family fibration $\begin{smallmatrix}\mathbf{Fam(Sets)}\\\downarrow\\\mathbf{Sets}\end{smallmatrix}$. This fibration is a (split) CCompC: it has full comprehension plus products and coproducts along all morphisms in the base category, so certainly along projections π_X. Explicitly, for a family $X = (X_i)_{i\in I}$ over I and a family $Y = (Y_\alpha)_{\alpha\in\coprod_{i\in I} X_i}$ over $\{X\} = \coprod_{i\in I} X_i$ we have the standard formulas:

$$
\begin{aligned}
\textstyle\prod_X(Y)_i &= \{f\colon X_i \to \bigcup_{x\in X_i} Y_{(i,x)} \mid \forall x \in X_i.\, f(x) \in Y_{(i,x)}\} \\
\textstyle\coprod_X(Y)_i &= \{\langle x, y\rangle \mid x \in X_i \text{ and } y \in Y_i\}.
\end{aligned}
$$

(Interestingly, no equality is needed to define these adjoints along dependent projections, whereas one does need equality for adjoints to arbitrary substitution functors u^*, see the proof of Lemma 1.9.5. These dependent products and coproducts are natural extensions of the simple products and coproducts along Cartesian projections, see Lemma 1.9.2)

The above coproduct $\coprod$ is strong, since there is a (canonical) isomorphism

$$
\begin{aligned}
\{\textstyle\coprod_X(Y)\} &= \{\langle i, \langle x, y\rangle\rangle \mid i \in I, x \in X_i \text{ and } y \in Y_{(i,x)}\} \\
&\cong \{\langle\langle i, x\rangle, y\rangle \mid i \in I, x \in X_i \text{ and } y \in Y_{(i,x)}\} \\
&= \{\langle \alpha, y\rangle \mid \alpha \in \textstyle\coprod_{i\in I} X_i \text{ and } y \in Y_\alpha\} \\
&= \{Y\}.
\end{aligned}
$$

Instead of modelling dependent types as set-indexed-sets, one can also take objects of a category indexed by themselves, as formalised by a codomain fibration. The latter is a (non-split) CCompC if and only if the underlying category is locally Cartesian closed, see Theorem 10.5.5 (ii) below.

As mentioned after Lemma 10.4.9, terms of type $X \in \mathbb{E}_I$ in a comprehension category with unit $\begin{smallmatrix}\mathbb{E}\\\downarrow p\\\mathbb{B}\end{smallmatrix}$ can be identified with vertical maps

$1(I) \rightarrow X$. We sketch the interpretation of the introduction and elimination rules for Π and Σ in a CCompC as above. For Π-introduction, assume a term $I, x: X \vdash f: Y$, given as a vertical map $\pi_X^*(1(I)) \cong 1\{X\} \rightarrow Y$. Its transpose yields the abstraction map $\lambda_X(f): 1(I) \rightarrow \prod_X(Y)$. And for Π-elimination, we assume terms $g: 1(I) \rightarrow \prod_X(Y)$ and $h: 1(I) \rightarrow X$. Then we get a transpose $\pi_X^*(1(I)) \rightarrow Y$ of f, and also a transpose $h^\vee: I \rightarrow \{X\}$ of h across $(1 \dashv \{-\})$. These can be combined into an application term $gh: 1(I) \cong (h^\vee)^* \pi_X^*(1(I)) \rightarrow (h^\vee)^*(Y) = Y(h)$.

For sum-introduction, assume maps $f: 1(I) \rightarrow X$ and $g: 1(I) \rightarrow (f^\vee)^*(Y)$. By using the unit of the adjunction $(\coprod_X \dashv \pi_X^*)$ at Y we get a tuple term $\langle f, g \rangle: 1(I) \rightarrow (f^\vee)^*(Y) \rightarrow (f^\vee)^* \pi_X^* \coprod_X(Y) \cong \coprod_X(Y)$. And for (strong) Σ-elimination, assume an object $Z \in \mathbb{E}$ over $\{\coprod_X(Y)\}$ with a term $I, x: X, y: Y \vdash h: Z(\langle x, y \rangle)$. This h can be identified with a map $\overline{h}: \{Y\} \rightarrow \{Z\}$ in $\mathbb{B}$ with $\pi_Z \circ \overline{h} = \kappa: \{Y\} \xrightarrow{\cong} \{\coprod_X(Y)\}$. Hence $\overline{h} \circ \kappa^{-1}$ is a section of π_Z, and thus yields a term $1(\coprod_X(Y)) \rightarrow Z$, as required. Further details are left to the reader. More elaborate studies of the categorical interpretation of (first and higher order) dependent type theory may be found in [319, 134].

Notice that we do not include equality in the definition of a CCompC. We shall have more to say about equality towards the end of this section.

First a basic result.

10.5.4. Proposition. *If* $\mathcal{P}: \mathbb{E} \rightarrow \mathbb{B}^\rightarrow$ *is a closed comprehension category, then its underlying fibration* $p = \text{cod} \circ \mathcal{P}: \mathbb{E} \rightarrow \mathbb{B}$ *is Cartesian closed. The fibred Cartesian product and exponent are given by:*

$$X \times Y = \coprod_X(\pi_X^*(Y)) \qquad and \qquad X \Rightarrow Y = \prod_X(\pi_X^*(Y)).$$

In particular, the underlying fibration of a CCompC is locally small, see Corollary 10.4.11. Hence it is small if and only if it has a generic object, by Corollary 9.5.6. See Exercise 10.6.3 for an example of a small CCompC.

Proof. In the fibre over $I \in \mathbb{B}$ one has:

$$\mathbb{E}_I(Z, X \times Y) \cong \mathbb{B}/I\left(\pi_Z, \pi_{\coprod_X \pi_X^*(Y)}\right) \qquad \text{by fullness}$$

$$\cong \mathbb{B}/I\left(\pi_Z, \coprod_{\pi_X}(\pi_{\pi_X^*(Y)})\right) \qquad \text{since } \coprod \text{ is strong}$$

$$\cong \mathbb{B}/I\left(\pi_Z, \coprod_{\pi_X} \pi_X^*(\pi_Y)\right) \qquad \text{by Lemma 10.4.4}$$

$$\cong \mathbb{B}/I\left(\pi_Z, \pi_X \times \pi_Y\right)$$

$$\cong \mathbb{B}/I\left(\pi_Z, \pi_X\right) \times \mathbb{B}/I\left(\pi_Z, \pi_Y\right)$$

$$\cong \mathbb{E}_I(Z, X) \times \mathbb{E}_I(Z, Y).$$

(Actually, it can be shown that the object $X \times Y = \coprod_X(\pi_X^*(Y))$ is a product without assuming that the coproduct $\coprod$ is strong, see Exercise 10.5.4 (ii).)

For fibred exponents we have adjoint correspondences:

$$\frac{\frac{\coprod_X \pi_X^*(Z) = Z \times X \longrightarrow Y}{\pi_X^*(Z) \longrightarrow \pi_X^*(Y)}}{Z \longrightarrow \prod_X \pi_X^*(Y) = X \Rightarrow Y} \qquad \qquad \square$$

The above categorical descriptions of the binary product $\times$ and the exponent $\Rightarrow$ coincide with the syntactic definitions $\sigma \times \tau = \Sigma x{:}\sigma.\, \tau$ and $\sigma \to \tau = \Pi x{:}\sigma.\, \tau$, for x not free in τ, given in Example 10.1.2. By a standard argument (as in the proof of Lemma 1.9.12) one can show that in presence of these fibred exponents $\Rightarrow$, the coproduct $\coprod$ and equality Eq in a CCompC automatically satisfy the Frobenius property.

Next we consider the three main (general) examples of comprehension categories. The first two involve what we have called the 'type theoretic' fibrations and the third one the 'logic' fibration.

10.5.5. Theorem. *Let $\mathbb{A}$ be a category with finite products and $\mathbb{B}$ a category with finite limits. The simple fibration $\begin{smallmatrix} s(\mathbb{A}) \\ \downarrow \\ \mathbb{A} \end{smallmatrix}$, the codomain fibration $\begin{smallmatrix} \mathbb{B}^{\to} \\ \downarrow \\ \mathbb{B} \end{smallmatrix}$ and the subobject fibration $\begin{smallmatrix} \mathrm{Sub}(\mathbb{B}) \\ \downarrow \\ \mathbb{B} \end{smallmatrix}$ are all fibrations admitting full comprehension with strong coproducts. Moreover:*

(i) $\begin{smallmatrix} s(\mathbb{A}) \\ \downarrow \\ \mathbb{A} \end{smallmatrix}$ *is a CCompC if and only if $\mathbb{A}$ is a CCC;*

(ii) $\begin{smallmatrix} \mathbb{B}^{\to} \\ \downarrow \\ \mathbb{B} \end{smallmatrix}$ *is a CCompC if and only if $\mathbb{B}$ is a LCCC;*

(iii) $\begin{smallmatrix} \mathrm{Sub}(\mathbb{B}) \\ \downarrow \\ \mathbb{B} \end{smallmatrix}$ *is a CCompC if and only if it is a fibred CCC.*

The three points of this result bring a number of different notions under a single heading. In particular, points (i) and (ii) show that finite products and exponents are related like finite limits and local exponentials (exponents in slices).

Proof. In the previous section we have already seen that all three fibrations admit full comprehension. The codomain and subobject fibrations have strong sums, simply by composition. A coproduct functor $\coprod_{(I,X)}$, left adjoint to $\pi_{I,X}^*$, for the simple fibration $\begin{smallmatrix} s(\mathbb{A}) \\ \downarrow \\ \mathbb{A} \end{smallmatrix}$ along the projection $\pi_{I,X}\colon I \times X \to I$ is given by

$(I \times X, Y) \mapsto (I, X \times Y)$. These coproducts are strong since

$$
\begin{aligned}
\{\textstyle\coprod_{(I,X)}(I \times X, Y)\} &= \{(I, X \times Y)\} \\
&= I \times (X \times Y) \\
&\cong (I \times X) \times Y \\
&= \{(I \times X, Y)\}.
\end{aligned}
$$

(i) In case $\mathbb{A}$ is Cartesian closed we get a product functor $\prod_{(I,X)}$ along $\pi: I \times X \to I$ by $(I \times X, Y) \mapsto (I, X \Rightarrow Y)$. Then, writing the fibre $s(\mathbb{A})_I$ as the simple slice $\mathbb{A}/\!\!/ I$, we get

$$
\begin{aligned}
s(\mathbb{A})_I\big((I, Z), \textstyle\prod_{(I,X)}(I \times X, Y)\big) &= \mathbb{A}/\!\!/I\big(Z, \, X \Rightarrow Y\big) \\
&= \mathbb{A}\big(I \times Z, \, X \Rightarrow Y\big) \\
&\cong \mathbb{A}\big((I \times X) \times Z, \, Y\big) \\
&= \mathbb{A}/\!\!/(I \times X)\big(Z, \, Y\big) \\
&= s(\mathbb{A})_{I \times X}\big(\pi^*(I, Z), \, (I \times X, Y)\big).
\end{aligned}
$$

In the reverse direction, we have, by the previous proposition, that if the simple comprehension category $s(\mathbb{A}) \to \mathbb{A}^{\to}$ is closed, then the underlying simple fibration $\begin{smallmatrix} s(\mathbb{A}) \\ \downarrow \\ \mathbb{A} \end{smallmatrix}$ is Cartesian closed. In particular, the fibre $s(\mathbb{A})_1 \cong \mathbb{A}/\!\!/1 \cong \mathbb{A}$ over the terminal object 1 is Cartesian closed.

(ii) By Proposition 1.9.8, since the induced comprehension category is the identity functor $\mathbb{B}^{\to} \to \mathbb{B}^{\to}$, which has all maps in $\mathbb{B}$ as projections.

(iii) The (only if)-part follows from the previous proposition. For the converse, assume subobjects $m: X \rightarrowtail I$ and $n: Y \rightarrowtail X$. The product subobject $\prod_m(n)$ of I may be defined as $m \supset (m \circ n)$, where $\supset$ is the exponent in the fibre over I. For a mono $k: Z \rightarrowtail X$, we get appropriate correspondences:

$$
\cfrac{k \;\leq\; \textstyle\prod_m(n) = m \supset (m \circ n)}{\cfrac{m \circ m^*(k) = m \wedge k \;\leq\; (m \circ n)}{m^*(k) \;\leq\; n}}
$$

The latter holds because m is monic. $\qquad\qquad\qquad\qquad\qquad\square$

10.5.6. Term model. Assume a dependent calculus with unit type $1: \mathsf{Type}$ and with dependent products Π and strong sums Σ. In Example 10.4.8 we described how the structure of contexts in DTT leads to a full comprehension category with unit $\begin{smallmatrix} \mathbb{T} \\ \downarrow \\ \mathbb{C} \end{smallmatrix}$ given by types $(\Gamma \vdash \sigma: \mathsf{Type}) \in \mathbb{T}$ over their contexts $\Gamma \in \mathbb{C}$. Here we show that the type theoretic products Π and strong sums Σ make this fibration into a closed comprehension category.

For a type $\Gamma \vdash \sigma$: Type we have to produce products and strong coproducts along the associated projection map $\pi : (\Gamma, x : \sigma) \to \Gamma$ in $\mathbb{C}$. Essentially this is already done in Proposition 10.3.3, but here we give a reformulation in terms of comprehension categories: this means that these product and coproduct functors do not act on the projections (*i.e.* display maps) induced by types, but on the types themselves. Explicitly, these functors are given by

$$\text{product } \textstyle\prod : \quad \big(\Gamma, x : \sigma \vdash \tau : \text{Type}\big) \;\mapsto\; \big(\Gamma \vdash \Pi x : \sigma. \tau : \text{Type}\big)$$
$$\text{coproduct } \textstyle\coprod : \quad \big(\Gamma, x : \sigma \vdash \tau : \text{Type}\big) \;\mapsto\; \big(\Gamma \vdash \Sigma x : \sigma. \tau : \text{Type}\big).$$

The mate correspondences for dependent product Π and sum Σ (see Exercise 10.3.2) show that these assignments yield right and left adjoints to weakening π^*. The categorial requirement of *strong* coproducts is satisfied, since the canonical pairing morphism

$$\big(\Gamma, x : \sigma, y : \tau\big) \xrightarrow{\;\;\kappa = (\vec{v}, \langle x, y \rangle)\;\;} \big(\Gamma, z : \Sigma x : \sigma. \tau\big)$$

is an isomorphism. Its inverse is the sequence $(\vec{v}, \pi z, \pi' z)$, which uses first and second projections $\pi z : \sigma$ and $\pi' z : \tau[\pi z / x]$, typical for strong sums see Proposition 10.1.3 (i). In this way we get a term model closed comprehension category.

One can extend this example with equality, and show that the type theoretic formulations used in Section 10.1 lead to the appropriate categorical structure. Assume therefore equality types Eq in the calculus. For a type $\Gamma \vdash \sigma$: Type there is a diagonal morphism $\delta : (\Gamma, x : \sigma) \to (\Gamma, x : \sigma, x' : \sigma)$ in $\mathbb{C}$, by $(\vec{v}, x, x)$, and an associated contraction functor δ^*, which maps

$$\big(\Gamma, x : \sigma, x' : \sigma \vdash \tau : \text{Type}\big) \quad \text{to} \quad \big(\Gamma, x : \sigma \vdash \tau[x/x'] : \text{Type}\big).$$

This functor δ^* has a left adjoint, namely

$$\big(\Gamma, x : \sigma \vdash \rho : \text{Type}\big) \;\mapsto\; \big(\Gamma, x : \sigma, x' : \sigma \vdash \text{Eq}_\sigma(x, x') \times \rho : \text{Type}\big).$$

The adjunction requires a bijective correspondence between terms P and Q in

$$\frac{\Gamma, x : \sigma, x' : \sigma, z : \text{Eq}_\sigma(x, x') \times \rho \vdash P : \tau}{\Gamma, x : \sigma, w : \rho \vdash Q : \tau[x/x']} \quad \text{(Eq-mate)}$$

It is given by

$$P \;\mapsto\; P[x/x', \langle \mathrm{r}, w \rangle / z]$$
$$Q \;\mapsto\; Q[\pi' z / w] \text{ with } x' = x \text{ via } \pi z.$$

This equality is strong in the syntactic sense if and only if it is strong in the categorical sense (as in Definition 10.5.2). This can be seen as follows. In the category $\mathbb{C}$ of contexts there is the canonical map

$$\left(\Gamma, x\colon \sigma, w\colon \rho\right) \xrightarrow{\;\;\kappa = (\vec{v}, x, x, \langle \mathsf{r}, w\rangle)\;\;} \left(\Gamma, x\colon \sigma, x'\colon \sigma, z\colon \mathrm{Eq}_\sigma(x, x') \times \rho\right)$$

An inverse of κ should be a sequence $\mu = (\vec{M}, P, Q)$ of terms for which there are conversions—in context $\Gamma, x\colon \sigma, x'\colon \sigma, z\colon \mathrm{Eq}_\sigma(x, x') \times \rho$—of the form:

$$M_i = v_i, \qquad P = x = x', \qquad \mathsf{r} = \pi z, \qquad Q = \pi' z.$$

Hence the presence of such an isomorphism μ leads to the strong equality conversions as in Proposition 10.1.3 (ii). And conversely, these conversions tell us how to define an inverse μ if equality is very strong.

In the presence of fibred equalisers—or equivalently, strong equality types, see Theorem 10.5.10 below—products $\prod$ in a CCompC can be obtained from exponents and strong coproducts $\coprod$, following a standard formula (used earlier for LCCCs). This is the content of the following result. We only sketch the proof, since the details are not so interesting.

10.5.7. Proposition. *Let $\begin{smallmatrix}\mathbb{E}\\\downarrow p\\\mathbb{B}\end{smallmatrix}$ a fibration with fibred finite limits, full comprehension, strong coproducts and a terminal object in its base category. Then p is a fibred CCC if and only if it is a CCompC (actually a CCompC with strong equality, by Theorem 10.5.10 below).*

Proof. We shall show how to define products $\prod$ from exponents $\Rightarrow$. For objets $X \in \mathbb{E}$ over $I \in \mathbb{B}$ and $Y \in \mathbb{E}$ over $\{X\}$, form $\prod_X(Y)$ as domain of the equaliser

$$\prod_X(Y) \rightarrowtail (X \Rightarrow \coprod_X(Y)) \underset{(X \Rightarrow \mathsf{fst})}{\overset{\Lambda(\pi')}{\rightrightarrows}} (X \Rightarrow X)$$

where $\mathsf{fst}\colon \coprod_X(Y) \to X$ is the first projection, obtained as unique map with $\{\mathsf{fst}\} = \pi_Y \circ \kappa^{-1}\colon \{\coprod_X(Y)\} \xrightarrow{\cong} \{Y\} \to \{X\}$, using fullness. $\square$

Notice that the formula for the product $\prod_X(Y)$ in this proof is used for the family example immediately after Definition 10.5.3, and also in the proof of Theorem 10.5.5 (iii).

We continue with examples of CCompCs.

10.5.8. PERs and omega-sets. We have mentioned set-indexed-sets as a model of type dependency. Of course we can also take ω-set-indexed-ω-sets and PER-indexed-PERs as denotations of dependent types since the categories ω-**Sets** and **PER** of ω-sets and PERs are locally Cartesian closed. Interestingly, there are equivalences of fibrations

$$
\begin{array}{ccc}
\mathrm{UFam}(\omega\text{-}\mathbf{Sets}) & & \omega\text{-}\mathbf{Sets}^{\rightarrow} \\
\downarrow & \simeq & \downarrow \\
\omega\text{-}\mathbf{Sets} & & \omega\text{-}\mathbf{Sets}
\end{array}
\quad \text{and} \quad
\begin{array}{ccc}
\mathrm{UFam}(\mathbf{PER}) & & \mathbf{PER}^{\rightarrow} \\
\downarrow & \simeq & \downarrow \\
\mathbf{PER} & & \mathbf{PER}
\end{array}
$$

(see Propositions 1.4.7 and 1.5.3) and we obtain split CCompCs on the left. Further instances of split CCompCs are the fibrations

$$
\begin{array}{ccc}
\mathrm{UFam}(\mathbf{PER}) & \mathrm{UFam}(\mathbf{PER}) & \mathrm{UFam}(\omega\text{-}\mathbf{Sets}) \\
\downarrow & \downarrow & \downarrow \\
\omega\text{-}\mathbf{Sets} & \mathbf{Eff} & \mathbf{Eff}
\end{array}
$$

We shall sketch some details of the first and third example.

The fibration $\begin{array}{c}\mathrm{UFam}(\mathbf{PER})\\\downarrow\\\omega\text{-}\mathbf{Sets}\end{array}$ of PERs over ω-sets admits full comprehension, with right adjoint to the terminal object functor given by

$$
R = (R_i)_{i \in (I,E)} \longmapsto \{R\} = \coprod_{i \in I} \mathbb{N}/R_i, \text{ with } E(i, [n]_{R_i}) = E(i) \wedge [n]_{R_i}.
$$

For a second collection of PERs $S = (S_\alpha)_{\alpha \in \{R\}}$ we have product and coproduct

$$
\begin{aligned}
\textstyle\prod_R(S)_i &= \{(k, k') \mid \forall n, n'.\, nR_i n' \text{ implies } k \cdot nS_{(i,[m])} k' \cdot n'\} \\
\textstyle\coprod_R(S)_i &= \{(\langle n, m\rangle, \langle n', m'\rangle) \mid nR_i n' \text{ and } mS_{(i,[n])} m'\}.
\end{aligned}
$$

These are special cases of the formulas for quantification along arbitrary maps in the proof of Lemma 1.9.6. This coproduct is strong since

$$
\begin{aligned}
\{\textstyle\coprod_R(S)\} &= \{(i, [k]) \mid i \in I \text{ and } k \in |\textstyle\coprod_R(S)_i|\} \\
&= \{(i, [\langle m, n\rangle]) \mid i \in I, n \in |R_i| \text{ and } m \in |S_{(i,[n])}|\} \\
&\cong \{((i, [n]), [m]) \mid i \in I, n \in |R_i| \text{ and } m \in |S_{(i,[n])}|\} \\
&= \{S\}.
\end{aligned}
$$

We turn to the fibration $\begin{array}{c}\mathrm{UFam}(\omega\text{-}\mathbf{Sets})\\\downarrow\\\mathbf{Eff}\end{array}$, as introduced in Section 6.3. The functor $\{-\}\colon \mathrm{UFam}(\omega\text{-}\mathbf{Sets}) \to \mathbf{Eff}$, right adjoint to the terminal object functor, is described in the first few lines of the proof of Proposition 6.3.3. For a family of ω-sets $X = (X_{[i]})_{[i] \in \Gamma(I,\approx)}$ over $(I, \approx)$ and a family $Y = (Y_\alpha)$ over $\Gamma\{X\} = \coprod_{[i] \in \Gamma(I,\approx)} X_{[i]}$, we have a (strong) coproduct

$$
\textstyle\coprod_X(Y)_{[i]} = \{(x, y) \mid x \in X_{[i]} \text{ and } y \in Y_{([i],x)}\}.
$$

Products $\prod$ are obtained by the previous proposition. It applies since the fibration $\begin{array}{c}\mathrm{UFam}(\omega\text{-}\mathbf{Sets})\\\downarrow\\\mathbf{Eff}\end{array}$ has fibred finite limits and exponents by a pointwise construction.

10.5.9. Topos models. Let $\mathbb{B}$ be a topos. By Proposition 5.4.7 $\mathbb{B}$ is locally Cartesian closed, so the codomain fibration on $\mathbb{B}$ is a (non-split) CCompC by Theorem 10.5.5 (ii). And by point (iii) in the same result, the subobject fibration on $\mathbb{B}$ is a (split poset) CCompC. What we will show next is that the codomain fibration on $\mathbb{B}$ is equivalent to a *split* fibration $\begin{smallmatrix}\mathcal{F}(\mathbb{B})\\\downarrow\\\mathbb{B}\end{smallmatrix}$, and that the latter fibration is a split CCompC. In the codomain fibration one has display indexing and substitution by pullback, whereas in the new split fibration one has pointwise indexing and substitution by composition. The construction comes from [154], and generalises the idea of inclusions followed by projections as display maps, as used in [46] in a set-theoretic setting.

The total category $\mathcal{F}(\mathbb{B})$ of our split fibration arises as follows.

objects	triples (I, X, φ) where $\varphi \colon I \times X \to \Omega$ is a map in $\mathbb{B}$. For each such object we choose a corresponding mono $m_\varphi \colon \{\varphi\} \rightarrowtail I \times X$, and write $\pi_\varphi = \pi \circ m_\varphi \colon \{\varphi\} \to I$.
morphisms	$(I \times X \xrightarrow{\varphi} \Omega) \to (J \times Y \xrightarrow{\psi} \Omega)$ are morphisms $\pi_\varphi \to \pi_\psi$ in $\mathbb{B}^\to$. They consist of a pair of maps $u \colon I \to J$ and $f \colon \{\varphi\} \to \{\psi\}$ with $\pi_\psi \circ f = u \circ \pi_\varphi$.

By construction there is a full and faithful functor $\mathcal{F}(\mathbb{B}) \to \mathbb{B}^\to$. Postcomposition with the codomain functor yields a fibration $\mathcal{F}(\mathbb{B}) \to \mathbb{B}$, which sends an object $(I \times X \xrightarrow{\varphi} \Omega)$ to I. It is a split fibration since for such an object φ over I and a morphism $u \colon J \to I$ one gets a reindexed object $u^*(\varphi) = \varphi \circ u \times \mathrm{id}$ over J by composition. The idea is to think of $\varphi \colon I \times X \to \Omega$ as a family of I-indexed sets $(X_i)_{i \in I}$, given as $X_i = \{x \in X \mid \varphi(i, x) = \mathbf{true}\}$. The functor $\mathcal{F}(\mathbb{B}) \to \mathbb{B}^\to$ given by $\varphi \mapsto \pi_\varphi$ is a fibred equivalence: in the reverse direction a family $\left(\begin{smallmatrix}Y\\\downarrow\psi\\J\end{smallmatrix}\right)$ is sent to the characteristic map $J \times Y \to \Omega$ of the mono $\langle \psi, \mathrm{id}\rangle \colon Y \rightarrowtail J \times Y$. A (split) terminal object functor $1 \colon \mathbb{B} \to \mathcal{F}(\mathbb{B})$ is obtained by $J \mapsto (\mathbf{true} \circ \pi' \colon J \times 1 \to \Omega)$; it is left adjoint to $\varphi \mapsto \{\varphi\}$.

We turn to dependent coproduct and product. Assume therefore objects $\varphi \colon I \times X \to \Omega$ over I and $\psi \colon \{\varphi\} \times Y \to \Omega$ over $\{\varphi\}$. We have to produce objects $\coprod_\varphi(\psi)$ and $\prod_\varphi(\psi)$ over I. The coproduct $\coprod_X(Y)$ is obtained as characteristic map in

$$
\begin{array}{ccccccc}
\{\psi\} & \overset{m_\psi}{\rightarrowtail} & \{\varphi\} \times Y & \overset{m_\varphi \times \mathrm{id}}{\rightarrowtail} & (I \times X) \times Y & \overset{\cong}{\to} & I \times (X \times Y) \\
\downarrow & \llcorner & & & & & \vdots\; \Big\downarrow \coprod_\varphi(\psi) \\
1 & \rightarrowtail & & \underset{\mathbf{true}}{\longrightarrow} & & & \Omega
\end{array}
$$

Then, by construction the projection $\pi_{\coprod_\varphi(\psi)}$ of this coproduct equals $\pi_\varphi \circ \pi_\psi$, so that we have a strong coproduct.

For the product $\prod_\varphi(\psi)$ we first form the maps $\widehat{\varphi}$, $\widetilde{\psi}$, using the partial map classifier in $\mathbb{B}$ (see Proposition 5.4.5):

$$
\begin{array}{ccc}
\{\varphi\} \overset{m_\varphi}{\rightarrowtail} I \times X & \qquad \{\psi\} \overset{m_\psi}{\rightarrowtail} \{\varphi\} \times Y \overset{m_\varphi \times \eta_Y}{\rightarrowtail} (I \times X) \times \bot Y \\
\Vert \quad \lrcorner \qquad \downarrow \widehat{\varphi} & \qquad \Vert \quad \lrcorner \qquad\qquad\qquad\qquad \downarrow \widetilde{\psi} \\
\{\varphi\} \overset{}{\underset{\eta_\varphi}{\rightarrowtail}} \bot\{\varphi\} & \qquad \{\psi\} \overset{}{\underset{\eta_\psi}{\longrightarrow}} \bot\{\psi\}
\end{array}
$$

Together with auxiliary maps

$$
\begin{aligned}
\alpha &= \langle \pi \times \mathrm{id}, \mathrm{ev} \circ \pi' \times \mathrm{id} \rangle : (I \times (X \Rightarrow \bot Y)) \times X \longrightarrow (I \times X) \times \bot Y \\
\beta &= \Lambda(\bot(\pi' \circ m_\varphi) \circ \widehat{\varphi}) \circ \pi : I \times (X \Rightarrow \bot Y) \longrightarrow (X \Rightarrow \bot X) \\
\gamma &= \Lambda(\bot(\pi' \circ m_\varphi \circ \pi_\psi) \circ \widetilde{\psi} \circ \alpha) : I \times (X \Rightarrow \bot Y) \longrightarrow (X \Rightarrow \bot X).
\end{aligned}
$$

Finally, one obtains the product object $\prod_\varphi(\psi) : I \times (X \Rightarrow \bot Y) \to \Omega$ as characteristic map of the equaliser $\{\prod_\varphi(\psi)\} \rightarrowtail I \times (X \Rightarrow \bot Y)$ of β, γ. It yields a split version of the usual dependent product in LCCCs.

We conclude this section with a characterisation of strong equality in CCompCs in terms of fibred equalisers. It shows that strong equality is quite natural, from a categorical perspective.

10.5.10. Theorem. (i) *A CCompC has strong equality if and only if its underlying fibration has fibred equalisers.*

(ii) *And in this situation, the fibration is a fibred LCCC.*

With this result it is easy to see that the CCompCs in Example 10.5.8 of families of PERs and of ω-sets have strong equality, because the underlying fibrations have fibred equalisers, obtained pointwise from equalisers in the categories of PERs and of ω-sets.

The link between locally Cartesian closed categories and dependent type theory with unit, product and strong sum and equality was uncovered in [306]. There, only empty contexts are used, whereas here, LCCC-structure is obtained in every context (and hence in fibred form).

Proof. (i) Let $\begin{smallmatrix}\mathbb{E}\\\downarrow p\\\mathbb{B}\end{smallmatrix}$ be a CCompC. Assume it has strong equality. For parallel maps $f, g : X \rightrightarrows Y$ in a fibre, say over $I \in \mathbb{B}$, consider the mediating map $\langle\{f\}, \{g\}\rangle : \{X\} \to \{\pi_Y^*(Y)\}$ in $\mathbb{B}$ for the pullback of π_Y against itself. Put $E(f, g) = \langle\{f\}, \{g\}\rangle^*(\mathrm{Eq}_Y(1)) \in \mathbb{E}_{\{X\}}$ and $\mathrm{Eq}(f, g) = \coprod_X E(f, g) \in \mathbb{E}_I$, with first projection $\mathsf{fst}\colon \coprod_X E(f,g) \to X$. We claim this is the equaliser of f and g in $\mathbb{E}_I$. Consider therefore the diagram

$$
\begin{array}{ccccccc}
\{\textstyle\coprod_X E(f,g)\} & \overset{\cong}{\underset{\kappa}{\leftarrow}} & \{E(f,g)\} & \longrightarrow & \{Y\} & \overset{\cong}{\longrightarrow} & \{\mathrm{Eq}_Y(1)\} \\
& {\scriptstyle\{\mathsf{fst}\}}\searrow & \;\;\downarrow{\scriptstyle\pi}\;\lrcorner & & \downarrow{\scriptstyle\delta_Y}\;\;\;\nearrow{\scriptstyle\pi} & & \\
& & \{X\} & \xrightarrow[\;\langle\{f\},\{g\}\rangle\;]{} & \{\pi_Y^*(Y)\} & &
\end{array}
$$

In this situation it follows that $\{f\} \circ \{\mathsf{fst}\} = \{g\} \circ \{\mathsf{fst}\}$, and so $f \circ \mathsf{fst} = g \circ \mathsf{fst}$. If also $h\colon Z \to X$ in $\mathbb{E}_I$ satisfies $f \circ h = g \circ h = k$, say, then the maps $\{h\}\colon \{Z\} \to \{X\}$ and $\{k\}\colon \{Z\} \to \{Y\}$ induce a unique mediating map $\{Z\} \dashrightarrow \{\coprod_X E(f,g)\}$ in the diagram, and thus a mediating map $Z \dashrightarrow \coprod_X E(f,g)$ in $\mathbb{E}_I$ by fullness.

    In the reverse direction we assume fibred equalisers, and write for an object $X \in \mathbb{E}$, $X' = \pi_X^*(X) \in \mathbb{E}_{\{X\}}$ and $\theta_X\colon 1\{X\} \to X'$ for the vertical part of the counit component $\varepsilon_X\colon 1\{X\} \to X$. Similarly, we write $X'' = \pi_{X'}^*(X')$ and $\theta_{X'}$ for the vertical part of $\varepsilon_{X'}$. It is not hard to see that the transpose $\{X\} \to \{X'\}$ of θ_X is the diagonal δ_X.

    Let $E(X) \in \mathbb{E}$ be the domain of the following equaliser in the fibre over $\{X'\}$.

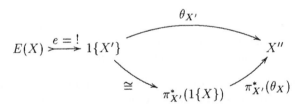

With some effort one can show that there is an isomorphism

$$
\begin{array}{ccc}
\{E(X)\} & \xrightarrow[\;\cong\;]{\pi_{X'}\,\circ\,\pi_{E(X)}} & \{X\} \\
{\scriptstyle\pi_{E(X)}}\searrow & & \swarrow{\scriptstyle\delta_X} \\
& \{X'\} &
\end{array}
$$

so that diagonals occur as projections. Then we can define an equality functor $\mathrm{Eq}_X\colon \mathbb{E}_{\{X\}} \to \mathbb{E}_{\{X'\}}$ by $\mathrm{Eq}_X(Y) = \pi_{X'}^*(Y) \times E(X)$. We get the appropriate adjoint correspondences:

$$\frac{\frac{\mathrm{Eq}_X(Y) = \pi^*(Y) \times Z \longrightarrow Z}{E(X) \longrightarrow \pi^*(Y) \Rightarrow Z}}{\frac{\delta_X \cong \pi_{E(X)} \longrightarrow \pi_{\pi^*(Y) \Rightarrow Z}}{\frac{1 \longrightarrow \delta_X^*(\pi^*(Y) \Rightarrow Z) \cong Y \Rightarrow \delta_X^*(Z)}{Y \longrightarrow \delta_X^*(Z)}}}$$

And this equality is strong since we have isomorphisms in $\mathbb{B}/\{X\}$:

$$
\begin{aligned}
\pi_{\mathrm{Eq}_X(Y)} &\cong \pi_{\pi_{X'}^*(Y)} \times \pi_{E(X)} \\
&\cong \pi_{X'}^*(\pi_Y) \times \delta_X \\
&\cong \delta_X^* \pi_{X'}^*(\pi_Y) \circ \delta_X \\
&\cong \pi_Y \circ \delta_X.
\end{aligned}
$$

(ii) The LCCC-structure is obtained from equivalences $(\mathbb{E}_I)/X \simeq \mathbb{E}_{\{X\}}$ for $X \in \mathbb{E}_I$, which arise as follows. In one direction, an object $Y \in \mathbb{E}_{\{X\}}$ is mapped to the first projection $\mathsf{fst} \colon \coprod_X(Y) \to X$. This yields a full and faithful functor. In the other direction, one maps a vertical family $\begin{pmatrix} Z \\ \downarrow \varphi \\ X \end{pmatrix}$ to the domain $D(\varphi)$ of the following equaliser.

$$D(\varphi) \rightarrowtail \pi_X^*(Z) \underset{\theta_X \,\circ\, !}{\overset{\pi_X^*(\varphi)}{\rightrightarrows}} \pi_X^*(X) \qquad\qquad \square$$

Exercises

10.5.1. Show that the fibration $\begin{smallmatrix} \mathbf{Fam(Sets)} \\ \downarrow \\ \mathbf{Cat} \end{smallmatrix}$ of presheaves (see Example 1.10.3 (ii)) is a closed comprehension category with strong equality.
[Comprehension for this fibration is described in [193]. For every functor $F \colon \mathbb{A} \to \mathbb{B}$ in the base category the reindexing functor F^* has both a left and a right adjoint (by left and right Kan extension), but Beck-Chevalley does not hold in general. It does hold along the *dependent* projections and diagonals.]

10.5.2. Describe the interpretation of the *weak* sum elimination rule, in a full comprehension category with unit and coproducts.

10.5.3.   Assume a category $\mathbb{C}$ with terminal object 1 and arbitrary coproducts $\coprod_{i \in I} X_i$.

   (i)   Show that if the global sections functor $\mathbb{C}(1, -) \colon \mathbb{C} \to \mathbf{Sets}$ preserves coproducts then the family comprehension category $\mathrm{Fam}(\mathbb{C}) \to \mathbf{Sets}^{\to}$ has strong coproducts.

   (ii) Suppose that coproducts in $\mathbb{C}$ are disjoint and universal. Prove that the functor $\mathbb{C}(1, -) \colon \mathbb{C} \to \mathbf{Sets}$ preserves coproducts if and only if the terminal object 1 is **indecomposable** (*i.e.* if $1 \cong \coprod_{i \in I} X_i$ then $1 \cong X_i$ for some $i \in I$).

10.5.4.   Let $\mathcal{P} \colon \mathbb{E} \to \mathbb{B}^{\to}$ be a comprehension category with unit and coproducts (*i.e.* with the underlying fibration $p = \mathrm{cod} \circ \mathcal{P}$ having $\mathcal{P}$-coproducts).

   (i)   Show that if p admits *full* comprehension, then it has fibred finite products—with $X \times Y = \coprod_X (\pi_X^* (Y))$ as in Proposition 10.5.4.

   (ii) Use (i) to show that Frobenius holds automatically for $\coprod$, in case comprehension is full.

   (iii) Prove that p admits full comprehension if and only if the canonical maps $\coprod_X (1\{X\}) \to X$ are isomorphisms. That is, if and only if the counit components $\varepsilon_X \colon 1\{X\} \to X$ are opcartesian.

   (iv) Show that the following statements are equivalent.

      (a) The coproducts $\coprod$ are strong.

      (b) The functors $\kappa^* \colon \mathbb{E}_{\{\coprod_X (Y)\}} \to \mathbb{E}_{\{Y\}}$ induced by the canonical maps $\kappa \colon \{Y\} \to \{\coprod_X (Y)\}$ are full and faithful.

      (c) The maps $1\kappa \colon 1\{Y\} \to 1\{\coprod_X (Y)\}$ are opcartesian.

10.5.5.   (From [154]) For a closed comprehension category $\mathcal{P} \colon \mathbb{E} \to \mathbb{B}^{\to}$, form the fibration $\{-\}^*(p)$ by change-of-base in:

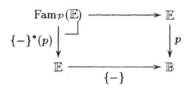

   (i)   Prove that $\{-\}^*(p)$ is again a CCompC, with as projections the vertical maps $\mathrm{fst} \colon \coprod_X (Y) \to X$. It may be seen as a CCompC over p.

   (ii) Conclude from (the proof of) Theorem 10.5.10 (ii) that the induced functor $\mathrm{Fam}_\mathcal{P}(\mathbb{E}) \to V(\mathbb{E})$ is an equivalence if $\mathcal{P} \colon \mathbb{E} \to \mathbb{B}^{\to}$ has strong equality.

10.5.6.   Consider a comprehension category $\mathcal{P} \colon \mathbb{E} \to \mathbb{B}^{\to}$ whose underlying fibration has fibred coproducts $+$. We call these **strong fibred coproducts** if for each pair of objects $X, Y \in \mathbb{E}$ in the same fibre, the induced tuple of reindexing functors

$$\mathbb{E}_{\{X+Y\}} \xrightarrow{\quad \langle \{\kappa\}^*, \{\kappa'\}^* \rangle \quad} \mathbb{E}_{\{X\}} \times \mathbb{E}_{\{Y\}}$$

is full and faithful.

(i) Investigate what this means in the term model comprehension category $\mathbb{T} \to \mathbb{C}^{\to}$.

(ii) Prove that for a codomain fibration $\begin{smallmatrix}\mathbb{B}^{\to}\\\downarrow\\\mathbb{B}\end{smallmatrix}$ the following statements are equivalent.

   (a) The category $\mathbb{B}$ has universal coproducts $+$.

   (b) The fibration $\begin{smallmatrix}\mathbb{B}^{\to}\\\downarrow\\\mathbb{B}\end{smallmatrix}$ has fibred coproducts $+$.

   (c) The identity comprehension category $\mathbb{B}^{\to} \to \mathbb{B}^{\to}$ has strong fibred coproducts $+$.

   Hence fibred coproducts $+$ are automatically strong in a codomain fibration.

(iii) Assume strong coproducts $+$, and give categorical proofs of the isomorphisms

$$\coprod_{X+Y}(Z) \cong \coprod_X\{\kappa\}^*(Z) + \coprod_Y\{\kappa'\}^*(Z)$$
$$\prod_{X+Y}(Z) \cong \prod_X\{\kappa\}^*(Z) \times \prod_Y\{\kappa'\}^*(Z)$$

from Exercise 10.1.7.

10.6 Domain theoretic models of type dependency

In this section we describe two domain theoretic examples of closed comprehension categories (forming models of DTT with $1, \prod$ and strong $\coprod$, see the previous section). The first example is based on directed complete partial orders (dcpos), and the second one on closures (in $P\omega$). In the first example, a type-indexed-type will be a family of dcpos *continuously* indexed by an index dcpo. Such a family will be understood as a functor from the index dcpo to a category of dcpos which preserves filtered colimits[1]. We simply use dcpos to construct a model, but one can also work with Scott domains, see [57, 245] (and Exercise 10.6.2).

In the second example, a type-indexed-type will be a closure-indexed-closure. The latter will simply be a morphism in the category of closures from the index closure to a distinguished object Ω, which may be understood as the closure of all closure—or as the type of all types. It is a model of higher order dependent type theory and forms a bridge between this chapter and the next one. This model is based on [302, 15].

---

[1] The idea of using such continuous functors, together with dependent product and coproduct as defined below, goes back to Plotkin's "Pisa Notes" (1978). Here we only put these ideas in the present categorical framework.

In both these examples the detailed verifications that everything works appropriately are quite lengthy, and for the most part left to the reader. More information about domain theoretic models may be found in [329, 330, 148, 57, 245, 61].

We start with some order-theoretic preliminaries. Recall that a directed complete partial order is a poset in which every directed subset has a join. Categorically, this means that the poset, as a category, has filtered colimits. A function between dcpos is called (Scott-)continuous if it is monotone and preserves directed joins. This yields a category, for which we write **Dcpo**. It is Cartesian closed, with singleton poset as terminal object, product of the underlying sets with componentwise order as Cartesian product, and with set of continuous functions with pointwise order as exponent. We shall use a second category of dcpos, namely $\mathbf{Dcpo}^{\mathrm{EP}}$. It has as objects dcpos, and as morphisms $f: A \to B$ pairs $f = (f^e, f^p)$ of continuous functions $f^e: A \to B$ and $f^p: B \to A$ which satisfy $f^p \circ f^e = \mathrm{id}$ and $f^e \circ f^p \leq \mathrm{id}$. One calls f^e an embedding, and f^p a projection.

Let $\phi: A \to \mathbf{Dcpo}^{\mathrm{EP}}$ be a functor with a dcpo A as domain. For an inequality $a_1 \leq a_2$ in A, we write the corresponding morphism $\phi(a_1) \to \phi(a_2)$ in $\mathbf{Dcpo}^{\mathrm{EP}}$ as an embedding and projection pair:

$$\phi(a_1 \leq a_2)^e : \phi(a_1) \to \phi(a_2) \qquad \text{and} \qquad \phi(a_1 \leq a_2)^p : \phi(a_2) \to \phi(a_1).$$

Such a functor $\phi: A \to \mathbf{Dcpo}^{\mathrm{EP}}$ will be called **continuous** if it preserves filtered colimits. A classical result in this area, see *e.g.* [3], is that continuity of ϕ amounts to the following requirement: every directed join $a = \bigvee_{i \in I} a_i$ in A yields a directed join

$$\left(\bigvee_{i \in I} \phi(a_i \leq a)^e \circ \phi(a_i \leq a)^p \right) = \mathrm{id}$$

in the dcpo of continuous functions $\phi(a) \to \phi(a)$. We shall work as if this condition defines continuity for ϕ.

10.6.1. Definition. Consider the following functor $\mathbf{Dcpo}^{\mathrm{op}} \to \mathbf{Cat}$. Its fibre category over $A \in \mathbf{Dcpo}$ has

objects   continuous functors $\phi: A \to \mathbf{Dcpo}^{\mathrm{EP}}$ with A as domain.

morphisms $(A \xrightarrow{\phi} \mathbf{Dcpo}^{\mathrm{EP}}) \to (A \xrightarrow{\psi} \mathbf{Dcpo}^{\mathrm{EP}})$ are families

$$f = \left(\phi(a) \xrightarrow{f_a} \Psi(a) \right)_{a \in A}$$

of continuous functions f_a, satisfying the following two

requirements.

(1) If $a_1 \leq a_2$ in A, then

$$\phi(a_1 \leq a_2)^e \circ f_{a_1} \circ \phi(a_1 \leq a_2)^p \leq f_{a_2}.$$

(2) If $a = \bigvee_{i \in I} a_i$ is a directed join in A, then f_a can be written as a directed join:

$$f_a = \bigvee_{i \in I} \phi(a_i \leq a)^e \circ f_{a_i} \circ \phi(a_i \leq a)^p.$$

The identity on an object $\phi \colon A \to \mathbf{Dcpo}^{\mathrm{EP}}$ over $A \in \mathbf{Dcpo}$ is the collection of identity functions on $\phi(a)$, for $a \in A$. This is an appropriate morphism by the continuity condition for ϕ. And composition of $\phi \xrightarrow{f} \psi \xrightarrow{g} \chi$ is given by the collection $(g_a \circ f_a)_{a \in A}$. Thus we get a category (over A).

Reindexing is done by composition: for a continuous function $u \colon B \to A$ in $\mathbf{Dcpo}$ we get a substitution functor by

$$\phi \mapsto \phi \circ u \qquad \text{and} \qquad \left(f_a \right)_{a \in A} \mapsto \left(f_{u(b)} \right)_{b \in B}.$$

The resulting split fibration of continuous families of dcpos over dcpos will be written as $\begin{smallmatrix} \mathrm{CFam}(\mathbf{Dcpo}) \\ \downarrow \\ \mathbf{Dcpo} \end{smallmatrix}$.

10.6.2. Lemma. *The fibration* $\begin{smallmatrix} \mathrm{CFam}(\mathbf{Dcpo}) \\ \downarrow \\ \mathbf{Dcpo} \end{smallmatrix}$ *has a (split) terminal object functor* $1 \colon \mathbf{Dcpo} \to \mathrm{CFam}(\mathbf{Dcpo})$ *with* $1(B) \colon B \to \mathbf{Dcpo}^{\mathrm{EP}}$ *sending every element* $b \in B$ *to a singleton poset. This functor* 1 *has a right adjoint* $\{-\}$*, which maps a continuous functor* $\phi \colon A \to \mathbf{Dcpo}^{\mathrm{EP}}$ *to the "Grothendieck completion"*

$$\{\phi\} = \{(a, x) \mid a \in A \text{ and } x \in \phi(a)\}$$

with order

$$(a_1, x_1) \leq (a_2, x_2) \ \Leftrightarrow \ a_1 \leq a_2 \text{ in } A \text{ and } \phi(a_1 \leq a_2)^e(x_1) \leq x_2 \text{ in } \phi(a_2).$$

The resulting functor $\mathrm{CFam}(\mathbf{Dcpo}) \to \mathbf{Dcpo}^{\to}$ *mapping* ϕ *over* A *to the first projection* $\pi_\phi \colon \{\phi\} \to A$ *is full and faithful, so that the fibration* $\begin{smallmatrix} \mathrm{CFam}(\mathbf{Dcpo}) \\ \downarrow \\ \mathbf{Dcpo} \end{smallmatrix}$ *admits full comprehension.*

Proof. It is easy to see that $1 \colon \mathbf{Dcpo} \to \mathrm{CFam}(\mathbf{Dcpo})$ is a terminal object functor, so we concentrate on its right adjoint. For $\phi \colon A \to \mathbf{Dcpo}^{\mathrm{EP}}$ the set $\{\phi\} = \{(a, x) \mid a \in A \text{ and } x \in \phi(a)\}$ is a dcpo, since for a directed collection (a_i, x_i) in $\{\phi\}$ one can compute the join as:

$$\bigvee_i (a_i, x_i) = (a, \bigvee_i \phi(a_i \leq a)^e(x_i)), \quad \text{where} \quad a = \bigvee_i a_i.$$

The first projection function $\pi_\phi \colon \{\phi\} \to A$ is clearly continuous. For a morphism $f \colon \phi \to \psi$ over A we get a function $\{f\} \colon \{\phi\} \to \{\psi\}$ by $(a, x) \mapsto (a, f_a(x))$. This $\{f\}$ is continuous function: for a directed collection $(a_i, x_i)_{i \in I}$ in $\{\phi\}$, the join is (a, x) where $a = \bigvee_{i \in I} a_i$ and $x = \bigvee_{i \in I} \phi(a_i \leq a)^e(x_i)$ and so we have

$$\{f\}(a, x) = (a, f_a(x)) \overset{(*)}{=} (a, \textstyle\bigvee_{i \in I} \phi(a_i \leq a)^e f_{a_i}(x_i)) = \textstyle\bigvee_{i \in I} \{f\}(a_i, x_i)$$

where the equation $(*)$ holds because

$$
\begin{aligned}
f_a(x) &= \bigvee_{j \in I} \phi(a_j \leq a)^e f_{a_j} \phi(a_j \leq a)^p(x) \\
&= \bigvee_{j \in I} \bigvee_{i \in I} \phi(a_j \leq a)^e f_{a_j} \phi(a_j \leq a)^p \phi(a_i \leq a)^e(x_i) \\
&= \bigvee_{i \in I} \phi(a_i \leq a)^e f_{a_i} \phi(a_i \leq a)^p \phi(a_i \leq a)^e(x_i) \\
&= \bigvee_{i \in I} \phi(a_i \leq a)^e f_{a_i}(x_i).
\end{aligned}
$$

where the two joins over I can be combined into a single join, because they are directed joins.

The resulting functor $\mathrm{CFam}(\mathbf{Dcpo})_A \to \mathbf{Dcpo}/A$ between fibres is obviously faithful. And it is full because if we have a continuous function $u \colon \{\phi\} \to \{\psi\}$ with $\pi_\psi \circ u = \pi_\phi$, then there is a morphism $f \colon \phi \to \psi$ over A with $f_a(x) = \pi' u(a, x)$. Then obviously $\{f\} = u$. We leave it to the reader to verify that this collection $f = (f_a)_{a \in A}$ is indeed a morphism $\phi \to \psi$ over A. $\qquad\square$

10.6.3. Lemma. *The comprehension category with unit* $\begin{smallmatrix} \mathrm{CFam}(\mathbf{Dcpo}) \\ \downarrow \\ \mathbf{Dcpo} \end{smallmatrix}$ *admits strong coproducts. For a family* $\phi \colon A \to \mathbf{Dcpo}^{\mathrm{EP}}$ *over* A, *and another family* $\psi \colon \{\phi\} \to \mathbf{Dcpo}^{\mathrm{EP}}$ *over* $\{\phi\}$, *we get a coproduct* $\coprod_\phi(\psi) \colon A \to \mathbf{Dcpo}^{\mathrm{EP}}$ *by*

$$a \mapsto \{\psi(a, -)\} = \{\langle x, y \rangle \mid x \in \phi(a) \text{ and } y \in \psi(a, x)\},$$

ordered by

$$(x_1, y_1) \leq (x_2, y_2) \iff x_1 \leq x_2 \text{ and } \psi((a, x_1) \leq (a, x_2))^e(y_1) \leq y_2.$$

Proof. For $a_1 \leq a_2$ in A we have to define a pair of continuous functions

$$\{\psi(a_1, -)\} \xrightarrow[\coprod_\phi(\psi)(a_1 \leq a_2)^p]{\coprod_\phi(\psi)(a_1 \leq a_2)^e} \{\psi(a_2, -)\}$$

They are given by

$$\coprod_\phi(\psi)(a_1 \le a_2)^e(x,y)$$
$$= \langle \phi(a_1 \le a_2)^e(x), \psi(a_1, \phi(a_1 \le a_2)^e(x)) \le (a_2, x))^e(y) \rangle$$
$$\coprod_\phi(\psi)(a_1 \le a_2)^p(x,y)$$
$$= \langle \phi(a_1 \le a_2)^p(x), \psi((a_1, x) \le (a_2, \phi(a_1 \le a_2)^p(x))^p(y) \rangle.$$

By some lengthy computations one verifies that these functions form an embedding-projection pair and make $a \mapsto \coprod_\phi(\psi)(a)$ a continuous functor.

For a continuous functor $\chi \colon A \to \mathbf{Dcpo}^{EP}$ over A, there is a bijective correspondence between vertical morphisms:

$$\frac{\coprod_\phi(\psi) \xrightarrow{\;f\;} \chi \quad \text{over } A}{\Psi \xrightarrow[\;g\;]{} \pi_\phi^*(\chi) \quad \text{over } \{\phi\}}$$

which is described as follows. Starting with a family $f = (f_a)_{a \in A}$ over A one gets a family $\overline{f}$ over $\{\phi\}$ by

$$\overline{f}_{(a,x)}(y) = f_a(x,y).$$

And in the reverse direction, starting with $g = (g_{(a,x)})_{(a,x) \in \{\phi\}}$ one takes

$$\overline{g}_a(x,y) = g_{(a,x)}(y).$$

These operations produce appropriate new families and are obviously each others inverses. It remains to show that these coproducts are strong. But it is not hard to see that the canonical map

$$\{\psi\} \longrightarrow \{\coprod_\phi(\psi)\} \qquad \text{given by} \qquad ((a,x),y) \mapsto (a,(x,y))$$

is an (order) isomorphism. $\qquad\qquad\qquad\qquad\qquad\qquad\qquad\qquad\qquad\qquad\qquad\qquad\square$

10.6.4. Lemma. *The comprehension category* $\begin{smallmatrix} \mathbf{CFam(Dcpo)} \\ \downarrow \\ \mathbf{Dcpo} \end{smallmatrix}$ *admits products: for families* $\phi \colon A \to \mathbf{Dcpo}^{EP}$ *and* $\psi \colon \{\phi\} \to \mathbf{Dcpo}^{EP}$, *there is a product family* $\prod_\phi(\psi) \colon A \to \mathbf{Dcpo}^{EP}$ *by*

$$a \mapsto \{h \colon \phi(a) \to \coprod_\phi(\psi)(a) \mid h \text{ is continuous and } \pi \circ h = \mathrm{id}\},$$
$$\text{ordered pointwise.}$$

Proof. The action of the product $\prod_\phi(\psi)$ on a morphism $a_1 \le a_2$ in A is given by the following embedding-projection pair.

$$\prod_\phi(\psi)(a_1 \le a_2)^e(h)$$
$$= \lambda z \in \phi(a_2). \langle z, \psi((a_1, \phi(a_1 \le a_2)^p(z)) \le (a_2, z))^e(\pi'h(\phi(a_1 \le a_2)^p(z)))\rangle$$
$$\prod_\phi(\psi)(a_1 \le a_2)^p(k)$$
$$= \lambda x \in \phi(a_1). \langle x, \psi((a_1, x) \le (a_2, \phi(a_1 \le a_2)^e(x))^p(\pi'k(\phi(a_1 \le a_2)^e(x)))\rangle.$$

This functor $\prod_\phi(-)$ is right adjoint to the weakening functor $\pi_\phi^*(-)$, by the adjoint correspondence

$$
\begin{array}{c}
\chi \xrightarrow{\ \ f\ \ } \prod_\phi(\psi) \quad \text{over } A \\
\rule{5cm}{0.4pt} \\
\pi_\phi^*(X) \xrightarrow[\ \ g\ \ ]{} \psi \quad \text{over } \{\phi\}
\end{array}
$$

described as follows. Given a family $f = (f_a)$ over A, define $\overline{f}_{(a,x)}(y) = \pi'f_a(y)(x) \in \psi(a, x)$. And conversely, given a family $g = (g_{(a,x)})$ over $\{\phi\}$ one puts $\overline{g}_a(y) = \lambda x. \langle x, g_{(a,x)}(y)\rangle \in \prod_\phi(\psi)(a)$. $\qquad\square$

By collecting the above lemmas we get the main result.

10.6.5. Theorem. *The fibration* $\begin{array}{c}\text{CFam}(\mathbf{Dcpo})\\ \downarrow \\ \mathbf{Dcpo}\end{array}$ *of "dcpos continuously indexed by dcpos" is a closed comprehension category.* $\qquad\square$

Closures indexed by closures

In the remainder of this section we sketch how the so-called closure model from [302, 15] fits in a categorical framework of closed comprehension categories. Let $(P\omega, \subseteq)$ be the complete lattice of subsets of the set of natural numbers ω ($= \mathbb{N}$). The set of (Scott-)continuous functions $[P\omega \to P\omega]$ comes equipped with continuous maps $F: P\omega \to [P\omega \to P\omega]$ and $G: [P\omega \to P\omega] \to P\omega$ for application and abstraction, satisfying $F \circ G = \text{id}$ and $G \circ F \ge \text{id}$. This makes it a (non-extensional, additive) model of the untyped λ-calculus, see [13]. As usual, we write $x \cdot y$ for $F(x)(y)$ and $\lambda x \ldots$ for $G(\lambda x \ldots)$. Further, we use that there is a continuous surjective pairing $[-, -]: P\omega \times P\omega \to P\omega$ with continuous projections π, π'.

A **closure** is an element $a \in P\omega$ satisfying $a \circ a = a \ge I$, where $a \circ a = \lambda x. a \cdot (a \cdot x)$ and $I = \lambda x. x$. Closures form a category **Clos** by the stipulating that a morphism $u: a \to b$ between closures is an element $u \in P\omega$ satisfying $b \circ u \circ a = u$ (or equivalently, $b \circ u = u$ and $u \circ a = u$). One easily verifies that **Clos** is a Cartesian closed category (see [302]), with terminal object $1 = \lambda x. \omega$, product $a \times b = \lambda x. [a \cdot \pi x, b \cdot \pi'x]$ and exponent $a \Rightarrow b = \lambda x. b \circ x \circ a$. For

$a \in \mathbf{Clos}$ we write $\mathrm{Im}(a) = \{a \cdot x \mid x \in P\omega\}$; then $\mathrm{Im}(a) = \{x \in P\omega \mid a \cdot x = x\}$ and $\mathrm{Im}(a \Rightarrow b) = \mathbf{Clos}(a, b)$.

A crucial result is the existence of a closure Ω with $\mathrm{Im}(\Omega) = \mathrm{Obj}\,\mathbf{Clos}$, so that $a \in P\omega$ is a closure if and only if $\Omega \cdot a = a$ (see [15, Theorem 1.12], where this result is attributed to Martin-Löf, Hancock and D. Scott independently). This gives us the possibility to define a split fibration $\begin{smallmatrix} \mathrm{Fam}(\mathbf{Clos}) \\ \downarrow \\ \mathbf{Sets} \end{smallmatrix}$ of 'closure-indexed closures'. Objects of the total category $\mathrm{Fam}(\mathbf{Clos})$ are arrows $X: a \to \Omega$ in $\mathbf{Clos}$. An arrow $(X: a \to \Omega) \to (Y: b \to \Omega)$ is a pair (u, α) with $u: a \to b$ a morphism in the base category $\mathbf{Clos}$ and $\alpha \in P\omega$ is an 'a-indexed family of morphisms'. The latter means that $\alpha \circ a = \alpha$ and $\alpha \cdot z: X \cdot z \to Y \cdot (u \cdot z)$ is a morphism in $\mathbf{Clos}$ (for each $z \in P\omega$). Here we use that $X \cdot z$ is an element of $\mathrm{Im}(\Omega) = \mathrm{Obj}\,\mathbf{Clos}$. The functor $\mathrm{Fam}(\mathbf{Clos}) \to \mathbf{Clos}$ sending $(X: a \to \Omega)$ to a is then a split fibration. It has a fibred terminal object via a functor $1: \mathbf{Clos} \to \mathrm{Fam}(\mathbf{Clos})$ by

$$a \mapsto (\lambda xy.\,\omega: a \to \Omega).$$

A right adjoint $\{-\}: \mathrm{Fam}(\mathbf{Clos}) \to \mathbf{Clos}$ to 1 is described by

$$(X: a \to \Omega) \mapsto \lambda z.\,[a \cdot \pi z, X \cdot \pi z \cdot \pi' z].$$

It is not hard to check that we get a fibration with full comprehension. It further has products and strong coproducts (along the induced projections $\pi_X: \{X\} \to a$): for an object $X: a \to \Omega$ over $a \in \mathbf{Clos}$ and an object $Y: \{X\} \to \Omega$ over $\{X\}$ one defines coproduct and product objects $\coprod_X(Y), \prod_X(Y): a \rightrightarrows \Omega$ over a by

$$\coprod_X(Y) = \lambda zv.\,[X \cdot z \cdot \pi v, Y \cdot [a \cdot z, X \cdot z \cdot \pi v] \cdot \pi'v]$$
$$\prod_X(Y) = \lambda zvw.\,Y \cdot [a \cdot z, X \cdot z \cdot w] \cdot (v \cdot (X \cdot z \cdot w)).$$

This yields a (split) closed comprehension category.

Exercises

10.6.1. (i) For continuous functors $\phi, \psi: A \to \mathbf{Dcpo}^{\mathrm{EP}}$, show that the product and exponent in the fibre over A, which are by Proposition 10.5.4 described by the formulas

$$\phi \times \psi = \coprod_\phi(\pi_\phi^*(\psi)) \qquad \text{and} \qquad \phi \Rightarrow \psi = \prod_\phi(\pi_\phi^*(\psi))$$

are pointwise the Cartesian product and exponent of $\mathbf{Dcpo}$.

    (ii) Check that Beck-Chevalley holds for the dependent coproducts $\prod_\phi(\psi)$ and products $\coprod_\phi(\psi)$ in Lemmas 10.6.3 and 10.6.4.

10.6.2. A Scott-domain is a bounded complete algebraic dcpo. Write $\mathbf{SD} \hookrightarrow \mathbf{Dcpo}$ for the full subcategory of Scott-domains (and Scott-continuous functions).

For a Scott-domain A a functor $\phi \colon A \to \mathbf{SD}^{\mathrm{EP}}$ will be called continuous if it is continuous as a family of dcpos, *i.e.* if it satisfies the condition mentioned in the beginning of this section. Such a functor ϕ may be seen as a continuous family of Scott-domains, indexed by a Scott-domain. Prove that the CFam($-$) fibration for Scott-domains also yields a closed comprehension category, essentially by checking that for a continuous family $\phi \colon A \to \mathbf{SD}^{\mathrm{EP}}$

(i)   the Grothendieck completion $\{\phi\}$ is again a Scott-domain;

(ii) the dcpo $\prod_\phi(\psi)(a)$ in Lemma 10.6.4 is also a Scott-domain, for $\psi \colon \{\phi\} \to \mathbf{SD}^{\mathrm{EP}}$ a continuous family over ϕ.

10.6.3. (i)   Show that the fibration $\begin{smallmatrix} \mathrm{Fam}(\mathbf{Clos}) \\ \downarrow \\ \mathbf{Sets} \end{smallmatrix}$ of closure-indexed-closures has a split generic object (and thus that it is a small fibration).

(ii) Explain that, as a result of (i), one has a model in which the axiom $\vdash \mathsf{Type} \colon \mathsf{Type}$ holds.

[A dependent calculus with such a type of all types is inconsistent in the sense that every type is inhabited. This follows from Girard's paradox, see Exercise 11.5.3.]

(iii) Show that it is a $\lambda\omega$-fibration. (As such, it is described in [307].)

(iv) Consider the category **Clos** with objects

$$\Omega_0 = \Omega$$
$$\Omega_1 = \lambda x. \left[\Omega \cdot (\pi_1 x), \ \Omega \cdot (\pi_2 x), \ \Omega \cdot (\pi_2 x) \circ \pi_3 x \circ \Omega \cdot (\pi_1 x) \right]$$

where we have used $[\_, \_, \_]$ for the 3-tuple with projections π_1, π_2, π_3. Show that we get an internal category $(\Omega_1 \rightrightarrows \Omega_0)$ in **Clos** and that its externalisation is the above fibration of closure-indexed-closures.

Chapter 11

Higher order dependent type theory

In this final chapter several lines come together: the various logics and type theories that we have so far studied in isolation, are now combined into several powerful higher order dependent type theories. These type theories will be introduced as suitable combinations of earlier type theories. And correspondingly, their categorical semantics will be described via suitable combinations of structures—notably fibrations and comprehension categories—that we used earlier for the component type theories. We focus mostly on these modular aspects of higher order dependent type theory, and, accordingly, we leave many of the details of the syntax implicit. On the categorical side, the double rôle of comprehension categories—on the one hand as domains of quantification, see Section 9.3, and on the other as models of dependent type theory, see Section 10.4—is crucial in achieving this modularity.

We will consider three systems of higher order dependent type theory.

- Higher order predicate logic over *dependent* type theory; often, it will be called dependent predicate logic, with DPL as abbreviation.
- Polymorphic dependent type theory (PDTT); this may be seen as a propositions-as-types extension of DPL.
- Full higher order dependent type theory (FhoDTT), which, in a more drastic manner, is also a propositions-as-types extension of DPL.

We start this chapter with higher order dependent predicate logic (DPL). It may be contrasted with ordinary or simple (first or higher order) predicate logic, which is predicate logic over *simple* type theory (abbreviated as SPL, see Chapter 4), and also with polymorphic predicate logic PPL (discussed in

Section 8.6). Characteristic for DPL is that for a type σ,

$$\text{term variables } x\colon\sigma \text{ may occur both in } \begin{cases} \text{predicates } \varphi(x)\colon \mathsf{Prop} \\ \text{and in} \\ \text{types } \tau(x)\colon \mathsf{Type}. \end{cases}$$

We will consider the higher order version of DPL, with the axiom $\mathsf{Prop}\colon\mathsf{Type}$, but of course, one may also choose to use a first order logic over dependent type theory. This dependent predicate logic is quite natural and expressive, as will be argued below. It forms the basis for the proof assistant PVS see [242, 241].

The other two systems PDTT and FhoDTT of higher order dependent type theory are obtained by extending this *logic* over dependent type theory to a *type theory* over dependent type theory, in the propositions-as-types manner, by introducing explicit proof-objects (or terms) in type theory for derivations in logic. Instead of 'propositions over types' as in logic we shall talk about 'types over kinds' in type theory. In this situation we assume (like in the logic) that for a kind $A\colon\mathsf{Kind}$, we may have

$$\text{variables } \alpha\colon A \text{ occurring in } \begin{cases} \text{types } \sigma(\alpha)\colon \mathsf{Type} \\ \text{and in} \\ \text{kinds } B(\alpha)\colon \mathsf{Kind}, \end{cases}$$

so that we have both types and kinds depending on kinds. We then consider two possible extensions: one may additionally have

(1) types depending on types: variables $x\colon\sigma$, for $\sigma\colon\mathsf{Type}$, occurring in types $\tau(x)\colon\mathsf{Type}$;
(2) kinds depending on types: variables $x\colon\sigma$, for $\sigma\colon\mathsf{Type}$, occurring in kinds $A(x)\colon\mathsf{Kind}$.

In the first case one gets what we call **polymorphic dependent type theory** (PDTT). Its basis is formed by sequents of kinds A and types σ of the form

$$\alpha_1\colon A_1,\ldots,\alpha_n\colon A_n \mid x_1\colon\sigma_1,\ldots,x_m\colon\sigma_m \vdash \sigma_{m+1}\colon \mathsf{Type}$$

like in polymorphic type theory, with the addition that both kind and type contexts contain dependent kinds and types: kinds A_{i+1} may contain variables $\alpha_1,\ldots,\alpha_i$ in kinds and types σ_{j+1} may contain variables $\alpha_1,\ldots,\alpha_n$ ranging over kinds (as in polymorphic 'simple' type theory), and also variables $x_1,\ldots,x_j$ ranging over types. Hence, polymorphic dependent type theory adds both "kind dependency" and "type dependency" to polymorphic simple type theory.

In the second case with kinds additionally depending on types one looses the possibility to separate kind and type contexts, so that one gets sequents of the form

$$x_1\!:\!C_1, \ldots, x_n\!:\!C_n \vdash D\!:\mathsf{Kind/Type}$$

where $C_i\!:\!\mathsf{Kind/Type}$, *i.e.* where C_i is either a type or a kind. This yields systems like the Calculus of Constructions of Coquand and Huet [58]. Type theories like in (1), which capture dependent type theory over dependent kind theory have been proposed as HML by Moggi [230] and as the theory of predicates by Pavlović [252, 253]. Moggi precludes kinds depending on types in HML, because he wishes to use HML as a rudimentary programming language with a compile-time part of kinds which is independent of a run-time part of types. The motivation of Pavlović comes from logic: he thinks of types as propositions and of kinds as sets and does not want sets to depend on proofs (*i.e.* on inhabitants of propositions). Both arguments make good sense and form the basis for sensible type theories.

In this chapter we will be increasingly blurring the distinction between type theory and category theory, assuming that the reader is sufficiently prepared by the previous chapters. Also, we shall be rather sketchy in describing particular type theories, mainly because we see them as modular composites of other systems that we described earlier in greater detail. Here we mostly put emphasis on the way in which these components are put together. The "dependency relation" between syntactic categories (like Type and Kind) will be of crucial importance in such combinations, see the beginning of Section 11.5.

This chapter starts with an introduction to (the syntax of) dependent predicate logic. This syntactical material is immediately organised in a term model. It suggests the underlying categorical structure, which will be elaborated in Section 11.2. The dependent version of polymorphic type theory is studied in Section 11.3. It prompts a detailed investigation of both syntactical and categorical aspects of different sum types and equality types. We identify weak, strong and very strong versions of these sum and equality types, by distinguishing between certain dependencies in the elimination rules. These dependencies in type theory are related to indexing in category theory. They are described systematically in the beginning of Section 11.5. The remainder of this section is devoted to the syntax of full higher order dependent type theory (FhoDTT). In the subsequent Section 11.6 we describe various models of this FhoDTT, including different PER-models. And the final Section 11.7 will elaborate on the special model consisting of PERs in the effective topos **Eff**. It focuses on the (weak) completeness of the fibrations of families of PERs and of ω-sets over **Eff**. The "stack completions" of these fibrations will be identified as complete fibrations (of separated families and of separated families orthogonal to $\nabla 2 \in$ **Eff** respectively).

11.1 Dependent predicate logic

In this section we will first introduce (higher order) predicate logic over dependent type theory, or dependent predicate logic (DPL), for short, together with some motivating examples (mostly involving subset and quotient types). Then we sketch the syntax for such a logic. We put special emphasis on the organisation of (type and proposition) contexts in this logic, since this is what determines the underlying fibred structure. Towards the end of this section we shall describe a term model of DPL. Its categorical structure will be further investigated in the next section.

As already mentioned in the introduction to this chapter, the key ingredient of dependent predicate logic is that term variables $x : \sigma$ in $\sigma : \mathsf{Type}$ may occur both in propositions $\varphi(x) : \mathsf{Prop}$ (as in ordinary, or simple, predicate logic SPL), and in types $\tau(x) : \mathsf{Type}$ (as in dependent type theory DTT). This means that we have formation sequents in DPL of the form:

$$x_1 : \sigma_1, \ldots, x_n : \sigma_n \vdash \tau : \mathsf{Type} \qquad \text{and} \qquad x_1 : \sigma_1, \ldots, x_n : \sigma_n \vdash \varphi : \mathsf{Prop}$$

in which the term variables $\vec{x} : \vec{\sigma}$ may occur both in τ and in φ. But there are also entailment sequents of the form:

$$x_1 : \sigma_1, \ldots, x_n : \sigma_n \mid \varphi_1, \ldots, \varphi_m \vdash \psi$$

expressing that the proposition ψ follows from the premises $\varphi_1, \ldots, \varphi_m$, in the (dependent) type context of term variables $\vec{x} : \vec{\sigma}$. As before, we call $\vec{\varphi}$ the proposition context. In this sequent it is understood the φ_i and ψ are propositions in context $\vec{x} : \vec{\sigma}$.

In the higher order version of dependent predicate logic we shall be using an axiom $\vdash \mathsf{Prop} : \mathsf{Type}$. As a result, we can also have proposition variables $\alpha : \mathsf{Prop}$ occurring in types and propositions—and we may also (impredicatively) quantify over these, like in $\exists \alpha : \mathsf{Prop}.\, \varphi(\alpha)$ and $\forall \alpha : \mathsf{Prop}.\, \varphi(\alpha)$.

One of the key advantages of dependent (over simple) predicate logic is that it allows us to make full use of subset types and quotient types—introduced in simple predicate logic in the earlier sections 4.6 and 4.7. For example, in forming (dependent!) types

$$\frac{p : \mathsf{N}, n : \mathsf{N} \vdash n < p : \mathsf{Prop}}{p : \mathsf{N} \vdash \mathsf{Nat}(p) \stackrel{\mathrm{def}}{=} \{n : \mathsf{N} \mid n < p\} : \mathsf{Type}}$$

of natural numbers below p. Similarly, with quotients one can form the dependent type (or group) $\mathsf{Z}/p\mathsf{Z}$ of integers modulo p, in:

$$\frac{p : \mathsf{Z}, x : \mathsf{Z}, y : \mathsf{Z} \vdash x \sim_p y \stackrel{\mathrm{def}}{=} \exists z : \mathsf{Z}.\, (x - y) = z \cdot p : \mathsf{Prop}}{p : \mathsf{Z} \vdash \mathsf{Z}/p\mathsf{Z} \stackrel{\mathrm{def}}{=} \mathsf{Z}/\sim_p : \mathsf{Type}}$$

What happens here is that subset types $\{n: \mathsf{N} \mid n < p\} \rightarrowtail \mathsf{N}$ and quotient types $\mathsf{Z} \twoheadrightarrow \mathsf{Z}/p\mathsf{Z}$ are formed which depend on a term variable p. This is because p is occurring free in the propositions which give rise to these subsets and quotients. The natural setting for these features is a logic over dependent types.

In such a logic one also has convenient ways of expressing results like

$$p: \mathsf{N}, f: \mathsf{Nat}(p) \rightarrow \mathsf{Nat}(p) \mid \mathsf{injective}(f) \vdash \mathsf{surjective}(f)$$

saying (in essence) that every injective endofunction on a finite set is surjective. A similar example is the following standard result in topology.

$$n: \mathsf{N}, a: P(\mathsf{R}^n) \mid \mathsf{compact}(a) \vdash \mathsf{closed}(a) \wedge \mathsf{bounded}(a).$$

Another argument supporting the naturality of this dependent predicate logic is that it actually is an expressive version of the internal language of a topos. There are various ways to describe such a language. In most topos theoretic texts it is presented as a higher order predicate logic over *simple* type theory (see *e.g.* [186]). Phoa [262] describes this internal language explicitly as a logic over *dependent* type theory. Of course, this extended language can be translated back into the logic over simple type theory, but its advantage lies in its additional flexibility and expressiveness (see the examples above). Such dependent predicate logic also occurs in [165] and in [242, 241].

The basis for the syntax of DPL is formed by the rules for dependent type theory in Section 10.1. So we shall use dependent product types $\Pi x: \sigma. \tau$, strong dependent sum types $\Sigma x: \sigma. \tau$ and a unit type 1. We exclude the equality type $\mathsf{Eq}_\sigma(x, x')$, because equality will be dealt with at the propositional level. There is a special type $\mathsf{Prop}: \mathsf{Type}$ of propositions, such that propositions occur as inhabitants of Prop. This gives us higher order logic. As already mentioned, in DPL there are entailment sequents of the following form.

$$\overbrace{x_1: \sigma_1, \ldots, x_n: \sigma_n}^{\substack{\text{dependent type} \\ \text{context}}} \mid \underbrace{\varphi_1, \ldots, \varphi_m}_{\substack{\text{ordinary} \\ \text{proposition context}}} \vdash \psi$$

We shall often abbreviate such type contexts as $\Gamma = (x_1: \sigma_1, \ldots, x_n: \sigma_n)$ and proposition contexts as $\Theta = (\varphi_1, \ldots, \varphi_m)$. The logical rules for dependent predicate logic (DPL) have the same form as for simple predicate logic (SPL) in Figure 4.1 (on page 225), except that the type contexts Γ are to be understood as contexts in dependent type theory. For example, for the universal quantifier $\forall$ we have the formation rule

$$\frac{\Gamma, x\!:\!\sigma \vdash \varphi\!:\!\mathsf{Prop}}{\Gamma \vdash \forall x\!:\!\sigma.\,\varphi\!:\!\mathsf{Prop}}$$

with introduction and elimination rules

$$\frac{\Gamma, x\!:\!\sigma \mid \Theta \vdash \varphi}{\Gamma \mid \Theta \vdash \forall x\!:\!\sigma.\,\varphi}\ (x \text{ not in } \Theta) \qquad \frac{\Gamma \vdash M\!:\!\sigma \qquad \Gamma \mid \Theta \vdash \forall x\!:\!\sigma.\,\varphi}{\Gamma \mid \Theta \vdash \varphi[M/x]}$$

So the additional type dependency in DPL has no influence on the form of the rules: we can use the same logical rules in simple and dependent predicate logic. But notice that in DPL we cannot form

$$\frac{\Gamma, x\!:\!\sigma, \Delta \vdash \varphi\!:\!\mathsf{Prop}}{\Gamma, \Delta \vdash \forall x\!:\!\sigma.\,\varphi\!:\!\mathsf{Prop}}$$

unless x does not occur free in Δ: then we can exchange $x\!:\!\sigma$ and Δ (according to the exchange rule in dependent type theory, so that the earlier mentioned formation rule can be used). This complication is due to the fact that in dependent type theory the order of the variable declarations in a type context is important—whereas in simple type theory it is not.

As already mentioned, the axiom $\vdash \mathsf{Prop}\!:\!\mathsf{Type}$ provides us with higher order logic in which one can quantify over propositions and over predicates, like in the (closed) propositions

$$\forall \alpha\!:\!\mathsf{Prop}.\,\alpha \supset \alpha \qquad \text{and} \qquad \forall \alpha\!:\!\mathsf{Prop}.\,\forall p\!:\!\alpha \to \mathsf{Prop}.\,\exists x\!:\!\alpha.\,px.$$

And like in higher order predicate logic over simple type theory, there is an "extensionality of entailment" rule:

$$\frac{\Gamma \vdash P, Q\!:\!\sigma \to \mathsf{Prop} \qquad \Gamma, x\!:\!\sigma \mid \Theta, Px \vdash Qx \qquad \Gamma, x\!:\!\sigma \mid \Theta, Qx \vdash Px}{\Gamma \mid \Theta \vdash P =_{\sigma \to \mathsf{Prop}} Q}$$

The main novelty in dependent predicate logic is that subset and quotient types can be exploited in full generality. This makes DPL (instead of SPL) the natural logic for these type constructors. They now have formation rules

$$\frac{\Gamma, x\!:\!\sigma \vdash \varphi(x)\!:\!\mathsf{Prop}}{\Gamma \vdash \{x\!:\!\sigma \mid \varphi(x)\}\!:\!\mathsf{Type}} \qquad \frac{\Gamma, x\!:\!\sigma, y\!:\!\sigma \vdash R(x, y)\!:\!\mathsf{Prop}}{\Gamma \vdash \sigma/R\!:\!\mathsf{Type}}$$

in which we can have a proper type context Γ—which was required to be empty in Section 4.6 in order to stay within simple type theory. The associated introduction and elimination rules are formally as in Section 4.6, but one should read the type contexts as containing dependent types.

As an example, for an integer $p\!:\!\mathsf{Z}$ we shall describe the quotient group $\mathsf{Z}/p\mathsf{Z}$ of integers modulo p in some detail. Here we assume that Z is the ring of integers with function symbols $(0, +, -, 1, \cdot)$ for addition and multiplication, satisfying the usual equations.

One forms the quotient Z/pZ from the (equivalence) relation $\sim_p$ on Z defined as

$$p: Z, x, y: Z \vdash x \sim_p y \stackrel{\text{def}}{=} \exists z: Z. (x - y) =_Z z \cdot p : \mathsf{Prop}.$$

So that we can form the dependent type

$$p: Z \vdash Z/pZ \stackrel{\text{def}}{=} Z/\sim_p : \mathsf{Type}.$$

It comes equipped with the canonical map

$$p: Z, x: Z \vdash [x]_p : Z/pZ.$$

Now we can put

$$p: Z \vdash 0_p \stackrel{\text{def}}{=} [0]_p : Z/pZ.$$

for a (new) neutral element. Inverse $-_p$ and addition $+_p$ operations on Z/pZ can be defined via representatives:

$$p: Z, a: Z/pZ \vdash -_p a \stackrel{\text{def}}{=} \text{pick } x \text{ from } a \text{ in } [-x] : Z/pZ$$

$$p: Z, a, b: Z/pZ \vdash a +_p b \stackrel{\text{def}}{=} \text{pick } x, y \text{ from } a, b \text{ in } [x + y] : Z/pZ.$$

In this way the type Z/pZ of integers modulo p becomes an Abelian group.

In the remainder of this section we show how the context structure of dependent predicate logic gives rise to a term model constellation of (split) fibred categories

where

- $\mathbb{C}$ is the category of (dependent) type contexts Γ, as introduced in Section 10.3.
- $\mathbb{T}$ is the category of dependent-types-in-context $\Gamma \vdash \sigma: \mathsf{Type}$, fibred over $\mathbb{C}$ via $(\Gamma \vdash \sigma: \mathsf{Type}) \mapsto \Gamma$. This gives a closed comprehension category $\mathbb{T} \to \mathbb{C}^{\to}$, see Example 10.5.6.
- $\mathbb{P}$ is the category of propositions-in-dependent-type-contexts $\Gamma \vdash \varphi: \mathsf{Prop}$. A morphism $(\Gamma \vdash \varphi: \mathsf{Prop}) \to (\Delta \vdash \psi: \mathsf{Prop})$ in this category consists of a context morphism $\vec{M}: \Gamma \to \Delta$ in $\mathbb{C}$ for which one can derive $\Gamma \mid \varphi \vdash \psi(\vec{M})$ in DPL. This category $\mathbb{P}$ is then fibred over $\mathbb{C}$ via $(\Gamma \vdash \varphi: \mathsf{Prop}) \mapsto \Gamma$.

We see that since both propositions and types depend on (*i.e.* are indexed by) types, we get two fibrations $\frac{\mathbb{P}}{\mathbb{C}}$ and $\frac{\mathbb{T}}{\mathbb{C}}$ of categories of propositions and of types over contexts of types. These fibrations are related in the following way.

(1) The fibration $\overset{\mathbb{P}}{\underset{\mathbb{C}}{\downarrow}}$ of propositions has products and coproducts with respect to the comprehension category $\mathbb{T} \to \mathbb{C}^{\to}$ of types (see Section 9.3 for what this means precisely). Notice that this involves quantification along *dependent* projections $\pi: (\Gamma, x{:}\sigma) \to \Gamma$. But the pattern is the same as in *simple* predicate logic, where the fibration of propositions admits quantification with respect to the *simple* comprehension category of types (with Cartesian projections only).

(2) The fibration $\overset{\mathbb{P}}{\underset{\mathbb{C}}{\downarrow}}$ of propositions has a (split) generic object $(\alpha{:}\mathsf{Prop} \vdash \alpha{:}\mathsf{Prop}) \in \mathbb{P}$ over the singleton context $(\alpha{:}\mathsf{Prop}) \in \mathbb{C}$. And the latter is the domain of the projection $(\alpha{:}\mathsf{Prop}) \to ()$ in $\mathbb{C}$ induced by the closed type $(\vdash \mathsf{Prop}{:}\mathsf{Type}) \in \mathbb{T}$.

We shall elaborate on the categorical aspects of quantification ($\forall$ and $\exists$) of propositions over dependent types in DPL, described via quantification in terms of comprehension categories (as mentioned in (1)). For each dependent type $(\Gamma \vdash \sigma{:}\mathsf{Type}) \in \mathbb{T}$ over $\Gamma \in \mathbb{C}$ we have a dependent projection $\pi: (\Gamma, x{:}\sigma) \to \Gamma$. This map in the base category $\mathbb{C}$ induces a weakening functor $\pi^*: \mathbb{T}_\Gamma \to \mathbb{T}_{(\Gamma, x{:}\sigma)}$ acting on types, and also a weakening functor $\pi^*: \mathbb{P}_\Gamma \to \mathbb{P}_{(\Gamma, x{:}\sigma)}$ acting on propositions. The latter will be of interest here; it maps

$$\left(\Gamma \vdash \varphi{:}\mathsf{Prop}\right) \mapsto \left(\Gamma, x{:}\sigma \vdash \varphi{:}\mathsf{Prop}\right)$$

by adding a dummy variable $x{:}\sigma$. In DPL (like in SPL) the rules for existential $\exists x{:}\sigma. (-)$ and universal $\forall x{:}\sigma. (-)$ quantification can be reformulated as 'mate' rules (see Lemma 4.1.8):

$$\frac{\Gamma \mid \varphi \vdash \forall x{:}\sigma. \psi}{\Gamma, x{:}\sigma \mid \varphi \vdash \psi} \qquad\qquad \frac{\Gamma \mid \exists x{:}\sigma. \psi \vdash \varphi}{\Gamma, x{:}\sigma \mid \psi \vdash \varphi}$$

where φ below the lines is really $\pi^*(\varphi)$. This shows that we have adjunctions

$$\exists x{:}\sigma. (-) \dashv \pi^* \dashv \forall x{:}\sigma. (-)$$

The Beck-Chevalley condition holds because substitution $[L/-]$ of terms L in propositions $\forall x{:}\sigma. \psi$ and $\exists x{:}\sigma. \psi$ commutes appropriately with $\forall$ and $\exists$, as in Section 4.1. Thus the fibration $\overset{\mathbb{P}}{\underset{\mathbb{C}}{\downarrow}}$ has products and coproducts with respect to the comprehension category $\mathbb{T} \to \mathbb{C}^{\to}$.

The structure of these two fibrations of types and of propositions over contexts will be studied more systematically in the next section.

Exercises

11.1.1. (i) Consider the quotient group Z/pZ described above, and prove that there are conversions:

$$p: Z, a: Z/pZ \vdash a = \text{pick } x \text{ from } a \text{ in } [x + p]_p : Z/pZ$$
$$p: Z, a: Z/pZ \vdash a = \text{pick } x \text{ from } a \text{ in } [x - p]_p : Z/pZ.$$

(ii) And derive:

$$p: Z, a: Z/pZ \mid \top \vdash \exists x: Z. \, a = [x]_p.$$

11.1.2. Describe equality propositions $\Gamma, x: \sigma, x': \sigma \vdash x =_\sigma x': \text{Prop}$, for $\Gamma \vdash \sigma: \text{Type}$ in DPL (with rules as in SPL) as equality for the (term model) fibration $\begin{smallmatrix} \mathbb{P} \\ \downarrow \\ \mathbb{C} \end{smallmatrix}$ of propositions with respect to the comprehension category $\mathbb{T} \to \mathbb{C}^\to$ of types.

11.2 Dependent predicate logic, categorically

In this section we describe a fibred categorical structure capturing dependent predicate logic, essentially by suitably combining the components of this logic. We then proceed to describe dependent subset types and quotient types in this setting.

We have seen in Chapters 4 and 10 that

(1) Predicate logic over simple type theory (SPL) is described by a preorder fibration

$$\begin{array}{c} \mathbb{D} \\ \downarrow q \\ \mathbb{B} \end{array}$$

where $\mathbb{B}$ is the category of type contexts (or just types, considered as singleton contexts, if we use finite products of types).

(2) Dependent type theory (DTT) is captured by a (closed) comprehension category

$$\mathbb{E} \xrightarrow{\;\mathcal{P}\;} \mathbb{B}^\to$$
$$p \searrow \quad \swarrow \text{cod}$$
$$\mathbb{B}$$

where $\mathbb{B}$ is the category of (dependent) type contexts.

Combining these we get a structure for dependent predicate logic (DPL).

It looks as follows,

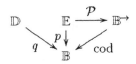

where $\begin{smallmatrix}\mathbb{D}\\\downarrow q\\\mathbb{B}\end{smallmatrix}$ is a preorder fibration and $\mathcal{P}\colon \mathbb{E} \to \mathbb{B}^{\to}$ is a comprehension category. The type formers $\Sigma, \Pi, 1$ and proposition formers $\exists, \forall, =$ and $\top, \wedge, \bot, \vee, \supset$ of DPL are then incorporated by imposing the following additional conditions on this structure:

(1) $\mathcal{P}$ is a closed comprehension category;

(2) q is a fibred bicartesian closed preorder fibration.

(3) q has $\mathcal{P}$-products $\forall$, $\mathcal{P}$-coproducts $\exists$ and $\mathcal{P}$-equality Eq.

For the higher order axiom $\vdash$ **Prop**: **Type** of DPL we further impose:

(4) for higher order: there is a closed type $\Omega \in \mathbb{E}_1$ in the fibre over the terminal object $1 \in \mathbb{B}$, such that the fibration q of propositions has a generic object $T \in \mathbb{D}$ above $\{\Omega\} = \mathrm{dom}(\mathcal{P}\Omega)$.

11.2.1. Definition. We shall call a structure $\begin{smallmatrix}\mathbb{D}\\\downarrow\\\mathbb{B}\end{smallmatrix}$ with $\mathbb{E} \to \mathbb{B}^{\to}$ as above, satisfying requirements (1) – (4) a **DPL-structure**, with 'DPL' for 'dependent predicate logic'.

It may be clear that the term model of DPL consisting of a fibration $\begin{smallmatrix}\mathbb{P}\\\downarrow\\\mathbb{C}\end{smallmatrix}$ and a comprehension category $\mathbb{T} \to \mathbb{C}^{\to}$ as discussed at the end of the previous section, is an instance of such a DPL-structure.

Our main examples of higher order fibrations:

$\begin{smallmatrix}\mathrm{Fam}(A)\\\downarrow\\\mathbf{Sets}\end{smallmatrix}$ for A a frame, $\qquad$ $\begin{smallmatrix}\mathrm{Sub}(\mathbb{B})\\\downarrow\\\mathbb{B}\end{smallmatrix}$ for $\mathbb{B}$ a topos, $\qquad$ $\begin{smallmatrix}\mathrm{UFam}(P\mathbb{N})\\\downarrow\\\mathbf{Sets}\end{smallmatrix}$

are in fact DPL-structures of the form,

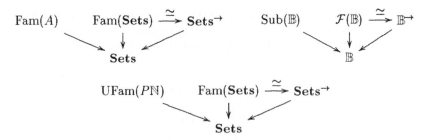

where we have chosen to present the comprehension categories in split form (with $\begin{smallmatrix}\mathcal{F}(\mathbb{B})\\\downarrow\\\mathbb{B}\end{smallmatrix}$ the split presentation of the codomain fibration of a topos $\mathbb{B}$, see

Example 10.5.9). These examples all arise (essentially) from the second point in the following result.

11.2.2. Proposition. *Let* $\begin{smallmatrix} \mathbb{D} \\ \downarrow \\ \mathbb{B} \end{smallmatrix}$ *be a higher order fibration.*

(i) *It forms part of a "simple" DPL-structure*

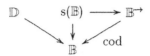

(ii) *If the base category* $\mathbb{B}$ *is a locally Cartesian closed category and the induced adjoints* $\coprod_u \dashv u^* \dashv \prod_u$ *in* q *(see Example 4.3.7) satisfy Beck-Chevalley, then we have a DPL-structure*

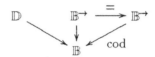

Proof. (i) Since the projections and diagonals of the simple comprehension category $s(\mathbb{B}) \to \mathbb{B}^{\to}$ are the Cartesian projections and diagonals, the higher order fibration has by definition quantification with respect to this comprehension category. The generic object condition holds, since the fibre over $1 \in \mathbb{B}$ of the simple fibration is isomorphic to $\mathbb{B}$.

(ii) Validity of the Beck-Chevalley condition ensures that the fibration $\begin{smallmatrix} \mathbb{D} \\ \downarrow \\ \mathbb{B} \end{smallmatrix}$ has products and coproducts with respect to the identity comprehension category $\mathbb{B}^{\to} \to \mathbb{B}^{\to}$ (see also Example 9.3.6). $\qquad\square$

This result tells us how to obtain DPL-structures from higher order fibrations. There is also a way to extract a higher order fibration from a DPL-structure, see Exercise 11.2.4 below.

In the remainder of this section we define what it means for a DPL-structure to admit (dependent) subset and quotient types. These definitions are important in the light of our claim that DPL is the natural logic for subset and quotient types, where these type constructors can be used in full generality (see the example of the quotient group Z/pZ of integers modulo p in the previous section). As examples of DPL-structures with dependent subset and quotient types we shall only discuss term models and topos models.

11.2.3. Definition. Consider a DPL-structure

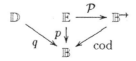

as introduced above. Then we can form the diagram

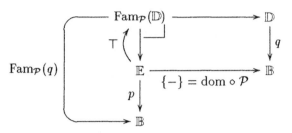

where $\mathrm{Fam}_{\mathcal{P}}(q)$ is defined as the composite $\mathrm{Fam}_{\mathcal{P}}(\mathbb{D}) \to \mathbb{E} \to \mathbb{B}$. The (fibred) terminal object functor $\top\colon \mathbb{E} \to \mathrm{Fam}_{\mathcal{P}}(\mathbb{D})$ is induced by the terminal object functor $\top\colon \mathbb{B} \to \mathbb{D}$ to q, namely as $X \mapsto (X, \top\{X\})$. In this situation, we say the DPL-structure has (**dependent**) **subset types** if there is a *fibred* right adjoint $\{-\}$ to $\top$ in the situation:

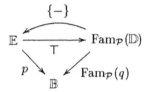

Such an adjoint induces a fibred projection functor $\mathrm{Fam}_{\mathcal{P}}(\mathbb{D}) \to V(\mathbb{E})$ over $\mathbb{E}$, like in Definition 4.6.1—where $V(\mathbb{E}) \hookrightarrow \mathbb{E}^{\to}$ is the full subcategory of vertical maps. We shall say that we have **full** dependent subset types if this functor is full and faithful.

(In the diagrams above we use the "Fam" notation because of the analogy with the construction in Exercise 1.9.11.)

The above right adjoint $\{-\}$ has to be a fibred one over p, because the type context Γ remains the same in the formation rule for dependent subset types:

$$\frac{\Gamma, x\colon \sigma \vdash \varphi\colon \mathsf{Prop}}{\Gamma \vdash \{x\colon \sigma \mid \varphi\}\colon \mathsf{Type}}$$

The sequent above the line yields an object $(\sigma, \varphi) \in \mathbb{E} \times_{\mathbb{B}} \mathbb{D} = \mathrm{Fam}_{\mathcal{P}}(\mathbb{D})$ over $\Gamma \in \mathbb{B}$. And also $\{x\colon \sigma \mid \varphi\} \in \mathbb{E}$ has to live above $\Gamma \in \mathbb{B}$.

We briefly describe this fibred subset adjunction in the term model $\begin{smallmatrix} \mathbb{P} \\ \downarrow \\ \mathbb{C} \end{smallmatrix}$ with $\mathcal{P}\colon \mathbb{T} \to \mathbb{C}^{\to}$ from the previous section. Notice that the auxiliary category $\mathrm{Fam}_{\mathcal{P}}(\mathbb{P})$ has:

objects   pairs consisting of a dependent type $\Gamma \vdash \sigma\colon \mathsf{Type}$ and a predicate $\Gamma, x\colon \sigma \vdash \varphi\colon \mathsf{Prop}$ on σ.

morphisms   from $\langle \Gamma \;\vdash\; \sigma\colon \mathsf{Type}, \; \Gamma, x\colon \sigma \;\vdash\; \varphi\colon \mathsf{Prop}\rangle$ to $\langle \Delta \;\vdash\; \tau\colon \mathsf{Type},$
$\Delta, y\colon \tau \;\vdash\; \psi\colon \mathsf{Prop}\rangle$ consist of a morphism $\vec{M}\colon \Gamma \to \Delta$ of contexts together with a term $\Gamma, x\colon \sigma \vdash N\colon \tau$ such that one can derive

$$\Gamma, x\colon \sigma \mid \varphi \vdash \psi(\vec{M}, N).$$

The terminal object functor $\mathbb{T} \to \mathrm{Fam}_{\mathcal{P}}(\mathbb{P})$ is then given by

$$\big(\Gamma \vdash \sigma\colon \mathsf{Type}\big) \mapsto \big(\Gamma, x\colon \sigma \vdash \mathsf{T}\colon \mathsf{Prop}\big).$$

A right adjoint to this functor T in the fibre over Γ involves a bijective correspondence between terms M and N in:

$$\frac{\big(\Gamma, x\colon \sigma \vdash \mathsf{T}\colon \mathsf{Prop}\big) \overset{M}{\longrightarrow} \big(\Gamma, y\colon \tau \vdash \psi\colon \mathsf{Prop}\big)}{\big(\Gamma \vdash \sigma\colon \mathsf{Type}\big) \underset{N}{\longrightarrow} \big(\Gamma \vdash \{y\colon \tau \mid \psi\}\colon \mathsf{Type}\big)}$$

That is, in:

$$\frac{\Gamma, x\colon \sigma \vdash M\colon \tau \quad\text{with}\quad \Gamma, x\colon \sigma \mid \mathsf{T} \vdash \psi[M/y]}{\Gamma, x\colon \sigma \vdash N\colon \{y\colon \tau \mid \psi\}}$$

The correspondence is given by

$$M \;\mapsto\; \mathsf{i}(M)$$
$$N \;\mapsto\; \mathsf{o}(N),$$

using the 'i' for 'in' and 'o' for 'out' as in the introduction and elimination rules for subset types in simple predicate logic (in Section 4.6).

11.2.4. Proposition. *For a topos* $\mathbb{B}$, *the associated DPL-structure*

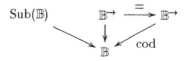

always has full dependent subset types. (For convenience we state this result for the the non-split presentation of this model.)

Proof. Consider the situation

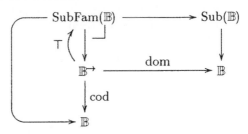

The category $\mathrm{SubFam}(\mathbb{B})$ thus has subobjects $Y \rightarrowtail X \longrightarrow I$ of families $X \longrightarrow I$ as objects. The terminal object functor $\mathbb{B}^{\rightarrow} \to \mathrm{SubFam}(\mathbb{B})$ maps

$$\left(Z \longrightarrow I \right) \;\longmapsto\; \left(Z \overset{\mathrm{id}}{\rightarrowtail} Z \longrightarrow I \right).$$

And it has a right adjoint over I, by composition:

$$\frac{\left(Z \overset{\mathrm{id}}{\rightarrowtail} Z \overset{\psi}{\longrightarrow} I \right) \overset{(f,g)}{\longrightarrow} \left(Y \overset{m}{\rightarrowtail} X \overset{\varphi}{\longrightarrow} I \right)}{\left(Z \overset{\psi}{\longrightarrow} I \right) \underset{g}{\longrightarrow} \left(Y \overset{\varphi \circ m}{\longrightarrow} I \right)}$$

since in a commuting diagram:

$$
\begin{array}{ccc}
Z & \overset{g}{\longrightarrow} & Y \\
& & \\
\mathrm{id}\searrow & f & \nearrow m \\
& Z \overset{f}{\longrightarrow} X & \\
& \psi\searrow \quad \swarrow\varphi & \\
& I &
\end{array}
$$

the map f is determined as composite $m \circ g$.

The associated projection functor $\mathrm{SubFam}(\mathbb{B}) \longrightarrow V(\mathbb{B}^{\rightarrow}) = \mathbb{B}^{\rightarrow \rightarrow}$ is

$$\left(Y \rightarrowtail X \longrightarrow I \right) \;\longmapsto\; \left(\begin{array}{c} Y \rightarrowtail X \\ \searrow \swarrow \\ I \end{array} \right).$$

It is obviously full and faithful. □

We turn to quotient types in DPL. Recall that for a comprehension category $\mathcal{P} \colon \mathbb{E} \to \mathbb{B}^{\rightarrow}$ we usually write $\{-\} \colon \mathbb{E} \to \mathbb{B}$ for $\mathrm{dom} \circ \mathcal{P}$. Let us write here $\{\{-\}\}$

for cod ∘ $\delta : \mathbb{E} \to \mathbb{B}$, where $\delta(X)$ is the diagonal map used in Definition 9.3.5 to define equality. Thus $\{\!\{-\}\!\}$ maps an object $X \in \mathbb{E}$ to

$$\{\!\{X\}\!\} = \{\mathcal{P}X^*(X)\} \longrightarrow \{X\}$$

Type theoretically, $\{\!\{-\}\!\}$ maps a dependent type $\Gamma \vdash \sigma \colon \mathsf{Type}$ to the context $(\Gamma, x \colon \sigma, x' \colon \sigma)$ that extends Γ with two variables of type σ.

11.2.5. Definition. For a DPL-structure

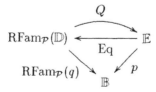

consider the category of relations, obtained by change-of-base in:

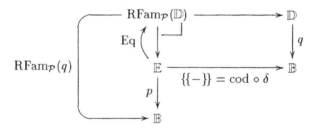

where $\mathrm{RFam}_{\mathcal{P}}(q)$ is the composite $\mathrm{RFam}_{\mathcal{P}}(\mathbb{D}) \to \mathbb{E} \to \mathbb{B}$. The (fibred) equality object functor $\mathrm{Eq} \colon \mathbb{E} \to \mathrm{RFam}_{\mathcal{P}}(\mathbb{D})$ is induced by the equality Eq in q (with respect to $\mathcal{P}$), namely as $X \mapsto (X, \mathrm{Eq}(1\{X\}))$. We say the DPL-structure has **(dependent) quotient types** if there is a *fibred* left adjoint Q to Eq in the situation:

$$\mathrm{RFam}_{\mathcal{P}}(\mathbb{D}) \underset{\mathrm{Eq}}{\overset{Q}{\rightleftarrows}} \mathbb{E}$$

Such an adjoint induces a "canonical quotient map" functor $\mathrm{RFam}_{\mathcal{P}}(\mathbb{D}) \to V(\mathbb{E})$ commuting with the domain functor dom: $V(\mathbb{E}) \to \mathbb{E}$, as in Proposition 4.8.5. We shall say that we have **full (or effective)** dependent quotient types if this functor is full and faithful, when restricted to equivalence relations.

In the type theoretic example with $\begin{smallmatrix}\mathbb{P}\\\downarrow\\\mathbb{C}\end{smallmatrix}$ and $\mathcal{P}\colon \mathbb{T} \to \mathbb{C}^{\to}$ as in the previous section, the category $\mathrm{RFam}_{\mathcal{P}}(\mathbb{P})$ has

objects	dependent types $\Gamma \vdash \sigma\colon \mathsf{Type}$ together with a relation $\Gamma, x\colon \sigma, x'\colon \sigma \vdash R(x, x')\colon \mathsf{Prop}$ on σ.
morphisms	from $(\Gamma, x\colon \sigma, x'\colon \sigma \vdash R(x, x')\colon \mathsf{Prop})$ to $(\Delta, y\colon \tau, y'\colon \tau \vdash S(y, y')\colon \mathsf{Prop})$ consist of a morphism $\vec{M}\colon \Gamma \to \Delta$ of contexts together with a term $\Gamma, x\colon \sigma \vdash N\colon \tau$ such that one can derive

$$\Gamma, x\colon \sigma, x'\colon \sigma \mid R(x, x') \vdash S(\vec{M}, N(x), N(x')).$$

The equality functor $\mathbb{T} \to \mathrm{RFam}_{\mathcal{P}}(\mathbb{P})$ sends a type $\Gamma \vdash \tau\colon \mathsf{Type}$ to the equality relation $\Gamma, y\colon \tau, y'\colon \tau \vdash y =_{\tau} y'\colon \mathsf{Prop}$ on τ. Quotient types in dependent predicate logic provide a left adjoint, since they induce a bijective correspondence (over Γ) between terms N and M in

$$\frac{\big(\Gamma, x\colon \sigma, x'\colon \sigma \vdash R(x, x')\colon \mathsf{Prop}\big) \xrightarrow{\ N\ } \big(\Gamma, y\colon \tau, y'\colon \tau \vdash y =_{\tau} y'\colon \mathsf{Prop}\big)}{\big(\Gamma \vdash \sigma/R\colon \mathsf{Type}\big) \xrightarrow[\ M\ ]{} \big(\Gamma \vdash \tau\colon \mathsf{Type}\big)}$$

i.e. in

$$\frac{\Gamma, x\colon \sigma \vdash N\colon \tau \quad \text{with} \quad \Gamma, x\colon \sigma, x'\colon \sigma \mid R(x, x') \vdash N(x) =_{\tau} N(x')}{\Gamma, a\colon \sigma/R \vdash M\colon \tau}$$

This correspondence is given by

$$N(x) \mapsto (\text{pick } x \text{ from } a \text{ in } N(x)) \qquad \text{and} \qquad M(a) \mapsto M[[x]_R/a].$$

11.2.6. Proposition. *Every topos—as a DPL-structure—has full dependent quotient types.*

Proof. Let $\mathbb{B}$ be our topos. Consider the relevant change-of-base situation

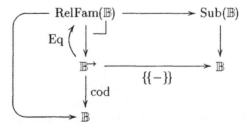

where the functor $\{\{-\}\}$ maps a family $X \to I$ to the domain $X \times_I X$ of its kernel pair $X \times_I X \rightrightarrows X$. An object of $\mathrm{RelFam}(\mathbb{B})$ is then a family $\varphi \colon X \to I$ together with a mono $\langle r_0, r_1 \rangle \colon R \rightarrowtail X \times_I X$ with $\varphi \circ r_0 = \varphi \circ r_1$. The equality functor $\mathbb{B}^{\to} \to \mathrm{RelFam}(\mathbb{B})$ sends a family $Y \to I$ to the (vertical) diagonal $Y \rightarrowtail Y \times_I Y$ on $Y \to I$. For a relation $\langle r_0, r_1 \rangle \colon R \rightarrowtail X \times_I X$ over $\varphi \colon X \to I$ we can form the coequaliser $c \colon X \twoheadrightarrow X/R$ in

$$
\begin{array}{c}
R \underset{r_1}{\overset{r_0}{\rightrightarrows}} X \xrightarrow{\ c\ } X/R \\
\varphi \searrow \quad \downarrow \varphi/R \\
I
\end{array}
$$

so that we get a new family φ/R over I. There is then a bijective correspondence (over I):

$$
\cfrac{\left(R \rightarrowtail X \times_I X \text{ over } X \xrightarrow{\varphi} I \right) \xrightarrow{\ (f,g)\ } \left(Y \rightarrowtail Y \times_I Y \text{ over } Y \xrightarrow{\psi} I \right)}{\left(X/R \xrightarrow{\varphi/R} I \right) \xrightarrow[h]{} \left(Y \xrightarrow{\psi} I \right)}
$$

since there is a map $f \colon X \to Y$ satisfying $\psi \circ f = \varphi$ with $g \colon R \dashrightarrow Y$ in the diagram below on the left, if and only if there is a map $h \colon X/R \to Y$ in the diagram on the right:

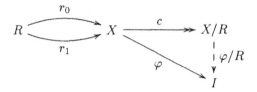

Fullness is left as an exercise below.   □

Exercises

11.2.1. Prove that dependent subset projections are monos, and similarly that dependent quotient maps are epis—in analogy with Lemmas 4.6.2 (i) and 4.8.2 (ii).

11.2.2. Consider DPL with dependent subset and quotient types. Give explicit descriptions in the term model $\overset{\mathbb{P}}{\underset{\mathbb{C}}{\downarrow}}$ with $\mathcal{P}\colon \mathbb{T} \to \mathbb{C}^{\to}$ of:

(i) the subset functor $\mathrm{Fam}_{\mathcal{P}}(\mathbb{P}) \to \mathbb{T}$

(ii) the quotient functor $\mathrm{RFam}_{\mathcal{P}}(\mathbb{P}) \to \mathbb{T}$

and show that these are fibred functors.

11.2.3. (i) Prove that the assignment $(\varphi, R) \mapsto \varphi/R$ defined in the proof of Proposition 11.2.6 yields a fibred functor $\mathrm{RelFam}(\mathbb{B}) \to \mathbb{B}^{\to}$.

[Remember that pullback functors preserve colimits in a topos.]

(ii) Prove fullness of dependent quotient types in this situation.

(iii) Formulate a Frobenius property for dependent quotient types, and show that it holds automatically in topos models. Explain also why.

11.2.4. Consider a DPL-structure $\overset{\mathbb{D}}{\underset{\mathbb{B}}{\downarrow}}q$ with $\mathcal{P}\colon \mathbb{E} \to \mathbb{B}^{\to}$ as described in the beginning of this section, and form by change-of-base:

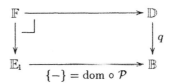

where $\mathbb{E}_1$ is the fibre category over the terminal object $1 \in \mathbb{B}$.

(i) Show that $\overset{\mathbb{F}}{\underset{\mathbb{E}_1}{\downarrow}}$ is a higher order fibration.

(ii) Check that if the DPL-structure has dependent (full) subset / quotient types, then $\overset{\mathbb{F}}{\underset{\mathbb{E}_1}{\downarrow}}$ has simple (full) subset / quotient types.

11.3 Polymorphic dependent type theory

Simple type theory is a 'propositions-as-types' extension of (constructive) propositional logic, via a Curry-Howard correspondence between inhabitation in type theory and derivability in logic, see Section 2.3. Models of propositional logic are Heyting algebras, which are poset (or preorder) Cartesian closed categories (CCCs). And these CCCs are models of simple type theory. Similarly, there are certain fibred categories for simple (higher order) predicate logic (SPL), which are preorder versions of the fibred categories for polymorphic type theory (PTT), see Section 8.1. The additional structure in the fibres of polymorphic type theory corresponds to derivations in predicate logic. An obvious next step is then to consider similar propositions-as-types extensions of dependent predicate logic (DPL), as described in the previous two sections. Then one looks for DPL-structures like in Definition 11.2.1, where the fibred preorders for propositional logic are replaced by proper fibre categories. The type theory obtained by extending dependent predicate logic in such a fashion will be called **polymorphic dependent type theory** (PDTT). The system HML of [230] and the calculus/theory of predicates of [252, 253] are of this kind. In this section we sketch the syntax of this polymorphic dependent type theory, and investigate its categorical semantics. Of special interest will be the description of quantification and generic objects via change-of-base.

Actually, in moving from DPL to DPTT we do not only replace a fibred preorder by a proper fibration, but by a comprehension category, so that we get a *dependent* type theory.

The syntax of PDTT will be very much like the syntax of PTT: we replace the logic of Prop's over Type's in SPL and DPL by a type theory with Type's over Kind's in PTT and PDTT—so that one can really read propositions as types in comparing logics and type theories. An additional advantage of using different syntactic universes Type and Kind in type theory is that it makes it still possible at some later stages to add an extra logical level by adding a level of Prop's. Thus we will have kinds and types in this polymorphic dependent type theory, written as:

$$\vdash A \colon \mathsf{Kind} \qquad \text{and} \qquad \vdash \sigma \colon \mathsf{Type}$$

both in appropriate contexts. In PDTT it is allowed that variables $\alpha \colon A$ inhabiting kinds occur both in kinds and in types, but variables $x \colon \sigma$ inhabiting types can only occur in types. (This restriction will disappear in Section 11.5.) As a result, in PDTT—like in PTT—one can (still) separate contexts,

$$\underbrace{\alpha_1 \colon A_1, \ldots, \alpha_n \colon A_n}_{\substack{\text{dependent kind} \\ \text{context}}} \mid \underbrace{x_1 \colon \sigma_1, \ldots, x_m \colon \sigma_m}_{\substack{\text{dependent} \\ \text{type context}}} \vdash M \colon \sigma_{m+1}$$

into a kind context followed by a type context. We often write these sequents as $\Xi \mid \Gamma \vdash M \colon \sigma_{m+1}$. Notice that in kind A_i the variables $\alpha_1, \ldots, \alpha_{i-1}$ may occur. And in type σ_j one may have free variables $\alpha_1, \ldots, \alpha_n, x_1, \ldots, x_{j-1}$.

Since we have both "kind-dependency" and "type-dependency", there are two ways of extending contexts (via comprehension):

$$\Xi \vdash B \colon \mathsf{Kind} \quad \text{yields an extended kind context} \quad \Xi, \beta \colon B$$
$$\Xi \mid \Gamma \vdash \sigma \colon \mathsf{Type} \quad \text{yields an extended type context} \quad \Xi \mid \Gamma, x \colon \sigma.$$

And indeed, the corresponding categorical structures will involve two comprehension categories: one for kinds and one for types.

(One can also consider type theories with such dependency only at the level of kinds, or only at the level of types, but we skip these intermediate versions.

They are captured by the categorical structures below in which one of the comprehension categories is "simple", see towards the end of this section.)

The features that we consider for these polymorphic dependent calculi are the following ones.

(1) Dependent product $\Pi\alpha: A.\ B$ and strong sum $\Sigma\alpha: A.\ B$ of kinds over kinds, plus a singleton kind $\vdash 1: \mathsf{Kind}$.
(2) Dependent product $\Pi x: \sigma.\ \tau$ and strong sum $\Sigma x: \sigma.\ \tau$ of types over types, together with a singleton type $\vdash 1: \mathsf{Type}$.
(3) Polymorphic product $\Pi\alpha: A.\ \sigma$ and sum $\Sigma\alpha: A.\ \sigma$ of types over kinds.
(4) A higher order axiom $\vdash \mathsf{Type}: \mathsf{Kind}$ together with the stipulation that the types in empty context:

$$\Xi \mid \emptyset \vdash \sigma: \mathsf{Type} \qquad \text{are the terms} \qquad \Xi \vdash \sigma: \mathsf{Type}$$

in this kind Type.

Notice that the three products and sums Π, Σ in (1), (2) and (3) are all different; they describe quantification of kinds over kinds, of types over types, and of types over kinds. In syntax it is custom to write the same symbols Π, Σ for these three cases, but categorically, these three forms of quantification are captured by three completely different adjunctions.

The rules for dependent products and sums of kinds over kinds and of types over types in (1) and (2) are as in dependent type (or kind) theory, see Section 10.1. The rules for polymorphic quantification of types over kinds are as in polymorphic type theory (see Section 8.1), except that the kind and type contexts are now 'dependent'. But this does not affect the form of the rules. What it means to have *strong* polymorphic sums $\Sigma\alpha: A.\ \sigma$ of types over kinds will be explained in the next section.

We turn to a categorical description of polymorphic dependent type theory (PDTT). Recall from the previous section that we describe dependent predicate logic via a "DPL-structure" of the form

$$\mathbb{D} \qquad \mathbb{E} \xrightarrow{\;\mathcal{P}\;} \mathbb{B}^{\rightarrow}$$
$$q \searrow \quad {\scriptstyle p} \downarrow \quad \swarrow \text{cod}$$
$$\mathbb{B}$$

We now wish to extend the (preorder) propositional part $\frac{\mathbb{D}}{\mathbb{B}}$ to a full type theory. Not just by allowing that $\frac{\mathbb{D}}{\mathbb{B}}$ is non-preordered (so that we get a simple type theory), but we wish to allow dependent types here. Thus $\mathbb{D}$ must be replaced by a (closed) comprehension category $\mathbb{D} \to \mathbb{A}^{\to}$ in a situation:

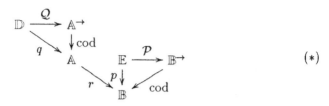

$$(*)$$

where the structure of $\mathbb{D} \to \mathbb{A}^{\to}$ is vertical with respect to the fibration $\begin{smallmatrix} \mathbb{A} \\ \downarrow r \\ \mathbb{B} \end{smallmatrix}$.
Type theoretically, this functor r describes the category $\mathbb{A}$ of kind-and-type-contexts $\Xi \mid \Gamma$ fibred over the category $\mathbb{B}$ of kind contexts Ξ. The objects
of $\mathbb{D}$ are then the types-in-context $\Xi \mid \Gamma \vdash \sigma:$ Type and the objects of $\mathbb{E}$ are
the kinds-in-context $\Xi \vdash A:$ Kind. The two projection functors $\mathcal{P}$ and $\mathcal{Q}$ map
a kind to its associated projection between kind contexts, and a type to its
projection between type contexts. In detail:

$$(\Xi \vdash A: \mathsf{Kind}) \overset{\mathcal{P}}{\longmapsto} ((\Xi, \alpha: A) \overset{\pi}{\longrightarrow} \Xi)$$
$$(\Xi \mid \Gamma \vdash \sigma: \mathsf{Type}) \overset{\mathcal{Q}}{\longmapsto} ((\Xi \mid \Gamma, x: \sigma) \overset{\pi}{\longrightarrow} (\Xi \mid \Gamma)).$$

There are the following four non-trivial points to be clarified about a structure $(*)$ as above.

(1) What does it mean that $\mathcal{Q}: \mathbb{D} \to \mathbb{A}^{\to}$ is a closed comprehension category
    "over r" (*i.e.* that it is vertical with respect to r)?
(2) How does one capture polymorphic quantification $\Pi\alpha: A.\,\sigma$ and $\Sigma\alpha: A.\,\sigma$
    of types over kinds in such a situation? The problem is that one cannot
    simply require that q has quantification with respect to $\mathcal{P}$ (as described in
    Section 9.3), since q and $\mathcal{P}$ have different base categories.
(3) What is the (categorical) rôle of the higher order axiom $\vdash$ Type: Kind?
(4) What does it mean in such a situation that polymorphic sums are strong?
    An answer to this last question will be postponed until the next section—
    where it will be shown in Proposition 11.4.3 that the polymorphic sums in
    PDTT are automatically strong.

We shall address the questions (1)–(3).

(1) As mentioned above, from a type theoretic perspective, a $\mathcal{Q}$-projection is
    a context projection $(\Xi \mid \Gamma, x: \sigma) \to (\Xi \mid \Gamma)$ for a type $\Xi \mid \Gamma \vdash \sigma:$ Type.
    This projection is between different type contexts in the same kind context Ξ. That is, it may be seen as a vertical projection $(\Gamma, x: \sigma) \to \Gamma$ in
    the fibre of $\mathbb{A}$ over the kind context $\Xi \in \mathbb{B}$. What we thus mean by re-

quiring that $Q: \mathbb{D} \to \mathbb{A}^{\to}$ is a comprehension category over r is that the Q-projections are r-vertical. Equivalently, $Q: \mathbb{D} \to \mathbb{A}^{\to}$ restricts to a functor $Q: \mathbb{D} \to V(\mathbb{A})$—where $V(\mathbb{A}) \hookrightarrow \mathbb{A}^{\to}$ is the full subcategory of vertical maps. This comprehension category Q is then "closed over r" if it is closed in the usual sense (see Section 10.5).

(In case Q is a comprehension category with unit, then this verticality of the Q-projections is equivalent to verticality of the counit of the comprehension adjunction $(1 \dashv \{-\})$, see Exercise 11.3.1 below.)

(2) For the polymorphic sum $\Sigma \alpha: A.\sigma$ and product $\Pi \alpha: A.\sigma$ of types over kinds one additionally requires the following. Since $r: \mathbb{A} \to \mathbb{B}$ is a fibration, one can transform the comprehension category $\mathcal{P}: \mathbb{E} \to \mathbb{B}^{\to}$ with basis $\mathbb{B}$ by change-of-base (as in Lemma 9.3.10) into a lifted comprehension category $r^*(\mathcal{P}): \mathbb{A} \times_{\mathbb{B}} \mathbb{E} \to \mathbb{A}^{\to}$ with basis $\mathbb{A}$. Then we simply say that the fibration $\begin{smallmatrix} \mathbb{D} \\ \downarrow q \\ \mathbb{A} \end{smallmatrix}$ of types has coproducts / products with respect to this lifted comprehension category $r^*(\mathcal{P})$.

We check in some detail that this captures the rules of polymorphic sum and product in a term model:

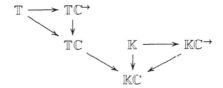

where $\mathbb{KC}$ and $\mathbb{TC}$ are categories of kind-contexts Ξ (and context morphisms), and of kind-and-type-contexts $\Xi \mid \Gamma$ respectively. A morphism $(\Xi \mid \Gamma) \to (\Xi' \mid \Gamma')$ in $\mathbb{TC}$ consists of two sequences of context morphisms $\vec{M}: \Xi \to \Xi'$ (in $\mathbb{KC}$) and $\vec{N}: \Gamma \to \Gamma'(\vec{M})$ (in kind context Ξ). Then $\mathbb{K}$ is the category of kinds-in-context $\Xi \vdash A: \mathsf{Kind}$ fibred over their contexts, with projection functor $\mathbb{K} \to \mathbb{KC}^{\to}$ sending a kind-in-context $\Xi \vdash A: \mathsf{Kind}$ to the projection $\pi: (\Xi, \alpha: A) \to \Xi$ in $\mathbb{KC}$. And $\mathbb{T}$ is the category of types-in-kind-and-type-context $\Xi \mid \Gamma \vdash \sigma: \mathsf{Type}$ fibred over their (kind plus type) contexts. It comes with a projection functor $\mathbb{T} \to \mathbb{TC}^{\to}$ which sends a type $\Xi \mid \Gamma \vdash \sigma: \mathsf{Type}$ to the vertical projection $\pi: (\Xi \mid \Gamma, x: \sigma) \to (\Xi \mid \Gamma)$ in $\mathbb{TC}$ over $\Xi \in \mathbb{KC}$.

The lifted comprehension category $\mathbb{TC} \times_{\mathbb{KC}} \mathbb{K} \longrightarrow \mathbb{TC}^{\to}$ in this situation is:

$$\langle \Xi \mid \Gamma, \Xi \vdash A: \mathsf{Kind} \rangle \longmapsto \begin{pmatrix} (\Xi, \alpha: A \mid \Gamma) \\ \downarrow \pi_A \\ (\Xi \mid \Gamma) \end{pmatrix}$$

(See Lemma 9.3.10.) The associated weakening functor is:

$$\mathbb{T}_{(\Xi \mid \Gamma)} \xrightarrow{\qquad \overline{\pi_A}^* \qquad} \mathbb{T}_{(\Xi, \alpha : A \mid \Gamma)}$$

$$(\Xi \mid \Gamma \vdash \tau : \mathsf{Type}) \longmapsto (\Xi, \alpha : A \mid \Gamma \vdash \tau : \mathsf{Type}).$$

Left and right adjoints to $\overline{\pi_A}^*$ thus involve correspondences

$$\frac{\Xi \mid \Gamma, z : \Sigma\alpha : A.\, \sigma \vdash M : \tau}{\Xi, \alpha : A \mid \Gamma, x : \sigma \vdash N : \tau} \qquad\qquad \frac{\Xi \mid \Gamma \vdash M : \Pi\alpha : A.\, \tau}{\Xi, \alpha : A \mid \Gamma \vdash N : \tau}$$

These are given by the standard introduction and elimination terms for polymorphic quantification: for Σ by

$$M(z) \mapsto M[\langle \alpha, x \rangle / z] \qquad \text{and} \qquad N(\alpha, x) \mapsto \mathsf{unpack}\; z \;\mathsf{as}\; \langle \alpha, x \rangle \;\mathsf{in}\; N$$

and for Π by

$$M \mapsto M\alpha \qquad \text{and} \qquad N \mapsto \lambda\alpha : A.\, N.$$

(3) What remains to be clarified is the rôle of the higher order axiom $\vdash \mathsf{Type} : \mathsf{Kind}$. In the type theory, terms $\Xi \vdash \sigma : \mathsf{Type}$ inhabiting this kind Type are the same as types $\Xi \mid \emptyset \vdash \sigma : \mathsf{Type}$ in the empty type context, (*i.e.* in the fibre of $\mathbb{T}$ over the terminal object 1Ξ in the fibre over Ξ). What one needs is a generic object for these types in empty type context.

In general, we consider the structure $(*)$ as above, and form the fibration $1^*(q)$ of types in empty type context by change-of-base along the terminal object functor $1 \colon \mathbb{B} \to \mathbb{A}$,

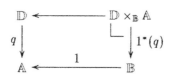

Then we can require that there is an object $\Omega \in \mathbb{E}$ in the fibre over the terminal object $1 \in \mathbb{B}$, such that this fibration $1^*(q)$ has a generic object over $\{\Omega\} = \mathrm{dom}(\mathcal{P}\Omega) \in \mathbb{B}$.

The pattern to describe generic objects in such multi-indexed structures is thus as for quantification (in the previous point): we first make the base categories match via change-of-base. Then we apply standard definitions.

11.3.1. Definition. A structure $(*)$ on page 665 (with two comprehension categories $\mathbb{E} \to \mathbb{B}^{\to}$, $\mathbb{D} \to \mathbb{A}^{\to}$ and a fibration $\begin{smallmatrix} \mathbb{A} \\ \downarrow \\ \mathbb{B} \end{smallmatrix}$) will be called a **PDTT-structure** if it satisfies the requirements as explained in the above points (1) – (3).

We leave it to the reader to describe a term model PDTT-structure in further detail. More examples are obtained via the following auxiliary result.

11.3.2. Lemma. *Let* $\begin{smallmatrix}\mathbb{E}\\\downarrow p\\\mathbb{B}\end{smallmatrix}$ *be a fibration with fibred Cartesian products* $\times$, *and let* $Q \colon \mathbb{D} \to \mathbb{B}^{\to}$ *be a comprehension category. This* Q *can be lifted along* p *so that we get the following situation.*

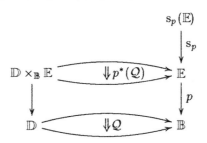

Then:

(i) *If* p *has* Q-*products (resp.* Q-*coproducts satisfying Frobenius), then the simple fibration* s_p *on* p *has* $p^*(Q)$-*products (resp.* $p^*(Q)$-*coproducts satisfying Frobenius).*

(ii) *If* p *is the fibration underlying a comprehension category* $\mathcal{P} \colon \mathbb{E} \to \mathbb{B}^{\to}$, *and we form the fibration* $\begin{smallmatrix}\mathrm{Fam}\mathcal{P}(\mathbb{E})\\\downarrow\{-\}^*(p)\\\mathbb{E}\end{smallmatrix}$ *by pullback of* p *along* $\{-\} = $ dom $\circ$ $\mathcal{P} \colon \mathbb{E} \to \mathbb{B}$ *(like in Exercise 10.5.5), we similarly get: if* p *has* Q-*products/coproducts, then* $\{-\}^*(p)$ *has* $p^*(Q)$-*products/coproducts.*

Proof. (i) We only do the case of coproducts. Assume therefore that p has coproducts $(\coprod_A \dashv \pi_A^*)$ along the projections $\pi_A = Q(A)$, satisfying Frobenius. For objects $A \in \mathbb{D}$ and $X \in \mathbb{E}$ with $qA = pX$ we get a projection map $p^*(Q)(A, X) = \overline{\pi_A}(X)$ in $\mathbb{E}$, see Lemma 9.3.10. It induces a weakening functor

$$s_p(\mathbb{E})_X \xrightarrow{\ \left(\overline{\pi_A}(X)\right)^*\ } s_p(\mathbb{E})_{\pi_A^*(X)} \qquad \text{by} \qquad Y \longmapsto \pi_A^*(Y).$$

Hence we have in the reverse direction $Z \mapsto \coprod_A(Z)$, with:

$$
\begin{aligned}
s_p(\mathbb{E})_{\pi_A^*(X)}\left(Z, \left(\overline{\pi_A}(X)\right)^*(Y)\right) &= \mathbb{E}_{\{A\}}\left(\pi_A^*(X) \times Z,\ \pi_A^*(Y)\right)\\
&\cong \mathbb{E}_{qA}\left(\coprod_A(\pi_A^*(X) \times Z),\ Y\right)\\
&\cong \mathbb{E}_{qA}\left(X \times \coprod_A(Z),\ Y\right)\\
&\doteq s_p(\mathbb{E})_X\left(\coprod_A(Z),\ Y\right).
\end{aligned}
$$

(ii) For $A \in \mathbb{D}$ and $X \in \mathbb{E}$, write $\gamma = \gamma_{(A,X)} \colon \{\pi_A^*(X)\} \overset{\cong}{\Rightarrow} \{\pi_X^*(A)\}$ for the mediating isomorphism between the following two pullback diagrams—both arising as in Lemma 10.4.4.

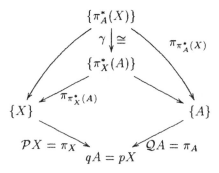

We now have a weakening functor for $\{-\}^*(p)$ induced by $p^*(\mathcal{Q})(A, X)$ between fibre categories:

$$\mathrm{Fam}_{\mathcal{P}}(\mathbb{E})_X = \mathbb{E}_{\{X\}} \longrightarrow \mathbb{E}_{\{\pi_A^*(X)\}} = \mathrm{Fam}_{\mathcal{P}}(\mathbb{E})_{\pi_A^*(X)}$$

namely $Y \mapsto \{\overline{\pi_A}(X)\}^*(Y)$. It has left and right adjoints by

$$Z \mapsto \coprod_{\pi_X^*(A)}(\gamma^*(Z)) \quad \text{and} \quad Z \mapsto \prod_{\pi_X^*(A)}(\gamma^*(Z))$$

where $\coprod$ and $\prod$ are the assumed ($\mathcal{Q}$-) coproducts and products of p. □

11.3.3. Proposition. (i) *A higher order fibration* $\overset{\mathbb{E}}{\underset{\mathbb{B}}{\downarrow}} p$ *gives rise to a "simple" PDTT-structure as on the left below.*

(ii) *If* $\overset{\mathbb{E}}{\underset{\mathbb{B}}{\downarrow}} p$ *is a fibred LCCC on a base category that is an LCCC, with additionally a generic object and coproducts and products along arbitrary maps in* $\mathbb{B}$, *then we get a PDTT-structure as on the right.*

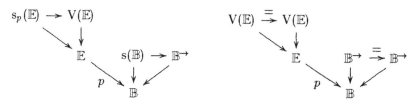

Proof. (i) The two simple comprehension categories $s(\mathbb{B}) \to \mathbb{B}^{\to}$ and $s_p(\mathbb{E}) \to V(\mathbb{E}) \hookrightarrow \mathbb{E}^{\to}$ are closed because of the CCC-structure in the basis, and in the fibres. And since p admits simple quantification, the simple fibration s_p on p admits quantification with respect to the lifted (along p)

simple comprehension category on $\mathbb{B}$. This follows from (i) in the previous lemma.

(ii) The LCCC-structure in the basis and in the fibres yields two closed comprehension categories, and the third form of quantification results from (ii) in the above lemma; it applies by Exercise 10.5.5 (ii).   □

The first point is a (type theoretic) analogue of Proposition 11.2.2 allowing us to transform $\lambda\omega$-fibrations into "simple" PDTT-structures (Exercise 11.3.2 below presents the reverse construction). The second point applies especially to the fibration $\begin{smallmatrix} \text{UFam}(\mathbf{PER}) \\ \downarrow \\ \omega\text{-Sets} \end{smallmatrix}$ of PERs over ω-sets—which is a fibred LCCC by Theorem 10.5.10. It does not apply to the fibration $\begin{smallmatrix} \text{UFam}(\mathbf{PER}) \\ \downarrow \\ \mathbf{Eff} \end{smallmatrix}$ of PERs over the effective topos, since this fibration does not have coproducts and products along all maps, see Section 11.7.

A minimal example

In the remainder of this section we shall further elaborate on the description of $\lambda\omega$-fibrations as special kind of PDTT-structures (as in (i) in the previous result). In particular, we focus on the use of "CT-structures" to describe models without all the type and kind formers. Recall from Section 2.4 that simple comprehension categories of the form $s(T) \to \mathbb{B}^{\to}$, for $T \subseteq \text{Obj}\,\mathbb{B}$ a CT-structure, can be used to describe models of simple type theory with exponent types $\to$ but without assuming Cartesian product types $\times$. Below we shall make similar use of these CT-structures to describe an ideal model—extending Example 2.4.9—for a version of second order polymorphic type theory with exponent $\to$ and polymorphic product and sum Π, Σ as only type constructors. The underlying categorical structure is as for PDTT-structures, but not all the operations are present. We present this ideal model as another example in which quantification and a generic object are described in PDTT-style via change-of-base.

More information on ideal models may be obtained from [206, 216, 155]. The construction below is based on [206].

11.3.4. Example. Let D be a reflexive dcpo $D \cong [D \to D]$ as in Example 2.4.9, and write $\mathcal{I}_D$ for the set of ideals $I \subseteq D$. Ordered by inclusion, this set forms a complete lattice, because ideals are closed under arbitrary intersections. Our aim is to construct a fibration $\begin{smallmatrix} \mathbb{E} \\ \downarrow p \\ \mathbb{B} \end{smallmatrix}$ together with two simple comprehension categories:

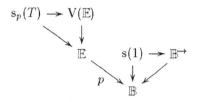

where the fibration of types $\overset{\mathrm{s}_p(T)}{\underset{\mathbb{E}}{\downarrow}}$ arising from $T \subseteq \mathrm{Obj}\,\mathbb{E}$ admits

- exponent types, in the sense that the simple comprehension category $\mathrm{s}_p(T) \to \mathbb{E}^{\to}$ has products with respect to itself.
- polymorphic products and coproducts with respect to the lifted comprehension category $\mathrm{s}(1) \to \mathbb{B}^{\to}$, where $1 \in \mathrm{Obj}\,\mathbb{B}$ yields a singleton CT-structure.
- a split generic object living over $1 \in \mathbb{B}$ for the fibration obtained by change-of-base of $\overset{\mathrm{s}_p(T)}{\underset{\mathbb{E}}{\downarrow}}$ along the terminal object functor $1 \colon \mathbb{B} \to \mathbb{E}$ of p.

In this ideal model we thus have many of the properties of a (simple) PDTT-structure.

The base category $\mathbb{B}$ has:

    objects   natural numbers $n \in \mathbb{N}$.

    morphisms $n \to m$ are m-tuples $u = (u_1, \dots u_m)$ of functions $u_i \colon \mathcal{I}_D^n \to \mathcal{I}_D$.

In this standard construction we get $0 \in \mathbb{B}$ as terminal object, and $n + m \in \mathbb{B}$ as Cartesian product of $n, m \in \mathbb{B}$. The object $1 \in \mathbb{B}$ gives rise to a simple comprehension category $\mathrm{s}(1) \to \mathbb{B}^{\to}$ with maps $\pi \colon n + 1 \to n$ as projections, see Definition 1.3.3.

    On top of $\mathbb{B}$ we construct a split fibration $\overset{\mathbb{E}}{\underset{\mathbb{B}}{\downarrow}}p$ with as fibre $\mathbb{E}_n$ over $n \in \mathbb{B}$ the category with

    objects   sequences $\langle X_1, \dots, X_k \rangle$ of maps $X_i \colon n \to 1$ in $\mathbb{B}$; these are in fact maps $n \to k$ in $\mathbb{B}$.

    morphisms $\langle X_1, \dots, X_k \rangle \to \langle Y_1, \dots, Y_\ell \rangle$ are ℓ-tuples $(f_1, \dots, f_\ell)$ of continuous functions $f_j \colon D^n \to D$ satisfying

$$\forall \vec{I} \in \mathcal{I}_D^n. \forall x_1 \in X_1(\vec{I}). \cdots \forall x_k \in X_k(\vec{I}).$$
$$f_j(x_1, \dots, x_k) \in Y_j(\vec{I}).$$

Reindexing is done by pre-composition, and fibred finite products are obtained by concatenation. The terminal object functor $1: \mathbb{B} \to \mathbb{E}$ sends $n \in \mathbb{B}$ to the empty sequence $n \to 0$ over n.

We choose the "set of types" $T \subseteq \mathrm{Obj}\,\mathbb{E}$ to be the set of singleton sequences $X: n \to 1$ in $\mathbb{B}$. Clearly it is closed under substitution, so it gives rise to a simple fibration $\begin{smallmatrix} s_p(T) \\ \downarrow \\ \mathbb{E} \end{smallmatrix}$ over p, as in Exercise 9.4.3. By change-of-base along the terminal object functor $1: \mathbb{B} \to \mathbb{E}$ we single out the maps $X: n \to 1$, for which the identity map $1 \to 1$ is a split generic object.

We have a simple comprehension category $s_p(T) \to \mathbb{E}^{\to}$ over p (see Exercise 9.4.3) with projections

$$(\langle X_1, \ldots, X_k \rangle, X) \mapsto \left(\begin{array}{c} \langle X_1, \ldots, X_k, X \rangle \\ \downarrow \pi \\ \langle X_1, \ldots, X_k \rangle \end{array} \right).$$

This simple fibration over p has products along these projections, since for an object $Y \in s_p(T)$ over $\langle X_1, \ldots, X_k, X \rangle$ we can define

$$\prod_X (Y)(\vec{I}) = \{ y \in D \mid \forall x \in X(\vec{I}).\, y \cdot x \in Y(\vec{I}) \}$$

over $\langle X_1, \ldots, X_k \rangle$. Then one easily checks the correspondences

$$\frac{\pi^*(Z) \longrightarrow Y \quad \text{over } \langle X_1, \ldots, X_k, X \rangle}{Z \longrightarrow \prod_X (Y) \quad \text{over } \langle X_1, \ldots, X_k \rangle}$$

which are pointwise as in Example 2.4.9. These simple products $\prod_X (Y)$ correspond to exponents $X \to Y$ of types.

The simple fibration $\begin{smallmatrix} s_p(T) \\ \downarrow \\ \mathbb{E} \end{smallmatrix}$ over p also has polymorphic products and coproducts with respect to the p-lifting of $s(1) \to \mathbb{B}^{\to}$: for a projection $\pi: n+1 \to n$ in $\mathbb{B}$ and an object $\langle X_1, \ldots, X_k \rangle \in \mathbb{E}$ over n we have a lifting in $\mathbb{E}$,

$$\langle X_1 \circ \pi, \ldots, X_k \circ \pi \rangle \xrightarrow{\quad \overline{\pi} \quad} \langle X_1, \ldots, X_k \rangle$$

Then, for an object $Y \in s_p(T)$ over $\langle X_1 \circ \pi, \ldots, X_k \circ \pi \rangle \in \mathbb{E}$ over $n+1 \in \mathbb{B}$, we define

$$\prod_n (Y)(\vec{I}) = \bigcap_{J \in \mathcal{I}_D} Y(\vec{I}, J) \qquad \text{and} \qquad \coprod_n (Y)(\vec{I}) = \bigvee_{J \in \mathcal{I}_D} Y(\vec{I}, J).$$

Then there are simply identities between:

$$
\frac{\overline{\pi}^*(Z) \xrightarrow{\ g\ } Y \quad \text{over } \pi^*(\vec{X})}{Z \xrightarrow[g]{} \prod_n(Y) \quad \text{over } \vec{X}}
\qquad
\frac{Y \xrightarrow{\ f\ } \overline{\pi}^*(Z) \quad \text{over } \pi^*(\vec{X})}{\coprod_n(Y) \xrightarrow[f]{} Z \quad \text{over } \vec{X}}
$$

The product correspondence is obvious, so we check the case of coproducts. For convenience we write boldface I for the sequence $\vec{I}$.

$$f\colon Y \longrightarrow \overline{\pi}^*(Z) \quad \text{over } \pi^*(\vec{X})$$

$\Leftrightarrow \forall I, J \in \mathcal{I}_D^{n+1}.\, \forall \vec{x} \in \overrightarrow{X(I)}.\, \forall y \in Y(I, J).\, f(\vec{x}, y) \in Z(I)$

$\Leftrightarrow \forall I \in \mathcal{I}_D^{n}.\, \forall \vec{x} \in \overrightarrow{X(I)}.\, \bigcup_{J \in \mathcal{I}_D} Y(I, J) \subseteq \{ y \mid f(\vec{x}, y) \in Z(I) \}$

$\Leftrightarrow \forall I \in \mathcal{I}_D^{n}.\, \forall \vec{x} \in \overrightarrow{X(I)}.\, \bigvee_{J \in \mathcal{I}_D} Y(I, J) \subseteq \{ y \mid f(\vec{x}, y) \in Z(I) \}$

    since $\{ y \mid f(\vec{x}, y) \in Z(I) \}$ is an ideal

$\Leftrightarrow \forall I \in \mathcal{I}_D^{n}.\, \forall \vec{x} \in \overrightarrow{X(I)}.\, \forall y \in \bigvee_{J \in \mathcal{I}_D} Y(I, J).\, f(\vec{x}, y) \in Z(I)$

$\Leftrightarrow f\colon \coprod_n(Y) \longrightarrow Z \quad \text{over } \vec{X}.$

This concludes our brief investigation of polymorphic dependent type theory. In the next section we will investigate strength of sums Σ (and of equality Eq) in such type theories. There we will see that the weak polymorphic sums that we have used in PDTT behave like strong ones—because of the presence of strong sums of types over types (see Proposition 11.4.3 in particular).

Exercises

11.3.1. Consider a fibration $\begin{smallmatrix} \mathbb{A} \\ \downarrow r \\ \mathbb{B} \end{smallmatrix}$ and a comprehension category with unit $\mathcal{Q}\colon \mathbb{D} \to \mathbb{A}^{\to}$ —given by a right adjoint $\{-\}$ to the terminal object functor 1 of the underlying fibration $q = \mathrm{cod} \circ \mathcal{Q}\colon \mathbb{D} \to \mathbb{A}$. Show that the following are equivalent.

(i) The functor $\mathcal{Q}\colon \mathbb{D} \to \mathbb{A}^{\to}$ restricts to vertical maps in $\mathbb{D} \to V(\mathbb{A}) \hookrightarrow \mathbb{A}^{\to}$.

(ii) The counit of the comprehension adjunction $(1 \dashv \{-\})$ is vertical.

(iii) The adjunction $(1 \dashv \{-\})$ is a fibred one in a situation:

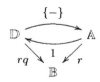

11.3.2. Consider a PDTT-structure $(*)$ as on page 665, and form the fibration $\begin{smallmatrix} \mathbb{F} \\ \downarrow \\ \mathbb{E}_1 \end{smallmatrix}$ by change-of-base in:

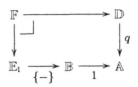

where $\mathbb{E}_1$ is the fibre category over the terminal object $1 \in \mathbb{B}$.

(i) Prove that $\begin{smallmatrix} \mathbb{F} \\ \downarrow \\ \mathbb{E}_1 \end{smallmatrix}$ is a $\lambda\omega$-fibration.

(ii) Check that first transforming a $\lambda\omega$-fibration into a simple PDTT-structure (as in Proposition 11.3.3 (i)), and then turning it back into a $\lambda\omega$-fibration returns the original fibration.

11.4 Strong and very strong sum and equality

In the previous section we described polymorphic dependent type theory (PDTT) in which types are indexed both by kinds and by types, and kinds only by kinds, and in which all three forms of quantification exist: dependent products and sums of kinds over kinds, of types over types and polymorphic products and sums of types over kinds. Strong sums $\Sigma\alpha: A.\, B$ of kinds over kinds and $\Sigma x: \sigma.\, \tau$ of types over types are as in Section 10.1 with first and second projections π, π'. What was left unexplained was the precise nature of strong versions of polymorphic sums $\Sigma\alpha: A.\, \sigma$ of types over kinds. It turns out that in general there are two versions, which we shall call 'strong' and 'very strong'. These will be discussed in the present section, together with similar 'strong' and 'very strong' versions of equality types. This material is based on joint work with Streicher.

We shall thus distinguish 'weak', 'strong' and 'very strong' sums $\Sigma\alpha: A.\, \sigma$ of types over kinds. The differences between these three versions involve certain dependencies in the elimination rules (such dependencies will be described more systematically in the next section). The formation and introduction rules are the same in all three cases, namely:

$$\frac{\Xi \vdash A: \mathsf{Kind} \qquad \Xi, \alpha: A \vdash \sigma: \mathsf{Type}}{\Xi \vdash \Sigma\alpha: A.\, \sigma: \mathsf{Type}} \qquad \frac{\Xi \vdash M: A \qquad \Xi \mid \Gamma \vdash N: \sigma[M/\alpha]}{\Xi \mid \Gamma \vdash \langle M, N \rangle: \Sigma\alpha: A.\, \sigma: \mathsf{Type}}$$

The only requirement so far on the ambient type theory is that variables $\alpha: A$ in kinds may occur in types $\sigma(\alpha): \mathsf{Type}$. The **weak** elimination rule is as follows.

$$\frac{\Xi \vdash \rho : \mathsf{Type} \qquad \Xi, \alpha : A \mid \Gamma, x : \sigma \vdash Q : \rho}{\Xi \mid \Gamma, z : \Sigma\alpha : A.\, \sigma \vdash (\mathsf{unpack}\ z\ \mathsf{as}\ \langle \alpha, x \rangle\ \mathsf{in}\ Q) : \rho} \text{(weak)}$$

This is the elimination rule for polymorphic sums of types over kinds as used earlier in Section 8.1 in (simple) polymorphic type theory.

Stronger versions of this rule are obtained by allowing extra freedom at the position "$\rho : \mathsf{Type}$" in the first assumption of this rule. For **strong** sums one allows the above type $\rho : \mathsf{Type}$ to contain a variable z of the sum type $\Sigma\alpha : A.\, B$. For these kind of strong sums one thus needs a type theory in which types may depend on types, like in polymorphic *dependent* type theory (but not in polymorphic *simple* type theory). The strong elimination rule then takes the following form.

$$\frac{\Xi \mid \Gamma, z : \Sigma\alpha : A.\, \sigma \vdash \rho : \mathsf{Type} \qquad \Xi, \alpha : A \mid \Gamma, x : \sigma \vdash Q : \rho[\langle \alpha, x \rangle / z]}{\Xi \mid \Gamma, z : \Sigma\alpha : A.\, \sigma \vdash (\mathsf{unpack}\ z\ \mathsf{as}\ \langle \alpha, x \rangle\ \mathsf{in}\ Q) : \rho} \text{(strong)}$$

In a next step, the "very strong" sums allow elimination with respect to types $\rho : \mathsf{Type}$ as above, and additionally with respect to kinds $\rho : \mathsf{Kind}$. And in both cases the term variable $z : \Sigma\alpha : A.\, \sigma$ may occur in ρ. We have not yet seen type theories where kinds may depend on (be indexed by) types, but we shall encounter them explicitly in the next section. One of the most important aspects of such calculi is that one can no longer separate kind and type contexts: there are contexts $\Gamma = (x_1 : C_1, \ldots, x_n : C_n)$ where each C_i is either a kind or a type (in the preceding context). We write this as

$$x_1 : C_1, \ldots, x_{i-1} : C_{i-1} \vdash C_i : \mathsf{Kind/Type}.$$

The **very strong** sum elimination rule has the following form.

$$\frac{\Gamma, z : \Sigma x : A.\, \sigma \vdash C : \mathsf{Type/Kind} \qquad \Gamma, \alpha : A, x : \sigma \vdash Q : C[\langle \alpha, x \rangle / z]}{\Gamma, z : \Sigma\alpha : A.\, \sigma \vdash (\mathsf{unpack}\ z\ \mathsf{as}\ \langle \alpha, x \rangle\ \mathsf{in}\ Q) : C} \left(\begin{smallmatrix} \text{very} \\ \text{strong} \end{smallmatrix} \right)$$

In all three cases the conversion rules are the same:

$$\mathsf{unpack}\ \langle M, N \rangle\ \mathsf{as}\ \langle \alpha, x \rangle\ \mathsf{in}\ Q\ =\ Q[M/\alpha, N/x] \qquad (\beta)$$
$$\mathsf{unpack}\ P\ \mathsf{as}\ \langle \alpha, x \rangle\ \mathsf{in}\ Q[\langle \alpha, x \rangle / z]\ =\ Q[P/z] \qquad (\eta).$$

These conversions are the same as for sums in polymorphic and in dependent type theory (so that one can derive the commutation conversion as in Exercise 10.1.4).

11.4.1. Remark. The above rules are presented for two syntactic categories Kind and Type. But we wish to include in our expositions the possibility that they coincide, *i.e.* that Kind = Type. Then there is no distinction anymore between strong and very strong sums; and these are then the same as the strong sums we described in dependent type theory in Section 10.1. The distinction strong/very strong is thus only of interest in situations where Kind and Type are different syntactic universes.

In Proposition 10.1.3 (i) we saw that the strong sums (of types over types) in dependent type theory can equivalently be described with a first and second projection map. The same result, with the same proof, can be obtained in the present more general setting for *very strong* sums.

11.4.2. Proposition. *The very strong sum-elimination rule can equivalently be formulated with first and second projection rules:*

$$\frac{\Gamma \vdash P \colon \Sigma\alpha \colon A.\, \sigma}{\Gamma \vdash \pi P \colon A} \qquad \frac{\Gamma \vdash P \colon \Sigma\alpha \colon A.\, \sigma}{\Gamma \vdash \pi' P \colon \sigma[\pi P/\alpha]}$$

with conversions $\pi\langle M, N\rangle = M$, $\pi'\langle M, N\rangle = N$ *and* $\langle \pi P, \pi' P\rangle = P$. $\qquad\square$

We emphasise that such a result cannot be obtained for strong sums, because for the first projection $\pi P = $ unpack P as $\langle\alpha, x\rangle$ in α one needs elimination with respect to kinds.

The next result (from [166]) is somewhat surprising.

11.4.3. Proposition. *In the presence of strong dependent sums* $\Sigma x \colon \tau.\, \rho$ *of types over types, polymorphic sums* $\Sigma\alpha \colon A.\, \sigma$ *of types over kinds are automatically strong.*

Proof. We shall use first and second projections π, π' for the dependent sums of types over types. Assume a type and term:

$$\Xi \mid \Gamma, z \colon \Sigma\alpha \colon A.\, \sigma \vdash \rho \colon \mathsf{Type} \qquad \Xi, \alpha \colon A \mid \Gamma, x \colon \sigma \vdash Q \colon \rho[\langle\alpha, x\rangle/z].$$

Using the "weak" unpack-terms and the projections π, π' we produce a "strong" unpack-term

$$\Xi \mid \Gamma, z \colon \Sigma\alpha \colon A.\, \sigma \vdash \underline{\mathsf{unpack}}\ z\ \underline{\mathsf{as}}\ \langle\alpha, x\rangle\ \underline{\mathsf{in}}\ Q \colon \rho.$$

as follows. Consider the combined sum-type

$$\Xi \mid \Gamma \vdash \rho' \stackrel{\mathrm{def}}{=} \Sigma z \colon (\Sigma\alpha \colon A.\, \sigma).\, \rho \colon \mathsf{Type}$$

together with the term

$$\Xi, \alpha \colon A \mid \Gamma, x \colon \sigma \vdash Q' \stackrel{\mathrm{def}}{=} \langle\langle\alpha, x\rangle, Q\rangle \colon \rho'.$$

The weak elimination rule then gives a term

$$\Xi \mid \Gamma, z \colon \Sigma\alpha \colon A.\, \sigma \vdash \mathsf{unpack}\ z\ \mathsf{as}\ \langle\alpha, x\rangle\ \mathsf{in}\ Q' \colon \rho'.$$

Hence we can put as our required term

$$\underline{\mathsf{unpack}}\ z\ \underline{\mathsf{as}}\ \langle\alpha, x\rangle\ \underline{\mathsf{in}}\ Q \stackrel{\mathrm{def}}{=} \pi'\big(\mathsf{unpack}\ z\ \mathsf{as}\ \langle\alpha, x\rangle\ \mathsf{in}\ Q'\big).$$

It is of type ρ, since

$$\pi\big(\text{unpack } z \text{ as } \langle \alpha, x \rangle \text{ in } Q'\big)$$
$$= \text{ unpack } \pi\big(\text{unpack } \langle \beta, y \rangle \text{ as } \langle \alpha, x \rangle \text{ in } Q'\big) \text{ as } \langle \beta, y \rangle \text{ in } z$$
$$= \text{ unpack } \pi Q'[\beta/\alpha, y/x] \text{ as } \langle \beta, y \rangle \text{ in } z$$
$$= \text{ unpack } \langle \beta, y \rangle \text{ as } \langle \beta, y \rangle \text{ in } z$$
$$= z.\qquad\qquad\qquad\qquad\qquad\qquad\qquad\qquad\qquad\qquad\square$$

This result tells us for example that in polymorphic dependent type theory in the previous section, the polymorphic sum $\Sigma\alpha\colon A.\,\sigma$ of types over kinds is automatically strong. Thus we were not negligent in not mentioning the requirement that polymorphic sums in PDTT are strong, because this holds automatically. And if we wish to use a polymorphic dependent calculus with *weak* sums $\Sigma\alpha\colon A.\,\sigma$ of types over kinds, then we are forced to use also weak sums of types over types. (Remember that very strong polymorphic sums cannot be used in PDTT because kinds do not depend on types.)

We turn to the categorical description of strong and very strong sums. We formulate these notions first in a situation of two comprehension categories with the same base category, where one describes dependent kinds and the other dependent types. This will later enable us to say what strong polymorphic sums are in a PDTT-structure via lifting of comprehension categories. Recall from Definition 10.5.2 that the categorical formulation of strength for coproducts involves canonical maps $\kappa\colon\{Y\} \to \{\coprod_X(Y)\}$. These will be used again in the present situation. Briefly, coproducts are strong when these maps are orthogonal (to the kind-projections), and very strong when they are isomorphisms. This orthogonality requirement comes from [148].

11.4.4. Definition. Take two comprehension categories $\mathbb{D} \xrightarrow{\;Q\;} \mathbb{B}^{\to} \xleftarrow{\;P\;} \mathbb{E}$ on the same base category $\mathbb{B}$.

(i) We say that Q has **strong** P-coproducts if, first of all, the underlying fibration $q = \mathrm{cod} \circ Q\colon\mathbb{D} \to \mathbb{B}$ has P-coproducts in the ordinary sense; that is, if for all objects $X \in \mathbb{E}$, the weakening functors $P(X)^* = \pi_X^*$ between the fibres of q have left adjoints $\coprod_X$ satisfying Beck-Chevalley. And second, if the induced canonical maps $\kappa = \kappa_{(X,A)}\colon\{A\} \to \{\coprod_X(A)\}$ are orthogonal to all Q-projections. This means: for every object $B \in \mathbb{D}$ and commuting rectangle in $\mathbb{B}$,

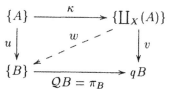

there is a unique 'diagonal-fill-in' w making the two triangles commute: $w \circ \kappa = u$ and $QB \circ w = v$. (Note that $\{-\}$ in this diagram is dom $\circ$ Q.)

(ii) And we say that Q has **very strong** $\mathcal{P}$-coproducts in case q has $\mathcal{P}$-coproducts in such a way that all these canonical maps $\kappa \colon \{A\} \to \{\coprod_X(A)\}$ in $\mathbb{B}$ are isomorphisms.

Recall that the canonical map $\kappa \colon \{A\} \to \{\coprod_X(A)\}$ arises by applying $\{-\} =$ dom $\circ$ $Q \colon \mathbb{D} \to \mathbb{B}$ to the (opcartesian) composite

$$
\begin{array}{c}
A \\
\eta \downarrow \\
\pi_X^* \coprod_X(A) \longrightarrow \coprod_X(A)
\end{array}
$$

Further, notice that when the comprehension categories $\mathcal{P}$ and $\mathcal{Q}$ happen to coincide (*i.e.* when Kind = Type), there is no difference between strong and very strong coproducts. The inverse of κ required for the implication (strong) $\Rightarrow$ (very strong) is obtained in the following diagram.

$$
\begin{array}{ccc}
\{A\} & \xrightarrow{\ \kappa\ } & \{\coprod_X(A)\} \\
\| & & \downarrow{\scriptstyle \pi\coprod_X(A)} \\
\{A\} & \xrightarrow[\ \pi_A\ ]{} & qA
\end{array}
$$

By uniqueness of fill-in maps, this morphism $\{\coprod_X(A)\} \to \{A\}$ is then also right-sided inverse of κ. This observation is in line with our earlier remark that in dependent type theory there is no difference between strong and very strong sums (of types over types).

Since projections are closed under pullback, there is the following reformulation of orthogonality—which is sometimes more convenient. The proof is easy and left to the reader.

11.4.5. Lemma. *The canonical map $\kappa \colon \{A\} \to \{\coprod_X(A)\}$ in the above definition is orthogonal to all $\mathcal{Q}$-projections if and only if: for every object $B \in \mathbb{D}$ over $\{\coprod_X(A)\} \in \mathbb{B}$ and morphism $u \colon \{A\} \to \{B\}$ in a commuting (outer) diagram*

$$
\begin{array}{ccc}
\{A\} & \xrightarrow{\ \kappa\ } & \{\coprod_X(A)\} \\
u \downarrow & {\scriptstyle \widehat{u}} & \| \\
\{B\} & \xrightarrow[\ \pi_B\ ]{} & \{\coprod_X(A)\} = qB
\end{array}
$$

there is a unique $\widehat{u}\colon \{\coprod_X(A)\} \dashrightarrow \{B\}$ *with* $\pi_B \circ \widehat{u} = \mathrm{id}$ *and* $\widehat{u} \circ \kappa = u$.   □

The above description of strong and very strong coproducts applies to a situation with two comprehension categories with the *same* base categories. This will be the case in the next section, but the two comprehension categories used for PDTT-structures in the previous section have *different* base categories (connected via a fibration, see the diagram below). In such a situation we can still say when coproducts are strong by suitably lifting the comprehension category of kinds (as we did to define polymorphic quantification).

11.4.6. Definition. Consider two comprehension categories $\mathcal{P}, \mathcal{Q}$ and a fibration r in the following diagram.

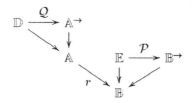

Then we say that $\mathcal{Q}$ has **strong coproducts** with respect to $\mathcal{P}$ in case $\mathcal{Q}$ has strong coproducts with respect to the lifted comprehension category $r^*(\mathcal{P})$.

All PDTT-structures have strong (polymorphic) coproducts of this kind by virtue of Proposition 11.4.3. Below we describe the orthogonality condition in a term model, and we leave it as an exercise to the reader to check that polymorphic coproducts are strong in the ideal model from Example 11.3.4.

11.4.7. Example. We elaborate the details of strong polymorphic sums $\Sigma\alpha\colon A.\,\sigma$ in a term model of PDTT (as described in the previous section) in relation to the above orthogonality condition. A lifted projection associated with a kind $\Xi \vdash A\colon \mathsf{Kind}$ in the category $\mathbb{TC}$ of type-and-kind-contexts at $\Xi \mid \Gamma$ is:

$$(\Xi, \alpha\colon A \mid \Gamma) \xrightarrow{\;\;\overline{\pi} = (\vec{\beta},\, \vec{v})\;\;} (\Xi \mid \Gamma)$$

of variables $\vec{\beta}$ in Ξ and $\vec{v}$ in Γ. For a type $\Xi, \alpha\colon A \mid \Gamma \vdash \sigma\colon \mathsf{Type}$ we assume that we can form the polymorphic sum $\Xi \mid \Gamma \vdash \Sigma\alpha\colon A.\,\sigma\colon \mathsf{Type}$. It is strong according to Lemma 11.4.5 if for each type

$$\Xi \mid \Gamma, z\colon \Sigma\alpha\colon A.\,\sigma \vdash \rho\colon \mathsf{Type} \quad \text{with term} \quad \Xi, \alpha\colon A \mid \Gamma, x\colon \sigma \vdash Q\colon \rho[\langle \alpha, x\rangle / z]$$

in a commuting (outer) diagram

$$
\begin{array}{ccc}
(\Xi, \alpha\colon A \mid \Gamma, x\colon \sigma) & \xrightarrow{\ \kappa = (\vec{\beta}, (\vec{v}, \langle \alpha, x \rangle))\ } & (\Xi \mid \Gamma, z\colon \Sigma\alpha\colon A.\,\sigma) \\[2pt]
{\scriptstyle (\vec{\beta},(\vec{v},\langle \alpha, x\rangle, Q))} \Big\downarrow & \quad \underset{\longleftarrow}{\dashrightarrow} \;\; {\scriptstyle (\vec{\beta},(\vec{v},z,\widehat{Q}))} & \Big\| \\[6pt]
(\Xi \mid \Gamma, z\colon \Sigma\alpha\colon A.\,\sigma, y\colon \rho) & \xrightarrow[\ \pi = (\vec{\beta}, (\vec{v}, z))\ ]{} & (\Xi \mid \Gamma, z\colon \Sigma\alpha\colon A.\,\sigma)
\end{array}
$$

there is a unique diagonal as indicated. This means that there is a term $\Xi \mid \Gamma, z\colon \Sigma\alpha\colon A.\,\sigma \vdash \widehat{Q}\colon \rho$ subject to the conversion $\widehat{Q}[\langle \alpha, x \rangle / z] = Q$. But then

$$
\widehat{Q} = \widehat{Q}[z/z] = \mathsf{unpack}\ z\ \mathsf{as}\ \langle \alpha, x \rangle\ \mathsf{in}\ \widehat{Q}[\langle \alpha, x \rangle / z] = \mathsf{unpack}\ z\ \mathsf{as}\ \langle \alpha, x \rangle\ \mathsf{in}\ Q.
$$

Hence we have the strong elimination rule as formulated in the beginning of this section.

Strong and very strong equality types

In the remainder of this section we have a brief look at strong and very strong versions of equality types. Since like sums, equality also involves left adjoints—not to weakening functors π^* but to contraction functors δ^*—a similar analysis applies.

For clarity, we are discussing "polymorphic" equality types over kinds here, with standard formation and introduction rules

$$
\frac{\Xi \vdash A\colon \mathsf{Kind}}{\Xi, \alpha\colon A, \beta\colon A \vdash \mathrm{Eq}_A(\alpha, \beta)\colon \mathsf{Type}}
\qquad\qquad
\frac{\Xi \vdash A\colon \mathsf{Kind}}{\Xi, \alpha\colon A \mid \emptyset \vdash \mathsf{r}_A(\alpha)\colon \mathrm{Eq}_A(\alpha, \alpha)}
$$

where we usually write r for $\mathsf{r}_A(\alpha)$ when confusion is unlikely.

One distinguishes three equality elimination rules, much like for sums: weak, strong and very strong ones. The difference lies again in the dependencies that one may have. The **weak** equality elimination rule is as follows.

$$
\frac{\Xi, \alpha\colon A, \beta\colon A \vdash \sigma\colon \mathsf{Type} \qquad \Xi, \alpha\colon A \mid \Gamma[\alpha/\beta] \vdash Q\colon \sigma[\alpha/\beta]}{\Xi, \alpha\colon A, \beta\colon A \mid \Gamma, z\colon \mathrm{Eq}_A(\alpha, \beta) \vdash (Q\ \mathsf{with}\ \beta = \alpha\ \mathsf{via}\ z)\colon \sigma} \ (\mathrm{weak})
$$

It is as in Section 8.1 for simple polymorphic type theory. If one uses a type theory where types may depend on types (like in a polymorphic dependent type theory), then the type $\Xi, \alpha\colon A, \beta\colon A \mid \Gamma \vdash \sigma\colon \mathsf{Type}$ in the first premise above, is allowed to contain an additional term variable $z\colon \mathrm{Eq}_A(\alpha, \beta)$, so that one can formulate the **strong** elimination rule as:

$$
\frac{\Xi, \alpha\colon A, \beta\colon A \mid \Gamma, z\colon \mathrm{Eq}_A(\alpha, \beta) \vdash \sigma\colon \mathsf{Type} \qquad \Xi, \alpha\colon A \mid \Gamma[\alpha/\beta] \vdash Q\colon \sigma[\alpha/\beta, \mathsf{r}/z]}{\Xi, \alpha\colon A, \beta\colon A \mid \Gamma, z\colon \mathrm{Eq}_A(\alpha, \beta) \vdash (Q\ \mathsf{with}\ \beta = \alpha\ \mathsf{via}\ z)\colon \sigma} \quad (\mathrm{strong})
$$

Finally in the very strong equality elimination rule one allows elimination not only with respect to types σ as above, but also with respect to kinds. This only makes sense in type theories where kinds may depend on types, *i.e.* where kinds may contain term variables (inhabiting types). And in such a type theory one cannot separate kind and type contexts anymore, so that the **very strong** elimination rule takes the following form.

$$\frac{\Gamma, \alpha\colon A, \beta\colon A, z\colon \mathsf{Eq}_A(\alpha, \beta) \vdash C\colon \mathsf{Type/Kind} \qquad \Gamma, \alpha\colon A \vdash Q\colon C[\alpha/\beta, \mathsf{r}/z]}{\Gamma, \alpha\colon A, \beta\colon A, z\colon \mathsf{Eq}_A(\alpha, \beta) \vdash (Q \text{ with } \beta = \alpha \text{ via } z)\colon C} \text{ (very strong)}$$

In all three cases the conversions are the same, namely:

$$Q \text{ with } \alpha = \alpha \text{ via } \mathsf{r} = Q \qquad (\beta)$$
$$Q[\alpha/\beta, \mathsf{r}/z] \text{ with } \beta = \alpha \text{ via } P = Q[P/z] \qquad (\eta).$$

The next two results are analogues of Propositions 11.4.2 and 11.4.3 for (strong) sums.

11.4.8. Proposition. *The very strong equality-elimination rule can equivalently be formulated by the following two rules.*

$$\frac{\Xi \vdash P\colon \mathsf{Eq}_A(M, N)}{\Xi \vdash M = N\colon A} \qquad \frac{\Xi \vdash P\colon \mathsf{Eq}_A(M, N)}{\Xi \vdash P = \mathsf{r}\colon \mathsf{Eq}_A(M, N)} \qquad \square$$

The proof of this result is the same as in dependent type theory (see Proposition 10.1.3 (ii)). It tells us that equality is very strong if and only if internal equality (inhabitation of $\mathsf{Eq}_A(M, N)$) and external equality (conversion $M = N\colon A$) are the same. In models of higher order dependent type theories *very strong* polymorphic equality occurs most frequently. But the version of polymorphic sum that one most often finds is the *strong* sum. This is because the very strong sums lead to "Girard's paradox" in the presence of the higher order axiom $\vdash \mathsf{Type}\colon \mathsf{Kind}$, see Exercise 11.5.3 in the next section.

11.4.9. Proposition. *Consider a type theory where types depend both on kinds and on types. In presence of strong dependent sums of types over types, polymorphic equality $\mathsf{Eq}_A(\alpha, \beta)\colon \mathsf{Type}$ for $A\colon \mathsf{Kind}$ is automatically strong.*

Proof. We proceed exactly as in the proof of Proposition 11.4.3. Assume a type $\Xi, \alpha\colon A, \beta\colon A \mid \Gamma, z\colon \mathsf{Eq}_A(\alpha, \beta) \vdash \sigma\colon \mathsf{Type}$ with a term $\Xi, \alpha\colon A \mid \Gamma[\alpha/\beta] \vdash Q\colon \sigma[\alpha/\beta, \mathsf{r}/z]$ and write

$$\Xi, \alpha\colon A, \beta\colon A \mid \Gamma \vdash \sigma' \stackrel{\text{def}}{=} \Sigma z\colon \mathsf{Eq}_A(\alpha, \beta). \sigma \colon \mathsf{Type}$$
$$\Xi, \alpha\colon A \mid \Gamma[\alpha/\beta] \vdash Q' \stackrel{\text{def}}{=} \langle \mathsf{r}, Q \rangle \colon \sigma'[\alpha/\beta].$$

Then we get by weak equality elimination, the term

$$\Xi, \alpha\colon A, \beta\colon A \mid \Gamma, z\colon \mathsf{Eq}_A(\alpha, \beta) \vdash (Q \text{ with } \beta = \alpha \text{ via } z)\colon \sigma'.$$

The strong elimination term that we seek is now obtained via second projection:

$$Q \text{ with } \beta = \alpha \text{ via } z \stackrel{\text{def}}{=} \pi'(Q' \text{ with } \beta = \alpha \text{ via } z).$$

This term is of the required type σ, since

$$\pi(Q' \text{ with } \beta = \alpha \text{ via } z)$$
$$= \pi(Q' \text{ with } \beta = \alpha \text{ via } z)[\alpha/\beta, \mathsf{r}/z] \text{ with } \beta = \alpha \text{ via } z$$
$$= \pi(\langle \mathsf{r}, Q[\alpha/\beta, \mathsf{r}/z]\rangle \text{ with } \alpha = \alpha \text{ via } \mathsf{r}) \text{ with } \beta = \alpha \text{ via } z$$
$$= \mathsf{r} \text{ with } \beta = \alpha \text{ via } z$$
$$= z[\alpha/\beta, \mathsf{r}/z] \text{ with } \beta = \alpha \text{ via } z$$
$$= z. \qquad\qquad \square$$

What is still lacking is a categorical description of strong and very strong equality. This follows the same pattern as for strong and very strong sums in Definition 11.4.4.

11.4.10. Definition. Assume two comprehension categories $\mathbb{D} \xrightarrow{\;Q\;} \mathbb{B}^{\rightarrow} \xleftarrow{\;P\;} \mathbb{E}$ on the same basis.

(i) We say that Q has **strong** P-equality if the underlying fibration $q = \mathrm{cod} \circ Q\colon \mathbb{D} \to \mathbb{B}$ has P-equality (via adjunctions $\mathrm{Eq}_X \dashv \delta_X^*$ plus Beck-Chevalley, where δ_X is the P-diagonal associated with $X \in \mathbb{E}$) in such a way that the induced canonical maps $\kappa = \kappa_{(X,A)}\colon \{A\} \to \{\mathrm{Eq}_X(A)\}$ are orthogonal to all Q-projections. This means: for every object $B \in \mathbb{D}$ and commuting rectangle in $\mathbb{B}$,

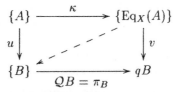

there is a unique diagonal making everything in sight commute.

(ii) And we say that Q has **very strong** P-equality in case q has P-equality in such a way that all these canonical maps $\{A\} \to \{\mathrm{Eq}_X(A)\}$ in $\mathbb{B}$ are isomorphisms.

This description of very strong equality captures the one used earlier in Proposition 4.9.3 in predicate logic as a special case. In Exercise 11.4.5 below it will turn out to be sufficient to have the Q-projection of "equality at 1" isomorphic to a P-diagonal to get very strong equality (in the presence of fibred CCC-structure). The essence of very strong equality is then that P-diagonals occur as Q-projections, as in the triangle below. This is often useful, see the examples in the subsequent Exercise 11.4.6.

Exercises

11.4.1. (i) Show that the elimination term $\underline{\text{unpack}}\ z\ \underline{\text{as}}\ \langle\alpha, x\rangle\ \underline{\text{in}}\ Q$ constructed in the proof of Proposition 11.4.3 comes with the appropriate conversions.

     (ii) Do the same for the term $Q\ \underline{\text{with}}\ \beta = \alpha\ \underline{\text{via}}\ z$ in the proof of Proposition 11.4.9.

11.4.2. Show that if a simple comprehension category $s(T) \to \mathbb{B}^{\to}$ has coproducts with respect to a comprehension category $\mathcal{P}\colon \mathbb{E} \to \mathbb{B}^{\to}$, then these coproducts are automatically strong.

[The underlying reason is that in simple comprehension categories there is no real type dependency so there is no difference between weak and strong sums.]

Conclude that the polymorphic coproducts $\coprod_n$ in the ideal model from Example 11.3.4 are strong.

11.4.3. Assume in Definition 11.4.10 that the comprehension categories $\mathcal{P}$ and $\mathcal{Q}$ are the same. Prove (categorically) that there is then no difference between strong and very strong equality.

11.4.4. Investigate the categorical description of strong equality in the term model of a polymorphic dependent calculus (like for sums in Example 11.4.7).

11.4.5. Let $\mathbb{D} \xrightarrow{\mathcal{Q}} \mathbb{B}^{\to} \xleftarrow{\mathcal{P}} \mathbb{E}$ be comprehension categories, where $\mathcal{Q}$ is full and has a unit $1\colon \mathbb{B} \to \mathbb{D}$. Recall (*e.g.* from Section 3.4) that we are often mostly interested in "equality at 1". This is the subject of the present exercise: we say $\mathcal{Q}$ has **pre-equality** with respect to $\mathcal{P}$ if for every object $X \in \mathbb{E}$ there is an opcartesian map $\underline{X}\colon 1\{X\} \to \text{Eq}(X)$ in $\mathbb{D}$ above the ($\mathcal{P}$-)diagonal δ_X, together with a Beck-Chevalley condition (in the style of Exercise 9.3.6).

Assume for each $X \in \mathbb{E}$ there is an object $\text{Eq}(X) \in \mathbb{D}$ over $\{\mathcal{P}(X)^*(X)\}$ with a map φ_X in a commuting triangle

$$
\begin{array}{ccc}
\{X\} & \xrightarrow{\ \varphi_X\ } & \{\text{Eq}(X)\} \\
& {\scriptstyle \delta_X}\searrow & \downarrow{\scriptstyle \mathcal{Q}(\text{Eq}(X))} \\
& \{\mathcal{P}(X)^*(X)\} &
\end{array}
$$

     (i) Show that if φ_X is orthogonal to all $\mathcal{Q}$-projections, then its transpose $1\{X\} \to \text{Eq}(X)$ is opcartesian over δ_X, so that $\mathcal{Q}$ has strong pre-equality.

     (ii) Prove that if each φ_X is an isomorphism, then $\mathcal{Q}$ has very strong pre-equality.

     (iii) Assume additionally that $q = \text{cod} \circ \mathcal{Q}$ is a fibred CCC. Prove that the definition

$$\text{Eq}_X(A) = \text{Eq}(X) \times \mathcal{P}\big(\mathcal{P}(X)^*(X)\big)^*(A)$$

yields that $\mathcal{Q}$ has (ordinary) very strong equality.

11.4.6. (i) Conclude from (ii) in the previous exercise that the comprehension categories of subobjects and of regular subobjects have very strong pre-equality with respect to every comprehension category.

(ii) And conclude also that the comprehension category UFam(**PER**) $\to$ ω-**Sets**$^\to$ of PERs over ω-sets has very strong equality with respect to UFam(ω-**Sets**) $\overset{\cong}{\to}$ ω-**Sets**$^\to$. Check that the same holds over **Eff**.

11.5 Full higher order dependent type theory

We come to the last type theory that will be discussed in this book. It is in a sense an extension of polymorphic dependent type theory (PDTT): it allows kinds to "depend on" types: kinds $A(x)$: Kind may contain a variable $x: \sigma$ inhabiting a type σ: Type (which is forbidden in PDTT, see the previous section). Since this new type theory allows all possible dependencies between Type and Kind—in a sense to be made precise below—we shall call it **full higher order dependent type theory**, abbreviated as FhoDTT. It is based on the Calculus of Constructions of [58], and can be seen as another combination (besides PDTT) of polymorphic and dependent type theory. Extensions of FhoDTT have been implemented in the proof tools COQ and LEGO, see the end of this section.

We start below by discussing some of the syntactic aspects of FhoDTT, and devote the next section to the categorical semantics of FhoDTT. For proof theoretic investigations we refer to [55, 92, 90, 22]. We first introduce a general notion of "dependency" in type theory, so that we can clearly characterise the type theory of FhoDTT among the many type theories that we have seen so far. These type dependencies actually form the basis for the classification of type theories in this book, see also [154, 163]. Basically, they determine the (indexed) categorical structures underlying the various type theories. In the second half of this section we describe and investigate FhoDTT in some detail. We put particular emphasis on a reflection between types and kinds, which results from the presence of sums Σ and unit types and kinds 1.

Dependencies in type theory

Remember from Section 11.3 that in polymorphic dependent type theory (PDTT) there are types depending both on kinds and on types, and kinds depending only on kinds. In the new type theory FhoDTT there are all four (combinatorially possible) dependencies between types and kinds. These observations prompt a more detailed investigation of such dependencies.

We will formulate these dependencies abstractly. Consider therefore two syntactic categories (or "universes", or "sorts") s_1, s_2 in a specific type theory—for example, $s_1 = \mathsf{Kind}$, $s_2 = \mathsf{Type}$. We then say that

s_2 **depends on** s_1, which will be written as $s_2 \succ s_1$,

if there are derivable sequents

$$\Gamma \vdash A \colon s_1 \qquad \text{and} \qquad \Gamma, x \colon A \vdash B(x) \colon s_2$$

in this type theory, with $x \colon A$ (actually) occurring free in $B(x)$. This means that there are "s_2-types" $B(x) \colon s_2$ containing variables $x \colon A$ inhabiting an "s_1-type" $A \colon s_1$. Put differently, $s_2 \succ s_1$ means that we may have

s_1-indexed s_2's as in the example $\left(B(x) \colon s_2\right)_{x \colon A}$ for $A \colon s_1$.

This last formulation is easiest to remember, so we repeat it explicitly.

11.5.1. Definition. In a type theory with universes s_1, s_2 we put

$s_2 \succ s_1 \Leftrightarrow$ there are (well-formed) expressions $A \colon s_1, B(x) \colon s_2$

forming an indexed collection

$$\left(B(x) \colon s_2\right)_{x \colon A \colon s_1}.$$

where x is actually free in B. And in that case we say: s_2 **depends on** s_1, or: s_2 **is indexed by** s_1.

For example, in polymorphic type theory PTT there are types depending on kinds (*i.e.* $\mathsf{Type} \succ \mathsf{Kind}$), typically in:

$$\left((\alpha \to \alpha) \colon \mathsf{Type}\right)_{\alpha \colon \mathsf{Type} \colon \mathsf{Kind}}.$$

This dependency $\mathsf{Type} \succ \mathsf{Kind}$ is characteristic of PTT. And in dependent type theory there are term variables occurring in types (amounting to $\mathsf{Type} \succ \mathsf{Type}$), as in the example of the type $\mathsf{NatList}(n)$ of lists (of natural numbers) of length n:

$$\left(\mathsf{NatList}(n) \colon \mathsf{Type}\right)_{n \colon \mathsf{N} \colon \mathsf{Type}}.$$

The dependency $\mathsf{Type} \succ \mathsf{Type}$ is typical for DTT. Figure 11.1 gives an overview of the dependencies that we have seen so far.

The first three type theories STT, PTT and DTT in this table can be seen as basic building blocks. The last two PDTT and FhoDTT are combinations. As is apparent in this table, the additional dependence in FhoDTT with respect to PDTT is the dependence of kinds on types. Under a propositions-as-types reading this becomes the dependence of types (or sets) on propositions. One can think of examples here—*e.g.* the set $\mathsf{steps}(p)$ of derivation steps in a proof $p \colon \sigma$ of proposition σ—but the naturality of this dependency is debatable (see also [230, 252, 253] for further discussion).

name	abbreviation	universes	dependencies
Simple Type Theory	STT	Type	—
Dependent Type Theory	DTT	Type	$\mathsf{Type} \succ \mathsf{Type}$
Polymorphic Type Theory	PTT	Type, Kind	$\mathsf{Type} \succ \mathsf{Kind}$
Polymorphic Dependent Type Theory	PDTT	Type, Kind	$\mathsf{Type} \succ \mathsf{Kind}$ $\mathsf{Type} \succ \mathsf{Type}$ $\mathsf{Kind} \succ \mathsf{Kind}$
Full higher order Dependent Type Theory	FhoDTT	Type, Kind	$\mathsf{Type} \succ \mathsf{Kind}$ $\mathsf{Kind} \succ \mathsf{Type}$ $\mathsf{Type} \succ \mathsf{Type}$ $\mathsf{Kind} \succ \mathsf{Kind}$

Fig. 11.1. Dependencies in various type theories

Taking dependencies as a starting point in the classification of type theories comes from [154, 163]. There, a collection of universes carrying a transitive relation $\succ$ of dependency is called a **setting**. The setting of a type theory determines the basic categorical structure of the type theory: $s_2 \succ s_1$ means that s_2 is fibred over s_1, since the s_2-types may be indexed by s_1-terms, as in $\left(B(x) : s_2 \right)_{x : A : s_1}$. This correspondence between dependency in type theories and indexing in category theory is the basis for all the categorical structures that we describe.

The view taken in this book is that a logic is always a logic over some type theory. We have explicitly studied predicate logic over STT in chapter 4, over PTT in Section 8.6 and also over DTT in Section 11.1. One can go a step further and consider logics over PDTT and over FhoDTT. Categorically this involves putting a suitable (preorder) fibration on top of a PDTT-/FhoDTT-structure. The above table can thus be extended with various logics. They can be described with an additional sort Prop, with typical dependency $\mathsf{Prop} \succ \mathsf{Type}$. It arises from predicates $\varphi(x) : \mathsf{Prop}$ which are indexed by types $\sigma : \mathsf{Type}$, as in:

$$\left(\varphi(x) : \mathsf{Prop} \right)_{x : \sigma : \mathsf{Type}}.$$

We conclude this excursion on dependencies with a review of the different settings in which the various sum and equality types (weak, strong, very strong) from the previous section are described. In a type theory with two sorts s_1, s_2 with dependency $s_2 \succ s_1$ it makes sense to consider products, sums and equality "s_2 over s_1", with the following formation rules.

$$\frac{\Gamma \vdash C : s_1 \qquad \Gamma, x : C \vdash D : s_2}{\Gamma \vdash \Pi x : C. D : s_2} \qquad\qquad \frac{\Gamma \vdash C : s_1 \qquad \Gamma, x : C \vdash D : s_2}{\Gamma \vdash \Sigma x : C. D : s_2}$$

$$\frac{\Gamma \vdash C : s_1}{\Gamma, x, x' : C \vdash \mathrm{Eq}_C(x, x') : s_2}$$

Sometimes we refer to these as (s_1, s_2)-products/sums/equality, when we wish to explicitly mention the relevant sorts.

The setting in which it makes sense to consider weak/strong/very strong (s_1, s_2) sum or equality types is summarised in the following table.

(s_1, s_2)-sum/equality	weak	strong	very strong
required dependency	$s_2 \succ s_1$	$s_2 \succ s_1$ $s_2 \succ s_2$	$s_2 \succ s_1$ $s_2 \succ s_2$ $s_1 \succ s_2$

It can be explained as follows. The weak elimination rule for (s_1, s_2)-sums is:

$$\frac{\Gamma \vdash B : s_2 \qquad \Gamma, x : C, y : D \vdash Q : B}{\Gamma, z : \Sigma x : C. D \vdash Q : B}$$

In the strong elimination rule one allows the variable $z : \Sigma x : C. D$, inhabiting the sum $\Sigma x : C. D$ of sort s_2, to occur free in $B : s_2$. This requires "s_2-type dependency" $s_2 \succ s_2$. In the very strong rule one additionally allows B to be of sort s_1. Since B may contain a variable $y : D$, where D is of sort s_2, this requires the dependency $s_1 \succ s_2$. The same analysis applies to equality types.

We see that the language of dependencies makes it easier to explain the differences between the various forms of sum and equality types. Also, in combination with Figure 11.1, it is now easy to see in which type theories it makes sense to consider, for example, strong (Kind, Type)-sums. Similarly, this language can be used in the description of constants, see [163] for more details.

Syntactical aspects of full higher order dependent type theory

In the remainder of this section we concentrate on a single type theory, namely on full higher order dependent type theory (FhoDTT). The starting point in the description of FhoDTT is the stipulation that there are two sorts Type, Kind and all (four) dependencies Type $\succ$ Kind, Kind $\succ$ Type, Type $\succ$ Type and Kind $\succ$ Kind between them. As a result, one cannot separate kind and type contexts, like in polymorphic dependent type theory: since variables inhabiting types may occur in kinds and variables inhabiting kinds may occur in types, a context is a sequence of variable declarations for types and for kinds. There are then sequents of the form

$$x_1\colon C_1, \ldots, x_n\colon C_n \vdash D\colon \mathsf{Type/Kind}$$

where $x_1\colon C_1, \ldots, x_i\colon C_i \vdash C_{i+1}\colon \mathsf{Type/Kind}$ for each $i < n$. The type and kind forming operations that we use in FhoDTT are

higher order axiom	$\vdash \mathsf{Type}\colon \mathsf{Kind}$
units for type and kind	$\vdash 1_T\colon \mathsf{Type}$ and $\vdash 1_K\colon \mathsf{Kind}$
with (sole) inhabitants	$\vdash \langle\rangle_T\colon 1_T$ and $\vdash \langle\rangle_K\colon 1_K$
products	$\Pi x\colon C.\, D$ for $C, D\colon \mathsf{Type/Kind}$.
strong sums	$\Sigma x\colon C.\, D$ for $C\colon \mathsf{Kind}, D\colon \mathsf{Type}$.
very strong sums	$\Sigma x\colon C.\, D$ for $C\colon \mathsf{Type}, D\colon \mathsf{Type}; C\colon \mathsf{Type}, D\colon \mathsf{Kind};$
	$C\colon \mathsf{Kind}, D\colon \mathsf{Kind}$.

So formally we have four different products and four different sums: types over types (Type, Type), kinds over types (Type, Kind), kinds over kinds (Kind, Kind) and types over kinds (Kind, Type). The first three sums are very strong, and the last one is only strong. (Recall, that there is actually no difference between strong and very strong for (Type, Type)-sums and (Kind, Kind-sums.)

In principle, these (Kind, Type)-sums can be very strong as well (since we have the dependency Kind $\succ$ Type in FhoDTT needed for the very strong elimination rule to make sense). But having very strong (Kind, Type)-sums has some detrimental effects. We will see below (in Corollary 11.5.4) that it results in an equivalence of types and kinds. This effectively gives us a type theory with a type of all types (Type: Type), and in such a type theory every type is inhabited. The result is known as "Girard's paradox". The original source is [95]; but many variations exist, see *e.g.* [54, 272, 153, 141] (and Exercise 11.5.3 below). The fact that all types are inhabited does not trivialise the type theory, but it makes it unusable in a propositions-as-types scenario, because it then means that each proposition has a proof.

It makes sense to consider "weak FhoDTT" with weak (Kind, Type)-sums, instead of strong ones, as required above. Proposition 11.4.3 then forces the (Type, Type)-sums to be weak as well. Extensional PERs (ExPERs) over ω-sets in Example 11.6.7 in the next section form a model of weak FhoDTT (but not of ordinary FhoDTT).

We shall see that the sums (and units) in FhoDTT give rise to a "reflection" Type $\leftrightarrows$ Kind between types and kinds. In order to make this categorically precise we have to use a term model of FhoDTT, consisting of two fibrations of types and of kinds, both over the same base category of (type and kind) contexts. This base category $\mathbb{C}$ consists of contexts $\Gamma = (x_1 : C_1, \ldots, x_n : C_n)$ in FhoDTT—with C_i: Type or C_i: Kind—and context morphisms (sequences of terms) between them. A category $\mathbb{T}$ of types over $\mathbb{C}$ has types-in-context $\Gamma \vdash \sigma$: Type as objects over $\Gamma \in \mathbb{C}$. A morphism $(\Gamma \vdash \sigma : \text{Type}) \to (\Delta \vdash \tau : \text{Type})$ in $\mathbb{T}$ consists of a context morphism $\vec{M} : \Gamma \to \Delta$ in $\mathbb{C}$ together with a term $\Gamma, x : \sigma \vdash N : \tau(\vec{M})$. There is a (full split) comprehension category $\mathbb{T} \to \mathbb{C}^{\to}$ mapping a type $\Gamma \vdash \sigma$: Type to the associated projection $\pi : (\Gamma, x : \sigma) \to \Gamma$. Similarly there is a category $\mathbb{K}$ of kinds-in-context $\Gamma \vdash A$: Kind as objects over $\Gamma \in \mathbb{C}$. And the (full split) comprehension category $\mathbb{K} \to \mathbb{C}^{\to}$ sends a kind $\Gamma \vdash A$: Kind to its context projection $\pi : (\Gamma, \alpha : A) \to \Gamma$. This leads to a situation:

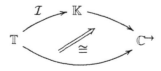

It will turn out that the very strong (Type, Kind)-sums and the (weak) (Kind, Type)-sums of (weak) FhoDTT yield a fibred reflection $\mathbb{T} \leftrightarrows \mathbb{K}$ between types and kinds. This is the content of the next two propositions.

11.5.2. Proposition. (i) *In the presence of a unit kind 1_K: Kind and weak* (Type, Kind)-*sums one can define a fibred functor* $\mathcal{I} : \mathbb{T} \to \mathbb{K}$ *from types to kinds by*

$$(\Gamma \vdash \sigma : \text{Type}) \mapsto (\Gamma \vdash \Sigma x : \sigma. 1_K : \text{Kind}).$$

(ii) *If* (Type, Kind)-*sums are very strong, then this functor* $\mathcal{I} : \mathbb{T} \to \mathbb{K}$ *is full and faithful and commutes up-to-isomorphism with the type- and kind-projection functors:*

$$\mathbb{T} \xrightarrow{\mathcal{I}} \mathbb{K} \quad \cong \quad \mathbb{C}^{\to}$$

Proof. (i) Notice that for a (vertical) morphism $\Gamma, x : \sigma \vdash M : \tau$ from $(\Gamma \vdash$

σ: Type) to $(\Gamma \vdash \tau$: Type) in $\mathbb{T}$ we can define $\mathcal{I}(M)$ as the term

$$\Gamma, z\colon \Sigma x\colon \sigma. 1_K \vdash \text{unpack } z \text{ as } \langle x, w \rangle \text{ in } \langle M(x), w \rangle \colon \Sigma y\colon \tau. 1_K.$$

(ii) If our (Type, Kind)-sums are very strong, then the canonical map

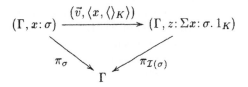

is invertible with as inverse $(\vec{v}, \pi z)$. One now easily concludes that $\mathcal{I}\colon \mathbb{T} \to \mathbb{K}$ is full and faithful:

$$\mathbb{T}_\Gamma\big(\sigma, \tau\big) \cong \mathbb{C}/\Gamma\big(\pi_\sigma, \pi_\tau\big) \cong \mathbb{C}/\Gamma\big(\pi_{\mathcal{I}(\sigma)}, \pi_{\mathcal{I}(\tau)}\big) \cong \mathbb{K}_\Gamma\big(\mathcal{I}(\sigma), \mathcal{I}(\tau)\big),$$

using that the comprehension categories of types and of kinds are full. $\square$

The reverse of (ii) also holds (under an extra assumption), see Exercise 11.5.1 below.

11.5.3. Proposition. *In the presence of very strong* (Type, Kind)*-sums and weak* (Kind, Type) *sums there is a fibred reflection*

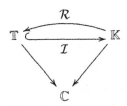

where $\mathcal{I}$ is the full and faithful functor resulting from (ii) in the previous proposition, and its left adjoint $\mathcal{R}$ is obtained from (i).

Proof. What we need for $\Gamma \vdash A$: Kind and $\Gamma \vdash \sigma$: Type is an adjointness correspondence between terms M and N in

$$\frac{\Gamma, z\colon \Sigma \alpha\colon A. 1_T \vdash M(z)\colon \sigma}{\Gamma, \alpha\colon A \vdash N\colon \Sigma x\colon \sigma. 1_K}$$

It is given by

$$M \;\mapsto\; \overline{M}(\alpha) = \langle M[\langle \alpha, \langle \rangle_T \rangle / z], \langle \rangle_K \rangle$$
$$N \;\mapsto\; \overline{N}(z) = \text{unpack } z \text{ as } \langle \alpha, x \rangle \text{ in } \pi N(\alpha).$$

Then:

$$\overline{\overline{M}}(z) \;=\; \text{unpack } z \text{ as } \langle \alpha, x \rangle \text{ in } M[\langle \alpha, \langle \rangle_T \rangle / z]$$
$$=\; \text{unpack } z \text{ as } \langle \alpha, x \rangle \text{ in } M[\langle \alpha, x \rangle / z]$$
$$=\; M.$$

$$\overline{\overline{N}}(\alpha) \;=\; \langle \text{unpack } \langle \alpha, \langle \rangle_T \rangle \text{ as } \langle \alpha, x \rangle \text{ in } \pi N(\alpha), \langle \rangle_K \rangle$$
$$=\; \langle \pi N(\alpha), \langle \rangle_K \rangle$$
$$=\; \langle \pi N(\alpha), \pi' N(\alpha) \rangle$$
$$=\; N(\alpha). \qquad \qquad \square$$

11.5.4. Corollary. *In FhoDTT with very strong* (Kind, Type) *sums there is an equivalence* $\mathbb{T} \simeq \mathbb{K}$ *of types of kinds.*

Proof. With very strong (Kind, Type)-sums the left adjoint $\mathcal{R} \colon \mathbb{K} \to \mathbb{T}$ in the previous proposition is also full and faithful, so that we get an equivalence $\mathbb{T} \xrightarrow{\;\simeq\;} \mathbb{K}$, as stated. $\qquad \square$

Polymorphic dependent type theory FhoDTT as sketched in this section was first formulated by Coquand and Huet [58] as the Calculus of Constructions, see also [14]. It was introduced as an expressive combination of polymorphic and dependent type theory. Actual representation and verification in the Calculus of Constructions turned out to be problematic (notably because of problems with inductively defined types, see [321]). This gave rise to two extensions, with additional universes for separating logic from data. Both these extensions have been implemented in proof tools.

(i) The Calculus of Inductive Definitions [59, 248] forms the basis for the proof tool COQ [20]. The emphasis in the use of COQ lies on program abstraction, via a duplication of the structure of the Calculus of Constructions (separating programming and logic). There is a facility for extracting executable ML programs from suitable terms, see [249, 31]. (See also [276] for a comparably duplicated version of second order polymorphic type theory, for similar purposes.)

(ii) The Extended Calculus of Constructions (ECC) adds to the Calculus of Constructions an infinite hierarchy of kinds $(\mathsf{Kind}_i)_{i \in \mathbb{N}}$ with inclusions $\mathsf{Type} \subseteq \mathsf{Kind}_0 \subseteq \mathsf{Kind}_1 \subseteq \cdots$, see [201, 202]. There is an ω-set based semantics, using an infinite sequence of inaccessible cardinals to interpret the Kind_i, see [200]. This hierarchy of kinds facilitates the formalisation of abstract mathematics. The LEGO system is a proof-assistant based on ECC, extended with inductive types, see [203].

Exercises

11.5.1. Assume that the (Type, Kind)-sums in Proposition 11.5.2 (i) are strong. Show then that if the induced functor $\mathcal{I}\colon \mathbb{T} \to \mathbb{K}$ is full and faithful, then the (Type, Kind)-sums are very strong.

11.5.2. Consider the full and faithful functor $\mathcal{I}\colon \mathbb{T} \to \mathbb{K}$ from types to kinds, arising from very strong (Type, Kind)-sums in Proposition 11.5.2 (ii). Prove that one has strong (Type, Type)-sums if and only if "$\mathcal{I}$ preserves sums", in the sense that for $\Gamma \vdash \sigma\colon \mathsf{Type}$ and $\Gamma, x\colon \sigma \vdash \tau\colon \mathsf{Type}$ the canonical term P in:

$$\Gamma, z\colon \Sigma\alpha\colon \mathcal{I}(\sigma).\, \mathcal{I}(\tau)[\pi\alpha/x] \vdash P\colon \mathcal{I}(\Sigma x\colon \sigma.\, \tau)$$

is invertible.

11.5.3. We sketch a version of Girard's paradox in FhoDTT with very strong (Kind, Type)-sums, mimicking Mirimanoff's paradox (see [153] for details) about the non-existence of a set Ω of well-founded sets (problem: is Ω itself well-founded?). Type theoretically, one takes Ω to be

$$\Omega = \Sigma\alpha\colon \mathsf{Type}.\, \Sigma <\colon \alpha \to \alpha \to \mathsf{Type}.\, \mathrm{WF}(\alpha, <),$$

where $\mathrm{WF}(\alpha, <)$ is the type

$$\Pi p\colon \alpha \to \mathsf{Type}.\, \big(\Sigma x\colon \alpha.\, px \times \Pi x\colon \alpha.\, [px \to \Sigma y\colon \alpha.\, (py \times y < x)]\big) \to \bot.$$

where $\bot = (\Pi\alpha\colon \mathsf{Type}.\, \alpha)\colon \mathsf{Type}$. The inhabitants of Ω are thus triples $\langle \alpha, \langle <_\alpha, q_\alpha\rangle\rangle$ consisting of a type α with an ordering $<_\alpha$, and a proof-term q_α witnessing that $<_\alpha$ on α is well-founded. The next step is to define an ordering $<_\Omega$ on Ω. Informally, one says that $\langle \alpha, \langle <_\alpha, q_\alpha\rangle\rangle$ is below $\langle \beta, \langle <_\beta, q_\beta\rangle\rangle$ if there is an order preserving function $\alpha \to \beta$, which stays below a certain point in β.

$$<_\Omega\ =\ \lambda u\colon \Omega.\, \lambda v\colon \Omega.\, \Sigma f\colon \alpha \to \beta.\, \Sigma z\colon \beta.$$

$$\text{unpack } u, v \text{ as } \langle\langle \alpha, \langle <_\alpha, q_\alpha\rangle\rangle, \langle \beta, \langle <_\beta, q_\beta\rangle\rangle,\rangle \text{ in}$$

$$\big(\Pi x\colon \alpha.\, \Pi y\colon \alpha.\, x <_\alpha y \to fx <_\beta fy\big) \times \big(\Pi x\colon \alpha.\, fx <_\beta z\big).$$

(i) Check that in order to define $u\ <_\Omega\ v\colon \mathsf{Type}$ for $u, v\colon \Omega$ we need sum-elimination with respect to $\mathsf{Type}\colon \mathsf{Kind}$—and thus very strong (Kind, Type)-sums.

(ii) Construct:

    (1) a proof-term $q_\Omega\colon \mathrm{WF}(\Omega, <_\Omega)$;

    (2) a proof-term $r\colon \underline{\Omega} <_\Omega \underline{\Omega}$, where $\underline{\Omega} = \langle \Omega, \langle <_\Omega, q_\Omega\rangle\rangle\colon \Omega$;

    (3) a proof-term $s\colon \bot$ inhabiting *falsum*, using (1) and (2).

11.6 Full higher order dependent type theory, categorically

In this section we describe appropriate fibred categories for full higher order dependent type theory (FhoDTT), and consider some examples. First of all, we present two 'degenerate' examples: one—resulting from an arbitrary topos—in which the fibration of types is a poset. Then there are no terms (or proof-objects) between types, and the model is 'logical' instead of 'type theoretic', in the sense that provability is modelled, and not proofs (under a propositions-as-types reading). The other degenerate example involves the fibration of closure-indexed-closures, as described at the end of Section 10.6. It is a model for the axiom ⊢ Type: Type, and so we can turn it into a model of polymorphic dependent type theory with Type = Kind. In this situation, Girard's paradox (Exercise 11.5.3) applies, so that every type is inhabited. This can also be seen as a form of (logical) degeneracy.

Next we present three realisability examples, involving PERs over ω-**Sets**, PERs over **Eff**, and 'extensional PERs' over ω-**Sets**. The latter is a model of weak-FhoDTT, in which the types are closed under weak dependent sums Σ, but not under strong ones, see [320]. We conclude with some generalities about "FhoDTT-structures".

As we already emphasised in the previous section, the main difference between full higher order dependent type theory (FhoDTT) and polymorphic dependent type theory (PDTT) is that in FhoDTT kinds may depend on types (*i.e.* Kind ≻ Type, in the notation from the previous section). As a result, one cannot separate kind- and type-contexts in FhoDTT, like in PDTT. This has consequences for the corresponding categorical structure: one does not have two base categories of contexts like in PDTT (namely $\mathbb{B}$ and $\mathbb{A}$ in the diagram (∗) on page 665), but a single base category of (kind- and type-) contexts (like in the base category $\mathbb{C}$ in the term model described before Proposition 11.5.2 in the previous section). Moreover, there will be a (fibred) reflection Type ⇆ Kind between types and kinds resulting from the sums in FhoDTT (see Proposition 11.5.3). Thus we shall combine a comprehension category $\mathbb{E} \to \mathbb{B}^{\to}$ for kinds with a reflection $\mathbb{D} \leftrightarrows \mathbb{E}$ of types in kinds, like in:

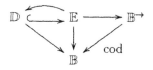

The functor $\mathbb{D} \leftarrow \mathbb{E}$ from kinds to types is the left adjoint to the inclusion, as in Proposition 11.5.3. Actually taking this reflection $\mathbb{E} \leftrightarrows \mathbb{D}$ as primitive greatly simplifies matters, since it induces much of the structure on types (see Lemma 9.3.9, Exercise 9.3.9 and the table below).

11.6.1. Definition. A structure

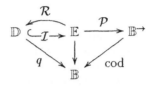

is called a **weak FhoDTT-structure** if

- $\mathcal{P}: \mathbb{E} \to \mathbb{B}^{\to}$ is a closed comprehension category (of kinds);
- $q = \text{cod} \circ \mathcal{P} \circ \mathcal{I}$ is a fibration (of types), and $\mathbb{D} \leftrightarrows \mathbb{E}$ is a fibred reflection;
- there is an object $\Omega \in \mathbb{E}$ over the terminal object $1 \in \mathbb{B}$, such that the fibration q has a generic object over $\{\Omega\} = \text{dom}\mathcal{P}(\Omega) \in \mathbb{B}$.

In such a situation the composite $\mathcal{Q} = \mathcal{P}\mathcal{I}: \mathbb{D} \to \mathbb{B}^{\to}$ is a comprehension category, which, by the reflection, has a unit $\mathcal{R}1: \mathbb{B} \to \mathbb{D}$—where $1: \mathbb{B} \to \mathbb{E}$ is the unit of $\mathcal{P}$. Further, the required four forms of coproduct—kinds/types over kinds/types—exist in such a structure, according to the following table.

coproducts	are present because
(very) strong (Kind, Kind)	$\mathcal{P}$ is a CCompC
strong (Type, Kind)	$\mathcal{Q}$-projections are $\mathcal{P}$-projections
weak (Kind, Type)	of the reflection $\mathbb{D} \leftrightarrows \mathbb{E}$
weak (Type, Type)	of the reflection and because $\mathcal{Q}$-projections are $\mathcal{P}$-projections

A similar table applies to products $\prod$ instead of coproducts $\coprod$. The last two coproducts (of types over kinds and over types) need not be strong, as will be shown in Example 11.6.7 below. Therefore we have the following addition.

11.6.2. Definition. A **(strong) FhoDTT-structure** is a weak FhoDTT-structure as above where the comprehension category $\mathcal{Q} = \mathcal{P}\mathcal{I}: \mathbb{D} \to \mathbb{B}^{\to}$ of types is closed.

(This amounts to the requirement that $\mathcal{Q}$ has strong $\mathcal{Q}$-coproducts, or, equivalently by Proposition 11.4.3, that $\mathcal{Q}$ has strong $\mathcal{P}$-coproducts.)

Since the situation with strong sums is most common, we often omit 'strong' and just say 'FhoDTT-structure' for 'strong FhoDTT-structure'. Also, we shall call a (weak or strong) FhoDTT-structure **split** if all of its structure is split. Below we follow our usual preference for presenting examples in split form.

11.6.3. Example (Toposes). Let $\mathbb{B}$ be a topos. Then $\mathbb{B}$ is a regular category, so that there is a fibred reflection $\text{Sub}(\mathbb{B}) \leftrightarrows \mathbb{B}^{\rightarrow}$, as in Theorem 4.4.4. The reflector $\mathbb{B}^{\rightarrow} \to \text{Sub}(\mathbb{B})$ sends an arbitrary morphism to its monic part, using the epi-mono factorisation in a topos. And since monos are closed under composition, $\text{Sub}(\mathbb{B}) \to \mathbb{B}^{\rightarrow}$ is a closed comprehension category. We thus get a FhoDTT-structure of the form:

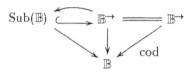

Recall from Example 10.5.9 that the codomain fibration $\begin{smallmatrix} \mathbb{B}^{\rightarrow} \\ \downarrow \\ \mathbb{B} \end{smallmatrix}$ of a topos $\mathbb{B}$ is equivalent to the split fibration $\begin{smallmatrix} \mathcal{F}(\mathbb{B}) \\ \downarrow \\ \mathbb{B} \end{smallmatrix}$ of families $I \times X \to \Omega$ in $\mathbb{B}$ (with reindexing given by composition, instead of by pullback). We may thus present the above structure also in split form as

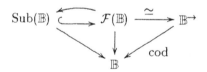

These structures are degenerate models of FhoDTT, since the subobject fibration $\begin{smallmatrix} \text{Sub}(\mathbb{B}) \\ \downarrow \\ \mathbb{B} \end{smallmatrix}$ for types is a poset. Thus all structure of terms between types is destroyed.

11.6.4. Example (Closures). Recall from the last part of Section 10.6 the closed comprehension category $\begin{smallmatrix} \text{Fam}(\mathbf{Clos}) \\ \downarrow \\ \mathbf{Clos} \end{smallmatrix}$ of closure-indexed-closures. It admits a type of all types (*i.e.* Type: Type): there is the universal closure $\Omega \in \mathbf{Clos}$ with the set of closures as image. Formally, the family $\text{id}_\Omega \colon \Omega \to \Omega$ over Ω is split generic object for $\begin{smallmatrix} \text{Fam}(\mathbf{Clos}) \\ \downarrow \\ \mathbf{Clos} \end{smallmatrix}$: for every family $X \colon a \to \Omega$ in Fam($\mathbf{Clos}$) over $a \in \mathbf{Clos}$ there is a unique map $a \dashrightarrow \Omega$ in the base category $\mathbf{Clos}$, namely X itself, which yields $X = X^*(\text{id}_\Omega)$. And we can see Ω as the domain of a family $\underline{\Omega} \colon 1 \to \Omega$ in Fam($\mathbf{Clos}$) over the terminal object $1 \in \mathbf{Clos}$, so that $\{\underline{\Omega}\} \cong \Omega$. All told, we have a FhoDTT-structure

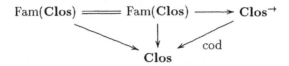

in which the fibrations of types and of kinds are the same: this is a degenerate model with Type = Kind.

We turn our attention to realisability models of FhoDTT. We discuss successively: PERs over ω-Sets, PERs over **Eff** and ExPERs over ω-Sets. The first two of these examples are obtained by combining various results on PERs from previous sections.

11.6.5. Example (PERs over ω-Sets). To start with, recall from Example 1.8.7 (ii) that the reflection **PER** $\leftrightarrows$ ω-**Sets** lifts to a fibred reflection UFam(**PER**) $\leftrightarrows$ UFam(ω-**Sets**) over ω-**Sets**. This gives us a FhoDTT-structure

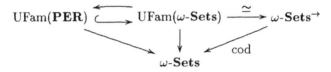

in which kinds are ω-set-indexed families of ω-sets, and types are ω-set indexed families of PERs. Since the fibration of PERs over ω-sets (on the left) is the externalisation of the internal category **PER** in ω-**Sets**, it has a split generic object, given by the set of PERs, considered as object $\mathbf{PER}_0$ in ω-**Sets**. And this object comes from a family over 1 in UFam(ω-**Sets**). Finally, in Example 10.5.8 we saw that UFam(**PER**) $\rightarrow$ ω-**Sets**$^\rightarrow$ is a closed comprehension category. In particular, there are strong (Type, Type)-sums in the above FhoDTT-structure.

11.6.6. Example (PERs over **Eff**). The reflection **PER** $\leftrightarrows$ ω-**Sets** that we used in the previous example, also lifts to a reflection over the effective topos **Eff**, see Proposition 6.3.2 (iii). Hence we get a similar FhoDTT-structure over **Eff**.

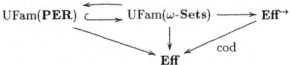

11.6.7. Example (ExPERs over ω-**Sets**). This example involves the (reflective) subcategory **ExPER** $\leftrightarrows$ **PER** of so-called 'extensional PERs', introduced in [81]. Streicher [320] shows that in this "submodel" example one does *not* have strong (Type, Type)-sums: it is a *weak* FhoDTT-structure.

We start with some preliminary definitions and results on extensional PERs. The definition of extensionality for PERs makes crucial use of the special PER $N_\perp$ (but see Exercise 11.6.1 below for an alternative definition):

$$N_\perp = \{(n, n') \mid n \cdot 0 \simeq n' \cdot 0\}.$$

where $\simeq$ means that either both sides are undefined, or both sides are defined and are equal. It is clear that $N_\perp$ is a PER. Its quotient set $\mathbb{N}/N_\perp$ is the union $\{[\Lambda x. \uparrow]\} \cup \{[\Lambda x. n] \mid n \in \mathbb{N}\}$; it can be identified with the set $\perp\mathbb{N} = 1 + \mathbb{N}$ of natural numbers with additional bottom element.

We now define a PER R to be **extensional** if the canonical map

$$R \longrightarrow N_\perp^{(N_\perp^R)} \qquad \text{mapping} \qquad x \longmapsto \lambda\alpha : N_\perp^R. \alpha(x)$$

is a regular mono in ω-**Sets**. (Recall from Lemma 5.3.8 (i) that regular subobjects in ω-**Sets** correspond to subsets of the carrier sets, with inherited existence predicate.) It follows that if $R \in$ **PER** is extensional, then two elements $x, y \in \mathbb{N}/R$ are equal if and only if $\alpha(x) = \alpha(y)$ for each $\alpha : R \to N_\perp$ in ω-**Sets**. It is easy to see that $N_\perp$ itself is extensional.

We write **ExPER** $\hookrightarrow$ **PER** for the full subcategory of extensional PERs. It is easy to construct a left adjoint to this inclusion: for an arbitrary PER S take the regular image in ω-**Sets** of the canonical map $S \to N_\perp^{(N_\perp^S)}$ as

$$S \xrightarrow{\ \eta_S\ } S' \overset{\text{regular mono}}{\rightarrowtail} N_\perp^{(N_\perp^S)}$$

Then, by the equivalence between (i) and (ii) in Exercise 11.6.1, S' is an ExPER. And every map $f : S \to R$ to an ExPER R, factors uniquely as $\overline{f} : S' \dashrightarrow R$ with $\overline{f} \circ \eta_S = f$ using the diagonal-fill-in:

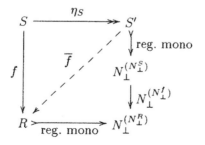

The ExPER model of polymorphic dependent type theory is easy to describe. The reflection **ExPER** $\leftrightarrows$ **PER** lifts to a fibred reflection UFam(**ExPER**) $\leftrightarrows$ UFam(**PER**) over ω-**Sets**. We thus get a weak FhoDTT-structure

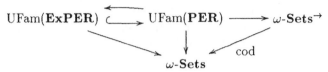

since the fibred reflection transports all the required structure from the closed comprehension category of PERs over ω-sets, to the fibration of ExPERs over ω-sets. In particular, this fibration of ExPERs is a fibred CCC, and so the category **ExPER** (the fibre over $1 \in \omega$-**Sets**) is a CCC.

In this example we have kinds as families of PERs over ω-sets. But since there is also a fibred reflection UFam(**PER**) $\leftrightarrows$ UFam(ω-**Sets**) over ω-**Sets**, we have by composition of fibred reflections, another example

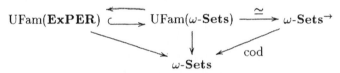

in which we have kinds as ω-set indexed families of ω-sets.

We now come to Streicher's counter example, showing that in this last situation one does not have strong (Type, Type)-coproducts (*i.e.* sums). Consider therefore the exponent ExPER $N_\perp^N$, where N is the PER of natural numbers. Over this object, define the family $S_f \in$ **ExPER** given for $f \in N_\perp^N$ by

$$S_f = \begin{cases} N_\perp & \text{if } \forall x.\, f(x) = 1 \\ \mathbb{N} \times \mathbb{N} & \text{else.} \end{cases}$$

where $\mathbb{N} \times \mathbb{N}$ is the terminal ExPER with singleton quotient set $\{\mathbb{N}\}$. We describe the coproduct of ExPERs over an ExPER as an ω-set:

$$X = \coprod_{f \in N_\perp^N} S_f$$
$$= \{(\lambda x.\, 1, z) \mid z \in 1 + \mathbb{N}\} \cup \{(f, \mathbb{N}) \mid f: \mathbb{N} \rightharpoonup \mathbb{N} \text{ part. rec., } f \neq \lambda x.\, 1\}$$

see Exercise 11.6.2. The carrier set of X can also be split into a union

$$X \cong X_1 \cup X_2$$

of sets of partial recursive functions:

$$X_1 = \{g: \mathbb{N} \rightharpoonup \mathbb{N} \text{ partial recursive} \mid \forall x \in \mathbb{N}.\, g(x+1) = 1\}$$
$$X_2 = \{g: \mathbb{N} \rightharpoonup \mathbb{N} \text{ partial recursive} \mid g(0) \uparrow \text{ and}$$
$$\exists x \in \mathbb{N}.\, (g(x+1) \neq 1 \text{ or } g(x+1) \uparrow)\}.$$

The value at 0 of $g \in X_1$ is thus used to give the value in the fibre S_f over the function $f(x) = g(x + 1)$. In the second case this value at 0 must be fixed, and divergence is chosen. The existence predicate E on this coproduct X is:

$$e \in E(g) \iff e \cdot 0 \simeq g(0) \text{ and } \forall x \in \mathbb{N}. \, e \cdot (x + 1) \simeq g(x + 1) = 1,$$
$$\text{for } g \in X_1;$$
$$e \in E(g) \iff \forall x \in \mathbb{N}. \, e \cdot (x + 1) \simeq g(x + 1),$$
$$\text{for } g \in X_2.$$

The outcome $e \cdot 0$ at 0 of a realiser $e \in E(g)$ for $g \in X_2$ is irrelevant since any two realisers of g (differing solely at 0) will be identified when X is considered as a PER. Thus, any code $n \in \mathbb{N}$ for a partial recursive function is a realiser for some function h_n in X, namely for

$$h_n = \begin{cases} \lambda y. \, e \cdot y & \text{if } \forall x. \, e \cdot (x + 1) = 1 \\ \lambda y. \begin{cases} \uparrow & \text{if } y = 0 \\ e \cdot y & \text{else} \end{cases} & \text{otherwise.} \end{cases}$$

Notice that if n_1 and n_2 are codes for the same partial recursive function, then they realise the same function in X.

If the fibration $\begin{smallmatrix} \mathbf{UFam}(\mathbf{ExPER}) \\ \downarrow \\ \omega\text{-}\mathbf{Sets} \end{smallmatrix}$ of ExPERs over ω-sets has strong coproducts, then by Exercise 11.5.2 this coproduct ω-set (X, E) must be an ExPER. This requires for $g_1, g_2 \in X$ that $g_1 = g_2$ if $\alpha(g_1) = \alpha(g_2)$ for all morphisms $\alpha \colon X \to N_\perp$ in ω-**Sets**. It turns out that this property fails, by the fact that the effective operations are continuous. This is the Myhill-Shepherdson Theorem from recursion theory, see *e.g.* [236, II.4].

Consider the following two functions in X.

$$g_1(x) = 1 \qquad \text{and} \qquad g_2(x) = \begin{cases} \uparrow & \text{if } x = 0 \\ 1 & \text{else.} \end{cases}$$

Then both g_1 and g_2 are in X_1, but obviously $g_1 \neq g_2$. We show that $\alpha(g_1) = \alpha(g_2)$ for each $\alpha \colon X \to N_\perp$ in ω-**Sets**. This means that X is not (isomorphic to) an ExPER.

Assume $\alpha \colon X \to N_\perp$ in ω-**Sets** is given, say tracked by $d \in \mathbb{N}$. Let t be a primitive recursive function—obtained via the *s-m-n* Theorem—such that for each $n, x \in \mathbb{N}$,

$$\varphi_{t(n)}(x) = d \cdot n \cdot 0.$$

We can now define an effective operation F by

$$F(\varphi_n) = \varphi_{t(n)}$$

since t is "extensional":

$$\varphi_{n_1} \simeq \varphi_{n_2} \Rightarrow n_1, n_2 \text{ are codes of the same function in } X$$
$$\Rightarrow d \cdot n_1 N_\perp d \cdot n_2$$
$$\Rightarrow d \cdot n_1 \cdot 0 \simeq d \cdot n_2 \cdot 0$$
$$\Rightarrow \varphi_{t(n_1)} \simeq \varphi_{t(n_2)}.$$

Assume now $\alpha(g_1) \neq \alpha(g_2)$ and thus $F(g_1) \neq F(g_2)$, say $F(g_1)(k) \neq F(g_2)(k)$. Without loss of generality, assume that $F(g_1)(k)$ is defined, and has value ℓ. By the Myhill-Shepherdson Theorem all effective operations are continuous, so there is a finite approximation $h_1 \subseteq g_1$ with $F(h)(k) = \ell$ for all $h \supseteq h_1$. We distinguish two cases:

- If $h_1(0) \uparrow$, then also $h_1 \subseteq g_2$, so that $F(g_2)(k) = \ell$. This is impossible.
- If $h_1(0) \downarrow$, one must have $h_1(0) = 1$. We take another finite function h_2, which acts like h_1, except on 0, where it is undefined. Any two realisers m_i for h_i (with $i = 1, 2$), are codes for the same function in X_2, so $d \cdot m_1 \simeq d \cdot m_2$. But then $F(h_1) = F(h_2)$, and since h_2 approximates g_2, we must have $F(g_2)(k) = \ell$ by monotonicity of F. Hence also this second case leads to a contradiction.

Thus we have shown that the coproduct X of ExPERs over an ExPER is not an ExPER. We may conclude that these ExPERs form a *weak* FhoDTT-structure.

11.6.8. Remark. Recall from Lemma 9.3.9 that products $\prod$ and coproducts $\coprod$ are "transported along a reflector". This ExPER example shows that strong coproducts are *not* transported: there is a fibred reflection UFam(**ExPER**) $\leftrightarrows$ UFam(ω-**Sets**) and the fibration of ω-**Sets** has strong coproducts, but the fibration of ExPERs does not.

These are the examples of FhoDTT-structures that we describe here. We should also mention the model (without sums) of Lamarche [185], which is an adaptation of the 'coherent domain' model of polymorphic type theory from [96]. Also there are the examples of Hyland and Pitts [148] involving particular kinds of Grothendieck toposes, namely 'algebraic' toposes (presheaf toposes on small categories with finite limits) and 'algebraic-localic' toposes (presheaf toposes on meet semi-lattices). The first of these examples is degenerate in the sense that it is a model with Type = Kind—and thus with Type: Type, using that there is an algebraic topos 'encoding' all algebraic toposes. And in the second case one uses algebraic localic toposes for types, and algebraic toposes for kinds. One then still has a kind of all kinds. Variations of the above model of PERs over ω-sets are used in [320, 321] for various independence results in FhoDTT. See also [312] for similar use of models based on combinatory algebras.

We conclude this section with some general points worth noticing about FhoDTT-structures.

11.6.9. Theorem. *Let* $\mathbb{D} \underset{I}{\overset{I}{\rightleftarrows}} \mathbb{E} \overset{P}{\to} \mathbb{B}^{\to}$ *be a weak FhoDTT-structures, as in Definition 11.6.1. Write* $Q = P\mathcal{I}: \mathbb{D} \to \mathbb{B}^{\to}$ *for the comprehension category of types, and* $q = \mathrm{cod} \circ Q: \mathbb{D} \to \mathbb{B}$ *for the associated fibration. Then*

(i) q *is a fibred CCC, and thus a locally small fibration;*

(ii) q *is a "full" small fibration: it is the externalisation of a full internal category* $\mathbf{C}$ *in* $\mathbb{B}$*; fullness is automatic because* Q *is a full and faithful (fibred) functor* $\mathrm{Fam}(\mathbf{C}) = \mathbb{D} \to \mathbb{B}^{\to}$.

Proof. (i) Cartesian closure follows as in Proposition 10.5.4, except that we have to remember that also the weak (Type, Type)-sums yield fibred Cartesian products, see Exercise 10.5.4 (i).

Since q is a thus a fibred CCC with comprehension, it is locally small by Corollary 10.4.11.

(ii) Let $U \in \mathbb{D}$ be the generic object of q over $C_0 = \{\Omega\} \in \mathbb{B}$. We then get an internal category $\mathbf{C} = \mathrm{Full}(U)$ in $\mathbb{B}$, as in Theorem 9.5.5, with a full and faithful fibred functor $\mathrm{Full}(U) \to \mathbb{D}$, provided the relevant pullbacks exist in the base category $\mathbb{B}$ to form the objects of composable tuples and triples. And this functor $\mathrm{Full}(U) \to \mathbb{D}$ is an equivalence like in Corollary 9.5.6 because q has a generic object.

The existence of the relevant pullbacks follows because one can always form pullbacks along the $\mathcal{P}$- and Q-projections. For example, the Cartesian product $C_0 \times C_0 \in \mathbb{B}$ is formed as the pullback

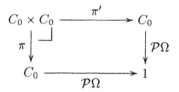

so that the Cartesian projections π and π' are $\mathcal{P}$-projections. The pair $\langle \partial_0, \partial_1 \rangle: C_1 \to C_0 \times C_0$ of domain and codomain maps of $\mathbf{C}$ is then the Q-projection

$$Q(\pi^*(U) \Rightarrow \pi'^*(U)).$$

This can be seen by combining the description of $\langle \partial_0, \partial_1 \rangle$ in the proof of Theorem 9.5.5 with the description of representing arrows via exponents $\Rightarrow$ underlying Corollary 10.4.11. Since $\partial_0 = \pi \circ \langle \partial_0, \partial_1 \rangle$ is a composite of $\mathcal{P}$- and

Q-projections, the pullback $u^*(\partial_0)$ exists in $\mathbb{B}$ for any map $u\colon I \to C_0$. This allows us to form the pullbacks C_2 and C_3 in $\mathbb{B}$ of composable tuples and triples (as in Definition 7.1.1).   □

This result emphasises the rôle of internal categories in (weak) FhoDTT-structures. In the special case where the fibration of kinds is a codomain fibration, the reflector from kinds to types makes this internal category complete. In effect, completeness is equivalent to existence of such a reflector. This result, due to [143, 76], will be presented next. It gives a particularly easy description of certain models of FhoDTT, like the PER and ExPER models over ω-sets. Such a structure is sometimes studied as a type theoretic generalisation of a topos, in which the (internal) preorder structure of the object Ω of propositions in a topos is replaced by a proper internal category of types, see *e.g.* [76, 255].

11.6.10. Theorem. *Let $\mathbb{B}$ be a locally Cartesian closed category, containing a full internal category* $\mathbf{C}$*, with full and faithful fibred functor* $\mathcal{I}\colon \mathrm{Fam}(\mathbf{C}) \to \mathbb{B}^{\to}$*. Then:* $\mathcal{I}$ *has a fibred left adjoint if and only if* $\mathbf{C}$ *is a small complete category in* $\mathbb{B}$ *and* $\mathcal{I}$ *is continuous.*

Under these equivalent conditions there is a weak FhoDTT-structure of the following form.

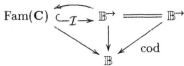

Since the codomain fibration of an LCCC is complete, the existence of a reflection obviously makes the internal category complete, so the interesting part of the statement concerns the construction of the reflector, assuming completeness. This may be done (like in [143]) via an application of a (fibred) adjoint functor theorem (see [47, 246]). But with the products $\prod$ one can define a coproduct, like in Exercise 8.1.5, yielding a weak left adjoint. What remains to be done then, is to make this into a real adjoint via a standard construction with equalisers (as used for example in the proof of [187, V, 6, Theorem 1] to turn a weak initial object into a real initial object). See [254] for a general analysis of the adjoint functor theorem in terms of such definable coproducts.

We shall present this construction in a type theoretic formulation, using some ad hoc notation for the inclusion $\mathcal{I}$ of types into kinds, and for equalisers of (parallel) terms between types. The inclusion $\mathcal{I}$ yields for a type $\Gamma \vdash \sigma\colon \mathsf{Type}$ a kind $\Gamma \vdash \mathcal{I}(\sigma)\colon \mathsf{Kind}$ in the same context. By fullness it comes with introduction and elimination rules:

$$\frac{\Gamma \vdash M : \sigma}{\Gamma \vdash i(M) : \mathcal{I}(\sigma)} \qquad\qquad \frac{\Gamma \vdash N : \mathcal{I}(\tau)}{\Gamma \vdash o(N) : \tau}$$

with 'i' and 'o' for 'in' and 'out'. The associated conversions are simply

$$o(i(M)) = M \qquad \text{and} \qquad i(o(N)) = N.$$

For types $\Gamma \vdash \sigma, \tau \colon \mathsf{Type}$ in the same context, and "parallel" terms $\Gamma, x \colon \sigma \vdash M_1, M_2 \colon \tau$ we shall use an "equaliser type" $\Gamma \vdash E(M_1, M_2) \colon \mathsf{Type}$, together with a term $\Gamma, z \colon E(M_1, M_2) \vdash e_{M_1, M_2} \colon \sigma$ satisfying

$$\Gamma, z \colon E(M_1, M_2) \vdash M_1[e_{M_1, M_2}/x] = M_2[e_{M_1, M_2}/x] \colon \sigma.$$

Further, we translate the familiar universal property of equalisers into this type theoretic language as follows. For each term $\Gamma, \Delta \vdash N \colon \sigma$ satisfying $\Gamma, \Delta \vdash M_1[N/x] = M_2[N/x] \colon \sigma$, there is a unique term $\Gamma, \Delta \vdash \overline{N} \colon E(M_1, M_2)$ satisfying $\Gamma, \Delta \vdash e_{M_1, M_2}[\overline{N}/z] = N \colon \sigma$.

Proof. For a kind $\Gamma \vdash A \colon \mathsf{Kind}$ we construct a type $\Gamma \vdash \mathcal{R}(A) \colon \mathsf{Type}$, such that for a type $\Gamma \vdash \sigma \colon \mathsf{Type}$ we get a bijective correspondence between terms M and N in:

$$\frac{\Gamma, \alpha \colon A \vdash M \colon \mathcal{I}(\sigma)}{\Gamma, x \colon \mathcal{R}(A) \vdash N \colon \sigma} \tag{$*$}$$

yielding the required adjunction $\mathcal{R} \dashv \mathcal{I}$. We construct $\mathcal{R}(A)$ by first constructing a weak left adjoint $\mathcal{R}_{\mathrm{w}}(A)$, together with two equaliser types. We put

$$\Gamma \vdash \mathcal{R}_{\mathrm{w}}(A) \stackrel{\mathrm{def}}{=} \Pi\beta \colon \mathsf{Type}. \, (\Pi\alpha \colon A. \, \beta) \to \beta \colon \mathsf{Type}$$

$$\Gamma \vdash P \stackrel{\mathrm{def}}{=} \lambda\alpha \colon A. \, \lambda\beta \colon \mathsf{Type}. \, \lambda g \colon (\Pi\alpha \colon A. \, \beta). \, g \cdot \alpha \colon \Pi\alpha \colon A. \, \mathcal{R}_{\mathrm{w}}(A)$$

$$\Gamma, f \colon \mathcal{R}_{\mathrm{w}}(A) \to \mathcal{R}_{\mathrm{w}}(A) \vdash P' \stackrel{\mathrm{def}}{=} \lambda\alpha \colon A. \, f \cdot (P \cdot \alpha) \colon \Pi\alpha \colon A. \, \mathcal{R}_{\mathrm{w}}(A).$$

By weakening we can also put P in the same context as P', so that we can form their equaliser; for convenience we use the following abbreviations.

$$\Gamma \vdash E \stackrel{\mathrm{def}}{=} E(P, P') \colon \mathsf{Type} \qquad\qquad \text{with canonical term:}$$

$$\Gamma, z \colon E \vdash e \stackrel{\mathrm{def}}{=} e_{P,P'} \colon \mathcal{R}_{\mathrm{w}}(A) \to \mathcal{R}_{\mathrm{w}}(A) \qquad \text{satisfying:}$$

$$\Gamma, z \colon E \vdash P = \lambda\alpha \colon A. \, e \cdot (P \cdot \alpha) \colon \Pi\alpha \colon A. \, \mathcal{R}_{\mathrm{w}}(A).$$

We also consider the following two terms with the intended reflector $\mathcal{R}(A)$ as their equaliser.

$$\Gamma, y : \mathcal{R}_w(A) \vdash Q \overset{\text{def}}{=} \lambda z : E. \, y : E \to \mathcal{R}_w(A)$$

$$\Gamma, y : \mathcal{R}_w(A) \vdash Q' \overset{\text{def}}{=} \lambda z : E. \, e \cdot y : E \to \mathcal{R}_w(A)$$

$$\Gamma \vdash \mathcal{R}(A) \overset{\text{def}}{=} E(Q, Q') : \mathsf{Type} \qquad\qquad \text{with canonical term:}$$

$$\Gamma, x : \mathcal{R}(A) \vdash c \overset{\text{def}}{=} e_{Q,Q'} : \mathcal{R}_w(A) \qquad\qquad \text{satisfying:}$$

$$\Gamma, x : \mathcal{R}(A) \vdash \lambda z : E. \, c = \lambda z : E. \, e \cdot c : E \to \mathcal{R}_w(A).$$

Notice that in this situation we have a conversion

$$\begin{aligned}
\Gamma, \alpha : A \vdash Q[(P \cdot \alpha)/y] &= \lambda z : E. \, P \cdot \alpha \\
&= \lambda z : E. \, e \cdot (P \cdot \alpha) \\
&= Q'[(P \cdot \alpha)/y] : E \to \mathcal{R}_w(A)
\end{aligned}$$

so that we get a unique term

$$\Gamma, \alpha : A \vdash \overline{P} : \mathcal{R}(A) \qquad \text{with} \qquad \Gamma, \alpha : A \vdash c[\overline{P}/x] = P \cdot \alpha : \mathcal{R}_w(\Lambda).$$

We claim that there is then a correspondence $(*)$ as in the beginning of the proof: given terms $\Gamma, \alpha : A \vdash M : \mathcal{I}(\sigma)$ and $\Gamma, x : \mathcal{R}(A) \vdash N : \sigma$ we define as transposes:

$$\Gamma, x : \mathcal{R}(A) \vdash M^\wedge \overset{\text{def}}{=} c \cdot \sigma \cdot (\lambda \alpha : A. \, \mathsf{o}(M)) : \sigma$$

$$\Gamma, \alpha : A \vdash N^\vee \overset{\text{def}}{=} \mathsf{i}\big(N[\overline{P}/x]\big) : \mathcal{I}(\sigma).$$

Then it is easy to see that

$$\begin{aligned}
M^{\wedge\vee} &= \mathsf{i}\big(c[\overline{P}/x] \cdot \sigma \cdot (\lambda \alpha : A. \, \mathsf{o}(M))\big) \\
&= \mathsf{i}\big(P \cdot \alpha \cdot \sigma \cdot (\lambda \alpha : A. \, \mathsf{o}(M))\big) \\
&= \mathsf{i}\big((\lambda \alpha : A. \, \mathsf{o}(M)) \cdot \alpha\big) \\
&= \mathsf{i}\big(\mathsf{o}(M)\big) \\
&= M.
\end{aligned}$$

It is more complicated to show $N^{\vee\wedge} = N$. We form the equaliser

$$\Gamma \vdash D \overset{\text{def}}{=} E(N, N^{\vee\wedge}) : \mathsf{Type} \qquad \text{with canonical term:}$$

$$\Gamma, w : D \vdash d \overset{\text{def}}{=} e_{N, N^{\vee\wedge}} : \mathcal{R}(A) \quad \text{satisfying:}$$

$$\Gamma, w : D \vdash N[d/x] = N^{\vee\wedge}[d/x].$$

Now it is easy to show that $N^{\vee\wedge}[\overline{P}/x] = N[\overline{P}/x]$, so that there is a unique term

$$\Gamma, \alpha : A \vdash \overline{\overline{P}} : D \qquad \text{with} \qquad \Gamma, \alpha : A \vdash d[\overline{\overline{P}}/w] = \overline{P} : \mathcal{R}(A).$$

We abbreviate

$$\Gamma \vdash L \stackrel{\mathrm{def}}{=} \lambda y \colon \mathcal{R}_{\mathrm{w}}(A).\, c[d/x][(y \cdot D \cdot (\lambda \alpha \colon A.\, \overline{\overline{P}}))/w] \colon \mathcal{R}_{\mathrm{w}}(A) \to \mathcal{R}_{\mathrm{w}}(A)$$

and check that

$$
\begin{aligned}
P'[L/f] &= \lambda \alpha \colon A.\, L \cdot (P \cdot \alpha) \\
&= \lambda \alpha \colon A.\, c[d/x][(P \cdot \alpha \cdot D \cdot (\lambda \alpha \colon A.\, \overline{\overline{P}}))/w] \\
&= \lambda \alpha \colon A.\, c[d/x][\overline{\overline{P}}/w] \qquad \text{by definition of } P \\
&= \lambda \alpha \colon A.\, c[\overline{P}/x] \\
&= \lambda \alpha \colon A.\, P \cdot \alpha \\
&= P \\
&= P[L/f].
\end{aligned}
$$

We thus get a unique term $\Gamma \vdash \overline{L} \colon E$ with $\Gamma \vdash e[\overline{L}/z] = L \colon \mathcal{R}_{\mathrm{w}}(A) \to \mathcal{R}_{\mathrm{w}}(A)$, because e is the equaliser term of P, P'. For $d' = d[((c \cdot D \cdot (\lambda \alpha \colon A.\, \overline{\overline{P}}))/w]$ we have

$$
\begin{aligned}
c[d'/x] &= c[d/x][(c \cdot D \cdot (\lambda \alpha \colon A.\, \overline{\overline{P}}))/w] \\
&= L \cdot c \\
&= e[\overline{L}/z] \cdot c \\
&= (\lambda z \colon E.\, e \cdot c) \cdot \overline{L} \\
&= (\lambda z \colon E.\, c) \cdot \overline{L} \\
&= c.
\end{aligned}
$$

Hence there are two terms $\Gamma, x \colon \mathcal{R}(A) \vdash x, d' \colon \mathcal{R}(A)$ with $c[d'/x] = c[x/x]$. By uniqueness—with respect to the equaliser $\mathcal{R}(A)$—they must be equal: $d' = x$. But then we can finally conclude that $N = N^{\vee \wedge}$ since

$$
\begin{aligned}
N &= N[d'/x] \\
&= N[d/x][(c \cdot D \cdot (\lambda \alpha \colon A.\, \overline{\overline{P}}))/w] \\
&= N^{\vee \wedge}[d/x][(c \cdot D \cdot (\lambda \alpha \colon A.\, \overline{\overline{P}}))/w] \\
&= N^{\vee \wedge}[d/x] \\
&= N^{\vee \wedge}.
\end{aligned}
$$
$\qquad\qquad\square$

We have described some of the essential aspects of categorical structures for full higher order dependent type theory. In the next, final section we shall further investigate one particular example of such a structure, namely PERs over **Eff** from Example 11.6.6.

Exercises

11.6.1. Prove that for a PER R the following statements are equivalent:

(i) R is extensional: the canonical map $R \to (N_\perp)^{((N_\perp)^R)}$ is a regular mono in ω-**Sets**;

(ii) there is an ω-set $A = (A, E)$ with a regular mono $R \rightarrowtail (N_\perp)^A$ in ω-**Sets**;

(iii) R is extensional in the sense of [81]: there is a "base" $B \subseteq \mathbb{N}$ such that for all $n \in |R|$ and $m \in \mathbb{N}$

$$n R m \quad \Leftrightarrow \quad n, m \text{ are "co-extensional mod } B"$$
$$\Leftrightarrow \quad \forall k \in B. n \cdot k \simeq m \cdot k.$$

11.6.2. Recall Exercise 1.2.10 and check that the homset of maps $N \to N_\perp$ in ω-**Sets** can be identified with the set of partial recursive functions $\mathbb{N} \rightharpoonup \mathbb{N}$.

11.6.3. Show that one can obtain a $\lambda\omega$-fibration r from a FhoDTT-structure $\mathbb{D} \underset{I}{\overset{I}{\rightleftarrows}} \mathbb{E} \overset{P}{\to} \mathbb{B}^{\to}$, by forming the fibration r via change-of-base in:

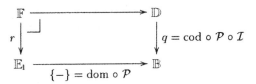

(Where $\mathbb{E}_1$ is the fibre over the terminal object $1 \in \mathbb{B}$.)

11.6.4. (i) Show that every FhoDTT-structure $\mathbb{D} \overset{I}{\rightleftarrows} \mathbb{E} \overset{P}{\to} \mathbb{B}^{\to}$ can be transformed into a PDTT-structure, by taking

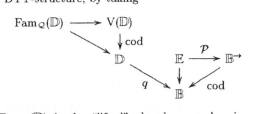

where $\mathrm{Fam}_Q(\mathbb{D})$ is the "lifted" closed comprehension category over $q = \mathrm{cod} \circ Q$, obtained as in Exercise 10.5.5 from $Q = PI$.
[*Hint.* Remember Lemma 11.3.2 (ii).]

(ii) Show that turning a FhoDTT-structure first into a PDTT-structure (as above) and then into a $\lambda\omega$-fibration (as in Exercise 11.3.2) yields the same result as turning the FhoDTT-structure directly into a $\lambda\omega$-fibration (as in the previous exercise).

[In [166] one can find how a $\lambda\omega$-fibration can be turned into a FhoDTT-structure. The construction is based on an idea from [230], and is rather complicated; therefore it will not be reproduced here.]

11.7 Completeness of the category of PERs in the effective topos

In this final section we shall have a closer look at the sense in which the internal category of PERs in the effective topos **Eff** is complete. It turns out not to be complete in the usual sense (see Definition 1.9.11) since the Beck-Chevalley condition does not hold for the product functors $\prod_F$, right adjoint to reindexing F^* along an arbitrary map F in **Eff**. A (non-trivial) counterexample may be found in [150, Proposition 7.5]. It forms the starting point for further investigations, leading to the characterisation that PERs in **Eff** are "weakly" complete. In this section we sketch some of the basic results in this direction: we assemble material from [143, 150, 41] to show that the fibrations of PERs and of ω-sets over **Eff** are "weakly complete", by describing their "stack completions" as the complete fibrations respectively of separated families, and of separated families which are orthogonal to $\nabla 2 \in$ **Eff**. The idea to describe the (weak) completeness of PERs via orthogonality is due to Freyd, see also Remark 8.3.6 (i) (b).

From a type theoretic perspective, this failure of Beck-Chevalley means that PERs in **Eff** do not form a FhoDTT-structure like in Theorem 11.6.10 (a complete internal category in an LCCC). Hence we do not get a model of full higher order dependent type theory (FhoDTT) with types as families of PERs and kinds as families of objects of **Eff**, both indexed over **Eff**. But PERs in **Eff** are complete enough to form a model of FhoDTT with kinds interpreted as (families of) ω-sets (as we have seen in Example 11.6.6). Then one only requires adjoints to certain weakening functors, induced by (generalised) projections given by kinds. Hence PERs in **Eff** can be used to give a model of FhoDTT.

The investigations below involve 'stacks'. These can be understood as generalisations of sheaves, which may have arbitrary categories as fibres, and not just discrete categories, *i.e.* sets. Thus stacks generalise sheaves like indexed categories generalise presheaves. In most general form, a stack consists of a "continuously" indexed category, over a base category equipped with a Grothendieck topology. See [100] or [168] for more information and references. Here we only need stacks with respect to the regular epi topology, as in Example 5.6.3 (ii). (Recall that in a topos every epi is regular, see Corollary 5.5.5.) Moreover, we restrict ourselves to full subfibrations of a codomain fibration $\begin{smallmatrix} \mathbb{B}^{\rightarrow} \\ \downarrow \\ \mathbb{B} \end{smallmatrix}$ for $\mathbb{B}$ a topos. Such a full subfibration can be identified with a collection $\mathcal{D} \subseteq \mathrm{Arr}\,\mathbb{B}$ of "display maps" of $\mathbb{B}$, which is closed under pullback: if a family φ is in $\mathcal{D}$, then each pullback $u^*(\varphi)$ of φ is again in $\mathcal{D}$. As before, we write $\begin{smallmatrix} \mathcal{D}^{\rightarrow} \\ \downarrow \\ \mathbb{B} \end{smallmatrix}$ for the associated fibration of families in $\mathcal{D}$.

11.7.1. Definition. Consider a full subfibration $\begin{smallmatrix} \mathcal{D}^{\rightarrow} \\ \downarrow \\ \mathbb{B} \end{smallmatrix}$ of the codomain fibration $\begin{smallmatrix} \mathbb{B}^{\rightarrow} \\ \downarrow \\ \mathbb{B} \end{smallmatrix}$ of a topos $\mathbb{B}$, given by a collection of display maps $\mathcal{D} \subseteq \operatorname{Arr} \mathbb{B}$.

(i) This subfibration is called a **stack** (with respect to the regular epi topology on $\mathbb{B}$) if for each pullback square

$$
\begin{array}{ccc}
Y & \longrightarrow & X \\
\psi \downarrow & \lrcorner & \downarrow \varphi \\
J & \xrightarrow{\;\;u\;\;} & I
\end{array}
$$

with $u \colon J \twoheadrightarrow I$ a (regular) epi, one has

$$
\psi \in \mathcal{D} \Rightarrow \varphi \in \mathcal{D}.
$$

(ii) The **stack completion** of the subfibration induced by $\mathcal{D}$ is given by the collection $\overline{\mathcal{D}}$ of display maps containing those families $\left(\begin{smallmatrix} X \\ \downarrow \varphi \\ I \end{smallmatrix} \right)$ for which there is an epi $u \colon J \twoheadrightarrow I$ in $\mathbb{B}$ with $u^*(\varphi) \in \mathcal{D}$.

11.7.2. Definition (See [39, 150]). A full subfibration $\begin{smallmatrix} \mathcal{D}^{\rightarrow} \\ \downarrow \\ \mathbb{B} \end{smallmatrix}$ of $\begin{smallmatrix} \mathbb{B}^{\rightarrow} \\ \downarrow \\ \mathbb{B} \end{smallmatrix}$ is called **weakly complete** if its stack completion $\begin{smallmatrix} \overline{\mathcal{D}}^{\rightarrow} \\ \downarrow \\ \mathbb{B} \end{smallmatrix}$ is a complete fibration.

The difference between weak completeness and ordinary completeness is that in the weak case it is true in the internal language that limits exist, whereas for ordinary completeness these limits are given to us by external functors. It is like for ordinary categories where one may have binary products $\times$ given 'weakly' as

for every pair X, Y of objects there is a product diagram $X \leftarrow X \times Y \rightarrow Y$

and 'ordinarily' as

there is a right adjoint $\times$ to the diagonal functor.

One needs the Axiom of Choice in the meta-theory to establish the equivalence of these two descriptions. But within the universe given by a (subobject fibration of a) topos, this axiom may fail.

We can be more explicit if the display maps in $\mathcal{D}$ come from an internal category $\mathbf{C} = (C_1 \rightrightarrows C_0)$ in $\mathbb{B}$. For objects $X, Y \colon I \rightrightarrows C_0$ over I, existence of their Cartesian product in the internal language is

$$\forall i\colon I.\, S(i) \quad \text{where} \quad \begin{cases} S(i) \equiv \exists P\colon C_0.\, \exists \pi\colon P \to X_i.\, \exists \pi'\colon P \to Y_i.\, T(i, P, \pi, \pi') \\ \quad \text{with} \\ T(i, P, \pi, \pi') \equiv \text{``}X_i \xleftarrow{\pi} P \xrightarrow{\pi'} Y_i \text{ is a product diagram''} \end{cases}$$

If $\forall i\colon I.\, S(i)$ holds we get a situation

$$
\begin{array}{ccc}
T & \longrightarrow\!\!\!\!\to & S \\
\downarrow & & \downarrow{\scriptstyle \cong} \\
I \times (C_0 \times C_1 \times C_1) & \xrightarrow{\quad \pi \quad} & I
\end{array}
$$

where $T \twoheadrightarrow S \rightarrowtail I$ is the image factorisation of the composite $T \rightarrowtail I \times (C_0 \times C_1 \times C_1) \to I$. We thus get an epi $e\colon T \twoheadrightarrow I$ such that $e^*(X)$ and $e^*(Y)$ have a (canonically given) product. This is in essence what is expressed by the above definition of weak completeness.

Our first aim is to show that the fibration $\begin{smallmatrix} \text{UFam}(\omega\text{-}\mathbf{Sets}) \\ \downarrow \\ \mathbf{Eff} \end{smallmatrix}$ of ω-**Sets** over **Eff** is weakly complete, by showing that its stack completion is the fibration $\begin{smallmatrix} \text{FSep}(\mathbf{Eff}) \\ \downarrow \\ \mathbf{Eff} \end{smallmatrix}$ of separated families for the double negation nucleus on **Eff**. The latter is always a stack, which is complete as a fibration. This can be established in greater generality.

11.7.3. Lemma. *Let $\mathbb{B}$ be a topos. A full subfibration $\begin{smallmatrix} \mathcal{D}^{\to} \\ \downarrow \\ \mathbb{B} \end{smallmatrix}$ of $\begin{smallmatrix} \mathbb{B}^{\to} \\ \downarrow \\ \mathbb{B} \end{smallmatrix}$ which is definable is a stack.*

As a result, for a nucleus $\mathbf{j}$ in $\mathbb{B}$ the fibrations $\begin{smallmatrix} \text{FSep}_{\mathbf{j}}(\mathbb{B}) \\ \downarrow \\ \mathbb{B} \end{smallmatrix}$ and $\begin{smallmatrix} \text{FSh}_{\mathbf{j}}(\mathbb{B}) \\ \downarrow \\ \mathbb{B} \end{smallmatrix}$ of families of separated objects and sheaves are stacks.

Proof. Assume a pullback square

$$
\begin{array}{ccc}
Y & \longrightarrow & X \\
{\scriptstyle \psi}\downarrow & \lrcorner & \downarrow{\scriptstyle \varphi} \\
J & \xrightarrow[\quad u \quad]{} & I
\end{array}
$$

in which u is an epi and ψ is in $\mathcal{D}$. Since $\mathcal{D}$ is definable, there is a mono $m\colon I' \rightarrowtail I$ with $\varphi' = m^*(\varphi) \in \mathcal{D}$. By the universal property of φ' we get a

unique map $v\colon J \to I'$ such that

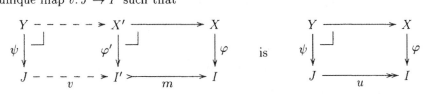

is

This makes m an epi, and hence an isomorphism. Thus φ is in $\mathcal{D}$.

The second part of the lemma follows because the fibrations of families of separated objects and of families of sheaves are definable, see Proposition 9.6.9. □

11.7.4. Lemma. *A fibration* $\begin{smallmatrix} \mathrm{FSep}_j(\mathbb{B}) \\ \downarrow \\ \mathbb{B} \end{smallmatrix}$ *of* **j**-*separated families in a topos* $\mathbb{B}$, *is always complete.*

Proof. It is easy to see that separated families are closed under finite limits. Closure under products $\prod$ follows from Exercise 5.7.6 (ii). □

Before we can analyse the situation in **Eff**, we need the following result; it says that every object in **Eff** has a separated cover.

11.7.5. Lemma. *For every object* $(I, \approx) \in$ **Eff** *there is an* ω-*set* $(I', E) \in$ ω-**Sets** *with an epi* $(I', E) \twoheadrightarrow (I, \approx)$ *in* **Eff**.

Proof. Take
$$I' = \{(i, n) \mid i \in I \text{ and } n \in E(i) = |i \approx i|\}$$
with existence predicate $E(i, n) = \{n\}$, and thus as equality
$$|(i, n) \approx (i', n')| = \begin{cases} \{n\} & \text{if } i = i' \text{ and } n = n' \\ \emptyset & \text{otherwise.} \end{cases}$$
There is then an epi $P\colon (I', E) \twoheadrightarrow (I, \approx)$ with $P(i, n, i') = |i \approx i'| \wedge \{n\}$. □

Notice that taking such separated covers is not functorial.

11.7.6. Theorem. *The fibration* $\begin{smallmatrix} \mathrm{UFam}(\omega\text{-}\mathbf{Sets}) \\ \downarrow \\ \mathbf{Eff} \end{smallmatrix}$ *considered as a full subfibration of the codomain fibration* $\begin{smallmatrix} \mathbf{Eff}^{\rightarrow} \\ \downarrow \\ \mathbf{Eff} \end{smallmatrix}$ *(via Proposition 6.3.4), has the fibration* $\begin{smallmatrix} \mathrm{FSep}(\mathbf{Eff}) \\ \downarrow \\ \mathbf{Eff} \end{smallmatrix}$ *of double negation separated families as its stack completion. Hence families of* ω-*sets over* **Eff** *form a weakly complete fibration.*

Proof. Every $(I, \approx)$-indexed family $(X_{[i]}, E_{[i]})_{[i] \in \Gamma(I, \approx)}$ of ω-sets yields a separated family $\begin{pmatrix} (\{X\}, \approx) \\ \downarrow \\ (I, =) \end{pmatrix}$ as in Proposition 6.3.4. And since separated fam-

ilies form a stack, they also contain the stack completion of such families
$$\left(\begin{array}{c} (\{X\},\approx) \\ \downarrow \\ (I,\approx) \end{array} \right).$$

Conversely, assume $\left(\begin{array}{c} (X,\approx) \\ \downarrow \varphi \\ (I,\approx) \end{array} \right)$ is a separated family. There is by the pre-

vious lemma an ω-set (I', E) together with an epi $P\colon (I', E) \twoheadrightarrow (I,\approx)$. Taking
the pullback along P yields another separated family, call it φ', over an ω-set.
By Proposition 6.3.3, this family φ' corresponds to a family in $\begin{array}{c} \text{UFam}(\omega\text{-Sets}) \\ \downarrow \\ \textbf{Eff} \end{array}$
Hence the original family φ is in the stack completion of this fibration. $\square$

We turn to the completeness of the fibration $\begin{array}{c} \text{UFam}(\textbf{PER}) \\ \downarrow \\ \textbf{Eff} \end{array}$ of families of PERs
over **Eff**. To make things easier we first investigate the non-fibred situation.
Below, a special rôle is played by the two-element set $2 = \{0, 1\} \in \textbf{Sets}$ and
by its image $\nabla 2 \in \textbf{Eff}$. We shall be especially interested in objects in **Eff**
'orthogonal' to $\nabla 2$ (and in families of these).

11.7.7. Definition. Let A be a fixed object in an arbitrary Cartesian closed
category $\mathbb{B}$. One calls an object $X \in \mathbb{B}$ **orthogonal to** A if every map $A \to X$
is constant. More precisely, if the canonical map $X \to (A \Rightarrow X)$—obtained as
exponential transpose of the projection $X \times A \to X$—is an isomorphism.

This concept extends to families: $\left(\begin{array}{c} X \\ \downarrow \\ I \end{array} \right)$ is orthogonal to $A \in \mathbb{B}$ if $\left(\begin{array}{c} X \\ \downarrow \\ I \end{array} \right)$ is

orthogonal to $I^*(A) = \left(\begin{array}{c} I \times A \\ \downarrow \pi \\ I \end{array} \right)$ in the slice category $\mathbb{B}/I$.

Orthogonality is related to uniformity: recall that an object U is uniform
with respect to N if every total relation $R \subseteq U \times N$ is constant: for some
$n\colon N$, $R(u, n)$ for all $u\colon U$. Such an object U is then clearly orthogonal to N.

The next result relates objects orthogonal to $\nabla 2$ and modest sets (or PERs)
in **Eff**. It shows a certain formal resemblance with connectivity in topological
spaces. We refer to the discussion in Remark 8.3.6 for the reason why this
result yields sufficient completeness for polymorphic (simple and dependent)
type theories. Instead of $\nabla 2$ one can also use the subobject classifier $\Omega \in \textbf{Eff}$,
see [150, Proposition 6.1].

11.7.8. Proposition (Freyd). *Consider an ω-set $(I, E) \in \textbf{Eff}$. Then (I, E)
is a modest set if and only if it is orthogonal to $\nabla 2 \in \textbf{Eff}$.*

Proof. Suppose (I, E) is modest and let $f\colon \nabla 2 \to (I, E)$ be tracked by e.
Write $i_0 = f(0)$ and $i_1 = f(1)$. Since $0 \in \mathbb{N} = E_{\nabla 2}(0) \cap E_{\nabla 2}(1)$, we get
$e \cdot 0 \in E_I(i_0) \cap E_I(i_1)$. Thus $i_0 = i_1$ and f is constant.

Conversely, assume (I, E) is orthogonal to $\nabla 2$, and let $i_0, i_1 \in I$ be given with $n \in E_I(i_0) \cap E_I(i_1)$. There is then a morphism $f \colon \nabla 2 \to (I, E)$ by $f(0) = i_0$ and $f(1) = i_1$, which is tracked by $\Lambda x . n$. But since f must be constant, we get $i_0 = i_1$. □

We would like to extend this result to families of PERs. But first we state the completeness of a full subfibration of orthogonal families.

11.7.9. Lemma. *Let $\mathbb{B}$ be a topos with a nucleus $\mathbf{j}$ and a fixed object $A \in \mathbb{B}$. We write $\mathrm{Orth}(A) \hookrightarrow \mathbb{B}^{\to}$ for the full subcategory of families that are orthogonal to A and $\mathrm{SepOrth}_{\mathbf{j}}(A) \hookrightarrow \mathrm{Orth}(A)$ for the full subcategory of $\mathbf{j}$-separated orthogonal families.*

(i) The codomain functors $\mathrm{Orth}(A) \to \mathbb{B}$ and $\mathrm{SepOrth}_{\mathbf{j}}(A) \to \mathbb{B}$ are fibrations.

(ii) These fibrations $\begin{array}{c} \mathrm{Orth}(A) \\ \downarrow \\ \mathbb{B} \end{array}$ *and* $\begin{array}{c} \mathrm{SepOrth}_{\mathbf{j}}(A) \\ \downarrow \\ \mathbb{B} \end{array}$ *are complete.*

(iii) And they are both definable—and hence stacks by Lemma 11.7.3.

Often, the subscript $\mathbf{j}$ in $\mathrm{SepOrth}_{\mathbf{j}}(A)$ will be ommitted if the nucleus $\mathbf{j}$ is clear from the context.

Proof. (i) For a family $\left(\begin{array}{c} X \\ \downarrow \varphi \\ I \end{array} \right)$ orthogonal to $A \in \mathbb{B}$ and a morphism $u \colon J \to I$, the pullback $u^*(\varphi)$ is again orthogonal to A, since: if $\varphi \cong I^*(A) \Rightarrow \varphi$ then

$$u^*(\varphi) \cong u^*(I^*(A) \Rightarrow \varphi) \cong u^* I^*(A) \Rightarrow u^*(\varphi) \cong J^*(A) \Rightarrow u^*(\varphi)$$

making $u^*(\varphi)$ orthogonal to $J^*(A)$ in $\mathbb{B}/J$.

Since separated families are closed under pullback, the codomain functor $\mathrm{SepOrth}_{\mathbf{j}}(A) \to \mathbb{B}$ is also a fibration.

(ii) By Lemma 11.7.4 it suffices to prove that the fibration of orthogonal families is complete. Fibrewise (finite) completeness is easy: a map $f \colon I^*(A) \to \varphi = \varprojlim_i \varphi_i$ is determined by the composites $I^*(A) \xrightarrow{f} \varphi \xrightarrow{\pi_i} \varphi_i$, which must be constant.

So what remains is closure under products $\prod_u$. Assume $u \colon I \to J$ in $\mathbb{B}$ and a family $\left(\begin{array}{c} X \\ \downarrow \varphi \\ I \end{array} \right)$. Maps $J^*(A) \to \prod_u(\varphi)$ in $\mathbb{B}/I$ then correspond by transposition to maps $I^*(A) \cong u^* J^*(A) \to \varphi$ in $\mathbb{B}/I$, which are constant. Hence the former are constant.

(iii) For a family φ in $\mathbb{B}/I$, write $\widehat{\varphi}$ for the canonical map $\varphi \to (I^*(A) \Rightarrow \varphi)$ and form the mono $\{\widehat{\varphi} \in \mathrm{vIso}\} \rightarrowtail I$, using that (vertical) isomorphisms are definable—since the codomain fibration on $\mathbb{B}$ is locally small, see Lemma 9.6.7. For a morphism $u \colon J \to I$ we get: $u^*(\varphi)$ is orthogonal to A if and only u

factors through $\{\hat{\varphi} \in vIso\} \rightarrowtail I$. This shows that orthogonality is definable. The families which are both orthogonal and separated are then also definable, see Exercise 9.6.3. □

11.7.10. Theorem. *The fibration* $\begin{smallmatrix} \mathrm{UFam}(\mathbf{PER}) \\ \downarrow \\ \mathbf{Eff} \end{smallmatrix}$ *considered as a full subfibration of the codomain fibration* $\begin{smallmatrix} \mathbf{Eff}^{\rightarrow} \\ \downarrow \\ \mathbf{Eff} \end{smallmatrix}$ *has the fibration* $\begin{smallmatrix} \mathrm{SepOrth}(\nabla 2) \\ \downarrow \\ \mathbf{Eff} \end{smallmatrix}$ *of separated families that are orthogonal to* $\nabla 2 \in \mathbf{Eff}$ *as its stack completion. Hence families of PERs over* $\mathbf{Eff}$ *form a weakly complete fibration.*

Proof. In essence we reproduce the argument underlying Proposition 11.7.8 for families of PERs. In one direction, an $(I, \approx)$-indexed family $R = (R_{[i]}, E_{[i]})_{[i] \in \Gamma(I, \approx)}$ of PERs gives rise to a separated family $\begin{pmatrix} (\{R\}, \cong) \\ \downarrow \pi_R \\ (I, \approx) \end{pmatrix}$ via Proposition 6.3.4 with

$$\{R\} = \{(i, [n]) \mid i \in I \text{ with } E(i) \neq \emptyset \text{ and } [n] \in \mathbb{N}/R_{[i]}\}$$

and

$$|(i, [n]) \cong (i', [n'])| = \begin{cases} |i \approx i'| \wedge [n] & \text{if } [i] = [i'] \text{ and } nR_{[i]}n' \\ \emptyset & \text{otherwise.} \end{cases}$$

This family is then orthogonal to $\nabla 2$: consider a morphism F in a commuting triangle

$$(I, \approx) \times \nabla 2 \xrightarrow{\quad F \quad} (\{R\}, \cong)$$
$$\pi \searrow \quad \swarrow \pi_R$$
$$(I, \approx)$$

There are then codes $a, b, c, d, e \in \mathbb{N}$ such that for each $i \in I$ with $n \in E(i)$ we have

$$a \cdot n \in F(i, 0, i_0, [n_0]) \cap F(i, 1, i_1, [n_1])$$
$$\text{for certain } i_0, i_1 \in I \text{ and } [n_0] \in \mathbb{N}/R_{[i_0]}, [n_1] \in \mathbb{N}/R_{[i_1]}$$
$$b \cdot n \in E(i_0, [n_0]) \cap E(i_1, [n_1]).$$

This yields a first projection $\mathbf{p}(b \cdot n) \in E(i_0) \cap E(i_1)$, so that $[i_0] = [i_1]$, and a second projection $\mathbf{p}'(b \cdot n) \in [n_0] \cap [n_1]$, so that $n_0 R_{[i_0]} n_1$. We also have

$$b \cdot n \in \pi_R(i_0, [n_0], i_0) \cap \pi_R(i_1, [n_1], i_1)$$
$$c \cdot n \in (\pi_R \circ F)(i, 0, i_0) \cap (\pi_R \circ F)(i, 1, i_1)$$
$$d \cdot n \in |i \approx i_0| \cap |i \approx i_1|$$
$$e \cdot n \in |(i_0, n_0) \cong (i_1, n_1)|.$$

This shows that F is constant.

Conversely, assume $\begin{pmatrix} (X,\approx) \\ \downarrow\varphi \\ (I,\approx) \end{pmatrix}$ is a separated family orthogonal to $\nabla 2$. There is by Lemma 11.7.5 an ω-set (I', E) together with an epi $P\colon (I', E) \twoheadrightarrow (I, \approx)$. Taking the pullback along P yields a family $\begin{pmatrix} (Y,E) \\ \downarrow\psi \\ (I',E) \end{pmatrix}$ which is separated and orthogonal to $\nabla 2$ over an ω-set. By Proposition 6.3.3, this family ψ corresponds to a family of ω-sets $(Y_{(i,n)}, E_{(i,n)})$, indexed by $(i, n) \in I'$ (as described in the proof of Lemma 11.7.5). Thus for each $(i, n) \in I'$ we have an (outer) pullback square

$$
\begin{array}{ccccc}
(Y_{(i,n)}, E_{(i,n)}) & \longrightarrow & (Y, E) & \longrightarrow & (X, \approx) \\
\downarrow & \lrcorner & \psi\downarrow & \lrcorner & \downarrow\varphi \\
1 & \xrightarrow{\;\;(i,n)\;\;} & (I', E) & \xrightarrow{\;\;P\;\;} & (I, \approx)
\end{array}
$$

showing that $(Y_{(i,n)}, E_{(i,n)})$ is orthogonal to $\nabla 2$. But since we already know that it is separated, it is a modest set by Proposition 11.7.8 (and hence comes from a PER). Hence ψ is isomorphic to a family of PERs over (I', E). Thus the family φ that we started from is in the stack completion of $\begin{array}{c}\mathbf{UFam(PER)}\\\downarrow\\\mathbf{Eff}\end{array}$. $\square$

In conclusion, there is a diagram of categories and functors over **Eff**,

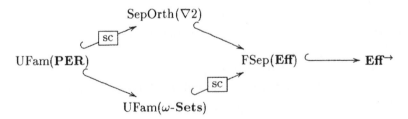

where 'sc' stands for 'stack completion'.

Exercises

11.7.1. Let $\mathcal{D}$ be a collection of display maps (closed under pullback) in a topos $\mathbb{B}$, with $\overline{\mathcal{D}}$ as its stack completion. Show that for $u\colon I \to J$ in $\mathbb{B}$ the pullback functor $u^*\colon \mathbb{B}/J \to \mathbb{B}/I$ restricts to $u^*\colon \overline{\mathcal{D}}/J \to \overline{\mathcal{D}}/I$.

11.7.2. Let $\mathbb{B}$ be a topos with nucleus j and object $A \in \mathbb{B}$. Prove that the fibrations associated with the following collections of morphisms in $\mathbb{B}$ are closed comprehension categories.

(i)   Separated families.

(ii) Families of sheaves.

(iii) Families orthogonal to A.

(iv) Separated families orthogonal to A.

[Essentially, one only needs to show that these collections are closed under composition.]

References

[1] M. Abadi and L. Cardelli. *A Theory of Objects*. Monographs in Comp. Sci. Springer, 1996.

[2] M. Abadi, L. Cardelli, and P.-L. Curien. Formal parametric polymorphism. *Theor. Comp. Sci.*, 121:9–58, 1993.

[3] S. Abramsky and A. Jung. Domain theory. In S. Abramsky, Dov M. Gabbai, and T.S.E. Maibaum, editors, *Handbook of Logic in Computer Science*, volume 3, pages 1–168. Oxford Univ. Press, 1994.

[4] Th. Altenkirch. *Constructions, Inductive Types and Strong Normalization*. PhD thesis, Univ. Edinburgh, 1993. Techn. rep. LFCS-93-279.

[5] Th. Altenkirch, M. Hofmann, and Th. Streicher. Categorical reconstruction of a *reduction free* normalization proof. In D.H. Pitt, D.E. Rydeheard, and P.T. Johnstone, editors, *Category Theory and Computer Science*, number 953 in Lect. Notes Comp. Sci., pages 182–199. Springer, Berlin, 1995.

[6] M.A. Arbib and E.G. Manes. Parametrized data types do not need highly constrained parameters. *Inf. & Contr.*, 52:139–158, 1982.

[7] A. Asperti. *Categorical Topics in Computer Science*. PhD thesis, Univ. Pisa, 1985. Techn. Rep. 7/90.

[8] A. Asperti and S. Martini. Categorical models of polymorphism. *Inf. & Comp.*, 99:1–79, 1992.

[9] K. Baclawski, D. Simovici, and W. White. A categorical approach to database semantics. *Math. Struct. in Comp. Sci.*, 4:147–183, 1994.

[10] E.S. Bainbridge. *A unified minimal realization theory with duality*. PhD thesis, Univ. Michigan, Ann Arbor, 1972. Techn. rep. 140, Dep. of Comp. and Comm. Sci.

[11] E.S. Bainbridge, P.J. Freyd, A. Scedrov, and P.J. Scott. Functorial polymorphism. *Theor. Comp. Sci.*, 70(1):35–64, 1990. Corrigendum in *Theor. Comp. Sci.* 71(3):431, 1990.

[12] R. Banach. Term graph rewriting and garbage collection using opfibrations. *Theor. Comp. Sci.*, 131(1):29–94, 1994.

[13] H.P. Barendregt. *The Lambda Calculus. Its Syntax and Semantics*. North-Holland, Amsterdam, 2nd rev. edition, 1984.

[14] H.P. Barendregt. Lambda calculi with types. In S. Abramsky, Dov M. Gabbai, and T.S.E. Maibaum, editors, *Handbook of Logic in Computer Science*, volume 2, pages 117–309. Oxford Univ. Press, 1992.

[15] H.P. Barendregt and A. Rezus. Semantics for classical AUTOMATH and related systems. *Inf. & Contr.*, 59:127–147, 1983.

[16] M. Barr. Fixed points in cartesian closed categories. *Theor. Comp. Sci.*, 70:65–72, 1990.

[17] M. Barr, P.A. Grillet, and D.H. van Osdol. *Exact Categories and Categories of Sheaves*. Number 236 in Lect. Notes Math. Springer, Berlin, 1971.

[18] M. Barr and Ch. Wells. *Toposes, Triples and Theories*. Springer, Berlin, 1985.

[19] M. Barr and Ch. Wells. *Category Theory for Computing Science*. Prentice Hall, 1990.

[20] B. Barras, S. Boutin, C. Cornes, J. Courant, J.-Chr. Filliâtre, E. Giménez, H. Herbelin, G. Huet, C. Muñoz, C. Murthy, C. Parent, C. Paulin-Mohring, A. Saïbi, and B. Werner. The Coq Proof Assistant User's Guide Version 6.1. Technical Report 203, INRIA Rocquencourt, France, May 1997.

[21] G. Barthe. Extensions of pure type systems. In M. Dezani-Ciancaglini and G. Plotkin, editors, *Typed Lambda Calculi and Applications*, number 902 in Lect. Notes Comp. Sci., pages 16–31. Springer, Berlin, 1995.

[22] G. Barthe. The relevance of proof-irrelevance. In K. Larsen, S. Skyum, and G. Winskel, editors, *International Colloquium on Automata, Languages and Programming*, number 1443 in Lect. Notes Comp. Sci., pages 755–768. Springer, Berlin, 1998.

[23] M.J. Beeson. *Foundations of Constructive Mathematics*. Springer, Berlin, 1985.

[24] J.L. Bell. *Toposes and Local Set Theories. An Introduction*. Number 14 in Logic Guides. Oxford Science Publ., 1984.

[25] G. Bellé and E. Moggi. Typed intermediate languages for shape analysis. In Ph. de Groote and J.R. Hindley, editors, *Typed Lambda Calculi and Applications*, number 1210 in Lect. Notes Comp. Sci., pages 11–29. Springer, Berlin, 1997.

[26] R. Belluci, M. Abadi, and P.-L. Curien. A model for formal parametric polymorphism: Per interpretation for system ∇. In M. Dezani-Ciancaglini and G. Plotkin, editors, *Typed Lambda Calculi and Applications*, number 902 in Lect. Notes Comp. Sci., pages 32–46. Springer, Berlin, 1995.

[27] J. Bénabou. Fibrations petites et localement petites. *C. R. Acad. Sc. Paris*, 281:A897–A900, 1975.

[28] J. Bénabou. Théories relatives à un corpus. *C. R. Acad. Sc. Paris*, 281:A831–A834, 1975.

[29] J. Bénabou. Fibered categories and the foundations of naive category theory. *Journ. Symb. Logic*, 50(1):10–37, 1985.

[30] J. Bénabou and J. Roubaud. Monades et descente. *C. R. Acad. Sc. Paris*, 270:A96–A98, 1970.

[31] S. Berardi. An application of PER models to program extraction. *Math. Struct. in Comp. Sci.*, pages 309–331, 1993.

[32] I. Bethke. *Notes on Partial Combinatory Algebras*. PhD thesis, Univ. Amsterdam, 1988.

[33] L. Birkedal, A. Carboni, G. Rosolini, and D.S. Scott. Type theory via exact categories. extended abstract. In *Logic in Computer Science*, pages 188–198. IEEE, Computer Science Press, 1998.

[34] G. Birkhoff and J.D. Lipson. Heterogenuous algebras. *Journ. Combinatorial Theory*, 8:115–133, 1970.

[35] C. Böhm and A. Berarducci. Automatic synthesis of typed λ-programs on term algebras. *Theor. Comp. Sci.*, 39:135–154, 1985.

[36] F. Borceux. *Handbook of Categorical Algebra*, volume 50, 51 and 52 of *Encyclopedia of Mathematics*. Cambridge Univ. Press, 1994.

[37] V. Breazu-Tannen and Th. Coquand. Extensional models for polymorphism. *Theor. Comp. Sci.*, 59:85–114, 1988.

[38] K.B. Bruce, A.R. Meyer, and J.C. Mitchell. The semantics of second-order lambda calculus. *Inf. & Comp.*, 85:76–134, 1990.

[39] M. Bunge and R. Paré. Stacks and equivalence of indexed categories. *Cah. de Top. et Géom. Diff.*, XX:404–436, 1979.

[40] R. Burstall and J. McKinna. Deliverables: an approach to program development in the calculus of constructions. In G. Huet and G. Plotkin, editors, *Logical Frameworks*, pages 113–121. Cambridge Univ. Press, 1991.

[41] A. Carboni, P.J. Freyd, and A. Scedrov. A categorical approach to realizability and polymorphic types. In M. Main, A. Melton, M. Mislove, and D. Schmidt, editors, *Mathematical Foundations of Programming Language Semantics*, number 298 in Lect. Notes Comp. Sci., pages 23–42. Springer, Berlin, 1988.

[42] A. Carboni, S. Lack, and R.F.C. Walters. Introduction to extensive and distributive categories. *Journ. Pure & Appl. Algebra*, 84(2):145–158, 1993.

[43] L. Cardelli and G. Longo. A semantic basis for Quest. *Journ. Funct. Progr.*, 1:417–458, 1991.

[44] L. Cardelli and P. Wegner. On understanding types, data abstraction and polymorphism. *ACM Comp. Surv.*, 4:471–522, 1985.

[45] J. Cartmell. *Generalized algebraic theories and contextual categories*. PhD thesis, Univ. Oxford, 1978.

[46] J. Cartmell. Generalised algebraic theories and contextual categories. *Ann. Pure & Appl. Logic*, 32:209–243, 1986.

[47] J. Celeyrette. *Catégories internes et fibrations*. PhD thesis, Univ. Paris-Nord, 1975.

[48] M. Cerioli and J. Meseguer. May I borrow your logic? (transporting logical structures along maps). *Theor. Comp. Sci.*, 173:311–347, 1997.

[49] A. Church. A formulation of the simple theory of types. *Journ. Symb. Logic*, 5:56–68, 1940.

[50] J.R.B. Cockett. List-arithmetic distributive categories: locoi. *Journ. Pure & Appl. Algebra*, 66:1–29, 1990.

[51] J.R.B. Cockett. Introduction to distributive categories. *Math. Struct. in Comp. Sci.*, 3:277–307, 1993.

[52] J.R.B. Cockett and D. Spencer. Strong categorical datatypes I. In R.A.G. Seely, editor, *Category Theory 1991*, number 13 in CMS Conference Proceedings, pages 141–169, 1992.

[53] J.R.B. Cockett and D. Spencer. Strong categorical datatypes II: A term logic for categorical programming. *Theor. Comp. Sci.*, 139:69–113, 1995.

[54] Th. Coquand. An analysis of Girard's paradox. In *Logic in Computer Science*, pages 227–236. IEEE, Computer Science Press, 1986.

[55] Th. Coquand. Metamathematical investigations of a calculus of constructions. In P. Odifreddi, editor, *Logic and computer science*, pages 91–122. Academic Press, London, 1990. The APIC series, vol. 31.

[56] Th. Coquand and Th. Ehrhard. An equational presentation of higher order logic. In D.H. Pitt, A. Poigné, and D.E. Rydeheard, editors, *Category and Computer Science*, number 283 in Lect. Notes Comp. Sci., pages 40–56. Springer, Berlin, 1987.

[57] Th. Coquand, C. Gunter, and G. Winskel. Domain theoretic models of polymorphism. *Inf. & Comp.*, 81:123–167, 1989.

[58] Th. Coquand and G. Huet. The calculus of constructions. *Inf. & Comp.*, 76(2/3):95–120, 1988.

[59] Th. Coquand and Ch. Paulin. Inductively defined types. In P. Martin-Löf and G. Mints, editors, *COLOG 88 International conference on computer logic*, number 417 in Lect. Notes Comp. Sci., pages 50–66. Springer, Berlin, 1988.

[60] R.L. Crole. *Programming metalogics with a fixpoint type.* PhD thesis, Univ. Cambridge, 1992. Comp. Lab. Techn. Rep. 247.

[61] R.L. Crole. *Categories for Types.* Cambridge Mathematical Textbooks. Cambridge Univ. Press, 1993.

[62] P.-L. Curien. Alpha-conversion, conditions on variables and categorical logic. *Studia Logica*, XLVIII 3:319–360, 1989.

[63] P.-L. Curien. *Categorical Combinators, Sequantial Algorithms and Functional Programming.* Progress in Theor. Comp. Sci. Birkhäuser, Boston, 1993.

[64] P.-L. Curien and R. di Cosmo. A confluent reduction for the lambda-calculus with surjective pairing and terminal object. *Journ. Funct. Progr.*, 6(2):299–327, 1996.

[65] H.B. Curry and R. Feys. *Combinatory Logic.* North-Holland, Amsterdam, 1958.

[66] N.J. Cutland. *Computability.* Cambridge Univ. Press, 1980.

[67] D. van Dalen. *Logic and structure.* Springer, Berlin, 2nd edition, 1983.

[68] L. Damas and R. Milner. Principal type-schemes for functional programs. In *Principles of Programming Languages.* ACM Press, 1982.

[69] B.A. Davey and H.A. Priestley. *Introduction to Lattices and Order.* Math. Textbooks. Cambridge Univ. Press, 1990.

[70] P. Dybjer. Inductive sets and families in Martin-Löf's type theory and their set-theoretic semantics. In G. Huet and G. Plotkin, editors, *Logical Frameworks*, pages 280–306. Cambridge Univ. Press, 1991.

[71] P. Dybjer. Inductive families. *Formal Aspects of Comp.*, 6:440–465, 1994.

[72] P. Dybjer. Internal type theory. In S. Berardi and M. Coppo, editors, *Types for Proofs and Programs*, number 1158 in Lect. Notes Comp. Sci., pages 120–134. Springer, Berlin, 1996.

[73] P. Dybjer. Representing inductively defined sets by wellorderings in Martin-Löf's type theory. *Theor. Comp. Sci.*, 179:329–335, 1997.

[74] Th. Ehrhard. A categorical semantics of constructions. In *Logic in Computer Science*, pages 264–273. IEEE, Computer Science Press, 1988.

[75] Th. Ehrhard. *Une sémantique catégorique des types dépendants: Application au Calcul des Constructions.* PhD thesis, Université Paris VII, 1988.

[76] Th. Ehrhard. Dictoses. In D.H. Pitt, A. Poigné, and D.E. Rydeheard, editors, *Category Theory and Computer Science*, number 389 in Lect. Notes Comp. Sci., pages 213–223. Springer, Berlin, 1989.

[77] H. Ehrig and B. Mahr. *Fundamentals of Algebraic Specification I: Equations and Initial Semantics.* Number 6 in EATCS Monographs. Springer, Berlin, 1985.

[78] R.L. Constable *et al. Implementing Mathematics with the Nuprl Proof Development System.* Prentice Hall, 1986.

[79] M.P. Fiore. A coinduction principle for recursive data types based on bisimulation. *Inf. & Comp.*, 127(2):186–198, 1996.

[80] M.P. Fourman and D.S. Scott. Sheaves and logic. In M.P. Fourman and C.J. Mulvey D.S. Scott, editors, *Applications of Sheaves*, number 753 in Lect. Notes Math., pages 302–401. Springer, Berlin, 1979.

[81] P. Freyd, Ph. Mulry, G. Rosolini, and D. Scott. Extensional PERs. *Inf. & Comp.*, 98(2):211–227, 1992.

[82] P.J. Freyd, editor. *Abelian Categories: An Introduction to the Theory of Functors.* Harper and Row, New York, 1964.

[83] P.J. Freyd. Aspects of topoi. *Bull. Austr. Math. Soc.*, 7:1–76 and 467–480, 1972.

[84] P.J. Freyd. Structural polymorphism. *Theor. Comp. Sci.*, 115:107–129, 1993.

[85] P.J. Freyd and A. Scedrov. *Categories, Allegories.* Number 39 in Math. Library. North-Holland, Amsterdam, 1990.

[86] Y. Fu. *Topics in Type Theory.* PhD thesis, Univ. Manchester, 1992. Techn. rep. 92-11-1.

[87] Y. Fu. Categorical properties of logical frameworks. *Math. Struct. in Comp. Sci.*, 7:1–47, 1997.

[88] J.R. Funk. *Descent for Cocomplete Categories.* PhD thesis, McGill Univ. Montreal, 1990.

[89] Ph. Gardner. *Representing Logics in Type Theory.* PhD thesis, Univ. Edinburgh, 1992. Tech. Rep. 93–92.

[90] J.H. Geuvers. The Church-Rosser property for $\beta\eta$-reduction in typed lambda calculi. In *Logic in Computer Science*, pages 453–460. IEEE, Computer Science Press, 1992.

[91] J.H. Geuvers. *Logics and Type Systems.* PhD thesis, Univ. Nijmegen, 1993.

[92] J.H. Geuvers and M.-J. Nederhof. A modular proof of strong normalization for the calculus of constructions. *Journ. Funct. Progr.*, 1(2):155–189, 1991.

[93] S. Ghilardi and G.C. Meloni. Modal and tense predicate logic: models in presheaves and categorical conceptualization. In F. Borceux, editor, *Categorical Algebra and its Applications*, number 1348 in Lect. Notes Math., pages 130–142. Springer, Berlin, 1988.

[94] J.-Y. Girard. Une extension de l'interprétation de Gödel à l'analyse et son application à l'élimination des coupures dans l'analyse et la théorie des types. In J.E. Fenstad, editor, *Proceedings of the 2nd Scandinavian Logic Symposium*, pages 63–92, Amsterdam, 1971. North-Holland.

[95] J.-Y. Girard. *Interprétation fonctionelle et élimination des coupures dans l'arithmétique d'ordre supérieur.* PhD thesis, Université Paris VII, 1972.

[96] J.-Y. Girard. The system F of variable types, 15 years later. *Theor. Comp. Sci.*, 45:159–192, 1986.

[97] J.-Y. Girard. Linear logic. *Theor. Comp. Sci.*, 50:1–102, 1987.

[98] J.-Y. Girard. *Proofs and types.* Number 7 in Tracts in Theor. Comp. Sci. Cambridge Univ. Press, 1989.

[99] J.-Y. Girard, A. Scedrov, and P.J. Scott. Normal forms and cut-free proofs as natural transformations. In Y.N. Moschovakis, editor, *Logic from Computer Science*, number 21 in Math. Sci. Research Inst. Publ., pages 217–241. Springer, 1992.

[100] J. Giraud. *Cohomologie non abélienne.* Springer, Berlin, 1971.

[101] J.A. Goguen, J. Thatcher, and E. Wagner. An initial algebra approach to the specification, correctness and implementation of abstract data types. In R. Yeh, editor, *Current Trends in Programming Methodology*, pages 80–149. Prentice Hall, 1978.

[102] R. Goldblatt. *Topoi. The Categorial Analysis of Logic.* North-Holland, Amsterdam, 2nd rev. edition, 1984.

[103] R. Goldblatt. *Mathematics of Modality.* CSLI Lecture Notes 43, Stanford, 1993.

[104] M.J.C. Gordon and T.F. Melham. *Introduction to HOL: A theorem proing environment for higher order logic.* Cambridge Univ. Press, 1993.

[105] J.W. Gray. Fibred and cofibred categories. In *Proc. Conf. on Categorical Algebra.*

LaJolla 1965, pages 21–83. Springer, Berlin, 1966.

[106] J.W. Gray. The categorical comprehension scheme. In P. Hilton, editor, *Category Theory, Homology Theory and their Applications III*, number 99 in Lect. Notes Math., pages 242–312. Springer, Berlin, 1969.

[107] A. Grothendieck. Catégories fibrées et descente (Exposé VI). In A. Grothendieck, editor, *Revêtement Etales et Groupe Fondamental (SGA 1)*, number 224 in Lect. Notes Math., pages 145–194. Springer, Berlin, 1970.

[108] C.A. Gunter. *Semantics of Programming Languages. Structures and Techniques.* The MIT Press, Cambridge, MA, 1992.

[109] C.A. Gunter and J.C. Mitchell, editors. *Theoretical Aspects of Object-Oriented Programming. Types, Semantics and Language Design.* The MIT Press, Cambridge, MA, 1994.

[110] T. Hagino. *A categorical programming language.* PhD thesis, Univ. Edinburgh, 1987. Techn. Rep. 87/38.

[111] T. Hagino. A typed lambda calculus with categorical type constructors. In D.H. Pitt, A Poigné, and D.E. Rydeheard, editors, *Category and Computer Science*, number 283 in Lect. Notes Comp. Sci., pages 140–157. Springer, Berlin, 1987.

[112] T. Hagino. Codatatypes in ML. *Journ. Symb. Computation*, 8:629–650, 1989.

[113] K. Hanna, N. Daeche, and G. Howells. Implementation of the Veritas design logic. In V. Stavridou, T.F. Melham, and R.T. Boute, editors, *Theorem Provers in Circuit Design*, IFIP Transactions A, pages 77–94. North Holland, 1992.

[114] K. Hanna, N. Daeche, and M. Longley. Specification and verification using dependent types. In *Trans. on Softw. Eng. 9, number 16*, pages 949–964, 1990.

[115] R. Harper, F. Honsell, and G.D. Plotkin. A framework for defining logics. *Journ. ACM*, 40(1):143–184, 1992.

[116] M. Hasegawa. Decomposing typed lambda calculus into a couple of categorical programming languages. In D.H. Pitt, D.E. Rydeheard, and P.T. Johnstone, editors, *Category Theory and Computer Science*, number 953 in Lect. Notes Comp. Sci., pages 200–219. Springer, Berlin, 1995.

[117] R. Hasegawa. Parametricity of extensionally collapsed models of polymorphism and their categorical properties. In T. Ito and A.R. Meyer, editors, *Theoretical Aspects of Computer Software*, number 526 in Lect. Notes Comp. Sci., pages 495–512. Springer, Berlin, 1991.

[118] R. Hasegawa. Categorical data types in parametric polymorphism. *Math. Struct. in Comp. Sci.*, 4:71–109, 1994.

[119] S. Hayashi. Adjunction of semifunctors: categorical structures in nonextensional lambda calculus. *Theor. Comp. Sci.*, 41:95–104, 1985.

[120] S. Hayashi. Logic of refinement types. In H. Barendregt and T. Nipkow, editors, *Types for Proofs and Programs*, number 806 in Lect. Notes Comp. Sci., pages 108–126. Springer, Berlin, 1994.

[121] L. Henkin, J.D. Monk, and A. Tarski. *Cylindric Algebras.* North-Holland, Amsterdam, 1971/1985. 2 volumes.

[122] U. Hensel. *Proof Principles for Categorical Datatypes.* PhD thesis, Univ. of Dresden, Germany, 1998.

[123] U. Hensel, M. Huisman, B. Jacobs, and H. Tews. Reasoning about classes in object-oriented languages: Logical models and tools. In Ch. Hankin, editor, *European Symposium on Programming*, number 1381 in Lect. Notes Comp. Sci., pages 105–121. Springer, Berlin, 1998.

[124] U. Hensel and B. Jacobs. Proof principles for datatypes with iterated recursion. In

E. Moggi and G. Rosolini, editors, *Category Theory and Computer Science*, number 1290 in Lect. Notes Comp. Sci., pages 220–241. Springer, Berlin, 1997.

[125] C. Hermida. *Fibrations, Logical Predicates and Indeterminates*. PhD thesis, Univ. Edinburgh, 1993. Techn. rep. LFCS-93-277. Also available as Aarhus Univ. DAIMI Techn. rep. PB-462.

[126] C. Hermida. On fibred adjunctions and completeness for fibred categories. In H. Ehrig and F. Orejas, editors, *Recent Trends in Data Type Specification*, number 785 in Lect. Notes Comp. Sci., pages 235–251. Springer, Berlin, 1994.

[127] C. Hermida. Some properties of Fib as a fibred 2-category. *Journ. Pure & Appl. Algebra*, 1998, to appear.

[128] C. Hermida and B. Jacobs. An algebraic view of structural induction. In L. Pacholski and J. Tiuryn, editors, *Computer Science Logic 1994*, number 933 in Lect. Notes Comp. Sci., pages 412–426. Springer, Berlin, 1995.

[129] C. Hermida and B. Jacobs. Fibrations with indeterminates: Contextual and functional completeness for polymorphic lambda calculi. *Math. Struct. in Comp. Sci.*, 5:501–531, 1995.

[130] C. Hermida and B. Jacobs. Structural induction and coinduction in a fibrational setting. *Inf. & Comp.*, 1998, to appear.

[131] C. Hermida and J. Power. Fibrational control structures. In I. Lee and S.A. Molka, editors, *Concur'95: Concurrency Theory*, number 962 in Lect. Notes Comp. Sci., pages 117–129. Springer, Berlin, 1995.

[132] M. Hofmann. Elimination of extensionality and quotient types in Martin-Löf type theory. In H. Barendregt and T. Nipkow, editors, *Types for Proofs and Programs*, number 806 in Lect. Notes Comp. Sci., pages 166–190. Springer, Berlin, 1994.

[133] M. Hofmann. *Extensional concepts in intensional type theory*. PhD thesis, Univ. Edinburgh, 1995. Techn. rep. LFCS-95-327.

[134] M. Hofmann. On the interpretation of type theory in locally cartesian closed categories. In L. Pacholski and J. Tiuryn, editors, *Computer Science Logic 1994*, number 933 in Lect. Notes Comp. Sci., pages 427–441. Springer, Berlin, 1995.

[135] M. Hofmann. A simple model for quotient types. In M. Dezani-Ciancaglini and G. Plotkin, editors, *Typed Lambda Calculi and Applications*, number 902 in Lect. Notes Comp. Sci., pages 216–234. Springer, Berlin, 1995.

[136] M. Hofmann. Conservativity of equality reflection over intensional type theory. In S. Berardi and M. Coppo, editors, *Types for Proofs and Programs*, number 1158 in Lect. Notes Comp. Sci., pages 153–164. Springer, Berlin, 1996.

[137] M. Hofmann. Syntax and semantics of dependent types. In P. Dybjer and A. Pitts, editors, *Semantics of Logics of Computation*, pages 79–130. Cambridge Univ. Press, 1997.

[138] M. Hofmann and B.C. Pierce. A unifying type-theoretic framework for objects. *Journ. Funct. Progr.*, 5(4):593–635, 1995.

[139] M. Hofmann and Th. Streicher. A groupoid model refutes uniqueness of identity proofs. In *Logic in Computer Science*, pages 208–212. IEEE, Computer Science Press, 1994.

[140] W.A. Howard. The formulae-as-types notion of construction. In J.R. Hindley and J.P. Seldin, editors, *To H.B Curry: Essays on Combinatory Logic, Lambda Calculus and Formalism*, pages 479–490. Academic Press, New York and London, 1980.

[141] A.J.C. Hurkens. A simplification of Girard's paradox. In M. Dezani-Ciancaglini and G. Plotkin, editors, *Typed Lambda Calculi and Applications*, number 902 in Lect. Notes Comp. Sci., pages 266–278. Springer, Berlin, 1995.

[142] J.M.E. Hyland. The effective topos. In A.S. Troelstra and D. van Dalen, editors, *The L.E.J. Brouwer centenary symposium*, pages 165–216. North-Holland, Amsterdam, 1982.

[143] J.M.E. Hyland. A small complete category. *Ann. Pure & Appl. Logic*, 40:135–165, 1988.

[144] J.M.E. Hyland. First steps in synthetic domain theory. In A. Carboni, M.C. Pedicchio, and G. Rosolini, editors, *Como Conference on Category Theory*, number 1488 in Lect. Notes Math., pages 131–156. Springer, Berlin, 1991.

[145] J.M.E. Hyland, P.T. Johnstone, and A.M. Pitts. Tripos theory. *Math. Proc. Cambridge Phil. Soc.*, 88:205–232, 1980.

[146] J.M.E. Hyland and E. Moggi. The *S*-replete construction. In D.H. Pitt, D.E. Rydeheard, and P.T. Johnstone, editors, *Category Theory and Computer Science*, number 953 in Lect. Notes Comp. Sci., pages 96–116. Springer, Berlin, 1995.

[147] J.M.E. Hyland and C.-H.L. Ong. Modified realizability toposes and strong normalization proofs. In M. Bezem and J.F. Groote, editors, *Typed Lambda Calculi and Applications*, number 664 in Lect. Notes Comp. Sci., pages 179–194. Springer, Berlin, 1993.

[148] J.M.E. Hyland and A.M. Pitts. The theory of constructions: categorical semantics and topos-theoretic models. In J. Gray and A. Scedrov, editors, *Categories in Computer Science and Logic*, number 92 in AMS Contemp. Math., pages 137–199, Providence, 1989.

[149] J.M.E. Hyland, E.P. Robinson, and G. Rosolini. Algebraic types in PER models. In M. Main, A. Melton, M. Mislove, and D. Schmidt, editors, *Mathematical Foundations of Programming Language Semantics*, number 442 in Lect. Notes Comp. Sci., pages 333–350. Springer, Berlin, 1990.

[150] J.M.E. Hyland, E.P. Robinson, and G. Rosolini. The discrete objects in the effective topos. *Proc. London Math. Soc.*, 60:1–36, 1990.

[151] A. Islam and W. Phoa. Categorical models of relational databases I: fibrational formulation, schema integration. In M. Hagiya and J.C. Mitchell, editors, *Theoretical Aspects of Computer Science*, number 789 in Lect. Notes Comp. Sci., pages 618–641. Springer, Berlin, 1994.

[152] J.A Goguen and R. Burstall. Institutions: Abstract model theory for specification and programming. *Journ. ACM*, 39(1):95–146, 1992.

[153] B. Jacobs. The inconsistency of higher order extensions of Martin-Löf's type theory. *Journ. Phil. Logic*, 18:399–422, 1989.

[154] B. Jacobs. *Categorical Type Theory*. PhD thesis, Univ. Nijmegen, 1991.

[155] B. Jacobs. Semantics of the second order lambda calculus. *Math. Struct. in Comp. Sci.*, 1(3):327–360, 1991.

[156] B. Jacobs. Simply typed and untyped lambda calculus revisited. In M.P. Fourman, P.T. Johnstone, and A.M. Pitts, editors, *Applications of Categories in Computer Science*, number 177 in LMS, pages 119–142. Cambridge Univ. Press, 1992.

[157] B. Jacobs. Comprehension categories and the semantics of type dependency. *Theor. Comp. Sci.*, 107:169–207, 1993.

[158] B. Jacobs. Semantics of lambda-I and of other substructure lambda calculi. In M. Bezem and J.F. Groote, editors, *Typed Lambda Calculi and Applications*, number 664 in Lect. Notes Comp. Sci., pages 195–208. Springer, Berlin, 1993.

[159] B. Jacobs. Mongruences and cofree coalgebras. In V.S. Alagar and M. Nivat, editors, *Algebraic Methodology and Software Technology*, number 936 in Lect. Notes Comp. Sci., pages 245–260. Springer, Berlin, 1995.

[160] B. Jacobs. Parameters and parametrization in specification using distributive categories. *Fund. Informaticae*, 24(3):209–250, 1995.

[161] B. Jacobs. Subtypes and bounded quantification from a fibred perspective. In S. Brookes, M. Main, A. Melton, and M. Mislove, editors, *Mathematical Foundations of Program Semantics*, number 1 in Elect. Notes in Theor. Comp. Sci. Elsevier, Amsterdam, 1995.

[162] B. Jacobs. Objects and classes, co-algebraically. In B. Freitag, C.B. Jones, C. Lengauer, and H.-J. Schek, editors, *Object-Orientation with Parallelism and Persistence*, pages 83–103. Kluwer Acad. Publ., 1996.

[163] B. Jacobs. On cubism. *Journ. Funct. Progr.*, 6:379–391, 1996.

[164] B. Jacobs. Invariants, bisimulations and the correctness of coalgebraic refinements. In M. Johnson, editor, *Algebraic Methodology and Software Technology*, number 1349 in Lect. Notes Comp. Sci., pages 276–291. Springer, Berlin, 1997.

[165] B. Jacobs and T. Melham. Translating dependent type theory into higher order logic. In M. Bezem and J.F. Groote, editors, *Typed Lambda Calculi and Applications*, number 664 in Lect. Notes Comp. Sci., pages 209–229. Springer, Berlin, 1993.

[166] B. Jacobs, E. Moggi, and Th. Streicher. Relating models of impredicative type theories. In D.H. Pitt et al., editor, *Category and Computer Science*, number 530 in Lect. Notes Comp. Sci., pages 197–218. Springer, Berlin, 1991.

[167] B. Jacobs and J. Rutten. A tutorial on (co)algebras and (co)induction. *EATCS Bulletin*, 62:222–259, 1997.

[168] G. Janelidze and W. Tholen. Facets of descent, I. *Appl. Categorical Struct.*, 2:245–281, 1994.

[169] P.T. Johnstone. *Topos Theory*. Academic Press, London, 1977.

[170] P.T. Johnstone. *Stone Spaces*. Number 3 in Cambridge Studies in Advanced Mathematics. Cambridge Univ. Press, 1982.

[171] P.T. Johnstone. Fibrations and partial products in a 2-category. *Appl. Categorical Struct.*, 1:141–179, 1993.

[172] P.T. Johnstone. Cartesian monads on a topos. *Journ. Pure & Appl. Algebra*, 116:199–220, 1997.

[173] P.T. Johnstone and R. Paré, editors. *Indexed Categories and their Applications*. Number 661 in Lect. Notes Math. Springer, Berlin, 1978.

[174] A. Joyal and M. Tierney. An extension of the Galois theory of Grothendieck. *Memoirs of the AMS*, 51(309-4), 1984.

[175] S. Kasangian, G.M. Kelly, and F. Rossi. Cofibrations and the realization of nondeterministic automata. *Cah. de Top. et Géom. Diff.*, XXIV:23–46, 1983.

[176] G.M. Kelly and R. Street. Review of the elements of 2-categories. In G.M. Kelly, editor, *Proc. Sydney Category Theory Seminar 1972/1973*, number 420 in Lect. Notes Math., pages 75–103. Springer, Berlin, 1974.

[177] R.E. Kent. The metric closure powerspace construction. In M. Main, A. Melton, M. Mislove, and D. Schmidt, editors, *Mathematical Foundations of Programming Language Semantics*, number 442 in Lect. Notes Comp. Sci., pages 173–199. Springer, Berlin, 1990.

[178] S.C. Kleene. On the interpretation of intuitionistic number theory. *Journ. Symb. Logic*, 10:109–124, 1945.

[179] A. Kock. Algebras for the partial map classifier monad. In A. Carboni, M.C. Pedicchio, and G. Rosolini, editors, *Como Conference on Category Theory*, number 1488 in Lect. Notes Math., pages 262–278. Springer, Berlin, 1991.

[180] A. Kock. Monads for which structures are adjoint to units. *Journ. Pure & Appl.*

Algebra, 104:41–59, 1995.

[181] K. Koymans. *Models of the Lambda Calculus*. PhD thesis, Univ. Utrecht, 1984. Also available as: CWI Tracts 9, Amsterdam.

[182] S.A. Kripke. Semantical analysis of intuitionistic logic. In J. Crossley and M.A.E. Dummett, editors, *Formal Systems and Recursive Functions*, pages 92–130, Amsterdam, 1965. North-Holland.

[183] Y. Lafont. *Logiques, Catégories et Machines*. PhD thesis, Univ. Paris VII, 1988.

[184] Y. Lafont and Th. Streicher. Game semantics for linear logic. In *Logic in Computer Science*, pages 43–50. IEEE, Computer Science Press, 1991.

[185] F. Lamarche. *Modelling Polymorphism with Categories*. PhD thesis, McGill Univ., Montréal, 1991.

[186] J. Lambek and P.J. Scott. *Introduction to higher order Categorical Logic*. Number 7 in Cambridge Studies in Advanced Mathematics. Cambridge Univ. Press, 1986.

[187] S. Mac Lane. *Categories for the Working Mathematician*. Springer, Berlin, 1971.

[188] S. Mac Lane and I. Moerdijk. *Sheaves in Geometry and Logic. A First Introduction to Topos Theory*. Springer, New York, 1992.

[189] S. Mac Lane and R. Paré. Coherence for bicategories and indexed categories. *Journ. Pure & Appl. Algebra*, 37:59–80, 1985.

[190] S. Lang. *Algebra*. Addison Wesley, 2nd rev. edition, 1984.

[191] F.W. Lawvere. Functorial semantics. *Proc. Nat. Acad. Sci. USA*, 50:869–872, 1963.

[192] F.W. Lawvere. Ajointness in foundations. *Dialectica*, 23:281–296, 1969.

[193] F.W. Lawvere. Equality in hyperdoctrines and comprehension scheme as an adjoint functor. In A. Heller, editor, *Applications of Categorical Algebra*, pages 1–14, Providence, 1970. AMS.

[194] F.W. Lawvere. Metric spaces, generalized logic, and closed categories. *Seminario Matematico e Fisico. Rendiconti di Milano*, 43:135–166, 1973.

[195] F. Leclerc and Ch. Paulin-Mohring. Programming with streams in Coq. A case study: the sieve of Eratosthenes. In H. Barendregt and T. Nipkow, editors, *Types for Proofs and Programs*, number 806 in Lect. Notes Comp. Sci., pages 191–212. Springer, Berlin, 1994.

[196] D. Leivant. Reasoning about functional programs and complexity classes associated with type discipline. In *Found. Comp. Sci.*, pages 460–469. IEEE, 1983.

[197] R. Loader. Equational theories for inductive types. *Ann. Pure & Appl. Logic*, 84:175–217, 1997.

[198] J.R. Longley. *Realizability Toposes and Language Semantics*. PhD thesis, Edinburgh Univ., 1994.

[199] G. Longo and E. Moggi. Constructive natural deduction and its 'ω-set' interpretation. *Math. Struct. in Comp. Sci.*, 1(2):215–254, 1991.

[200] Z. Luo. ECC the Extended Calculus of Constructions. In *Logic in Computer Science*, pages 386–395. IEEE, Computer Science Press, 1989.

[201] Z. Luo. Program specification and data refinement in type theory. *Math. Struct. in Comp. Sci.*, 3(3):333–363, 1993.

[202] Z. Luo. *Computation and Reasoning. A Type Theory for Computer Science*. Clarendon Press, Oxford, 1994.

[203] Z. Luo and R. Pollack. LEGO proof development system: User's manual. Techn. rep. LFCS-92-211, Univ. Edinburgh, 1992.

[204] Q. Ma and J. Reynolds. Types, abstraction, and parametric polymorphism, Part 2. In M. Mislove S. Brookes, M. Main and D. Schmidt, editors, *Mathematical Foundations of Program Semantics*, number 598 in Lect. Notes Comp. Sci., pages 1–40. Springer,

Berlin, 1992.
[205] D.B. MacQueen. Using dependent types to express modular structure. In *Principles of Programming Languages*, pages 277–286. ACM Press, 1986.
[206] D.B. MacQueen, R. Sethi, and G. Plotkin. An ideal model for recursive types. *Inf. & Contr.*, 71:95–130, 1986.
[207] L. Magnusson and B. Nordström. The ALF proof editor and its proof engine. In H. Barendregt and T. Nipkow, editors, *Types for Proofs and Programs*, number 806 in Lect. Notes Comp. Sci., pages 213–237. Springer, Berlin, 1994.
[208] M. Makkai. Duality and definability in first order logic. *Memoirs of the AMS*, 105(Number 503(4)), 1993.
[209] M. Makkai. The fibrational formulation of intuitionistic predicate logic I: completeness according to Gödel, Kripke, and Läuchli. Part 1. *Notre Dame Journ. Formal Log.*, 34(3):334–377, 1993.
[210] M. Makkai. The fibrational formulation of intuitionistic predicate logic I: completeness according to Gödel, Kripke, and Läuchli. Part 2. *Notre Dame Journ. Formal Log.*, 34(4):471–499, 1993.
[211] M. Makkai and G.E. Reyes. *First Order Categorical Logic*. Number 611 in Lect. Notes Math. Springer, Berlin, 1977.
[212] E.G. Manes. *Algebraic Theories*. Springer, Berlin, 1974.
[213] P. Martin-Löf. An intuitionistic theory of types: predicative part. In H.E. Rose and J.C. Shepherson, editors, *Logic Colloquium '73*, pages 73–118, Amsterdam, 1975. North-Holland.
[214] P. Martin-Löf. Constructive mathematics and computer programming. In L.C. Cohen, J. Los, , H. Pfeiffer, and K.P. Podewski, editors, *Logic, Methodology and the Philosophy of Science VI*, pages 153–179. North Holland, 1982.
[215] P. Martin-Löf. *Intuitionistic Type Theory*. Bibliopolis, Napoli, 1984.
[216] S. Martini. An interval model for second order lambda calculus. In D.H. Pitt, A. Poigné, and D. Rydeheard, editors, *Category Theory and Computer Science*, number 283 in Lect. Notes Comp. Sci., pages 219–237. Springer, Berlin, 1987.
[217] J.H. McKinna. *Deliverables: a Categorical Approach to Program Development in Type Theory*. PhD thesis, Univ. Edinburgh, 1992. Techn. rep. LFCS-92-247.
[218] C. McLarty. *Elementary Categories, Elementary Toposes*. Number 21 in Logic Guides. Oxford Science Publ., 1992.
[219] N.P. Mendler, P. Panangaden, P.J. Scott, and R.A.G. Seely. A logical view of concurrent constraint programming. *Nordic Journ. Comput.*, 2:181–220, 1995.
[220] J. Meseguer. Relating models of polymorphism. In *Principles of Programming Languages*, pages 228–241. ACM Press, 1989.
[221] A.R. Meyer. What is a model of the lambda calculus? *Inf. & Contr.*, 52:87–122, 1982.
[222] R. Milner. A theory of type polymorphism in programming. *Journ. Comp. Softw. Syst.*, 17:348–375, 1978.
[223] R. Milner and M. Tofte. *Commentary on Standard ML*. The MIT Press, Cambridge, MA, 1990.
[224] R. Milner, M. Tofte, and R. Harper. *The Definition of Standard ML*. The MIT Press, Cambridge, MA, 1991.
[225] J.C. Mitchell. A type-inference approach ro reduction properties and semantics of polymorphic expressions (summary). In *ACM Conf. on LISP and Funct. Progr.*, pages 308–319. ACM Press, 1986.
[226] J.C. Mitchell and E. Moggi. Kripke style models for typed lambda calculus. *Ann.*

  Pure & Appl. Logic, 51(1/2):99–124, 1991.

[227] J.C. Mitchell and G.D. Plotkin. Abstract types have existential type. *ACM Trans. on Progr. Lang. and Systems*, 10(3):470–502, 1988.

[228] John C. Mitchell. *Foundations of Programming Languages*. The MIT Press, Cambridge, MA, 1996.

[229] J.L. Moens. *Caractérisation des topos de faisceaux sur un site interne à un topos*. PhD thesis, Univ. Cath. de Louvain-la-Neuve, 1982.

[230] E. Moggi. A category-theoretic account of program modules. *Math. Struct. in Comp. Sci.*, 1(1):103–139, 1991.

[231] R.P. Nederpelt, J.H. Geuvers, and R.C. de Vrijer, editors. *Selected papers on Automath*, Amsterdam, 1994. North-Holland.

[232] B. Nordström, K. Peterson, and J.M. Smith. *Programming in Martin-Löf's Type Theory: an introduction*. Number 7 in Logic Guides. Oxford Science Publ., 1990.

[233] A. Obtułowicz. Functorial semantics of type-free λ-$\beta\eta$ calculus. In Karpinsky, editor, *Fundamentals of Computation Theory*, number 56 in Lect. Notes Comp. Sci., pages 302–307. Springer, Berlin, 1977.

[234] A. Obtułowicz. Categorical and algebraic aspects of Martin-Löf type theory. *Studia Logica*, XLVIII 3:299–317, 1989.

[235] A. Obtułowicz and A. Wiweger. Categorical, functorial and algebraic aspects of the type-free lambda calculus. In *Univ. Algebra and Appl.*, number 9 in Banach Center Publ., pages 399–422, Warsaw, 1982.

[236] P. Odifreddi. *Classical Recursion Theory*. North-Holland, Amsterdam, 1989.

[237] P.W. O'Hearn and R.D. Tennent. Relational parametricity and local variables (preliminary report). In *Principles of Programming Languages*, pages 171–184. ACM Press, 1992.

[238] C.-H.L. Ong and E. Ritter. A generic strong normalization argument: application to the calculus of constructions. In E. Börger, Y. Gurevich, and K. Meinke, editors, *Computer Science Logic 1993*, number 832 in Lect. Notes Comp. Sci., pages 261–279. Springer, Berlin, 1994.

[239] J. van Oosten. *Exercises in Realizability*. PhD thesis, Univ. Amsterdam, 1991.

[240] J. van Oosten. Axiomatizing higher order Kleene realizability. *Ann. Pure & Appl. Logic*, 70:87–111, 1994.

[241] S. Owre, S. Rajan, J.M. Rushby, N. Shankar, and M. Srivas. PVS: Combining specification, proof checking, and model checking. In R. Alur and T.A. Henzinger, editors, *Computer Aided Verification*, number 1102 in Lect. Notes Comp. Sci., pages 411–414. Springer, Berlin, 1996.

[242] S. Owre, J.M. Rushby, N. Shankar, and F. von Henke. Formal verification for fault-tolerant architectures: Prolegomena to the design of PVS. *IEEE Trans. on Softw. Eng.*, 21(2):107–125, 1995.

[243] V.C.V de Paiva. The dialectica categories. In J. Gray and A. Scedrov, editors, *Categories in Computer Science and Logic*, number 92 in AMS Contemp. Math., pages 47–62, Providence, 1989.

[244] V.C.V. de Paiva. A dialectica-like model of linear logic. In D.H. Pitt, A. Poigné, and D.E. Rydeheard, editors, *Category Theory and Computer Science*, number 389 in Lect. Notes Comp. Sci., pages 341–356. Springer, Berlin, 1989.

[245] E. Palmgren and V. Stoltenberg-Hansen. Domain interpretations of Martin-Löf's partial type theory. *Ann. Pure & Appl. Logic*, 48:135–196, 1990.

[246] R. Paré and D. Schumacher. Abstract families and the adjoint functor theorems. In P.T. Johnstone and R. Paré, editors, *Indexed Categories and their Applications*,

number 661 in Lect. Notes Math., pages 1–125. Springer, Berlin, 1978.

[247] Ch. Paulin-Mohring. *Extraction des Programmes dans le Calcul des Constructions.* PhD thesis, Université Paris VII, 1989.

[248] Ch. Paulin-Mohring. Inductive definitions in the system Coq. Rules and properties. In M. Bezem and J.F. Groote, editors, *Typed Lambda Calculi and Applications*, number 664 in Lect. Notes Comp. Sci., pages 328–345. Springer, Berlin, 1993.

[249] Ch. Paulin-Mohring and B. Werner. Synthesis of ML programs in the system Coq. *Journ. Symb. Computation*, 5-6:607–640, 1993.

[250] L.C. Paulson. *Isabelle: A Generic Theorem Prover.* Number 828 in Lect. Notes Comp. Sci. Springer, Berlin, 1994.

[251] L.C. Paulson. *ML for the Working Computer Scientist.* Cambridge Univ. Press, 2nd rev. edition, 1996.

[252] D. Pavlović. *Predicates and Fibrations.* PhD thesis, Univ. Utrecht, 1990.

[253] D. Pavlović. Constructions and predicates. In D.H. Pitt et al., editor, *Category and Computer Science*, number 530 in Lect. Notes Comp. Sci., pages 173–196. Springer, Berlin, 1991.

[254] D. Pavlović. A logical view of the adjoint functor theorem. In R.A.G. Seely, editor, *Category Theory 1991*, number 13 in CMS Conference Proceedings, pages 361–366, 1992.

[255] D. Pavlović. On completeness and cocompleteness in and around small categories. *Ann. Pure & Appl. Logic*, 74:121–152, 1995.

[256] D. Pavlović. Maps II: chasing proofs in the Lambek-Lawvere logic. *Journ. of the IGPL*, 4(2):159–194, 1996.

[257] J. Penon. Algèbre de catégories — catégories localement internes. *C. R. Acad. Sc. Paris*, 278:A1577–A1580, 1974.

[258] W.K.-S. Phoa. Relative computability in the effective topos. *Math. Proc. Cambridge Phil. Soc.*, 106:419–422, 1989.

[259] W.K.-S. Phoa. Effective domains and intrinsic structure. In *Logic in Computer Science*, pages 366–377. IEEE, Computer Science Press, 1990.

[260] W.K.-S. Phoa. *Domain Theory in Realizability Toposes.* PhD thesis, Univ. Cambridge, 1991. Extended version is available as Edinburgh Techn. Rep. CST-82-91.

[261] W.K.-S. Phoa. Building domains from graph models. *Math. Struct. in Comp. Sci.*, 2:277–299, 1992.

[262] W.K.-S. Phoa. An introduction to fibrations, topos theory, the effective topos and modest sets. Technical Report LFCS-92-208, Edinburgh Univ., 1992.

[263] W.K.-S. Phoa. Using fibrations to understand subtypes. In M.P. Fourman, P.T. Johnstone, and A.M. Pitts, editors, *Applications of Categories in Computer Science*, number 177 in LMS, pages 239–257. Cambridge Univ. Press, 1992.

[264] W.K.-S. Phoa. From term models to domains. *Inf. & Comp.*, 109:211–255, 1994.

[265] W.K.-S. Phoa and M. Fourman. A proposed semantics for pure ML. In W. Kuich, editor, *International Colloquium on Automata, Languages and Programming*, number 623 in Lect. Notes Comp. Sci., pages 533–544. Springer, Berlin, 1992.

[266] B.C. Pierce and D.N. Turner. Simple type theoretic foundation for object-oriented programming. *Journ. Funct. Progr.*, 4(2):207–247, 1994.

[267] A.M. Pitts. *The Theory of Triposes.* PhD thesis, Univ. Cambridge, 1981.

[268] A.M Pitts. Polymorphism is set theoretic, constructively. In D.H. Pitt, A. Poigné, and D.E. Rydeheard, editors, *Category and Computer Science*, number 283 in Lect. Notes Comp. Sci., pages 12–39. Springer, Berlin, 1987.

[269] A.M. Pitts. Non-trivial power types can't be subtypes of polymorphic types. In *Logic*

in Computer Science, pages 6–13. IEEE, Computer Science Press, 1989.

[270] A.M. Pitts. A co-induction principle for recursively defined domains. *Theor. Comp. Sci.*, 124(2):195–219, 1994.

[271] A.M. Pitts. Categorical logic. In S. Abramsky, Dov M. Gabbai, and T.S.E. Maibaum, editors, *Handbook of Logic in Computer Science*, volume 6. Oxford Univ. Press, to appear.

[272] A.M. Pitts and P. Taylor. A note on Russell's paradox in locally cartesian closed categories. *Studia Logica*, XLVIII 3:377–387, 1989.

[273] G. Plotkin and M. Abadi. A logic for parametric polymorphism. In M. Bezem and J.F. Groote, editors, *Typed Lambda Calculi and Applications*, number 664 in Lect. Notes Comp. Sci., pages 361–375. Springer, Berlin, 1993.

[274] G.D. Plotkin. Set-theoretical and other elementary models of the λ-calculus. *Theor. Comp. Sci.*, 121:351–409, 1993.

[275] A. Poigné. On specifications, theories, and models with higher types. *Inf. & Comp.*, 68:1–46, 1986.

[276] E. Poll. *A Programming Logic based on Type Theory*. PhD thesis, Techn. Univ. Eindhoven, 1994.

[277] E. Poll, C. Hemerik, and H.M.M. Ten Eikelder. CPO-models for second order lambda calculus with recursive types and subtyping. *Inf. Théor. et Appl.*, 27(3):221–260, 1993.

[278] A.J. Power. An abstract formulation for rewriting systems. In D.H. Pitt, A. Poigné, and D.E. Rydeheard, editors, *Category Theory and Computer Science*, number 389 in Lect. Notes Comp. Sci., pages 300–312. Springer, Berlin, 1989.

[279] J. Power and H. Thielecke. Environments, continuation semantics and indexed categories. In M. Abadi and T. Ito, editors, *Theoretical Aspects of Computer Software*, number 1281 in Lect. Notes Comp. Sci., pages 391–414. Springer, Berlin, 1997.

[280] D. Prawitz. *Natural Deduction*. Almqvist and Wiksell, Uppsala, 1965.

[281] H. Rasiowa and R. Sikorsky. *The Mathematics of Metamathematics*. PWN, Warsaw, 1963.

[282] H. Reichel. *Initial Computability, Algebraic Specifications, and Partial Algebras*. Number 2 in Monographs in Comp. Sci. Oxford Univ. Press, 1987.

[283] H. Reichel. An approach to object semantics based on terminal co-algebras. *Math. Struct. in Comp. Sci.*, 5:129–152, 1995.

[284] G.R. Reyes. From sheaves to logic. In A. Daigneault, editor, *Studies in Algebraic Logic*, number 9 in MAA Studies in Math., pages 143–204, 1974.

[285] J.C. Reynolds. Towards a theory of type structure. In *Programming Symposium*, number 19 in Lect. Notes Comp. Sci., pages 408–425. Springer, Berlin, 1974.

[286] J.C. Reynolds. Types, abstraction and parametric polymorphism. In R.E. A. Mason, editor, *Information Processing 83*, pages 513–523. IFIP, Elsevier Sci. Publ. (North-Holland), 1983.

[287] J.C. Reynolds. Polymorphism is not set-theoretic. In G. Kahn, D.B. MacQueen, and G.D. Plotkin, editors, *Semantics of Data Types*, number 173 in Lect. Notes Comp. Sci., pages 145–156. Springer, Berlin, 1984.

[288] J.C. Reynolds and G.D. Plotkin. On functors expressible in the polymorphic typed lambda calculus. *Inf. & Comp.*, 105:1–29, 1993.

[289] E. Ritter. *Categorical abstract machines for higher order typed lambda calculi*. PhD thesis, Univ. Cambridge, 1993. Comp. Lab. Techn. Rep. 297.

[290] E. Ritter. Categorical abstract machines for higher-order lambda calculi. *Theor. Comp. Sci.*, 136(1):125–162, 1994.

[291] E.P. Robinson. Parametricity as isomorphism. *Theor. Comp. Sci.*, 136:163–181, 1994.

[292] E.P. Robinson and G. Rosolini. Colimit completions and the effective topos. *Journ. Symb. Logic*, 55:678–699, 1990.

[293] E.P. Robinson and G. Rosolini. Reflexive graphs and parametric polymorphism. In *Logic in Computer Science*, pages 364–371. IEEE, Computer Science Press, 1994.

[294] H. Rogers. *Theory of Recursive Functions and Effecitve Computability*. MIT Press, Cambridge, MA, 1967.

[295] R. Rosebrugh and R.J. Wood. Relational databases and indexed categories. In R.A.G. Seely, editor, *Category Theory 1991*, number 13 in CMS Conference Proceedings, pages 391–407, 1992.

[296] G. Rosolini. *Continuity and effectiveness in topoi*. PhD thesis, Univ. Oxford, 1986.

[297] G. Rosolini. About modest sets. *Int. Journ. Found. Comp. Sci.*, 1:341–353, 1990.

[298] J. Rutten and D. Turi. On the foundations of final semantics: non-standard sets, metric spaces and partial orders. In J.W. de Bakker, W.P. de Roever, and G. Rozenberg, editors, *Semantics: Foundations and Applications*, number 666 in Lect. Notes Comp. Sci., pages 477–530. Springer, Berlin, 1993.

[299] J. Rutten and D. Turi. Initial algebra and final coalgebra semantics for concurrency. In J.W. de Bakker, W.P. de Roever, and G. Rozenberg, editors, *A Decade of Concurrency*, number 803 in Lect. Notes Comp. Sci., pages 530–582. Springer, Berlin, 1994.

[300] A. Scedrov. A guide to polymorphic types. In P. Odifreddi, editor, *Logic and computer science*, pages 387–420. Academic Press, London, 1990. The APIC series, vol. 31.

[301] D.S. Scott. Continuous lattices. In F.W. Lawvere, editor, *Toposes, Algebraic Geometry and Logic*, number 274 in Lect. Notes Math., pages 97–136. Springer, Berlin, 1972.

[302] D.S. Scott. Data types as lattices. *SIAM Journ. Comput.*, 3:523–587, 1976.

[303] D.S. Scott. Lambda calculus: some models some philosophy. In J. Barwise, H.J. Keisler, and K. Kunen, editors, *The Kleene Symposium*, pages 223–266, Amsterdam, 1980. North-Holland.

[304] D.S. Scott. Relating theories of the λ-calculus. In J.R. Hindley and J.P. Seldin, editors, *To H.B. Curry: Essays on Combinatory Logic, Lambda Calculus and Formalism*, pages 403–450, New York and London, 1980. Academic Press.

[305] R.A.G Seely. Hyperdoctrines, natural deduction and the Beck condition. *Zeits. Math. Log. Grundl. Math.*, 29:33–48, 1983.

[306] R.A.G. Seely. Locally cartesian closed categories and type theories. *Math. Proc. Cambridge Phil. Soc.*, 95:33–48, 1984.

[307] R.A.G. Seely. Categorical semantics for higher order polymorphic lambda calculus. *Journ. Symb. Logic*, 52:969–989, 1987.

[308] R.A.G. Seely. Modelling computations—a 2-categorical approach. In *Logic in Computer Science*, pages 65–71. IEEE, Computer Science Press, 1987.

[309] M.B. Smyth and G.D. Plotkin. The category theoretic solution of recursive domain equations. *SIAM Journ. Comput.*, 11:761–783, 1982.

[310] E.W. Stark. Dataflow networks are fibrations. In D.H. Pitt et al., editor, *Category and Computer Science*, number 530 in Lect. Notes Comp. Sci., pages 261–281. Springer, Berlin, 1991.

[311] R. Statman. Logical relations and the typed lambda calculus. *Inf. & Contr.*, 65:85–97, 1985.

[312] M. Stefanova and H. Geuvers. A simple model construction for the calculus of constructions. In S. Berardi and M. Coppo, editors, *Types for Proofs and Programs*,

number 1158 in Lect. Notes Comp. Sci., pages 249–264. Springer, Berlin, 1996.

[313] C. Stirling. Modal and temporal logics. In S. Abramsky, Dov M. Gabbai, and T.S.E. Maibaum, editors, *Handbook of Logic in Computer Science*, volume 2, pages 477–563. Oxford Univ. Press, 1992.

[314] R. Street. The formal theory of monads. *Journ. Pure & Appl. Algebra*, 2:149–169, 1972.

[315] R. Street. Fibrations and Yoneda's lemma in a 2-category. In G.M. Kelly, editor, *Proc. Sydney Category Theory Seminar 1972/1973*, number 420 in Lect. Notes Math., pages 104–133. Springer, Berlin, 1974.

[316] R. Street. Cosmoi of internal categories. *Trans. Am. Math. Soc.*, 258(2):271–317, 1980.

[317] R. Street. Fibrations in bicategories. *Cah. de Top. et Géom. Diff.*, XXI-2:111–160, 1980.

[318] Th. Streicher. *Correctness and Completeness of a Categorical Semantics of the Calculus of Constructions*. PhD thesis, Univ. Passau, 1989. Techn. Rep. MIP - 8913.

[319] Th. Streicher. *Semantics of Type Theory. Correctness, Completeness and Independence results*. Progress in Theor. Comp. Sci. Birkhäuser, Boston, 1991.

[320] Th. Streicher. Dependence and independence results for (impredicative) calculi of dependent types. *Math. Struct. in Comp. Sci.*, 2(1):29–54, 1992.

[321] Th. Streicher. Independence of the induction principle and the axiom of choice in the pure calculus of constructions. *Theor. Comp. Sci.*, 103(1):395–408, 1992.

[322] Th. Streicher. *Investigations into intensional type theory*. Habil. Thesis, Ludwig Maximilian Univ. München, 1993.

[323] G. Sundholm. Proof theory and meaning. In D. Gabbay and F. Guenthner, editors, *Handbook of Philosophical Logic*, volume 3, pages 471–506, Dordrecht, 1984. Reidel.

[324] G. Sundholm. Constructive generalized quantifiers. *Synthese*, 79:1–12, 1989.

[325] W.W. Tait. A realizability interpretation of the theory of species. In *Logic Colloquium. Symposium on Logic held at Boston 1972 - 1973*, number 453 in Lect. Notes Math., pages 240–251. Springer, Berlin, 1975.

[326] I. Takeuti. An axiomatic system of parametricity. In Ph. de Groote and J.R. Hindley, editors, *Typed Lambda Calculi and Applications*, number 1210 in Lect. Notes Comp. Sci., pages 354–372. Springer, Berlin, 1997.

[327] A. Tarlecki, R.M. Burstall, and J.A. Goguen. Some fundamental algebraic tools for the semantics of computation: Part 3. Indexed categories. *Theor. Comp. Sci.*, 91:239–264, 1991.

[328] A. Tarski. Der Aussagenkalkül und die Topologie. *Fundam. Math.*, 31:103–134, 1938.

[329] P. Taylor. Internal completeness of categories of domains. In D.H. Pitt, S. Abramsky, A. Poigné, and D.E. Rydeheard, editors, *Category Theory and Computer Programming*, number 240 in Lect. Notes Comp. Sci., pages 449–465. Springer, Berlin, 1985.

[330] P. Taylor. *Recursive domains, indexed category theory and polymorphism*. PhD thesis, Univ. Cambridge, 1986.

[331] P. Taylor. The fixed point property in synthetic domain theory. In *Logic in Computer Science*, pages 152–371. IEEE, Computer Science Press, 1991.

[332] P. Taylor. *Practical Foundations*. Cambridge Univ. Press, 1998, to appear.

[333] A.S. Troelstra. *Metamathematical Investigation of Intuitionistic Arithmetic and Analysis*. Number 344 in Lect. Notes Math. Springer, Berlin, 1973.

[334] A.S. Troelstra. On the syntax of Martin-Löfs type theories. *Theor. Comp. Sci.*, 51:1–26, 1987.

[335] A.S. Troelstra and D. van Dalen. *Constructivism in Mathematics. An Introduction*.

North-Holland, Amsterdam, 1988. 2 volumes.

[336] D. Čubrić. On the semantics of the universal quantifier. *Ann. Pure & Appl. Logic*, 87:209–239, 1997.

[337] B. Veit. A proof of the associated sheaf theorem by means of categorical logic. *Journ. Symb. Logic*, 46:45–55, 1981.

[338] S. Vickers. *Topology Via Logic*. Number 5 in Tracts in Theor. Comp. Sci. Cambridge Univ. Press, 1989.

[339] F.-J. de Vries. *Type Theoretical Topics in Topos Theory*. PhD thesis, Univ. Utrecht, 1989.

[340] Ph. Wadler. Theorems for free! In *Funct. Progr. & Comp. Architecture*, pages 347–359. ACM Press, 1989.

[341] R.F.C. Walters. *Categories and Computer Science*. Carslaw Publications, Sydney, 1991. Also available as: Cambridge Computer Science Text 28, 1992.

[342] R.F.C. Walters. An imperative language based on distributive categories. *Math. Struct. in Comp. Sci.*, 2:249–256, 1992.

[343] W. Wechler. *Universal Algebra for Computer Scientists*. Number 25 in EATCS Monographs. Springer, Berlin, 1992.

[344] G. Winskel. A compositional proof system on a category of labelled transition systems. *Inf. & Comp.*, 87:2–57, 1990.

[345] G. Winskel and M. Nielsen. Models for concurrency. In S. Abramsky, Dov M. Gabbai, and T.S.E. Maibaum, editors, *Handbook of Logic in Computer Science*, volume 4, pages 1–148. Oxford Univ. Press, 1995.

[346] G.C. Wraith. A note on categorical datatypes. In D.H. Pitt, A. Poigné, and D.E. Rydeheard, editors, *Category Theory and Computer Science*, number 389 in Lect. Notes Comp. Sci., pages 118–127. Springer, Berlin, 1989.

Notation Index

Subject Index

743

Printed and bound by CPI Group (UK) Ltd, Croydon, CR0 4YY

08/05/2025

01864930-0001